JACOB PALIS –
Selected Works

Jacob Palis

Photo by Ana Paula Rodrigues

JACOB PALIS – Selected Works

Edited by

Welington de Melo

impa
instituto nacional de
matemática
pura e aplicada

SBM

Springer

Jacob Palis
IMPA
Rio de Janeiro
Brazil

Editor
Welington de Melo
IMPA
Rio de Janeiro
Brazil

ISBN 978-3-319-06967-8

Springer Cham Heidelberg New York Dordrecht London

Library of Congress Control Number: 2012954381

Math. Subj. Classification (2010): 37-XX, 34-XX, 58-XX

Printed on acid-free paper

Springer is part of Springer Science+Business Media (www.springer.com)

Selected Works of Outstanding Brazilian Mathematicians

Mathematics in Brazil is a very young endeavor. Regular research activities and graduate studies go back to the 1950s, when the biannual Brazilian Mathematical Colloquium was launched and Brazil joined the International Mathematical Union (IMU). The Instituto de Matemática Pura e Aplicada (IMPA) was founded in 1952. Creation of the SBM, Sociedade Brasileira de Matemática, would follow, in 1969.

Half a century later, Brazil has reached Group IV in the IMU's ranking and is responsible for about 1.8 % of the world's mathematical production (according to MathSciNet). The Brazilian mathematical community is growing steadily, and has earned international reputation in several fields of research. While many great challenges remain ahead, one can certainly be proud of what has been achieved in such a short time.

This progress was largely due to a number of leading mathematicians who, in addition to their own fundamental contributions to Mathematics, created genuine "schools" around them that raised entire research fields in Brazil to the best international standards. Through this collection, the SBM and IMPA aim to highlight their scientific trajectories and to pay tribute to their pioneering efforts.

Editors
Hilário Alencar
César Camacho
André Nachbin
Marcelo Viana

Preface

This volume presents a selection of Jacob Palis' mathematical contributions and reflects their fundamental impact on the theory of dynamical systems, starting with his Ph.D. thesis and ending with papers on what is widely known as the Palis Conjecture.

I had the privilege to be his first student and to participate in the seminars he organized in 1970–1971 to prepare Brazilian mathematicians for the International Conference on Dynamical Systems held in Salvador, Brazil in July 1971. Most of the prominent mathematicians in the area attended the conference and many of them visited IMPA for a longer period, a testament to the wonderful atmosphere Palis had created at IMPA at the time. This marked the beginning of his strong leadership role in Brazilian mathematics, which soon expanded to Latin America and to many other countries around the world when he became involved with the International Mathematical Union—of which he was first Secretary and later President—and with the Academy of Sciences for the Developing World (TWAS), of which he was also President.

This volume opens with a brief overview of Jacob Palis' scientific life that is a minor modification of a text by Marcelo Viana, one of Palis' former Ph.D. students.

To date, Jacob Palis has supervised 41 Ph.D. dissertations. A complete list of his students, many of whom have since gone on to become prominent mathematicians in their own right, can be found at the end of this volume. A full list of his publications, which includes two books that profoundly influenced several generations of mathematicians all over the world, is also included.

Rio de Janeiro *Welington de Melo*

Acknowledgments

I would like to take this opportunity to thank, first of all, my family for their constant and unsurpassed support. Also, my Ph.D. advisor, Steve Smale, my colleagues and students at IMPA, as well as esteemed colleagues abroad, most especially Jean-Christophe Yoccoz, for the wealth of inspiration he has given me throughout my scientific career. Special thanks go to Welington de Melo for kindly and patiently editing the present collection of Selected Papers. Finally, I wish to dedicate my scientific work and achievements to my wife Suely, my children and grandchildren.

Jacob Palis

Contents

On the Scientific Life of Jacob Palis xv
Welington de Melo

On Morse–Smale Dynamical Systems 1
J. Palis

Structural Stability Theorems 21
J. Palis and S. Smale

A Note on Ω-Stability 31
J. Palis

Neighborhoods of Hyperbolic Sets 33
M. Hirsch, J. Palis, C. Pugh and M. Shub

Hyperbolic Nonwandering Sets on Two-Dimensional Manifolds 47
S. Newhouse and J. Palis

The Topology of Holomorphic Flows with Singularity 57
C. Camacho, N. H. Kuiper and J. Palis

Topological Equivalence of Normally Hyperbolic Dynamical Systems . . 91
J. Palis and F. Takens

Moduli of Stability and Bifurcation Theory 103
J. Palis

Characterising Diffeomorphisms with Modulus of Stability One 109
W. de Melo, J. Palis and S. J. van Strien

Bifurcations and Stability of Families of Diffeomorphisms 129
S. Newhouse, J. Palis and F. Takens

Stability of Parametrized Families of Gradient Vector Fields 197
J. Palis and F. Takens

A Note on the Inclination Lemma (λ-Lemma) and Feigenbaum's Rate of Approach 237
J. Palis

Cycles and Measure of Bifurcation Sets for Two-Dimensional Diffeomorphisms 243
J. Palis and F. Takens

Hyperbolicity and the Creation of Homoclinic Orbits 269
J. Palis and F. Takens

On the C^1 Ω-Stability Conjecture 307
J. Palis

Centralizers of Anosov Diffeomorphisms on Tori 313
J. Palis and J.-C. Yoccoz

Homoclinic Tangencies for Hyperbolic Sets of Large Hausdorff Dimension 323
J. Palis and J.-C. Yoccoz

High Dimension Diffeomorphisms Displaying Infinitely Many Periodic Attractors 369
J. Palis and M. Viana

On the Arithmetic Sum of Regular Cantor Sets 413
J. Palis and J.-C. Yoccoz

A Global View of Dynamics and a Conjecture on the Denseness of Finitude of Attractors 431
J. Palis

Fers à cheval non uniformément hyperboliques engendrés par une bifurcation homocline et densité nulle des attracteurs. (French) [Non-Uniformly Hyperbolic Horseshoes Generated by Homoclinic Bifurcations and Zero Density of Attractors] 445
J. Palis and J.-C. Yoccoz

Homoclinic Tangencies and Fractal Invariants in Arbitrary Dimension . 451
C. G. Moreira, J. Palis and M. Viana

A Global Perspective for Non-Conservative Dynamics 457
J. Palis

Non-Uniformly Hyperbolic Horseshoes Arising From Bifurcations of Poincaré Heteroclinic Cycles 481
J. Palis and J.-C. Yoccoz

List of Publications of Jacob Palis 699

List of Ph.D. Students of Jacob Palis at IMPA 707

Acknowledgments 709

On the Scientific Life of Jacob Palis

Jacob Palis Jr. was born in Uberaba, in the Brazilian state of Minas Gerais, on March 15, 1940, being the eighth and last child of Jacob and Sames Palis. At the age of 16, he moved to Rio de Janeiro, the national capital at the time, where he graduated with a degree in Engineering from the University of Brazil in December 1962, receiving the university's Best Graduating Student prize.

Deciding to pursue high-level scientific research in mathematics and greatly motivated by Mauricio Peixoto, Elon Lima and Lindolpho de Carvalho Dias, whom he first met at the Engineering School and then IMPA, in 1964 he went on to join the University of California at Berkeley. In 1967, he completed his Ph.D. degree under the supervision of Steve Smale, a 1966 Fields Medalist.

In his doctoral thesis, Palis proved that gradient-like (or Morse-Smale) dynamical systems in lower dimensions are stable, i.e., that their orbit structure remains qualitatively the same under small perturbations of the evolution law. This remarkable finding, which provided the first class of stable systems existing on any smooth configuration space (manifold), brought to maturity the theory of structural stability first introduced by Andronov and Pontryagin some three decades before. Equally important, the methods of proof embodied a new geometric approach that would fundamentally shape subsequent developments in the field: he created the notion of stable foliations being partially subfoliated to include those of critical points (or isolated periodic motions) of higher indices where they accumulate upon.

Soon thereafter, in a joint work with Smale, the result of his thesis was extended to all dimensions and the authors formulated the famous Stability Conjectures, which proposed precise conditions for a dynamical system to be stable, or stable restricted to its limit-set. These conjectures remained a major topic of research in the area until they were solved some twenty years later, most remarkably by Mañé, one of Palis' first doctoral students, in the case of global stability, and by Palis himself in the case of limit-set stability (Publ. Math. IHES). Both results were established for discrete-time systems. Important partial progress had been obtained, among others, by de Melo, another of his doctoral students, whose thesis presented a considerable extension of Palis' geometric approach (Invent. Math). The conjectures for flows were later proved by Hayashi (Annals of Math). The solution of the Palis-Smale Stability Conjectures remains, without a doubt, one of the most beautiful achievements in dynamical systems.

In 1968, Palis returned to Rio de Janeiro to pursue a career at the Instituto de Matemática Pura e Aplicada (IMPA), rapidly becoming part of the soul of and a main driving force at the institution. His influence and leadership were fundamental in making IMPA one of the finest scientific centers in the developing world and a symbol of excellence in mathematics recognized around the globe.

Upon his return to Brazil, he broadened his research goals considerably. While stability would remain close to his heart, the 1970s saw him make fundamental and pioneering contributions to the theory of bifurcations. He and his colleagues, most notably Newhouse and Takens, developed a very successful approach based on considering parametrized families of dynamical systems obtained by deformation of a stable one, and analyzing the first bifurcation parameter for which stability breaks down. Also, around this time, he proved with F. Takens that most parametrized families of gradient-like vector fields are stable. The proof, published in the Annals of Mathematics, is an impressive demonstration of the power of the geometric method.

Another relevant front in his work was the use of certain smooth invariants for topological equivalence of dynamical systems. These invariants are very efficient tools for the classification of dynamical systems and have been used by several authors in contexts that greatly expand on the original, e.g. in the classification of linear holomorphic vector fields. He presented this work in 1978 in an invited address at the International Congress of Mathematicians (ICM), held every four years by the International Mathematical Union (IMU).

In the early eighties, Palis became increasingly interested in another branch of research where he would make another and one of the most lasting impacts in dynamics in the last two to three decades: the unfolding of homoclinic tangencies. Unfolding homoclinic tangencies are a major mechanism for complex dynamical behavior, discovered by the great French mathematician Henri Poincaré at the end of the nineteenth century and studied by several important dynamicists, like Birkhoff, Cartright-Littlewood and Smale. The work of Palis and his outstanding collaborators, especially Newhouse, Takens, Viana, and Yoccoz, who is also a Fields Medal laureate, showed that the creation of such dynamics through homoclinic bifurcations is accompanied by a remarkable variety of other complex dynamical changes.

Indeed, going beyond that conclusion, Palis formulated a series of conjectures to the effect that homoclinic bifurcations are the key mechanism underlying global instabilities of the dynamical behavior. These Palis conjectures have been a central topic of research in the area in the last decade or so, being a key object of study for mathematicians in France, Japan, China, the United States, England, Brazil and other countries in Latin America. They have generated a remarkable amount of scientific activity, as can be seen by a dozen plenary and invited lectures at the last ICMs, especially those in Zurich, Berlin and Beijing.

Palis' collaboration with Yoccoz, which now includes nine scientific papers, started in the late 1980s with a series of articles in which they solved the problem of centralizers for hyperbolic dynamical systems by proving that the majority of such systems admit only trivial smooth symmetries. These results, as well as their work on homoclinic bifurcations published in the famous journal Acta Mathematica, were explicitly cited on the occasion of Yoccoz being awarded the Fields Medal in 1994.

One of Palis' greatest contributions in the study of homoclinic tangencies is his revelation of the fundamental role played by fractal dimensions in connection with the frequency of dynamical bifurcations. Fractal dimensions were first introduced in the 19th century by the German mathematician Hausdorff, and received considerable attention in the late 1970s when the development of personal computers led Mandelbrot, Feigenbaum and other scientists to the discovery of many important objects and phenomena with a fractal nature.

In Palis' own work, particularly in the joint papers with Takens, Viana, and Yoccoz published in key journals such as the Annals of Mathematics, Inventiones Mathematicae, and Acta Mathematica, fractal dimensions interact in a completely new way, determining in a very precise sense the frequency of stable dynamical behavior. In fact, inspired by these discoveries, Palis formulated a deep conjecture relating the structure of the arithmetic difference of fractal (Cantor) sets to their fractal dimensions. This difficult problem was solved only a few years later by Moreira, another of Palis' students, together with Yoccoz, in a paper that was also published in the Annals of Mathematics.

Palis has an equally impressive record as a mentor to and educator of talented young mathematicians. To date he has supervised 41 doctoral students from Argentina, Chile, Iran, Italy, Mexico, Peru, Portugal, Spain, Uruguay, Venezuela and Brazil, many of whom have since achieved worldwide renown in their fields of research. In particular, five of his Ph.D. students were invited as speakers at the International Congress of Mathematicians. All of them have received the TWAS award in mathematics, and two have been awarded the recently created Ramanujan Prize by the International Center of Theoretical Physics. Also, many of his students are among the most active mathematicians in their countries, a fact that has greatly contributed to the dissemination of high-quality research in various corners of the globe. As of today, his mathematical "family tree" lists 180 Ph.D. descendants.

Further, Palis' role in the remarkable advancement of mathematics in Brazil, Latin America and the world at large over the last several decades can hardly be overemphasized. He was a founder and the first Scientific Coordinator of UMALCA, the Mathematical Union for Latin America and the Caribbean, which was created in Rio de Janeiro in 1995. Under his inspiration the Union became a model collaboration network, promoting the development of mathematics and the mobility of young researchers in the region. Palis was also the Director of the National Institute for Pure and Applied Mathematics (IMPA), Rio de Janeiro, for ten years (1993–2003), a period during which the Institute attained and consolidated a position as the leading research center it is today.

In the period 2001–2005, together with Phillip Griffiths, he played a fundamental part in the launching of the Millennium Science Institutes, a major project for a remarkable advancement of science in Brazil.

More globally, Palis has been active at the highest level of several international institutions. He was a member of the Executive Committee of the International Mathematical Union (IMU) for an unprecedented six consecutive terms, from 1983 through 2006, including eight years as Secretary General and four years as President. His term as President of the IMU was distinctly marked by an opening of the Union towards other disciplines, as well as by a focus on bringing mathematical development to all parts of the world, with a special emphasis on developing countries. Equally important, he led the IMU to provide full support to the government and scientific community of Norway in the creation of the Abel Prize in order to immortalize the greatest mathematicians of our time, and without the age restrictions that apply e.g. for the Fields Medal.

These same policies drove Palis' actions as a member from the early 1990's through 2005 and then as Chair of the Scientific Council of the International Center for Theoretical Physics in Trieste for 2 years, during which he helped to create the Ramanujan Prize for young mathematical researchers from developing countries. He also made enormous contributions to both the Academy of Sciences for the Developing World (TWAS) as a former Secretary General and now its President, and to the Brazilian Academy of Sciences.

All his Ph.D. supervision duties and contributions to science worldwide have by no means hampered Palis' great capacity to remain at the forefront of research, and be a permanent source of ideas on dynamical systems. By the late 1980's, new fundamental developments were taking place in the study of chaotic systems. Foremost among them was the first rigorous proof, by the Swedish mathematicians Benedicks and Carleson, of the existence of strange attractors in the Hénon model of planar transformations. Immediately upon their announcement, and long before the actual paper appeared, Palis conjectured that the same conclusion should remain true in the much more general context of homoclinic bifurcations. This conjecture was fully confirmed by Mora and Viana, Palis' doctoral students at the time, in a paper that was published in Acta Mathematica and played a crucial part in the substantial progress that followed.

These and other developments further consolidated the conclusion that Smale's uniform hyperbolicity, though an important element, is hardly sufficient for describing "most" dynamical systems. Indeed, during the 1970s it became commonly accepted that such a broad picture might not even be possible. However, by 1995, Palis had formulated a bold series of ideas and conjectures that encompassed a global view of dynamics, with a much more probabilistic flavor than had been attempted before. The Palis Program, as it is often called, roughly states that most systems should display only finitely many attractors, which trajectories accumulate upon in the future. It has been remarkably successful in the case of low dimensional systems, such as transformations of the line, where its completion may be envisaged for the near future. Good progress has also been made regarding the general case. In fact, the program has already played a fundamental part in guiding current research and setting exciting partial goals, and will continue to do so in years to come. It may also be applicable to other branches of science, such as turbulence molded by evolution equations, at least for complete solutions that accumulate upon a finite dimension space in the future.

Palis himself has also made important contributions in this direction, especially through his new and extremely promising joint project with Yoccoz, in which they study the formation of "strange horseshoes" in the unfolding of homoclinic tangencies. Their recent (2010) and long paper, extending over 200 pages, advances the theory of strange chaotic dynamics to a new level of sophistication and precision, and is bound to become another landmark work in the field. It has appeared by invitation in the highly prestigious journal Publications Mathématiques de l'IHÉS (Inst. des Hautes Etudes. Scient.).

Over the course of his career, Palis has received a remarkable list of awards and distinctions for scientific merit, including nine honorary doctorates and professorships from universities in six different countries and the Chinese Academy of Sciences. He is or has been a member of 16 Academies of Sciences, including the US National Academy of Sciences; the French, Russian, Norwegian, Portuguese Lisbon, Italian Lincei, Germany Leopoldina, Chinese and Indian Academies of Sciences, and has received high honors in Brazil and abroad, like the Brazilian Grand-Croix National Order of Scientific Merit, the French Légion d'Honneur and the Solomon Lefschetz Medal for the promotion of mathematics in America, besides invitations to highly prestigious meetings and distinguished special lectures. The list also includes major Science Prizes in Brazil and Mexico from the Organization of American States, Academy of Sciences for the Developing World and Trieste, Accademia dei Lincei, and most recently the coveted Balzan Prize, which also provides financial support for a large research project for five years. The project, coordinated by Palis and Yoccoz, will involves the participation of excellent mathematicians from various countries.

Welington de Melo
Marcelo Viana

Topology, Vol. 8, pp. 385—405. Pergamon Press, 1969. Printed in Great Britain

ON MORSE–SMALE DYNAMICAL SYSTEMS

J. PALIS †

(*Received* 15 *September* 1968)

INTRODUCTION

WE CONSIDER here dynamical systems on a compact C^{∞} manifold without boundary, which are C^r actions either of Z or R on M, $r \geq 1$. From the point of view of orbit structure, there is a natural identification of these actions with the sets of C^r diffeomorphisms and C^r vector fields of M (see [13]). These sets with the uniform C^r topology are denoted by Diff(M) and $\chi(M)$.

In [10] Smale defined a subset $S(M)$ of Diff(M), called now Morse–Smale diffeomorphisms, and conjectured that $S(M)$ is open and each of its elements is structurally stable. The main purpose of this paper is to prove the first part of this conjecture and the second part when dim $M \leq 3$. The correspondent results hold for the vector field case and then they extend results of [7].

For $f \in$ Diff(M) we denote by $\Omega(f)$ the set of nonwandering points of f and by Per(f) the set of periodic points of f. A point $x \in$ Per(f) of period n is called hyperbolic if $(Df^n)_x$ has its spectrum disjoint from the unit circle in the complex plane. In this case we can define the stable and unstable manifold of x, $W^s(x)$ and $W^u(x)$. We call $s = \dim W^s(x)$ the stable index of x. For $s = \dim M$, $0 < s < \dim M$, $s = 0$, x is called a sink (attractor), saddle, source respectively. We refer the reader to [13] for a detailed presentation of these concepts and a general background to the subject.

Let $S(M)$ be the subset of Diff(M) defined by: $f \in S(M)$ iff

i) $\Omega(f)$ is finite (this implies $\Omega(f) =$ Per(f)),
ii) all points in Per(f) are hyperbolic,
iii) for any pair of points $x, y \in$ Per(f), $W^s(x)$ and $W^u(y)$ have a transversal intersection.

The analogue of $S(M)$ for the vector field case is the set $\mathscr{S}(M)$ defined by: $X \in \mathscr{S}(M)$ iff

i) $\Omega(X)$ is the union of a finite number of fixed points $x_1, \ldots, x_m$ and closed orbits $\gamma_1, \ldots, \gamma_n$ of X,
ii) the x_i, γ_j are all hyperbolic,
iii) the stable and unstable manifolds of the x_i, γ_j have a transversal intersection.

† Partially supported by C.N. Pq., Brazil.

For $f \in S(M)$ its periodic orbits are partially ordered by the relation: $\mathcal{O}(x) \leq \mathcal{O}(y)$ iff $W^s(x) \cap W^u(y) \neq \varnothing$. The periodic orbits of f with this partial order structure is called the phase-diagram $D(f)$. By a diagram isomorphism we mean a map $\rho: D(f) \to D(g)$ which is bijective, index and order preserving.

THEOREM. *$S(M)$ is open in* Diff(M). *If $f \in S(M)$ then its phase-diagram is stable (up a diagram isomorphism) under C^r small pertubations.*

An element $f \in \text{Diff}(M)$ is called structurally stable if there is a neighborhood $N(f)$ in Diff(M) such that for each $g \in N(f)$ there exists a homeomorphism h of M satisfying $hf(x) = gh(x)$ for all $x \in M$. Similarly, $X \in \chi(M)$ is structurally stable if there is a neighborhood $N(X)$ in $\chi(M)$ such that for each $Y \in N(X)$ there exists a homeomorphism of M sending trajectories of χ into those of Y.

If x is a hyperbolic fixed point of $f \in \text{Diff}(M)$, then f is locally topologically conjugate to $(Df)_x$. This is due to Hartman [3] and Grobman [2]. We give here a new proof of this result, which is related to the proof of the following

THEOREM. *If $f \in S(M)$ and* dim $M \leq 3$ *then f is structurally stable.*

The correspondent results hold for the vector fields in $\mathcal{S}(M)$.

In [11] it was shown that given any Riemannian metric on M, $\mathcal{S}(M) \cap G(M)$ is an open and dense subset of $G(M)$, where $G(M)$ is the set of all gradient vector fields on M. Thus in any manifold $\mathcal{S}(M)$ is far from being empty. Also if $X \in \mathcal{S}(M) \cap G(M)$ then its induced diffeomorphism at time $t = 1$ $X_{t=1} \in S(M)$. Using this and the above result we conclude that any manifold M with dim $M \leq 3$ has structurally stable diffeomorphisms and vector fields.

In the last section we analyze the problem of embedding a diffeomorphism of M in a topological flow. It is well known that there are diffeomorphisms near the identity map 1 on M that do not embed in smooth flows. We show here that in any neighborhood of 1 in Diff(M) there is an open set of diffeomorphisms that do not embed in topological flows.

This paper originated from the author's doctoral thesis at the University of California under the guidance of S. Smale. I wish to thank M. Hirsch, N. Kopell, I. Kupka, E. Lima, M. Peixoto, C. Pugh, M. Shub, J. Sotomayor, R. Thom and especially S. Smale for many helpful conversations.

§1

The main goal of this section is to prove that the set of Morse–Smale diffeomorphisms $S(M)$ is open in Diff(M). We consider a set $R(M) \supset S(M)$, which is open in Diff(M) and then we show that $S(M)$ is open in $R(M)$.

We start by discussing the phase-diagram of these elements of Diff(M). The basic reference is [10].

Let $f \in \text{Diff}(M)$ and P be a hyperbolic fixed point of f. We denote by $LS(P)$ and $LU(P)$ the local stable and unstable manifold of P. Let B^s be a cell neighborhood of P in $LS(P)$, such that $f(\partial B^s) \subset \text{Int } B^s$. The existence of such a cell B^s follows from the fact that $f/LS(P)$

is a contraction. The embedded annulus in $LS(P)$ whose boundaries are $\partial B^s, f(\partial B^s)$ is called a fundamental domain $G^s(P)$ of $W^s(P)$. We have $W^u(P) = \bigcup_{n\in\mathbf{Z}} f^n(G^s(P)) \cup \{P\}$. Any neighborhood $N^s(P)$ of $G^s(P)$ in M, disjoint from $LU(P)$, is called a fundamental neighborhood associated with $W^s(P)$.

Dually we can define $G^u(P)$ and $N^u(P)$.

If $P \in \mathrm{Per}(f)$ is hyperbolic with period $n > 1$, we can define $G^s(P)$, $N^s(P)$ as before by considering the map f^n for which P is fixed. In general, a description of f^n in a small neighborhood of P is transferred, via iteration by f, to small neighborhoods of the points in the orbit $\mathcal{O}(P)$. This observation will be used throughout this paper.

Let W_1, W_2 be submanifolds of M with i_1, i_2 being their inclusion maps. We say that W_2 is ε C^1-close to W_1 is there is a diffeomorphism $\gamma : W_1 \to W_2$ such that $|i_1 - i_2 \circ \gamma|_{C^1} < \varepsilon$.

(1.1) LEMMA (λ-lemma). *Suppose P is a hyperbolic fixed point for $f \in \mathrm{Diff}(M)$, dim $W^u(P) = r$, $0 < r < m$, and N is a 1–1 immersed manifold in M, invariant under f and having a point Q of transversal intersection with $W^s(P)$. For any given cell neighborhood B^r of P in $W^u(P)$ and any $\varepsilon > 0$, there exists an r-cell in N, ε C^1-close to B^r.*

Proof. First we notice that it is enough to prove the lemma when dim $N = r$, since in any case we can restrict the argument to a small r-cell in N transversal to $W^s(P)$ at Q and its iterates by f. Also, since $W^s(P)$ and N are invariant under f, we can reduce the question to a local one around P.

From the stable manifold theorem [3], there exists a neighborhood $V = LS(P) \times LU(P)$ of P, where f can be expressed by

$$f(x_1, x_2) = (L_s x_1 + X_1(x_1, x_2), L_u x_2 + X_2(x_1, x_2)) \tag{I}$$

where $(L_s, L_u) = (Df)_P$, $x_1 \in LS(P)$, $x_2 \in LU(P)$), and

$$\|L_s\| < a < 1,\ \|L_u^{-1}\| < \frac{1}{b} < 1,\ \frac{\partial X_1}{\partial x_2}\bigg|_{LU(P)} = \frac{\partial X_2}{\partial x_1}\bigg|_{LS(P)} = 0. \tag{II}$$

Let $k = \max_{C^1 V} \left\| \frac{\partial X_i}{\partial x_j} \right\|$, $1 \le i, j \le 2$. If V is small enough we can set

$$0 < k < 1,\ b_1 = b - k > 1,\ a_1 = a + k < 1,\ k < \frac{(b-1)^2}{4}. \tag{III}$$

As we pointed out above, we can assume $B^r \subset V$ and $Q \in V$. Let v_0 be any unit vector in $(TN)_Q$. With respect to the splitting $V = LS(P) \times LU(P)$, we can write $v_0 = (v_0{}^s, v_0{}^u)$. We call the inclination of v_0, λ_0 :

$$\lambda_0 = \frac{\|v_0{}^s\|}{\|v_0{}^u\|}, \qquad \|v_0{}^u\| \ne 0.$$

Set $Q_n = f^n(Q)$ and $v_n = (Df)_{Q_{n-1}}(v_{n-1})$. Using (II) and (III) we have

$$\lambda_n \le \frac{\lambda_0}{b_1{}^n} + k \sum_{i=1}^{n} \frac{1}{b_1{}^i}.$$

or

$$\lambda_n \le \frac{\lambda_0}{b_1{}^n} + \frac{k}{b_1 - 1} \tag{IV}$$

So there exists $n_0 \in Z_+$ such that

$$\lambda_n \le \frac{b_1 - 1}{4} \quad \text{for} \quad n \le n_0 . \tag{V}$$

Now, choose a product neighborhood $V_1(P) = \delta I^s \times B^r$, $\delta < \varepsilon$, $V_1 \subset V$ and in $Cl V_1$

$$k_1 = \max_{ClV_1} \left\| \frac{\partial X_1}{\partial x_2} \right\| < \min(\varepsilon, k).$$

Also, choose a small cell neighborhood D^r of $Q_{n_0} = f^{n_0}(Q)$ in N, such that the inclination of any unit vector in $(TN)_R$, $R \in D^r$, satisfies:

$$\lambda \le \frac{b_1 - 1}{2}. \tag{VI}$$

By considering the positive iterates of D^r restricted to V_1 and using the same notation as above, for $b_2 = \dfrac{b_1 + 1}{2}$ we get

$$\lambda_{n+n_0} \le \frac{\lambda_{n_0}}{b_2{}^n} + \frac{k_1}{b_2 - 1} \tag{VII}$$

which is an expression similar to (IV). If $n \in Z_+$ is large enough, we get

$$\lambda_{n+n_0} \le \varepsilon\left(1 + \frac{1}{b_2 - 1}\right)$$

in (VII), which implies the lemma.

We notice that the above estimates hold in Banach spaces. Also, the condition that P be hyperbolic can be somewhat relaxed if we consider the strong unstable and central stable manifolds of P instead of $W^u(P)$, $W^s(P)$ and require that the point Q has its positive orbit coverging to P.

As immediate consequences of lemma (1.1) we have:

(1.2) COROLLARY. *Let $P \in \text{Per}(f)$ be hyperbolic and let $N^s(P)$ be a fundamental neighborhood associate with $W^s(P)$. Then $\bigcup_{n \in z} f^n(N^s(P)) \supset V - LU(P)$, where V is a neighborhood of P in M.*

(1.3) COROLLARY. *Let $P_i \in \text{Per}(f)$, $1 \le i \le 3$, be all hyperbolic. If $W^u(P_1)$ and $W^s(P_2)$, $W^u(P_2)$ and $W^s(P_3)$ have points Q_1, $Q_2 \notin \text{Per}(f)$ of transversal intersection* then *$W^u(P_1)$ and $W^s(P_3)$ also have a point $Q_3 \notin \text{Per}(f)$ of transversal intersection.*

(1.4) COROLLARY. *Let $P \in \text{Per}(f)$ be hyperbolic. If $W^u(P) \cap W^s(P) \neq \mathcal{O}(P)$, where $\mathcal{O}(P)$ is the orbit of P, then $\Omega(f)$ is not finite.*

Proof. In fact $W^u(P) \cap W^s(P) \subset \Omega(f)$. So if $Q \in W^u(P) \cap W^s(P)$ and $q \notin \mathcal{O}(P)$ then $Q \notin \text{Per}(f)$ and $\bar{\mathcal{O}}(Q) \subset \Omega(f)$.

Unless otherwise stated, we will be considering in this section the subset $R(M)$ of Diff(M) defined by: $f \in R(M)$ iff

i) $\Omega(f)$ is finite (this implies $\Omega(f) = \text{Per}(f)$),
ii) all points in Per(f) are hyperbolic,
iii) for $x, y \in \text{Per}(f)$ if $W^s(x) \cap W^u(y) \neq \emptyset$ then $W^s(x)$ and $W^u(y)$ have a point of transversal intersection.

Clearly the set of Morse–Smale diffeomorphisms $S(M)$ is contained in $R(M)$. For $f \in R(M)$, since $\Omega(f)$ is finite we have $M = \bigcup_i W^s(P_i) = \bigcup_i W^u(P_i)$, $P_i \in \text{Per}(f)$. Consequently there is at least one sink and one source in Per(f).

(1.5) Lemma (Smale). *Let* $Cl\,W^u(Q) \cap W^u(P) \neq \emptyset$, $P, Q \in \text{Per}(f)$. *There is a sequence* $P_1, \ldots, P_n \in \text{Per}(f)$, $P_1 = P$, $P_n = Q$ *such that* $W^s(P_i) \cap W^u(P_{i+1}) \neq \emptyset$ *for* $1 \leq i \leq n-1$.

Proof. P can not be a source. If P is a sink it is enough to consider the pair P, Q. So let P be a saddle point. There is a point $x_1 \notin \mathcal{O}(P)$ such that $x_1 \in W^s(P) \cap Cl\,W^u(Q)$. Otherwise we can construct a fundamental neighborhood $N^s(P)$ so that $N^s(P) \cap Cl\,W^u(Q) = \emptyset$. Since $\bigcup_n f^n(N^s(P)) \cup LU(P)$, $n \in Z$, contains a neighborhood of P we have $Cl\,W^u(Q) \cap LU(P) = \emptyset$, contradicting the assumption. Now $x_1 \in W^u(P_2)$ for some $P_2 \in \text{Per}(f)$, so that $W^s(P) \cap W^u(P_2) \neq \emptyset$ and $Cl\,W^u(Q) \cap W^u(P_2) \neq \emptyset$. If $P_2 = Q$ we are done. If not we can repeat the argument and thus get a sequence $P_1, P_2 \ldots$ as above with $P_1 = P$. Notice that in this sequence $P_i \neq P_j$ for $i \neq j$ since otherwise $W^s(P_i) \cap W^u(P_i) \neq \mathcal{O}(P_i)$ and from (1.4) $\Omega(f)$ would not be finite. Since Per(f) is finite we must reach Q with this process and thus proving the lemma.

(1.6) Proposition (Smale). *Let* $f \in R(M)$. *Then for each* $P \in \text{Per}(f)$, $W^s(P)$ *and* $W^u(P)$ *are (embedded) submanifolds of* M. *The periodic orbits of* f *are partially ordered by*: $\mathcal{O}(P) \leq \mathcal{O}(Q)$ *iff* $Cl\,W^u(Q) \cap W^u(P) \neq \emptyset$.

Proof. Both statements follow from (1.3), (1.4) and (1.5), for they imply

$$Cl\,W^u(Q) \cap W^u(P) \neq \emptyset \text{ iff } W^u(Q) \cap W^s(P) \neq \emptyset.$$

Thus this partial order can equivalently be defined by: $\mathcal{O}(P) \leq \mathcal{O}(Q)$ iff $W^s(P) \cap W^u(Q) \neq \emptyset$. Consequently for each $P \in \text{Per}(f)$ there is a fundamental neighborhood $N^S(P)$ such that for $Q \in \text{Per}(f)$ if $W^s(P) \cap W^u(Q) = \emptyset$ then $N^s(P) \cap W^u(Q) = \emptyset$. From now on in this paper we will consider only such fundamental neighborhoods.

For $P, Q \in \text{Per}(f)$, a chain connecting Q to P in the phase-diagram of f is a sequence $P_0, \ldots, P_n$ with each $P_i \in \text{Per}(f)$, $P_i \notin \mathcal{O}(P_{i+1})$, $P_1 = P$ and $P_n = Q$, such that $W^s(P_i) \cap W^u(P_{i+1}) \neq \emptyset$. In this case n is called the length of the chain.

(1.7) *Definition.* For $P, Q \in \text{Per}(f)$, Q is said to have k behavior relative to P, beh$(Q \mid P) = k$, if the maximum length of chains connecting Q to P is k, $k \in Z_+$. We complete by setting beh$(Q \mid P) = 0$ if $W^s(P) \cap W^u(Q) = \emptyset$. Thus beh$(Q \mid P)$ indicates how P and Q are related in the phase-diagram of f.

Notice that if $\mathrm{beh}(Q|P) = 1$ and $N^s(P)$ is a closed fundamental neighborhood associate with $W^s(P)$ then $W^u(Q)$ is closed in $N^s(P)$. Also, since $W^u(P)$ has the induced topology, there is $n_0 \in Z_+$ such that for $n \geq n_0, f^n(G^u(Q)) \cap N^s(P) = \varnothing$, where $G^u(Q)$ is a fundamental domain of $W^u(Q)$.

Examples. We give now examples of Morse–Smale diffeomorphisms showing different types of behavior. The following pictures illustrate the main features of the diffeomorphisms.

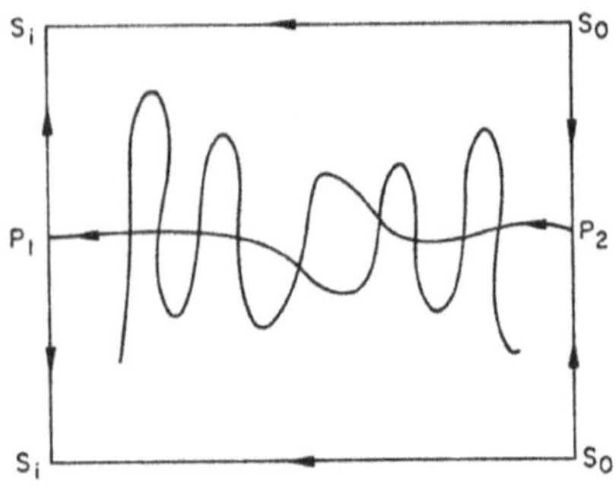

beh$(P_2/P_1) = 1$

Figure 1.

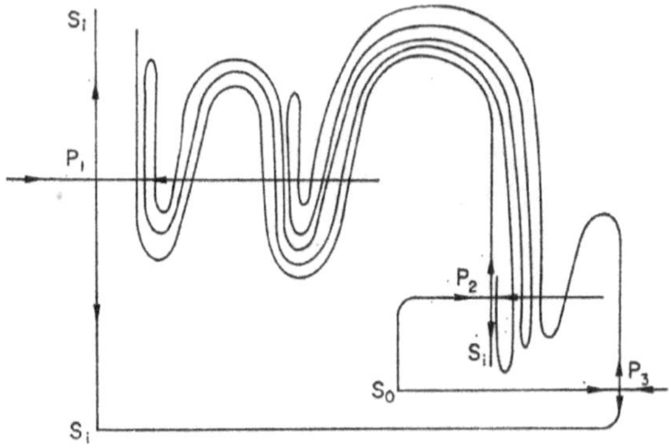

beh$(P_2/P_1) = 1$, beh$(P_3/P_1) = 2$

Figure 2.

Both examples can be completed to Morse–Smale diffeomorphisms of the sphere S^2. For the first, take four copies of the above and match them along the edges, forming the vertical faces of a cube. Then put a saddle point in center of the top and bottom faces. For the second example one has only to add a source outside the picture. Also, we can construct in any compact two dimensional manifold a Morse–Smale diffeomorphism with an arbitrarily high behavior $n \in Z_+$ in its phase-diagram as follows. We start with a Morse–Smale vector field X with no closed orbits. Let S be a sink for X and let D be a disc around S, ∂D transversal to X. We redefine X in D so that

i) $\Omega(X)$ is a finite union of critical points, all hyperbolic,

ii) X has in D $n+1$ saddles $P_0, \ldots, P_n$. For each P_i one of the components of $W^s(P_i) - P_i$ coincides with one of the components of $W^u(P_{i+1}) - P_{i+1}$.

Let f be the diffeomorphism induces by X at time $t = 1$. Taking fundamental neighborhoods $N^s(P_i)$, we can make a series of small C^r pertubations each having support in one of the $N^s(P_i)$, such that if g is the resulting diffeomorphism then $\Omega(g) = \Omega(f)$ and $W^s(P_i, g)$ and $W^u(P_{i+1,g})$ have a non-empty transversal intersection. If necessary, with a further C^r small perturbation outside D, we can set all the stable and unstable manifolds of $\Omega(g)$ in general position. We thus get a Morse–Smale diffeomorphism g_1 so that $P_0, P_n \in \Omega(g_1)$ and $\mathrm{beh}(P_n | P_0) = n$.

See [12] for the proof of the following:

(1.8) Lemma. *Let $P \in \mathrm{Per}(f)$ be hyperbolic of period n and let U be a neighborhood of P with no periodic points of period less than or equal to n. Let B^s, B^u be cell neighborhoods of P*

in $LS(P)$, $LU(P)$ and $B^s \times B^u \subset U$. There is a neighborhood $N(f)$ in Diff(M) *and a continuous map $\rho : N(f) \to$* Diff(M) *such that*

i) $\rho(f) = 1$,

ii) for each $g \in N(f)$, $P^* = \rho(g)(P)$ is a hyperbolic period point of g in U, P^* has the same period and stable index as P and is the only periodic point of g in U with period less than or equal to n,

iii) $\rho(g)(B^s)$, $\rho(g)(B^u)$ are cell neighborhoods of P^* in $LS(P^*)$, $LU(P^*)$.

When Per(f) is a finite hyperbolic, we get from (1.8) a well defined map $\rho(g) : \text{Per}(f) \to \text{Per}(g)$, $g \in N(f)$.

An element $f \in \text{Diff}(M)$ is called Ω-stable if there is a neighborhood $N(f)$ in Diff(M) such that for each $g \in N(f)$ the maps $f/\Omega(f)$ and $g/\Omega(g)$ are topologically conjugate.

(1.9) Theorem. *The set $R(M)$ is open in* Diff(M). *If $f \in R(M)$ then there is aood neighborh $N(f)$ such that for each $g \in N(f)$ the map $\rho(g) : \text{Per}(f) \to \text{Per}(g)$ defined above is a diagram isomorphism. In particular f is Ω-stable.*

Proof. Let $f \in R(M)$ and let $N_1(f)$ be such that the map $\rho(g) : \text{Per}(f) \to \text{Per}(g)$ is well defined in $N_1(f)$. We claim that there exist a neighborhood $N(f) \subset N_1(f)$ and a neighborhood V of the sinks, sources, stable and unstable manifolds of the saddles of f such that $\Omega(g) \cap V = \rho(g)(\text{Per}(f))$, for $g \in N(f)$. First we construct a neighborhood $N(f)$ and a neighborhood V^u of the sinks and unstable manifolds of the saddles of f such that $\Omega(g) \cap V^u = \rho(g)(\text{Per}(f) \cap V^u)$, for $g \in N(f)$. We proceed by induction on the phase-diagram of f.

For each sink S_i of f, choose a neighborhood $V_0(S_i) \subset W^s(S_i)$ and $\varepsilon_0(S_i) > 0$ such that if $|g - f|_{C^r} < \varepsilon_0(S_i)$ then $V_0 \subset W^s(S_i^*)$, where $S_i^* = \rho(g)(S_i)$. This follows from lemma (1.8). To simplify the notation let $V_0 = \bigcup_i V_0(S_i)$ and $\varepsilon_0 = \min\{\varepsilon_0(S_i) \mid S_i \text{ sink of } f\}$. In V_0 the condition $\Omega(g) \cap V_0 = \rho(g)(\text{Per}(f) \cap V_0)$ is readily verified.

Suppose now that we have constructed V_k, ε_k corresponding to the points in Per(f) whose behavior with respect to the sinks of f is less than or equal to k, so that $\Omega(g) \cap V_k = \rho(g)(\text{Per}(f) \cap V_k)$ for $|g - f|_{C^r} < \varepsilon_k$. Let P_k be a point next to these in the phase-diagram of f and let $N^u(P_{k+1})$ be a closed fundamental neighborhood. We assert that there exists $n_0 \in Z_+$ such that for each $x \in N^u(P_{k+1})$, $f^n(x) \in V_k$ for some n, $1 \le n \le n_0$. In fact, let P_k be next to P_{k+1} in the phase-diagram and take an open fundamental neighborhood $N^s(P_k) \subset V_k$. Since $W^s(P_k)$ is closed in $N^u(P_{k+1})$ there exists $n_1 \in Z_+$ such that $\bigcup_i f^{-1}(N^s(P_k))$, $1 \le {}_i \le n_1$, contains a neighborhood T_k of $W^s(P_k) \cap N^u(P_{k+1})$. Similar neighborhoods can be constructed for the stable manifolds of all points next to P_{k+1} in the phase-diagram. Let us denote the union of these neighborhoods simply by T_k. Let now P_{k-1} be such that beh$(P_{k+1} \mid P_{k-1}) = 2$. In $N^u(P_{k+1}) - T_k$, $W^s(P_{k-1})$ is closed. Thus we can repeat the argument for $W^s(P_{k-1})$ finding $n_2 \in Z_+$ as above and so on. Taking $n_0 = \max(n_i)$ the assertion is proved. Now let U be a neighborhood of P_{k+1} such that $U \supset N^u(P_{k+1})$ and for each $x \in U - LS(P_{k+1})$, $f^n(x) \in N^u(P_{k+1})$ for some $n \ge 0$. If $\varepsilon_{k+1} > 0$, is small enough the same is true for g, $|g - f|_{C^r} < \varepsilon_{k+1}$ and each $x \in U - LS(P_{k+1}{}^*)$. This follows from lemma (1.8)

Also for each $x \in N^u(P_{k+1})$, $g^n(x) \in V_k$ for some n, $1 \leq n \leq n_0$. The induction step of the assertion is proved by taking $V_{k+1} = V_k \bigcup_i f^n(N^u(P_{k+1})) \cup U$ for $1 \leq n \leq n_0$ and $\varepsilon_{k+1} < \varepsilon_k$. Let now $V^u = \bigcup_k V_k$ and $\varepsilon^u = \min_k \varepsilon_k$. Dually, we can construct V^s, ε^s. Our claim is proved by setting $V = V^s \cup V^u$ and $\varepsilon = \min(\varepsilon^s, \varepsilon^u)$.

Now, for g C^0 near f, $\Omega(g)/M - V = \emptyset$. In fact, $M - \operatorname{Int} V$ is compact and contained in the stable manifolds of the sinks of f.

Thus for g near f in Diff(M) the map $\rho(g): \operatorname{Per}(f) \to \operatorname{Per}(g)$ is onto. Moreover, the above construction shows that if $W^s(P_i) \cap W^u(P_j) = \emptyset$ then $W^s(P_i{}^*) \cap W^u(P_j{}^*) = \emptyset$. And since the property of $W^s(P_i)$ and $W^u(P_j)$ having a point of transversal intersection is clearly an open one in Diff(M), we conclude that $\rho(g)$ is a diagram isomorphism and thus the proof of the theorem.

(1.10) COROLLARY. *The set of Morse–Smale diffeomorphisms $S(M)$ is dense in $R(M)$.*

Proof. In [4], [12] it is shown that any diffeomorphism of M can be approximated by one having all of its periodic points hyperbolic and their stable and unstable manifolds in general position. Since $R(M)$ is open, (1.10) follows.

(1.11) LEMMA. *Let f be a Morse–Smale diffeomorphism of M and let $P \in \operatorname{Per}(f)$ be of unstable index r. Fix a cell neighborhood B^r of P in $LU(P)$. There exists a neighborhood V of P, $V = \delta I^{m-r} \times B^r$, and a neighborhood $N(f)$ in* Diff(M) *such that for any $g \in N(f)$, $W^u(P_i{}^*) \cap V$, with the fibers C^1 close to B^r. As usual, $P_i{}^* = \rho(g)(P_i)$, ρ being the map defined in lemma* (1.8).

Proof. First we prove the assertion for the invariant manifolds of Per(f). We proceed by induction on the phase-diagram of f. If P is next to a source, there is nothing to be proved. Suppose the assertion is true for all periodic points P_i such that $W^u(P_i) \cap W^s(P) \neq \emptyset$, $P_i \notin \mathcal{O}(P)$. Let $N^s(P)$ be a closed fundamental neighborhood associate with $W^s(P)$ and let $P_1 \in \operatorname{Per}(f)$ be such that $\operatorname{beh}(P_1 \mid P) = 1$. So $L_1 = W^u(P_1) \cap N^s(P) \cap W^s$ is compact. From the transversality assumption, the normal bundle of L_1, restricted to a small neighbohood V_1 of L_1 in $W^u(P_1)$, yields a r-cell fiber bundle of $W^u(P_1) \cap V_1$ over L_1, the cells having maximal inclination say less than λ_1. The other periodic points of f that have 1 behavior relative to P are treated similarly and they will be denoted simply by P_1. Let now $P_2 \in \operatorname{Per}(f)$ be such that $\operatorname{beh}(P_2 \mid P) = 2$. Since from the induction assumption, the lemma was supposed to be true for P_1, we get immediately that for a small tubular neighborhood T_1 of $W^u(P_1)$ in $N^s(P)$, $W^u(P_2) \cap T_1$ also fibers over its intersection with $W^s(P)$, with the fibers having maximal inclination less than $2\lambda_1$. In $N^s(P) - \operatorname{Int} T_1$, $W^u(P_2)$ is compact, so we can repeat the argument, getting eventually a new fundamental neighborhood $N^s(P)$ and a new upper bound for the inclination of the fibers, say λ_2 and so on until we exhaust all the periodic points P_i such that $W^u(P_i) \cap W^s(P) \neq \emptyset$. Applying now lemma (1.1), we finish the proof for Per(f).

The result can be extended to a small neighborhood of f in Diff(M) in view of theorem (1.9), lemma (1.8) and the uniform estimates of lemma (1.1).

From theorem (1.10) and lemma (1.11) we have

(1.12) THEOREM. *The set of Morse–Smale diffeomorphisms of M is open in* Diff(M). *Moreover, if f is Morse–Smale, then its phase-diagram is stable (up a diagram isomorphism) under small C^r perturbations.*

§2

Here we discuss the (local) stability of $f \in \text{Diff}(M)$ in a neighborhood of a point $P \in \text{Per}(f)$ which is hyperbolic. Then we prove the (global) stability of the Morse–Smale diffeomorphisms when $\dim M \leq 3$.

(2.1) THEOREM. *Let $f: U(P) \to M$ be a local C^r diffeomorphism, $r \geq 1$ and $U(P)$ a neighborhood of P in M, having P as a hyperbolic fixed point. There exists a local homeomorphism $h: V \to M$ such that $hL(x) = fh(x)$ for all $x \in V$, where $L = (Df)_P$ and V is a neighborhood of the origin in the tangent space of M at P.*

Notice that, by taking a neighborhood chart of P in M and making the obvious identifications, the question can be reduced to a similar one in R^m, where $m = \dim M$. The extension of (2.1) to the case when P is a hyperbolic periodic point is immediate.

The result is due to Hartman [3] and Grobman [2].

In order to prove theorem (2.1) we need the following:

(2.2) LEMMA. *For any given $\varepsilon > 0$, there is a neighborhood $B^s \times B^u$ of P, $B^s \subset LS(P)$ and $B^u \subset LU(P)$, and a diffeomorphism $h_s: V \to LS(P) \times LU(P)$, $V \supset B^s \times B^u$, such that*

i) $|h^s - Id|_{C^1} < \varepsilon$,
ii) $h^s = Id$, $h_s \circ f = L$ *on* $\partial b B^s \times B^u$.

Here Id stands for the identity map.

Proof. Fix a cell neighborhood $B_1(P)$ and let $C^1(B_1, R^m)$ be the set of C^1 maps of B_1 into R^m with the uniform C^1 topology. Let $U_1 = B_1 \cap L(B_1)$ and U_2 be the complement of a neighborhood of U_1 in R^m, so that we can choose $B_1{}^s \times B_2{}^u \subset B_1 \cap L^{-1}(B_1)$ with $\partial B_1{}^s \times B_1{}^u \subset U_2$. Let $\alpha: R^m \to R$ be a C^∞ function with uniformly bounded derivatives and $\alpha/U_1 = 1$, $\alpha/U_2 = 0$ and let $\gamma_n: R^m \to R^m$ be the map $\gamma_n(z) = \dfrac{1}{n} z$, $n \in Z_+$. Setting $f = L + \Phi$, define for each $n \in Z_+$ $h_n = \alpha k_n + (1-\alpha)Id: B_1 \to R^m$, where $k_n = \text{Id} + \Phi \circ L^{-1} \circ \gamma_n: B_1 \to R^m$. For large n, h_n is near Id in $C^1(B_1, R^m)$ and thus a diffeomorphism of B_1 into R^m. Finally, let $h^*: B \to R^m$ be defined by $h^*(z) = \gamma_n\, h_n(nz)$, where $B = \gamma_n(B_1)$. For large n, $|h^* - Id|_{C^1} < \varepsilon$ and $|(h^*)^{-1} - Id|_{C^1} < \varepsilon$. One verifies directly that on $\partial B^s \times B^u$, $h^* = Id$ and $h^* \circ L = f$, where $B^s = \gamma_n(B_1{}^s)$ and $B^u = \gamma_n(B_1{}^u)$. Setting $h_s = (h^*)^{-1}$, the lemma is proved. Notice that $B^s \subset LS(P)$ and $B^u \subset LU(P)$ are invariant under h_s and h_s is C^r.

Proof of (2.1). Let $U = B^s \times B^u$, h_s as in lemma (2.2) for some $\varepsilon > 0$. We construct a homeomorphism φ of U such that

i) φ is C^r off B^s and B^u,
ii) $\varphi/B^s = Id$ and $\varphi/B^u = Id$,
iii) $\varphi \circ f \circ \varphi^{-1} = f_s \times f_u$, where $f_s = f/B^s$ and $f_u = f/B^u$.

From this the theorem will follow easily.

Let us construct $\varphi = (\varphi_s, \varphi_u)$, where $\varphi_s = \pi_s \circ \varphi$, $\varphi_u = \pi_u \circ \varphi$ and π_s, π_u are the projections $B^s \times B^u \to B^s$, $B^s \times B^u \to B^u$. Let $D^s = \partial B^s \times B^u$. Since $h_s \circ f = L$ on D^s and $L_s = L/B^s$ is a contraction we have $f(D^s) \cap D^s = \emptyset$ and $f(D^s)$ separates $D^s \times D^u$ into two disjoint components. Thus we can define

$V_0 =$ the region of $B^s \times B^u$ bounded by D^s and $f(D^s)$, and by induction

$V_n =$ the region of $B^s \times B^u$ bounded by $f^n(D^s)$ and $f^{n+1}(D^s)$.

Clearly, $\bigcup_{n=0}^{\infty} V_n = B^s \times B^u - \{P\} \times B^u$. We define $\varphi_s/V_0 = hs^{-1} \circ \pi_s \circ h_s$, $\varphi_s/V_n = f^n \circ \varphi_s \circ f^{-n}$ and $\varphi_s(B^u) = P$. φ_s is well defined. To prove this it is enough to assure that φ_s is well defined on the boundaries of the V_n. This follows from the fact that on D^s, $h_s = Id$ and $h_s \circ f = L$. Dually, we can define φ_u by considering f^{-1} and a diffeomorphism h_u similar to the h_s of lemma (2.2). From its construction it is immediate that φ satisfies conditions (i) to (iii) above. It remains to show that φ is a homeomorphism.

φ is one to one and onto $B^s \times B^u$. Taking ε, used to define h_s and h_u in lemma (2.2), small enough it follows from lemma (1.1) that for each $z \in B^s$, $\varphi_s^{-1}(z)$ is C^1 close to B^s and for each $z_1 \in B^u$, $\varphi_u^{-1}(z_1)$ is C^1 close to B^s. Since B^s and B^u intersects in a unique point, the same is true for $\varphi_s^{-1}(z)$, $\varphi_u^{-1}(z_1)$ for $z \in B^s$, $z_1 \in B^u$. This implies the assertion.

φ is continuous. From its definition φ is C^r except possibly at B^s, B^u. Let us check the continuity of φ on $B^u - \{P\}$. Let $z_n \to z \in B^u - \{P\}$. Since φ_u is clearly continuous on $B^u - \{P\}$, $\varphi_u(z_n) \to \varphi_u(z) = z$. So it is enough to show that $|\varphi_s(z_n)| \to 0$. For that we can assume $z_n \notin B^u$ and thus for each z_n there is $n_i > 0$ and $y_{n_i} \in V_0$ such that $f^{n_i}(y_{n_i}) = z_n$. So $\varphi_s(z_n) = f^{n_i}\varphi_s(y_{n_i})$ and from $z_n \to z \in B^u$ we have $n_i \to \infty$. The assertion now follows from the fact that f/B^s is a contraction. The proof of the continuity of φ on B^s is similar.

To finish the proof of the theorem, we notice that $f_s = f/B^s$ and $L_s = L/B^s$ are contractions. Moreover, if $G^s(f)$ and $G^s(L)$ are the fundamental domains with boundaries ∂B^s, $f_s(\partial B^s)$ and ∂B^s, $L_s(\partial B^s)$ respectively, then $h_s : G^s(f) \to G^s(L)$ satisfies $h_s f_s(z) = L_s h_s(z)$ for $z \in \partial B^s$. Thus we can redefine h_s in all of B^s by $h_s(z) = L_s{}^n h_s f_s{}^{-n}(z)$ for $f_s{}^{-n}(z) \in G^s(f)$ and $h_s(P) = P$. Similarly, we can redefine h_u in B^u so that $h_u f_u = L_u h_u$. Setting $h = h_s \times h_u$ and using that $\varphi f \varphi^{-1} = f_s \times f_u$ we get $h\varphi f \varphi^{-1} = Lh$ on $B^s \times B^u$. This shows that $h\varphi$ is a conjugacy between f and L on $B^s \times B^u$. The theorem is proved.

An immediate consequence of theorem (2.1) and lemma (1.8) is

(2.3) Theorem. (Local Stability). *Let $P \in$ Per(f) be hyperbolic. There exist a neighborhood V of P and a neighborhood $N(f)$ in* Diff(M) *such that for each $g \in N(f)$ there is a local homeomorphism $h : V \to M$ satisfying $hf = gh$ on V.*

We now pursue a global version of theorem (2.3) for a Morse–Smale diffeomorphism.

All submanifolds of M considered here will have the induced topology.

(2.4) *Definition.* Let P be a fixed point of f. A neighborhood $T^u(P)$ is called a $\mathscr{F}$-neighborhood if T^u is trivially fibered over $B^s = T^u \cap W^s(P)$, B^s a cell neighborhood of P in $W^s(P)$, the typical fiber being $B^u = T^u \cap W^u(P)$, B^u a cell neighborhood of P in $W^u(P)$. Moreover, if π is the projection $\pi : T^u \to B^s$ then

i) T^u is f-invariant, which means $f(\pi^{-1}(y)) \supset \pi^{-1}(f(y))$ for $y \in B^s$,
ii) for each $y \in B^s$, $\pi^{-1}(y)$ is a C^r submanifold of M, π is continuous on T^u and C^r on each fiber $\pi^{-1}(y)$,
iii) the fibers $\pi^{-1}(y)$, $y \in B^s$, are C^1 uniformly close to B^u in the following sense: for any $\varepsilon > 0$ there is a neighborhood $U(P) \subset B^s$ so that $\pi^{-1}(y)$ is ε C^1 close to B^u for all $y \in U$.

We denote the space of fibers $\{\pi^{-1}(y) \mid y \in B^s\}$ above by $\mathscr{F}^u$.

As usual, if P is periodic of period n, then we can define a $\mathscr{F}$-neighborhood of the orbit of P via itteration by f of a $\mathscr{F}$-neighborhood of P for f^n.

(2.5) *Remark.* Let $T^u(P)$ be a $\mathscr{F}$-neighborhood of P, $\pi : T^u \to B^s = T^u \cap W^s(P)$ being its projection map. If $B_1{}^s \subset B^s$ is a cell-neighborhood of P in $W^s(P)$, then the restriction $\pi : \pi^{-1}(B_1{}^s) \to B_1{}^s$ defines a $\mathscr{F}$-neighborhood of P. If $B_1{}^u \supset B^u = T^u \cap W^u(P)$ is a cell neighborhood of P in $W^u(P)$, then via itteration by f we get a $\mathscr{F}$-neighborhood $T_1{}^u(P)$ such that $T_1{}^u \cap W^u(P) = B_1{}^u$.

(2.6) *Definition.* A system of $\mathscr{F}$-neighborhoods $\{T^u(P) \mid P \in \text{Per}(f)\}$ is called admissible if

i) for each $P \in \text{Per}(f)$, there exists a fundamental neighborhood $G^s(P) \subset T^u(P)$ such that if $W^u(P_i) \cap W^s(P) \neq \varnothing$, i.e., $\text{beh}(P_i \mid P) > 0$, then $W^u(P_i) \cap G^s(P) \subset \bigcup_k T^u(P_k)$, the P_k being such that $\text{beh}(P_i \mid P_k) \geq 0$,
ii) for $e_i \in \mathscr{F}^u(P_i)$ and $e_j \in \mathscr{F}^u(P_j)$, if $e_i \cap e_j \neq \varnothing$ then $e_i \subset e_j$ or vice-versa.

(2.7) Lemma. (Stability Lemma). *Let $f \in S(M)$. Then f is structurally stable if there exists a neighborhood $N(f)$ in* Diff(M) *such that*

1) *for each $g \in N(f)$, the map $\rho(g) : \text{Per}(f) \to \text{Per}(g)$ defined in* (1.8) *is a diagram isomorphism,*
2) *each $g \in N(f)$ has an admissible system of $\mathscr{F}$-neighborhoods, $\{T^u(P^*) \mid P^* \in \text{Per}(g)\}$, with their space of fibers denoted by $\mathscr{F}^u(P^*)$ and $P^* = \rho(g)(P)$, $P \in \text{Per}(f)$,*
3) *for each $P_i \in \text{Per}(f)$ there exists a conjugacy $h_i : LS(P_i) \to LS(P_i^*)$ between $f/LS(P)$ and $g/LS(P_i^*)$ such that if $h_i^* : \mathscr{F}^u(P_i) \to \mathscr{F}^u(P_i)$ is the map induced by h_i on $\mathscr{F}^u(P_i)$ then h_i^* is admissible, i.e. for $e_i \in \mathscr{F}^u(P_i)$ and $e_j \in \mathscr{F}^u(P_j)$ if $e_i \subset e_j$ then $h_i^*(e_i) \subset h_j^*(e_j)$.*

Proof. Define $H(z) = g^{-n}h_i f^n(z)$ for $f^n(z) \in LS(P_i)$. Since $M = \bigcup_{i,n} f^n(LS(P_i))$, H is a well defined map $H : M \to M$, with an inverse $H^{-1}(z) = f^{-n}h_i^{-1}g^n(z)$ for $g^n(z) \in LS(P_i^*)$ and $Hf = gH$ on M. Thus, if H and H^{-1} are continuous then H is a conjugacy between f and g for $g \in N(f)$ and so f is structurally stable. Let us show that H is continuous. It is enough to show that H is continuous at $W^s(P_i)$ in a neighborhood of P_i, for each $P_i \in \text{Per}(f)$. So let $z_n \to z$, $z_n \in T^u(P_i)$ and $z \in T^u(P_i) \cap LS(P_i)$. Let $z_n \in e_n \in \mathscr{F}^u(P_i)$ and $z \in e \in \mathscr{F}^u(P_i)$. From condition (3) above, $H(z_n) \in h_i^*(e_n)$. Also from (3) we have that the set of accumulation points of $\{H(z_n)\}$ is contained in $LS(P_i^*)$. Thus, $H(z_n) \to h_i^*(e) \cap LS(P_i^*) = h_i(z) = H(z)$, proving the assertion. Similarly, H^{-1} is continuous and so the lemma is proved.

In the final part of this section we show that if $f \in S(M)$ and dim $M \leq 3$ then the conditions of lemma (2.7) can be fulfilled and thus f is structurally stable.

The next proposition was shown to me by M. Shub from lectures of A. Haefliger. See [1] for the concepts appearing here.

(2.8) PROPOSITION. *Let N be a compact manifold possibly with corners, A be a submanifold of ∂N and G be a compact manifold with boundary. Consider an embedding $J: A \to \partial G$ and let $\mathrm{Emb}_{J,A}(N,G)$ be the subset of embeddings of N in G such that $g \in \mathrm{Emb}_{J,A}(N,G)$ iff $g^{-1}(\partial N) = A$, $g/A = J$ and g is transversal to ∂G on A. Let $\mathrm{Diff}_{\partial G}(G)$ be the subset of $\mathrm{Diff}(G)$ whose elements are the identity map on ∂G. For any given $g \in \mathrm{Emb}_{J,A}(N,G)$ the map $\eta : \mathrm{Diff}_{\partial G}(G) \to \mathrm{Emb}_{J,A}(N,G)$, defined by $\eta(H) = H \circ g$, is a local trivial fibration.*

The proof of (2.8) follows essentially from [6] (see also [5]) and a lemma due to Seeley [9].

If N in (2.8) is a submanifold of G and φ_t is an isotopy of N leaving fixed all points in $N \cap \partial G$ then φ_t can be extended to an isotopy of G leaving fixed all points in ∂G and outside a compact neighborhood of N.

(2.9) LEMMA. *Let $f \in S(M)$ and dim $M \leq 3$. There exists a system $\{T^u(P) \mid P \in \mathrm{Per}(f)\}$ of $\mathscr{F}$-neighborhoods such that*

1) *$\{T^u(P) \mid P \in \mathrm{Per}(f)\}$ is admissible,*
2) *there exist fundamental domains $G^s(P)$ in $T^u(P)$ and diffeomorphisms $\psi_i : T^u(P_i) \cap V_i \to (W^u(P_i) \cap V_i) \times \delta_i I^{s_i}$, where $V_i = G^s(P) - \bigcup_k \mathrm{Int}\, T^u(P_k)$, $\mathrm{beh}(P_i \mid P_k) > 0$, $\delta_i > 0$ and $s_i = \dim W^s(P_i)$. Moreover, if $x \in W^u(P_i) \cap V_i \cap \partial T^u(P_k)$ then $\psi_k \psi_i^{-1}(x \times \delta_i I^{s_i}) \subset y \times \delta_k I^{s_k}$ for some $y \in W^u(P_k)$. If $x \in W^u(P_i) \cap V_i \cap \partial G^s(P)$ then $\psi_i^{-1}(x \times \delta_i I^{s_i}) \subset \partial G^s(P)$.*

Notice that (2) above express the triviality of the normal bundle of $W^u(P_i) \cap V_i$ in $G^s(P)$ and how these normal bundles fit together.

Proof. We use induction on the phase-diagram of f. For the sources of f, we simply take cells in their unstable manifolds to be their $\mathscr{F}$-neighborhoods.

Let $P \in \mathrm{Per}(f)$ and $T^u(P_i)$ constructed satisfying (1) and (2) above for all $P_i \in \mathrm{Per}(f)$ such that $\mathrm{beh}(P_i \mid P) > 0$.

We first assume dim $W^s(P) = 2$ and dim $M = 3$, which turns out to be the most interesting case. Consider $G^s(P) \subset LS(P)$. From (2.5) we may assume $UT^u(P_i)$ to contain a neighborhood of $\cup W^u(P_i) \cap G^s(P)$. We show first that we may take $G^s(P)$ so that condition (2) is satisfied. Then we construct an admissible $\mathscr{F}$-neighborhood $T^u(P) \supset G^s(P)$.

Let P_1 be next to P in the phase-diagram of f, i.e. $\mathrm{beh}(P_1 \mid P) = 1$. We have two cases:

a) dim $W^u(P_1) = 1$. In this case we can take $T^u(P_1) \cap \partial G^s(P) = \varnothing$, since $W^u(P_1) \cap G^s(P)$ is made up of a finite number of points. Condition (2) is then trivially true.

b) dim $W^u(P_1) = 2$. We can take $\partial G^s(P)$ to be transversal to $L_1 = W^u(P_1) \cap G^s(P)$ L_1. is a compact one dimensional submanifold of $G^s(P)$, $\partial L_1 \subset \partial G^s(P)$. Each component of L_1 is diffeomorphic either to I^1 or S^1. Thus there is a diffeomorphism $\psi_1 : T^u(P_1) \cap G^s(P) \to$

$L_1 \times \delta_1 I^1$ for some $\delta_1 > 0$, such that if $x \in W^u(P_1) \cap \partial G^s(P)$ then $\psi_1^{-1}(x \times \delta_1 I^1) \subset \partial G^s(P)$ as required by condition (2).

For the other periodic points next to P in the phase-diagram we proceed as in (a) or (b) above. In fact, their $\mathscr{F}$-neighborhoods can be taken to be pairwise disjoint.

Let P_2 be such that $\text{beh}(P_2 | P) = 2$. When dim $W^u(P_2) = 1$ we can take $T^u(P_2) \cap T^u(P_k) = \varnothing$ for $\text{beh}(P_2 | P_k) \geq 0$. When dim $W^u(P_2) = 2$ then $L_2 = W^u(P_2) \cap V_2$, where $V_2 = G^s(P) - \bigcup_k \text{Int } T^u(P_k)$ for $\text{beh}(P_2 | P_k) \geq 0$, is a compact one dimensional submanifold of $G^s(P)$. In either case we are reduced to a situation similar to (a) or (b) above. The same is true for points with higher behavior relative to P, which implies condition (2) of the lemma.

We now construct an admissible $T^u(P) \supset G^s(P)$. We may assume P to be a fixed point of f, for otherwise we proceed as in the definition of a $\mathscr{F}$-neighborhood of a periodic point. Let $\partial G^s(P) = S_E \cup S_I$, $f(S_E) = S_I$. First we notice that a neighborhood U of $G^s(P)$ can be fibered over $U \cap LS(P)$, such that if π_1 is the projection $\pi_1 : U \to U \cap LS(P)$, then π_1 is C^r and $\pi_1 f = f\pi_1$ in a neighborhood of S_E. This is done as follows. Let $\pi_s : LS(P) \times LU(P) \to LS(P)$ be the natural projection, $C_E = G^s(P) \times \delta I^1$ for some $\delta > 0$ and $C_I = f(C_E)$. Let φ_t be the C^r isotopy in a neighborhood U_1 of C_I, which is parallel to $LS(P)$, it takes at time $t = 1$, $f(\pi_s^{-1}(f^{-1}(y)))$ into $\pi_s^{-1}(y)$ for $y \in U_1 \cap LS(P)$ and it is defined proportional to arc length. φ_t can be extended to all of M leaving fixed all points in $LS(P)$ and outside a neighborhood of C_I. Now define $\pi_1 = \pi_s \circ \varphi_1$ in some neighborhood of $G^s(P)$.

From the C^r product structure given by the ψ_i of condition (2), we get a topological product $\psi_i : T^u(P_i) \cap V_i \to (W^u(P_i) \cap V_i) \times \delta_i I^{s_i}$, such that for $t \in \delta_i I^{s_i}$, $\psi_i^{-1}(W^u(P_i) \times t) \subset e_t$ for some $e_t \in \mathscr{F}^u(P_i)$. This is a consequence of the fact that in $T^u(P_i)$ the fibers $e_t \in \mathscr{F}^u(P_i)$ are C^1 close to $W^u(P_i)$. Thus in some neighborhood of $G^s(P)$, $N = \bigcup_i T^u(P_i)$, P_i periodic saddles of f, is a fiber bundle over $N \cap LS(P)$ with typical fiber I^1. If $\pi : N \to N \cap LS(P)$ is its projection map then, from condition (2) and f-invariances of $\mathscr{F}^u(P_i)$, $\pi f = f\pi$ in a neighborhood of C_E. We claim that there exists a neighborhood U_0 of $G^s(P)$, which fibers over $U_0 \cap LS(P)$ with projection map $\pi_0 : U_0 \to U_0 \cap LS(P)$ such that $\pi_0/N = \pi$, π_0 is C^r off N and $\pi_0 f = f\pi_0$ in some neighborhood of C_E. We proceed as follows. ∂N is a two dimensional submanifold of M possibly with corners. But from condition (2) above these corners do not occur in a neighborhood of $\partial G^s(P)$. We define an isotopy ϕt of ∂N in a neighborhood of C_E such that ϕ_t is parallel to $LS(P)$, it takes at time $t = 1$ $\pi^{-1}(y)$ into $\pi_1^{-1}(y)$ for $y \in \partial N \cap LS(P)$ and it is defined proportional to arc length. ϕ_t can be extended to a neighborhood of C_E. This induces a isotopy in a neighborhood of C_I, defined by $f\phi_t f^{-1}$. The union of these isotopies can now be extended to all of M, leaving fixed all points in $LS(P)$. Define $\pi_2 = \pi_1 \circ \phi_1 : U_2 \to U_2 \cap LS(P)$, where U_2 is a neighborhood of $G^s(P)$. Clearly $\pi_2 f = f\pi_2$ near C_E. Finally, define an isotopy Φ_t of all ∂N similar to the previous ones, but taking at time $t = 1$ $\pi^{-1}(y)$ into $\pi_2^{-1}(y)$ for $y \in \partial N \cap LS(P)$. Since $\pi_2/\partial N = \pi$ near $\partial G^s(P)$, Φ_t leaves fixed all points of ∂N near $\partial G^s(P)$. Then from proposition (2.8), Φ_t can be extended to all of M, leaving fixed all points in $LS(P)$ and in a neighborhood of $\partial G^s(P)$. Thus we get a projection $\pi_3 = \pi_2 \circ \Phi_1$, which extends $\pi/\partial N$ and $\pi_3 f = f\pi_3$ near C_E. Our assertion is proved by setting $\pi_0 = \pi/N \cup \pi_3/U_0 - N$, where U_0 is some neighborhood of $G^s(P)$.

We now redefine π_0 in a neighborhood of P by setting $\pi_0^{-1}(y) = f^n \pi_0^{-1}(f^{-n}(y))$ for $f^{-n}(y) \in G^s(P)$. Using lemma (1.1), we have that π_0 defines an $\mathscr{F}$-neighborhood of P which is admissible.

When dim $W^s(P) = 1$, the induction step in proving conditions (1) and (2) of the lemma is much simpler. We can take all $T^u(P_i)$, P_i periodic saddles of f with beh$(P_i | P) > 0$, pairwise disjoint and disjoint from $\partial G^s(P)$. Thus condition (2) is trivially true. For condition (1) we proceed as in the previous case. Using the same notation, we notice that here the extension of the projection $\pi : N \to N \cap LS(P)$ is much easier.

To conclude the lemma we must prove condition (2) when dim $M = 3$ and P is a sink. If dim $W^u(P_i) = 1$, we can take $\partial G^s(P)$ transversal to $W^u(P_i)$ and condition (2) is immediate. If dim $W^u(P_i) = 2$, we take $\partial G^s(P)$ transversal to $W^u(P_i)$ and to $W^u(P_i) \cap \partial T^u(P_k)$ for beh$(P_i | P_k) > 0$. Thus $L_2 = W^u(P_i) \cap V_i$ is a compact two dimensional submanifold of $G^s(P)$ possibly with corners. Via iteration by f^{-1}, the question can be considered in a fundamental domain $G^u(P_i) \subset LU(P_i)$. Condition (2) can then be verified in the same way as condition (1) for the case dim $W^s(P) = 2$ and dim $M = 3$, treated above. This finishes the proof of the lemma.

(2.10) *Remark.* Let $\{T^u(P_i)\}$ be an admissible system of $\mathscr{F}$-neighborhoods for Per(f), with projection maps $\pi_i : T^u(P_i) \to T^u(P_i) \cap W^s(P_i)$, as constructed in lemma (2.9). Using lemmas (1.1) and (1.8) at each stage of this construction in (2.9), we get for each g in some neighborhood $N(f)$ in Diff(M) an admissible system $\{T^u(P_i^*)\}$ of $\mathscr{F}$-neighborhoods for Per (g), with projection maps $\pi_i^* : T^u(P_i^*) \to T^u(P_i^*) \cap W^s(P_i^*)$ such that

1) $\partial T^u(P_i^*) = \partial T^u(P_i)$ for each index i,
2) π_i^* is C^r close to π on $U \cap \partial(\bigcup_j T^u(P_j))$, where the P_j are saddle points with beh$(P_j | P_i) > 0$ and U is a neighborhood of a fundamental domain $G^s(P_i) \subset T^u(P_i)$.

(2.11) Lemma. *Let P be a fixed point of f and $G^s(P)$ be a fundamental domain of $W^s(P)$, $\partial G^s(P) = S_1 \cup S_E$ with $S_I = f(S_E)$. Let N be a compact submanifold of $G^s(P)$ possibly with corners, such that N is f-invariant, i.e. $N \cap S_I = f(N \cap S_E)$, N is transversal to $G^s(P)$, ∂N separates $G^s(P)$ in two disjoint components and $\partial N \cap S_E$ is disjoint from $f^{-1}(\partial N \cap S_I)$. Suppose that for each g in some neighborhood of f in* Diff(M) *we have a homeomorphism h_g of N into $G^s(P^*)$ such that $h_g f = g h_g$ on $\partial N \cap S_E$ and the correspondence $g \to h_g | \partial N$ is continuous at f and $h_f = 1$ on ∂N. Then for g near f in* Diff(M), *h_g can be extended to a homeomorphism H_g of $G^s(P)$ onto $G^s(P^*)$ such that $H_g f = g H_g$ on S_E, H_g being C^r small in $G^s(P) - N$.*

Here, we assume $\partial N \cap S_E$ and $f^{-1}(\partial N \cap S_I)$ disjoint because N might not be f^{-1}-invariant. The same result holds for $P \in$ Per(f) : if n is the period of P, one has to consider the map f^n.

Proof. From lemma (1.8) we have a diffeomorphism $\rho : G^s(P) \to G^s(P^*)$, C^r small if g is near f in Diff(M). Let i_E, i_I be the inclusions of $\partial N \cap S_E$, $\partial N \cap S_I$ in $G^s(P)$. Then i_E and $\rho^{-1} h_g$, $f^{-1} i_I$ and $\rho^{-1} g^{-1} h_g i_1$ are C^r close embeddings of $\partial N \cap S_E$, $\partial N \cap S_I$ in S_E. From (2.8) there is a C^r small diffeormorphism η_E of $LS(P)$ such that $\eta_E i_E = \rho^{-1} h_g i_E$, $\eta_E f^{-1} i_I = \rho^{-1} g^{-1} h_g i_I$ and η_E leaves fixed all points outside a small neighborhood of S_E. Consider now the two close embeddings fi_E and $\rho^{-1}g\ \rho\ \eta_E i_E$ of S_E in $LS(P)$. There is a C^r diffeomorphism η_I of $LS(P)$ such that $\eta_I f i_E = \rho^{-1}g\ \eta_E$ and η_I leaves fixed all points outside a small neighborhood of S_I. Let $\eta = \eta_I\eta_E$. We have $\rho\eta f = g\rho\eta$ on S_E and $\rho\eta = h_g$ on $\partial N \cap \partial G^s(P)$. Finally consider the C^r close embeddings η and $\rho^{-1}h_g$ of ∂N in $G^s(P)$. Since ∂N is transversal to $\partial G^s(P)$ and $\eta = \rho^{-1}h_g$ on $\partial N \cap \partial G^s(P)$, there is from (2.8) a diffeomorphism η^* of $G^s(P)$ such that $\eta^*\eta = \rho^{-1}h_g$ on ∂_N and η^* fixes all points in $\partial G^s(P)$. Therefore, $H_g = h_g/N \cup \rho\eta^*/G^s(P) - N$ satisfies all the required conditions. The lemma is proved.

(2.12) THEOREM. *A Morse–Smale diffeomorphism f of M, $\dim M \leq 3$, is structurally stable.*

Proof. It is enough to show that conditions (1) to (3) of lemma (2.7) can be fulfilled. Condition (1) follows from (1.9): for g near f in Diff(M) we have a diagram isomorphism $\rho(g)\colon \mathrm{Per}(f) \to \mathrm{Per}(g)$. As before, we denote $P^* = \rho(g)(P)$.

Let $\{T^u(P_i)\}$, $\{T^u(P_i^*)\}$ and $G^s(P)$ be defined as in (2.10). We now construct homeomorphisms $h_i\colon LS(P_i) \to LS(P_i^*)$ satisfying condition (3) of (2.7). We will require also $h_i/G^s(P) \cap \partial(\bigcup_j T^u(P_j))$, where the P_j are saddle points with $\mathrm{beh}(P_j\,|\,P_i) > 0$, to be C^r small.

If P_i is a source, h_i is defined simply by $H(P_i) = P_i^*$. Suppose h_j has been constructed as required above for all $LS(P_j)$ such that $\mathrm{beh}(P_j\,|\,P_i) > 0$. Let P_{i+1} be such that $\mathrm{beh}(P_{i+1}\,|\,P_i) = 1$. From lemma (1.8) we get a C^r small diffeomorphism $\tau_1\colon W^u(P_{i+1}) \cap G^s(P_i) \to W^u(P^*_{i+1}) \cap G^s(P_i^*)$ such that $\tau_1 f = g\tau_1$ on the exterior boundary S_E of $G^s(P_i)$, for g near f in Diff(M). On the other hand, as in lemma (2.9) we have a product structure.

$$\hat{\psi}_{i+1}\colon W^u(P_{i+1}) \cap G^s(P_i) \to (W^u(P_{i+1}) \cap G^s(P_i)) \times \delta I^{s_{i+1}}$$

so that for $t \in \delta I^{s_{i+1}}, \hat{\psi}^{-1}(W^u(P_{i+1}) \times t) \subset e_t$ for som $e_t \in \mathscr{F}^u(P_{i+1})$. Let $\sum^u(P_{i+1}) = \{(W^u(P_{i+1}) \cap G^s(P_i)) \times t \,|\, t \in \delta I^{s_{i+1}}$ If $h_{i+1}^*\colon \mathscr{F}^u(P_{i+1}) \to \mathscr{F}^u(P_i)$ is the map induced by h_{i+1} on $\mathscr{F}^u(P_{i+1})$, then h_{i+1}^* induces a map $\mu_1\colon \sum^u(P_{i+1}) \to \sum^u(P_{i+1}^*)$ defined by $\mu_1(e_t \cap G^s(P)) = h_{i+1}(e_t) \cap G^s(P^*)$. Thus the map $h_i^*\colon T^u(P_{i+1}) \cap G^s(P_i) \to T^u(P_{i+1}^*) \cap G^s(P_i^*)$ satisfies $h_i f = gh_i$ on S_E and h_i is C^r small on $G^s(P_i) \cap A_{i+1}$, where $A_{i+1} = \partial T^u(P_{i+1}) - \bigcup_k \mathrm{Int}\, T^u(P_k)$ for saddles P_k with $\mathrm{beh}(P_k\,|\,P_i) > 0$.

We now use induction on the behavior of the periodic saddles P_j relative to P_i in order to extend h_i to $h_i\colon \bigcup_j T^u(P_j) \cap G^s(P_i) \to \bigcup_j T^u(P_j^*) \cap G^s(P_i^*)$ such that $h_i f = gh_i$ on S_E and h_i is C^r small on $\partial(\bigcup_j T^u(P_j)) \cap G^s(P_i)$. The induction step is done by using again (2.9) and the argument is the same as above. When we reach the sources of f in this process, then lemma (2.11) allow us to extend h_i to h_i: $G^s(P_i) \to G^s(P_i^*)$ satisfying condition (3) of lemma (2.7) on $G^s(P)$ and in particular $h_i f = gh_i$ on S_E. Finally, we define h on all of $LS(P)$ by $h_i(z) = g^n h_i f^{-n}(z), f^{-n}(z) \in G^s(P)$ and $h_i(P_i) = P_i^*$. Since h_i satisfy all the required conditions, the theorem is proved.

§3

Here we consider the extension of the results of the previous sections to the set of Morse–Smale vector fields $\mathscr{S}(M) \subset \chi(M)$. The proofs in this case are similar to those of the previous sections, once some of the concepts given for diffeomorphisms are adapted for vector fields.

For $X \in \chi(M)$, we denote its induced flow by X_t. We recall that a singularity x of X is hyperbolic if x is a hyperbolic fixed point for $X_{t=1}$, the diffeomorphism induced by X at time $t = 1$. A closed orbit γ of X is called hyperbolic if for $x \in \gamma$, the Poincaré transformation $f: U \to S$ has x as a hyperbolic fixed point. Here S is a cross section of γ at x and U is a neighborhood of x in S. We denote the set of singularities (fixed points) and closed orbits of X by $\Lambda(X)$. When $\alpha \in \Lambda(X)$ is hyperbolic we can define the stable and unstable manifolds of α [13], denoted by $W^s(\alpha)$ and $W^u(\alpha)$. The local stable and unstable manifolds of α are denoted by $LS(\alpha)$, $LU(\alpha)$.

(3.1) LEMMA. *Let $\alpha \in \Lambda(X)$ be hyperbolic. Suppose N is a one-to-one immersed manifold in M having a point of transversal intersection with $W^s(\alpha)$ and being invariant under X_t, i.e. $x \in N$ implies $X_t(x) \in N$ for all $t \in R$. If α is a fixed point, let B be a cell neighborhood of α in $W^u(\alpha)$. If α is a closed orbit, let B be a cell neighborhood of α in $W^u(\alpha) \cap S$, S a cross section of γ at x. Given $\varepsilon > 0$, there exists a submanifold of N, ε close to B in the C^1 sense.*

Proof. If α is a fixed point, consider the time one map $X_{t=1}$, for which x is hyperbolic. If α is a closed orbit, consider the Poincaré transformation $f: U \to S$ for which x is hyperbolic, U a neighborhood of x in S. Since N is invariant under X_t, N has a point of transversal intersection with $W^s(\alpha)$ in U. In either case, the result follows from lemma (1.1).

(3.2) PROPOSITION. (Smale). *Let $X \in \mathscr{S}(M)$. Then*

1) *for any $\alpha \in \Lambda(M)$, $W^s(\alpha)$ and $W^u(\alpha)$ have the topology induced by M.*
2) *$\Lambda(M)$ has a partial order structure defined by: for $\alpha_1, \alpha_2 \in \Lambda(M)$, $\alpha_1 \leq \alpha_2$ if and only if $W^u(\alpha_2) \cap W^s(\alpha_1) \neq \emptyset$.*

Proof. It follows from lemma (3.1) and the fact that $M = \bigcup_i W^s(\alpha_i) = \bigcup_i W^u(\alpha_i)$, $\alpha_i \in \Lambda(X)$, using the same argument as in the proof of (1.5) and (1.6).

$\Lambda(M)$ with this partial order structure is called the phase-diagram of X.

See [13] for the proof of the following analogue of lemma (1.8).

(3.3) LEMMA. *Let $\alpha \in \Lambda(X)$ be hyperbolic of period ω (if α is a fixed point we consider $\omega = 0$). Let U be a neighborhood of α in M containing no element of $\Lambda(X)$ of period less than 2ω, besides α. There exists a neighborhood $N(X)$ in $\chi(M)$ and a continuous map $\rho : N(X) \to$ Diff(M) such that*

1) *$\rho(X) = 1$, the identity map on M,*
2) *$\alpha^* = \rho(Y)(\alpha)$ is the unique element of $\Lambda(Y)$ in U of period less than 2ω, α^* is hyperbolic and has the same stable index as α,*
3) *$LS(\alpha), LU(\alpha)$ are isotopic to $LS(\alpha^*), LU(\alpha^*)$ and the isotopy depends continuously on $Y \in N(X)$.*

(3.4) *Definition.* Let $X \in \chi(M)$ and $\alpha \in \Lambda(M)$ be hyperbolic. Let B be a neighborhood of α in $W^s(\alpha)$ whose boundary ∂B is transversal to X. We call $G^s(\alpha) = \partial B$ a fundamental domain of $W^s(\alpha)$. Any cross section of X containing $G^s(\alpha)$ and transversal to $W^s(\alpha)$ is called a fundamental neighborhood $N^s(\alpha)$ associate with $W^s(\alpha)$. We have $W^s(\alpha) = \bigcup_{t \in R} X_t(G^s(\alpha)) \cup \{\alpha\}$ and $\bigcup_{t \geq 0} X_t(N^s(\alpha)) \cup W^u(\alpha)$ contains a neighborhood of α in M.

(3.5) THEOREM. *The set $\mathscr{S}(M)$ is open in $\chi(M)$. Also if $X \in \mathscr{S}(M)$ its phase-diagram is stable (up a diagram isomorphism) under C^r small perturbations.*

Proof. Using fundamental domains and neighborhoods as in (3.4), the proof of (1.9) goes through here to show that the phase-diagram of $X \in \mathscr{S}(M)$ is stable. The openess of the transversality condition, i.e. for Y in some neighborhood of X in $\chi(M)$ any pair of elements $\alpha_i, \alpha_j \in \Lambda(Y)$ have their unstable and stable manifolds transversal to each other, is proved by induction on the phase-diagram of X using lemmas (3.1) and (3.3).

Let us now define $\mathscr{F}$-neighborhoods of the elements of $\Lambda(X)$ for $X \in \mathscr{S}(M)$. If $\alpha \in \Lambda(X)$ the definition is the same as that of (2.4), the invariance condition being under X_t for $t \geq 0$. If α is a closed orbit with period ω, we first assume that there is a cross section S of γ at $x \in \gamma$ which is invariant in the following sense: there is a neighborhood U of x at S such that $X_{t=w}(U) \subset S$. Let $f: U \to S$ be the induced Poincaré map and $T^u(x,f)$ be an $\mathscr{F}$-neighborhood of x in U with projection map $\pi: T^u(x,f) \to T^u(x,f) \cap W(x,f)$. We then define an $\mathscr{F}$-neighborhood of α as $T^u(\alpha) = \bigcup_t X_t(T^u(x,f))$ for $0 \leq t \leq \omega$ with projection map $\pi_0: T^u(\alpha) \to T^u(\alpha) \cap W^s(\alpha)$ defined by $\pi_0(z) = X_t \pi X_{-t}(z)$ for $X_{-t}(z) \in T^u(x,f)$. We now show that in order to prove the structural stability of $X \in \mathscr{S}(M)$, we may assume the closed orbits of all vector fields in some neighborhood of X to have invariant cross sections. In fact, let $\gamma_1, \ldots, \gamma_m$ be the closed orbits of X with periods $\omega_1, \ldots, \omega_n$ and cross sections $S_1, \ldots, S_m$, respectively. By a local change of parametrization on their induced flows near $\gamma_1, \ldots, \gamma_m$, we get a continuous map $\mu: N(X) \to \mathscr{S}(M)$ of a neighborhood $N(X) \subset \mathscr{S}(M)$ ($\mathscr{S}(M)$ is open in $\chi(M)$) such that for each $Y \in N(X)$

1) Y and $\mu(Y)$ have the same trajectories and they are topologically equivalent, i.e., there is a homeomorphism of M sending trajectories of Y into those of $\mu(Y)$,
2) the closed orbits $\gamma_1^*, \ldots, \gamma_m^*$ of $\mu(Y)$ have periods $\omega_1, \ldots, \omega_m$ and $S_1, \ldots, S_m$ as invariant cross sections. In fact, we may assume that for each $Y \in N(X)$ all periodic orbits of $\mu(Y)$ have the same period of $w = 1$.

The definition of an admissible set of $\mathscr{F}$-neighborhoods is the same as in (2.6).

(3.6) LEMMA. (Stability Lemma). *Let $X \in \mathscr{S}(M)$. Then X is structurally stable if there exists a neighborhood $N(X)$ in $\mathscr{S}(M)$ such that*

1) *for each $Y \in N(X)$, the map $\rho(Y): \Lambda(X) \to (Y)$ defined in* (3.3) *is a diagram isomorphsim,*
2) *each $Y \in N(X)$ has an admissible system of $\mathscr{F}$-neighborhoods, $\{T^u(\alpha^*) \mid \alpha^* \in \Lambda(Y)\}$, with their space of fibers denoted by $\mathscr{F}^u(\alpha^*)$ and $\alpha^* = \rho(Y)(\alpha)$ for $\alpha \in \Lambda(X)$,*
3) *for each $\alpha_i \in \Lambda(X)$ there is a conjugacy $h_i: LS(\alpha) \to LS(\alpha^*)$ between $X_t/LS(\alpha)$ and $Y_t/LS(\alpha^*)$, i.e. $h_i X_t = Y_t h_i$ on $LS(\alpha_i)$, such that if $h_i: \mathscr{F}^u(\alpha_i) \to \mathscr{F}^u(\alpha_i^*)$ is its induced map on $\mathscr{F}^u(\alpha_i)$ then h_i^* is admissible in the following sense: for $e_i \in \mathscr{F}^u(\alpha_i)$ and $e_j \in \mathscr{F}^u(\alpha_j)$ if $e_i \subset e_j$ then $h_i^*(e_i) \subset h_i^*(e_j)$.*

Proof. Define $H(z) = Y_{-t}h_i X_t(z)$ for $X_t(z) \in LS(\alpha_i)$. Since $M = \bigcup_{i,t} X_t(LS(\alpha_i))$, H is a well defined map on M, with an inverse $H^{-1}(z) = X_{-t}h_i^{-1}Y_t(z)$ for $Y_t(z) \in LS(\alpha_i^*)$. Clearly $HX_t = Y_t H$ on M, for each $t \in R$. Thus the lemma will follow if H, H^{-1} are continuous. But this can be proved in the same way as in (2.7).

The proof that conditions (1) to (3) of lemma (3.6) can be fulfilled when $\dim M \leq 3$ is very similar to that of the diffeomorphism case. We just remark that, in the construction of an admissible system of $\mathscr{F}$-neighborhoods, for a fixed point one argues with a fundamental domain of its stable manifold (definition (3.4)) and for a closed orbit one argues with the Poincaré transformation of an invariant cross section, as in the diffeomorphism case.

Thus, we have

(3.7) THEOREM. *A Morse–Smale vector field on M, $\dim M \leq 3$, is structurally stable.*

§4

In this section we discuss the question of embedding a diffeomorphism of M in a topological flow. A topological flow of M is homomorphism $\varphi : R \to \text{Homeo}(M)$ such that the induced map $\varphi^* : R \times M \to M$, defined by $\varphi^*(t, x) = \varphi(t)(x)$, is continuous. Here R is the real line with its additive Lie group structure and Homeo(M) is the group of homeomorphisms of M.

We will be restricted to the following Baire subset of Diff(M):

i) all points in Per(f) are hyperbolic,

ii) for each pair of points $P, Q \in \text{Per}(f)$, $W^s(P)$ and $W^u(Q)$ are in general position,

iii) $Cl\,\text{Per}(f) = \Omega(f)$.

A diffeomorphism satisfying conditions (i) and (ii) above (see [4] and [12]) is called Kupka–Smale. The Baire property of $B(M)$ is in [8].

(4.1) *Definition.* Let $f \in B(M)$. We say that there exists a heteroclinic phenomenon associated with f if for a pair of points $P, Q \in \text{Per}(f)$, $P \notin \mathcal{O}(Q)$, $W^s(P) \cap W^u(Q)$ has a closed non-empty component.

If $\dim W^s(P) = \dim W^s(Q)$ in the above definition, then $W^s(P) \cap W^u(Q)$ is made up of points called heteroclinic. We shall give later an example of heteroclinic phenomenon with $\dim W^s(P) \neq \dim W^s(Q)$.

(4.2) THEOREM. *Suppose $f \in B(M)$ embeds in a topological flow of M. Then:*

1) $\text{Per}(f) = \text{Fix}(f)$

2) *f restricted to each invariant manifold of its periodic points is orientation preserving,*

3) *there is no heteroclinic phenomenon associated to f. When* $\dim M = 2$, *these conditions are also sufficient for $f \in B(M)$ to embed in a topological flow.*

(4.3) *Remark.* For $f \in B(M)$, condition (1) above implies that f is Morse–Smale.

Proof of (4.2). Suppose $f \in B(M)$ embeds in a topological flow φ_t. By a change of parametrization if necessary, we can assume $f = \varphi_1$. If $P \in \text{Per}(f)$ has period $n > 1$, then the orbit γ of φ_t through P is closed. Also, all points in γ will be periodic for f with period n. But this contradicts the fact that the periodic points of $f \in B(M)$ with period less or equal to n are isolated. Thus, $\mathrm{Per}(f) = \mathrm{Fix}(f)$. Since for $P \in \mathrm{Per}(f)$, $W^s(P)$ and $W^u(P)$ are invariant under φ_t and φ_t, $0 \le t \le 1$, is a homotopy between Id_M and f, condition (2) is proved. If K is a closed non-empty component of $W^s(P) \cap W^u(Q)$ for $P, Q \in \mathrm{Per}(f)$ and $P \notin \mathcal{O}(Q)$, then K must be invariant under φ_t. In particular K must be invariant under $\varphi_1 = f$, which is impossible since $K \subset W^s(P)$. Thus, the first part of the theorem is proved.

Let us assume now that $f \in B(M)$ satisfies conditions (1) to (3) and $\dim M = 2$. We are going to construct a topological flow X_t in M such that $X_{t=1} = f$. Let S_a be the set of saddle points of f. We treat first the case $S_a \neq \emptyset$. From (2.1), we may assume f to be linear in a neighborhood of each of its fixed points. This and conditions (1) to (3) above imply the existence of coordinate neighborhoods $U_i = LS(P_i) \times LU(P_i)$ of each $P_i \in S_a$, such that:

a) $\bigcup_{n \in Z} f^n(U_i) \cap \bigcup_{n \in Z} f^n(U_j) = \emptyset$ for $P_i \neq P_j$.

b) $f(x_1, x_2) = (\lambda_1 x_1, \lambda_2 x_2)$, $0 < \lambda_1 < 1$ and $\lambda_2 > 1$ $x_1 \in LS(P_i)$ and $x_2 \in LU(P_i)$.

Let $X_t(x_1, x_2) = (\lambda_1{}^t x_1, \lambda_2{}^t x_2)$ be the local one parameter group generated by f and let X be the associated vector field. We can extend X_t, and thus X, to all of $\bigcup_{n \in Z} f^n(U_i)$ by:

$$X_t(y) = f^n X_t f^{-n}(y), f^{-n}(y) \in U_i \qquad (*)$$

Consider now an embedded disc D around each sink P of f, such that $f(\partial D) \subset \mathrm{Int}\, D$ and ∂D is transversal to X. Thus, X is defined on a finite number of disjoint strips of $G^s(P)$, where $G^s(P)$ is the closed annulus bounded by ∂D and $f(\partial D)$. The extension of X to all of $G^s(P)$, in such a way that $X_{t=1} = f$, can be done according to a general procedure introduced by Smale in [12]. In fact, this question is reduced to the following one. Let I_1, I_2 be copies of $[0, 1]$, let $f: I_1 \times \{0\} \to I_1 \times \{1\}$ be an orientation preserving diffeomorphism and let $X = \partial/\partial x_2$ restricted to $\{0\} \times I_2 \cup \{1\} \times I_2$. We want to extend X to all of $I_1 \times I_2$ such that $X_{t=1} = f$. To do this, we notice that identifying, in $I_1 \times I_2$, $(x, 0)$ with $f(x, 0)$ we get a manifold N diffeomorphic to $I_1 \times S^1$, say through λ. Parametrizing $I_1 \times S^1$ by $\{(t, \theta) \mid 0 \le t \le 1, 0 \le \theta \le 2\pi\}$, the pull back by λ of $\partial/\partial\theta$ gives the desired vector field X on $I_1 \times I_2$.

Having defined the flow on fundamental domains of the sinks of f, we can extend it to all of M by saturation through f, as in $(*)$ above, and define the flow to be constant at the sinks and sources of f. The continuity of the flow in these points is a consequence of the fact that $f(f^{-1})$ restricted to a neighborhood of a sink (source) of f is a contraction. This concludes the proof of the theorem when the set of saddle points of f S_a is not empty. If $S_a = \emptyset$, there must be a unique sink and source for f and M must be the sphere S^2. The construction of the flow in this case can also be done by the previous method. This completes the proof of the theorem.

(4.4) *Remark*. By making small C^r perturbations of $\partial/\partial\theta$ in $I_1 \times S^1$ above, we see that f actually embeds in uncountable many flows.

(4.5) Corrollary. *For any neighborhood U of* 1 *in* Diff(M) *there exists an open set $V \subset U$ such that no element of V embeds in a topological flow.*

Proof. Let φ_t be a C^r flow in M with a closed orbit γ. Let $x \in \gamma$ and let $t_0 \in R$ be such that $\varphi_{t_0} \in U$ and x is a periodic point for φ_{t_0} of period $n > 1$, for some $n \in Z_+$. From [12] there exists g arbitrarily close to φ_{t_0} in Diff(M), such that x is a hyperbolic periodic point for g of period n. Also there exists a neighborhood V_1 of g in Diff(M), with the property that each element of V_1 has a hyperbolic periodic point of period $n > 1$. Thus, by theorem (4.2) no elements of V_1 embeds in a topological flow. Take V to be an open neighborhood of g in Diff(M), $V \subset U \cap V_1$. The corollary is proved.

We give now an example of an heteroclinic phenomenon, more general than the ones furnished by heteroclinic points. We start with a Morse–Smale diffeomorphism f of S^2, having at least one saddle point. Let S^3 be the suspension of S^2 and extend f to be Morse–Smale in S^3, having the north and south poles as sinks. Let g be this extension of f. There will be two saddle points for g, say P_1 and P_2 such that dim $W^u(P_2) =$ dim $W^s(P_1) = 2$ and $W^u(P_2) \cap W^s(P_1)$ is non-empty and made up of at most two connected components L_i, each homeomorphic to the real line. Let $N^s(P_1)$ be a fundamental neighborhood and let D be a small embedded disc in $W^u(P_2) \cap \operatorname{Int} N^s(P_1)$, $D \cap W^s(P_1) = \emptyset$. Finally, let η be an ambient diffeomorphism of S^3 such that $\gamma = Id$ outside $N^s(P_1)$ and on ∂D, γ/D is transversal to $W^s(P_1)$, $\gamma(D) \cap W^s(P_1) \neq \emptyset$ and $\gamma(D) \cap L_i = \emptyset$. Define $g = \eta g$. Clearly, g is Morse–Smale and $\operatorname{Per}(g) = \operatorname{Per}(g)$. Moreover, $W^u(P_2, g) \cap W^s(P_1, g)$ is made up of embedded lines and circles.

The generalization of (4.2), for higher dimensions of M, is certainly related to the obstruction to extending vector fields defined in some subsets of M. This generalization amounts to the following.

Problem. What conditions, besides those of (4.2), should $f \in B(M)$ satisfy in order to embed in a topological flow.

REFERENCES

1. A. Douady: Variétés à bords anguleau et voisinages tubilaires, *Sémin. Henri Cartan*, 1961/62, No. 1.
2. D. Grobman: Homeomorphisms of systems of differential equations, *Dokl. Akad. Nauk.* **128**(1965), 880–881.
3. P. Hartmann: *Ordinary Differential Equations*, Wiley, 1964.
4. I. Kupka: Contribution à la théorie des champs génériques, *Contr. diffl. Equat.* **2** (1963), 457–484.
5. E. Lima: On the local triviality of the restriction map for embeddings. *Comment. math. helvet* **38** (1964), 163–164.
6. R. Palais: Local triviality of the restriction map for embeddings. *Comment. math. helvet.* **34** (1960), 305–312.
7. M. Peixoto: Structural stability on two-dimensional manifolds, *Topology* **1** (1962), 101–120.
8. C. Pugh: An improved closing lemma and a general density theorem, *Am. J. Math.* **89** (1967), 1010 1021.
9. R. Seeley: Extension of C^∞ functions defined in a half space, *Proc. Am. math. Soc.* **15** (1964), 625–626.
10. S. Smale: Morse inequalities for a dynamical system, *Bull. Am. math. Soc.* **66** (1960), 43–49.
11. S. Smale: On gradient dynamical systems, *Ann. Math.* **74** (1961), 199–206.
12. S. Smale: Stable manifolds for differential equations and diffeomorphisms, *Annali Scu. norm. sup., Pisa* **XVII** (1963) 97–116.
13. S. Smale: Differentiable dynamic systems, *Bull. Am. math. Soc.* (Nov. 1967).

I.M.P.A., Rio de Janeiro

University of California, Berkeley

STRUCTURAL STABILITY THEOREMS

J. PALIS AND S. SMALE

1. We prove here that a class of diffeomorphisms, i.e. differentiable automorphisms, introduced in [3] and studied in [2] are structurally stable. Since every compact manifold has such diffeomorphisms, this implies a positive answer to the well-known question as to whether every manifold has a structurally stable diffeomorphism.

These results are proved in [2] for manifolds with dimension ≤ 3.

We show also that the similar theorems for differential equations are true.

The precise results are as follows. Let M be a compact C^∞ manifold without boundary, and for $r \geq 1$ let Diff (M) be the set of C^r diffeomorphisms of M with the uniform C^r topology. For $f \in$ Diff (M) we denote by $\Omega(f)$ the set of nonwandering points of f (see [5] for elementary definitions and a general background). Consider $f \in$ Diff (M) which satisfies

(1.1)
(a) $\Omega(f)$ is finite,
(b) $\Omega(f)$ is hyperbolic,
(c) Transversality Condition.

Condition (a) implies that $\Omega = \Omega(f)$ consists of periodic points. Then (b) means that if $x \in \Omega$ and $f^m(x) = x$, for some $m > 0$, then the derivative $Df^m(x): T_x(M) \to T_x(M)$ has its eigenvalues not equal to one in absolute value. From condition (b) one has defined stable and unstable manifolds for each $x \in \Omega$ denoted by $W^s(x)$, $W^u(x)$ respectively. Then condition (c) means that for each $x, y \in \Omega$, $W^s(x)$, $W^u(y)$ have transversal intersection in M.

A diffeomorphism f Diff (M) is said to be *structurally stable* if there exists a neighborhood $N(f) \subset$ Diff (M) such that if $g \in N(f)$, then f and g are topologically conjugate, i.e. there exists a homeomorphism $h: M \to M$ satisfying $hf = gh$.

(1.2) MAIN THEOREM. *If $f \in$ Diff(M) satisfies* (1.1) *then f is structurally stable.*

Using [4], we have the following corollaries:

(1.3) COROLLARY. *The gradient dynamical systems on a compact Riemannian manifold M which are structurally stable form a dense open set among all gradient dynamical systems on M. Moreover, a gradient dynamical system is structurally stable iff its induced diffeomorphism at time $t = 1$ satisfies* (1.1).

(1.4) COROLLARY. *Any compact manifold admits a structurally stable diffeomorphism.*

Originally published in: *Global Analysis (Proc. Sympos. Pure Math., Vol. XIV, Berkeley, Calif., 1968), 223–231,* © American Mathematical Society, Providence, R.I. 1970
Reprint by Springer-Verlag Berlin Heidelberg 2014 with kind permission

This paper is mainly devoted to proving (1.2). §5 gives the corresponding results for differential equations and poses a question related to the main result in this paper.

Conversations with M. Hirsch and C. Pugh have been very helpful in preparing this paper.

2. In this section we first review the needed background results from [2] and [3]. We then construct for f satisfying (1.1) tubular families of its stable manifolds, as defined below.

Let f satisfy (1.1) and $\Omega = \Omega(f)$. If Ω_k is a periodic orbit of Ω, let $W^s(\Omega_k) = \bigcup_{p\in\Omega_k} W^s(p)$ and $W^u(\Omega_k) = \bigcup_{p\in\Omega_k} W^u(p)$. Then from [3] (see also [2] and [5]):

(2.1) THEOREM. *The Ω_k are partially ordered by: $\Omega_i \le \Omega_k$ iff $W^s(\Omega_i) \cap W^u(\Omega_k) \ne \emptyset$. For all $1 \le k \le m$, $W^s(\Omega_k)$ and $W^u(\Omega_k)$ are properly imbedded submanifolds of M.*

We now choose once and for all a simple ordering consistent with this partial ordering, so that $\Omega_1 \le \Omega_2 \le \Omega_3 \le \ldots \le \Omega_{m-1} \le \Omega_m$, the Ω_k being all the periodic orbits of f. Thus Ω_1 (and perhaps Ω_2) is a sink (or attractor) and Ω_m is a source.

Submanifolds are always supposed to be properly imbedded and without boundary unless so stated. By the general theory for $p \in \mathrm{Per}(f)$, $W^s(p)$ is a submanifold of M.

(2.2) DEFINITION (TUBULAR FAMILY OF $W^s(p)$). A *tubular family* T of $W^s(p) = T_0$ is a collection of disjoint C^r submanifolds $\{T_y\}$ of M indexed by y in an open neighborhood N of p in $W^u(p)$ with the following properties.

(1) $V = V(T_0) = \bigcup_{y\in N} T_y$ is an open set of M containing T_0.

(2) $T_0 = T_p$.

(3) T_y intersects N transversally in the single point y.

Finally

(4) the map $V \to N$ which sends T_y into y is continuous; the section s which sends $x \in T_y$ into the tangent space of T_y at x is a continuous map from V into the Grassmann bundle over V.

Note that s in (4) is not required to be smooth.

If Ω_k is a periodic orbit of f satisfying (1.1), then we will speak of a tubular family for $W^s(\Omega_k)$. This is simply a tubular family $T = \{T_x\}$ for $W^s(p)$, some $p \in \Omega_k$ together with the submanifolds $\{f^i(T_x)\}_{i=0,\ldots,n-1}$ n = period Ω_k.

A tubular family $T = \{T_y\}$ of $W^s(p)$ is called *invariant* if $f^{-n}T_y = T_{f^{-n}(y)}$ where n is the period of p. In this case the corresponding tubular family of $W^s(\Omega_k)$ will also be called invariant.

A *system* of tubular families for f is a set of tubular families $T^k = \{T^k_x\}$, one for each Ω_k. The system will be called *compatible* when it meets the condition, if $T^i_x \cap T^j_y \ne \emptyset$, then one submanifold contains the other.

The main goal of this section is to prove that $f \in \mathrm{Diff}(M)$ satisfying (1.1) has a compatible system of invariant tubular families.

A standard result in Differential Topology is:

(2.3) LEMMA. *Let N be a C^r manifold possibly with boundary and $I = [0, 1]$. There exist a neighborhood $U(\Delta)$ of the diagonal Δ in $N \times N$ and a C^r function $\theta: U(\Delta) \times I \to N$ such that $\theta(x, y, 0) = x$, $\theta(x, y, 1) = y$, $\theta(x, x, t) = x$ for all $t \in I$, and if $x, y \in \partial N$, then $\theta(x, y, t) \in \partial N$.*

PROOF. By a theorem of Whitney, choose a C^∞ structure on N compatible with its C^r structure and define θ via geodesics in some C^∞ Riemannian metric, for which ∂N is totally geodesic. Thus $\theta(x, y, t)$ is the point on the unique geodesic joining x and y at "time t from x."

A C^r retraction $r \geq 0$ of an open set $U \subset M$ into a closed submanifold B of U is a C^r map $p: U \to B$ which is the identity on B.

(2.4) LEMMA (C^r RETRACTION LEMMA). *Let M be a C^∞ manifold, B a closed C^r submanifold. Let A be a compact set in B, U_0 a neighborhood of A in M and $r_0: U_0 \to B$ a C^r retraction onto $U_0 \cap B$. Then there exists a neighborhood U of B and a retraction $r: U \to B$ such that $r/U'_0 = r_0/U'_0$, U'_0 a neighbourhood of A.*

Furthermore if B has boundary, the above holds if $A \subset \text{int}\, B$ and U is a tubular neighborhood "with boundary" ∂U, fibered by r over ∂B.

PROOF. The tubular neighborhood theorem yields a C^r retraction $\pi: U_1 \to B$ where U_1 is a neighborhood of B. Let ϕ be a C^∞ "bump function", i.e. ϕ is 1 on a neighborhood of A, with support of ϕ contained in U_0 and $0 \leq \phi \leq 1$.

Let θ be the "geodesic function" on B by the previous lemma. Then define $r(x) = \pi(x)$ if $x \notin U_0$ and $\theta(\pi(x), r_0(x); \phi(x))$ if $x \in U_0$. This proves (2.4).

Having seen this basic extension process, we now prove a parameterized version of it in the next lemma, which we explicitly use in constructing tubular families.

(2.5) LEMMA *Let $T = \{T_x\}$ be a tubular family for $W^s(\Omega_i)$ and $T_0 = \bigcup_{x \in \Omega_i} T_x$. Let W be a C^r submanifold meeting T_0 transversally and A be a compact set in W. Suppose that $r_0: U_0 \to W$ is a given continuous retraction of a neighborhood U_0 of $(\partial T_0) \cap A$ (∂T_0 the point set boundary) which is*

(i) *C^r on each T_x,*

(ii) *$y \in T_x \cap W \cap U$ implies $r_0^{-1}(y) \subset T_x$ and*

(iii) *the family $\{r^{-1}(y)\}$ satisfies the Grassmannian continuity condition of* (2.2)–4.

Then there exists a continuous retraction $r: U \to W$, U a neighborhood of $\bar{T}_0 \cap A$ such that (i), (ii) *and* (iii) *hold for r and r restricted to some neighborhood of* $(\partial T_0) \cap A$ *is r_0.*

(Here $\bar{T}_0$ is the closure of T_0 in M.)

We indicate how we apply this crucial lemma in the construction of tubular families. We take $W = W^u(p)$, $p \in \Omega_k$, period $\Omega_k = n$. A fundamental domain D for f^{-n} on $W^u(p)$ is a subset of the form closure $(N - f^n(N))$ where N is a compact ball in $W^u(p)$ such that $f^{-n}(N) \subset \text{int}\, N$. Then $\partial D = S_E \cup S_I$, a disjoint union of two spheres, an exterior one S_E, an interior one S_I with $f^{-n}(S_E) = S_I$. In the applications, A will be either S_E or D, and i will be less than k.

This will be carried out in (2.6). Now we prove (2.5).

PROOF. First assume Ω_i is a point p. Since the question is restricted to a neighborhood of A, we may assume W to be compact and $A \subset \operatorname{int} W$. Let U_1 be an open neighborhood of $\partial T_0 \cap W$ such that $\bar{U}_1 \cap A \subset U_0 \cap A$. Then choose a C^r retraction $F: U(T_0) \to T_0$ defined off U_1 and such that $F(W) \subset W \cap T_0$. This can be done using a tubular neighborhood of T_0 which extends a tubular neighborhood of $W \cap T_0$ in W. Let $F_x = F/T_x$ so $F_x: T_x \to T_0$ is C^r for each $x \in N$, where N, the index set for T, is a neighborhood of Ω_i in $W^u(\Omega_i)$. Then N can be chosen small enough so that for each $x \in N$, F_x is invertible on neighborhood $T_0 \cap A$.

Choose a C^r retraction $\pi_0: T_0 \cap V \to T_0 \cap W$ with V some neighborhood of $T_0 \cap A$. Then define for $x \in N$ a C^r retraction $\pi_x: T_x \cap V \to T_x \cap W$ by $\pi_x = F_x^{-1} \circ \pi_0 \circ F_x$. This makes sense off U_1.

Also, off U_1 define via (2.3) a "geodesic function" $\mathcal{O}_0$ for $T_0 \cap W$ and then let $\mathcal{O}_x$ be the geodesic function on $T_x \cap W$ off U_1 induced by the map $F_x: T_x \to T_0$.

Now we follow the proof of (2.4) to obtain our desired retraction r. Thus let ϕ be a C^∞ bump function which is 1 on U_1, has support in U_0 and $0 \leqq \Phi \leqq 1$. Define $r = r_0$ on U_1, $r(z) = \pi_x(z)$ if $z \in T_x$-supp Φ znd $r(z) = \theta_x(\pi_x z, r_0 z, \Phi(z))$ otherwise.

Thus r is defined in some neighborhood U of $T_0 \cap A$ and one checks directly that r has the desired properties.

In the general case $T_0 = W^s(\Omega_i)$, proceed as above for each component of T_0. This proves (2.5).

(2.6) TUBULAR FAMILY THEOREM. *Let $f \in \operatorname{Diff}(M)$ satisfy (1.1) and let $\Omega(f) = \bigcup \Omega_k$, where the Ω_k are the periodic orbits of $\Omega(f)$ simply ordered as above. There exists a compatible, invariant system of tubular families $\{T^k\}$, each T^k being a tubular family of $W^s(\Omega_k)$. Thus $T_0^k = W^s(\Omega_k)$.*

Moreover, the map $\pi: \bigcup_{x \in N_k} T_x^k \to W^u(\Omega_k)$, defined by $\pi(T_x^k) = x = T_x^k \cap W^u(\Omega_k)$ is C^r on T_x^i, each x and $1 \leq i \leq k$.

PROOF. We sometimes let T^k be the union $\bigcup_{x \in N_k} T_x^k$, restricting N_k when necessary. Proceeding inductively define T^1 as simply $\{W^u(p) | p \in \Omega_1\}$. Now assume $T^1, \ldots, T^{k-1}$ constructed as in the statement of the theorem. Let us construct T^k. For $p \in \Omega_k$ of period m, let D be a fundamental domain with $\partial D = S_E \cup S_I$. We will define a continuous retraction $\pi: U(D) \to W^u(p)$ using a second induction on $T^i \cap U(D)$ for $i = k-1, \ldots, 1$, such that $\pi f^{-n} = f^{-n} \pi$ on a neighborhood of S_E, conditions (i), (ii) of (2.5) are met for each T^i, $1 \leq i \leq k-1$ and $\{\pi^{-1}(y)\}$ satisfy (2.5)–(iii).

Once this is done, the desired T^k is obtained as follows. Let N be the neighborhood of $p \in \Omega_k$ in $W^u(p)$, with $\partial N = S_E$. For each $x \in N$ there is a unique point $y \in D - S_E$ and an integer $j \geq 0$ so that $x = f^{-nj}(y)$. We then set $T_x^k = f^{-nj}(\pi^{-1}(y))$ and $T_p^k = W^s(p)$. $\{T_x^k\}_{x \in N}$ is a tubular family of $W^s(p)$. This follows from Lemma 1.1 of [2]. It amounts to seeing that $\bigcup_{x \in N} T_x^k$ is open (which also follows from Hartman's Theorem) and the continuity condition (2.2)–4 is met. This continuity is also a consequence of basic known linear estimates.

We now construct the retraction $\pi: U(D) \to W^u(p)$ as required above. To give the idea, the retraction is constructed in the following order. From

$U((\partial T_0^{k-1}) \cap S_E)$ it is extended to $U(T_0^{k-1} \cap S_E)$. This induces via f^{-n} the retraction on $U(T_0^{k-1} \cap S_I)$, which is now extended from $U((T_0^{k-1}) \cap S_E) \cup U(T_0^{k-1} \cap S_I)$ to $U(T_0^{k-1} \cap D)$. Then we extend it from $U((\partial T_0^{k-2}) \cap S_E)$ to $U(T_0^{k-2} \cap S_E)$. This induces the retraction on $U((T_0^{k-2}) \cap S_E)$ to $U(T_0^{k-2} \cap S_E)$. This induces the retraction on $U(T_0^{k-2} \cap S_I)$ and we extend it from $U(T_0^{k-1} \cap D) \cup U(T_0^{k-2} \cap S_E) \cup U(T_0^{k-2} \cap S_I)$ to $U(T_0^{k-2} \cap D)$. We continue with T_0^{k-3} in the same manner.

More precisely let $\pi : T^{k-1} \cap U(S_E) \to W^u(p)$ given by (2.5). For that we take in (2.5) A to be S_E, $W = W^u(p)$, T to be T^{k-1}, using the fact that $T^{k-1} \cap D$ is compact and $\partial T_0 \cap A = \varnothing$. Since T^{k-1} is invariant and $f^{-n}(S_E) = S_I$, the composition $f^{-n}\pi$ defines a retraction of $T^{k-1} \cap U(S_I)$ into $W^u(p)$, denoted also by π. Here $U(S_I) = f^{-n}(U(S_E))$. Now apply again (2.5) with $A = D$ to extend π to the desired retraction of $T^{k-1} \cap U(D)$ into $T^{k-1} \cap D$.

We proceed in the same way to extend π to $T^{k-2} \cap U(D)$, using again (2.5) and the fact that $(\partial T_0^{k-2}) \cap D \subset (T_0^{k-1} \cup T_0^k) \cap D$. Continuing this process through $T^{k-2}, T^{k-3}, \ldots, T^1$ yields the desired π and hence (2.6).

(2.7) Remark. Via iteration by f, the system $\{T^k\}$ obtained in (2.6) can be extended to be also f invariant. Such a system will be denoted by $\{T^{k,s}\}$, while $\{T^{k,u}\}$ will denote a similar system of tubular families of the $W^u(\Omega_k)$. $T^{k,s}(x)$, $T^{k,u}(x)$ will stand for the element of the family $T^{k,s}$, $T^{k,u}$ containing the point x.

(2.8) Remark. The set of diffeomorphisms satisfying (1.1) is open in Diff(M) [2]. Let $\{T^{k,s}\}$, $\{T^{k,u}\}$ be as above. There exists a neighborhood N of f such that for each $g \in N$ we can construct compatible, g and g^{-1} invariant systems of tubular families $\{T_g^{k,s}\}$, $\{T_g^{k,u}\}$ of the $W^s(\Omega_k(g))$, $W^u(\Omega_k(g))$ where $\{T_f^{k,s}\} = T^{k,s}$ and $\{T_f^{k,u}\} = \{T^{k,u}\}$. Moreover we may assume that the maps $\phi_1(g) = S_g^{k,s}$ and $\phi_2(g) = S_g^{k,u}$, $g \in N$, are continuous, at f on compact parts of the $T^{k,s}$, $T^{k,u}$ where $S_g^{k,s}$, $S_g^{k,u}$ are the sections on the Grassman bundle of (2.2)–4 defined for $T_g^{k,s}$, $T_g^{k,u}$. This can be checked at each stage of the proof of the Tubular Family Theorem, using the same uniform estimates on linear maps mentioned there. The tubular families that appear in the following sections are the ones given by the maps ϕ_1, ϕ_2 above.

3. Let N be a neighborhood of f in Diff(M). For each $g \in N$, gf^{-1} is called a perturbation of f. The perturbation is said to be localized at an open set U of M if it is 1 on $M - U$, where 1 stands for the identity map.

The main purpose of this section is to show that perturbations of f near 1 can be decomposed into simple ones. That is, into localized perturbations leaving invariant the tubular family structure introduced in the previous section.

We denote by supp$f$ the closure of the set where $f \neq 1$. The following lemma was detailed to us by M. Hirsch.

(3.1) Lemma. *Let $\{U_k | 1 \leq k \leq n\}$ be an open cover of M and N be a neighborhood of* 1. *There exists a neighborhood N_1 of* 1 *such that if $f \in N_1$ then f can be factored as $f = f_n \circ \ldots \circ f_1$, where $f_i \in N$, and* supp$f_i \subset U_i$.

Proof. First choose N_1 so that if $f \in N_1$ then there is an isotopy $F : M \times I \to M$, $F_0 = 1$ and $F_1 = f$. Let now $\{\lambda_i : M \to R | \text{supp}\, \lambda_i = \overline{W}_i \subset U_i\}$ be a partition of unity subordinate to $\{U_i\}$ and define $\mu_i : M \to M \times I$ as $\mu_i = (1, \lambda_1 + \ldots + \lambda_i)$.

For N_1 small, F is close to the projection map $\pi: M \times I \to M$ and the maps $g_i = F_0\mu_i$ are close to 1. Thus $g_i \in \text{Diff}(M)$, $g_i \circ g_{i-1}^{-1} \in N$ and $g_{i-1}^{-1}(M - V_i) \subset M - \bar{W}_i$. Here V_i is open, $\bar{W}_i \subset V_i$ and $\bar{V}_i \subset U_i$. Since $g_i = g_{i-1}$ on $M - W_i$ and $g_n = f$, setting $f_i = g_i \circ g_{i-1}^{-1}$ we have $f = f_n \circ \ldots \circ f_1$. The lemma is proved.

(3.2) LEMMA. *Let $f \in \text{Diff}(M)$ satisfy* (1.1). *There exist open neighborhoods N of f, $U_k = U_k(\Omega_k)$ and $J > 0$ such that*

(a) $\{f^j(U_k)\}_{0 \le j < J, 1 \le k \le m}$ *is a covering of M,*

(b) *for each $g \in N$, k as above and $0 \le j \le J$, $f^j(U_k) \subset T_f^{k,s} \cap T_f^{k,u} \cap T_g^{k,s} \cap T_g^{k,u}$,*

(c) *$T^{k,s}(x)$ and $T^{k,u}(y)$ intersect transversally in a unique point for each $x, y \in f^j(U_k)$. Here $T^{k,s}$ stands either for $T_f^{k,s}$ or $T_g^{k,s}$ and $T^{k,u}$ for $T_f^{k,u}$ or $T_g^{k,u}$.*

REMARK. We assume that N is such that $T_g^{k,s}$ is defined by (2.8).

PROOF. $T_f^{k,s}$, $T_f^{k,u}$ are tubular families of $W^s(\Omega_k)$, $W^u(\Omega_k)$. Thus for each k we can choose $U_k = U_k(\Omega_k)$ so that (b) and (c) are true for $j = 0$, $\bar{U}_k$ and $T_f^{k,s}$, $T_f^{k,u}$. Since $\Omega(f) = \bigcup \Omega_k$ and M is compact, there exists $J > 0$ so that $\{f^j(U_k)\}_{0 \le j < J, 1 \le k \le m}$ is a covering of M. $T_f^{k,s}$, $T_f^{k,u}$ being invariant (a) and (b) are true for $0 \le j \le J$, $f^j(\bar{U}_k)$ and $T_f^{k,s}$, $T_f^{k,u}$. Now $\bigcup_{0 \le j \le J} f^j(\bar{U}_k)$ being compact, (2.8) implies the existence of a neighborhood N as required by the lemma. This proves (3.2).

(3.3) LEMMA. *Let $\{f^j(U_k)\}$ be the covering of M and let N be the neighborhood of f as in* (3.2). *If $g \in N$ and* $\text{supp}\, gf^{-1} \subset f^j(U_k)$ *for fixed j, k then there are homeomorphisms $\eta^s, \eta^u: M \to M$ so that for all $1 \le i \le m$ and $x \in M$*

(a) *$\eta^s(T_f^{i,s}(x)) = T_f^{i,s}(x)$ and $\eta^s = 1$ on $M - U$.*

(b) *$\eta^u(T_g^{i,u}(x)) = T_g^{i,u}(x)$ and $\eta^u = 1$ on $M - U$.*

(c) *$\eta^u\eta^s f(x) = g(x)$.*

Here $U = f^j(U_k) \cup f^{j+1}(U_k)$.

PROOF. Since the systems of tubular families considered here are compatible, it is enough to construct η^s, η^u satisfying (a) to (c) above for $i = k$. In this case (3.2) implies that the map $\eta^s: U \to U$, $\eta^s(x) = T_f^{k,s}(x) \cap gT_g^{k,u}(f^{-1}(x))$, is well defined and is, in fact, a homeomorphism. The same is true for $\eta^u: U \to U$, defined by $\eta^u(x) = gT_f^{k,s}(f^{-1}(x)) \cap T_g^{k,u}(x)$. Since $\text{supp}\, gf^{-1} = \bar{V} \subset f^j(U_k)$, we have $g = f$ $g^{-1} = f^{-1}$ on $M - W$, where $W = V \cup f(V)$. Consequently $\eta^s, \eta^u = 1$ on $U - W$ and so they can be extended to be 1 on $M - U$.

From the definitions of η^s and η^u, $\eta^s(T_f^{k,s}(x)) = T_f^{k,s}(x)$ and $\eta^u(T_g^{k,u}(x)) = T_g^{k,u}(x)$ for all $x \in M$.

For $x \in U$, $\eta^u\eta^s f(x) = g(T_f^{k,s}(x) \cap T_g^{k,u}(x))$ and so $\eta^u\eta^s f(x) = g(x)$ on U. Since this is true on $M - U$, we have $\eta^u\eta^s f = g$ on M.

Thus conditions (a) to (c) of the lemma are met by η^s, η^u and hence (3.3) is proved.

4. In this section we prove the structural stability of diffeomorphisms satisfying (1.1).

(4.1) THEOREM. *Let $f \in \text{Diff}(M)$ satisfy* (1.1). *Then f is structurally stable.*

PROOF. Let $\{f^j(U_k)\}$ be the covering of M as in (3.2). If g is near f, then from

5. We consider here flows or vector fields on M and extend the previous results to this case.

Let χ be the set of C^r vector fields on M, with the uniform C^r topology. This makes χ into a Banach space. For each $X \in \chi$ we denote by X_t its induced flow. A fixed (critical) point of X is called hyperbolic if x is a hyperbolic fixed point for $X_{t=1}$. In a neighborhood of a closed orbit γ, X is the suspension of a diffeomorphism f defined in a cross section S of γ through $x \in \gamma$. We call γ hyperbolic if f has x as a hyperbolic fixed point.

We now define the analogue of the diffeomorphisms satisfying (1.1) as the vector fields X such that

(5.1)
(a) $\Omega(X)$ is the union of a finite number of fixed points $x_1, \ldots, x_m$ and a finite number of closed orbits $\gamma_1, \ldots, \gamma_n$ of X,
(b) the x_i, γ_j are all hyperbolic,
(c) the stable and unstable manifolds of the x_i, γ_j have transversal intersection.

X, $Y \in \chi$ are called *equivalent* if there is a homeomorphism of M sending trajectories of X into trajectories of Y. X is called structurally stable if there exists a neighborhood N of X in χ such that each $Y \in N$ is equivalent to X.

(5.2) THEOREM. *If $X \in \chi$ satisfies* (5.1) *then X is structurally stable.*

To prove (5.2), we first indicate how to carry over to this case the corresponding concepts and results of §§2 and 3. The only novelty here is the concept of tubular families for the stable and unstable manifolds of the closed orbits of the flow.

Let $X \in \chi$ satisfy (5.1). For the fixed points of X, we define tubular families of their stable and unstable manifolds as in (2.2). Let now γ be a closed orbit of X with period π (not necessarily 3.14 . .!) and S be a cross-section of γ at $x \in \gamma$. S is called invariant if $X_{t=\pi}(U) \subset S$, where U is a neighborhood of x in S. In this case we define a tubular family of $W^s(\gamma)$ as follows. Let $f: U \to S$ be the restriction of X_π and let $\{T_f^s(y)\}_{y \in U}$ be an invariant (under f^{-1}) tubular family of $W_f^s(x)$. The tubular family of $W^s(\gamma)$ is now defined by $T^s(X_t(y)) = X_t(T_f^s(y))$, for $y \in U$ and all t.

We show that in proving (5.2) we may assume X, and in fact all vector fields in some neighborhood of X, to have invariant cross-sections for its closed orbits. Let γ be a closed orbit of X with period π, S be a cross-section of γ at $x \in \gamma$ and $z \in \gamma$, $z \neq x$. From simple properties of flows near a closed orbit (e.g. [**1**, p. 251]), for each Y in some neighborhood N of X we get a positive C^r function $\rho(Y): M \to R$ so that $(\rho(Y)Y)_{t=\pi}(U) \subset S$ and $\rho(Y) = 1$ on $M - V$, where U is a neighborhood of x in S and V a neighborhood of z in M. Moreover, the correspondence $Y \in N \to \rho(Y)$ is continuous. We now repeat the same procedure for each closed orbit of X and notice that the set of vector fields satisfying (5.1) is open in χ [**2**], this yielding in particular a natural one-one correspondence between the fixed points and closed orbits of X and those of Y, Y near X. Thus, taking the neighborhood N of X small enough, we get a continuous map $\mu: N \to \chi$ such that for each $Y \in N$, $\mu(Y)$ satisfies (5.1), it is equivalent to Y with the same trajectories as Y

(3.1) $gf^{-1} = f_n \circ \dots \circ f_1$, each f_i near 1 and $\operatorname{supp} f_i \subset f^j(U_k)$ for some j, k where $0 \le j < J$, $1 \le k \le m$. Thus from the openness of the diffeomorphisms satisfying (1.1), we have that $f_i \circ \dots \circ f_1 \circ f$ also satisfy (1.1). The result will follow by proving the conjugacy between $f_i \circ \dots \circ f_1 \circ f$ and $f_{i-1} \circ \dots \circ f_1 \circ f$. Thus we may assume that g is of the form $g = f_i \circ f$, f_i as above. But for g in this form, we have $g = \eta^u \circ \eta^s \circ f$, η^s and η^u being the homeomorphisms given by (3.2). Since η^s and η^u preserves, respectively, each element of $\{T_f^{k,s}\}$ and $\{T_g^{k,u}\}$, it is enough to prove that $\eta^s f$ and f are conjugate.

Let $d = \eta^s f$. We define a conjugacy h between f and d as follows:

(4.2) $h(x) = \lim_{n\to\infty} d^n f^{-n}(x)$.

This amounts to

(a) $h(x) = x$ for $x \in M - \bigcup_{m\ge 0} f^m(U_k)$,

(b) $h(x) = d^n f^{-n}(x)$ for $x \in \bigcup_{m\ge 0} f^m(U_k)$ and $f^{-n}(x) \in M - \bigcup_{m\ge 0} f^m(U_k)$,

(c) $h(x) = T^{k,s}(x) \cap W^u(\Omega_k(g))$ for $x \in W^u(\Omega_k)$.

To prove this assertion it is enough to show that $h(x) = T^{k,s}(x) \cap W^u(\Omega_k(g))$ for $x \in W^u(\Omega_k) \cap U_k$. Since $d = \eta^s f$ and $\eta^s(T^{k,s}(x)) = T^{k,s}(x)$, $d^n f^{-n}(T^{k,s}(x)) = T^{k,s}(x)$ and thus $h(T^{k,s}(x)) = T^{k,s}(x)$. Also, since $d = (\eta^u)^{-1} g$ the set of limit points of the the sequences $\{d^n(y) | y \in U_k\}$ is contained in $\bigcup_{i\le k} W^u(\Omega_i(g))$. So if K is the set of limit points of the sequences $\{d^n f^{-n}(y) | y \in U_k\}$ then $K \subset \bigcup_{i\le k} W^u(\Omega_i(g)) \cap U_k = W^u(\Omega_k(g)) \cap U_k$. This implies $\lim_{n\to\infty} d^n f^{-n}(x) = T^{s,k}(x) \cap W^u(\Omega_k(g))$ and hence the assertion.

We have to show that h is well defined, $hf(x) = dh(x)$ for all $x \in M$ and h is a homeomorphism of M.

That h is well defined is clear, since it is well defined on $M - W^u(\Omega_k)$ and for each $x \in W^u(\Omega_k)$ there is a unique point of intersection $T^{k,s}(x) \cap W^u(\Omega_k(g))$.

For all $x \in M$, $dhf^{-1}(x) = \lim_{n\to\infty} d^{n+1} f^{-(n+1)}(x) = h(x)$. This proves $dh = hf$.

The continuity of h follows from its continuity on $W^u(\Omega_i)$, $i \le k$. Let us show that h is continuous on $W^u(\Omega_k)$. It is enough to do so on $W^u(\Omega_k) \cap U_k$. Let $x_m \in U_k$ be a sequence so that $x_m \to x \in W^u(\Omega_k) \cap U_k$. As we showed above $h(T^{k,s}(x_m)) = T^{k,s}(x_m)$. Consider now the following two cases. If $x_m \in W^u(\Omega_k)$ then $f^{-n}(x_m) \in U_k$ for all $n \ge 0$. If $x_m \notin W^u(\Omega_k)$ there exist $y_m \in f^{-1}(U_k) - U_k$, $n_m \ge 0$ so that $f^{-n_m}(x_m) = y_m$. Thus $n_m \to \infty$ as $m \to \infty$ and $h(x_m) = d^{n_m}(y_m)$. As before the set of limit points of the sequences $\{d^n(y) | y \in U_k \cup f^{-1}(U_k)\}$ is contained in $\bigcup_{i\le k} W^u(\Omega_i(g))$. Thus by combining the two cases if necessary, the set of limit points of $h(x_m)$ is contained in $\bigcup_{i\le k} W^u(\Omega_i(g)) \cap U_k = W^u(\Omega_k(g)) \cap U_k$.

Therefore $h(x_m) \to T^{k,s}(x) \cap W^u(\Omega_k(g))$ or $h(x_m) \to h(x)$ and thus h is continuous on $W^u(\Omega_k)$.

The proof of the continuity of h on $W^u(\Omega_i)$, $i < k$ is similar. For from the compatibility of $\{T^{i,s}\}_{1\le i\le m}$, $h(T^{i,s}(x)) = T^{i,s}(x)$ for $i \le k$ since this is true for $T^{k,s}$. Also if $x_m \to x \in W^u(\Omega_i)$, then the set of limit points of $h(x_m)$ is contained in $W^u(\Omega_i(g))$. From these two facts the assertion follows. Thus h is continuous.

Finally, to show that h is a homeomorphism we consider the map $h_1 = \lim_{n\to\infty} f^n d^{-n}(x)$. As before h_1 is well defined and continuous. Since $f^n d^{-n} d^n f^n = 1$ for all n, $h_1 h = 1$. Similarly $hh_1 = 1$ and hence h is a homeomorphism of M. This concludes the proof of the theorem.

and its closed orbits have the same period and invariant cross-sections as the correspondent ones for $\mu(X)$.

The definition of a compatible, invariant system of tubular families is the same as before, the invariance being considered for the maps X_t, for all t. As in (2.6) one can prove the existence of compatible, invariant systems of tubular families of the $W^s(x_i)$, $W^s(\gamma_j)$ and $W^u(x_i)$, $W^u(\gamma_j)$. Again, it is enough to do so for vector fields equivalent to X and we point out that instead of the neighborhood of the fundamental domain considered in (2.6), we take here a cross-section S to the flow whose trajectories together with $W^s(x_i)$ (or $W^s(\gamma_j)$) form a neighborhood of $W^s(x_i)$ (or $W^s(\gamma_j)$).

The translation to flows of the results of §3 is immediate, once we consider flows corresponding to the maps $\eta^s f$ and $(\eta^u)^{-1}g$ in Lemma (3.3).

The proof of (5.2) becomes now very much the same as that of (4.1). We just remark that when X_t has no closed orbits, (5.2) provides for Y_t near X_t a homeomorphism that not only sends trajectories of X_t into those of Y_t, but it is in fact a conjugacy between the flows. This is the case for gradient flows satisfying (5.1).

Using (4.1) (or (5.2) for the corresponding case for flows) and a converse known for some time, we have:

(5.3) COROLLARY. *Let $f \in \mathrm{Diff}(M)$, with $\Omega(f)$ being finite. Then f is structurally stable if and only if f satisfies* (1.1).

This gives a good characterization of structural stability in the Ω-finite case and leads us to the question of finding in the general case a similar characterization.

Referring the reader to [5] for the concepts appearing here, we pose as a possible answer to this question the following

CONJECTURE. $f \in \mathrm{Diff}(M)$ is structurally stable if and only if f satisfies

(a) AXIOM A. $\Omega(f)$ is hyperbolic and the set of periodic points of f is dense in $\Omega(f)$.

(b) STRONG TRANSVERSALITY CONDITION. For all $x, y \in \Omega(f)$, $W^s(x)$ and $W^u(y)$ have a transversal intersection.

REFERENCES

1. P. Hartman, *Ordinary differential equations*, Wiley, New York, 1964.
2. J. Palis, *On Morse-Smale dynamical systems*, Topology (to appear).
3. S. Smale, *Morse inequalities for a dynamical system*, Bull. Amer. Math. Soc. **66** (1960), 43–49.
4. ———, *On gradient dynamical systems*, Ann. of Math. (2) **74** (1961), 199–206.
5. ———, *Differentiable dynamical systems*, Bull. Amer. Math. Soc. **73** (1967), 747–817.

I.M.P.A., RIO DE JANEIRO
UNIVERSITY OF CALIFORNIA, BERKELEY

A NOTE ON Ω-STABILITY

J. PALIS

Recently the following important result in Dynamical Systems has been proved by Smale [4]: if a diffeomorphism f of a compact C^∞ manifold M satisfies Axiom A and $\Omega(f)$ has the no cycle property then f is Ω-stable. We show in this note that if f satisfies Axiom A and is Ω-stable then $\Omega(f)$ has the no cycle property.

Here $\Omega(f)$ stands for the nonwandering set of f. We denote by Diff(M) the set of C^r diffeomorphisms of M with the uniform C^r topology, $0 < r$. We call $f \in \text{Diff}(M)$ Ω-stable if there exists a neighborhood N of f in Diff(M) such that for each $g \in N$, $g/\Omega(g)$ is topologically conjugate to $f/\Omega(f)$.

(1) DEFINITION. We say that $f \in \text{Diff}(M)$ satisfies Axiom A if

(i) $\Omega(f)$ has a hyperbolic structure,

(ii) the set of periodic points is dense in $\Omega(f)$.

We refer the reader to [3] for a detailed presentation of these concepts.

From the Spectral Decomposition Theorem [3], if f satisfies Axiom A then $\Omega = \Omega(f)$ can be written as $\Omega = \Omega_0 \cup \ldots \cup \Omega_k$, where the Ω_i are closed invariant sets and f is topologically transitive on each of them. These Ω_i are called basic sets for f. For any closed hyperbolic set K one can define [1] its stable and unstable manifolds, denoted by $W^s(K)$ and $W^u(K)$.

(2) DEFINITION. Let $f \in \text{Diff}(M)$ satisfy Axiom A. An n-cycle ($n \geq 1$) on $\Omega(f)$ is a sequence of basic sets $\Omega_0, \ldots, \Omega_n, \Omega_{n+1}$ (reordering the indices if necessary) such that $W^s(\Omega_i) \cap W^u(\Omega_{i+1}) \neq \varnothing$, $\Omega_0 = \Omega_{n+1}$ and otherwise $\Omega_j \neq \Omega_k$ for $j \neq k$. When there is no n-cycle ($n \geq 1$) on $\Omega(f)$ we say that $\Omega(f)$ has the no cycle property.

(3) THEOREM. *Let $f \in \text{Diff}(M)$ satisfy Axiom* A *and be Ω-stable. Then $\Omega = \Omega(f)$ has the no cycle property.*

PROOF. The theorem will follow by showing that if there is an n-cycle on Ω, $n \geq 1$, then f is not Ω-stable.

We prove first a weaker assertion: if there is an n-cycle on Ω for $n > 1$ then f is not Ω-stable or there exists g near f in Diff(M) such that $g/\Omega = f/\Omega$ and Ω is the union $\Omega_0 \cup \ldots \cup \Omega_k$ of hyperbolic sets for g with an $(n-1)$-cycle. We then show that this last alternative also implies the non-Ω-stability of f.

Let $\Omega_0, \ldots, \Omega_n, \Omega_{n+1}$ be a cycle on $\Omega = \Omega(f)$. From $W^s(\Omega_i) = \bigcup_{x \in \Omega_i} W^s(x)$ and $W^u(\Omega_i) = \bigcup_{x \in \Omega_i} W^u(x)$ [4], we have $W^s(x_i) \cap W^u(x^*_{i+1}) \neq \varnothing$ for some $x_i, x_i^* \in \Omega_i$ and $0 \leq i \leq n$. We claim that at least for one of the indices i, $\dim W^s(x_i) + \dim W^u(x^*_{i+1}) \geq \dim M$. For otherwise, if $r = \sum_{i=0}^n (\dim W^s(x_i) + \dim W^u(x^*_{i+1}))$ then $r < n \dim M$. But since $\Omega_{n+1} = \Omega_0$ and $\dim W^s(x_i) + \dim W^u(x^*_i) = \dim M$ we have $r = n \dim M$, thus reaching a contradiction. Let $y \in W^s(x_j) \cap W^u(x^*_{j+1})$, the index j chosen as above. Clearly y is a wandering point for f. Thus after a small

Originally published in: *Global Analysis (Proc. Sympos. Pure Math., Vol. XIV, Berkeley, Calif., 1968), 221–222,* © American Mathematical Society, Providence, R.I. 1970
Reprint by Springer-Verlag Berlin Heidelberg 2014 with kind permission

local pertubation of f near y we get f_i near f in Diff(M) so that $f_1/\Omega = f/\Omega$, $W^s(x_j, f_1)$ and $W^u(x^*_{j+1}, f_1)$ have a point of transversal intersection and $W^s(x_i, f_1) \cap W^u(x^*_{i+1}, f_1) \neq \emptyset$ for $0 \leq i \leq n$. From the density of the periodic points in Ω we may assume that x_j, x^*_{j+1} are periodic points of f (and thus of f_1). Now, $f_1/\Omega_{j+1} = f/\Omega_{j+1}$ is topologically transitive and so we may further assume $x^*_{j+1} = x_{j+1}$. Furthermore if $z \in W^s(x_{j+1}, f_1) \cap W^u(x^*_{j+2}, f_1)$ is nonwandering for f_1 then $f_1/\Omega(f_1)$ is not conjugate to f/Ω, which implies that f is not Ω-stable. Suppose z is wandering for f_1. Then, since Closure $W^s(x_j, f_1) \supset W^s(x_{j+1}, f_1)$ [2], by a small local pertubation of f_1 near z we get g near f in Diff(M) with $g/\Omega = f/\Omega$ and $W^s(x_j, g) \cap W^u(x^*_{j+2}, g) \neq \emptyset$, thus producing an $(n-1)$-cycle on $\Omega \subset \Omega(g)$ and hence proving our assertion.

The proof of the theorem goes now by induction on the length of the cycle and the fact that if Ω has a 1-cycle then f is not Ω-stable. For as above after a small local pertubation of f, say to f_1, $W^s(x_1, f_1)$ and $W^u(x^*_2, f_1)$ has a point y of transversal intersection. Then y is nonwandering for f_1 and using that f/Ω_i is topologically transitive we conclude that f and f_1 are not Ω-conjugate and thus f is not Ω-stable. This finishes the proof of the theorem.

From [2] and this Theorem we get

(4) Corollary. *Let $f \in$ Diff(M) and $\Omega(f)$ be finite. Then f is Ω-stable iff $\Omega(f)$ is hyperbolic and has the no cycle property.*

For the general case, Smale posed in [5] the following

(5) Conjecture. *Let $f \in$ Diff(M). Then f is Ω-stable iff f satisfies Axiom* A *and $\Omega(f)$ has the no cycle property.*

From the Ω-stability Theorem [4] and the above Theorem it follows that the remaining question to consider in the above conjecture is whether Ω-stability implies Axiom A.

References

1. M. Hirsch and C. Pugh, *Stable manifolds and hyperbolic sets*, these Proceedings, vol. 14.
2. J. Palis, *On Morse-Smale dynamical systems*, Topology (to appear).
3. S. Smale, *Differentiable dynamical systems*, Bull. Amer. Math. Soc. 73 (1967), 747–817.
4. ———, *The Ω-stability theorem*, these Proceedings. vol. 14.
5. ———, *Global stability questions in dynamical systems*, Symposium on Global Analysis, Washington, D.C., April 1968.

University of California, Berkeley
IMPA, Rio de Janeiro

Inventiones math. 9, 121 – 134 (1970)

Neighborhoods of Hyperbolic Sets

M. Hirsch, J. Palis, C. Pugh, and M. Shub (Univ. of Warwick)

§ 1. Introduction

In this paper we study the asymptotic behavior of points near a compact hyperbolic set of a C^r diffeomorphism ($r \geqq 1$) $f\colon M \to M$, M being a compact manifold. The purpose of our study is to complete the proof of Smale's Ω-stability Theorem by demonstrating (2.1), (2.4) of [6].

Ω denotes the set of non-wandering points for f. Smale's Axiom A requires [5]:

(a) Ω has a hyperbolic structure,

(b) the periodic points are dense in Ω.

Hyperbolic structure, the stable manifold of Ω, and fundamental neighborhoods are discussed in §§ 2 and 5.

The result of [6] proved here is:

If f obeys Axiom A *then there exists a proper fundamental neighborhood V for the stable manifold of Ω such that the union of the unstable manifold of Ω and the forward orbit of V contains a neighborhood of Ω in M.*

As a consequence we have:

If f obeys Axiom A *then any point whose orbit stays near Ω is asymptotic with a point of Ω.*

Section 8 of the mimeographed version of [1] contains a generalization of the above results with an incorrect proof. A correct generalization is:

(1.1) Theorem. *If Λ is a compact hyperbolic set then $W^u(\Lambda) \cup 0_+ V$ contains a neighborhood U of Λ, where V is any fundamental neighborhood for $W^s(\Lambda)$ and $0_+ V = \bigcup_{n \geqq 0} f^n(V)$. If Λ has local product structure then a proper fundamental neighborhood may be found and any point whose forward orbit lies in U is asymptotic with some point of Λ.*

Theorem (1.1) is proved in § 5, local product structure is discussed in [5] and in § 2. In § 7 we prove the analogous theorems for flows.

Here is an example, due to Bowen, of a compact hyperbolic set Λ which does not have local product structure, has no proper fundamental neighborhood and for which there are points asymptotic to Λ without being asymptotic with any point of Λ.

Originally published in: *Invent. Math. 9 (1969/1970), 121–134,*

Consider the Cantor set C as the space of sequences of zeros and ones with a decimal point:

$$x=(\dots x_{-2}\,x_{-1}\,.\,x_0\,x_1\dots) \qquad x_j=0 \text{ or } 1.$$

The elements of C can be thought of as maps $\mathbf{Z}\to\{0,1\}$. The compact open topology makes C a metrizable compact space. A distance is given by

$$d(x,y)=\sum_{-\infty}^{\infty}\frac{|x_i-y_i|}{2^{|i|}}.$$

The shift map $\sigma\colon C\to C$ moves the decimal point one space to the right:

$$\sigma(\dots x_{-1}\,.\,x_0\,x_1\dots)=(\dots x_{-1}\,x_0\,.\,x_1\dots).$$

σ is a homeomorphism. By [4] (C,σ) is conjugate to the restriction of a diffeomorphism $f\colon S^2\to S^2$ to a certain compact invariant hyperbolic set K. Hence the topological properties of C reflect themselves exactly in K.

Consider the subset of C, and correspondingly of K,

$$\Lambda=\{x\in C\colon \text{any finite maximal string of 0's is of even length}\}.$$

Clearly Λ is a compact σ-invariant subset of C and the periodic points are dense in Λ. However, there are points $c\in C$ such that

$$d(\sigma^n c,\Lambda)\to 0 \qquad \text{as } n\to\pm\infty$$

$$d(\sigma^n c,\sigma^n x)\not\to 0 \qquad \text{as } n\to\pm\infty \text{ for any } x\in\Lambda.$$

That is, there are points of C tending to Λ but not asymptotic with any point of Λ. Even worse, there are periodic points of $C-\Lambda$ whose entire orbits lie arbitrarily close to Λ. This behavior is opposite to that of (1.1).

Define such c as follows. Put odd maximal strings of zeros of increasing length on both sides of the decimal point

$$\dots 1000001000101\,.\,1010001000001\dots.$$

Given any $m>0$, $\sigma^n c$ has at most one entry of 1 in $[-m,m]$ for large enough $|n|$. The sequence x with zeros except at this entry belongs to Λ and

$$d(\sigma^n c,x)\leqq\sum_{|i|\geqq m}1/2^{|i|}=1/2^{m-1}$$

which is arbitrarily small when m is large. Hence $\sigma^n c\to\Lambda$. On the other hand, for any $x\in C$ if $d(\sigma^n c,\sigma^n x)\to 0$ as $|n|\to\infty$ then

$$x=c \text{ to the right of some entry if } n\to\infty$$

$$x=c \text{ to the left of some entry if } n\to-\infty.$$

Either way, x can not belong to Λ.

A periodic point $p \in C$ is a sequence which endlessly repeats a finite block. If this block is of the form $10 \dots 0$ with $2m+1$ zeros then the orbit of p lies within a distance $1/2^{m-1}$ of Λ by the same reasoning as before. Thus, the existence of periodic points as claimed above is clear.

§ 2. Local Product Structure

Several notions and results from [1, 5] should be recalled.

A compact set $\Lambda \subset M$ is hyperbolic for the diffeomorphism $f: M \to M$ if $f\Lambda = \Lambda$ and Tf leaves invariant a continuous splitting $T_\Lambda M = E^s \oplus E^u$, expanding E^u and contracting E^s. That is

$$|Tf(v)| \leqq \tau |v| \qquad \text{if } v \in E^s$$
$$|Tf(v)| \geqq \tau^{-1} |v| \qquad \text{if } v \in E^u$$

for some constant τ, $0 < \tau < 1$, and some Riemannian metric on M. The constant τ is called the skewness.

Through such a Λ pass families of smooth unstable and stable manifolds tangent to E^u, E^s at Λ [1]. The unstable manifold of size ε through $p \in \Lambda$ is called $W_\varepsilon^u(p)$, the stable one $W_\varepsilon^s(p)$. This "size ε" refers to the radius measured in the tangent space at p:

$$W_\varepsilon^u(p) = \exp_p(\text{graph } g_p)$$

where $g_p: E_p^u(\varepsilon) \to E_p^s(\varepsilon)$ is a smooth map whose graph has slope $\leqq 1$ with $g_p(0) = 0$, $Tg_p(0) = 0$. $E_p^u(\varepsilon)$, $E_p^s(\varepsilon)$ are the ε-discs in E_p^u, E_p^s. Similarly for $W_\varepsilon^s(p)$.

The families $W_\varepsilon^u = \{W_\varepsilon^u(p) \mid p \in \Lambda\}$, $W_\varepsilon^s = \{W_\varepsilon^s(p) \mid p \in \Lambda\}$ are overflowing invariant in the following sense: $f^{-1} W_\varepsilon^u(p) \subset W_\varepsilon^u(f^{-1} p)$ and $f W_\varepsilon^s(p) \subset W_\varepsilon^s(fp)$. They are expanding from Λ and contracting to Λ in the sense that

$$\bigcap_{n \geqq 0} f^{-n} W_\varepsilon^u = \Lambda = \bigcap_{n \geqq 0} f^n W_\varepsilon^s$$

for $W_\varepsilon^u = \bigcup_{p \in \Lambda} W_\varepsilon^u(p)$ and $W_\varepsilon^s = \bigcup_{p \in \Lambda} W_\varepsilon^u(p)$. We call $W_\varepsilon^u = W_\varepsilon^u \Lambda$ the local unstable manifold of Λ. Similarly, $W_\varepsilon^s = W_\varepsilon^s \Lambda$ is called the local stable manifold of Λ. Notice that this terminology differs from that of Smale in [5].

If p is a hyperbolic fixed point (that is $\Lambda = p$ in the preceding discussion) and V is a local submanifold transverse to its stable manifold then locally f^n presses V toward the unstable manifold of p, $W^u(p)$. If V has the same dimension as $W^u p$ then as $n \to \infty$, $f^n V \to W^u(p)$ in the C^1

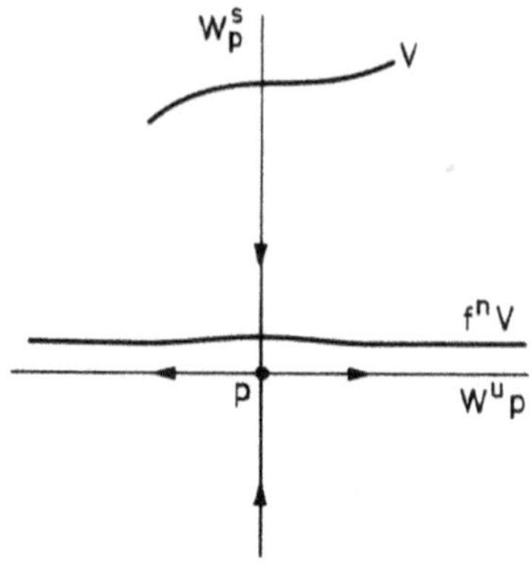

Fig. 1. f^n flattens V toward $W^u p$

sense, locally. This is the content of the λ-lemma [3] and is illustrated in Fig. 1.

A direct consequence of the λ-lemma is the

(2.1) *Cloud Lemma. If $f\colon M \to M$ is a diffeomorphism and p, q are hyperbolic periodic points with $W^u(p) \pitchfork W^s(q) \neq \emptyset$, $W^s(p) \pitchfork W^u(q) \neq \emptyset$ then these points of intersection are non-wandering.*

Proof [5]. See Fig. 2 for n so large that p, q are fixed points of f^n. The set U is a neighborhood of $x \in W^u(q) \cap W^s(p)$. We have drawn some iterates $f^{nk} U$ and shown how they must re-intersect U eventually, by the λ-lemma.

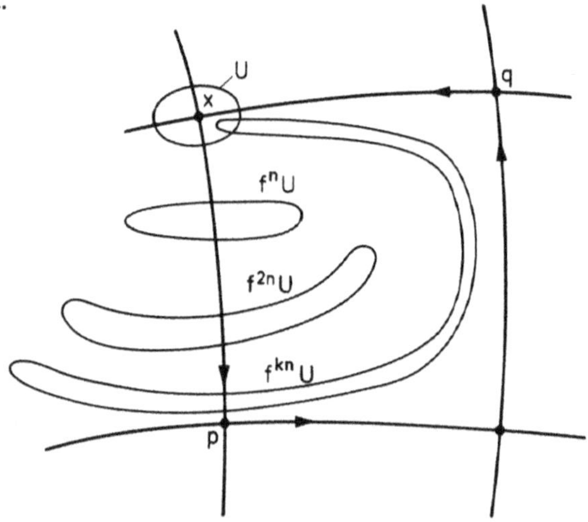

Fig. 2. f^{kn} U re-intersects U

Definition. A hyperbolic set Λ has local product structure if, for some $\varepsilon > 0$,

$$W_\varepsilon^u(p) \cap W_\varepsilon^s(p') \subset \Lambda$$

for all $p, p' \in \Lambda$.

(2.2) *Local Product Structure Theorem. If f obeys Axiom* A *then f has local product structure on Ω.*

Proof. Let $\varepsilon > 0$ be small enough that the 3ε-local stable and unstable manifolds through points of Ω are given by the stable manifold theory of [1]. Let $x, x' \in \Omega$ have $y \in W^s_\varepsilon(x) \cap W^s_\varepsilon(x')$. By [1] the intersection is transverse, consists of a single point, and the same is true of the intersection $W^s_{2\varepsilon}(x) \cap W^u_{2\varepsilon}(x')$. (See Fig. 3.)

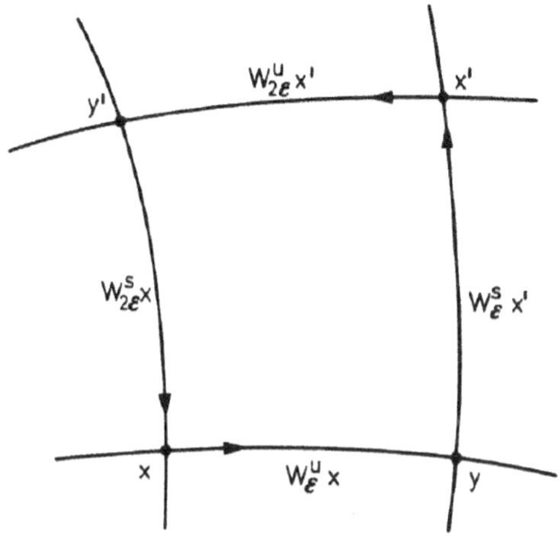

Fig. 3. Local product structure

Approximate x, x' by periodic points p, p'. By continuity of the stable and unstable manifolds, the intersections

$$W^u_{3\varepsilon}(p) \cap W^s_{3\varepsilon}(p'), \qquad W^s_{3\varepsilon}(p) \cap W^u_{3\varepsilon}(p')$$

continue to be transverse and to consist of single points, q, q' (see Fig. 4). By the Cloud Lemma, $q, q' \in \Omega$. As q is arbitrarily near y and Ω is a closed set, y belongs to Ω.

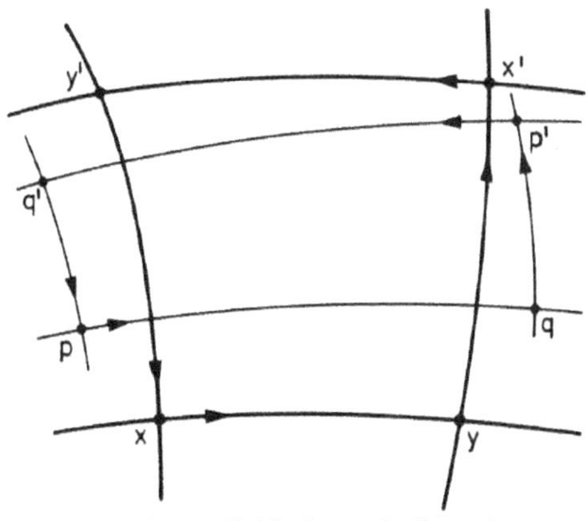

Fig. 4. Stable and unstable manifolds for periodic p, p' approximating x, x'

Corollary of the Proof. If Λ is any hyperbolic set isolated from $\Omega-\Lambda$ and the periodic points are dense in Λ then f has local product structure at Λ.

The following key lemma shows how local product structure simplifies the topology of $W_\varepsilon^s=\bigcup_{p\in\Lambda} W_\varepsilon^s(p)$.

(2.3) Lemma. *W_δ^s is a neighborhood of Λ in W_ε^s if $0<\delta\leqq\varepsilon$, ε is small, and Λ has local product structure.*

Proof. Suppose, on the contrary, that there existed a sequence of points $x_n\in W_\varepsilon^s(p_n)-W_\delta^s$ converging to some $p\in\Lambda$. As in (2.2), when ε is small this implies that $W_{2\varepsilon}^s(p_n)\pitchfork W_{2\varepsilon}^u(p)$ in a single point, say q_n. By local product structure $q_n\in\Lambda$ and $d(x_n,q_n)\to 0$. Thus $x_n\in W_\delta^s(q_n)$ when n is large, contradicting our assumption.

§3. Fundamental Domains

A fundamental domain for W_ε^s is a compact set $D\subset W_\varepsilon^s$ such that

$$W_\varepsilon^s-\Lambda\subset 0_+ D$$

where $0_+ D=\bigcup_{n\geqq 0} f^n D$, i.e. the forward orbit of D. A fundamental domain for W_ε^u is a compact set $D\subset W_\varepsilon^u$ such that

$$W_\varepsilon^u-\Lambda\subset 0_- D$$

where $0_- D=\bigcup_{n\geqq 0} f^{-n} D$. If in addition D is disjoint from Λ, we call D a *proper* fundamental domain.

(3.1) Lemma. *The set $D^s=Cl(W_\varepsilon^s-f W_\varepsilon^s)$ is a fundamental domain for W_ε^s and $D^u=Cl(W_\varepsilon^u-f^{-1} W_\varepsilon^u)$ is one for W_ε^u.*

Proof. $0_+ D^s\supset W_\varepsilon^s-\bigcap_{n\geqq 0} f^n W_\varepsilon^s=W_\varepsilon^s-\Lambda$. The proof for D^u is similar.

(3.2) Lemma. *If Λ has local product structure then D^s, D^u are proper fundamental domains.*

Proof. There is a δ, $0<\delta<\varepsilon$, such that $f W_\varepsilon^s\supset W_\delta^s$. By (2.3), W_δ^s is a neighborhood of Λ in W_ε^s. Therefore $D^s=Cl(W_\varepsilon^s-f W_\varepsilon^s)$ is disjoint from Λ. Similarly for D^u.

Question. Do the following conditions imply the existence of proper fundamental domain for $W^s\Lambda$:

(a) the periodic points are dense in Λ,

(b) there exists a neighborhood U of Λ such that if $x\in U$ and $0_+(x)\subset U$ then $\omega(x)\subset\Lambda$, where $\omega(x)$ is the ω-limit set of x.

§4. Semi Invariant Disc Families

In this section we state and prove the basic technical theorem required for (1.1).

The space $\mathrm{Emb}(D^u, M)$ of all embeddings of the closed u-disc into M may be thought of as a fiber bundle over M. The fiber at $x \in M$ is the set of all embeddings $e\colon D^u \to M$ such that $e(0)=x$. We put the uniform topology on $\mathrm{Emb}(D^u, M)$.

Definition. If $a\colon X \to \mathrm{Emb}(D^u, M)$ is a continuous section over $X \subset M$ then $\mathscr{A} = \{\text{image } a(x) \mid x \in X\}$ is a (continuous) u-disc family through X.

Clearly $W_\delta^u = \{W_\delta^u(p) \mid p \in \Lambda\}$ is a u-disc family for any small $\delta > 0$.

The following lemma may be thought of as a type of Inverse Function Theorem for certain u-disc families. By $R^m(\varepsilon)$ we mean the disc of radius ε in R^m, centered at the origin.

(4.1) Lemma. *If $u+s=m$ and $\mathscr{A} = \{A(y) \mid y \in R^s(\delta)\}$ is a u-disc family through $0 \times R^s(\delta) \subset R^m$ with $A(y) = \operatorname{graph} g_y$ for $g_y\colon R^u(\varepsilon) \to R^s$, $g_y(0) = y$ then $\bigcup_{y \in R^s} A(y)$ is a neighborhood of 0 in R^m.*

Proof. Continuity of $\mathscr{A}$ implies that $g\colon (x, y) \mapsto (x, g_y\, x)$ is a continuous map from $B = R^u(\varepsilon) \times R^s(\delta)$ into R^m. It suffices to prove that $g \mid \partial B$ is homotopic to the inclusion map $\partial B \hookrightarrow R^m - 0$. Such a homotopy is given by

$$G_t(x, y) = \big(x, (1-t)\, g_y + t\, y\big).$$

The curves $G_t(x, y)$ never pass through 0 when $(x, y) \in \partial B$ since $x=0$ implies $g_y(0) = y$ and so $G_t(x, y) = (0, y) \neq 0$. Thus, (4.1) is proved.

Now we may proceed to the main theorem of this section.

(4.2) Theorem. *For any small $\delta > 0$ there is a u-disc family $\tilde{W}_\delta^u$ through N, a neighborhood of Λ in M, which reduces to W_δ^u at Λ and is semi-invariant in the sense that*

$$\tilde{W}_\delta^u(f x) \subset f\, \tilde{W}_\delta^u(x) \qquad \text{for any } x \in N \cap f^{-1} N.$$

Moreover for any $p \in \Lambda$, $\bigcup_{y \in W_\varepsilon^s p} \tilde{W}_\delta^u(y)$ is a neighborhood of p in M.

Remark. If f is C^r we can make $\tilde{W}_\delta^u$ a continuous family of C^r u-discs.

Remark. We cannot expect any sort of uniqueness for $\tilde{W}_\delta^u$, as simple examples show with Λ taken to be one point. It is unknown whether $\tilde{W}_\delta^u$ can be chosen to foliate a neighborhood of Λ.

The proof of (4.2) is similar to the existence of $\{W_\varepsilon^u(p) \mid p \in \Lambda\}$ in [1]. All estimates needed here were proved in [1]. We must deal with graph transforms induced by maps of one space to another instead of to itself. We outline what must be re-interpreted.

Let $E=E_1\times E_2$, $F=F_1\times F_2$ be Banach spaces, equiped with the product norms

$$|(x_1,x_2)|=\max(|x_1|,|x_2|).$$

Define $\mathscr{G}_\varepsilon(E)$ to be the space of maps $g\colon E_1(\varepsilon)\to E_2$ such that

$$g(0)=0, \qquad L(g)\leqq 1$$

where L denotes the Lipschitz constant and $E(\varepsilon)$ denotes the closed ball of radius ε. The metric in $\mathscr{G}_\varepsilon(E)$ is $|g-g'|=\sup_x |g\,x-g'\,x|$ and is complete. Define $\mathscr{G}_\varepsilon(F)$ similarly.

Let $f\colon E(\varepsilon)\to F$, $f(0)=0$ be a map. It might happen that for every $g\in\mathscr{G}_\varepsilon(E)$ there exists $h\in\mathscr{G}_\varepsilon(F)$ such that

$$\text{graph } h\subset f(\text{graph } g).$$

If so, h is uniquely determined by g and we call h the graph transform of g by f,

$$h=\Gamma_f\, g.$$

The next theorem was proved in [1, §4].

(4.3) Theorem. *Let $T\colon E\to F$ be an isomorphism sending E_1 onto F_1, E_2 onto F_2 when $E_1\times E_2=E$, $F_1\times F_2=F$ as above. Suppose $\|T_2\|\leqq\tau$, $\|T_1^{-1}\|\leqq\tau$, $0<\tau<1$ for $T_i=T|E_i$, $i=1,2$. If $f\colon E(\varepsilon)\to F$ is sufficiently near $T|E(\varepsilon)$ then $\Gamma_f\colon\mathscr{G}_\varepsilon(E)\to\mathscr{G}_\varepsilon(F)$ is defined and has $L(\Gamma_f)<1$. The precise condition on f is*

$$f(0)=0 \quad \textit{and} \quad L(f-T)<\frac{1-\tau}{1+\tau}.$$

Before proving (4.2) we need a simple extension lemma.

(4.4) Lemma. *Let E, M be manifolds and $\pi\colon E\to M$ a fiber bundle. If $s\colon X\to E$ is a section over a closed $X\subset M$ then s extends to $\bar{s}\colon N\to E$, a section over a neighborhood of X.*

Proof. Since E is an ANR (absolute neighborhood rectract) and M is a normal space, s extends to a map $s_0\colon N_0\to E$ when N_0 is a neighborhood of X in M. Let d be a metric on M. Choosing small neighborhoods $N\subset N_0$ of X makes the function $d(\pi\, s_0\, x, x)$ small if $x\in N_0$. Because M is an ANR, this makes $\pi\, s_0|N\colon N\to M$ homotopic, rel X, to the inclusion $i_N\colon N\hookrightarrow M$. The Covering Homotopy Theorem [7] shows that $s_0|N$ is homotopic, rel X, to a map $\bar{s}\colon N\to E$ such that $\pi\,\bar{s}=i_N$. This proves the lemma.

Proof of (4.2). Let the skewness of $Tf|T_\Lambda M$ be $<\tau$. By (4.4) there exists a neighborhood N_0 of Λ and extensions $\bar{E}^u$, $\bar{E}^s$ of the bundles E^u, E^s to N_0 for they are sections of the Grassmann bundles $G^u(M)$, $G^s(M)$ restricted to Λ. Taking a smaller N_0 if necessary we have $\bar{E}^u\oplus\bar{E}^s=T_{N_0}M$ since $E^u\oplus E^s=T_\Lambda M$.

Applying (4.4) again we can find a "connector" θ, that is a continuous family of isomorphisms $\theta(x,x')\colon M_x \to M_{x'}$ defined for all $(x,x')\in M\times M$ sufficiently near the diagonal Δ of $\Lambda\times\Lambda$ such that

$$\theta(x,x)=I_x, \qquad \theta(x,x')\bar{E}^u_x=\bar{E}^u_{x'}, \qquad \theta(x,x')\bar{E}^s_x=\bar{E}^s_{x'}.$$

In fact $GL(\bar{E}^u)$ is a bundle over $N_0\times N_0$ whose fiber over (x,x') is the space of all isomorphisms $\bar{E}^u_x \to \bar{E}^u_{x'}$ and the map $(x,x)\to I_x$ is a section of $GL(\bar{E}^u)$ over Δ. By (4.4) this section extends giving a section θ^u. Similarly we get a section θ^s. Combining the two as $\theta^u\oplus\theta^s$ we get the connector θ.

Of course $\bar{E}^u\oplus\bar{E}^s$ is not likely to be Tf-invariant. Let $(Tf)_x\colon M_x\to M_{fx}$ be defined, with respect to this splitting, by the matrix

$$(Tf)_x=\begin{pmatrix} A_x & B_x \\ C_x & K_x \end{pmatrix}$$

for $x\in f^{-1}N_0\cap N_0$. Consider the linear map $T_x\colon M_x\to M_{fx}$ defined on $f^{-1}N_0\cap N_0$ by

$$T_x=\begin{pmatrix} A_x & 0 \\ 0 & K_x \end{pmatrix}.$$

At Λ the entries B and C vanish and $T=Tf$.

For any $N\subset f^{-1}N_0\cap N_0$ the map $T\colon T_N M\to T_{fN}M$ defined by $T|M_x=T_x$ is a bundle isomorphism leaving $\bar{E}^u\oplus\bar{E}^s$ invariant and equal Tf at Λ.

Let $\bar{f}=\exp^{-1}\circ f\circ\exp\colon M(\delta)\to TM$, where $M(\delta)=\bigcup_{x\in M} M_x(\delta)$. For small enough δ and a small enough neighborhood $N\subset N_0$ of Λ, $\bar{f}$ is well defined and

$$\sup_{x\in N} L((\bar{f}-T)\,|\,M_x(\delta))=\xi$$

satisfies the inequality of (4.3), $\xi<(1-\tau)/(1+\tau)$. For when $s\to 0$ and $N\to\Lambda$ we are merely seeing how well Tf approximates $\bar{f}$ at Λ. By (4.3) $\Gamma_{\bar{f}}\colon \mathcal{G}_\delta(T_N M)\to\mathcal{G}_\delta(T_{fN}M)$ is defined and has Lipschitz constant <1.

We may assume N chosen to be a smooth manifold with boundary. If V is a neighborhood of fN with $V\subset N_0$ and $\rho\colon V\to fN$ a retraction then $\theta(\rho x, x)$ will be defined for V small enough and ρ near enough the inclusion $i_V\colon V\hookrightarrow M$.

Define $h\colon V\to N$ by $h=f^{-1}\circ\rho$. For each $x\in V$ define $(\theta,\bar{f})_x\colon M_{hx}(\delta)\to M_x$ by

$$(\theta,\bar{f})_x=\theta(\rho x, x)\circ\bar{f}\,|\,M_{hx}(\delta)$$

and define $(\theta,T)_x\colon M_{hx}\to M_x$ by

$$(\theta,T)_x=\theta(\rho x,x)\circ T_{hx}.$$

Since θ is continuous, and equals I_x at Λ, $(\theta, T)_x$ may be made arbitrarily near $Tf|T_\Lambda M$ by taking N small, V near N, and ρ near i_V. That is

$$\sup_{x\in V} L((\theta,\bar{f})_x-(\theta, T)_x)=\bar{\xi}$$

can be made near ξ and in particular can be made $<(1-\tau)(1+\tau)$. But (θ, T) preserves $\bar{E}^u$ and $\bar{E}^s$ because both θ and T do. Taking N small enough assures (θ, T) has skewness $\leqq\tau$ because $Tf|T_\Lambda M$ has skewness $<\tau$. Hence we may apply (4.3) to conclude that $(\theta,\bar{f})$ defines a graph-transform $\Gamma_{(\theta,\bar{f})}: \mathscr{G}_\delta(T_N M)\to\mathscr{G}_\delta(T_V M)$ with Lipschitz constant <1.

Let $\varphi: M\to[0, 1]$ be continuous with $\varphi=0$ off V and $\varphi=1$ on fN. Define $F: \mathscr{G}_\delta(T_N M)\to\mathscr{G}_\delta(T_N M)$ by sending $g\in\mathscr{G}_\delta(T_N M)$ onto the element of $\mathscr{G}_\delta(T_N M)$ which, when restricted to $\bar{E}^u_x(\delta)$ equals

$$\begin{array}{ll} 0 & \text{if } x\in N-V, \\ \varphi(x)\,\Gamma_{(\theta,\bar{f})_x}(g) & \text{if } x\in V\cap N. \end{array}$$

If $x\in\partial V\cap N$, we know that $\varphi(x)=0$ and so the two definitions agree. Note that $\Gamma_{(\theta,\bar{f})_x}(g)$ is a map $\bar{E}^u_x(\delta)\to\bar{E}^s_x(\delta)$ and so our definition of F is well stated. Also note that F is *not* globally a graph transform: it does not cover a map $N\to N$.

Clearly F carries $\mathscr{G}_\delta(T_N M)$ into itself because $\Gamma_{(\theta,f)}$ carries $\mathscr{G}_\delta(T_N M)$ into $\mathscr{G}_\delta(T_V M)$ and multiplication by φ, $0\leqq\varphi\leqq 1$, can only help. In the computation of the Lipschitz size of the map $F(g)_x: \bar{E}^u_x(\delta)\to\bar{E}^s_x(\delta)$, φ is constant. For the same reason, $L(F)\leqq L(\Gamma_{(\theta,f)})<1$. Hence F has a unique fixed point, say g^*.

Put $\tilde{W}^u_\delta(x)=\exp_x(\text{graph}(g^*|M_x(\delta)))$, where $x\in f^{-1}N\cap N$. For such x, it is clear that $f\tilde{W}^u_\delta(x)\supset\tilde{W}^u_\delta(fx)$ by the definition of F. On Λ we have just gone through the construction of W^u_δ in [1] again and so $\tilde{W}^u_\delta(p)=W^u_\delta(p)$ by uniqueness of $\{W^u_\delta(p)|p\in\Lambda\}$. Thus $\tilde{W}^u_\delta$ is a semi-invariant family of u-discs extending W^u_δ. It remains to show that $\bigcup_{y\in W^s_\varepsilon(p)}\tilde{W}^u_\delta(y)$ is a neighborhood of p for $p\in\Lambda$.

There is a diffeomorphism $j: M_p(\varepsilon)\to M_p(\varepsilon)$ such that $(Tj)_p=I_p$ and $j(\exp_p^{-1}(W^s_\varepsilon p))=E^s_p(\varepsilon)$, $j(\exp_p^{-1}(W^u_\varepsilon p))=E^u_p(\varepsilon)$. So

$$\{j\circ\exp^{-1}(\tilde{W}^u_\delta y)|y\in W^s_\mu p\}$$

is a u-disc family through $E^s_p(\mu)$. Its radius may be slightly less than δ, but it is greater than $\delta/2$, say, if μ is small enough. Also, for small δ these u-discs will be graphs of maps $E^u_p(\delta/2)\to E^s_p(\varepsilon)$ having slope $\leqq 2$. By (4.3) this u-disc family includes a neighborhood of p in $M_p(\varepsilon)$ and hence its $\exp_p\circ j^{-1}$ image $\{\tilde{W}^u_\delta(y)|y\in W^s_\mu p\}$ contains a neighborhood of p in M, since $\exp_p\circ j^{-1}$ is a local diffeomorphism at p. This proves (4.2).

Remark. In a similar manner we could obtain an invariant family of linear graphs $G_x: \bar{E}^u_x \to \bar{E}^s_x$. These determine a subbundle of $T_N M$ semi-invariant under Tf, and thus a semi-invariant extension of $E^s \oplus E^u$ to $\tilde{E}^u \oplus \tilde{E}^s = T_N M$. Proceeding with the construction of $\tilde{W}^u_\delta$ we find that $\tilde{W}^u_\delta(y)$ is tangent to $\tilde{E}^u_y$ at y for y near Λ. In fact, it can be shown that $\tilde{W}^u_\delta(y)$ is C^r when f is C^r by imitating the proof of smoothness of unstable manifolds in either [1] or [2]. In this way one can prove.

(4.5) Theorem. *The semi-invariant family* $\tilde{W}^u_\delta$ *of* (4.2) *can be chosen to be a continuous family of* C^r *u-discs when* f *is* C^r.

§ 5. Fundamental Neighborhoods and the Proof of the Main Theorem

As before, let Λ be a hyperbolic set for $f \in \mathrm{Diff}(M)$. A *fundamental neighborhood* for W^s_ε is a compact neighborhood of a fundamental domain of W^s_ε. Similarly for W^u_ε.

The following theorem was stated in § 8 of the mimeographed version of [1] but was proved incorrectly.

(5.1) Theorem. *If* V *is a fundamental neighborhood for* W^s_ε *then* $W^u_\varepsilon \cup 0_+ V$ *is a neighborhood of* Λ *in* M.

From (5.1) we can finish the proof of (1.1). By (3.1), (3.2) it suffices to prove the following

(5.2) Theorem. *If* $W^u_\varepsilon \Lambda$ *has a proper fundamental domain* D *then* Λ *has a neighborhood* U *such that if* $0_+ x \subset U$ *then* $x \in W^s_\varepsilon(p)$ *for some* $p \in \Lambda$.

Proof. Let V be a compact neighborhood of D disjoint from Λ. Set

$$U = (W^s_\varepsilon \cup 0_- V) - V$$

where $0_- V = \bigcup_{n \geqq 0} f^{-n} V$. By (5.1) applied to f^{-1}, U is a neighborhood of Λ. If $f^n(x) \in U$ for all $n \geqq 0$ then $x \notin 0_- V$ for otherwise $f^n(x) \in V$ for some $n \geqq 0$, but $f^n(x) \in U \subset M - V$ which is a contradiction. Therefore $x \in W^s_\varepsilon$, proving the theorem.

Proof of (5.1). Let D be a fundamental domain for W^s_ε and let V be a neighborhood of it. We must show that $W^u_\varepsilon \cup 0_+ V$ is a neighborhood of Λ in M.

Consider the semi-invariant u-disc family $\tilde{W}^u_\delta$ constructed in (4.2) where δ has been chosen so small that $\tilde{W}^u_\delta(y) \subset V$ for any $y \in D$ and $0 < \delta \leqq \varepsilon$.

Clearly

$$W^u_\varepsilon \cup 0_+ V \supset W^u_\varepsilon \cup \bigcup_{y \in D} \tilde{W}^u_\delta(0_+ y)$$

by the semi-invariance of $\tilde{W}_\delta^u$. But $W_\varepsilon^s = \Lambda \cup \bigcup_{y \in D} (0_+ y)$ and so

$$W_\varepsilon^u \cup \bigcup_{y \in D} \tilde{W}_\delta^u(0_+ y) \supset \bigcup_{y \in W_\varepsilon^s} \tilde{W}_\delta^u(y).$$

But for each $p \in \Lambda$, $\bigcup_{y \in W_\varepsilon^s} \tilde{W}_\delta^u(y) \supset \bigcup_{y \in W_\varepsilon^s p} \tilde{W}_\delta^u(y)$ which is a neighborhood of p in M by (4.2). Hence $W_\varepsilon^u \cup 0_+ V$ is a neighborhood of Λ and (5.1) is proved.

§ 6. The Pseudo-Hyperbolic Case

The goal of this section is to generalize (4.2) to pseudo hyperbolic invariant sets of diffeomorphisms. We will apply this theorem to extend to flows the main results of the earlier sections about diffeomorphisms.

Recall that a compact subset Λ of M is pseudo hyperbolic for $f \in \mathrm{Diff}(M)$ if $f\Lambda = \Lambda$ and Tf leaves invariant a continuous spitting $T_\Lambda M = E_1 \oplus E_2$ such that $\|(Tf)^{-1} | E_1\| < k$, $\|Tf | E_2\| < \ell$, and $k\ell < 1$. For $f \in \mathrm{Diff}^r(M)$, $r \geqq 1$, Theorem 3B.5 of [2] says

(6.1) Theorem. *If $k < 1$ then there exists $\varepsilon > 0$ and a unique f-invariant C^r regular family $W_{1,\varepsilon} = \{W_{1,\varepsilon}(p) : p \in \Lambda\}$ of submanifolds of size ε such that $W_{1,\varepsilon}(p)$ is tangent to $E_1(p)$ for each $p \in \Lambda$. The manifold $W_{1,\varepsilon}(p)$ is characterized by $x \in W_{1,\varepsilon}(p)$ if and only if*

$$d(f^{-n} x, f^{-n} p) \leqq \varepsilon \qquad \textit{for all } n \geqq 0$$

$$\lim_{n \to \infty} \frac{d(f^{-n} x, f^{-n} p)}{k^n} = 0.$$

We extend W_1 to a disc family through a neighborhood of Λ as in § 4.

(6.2) Theorem. *If Λ is as in (6.1) then for any small $\delta > 0$ there is a disc family $\tilde{W}_{1,\delta}$ through N, a neighborhood of Λ, which reduces to $W_{1,\delta}$ at Λ and is semi-invariant in the sense that*

$$\tilde{W}_{1,\delta}(fx) \subset f\tilde{W}_{1,\delta}(x) \qquad \textit{for } x \in N \cap f^{-1} N.$$

Proof. The proof of (6.2) is essentially the same as that of (4.2), except that a new complete metric on the function spaces $\mathcal{G}_\varepsilon(E)$, $\mathcal{G}_\varepsilon(F)$, etc. must be used. It was introduced in [2, § 3B] and is

$$|g - g'|_* = \sup_{x \neq 0} |g x - g' x| / |x|.$$

The estimates demonstrating the analogue of (4.3) in the pseudo hyperbolic case with $k < 1$ are contained in [2, § 3B] and are similar to those in [1]. Once this analogue of (4.3) is proved, the proof of (6.2) is exactly the same as that of (4.2).

§ 7. Flows

Let $\{\varphi_t\}$ be a C^r flow on M, $r \geqq 1$. A compact invariant subset $\Lambda \subset M$ is said to be hyperbolic for $\{\varphi_t\}$ if, for every $t>0$, $T\varphi_t$ leaves invariant a continuous splitting

$$T_\Lambda M = E^u \oplus E^\varphi \oplus E^s$$

expanding E^u and contracting E^s, where E^φ is the tangent bundle to the orbits of the flow. If X is the vector field generating the flow then E^φ_x is the 0-dimensional or 1-dimensional space spanned by X_x. The equality $(T\varphi_t)_x(X_x) = X_{\varphi_t x}$ is valid for any smooth flow $\{\varphi_t\}$, so invariance of E^φ is automatic. In [5] and [2] details and equivalent definitions are given.

Note that Λ is a pseudo hyperbolic set for φ_t, $t \neq 0$. If $t>0$ then $E_1 = E^u$ and $E_2 = E^\varphi \oplus E^s$. If $t<0$ then $E_1 = E^s$ and $E_2 = E^u \oplus E^\varphi$. In either case we can choose $k<1$ where $\|(T\varphi_t | E_1)^{-1}\| < k$.

Applying (6.1) to φ_1, we get a unique φ_1-invariant family through Λ, $W^u_\varepsilon = \{W^u_\varepsilon p \mid p \in \Lambda\}$ tangent to E^u. By uniqueness, as in [2], W^u_ε is locally φ_t invariant for all t. Applying (6.1) to φ_{-1}, we get a unique φ_{-1} invariant family through Λ, $W^s_\varepsilon = \{W^\varepsilon_\varepsilon p : p \in \Lambda\}$ tangent to E^s and W^s_ε is φ_t-invariant for all t. For any orbit $\mathcal{O} \subset \Lambda$ we put

$$W^u_\varepsilon \mathcal{O} = \bigcup_{p \in \mathcal{O}} W^u_\varepsilon p, \qquad W^s_\varepsilon \mathcal{O} = \bigcup_{p \in \mathcal{O}} W^s_\varepsilon p.$$

(7.1) Theorem. *The* $\{W^u_\varepsilon \mathcal{O}\}$, $\{W^s_\varepsilon \mathcal{O}\}$ *form invariant families of* C^r *manifolds. The families* $\{W^u_\varepsilon p \mid p \in \mathcal{O}\}$, $\{W^s_\varepsilon p : p \in \mathcal{O}\}$ *form* C^r *fibrations of* $W^u_\varepsilon \mathcal{O}$, $W^s_\varepsilon \mathcal{O}$.

Proof. Since $\varphi_t W^u_\varepsilon p \supset W^u_\varepsilon(\varphi_t p)$ for $t>0$ and $\varphi_t W^s_\varepsilon p \supset W^s_\varepsilon(\varphi_t p)$ for $t<0$, (7.1) is a direct consequence of (6.1).

We say that Λ has local product structure if

$$W^s_\varepsilon \cap W^u_\varepsilon = \Lambda$$

for some $\varepsilon>0$. (2.3) and (3.1) generalize at once to

(7.2) Proposition. *Let* $\{\varphi_t\}$ *be a flow satisfying Axiom A′ [5], i.e. the nonwandering set* Ω *is hyperbolic and the periodic orbits are dense in* Ω. *Then* Ω *has local product structure.*

(7.3) Lemma. *If* $t>0$ *and* Λ *is hyperbolic for* $\{\varphi_t\}$ *then the set* $D^s = Cl(W^s_\varepsilon - \varphi_t W^s_\varepsilon)$ *is a fundamental domain for* $W^s_\varepsilon \Lambda$. *If* Λ *has local product structure,* D^s *is proper.*

We also have

(7.4) Theorem. *If* $D \subset W^s_\varepsilon \Lambda$ *is a fundamental domain for some* φ_τ, $\tau>0$, *and* V *is a neighborhood of* D. *Then the set*

$$W^u_\varepsilon \Lambda \cup 0_+ V$$

is a neighborhood of Λ *in* M.

10*

Proof. Extend the disc family W_ε^u to a disc family $\tilde{W}_\varepsilon^u$, semi-invariant by φ_τ, through a neighborhood of Λ. This can be done by (6.2). Then apply (4.1) as in (4.2). Since $0_+ V = \bigcup_{t \geq 0} \varphi_t V \supset \bigcup_{n \geq 0} \varphi_{n\tau} V$, the theorem is proved.

(7.2, 7.3, 7.4) provide the analogue of (1.1) for flows.

Remark. (7.4) holds for normally hyperbolic foliations as defined in [2]. The proof is the same.

References

1. Hirsch, M., Pugh, C.: Stable manifolds and hyperbolic sets. Proceedings of the A.M.S. Summer Institute on Global Analysis (Berkeley, 1967) (to appear).
2. — — Shub, M.: Invariant manifolds (to appear).
3. Palis, R.: On Morse-Smale dynamical systems. Topology **8**, No. 4, 385–405 (1969).
4. Smale, S.: Diffeomorphisms with many periodic points. Differential and combinatorial topology, pp. 63–80. Princeton, N. J.: Princeton Univ. Press 1965.
5. Smale: Differentiable dynamical systems. Bull. of A.M.S. **73**, No. 6, 747–817 (1967).
6. — The Ω-stability theorem. Proceedings of the A.M.S. Summer Institute on Global Analysis (Berkeley, 1967) (to appear).
7. Steenrod, N.: The topology of fiber bundles. Princeton, N. J.: Princeton Univ. Press 1951.

M. Hirsch
Math. Dept.
University of California
Berkeley, Calif. 94720
USA

C. Pugh, M. Shub
Institut des Hautes Etudes Scientifiques
35, Route de Chartres
F-91 Bures-sur-Yvette

J. Palis
Instituto de Matemática Pura e Appliada
rua Luis de Camoes 68
Rio de Janeiro 58
GB Brasil

(Received October 1, 1969)

Reprinted from:
DYNAMICAL SYSTEMS
© 1973
ACADEMIC PRESS, INC., NEW YORK AND LONDON

Hyperbolic Nonwandering Sets on Two-Dimensional Manifolds†

S. NEWHOUSE and J. PALIS
Instituto de Matemática Pura e Aplicada
Rio de Janeiro, Brasil

We are concerned here with diffeomorphisms of a compact manifold M whose nonwandering sets are hyperbolic.

Let f be such a diffeomorphism and Ω its nonwandering set. Consider the following questions:

(a) Are the periodic orbits of f dense in Ω?
(b) Can f be approximated by an Ω-stable diffeomorphism?

We show in this chapter that both questions have a positive answer when M is a closed two-dimensional manifold.

The first question was suggested by Smale in [7]. Related to it is Anosov's *closing lemma*: The periodic orbits are dense in the nonwandering set of f/Ω, f being as above. Also, one should notice that if f is an Anosov diffeomorphism, then the periodic points are dense in the nonwandering set. This follows, for instance, from the structural stability of f and Pugh's *closing lemma*.

With respect to the second question, if f has the no-cycle property, then f is Ω-stable [8], [3]. On the other hand, if there is a cycle, one can radically change the nonwandering set by small perturbations, so f is not Ω-stable [5]. What we do here is to break all existing cycles without changing Ω to achieve an Ω-stable diffeomorphism.

† AMS (MOS) 1970 Subject Classification: 34C35, 34C40.

Originally published in: *Dynamical systems (Proc. Sympos., Univ. Bahia, Salvador, 1971), 293–301,* © Academic Press Inc., New York and London 1973
Reprint by Springer-Verlag Berlin Heidelberg 2014 with kind permission

We now set some notation, basic definitions, and the precise statements of our results.

Let $\mathrm{Diff}^r(M)$ be the set of C^r diffeomorphisms of M with the C^r topology, $r \geq 1$. For $f \in \mathrm{Diff}^r(M)$, we denote by $P = P(f)$ the set of periodic points and by $\Omega = \Omega(f)$ the nonwandering set of f. We say that $\Omega = \Omega(f)$ is hyperbolic if there exist a continuous splitting $T_\Omega M = E^s \oplus E^u$, a riemannian norm $|\cdot|$ on TM, and a constant $0 < \lambda < 1$ such that $|Tf(v)| \leq \lambda |v|$ for $v \in E^s$ and $|Tf^{-1}(v)| \leq \lambda |v|$ for $v \in E^u$. The diffeomorphism $g \in \mathrm{Diff}^r(M)$ is called Ω-stable if there exists a neighborhood N of g in $\mathrm{Diff}^r(M)$ such that for any $\tilde{g} \in N$ there is a homeomorphism $h: \Omega(g) \to \Omega(\tilde{g})$ satisfying the condition $hg(x) = \tilde{g}h(x)$ for all $x \in \Omega(g)$.

Assume M is a compact manifold without boundary and $\dim M = 2$.

THEOREM A For $f \in \mathrm{Diff}^r(M)$, if $\Omega(f)$ is hyperbolic, then $P(f)$ is dense in $\Omega(f)$. That is, f satisfies Smale's Axiom A.

THEOREM B For $f \in \mathrm{Diff}^r(M)$, if $\Omega(f)$ is hyperbolic, then f can be approximated in $\mathrm{Diff}^r(M)$ by an Ω-stable diffeomorphism g with $\Omega(g) = \Omega(f)$.

The starting point for proving these results is the analogue of the spectral decomposition theorem [7] for limit sets as considered in [3].

Let $f \in \mathrm{Diff}^r(M)$. For $x \in M$, define

$$\omega(x) = \{y \in M : \text{there exists a sequence of integers } n_i \to \infty \text{ such that } \lim f^{n_i}(x) = y\}.$$

Dually, one defines $\alpha(x)$. Let

$$\omega(f) = \bigcup_{x \in M} \omega(x) \qquad \text{and} \qquad \alpha(f) = \bigcup_{x \in M} \alpha(x).$$

The limit set $L = L(f)$ of f is the closure of $\omega \cup \alpha$.

If L is hyperbolic, then $\bar{P} = L$, and L can be written as the disjoint union $L = L_1 \cup \cdots \cup L_n$, each L_i being closed, invariant by f, and topologically transitive. The periodic points are dense in each L_i, and L_i has a local product structure, i.e., there is an $\varepsilon > 0$ such that $W_\varepsilon^s(L_i) \cap W_\varepsilon^u(L_i) \subset L_i$([3], [7]).

The L_i are called basic limit sets.

Notice that

$$M = \bigcup_i W^s(L_i) = \bigcup_i W^u(L_i),$$

and using [2]

$$W^s(L_i) = \bigcup_{x\in L_i} W^s(x), \qquad W^u(L_i) = \bigcup_{x\in L_i} W^u(x).$$

W^s and W^u stand, respectively, for the stable and unstable manifolds of sets involved. By $W^s_{loc}(x)$ and $W^u_{loc}(x)$, $x \in L$, we mean $W_\varepsilon^s(x)$ and $W_\varepsilon^u(x)$, for some small $\varepsilon > 0$ as defined in Section 2 of [2] (see also [1]). If $\dim W^s(x) = 1$, define $W_+^s(x)$ and $W_-^s(x)$ to be the components of $W^s_{loc}(x) - \{x\}$.

From now on, we assume $\dim M = 2$ and $f \in \mathrm{Diff}^r(M)$ with $\Omega = \Omega(f)$ hyperbolic.

PROPOSITION 1 There exists a finite subset $F^s \subset P$ such that if $x \in L$, $\dim W^s(x) = 1$ and x is not a limit point of both $W_+^s(x) \cap L$ and $W_-^s(x) \cap L$, then $x \in W^u(p)$, $p \in F^s$.

Proof Let $\varepsilon > 0$ be such that no arc $V \subset W^s_{loc}(x)$ with both boundary points in L and length less than ε contains x in its interior.

We first show that $x \in W^u(p)$, for some $p \in P$. Suppose not. Since $x \in L$, $x \in L_i$ for some basic limit set L_i. Let $y \in \alpha(x)$ and n_i a sequence of integers such that $n_i \to \infty$ and $f^{-n_i}(x) \to y$. Clearly $y \in L_i$ and using the local product structure of L_i we get an integer $n > 0$ and an arc $(z, w) \subset f^{-n}W^s_{loc}(x)$ of length smaller than ε containing $f^{-n}(x)$ in its interior, z and w in L_i. Thus we reach a contradiction, since $f^n(z)$, $f^n(w) \in L_i$, and the arc $(f^n(z), f^n(w)) \subset W^s_{loc}(x)$ has length less than ε and contains x in its interior.

A similar argument shows that the subset of P for which there are points in their unstable manifolds satisfying the hypotheses of the proposition is finite.

The same fact is true interchanging stable and unstable manifold in the above proof. This yields a corresponding finite subset $F^u \subset P$. We denote by F the union of F^s and F^u.

We now define basic regions and a partial ordering relation on them. This will play a key role in proving the main results in the paper.

Let $p \in F \cap L_i$ with period n, L_i being a basic limit set. Clearly, p is a saddle point. Let $U, f(U), \ldots, f^{n-1}(U)$ be a pairwise disjoint small open disks containing the orbit $O(p)$ such that in $U' = \bigcup_{k=0}^{k=n-1} f^k(U)$. $W^s_{loc}(O(p))$ and $W^u_{loc}(O(p))$ are well defined and $U' \cap L_j = \varnothing$ for $j \neq i$, In U' we distinguish the open regions $R_1, \ldots, R_m$ bounded by $W^s_{loc}(O(p))$ and $W^u_{loc}(O(p))$ such that $U' \cap f^l(R_k) \subset R_k$ for $0 \leq l \leq n$. The R_k

are called basic regions. For instance, if $(Tf^n)_p$ has positive eigenvalues then there are four basic regions R_1, R_2, R_3, R_4 associated to $O(p)$, each one has n connected components corresponding to the open quadrants defined in U' by $W^s_{loc}(O(p))$ and $W^u_{loc}(O(p))$. If one eigenvalue of $(Tf^n)_p$ is negative, then there are two basic regions, each having $2n$ connected components.

Notice that $W^s_{loc}(O(p)) \cap R_k = \emptyset$ and $W^u_{loc}(O(p)) \cap R_k = \emptyset$ for each k.

For a basic limit set L_j, which is not a periodic orbit in F, the basic region is defined simply as a neighborhood R of L_j, where $W^s_{loc}(L_j)$ and $W^u_{loc}(L_j)$ are well defined and $R \cap L_r = \emptyset$ for $r \neq j$.

Let $R_1, R_2, \ldots, R_n$ be the basic regions considered above. In $\bar{R}_i$ we define

$$R_i^s = W^s_{loc}(O(p)) \cap \bar{R}_i - L,$$
$$R_i^u = W^u_{loc}(O(p)) \cap \bar{R}_i - L,$$

or

$$R_i^s = W^s_{loc}(L_i) - \{W^s(F \cap L_i) \cup L_i\},$$
$$R_i^u = W^u_{loc}(L_i) - \{W^u(F \cap L_i) \cup L_i\},$$

according to whether R_i is associated to $O(p) \subset F$ or to a basic limit set L_i not contained in F.

Notice that the correspondences $R_i \to R_i^s$ and $R_i \to R_i^u$ may be two-to-one for periodic orbits in F.

We also globalize these "local components" of stable and unstable manifolds by setting

$$\tilde{R}_i^s = \bigcup_0^\infty f^{-n}(R_i^s),$$

$$\tilde{R}_i^u = \bigcup_0^\infty f^n(R_i^u).$$

Given two basic regions R_i and R_j, we say that R_i almost visits R_j if there exist sequences $x_n \to x$, $x_n \in R_i$ and $x \in R_i^s$, and $m_n \to \infty$, such that $f^{m_n}(x_n) \to y$, $f^{m_n}(x_n) \in R_j$, and $y \in R_j^s$. In this case, we also say that x almost visits y in R_j from R_i.

LEMMA 2 Let R_i and R_j be basic regions such that R_i almost visits R_j. Then, given any open set V meeting R_i^s, we have $R_j^u \subset \overline{\bigcup_{n>0} f^n(V)}$.

Proof We construct a sequence of basic regions $R_{i_0} = R_i, R_{i_1}, \ldots, R_{i_l} = R_j$, and a sequence of points $x_0, x_1, \ldots, x_l$ such that

(a) R_{i_k} almost visits $R_{i_{k+1}}$ for $0 \le k < l$;
(b) Either $x_{k+1} \in R^u_{i_k} \cap \tilde{R}^s_{i_{k+1}}$ or $x_{k+1} \in W^u(q) \cap \tilde{R}^s_{i_{k+1}}$, where $q \in F \cap L_k$ and R_{i_k} is associated to some $O(p) \subset F \cap L_k$ or to L_k, $0 \le k < l$;
(c) For any open set V_k meeting $R^s_{i_k}$, $R^u_{i_p} \subset \overline{\bigcup_{n\ge 0} f^n(V_k)}$ for $0 \le k \le p \le l$.

Such a sequence, whose existence will be proved by induction, implies the lemma.

Since R_i almost visits R_j, we have sequences $x_n \to x_0$, $x_n \in R_i$ and $x_0 \in R_i{}^s$, and $f^{m_n}(x_n) \to y$, with $m_n \to \infty$, $f^{m_n}(x_n) \in R_j$, and $y \in R_j{}^s$. Take $R_{i_0} = R_i$. Let U be a small neighborhood of L_i, R_i being associated to L_i or to $O(p) \subset F \cap L_i$. From [2], there exist $x_1 \in W^u_{\text{loc}}(L_i) - L_i$ and a sequence $\varrho_n < m_n$ such that $\varrho_n \to \infty$, $f^j(x_n) \in U$, for $0 \le j \le \varrho_n$, and $f^{\varrho_n}(x_n) \to x_1$. Either $x_1 \in W^u(L_i) - W^u(F \cap L_i)$ or $x_1 \in W^u(q)$ for some $q \in F \cap L_i$. Now we can find an index i_1 such that $x_1 \in \tilde{R}^s_{i_1}$, R_i almost visits R_{i_1} and R_{i_1} almost visits R_j, and given any neighborhood V meeting $R_i{}^s$, $R^u_{i_1} \subset \overline{\bigcup_{n\ge 0} f^n(V)}$. In the case $x_1 \in W^u(L_i) - W^u(F \cap L_i)$, this is easily checked since $W^u(L_i) \subset \overline{\bigcup_{n\ge 0} f^n(V)}$ and $R_\nu{}^u \subset \overline{W^u(L_i)}$ for the possible indices ν such that $x_1 \in \tilde{R}_\nu{}^s$. If $x_1 \in W^u(F \cap L_i)$, then $W^u(x_1) \subset \overline{\bigcup_{n\ge 0} f^n(V)}$. Of course, $W^u(x_1)$ and $W^s(x_1)$ might meet transversely at x_1 or not. Analyzing the types of possible intersections of $W^u(x_1)$ and $W^s(x_1)$ at x_1 and using the λ-lemma [4], one can choose R_{i_1} with $x_1 \in \tilde{R}^s_{i_1}$ as claimed. Once x_k, R_{i_k} have been chosen, a similar construction yields a point x_{k+1} and a region $R_{i_{k+1}}$ with $x_{k+1} \in \tilde{R}^s_{i_{k+1}}$, so that conditions (a)–(c) above hold for the sequence $R_{i_0}, \ldots, R_{i_{k+1}}$.

Finally, we claim that we must reach R_j with this process. In fact, the set of basic regions is finite, so the only possible obstruction to that would be the existence of cycles in the sequence, i.e., sequences $R_{i_m}, R_{i_{m+1}}, \ldots, R_{i_n}$ as above with $R_{i_m} = R_{i_n}$. But if this happens, condition (c) implies that all the points $x_k \in \tilde{R}^s_{i_k}$, $m \le k < n$, are nonwandering. Since Ω is hyperbolic, $W^u(x_k)$ and $W^s(x_k)$ meet transversely at x_k. This implies that each x_k is in the closure of the set of transversal homoclinic points of the periodic points. It now follows from [6] that $x_k \in \bar{P} = L$, contradicting the construction of x_k. Thus the lemma is proved.

For basic regions R_i and R_j, define $R_i \ge R_j$ if either $R_i = R_j$ or R_i almost visits R_j.

As an immediate consequence of Lemma 2 we have

THEOREM 3 The relation $\ge$ defines a partial ordering on the set of basic regions.

We can now prove Theorem A.

Proof of Theorem A Since $\bar{P} = L$, it is enough to prove $\Omega = L$. Suppose there exists $x \in \Omega - L$. Then $x \in \tilde{R}_i^{s} \cap \tilde{R}_j^{u}$ for basic regions R_i, R_j, and R_i almost visits R_j. We now proceed as in Lemma 2 constructing a sequence of basic regions $R_{i_0} = R_i, R_{i_1}, \ldots, R_{i_l} = R_j$, and a sequence of points $x_0 = x, x_1, \ldots, x_l$ such that

(a) R_{i_k} almost visits $R_{i_{k+1}}$, for $0 \leq k < l$;
(b) Either $x_{k+1} \in R_{i_k}^{u} \cap \tilde{R}_{i_{k+1}}^{s}$ or $x_{k+1} \in W^{u}(q) \cap \tilde{R}_{i_{k+1}}^{s}$, where $q \in F \cap L_k$ and R_{i_k} is associated to some $O(p) \subset F \cap L_k$ or to L_k, $0 \leq k < l$;
(c) For any open set V_k meeting $R_{i_k}^{s}$, $R_{i_p}^{u} \subset \overline{\bigcup_{n\geq 0} f^n(V_k)}$ for $0 \leq k \leq p \leq l$.

Also, since $x \in \Omega$ and Ω is hyperbolic,

(d) $W^{u}(x)$ and $W^{s}(x)$ meet transversely at x.

Thus taking any neighborhood V of x_1, $\tilde{R}_{i_l}^{u} \subset \bigcup_{n>0} f^n(V)$, and using (d), $\overline{\bigcup_{n>0} f^n(V)} \cap V \neq \emptyset$, which implies that $x_1 \in \Omega$. Again since Ω is hyperbolic, $W^{u}(x_1)$ and $W^{s}(x_1)$ meet transversely at x_1. Repeating the argument, it follows that each x_k is nonwandering and consequently $W^{u}(x_k)$ and $W^{s}(x_k)$ meet transversely at x_k, $0 \leq k < l$. Again, from [6], $x \in \bar{P} = L$, which contradicts the assumption that $x \in \Omega - L$. Therefore, $\Omega = L$ and so Theorem A is proved.

As a consequence of Theorem A, we have that each basic limit set L_i in the spectral decomposition theorem corresponds, in fact, to a basic set Ω_i as defined by Smale in [7]. We keep the notation $\Omega = L = L_1 \cup \cdots \cup L_n$.

As in [3], define the relation $\geq$ on the basic sets by $L_i \geq L_j$ if there is a sequence $L_{i_0} = L_i, L_{i_1}, \ldots, L_{i_l} = L_j$ such that $\overline{W^{u}(L_{i_k})} \cap W^{s}(L_{i_{k+1}}) \neq \emptyset$ for $0 \leq k < l$. A corresponding equivalence relation is defined on the $\{L_i\}$ by $L_i \sim L_j$ if $L_i \geq L_j$ and $L_j \geq L_i$. Let $\{\gamma_1, \ldots, \gamma_m\}$ be the distinct equivalence classes of $\{L_i\}$ under $\sim$. The $\{\gamma_i\}$ are partially ordered by $\gamma_i \geq \gamma_j$ if there exist $L_i \in \gamma_i$ and $L_j \in \gamma_j$ such that $L_i \geq L_j$. In what follows, the union of the basic sets in a class γ_i will also be denoted by γ_i.

We now relate these notions with basic regions and stable and unstable manifolds of basic regions.

A component of the class γ is a global unstable manifold $\tilde{R}_k^{u}$ of the basic region R_k, where R_k is associated to a periodic orbit in $F \cap \gamma$. The component $\tilde{R}_k^{u}$ is free if $\tilde{R}_k^{u} \subset \bigcup_{\gamma_j<\gamma} W^{s}(\gamma_j)$.

LEMMA 4 Suppose that $\tilde{R}_k{}^{\mathrm{u}} \subset W^{\mathrm{u}}(\gamma)$ and $\tilde{R}_k{}^{\mathrm{u}} \cap W^{\mathrm{s}}(\gamma) \neq \emptyset$ for some equivalence class γ. Then there are basic regions associated to γ, $R_{k_0} = R_k > R_{k_1} > \cdots > R_{k_n}$ and $R_{k_{n+1}}, R_{k_{n+2}}, \ldots, R_{k_{n+m}}$ such that

(a) $R_{k_n} > R_{k_{n+r}}$ for $1 \leq r \leq m$;

(b) $\tilde{R}^{\mathrm{u}}_{k_{n+r}}$ is free for $1 \leq r \leq m$;

(c) $\tilde{R}^{\mathrm{u}}_{k_n} \subset \bigcup_{r=1}^{m} W^{\mathrm{s}}(O(p_r)) \cup \bigcup_{\gamma_j < \gamma} W^{\mathrm{s}}(\gamma_j)$, and

$$(\overline{\tilde{R}^{\mathrm{u}}_{k_n}} - (\tilde{R}^{\mathrm{u}}_{k_n} \cup L_{k_n})) \cap W^{\mathrm{u}}(\gamma) \subset \bigcup_{r=1}^{m} \tilde{R}^{\mathrm{u}}_{k_{n+r}} \cup O(p_r),$$

where L_{k_n} is the basic set to which R_{k_n} is associated, and $O(p_r)$ is the periodic orbit of $F \cap \gamma$ to which $R_{k_{n+r}}$ is associated.

(d) $\tilde{R}^{\mathrm{u}}_{k_{n+r}}$ is connected, for $1 \leq r \leq m$.

Proof The lemma is proved by induction on the number of basic regions involved. Suppose $R_{k_0} = R_k > R_{k_1} > \cdots > R_{k_j}$ has been defined. If there are no $R_{k_{j+1}}, \ldots, R_{k_{j+s}}$ such that conditions (a)–(d) are satisfied, we can proceed to construct $R_{k_{j+1}}$ such that

$$R_{k_0} = R_k > R_{k_1} > \cdots > R_{k_j} > R_{k_{j+1}}.$$

Since the number of basic regions is finite and by Theorem 3 they are partially ordered by $>$, eventually we get a sequence of basic regions as desired.

COROLLARY 5 In the previous lemma the basic region R_{k_n} is associated to a periodic orbit in $F \cap \gamma$.

From Lemma 4 and Corollary 5, it follows that if each class γ_i has only free components, then f has the no-cycle property and thus is Ω-stable ([3, 8]). Theorem B will be proved through a finite number of small perturbations that keep the nonwandering set the same and increase the number of free components.

We now restrict our attention to the periodic orbits $O(p_1), \ldots, O(p_r)$ that occur in Lemma 4.

Let γ be a $\sim$-class consisting of more than one basic set. Let R_{i_1} and R_{i_2} be basic regions associated to a periodic orbit $O(p)$ in $F \cap L_i$, $L_i \in \gamma$, so that $\tilde{R}^{\mathrm{u}}_{i_1} = \tilde{R}^{\mathrm{u}}_{i_2}$ is free. Suppose R_{i_3} and R_{i_4} are the other basic regions associated to $O(p)$, R_{i_3} adjacent to R_{i_1} and R_{i_4} adjacent to R_{i_2}, i.e., $\tilde{R}^{\mathrm{s}}_{i_1} = \tilde{R}^{\mathrm{s}}_{i_3}$ and $\tilde{R}^{\mathrm{s}}_{i_2} = \tilde{R}^{\mathrm{s}}_{i_4}$. Notice that possibly $i_1 = i_2$ and

$i_3 = i_4$. For $\varepsilon > 0$ small and $\delta < \varepsilon/2$, let $G = G' \cap W_\varepsilon^s(O(p))$, where

$$G' = \overline{W_\delta^s(L_i) - f W_\delta^s(L_i)},$$

i.e., G' is a proper fundamental domain for L_i [2].

LEMMA 6 There exists a neighborhood N of G such that for any $y \in M - \Omega$ and $x', x'' \in N \cap \{\bar{R}_{i_1} \cup \bar{R}_{i_3}\}$ if x' almost visits y from R_{i_3}, then y can only visit x'' in R_{i_1}.

The same is true interchanging R_{i_1} and R_{i_3} by R_{i_2} and R_{i_4}.

Proof If the lemma were false, choosing appropriate convergent subsequences, we could find $x', x'' \in R_{i_3}^s \cap N$ and $y \in M - \Omega$ such that x' almost visits y from R_{i_3} and y almost visits x'' in R_{i_3}. Since $y \notin \Omega = L$, we can choose R_{j_1} and R_{j_2} such that $\tilde{R}_{j_1}^s \cap \overline{\tilde{R}_{j_2}^u} \ni y$, R_{j_1} almost visits R_{i_3} and R_{i_3} almost visits R_{j_2}. Therefore, R_{j_1} almost visits R_{j_2} by Theorem 3. Moreover, constructing a sequence as in Lemma 2, we see that $V \cap \bigcup_{n>0} f^n(V) \neq \emptyset$ for any neighborhood V of y. Thus $y \in \Omega$, contradicting the assumption. Lemma 6 is proved.

Notice that if we perform a perturbation of f with support in N that creates a new nonwandering point $y \in M$, then there must exist a pair of points x', x'' as above.

Proof of Theorem B The theorem is proved by making a sequence of approximations, each increasing the number of free components, without changing Ω.

We can take on $\{\gamma_i\}$ a linear ordering compatible with $\geq$, so that $\gamma_n \geq \gamma_{n-1} \geq \cdots \geq \gamma_1$. Suppose all components of γ_j are free for $\gamma_j \leq \gamma_{i-1}$, so each class γ_j is made of a unique basic set. Suppose $\tilde{R}_k{}^u$

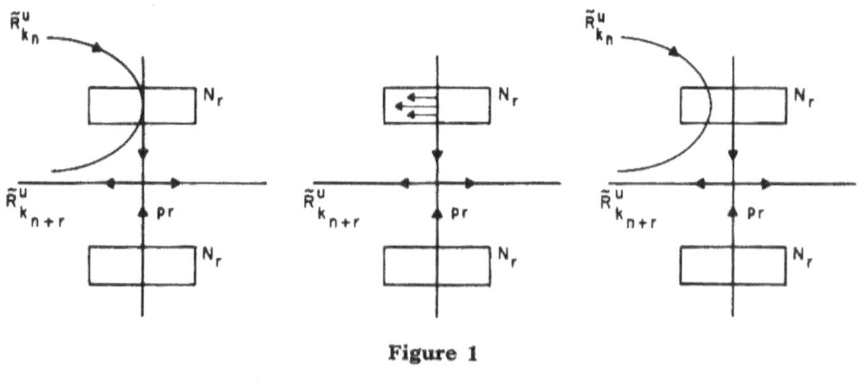

Figure 1

is associated to the class γ_i and $\tilde{R}_k^u \cap W^s(\gamma_i) \neq \emptyset$. From Lemma 4, there are regions $R_{k_n}, R_{k_{n+1}}, \ldots, R_{k_{n+m}}$ such that $R_k > R_{k_i}$ for $n \leq i \leq m + n$, and (a)–(d) in the lemma are satisfied.

Using Lemma 6, we can perform an arbitrarily small C^r perturbation of f such that Ω remains the same, the free components of γ_j remain free for $j \leq i$, and $\tilde{R}_{k_n}^u$ becomes free. This perturbation has support in the fundamental neighborhoods N_r, $1 \leq r \leq m$, defined in Lemma 6 and is taken so small as to be disjoint from all free components of γ_j, for $j \leq i$. For each r, $1 \leq r \leq m$, the perturbation is the time-one map of a vector field parallel to $\tilde{R}_{k_{n+r}}^u$ with the same "orientation," as described in Figure 1.

In this way, we achieve a diffeomorphism g, C^r near f, such that $\Omega(g) = \Omega(f)$ and all the components of the basic sets for g are free. Therefore g has the no-cycle property and by [8] is Ω-stable.

References

[1] M. Hirsch and C. Pugh, Stable manifolds for hyperbolic sets, *Global Anal. Proc. Symp. Pure Math.* **14**, pp. 133–163. Amer. Math. Soc., Providence, Rhode Island, 1970.

[2] M. Hirsch, J. Palis, C. Pugh, and M. Shub, Neighborhoods of hyperbolic sets, *Inventiones Math.* **9** (1970), 212–234.

[3] S. Newhouse, Hyperbolic limit sets, *Trans. Amer. Math. Soc.* **167** (1972), pp. 125–150.

[4] J. Palis, On Morse–Smale dynamical systems, *Topology* **8** (1969), 385–405.

[5] J. Palis, A note on Ω-stability, *Global analysis, Proc. Symp. Pure Math.* 1970, **14**, pp. 221–222. Amer. Math. Soc., Providence, Rhode Island, 1970.

[6] S. Smale, Diffeomorphisms with many periodic points, "Differential and Combinatorial Topology," Princeton Univ. Press, Princeton, New Jersey, 1965, pp. 63–80.

[7] S. Smale, Differentiable dynamical systems, *Bull. Amer. Math. Soc.* **73** (1967), 747–817.

[8] S. Smale, The Ω-stability theorem, Global analysis, *Proc. Symp. Pure Math.* **14**, Amer. Math. Soc., Providence, Rhode Island, 1970.

THE TOPOLOGY OF HOLOMORPHIC FLOWS WITH SINGULARITY

by CESAR CAMACHO (1), NICOLAAS H. KUIPER (2) *and* JACOB PALIS (1)

Prologue and summary (3)

The integrals of the differential equations defined by a holomorphic vector field F on a complex manifold are complex curves parametrized by **C**. The corresponding action of **C** is called a holomorphic flow and the complex curves are its orbits. These orbits, in general two-dimensional real surfaces, form a foliation $\mathcal{F}(\mathrm{F})$ with singularities at the zeroes of the vector field F. We study the topology of such foliations $\mathcal{F}(\mathrm{F})$, in particular near a singularity. A simple example on $\mathbf{C}^2$, which is rather general from the point of view of topology as we will see later, is given by the differential equations in complex numbers:

$$\frac{dz_1}{d\mathrm{T}}=-z_1, \quad \frac{dz_2}{d\mathrm{T}}=iz_2,$$

with solution in $\mathrm{T}=u+iv$:

$$z_1=e^{-u-iv}w_1, \quad z_2=e^{-v+iu}w_2,$$

through the point (w_1, w_2). The solutions are real two-dimensional leaves of a foliation $\mathcal{F}$ with singularity at $o \in \mathbf{C}^2$. Special leaves (topologically cylinders) are the *coordinate axes* $(z_1 \neq o = z_2)$ and $(z_2 \neq o = z_1)$. We see that every other leaf is transversal to $|z_1|=r$, to $|z_2|=r$, and to the "sphere" $\sup_j |z_j|=r$ for $r>o$. It is topologically a cone with (deleted) top at $o \in \mathbf{C}^2$. Starting from any point (w_1, w_2), it is seen to *wrap around the* z_1*-axis while converging to it* for $u=o$, $v \to \infty$:

$$z_1=e^{-iv}w_1, \quad z_2=e^{-v}w_2,$$

---

(1) Instituto de Matemática Pura et Aplicada, Rio de Janeiro.
(2) Institut des Hautes Études Scientifiques, Bures-sur-Yvette.
This research was partially supported by CNPQ (Brásil) and CNRS (France).
(3) The results of chapter I and II were announced by M. Jean Leray at the meeting of the Academie des Sciences of May 3, 1976 [2]. After completing this paper we learned that theorem 1 was also found in USSR. See Ilyashenko [15].
The authors are greatful for important critical remarks of the referee.

Originally published in: *Publ. Math. Inst. Hautes Études Sci. no. 48 (1978), 5–38,*

and that it *wraps around the z_2-axis while converging to it* for $v=0$, $u\to\infty$:

$$z_1=e^{-u}w_1, \quad z_2=e^{iu}w_2.$$

Such leaves that wrap converging along two coordinate axes will be called *Poincaré leaves*, also in the more general case of linear differential equations, in normal form, on $\mathbf{C}^m$:

$$\frac{dz_j}{d\mathrm{T}}=\lambda_j z_j, \quad z_j=e^{\lambda_j \mathrm{T}}w_j, \quad j=1, \ldots, m, \quad \mathrm{T}\in\mathbf{C}.$$

If $m\geq 3$ a different kind of leaf to be called a *Siegel leaf* may arise. A Siegel leaf is a closed embedding of $\mathbf{C}$ in $\mathbf{C}^m$ with minimal distance $\|\zeta\|=\rho>0$ to the origin $0\in\mathbf{C}^m$ at a point ζ. If we fix ζ, then the points in the leaf at distance $r>\rho$ from the origin form an embedded circle, because the distance to 0 has at most one critical point (ζ) on a leaf. For decreasing $\rho>0$, moving ζ to $0\in\mathbf{C}^m$, but keeping $r=1$ fixed, we have the curious phenomenon that an increasing portion (with respect to length) of the embedded circle is very near to the axes and a point moving on the circle wraps around an axis a finite number of times, before going to the next. The finite number for the j-th axis can be defined as the number n_j of intersection points of the leaf with a small transversal section to the j-th axis. It now happens that the sequence of ratios $n_j/\sum_k n_k$ for $j=1, \ldots, m$, has as accumulation points as $\zeta\to 0\in\mathbf{C}^m$, that is as $\sum_k n_k\to\infty$, exactly the set of sequences of non-negative numbers (measures) $c_1, \ldots, c_m$:

$$\Delta=\{c_1\geq 0, \ldots, c_m\geq 0 : \sum_j c_j=1, \ \sum_j c_j\lambda_j^{-1}=0\}.$$

Δ is a topological invariant of the foliation $\mathscr{F}$ and it is the only one under the assumption that no two of $\lambda_1, \ldots, \lambda_m$ are linearly dependent over $\mathbf{R}$. This is theorem I of chapter I. Note that Δ is empty in case 0 is not in the convex hull $\mathscr{H}(\lambda_1, \ldots, \lambda_m)$ of $\lambda_1, \ldots, \lambda_m$ in $\mathbf{C}$.

Let the foliation $\mathscr{F}$ be a member of a family Φ, a topological space with a linear or at least a differentiable structure. $\mathscr{F}$ is *called topologically stable of codimension $\leq d$* (or just *stable* in case $d=0$) in Φ, if all members in some neighborhood U of $\mathscr{F}$ in Φ are completely classified up to homeomorphism by d linear or differentiable real functions.

Theorem I (chapter I) can now be expressed as follows: Let Φ_{L} be the set of foliations coming from linear vector fields on $\mathbf{C}^m$ with $i\neq j\Rightarrow\lambda_i\notin\mathbf{R}\lambda_j$. The foliation $\mathscr{F}$ is stable, respectively stable of codimension $2m-4$, in Φ_{L}, in case $0\notin\mathscr{H}(\lambda_1, \ldots, \lambda_m)$, respectively $0\in\mathscr{H}(\lambda_1, \ldots, \lambda_m)$.

Chapter II gives an application of the linear theory of chapter I to *holomorphic flows on the complex projective space* $\mathrm{P}=\mathbf{C}\mathrm{P}(m)$. As we recall and prove again in § 8 every vector field F on $\mathbf{C}\mathrm{P}(m)$ arises naturally from a linear vector field $\sigma : \mathbf{C}^{m+1}\to\mathbf{C}^{m+1}$ as the quotient by the action of $\mathbf{C}^*=\mathbf{C}-\{0\}$ by scalar multiplication. At the $m+1$ singular points in P, $\mathscr{F}$(F) has the topological invariants of chapter I. This gives the complete topological classification of the foliations of such flows: *Assuming that no three of the eigenvalues $\lambda_0, \ldots, \lambda_m$ of σ are collinear in the complex plane $\mathbf{C}$, $\mathscr{F}$ is stable, respectively stable of codimension $2m-4$, in case $\mathscr{H}(\lambda_0, \ldots, \lambda_m)$ is an $(m+1)$-gon, respectively an m-gon in $\mathbf{C}$. If* *more than one among* $\lambda_0, \ldots, \lambda_m$ *are inside* $\mathcal{H}(\lambda_0, \ldots, \lambda_m)$, *then the topological classification coincides with the classification under projective transformations taken together with complex conjugation of* $\mathbf{C}$, *and* $\mathcal{F}$ *is "stable of codimension* $2m-2$*".*

This is theorem II. We recall that holomorphic vector fields are rare on algebraic smooth varieties that are different from the complex projective spaces $\mathbf{CP}(m)$. For a precise statement see Lieberman [8].

Consider now the larger class Φ of foliations of all holomorphic vector fields with an isolated singularity at $0 \in \mathbf{C}^m$:

$$\frac{dz}{dT} = F(z) = \sigma z + R(z) \in \mathbf{C}^m, \qquad z \in \mathbf{C}^m, \qquad F(0) = 0 \tag{1}$$

$$\sigma z = (DF_0)(z),$$

σ has eigenvalues $\lambda_1, \ldots, \lambda_m$.

The problem of finding a *holomorphic local equivalence* between F and σ was considered by Poincaré [11] and Siegel [13], see also [1], [4], [12]:

Theorem of Poincaré. — Assume that $i \neq j \Rightarrow \lambda_i \notin \mathbf{R}\lambda_j$ and $0 \notin \mathcal{H}(\lambda_1, \ldots, \lambda_m)$. If no relation:

$$\lambda_j = \sum_{i=1}^{m} k_i \lambda_i, \qquad k_i \in \mathbf{Z}_+, \qquad \sum_{i=1}^{m} k_i \geq 2, \qquad j = 1, \ldots, m,$$

holds, then F is holomorphically equivalent to σ near $0 \in \mathbf{C}^m$.

Theorem of Siegel. — Assume that $i \neq j \Rightarrow \lambda_i \notin \mathbf{R}\lambda_j$ and $0 \in \mathcal{H}(\lambda_1, \ldots, \lambda_m)$. Then for almost all $\Lambda = \{\lambda_1, \ldots, \lambda_m\}$, with respect to Lebesgue measure, F is holomorphically equivalent to σ near $0 \in \mathbf{C}^m$.

From these theorems we obtain easily in Chapter I, § 7, the characterization for local stability of F near $0 \in \mathbf{C}^m$:

Corollary. — F *is stable (of codimension zero) if and only if* $i \neq j \Rightarrow \lambda_i \notin \mathbf{R}\lambda_j$ *and* $0 \notin \mathcal{H}(\lambda_1, \ldots, \lambda_m)$.

The sufficiency of this condition is *Guckenheimer's* [5] *stability theorem*: Any two foliations $\mathcal{F}(F)$ and $\mathcal{F}(F')$ of vector fields F and F' with singularity at $0 \in \mathbf{C}^m$, and with spectra Λ of DF_0 and Λ' of DF'_0 in the Poincaré domain, are locally homeomorphic.

In chapters III and IV we study the local problem for the Siegel case:

$$0 \in \mathcal{H}(\lambda_1, \ldots, \lambda_m), \qquad i \neq j \Rightarrow \lambda_i \notin \mathbf{R}\lambda_j.$$

We conjecture that the foliation $\mathcal{F}(F)$ near the isolated singularity of F at $0 \in \mathbf{C}^m$ is homeomorphic to the foliation $\mathcal{F}(\sigma)$ of its linear part $\sigma = (DF)_0$. We prove this for $m = 3$ in chapter III (theorem III) and find therefore with theorem I: *If* $0 \in \mathcal{H}(\lambda_1, \lambda_2, \lambda_3)$, *then (the germ at* 0 *of)* $\mathcal{F}$ *is stable of codimension two in the space* Φ *of (germs at zero of) foliations of holomorphic vector fields with singularity at zero.* This theorem is rather different from classical results concerning *holomorphic equivalence* to linear or other normal forms, in which " small " and " zero divisors " play an important role (Poincaré, Siegel and others. Compare Brjuno [1]).

In chapter IV (and chapter III, § 11 and § 12 for $m=3$) we give a *weak normal form* for any F (see (1)) by proving the existence of a *holomorphic change of coordinates* after which the remainder $R(z)$ belongs to a specific simple class. In this weak normal form the union of all Poincaré leaves for $\mathcal{F}(F)$ is already in the same stratified union V of linear subspaces as for the corresponding linear case $\mathcal{F}(\sigma)$. This is a first step in the proof of our conjecture for $m\geq 4$, which we hope to give in another paper (1).

I. — LINEAR FLOWS

1. Introduction and main theorem.

Let $\mathcal{F}(\sigma)$ be the holomorphic foliation or flow with singularity at 0, defined by the vector field $F(z)$ in $\mathbf{C}^m$:

$$\frac{dz}{dT}=F(z)=\sigma z\in\mathbf{C}^m, \quad T\in\mathbf{C}, \quad \sigma\in GL(m, \mathbf{C}) \tag{2}$$

with real two-dimensional leaves:

$$z=e^{\sigma T}w, \quad w\in\mathbf{C}^m. \tag{3}$$

Set:

$$\text{spectrum } \sigma=\Lambda=\{\lambda_1, \dots, \lambda_m\}\subset\mathbf{C} \tag{4}$$

$$\text{spectrum } 2\pi i\sigma^{-1}=\hat{\Lambda}=\{\hat{\lambda}_1, \dots, \hat{\lambda}_m\}$$

$$\hat{\lambda}_j=2\pi i\lambda_j^{-1}. \tag{5}$$

The *equivalence class* of $\hat{\Lambda}\subset\mathbf{C}$ *under the natural action of* $GL(2, \mathbf{R})$ in $\mathbf{C}=\mathbf{R}^2$ is denoted:

$$\eta(\sigma). \tag{6}$$

In § 2 we give the easy proof of the

Pre-theorem. — If σ is diagonal, then the topology of $\mathcal{F}(\sigma)$ is completely determined by $\eta(\sigma)$.

Already in case $m=2$ equality and even real dependence of two eigenvalues of σ complicates the topology of $\mathcal{F}(\sigma)$ very much. We therefore *assume* that any two eigenvalues are independent over $\mathbf{R}$:

$$i\neq j \Rightarrow \lambda_i\notin\mathbf{R}\lambda_j, \quad i,j=1, \dots, m. \tag{7}$$

---

(1) *Added in proof* (May 1978): For a non linear flow F with singularity at $0\in\mathbf{C}^m$, we can now define the topological invariant Δ, and it depends, in the same way as before, only on the linear part of F at 0. This is necessary but not sufficient to prove the conjecture also for $n\geqslant 4$. Dumortier and Roussarie [17] have important related results on linearization.

The *convex hull* of $\Lambda=\{\lambda_1, \ldots, \lambda_m\}$ in $\mathbf{C}$ is denoted $\mathscr{H}(\Lambda)$. The open set of unordered m-tuples $\{\Lambda : (7)\}$ consists of a connected component, the *Poincaré domain* $\{\Lambda : 0\notin\mathscr{H}(\Lambda)\}$, and its complement, the *Siegel domain* [1]:

$$\{\Lambda : 0\in\mathscr{H}(\Lambda)\}.$$

$\eta(\sigma)$ is topologically irrelevant in the case of the Poincaré domain (Guckenheimer) as we prove again (for later applications) by an explicit homeomorphism in § 6.

For the Siegel case the situation is different and we have (§ 5):

Main Theorem 1. — If the spectrum of σ *lies in the Siegel domain* ((7) *and* $0\in\mathscr{H}(\Lambda)$), *then* $\eta(\sigma)$ *is a topological invariant. It determines and is determined by the topology of the foliation* $\mathscr{F}(\sigma)$.

2. Proof of the pre-theorem.

In suitable coordinates, (1) (2) is expressed by:

$$dz_j=\lambda_j z_j d\mathrm{T}, \qquad z_j=e^{\lambda_j \mathrm{T}}w_j=e^{\lambda_j(\mathrm{T}+\mathrm{C}_j)}, \qquad j=1, \ldots, m. \tag{8}$$

For a given diagonal σ, and analogously for σ', recall that $\hat{\lambda}_j=2\pi i\lambda_j^{-1}$. We assume first $\hat{\lambda}'_j=g\hat{\lambda}_j$, $j=1, \ldots, m$, $g\in\mathrm{GL}^+(2, \mathbf{R})$. The *homeomorphism*:

$$h : (\mathbf{C}^m, \mathscr{F}(\sigma)) \to (\mathbf{C}^m, \mathscr{F}(\sigma'))$$

required for the pre-theorem is then defined as follows:

If $\qquad z_j(z)=e^{\lambda_j \mathrm{T}_j}$ respectively 0,

then $\qquad z_j(h(z))=e^{\lambda'_j \mathrm{T}'_j}$ respectively 0, $\quad$ where $\quad \mathrm{T}'_j=g\mathrm{T}_j$.

This is well defined because T_j is determined modulo $\hat{\lambda}_j$ and T'_j modulo $\hat{\lambda}'_j$. Moreover the image of an (any) $\mathscr{F}(\sigma)$-leaf (8) is the set:

$$e^{\lambda'_j(g\mathrm{T}+g\mathrm{C}_j)}=e^{\lambda'_j \mathrm{T}'}w'_j,$$

and this is an $\mathscr{F}(\sigma')$-leaf.

The non-oriented elements [2] g of $\mathrm{GL}(2, \mathbf{R})$ are realized by composing with one of them *e.g.* complex conjugation of $\mathbf{C}=\mathbf{R}^2$. It sends $(\lambda_1, \ldots, \lambda_m)$ into $(\bar{\lambda}_1, \ldots, \bar{\lambda}_m)$. The required homeomorphism of $\mathbf{C}^m$ is given by complex conjugation:

$$h : (z_1, \ldots, z_m) \mapsto (\bar{z}_1, \ldots, \bar{z}_m).$$

Remark. — If $z_j(z)=e^{u+iv}$, then $z_j(h(z))=e^{\alpha_j u+i(\beta_j u+v)}$ for some real constants α_j, β_j and the mapping h:

$$\begin{cases} |z_j(h(z))|=|z_j(z)|^{\alpha_j} \\ \arg z_j(h(z))=\arg z_j(z)+\beta_j \ln|z_j(z)| \end{cases} \tag{9}$$

---

(1) (Replacing λ_j by $\lambda_j/|\lambda_j|$, we easily see that the Siegel domain is connected for $m=3$ and 4, and it has three components for $m=5$). A more complicated definition of Siegel domain is customary in the theory of holomorphic equivalence.

(2) J.-P. Françoise drew our attention to this case which we had overlooked.

produces in the j-th coordinate axis $\{z : z_k = 0 \text{ for } k \neq j\}$ a *spiraling homeomorphism* (9) for $z_j \to 0$ or ∞, with the unit circle $|z_j| = 1$ pointwise fixed. It leaves invariant each of the manifolds $z_j = 0$ and $|z_j| = 1$, as well as the piecewise smooth $(2m-1)$-sphere:

$$S = \{z : \sup_j |z_j| = 1\}. \tag{10}$$

Remark. — h preserves the additive group action of $\mathbf{C} = \mathbf{R}^2$ (see T in (8)).

3. The foliation on W, the union of the Siegel leaves, is stable.

We assume (7). The real function $||z||^2 = \sum_j z_j \bar{z}_j$ has a critical value on a leaf (8) at a point z if and only if:

$$0 = d\sum_j z_j\bar{z}_j = \sum_j (z_j\bar{\lambda}_j\bar{z}_j d\bar{T} + \bar{z}_j\lambda_j z_j dT)$$
$$= \sum_j z_j\bar{z}_j(\lambda_j dT + \bar{\lambda}_j d\bar{T}) = 0 \quad \text{for} \quad dT \in \mathbf{C}$$

$$dT = 1 \quad \text{yields} \quad \sum_j z_j\bar{z}_j(\lambda_j + \bar{\lambda}_j) = 0$$

$$dT = i \quad \text{yields} \quad \sum_j z_j\bar{z}_j(\lambda_j - \bar{\lambda}_j) = 0.$$

The union M of the 0-nearest points, $z \neq 0$, has therefore the equation:

$$M : \sum_j z_j\bar{z}_j\lambda_j = 0, \qquad z \neq 0. \tag{11}$$

No leaf has two (or more) critical points and every critical value is a minimum, because for any $T_1 \neq T_2$ the *real* function:

$$t \mapsto \sum_j |z_j|^2, \qquad z_j = e^{\lambda_j(tT_1 + (1-t)T_2)} w_j, \qquad j = 1, \ldots, m,$$

is a *sum of real exponential functions* in $t \in \mathbf{R}$, hence concave. A leaf with a minimum is called a *Siegel leaf*. It is a closed embedding of $\mathbf{C}$ and can be characterised by its critical point $\zeta = (\zeta_1, \ldots, \zeta_m)$ in M. *The union* W *of all Siegel leaves is therefore the total space of a trivial bundle,* $W = M \times \mathbf{C} \to M$, *embedded in* $\mathbf{C}^m$ *by*:

$$z_j = \zeta_j e^{\lambda_j T}, \quad j = 1, \ldots, m; \quad (\zeta, T) \in M \times \mathbf{C}$$

with base space M.

M is seen to be a *manifold* by putting in (11):

$$z_j = x_j + iy_j, \quad \lambda_j = \mu_j + i\nu_j, \quad M : \sum_j (x_j^2 + y_j^2)\mu_j = \sum_j (x_j^2 + y_j^2)\nu_j = 0,$$

and by calculating the tangent space:

$$(dx_1, \ldots, dy_m) : \sum_j (x_j dx_j + y_j dy_j)\mu_j = \sum_j (x_j dx_j + y_j dy_j)\nu_j = 0$$

with coefficient matrix of rank 2, because every determinant $\mu_j\nu_k - \mu_k\nu_j \neq 0$, for $j \neq k$ by (7). The manifold M is a *cone* with deleted top $0 \in \mathbf{C}^m$ over the compact manifold:

$$M(1) = \{z \in M : ||z||^2 = \sum_j z_j\bar{z}_j = 1\}.$$

10

From (11), where $c_j = z_j\bar{z}_j \geq 0$, we see that $0 \in \mathbf{C}$ is a weighted mean of the set of complex numbers $\Lambda = \{\lambda_1, \ldots, \lambda_m\}$. Hence Siegel leaves can only exist (and M is not empty) in the Siegel domain case $0 \in \mathscr{H}(\Lambda)$. In § 4 we will see that then:

$$W = \{z : 0 \in \mathscr{H}(\{\lambda_j : j \in J(z)\})\} \tag{12}$$

where:
$$J(z) = \{j : z_j(z) \neq 0\}. \tag{13}$$

It is *open dense* in $\mathbf{C}^m$.

The (abstract) *differentiable manifold* M depends by (11) continuously on Λ, and is therefore "*constant*" *on each component of the Siegel domain. Then also the topology of the restriction of the foliation*: $\mathscr{F}(\sigma) \mid W$ *is locally constant* (=stable), *and gives therefore no topological invariants.*

4. Geometry in the T-plane of a leaf.

We assume (7), use coordinates as in (8) but ordered in such a way that:

$$0 \leq \arg \hat{\lambda}_1 < \arg \hat{\lambda}_2 \ldots < \arg \hat{\lambda}_m < 2\pi. \tag{14}$$

The parameter T in the leaf of a point z is determined up to translations in $\mathbf{C} = \mathbf{R}^2$. The intersection of a leaf with the "ball" $B = \{z : \sup_j |z_j| \leq 1\}$ and with the manifolds $|z_j| = 1$, gives rise to interesting configurations in the T-plane of that leaf. We introduce the *configuration* $G = G(z)$ consisting of the half-planes (see (8) and fig. 1):

$$\alpha_j = \{T : |z_j| \leq 1\} \subset \mathbf{C}, \quad j \in J(z) = \{j : z_j \neq 0\}.$$

The boundary $\partial\alpha_j$ is a line parallel to and oriented by the vector $\hat{\lambda}_j$. We also define the convex disc:

$$D(z) = \bigcap_j \alpha_j \subset \mathbf{C}, \tag{15}$$

which represents the *intersection of the leaf with* B, *and its boundary, the oriented convex polygon* $C = C(z)$. Let $I(z) \subset J(z)$ be the set of indices j involving edges $\partial\alpha_j$ of $C(z)$. Let the edge on $\partial\alpha_j$ be between vertices T_{j_-} and T_j where j is the cyclic successor of j_- in $I(z)$. For later use we define $\tilde{n}_j \in \mathbf{R}$ by:

$$T_j = T_{j_-} + \tilde{n}_j \hat{\lambda}_j \tag{16}$$

$2\pi\tilde{n}_j$ is the increase of the argument of z_j from T_{j_-} to T_j. The real number $\tilde{n}_j$ differs from the integral number of those points on the edge $T_{j_-}T_j$ where z_j is real by at most one.

If the polygon C is bounded and if we set $\tilde{n}_j = 0$ for $j \notin I(z)$, then clearly (see (16) or fig. 1):

$$\sum_j \tilde{n}_j \hat{\lambda}_j = 0. \tag{17}$$

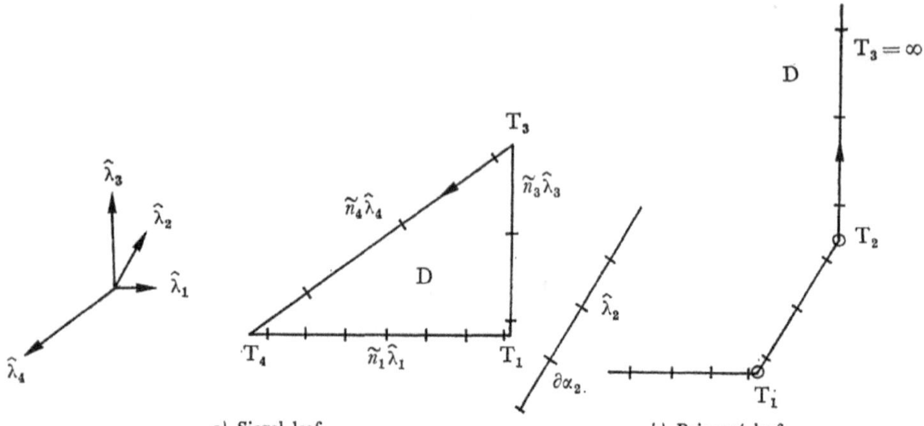

a) Siegel leaf *b)* Poincaré leaf

Fig. 1

The *complete configuration* $G^*(z)$ of the leaf of z consists of the set $G(z)$ of half planes α_j numbered by $j \in J(z)$ with oriented boundaries $\partial\alpha_j$, together with the set of those points (marked in fig. 1) on $\partial\alpha_j$ where z_j is real. $G^*(z)$ is to be considered *modulo translations of* $\mathbf{C}=\mathbf{R}^2$. The point $0 \in \mathbf{C}^m$ is represented by the empty configuration.

Lemma 1. — Every complete configuration that agrees with $\hat\Lambda=(\hat\lambda_1, \ldots, \hat\lambda_m)$ *determines a unique leaf. A point* $T \in \mathbf{C}$ *determines a unique point* z *in that leaf.*

Proof. — If the half-plane α_j is not in $G(z)$, then $z_j(z)=0$ on the leaf. If z_j is known at *some* point of the leaf (and z_j is known to be 1 at the marked points of $\partial\alpha_j$!) then the formulas (8) determine z_j at every other point T and for example at $T(z)$. So then $z=z(T)$ and its leaf are determined.

Remark. — By letting $g \in GL(2, \mathbf{R})$ with $g\hat\Lambda=\hat\Lambda'$ act on the T-plane $\mathbf{C}=\mathbf{R}^2$ and on all complete configurations with respect to $\hat\Lambda$ in $\mathbf{R}^2$, we obtain the homeomorphism h of § 2.

Lemma 2. — Assuming (7), *every leaf of* $\mathcal{F}(\sigma)$ *outside* $0 \in \mathbf{C}^m$, *is of one of the following kinds:*

- — A coordinate axis, *topologically a cylinder, in case the polygon* $C(z)$ *is one line. There are m axes.*
- — A Siegel leaf (*see* (13)), *a closed embedding of* $\mathbf{C}$ *in* $\mathbf{C}^m$, *with bounded or empty polygon* $C(z)$, *in case* $0 \in \mathcal{H}(\{\hat\lambda_j : j \in J(z)\})$.
- — A Poincaré leaf, *an embedding of* $\mathbf{C}$ *in* $\mathbf{C}^m$, *transversal to each " sphere "* $\sup_j |z_j|=r>0$, *with unbounded polygon* $C(z)$, *in case* $0 \notin \mathcal{H}(\{\hat\lambda_j : j \in J(z)\})$.

Proof. — First suppose $z_j \neq 0$ *for all* j, $0 \notin \mathcal{H}(\{\hat\lambda_1, \ldots, \hat\lambda_m\})$, $m \geq 2$. We then may assume:

$$0 \leq \arg\hat\lambda_1 < \arg\hat\lambda_2 < \ldots < \arg\hat\lambda_m < \pi \tag{18}$$

and the half planes α_j clearly have an unbounded intersection. Along any real vector $\mu \in \mathbf{C} = \mathbf{R}^2$ for which:

$$\arg \hat{\lambda}_m < \arg \mu < \pi$$

attached at any point in the T-plane, the *linear function on* $\mathbf{R}^2$, $\ln|z_j|$, *decreases for* $j = 1, 2, \ldots, m$. Then the leaf is transversal to every " sphere " $\sup|z_j| = r > 0$. Topologically, the leaf is a cone over its intersection $C(z)$ (homeomorphic to $\mathbf{R}$) with $S = \{z : \sup|z_j| = 1\}$. For $w \in \mathbf{R} = C(z)$ (see fig. 1 *b*) converging to $-\infty$ (resp. ∞) the first (resp. last) coordinate converges in absolute value to 1, and all others to zero. The point $w \in C(z)$ converges to the unit circle in the first (resp. last) axis. The leaf is called a *Poincaré leaf.*

The same argument applies to any z for which $0 \notin \mathscr{H}(\{\hat{\lambda}_j : j \in J(z)\})$ in case $J(z)$ contains at least two indices. We then restrict the argument to the coordinates z_j for which $j \in J(z)$.

There remains the case where $0 \in \mathscr{H}(\{\hat{\lambda}_j : j \in J(z)\})$. Then $C(z)$ is either a bounded polygon or empty. In both cases $\{T : \sup_j|z_j| \leq N\}$ is for large $N > 0$ a compact convex set on the T-plane and $||z||$ has a minimum in the interior. So the leaf is a *Siegel leaf* by the definition in § 3.

As announced in (12) we have:

$$W = \{z \in \mathbf{C}^m : 0 \in \mathscr{H}(\{\hat{\lambda}_j : j \in J(z)\})\}.$$

An immediate corollary of lemma 1 is (see fig. 1):

Lemma 3. — The leaf of z, given Λ, is completely determined by the following " coordinates ":

1) $J(z)$, $I(z)$ *and* $\tilde{n}_j$ *for* $j \in I(z)$.
2) *The maximum* $e^{-\beta_s} \leq 1$ *of* $|z_s(T)|$ *for* $T \in C(z)$, $s \in J(z) \setminus I(z)$. *This equals* $|z_s(T_j)|$ *for* $j_- < s < j$ *in cyclic order* $j_-, j \in I(z)$.
3) *The argument* $\varphi_s = \arg z_s$, *at the vertex* $T_j \in C$, *for* $j_- < s \leqslant j$.

In the case of a Siegel leaf, $0 \in \mathscr{H}(\{\hat{\lambda}_j : j \in J\})$, all these, $I \subset J$, $\tilde{n}_j$, β_s, $\varphi_s \bmod 2\pi$, can be chosen arbitrarily, but for the condition:

$$\sum_j \tilde{n}_j \hat{\lambda}_j = 0.$$

5. The topological invariant $\eta(\sigma) = \Delta$ in the Siegel domain case.

We prove theorem 1, knowing the pre-theorem, by giving a *topological description of the* $(m-3)$*-dimensional convex polytope*:

$$\Delta = \{(c_1, \ldots, c_m) : c_j \geq 0 \ \forall j, \quad \sum_j c_j = 1, \quad \sum_j c_j \hat{\lambda}_j = 0\} \subset \mathbf{R}^m. \tag{19}$$

A sequence of weights $c_1, \ldots, c_m$ in (19) which makes 0 the barycenter of $\hat{\Lambda}$ is invariant under the action of $GL(2, \mathbf{R})$ on $\hat{\Lambda}$. Vice versa Δ determines $\hat{\Lambda}$ modulo that action.

To see this take $\hat{\lambda}'_1 = 1$, $\hat{\lambda}'_2 = i$, and determine $\hat{\lambda}'_j$ by taking $c_k = 0$ for $k \neq 1, 2$ or j in (19). Hence Δ is *equivalent to* $\eta(\sigma)$.

Let S_j be a small section transversal to the foliation $\mathscr{F}(\sigma)$ at a point p_j on the j-th axis and $n_j = n_j(L)$ the number of intersection points with some Siegel leaf L. We now define Δ' as the closure in $\mathbf{R}^m$ of the set of m-tuples $(c_1, \ldots, c_m)$ of positive numbers, for which there exists a sequence of Siegel leaves L_α $(\alpha = 1, 2, \ldots)$ such that for $\alpha \to \infty$:

$$n_j \to \infty \quad \text{and} \quad \lim \frac{n_j}{\sum_k n_k} = c_j \text{ for all } j \tag{20}$$

The definition of Δ' is purely topological. We prove that for every choice of S_j, p_j:

Lemma 4. — $\Delta' = \Delta$.

Proof. — Under holonomic transport of S_j with respect to the foliation, the intersection numbers with any leaf remain constant. After such transport along a curve in the j-th axis from p_j to the point with coordinates $z_j = 1$, $z_k = 0$ for $k \neq j$, we may assume for some $0 < \delta < 1$:

$$S_j(\delta) \subset S_j \subset S_j(1) \subset S = \{z : \sup_k |z_k| = 1\}$$

where:
$$S_j(\delta) = \{z : z_j = 1, \quad |z_k| < \delta \leq 1 \quad \text{for} \quad k \neq j\}. \tag{21}$$

The Siegel leaf of z meets $S_j(\delta)$ and S_j inside S, hence in the convex polygon $C(z)$ in the T-plane and *in marked points of* $\partial\alpha_j$. Because the real functions $\ln|z_k| = \mathrm{Re}\,\lambda_k(T + c_k)$ on the T-plane $\mathbf{R}^2$ with level lines parallel to $\hat{\lambda}_k$, have constant gradients, no two of which are $\mathbf{R}$-linearly dependent by (7), there exists for any $\delta > 0$ a *number* $K > 0$ such that:

$$|\ln|z_k(T + t\hat{\lambda}_j)| - \ln|z_k(T)|| > |\ln \delta|$$

for $t > K$, for all k and $j \neq k$. In particular in the edge $T_{j-}T_j$ (see fig. 2) of the polygon $C(z)$ (on which $|z_k| \leq 1$ for all k) we see that S_j and

$$S_j(\delta) : |z_j| = 1, \quad |z_k(T)| < \delta < 1 \quad \text{for} \quad k \neq j \tag{22}$$

contain all points:

$$T = T_{j-} + t\hat{\lambda}_j, \quad K < t < \tilde{n}_j - K. \tag{22'}$$

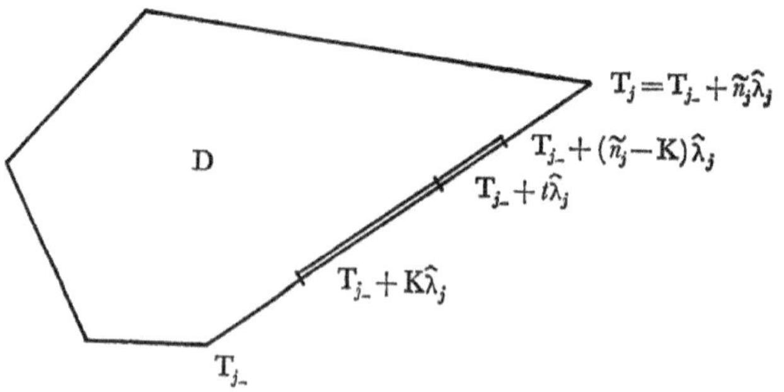

FIG. 2

Counting intersection points (marked on $\partial\alpha_j$) of a Siegel leaf L_α we see from (21):

$$n_j(S_j(\delta)) \leq n_j(S_j) = n_j \leq n_j(S_j(1)) \leq \tilde{n}_j + 1.$$

From (22) we read that all marked points on the interval (22') belong to $S_j(\delta)$. There remain at most $2(K+1)$ other marked points between $T_{j\_}$ and T so that:

$$\tilde{n}_j - n_j(S_j(\delta)) \leq 2K+3.$$

Hence *for all Siegel leaves*:

$$|\tilde{n}_j - n_j| \leq 2K+3. \tag{23}$$

But by (17) $\sum_j \tilde{n}_j \hat{\lambda}_j = 0$.

Then:
$$\sum_j \frac{\tilde{n}_j}{\sum_k \tilde{n}_k} \hat{\lambda}_j = 0.$$

From (20) and (23) we obtain:

$$\sum_j c_j \hat{\lambda}_j = 0.$$

We have proved $\Delta' \subset \Delta$. If we take any $(c_1, \ldots, c_n)$ in Δ with $c_j > 0$ for all j, then there is by lemmas 1 and 3 a Siegel leaf L_α with:

$$\tilde{n}_j = \alpha c_j, \quad j = 1, \ldots, m.$$

For $\alpha = 1, 2, 3, \ldots, \alpha \to \infty$ we have:

$$\tilde{n}_j / \sum_k \tilde{n}_k = c_j, \quad \text{hence} \quad \lim_{\alpha\to\infty} n_j / \sum_k n_k = c_j$$

by (23). Consequently $\Delta' = \Delta$ and theorem I is proved.

6. An explicit homeomorphism in the case of the Poincaré domain.

Here we assume (7), (8), $0 \notin \mathcal{H}(\hat{\Lambda})$:

$$0 \leq \arg\hat{\lambda}_1 < \arg\hat{\lambda}_2 \ldots < \arg\hat{\lambda}_m < \arg\mu < \pi. \tag{24}$$

Every leaf that is not an axis is a Poincaré leaf, meeting S in a curve that is represented in the T-plane by an unbounded convex polygon C. It has at least one vertex and is transversal to the constant vector field μ. The T-plane of a leaf is then naturally a product:

$$T = T_0 + s\mu, \quad T_0 \in C, \quad s \in \mathbf{R}.$$

Taking all these products together we write $\mathcal{F}(\sigma)$ as the product of a 1-foliation $\mathcal{F}_1(\sigma) = \mathcal{F}(\sigma) \cap S$ and $\mathbf{R}$, by the formulas for $0 \neq w \in \mathbf{C}^m$, $z \in S$, $s \in \mathbf{R}$:

$$w_j = e^{\lambda_j s\mu} z_j, \quad j = 1, \ldots, m. \tag{25}$$

Let σ' fulfil the same conditions as σ. In order to define a homeomorphism:

$$h : \mathcal{F}(\sigma) \to \mathcal{F}(\sigma')$$

it suffices by the last remark and (25) to define the restriction:

$$h_1 = h|S : \mathscr{F}_1(\sigma) \to \mathscr{F}_1(\sigma')$$

of h to S. The map h_1 will induce a map h_L from the set of leaves of $\mathscr{F}_1(\sigma)$ onto the set of leaves of $\mathscr{F}_1(\sigma')$. We begin with the definition of h_L. Recall lemma 1, § 4 saying that for given Λ the leaves (except axes) are 1-1-represented by complete configurations G^* modulo translation. We define h_L by claiming that it is expressed by the identity in terms of the " coordinates " of lemma 3, § 4. This does not work for the m axes. We let h_L map each axis onto itself. We now examine this definition of h_L in detail.

Equality of the first sets of " coordinates " in lemma 3 has the following consequences: $J' = J$ gives the invariance of (the union of all leaves in) $z_j = 0$ for $j = 1, \ldots, m$.

If C and C′ are convex polygons corresponding with a leaf γ and its image leaf $h_L(\gamma)$, then $I' = I$ implies that the same coordinates among $z_1, \ldots, z_m$ take absolute value *one* on edges of C and of C′. This determines a correspondence of edges.

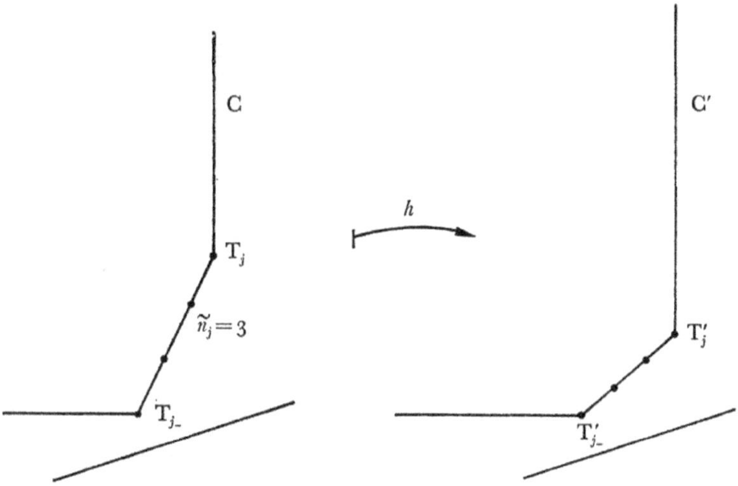

FIG. 3

$\tilde{n}'_j = \tilde{n}_j$ for $j \in I = I'$ determines, for given Λ, Λ', the lengths of the bounded edges of C′ of the leaf $h_L(\gamma)$ once those of C of the leaf γ are given. Therefore we now have obtained a one-one-correspondence between polygons C and C′ modulo translations, which correspondence must lift to h_L.

With equality of the second sets of " coordinates " in lemma 3, we obtain the necessary information on the absolute values of those coordinates z_j, $j \in J$, for which $j \notin I$, at certain vertices of C and C′:

$$e^{-\beta'_s} = |z_s(T'_{j_-})| = e^{-\beta_s} = |z_s(T_{j_-})|$$

for $j_- < s < j$, and j the successor of j_- in I.

With equality of the third sets of "coordinates" in lemma 3, we *complete the definition of* h_L because we obtain the necessary information on the arguments of the coordinates z_j of certain vertices of C and C':

$$\varphi'_s = \arg z_s(T'_j) = \varphi_s = \arg z_s(T_j)$$

for $j_- \leq s \leq j$, j successor of j_-.

On one hand $z_s(T_j)$ is not defined if $T_j = \infty \in \mathbf{C} \cup \infty$ but no ambiguities arise in case neither T_{j_-} nor T_j is ∞, because $\widetilde{n}'_j = \widetilde{n}_j$ implies:

$$\varphi'_j(T'_{j_-}) = \varphi_j(T_{j_-}) \Leftrightarrow \varphi'_j(T'_j) = \varphi_j(T_j).$$

Having obtained the map h_L we now define a point set bijection $h_1 : S \to S$, which is a lift of h_L, by assigning to the point $T = T_{j_-} + t\hat{\lambda}_j$ on the polygon C of a leaf γ of $\mathscr{F}_1(\sigma)$ the point $T' = T'_{j_-} + t\hat{\lambda}'_j$ on the polygon C' of the leaf $h_L(\gamma)$ of $\mathscr{F}_1(\sigma')$, and similarly in case $T_{j_-} = \infty$ with $T = T_j - t\hat{\lambda}_j$. In particular vertices of C go to vertices of C'. We define h_1 to be the identity map on each axis. It remains to prove that h_1 is continuous. Then also h_1^{-1} is continuous by interchange of $\mathscr{F}(\sigma)$ and $\mathscr{F}(\sigma')$.

Proof. — For a given Λ, the set of all leaves with a fixed set of nonzero coordinates J is homeomorphically represented by the set of all its complete configurations (lemma 1) in its natural topology. This space is also seen to be homeomorphically represented (embedded) by the following sets of "coordinates" of lemma 3:

$$\begin{aligned} e^{\widetilde{n}_s} z_s(T_j) &= e^{\widetilde{n}_s - \beta_s + i\varphi_s} \quad j_- < s \leq j \quad \text{for} \quad T_j \neq \infty \neq T_{j_-} \\ &= z_j(T_{j_-}) \quad \text{for} \quad T_j = \infty \\ &= z_j(T_j) \quad \text{for} \quad T_{j_-} = \infty; \end{aligned} \tag{26}$$

j is the successor of j_- in I.

Recall that $T_j \neq \infty$ is a point in the polygon (1-leaf) C, at which $|z_s|$ takes its maximal value ≤ 1. If this value is smaller than one, then $\widetilde{n}_s$ is automatically zero.

As h is the identity in these "coordinates" we conclude that *the restriction of* h to $\{z : J = \{j : z_j \neq 0\}\} \cap S$ induces a homeomorphism of the space of those 1-leaves in S, and then h_1 is a homeomorphism of that part of S onto itself as well.

The formulas (26) tell even more, because we can include the values $z_s = 0$ in the consideration and let s run through all indices between the first, j_b, and the last, j_e, of J. Therefore we can conclude that h_1 is a homeomorphism onto itself on each of the sets:

$$\Omega(j_b, j_e) = \{z : z_{j_b} \neq 0, \ z_{j_e} \neq 0, \ z_j = 0 \text{ for } j < j_b < j_e \text{ and } z_j = 0 \text{ for } j > j_e\} \cap S,$$

and in particular on the open dense set:

$$\Omega(1, m) = \{z : z_1 \neq 0, \ z_m \neq 0\} \cap S \subset S.$$

We next prove that h is also continuous at any point $w \notin \Omega(j_b, j_e)$, w not on an axis:

$$w = (w_{\mathrm{I}}; w_{\mathrm{II}}; w_{\mathrm{III}}) = (0; w_{\mathrm{II}}; 0) = (0, \ldots, 0; w_{j_b}, \ldots, w_{j_e}; 0, \ldots, 0)$$

$$w_{j_b} \neq 0, \quad w_{j_e} \neq 0, \quad j_b < j_e.$$

By the above consideration h is continuous at and near w, on the subspace defined by $z_{\mathrm{I}}=0$ and $z_{\mathrm{III}}=0$. So we can restrict our study of continuity to points:

$$z=(z_{\mathrm{I}};\,z_{\mathrm{II}};\,z_{\mathrm{III}})\quad \text{with}\quad z_{\mathrm{II}}=w_{\mathrm{II}}.$$

The point $w\in S$ is represented by some point $T(w)$ on the configuration $G(w)$. The configuration $G(z)$, in the T-plane of the leaf of z, is then obtained from $G(w)$ by adding half-planes α_j for the coordinates in z_{I} $(z_j : j<j_b)$ and half-planes for the coordinates in z_{III} $(z_j : j>j_e)$. Let $\|z-w\|=\delta$. If δ is small then all the new boundaries $\partial\alpha_j$ will meet $C(w)$ *far away* from its vertices and from $T(w)$ on either of the two unbounded edges (fig. 3). Then the point w and the point z are represented by the same point of $C(w)$. (See the equations for a leaf.) The 1-1-correspondence h_1 preserves this property of far-ness concerning the images $G'(h_1(w))$ and $G'(h_1(z))$. Moreover *vice versa* far-ness of the new half-plane boundaries $\partial\alpha'_j$ implies that $|z_j|$ is small for $j<j_b$ and $j>j_e$. Therefore continuity follows:

$$\|h(z)-h(w)\|=O(\delta).$$

By the equations (26), we find for any point $z\in S$ not on the j-th axis, but so that $|z_j(z)|=1$:

$$z_j(h(z))=z_j(z).$$

This identity relation is also the definition of h on the j-th axis. With a "far-away" argument concerning other coordinates, this proves continuity also at axis-points in S.

7. A corollary on stability admitting non linear perturbations as well.

Corollary 1. — *Let* F *be a holomorphic vector field in* $\mathbf{C}^m$, $F(0)=0$, *and let* $\sigma=DF_0$ *have the spectrum* $\{\lambda_1, \ldots, \lambda_m\}$. *Then* F *is locally stable (of codimension zero) near* $0\in\mathbf{C}^m$ *if and only if* $i\neq j \Rightarrow \lambda_i\notin\mathbf{R}\lambda_j$, *and* $0\notin\mathscr{H}(\lambda_1, \ldots, \lambda_m)$.

Proof. — If $0\in\mathscr{H}(\lambda_1, \ldots, \lambda_m)$, then we can approximate F, by Siegel's theorem ([13], [12]), by another vector field $\widetilde{F}$, $\widetilde{F}(0)=0$, which is holomorphically equivalent to its linear part $\widetilde{\sigma}=D\widetilde{F}_0$, and whose spectrum $\widetilde{\sigma}$ is in the Siegel domain. By theorem I $\widetilde{\sigma}$ is not stable, so F is not stable. On the other hand, if $i\neq j\Rightarrow\lambda_i\notin\mathbf{R}\lambda_j$ and $0\notin\mathscr{H}(\Lambda)$, Guckenheimer [2] proved that $\mathscr{F}(\sigma)$ meets every sphere

$$S_r:\ \|z\|^2=\sum_{j=1}^{m} z_j\bar{z}_j=r^2>0$$

transversally, hence in a real 1-foliation, and that the leaves are the orbits of a Morse-Smale vector field with m closed orbits. From the structural stability of these vector fields [10] follows the local stability of F, also under small non-linear perturbations. So it remains to show that whenever $0\notin\mathscr{H}(\Lambda)$ and two eigenvalues are dependent over $\mathbf{R}$ then F is not stable. Suppose $\lambda_2\in\mathbf{R}\lambda_1$, $0\notin\mathscr{H}(\lambda_1, \ldots, \lambda_m)$. Arbitrarily near to F we find F' with $\{\lambda'_1, \ldots, \lambda'_m\}$ in the Poincaré domain: $0\notin\mathscr{H}(\lambda'_1, \ldots, \lambda'_m)$,

$i \neq j \Rightarrow \lambda'_i \notin \mathbf{R}\lambda'_j$, *and* moreover obeying the conditions $\lambda'_j - \sum_{i=1}^{m} k_i \lambda'_i \neq 0$ for any non-negative integers $k_1, \ldots, k_m$. By Poincaré [11] F′ is locally holomorphically equivalent to its linear part. It has Poincaré leaves only except for the cylindrical coordinate axes. Arbitrarily near to F we also find F″ with $\{\lambda''_1, \ldots, \lambda''_m\}$ obeying the following conditions: $0 \in \mathscr{H}(\lambda''_1, \ldots, \lambda''_m)$, $\lambda''_2 = r\lambda''_1$, r rational, $\lambda''_i \notin \mathbf{R}\lambda''_j$ for $i \neq j$, $i \geq 2$, $j \geq 2$, and:

$$\lambda''_j - \sum_{i=1}^{m} k_i \lambda''_i \neq 0$$

for any non-negative integers $k_1, \ldots, k_m$. By Poincaré [11] F″ is locally holomorphically equivalent to its linear part σ''. But all leaves of σ'' in the linear subspace with equations $z_3 = z_4 = \ldots = z_m = 0$ are cylinders, so $\mathscr{F}(\mathrm{F}')$ and $\mathscr{F}(\mathrm{F}'')$ are not homeomorphic near zero, and F is not stable.

II. — HOLOMORPHIC FLOWS ON $\mathbf{CP}(m)$

8. Holomorphic flows on CP(m) arise from linear vector fields on $\mathbf{C}^{m+1}$.

Here we prove the (known)

Lemma. — *Every holomorphic vector field over* $\mathrm{P} = \mathbf{CP}(m)$ *originates naturally from a linear vector field on* $\mathbf{C}^{m+1}$ $(\sigma z \in \mathbf{C}^{m+1}$, $\sigma \in \mathrm{GL}(m+1, \mathbf{C}))$.

Proof. — Consider the embedding of the trivial one dimensional vector bundle over $\mathbf{C}_*^{m+1} = \mathbf{C}^{m+1} - \{0\}$ into the (trivial) tangent bundle, given by the following inclusion of total spaces:

$$\{(z, \mu z) : z \in \mathbf{C}_*^{m+1}, \mu \in \mathbf{C}\} \subset \mathbf{C}_*^{m+1} \times \mathbf{C}^{m+1}.$$

The first bundle has the section $\mu = 1$. This section, as well as the embedding, is invariant under the action of $\mathbf{C}^* = \mathbf{C} - \{0\}$:

$$\lambda . (z, w) = (\lambda z, \lambda w), \qquad \lambda \in \mathbf{C}^*.$$

The quotient is an embedding of vector bundles over P that can be completed in an exact sequence with the tangent bundle τ of P:

$$0 \to \theta \to \eta \to \tau \to 0.$$

θ is trivial with non zero section $(\mu = 1)$. Čech cohomology of P with coefficients in the sheaves of germs of sections of these bundles, gives rise to a long exact sequence that begins with groups of global cross sections $\mathrm{H}_0 = \Gamma$:

$$0 \to \Gamma(\mathrm{P}, \theta) = \mathbf{C} \to \Gamma(\mathrm{P}, \eta) (= \mathbf{C}^{(m+1)^2}, \text{ see below})$$

$$\xrightarrow{\iota} \Gamma(\mathrm{P}, \tau) \to \mathrm{H}_1(\mathrm{P}, \text{sheaf } \theta) = \mathrm{H}_{0,1}(\mathrm{P}, \mathbf{C}) = 0.$$

Hence ι is surjective onto the set of holomorphic vector fields $\Gamma(P, \tau)$. Each holomorphic section of $\Gamma(P, \eta)$ lifts to a holomorphic vector field $F(z)$ on C_*^{m+1} that is invariant under the action of $\mathbf{C}^*$:

$$\lambda F(z_0, \ldots, z_m) = F(\lambda z_0, \ldots, \lambda z_m).$$

Differentiation with respect to z_j yields

$$\lambda \partial_j F(z_0, \ldots, z_m) = \partial_j F(\lambda z_0, \ldots, \lambda z_m) . \lambda.$$

The holomorphic vector field

$$\partial_j F(\lambda z_0, \ldots, \lambda z_m) = \partial_j F(z_0, \ldots, z_m)$$

is bounded near $0 \in \mathbf{C}^{m+1}$, hence it extends over zero, with value $0 \in \mathbf{C}^{m+1}$. Then:

$$\partial_j F(z_0, \ldots, z_m) = \partial_j F(0, \ldots, 0) = \text{constant}.$$

$F(z)$ is linear and the lemma is proved.

9. The topological invariants.

Let $\mathscr{F}$ be the flow of a holomorphic vector field on the projective space $\mathbf{CP}(m)$, which comes from the linear vector field on $\mathbf{C}^{m+1}$:

$$\frac{dz}{dT} = \sigma z, \qquad z = e^{\sigma T} w, \qquad z, w \in \mathbf{C}^{m+1}, \qquad T \in \mathbf{C}. \tag{27}$$

As before $z = (z_0, \ldots, z_m)$ is a set of homogeneous coordinates for

$$\mathbf{CP}(m) = (\mathbf{C}^{m+1} - \{0\}) / \mathbf{C}^*.$$

The spectrum $\Lambda = \{\lambda_0, \ldots, \lambda_n\}$ of σ is a projective invariant of the flow, but should now be *considered modulo the group of all translations and similarities in* $\mathbf{C}$. (If we replace T by $\omega^{-1}T$, then $\Lambda \subset \mathbf{C}$ is multiplied by $\omega \in \mathbf{C}^*$, and if we replace $e^{\sigma T} w$ by $e^{\sigma T - \lambda T} w$, $\lambda \in \mathbf{C}$, this translates λ_j to $\lambda_j - \lambda$ for $j = 0, \ldots, m$.) If σ is diagonisable we have in preferred coordinates the flow $\mathscr{F}(\Lambda)$:

$$\frac{dz_j}{dt} = \lambda_j z_j, \qquad z_j = e^{\lambda_j T} w_j, \qquad j = 0, \ldots, m. \tag{28}$$

It has a singularity at each of the vertices of the coordinate simplex. Outside the coordinate hyperplane $z_k = 0$, we take $z_k = 1$ and non-homogeneous coordinates $z_j / z_k = z_j$ and we obtain the linear flow on $\mathbf{C}^m$:

$$\mathscr{F}_k = \mathscr{F}_k(\Lambda) : z_k = 1, \qquad z_j = e^{(\lambda_j - \lambda_k)T} w_j \quad j \neq k. \tag{29}$$

In order to have $\lambda_i - \lambda_k$ and $\lambda_j - \lambda_k$ real independent for every $i \neq j \neq k \neq i$ (condition (7)), we make the

Assumption. — *For* $\lambda_i, \lambda_j, \lambda_k \in \Lambda$, $i \neq j \neq k \neq i$, λ_i, λ_j, λ_k *are not collinear in the plane* $\mathbf{C}$. (30)

By theorem 1, $\mathscr{F}_k$ has only the non-trivial topological invariant $\eta(\sigma)=\Delta$ in case the spectrum $\Lambda_k=\{(\lambda_j-\lambda_k),\ j\neq k\}$ is in the Siegel domain, that is in case λ_k is in the interior of $\mathscr{H}(\Lambda)$, and no topological invariant otherwise. We now formulate:

Theorem II. — A complete set of topological invariants of a holomorphic flow $\mathscr{F}(\Lambda)$ (27) *on* $\mathbf{C}P(m)$, *under the general position assumption* (30), *consists of the topological invariants of chapter I at the* $m+1$ *singular points. In other words:*

A) *If the boundary* $\partial\mathscr{H}(\Lambda)$ *of the convex hull* $\mathscr{H}(\Lambda)$ *is an* $(m+1)$*-gon, then there are no topological invariants:* $\mathscr{F}$ *is stable.*

B) *If* $\partial\mathscr{H}(\Lambda)$ *is an m-gon with one eigenvalue, say* λ_k, *in the interior, then*

$$\{2\pi i(\lambda_j-\lambda_k)^{-1},\ j\neq k\}\subset \mathbf{C}$$

modulo action of $\mathrm{GL}(2,\mathbf{R})$, *is the only topological invariant.*

C) *If* $\mathscr{H}(\Lambda)$ *has at least two eigenvalues in its interior, then* Λ, *modulo translations, similarities and reflections in* $\mathbf{C}=\mathbf{R}^2$, *is a topological invariant. It is the only one because it clearly is the complete invariant of* $\mathscr{F}$ *under projective transformations and complex conjugation of* $\mathbf{C}P(m)$.

We first prove case C by determining the topological invariant. In § 10 we prove case A. We shall not elaborate on the proof of case B which goes along the same line as case A. For $m=2$ cases B and C do not occur, and case A was proved in [16].

For case C we assume (30) for Λ and Λ'. Let $h:\mathscr{F}(\Lambda)\to\mathscr{F}(\Lambda')$ be a homeomorphism of $\mathbf{C}P(m)$ onto itself sending leaves of $\mathscr{F}(\Lambda)$ onto leaves of $\mathscr{F}(\Lambda')$. It sends any singular point onto a singular point with the same local topological invariants. We may assume after projective transformation of $\mathscr{F}(\Lambda')$ in $\mathbf{C}P(m)$ that each of the $m+1$ singular points is invariant under h.

Let λ_0 and λ_1 be interior points of $\mathscr{H}(\Lambda)$. The corresponding singular points are then of Siegel type for $\mathscr{F}(\Lambda)$ and the same holds for their images under h which are singular points for $\mathscr{F}(\Lambda')$. Then λ_0' and λ_1' are also interior points of $\mathscr{H}(\Lambda')$. We can assume $\lambda_0=\lambda_0'=0$ and $\lambda_1=\lambda_1'=1$ by permitted changes of coordinates and parameters (translations and similarities).

By theorem 1, there exists $g_k\in\mathrm{GL}(2,\mathbf{R})$, $k=0,1$, such that for all j:

$$g_k(2\pi i(\lambda_j-\lambda_k)^{-1})=2\pi i(\lambda_j'-\lambda_k')^{-1}.$$

Hence if $x_k, y_k\in\mathbf{R}$, are such that:

$$(k=0)\qquad 2\pi i\lambda_3^{-1}=x_0 2\pi i\lambda_2^{-1}+y_0 2\pi i$$

and:

$$(k=1)\qquad 2\pi i(\lambda_3-1)^{-1}=x_1 2\pi i(\lambda_2-1)^{-1}+y_1 2\pi i(0-1)^{-1},$$

then the same equations hold for λ_2' and λ_3'. Elimination of λ_3 yields (for given x_0, y_0, x_1, y_1):

$$\lambda_3=\frac{\lambda_2}{x_0+y_0\lambda_2}=1+\frac{\lambda_2-1}{x_1+y_1(\lambda_2-1)}$$

which is a quadratic equation in λ_2 with real coefficients. If one solution is λ_2, then another is $\lambda_2' = \lambda_2$ or $\bar{\lambda}_2$ and it follows that g_1 is either the identity $(\lambda_2' = \lambda_2)$ or the reflection: complex conjugation $(\lambda_2' = \bar{\lambda}_2)$. The topology of $\mathscr{F}(\Lambda')$ is therefore determined by $\Lambda = \Lambda'$ modulo translations, similarities and reflections, and case C is proved. Observe that the foliation of $\Lambda' = \bar{\Lambda}$ is obtained from that of Λ by complex conjugation, a homeomorphism of $\mathbf{C}\mathrm{P}(m)$ onto itself.

10. Stable holomorphic flows on CP(m).

Here we prove case A of theorem II:

Theorem II A. — *The holomorphic flow* $\mathscr{F}(\Lambda)$ *on* $\mathbf{C}\mathrm{P}(m)$ *is stable in case* $\partial\mathscr{H}(\Lambda)$ *is a convex* $(m+1)$*-gon.*

Proof. — Let $\lambda_0, \lambda_1, \ldots, \lambda_m$ be cyclic successive vertices of $\mathscr{H}(\Lambda)$ (fig. 4). Let O_k be the singular point $\{z : z_j = 0,\ j \neq k,\ z_k = 1\} \in \mathbf{C}\mathrm{P}(m)$ and let $O_k O_\ell$ denote the "edge", a cylinder $\{z : z_j = 0 \text{ for } j \neq k, \ell,\ z_k \neq 0,\ z_\ell \neq 0\}$. The flow $\mathscr{F}_k$ on $z_k \neq 0$, as expressed in (29), has a singularity at O_k, and it is in the case of the Poincaré domain as described in § 6. Thus the leaf of a "general" point z (that is: $z_j \neq 0\ \forall j$) wraps around the axes (= "edges") $O_{k-1}O_k$ and $O_k O_{k+1}$ while converging to them. This

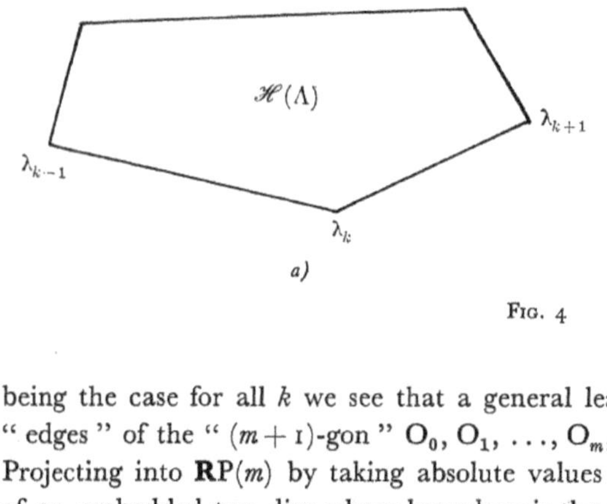

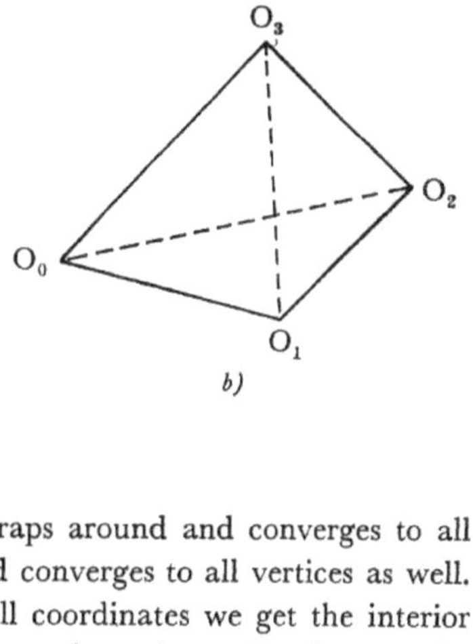

Fig. 4

being the case for all k we see that a general leaf wraps around and converges to all "edges" of the "$(m+1)$-gon" $O_0, O_1, \ldots, O_m$, and converges to all vertices as well. Projecting into $\mathbf{R}\mathrm{P}(m)$ by taking absolute values of all coordinates we get the interior of an embedded two-disc whose boundary is the ordinary $(m+1)$-gon $O_0, O_1, \ldots, O_m$ of the $\mathbf{R}$-coordinate simplex in $\mathbf{R}\mathrm{P}(m)$ (fig. 4 *b*).

In § 6 we saw that the topology of $\mathscr{F}_k$ is completely determined by the 1-flow in which it meets the "sphere" $S = S_k$; in homogeneous coordinates:

$$S_k : \{z : |z_k| = \sup_{j \neq k} |z_j|\}.$$

The intersection of a leaf with S_k is represented in the T-plane by a convex unbounded polygon:

$$C_k = \partial D_k \subset \mathbf{C}$$

boundary of the disc $D_k \subset \mathbf{C}$ that represents the intersection with

$$B_k = \{z : |z_k| = \sup_j |z_j|\} \subset \mathbf{C}P(m).$$

As $\bigcup_k B_k = \mathbf{C}P(m)$, therefore $\bigcup_k D_k = \mathbf{C}$ for a "general" leaf.

We now define the *graph* $GR = GR(z)$ *of a leaf* of a point $z \in \mathbf{C}P(m)$ as the union:

$$GR = \bigcup_k C_k = \{T \in \mathbf{C} : \exists k : |z_k| = \sup_{j \neq k} |z_j|\}.$$

In fig. 5 we give some example of graphs.

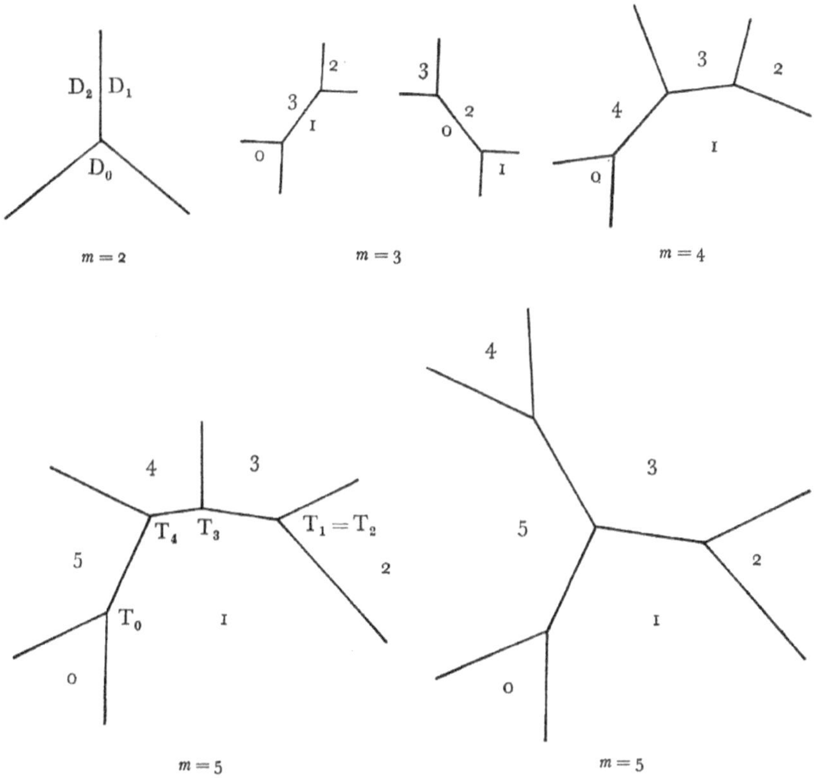

Fig. 5

The intersection $D_j \cap D_k \subset \mathbf{C}$ is *either an interval* parallel to the vector

$$\hat{\lambda}_{jk} = 2\pi i (\lambda_j - \lambda_k)^{-1}$$

with end points T_{1jk} and (see (29) and fig. 6):

$$T_{2jk} = T_{1jk} + \tilde{n}_{jk} \cdot 2\pi i(\lambda_j - \lambda_k)^{-1} \tag{31}$$

for some $0 < \tilde{n}_{jk} \leq \infty$, *or a point* and we put $\tilde{n}_{jk} = 0$, *or empty* and $\tilde{n}_{jk}$ is not defined. For cyclic successors k and $k+1$, $\tilde{n}_{k,k+1}$ is ∞.

Let T_k denote the endpoint of the infinite segment $D_k \cap D_{k+1}$ (fig. 5 and 6):

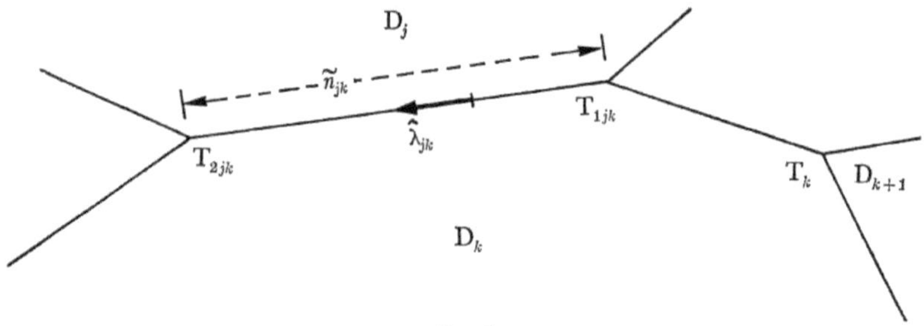

FIG. 6

We intersect the graph with a huge convex 2-disc which is then divided in $e_2 = m+1$ cells, and has e_0 vertices and e_1 edges, including $m+1$ vertices and $m+1$ edges on the boundary of the disc. The Euler characteristic of the disc is $1 = e_0 - e_1 + e_2$. In general every vertex is on three edges: $3e_0 = 2e_1$. Then the number of vertices is $e_0 = 2m$. Among these are $m-1$ vertices of GR. There are $e_1 = 3m$ edges, of which $m+1$ on the boundary of the disc and $m+1$ leading to this boundary. There remain $m-2$ bounded edges on GR giving rise to $m-2$ positive numbers $\tilde{n}_{jk}$, for $m-2$ specific pairs of indices j, k. Given this set, any $m-2$ positive numbers $\tilde{n}_{jk}$ yield up to translation a unique graph GR *compatible* with $\Lambda : D_j \cap D_k$ is parallel to $\hat{\lambda}_{jk}$. By admitting values $\tilde{n}_{jk} = 0$ for some of these index pairs we cover also the cases where more than three edges meet in a vertex.

For z such that $z_j \neq 0$ for all j, we know that in its leaf:

$$\left|\frac{z_{k+1}}{z_k}\right|(T_k) = 1 \quad \text{for} \quad k = 0, \ldots, m$$

and we define the argument φ_k by:

$$\left(\frac{z_{k+1}}{z_k}\right)(T_k) = e^{i\varphi_k}.$$

Lemma 5 *a.* — *Let* Λ *be given. The leaf of* $z \in \mathcal{S} = \bigcup_k S_k$, $z_j \neq 0\ \forall\ j$, *determines and is determined by the set of "coordinates"* $\{\tilde{n}_{jk}\}$, *a set of* $m-2$ *non negative numbers, for a specific set of index pairs* (j, k), *and the* m *arguments* $\varphi_j \bmod 2\pi$, $j = 0, \ldots, m-1$.

That the leaf z determines the "coordinates" is clear. Now suppose the "coordinates" given. Given Λ, the numbers $\tilde{n}_{jk}$ for a given k determine the convex disc D_k but for translations. We attach D_k to D_{k+1} along the common infinite edge for $k=0, 1, \ldots, m-1$. The finite sides fit also. We see that the $m-2$ numbers $\tilde{n}_{jk}$ determine the graph GR but for translations. Knowing φ_k, we know

$$z_{k+1}/z_k$$

at the point $T_k \in GR$. But in any other point $T \in GR$ we read from the formula:

$$(z_{k+1}/z_k)(T) = e^{(\lambda_{k+1}-\lambda_k)(T-T_k)}(z_{k+1}/z_k)(T_k).$$

So for every point $T \in GR$ we know without ambiguity $z_1/z_0, z_2/z_1, \ldots, z_m/z_{m-1}$, that is the set of non-zero homogeneous coordinates

$$(z_0, z_1, \ldots, z_m).$$

If *the point* T is on $D_j \cap D_k$ with vertex $T_{1jk} \in \mathbf{C}$, then T as well as the corresponding point $z \in \mathbf{C}^m$ can be *characterized by* $0 \leq t \leq \tilde{n}_{jk}$ for which

$$T = T_{1jk} + t2\pi i(\lambda_j - \lambda_k)^{-1}. \tag{32}$$

For a point z for which some (at most $m-2$) coordinates vanish, the same considerations apply to the remaining (at least three) non-zero coordinates, its leaf, its graph (with less domains D_k), etc. We get therefore:

Lemma 5. — *Let Λ be given. The leaf of a point $z \in \mathscr{S} = \bigcup_k S_k$ with $m'+1$ (at least three) non-zero coordinates determines and is determined by*:

$$J(z) = \{j : z_j \neq 0\},$$

$m'-2$ non negative numbers $\tilde{n}_{jk}$, and m' arguments φ_k mod 2π, $j, k \in J(z)$.

Now let Λ, Λ' be given and let $\partial\mathscr{H}(\Lambda)$ and $\partial\mathscr{H}(\Lambda')$ be convex $(m+1)$-gons. Define a 1-1 correspondence $h : \mathscr{S} \to \mathscr{S}$ by the identity in terms of the "coordinates" of lemma 5 and the coordinate t of (32) on $D_j \cap D_k$ outside the "edges", and by the ordinary identity map on $O_k O_\ell \cap \mathscr{S}$.

End of the proof. — Clearly h maps $S_k \subset \mathscr{S}$ onto itself. It is not exactly the same as the map h which we defined on S in § 6, but the same continuity arguments remain valid. So $h|S_k$ is a homeomorphism and it extends to a leaf preserving homeomorphism of B_k onto itself for each k (cone). This combines into the required homeomorphism:

$$h : (\mathbf{CP}(m), \mathscr{F}(\Lambda)) \to (\mathbf{CP}(m), \mathscr{F}(\Lambda')).$$

In case B of theorem II, we let $(\lambda_k =)\, \lambda_0 = 0$ be in the interior of $\mathscr{H}(\lambda_1, \ldots, \lambda_m)$ and we proceed as above. C_k is a convex polygon, unbounded for $k \neq 0$, bounded or empty in general in case $k=0$. The graph GR(z) of a leaf may therefore contain one cycle whose numbers $\tilde{n}_{j_0}$ then necessarily obey:

$$\sum_j \tilde{n}_{j_0} \lambda_j^{-1} = 0.$$

Apart from some special care concerning the case where C_0 is empty, the proof of II B follows the above pattern.

III. — NON LINEAR FLOWS NEAR A SINGULARITY IN DIMENSION $m=3$

11. A simple solution of a formal power series problem (m general).

We now start the study of the topology of a flow near a singularity at $0\in\mathbf{C}^m$, defined by:

$$\frac{dz}{dT}=F(z)=\sigma z+R(z)\in\mathbf{C}^m, \qquad z\in\mathbf{C}^m \tag{33}$$

where F is holomorphic, $F(0)=0$, $\sigma z=(DF)_0 z$ is the first term of the Taylor series of F and R is the rest. We assume again for the eigenvalues of σ:

$$i\neq j \Rightarrow \lambda_i\notin\mathbf{R}\lambda_j, \quad i,j=1,\ldots,m. \tag{7}$$

In suitable linear coordinates (33) is expressed as:

$$\frac{dz_j}{dT}=\lambda_j z_j+\varphi_j(z_1,\ldots,z_m), \quad j=1,\ldots,m \tag{34}$$

φ_j is considered as a convergent power series starting with terms of degree ≥ 2:

$$\varphi_j(z)=\sum_Q \varphi_{jQ}\, z^Q,$$

where $Q=(q_1,\ldots,q_m)$, $z^Q=z_1^{q_1}z_2^{q_2}\ldots z_m^{q_m}$ is of degree $\|Q\|=q_1+q_2+\ldots+q_m$ in $z_1,\ldots,z_m$.

Formal power series lemma 6 (known). — *There exists a unique solution in formal power series:*

$$z_j=w_j+\zeta_j(w_1,\ldots,w_m), \quad \zeta_j=\sum_{\|Q\|\geq 2}\zeta_{jQ}w^Q, \quad j=1,\ldots,m \tag{35}$$

which transforms (34) *into*

$$\frac{dw_j}{dT}=\lambda_j w_j+\psi_j(w_1,\ldots,w_m), \quad \psi_j=\sum_{\|Q\|\geq 2}\psi_{jQ}w^Q \tag{36}$$

where $\psi_{jQ}=0$ *in case* $\delta_{jQ}\overset{\text{def}}{=}q_1\lambda_1+\ldots+q_m\lambda_m-\lambda_j\neq 0$

and $\zeta_{jQ}=0$ *in case* $\delta_{jQ}=0$. (37)

Proof. — Substitution of (35) in (34) yields with (36):

$$\frac{dw_j}{dT}+\sum_k\frac{\partial\zeta_j}{\partial w_k}(\lambda_k w_k+\psi_k)=\lambda_j(w_j+\zeta_j)+\varphi_j(w_1+\zeta_1,\ldots,w_m+\zeta_m).$$

Substract (36) to get:

$$\sum_k\frac{\partial\zeta_j}{\partial w_k}\lambda_k w_k-\lambda_j\zeta_j+\psi_j=\varphi_j(w_1+\zeta_1,\ldots)-\sum_k\frac{\partial\zeta_j}{\partial w_k}\psi_k. \tag{38}$$

As $\sum_k \left(\frac{\partial}{\partial w_k} w^Q\right) \lambda_k w_k - \lambda_j w^Q = \delta_{jQ} w^Q$, we find the equivalent equations:

$$\sum_Q (\delta_{jQ} \zeta_{jQ} + \psi_{jQ}) w^Q = \varphi_j(w_1 + \zeta_j, \ldots) - \sum_k \frac{\partial \zeta_j}{\partial w_k} \psi_k. \tag{38}$$

The terms on the right hand side of degree $\|Q\| = n$ do not involve coefficients of terms of ζ_j, ψ_j, $j = 1, \ldots, m$, of degrees $\geq n$. By (37) we compute unique values ζ_{jQ} and ψ_{jQ} for $\|Q\| = n$, once we know them for $\|Q\| < n$.

The unique power series ζ_j and ψ_j satisfy (38). From (38) with (34) and (35) we can deduce the equations $(j = 1, \ldots, m)$:

$$W_j + \sum_{k=1}^{m} \frac{\partial \zeta_j}{\partial w_k} W_k = 0 \quad \text{for} \quad W_i = \frac{dw_i}{dT} - \lambda_i w_i - \psi_i.$$

The determinant of the coefficient matrix is near one for $(w_1, \ldots, w_m)$ near $(0, 0, \ldots, 0)$, so that (36) holds:

$$W_j = 0, \quad j = 1, \ldots, m.$$

Our unique power series give solutions indeed, and lemma 9 is proved.

12. Holomorphic normal forms for m = 3.

First normal form: Lemma 7. — *If* $m = 3$, $0 \in \mathcal{H}(\lambda_1, \lambda_2, \lambda_3)$, (7), *then there exists a holomorphic change of coordinates* (35) *near* $0 \in \mathbf{C}^3$, *transforming* (34) *into the (not unique) normal form*:

$$\frac{dw_j}{dT} = \lambda_j w_j + w_1 w_2 w_3 \chi_j, \quad j = 1, 2, 3 \tag{39}$$

$\chi_j = \chi_j(w_1, w_2, w_3)$ *holomorphic near* $0 \in \mathbf{C}^3$.

Proof. — Because $0 \in \mathcal{H}(\lambda_1, \lambda_2, \lambda_3)$, we conclude from geometry in the plane $\mathbf{C}$ that if $\delta_{1Q} = (q_1 - 1)\lambda_1 + q_2\lambda_2 + q_3\lambda_3$, $|\delta_{1Q}| < \delta$ and $\delta > 0$ small, then

$$q_1 \geq 2, \quad q_2 \geq 1, \quad q_3 \geq 1$$

and similarly for $j = 2, 3$. The *ideal* Ψ generated by the polynomial $w_1 w_2 w_3$ contains therefore among others *all* polynomials $w^Q = w_1^{q_1} w_2^{q_2} w_3^{q_3}$ for which $|\delta_{jQ}| < \delta$ for some j.

As in § 11 there is a formal power series solution (35) transforming (34) into (36), but now, instead of (37), such that

$$\zeta_{jQ} = 0 \quad \text{if} \quad w^Q \in \Psi, \qquad \psi_{jQ} = 0 \quad \text{if} \quad w^Q \notin \Psi \tag{40}$$

because all *small divisors* $|\delta_{jQ}| < \delta$ *(in particular zero divisors) are avoided in the computation of* ζ_j *from* (38). In order to prove that ζ_j is convergent near $0 \in \mathbf{C}^3$, we use the following notations concerning power series ξ. The series $\bar{\xi}$ (in Siegel's notation $\lceil \xi \rceil$, see [14]) is obtained from ξ by replacing each coefficient by its absolute value. $\bar{\bar{\xi}}$ is obtained from $\bar{\xi}$ by taking all arguments equal ($w_j = w$ for $j = 1, \ldots, m$):

$$\bar{\bar{\xi}}(w) = \bar{\xi}(w, w, \ldots, w).$$

We write (with Siegel) $\xi \prec \eta$ to express that $|\xi_Q| \leq |\eta_Q| = \eta_Q$ for all Q. Clearly if $\bar{\bar{\xi}}$ is convergent near $0 \in \mathbf{C}$, then $\bar{\xi}$ is convergent near $0 \in \mathbf{C}^m$, and then ξ is convergent near $0 \in \mathbf{C}^m$.

From (38) and $|\delta_{jQ}| \geq \delta$ we obtain:

$$\delta\bar{\zeta}_j \prec \sum_Q \delta_{jQ} |\zeta_{jQ}| w^Q + \bar{\psi} \prec \bar{\varphi}_j(w_1 + \bar{\zeta}_1, \ldots) + \sum_k \frac{\partial \bar{\zeta}_j}{\partial w_k} \bar{\psi}_k. \tag{41}$$

The power series at the extreme left has no terms in the ideal Ψ. So we can delete the last part in the form at extreme right which is in Ψ, and obtain:

$$\delta\bar{\zeta}_j \prec \bar{\varphi}_j(w_1 + \bar{\zeta}_1, w_2 + \bar{\zeta}_2, \ldots).$$

Hence
$$\sum_j \bar{\bar{\zeta}}_j \prec \delta^{-1} \sum_j \bar{\varphi}_j(w + \bar{\bar{\zeta}}_1 + \ldots + \bar{\bar{\zeta}}_m, w + \bar{\bar{\zeta}}_1 + \ldots + \bar{\bar{\zeta}}_m, \ldots). \tag{42}$$

We define a new power series $u = u(w) = \sum_{n \geq 1} u_n w^n$ by:

$$wu = \sum_j \bar{\bar{\zeta}}_j. \tag{43}$$

Recall that $\sum_j \varphi_j(w_1, \ldots, w_m)$ is given, convergent near 0, and it begins with terms of degree ≥ 2. Therefore $A_0 > 0$ and $A > 0$ exist such that:

$$\delta^{-1}. \sum_j \bar{\bar{\varphi}}_j(w) \prec \frac{A_0 w^2}{1 - Aw}. \tag{44}$$

(42), (43) and (44) yield

$$wu \prec \frac{A_0(w + wu)^2}{1 - A(w + wu)}$$

$$u \prec \frac{A_0 w(1 + u)^2}{1 - Aw(1 + u)} = A_0 w(1 + u)^2 \sum_{k=0}^{\infty} (Aw(1 + u))^k. \tag{45}$$

We compare (45) with the equation for the power series $v(w) = \sum_{n=1}^{\infty} v_n w^n$, $v(0) = 0$:

$$v = \frac{A_0 w(1 + v)^2}{1 - Aw(1 + v)} = A_0 w(1 + v)^2 \sum_{k=0}^{\infty} (Aw(1 + v))^k. \tag{46}$$

$v(w)$ is unique and convergent near $0 \in \mathbf{C}$ because

$$v_1 = \left(\frac{dv}{dw}\right)_{w=0} = A_0 \neq 0.$$

By choosing $A_0 > 0$ big enough we find $v_k > u_k > 0$ for the first non zero coefficient u_k of u. Then by induction with respect to n while reading and comparing (45) and (46) we obtain the majoration $v_n \geq u_n \geq 0$ for all n. Then u, $\bar{\bar{\zeta}}_j$, $\bar{\zeta}_j$ and ζ_j are convergent near 0 and lemma 9 is proved.

We push the normalisation further in the

Second normal form: Lemma 8. — *With the conditions of lemma* 7 *there exists a holomorphic change of coordinates* (35) *near* $0 \in \mathbf{C}^3$ *transforming* (34) *into the normal form*

$$\frac{dw_j}{dT} = \lambda_j w_j (1 + w_1 w_2 w_3 \chi_j) \quad j = 1, 2, 3. \tag{47}$$

Proof. — By lemma 7 we can assume for (34):

$$\frac{dz_j}{dT} = \lambda_j z_j + \varphi_j, \quad \varphi_j \in \Psi, \quad j = 1, 2, 3. \tag{48}$$

Again we formally solve:

$$\sum_Q (\delta_{jQ} \zeta_{jQ} + \psi_{jQ}) w^Q = \varphi_j(w_1 + \zeta_1, w_2 + \zeta_2, \dots) + \sum_k \frac{\partial \zeta_k}{\partial w_k} \psi_k \tag{35}$$

but now instead of (37) or (40) we claim

$$\zeta_{jQ} = 0 \quad \text{if} \quad w^Q \in \Psi_j, \qquad \psi_{jQ} = 0 \quad \text{if} \quad w^Q \notin \Psi_j$$

where $\Psi_j \subset \Psi$ is the ideal generated by $z_j z_1 z_2 z_3$. We can solve because $|\delta_{jQ}| > \delta > 0$ for $w^Q \notin \Psi_j$!

By induction with respect to $\|Q\| = n$ we see that the formal power series ζ_j belongs to Ψ, and for all Q, by construction, $\zeta_{jQ} \neq 0$ implies that $\zeta_{jQ} w^Q \notin \Psi_j$. For $j = 1$ for example we can therefore write

$$\zeta_1 = w_1 w_2 w_3 \xi_{23}(w_2, w_3)$$

$$\psi_k = w_k w_1 w_2 w_3 \chi_k(w_1, w_2, w_3)$$

and consequently for $k = 1, 2, 3$, $\frac{\partial \zeta_j}{\partial w_k} \psi_k$ has a factor $w_1^2 w_2^2 w_3^2$ and belongs to the ideal Ψ^2. So does $\frac{\partial \zeta_j}{\partial w_k} \psi_k$ for $j = 2, 3$. The power series ζ_j has no terms in $\Psi^2 \supset \Psi_j$. We can then repeat the arguments with the equations (41)-(46), neglecting terms in Ψ^2, and conclude the convergence of ζ_j $j = 1, 2, 3$ near $0 \in \mathbf{C}^3$. Lemma 10 is proved.

13. Local stability of codimension two.

Theorem III. — *The foliation* $\mathcal{F}(F)$ *defined by a holomorphic vector field* $F(z)$ *near* $0 \in \mathbf{C}^3$, *with singularity at* 0 *such that the set of eigenvalues* $\{\lambda_1, \lambda_2, \lambda_3\}$ *of* $(Df)_0 = \sigma$ *is in the Siegel domain, is locally homeomorphic to the foliation of its linear part* $\mathcal{F}(\sigma)$. *The invariant* Δ $(=\eta)$ *of chapter I characterises the topology completely.* $\mathcal{F}(F)$ *is locally stable of codimension two.*

By lemma 8 we can assume (34) in the form

$$\frac{dz_j}{dT}=\lambda_j z_j(1+z_1 z_2 z_3 \chi_j), \quad j=1, 2, 3. \tag{49}$$

The three coordinate planes are invariant and they already contain linear vector fields. There exist $0<\varepsilon_0<1$ and $K>0$ such that $\chi_j(z_1, z_2, z_3)$ is convergent and $|\chi|<K$ for $\sup_i|z_i|<\varepsilon_0$. Assume $\varepsilon_0 K<1$. By substituting $\varepsilon^3 z_j$ for z_j, $j=1, 2, 3$ with $0<\varepsilon<\varepsilon_0$, we obtain new equations instead of (49) with new functions χ_j for which we can assume convergence in the " unit ball " $\sup_j|z_j|\leq 1$ and moreover:

$$|\chi_j|<\varepsilon^8. \tag{50}$$

We will first construct a homeomorphism h_1 of S onto S, carrying the leaves of $\mathscr{F}_1(F)=\mathscr{F}(F)\cap S$ onto those of $\mathscr{F}_1(\sigma)=\mathscr{F}(\sigma)\cap S$, and which is the identity on $V_1=V\cap S$, where

$$V=\{z : z_1 z_2 z_3=0\}.$$

Our strategy will be to let h_1 preserve the strata of S and to let h_1 be identity on $\{z : |z_1|=|z_3|=1\}$ (see fig. 7).

The first part of theorem III will then follow from general considerations, and the last part is a consequence of theorem **I**.

For later estimation purposes we write the unit disc in **C**:

$$\theta=\{u\in\mathbf{C} : |u|\leq 1\} \tag{51}$$

and $f(\theta)=\{f(u) : u\in\theta\}$ for any function f of $u\in\mathbf{C}$.

We assume $\arg\hat\lambda_1<\arg\lambda_2<\arg\hat\lambda_3<\arg\hat\lambda_1+2\pi$ so that with $\hat\lambda_j=2\pi i\lambda_j^{-1}$:

$$\operatorname{Re}\lambda_2\hat\lambda_1>0, \qquad \operatorname{Re}\lambda_3\hat\lambda_1<0. \tag{52}$$

The leaves of $\mathscr{F}(\sigma)$ are transversal to S except at the points where $|z_1|=|z_2|=|z_3|=1$. The same holds for the slightly perturbed $\mathscr{F}(F)$. Any 1-leaf in S meets $|z_3|=|z_1|=1$ in at most one point $z(T_0)=(z_1(T_0), z_2(T_0), z_3(T_0))$ with parameter value T_0 say; it meets $|z_1|=|z_2|=1$ in $z(T_1)$; it meets $|z_2|=|z_3|=1$ in $z(T_2)$. This will be made clear in the following calculations. We start at $t=0$ from a point:

$$z(T_0)=(e^{i\alpha_1}, e^{i\alpha_2-N}, e^{i\alpha_3}), \qquad N>0. \tag{53}$$

We shall perform the calculations only in the special case $\alpha_1=\alpha_2=\alpha_3=0$, hence

$$z(T_0)=(1, e^{-N}, 1). \tag{53'}$$

The general case (53) is formally but not essentially more complicated. We follow its 1-leaf in S with respect to $\mathscr{F}(F)$ in $|z_1|=1$, substitute

$$z_1=e^{2\pi i t}, \quad t\geq 0 \tag{54}$$

in (49), and find:

$$2\pi i=\frac{d\ln z_1}{dt}=\lambda_1(1+z_1 z_2 z_3\chi_1)\frac{dT}{dt}.$$

If $\varepsilon > 0$ is small enough, depending only on $\lambda_1, \lambda_2, \lambda_3$, then (see (50) and (51))

$$\frac{dT}{dt} = \hat{\lambda}_1 (1 + z_1 z_2 z_3 \chi_1)^{-1} \in \hat{\lambda}_1 + \varepsilon^7 \sup |z_1 z_2 z_3| \theta \tag{55}$$

$$\frac{dT}{dt} \in \hat{\lambda}_1 + \varepsilon^7 \theta. \tag{56}$$

The sup is here over the segment from T_0 to T_1 where, as we see below in (57), $|z_2|$ increases up to its value 1 at T_1 at which we set $t = \tilde{n}_1$. Similarly $|z_3|$ decreases between T_0 and T_1. In the T-plane the curve from T_0 to T_1 is for small ε almost parallel to $\hat{\lambda}_1$ by (56). See fig. 7.

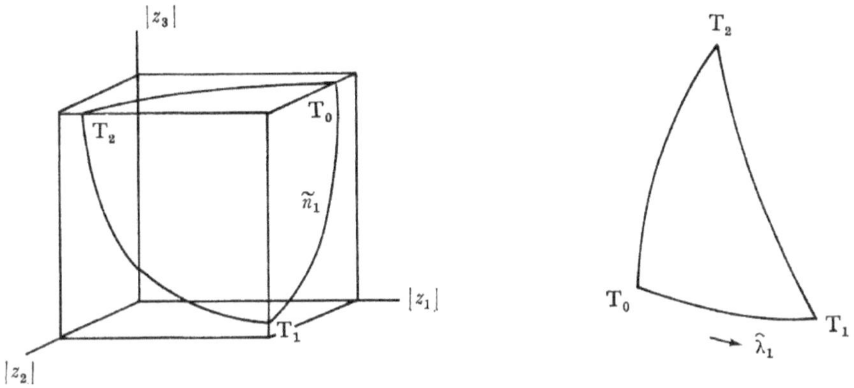

FIG. 7

Next we use (55) and the equation (49) for z_2 to get, for small $\varepsilon > 0$:

$$\frac{d \ln z_2}{dt} = \lambda_2 (1 + z_1 z_2 z_3 \chi_2) \frac{dT}{dt} \in \lambda_2 \hat{\lambda}_1 + \varepsilon^6 \sup |z_1 z_2 z_3| \theta \subset \lambda_2 \hat{\lambda}_1 + \varepsilon^6 \theta. \tag{57}$$

We integrate:

$$\ln z_2 \in -N + (\lambda_2 \hat{\lambda}_1 + \varepsilon^6 \sup |z_1 z_2 z_3| \theta) t \subset -N + (\lambda_2 \hat{\lambda}_1 + \varepsilon^6 \theta) t \tag{58}$$

$$\ln |z_2| < -N + (\operatorname{Re} \lambda_2 \hat{\lambda}_1 + \varepsilon^6) t \stackrel{\text{def}}{=} -N + \gamma_3 t, \qquad \gamma_3 > 0. \tag{59}$$

Analogously for z_3:

$$\ln z_3 \in (\lambda_3 \hat{\lambda}_1 + \varepsilon^6 \sup |z_1 z_2 z_3| \theta) t \subset (\lambda_3 \hat{\lambda}_1 + \varepsilon^6 \theta) t \tag{60}$$

$$\ln |z_3| < (\operatorname{Re} \lambda_3 \hat{\lambda}_1 + \varepsilon^6) t \stackrel{\text{def}}{=} -\gamma_2 t, \qquad \gamma_2 > 0. \tag{61}$$

From (59) and (61) it follows that

$$\gamma_2 \ln |z_2| + \gamma_3 \ln |z_3| < -\gamma_2 N.$$

As $\ln|z_1|=1$, $\ln|z_2|\leq 0$, $\ln|z_3|\leq 0$:

$$\ln|z_1 z_2 z_3| = \ln|z_2| + \ln|z_3| < \max\left(-1, -\frac{\gamma_2}{\gamma_3}\right). N \stackrel{\text{def}}{=} -\gamma_4 N$$

$$\sup|z_1 z_2 z_3| < e^{-\gamma_4 N}. \tag{62}$$

$|z_2(T_1)|=1$ for $t=\widetilde{n}_1$, and substituting (62) in (58) gives an estimate for $\widetilde{n}_1$:

$$0=\ln|z_2(T_1)| \in -N + (\operatorname{Re}\lambda_2\widehat{\lambda}_1 + \varepsilon^6 e^{-\gamma_4 N}\theta)\widetilde{n}_1 \subset -N + \operatorname{Re}\lambda_2\widehat{\lambda}_1(1+\varepsilon^5 e^{-\gamma_4 N}\theta)\widetilde{n}_1.$$

Then for ε small enough:

$$\widetilde{n}_1 < \frac{2N}{\operatorname{Re}\lambda_2\widehat{\lambda}_1}, \quad \text{hence} \quad \left|\widetilde{n}_1 - \frac{N}{\operatorname{Re}\lambda_2\widehat{\lambda}_1}\right| \leq \varepsilon^5 e^{-\gamma_4 N}\operatorname{Re}\lambda_2\widehat{\lambda}_1\widetilde{n}_1 \leqslant \varepsilon^4 N e^{-\gamma_4 N}. \tag{63}$$

The corresponding answer for a 1-leaf of $\mathscr{F}(\sigma)$ *starting at the same point*

$$z'(T'_0)=z(T_0)=(1, e^{-N}, 1),$$

is obtained by putting $\varepsilon=0$. We use primes for the linear case $\mathscr{F}(\sigma)$:

$$\widetilde{n}'_1 = \frac{N}{\operatorname{Re}\lambda_2\widehat{\lambda}_1}. \tag{63'}$$

The difference is small for large N:

$$|2\pi(\widetilde{n}_1-\widetilde{n}'_1)| = |\ln z_1(T_1) - \ln z'_1(T'_1)| = |\arg z_1(T_1) - \arg z'_1(T'_1)| \leq \varepsilon^3 N e^{-\gamma_4 N}. \tag{64}$$

Substituting (62) and $t=\widetilde{n}_1$ in (58) gives:

$$\begin{aligned} &\ln z_2(T_1) \in -N + (\lambda_2\widehat{\lambda}_1 + \varepsilon^6 e^{-\gamma_4 N}\theta)\widetilde{n}_1 \\ &\ln z'_2(T'_1) = -N + \lambda_2\widehat{\lambda}_1\widetilde{n}'_1 \end{aligned} \tag{65}$$

Hence $\quad |\ln z_2(T_1) - \ln z'_2(T'_1)| = |\arg z_2(T_1) - \arg z'_2(T'_1)| \leq \varepsilon^2 N e^{-\gamma_4 N}. \quad (66)$

Substituting (62) and $t=\widetilde{n}_1$ in (60) gives:

$$\begin{aligned} &\ln z_3(T_1) \in (\lambda_3\widehat{\lambda}_1 + \varepsilon^6 e^{-\gamma_4 N}\theta)\widetilde{n}_1 \\ &\ln z'_3(T'_1) = \lambda_3\widehat{\lambda}_1\widetilde{n}'_1 = N\lambda_3\widehat{\lambda}_1/\operatorname{Re}\lambda_2\widehat{\lambda}_1 \end{aligned} \tag{67}$$

Hence $\quad |\ln z_3(T_1) - \ln z'_3(T'_1)| \leq \varepsilon^2 N e^{-\gamma_4 N}. \quad (68)$

We conclude from (64), (66) and (68) *that the mapping* $z(T_1) \to z'(T'_1)$ of

$$S \cap \{z : |z_2|=|z_1|=1, z_3 \neq 0\}$$

onto itself *tends to the identity map in* $z_3=0$ (*for* $N\to\infty$). This is equally true starting from (53) instead of (53'). The same calculations for the 1-leaf segments T_0T_2 in $|z_3|=1$ give the *analogous conclusions for the mapping* $z(T_2)\to z'(T'_2)$.

We now define the map

$$h_1 = (S, \mathscr{F}_1(F)) \to (S, \mathscr{F}_1(\sigma))$$

by the following conditions:

a) The restriction of h_1 to the union $V=\{z : z_1 z_2 z_3 = 0\}$ of the Poincaré leaves is the identity.

b) h_1 leaves invariant each stratum of the stratification of S by $|z_j|=1$, $j=1, 2, 3$.

c) On the stratum $|z_1|=|z_3|=1$, h_1 is the identity map. In our notation *c)* means $h_1(z(T_0))=z'(T'_0)=z(T_0)$. As every Siegel 1-leaf in S meets $|z_3|=|z_1|=1$, *c) determines for each Siegel $\mathscr{F}_1$(F)-leaf its image $\mathscr{F}_1(\sigma)$-leaf.* In view of *b)* we have by intersection:

$$h_1(z(T_1))=z'(T'_1) \quad \text{if} \quad |z_2|=|z_1|=1, \quad z_3\neq 0 \text{ at } z(T_1).$$

This agrees continuously with the identity at $z_3=0$ (see *a)*). Similarly for

$$h_1(z(T_2))=z'(T'_2).$$

d) A point on the edge T_0T_1 of an $\mathscr{F}_1$(F)-leaf with total z_1-argument-length $2\pi\tilde{n}_1$, is determined by a rotation number t, $0\leq t\leq\tilde{n}_1$, if we start from $z(T_0)$, or $\tilde{n}-t$ if we start from $z(T_1)$ (see (54)). It is analogous for the edges T_1T_2 and T_2T_0 by cyclic permutation of 1, 2 and $3=0$. The same applies to the linear case $\mathscr{F}(\sigma)$, which we continue to distinguish in the notation by primes. The action of h_1 on the points of an $\mathscr{F}_1$(F)-leaf onto its image $\mathscr{F}_1(\sigma)$-leaf, is given by proportional rotation numbers:

$$t'_j/t_j=\tilde{n}'_j/\tilde{n}_j, \quad j=1, 2, 3.$$

For $n\to\infty$ these quotients converge to 1. (64) tells this for $j=1$. For $j=3$ it follows by studying T_0T_2 instead of T_0T_1. The value $2\pi\tilde{n}_2$ is the z_2-argument difference between $z(T_0)$ and $z(T_2)$, which can be read also going along two edges via $z(T_1)$. The formulas then tell again that also $\tilde{n}'_2/\tilde{n}_2$ tends to 1 for $N\to\infty$.

If we keep $t=t_j$ or $\tilde{n}_j-t_j$ fixed for some j and let N go to ∞, then the map of the initial point in the $\mathscr{F}_1$(F)-leaf converges to the identity map of a point of a Poincaré-$\mathscr{F}_1$(F)-leaf in V. Therefore h_1 agrees continuously with the identity map on V, it is continuous as is its inverse. *Then h_1 is the desired homeomorphism.*

There remains to define with the help of h_1 a homeomorphism $h:\mathscr{F}(F)\to\mathscr{F}(\sigma)$ near $0\in\mathbf{C}^3$. Let U_ρ be the neighborhood of V in $B=\{z:\sup_j|z_j|\leq 1\}$ that is the union of all $\mathscr{F}$(F)-leaves in B at euclidean distance $(\|z\|^2=\sum_j|z|^2)$ smaller than some ρ with $0<\rho<\frac{1}{2}$ from $0\in\mathbf{C}^3$. The restriction $\mathscr{F}(F)|U_\rho$ is transversal to S and to $S_\tau=\{z:\|z\|=\tau\}$ for $\tau>\rho$. Recall that in each Siegel leaf of $\mathscr{F}(\sigma)$ the function $\|z\|$ has exactly one critical point, where $\|z\|$ has a minimum, and the critical points form together a manifold M. These two properties are stable under our small perturbations and hold equally well for $\mathscr{F}$(F). This can be seen also by a calculation of $d\sum_j z_j\bar{z}_j=0$. Call the manifold of nearest Siegel leaf points M^F. First observe that:

$$\mathscr{F}(F)\left|\left(U_\rho\cap\left\{z:\|z\|>\frac{1}{2}\right\}\right)\right.$$

is homeomorphic ($\overset{h}{\sim}$) to $\mathscr{F}_1(F)\times\left[\frac{1}{2}, 1\right]$. Extend this homeomorphism with the help of the $\|z\|$-gradient lines in each $\mathscr{F}$(F)-leaf in U_ρ to obtain a homeomorphism:

$$\mathscr{F}(F)|(U_\rho\setminus M^F\cup 0)\overset{h}{\sim}\mathscr{F}_1(F)\times(0, 1].$$

The union of the Siegel leaves in this foliation is a trivial 2-disc fibration with base space $U_\rho \cap M^F$ from which a cross-section $U_\rho \cap M^F$ is deleted. We recover it by compactifying each Siegel-leaf with one point, and we recover $\mathscr{F}(F)|U_\rho$ by compactifying the space so obtained with one more point, the origin $0 \in \mathbf{C}^3$. The foliation $\mathscr{F}(F)|U_\rho$ of $\mathscr{F}(F)$ near 0 is now up to topological equivalence completely determined by $\mathscr{F}_1(F)|(U_\rho \cap S)$, and the same holds for $\mathscr{F}(\sigma)$. Therefore the existence of h_1 carries with itself the existence of a homeomorphism of $\mathscr{F}(F)|U_\rho$ onto $\mathscr{F}(\sigma)|U'_\rho$ where U'_ρ is obtained from σ in the same way as U_ρ from F. Theorem III is proved.

IV

14. Holomorphic reduction to normal forms, $m \geq 3$.

We terminate this paper with the necessary preparation for a proof (that we hope to give later) of the stability of codimension $2m-4$ for Siegel domain type singularities of vector fields on $\mathbf{C}^m$, $m>3$ (see footnote (1) on page 8).

Theorem **(IV.1)**. — *Given, as before, near* $0 \in \mathbf{C}^m$, *the equations*

$$\frac{dz}{dT} = F : \frac{dz_j}{dT} = \lambda_j z_j + \varphi_j, \quad i \neq j \Rightarrow \lambda_i \notin \mathbf{R}\lambda_j, \quad i,j = 1, \ldots, m \qquad (7), (34)$$

we can assume after a suitable holomorphic change of coordinates that the union $V(F)$ *of the Poincaré leaves and the axes equals the unperturbed set*

$$V(\sigma) = V = \{z : \text{if } \{j_1, \ldots, j_r\} = \{j : z_j \neq 0\}, \text{ then } 0 \notin \mathscr{H}(\lambda_{j_1}, \ldots, \lambda_{j_r})\}.$$

This follows from the reduction to normal form:

Theorem **(IV.2)**. — *Assuming* (7), (34), *there is a holomorphic change of coordinates*

$$z_j = w_j + \zeta_j(w_1, \ldots, w_m) \qquad (35)$$

transforming (34) *into*

$$\frac{dw_j}{dT} = \lambda_j w_j + \psi_j + \chi_j \qquad (69)$$

such that $\psi_j(w_1, \ldots, w_m)$ *is in* Ψ, *the ideal in the ring of convergent power series generated by the monomials* $w_i w_k w_\ell$ *for which* $0 \in \mathscr{H}(\lambda_i, \lambda_k, \lambda_\ell)$,

$$\psi_j = \sum_{0 \in \mathscr{H}(\lambda_i, \lambda_k, \lambda_\ell)} w_i w_k w_\ell \zeta_{jik\ell},$$

and $\chi_j(w_1, \ldots, w_m)$ *is a sum of scalar multiples of terms, finite in number,*

$$w^Q = w_{j_1}^{q_1} w_{j_2}^{q_2}, \ldots, w_{j_r}^{q_r}, \quad q_1 > 0, \ldots, q_r > 0,$$

for which δ_{jQ} *is a Poincaré zero divisor:*

$$\delta_{jQ} = q_1 \lambda_{j_1} + \ldots + q_r \lambda_{j_r} - \lambda_j = 0 \qquad (70)$$

and there is an open half plane for some $\omega \in \mathbf{R}$:

$$\{\lambda \in \mathbf{C} : \operatorname{Re} e^{i\omega}\lambda > 0\} \tag{71}$$

to which $\lambda_{j_1}, \ldots, \lambda_{j_r}$ *and* λ_j *belong.*

Corollary. — *The theorem of Dulac* [4]. This is the special Poincaré domain case $0 \notin \mathscr{H}(\lambda_1, \ldots, \lambda_m)$, for which Ψ is empty, hence $\psi_j = 0$ in (69).

Proof of $V \subset V(F)$ *in theorem* (IV.1), *from* (IV.2).

We shall assume that the vector field F has the form (69) with the condition of theorem (IV.2). The set V is a finite union of maximal linear subspaces, and we first prove that any one of them, say V_0, is invariant under F. There is no loss of generality in assuming that V_0 is defined by equations

$$V_0 : z_{r+1} = \ldots = z_m = 0 \tag{72}$$

for some r, which implies that, for some ω,

$$\lambda_i \in \{\lambda : \operatorname{Re} e^{i\omega}\lambda > 0\}, \quad \text{for} \quad 1 \leq i \leq r. \tag{73}$$

We have to prove that, on V_0, (72) implies that

$$\frac{dz_j}{dT} = 0 \quad \text{for} \quad j \geq r+1.$$

To see this substitute (72) in the right hand side of:

$$\frac{dz_j}{dT} = \lambda_j z_j + \psi_j + \chi_j, \quad j \geq r+1.$$

Then $\lambda_j z_j = 0$; $\psi_j = 0$ because the definition of Ψ implies that every term in ψ_j contains one of the z_ℓ for which $\operatorname{Re} e^{i\omega}\lambda_\ell \leq 0$, that is $\ell > r$ and $z_\ell = 0$; $\chi_j = 0$ because if it has a nonzero term $cz^Q = cz_1^{q_1} \ldots z_r^{q_r}$ then $\operatorname{Re}(e^{i\omega}\delta_{jQ}) = \operatorname{Re}(e^{i\omega}(q_1\lambda_1 + \ldots + q_r\lambda_r - \lambda_j)) > 0$ and δ_{jQ} cannot be zero. Then the left hand side vanishes as well.

The nonlinear vector field F defines in the invariant part (73) of V a Poincaré domain vector field near 0. So this part, like any part near 0 of V, lies in V(F) and:

$$V \subset V(F).$$

We do not give the proof here of the stronger assertion:

$$V = V(F)$$

which is an elaborate calculation.

Proof of theorem (IV.2). — As in § 11 we can obtain a (unique) formal solution for ζ_j from equations (38) but now with the conditions

$$\begin{aligned} \psi_{jQ} &= 0 \quad \text{if} \quad w^Q \notin \Psi, \\ \zeta_{jQ} &= 0 \quad \text{if} \quad w^Q \in \Psi \quad \text{or if} \quad \delta_{jQ} = 0, \\ \chi_{jQ} &= 0 \quad \text{if} \quad w^Q \in \Psi \quad \text{or if} \quad \delta_{jQ} \neq 0. \end{aligned}$$

We have to prove that ζ_j is convergent near $0 \in \mathbf{C}^m$ for $j = 1, \ldots, m$.

We first prove the

Lemma 9. — *There exist* $n_0>0$ *and* $\delta>0$ *such that if*

$$w^Q\notin\Psi,\quad \delta_{jQ}\neq 0,\quad \|Q\|=q_1+\ldots+q_m\geq n_0,$$

then
$$|\delta_{jQ}|>\delta.\|Q\|.$$

As $w^Q=w_{j_1}^{q_1}\ldots w_{j_r}^{q_r}\notin\Psi$, $q_1>0,\ldots,q_r>0$, we conclude that $\lambda_{j_1},\ldots,\lambda_{j_r}$ are contained in some open half plane $\{\lambda:\mathrm{Re}\,e^{i\omega}\lambda>0\}$. If λ_j is *not in* that half plane, hence $\mathrm{Re}\,e^{i\omega}\lambda_j\leq 0$, then:

$$\begin{aligned}|\delta_{jQ}|>\mathrm{Re}\,e^{i\omega}\delta_{jQ}&=\mathrm{Re}\,e^{i\omega}(q_1\lambda_{j_1}+\ldots+q_r\lambda_{j_r}-\lambda_j)\\&\geq\mathrm{Re}\,e^{i\omega}(q_1\lambda_{j_1}+\ldots+q_r\lambda_{j_r})\\&\geq\|Q\|\min(\mathrm{Re}\,e^{i\omega}\lambda_{j_1},\ldots,\mathrm{Re}\,e^{i\omega}\lambda_{j_r})\overset{\mathrm{def}}{=}\|Q\|\delta_\omega>0.\end{aligned}$$

If λ_j is *in* that half plane, hence $\mathrm{Re}\,e^{i\omega}\lambda_j=C_{\omega j}>0$, then for $\|Q\|>n_0$ large:

$$\begin{aligned}|\delta_{jQ}|>\mathrm{Re}\,e^{i\omega}\delta_{jQ}&=\mathrm{Re}\,e^{i\omega}(q_1\lambda_{j_1}+\ldots+q_r\lambda_{j_r})-\mathrm{Re}\,e^{i\omega}\lambda_j\\&\geq\|Q\|\delta_\omega-C_{\omega j}>\|Q\|\left(\delta_\omega-\frac{C_{\omega j}}{n_0}\right).\end{aligned}$$

We need to consider only a finite number of half planes, that is of values of ω, and can choose $\delta>0$ small and n_0 big to satisfy lemma 9.

We make a preliminary change of coordinates by finite polynomials, to arrange that ζ_{jQ} will be 0 for $\|Q\|\leq n_0$, below. We can do this because we can do it in formal finite power series.

Next we proceed as in § 12. In order to prove that ζ_j is convergent, it suffices to prove the convergence of $\bar{\zeta}_j=\sum_Q|\zeta_{jQ}|.w^Q$, hence of $\bar{\bar{\zeta}}_j=\sum_{n\geq 2}\bar{\bar{\zeta}}_{jn}w^n$ (where $\bar{\bar{\zeta}}_j$ is defined as in § 12, which means $\bar{\bar{\zeta}}_{jn}=\sum_{\|Q\|=n}|\zeta_{jQ}|$), hence of $u=u(w)$ defined by (the factor n will be needed below!):

$$wu=\sum_j\sum_{n\geq n_0}n\bar{\bar{\zeta}}_{jn}w^n. \tag{74}$$

(With respect to notation we recall that: w^Q means $w_1^{q_1}\ldots w_m^{q_m}$, whereas w^n means the n-th power of one variable w.)

For the calculation to follow we also observe that if we let $w_1=w_2\ldots=w_m=w$, $\|Q\|=q_1+\ldots+q_m=n$, then $w_1^{q_1}w_2^{q_2}\ldots w_m^{q_m}=w^n$, and

$$\sum_k\frac{\partial}{\partial w_k}(w_1^{q_1}w_2^{q_2}\ldots w_m^{q_m})=\sum_k\frac{\partial}{\partial w_k}(w_1^{q_1}w_2^{q_2}\ldots w_m^{q_m})\frac{\partial w_k}{\partial w}=\frac{\partial w^n}{\partial w}=nw^{n-1} \tag{75}$$

For example:

$$\sum_{\|Q\|=n}\left(\overline{\overline{\sum_k\frac{\partial\zeta_{jQ}w^Q}{\partial w_k}}}\right)=n\bar{\bar{\zeta}}_{jn}w^n.w^{-1}.$$

We study the solutions of (38) modified according to (69):

$$\sum_Q \delta_{jQ}\zeta_{jQ}w^Q + \psi_j + \chi_j = \varphi_j(w_1+\zeta_1, \ldots, w_m+\zeta_m) - \sum_k \frac{\partial\zeta_j}{\partial w_k}(\psi_k+\chi_k).$$

We replace coefficients of all formal power series in this equation by their absolute values, delete all terms of the ideal Ψ that we see, use the lemma, and apply obvious majorations to obtain (compare § 11):

$$\delta\sum_Q \|Q\|.|\zeta_{jQ}|.w^Q \prec \sum_Q |\delta_{jQ}\zeta_{jQ}|w^Q$$

$$\prec \bar{\varphi}_j(w_1+\bar{\zeta}_1, \ldots, w_m+\bar{\zeta}_m) + \left(\sum_k \frac{\partial\zeta_j}{\partial w_k}\right)(\sum_l \bar{\chi}_l). \tag{76}$$

We sum (76) over j, substitute $w_1 = w_2 = \ldots = w_m = w$, use (74) and (75) and apply considerable majorations. We also use that, for some $A_0 > 0$ and $A > 0$,

$$\delta^{-1}\sum_j \bar{\bar{\varphi}}_j(w) \prec \frac{A_0 w^2}{1 - Aw},$$

and:

$$\overline{\overline{\sum_j\sum_k \frac{\partial\zeta_j}{\partial w_k}}} = \sum_{j,n} \frac{\partial\bar{\bar{\zeta}}_{jn}w^n}{\partial w} = wu.w^{-1} = u.$$

Then we find:

$$wu \prec \frac{A_0(w+wu)^2}{1 - A(w+wu)} + u(\sum_k \bar{\bar{\chi}}_k/\delta)$$

$$u \prec \frac{A_0 w(1+u)^2}{1 - Aw(1+u)} + u(\sum_k \bar{\bar{\chi}}_k/\delta w). \tag{77}$$

$\sum_k \bar{\bar{\chi}}_k$ is a (finite) polynomial in w, that starts with terms of degree ≥ 2. Now compare (77) with the equation for $v = \sum_1^\infty v_n w^n$:

$$v = \frac{A_0 w(1+v)^2}{1 - Aw(1+v)} + v(\sum_k \bar{\bar{\chi}}_k/\delta w) \tag{78}$$

which has a convergent solution near $w = 0$, because

$$\left(\frac{dv}{dw}\right)_{w=0} = A_0 \neq 0.$$

For $A_0 > 0$ big enough we obtain:

$$v_r > u_r, \quad r \geq n_0$$

where u_r is the first nonzero term of the power series u.

Then by comparing (77) and (78) we find by induction on $n: 0 \leq u_n < v_n$, and also u is convergent. Then ζ_j $(j = 1, \ldots, m)$ is convergent, and theorem (IV.2) is proved.

REFERENCES

[1] A. D. Brjuno, Analytical form of differential equations, *Trudy Moscow Math. Obšč (Trans. Moscow Math. Soc.)*, vol. 25 (1971), p. 131-288.

[2] C. Camacho, N. H. Kuiper, J. Palis, La topologie du feuilletage d'un champ de vecteurs holomorphe près d'une singularité, *C.R. Acad. Sc. Paris*, t. 282 A, p. 959-961.

[3] Topological properties of $\mathbf{R}^2$-actions are studied in:
C. Camacho, On $\mathbf{R}^k \times Z^l$-actions, *Proceed. Symp. on Dynamical Systems*, Salvador 1971, Ed. Peixoto, p. 23-70.
G. Palis, Linearly induced vector fields and $\mathbf{R}^2$-actions on spheres, to appear in *J. Diff. Geom.*
C. Camacho, Structural stability theorems for integrable differential forms on 3-manifolds, to appear in *Topology*.

[4] H. Dulac, Solutions d'un système d'équations différentielles dans le voisinage des valeurs singulières, *Bull. Soc. Math. France*, **40** (1912), 324-383.

[5] J. Guckenheimer, Hartman's theorem for complex flows in the Poincaré domain, *Compositio Math.*, **24** (1972), p. 75-82.

[6] A real analogue of theorem III is the classical Grobman-Hartman theorem:
P. Hartman, *Proc. AMS*, **11** (1960), p. 610-620.

[7] Topological properties of *real* linear flows on $\mathbf{R}^n$ are studied in:
N. H. Kuiper, *Manifolds Tokyo, Proceedings Int. Conference, Math. Soc. Japan* (1973), p. 195-204, and:
N. N. Ladis, *Differentialnye Uravnenya Volg.* (1973), p. 1222-1235.

[8] D. Lieberman, Holomorphic vector fields on projective varieties, *Proc. Symp. Pure Math.*, XXX (1976), 273-276.

[9] The invariant of chapter I goes back to an invariant in the study of stability in one parameter families of diffeomorphisms :
S. Newhouse, J. Palis, F. Takens, to appear. See also:
J. Palis, *A differentiable invariant of topological conjugacies and moduli of stability*, preprint IMPA.

[10] J. Palis, S. Smale, Structural stability theorems, *Global Analysis, Symp. Pure Math., AMS*, vol. XIV (1970), p. 223-231.

[11] H. Poincaré, Sur les propriétés des fonctions définies par les équations aux différences partielles, thèse, Paris, 1879 = *Œuvres complètes*, I, p. xcix-cv.

[12] H. Russmann, *On the convergence of power series transformations of analytic mappings near a fixed point into a normal form*, Bures-sur-Yvette, preprint I.H.E.S.

[13] C. L. Siegel, Über die Normalform analytischer Differentialgleichungen in der Nähe einer Gleichgewichtslösung, *Göttingen, Nachr. Akad. Wiss., Math. Phys. Kl.* (1952), p. 21-30.

[14] C. L. Siegel, J. Moser, *Celestial mechanics* (1971) (English edition of: C. L. Siegel, *Vorlesungen über Himmelsmechanik*, 1954, Springer Verlag).

[15] Ю. С. Ильяшенко ЗАМЕЧАНИЯ О ТОПОЛОГИИ ОСОБЫХ ТОЧЕК АНАЛИТИЧЕСКИХ ДИФФЕРЕНЦИАЛЬНЫХ УРАВНЕНИЙ В КОМПЛЕКСНОЙ ОБЛАСТИ И ТЕОРЕМА ЛАДИСА, *Функциональный анализ и его приложения*, Т. 11, ВЫН 2, 1977, 28-38.

[16] J. Guckenheimer, On holomorphic vector fields on $\mathbf{CP}(2)$, *An. Acad. Brasil. Cienc.*, **42** (1970), p. 415-420.

[17] F. Dumortier, R. Roussarie, Smooth linearization of germs of $\mathbf{R}^2$-actions and holomorphic vector fields, to appear.

Manuscrit reçu le 6 septembre 1977.

Topology Vol. 16, pp. 335–345. Pergamon Press, 1977. Printed in Great Britain

TOPOLOGICAL EQUIVALENCE OF NORMALLY HYPERBOLIC DYNAMICAL SYSTEMS

J. Palis and F. Takens

(Received 21 September 1976)

§1. INTRODUCTION

For more than a decade the concept of topological equivalence between vector fields, i.e. the existence of a homeomorphism of the ambient space carrying orbits of one vector field onto orbits of the other, has been playing a central role in the theory of Dynamical Systems. It is a natural concept to reflect when the vector fields have the same space of orbits or phase portrait. The main question studied in this context was to show that a large class of vector fields had under small perturbations the same topological type; such vector fields are called structurally stable. In the present paper we consider the problem of extending a topological equivalence defined on invariant submanifolds for vector fields to neighborhoods of these submanifolds. We were motivated to this problem by questions on structural stability as well as bifurcation theory; we will present applications to these topics. Similar results also hold for diffeomorphisms. These results are also useful for the topological classification of dynamical systems.

To be more precise, let X and X' be C^1 vector fields on smooth manifolds M and M' leaving invariant normally hyperbolic compact submanifolds $N \subset M$ and $N' \subset M'$. We prove here that a topological equivalence between $X|N$ and $X'|N'$, which can be lifted to bundle isomorphisms between the stable and unstable bundles, can be extended to neighborhoods of N in M and of N' in M'.

In the particular but important case when we start with a conjugacy, i.e. a topological equivalence between the flows $X_t|N$ and $X'_t|N'$ generated by X and X', the extension problem is simpler once we have the stable and unstable foliations as constructed for instance in Hirsch, Pugh and Shub[5]. This was considered by Pugh and Shub[9] in the context of normally linearizing the flow X_t in a neighborhood of N and in the local context by Shoshitaishvili[10]. Of course, all these results are extensions of the Grobman–Hartman Theorem for hyperbolic singularities[3], [4]. Historically, the idea of getting the equivalence from the invariant foliations was used by Anosov[1] in the globally hyperbolic case, by Palis[7] in a neighborhood of a hyperbolic singularity and in a somewhat more general way in [7], Palis and Smale[8] and Melo[6] for global cases where the invariant foliations (tubular families) are not uniquely determined.

The structure of the paper is as follows. In order to give a clearer view of the use of invariant foliations we start by showing the result in the case of a conjugacy between $X|N$ and $X'|N'$. This is done in §2. The main theorem concerning the extension of a topological equivalence is in §3. The results we discuss here have important applications to the unfoldings of singularities of several parameter families of vector fields as shown by Shoshitaishvili[10], Sotomayor[11] and others. For a presentation of this topic and more references see Arnold[2] and Takens[12]. We conclude that in order to obtain universal unfoldings of these singularities it is enough to analyse certain normally hyperbolic (center) submanifolds. This is in the last section together with an application to structural stability and the corresponding results for diffeomorphisms. The structural stability result is as follows: if N is normally hyperbolic then X is structurally stable in a neighborhood of N if $X|N$ is structurally stable. To be able to state such a fact we must clarify how we can consider $X|N$ a "C^1 vector field" although N is taken to be just C^1. This is done in the Appendix by introducing a compatible C^2 structure on N. A similar problem arises in the suspension construction for diffeomorphisms, which is also treated in the Appendix; we need this suspension to obtain from our theorems on vector fields from the corresponding results for diffeomorphisms.

This research was supported by C.N.Pq. (Brazil).

Originally published in: *Topology 16 (1977), no. 4, 335–345,*

We clarify that part of our motivation came from an incorrect statement in [12] saying that a weak version of our main theorem was implicitly contained in [9]. This mistake was pointed out by Melo and others. Fundamentally new ideas, as related in §3, were needed to deal with the set of singularities for $X|N$.

We now give some basic definitions and facts. Let X be a C^1 vector field on a riemannian manifold M. A submanifold $N \subset M$ is invariant if $X(x) \in T_xN$ for each $x \in N$. It is normally hyperbolic if,

(a) there is a continuous splitting of $T_N(M) = T(N) \oplus E^u(N) \oplus E^s(N)$ which is preserved by the flow X_t generated by X. That is, for every $t \in \mathbf{R}$, $(X_t)_* T(N) = T(N)$, $(X_t)_* E^u(N) = E^u(N)$ and $(X_t)_* E^s(N) = E^s(N)$, where $(X_t)_*$ indicates the derivative of X_t,

(b) there are numbers $0 < \lambda < 1 < \mu$ and $T > 0$ such that for any $v \in T(N)$ and $t > T$, $\|(X_t)_* v\| \in (e^{\lambda t}\|v\|, e^{\mu t}\|v\|)$; for any $v \in E^u(N)$ and $t > T$, $\|(X_t)_* v\| \geq e^{\mu t}\|v\|$; for any $v \in E^s(N)$ and $t > T$, $\|(X_t)_* v\| \leq e^{\lambda t}\|v\|$.

For N compact or $X|N$ with compact support there are C^0 locally invariant foliations over N, namely the strong stable $\mathscr{F}^s$ and strong unstable $\mathscr{F}^u$ foliations. Each leaf W_x^{ss}, $x \in N$, of $\mathscr{F}^s$ is C^1 and its dimension is the dimension of the fiber of $E^s(N)$. Similarly for the leaves of $\mathscr{F}^u$. They are invariant in the following sense: there is a neighborhood U of N such that if $S \cap U$ is a leaf of $\mathscr{F}^s$ and $t > 0$ then $X_t(S \cup U)$ is contained in a leaf of $\mathscr{F}^s$; if $S \cap U$ is a leaf of $\mathscr{F}^u$ and $t < 0$ then $X_t(S \cap U)$ is also contained in a leaf of $\mathscr{F}^u$. The foliations $\mathscr{F}^s$ and $\mathscr{F}^u$ are not in general unique, but the union of the leaves $W^s(N) = \bigcap_{x \in N} W_x^{ss}$ is uniquely characterized as the union of points in U whose positive orbits remain in a neighborhood $V \subset U$ of N; similarly for $W^u(N) = \bigcup_{x \in N} W_x^{uu}$. They are C^1 submanifolds of M and called the stable and unstable manfolds of N. We refer the reader to [5] for proofs of these facts.

A vector field X on N is topologically equivalent to a vector field X' on N' if there is a homeomorphism $h\colon N \to N'$ sending trajectories of X onto trajectories of X'. These vector fields are conjugate if there is a homeomorphism $h\colon N \to N'$ such that $hX_t = X'_t h$ for every $t \in \mathbf{R}$, where X_t and X'_t are the flows generated by X and X'.

§2. CONJUGACIES

THEOREM. *Let $h\colon N \to N'$ be a conjugacy between $X|N$ and $X'|N'$ that can be lifted to invertible linear bundle maps $h^s\colon E^s(N) \to E^s(N')$ and $h^u\colon E^u(N) \to E^u(N')$. We can extend h to a conjugacy H between X and X' from a neighborhood of N to a neighborhood of N'.*

Proof. The construction of H goes in two steps: first we extend h to a neighborhood of N in $W^u(N)$ and then to a neighborhood of N in M.

Step 1. Choose as in [9] a neighborhood U of N in $W^u(N)$ with smooth boundary ∂U such that ∂U is topologically transverse to X_t and intersects each leaf of the strong unstable foliation $\mathscr{F}^u$ transversely in a sphere; this sphere is radially homotopic to the standard sphere in the leaf. Let $\pi_u\colon U \to N$ be the projection of U on N along the leaves of $\mathscr{F}^u$. Let U' be a corresponding neighborhood of N' in $W^u(N')$ and $\pi'_u\colon U' \to N'$ the corresponding projection. Since h lifts to an invertible linear bundle map $h^u\colon E^u(N) \to E^u(N')$, there is a homomorphism $h_{\partial U}\colon \partial U \to \partial U'$ such that $h\pi_u = \pi'_u h_{\partial U}$ on ∂U. Now the conjugacy $h_u\colon U \to U'$ is determined by $h_u = h$ on N and $h_u X_{-t}(p) = X'_{-t} h_{\partial U}(p)$ for $p \in \partial U$ and $t \geq 0$. It is easy to check that h_u is a homeomorphism and by construction h_u is a conjugacy between X and X'.

Step 2. The extension H of h_u to a neighborhood of N in M is similar to the previous step once we observe the following facts. First, we can define extended strong stable foliations convering full neighborhoods of N in M and of N' in M'. To see this we consider $W^u(N)$, $W^u(N')$ as normally contracting submanifolds for X, X' on neighborhoods of N, N'. Let $V \supset U$ be a neighborhood of N in $W^u(N)$ with ∂V transverse to X [13]. We then define $\mathscr{F}^s$ over ∂U and simply extend it to $V - N$ via X_{-t} for all $t \geq 0$; that is, if $y = X_{-t}(x)$ for $x \in \partial V$ and $t \geq 0$ then $W_y^{ss} = X_{-t}(W_x^{ss})$. By the generalized λ-lemma [7], [9], $\mathscr{F}^s$ defined on $V - N$ extends over N. The construction for N', X' is the same. Second, since h lifts to invertible linear bundle maps $h^u\colon E^u(N) \to E^u(N')$ and $h^s\colon E^s(N) \to E^s(N')$, it follows that the conjugacy constructed in Step 1 $h_u\colon U \subset W^u(N) \to U' \subset W^u(N')$ lifts to an invertible linear bundle map $H^s\colon E^s(U) \to E^s(U')$. Using these facts we can proceed as in Step 1 to extend h_u to a conjugacy H between X and X' on neighborhoods of N in M and of N' in M'.

§3. EQUIVALENCES

THEOREM. *Let X and X' be C^1 vector fields on manifolds M and M' with compact normally hyperbolic invariant submanifolds N and N', respectively. Let h: $N \to N'$ be a topological equivalence between $X|N$ and $X'|N'$. If h lifts to invertible linear bundle maps h^u: $E^u(N) \to E^u(N')$, h^s: $E^s(N) \to E^s(N')$ and if both E^u and E^s are trivial on a neighborhood of the singular set, then there is a topological equivalence H between X and X' which extends h to a neighborhood of N.*

Remark. The condition on the triviality of E^u and E^s on a neighborhood of the critical set might not be necessary, but we need it for technical reasons. We point out that this condition is satisfied in many important applications like the ones we mentioned in the introduction. Also, it follows from our proof that the result remains true if we only assume that the supports of $X|N$ and $X'|N'$ are compact.

Proof. We first prove the theorem in a number of simpler cases:

Case 1. The theorem holds under the following extra assumptions:
(a) the unstable bundle is 0-dimensional and (b) X has no singularities on N (and hence X' also has no singularities on N').

Proof of Case 1. We choose a neighborhood S of N in M such that S has smooth boundary ∂S topologically transverse to X_t and ∂S intersects each leaf of the strong stable foliation in a sphere [9]. Let π_s be the projection of S on N collapsing each fiber of the strong stable foliation to its intersection with N. S' is a similar neighborhood of N' and π'_s the corresponding projection. Because h lifts to an invertible linear bundle map h^s: $E^s(N) \to E^s(N')$, there is a homoeomorphism $h_{\partial s}$: $\partial S \to \partial S'$ such that $h\pi_s = \pi'_s h_{\partial S}$ on ∂S. From this we obtain a homeomorphism H: $S \to S'$ as follows. If $s \in N$ then $H(s) = h(s)$. If $s \in S - N$ then $s = X_t(s_1)$ for some $s_1 \in \partial S$ and $t \geq 0$; we define $H(s)$ as the intersection of the positive X' orbit starting at $h_{\partial S}(s_1)$ with $(\pi'_s)^{-1}(h(\pi(s)))$. This last intersection may not be unique, namely if $\pi_s(s)$ lies on a periodic X orbit; however H is uniquely determined by continuity. Since H is clearly continuous on $S - N$, we have only to check the continuity of H along N. Let s_i, $i = 1, 2, \ldots$, be a sequence of points in $S - N$ converging to $s_\infty \in N$; we need to show that $H(s_i)$ has as limit $h(s_\infty)$. Since $\pi'_s H = H\pi_s$ and $H|N = h$, $\pi'_s H(s_i)$ converges to $h(s_\infty)$. So we have only to show that the distance of $H(s_i)$ to N' tends to zero. Let t_i, t'_i be such that $x_{-t_i}(s_i) \in \partial S$ and $X'_{-t'_i}(H(S_i)) \in \partial S'$. Since N and N' are normally contracting we have that s_i tends to N if and only if $t_i \to \infty$ and $H(s_i)$ tends to N' if and only if $t'_i \to \infty$. To show that $t'_i \to \infty$ we define a positive real function μ: $N \to \mathbf{R}$ by $X'_{\mu(n)}(h(n)) = h(X_1(n))$ for $n \in N$. The function μ is clearly continuous and let $c > 0$ be its minimum. If $X'_{t'}(h(n) = h(X_t(n))$ for $t > 0$, then $t' \geq c[t]$ when $[t]$ denotes the integral part of t. From this and the fact that H preserves the stable foliation it follows that $t'_i \to \infty$ and thus $H(s_i) \to h(s_\infty)$ proving the continuity of H. By construction H is one to one and sends trajectories of X onto trajectories of X'. This shows that H is a topological equivalence.

Case 2. The theorem holds when (a) the unstable bundle is zero-dimensional and (b) the stable bundle of N, and hence of N', is trivial.

Proof of Case 2. A neighborhood of N can be parametrized by $N \times V$, V an open neighborhood of the origin in some euclidean space. In what follows we use this parametrization; when necessary we assume that the riemannian metric corresponds to a fixed product metric on $N \times V$. From the Theorem in §2, we may assume that X is such that $X_t(n, v) = (X_t(n), e^{-\lambda t}v)$ for some $\lambda > 0$ and all $(n, v) \in N \times V$; λ is taken big enough so that we have normal hyperbolicity with respect to the product metric.

Next we want to construct a neighborhood S of N more or less like in Case 1; because of the possible presence of singularities we have to be more careful. Let $\Sigma \subset N$ be the set of singularities and for $\epsilon > 0$ let $S_\epsilon = \{(n, v); \|v\| \leq \epsilon$ and $\|v\| \leq \rho(n, \Sigma)\}$, where ρ denotes the distance defined by the riemannian metric. The lext lemma shows that for $\epsilon > 0$ small the positive orbits of $S_\epsilon - \Sigma$ move inside S_ϵ.

LEMMA 1. *There exists $\epsilon_0 > 0$ such that if $s \in S_\epsilon - \Sigma$ and $t > 0$ then $X_t(s) \in \operatorname{Int} S_\epsilon$, whenever $0 < \epsilon < \epsilon_0$.*

The proof of the lemma will be postponed to the end of this section. Similar considerations apply to the vector field X'. First, we may assume that X' is defined on a product neighborhood $N'\times V'$, where V' is a neighborhood of the origin in the same euclidean space as V. We also can take $X'_t(n',v')=(X_t(n'),e^{-\lambda t}v')$ with $\lambda>0$ and big enough so that X' is normally hyperbolic at N'. Define S'_ϵ as before and applying Lemma 1 one gets the corresponding ϵ'_0.

We now start to extend the topological equivalence $h\colon N\to N'$ as follows. Take $0<\epsilon<\min(\epsilon_0,\epsilon'_0)$ and define $h_\partial\colon \partial S_\epsilon\to\partial S'_\epsilon$ by $h_\partial(n,v)=(h(n),\sigma(n)v)$, where $\sigma(n)$ is the unique positive real number such that $(h(n),\sigma(n)v)\in\partial S'_\epsilon$. Since $h(\Sigma)=\Sigma'$, the singular set of X', $\sigma(n)$ is well defined whenever $n\not\in\Sigma$; otherwise $v=0$ and $h_\partial(n,0)=(h(n),0)$. Exactly like in Case 1 we can extend h_∂ and h to a topological equivalence $H^*\colon S_\epsilon\to S'_\epsilon$ so that H^* preserves the stable foliation, i.e. $H^*(n\times V)\cap S_\epsilon=(h(n)\times V')\cap S'_\epsilon$. Notice that if $x\in S_\epsilon-N$ then there is a unique $t\geq 0$ such that $X_{-t}(x)\in\partial S_\epsilon$. The proof that H is continuous at N is slightly more delicate here than in Case 1. This is shown as follows. Let s_i in S_ϵ be a sequence of points converging to $s\in N$ and we have to see that $H^*(s_i)$ converges to $H^*(s)$. As before, since H^* preserves the stable foliations it is enough to show that $H^*(s_i)$ tends to N'. So given $\delta>0$ we claim that $\rho(H^*(s_i),N')<\delta$ for all $i\geq i_0$ for some i_0. Let W'_δ be the δ-neighborhood of Σ' in N' and let $W_\delta=h^{-1}(W'_\delta)$. If s_i is such that $X_{-t}(s_i)\in\pi_s^{-1}(W_\delta)\cap S_\epsilon$ for some $t\geq 0$ then $\rho(H^*(s_i),N')<\delta$. In fact $\pi_s(X_{-t}(s_i))\in W_\delta$ and hence $\pi'_sH^*(X_{-t}(s_i))\in W'_\delta$. This means that $\rho(H^*(X_{-t}(s_i)),N')<\delta$. Since H^* preserves the orientation of the integral curves and X' contracts exponentially toward N', it follows that $\rho(H^*(s_i),N')<\delta$. Finally, if there is no $t\geq 0$ such that $X_{-t}(s_i)\in\pi_s^{-1}(W_\delta)\cap S_\epsilon$ we can apply the same argument of Case 1 since the orbit stays away from the singular set Σ. This concludes the proof that H^* is continuous.

The last step is to extend H^* to a topological equivalence H of a neighborhood of the singular set Σ. Together with S_ϵ and $\Sigma\times V$, we fill out such a neighborhood with the negative orbits of X starting at ∂S_ϵ. For points $(n,v)\in\partial S_\epsilon$ with $v\neq 0$, we recall that $H^*(n,v)=(h(n),\sigma(n)v)\in\partial S'_\epsilon$. So if $(\tilde n,\tilde v)=X_{-t}(n,v)$ for $(n,v)\in\partial S_\epsilon$, we define $H(\tilde n,\tilde v)=(n^*,\tilde v-v+\sigma(n)v)$. That is, $H(\tilde n,\tilde v)$ is in the orbit of $H^*(n,v)$ and it has V-coordinates $\tilde v-v+\sigma(n)v$. Finally, for $n\in\Sigma$ we set $H(n,v)=(h(n),v)$. From the construction, H sends trajectories of X onto trajectories of X', it is one to one and continuous away from $\Sigma\times V$. We proceed to show that H is continuous along $\Sigma\times V$. We will make use of the lemma below; its proof will be postponed to the end of this section.

Lemma 2. *Let $P\colon N\to N$ be defined as follows. For $n\in\Sigma$ we set $P(n)=n$. For $n\in N-\Sigma$ we take $v\neq 0$ so that $(n,v)\in\partial S_\epsilon$ and define $P(n)=n^*$, where n^* is such that $(n^*,\epsilon(v/\|v\|))$ is in the negative orbit of X trough (n,v). The map P is a homeomorphism.*

Remark. The map P is well defined because of the special form of the vector field X. A similar map P' can be constructed for X', N'.

Using this lemma we can conclude the continuity of H on $\Sigma\times V$. In fact, let $p_i=(n_i,v_i)$ be a sequence in $N\times V$ that converges to $p=(n,v)\in\Sigma\times V$. We may suppose that p_i is off both N and $\Sigma\times V$ for otherwise the statement is trivially true. Let q_i be the intersection of the positive orbit of p_i with ∂S_ϵ. Clearly $q_i\to\Sigma\subset N$ and so the V'-coordinate of $H(p_i)$ converges to v. Since $H(p)=(h(n),v)$, it remains to show that the N'-coordinate of $H(p_i)$ converges. We have to consider two possiblities. If $v=0$, then $q_i\to(n,0)$ because from the normal contraction hypotheses X has a relatively large V-component near Σ and thus X is uniformily transversal to ∂S_ϵ off Σ. Thus $q'_i=H(q_i)$ converges to $(h(n),0)$ for H is continuous on ∂S_ϵ. Applying the previous argument to X', Σ' and $\partial S'_\epsilon$ we conclude that $H(p_i)$ converges to $H(p)=(h(n),0)$. Suppose now that $v\neq 0$ and let r_i, r'_i be the intersection of the orbits of p_i, $p'_i=H(p_i)$ with $\tilde S_\epsilon=\{(n,v)\in N\times V;\ \|v\|=\epsilon\}$ and $\tilde S'_\epsilon=\{(n',v')\in N'\times V';\ \|v'\|=\epsilon\}$, respectively. Clearly p_i converges if and only if r_i converges and the same is true for p'_i, r'_i. Thus r_i converges and from Lemma 2 $q_i\in\partial S_\epsilon$ converges. But H is continuous on ∂S_ϵ and thus $q'_i=H(q_i)$ converges. Again Lemma 2 applied now to X' on $N'\times V'$ yields that the N'-coordinate of r'_i converges. Together with the previous argument on the convergence of the V'-coordinate, this implies that r'_i converges and so does $H(p_i)$. Therefore H is a topological equivalence between X and X' on neighborhoods of N and N'.

Case 3. The theorem holds when the unstable bundle is zero dimensional.

Proof. We just combine the proofs of Cases 1 and 2 as follows. In a neighborhood of the singular set we apply the methods of Case 2 using the fact that the stable bundle is trivial. Elsewhere we can apply the method of Case 1. There is no difficulty in the overlapping part: one takes a trivialization of the stable bundle near the singular set Σ induced by a given trivialization near Σ' through a bundle isomorphism h^s: $E^s \to E^{s'}$ which lifts the equivalence h: $N \to N'$.

Proof of the General Case. In view of Case 3, we can proceed similarly to the proof of the conjugacy case in §2. First we extend h to a topological equivalence h_u: $U \to U'$, where U is a neighborhood of N in $W^u(N)$ and U' is a neighborhood of N' in $W^u(N')$. Next we extend the stable foliation of N to a neighborhood $V \supset U$ in $W^u(N)$ and the stable foliation of N' to a neighborhood $V' \supset U'$ in $W^u(N')$. As before, since h lifts to bundle isomorphisms h^u: $E^u(N) \to E^u(N')$ and h^s: $E^s(N) \to E^s(N')$ then h_u lifts to a bundle isomorphism H^s: $E^s(U) \to E^s(U')$. Using these facts and considering U and U' as normally contracting in M and M', we can by Case 3 extend h_u to a topological equivalence H between X and X' in a neighborhood of N in M and of N' in M'. The proof of our main theorem is now complete.

Proof of Lemma 1. It is enough to show that, for small $\epsilon > 0$, small $t > 0$ and $s \in \partial S_\epsilon - \Sigma$, $X_t(s) \in \operatorname{Int} S_\epsilon$. For the points $s \in \partial S_\epsilon - \{(n, v); \|v\| = \rho(n, \Sigma) \le \epsilon\}$ it is clear from the product structure that the orbits move in. Let W_ϵ be the set of points $(n, v) \in \partial S_\epsilon - \Sigma$ such that the orbit of (n, v) does not enter in the interior of S_ϵ. If $0 < \epsilon' < \epsilon$ then $W_{\epsilon'} = W_\epsilon \cap S_{\epsilon'}$, so if W_ϵ is bounded away from Σ we are done. We will prove that W_ϵ is bounded away from Σ by contradiction. Let $q_i = (n_i, v_i)$ be a sequence in W_ϵ not bounded away from Σ. Since Σ is compact we can asume that q_i converges to $q = (n, 0) \in \Sigma$. Let $\rho_i = \rho(n_i, \Sigma)$, $K_i = \{k \in \Sigma;\ \rho(n_i, k) = \rho(n_i, \Sigma)\}$ and $\rho_i(t) = \rho(X_t(n_i), \Sigma)$. Then, for small $t > 0$,

$$\rho_i(t) = \rho_i + t \inf_{k \in K_i} \frac{\mathrm{d}}{\mathrm{d}t} \rho(X_t(n_i), k) + o(t)$$

$$= \rho_i + t \cdot \inf_{k \in K_i} \frac{\langle X(n_i), n_i - k\rangle}{\|(n_i - k)\|} + o(t).$$

To get the last expression we used local coordinates such that in $T_n N$ the Riemannian metric and the Euclidean metric given by the coordinates agree. The integral curve through (n_i, v_i) has the form $t \to (X_t(n_i), \mathrm{e}^{-\lambda t} v_i)$. Since $(n_i, v_i) \in \partial S_\epsilon - \Sigma$, $\|v_i\| = \rho_i$; since the integral curve through (n_i, v_i) does not enter in $\operatorname{Int} S_\epsilon$, we have that

$$\frac{\langle X(n_i), n_i - k_i\rangle}{\|n_i - k_i\|} \le -\lambda \|v_i\| \quad \text{for some} \quad k_i \in K_i.$$

It follows that $\langle (X(n_i)/\|n_i - k_i\|), (n_i - k_i)/\|n_i - k_i\|\rangle \le -\lambda$ for some $k_i \in K_i$. It is easy to see that $\lim k_i = n$; taking a subsequence if necessary we obtain that $\lim (n_i - k_i)/\|n_i - k_i\| = v$ for some unit vector $v \in T_n N$. And also $\lim (X(n_i)/\|n_i - k_i\|) = \lim (X(n_i) - X(k_i))/\|n_i - k_i\| = (\mathrm{d}X)_n(v)$. Then $(\mathrm{d}/\mathrm{d}t)\|(X_t)_* v\|_{t=0} \le -\lambda$, which is the required contradiction because we can not have a contraction along N as strong as the normal contraction.

Proof of Lemma 2. Clearly P is one to one, surjective and continuous outside Σ. We only have to show that P is continuous at Σ. Since N is normally contracting with normal contraction constant λ, there is $0 < \lambda' < \lambda$ such that $\|(X_t)_*(v)\| \ge \mathrm{e}^{-\lambda' t}\|v\|$ for all $v \in T(N)$ and $t > 0$. Given $q \in \Sigma$ and $\delta > 0$ we claim that there is $a > 0$ such that if $(n, v) \in \partial S_\epsilon - \Sigma$, $\rho(n, q) < a$ then the point $(P(n), \epsilon(v/\|v\|))$ in the negative orbit of (n, v) satisfies $\rho(P(n), q) < \delta$. To see this we choose for each (n, v) a point $q' \in \Sigma$ such that $\rho(n, q') = \rho(n, \Sigma) = b < a$. Then $\|v\| = b$ or else $\|v\| = \epsilon$ but then we are far from Σ. Since $X_{-t}(n, v) = (P(n), \epsilon(v/\|v\|))$, we have that $\mathrm{e}^{t\lambda} b = \epsilon$ or $t = (1/\lambda)(\ln \epsilon - \ln b)$. Next we take a curve of length b from (n, v) to (q', v) parallel to N and apply X_{-t} to it with t as above. Then its length is at most $\mathrm{e}^{\lambda' t} b$; but this is also an upper estimate for $\rho(P(n), q')$. It follows that $\rho(P(n), q') \le b + \epsilon b^{1-(\lambda'/\lambda)} \le a + \epsilon a^{1-(\lambda'/\lambda)}$. When $a \to 0$ we have that $\rho(P(n), q') \to 0$ because $0 < \lambda' < \lambda$. Also if $a \to 0$ clearly $\rho(q', q) \to 0$ and so $\rho(P(n), q) \to 0$. This proves that P is continuous at Σ and thus P: $N \to N$ is a homeomorphism as we claimed.

§4. APPLICATIONS

(a) Diffeomorphisms

In this section we want to show how to extend a given topological conjugacy defined on a normally hyperbolic invariant submanifold of a diffeomorphism to a neighborhood of the submanifold. We shall do this by first suspending the diffeomorphism and then applying our results on the extension of conjugacies for flows.

To be more precise, let M be a smooth manifold, φ: $M \to M$ a C^1 diffeomorphism and $N \subset M$ a compact normally hyperbolic invariant submanifold for φ. As in the case of flows, we have the splitting $T_N(M) = T(N) \oplus E^u(N) \oplus E^s(N)$; we denote by $d^u\varphi$, resp. $d^s\varphi$ the bundle isomorphisms obtained by restricting $d\varphi$ to $E^u(N)$, respectively $E^s(N)$. Let M' be another smooth manifold with C^1 diffeomorphism φ' and compact normally hyperbolic invariant manifold N' with stable and unstable bundle $E^s(N')$, $E^u(N')$, $E^u(N')$ respectively.

THEOREM. *Let φ: $M \to M$ and φ': $M' \to M'$ be as above. Let h: $N \to N'$ be a conjugacy between $\varphi|N$ and $\varphi'|N'$ that can be lifted to bundle isomorphisms h^u: $E^u(N) \to E^u(N')$, h^s: $E^s(N) \to E^s(N')$ such that h^u and $(d^u\varphi')^{-1} \circ h^u \circ d^u\varphi$* are homotopic as bundle isomorphisms covering h and the same holds for h^s and $(d^s\varphi')^{-1} \circ h^s \circ d^s\varphi$. Then h can be extended to a conjugacy defined on a neighborhood of N.

Proof. Let $\tilde{M}$, $\tilde{M}'$ be obtained by suspending the diffeomorphism φ, respectively φ' (see the Appendix). The corresponding vectorfields are denoted by X and X' and the invariant manifolds by $\tilde{N}$ and $\tilde{N}'$; X and X' may be regarded as C^1 vectorfields according to the Appendix. Recall that $\tilde{N} = N \times [0, 1]/(x, 1) = (\varphi(x), 0)$ and $\tilde{N}' = N' \times [0, 1]/(x', 1) = (\varphi'(x'), 0)$. From h we obtain the conjugacy $\tilde{h}$: $\tilde{N} \to \tilde{N}'$ between $\tilde{X}|\tilde{N}$ and $\tilde{X}'|\tilde{N}'$ by setting $\tilde{h}(x, t) = (h(x), t)$. Next we have to lift this conjugacy to bundle isomorphisms $\tilde{h}^u$ and $\tilde{h}^s$. Observe that $\tilde{N}$ and $\tilde{N}'$ are compact normally hyperbolic invariant manifolds for X, respectively X'. We only consider the construction of $\tilde{h}^u$; $\tilde{h}^s$ can be constructed in the same way.

From the suspension construction it follows that $E^u(\tilde{N}) = E^u(N) \times [0, 1]/(v, 1) = (d^u\varphi(v), 0)$. We define $\tilde{h}^u(v, 0)$ to be $(h^u(v), 0)$; then it follows that $\tilde{h}^u(v, 1) = ((d^u\varphi')^{-1} \circ h^u \circ d^u\varphi(v), 1)$. Because h^u and $(d^u\varphi')^{-1} \circ h^u \circ d^u\varphi$ are homotopic as bundle isomorphisms over h, there is a bundle isomorphism $\tilde{h}^u$: $E^u(\tilde{N}) \to E^u(\tilde{N}')$ covering $\tilde{h}$ and extending the previously defined part $\tilde{h}^u|N \times \{0\} = \tilde{h}^u|N \times \{1\}$.

Once we have these liftings $\tilde{h}^u$ and $\tilde{h}^s$, we can apply our result from §2 and obtain an extension of $\tilde{h}$ to a conjugacy $\tilde{H}$ which is defined on a neighborhood of $\tilde{N}$. To obtain the extension H of h to a neighborhood of N in M we define H as the composition $p \circ \tilde{H} \circ i$, where i: $M \to \tilde{M}$ is the injection $i(x) = (x, 0)$, $\tilde{H}$ is as above and p is a "projection" of a neighborhood of $M' \times \{0\}$ in $\tilde{M}'$ to M' by going along integral curves of X', i.e. p maps (x', t) to x' if $t \in [0, \epsilon)$ and to $\varphi'(x')$ if $t \in (1 - \epsilon, 1]$, $0 < \epsilon < (1/2)$.

(b) Structural stability

Let X be a C^1 vectorfield on a manifold M. If $K \subset M$, we say that X is *continuously (structurally) stable on K* if, for any neighborhood V of the cononical injection i: $K \to M$ in the C^0 topology there is a neighborhood U of X in the C^1 topology such that whenever $X' \in U$, there is an embedding $h \in V$, which maps connected components of integral curves of X in K to connected components of integral curves of X' in $h(K)$. If, in the above definition, we replace K by M, we have the usual definition of continuous stability.

For $N \subset M$ a normally hyperbolic invariant submanifold of X there is a C^2 structure on N, compatible with the given C^1 structure, which makes $X|N$ a C^1 vertorfield (see the Appendix). Hence it makes sense to consider the continuous stability of $X|N$.

THEOREM. *Let X be a C^1 vectorfield on M with $N \subset M$ a compact normally hyperbolic invariant submanifold. Then, if $X|N$ is continuously stable there is a neighborhood W of N in M such that X is continuously stable on W.*

Proof. If X' is a vectorfield which is C^1 near to X, X' has a normally hyperbolic invariant submanifold N' which is C^1 near N [5]. According to the Appendix, $X'|N'$ can then be considered as C^1 close to $X|N$. Hence there is an equivalence h: $N \to N'$ between $X|N$ and $X'|N'$ which is C^0 close to the canonical embedding $N \to M$. By [5], the splitting $T_{N'}(M) = T(N') \oplus E^u(N') \oplus E^s(N')$ is close to the splitting $T_N(M) = T(N) \oplus E^u(N) \oplus E^s(N)$ so there are liftings h^u: $E^u(N) \to E^u(N')$ and h^s: $E^s(N) \to E^s(N')$ of h to bundle isomorphisms. Now we can apply our result from §3 and the theorem follows.

Remark. It should be true that if X is continuously stable on a neighborhood of N then also $X|N$ is continuously stable. The proof of this is however obstructed by the following difficulty: if we take on N some C^2 structure as in the Appendix and then choose a vectorfield $\tilde{X}$ on N which is C^1 and C^1 close to X (with respect to this structure) then there is in general no extension of $\tilde{X}$ to a C^1 vectorfield on M. For diffeomorphisms these difficulties do not occur. Therefore we have

THEOREM. *Let φ: $M \to M$ be a C^1 diffeomorphism with compact normally hyperbolic invariant submanifold N. Then $\varphi|N$ is continuously stable if and only if there is a neighborhood W of N on which φ is continuously stable.*

(c) Local equivalences of vectorfields depending on parameters

We shall describe how our main theorem, with some modifications, can be applied to reduce certain local bifurcation problems to lower dimensional ones by "splitting off the hyperbolic part", see [2, 10, 12].

Let X_μ be a vectorfield on R^n which depends on a parameter $\mu \in \mathbf{R}^p$; we assume X_μ to be C^1 in $x \in \mathbf{R}^n$ and $\mu \in \mathbf{R}^p$. Let $X_0(0) = 0$ and let the derivative (in the $\mathbf{R}^n$ direction) of X_0 in 0 have u, s, c eigenvalues with real part >0, <0, $=0$ respectively. Then, according to the invariant manifold theory[5], there is a C^1 manifold W^c in $\mathbf{R}^n \times \mathbf{R}^p$, a center manifold of the vectorfield $X(x, \mu) = (X_\mu(x), 0)$ on $\mathbf{R}^n \times \mathbf{R}^p$, with the following properties:

(a) $(0, 0) \in W^c$;

(b) the dimension of W^c is $p + c$;

(c) for each $(x, \mu) \in W^c$, W^c is transversal to $\mathbf{R}^n \times \{\mu\}$ and $(X_\mu(x), 0)$ is tangent to $W_\mu{}^c \times \{\mu\} = W^c \cap (\mathbf{R}^n \times \{\mu\})$;

(d) all the eigenvalues of the derivative of $X_0|W_0{}^c$ in 0 have real part zero.

This means that, at least locally, W^c is a normally hyperbolic invariant submanifold of the vectorfield $X(x, \mu) = (X_\mu(x), 0)$ on $\mathbf{R}^n \times \mathbf{R}^p$.

Let X'_μ be another such vectorfield on $\mathbf{R}^n$ depending on $\mu \in \mathbf{R}^p$ with $X'_0(0) = 0$ and with derivative (in the $\mathbf{R}^n$ direction) of X'_0 in 0 having also u, s, c eigenvalues with real part respectively bigger, smaller or equal to zero and with center manifold W'^c.

Definition. For X_μ, X'_μ as above, X_μ is called locally equivalent, resp. locally equivalent on the center, with X'_μ if there is a homeomorphism h from a neighborhood of $(0, 0)$ in $\mathbf{R}^n \times \mathbf{R}^p$, resp. W^c, to a neighborhood of $(0, 0)$ in $\mathbf{R}^n \times \mathbf{R}^p$, resp. W'^c, such that:

(1) h maps orbits of X to orbits of X';

(2) for (x, μ), (x', μ) in the domain of h, the μ-coordinates of $h(x, \mu)$ and $h(x', \mu)$ are the same.

THEOREM. *For X_μ, X'_μ as above, X_μ and X'_μ are locally equivalent if X_μ and X'_μ are locally equivalent on the center.*

Proof. According to the result in §3 we can extend topological equivalences defined on normally hyperbolic invariant submanifolds. Since our center manifolds are at least locally normally hyperbolic we can in principle apply our theorem. There are however two observations to be made.

First, the center manifolds are only locally invariant manifolds; we need some modification of the vectorfields in order to make them complete. That is, we have to construct some $\tilde{X}_\mu$ such that $(x, \mu) \in W_\mu{}^c$ implies that $(\tilde{X}_\mu)_t \in W_\mu{}^c$ for all $t \in \mathbf{R}$. Of course $\tilde{X}_\mu$ should be locally equivalent with X_μ and W^c should by normally hyperbolic invariant for $\tilde{X}_\mu$.

Before starting the construction of $\tilde{X}$ we observe that in the proofs in §2 and §3 we also used incomplete (locally) invariant submanifolds and extended conjugacies or equivalence defined on them. Namely we considered the conjugacy or equivalence defined on $W^u(N)$ (or $W^s(N)$) and extended in the stable (unstable) direction. This however was possible, because in our extension constructions we need the invariant manifold only to be negatively complete (i.e. such that $X_t(x)$ is in the invariant manifold whenever $t < 0$ and x is in the invariant manifold) if we extend in the stable direction and positively complete if we extend in the unstable direction.

In the present case, W^c is in general neither positively nor negatively complete and we have to extend the equivalence both in the stable and in the unstable direction.

We write $\mathbf{R}^n$ as the direct sum of two subvector spaces M_1 and M_2, with dimension $s+u$, resp. c, such that both M_1 and M_2 are invariant subspaces of the derivative of X_0 in O and such that M_2 is the tangent space of $W_0^c = W^c \cap (\mathbf{R}^n \times \{0\})$. Near $(0,0)$, W^c has then the form $W^c = \{(m_1, m_2, \mu) \in M_1 \times M_2 \times \mathbf{R}^p | m_1 = F(m_2, \mu), (m_2, \mu) \in V\}$ for some C^1 map $F: M_2 \times \mathbf{R}^p \to M_1$ and a bounded neighborhood V of $(0,0)$ in $M_2 \times \mathbf{R}^p$. Next we take a non negative smooth function $\Psi: M_2 \times \mathbf{R}^p \to \mathbf{R}$, which is identically equal to 1 on a neighborhood of $(0,0)$ and such that support $(\Psi) = \bar{U} \subset V$, U an open subset of $M_2 \times \mathbf{R}^p$. Let π be the canonical projection of $\mathbf{R}^n \times \mathbf{R}^p$ onto $M_2 \times \mathbf{R}^p$ and define $\bar{X}_\mu$, for $(x, \mu) \in \pi^{-1}(V)$, to be the vectorfield $\bar{X}_\mu(m_1, m_2) = X_\mu(m_1, m_2) + (-1 + \Psi(m_2, \mu)) \cdot X_\mu(F(m_2, \mu), m_2)$. $\bar{X}_\mu$ has the following properties:

(a) $W^c \cap \pi^{-1}(V)$ is an invariant submanifold for $\bar{X}$, $\bar{X}(x, \mu) = \bar{X}_\mu(x)$;

(b) $\bar{X}|W^c$ has compact support, namely $\pi^{-1}(\bar{U}) \cap W^c$.

Even for $U \subset \bar{U} \subset V$ small, W^c does not have to be normally hyperbolic any more for $\bar{X}$; this is due to possible distorsions as a consequence of multiplying $X|W^c$ with $\Psi \circ \pi$. In order to make W^c again normally hyperbolic, we perform a second modification.

For this we choose a smooth non negative function $\bar{\Psi}: M_2 \times \mathbf{R}^p \to \mathbf{R}$ which is positive on a neighborhood of the complement in V of $\Psi^{-1}(1)$ and which is zero in a neighborhood of $(0,0)$. Set $\bar{X}_\mu(x) = \bar{X}_\mu^{(1)}(x) + \bar{X}_\mu^{(2)}(x)$ with $\bar{X}_\mu^{(1)}(x)$, $\bar{X}_\mu^{(2)}(x)$ in the M_1, resp. the M_2 direction. Now we define

$$\bar{\bar{X}}_\mu(m_1, m_2) = \bar{X}_\mu(m_1, m_2) + \lambda \cdot \bar{\Psi}(m_2, \mu) \cdot \{\bar{X}_\mu^{(1)}(m_1, m_2) - \bar{X}_\mu^{(1)}(F(m_2, \mu), m_2)\}.$$

For $\bar{\bar{X}}_\mu$ and W^c the above statements (a) and (b) still hold. Moreover, for λ sufficiently big, W^c is normally hyperbolic for $\bar{\bar{X}}$, at least if we use the following somewhat broader notion of normal hyperbolicity. Let M be a Riemannian manifold with C^1 vectorfield X. If N is an invariant manifold for X with support $(X|N)$ compact then we say that N is normally hyperbolic if there is an invariant splitting $T_N(M) = T(N) \oplus E^u(N) \oplus E^s(N)$, a positive constant t_0 and positive functions $\lambda, \mu: N \to \mathbf{R}$ with $\lambda > 1 > \mu$ such that for each $n \in N$:

(1) for each $v \in E_n^u(N)$, $\|\varphi_{t_0}(v)\| > \lambda(n) \cdot \|v\|$;

(2) for each $v \in E_n^s(N)$, $\|\varphi_{t_0}(v)\| < \mu(n) \cdot \|v\|$;

(3) for each $v \in T_n(N)$, $\mu(n) \cdot \|v\| < \|\varphi_{t_0}(v)\| < \lambda(n) \cdot \|v\|$.

For invariant manifolds in the above sense, stable and unstable foliations also exist [5], [9] and hence our results in §2 and §3 apply here.

If $h: W^c \to W'^c$ is the given equivalence between $X|W^c$ and $X'|W'^c$ we take $U' = h(U)$ (U as before; we may and do assume that U is in the domain of h). Now we carry out the same constructions for X' as we did for X and obtain $\bar{\bar{X}}'$. The map h is still an equivalence between $\bar{\bar{X}}|W^c$ and $\bar{\bar{X}}'|W'^c$, because on U, resp. U', X, resp. X', has only been multiplied with a positive function and on the complement of U, resp. U', in W^c, resp. W'^c, we have $\bar{\bar{X}}$, resp. $\bar{\bar{X}}'$, equal to zero. Also, by assumption the dimensions of the stable, resp. unstable bundles over W^c and W'^c are equal. Hence by our main theorem in §3 the equivalence $h: W^c \to W'^c$ can be extended to a neighborhood of W^c as an equivalence between $\bar{\bar{X}}$ and $\bar{\bar{X}}'$. Since X, resp. X', are, in a neighborhood of $(0,0)$, equal to X, resp. X', we have constructed an extension H which is a local equivalence between X and X'.

The second observation we have to make is that the extension H has to map points with equal μ coordinates to points with equal μ coordinates. That the construction in §3 can give such an extension follows from the fact that all the foliations can be choosen so that each leaf is contained in some $\mathbf{R}^n \times \{\mu\}$ and from the fact that each integral curve of $\bar{\bar{X}}$, $\bar{\bar{X}}'$ also is contained in some $\mathbf{R}^n \times \{\mu\}$.

REFERENCES

1. D. V. Anosov: Geodesic flows on compact Riemannian manifolds of negative curvature, *Proc. Steklov math. Inst.* **90** (1967).
2. V. I. Arnol'd: Lectures on bifurcation and versal families, *Russ. math. Survs* **27** (1972), 54–123.
3. D. Grobman: Homeomorphisms of systems of differential equations. *Dokl. Akad. Nauk SSSR* **128** (1959), 880–881.
4. P. Hartman: A lemma in the structural stability of differential equations, *Proc. Am. math. Soc.* **11** (1960), 610–620.
5. H. W. Hirsch, C. C. Pugh and M. Shub: Invariant manifolds, Lecture Notes in Mathematics, 583. Springer-Verlag (1977).
6. W. de Melo: Structural stability of diffeomorphisms on two-manifolds, *Invent. Math.* **21** (1973), 233–246.
7. J. Palis: On Morse–Smale dynamical systems, *Topology* **8** (1969), 385–404.

8. J. PALIS and S. SMALE: Structural stability theorems, Global Analysis, *Proc. Symposia pure Math.* **14** (1970) A.M.S.
9. C. C. PUGH and M. SHUB: Linearization of normally hyperbolic diffeomorphisms and flows, *Invent. Math.* **10** (1970), 187–198.
10. A. N. SHOSHITAISHVILI: Bifurcations of topological type at singular points of parametrized vectorfields, *Funct. Anal. appl.* **6** (1972), 169–170.
11. J. SOTOMAYOR: Generic one-parameter families of vectorfields on two-dimensional manifolds, *Publ. I.H.E.S.* **43** (1973), 5–46.
12. F. TAKENS, Singularities of vectorfields, *Publ. I.H.E.S.* **43** (1973), 47–100.
13. F. WILSON, The structure of the level surfaces of a Liapunov function, *J. Diff. Equations* **3** (1967), 323–329.

I.M.P.A.
Rua Luiz de Camões 68
Rio de Janeiro, Brazil

Mathematisch Instituut
Groningen, Holland

APPENDIX

1. Invariant manifolds

THEOREM. *Let M be a smooth manifold, X a C^1 vectorfield on M and $N \subset M$ a C^1 invariant manifold. Then there is a C^2 structure on N, compatible with the induced C^1 structure, such that $X|N$ is C^1 (with respect to this C^2 structure).*

Proof. Let $i: N \to M$ denote the canonical embedding. Take some C^2 structure on N; N with this C^2 structure, is denoted by $\tilde{N}$. Let $\pi: U \to N$, U a neighborhood of $i(N)$ in M, be a C^1 map of maximal rank such that $\pi i = id_N$. π induces a map $\tilde{\pi}: U \to \tilde{N}$ which is C^1; we approximate this map by a C^2 map $\tilde{\pi}'$: $U \to \tilde{N}$ such that for each $x \in i(N)$, $\ker(d\tilde{\pi}')(x)$ is transversal with respect to $T_x(N)$. Hence $\tilde{\pi}' \circ i$, restricted to N, has everywhere maximal rank and so induces a C^2 structure from $\tilde{N}$ on N; N with this C^2 structure is denoted by N^*. The C^2 structure of N^* is clearly compatible with the induced C^1 structure. We want to show that $X|N^*$ is C^1. This goes as follows:

Let $p \in i(N)$, $\tilde{p} = \tilde{\pi}' \circ i(p) \in \tilde{N}$ and let $x_1, \ldots, x_n$ be a C^2 local coordinate system on $\tilde{N}$ near $\tilde{p}$. Then we can find C^2 local coordinates $\tilde{x}_1, \ldots, \tilde{x}_m$ on M near p such that $\tilde{x}_1 = x_1 \circ \tilde{\pi}', \ldots, \tilde{x}_n = x_n \circ \tilde{\pi}'$. In these coordinates we have: $i(N)$ has the form $\{(\tilde{x}_1, \ldots, \tilde{x}_m) | \tilde{x}_{n+1} = f_{n+1}(\tilde{x}_1, \ldots, \tilde{x}_n), \ldots, \tilde{x}m = f_m(\tilde{x}_1, \ldots, \tilde{x}_n)\}$, with $f_{n+1}, \ldots, f_m$ C^1 functions; $X = \sum_{i=1}^{m} X_i(\tilde{x}_1, \ldots, \tilde{x}_m)(\partial/\partial \tilde{x}_i)$, with $X_1, \ldots, X_m$ C^1 functions; and hence

$$\tilde{\pi}'_*(X|i(N)) = \sum_{i=1}^{n} X_i(x_1, \ldots, x_n, f_{n+1}(x_1, \ldots, x_n), \ldots, f_m(x_1, \ldots, x_n))(\partial/\partial x_i)$$

with X_i and f_j as above. Hence $\tilde{\pi}'_*(X|i(N)))$ is a C^1 vectorfield on N^*.

Remark. For M, N, X as above and N compact, one has the following fact. If X' is C^1 close to X and N' is an invariant manifold for X' which is C^1 close to N, then, using the notation of the above proof,

(a) $\tilde{\pi}'|N'$: $N' \to \tilde{N}$ defines a C^2 structure on N;

(b) $(\tilde{\pi}'|N)_*X$ and $(\tilde{\pi}'|N')_*X'$ are two vectorfields which are C^1 close.

So, in a certain sense $X|N$ and $X'|N'$ are "C^1-close".

2. Suspensions

Let M be a smooth (C^∞) manifold and φ: $M \to M$ a C^1 diffeomorphism. The suspension of (M, φ) consists of a manifold $\tilde{M} = M \times [0, 1]/\sim$, where $\sim$ denotes the identification of $(x, 1)$ with $(\varphi(x), 0)$ for each $x \in M_0$, and a vectorfield X_φ on $\tilde{M}$ whose integral curves are of the form

$$t \mapsto \begin{cases} (x, t) \text{ for } t \in [0, 1] \\ (\varphi(x), t-1) \text{ for } t \in [1, 2] \\ (\varphi^{-1}(x), t+1) \text{ for } t \in [-1, 0] \\ \text{etc.} \end{cases}$$

The manifold $\tilde{M}$ has a canonical C^1 structure, so X_φ is only a C^0 vectorfield. The following fact has been used extensively in the literature on dynamical systems although we could not find a published proof (our proof was obtained in cooperation with C. C. Pugh). This proof will be presented here for the sake of completeness.

THEOREM. *For $\tilde{M}$ and X_φ as above, there is a C^∞ structure on $\tilde{M}$, which is compatible with the C^1 structure and with respect to which X_φ is C^1.*

For the proof we need the following:

LEMMA. *Let M be a smooth manifold. Then there is a C^1 neighborhood U of the identity in $\mathrm{Diff}^1(M)$ (with respect to the strong Whitney topology) such that for every $\Psi \in U$, there is a C^1 vectorfield Z_Ψ on $M \times [0,1]$ such that*

(1) $Z_\Psi(x,t) = (\partial/\partial t)$ for $t \in [0,(1/4)]$ and $t \in [(3/4),1]$;

(2) the time $t = 1$ map of Z_Ψ maps $(x,0)$ to $(\Psi(x),1)$.

Proof of the lemma. We take a complete and smooth Riemannian matric on M; exp: $T(M) \to M$ denotes the induced exponential map. It is well known that there is a (convex) neighborhood V of 0 in the space of C^1 vectorfields on M such that for each $X \in V$, exp X, defined by $x \mapsto \exp(X(x))$, is a C^1-diffeomorphism. $\{\exp X | X \in V\} = U$ is a C^1 neighborhood of the identity in $\mathrm{Diff}^1(M)$.

Let now $\Psi \in U$; choose $X \in V$ such that $\exp(X) = \Psi$. Furthermore let β: $[0,1] \to [0,1]$ be a smooth function with

$$\beta(t) = \begin{cases} 0 & \text{for } t \in [0,(1/4)] \\ 1 & \text{for } t \in [(3/4),1] \end{cases}.$$

Z_Ψ is now the vectorfield having the following integral curves:

$$t \mapsto (\exp(\beta(t) \cdot X(x)), t) \quad \text{for all} \quad x \in M.$$

These curves are smooth, so Z_Ψ is well defined. We only have to show that Z_Ψ is C^1 (Z_Ψ trivially satisfies conditions 1. and 2. in the Lemma). To do so, we write

$$Z_\Psi(x,t) = \tilde{Z}_\Psi(x,t) + \frac{\partial}{\partial t}$$

with

$$\tilde{Z}_\Psi(x,t) \in T_x(M).$$

We give $\tilde{Z}_\Psi$ as a composition of C^1 maps. So we write $\tilde{Z}_\Psi = (\mathrm{d}(\exp)) \circ \beta \circ \alpha$:

$$M \times [0,1] \xrightarrow{\alpha} M \times [0,1] \xrightarrow{\gamma} T(T(M)) \xrightarrow{\mathrm{d(exp)}} T(M),$$

where $\alpha(x,t) = ((\exp(\beta(t) \cdot X)^{-1}(x), t)$ and $\gamma(\tilde{x},t)$ is the vector in $T_{\beta(t)\cdot X(\tilde{x})}(T_{\tilde{x}}(M))$, obtained by parallel translation from $\beta'(t) \cdot X(\tilde{x}) \in T_{\tilde{x}}(M) \cong T_0(T_{\tilde{x}}(M))$. In other words, $\gamma(\tilde{x},t)$ is the tangent vector to the curve $t \mapsto \beta(t) \cdot X(\tilde{x}) \in T_{\tilde{x}}(M)$.

Proof of the Theorem. Let σ_1 be the C^∞ structure on $M \times [(0,(1/4)) \cup ((3/4),1])/\sim$ induced by the C^∞ structure on $M \times (-(1/4),+(1/4))$ through the map Σ_1: $M \times (-(1/4),+(1/4)) \to \tilde{M}$, defined by:

$$(x,t) \mapsto \begin{cases} (x,t) & \text{for } t \in [0,(1/4)) \\ (\varphi^{-1}(x), t+1) & \text{for } t \in (-(1/4),0] \end{cases}.$$

Next we choose a C^1 approximation Ψ: $M \to M$ of the identity such that

(1) Ψ is in the neighborhood U of the identity, described in the above lemma;

(2) $\varphi \cdot \Psi^{-1}$ is C^∞.

Let Z_Ψ be a vectorfield on $M \times [0,1]$ as constructed in the above Lemma. Define Σ_2: $M \times (0,1) \to \tilde{M}$ by

$$\Sigma_2(x,t) = \begin{cases} (x,t) \text{ for } x \in (0,(1/4)] \\ (\Psi^{-1}(x),t) \text{ for } x \in [(3/4),1) \\ (\Sigma_2)_*(Z_\Psi) = X_\varphi | M \times (0,1); \end{cases}$$

this is possible because of the properties of Z_Ψ. Let σ_2 be the C^∞ structure on $M \times (0,1) \subset \tilde{M}$

defined by Σ_2. Then σ_1 and σ_2 agree on their common domain (because $\varphi \circ \Psi^{-1}$ is C^∞) and hence define a C^∞ structure on $\tilde{M}$ which is compatible with the C^1 structure and makes X_φ a C^1 vectorfield.

Remark.

After this paper was written we become convinced that the main theorem is true without the assumption on the triviality of the stable and unstable bundles over the singular set. We thank M. Shub for a suggestion in this direction.

Proceedings of the International Congress of Mathematicians
Helsinki, 1978

Moduli of Stability and Bifurcation Theory

J. Palis

To Rufus Bowen, in memoriam

Bifurcation theory of dynamical systems goes as far back as Poincaré. One aims at describing the changes in the phase portrait (space of orbits) of a system depending on parameters when such parameters vary. This lecture concerns a line of recent developments in this direction, especially on bifurcations of one parameter families of vector fields and diffeomorphisms. A relevant role, in this context, is played by certain differentiable invariants of topological equivalence for systems that exhibit non-transversal saddle connections. Such invariants bring up, in a natural way, the notion of moduli of stability in dynamical systems. They also imply, as explained in one of the topics below, the existence of moduli of stability for holomorphic vector fields near a singularity.

We begin with some simple examples of such invariants. Let X, X' be vector fields of class C^2 of $\mathbf{R}^2$, each with two hyperbolic singularities of saddle type and trajectories γ and γ' connecting them. Let μ, μ' and λ, λ' be the eigenvalues of dX, dX' at the saddles along transversal directions to γ, γ'. Then **[13]**, there is a flow conjugacy between X and X' near γ and γ' if and only if $\mu/\lambda = \mu'/\lambda'$.

We recall that, for a flow conjugacy, we require a homeomorphism h such that $hX_t = X'_t h$, where X_t and X'_t for $t \in \mathbf{R}$ are the flows generated by X and X'. For a topological equivalence, we just require a homeomorphism sending orbits of X into orbits of X'.

We say that, up to conjugacy, the modulus of stability of X near γ as above is one, since there is one parameter family of different conjugacy classes for the vector fields C^2 near X. It follows that, up to topological equivalence, the modulus of stability of X near γ is zero.

This fact can be generalized to higher dimensions as follows. We now require

Originally published in: *Proceedings of the International Congress of Mathematicians (Helsinki, 1978), 835-839,* © Academia Scientiarum Fennica, Helsinki, 1980
Reprint by Springer-Verlag Berlin Heidelberg 2014 with kind permission

the connecting orbit γ to be a quasi-transversal (second order contact) intersection of the unstable manifold of a saddle p_1 and the stable of a saddle p_2, contained in the "weak" parts of both these submanifolds. We assume that the eigenvalues of dX at p_1 and p_2 are real. If $\mu_1, \mu_2, \dots, \mu_s$ are the negative eigenvalues of dX at p_1, we also assume that $\mu_i < \mu_1$, $2 \leq i \leq s$. Similarly, $\lambda_1 < \lambda_j$ for $2 \leq j \leq u$, where $\lambda_1, \lambda_2, \dots, \lambda_u$ are the positive eigenvalues of dX at p_2. Then, up to conjugacy, the modulus of stability of X near γ is one, given by the ratio μ_1/λ_1. Up to topological equivalence, it is zero.

Let us see the role of these invariants in the study of one parameter families of gradient flows. Let M be a smooth compact manifold and Grad (M) its C^∞ gradient vector fields, with the C^∞ topology. Consider the set of C^1 arcs $\varphi: I \to \text{Grad}(M)$, $I=[0, 1]$, with the C^1 topology. Two such arcs φ and φ' are conjugate, resp. equivalent, if there are a homeomorphism $\varrho: I \to I$ and a continuous map $h: I \to \text{Homeo}(M)$ such that, for each $s \in I$, $h(s)$ is a conjugacy, resp. a topological equivalence, between $\varphi(s)$ and $\varphi'(\varrho(s))$. An arc φ has finite modulus of stability if we can parametrize all equivalence classes of arcs near φ with finitely many real parameters; the minimum number of parameters is called the modulus of φ. Very recently, in a work still in production, Newhouse, Takens and the author proved the following result.

THEOREM [11]. *There is an open and dense subset B of one parameter families in* Grad (M) *such that if $\varphi \in B$ then the modulus of stability of φ, up to conjugacy, is finite. The modulus is given by the number of values $s \in I$ for which $\varphi(s)$ has a saddle connection.*

As a corollary we have

THEOREM. *There is an open and dense subset of stable one parameter families of gradient vector fields, up to topological equivalence.*

One of the main motivations for pursuing the ideas above was to obtain this result. It implies the stability of the simple arcs constructed by Newhouse–Peixoto connecting Morse–Smale vector fields, (see [14]). Note that, for $\dim M=2$, it is a consequence of the work of Sotomayor [16]. The following are, in my opinion, very pretty questions in this direction. Are the stable k-parameter families of gradient flows dense for $k>1$? If not, are the ones with finite modulus of stability dense?

There are analogous invariants for saddle connections of diffeomorphisms. Let f and f' be C^2 diffeomorphisms of R^2, each with two fixed saddles and orbits γ and γ' of parabolic contact between their unstable and stable manifolds. As before, we indicate by μ, μ' and λ, λ' the eigenvalues of df, df' at the saddles taken along transversal directions to these invariant submanifolds. Then, there exists a conjugacy between f and f' near γ and γ' if and only if $\log|\mu|/\log|\lambda| = \log|\mu'|/\log|\lambda'|$.

That such a relation among the eigenvalues is necessary for a conjugacy is in [13] and the converse in [7]. Earlier, Newhouse had pointed out that a conjugacy as above does not in general exist. We remark that a similar invariant can be obtained in higher dimensions [11]. Some global results for $\dim M=2$ are in [7]; for higher dimensions, Melo and the author believe it is possible to describe all Axiom A diffeomorphisms (see definition in [12]) with modulus of stability one. An interesting question is to obtain invariants of conjugacy for saddle connections with higher order contact. We also remark that all these facts and questions can be considered for saddle connections of closed orbits of vector fields. A particularly interesting case is discussed in the next topic.

We now describe a parallel development for holomorphic flows near a singularity. Let F be a holomorphic vector field near the origin 0 of C^n, with 0 as a singularity. Consider the real flow F induces by intersecting its orbits with a small sphere S^{2n-1} centered at the origin. We assume that any two eigenvalues of $dF(0)$ are independent over the reals. If the convex hull of these eigenvalues does not contain the origin of C, Guckenheimer showed that the real flow on S^{2n-1} is Morse–Smale, hence stable. It follows that F is locally stable. In the other case, the real flow on S^{2n-1} has pairs of closed orbits with nontransversal saddle connections, which persist under small perturbations of F. This fact implies the existence of moduli of stability: the modulus of stability of such F in C^n near the singularity is at least $2n-4$. For linear flows, it was shown in [2] and independently by Ladis (see [5]) that the modulus of stability is precisely $2n-4$. We conjecture that the same is true for non linear flows; for $n=3$ this is proved in [2]. Using the results for linear flows on C^{n+1}, a complete classification of holomorphic vector fields on the complex projective space CP^n is also carried out in [2].

Another interesting example of moduli is in [3], where the Lorenz attractor is proved to have modulus two of stability.

Let us mention some results of a global nature on bifurcations of diffeomorphisms. See [1], [12], [14], [17] for background, related accounts and references. In [9], [10] we studied how a family of diffeomorphisms depending on a parameter bifurcates from the Morse–Smale ones, these forming the best known class of stable diffeomorphisms. Consider C^1 families $\{\varphi_\mu\}$ of diffeomorphisms, $\mu\in[0, 1]$, such that φ_0 is Morse–Smale. Call $b=\inf\{\mu; \varphi_\mu$ is not Morse–Smale$\}$ the first bifurcation point. The problem is then to describe the phase portraits of φ_b and φ_μ, $b<\mu<b+\varepsilon$ for some $\varepsilon>0$. We assume the limit set $L(\varphi_b)$ to be a finite number of orbits. This is true in many examples and we conjecture that it is so for a Baire (residual) subspace A of arcs. We prove that A contains an open subset of all arcs, the remaining part of the conjecture being its density. The other results can be briefly summarized as follows. For a residual (generic) set of arcs, we have

(i) if $L(\varphi_b)$ has no cycles, then for some $\varepsilon>0$ φ_μ is Morse–Smale for μ in an open and dense subset of $(b, b+\varepsilon)$;

(ii) if $L(\varphi_b)$ is hyperbolic and has an equidimensional cycle, then for some

$\varepsilon > 0$ there are infinitely many distinct classes of stable φ_μ's with $L(\varphi_\mu)$ infinite, $\mu \in (b, b+\varepsilon)$.

In case (ii), the subset of $(b, b+\varepsilon)$ for which φ_μ is stable has a large measure relative to ε. However, on its complement there are residual subsets of small intervals for which φ_μ exhibits infinitely many sinks [**8**]. We remark that some of the above questions were also treated by Afraimovic and Silnikov (see [**12**] for references). In his forthcoming thesis [**6**], M. Levi shows that several of the above facts occur in a model, previously introduced by Levinson, of the forced van der Pol equation.

As for the stability of such arcs, a characterization is given in [**11**] of the stable ones up to and beyond the first bifurcation point. In doing so, we provide universal models for the unfoldings of the elementary bifurcations, the novelty being the surprising rigidity of the saddle-node case. From our previous discussion on moduli of stability, we conclude that arcs of diffeomorphisms going through nontransversal saddle connections are not stable. One of our results, proving the stability of a certain class of arcs, has been generalized by C. Robinson [**15**].

In our analysis, a specially interesting question is the stability of arcs of diffeomorphisms going through a saddle-node with cycle. In some cases, there appears an invariant circle and using Herman's results [**4**], it is shown that such arcs are not stable. In many other cases (in fact, we conjecture: in all other generic cases), nontransversal homoclinic orbits and horse-shoes appear. We succeeded in showing this fact for normally attracting (repelling) saddle-node. In this case, the saddle-node is contained in a normally attracting annulus. Through projection, the question is reduced to the study of a class of endomorphisms of degree one of the circle, with varying rotation "interval". Such a concept is similar to that of rotation number for a homeomorphism of the circle.

References

1. V. I. Arnold, *Lectures on bifurcations in versal families*, Russian Math. Surveys **27** (1972), 54—153.

2. C. Camacho, N. Kuiper and J. Palis, *The topology of holomorphic flows with singularity*, Inst. Hautes Études Sci. Publ. Math. **48** (1978), 1—38.; also, *La topologie du feuilletage d'un champ de vecteurs holomorphes près d'une singularité*, C. R. Acad. Sci. Paris **282** (1976) 959—961.

3. J. Guckenheimer and R. F. Williams, *Structural stability of Lorenz attractors*, Inst. Hautes Études Sci Publ. Math. (to appear).

4. M. Herman, *Sur la conjugaison différentiable des difféomorphismes du cercle à des rotations*, thesis, Orsay, 1976. Inst. Hautes Études Sci. Publ. Math. (to appear).

5. J. Ilyashenko, *Remarks on the topology of singular points of analytic differential equations in the complex domain and on the theorem of Ladis*, Functional Anal. Appl. **11** (1977), 28—38.

6. M. Levi, thesis, Courant Institute (to appear).

7. M. Melo, *Moduli of stability of two-dimensional diffeomorphisms*, Topology (to appear).

8. S. Newhouse, *The abundance of wild hyperbolic sets and non-smooth stable sets for diffeomorphisms*, Inst. Hautes Études Sci. Publ. Math. (to appear).

9. S. Newhouse and J. Palis, *Bifurcations of Morse-Smale dynamical systems*, Dynamical Systems, ed. M. Peixoto, Academic Press, New York, 1973, pp. 303—366.

10. ——, *Cycles and bifurcation theory*, Asterisque **31** (1976), 43—140.

11. S. Newhouse, J. Palis and F. Takens, *Stable families of dynamical systems* (to appear), also *Stable arcs of diffeomorphisms*, Bull. Amer. Math. Soc. **82** (1976), 499—502.

12. J. Palis, *Some developments on stability and bifurcations of dynamical systems*, Geometry and Topology, Lecture Notes in Math., vol. 597, ed., J. Palis and M. do Carmo, Springer-Verlag, Berlin and New York, 1977, pp. 495—509.

13. ——, *A differentiable invariant of topological conjugacies and moduli of stability*, Asterisque **51** (1978), 335—346.

14. M. Peixoto, *On bifurcations of dynamical systems*, Proc. Internat. Congress of Mathematicians, Vancouver (1974), pp. 315—319.

15. C. Robinson, *Global structural stability of a saddle-node bifurcation*, Trans. Amer. Math. Soc. **236** (1978), 155—171.

16. J. Sotomayor, *Generic one-parameter families of vector fields on two manifolds*, Inst. Hautes Études Sci. Publ. Math. **43** (1973), 5—46.

17. F. Takens, *Introduction to global analysis*, Comm. Math. Inst. Univ. Utrecht, 1973.

INSTITUTO MATEMATICA PURA E APPLICADA
RIO DE JANEIRO, BRAZIL

Characterising diffeomorphisms with modulus of stability one.

W. de Melo, J. Palis* & S.J. van Strien

In the theory of dynamical systems one tries to classify diffeomorphisms according to their orbit structure. One of the most useful equivalence relations in this context is topological conjugacy. In [P1] and [P.S.] it was found that the orbit structure of a diffeomorphism with finite limit set is stable under perturbations if it is "Morse-Smale".

Of course, not all diffeomorphisms are stable. For example, in dimension 2, we have that a diffeomorphism Φ_0 with a saddle-connection, as in figure 1(b) is not conjugate to the nearby diffeomorphisms without the saddle-connections (see figure 1(a) and 1(c)).

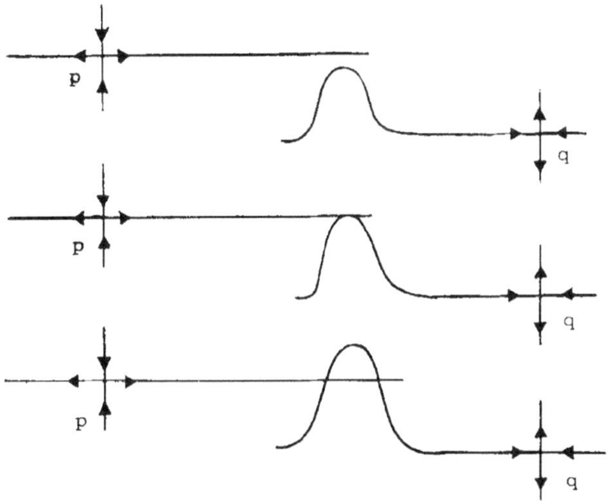

Figure 1(a) Φ_μ $\mu < 0$

Figure 1(b) Φ_0

Figure 1(c) Φ_μ $\mu > 0$

It is even not true that there are only three equivalence classes of diffeomorphisms near this saddle-connection diffeomorphism Φ_0. If we denote the contracting eigenvalue of Φ_0 at p by a and the expanding eigenvalue at q by b, then for diffeomorphisms as in figure 1(b), $P(\Phi) = \log|a|/\log|b|$ is a topological invariant, i.e., two diffeomorphisms Φ and $\tilde{\Phi}$ as in figure 1(b) can only be conjugate if $P(\Phi) = P(\tilde{\Phi})$. Such a condition is called a <u>modulus of stability</u>.

In this paper we shall give a characterisation of a large class of diffeomorphisms with modulus of stability one. In a previous paper, [M.P.], it was proved that these

* The second author gratefully acknowledges the financial support of the Stiftung Volkswagenwerk for a visit to the IHES, during which part of this work was developed.

Originally published in: *Dynamical systems and turbulence, Warwick 1980 (Coventry, 1979/1980), 266–285, Lecture Notes in Math. 898,* © Springer, Berlin Heidelberg, 1981
Reprint by Springer-Verlag Berlin Heidelberg 2014

diffeomorphisms, under certain conditions, have only one modulus. The conditions imposed on the diffeomorphisms are quite natural and in fact necessary ones as we show here.

§1. Statement of Results.

Let M be a compact C^∞ manifold without boundary and Diff(M) the set of C^∞ diffeomorphisms with the C^∞ topology. For Φ and $\tilde{\Phi}$ in Diff (M) we say that Φ and $\tilde{\Phi}$ are conjugate if there exists a homeomorphism h on M such that $h \circ \Phi = \tilde{\Phi} \circ h$. If each $\tilde{\Phi}$ sufficiently near Φ is conjugate to Φ then Φ is called stable. If the equivalence classes in small neighbourhoods of Φ can be parametrised by k real parameters then the modulus of stability is called k, (assume that k is chosen minimal). We also say that the modulus is zero if the number of equivalence classes is finite.

Let p be a hyperbolic periodic point for Φ. Then Φ has a strictly weakest contracting eigenvalue a at p, if the eigenvalue has multiplicity one and if for any other eigenvalue $\lambda \neq a$, a at p, with $|\lambda| < 1$ one has $|\lambda| < |a|$. In this case there is a unique invariant submanifold $W^{ss}(p)$, called the strong stable manifold, which is tangent to the eigenspace corresponding to eigenvalues with norm smaller that $|a|$, see [H.P.S.]. Moreover there is a uniquely defined foliation $F^{ss}(p)$ contained in $W^s(p)$ with smooth leaves such that $W^{ss}(p)$ is a leaf and such that Φ maps leaves to leaves (see [H. P.S.]). Similarly for the strictly weakest expanding eigenvalue.

We say that p is s-critical if there is some periodic point z such that $W^u(z)$ intersects some leaf of $F^{ss}(p)$ non-transversally.

When the non-wandering set $\Omega(\Phi)$ is finite, we say that $\Omega(\Phi)$ has no cycles if for any sequence $p_1, \ldots, p_n$ of distinct periodic orbits such that $W^s(p_i) \cap W^u(p_{i+1}) \neq \emptyset$, $1 \leq i \leq n-1$, we have that $W^s(p_n) \cap W^u(p_1) = \emptyset$.

We will consider the rather large class of diffeomorphisms with finite hyperbolic non-wandering set having no-cycles. This class is denoted by $\mathcal{A}$.

Definition (1.1). Let $\mathcal{B} \subset \mathcal{A}$ be the set of diffeomorphisms Φ such that :

(1) the stable and unstable manifolds of the periodic orbits of Φ are either transversal or quasi-transversal along their orbits of intersection. (See [N.P.T.] for the definition of quasi-transversality).

(2) for each pair of periodic orbits p, q such that $W^u(p)$ and $W^s(q)$ has an orbit of quasi-transversal intersection, the strictly weakest contracting eigenvalue at p and the strictly weakest expanding eigenvalue at q are defined;

(3) for each such pair of periodic points p and q, there are C^1 linearising coordinates near p and q (see [Ste]);

(4) let $W^{cu}(p)$ be an invariant C^1 manifold tangent at p to the direct sum of the expanding eigenspace and the weakest contracting eigenspace. These centre-unstable manifolds $W^{cu}(p)$ are not unique, but they are tangent to each other at $W^u(p)$. Therefore we can demand $W^s(q)$ to be transversal to $W^{cu}(p)$. Similarly, we assume that $W^u(p)$ is transversal to $W^{cs}(q)$.

We point out that $\mathcal{A}$-$\mathcal{B}$ is of codimension bigger than one in $\mathcal{A}$: most (open and dense) smooth one-parameter families of diffeomorphisms in $\mathcal{A}$ miss $\mathcal{A}$-$\mathcal{B}$. Our result really concerns diffeomorphisms in $\mathcal{A}$; we will restrict ourselves to $\mathcal{B}$ for technical reasons.

We can now state a characterisation of diffeomorphisms in $\mathcal{B}$ with modulus of stability one. Let $\mathcal{C} \subset \mathcal{B}$ be the subset of Φ such that :

(1) there is a pair of periodic orbits p and q such that $W^u(p)$ and $W^s(q)$ have a unique orbit of non-transversal intersection; along all other orbits the stable and unstable manifolds meet transversally;

(2) for such a pair of periodic orbits p and q, the strictly weakest contracting eigenvalue at p and the weakest expanding eigenvalue at q are real. Furthermore, p is not s-critical and q is not u-critical.

In [M-P] the following theorem was proved.

<u>Theorem.</u> If $\Phi \in \mathcal{C}$ then Φ has modulus of stability one.

Here we prove the converse.

Main Theorem. If $\Phi \in \mathcal{B}$ has modulus of stability one then $\Phi \in \mathcal{C}$.

We also show that the no-cycle condition is necessary for a diffeomorphism to have modulus of stability one.

The organisation of this paper is as follows. Let Φ be a diffeomorphism which has modulus of stability one.

In Section 2 we assume that (i) $W^u(p)$ and $W^s(q)$ have a quasi-transversal intersection, (ii) the weakest eigenvalues at p and q exist and are real. We then prove that $W^u(p)$ and $W^s(q)$ have only one orbit of quasi-transversal intersection, and furthermore that p is not s-critical and q is not u-critical. If any of these conditions is not satisfied, we exhibit at least two (real independent) conjugacy invariants and thus Φ has modulus of stability bigger than one.

In Section 3 we prove that the weakest eigenvalues must be real.

In Section 4 the proof of the main theorem is completed.

In Section 5 we prove that the no-cycle hypothesis is necessary.

Throughout this paper suppose that h is a conjugacy between two diffeomorphisms Φ and $\tilde{\Phi}$. We denote the analogue of $W^u(p)$ by $W^u(\tilde{p})$, of $W^s(q)$ by $W^s(\tilde{q})$ and so on.

Furthermore we adopt the following notation. Let c_i, d_i be two sequences of numbers. Then $c_i \lesssim d_i$ means that c_i/d_i has a supremum not bigger than one and $c_i \cong d_i$ means that c_i/d_i converges to one.

§2. Criticallity and number of orbits of non-transversal intersection.

In this section we suppose that

(i) Φ has periodic points p and q such that $W^u(p)$ and $W^s(q)$ have a quasi-transversal intersection in some orbit $O(r)$.

(ii) the strictly weakest contracting eigenvalue a at p exists and is real. Similarly for the weakest expanding eigenvalue b at q.

(iii) Furthermore suppose that conditions (3) and (4) of Definition (1.1) are satisfied.

§2.a. The strong-stable foliation is preserved.

Theorem (2.1). Let Φ and $\tilde{\Phi}$ be as above and h a conjugacy between them. Consider the restriction of h to $W^s(p)$; i.e., $h:W^s(p) \to W^s(\tilde{p})$. Then :

(1) The unique strong stable foliations in $W^s(p)$, $W^s(\tilde{p})$ are preserved by h; i.e., h sends leaves of $F^{ss}(p)$ onto leaves of $F^{ss}(\tilde{p})$.

(2) Suppose that $\frac{\log|a|}{\log|b|} \notin \mathbb{Q}$. In this case the induced map h_* on the fibre-space of these foliations is smooth except at $W^{ss}(p)$. It is, in fact, nearly uniquely defined in the following sense. If $a > 0$, take leaves F_1, F_2 of $F^{ss}(p)$ in the two components of $W^s(p) - W^{ss}(p)$. Then h_* is completely determined as soon as one chooses the image of two such leaves. For $a < 0$, the image of one leaf determines h_*. Similarly for the map induced on the space of leaves of $F^{uu}(q)$.

Remark. Take linearising coordinates $(y_1, \dots, y_n)$ near p, such that the y_1-axis is the eigenspace corresponding to the weakest contracting eigenvalue (i.e. a), and similarly for p. Then for $y \in W^s(p)$

$$\tilde{\pi}_1 \circ h(y) = \alpha \circ (\pi_1(y))^{\delta} ,$$

where α is some constant $\delta = \frac{\log|\tilde{a}|}{\log|a|}$ and $\pi_1(y_1, \dots, y_n) = y_1$.

For further use assume that $s = \dim W^s(q)$ and $u = \dim W^u(q)$.

Proof of Theorem (2.1). Assume that $a, b > 0$, otherwise take Φ^2 instead of Φ. We want to restrict our attention to a centre-unstable manifold $W^{cu}(p)$. The problem now is that

$W^{cu}(p)$ is not unique and it has no topological characterisation. Thus $h(W^{cu}(p))$ is a topological manifold which might not coincide with a C^1 centre-unstable manifold of p. But $h(W^u(p)) = W^u(\tilde{p})$ and so $h(W^{cu}(p))$ cannot be too far from $W^{cu}(\tilde{p})$. In fact we will show that $h(W^{cu}(p))$ lies in a cone tangent to $W^{cu}(\tilde{p})$ at $W^u(\tilde{p})$. First we will compare distances on M arising from different Riemannian metrics. Let $S \subset W$ be closed submanifolds of M and d be a distance function on M induced by a Riemannian metric. For $c > 0$ and $\tau > 1$ we consider the cone $C_d(S,W;c) = \{x \in M;\ d(x,W) \leq c(d(x,S))^\tau\}$. Since any two Riemannian matrics on M are equivalent it follows that they are cone-wise equivalent : if d' is another distance function there exist $c', c'' > 0$ such that $C_{d'}(S,W;c') \subset C_d(S,W;c) \subset C_{d'}(S,W;c'')$. The set $C_d(S,W;c)$ contains a neighbourhood of W-S and it is "tangent to W at S" in the following sense : if $\gamma:(-\varepsilon,\varepsilon) \to M$ is C^1 curve such that $\alpha(0) \in S$ and $\alpha(t) \in C_d(S,W;c)$ for values of t arbitrarily near 0, then $\alpha'(0) \in T_{\alpha(0)}W$.

Lemma 2.2. Let d_i, $i = 1,2$, be distance functions on M induced by Riemannian metrics $\langle\ ,\ \rangle^i$. Let $S \subset W$ be closed submanifolds of M such that S has codimension one in W. If $x \in S$ there exists a positive real number α such that for any sequence $x_n \in C_{d_1}(S,W;c)-S$ converging to x we have that $\dfrac{d_1(x_n,S)}{d_2(x_n,S)}$ converges to α.

Proof. Since it is a local problem we may assume that $M = \mathbb{R}^m$, S and W are subspaces of $\mathbb{R}^m$ and x is the origin. Let $\exp_y^i$ be the exponential mapping induced by the Riemannian metric $\langle\ ,\ \rangle^i$ at y. We have $\exp_y^i(v) = y + v + \varphi_i(y,v)$, where $\dfrac{\varphi_i(y,v)}{\|v\|}$ tends to zero as $\|v\| \to 0$. Here $\|\ \|$ is any norm in $\mathbb{R}^m$. For z near 0 there is a unique y_i near 0 such that $d_i(z,S) = d_i(z,y_i)$. Hence $v_i = (\exp_{y_i})^{-1}(z)$ is orthogonal to S with respect to the inner product $\langle\ ,\ \rangle^i_{y_i}$ and $d_i(z,S) = \|v_i\|^i_{y_i} = (\langle v_i, v_i\rangle^i_{y_i})^{1/2}$. Thus

$$z = y_1 + d_1(z,S)\frac{v_1}{\|v_1\|^1_{y_1}} + \varphi_1(y_1,v_2) = y_2 + d_2(z,S)\frac{v_2}{\|v_2\|^2_{y_2}} + \varphi_2(y_2,v_2)\,.$$

Let $e_i \in W$, $i = 1,2$, be such that $\langle e_i, e_i\rangle^i_0 = 1$, e_i is orthogonal to S with respect to the inner product $\langle\ ,\ \rangle^i_0$ and $\alpha = \langle e_1, e_2\rangle^1_0$ is positive. Since $\langle e_1, y_j\rangle^1_0 = 0$ for $j = 1,2$ we have

$$d_1(z,S)\,\langle e_1, \frac{v_1}{\|v_1\|^1_{y_1}}\rangle^1_0 + \langle e_1, \varphi_1(y_1,v_1)\rangle^1_0 =$$

$$= d_2(z,S)\,\langle e_1, \frac{v_2}{\|v_2\|^2_{y_2}}\rangle^1_0 + \langle e_1, \varphi_2(y_2,v_2)\rangle^1_0 .$$

Since

$$\frac{v_1}{\|v_1\|^1_{y_1}} \to \pm e_1, \qquad \frac{v_2}{\|v_2\|^2_{y_2}} \to \pm e_2, \qquad \frac{\varphi_i(y_i,v_i)}{\|v_i\|^i_{y_i}} \to 0$$

and $\frac{d_1(z,S)}{d_2(z,S)}$ is bounded it follows that $\frac{d_1(z,S)}{d_2(z,S)}$ converges to $\alpha = \langle e_1, e_2\rangle^1_0$. This finishes the proof of the lemma.

Proposition (2.3). Take a small compact neighbourhood V of r, and some centre-unstable manifolds $W^{cu}(p)$ and $\tilde{W}^{cu}(\tilde{p})$. Let d be a distance function on M induced by a Riemannian metric.

(1) For c sufficiently big and $\tau>1$ sufficiently near 1 the set $h(W^{cu}(p)) \cap V$ is contained in the cone $D^{cu}(\tilde{p}) = \{x \mid d(x, W^{cu}(\tilde{p})) \leq c.(d(x, W^{u}(\tilde{p})))^{\tau}\}$.

(2) Let $D^{cu}(p)$ and $D^{cs}(q)$ be cones similar to $D^{cu}(\tilde{p})$. Consider sequences $r^1_n, r^2_n \in D^{cu}(p)$, $r^1_n, r^2_n \to r$ with $\Phi^{-k_n}(r^{(i)}_n) \to s^{(i)}$, $s^{(i)} \in W^s(p) - \{p\}$. We claim that $s^{(i)} \in W^s(p)\backslash W^{ss}(p)$ and s^1, s^2 are both contained in the same leaf of $F^{ss}(p)$ if and only if

$$d(r^1_n, W^u(p)) \cong d(r^2_n, W^u(p)) .$$

For such a sequence r_n one has :

$$d(r_n, W^u(p)) \cong \sigma_p.(a)^{k_n}.|\pi_1(s)| ;$$

where $\pi_1(s)$ is the first coordinate of s (assuming that s is contained in the linearising neighbourhood near p as above) and $\sigma_p > 0$ is some "transition constant", depending only on d and r.

Proof. Take a full fundamental neighbourhood S of $W^s(p)$. Since we have that $h(W^{ss}(p)) = W^{ss}(\tilde{p})$, see [N.P.T.], we certainly have that $h(W^{cu}(p)) \cap W^{ss}(\tilde{p}) = \tilde{p}$. Therefore by compactness $h(W^{cu}(p)) \cap S$ is contained in the set $D^{cu}(\tilde{p}) = \{x : d(x, W^{cu}(\tilde{p})) \leq C_1.(d(x, W^u(\tilde{p})))^{\tau}\}$ for C_1 big. Since the metrics are cone-wise equivalent, we may assume that the metric defining the distance d is, in a neighbourhood of $\tilde{p}$, induced from the usual metric of R^n by the coordinate system linearising $\tilde{\Phi}$. Since the eigenvalue $\tilde{a}$ is strictly weaker than other contracting eigenvalues at $\tilde{p}$, $D^{cu}(\tilde{p})$ is positively invariant in a small neighbourhood U of $\tilde{p}$, provided $\tau > 1$ is sufficiently near 1. Therefore $h(W^{cu}(p)) \cap U \subset D^{cu}(\tilde{p})$. See figure 2. It is now easy to translate such a relation to any point in the orbit of r (taking a finite number of iterates by Φ and $\tilde{\Phi}$). This concludes part (1) of the proposition.

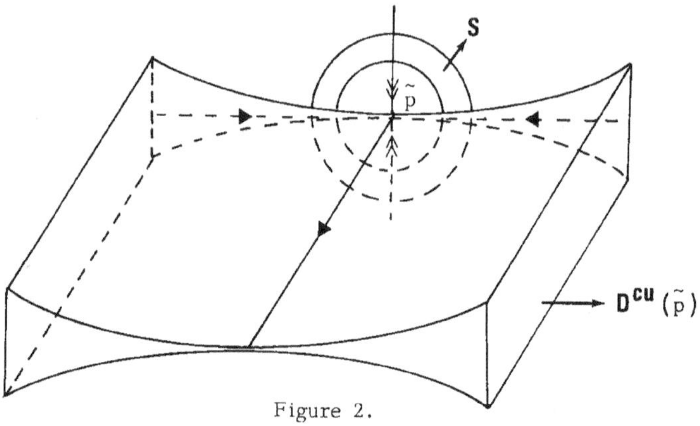

Figure 2.

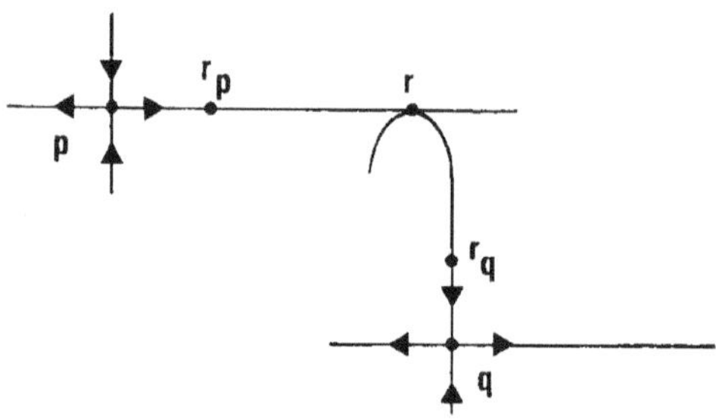

Figure 3.

For some $k, \ell \in \mathbb{N}$ we have that $r_p = \Phi^{-k}(r)$ and $r_q = \Phi^{\ell}(r)$ are contained in the linearising boxes near p, respectively q. In the linearising coordinates $(y_1, \ldots, y_n)$ let $\pi_1(y_1, \ldots, y_n) = y_1$, where the y_1-axis is the weakest contracting direction. Since $W^u(p)$ is a codimension one submanifold of $W^{cu}(p)$, it follows from Lemma (2.2) that, for any sequence $r_n \in D^{cu}(p)$, $r_n \to r$,

$$d(r_n, W^u(p)) \cong \sigma'_p |\pi_1(\Phi^{-k}(r_n))|$$

for some constant $\sigma'_p > 0$ independent of the sequence $\{r_n\}$. But since $(y_1,\ldots,y_n)$ are linearising coordinates

$$\pi_1(\Phi^{-k_n}(r_n)) = (a)^{k-k_n}.\pi_1(\Phi^{-k}(r_n)) .$$

Since $\Phi^{-k_n}(r_n) \to s$ this implies

$$d(r_n, W^u(p)) \cong \sigma_p.\pi_1(s).(a)^{k_n} .$$

Thus Proposition (2.3) is proved.

In the next proposition we prove that h preserves certain relations between the distance of points to $W^u(p)$ and $W^s(q)$. For this we need to use the assumption that $W^u(p)$ and $W^s(q)$ have a quasi-transversal intersection. This notion is intrinsically defined in [N.P.T.] and implies that there exist local coordinates $(x_1,\ldots,x_n)$ near r such that :

$$W^s(g) = \{x_1 = \ldots = x_{n-s} = 0\}$$

$$W^u(p) = \{x_{u+2} = \ldots = x_n = 0, \quad x_1 = Q(x_{n-s+1},\ldots,x_{u+1})\}$$

and $n-s+1 \leq u+2$.

Here Q is a non-degenerate quadratic function and if $n-s+1 = u+2$ one should read "$x_1 = 0$" instead of $x_1 = Q(\ldots)$". Furthermore, by assumption, $W^{cu}(p)$ is transversal to $W^s(q)$ and $W^{cs}(q)$ is transversal to $W^u(p)$. Therefore we can write

$$T_rW^s(q) = \langle\frac{\partial}{\partial x_{n-s+1}}, \ldots, \frac{\partial}{\partial x_n}\rangle ,$$

$$T_rW^u(p) = \langle\frac{\partial}{\partial x_2}, \ldots, \frac{\partial}{\partial x_{u+1}}\rangle$$

meaning that the subspaces on the left are generated by the vectors on the right hand side. Since $n-s+1 \leq u+2$ we have that, by the transversality assumptions, there exists a vector v at r such that

$$T_rW^{cs}(q) = \langle v,\frac{\partial}{\partial x_{n-s+1}}, \ldots, \frac{\partial}{\partial x_n}\rangle$$

$$T_r W^{cu}(p) = \langle v, \frac{\partial}{\partial x_2}, \ldots, \frac{\partial}{\partial x_{u+1}} \rangle .$$

Using this we will prove the next proposition. This proposition originates from [Str].

Proposition (2.4). (1) If the non-degenerate quadratic function Q has 0 as a saddle-point or if $n-s+1 = u+2$ then there exist distance functions $d, \tilde{d}$ on M and for each small $\alpha > 0$ a point $x^{\alpha} \in W^{cu}(p) \cap W^{cs}(q)$ such that

$$\alpha = d(x^{\alpha}, W^{u}(p)) \cong d(x^{\alpha}, W^{s}(q)) ,$$

and $\tilde{d}(h(x^{\alpha}), W^{u}(\tilde{p})) \cong \tilde{d}(h(x^{\alpha}), W^{s}(\tilde{q}))$.

(2) If the function Q has a maximum or a minimum (and $n-s+1 \le u+1$) then there exist a point $x_i^{\alpha} \in W^{cu}(p) \cap W^{cs}(q)$, for $i = 1, 2$, such that

$$\alpha = d(x^{\alpha}, W^{u}(p)) \cong d(x^{\alpha}, W^{s}(q)) ,$$
$$d(h(x_1^{\alpha}), W^{u}(p)) \le d(h(x_1^{\alpha}), W^{s}(q))$$

and

$$d(h(x_2^{\alpha}), W^{u}(p)) \ge d(h(x_2^{\alpha}), W^{s}(q)) .$$

Proof. As anywhere else in this section we have assumed that Φ and $\tilde{\Phi}$ are conjugate by the homeomorphism h. Now the index of the quadratic function Q determines the intersection pattern of $W^{s}(q)$ and $W^{u}(p)$. Therefore Q and $\tilde{Q}$ have the same index.

(1) First assume that $n-s+1 = u+2$. Let $E = W^{cu}(p) \cap W^{cs}(q)$ and $\tilde{E} = D^{cu}(\tilde{p}) \cap D^{cs}(\tilde{q})$. These sets contain one-dimensional curves, since $T_r W^{cu}(p) \cap T_r W^{cs}(q) = \langle v \rangle$. We may choose a Riemannian metric for which the vector v is orthogonal to $T_r W^{u}(p)$ and $T_r W^{s}(q)$. If d denotes the induced distance function, then $d(x_n, W^{s}(q)) \cong d(x_n, W^{u}(q))$ for any sequence $x_n \in E$ converging to r. Similarly we can choose a distance function $\tilde{d}$ so that $\tilde{d}(y_n, W^{s}(\tilde{q})) \cong \tilde{d}(y_n, W^{s}(\tilde{p}))$ for any sequence $y_n \in \tilde{E}$ converging to $\tilde{r}$. By Proposition (2.3) one has $h(E) \subset \tilde{E}$, which proves the statement in this case.

Now assume that $n-s+1 \le u+1$ and that Q and $\tilde{Q}$ have saddle points. Consider the set

$$E = (W^{cu}(p) \cap W^{cs}(q)) \cap \{Q(x_{n-s+1}, \ldots, x_{u+1}) = 0\}$$

and choose a distance function d such that $d(x, W^s(q)) \cong d(x, W^u(p))$ for $x \in E$. Let $E_\alpha = E \cap \{d(x, W^u(p)) = \alpha\}$. Since $T_r W^{cu}(p) \cap T_r W^{cs}(q)$ is equal to $\langle v, \frac{\partial}{\partial x_{n-s+1}}, \ldots, \frac{\partial}{\partial x_{u+1}} \rangle$, E_α is not empty for α small and this set is homeomorphic to a cone Similarly let $\tilde{E} = (D^{cu}(\tilde{p}) \cap D^{cs}(\tilde{q})) \cap \{\tilde{Q}(\tilde{x}_{n-s+1}, \ldots, \tilde{x}_{u+1}) = 0\}$ and $\tilde{d}$ be a distance function such that $\tilde{d}(x, W^s(\tilde{q})) \cong \tilde{d}(x, W^u(\tilde{q}))$ for $x \in \tilde{E}$.

The second step in the proof now is to show that, for α sufficiently small, $h(E_\alpha) \cap \tilde{E} \neq \emptyset$. From Proposition (2.3) it follows that $h(W^{cu}(p)) \subset D^{cu}(\tilde{p})$ so it suffices to show that if $h(E_\alpha)$ is contained in the set $\{\tilde{x} \mid \tilde{Q}(\tilde{x}) > 0\}$ then we get a contradiction. By assumption we had that Q has a saddle-point, i.e. we can write $\tilde{Q}(\tilde{x}_{n-s+1}, \ldots, \tilde{x}_{u+1}) = + \tilde{x}^2_{n-s+1} + \ldots + \tilde{x}^2_{\xi} - \tilde{x}^2_{\xi+1} - \ldots - \tilde{x}^2_{u+1}$ for some choice of the local coordinates $(\tilde{x}_1, \ldots, \tilde{x}_n)$. Let $V = \{\tilde{x}_1 = \ldots = \tilde{x}_{n-s} = 0, \tilde{x}_{\xi+1} = \ldots = \tilde{x}_{u+1} = \tilde{x}_{u+2} = \ldots = \tilde{x}_n = 0\}$ and $\text{proj}(\tilde{x}_1, \ldots, \tilde{x}_n) = (0, \ldots, 0, \tilde{x}_{n-s+1}, \ldots, \tilde{x}_\xi, 0, \ldots, 0)$. Now $E_0 = W^s(q) \cap W^u(p)$ and, since h is a conjugacy, $h(E_0) = W^s(\tilde{q}) \cap W^u(\tilde{p})$. Hence $\text{proj}(h(E_0))$ is a neighbourhood of 0 in V. This implies that, for α small, $\text{proj}(h(E_\alpha))$ contains a neighbourhood of 0. On the other hand, if $h(E_\alpha)$ is contained in the set $\{\tilde{x}; Q(\tilde{x}) > 0\}$ then $\text{proj}(h(E_\alpha))$ cannot contain $0 \in V$. This contradiction proves Proposition (2.4) (1). To prove the second part of the proposition we let $W = W^{cu}(p) \cap W^{cs}(q)$, $E \subset W$ be a C^1 curve through r transversal to $W^s(q)$. Let W_1 (resp. W_2) be the connected component of $W - W^u(p)$ (resp. $W - W^s(q)$) such that $W_1 \cap W^s(q)$ (resp. $W_2 \cap W^u(p)$) does not intersect a neighbourhood of r as in the figure 4.

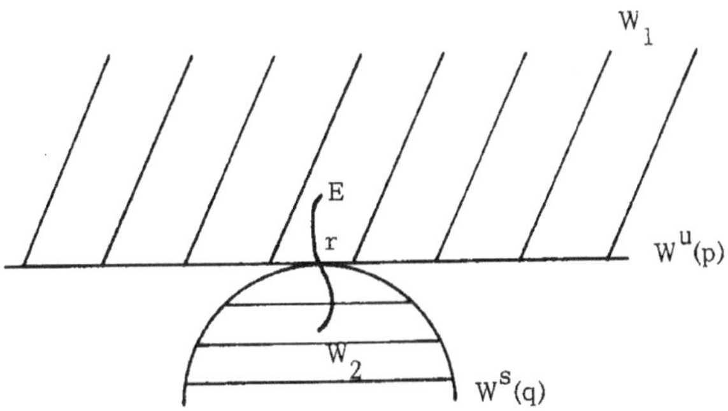

Figure 4.

Similarly we let $\tilde{W} = D^{cu}(\tilde{p}) \cap D^{cs}(\tilde{q})$ and $\tilde{W}_1$ (resp. $\tilde{W}_2$) be the connected component of $\tilde{W}-W^u(\tilde{p})$ (resp. $\tilde{W}-W^s(\tilde{q})$) such that $\tilde{W}_1 \cap W^s(\tilde{q})$ (resp. $\tilde{W}_2 \cap W^u(\tilde{p})$) does not intersect a small neighbourhood of $\tilde{r}$. Clearly $h(W_i) \subset \tilde{W}_i$ and if $\tilde{r}_n^i \to \tilde{r}$, $\tilde{r}_n^i \in \tilde{W}_i$ then

$$d(\tilde{r}_n^1, W^u(\tilde{p})) \leq d(\tilde{r}_n^1, W^s(\tilde{q}))$$
$$d(\tilde{r}_n^2, W^s(\tilde{q})) \leq d(\tilde{r}_n^2, W^u(\tilde{p})) \ .$$

Since E is transversal to $W^s(q)$ at r we have, for any sequence $r_n \in E$ converging to r,

$$d(r_n, W^s(q)) \cong d(r_n, W^u(p)) \ .$$

Choose $x_i^\alpha \in E \cap W_i$ such that $\alpha = d(x_i^\alpha, W^u(p))$. Clearly x_i^α satisfy condition (2) of Proposition (2.4). This concludes the proof of Proposition (2.4).

Conclusion of the Proof of Theorem (2.1).

We will prove the Theorem for the case that we are in the siutation of Proposition (2.4) (1). The other case is similar. Choose points s^1, s^2 in $W^s(p) - W^{ss}(p)$ in the same leaf F of $F^{ss}(p)$ and take two centre-unstable manifolds $W_1^{cu}(p)$, $W_2^{cu}(p)$ such that $s^i \in W_i^{cu}(p)$. Consider a sequence $r_n \to r$, $r_n \in W_1^{cu}(p) \cap W^{cs}(q)$, such that $\Phi^{-n}(r_n) \to s^1$. By Proposition (2.3), we have $\alpha_n = d(r_n, W^u(p)) \cong \sigma_p |a|^n \pi_1(s_1)$. For these α_n, we have from Proposition (2.4) that there are sequences $r_n^i \in W_i^{cu}(p) \cap W^{cs}(q)$ such that $\alpha_n = d(r_n^i, W^u(p)) \cong d(r_n^i, W^s(q))$ and $\tilde{d}(h(r_n^i), W^u(\tilde{p})) \cong \tilde{d}(h(r_n^i), W^s(\tilde{q}))$. But then, again by Proposition (2.3), we have $\Phi^{-n}(r_n^i) \to s^i$. Take now a subsequence k_n of intergers such that $\Phi^{k_n}(r_{k_n}^1) \to u \in W^u(q)$. By Proposition (2.3) $d(r_{k_n}^1, W^s(q)) \cong \sigma_q |b|^{-k_n} |\pi_1(u)|$. Since $d(r_{k_n}^1, W^s(q)) \cong d(r_{k_n}^1, W^u(p)) = d(r_{k_n}^2, W^u(p)) \cong d(r_{k_n}^2, W^s(q))$, we have that $d(r_{k_n}^2, W^s(q)) = \sigma_q |b|^{-k_n} |\pi_1(u)|$. But $r_k^2 \in W^{cs}(q)$ and thus by Proposition (2.3) we conclude that $\Phi^{k_n}(r_k^2)$ also converges to $u \in W^u(q)$. Thus $\tilde{\Phi}^{-n}(h(r_n^i)) \to h(s_i) \in W^s(\tilde{p})$ as well as $\tilde{\Phi}^{k_n}(h(r_{k_n}^i)) \to h(u) \in W^u(\tilde{q})$. From this and Proposition (2.3), we have that $\tilde{d}(h(r_{k_n}^1), W^s(\tilde{q})) \cong \tilde{d}(h(r_{k_n}^2), W^s(\tilde{q}))$ and from this, $\tilde{d}(h(r_{k_n}^1), W^u(\tilde{p})) \cong \tilde{d}(h(r_{k_n}^2), W^u(\tilde{p}))$. The last equivalence implies by Proposition (2.3) that $|\tilde{\pi}_1(h(s_1))| = |\tilde{\pi}_1(h(s_2))|$ and by continuity of h, $\tilde{\pi}_1(h(s_1)) = \tilde{\pi}_1(h(s_2))$. This finishes part (1) of Theorem (2.1). Let us prove part (2). Since $\frac{\log|a|}{\log|b|} \notin \mathbb{Q}$, we have as in [M] that for any given $u \in W^{cs}(q) \cap W^u(q)$ there exists a sequence of integers k_n such that $\Phi^{k_n}(r_n^1) \to u$. In fact this follows from the relation $\sigma_p|\pi_1(s_1)||a|^{k_n} \cong \sigma_q|\pi_1(u)||b|^{-k_n}$ obtained above. But then $\tilde{\sigma}_{\tilde{p}}|\tilde{\pi}_1(h(s_1))||\tilde{a}|^{k_n} \cong \tilde{\sigma}_{\tilde{p}}|\tilde{\pi}_1 h(u)||\tilde{b}|^{-k_n}$. But since $P(\Phi) = P(\tilde{\Phi})$, we can write $\tilde{a} = a^{\delta}$ and $\tilde{b} = b^{\delta}$ where $\delta = \frac{\log \tilde{a}}{\log a}$. Therefore, $\tilde{\pi}_1(h(u)) = K(\pi_1(u))^{\delta}$ for all $u \in W^{cs}(q) \cap W^u(p)$, where

$$K = \left(\frac{\sigma_q}{\sigma_p}\right)^{\delta} \frac{\tilde{\sigma}_{\tilde{p}}|\tilde{\pi}_1(h(s_1))|}{\tilde{\sigma}_{\tilde{q}}|\pi_1(s_1)|}.$$

This proves part (2) and the Remark following Theorem (2.1) for the induced map on the space of leaves of $F^{uu}(q)$. Similarly for the map induced on the space of leaves of $F^{ss}(p)$.

§2.b. Moduli due to orbits of non-transversal intersection.

Theorem (2.5). If $W^u(p)$ and $W^s(q)$ have k orbits of tangency (quasi-transversal) then there are at least k moduli.

Proof. According to the last formula of Section 2.a one has that the conjugacy h as a map on the set of leaves of $F^{ss}(p)$ determines the conjugacy on the set of leaves of $F^{uu}(q)$, for each tangency, namely $\tilde{\pi}_1(h(u)) = K_i(\pi_1(u))^{\delta}$ for $i = 1, 2, \ldots, k$. Therefore $K_i = K_1$ for all $1 \le i \le k$. This gives rise to k-1 new topological invariants.

§2.c. Moduli due to criticallity.

Suppose that Φ is as above and that z is a hyperbolic periodic point such that $W^u(z)$ intersects some leaf of $F^{ss}(p)$ non-transversally and furthermore that

(1) There exists a strictly weakest contracting eigenvalue c at z. In this section we assume c is real. In §3 we will deal with the case that c is complex.

(2) $W^u(z)$ intersects a leaf F of $F^{ss}(p)$ either transversally or quasi-transversally.

(3) $W^u(z)$ is transversal to $W^s(p)$ and $W^{cu}(z)$ is transversal to each leaf F of $F^{ss}(p)$, for which $W^u(z)$ is quasi-transversal to F.

Notice that the conditions (1), (2), (3) are open-dense conditions.

Theorem (2.6). Φ and $\tilde{\Phi}$ as above can only be conjugate if

(1) $$\frac{\log a}{\log b} = \frac{\log \tilde{a}}{\log \tilde{b}}$$

(2) $$c = \tilde{c}$$

Proof. The equation (1) is proved in [N.P.T.]. To prove equation (2) let us first look at the two-dimensional case :

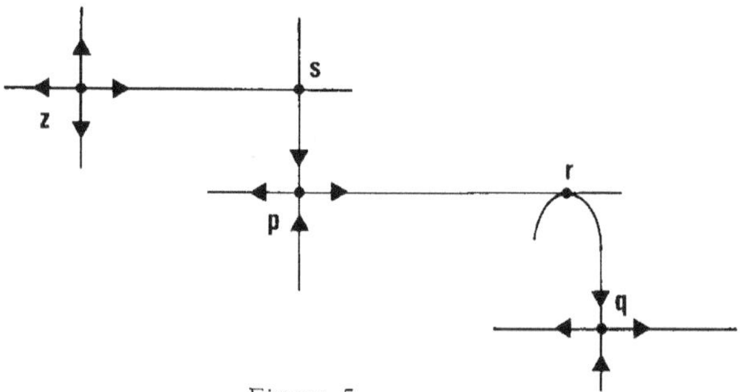

Figure 5.

Since it is proved, in Theorem (2.4) that the conjugacy $h|W^s(p)$ is differentiable near s, we have the following :

Take a sequence $x_n \in W^s(p)$, $x_n \to s$. Then for a proper choice of x_n one has that $\Phi^{-n}(x_n)$ converge to some point $t \in W^s(z)$. Therefore

$$d(x_n, s).\,|c|^{-n} \simeq \alpha$$

and $$d(h(x_n), h(s)).\,|\tilde{c}|^{-n} \simeq \beta$$

where α and β are positive real numbers.
Since $h/W^s(p)$ is differentiable near s, there exists $\gamma > 0$ such that

$$d(x_n, s) \cong \gamma d(h(x_n), h(s))$$

and therefore $|c| = |\tilde{c}|$. Clearly if Φ and $\tilde{\Phi}$ are conjugate then c and $\tilde{c}$ also have the same sign.

Now let us deal with the higher dimensional case, see figure 6, (but let $c \in \mathbb{R}$). The conjugacy preserves $W^u(z)$, the foliation $F^{ss}(p)$, the the topological type of the intersection of $W^u(z)$ and a leaf F of $F^{ss}(p)$. Therefore if s is a point where $W^u(z)$

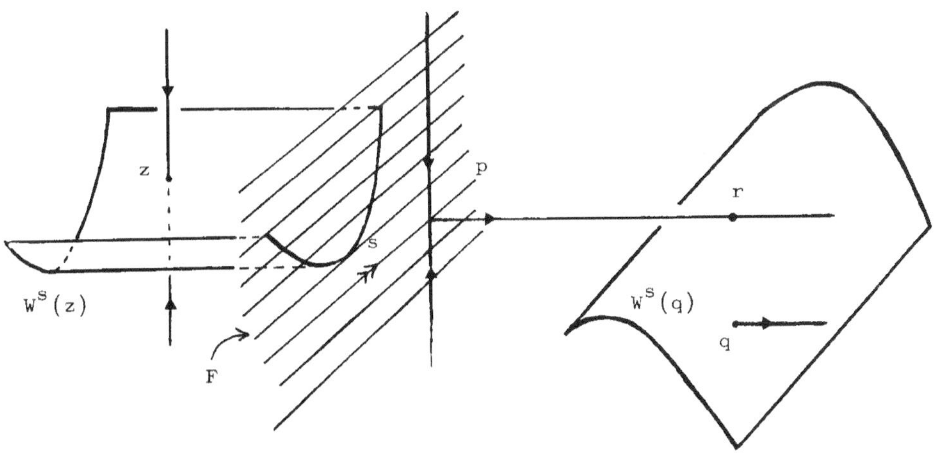

Figure 6

meets a leaf F of $F^{ss}(p)$ quasi-transversally then h(s) is also such a point. As in Proposition (2.4) we have to distinguish two cases. In the first case we have exactly as in Proposition (2.4) (1), a point $x^\alpha \in W^{cu}(z) \cap W^s(p)$ such that $d(x, W^u(z)) \cong d(x, F) = \alpha$ and $d(h(x), W^u(z)) \cong d(h(x), h(F))$, for each α sufficiently small. (We remark here that F has codimension one in $W^{cu}(z)$.) But since h is differentiable as a map on leaves of F^{ss}, outside $W^{ss}(p)$:

$$d(x, F) \cong \gamma d(h(x), h(F)), \quad \gamma > 0 .$$

Then the rest of the proof works exactly as in the two-dimensional case. If we are in the situation of Proposition (2.4) (2) we get two inequalities and, from them, we also get $c = \tilde{c}$.

§3. Moduli due to non-real eigenvalues.

§3.a. The eigenvalue a or b is non-real.

Suppose that $W^u(p)$ and $W^s(q)$ have a quasi-transversal intersection as above and that the weakest contracting eigenvalue a at p and the weakest expanding eigenvalue b at q exist. Furthermore suppose that either a or b (or both) are non-real. One may assume that $\theta = \arg(a)$, $\psi = \arg(b)$ are in the interval $[0, \pi]$.

Theorem (3.1). For conjugate diffeomorphisms Φ and $\tilde{\Phi}$ as above

(1) $$\frac{\log |a|}{\log |b|} = \frac{\log |\tilde{a}|}{\log |\tilde{b}|}$$

(2) $$\theta = \tilde{\theta} \text{ and } \psi = \tilde{\psi} .$$

Proof. Assume that $\psi \in (0, \pi)$ and take a point r where $W^u(p)$ and $W^s(q)$ intersect quasi-transversally. As before this implies that $n-s+1 \leq u+2$ and that there are local coordinates $(x_1, \ldots, x_n)$ near r such that :

$$W^s(q) = \{x_1 = \ldots = x_{n-s} = 0\} ,$$

and

$$W^u(p) = \{x_{u+2} = \ldots = x_n = 0,\ x_1 = Q(x_{n-s+1}, \ldots, x_{u+1})\}$$

where one should read $x_1 = 0$ if $n-s+1 = u+2$, and Q is quadratic. Furthermore we have assumed that $W^{cu}(p) \pitchfork W^u(q)$ and $W^{cs}(q) \pitchfork W^u(p)$. We distinguish two cases :

Case 1: $n-s+1 = u+2$.

In this case $T_r W^u(p) \cap T_r W^s(q) = 0$, and therefore for a small neighbourhood U of q,

$W^u(p) \cap U$ has countably many components which pile up at $W^s(q)$. The component of $W^u(p) \cap U$ which contains $\Phi^k(r)$ is called $W^u_k(p)$.

Assertion : If $(k_j.\psi)$ (mod 2π) has ℓ limit points in $[0,2\pi)$ then $W^u_{k_j}(p) \cap U$ accumulates to ℓ smooth codimension-one submanifolds of $W^u(q)$: $P_1,\ldots,P_\ell$. All these manifolds contain $W^{uu}(q)$.

Proof of Assertion : Take linearising coordinates $(y_1,\ldots,y_n)$ near q and let $\pi:U \to W^u(q)$ be the linear projection on $W^u(q)$ along $W^s(q)$. Since $T_rW^u(p) \cap T_rW^s(q) = 0$ and $\dim(T_rW^u(p) + T_rW^s(q)) = n-1$ one has that $\pi(W^u_k(p))$ is a smooth codimension-one submanifold of $W^u(q)$. Now in fact $\pi(W^u_k(p))$ is transversal to the two-dimensional eigenspace W^c corresponding to the weakest expanding eigenvalue b (b is non-real, so $\dim(W^c) = 2$). This last fact follows from the assumption $T_rW^u(p) \pitchfork T_rW^{cs}(q)$ and therefore $\pi(T_rW^u_k(p))$ is transversal to W^c as a subset of $W^u(q)$. But this certainly implies that $\pi(W^u_k(p))$ accumulates to the eigenspace of the strong-expanding eigenvalues $W^{uu}(q)$. Furthermore $\dim(T_rW^c \cap \pi(T_rW^u_k(p)) = 1$, and therefore the manifold $\pi(W^u_k(p))$ is rotated. From all this it follows that if $(k_i.\psi)$ (mod 2π) converges then $W^u_{k_i}(p)$ converges to a plane P in $W^u(q)$, which contains $W^{uu}(q)$. This finishes the proof of the assertion.

Now assume that k_j = integer part $(\frac{2\pi j}{\psi})$. Clearly then $W^u_{k_j}(p)$ converges to some manifold P in $W^u(q)$. If Φ and $\tilde{\Phi}$ are conjugate then $W^u_{k_j}(\tilde{p})$ must converge to the topological manifold h(P). But this is only possible if $k_j.\psi$ (mod 2π) converges to some constant. By the choice of the sequence k_j, we have that $\psi = \tilde{\psi}$.

Case 2: $n-s+1 < u+2$.

In this case $\dim(T_rW^u(p) \cap T_rW^s(q)) \geq 1$. Let N be a small cell-neighbourhood of r in U and let $W^u_k(p) = \Phi^k(W^u(p) \cap N) \cap U$. As before, assume that in U we have coordinates that linearise Φ and let $\pi:U \to W^u(q) \cap U$ be the corresponding projection along $W^s(q)$. We call a critical point of $\pi|W^u_k(p)$ a "fold" point and we say that $W^u(p)$ "folds" at $v \in W^u(q)$ if v is accumulated by "fold" (critical) points of $\pi|W^u_k(p)$, $k \in \mathbb{N}$. That is, v is a limit point of critical points of $\pi|W^u_{k_j}(p)$ for some sequence of integers $k_j \to \infty$.

Assertion: If $(k_j.\psi)$ (mod 2π) has ℓ limit points in $[0,2\pi)$ then $W^u_{k_j}(p)$ is "folded" at ℓ smooth codimension-one manifolds of $W^u(q)$:$P_1,\ldots,P_\ell$.

These "folding manifolds" are topologically characterised in the following way : For $v \in W^u(q)$ then $v \notin P_1 \cup \ldots \cup P_\ell$ if and only if there is a cell-neighbourhood V of v such that for each $v' \in W^u(q) \cap V$ there are arbitrarily small cell-neighbourhoods V' of v' in V such that for j sufficiently big $W^u_{k_j}(p) \cap V'$ and $W^u_{k_j}(p) \cap V$ are homeomorphic.

Proof of Assertion : Since $\dim(T_r W^u(p) + T_r W^s(q)) = n-1$ and $W^u(p)$ intersects $W^s(q)$ quadratically at r, one has that the set of critical values of $\pi | W^u_k(p)$ is a codimension-one submanifold of $W^u(q)$. As in Case 1 one can show that if $(\psi . k_j)$ (mod 2π) converges then the set of critical values of $\pi | W^u_{k_j}(p)$ converges to a plane P as $j \to \infty$. This plane P contains $W^{uu}(q)$ and has codimension one in $W^u(q)$. Using the normal form for the quasi-transversal intersection of $W^u(p)$ and $W^s(q)$ at r mentioned before Proposition (2.4), we get the desired topological characterisation of these "folding" manifolds.

As in Case 1, one can use the topological characterisation of the limit sets P to show that $\psi = \tilde{\psi}$.

§3.b. Criticallity in the complex case.

Now we deal with the criticallity condition. Suppose that $W^u(p)$ and $W^s(q)$ have a quasi-transversal intersection as above and that $\log|a|/|b|$ is irrational and a and b are real. Furthermore suppose that there is a hyperbolic periodic point z such that $W^u(z)$ intersects some leaf of $F^{ss}(p)$ non-transversally. We also suppose that conditions (2), (3) of §2.c are satisfied but now we assume that c is not real.
As before let $\theta = \arg(c)$ and assume that $\theta \in [0, \pi]$.

Theorem (3.2). For conjugate diffeomorphisms Φ, $\tilde{\Phi}$ as above we have :

(1) $$\frac{\log |a|}{\log |b|} = \frac{\log |\tilde{a}|}{\log |\tilde{b}|}$$

(2) $$\theta = \tilde{\theta} .$$

Proof. Let F be a leaf of $F^{ss}(p)$ such that F and $W^u(z)$ are non-transversal. By Theorem (2.1) the leaf F is mapped onto a similar leaf $\tilde{F}$ of $\tilde{F}^{ss}(p)$.

Since by assumption F is quasi-transversal to $W^u(z)$ and $W^{cu}(z)$ is transversal to F we can apply the same method as in Theorem (3.1) to the spaces $\Phi^{-k}(F)$. This will show that $\theta = \tilde{\theta}$.

§4. Conclusion of the proof of the main theorem.

Let $\Omega(\Phi)$ be finite, hyperbolic and without cycles. Suppose that for each pair of periodic orbits p and q such that $W^u(p)$ and $W^s(q)$ have a non-transversal intersection, conditions (1) to (4) of Definition (1.1) are met (i.e., let $\Phi \in \mathcal{B}$). If all stable and unstable manifolds meet transversally, then Φ is stable (it is Morse-Smale), see [P], [PS], and so Φ has modulus of stability zero. Thus if Φ has modulus of stability one, there must be a pair of periodic orbits p, q whose unstable and stable manifolds have an orbit of quasi-transversal intersection. According to §2, such a non-transversal orbit of intersection must be unique and according to §3 the relevant eigenvalues (strictly weakest contracting and expanding, resp.) must be real. Finally, again in §2, the non-criticallity condition is proved when these eigenvalues are real. This concludes the proof of our main result.

§5. The no-cycles condition.

Let $\Omega(\Phi)$ be finite and hyperbolic. Suppose that Φ had modulus of stability one.

<u>Theorem.</u> $\Omega(\Phi)$ has no cycles.

<u>Proof.</u> Suppose $\Omega(\Phi)$ has a cycle. Then we must have some orbit of non-transversal intersection of stable and unstable manifolds along the cycle for otherwise $\Omega(\Phi)$ would not be finite. In fact, we would have transversal homoclinic orbits. Now according to §2, there must be only one orbit γ of non-transversal intersection for otherwise Φ would have modulus of stability bigger than one. But this is again a contradiction because in this case $\gamma \in \Omega(\Phi)$ and γ is not periodic and thus $\Omega(\Phi)$ would not be finite.

References

[H.P.S.] M. Hirsch, C. Pugh & M. Shub : Invariant manifolds, Lecture Notes in Math., 583, Springer-Verlag, (1977).

[M] W. de Melo, Moduli of stability of two-dimensional diffeomorphisms, Topology 19, (1980), 9-21.

[M.P.] W. de Melo & J. Palis, Moduli of stability for diffeomorphisms, Conference at Northwestern Univ. on Global Theory of Dynamical Systems, Lecture Notes in Math., 819, Springer-Verlag, (1980), 318-339.

[N.P.T.] S. Newhouse, J. Palis & F. Takens, Stable families of diffeomorphisms, IMPA, preprint.

[P1] J. Palis, On Morse-Smale dynamical systems, Topology 8, (1969) 385-405.

[P2] J. Palis, A differentiable invariant of topological conjugacies and moduli of stability, Asterisque 51, (1978), 335-346.

[P.S.] J. Palis & S. Smale, Structural stability theorems, Proc. Symp. Pure Math. A.M.S. 14 (1970) 223-232.

[Ste] S. Sternberg, On the structure of local homeomorphisms of Euclidean n-spaces II, Amer. Journ. of Math. 80 (1958), 623-631.

[Str] S.J. van Strien, Saddle connections of arcs of diffeomorphisms, moduli of stability, this volume.

W. de Melo & J. Palis, Instituto de Matematica Pura e Aplicada, Rua Luiz de Camoes 68, Rio de Janeiro, R.J. Brazil.

S.J. van Strien, Mathematics Institute, University of Utrecht, Holland.

BIFURCATIONS AND STABILITY OF FAMILIES OF DIFFEOMORPHISMS

by S. NEWHOUSE, J. PALIS *and* F. TAKENS

Abstract

We consider one parameter families or arcs of diffeomorphisms. For families starting with Morse-Smale diffeomorphisms we characterize various types of (structural) stability at or near the first bifurcation point. We also give a complete description of the stable arcs of diffeomorphisms whose limit sets consist of finitely many orbits. Universal models for the local unfoldings of the bifurcating periodic orbits (especially saddle-nodes) are established, as well as several results on the global dynamical structure of the bifurcating diffeomorphisms. Moduli of stability related to saddle-connections are introduced.

CONTENTS

Chapter I. — **Introduction** 7

Chapter II. — **Local description of the elementary bifurcations** 13

1. Introduction 13
2. Center manifolds and periodic elementary bifurcations 13
3. The saddle-node 15
4. The flip 21
5. The Hopf point 22
6. Quasi-transversal intersections 23

Chapter III. — **Necessary conditions for stability of arcs** 26

1. The modulus condition (quasi-transversal intersection) 26
2. Necessary conditions for mild stability and stability 37
3. Endomorphisms of the circle 43
4. The saddle-node with 1-cycle in dimension 2 48
5. On the rigidity of the unfolding of the saddle-node 51

5

Originally published in: *Publ. Math. Inst. Hautes Études Sci. no. 57 (1983), 5–71,*

Chapter IV. — **Global stability** 55
1. Introduction 55
2. Local tubular families 55
3. Global tubular families 59
4. Stability 64

This research was carried out mainly at the Instituto de Matemática Pura e Aplicada, Rio de Janeiro. The first two authors were also supported by an agreement between CNPq (Brazil) and NSF (USA).

I. — INTRODUCTION

The study of the geometric structure of the orbits of dynamical systems (differential equations, flows, vector fields, or diffeomorphisms) defined on a manifold has been considered in many works since Poincaré and Liapunov. Two diffeomorphisms f and f' are said to have the same geometric structure if they are topologically conjugate, i.e. if there is a homeomorphism h from the domain of f to that of f' such that $hf = f'h$. Two flows or vector fields, are called topologically equivalent if there is a homeomorphism sending orbits of one system to orbits of the other; if, in addition, the homeomorphism preserves the flow parameter, we again say the systems are conjugate. In general terms, we aim at the classification of dynamical systems under conjugacy or topological equivalence. Since, however, much pathological behavior can occur, we must restrict ourselves to interesting special classes of systems. We shall be concerned here with systems having only mild recurrence; in particular, we shall frequently assume that their limit sets consist of only finitely many orbits.

The space of differentiable systems of class C^r, $r \geq 1$, has a natural topology given by uniform convergence of the first r derivatives. This is called the C^r topology. Given any equivalence relation E on the set of dynamical systems, one can define systems to be E-*stable* if they lie in the interior of their E-equivalence classes. When topological equivalence is used for E, an E-stable system is called structurally stable (or just stable). The stable diffeomorphisms and vector fields whose limit sets have finitely many orbits coincide with the Morse-Smale ones [24]. In fact, in this case the Birkhoff center [15] is finite because non-trivial recurrence implies uncountably many orbits in the limit set. Since one understands the structure of Morse-Smale systems pretty well, it is natural to consider one-parameter families of systems starting at a Morse-Smale one and to attempt to describe the structure of the elements of such families. In the present work we shall define three natural equivalence relations on these families, and we shall characterize their stable families in terms of geometric properties. In particular, we will characterize the stable one parameter families of diffeomorphisms whose elements have only finitely many orbits in their limit sets. This corresponds to the characterization of the Morse-Smale diffeomorphisms as the ones which are stable and have finitely many orbits in their limit sets. The results can, of course, be translated to certain classes of vector fields (those with global cross-sections). In [26a] related results are obtained for families of gradient vector fields.

We first present a preliminary description of our main results. Later, we shall give their precise statements.

In [19], [20], one studied how a one-parameter family of diffeomorphisms starting at a Morse-Smale one ceases to be stable (i.e. goes through a bifurcation) when the parameter evolves. For a generic family, the description given there is complete assuming that the diffeomorphism at the first bifurcation point has its limit set made of a finite number of orbits. It turns out that these orbits are all periodic except at most one. If the periodic orbits are all hyperbolic, then their stable and unstable manifolds meet transversally except along one orbit. In case one of the periodic orbits is not hyperbolic, this orbit must be an elementary bifurcation (a saddle-node, flip, or Hopf orbit), the other periodic orbits must be hyperbolic, and all stable and unstable manifolds meet transversally. The orbit structure of the diffeomorphism at the first bifurcation point will be basic for our results on the stability of these arcs.

Throughout this paper, M denotes a compact C^∞ manifold without boundary, and Diff(M) denotes the set of C^∞ diffeomorphisms of M. We let $\mathscr{P} = \mathscr{P}(M)$ be the space of C^∞ arcs of diffeomorphisms on M. That is, if I is the unit interval, then $\mathscr{P}(M)$ consists of C^∞ mappings $\Phi : M \times I \to M \times I$, such that $\Phi(m, \mu) = (\varphi_\mu(m), \mu)$ where $m \mapsto \varphi_\mu(m)$ is a C^∞ diffeomorphism for each $\mu \in I$. Elements of $\mathscr{P}$ will also be called one-parameter families or arcs of diffeomorphisms and will frequently be denoted by $\{\varphi_\mu\}$ or φ. We give Diff(M) and $\mathscr{P}$ the usual C^∞ topologies.

Let us consider three equivalence relations on the set $\mathscr{P}$ of one-parameter families of diffeomorphisms. We say two families are *topologically conjugate* if, modulo an orientation preserving homeomorphism of the interval I, each element of the first family is topologically conjugate to an element of the second family, and the conjugacy varies continuously with the parameter. If conjugacies exist but do not necessarily vary continuously, we say the families are *mildly conjugate*. Finally, if the elements of the families are topologically conjugate up to and including their first bifurcation points, we say they are *left conjugate*. The interiors of the equivalence classes of the preceding equivalence relations define, respectively, *stable*, *mildly stable*, and *left stable* arcs of diffeomorphisms.

For arcs $\{\varphi_\mu\}$ such that the limit set of φ_μ consists of finitely many orbits for each μ, we will give necessary and sufficient conditions for stability, and we will geometrically characterize left stability. For mild stability, our characterization is complete except for one case which we present as an interesting open question. In all cases, we exhibit necessary and sufficient conditions in terms of the orbit structures at the bifurcation points of the arcs. For arcs beginning at Morse-Smale diffeomorphisms we will first study the different types of stability for an interval in I containing the first bifurcation point in its interior.

Let us describe the results. To begin with, if, at the first bifurcation point, some stable and unstable manifolds meet non-transversally, then the arc is not even left stable. In fact, in this case we have at least a one-parameter family of different equivalence classes of arcs near the initial one. This corresponds to the existence of the modulus of stability as discussed in [27], [28]. In the other possible cases, when there is a saddle-node, a flip or a Hopf periodic orbit, the arc is left stable. When the arc goes through a Hopf orbit it is never mildly stable; this is due to the appearance of invariant circles with irrational rotations. On the other hand, we always get stability for arcs going through flip orbits. The case of a saddle-node orbit deserves a special discussion. First of all, it is much harder than in the flip case to prove the existence up to conjugacy of a universal model for its local unfolding. This is done in Chapter II. On the other hand, the restriction of such a conjugacy to the center manifold is surprisingly rigid. We discuss several applications of this fact in section 5 of Chapter III. Moreover, the strong stable and strong unstable foliations of the stable and unstable manifolds must be preserved by a conjugacy between two arcs going through saddle-nodes. Thus, criticallity or tangency of the invariant (stable, unstable) manifolds of other periodic orbits with respect to one of these foliations implies that the arc is not stable. Criticallity with respect to both foliations implies that the arc is not even mildly stable. Another crucial factor is the existence of cycles for the periodic orbits. When the arc goes through a saddle-node which is critical but not bicritical and has no cycles, then this arc is mildly stable but not stable. If there is a cycle, the arc is not stable and, we believe, it is not even mildly stable. We are able to prove this last statement for cycles of length bigger than one and for one-cycles when the saddle-node is normally attracting or repelling. This follows from the appearance of a non-transversal homoclinic orbit, which implies a non zero modulus of stability. To prove the existence of such a homoclinic orbit, we reduce the question to one-parameter families of endomorphisms of the circle and introduce a generalized notion of rotation number. Necessary conditions for the types of stability mentioned above are also established in Chapter III. The proof that, under these conditions, the arcs are stable, mildly or left stable is performed in the last chapter. There we use a suitable version of tubular families or foliations, some of them with singularities. Our constructions also provide a more elegant proof of the stability of Morse-Smale diffeomorphisms originally established in [24], [25]. Stability for arcs containing saddle-nodes was generalized by Robinson [31] to certain families starting at Axiom A diffeomorphisms.

Many of these results were announced in [21]; however, we mistakenly claimed to have characterized mild stability. As we mentioned above, it remains to prove that certain l-cycle cases are not mildly stable.

Let us now review some definitions and be more precise.

For $g \in \mathrm{Diff}(M)$, the *orbit* $o(x)$, of a point $x \in M$, is defined as $o(x) = \{g^n(x) \mid n \in \mathbf{Z}\}$. A point $y \in M$ is called a *limit point* of g if for some sequence $n_i \in \mathbf{Z}$ with $|n_i| \to \infty$, $\lim g^{n_i}(x) = y$. We denote by $L(g)$ the closure of the set of these limit points. A point $x \in M$ is a *periodic point* of g with period k if $g^k(x) = x$ and $g^\ell(x) \neq x$ for all $0 < \ell < k$; x is *hyperbolic* if $dg^k(x)$ has no eigenvalues on the unit circle. The stable, unstable sets or manifolds $W^s(x, g)$, $W^u(x, g)$, of a periodic point x are defined as

$$\{y \in M \mid \rho(g^n(y), g^n(x)) \to 0 \text{ for } n \to +\infty\}$$

and

$$\{y \in M \mid \rho(g^n(y), g^n(x)) \to 0 \text{ for } n \to -\infty\},$$

respectively, where ρ is a metric on M. If x is a hyperbolic periodic point of g, $W^s(x, g)$ and $W^u(x, g)$ are injectively immersed sub-manifolds of M. We say that a diffeomorphism $g \in \mathrm{Diff}(M)$ with finitely many periodic orbits has a *k-cycle* if there is a sequence of periodic orbits $\mathcal{O}(p_0), \ldots, \mathcal{O}(p_k)$ with $\mathcal{O}(p_0) = \mathcal{O}(p_k)$ and $\mathcal{O}(p_{i+1}) \subset \mathrm{closure}\,(W^u(\mathcal{O}(p_i)))$ for $0 \leq i < k$, and $\mathcal{O}(p_i) \neq \mathcal{O}(p_j)$ for $0 \leq i < j < k$, and if this sequence is maximal.

A diffeomorphism g is *Morse-Smale* if $L(g)$ is finite and hyperbolic and if all the intersections of stable and unstable manifolds are transverse. We denote the set of Morse-Smale diffeomorphisms by MS. This set is open [24] and each $g \in \mathrm{MS}$ is stable in the sense that any g' which is C^1 near g is conjugate to g [24], [25], i.e. there is a homeomorphism $h : M \to M$ so that $g'h = hg$.

For an arc $\{\varphi_\mu\} \in \mathcal{P}$ with $\varphi_0 \in \mathrm{MS}$, let $b = b(\varphi) = \inf\{\mu \in I \mid \varphi_\mu \notin \mathrm{MS}\}$. We always assume that $b(\varphi) < 1$. If $\{\varphi_\mu\}, \{\varphi'_\mu\} \in \mathcal{P}$, then we say that $(h, \{H_\mu\})$ is a *conjugacy* if $h : [0, 1] \to [0, 1]$ is a homeomorphism with $h(b(\varphi)) = b(\varphi')$, $H_\mu : M \to M$ is a conjugacy between φ_μ and $\varphi'_{h(\mu)}$ for all μ in some neighborhood of $[0, b(\varphi)]$, and H_μ depends continuously on μ. If H_μ does not necessarily depend continuously on μ, we say that $(h, \{H_\mu\})$ is a *mild conjugacy*, and if H_μ is only a conjugacy for $\mu \leq b(\varphi)$, not necessarily continuous in μ, then $(h, \{H_\mu\})$ is called a *left conjugacy*. Conjugacy, mild conjugacy and left conjugacy define equivalence relations in the set of those arcs in $\mathcal{P}$ which start in MS. An arc $\{\varphi_\mu\} \in \mathcal{P}$ is called stable, mildly stable or left stable if is an interior point of its corresponding equivalence class.

Now we come to the description of the class of arcs to which our results apply.

Definition. — $\mathcal{A} \subset \mathcal{P}$ is the subset of those $\{\varphi_\mu\} \in \mathcal{P}$ such that

1. $\varphi_0 \in \mathrm{MS}$;
2. $b = b(\varphi) = \inf\{\mu \in [0, 1] \mid \varphi_\mu \notin \mathrm{MS}\} < 1$;
3. the limit set of φ_b has finitely many orbits.

For the arcs of diffeomorphisms in $\mathcal{A}$ it is often useful to impose certain generic conditions on the first bifurcation. In order to describe them we need some more conditions.

Let x be a fixed point of a diffeomorphism $g \in \mathrm{Diff}(M)$. We call x *quasi-hyperbolic* if one of the following three conditions holds:

— $(dg)_x$ has one eigenvalue one, the other eigenvalues have norm different from 1 and there is a g-invariant curve α through x such that $g \mid \alpha$ has first but not second order contact with the identity at x; in this case, x is a saddle-node;

— $(dg)_x$ has one eigenvalue -1, the other eigenvalues have norm different from 1 and there is a g-invariant curve α through x such that $g^2 \mid \alpha$ has second but not third order contact with the identity at x; in this case, x is a flip.

— $(dg)_x$ has a pair $\lambda \neq \bar{\lambda}$ of eigenvalues on the unit circle, the other eigenvalues have norm different from 1 and there is a g-invariant surface α through x, tangent to the generalized eigenspace of the pair λ, $\bar{\lambda}$ at x and such that the 3-jet of g at x makes $g \mid \alpha$ an attractor or a repeller; in this case, x is a Hopf point.

For periodic orbits there is an analogous definition of quasi-hyperbolicity.

When $\{\varphi_\mu\} \in \mathscr{A}$, we say that $\{\varphi_\mu\}$ is *elementary* at its first bifurcation value b (or φ_b is elementary) if it fulfills one of the two following conditions: (1) all periodic points of φ_b are hyperbolic, there is one orbit of non-transversal intersections of a stable and an unstable manifold, and all other intersections of stable and unstable manifolds are transversal; or (2) there is one quasi-hyperbolic orbit, the other periodic orbits are hyperbolic and all intersections of stable and unstable manifolds are transverse.

In case x is a flip or saddle-node of φ_b we also require stable and unstable manifolds to be transversal to the strong stable and unstable manifold of x; the strong stable (resp. unstable) manifold consists of the points y such that the distance from x to $\varphi_b^i(y)$ (resp. $\varphi_b^{-i}(y)$) goes exponentially to zero (see also Chapter IV).

If $\{\varphi_\mu\} \in \mathscr{A}$, φ_b is elementary, and φ_b has a quasi-hyperbolic periodic orbit, then there are generic conditions one may impose on the dependence upon μ at these quasi-hyperbolic periodic points. Such conditions are described in Chapter II, § 3 (for the saddle-node), § 4 (for the flip) and § 5 (for the Hopf orbit). If these conditions are satisfied we say that the quasi-hyperbolic orbit *unfolds generically.*

Definition. — $\mathscr{B} \subset \mathscr{A}$ is defined to be the set of those arcs $\{\varphi_\mu\}$ in $\mathscr{A}$ for which φ_b is elementary and for which the quasi-hyperbolic periodic point of φ_b, if there is any, unfolds generically.

It can be shown [4], [19], [20], [36] that there is a residual subset $\mathscr{B}'$ in $\mathscr{P}$ such that $\operatorname{int}(\mathscr{B}) = \mathscr{A} \cap \mathscr{B}'$. We want to give necessary and sufficient conditions for arcs in $\mathscr{B}$ to be stable, mildly stable or left stable. For this we need to describe the notion of criticallity.

Let $g \in \operatorname{Diff}(M)$ and let x be a saddle-node of g. Then there is a unique foliation $\mathscr{F}^{ss}$ of $W^s(x, g)$ with smooth leaves such that the boundary of $W^s(x, g)$ is a leaf and such that g maps leaves to leaves; see [12]. $\mathscr{F}^{ss}$ is called the strong stable foliation. A corresponding foliation of $W^u(x, g)$, the strong unstable foliation, is denoted by $\mathscr{F}^{uu}$. We call x *s-critical* if there is some hyperbolic periodic point y of g such that $W^u(y, g)$ intersects some leaf of $\mathscr{F}^{ss}$ non-transversally; *u-critical* is defined similarly. Now, x is called *semi-critical* if it is either s- or u-critical, x is called *bi-critical* if it is both s- and u-critical, and x is called *non-critical* if it is not semi-critical.

Theorem. — *Let* $\{\varphi_\mu\}$ *be an arc in* $\mathscr{B}$.

1. $\{\varphi_\mu\}$ *is left stable if and only if all stable and unstable manifolds of* φ_b *intersect transversally;*
2. *if* $\{\varphi_\mu\}$ *is left stable, if the quasi-hyperbolic orbit is not a bi-critical saddle-node or a Hopf orbit and if* φ_b *has no cycles, then* $\{\varphi_\mu\}$ *is mildly stable;*

2'. *if* $\{\varphi_\mu\}$ *is mildly stable, then* $\{\varphi_\mu\}$ *is left stable, the quasi-hyperbolic orbit is not a bi-critical saddle-node or a Hopf orbit,* φ_b *has no cycles of length greater than* 1, *and* φ_b *has no non-critical* 1-*cycles;*

3. $\{\varphi_\mu\}$ *is stable if and only if* $\{\varphi_\mu\}$ *is left stable, the quasi-hyperbolic orbit of* φ_b *is not a semi-critical saddle-node or a Hopf orbit and* φ_b *has no cycles.*

This theorem, together with the remark at the end of section 2, Chapter III, and Theorem (4.4) in Chapter IV implies the following characterization of stable arcs of diffeomorphisms with limit sets consisting of finitely many orbits.

Theorem. — *Let* $\{\varphi_\mu\}$, $\mu \in [0, 1]$ *be an arc of diffeomorphisms such that the limit set of each* φ_μ *consists of finitely many orbits,* $\mu \in [0, 1]$. *Then* $\{\varphi_\mu\}$ *is stable if and only if there are only finitely many bifurcation values, say* $b_1, \ldots, b_s$, *in* $(0, 1)$ *and for each* $1 \leq i \leq s$, φ_{b_i} *has the following properties:*

— *all stable, strong stable, unstable, and strong unstable manifolds intersect transversally;*

— φ_{b_i} *has no cycles and has exactly one non-hyperbolic periodic orbit which is either a flip or a non-critical saddle-node; this non-hyperbolic periodic orbit unfolds generically.*

We acknowledge useful comments by several colleagues, especially W. de Melo, F. Przytycki, and C. Pugh.

II. — LOCAL DESCRIPTION OF THE ELEMENTARY BIFURCATIONS

1. Introduction

Let M be a smooth manifold. We shall consider smooth arcs of diffeomorphisms $\varphi_\mu : M \to M$, $\mu \in R$, i.e. such that the corresponding map $\Phi : M \times R \to M \times R$, defined by $\Phi(x, \mu) = (\varphi_\mu(x), \mu)$ is C^∞. For such an arc, there usually are points $(x, \mu) \in M \times R$ where φ_μ does not satisfy the Kupka-Smale conditions [13], [33] along the orbit of φ_μ through x. In this chapter we shall analyze the behavior of Φ near those points. Such a point is called an *elementary bifurcation* (point) of Φ (or φ_μ).

There are two types of elementary bifurcation points, namely those (x, μ) for which x is a non-hyperbolic periodic point of φ_μ, and those (x, μ) where x is a non-transversal point of intersection of stable and unstable manifolds of φ_μ (see [36]). Before we go into details we describe some facts concerning center manifolds in relation to the periodic elementary bifurcation points. Then a description of the types of periodic elementary bifurcations occurring in generic arcs is given. In subsequent sections all these types are analyzed. The saddle-node elementary bifurcation shows some exceptional and unexpected topological properties. In the final section, non-transversal intersections of stable and unstable manifolds are treated.

2. Center manifolds and periodic elementary bifurcations

Let $\{\varphi_\mu : M \to M\}$ be a smooth arc of diffeomorphisms having a periodic bifurcation at $(\bar{x}, \bar{\mu})$. We assume $\bar{x}$ to be a fixed point of $\varphi_{\bar{\mu}}$; if not we replace $\varphi_{\bar{\mu}}$ by $\varphi_{\bar{\mu}}^k$ where k is the period of $\bar{x}$. Let c be the number of eigenvalues of $(d\varphi_{\bar{\mu}})_{\bar{x}}$ of norm 1; since $(\bar{x}, \bar{\mu})$ is a bifurcation, $c \geq 1$.

From the theory of invariant manifolds [12], we conclude the existence of a " center manifold depending on μ ", namely a differentiable submanifold W^c of $M \times \mathbf{R}$ such that:

- $(\bar{x}, \bar{\mu}) \in W^c$;
- $\Phi(W^c) \cap W^c$ is open in W^c (and contains $(\bar{x}, \bar{\mu})$);
- the dimension of W^c is $c + 1$ and at each point $(x, \mu) \in W^c$, W^c is transversal with respect to $M \times \{\mu\}$;
- $d(\varphi_{\bar{\mu}} \mid W^c \cap (M \times \{\bar{\mu}\}))_{\bar{x}}$ has only eigenvalues of norm 1.

For any $k < \infty$ we may assume that W^c is C^k (but as k gets bigger, it may be necessary to take W^c smaller). However if in some neighborhood of $\bar{x}$ in $W^c_{\bar{\mu}} = W^c \cap (M \times \{\bar{\mu}\})$,

13

every orbit $\varphi^i_{\bar\mu}(x)$ tends to $\bar x$ for $i \to \pm\infty$, then W^c can be chosen so that $W^c_{\bar\mu}$ is C^∞; even in this last case, W^c can not be made C^∞.

We may consider $\Phi \mid W^c$ as a (local) arc of diffeomorphisms; the parameter being the restriction of μ to W^c. Also from the invariant manifold theory [26], [30] it follows that Φ near $(\bar x, \bar\mu)$, up to a " local conjugacy ", is completely determined by $\Phi \mid W^c$ and the normal data. The normal data consist of the numbers of eigenvalues of $(d\varphi_{\bar\mu})_{\bar x}$ with norm > 1, resp. < 1, and of the signs of the determinants of $(d\varphi_{\bar\mu})_{\bar x}$, restricted to the maximal invariant expanding, resp. contracting subspaces of $T_{\bar x}(M)$. We say that an arc $\Phi : M \times \mathbf{R} \to M \times \mathbf{R}$ is at $(\bar x, \bar\mu)$ *locally conjugate* to the arc $\bar\Phi : \bar M \times \mathbf{R} \to \bar M \times \mathbf{R}$ at (x, μ) if there is a homeomorphism (the local conjugacy) h from a neighborhood of $(\bar x, \bar\mu)$ to a neighborhood of (x, μ) such that

— $h \circ \Phi = \bar\Phi \circ h$ wherever defined;
— there is a local homeomorphism $h_{\mathbf{R}} : \mathbf{R} \to \mathbf{R}$, defined in a neighborhood of $\bar\mu$ such that the μ-coordinate of $h(x, \mu)$ equals $h_{\mathbf{R}}(\mu)$ wherever $h(x, \mu)$ is defined.

So in order to analyze Φ up to local conjugacy at $(\bar x, \bar\mu)$ it suffices to analyze $\Phi \mid W^c$ up to local conjugacy; this will be done in the following sections.

It should be pointed out that for Φ as above, one can choose invariant manifolds W^{cs} and W^{cu}, the center-stable and the center-unstable manifolds, for Φ, containing W^c such that the tangent space at $(\bar x, \bar\mu)$ is the direct sum of the tangent space of W^c at $(\bar x, \bar\mu)$ and the maximal invariant subspace of $T_{\bar x}(M)$ on which $(d\varphi_{\mu})_{\bar x}$ has only eigenvalues with norm smaller, resp. bigger, than 1. These invariant manifolds are in general not unique. In W^{cs} and W^{cu} one can choose invariant continuous foliations $\mathscr{F}^{ss}$ and $\mathscr{F}^{uu}$, the strong stable and the strong unstable foliations, such that the leaves are C^1, the tangent planes of the leaves depend continuously on the base-point in W^{cs}, resp. W^{cu}, and such that each leaf intersects W^c transversally (in W^{cs}, resp. W^{cu}) in one point. Also these foliations are not unique. For details on invariant manifolds and foliations see [12] and also Chapter IV of this paper. If the leaves of $\mathscr{F}^{ss}$, or $\mathscr{F}^{uu}$, have co-dimension one in $W^{cs}_\mu = W^{cs} \cap \{\mu\}$, or $W^{cu}_\mu = W^{cu} \cap \{\mu\}$, then, by [11], the foliation $\mathscr{F}^{ss}$, resp. $\mathscr{F}^{uu}$, is C^1.

Let Φ and $\bar\Phi$ be two arcs of diffeomorphisms with center manifolds W^c and $\bar W^c$, let $h : W^c \to \bar W^c$ be a local conjugacy between $\Phi \mid W^c$ and $\bar\Phi \mid \bar W^c$, and let the normal data of Φ and $\bar\Phi$ be equal. Choose invariant manifolds and foliations W^{cs}, W^{cu}, $\mathscr{F}^{ss}$ and $\mathscr{F}^{uu}$ for Φ, and $\bar W^{cs}$, $\bar W^{cu}$, $\bar{\mathscr{F}}^{ss}$ and $\bar{\mathscr{F}}^{uu}$ for $\bar\Phi$. From [26] it follows that there is an extension H of h to a local conjugacy between Φ and $\bar\Phi$ so that $H(W^{cs}) = \bar W^{cs}$, $H(W^{cu}) = \bar W^{cu}$. $H(\mathscr{F}^{ss}) = \bar{\mathscr{F}}^{ss}$ and $H(\mathscr{F}^{uu}) = \bar{\mathscr{F}}^{uu}$.

In the case of generic arcs Φ the only periodic bifurcations are those of the following three types (recall we assume $\bar x$ fixed):

1. $c = 1$, $(d\varphi_{\bar\mu})_{\bar x}$ has an eigenvalue 1 and the 2-jet of $\varphi_{\bar\mu} \mid W^c_{\bar\mu}$ at $\bar x$ is different from the 2-jet of the identity; in this case $(\bar x, \bar\mu)$ is called a saddle-node of Φ;

2. $c = 1$, $(d\varphi_{\bar{\mu}})_{\bar{x}}$ has an eigenvalue -1 and the 3-jet of $\varphi^2_{\bar{\mu}} \mid W^c_{\bar{\mu}}$ at $\bar{x}$ is different from the 3-jet of the identity; in this case we call $(\bar{x}, \bar{\mu})$ a flip of Φ;
3. $c = 2$, $(d\varphi_{\bar{\mu}})_{\bar{x}}$ has a pair of non-real complex conjugate eigenvalues on the unit circle and the 3-jet of $\varphi_{\bar{\mu}} \mid W^c_{\bar{\mu}}$ makes $\bar{x}$ an attractor or a repeller, we call these points Hopf points.

A periodic bifurcation will be called *elementary* if it is of type 1, 2, or 3 above and it unfolds generically with μ. This last condition will be explained in sections 3, 4, and 5.

Before we consider these cases in more detail, we make some more general remarks:

— in all these cases $W^c_{\bar{\mu}}$ is C^∞ (because near $\bar{x}$, each orbit $\varphi^i_{\bar{\mu}}(x)$ tends to $\bar{x}$ as $i \to +\infty$ or as $i \to -\infty$), but generically it is not possible to choose W^c to be C^∞ (we shall not use or prove this fact but see [37]).

This results described in this section, except those concerning saddle-nodes are all more or less well-known. Apart form the references in the various sections, one may consult [1], [4], [19], [20], [35], [36].

3. The saddle-node

First we consider arcs of diffeomorphisms $\{\varphi_\mu\}$ of $\mathbf{R}$ (with coordinate x) such that $\varphi_0(0) = 0$, $\frac{d}{dx}\varphi_0(0) = 1$, $\frac{d^2}{dx^2}\varphi_0(x) \neq 0$ and $\frac{d}{d\mu}\varphi_\mu(0) \neq 0$ (to simplify notation we took $\bar{\mu} = 0$ and $\bar{x} = 0$). Note that if $\bar{\Phi}$ is a generic arc of diffeomorphisms, then $\bar{\Phi}$, restricted to a center manifold of a saddle-node, has the above form. Without loss of generality, we shall assume that $\frac{d^2}{dx^2}\varphi_0(x) > 0$ and $\frac{d}{d\mu}\varphi_\mu(0) > 0$. We shall also assume that $\varphi_\mu(x)$, as a function of x and μ, is C^6 and that $\varphi_0(x)$ is a C^∞ function of x. Arcs of diffeomorphisms $\{\varphi_\mu\}$ satisfying the above conditions will be called (in this section) saddle-node arcs. We shall prove the following results.

Theorem (3.1). — *Any two saddle-node arcs are locally conjugate near* $(0, 0)$. *Moreover the conjugacy can be chosen to be continuously differentiable off the fixed point set.*

We recall that if $\{\varphi_\mu\}$ is a saddle-node arc, there is a unique C^∞ vector field X, defined on a neighborhood of 0 in $\mathbf{R}$, such that the time 1 map X_1 of X equals φ_0 [39].

Theorem (3.2). — *Let* $\{\varphi_\mu\}$ *and* $\{\bar{\varphi}_\mu\}$ *be two saddle-node arcs with corresponding vector fields* X *and* $\bar{X}$, i.e. *such that* $X_1 = \varphi_0$ *and* $\bar{X}_1 = \bar{\varphi}_0$. *Let h be a local conjugacy between* $\{\varphi_\mu\}$ *and* $\{\bar{\varphi}_\mu\}$. *Then* $h_- = h \mid \{x < 0, \mu = 0\}$ *and* $h_+ = h \mid \{x > 0, \mu = 0\}$ *are* C^∞ *and* $(h_\pm)_* X = \bar{X} \mid \{\pm x > 0\}$.

Remark (3.3). — The above theorem implies that the choice of the conjugacy along $\{\mu = 0\}$ is extremely restricted. Instead of the usual freedom to fix the conjugacy arbitrarily on a fundamental domain, we are here only free to fix h in two points: one in $\{x<0\}$ and one in $\{x>0\}$.

Let $\{\varphi_\mu\}$ be an arc of diffeomorphisms of an n-dimensional manifold M having a saddle-node fixed point at $(\bar{x}, \bar{\mu})$. Let $\Phi(x, \mu) = (\varphi_\mu(x), \mu)$. We say that $(\bar{x}, \bar{\mu})$ *unfolds generically* if for some (or any) center manifold W^c at $(\bar{x}, \bar{\mu})$, $\Phi \mid W^c$ is a saddle-node arc. Similarly, one can define generic unfolding of periodic saddle-nodes.

Theorem **(3.4)**. — *Let Φ be a smooth arc of diffeomorphisms of* M, *and let $(\bar{x}, \bar{\mu})$ be a saddle-node of Φ which unfolds generically. Let* $W^s = \{(x, \bar{\mu}) \mid \Phi^i(x, \bar{\mu}) \to (\bar{x}, \bar{\mu})$ *as* $i \to \infty\}$ (note that $W^s \subset W^{cs}$, that $\dim(W^s) = \dim(W^{cs}) - 1$ and that W^s has a boundary containing $(\bar{x}, \bar{\mu})$). *Then the strong stable foliation $\mathscr{F}^{ss}$, restricted to W^s, is unique and is preserved under any conjugacy.*

The proofs of these theorems occupy the rest of this section. As part of the proof of Theorem (3.1), we need to prove the corresponding theorem for vector fields.

Consider vector fields $X = X(x, \mu) \dfrac{\partial}{\partial x}$ on $\mathbf{R}^2$ with $X(0, 0) = 0$, $\dfrac{\partial}{\partial x} X(0, 0) = 0$, $\dfrac{\partial^2}{\partial x^2} X(0, 0) > 0$, $\dfrac{\partial}{\partial \mu} X(0, 0) > 0$ and which are at least C^2. These vector fields are called saddle-node fields. A saddle-node field can of course be considered as a one-parameter family of vector fields on $\mathbf{R}$; its time one map is a saddle-node arc (except for the differentiability); see the beginning of this section. Two saddle-node fields X and $\bar{X}$ are called *locally conjugate* if there is a homeomorphism h (compare section 2) from a neighborhood of $(0, 0)$ in $\mathbf{R}^2$ to another such neighborhood such that

— $h \circ X_t = \bar{X}_t \circ h$, whenever defined ($X_t$ stands for the time t map of the vector field X);

— there is a local homeomorphism $h_{\mathbf{R}} : \mathbf{R} \to \mathbf{R}$ such that the μ-coordinate of $h(x, \mu)$ equals $h_{\mathbf{R}}(\mu)$ whenever defined.

First we shall prove:

Theorem **(3.5)**. — *Any two saddle-node fields are locally conjugate by a conjugacy which is continuously differentiable with respect to x, in the complement of the singular set.*

Proof of Theorem (3.5). — Let X and $\bar{X}$ be two saddle-node fields.

For $X = X(x, \mu) \dfrac{\partial}{\partial x}$ we choose a box $U = \{(x, \mu) \mid |x| \le a, 0 \le \mu \le \varepsilon\}$ such that $X(x, \mu)$ and $\dfrac{\partial}{\partial \mu} X(x, \mu)$ are positive on $U \backslash (0, 0)$. Define $f : (0, \varepsilon] \to \mathbf{R}_+$ by $X_{f(\delta)}(-a, \delta) = (+a, \delta)$, or, equivalently, $f(\delta) = \int_{-a}^{+a} (X(x, \delta))^{-1} dx$. From this last formula it is clear that $\lim_{\delta \to 0} f(\delta) = +\infty$ and that the derivative $f'(\delta)$ is less than zero for $\delta \in (0, \varepsilon]$.

Let $\bar{U}$, $\bar{f}$, $\bar{a}$, $\bar{\varepsilon}$ be defined analogously for $\bar{X}$. Pick $0 < \alpha < \varepsilon$ so that

$f(\alpha) > \max(f(\varepsilon), \bar{f}(\bar{\varepsilon}))$ and take $0 < \bar{\alpha} < \bar{\varepsilon}$ so that $\bar{f}(\bar{\alpha}) = f(\alpha)$. Then there is a homeomorphism $\eta : [0, \alpha] \to [0, \bar{\alpha}]$ such that $f(\delta) = \bar{f}(\eta(\delta))$ for $0 < \delta \leq \alpha$. A conjugacy h from X to $\bar{X}$, restricted to $\{0 \leq \mu \leq \alpha\}$ is now determined by putting

$$h(-a, \mu) = (-\bar{a}, \eta(\mu));$$
$$h(a, \mu) = (\bar{a}, \eta(\mu));$$
$$h(0, 0) = (0, 0);$$

and requiring that $hX_t = \bar{X}_t h$. The continuity of h is automatic except at $(0, 0)$. We now prove continuity at $(0, 0)$. For any sequence (x_i, μ_i) in U, there are sequences s_i, t such that

$$X_{s_i}(a, \mu_i) = x_i \quad \text{and} \quad X_{t_i}(-a, \mu_i) = x_i.$$

$(x_i, \mu_i) \to (0, 0)$ if and only if $\mu_i \to 0$, $s_i \to -\infty$, and $t_i \to +\infty$. Hence h maps sequences converging to $(0, 0)$ to sequences converging to $(0, 0)$ which proves continuity.

The construction of $h \mid \{\mu < 0\}$ is easy. For example, take $h(-a, \mu) = (-\bar{a}, \mu)$, $h(0, \mu) = (0, \mu)$ and $h(a, \mu) = (\bar{a}, \mu)$ and extend with $hX_t = \bar{X}_t h$. On the complement of the set of singularities of X, $\frac{\partial h}{\partial x}$ clearly exists and equals $\bar{X}(h(x, \mu)) . (X(x, \mu))^{-1}$ which is continuous. This completes the proof of the theorem.

Let $\{\varphi_\mu\}$ be a saddle-node arc, and let X be the C^∞ vector field defined near 0 in $\mathbf{R}$ such that $X_1 = \varphi_0$. It would be useful if there were a C^∞ field $\hat{X}$ extending X to a neighborhood of $(0, 0)$ in $\mathbf{R}^2$ of the form $\hat{X}(x, \mu) = \tilde{X}(x, \mu) \frac{\partial}{\partial x}$ such that $\hat{X}(\cdot, \mu)_1 = \varphi_\mu(\cdot)$. Although we cannot find such an $\hat{X}$, lemma (3.6) provides us with a suitable substitute.

We say that a C^4 saddle-node field X is *adapted* to a saddle-node arc $\{\varphi_\mu\}$ if the function $g(x, \mu)$, defined by $(\varphi_\mu(x) + g(x, \mu), \mu) = X_1(x, \mu)$, vanishes along the x-axis and has at $(0, 0)$ its 4-jet equal to zero.

Lemma **(3.6).** — *For each saddle-node arc* $\{\varphi_\mu\}$ *there is an adapted saddle-node field* X.

Proof. — φ_0 is C^∞ and has, at $x = 0$, only a finite order of contact with the identity, so by [39], φ_0 embeds in a unique C^∞-vector field $\tilde{X} = \tilde{X}(x) \frac{\partial}{\partial x}$. In general [14], [40] if $\Psi : (\mathbf{R}^n, 0) \to (\mathbf{R}^n, 0)$ is a C^∞ diffeomorphism with all eigenvalues of $(d\Psi)_0$ equal to 1, there is a unique k-jet of a vector field $[Y]_k$, $k \geq 1$, such that any representative $Y \in [Y]_k$ satisfies

— the k-jets of Y_1 and Ψ agree at 0;
— the eigenvalues of dY at 0 are all equal to zero.

If we apply this to $\Phi : (x, \mu) \to (\varphi_\mu(x), \mu)$ at $(0, 0)$ we see that the corresponding 4-jet $[X]$ at $(0, 0)$ is uniquely determined and has a representative of the form $\hat{X}(x, \mu) \frac{\partial}{\partial x}$.

The restriction of this 4-jet to the x-axis equals the 4-jet of the previously constructed $\tilde{X}$ because of uniqueness of that 4-jet (which follows if we apply the above general statement to φ_0 at o). X is now obtained by taking $X = (\hat{X}(x, \mu) + \tilde{X}(x) - \hat{X}(x, 0)) \frac{\partial}{\partial x}$. The proof of (3.6) is complete.

The next lemma compares high iterates of a saddle-node arc with high iterates of the time one map of an adapted saddle-node field.

Lemma (3.7). — *Let* $\{\varphi_\mu\}$ *be a saddle-node arc and* $X = X(x, \mu) \frac{\partial}{\partial x}$ *an adapted saddle-node field which is at least* C^5. *Let* $U = \{(x, \mu) \mid 0 \leq \mu \leq \bar{\mu}, -a \leq x \leq a\}$ *be so that for* $(x, \mu) \in U \backslash (0, 0)$, $X(x, \mu) > 0$, $\varphi_\mu(x) > x$, $(\varphi_\mu(-a), \mu) \in U$ *and* $|g(x, \mu)| < \varphi_\mu(x) - x$. (Here $g(x, \mu)$ is the function defined just before Lemma (3.6). The fact that the last condition is satisfied for a and μ small enough follows from $|g(x, \mu)| = o(|(x, \mu)|^4)$ and $|\varphi_\mu(x) - x)| \geq k(\mu + x^2)$ on $\{\mu \geq 0\}$ for some constant k.)

Then there are constants C_1, C_2 *such that for any* $(x, \mu) \in U$, $i \in \mathbf{N}$, *with* $(\varphi^i_\mu(x), \mu) \in U$,

(i) $X_\alpha(x, \mu) = (\varphi^i_\mu(x), \mu)$ *for some* $\alpha \in \mathbf{R}$ *with* $|i - \alpha| \leq \mu . C_1$;

(ii) $|\log((d(\varphi^i_\mu)_x) X(x, \mu)) - \log(X(\varphi^i_\mu(x), \mu))| \leq \mu . C_2$.

Proof. — The above two estimates on (high) iterates of φ_μ are implied by the following two: there are constants K_1, K_2 such that for any $(x, \mu) \in U$ with $(\varphi_\mu(x), \mu) \in U$

(i)′ $X_\alpha(x, \mu) = (\varphi_\mu(x), \mu)$ for some α with

$$|1 - \alpha| \leq \mu . (\varphi_\mu(x) - x) . K_1;$$

(ii)′ $|\log((d\varphi_\mu)_x . X(x, \mu)) - \log(X(\varphi_\mu(x), \mu))| \leq \mu . (\varphi_\mu(x) - x) . K_2$.

Of course in (ii)′ we have to assume that $(x, \mu) \neq 0$, but since for $\mu = 0$ the whole lemma is trivial we shall assume, in what follows, that always $\mu > 0$. To show that (ii)′ really implies (ii) it suffices to observe that

$$\frac{d(\varphi^i_\mu)_x . (X(x, \mu))}{X(\varphi^i_\mu(x), \mu)} = \prod_{j=0}^{i-1} \frac{d(\varphi_\mu)_{\varphi^j_\mu(x)}(X(\varphi^j_\mu(x), \mu))}{X(\varphi^{j+1}_\mu(x), \mu)}.$$

In the following calculation, Δ will indicate that the formula in which it occurs is valid if Δ is replaced by some positive constant, i.e. independent of $(x, \mu) \in \{(x, \mu) \mid (x, \mu) \in U, \mu > 0, (\varphi_\mu(x), \mu) \in U\}$. From the various definitions we obtain

$$|X(x, \mu)| \geq \Delta . (x^2 + \mu^2)$$

and $$|g(x, \mu)| \leq \Delta . \mu . (x^2 + \mu^2)^2$$

(here one uses that X is C^5). Define $h(x, \mu)$ by $X_{1+h(x,\mu)}(x, \mu) = (\varphi_\mu(x), \mu)$. Since $h(x, \mu) = \int_\alpha^\beta [X(y, \mu)]^{-1} dy$, where $\alpha = X_1(x, \mu)$ and $\beta = \varphi_\mu(x)$, the above inequalities imply

$$|h(x, \mu)| \leq \Delta . \mu . (x^2 + \mu^2).$$

Now (i)′ follows from the obvious inequality

$$|\varphi_\mu(x) - x| \geq \Delta.(x^2 + \mu^2).$$

To prove (ii)′, we observe that from previous definitions it follows that

(a) the 4-jet of $(\Phi_* X) - X$ at (0, 0) is zero (since the 4-jet of X_1 and Φ agree at (0, 0); see Lemma (3.6))

(b) $\Phi_*(X) - X$ is zero along $\{\mu = 0\}$.

Hence

$$|(d\varphi_\mu)_x . X(x, \mu) - X(\varphi_\mu(x), \mu)| \leq \Delta.\mu.((\varphi_\mu(x))^2 + \mu^2)^2$$

and

$$|X(\varphi_\mu(x), \mu)| \geq \Delta.((\varphi_\mu(x))^2 + \mu^2),$$

so

$$\frac{|(d\varphi_\mu)_x . X(x, \mu) - X(\varphi_\mu(x), \mu)|}{|X(\varphi_\mu(x), \mu)|} \leq \Delta.\mu.((\varphi_\mu(x))^2 + \mu^2).$$

Since $(x, \mu) \mapsto (\varphi_\mu(x), \mu)$ is a diffeomorphism, $((\varphi_\mu(x))^2 + \mu^2) \leq \Delta(x^2 + \mu^2)$. From these inequalities it easily follows that

$$|\log((d\varphi_\mu)_x . X(x, \mu)) - \log(X(\varphi_\mu(x), \mu))| \leq \Delta.\mu.(x^2 + \mu^2) \leq \Delta.\mu.(\varphi_\mu(x) - x);$$

this proves (ii)′.

Proof of Theorem (3.1). — The proof of Theorem (3.1) will be obtained by showing that if $\{\varphi_\mu\}$ is a saddle-node arc and X an adapted vector field (of class C^5) then there is a local conjugacy $H(x, \mu) = (h_\mu(x), \mu)$ from $\{\varphi_\mu\}$ to the time one map X_1 of X. We shall also construct H so that $\frac{\partial H}{\partial x}$ exists and is continuous on the complement of the fixed-point set of Φ. Indeed, by Theorem (3.5) and Lemma (3.6), this implies (3.1).

We take $h_0 = H \mid \{\mu = 0\}$ to be the identity. Let

$$U = \{(x, \mu) \mid 0 \leq \mu \leq \bar{\mu}, \ -a \leq x \leq a\}$$

be as in Lemma (3.7). Take:

1. $h_\mu(-a) = -a$ for $0 \leq \mu \leq \bar{\mu}$ and hence
2. $(h_\mu(\varphi_\mu^n(-a)), \mu) = X_n(-a, \mu)$ whenever $(\varphi_\mu^n(-a), \mu) \in U$;
3. extend the definition of $h_\mu(x)$ to

$$\{(x, \mu) \mid 0 \leq \mu \leq \bar{\mu}, \ -a \leq x \leq \varphi_\mu(-a)\}$$

in such a way that it is C^1 and such that the extension, defined by

$$h_\mu(\varphi_\mu(x), \mu) = X_1(h_\mu(x), \mu)$$

is also differentiable in a neighborhood of $\{(\varphi_\mu(-a), \mu) \mid 0 \leq \mu \leq \bar{\mu}\}$. Now $H \mid U$ is uniquely determined; the continuity of H along $\{(x, \mu) \mid x \geq 0, \ \mu = 0\}$ follows from Lemma (3.7); also the fact that $\frac{\partial H}{\partial x}$ is continuous on $U \backslash (0, 0)$ follows from that same lemma.

The extension of H to $\{\mu \leq 0\}$ is simple: see the proof of Theorem (3.5).

Proof of Theorem (3.2). — Let $\{\varphi_\mu\}$ be a saddle-node arc and X an adapted saddle-node field, of class C^5, defined in a box U as in Lemma (3.7). We take $-a < x < 0$ and $0 < \bar{x} < a$ and show how $\{\varphi_\mu\}$ gives rise to a canonical homeomorphism $T_{\bar{x}x}$ from a neighborhood of x (in **R**) to a neighborhood of $\bar{x}$. These maps $T_{\bar{x}x}$ are called translations.

Let $\{x_i\}_{i\in\mathbf{N}}$ be any sequence converging to x; choose a sequence $\{\mu_i\}_{i\in\mathbf{N}}$ such that $\varphi^i_{\mu_i}(x_i)$ converges to $\bar{x}$ with $\mu_i \to 0$ as $i \to \infty$. To get such a sequence, choose $\mu_i \to 0$ as $i \to \infty$ such that $(X_i)(x_i, \mu_i) \to (\bar{x}, 0)$ and apply Lemma (3.7). Let x' be a point close to x; without loss of generality we may put $(x', 0) = (X_\alpha)(x, 0)$ for some $\alpha \in \mathbf{R}$. Then for any sequence $x'_i \to x'$, $\lim_{i\to\infty} \varphi^i_{\mu_i}(x'_i)$ (the same μ_i's as above) exists and equals $(X_\alpha)(\bar{x}, 0)$. This can be seen as follows: for some β_i, β'_i, α_i we have

$$\begin{aligned}(\varphi^i_{\mu_i}(x_i), \mu_i) &= X_{\beta_i}(x_i, \mu_i);\\ (\varphi^i_{\mu_i}(x'_i), \mu_i) &= X_{\beta'_i}(x'_i, \mu_i);\\ X_{\alpha_i}(x_i, \mu_i) &= (x'_i, \mu_i).\end{aligned}$$

By Lemma (3.7) $|i - \beta_i| \leq \mu_i . C_1$ and $|i - \beta'_i| \leq \mu_i . C_1$ for some constant C_1; because $x_i \to x$, $x'_i \to x'$ and $X_\alpha(x, 0) = (x', 0)$, $\alpha_i \to \alpha$. From this we conclude that if $X_{\tilde{\alpha}_i}(\varphi^i_{\mu_i}(x_i), \mu_i) = (\varphi^i_{\mu_i}(x'_i), \mu_i)$ then $\tilde{\alpha}_i = \alpha_i + \beta'_i - \beta_i$ so $\lim_{i\to\infty} \tilde{\alpha}_i = \alpha$. We define $T_{\bar{x}x}(x') = \lim_{i\to\infty} \varphi^i_{\mu_i}(x'_i)$. It is clear that the definition of the local homeomorphism $T_{\bar{x}x}$ is independent of X: X was only used to show that $\lim_{i\to\infty} \varphi^i_{\mu_i}(x'_i)$ exists. Observe that $(T_{\bar{x}x})_*(X) = X$. Now we extend the notion of translation. If x, $\bar{x}$ are on the same side of 0 and x' on the other side, we may define $T_{\bar{x}x} = T_{\bar{x}x'} \circ T_{x'x}$; this definition is independent (at least near x) of the choice of x'. Hence we see that the translations form the pseudo-group of all local diffeomorphisms of $(-a, +a)\backslash\{0\}$ which preserve $X \mid \{\mu = 0\}$.

If h is a local conjugacy between two saddle-node arcs $\{\varphi_\mu\}$ and $\{\bar{\varphi}_\mu\}$, it must also conjugate the translations defined by $\{\varphi_\mu\}$ with those of $\{\bar{\varphi}_\mu\}$. Let X, $\bar{X}$ be the smooth vector fields such that $X_1 = \varphi_0$ and $\bar{X}_1 = \bar{\varphi}_0$. For $x \neq 0$, $\bar{x} \neq 0$ and $z = X_t(x)$ near x, let $x' = h(x)$, $\bar{x}' = h(\bar{x})$, $z' = h(z)$ and $z' = \bar{X}_{t'}(x')$. Since h conjugates $T_{\bar{x}x}$ and $\bar{T}_{\bar{x}'x'}$, and $T_{\bar{x}x}(z) = X_t(\bar{x})$, $\bar{T}_{\bar{x}'x'}(z') = \bar{X}_{t'}(\bar{x}')$, we conclude that there is a (continuous) function $t' = \sigma(t)$ such that $hX_t h^{-1} = \bar{X}_{\sigma(t)}$. From the group property $X_{t_1} \circ X_{t_2} = X_{t_1+t_2}$ we deduce that $\sigma(t)$ is linear in t. Since $X_1 = \varphi_0$, $\bar{X}_1 = \bar{\varphi}_0$, we have $h \circ X_1 \circ h^{-1} = \bar{X}_1$ and hence $h \circ X_t \circ h^{-1} = \bar{X}_t$ for all t. This proves the theorem.

Proof of Theorem (3.4). — From Theorem (3.1) and invariant manifold theory (see Section 2 and [26]), it follows that Φ, near $(\bar{x}, \bar{\mu})$, is locally conjugate with $\tilde{\Phi}$ near (0, 0) where

$$\tilde{\Phi}(x_1, \ldots, x_m, \mu) = \left(x_1 + x_1^2 + \mu, \pm 2x_2, \ldots, \pm 2x_h, \pm \frac{1}{2}x_{h+1}, \ldots, \pm \frac{1}{2}x_m, \mu\right).$$

The stable manifold $\widetilde{W}^s$ of $\widetilde{\varphi}_0$ is $\{(x_1, \ldots, x_m) : x_1 \leq 0\}$. We take the strong stable foliation $\widetilde{\mathscr{F}}^{ss}$ of $\widetilde{\Phi}$ near (0, 0) to consist of the manifolds

$$\{(x_1, \ldots, x_m, \mu) : (x_1, \ldots, x_h, \mu) = \text{constant}\}.$$

In a neighborhood U of 0, we give a dynamical characterization of the leaves of $\widetilde{\mathscr{F}}^{ss} \cap \widetilde{W}^s$: $q_1, q_2 \in \widetilde{W}^s$ belong to the same leaf of $\widetilde{\mathscr{F}}^{ss}$ if and only if there are sequences $\{\mu_i\}_{i \in N}$, $\{q_1^i\}_{i \in N}$ and $\{q_2^i\}_{i \in N}$ such that

— $(q_1^i, \mu_i) \to (q_1, 0)$ and $(q_2^i, \mu_i) \to (q_2, 0)$,

— for each i, $\{\widetilde{\varphi}^j_{\mu_i}(q_1^i)\}_{j=0}^i$ and $\{\widetilde{\varphi}^j_{\mu_i}(q_2^i)\}_{j=0}^i$ are contained in U;

— $\lim(\widetilde{\varphi}^i_{\mu_i}(q_1^i), \mu_i) = \lim(\widetilde{\varphi}^i_{\mu_i}(q_2^i), \mu_i)$ and this limit is not the saddle-node point.

This follows from the special form of $\widetilde{\Phi}$.

Let $\mathscr{F}^{ss}$ be some strong stable foliation for Φ. The local conjugacy between Φ and $\widetilde{\Phi}$ mentioned before can be chosen to map $\mathscr{F}^{ss}$ to $\widetilde{\mathscr{F}}^{ss}$. This implies that $\mathscr{F}^{ss} \cap W^s$ also satisfies the above dynamical characterization. Hence, $\mathscr{F}^{ss} \cap W^s$ is unique and is preserved under conjugacies. Theorem (3.4) is proved.

Note that a local conjugacy of $\widetilde{\varphi}_0$, or $\varphi_{\widetilde{\mu}}$, does not have to respect the above foliation. This has to do with the fact that the dynamical characterization was only possible by using Φ on a full neighborhood of (0, 0) in $R^m \times R$.

4. The flip

Here we consider arcs $\{\varphi_\mu\}$ of diffeomorphisms on R such that $\varphi_0(0) = 0$, $(d\varphi_0)_0 = -1$ and such that the 3-jet of $(\varphi_0)^2$ at the origin differs from the 3-jet of the identity. Such arcs of diffeomorphisms are obtained by restricting a generic arc of diffeomorphisms, at a flip, to a center manifold. The origin is either a source or a sink of φ_0; in the following we shall assume it to be a sink of φ_0; the other case then occurs for φ_0^{-1} and is completely analogous. With a coordinate change of the form

$$\widetilde{x} = \widetilde{x}(x, \mu)$$
$$\widetilde{\mu} = \mu$$

we can put φ_μ in the form

$$\varphi_{\widetilde{\mu}}(\widetilde{x}) = -\widetilde{x} + \widetilde{x}^3 + \lambda(\widetilde{\mu}).\widetilde{x} + o(\widetilde{x}^4) + o(|\widetilde{\mu}|.\widetilde{x}^2)$$

where λ is a real function and $\lambda(0) = 0$. We shall say that the flip *unfolds generically* if $\frac{d\lambda}{d\widetilde{\mu}}(0) \neq 0$. This is, of course, a generic condition. In higher dimensions, we will use this terminology if $\frac{d\lambda}{d\widetilde{\mu}}(0) \neq 0$ on some center manifold. Similar considerations apply to periodic flips.

Now we return to the one dimensional case. We shall assume $\frac{d\lambda}{d\widetilde{\mu}}(0) > 0$; the other case can be reduced to this by replacing $\tilde{\mu}$ by $-\tilde{\mu}$. It is clear that the phase portrait looks as follows:

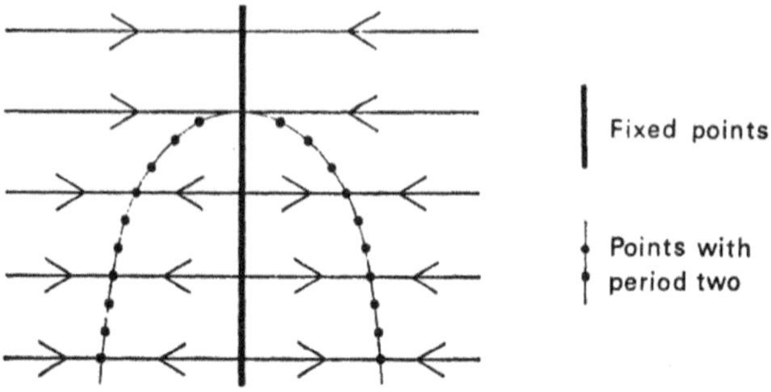

Fig. 1

The following theorem can be easily proved using the methods of proof used in Theorem (3.5) in the region $\{\mu \leq 0\}$:

Theorem (**4.1**). — *Any arc* $\{\varphi_\mu\}$ *of diffeomorphisms on* **R** *which is of the form* $\varphi_\mu(x) = -x + x^3 + \lambda(\mu).x + o(x^4) + o(|\mu|.x^2)$ *with* $\lambda'(0) > 0$ *is locally conjugate with* $\bar{\varphi}_\mu(x) = -x + x^3 + \mu x$.

5. The Hopf point

We consider arcs $\{\varphi_\mu\}$ of diffeomorphisms of $\mathbf{R}^2$ such that $\varphi_0(0) = 0$, $(d\varphi_0)_0$ has eigenvalues on the unit circle, but different from ± 1, and such that the 3-jet of φ_0 makes the origin an attractor or a repeller. We shall assume it to be an attractor; otherwise we consider φ_μ^{-1}. Up to a change of coordinates, the origin will be a fixed point of φ_μ if $|\mu|$ is small. Let $\lambda(\mu)$, $\bar{\lambda}(\mu)$ be the eigenvalues of $(d\varphi_\mu)_0$. For generic arcs one has $\frac{d}{d\mu}|\lambda(\mu)|_{\mu=0} \neq 0$, and we shall say, in this case, that the Hopf orbit *unfolds generically*. As for the saddle node and flip, we shall use the same terminology for periodic Hopf points if they unfold generically on center manifolds. In the following, we return to dimension two, and we shall assume that $\frac{d}{d\mu}|\lambda(\mu)|_{\mu=0} > 0$; otherwise we replace μ by $-\mu$. These arcs have been extensively studied; for references see [32].

From the fact that, for $\mu \leq 0$, the origin is an attractor of φ_μ one concludes:

Proposition (**5.1**). — *If* $\{\varphi_\mu\}$ *and* $\{\bar{\varphi}_\mu\}$ *are arcs of diffeomorphisms of* $\mathbf{R}^2$, *satisfying all the above requirements, then there is a continuous one parameter family of homeomorphisms* h_μ, $\mu \leq 0$, *from a neighborhood of* $0 \in \mathbf{R}^2$ *to a neighborhood of* $0 \in \mathbf{R}^2$, *such that* $h_\mu \circ \varphi_\mu = \bar{\varphi}_\mu \circ h_\mu$ *whenever defined*; i.e. $(x, \mu) \mapsto (h_\mu(x), \mu)$ *is a local conjugacy between* φ_μ *and* $\bar{\varphi}_\mu$ *on* $\mu \leq 0$.

A consequence of Proposition (5.1) is that arcs $\{\varphi_\mu\}$ as above are locally stable for $\mu \leq 0$. For $\mu > 0$ this is not the case.

Theorem **(5.2)**. — *Let* $\{\varphi_\mu\}$ *be a* C^∞ *arc of diffeomorphisms as above. Then in any* C^∞ *neighborhood of* $\{\varphi_\mu\}$ *there is an arc* $\{\bar{\varphi}_\mu\}$ *such that the arcs* $\{\varphi_\mu\}$, $\{\bar{\varphi}_\mu\}$ *are not conjugate, even not mildly conjugate.*

Proof. — We can first approximate the arc $\{\varphi_\mu\}$ by an arc $\{\tilde{\varphi}_\mu\}$ such that the eigenvalues of $(d\varphi_0)_0$ have the form $e^{2\pi i\alpha}$ with α irrational. Then we know [40] that $\tilde{\varphi}_\mu$ is rotationally symmetric in the formal sense. By this we mean that there is a smooth S^1 $(= \mathbf{R}/2\pi)$ action $\tilde{R}$ on $\mathbf{R}^2 \times \mathbf{R}$, differentiably conjugate with the usual action $R : R_\alpha(x_1, x_2, \mu) = (x_1.\cos\alpha + x_2.\sin\alpha, -x_1\sin\alpha + x_2\cos\alpha, \mu)$, such that for each α, $\tilde{R}_\alpha \circ \tilde{\Phi}$ and $\tilde{\Phi} \circ \tilde{R}_\alpha$ have the same ∞-jet in the origin. So with a second perturbation one can find an arc $\{\bar{\varphi}'_\mu\}$ such that the corresponding map $\bar{\Phi}'$ defined by $\bar{\Phi}'(x, \mu) = \bar{\varphi}'_\mu(x)$ commutes with the S^1 action $\tilde{R}$, at least on a small neighborhood of $(x = 0, \mu = 0)$.

One knows [32] that $\bar{\varphi}'_\mu$, for $\mu > 0$, has an invariant circle, say C_μ. Because $\bar{\Phi}'$ commutes with the smooth S^1 action (for μ small), $\bar{\varphi}'_\mu \mid C_\mu$ is differentiably conjugate to a rotation. Since there is a residual subset of the diffeomorphisms of S^1 no element of which is conjugate to a rotation, there is finally an approximation $\bar{\varphi}_\mu$ of $\bar{\varphi}'_\mu$ (in the C^∞ sense) such that $\bar{\varphi}_\mu$ also has C_μ as invariant circle but such that for some sequence $\{\mu_i\}_{i\in I}$, $\mu_i \to 0$ as $i \to \infty$, $\bar{\varphi}_\mu \mid C_\mu$ is not conjugate to a rotation.

It follows from the construction that $\{\bar{\varphi}_\mu\}$ and $\{\bar{\varphi}'_\mu\}$ are not conjugate and even not mildly conjugate near (0, 0). Hence φ_μ cannot be (mildly) conjugate to both $\{\bar{\varphi}_\mu\}$ and $\{\bar{\varphi}'_\mu\}$. This proves the theorem.

Remark **(5.3)**. — If $\{\varphi_\mu\}$ is an arc of diffeomorphisms on a manifold M of dimension $m > 2$ which has a Hopf point at $(\bar{x}, \bar{\mu})$ then the conclusion of Theorem (5.2) remains valid. In the construction of the C^∞ arcs $\{\bar{\varphi}'_\mu\}$, $\{\bar{\varphi}_\mu\}$, one has to add in this case a preliminary step, namely one has to modify $\{\varphi_\mu\}$ first so that it has a C^∞ centermanifold. Then the modifications, as described in the proof of (5.2) are carried out in that center manifold and extended to a neighborhood.

6. Quasi-transversal intersections

We consider a C^∞ arc of diffeomorphisms $\{\varphi_\mu : M \to M\}$ and assume that for some $\bar{\mu} \in \mathbf{R}$ and some $\bar{x} \in M$, there is a non-transversal intersection at $\bar{x}$ of a stable and an unstable manifold (of some periodic points) of $\varphi_{\bar{\mu}}$. In such a situation, if we denote the stable, resp. unstable, manifold by W^s, W^u, there is a canonical quadratic map $D : T_{\bar{x}}(W^s) \cap T_{\bar{x}}(W^u) \to T_{\bar{x}}(M)/T_{\bar{x}}(W^s) + T_{\bar{x}}(W^u)$, which is analogous to the intrinsic 2nd order derivative [3] and is defined as follows.

Let $i : T_{\bar{x}}(W^s) \cap T_{\bar{x}}(W^u) \to W^u$ be a smooth map such that $i(0) = \bar{x}$ and $(di)_0$ is the canonical injection; let π be a projection (*i.e.* $\pi^2 = \pi$) of a neighborhood U of $\bar{x}$

to a submanifold whose dimension equals the dimension of $T_{\bar{x}}(M)/(T_{\bar{x}}(W^s) + T_{\bar{x}}(W^u))$, and such that $\pi(\bar{x}) = \bar{x}$, $(d\pi)_{\bar{x}}(T_{\bar{x}}(W^s) + T_{\bar{x}}(W^u)) = 0$ and $\pi(W^s) = \bar{x}$; let $N = \pi(U)$. Then $T_{\bar{x}}(N)$ is canonically isomorphic with $T_{\bar{x}}(M)/(T_{\bar{x}}(W^s) + T_{\bar{x}}(W^u))$.

The map $\pi \circ i$ maps 0 to $\bar{x}$, $d(\pi \circ i)_0 = 0$; hence the second derivative $d^2(\pi \circ i)_0 : T_{\bar{x}}(W^s) \cap T_{\bar{x}}(W^u) \to T_{\bar{x}}(N)$ is a well defined quadratic map; we define D to be the composition of $d^2(\pi \circ i)_0$ with the canonical isomorphism

$$T_{\bar{x}}(N) \cong T_{\bar{x}}(M)/(T_{\bar{x}}(W^s) + T_{\bar{x}}(W^u)).$$

The maps D does not depend on the various choices.

It is not hard to show that, for generic arcs of diffeomorphisms $\{\varphi_\mu\}$, all the non-transversal intersection points $\bar{x}$ of stable and unstable manifolds of $\varphi_{\bar{\mu}}$, for $\bar{\mu} \in \mathbf{R}$, $\bar{x} \in M$, will satisfy (in the above terminology):

a) $\dim T_{\bar{x}}(M)/(T_{\bar{x}}(W^s) + T_{\bar{x}}(W^u)) = 1$;

b) D is non-degenerate.

Under these circumstances there are coordinates $x_1, \ldots, x_m$ on a neighborhood of $\bar{x}$ ($\bar{x}$ corresponding to 0) such that W^s and W^u locally have the following

Canonical form **(6.1)**

$$W^s = \{x_1 = \ldots = x_{m-s} = 0\}$$
$$W^u = \{x_{u+2} = \ldots = x_m = 0,\ x_1 = f(x_{m-s+1}, \ldots, x_{u+1})\}$$

where

$m = \dim(M)$;

$s = \dim(W^s)$;

$u = \dim(W^u)$;

f is a homogeneous non-degenerate quadratic function;

if $m - s + 1 > u + 1$ then one should read $x_1 = 0$.

Before proving (6.1), we note the following

c) $\max(s, u) < m$ by condition *(a)* above;

d) the vector $\left.\dfrac{\partial}{\partial x_1}\right|_{\bar{x}}$ is not in $T_{\bar{x}}(W^s) + T_{\bar{x}}(W^u)$;

e) the vectors $\left.\dfrac{\partial}{\partial x_{m-s+1}}\right|_{\bar{x}}, \ldots, \left.\dfrac{\partial}{\partial x_{u+1}}\right|_{\bar{x}}$ are in $T_{\bar{x}}(W^s) \cap T_{\bar{x}}(W^u)$.

Proof of (6.1). — Since $\dim(T_{\bar{x}}(W^s) + T_{\bar{x}}(W^u)) = m - 1$, we can choose a $(u + 1)$-manifold $\widetilde{W}^u$ containing W^u which is at $\bar{x}$ transversal to W^s. From this it clearly follows that one can choose coordinates $\bar{x}_1, \ldots, \bar{x}_m$ such that

$W^s = \{\bar{x}_1 = \ldots = \bar{x}_{m-s} = 0\}$;

$\widetilde{W}^u = \{\bar{x}_{u+2} = \ldots = \bar{x}_m = 0\}$;

W^u is tangent at $\bar{x}$ to $\{\bar{x}_1 = \bar{x}_{u+2} = \bar{x}_{u+3} = \ldots = \bar{x}_m = 0\}$.

Then, for some function $\bar{f}$

$$W^u = \{\bar{x}_{u+2} = \ldots = \bar{x}_m = 0,\ \bar{x}_1 = \bar{f}(\bar{x}_2, \ldots, \bar{x}_{u+1})\},$$
$$\bar{f}(0) = 0 \quad \text{and} \quad (d\bar{f})_0 = 0.$$

Now we replace $\bar{x}_1$ by the new coordinate

$$x_1 = \bar{x}_1 - \bar{f}(\bar{x}_2, \ldots, \bar{x}_{u+1}) + \bar{f}(0, \ldots, 0, \bar{x}_{m-s+1}, \ldots, \bar{x}_{u+1});$$

then we have

$$W^s = \{x_1 = \bar{x}_2 = \ldots = \bar{x}_{m-s} = 0\},$$
$$W^u = \{x_1 = \bar{f}(0, \ldots, 0, \bar{x}_{m-s+1}, \ldots, \bar{x}_{u+1}),\ \bar{x}_{u+2} = \ldots = \bar{x}_m = 0\},$$

the first derivative of $(\bar{x}_{m-s+1}, \ldots, \bar{x}_{u+1}) \mapsto \bar{f}(0, \ldots, 0, \bar{x}_{m-s+1}, \ldots, \bar{x}_{u+1})$ is zero and the second derivative is just D and hence has maximal rank. Now we can apply Morse's Lemma [3] and obtain a coordinate change of the form

$$x_j = \bar{x}_j \quad \text{for } 2 \leq j \leq m - s \quad \text{or} \quad u + 2 \leq j \leq m,$$
$$x_j = x_j(\bar{x}_{m-s+1}, \ldots, \bar{x}_{u+1}) \quad \text{for } m - s + 1 \leq j \leq u + 1,$$

for which we have

$$W^s = \{x_1 = \ldots = x_{m-s} = 0\},$$
$$W^u = \{x_1 = f(x_{m-s+1}, \ldots, x_{u+1}),\ x_{u+2} = \ldots = x_m = 0\},$$

with f homogeneous, quadratic and non-degenerate.

Remark **(6.2)**. — In suitable coordinates the generic unfolding of the quasi-transversal intersection puts the manifolds W^s_μ and W^u_μ in the form

$$W^s_\mu = \{x_1 = 0 = \ldots = x_{m-s} = 0\},$$
$$W^u_\mu = \{x_{u+2} = \ldots = x_m = 0,\ x_1 = f(x_{m-s+1}, \ldots, x_{u+1}) \pm (\mu - \bar{\mu})\}.$$

III. — NECESSARY CONDITIONS FOR STABILITY OF ARCS

In this chapter we shall obtain necessary conditions for the various kinds of stabilities of arcs we have defined. Let $I = [0, 1]$ and let $\mathscr{P}$ be the space of C^∞ arcs $\varphi : I \to \mathrm{Diff}^\infty(M)$ with the C^∞ topology. Elements of $\mathscr{P}$ will be denoted by φ, or by $\{\varphi_\mu\}$ when we wish to make the dependence on the parameter $\mu \in I$ explicit. Recall that $\mathscr{A} \subset \mathscr{P}$ is the subset of those arcs $\varphi \in \mathscr{P}$ such that

(1) $\varphi_0 \in \mathrm{MS}$,
(2) $b = b(\varphi) = \inf\{\mu \in I : \varphi_\mu \notin \mathrm{MS}\} < 1$,
(3) the limit set of φ_b consists of finitely many orbits.

A diffeomorphism f is called elementary if either

(a) there is exactly one quasi-hyperbolic periodic orbit, the other periodic orbits are hyperbolic, and all stable, strong stable, unstable, and strong unstable manifolds meet transversally; or
(b) all periodic orbits are hyperbolic, there is one quasi-transversal orbit of intersections of stable and unstable manifolds, and all other stable and unstable manifold intersections are transverse.

Let $\mathscr{B} \subset \mathscr{A}$ be the set of arcs $\varphi \in \mathscr{A}$ such that φ_b is elementary and the quasi-hyperbolic periodic orbit of φ_b, if it exists, unfolds generically.

We proceed to discuss necessary conditions for stability of arcs in $\mathscr{B}$.

1. The modulus condition (quasi-transversal intersection)

In this section we shall show that left stability of an arc $\{\varphi_\mu\}$ in $\mathscr{B}$ forces all stable and unstable manifolds of periodic points of φ_b to meet transversally. Before this, we consider the effect of a quasi-transversal orbit on topological conjugacy. The next theorem shows that generally such an orbit yields at least a one-parameter family of distinct topological types. We will refer to this phenomenon by saying that moduli occur. It should be noted that this occurs even in a locally isolated codimension one submanifold of the boundary of MS on the 2-sphere (hence, on any manifold of dimension larger than one).

For example, consider an MS diffeomorphism φ_0 on S^2 as in the next figure.

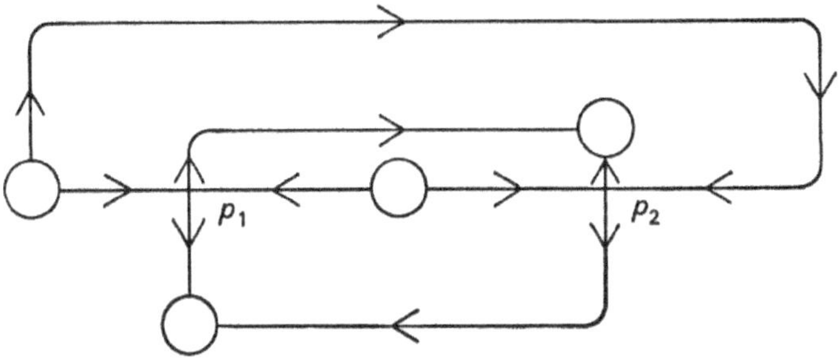

FIG. 2

The circles represent sources and sinks and there are two saddle fixed points p_1 and p_2. We choose a curve of C^2 diffeomorphisms $\{\varphi_\mu\}$, $0 \leq \mu \leq 1$, starting at φ_0 so that p_1 and p_2 remain fixed for each φ_μ, and $W^u(p_1, \varphi_{1/2})$ has a single orbit $o(x)$ of quasi-transversal intersections with $W^s(p_2, \varphi_{1/2})$ as in the next figure.

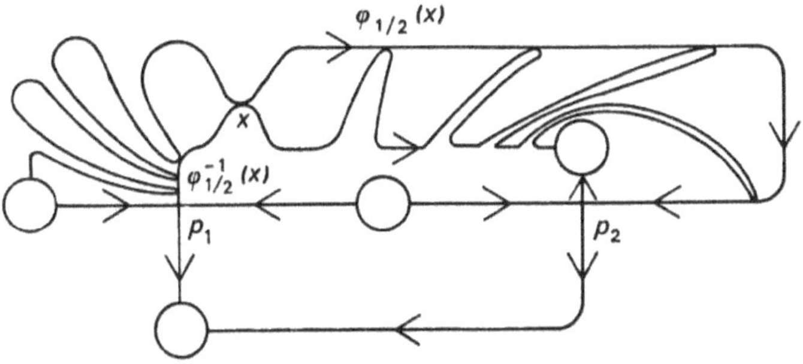

FIG. 3

This can be done so that φ_μ is in MS for $\mu \neq \frac{1}{2}$ and any perturbation $\{\varphi'_\mu\}$ of $\{\varphi_\mu\}$ has a unique bifurcation $b(\varphi')$ near $\frac{1}{2}$.

Let f be a diffeomorphism of M with a hyperbolic fixed point p. Let α^* be the largest modulus of the eigenvalues of $df(p)$ which are inside the unit circle, and let β^* be the smallest modulus of the eigenvalues of $df(p)$ which are outside the unit circle. If there is an eigenvalue α of $df(p)$ such that

(1) $|\alpha| = \alpha^*$,
(2) α has multiplicity one,
(3) any eigenvalue λ of $df(p)$ different from α or the complex conjugate $\bar{\alpha}$ satisfies $|\lambda| \neq \alpha^*$,

then we say that the weakest contracting eigenvalue α of f at p is defined. Similarly, if there is an eigenvalue β of multiplicity one of $df(p)$ such that $|\beta| = \beta^*$ and any eigenvalue $\lambda \neq \bar{\beta}, \beta$ satisfies $|\lambda| \neq \beta^*$, we say the weakest expanding eigenvalue β of f at p is defined.

Let p and q be hyperbolic fixed points of f such that the weakest contracting eigenvalue α at p and the weakest expanding eigenvalue β at q are defined. Let H_p be a C^1 invariant manifold containing $W^u(p)$ and tangent at p to the sum of the eigenspace of α and $T_pW^u(p)$, and let H_q be a C^1 invariant manifold containing $W^s(q)$ and tangent at q to the sum of the eigenspace of β and $T_qW^s(q)$. The existence of H_p and H_q is proved in [12]. They are not unique. However, the tangent bundle of H_p along $W^u(p)$ is unique, and so is that of H_q along $W^s(p)$.

Let r be a quasi-transversal intersection of $W^u(p)$ and $W^s(q)$. We say that r is a regular quasi-transversal intersection if $W^u(p)$ is transverse to H_q at r and $W^s(q)$ is transverse to H_p at r. This definition is independent of the choice of the manifolds H_p and Q_q because it depends only on TH_p, TH_q along $W^u(p)$, $W^s(q)$, respectively.

Note that part of the definition of a regular quasi-transversal intersection of $W^u(p)$ and $W^s(q)$ is that the weakest contracting and expanding eigenvalues α at p and β at q be defined. Note that for arcs φ in a residual subset of $\mathcal{P}$ all quasi-transversal orbits of φ_μ are regular.

Theorem (**1.1**). — *Let f, f' be C^2 diffeomorphisms. Let p and q be hyperbolic fixed points having an orbit $\{f^i(r)\}$ of regular quasi-transversal intersections of $W^u(p)$ and $W^s(q)$. Let α be the weakest contracting eigenvalue of f at p and β be the weakest expanding eigenvalue of f at q. Make analogous assumptions on f' concerning fixed points p', q', eigenvalues α', β', etc.*

If there is a conjugacy from f to f' defined on a neighborhood of the closure of $\{f^i(r)\}$ mapping p to p', q to q', and r to r', then

$$\frac{\log|\alpha|}{\log|\beta|} = \frac{\log|\alpha'|}{\log|\beta'|}.$$

Remark 1. — If p and q are periodic points of period $\tau(p)$ and $\tau(q)$ instead of fixed points, and n is the least common multiple of $\tau(p)$ and $\tau(q)$, the theorem can be applied to f^n. Doing this one obtains that if α is the weakest contracting eigenvalue of $f^{\tau(p)}$ at p and β is the weakest expanding eigenvalue of $f^{\tau(q)}$ at q and α', β' are the corresponding eigenvalues for f', then the existence of a conjugacy between f and f' implies

$$\frac{\log|\alpha|}{\log|\beta|} = \frac{\log|\alpha'|}{\log|\beta'|}.$$

2. Theorem (1.1) shows that quasi-transversal orbits lead to at least one dimensional invariants of topological conjugacy. It is interesting to ask what additional invariants in the presence of such orbits are sufficient to imply the existence of a topological conjugacy in various contexts. For some results in this direction, see [18], [28].

Proof of Theorem (1.1).

A) *The case* $m=2$; $m=\dim(M)$

In this case the dimensions of the stable and unstable manifolds are both one.

Replacing f, f' by f^2, f'^2, if necessary, we assume $\alpha, \beta, \alpha', \beta' > 0$. So we are in the following situation:

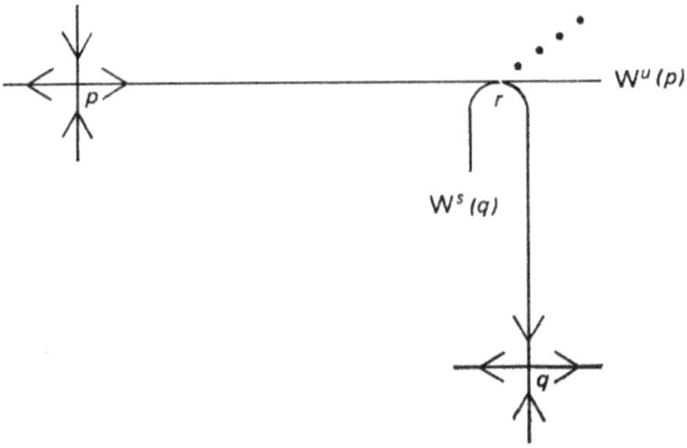

FIG. 4

We consider a sequence of points r_i converging to r but so that $r_i \notin (W^u(p) \cup W^s(q))$ for all i. By choosing a subsequence if necessary, we can arrange that there are sequences of integers $n_i \to \infty$, $m_i \to \infty$ with the property that $f^{-n_i}(r_i)$, $f^{m_i}(r_i)$, has a limit in $W^s(p) - p$, $W^u(q) - q$, respectively. Let $\rho(r_i, W^u(p))$, $\rho(r_i, W^s(q))$ denote the distance from r_i to $W^u(p)$, respectively $W^s(q)$, with respect to some Riemannian metric.

Because f is C^2, it is C^1 linearizable on $W^s(p)$ and $W^u(q)$ [8]. From this, we conclude that $\rho(r_i, W^u(p)) \sim \alpha^{n_i}$ and $\rho(r_i, W^s(q)) \sim \beta^{-m_i}$, where $\sim$ denotes equality up to a positive multiplicative factor, depending on i but uniformly bounded and bounded away from zero. It is clear from the picture, or rather from the normal forms in Chapter II, that a sequence r_i can be chosen so that $\rho(r_i, W^u(p)) \sim \rho(r_i, W^s(q))$. In that case we have

$$\frac{\log \alpha}{\log \beta} = -\lim_{i\to\infty} \frac{m_i}{n_i}.$$

Now we assume that there is a local conjugacy h, defined on a neighborhood of the closure of the orbit of r, as in the theorem. Let $h(r_i) = r'_i$. From the topology of the intersection of W^u and W^s and the position of the r'_i's (see figure) it follows that $\rho(r'_i, W^u(p')) \leq \rho(r'_i, W^s(q'))$; *i.e.*

$$\frac{\rho(r'_i, W^u(p'))}{\rho(r'_i, W^s(q'))} \leq 1.$$

Since $(f')^{-n_i}(r_i')$ and $(f')^{m_i}(r_i')$ must have a limit in $W^s(p') - \{p'\}$ resp. $W^u(q') - \{q'\}$, we conclude that $\rho(r_i', W^u(p')) \sim (\alpha')^{n_i}$ and $\rho(r_i', W^s(q')) \sim (\beta')^{-m_i}$. This, together with

$$\frac{\rho(r_i', W^u(p'))}{\rho(r_i', W^s(q'))} \leq 1$$

implies that $\dfrac{\log \alpha'}{\log \beta'} \leq - \lim\limits_{i\to\infty} \dfrac{m_i}{n_i} = \dfrac{\log \alpha}{\log \beta}$. Reversing the argument, one finds $\dfrac{\log \alpha'}{\log \beta'} \geq \dfrac{\log \alpha}{\log \beta}$ and hence $\dfrac{\log \alpha'}{\log \beta'} = \dfrac{\log \alpha}{\log \beta}$.

Observe that if $W^u(p)$ is transverse to $W^s(q)$ and $W^u(p')$ is transverse to $W^s(q')$, then we cannot conclude for any sequence $r_i \to r$ that $\dfrac{\rho(r_i', W^u(p'))}{\rho(r_i', W^s(q'))}$ is bounded. So we needed some of the properties of the " topology of the intersection ". In the next higher dimensional case we have to analyze this in detail (¹).

B) *The case* $\dim(W^u(p)) = \dim(W^u(q)) = m - 1$

As in case A we assume $\alpha, \beta, \alpha', \beta' > 0$ and we consider sequences $r_i \to r$ such that $f^{-n_i}(r_i)$ converges to a point in $W^s(p) - \{p\}$, and $f^{m_i}(r_i)$ converges to a point in $W^u(q) - \{q\}$ for some sequences $n_i, m_i \to \infty$. We also assume that one of the following three possibilities takes place (this can be obtained by taking a subsequence):

$$\frac{\rho(r_i, W^u(p))}{\rho(r_i, W^s(q))} \begin{cases} \to 0; \\ \text{is bounded and bounded away from zero (or } \sim 1); \\ \to \infty. \end{cases}$$

In these cases we find that

$$\frac{\log \alpha}{\log \beta} \begin{cases} \leq \liminf\limits_{i\to\infty}\left(-\dfrac{m_i}{n_i}\right); \\ = \lim\limits_{i\to\infty}\left(-\dfrac{m_i}{n_i}\right); \\ \geq \limsup\limits_{i\to\infty}\left(-\dfrac{m_i}{n_i}\right). \end{cases}$$

We assume that there is a local conjugacy h and denote $h(r_i)$ by r_i'. We assume also that $\dfrac{\rho(r_i', W^u(p'))}{\rho(r_i', W^s(q'))}$ either goes to zero, is ~ 1, or goes to ∞; in each of these cases we get a relation between $\dfrac{\log \alpha'}{\log \beta'}$ and " lim " $\left(-\dfrac{m_i}{n_i}\right)$ as above.

(¹) After the arguments for sections B, C, and D were written, a cleaner treatment was discovered by S. VAN STRIEN. This treatment uses arguments presented in the proof of Theorem (2.1) *in* [18*a*].

From this we claim that, if $\dfrac{\log \alpha}{\log \beta} < \dfrac{\log \alpha'}{\log \beta'}$, then, for each sequence $r_i \to r$ as above,

$$\frac{\rho(r_i, W^u(p))}{\rho(r_i, W^s(q))} \to 0 \quad \text{and/or} \quad \frac{\rho(r'_i, W^u(p'))}{\rho(r'_i, W^s(q'))} \to \infty.$$

Indeed suppose that $\dfrac{\log \alpha}{\log \beta} < \dfrac{\log \alpha'}{\log \beta'}$ and $r_i \to r$ as above. If $\dfrac{\rho(r'_i, W^u(p'))}{\rho(r'_i, W^s(q'))}$ does not go to ∞, then for some constant $C > 0$, we have

$$(\alpha')^{n_i}(\beta')^{m_i} \leq C.$$

This gives $n_i \log \alpha' + m_i \log \beta' \leq \log C$ or

$$\frac{\log \alpha'}{\log \beta'} \leq \frac{\log C}{n_i \log \beta'} - \frac{m_i}{n_i}.$$

If $\delta > 1$ is so that $\dfrac{\log \alpha + \log \delta}{\log \beta} = \dfrac{\log \alpha'}{\log \beta'}$, then

$$\frac{\log \alpha\delta}{\log \beta} \leq \frac{\log C}{n_i \log \beta} - \frac{m_i}{n_i},$$

or

$$n_i \log \alpha\delta + m_i \log \beta \leq \log C . \frac{\log \beta}{\log \beta'}.$$

This means that $(\alpha\delta)^{n_i}\beta^{m_i}$ is bounded, so $\alpha^{n_i}\beta^{m_i} \to 0$ as $i \to \infty$ which implies that

$$\frac{\rho(r_i, W^u(p))}{\rho(r_i, W^s(q))} \to 0.$$

This proves the claim. Similarly, for $\dfrac{\log \alpha}{\log \beta} > \dfrac{\log \alpha'}{\log \beta'}$ we have

$$\frac{\rho(r_i, W^u(p))}{\rho(r_i, W^s(q))} \to \infty \quad \text{and/or} \quad \frac{\rho(r'_i, W^u(p'))}{\rho(r'_i, W^s(q'))} \to 0.$$

We want to prove that this cannot happen for every sequence in case of a quasi-transversal intersection of $W^u(p)$ and $W^s(q)$ (and $W^u(p')$ and $W^s(q')$).

For this we take subsets A^s, A^u, A^0 of a neighborhood of r such that:

— $A^s \cup A^u \cup A^0$ is a neighborhood of r;

— if $r_i \to r$ is a sequence such that $\dfrac{\rho(r_i, W^u(p))}{\rho(r_i, W^s(q))} \to 0$, then $r_i \in A^u$ for i big;

— if $r_i \to r$ is a sequence such that $\dfrac{\rho(r_i, W^u(p))}{\rho(r_i, W^s(q))} \to \infty$, then $r_i \in A^s$ for i big;

— if $r_i \to r$ is a sequence such that $r_i \in A^0$ (and $r_i \notin W^u(p) \cup W^s(q)$) then

$$\frac{\rho(r_i, W^u(p))}{\rho(r_i, W^s(q))} \sim 1.$$

A'^u, A'^s and A'^0 are similar sets in a neighborhood of r'. From the above arguments it follows that if $\frac{\log\alpha}{\log\beta} < \frac{\log\alpha'}{\log\beta'}$, then there is a neighborhood U of r such that

$$h(U \cap (A^0 \cup A^s)) \subset A'^s$$

(and if $\frac{\log\alpha}{\log\beta} > \frac{\log\alpha'}{\log\beta'}$,

$$h(U \cap (A^0 \cup A^u)) \subset A'^u).$$

In order to derive a contradiction from this (namely from the assumption that $\frac{\log\alpha}{\log\alpha} < \frac{\log\alpha'}{\log\beta'}$) in the case of quasi-transversal intersection, we proceed to an explicit construction of A^u, A^s and A^0. From Chapter II we know that there are coordinates $x_1, \ldots, x_m$ in a neighborhood of r such that locally

$$W^s(q) = \{x_m = 0\}$$

and

$$W^u(p) = \{x_m = \xi(x_1, \ldots, x_{m-1})\}$$

where ξ is a homogeneous quadratic function. In these coordinates we take

$$A^s = \left\{ |x_m| \leq \tfrac{1}{8} |\xi(x_1, \ldots, x_{m-1})| \right\}$$

$$A^u = \left\{ |x_m - \xi(x_1, \ldots, x_{m-1})| \leq \tfrac{1}{8} |\xi(x_1, \ldots, x_{m-1})| \right\}$$

$$A^0 = \text{closure of the complement of } A^s \cup A^u.$$

Near r', $W^u(p')$ and $W^s(q')$ have the same form, so there is a diffeomorphism Ψ from a neighborhood of r to a neighborhood of r', mapping $W^u(p)$ to $W^u(p')$ and $W^s(q)$ to $W^s(q')$. We define $A'^s = \Psi(A^s)$, $A'^u = \Psi(A^u)$ and $A'^0 = \Psi(A^0)$.

It now follows that $\widetilde{h} = \Psi^{-1} \circ h$ is a local homeomorphism from a neighborhood of r to itself, inducing homeomorphisms in $W^u(p)$ and $W^s(q)$ and mapping $(A^0 \cup A^s) \cap U$ into A^s. Let $U^* \subset U$ be a subset of the form $U^* = \{(x_1, \ldots, x_m) \mid 0 < \Sigma x_i^2 < a^*\}$ and $U^{**} \supset \widetilde{h}(U^*)$ a set of the form $U^{**} = \{(x_1, \ldots, x_m) \mid 0 < \Sigma x_i^2 < a^{**}\}$. From the fact that the subsets $W^u(p)$, $W^s(q)$, A^u, A^s and A^0, restricted to U^* or U^{**} are all cone-formed, and the fact that $\widetilde{h}$ is a local homeomorphism inducing local homeomorphisms in $W^u(p)$ and $W^s(q)$ we conclude that the maps

$$U^* \cap W^s(q) \to U^{**} \cap W^s(q),$$
$$U^* - (W^u(p) \cap U^*) \to U^{**} - (W^u(p) \cap U^{**}),$$

induced by $\widetilde{h}$, induce isomorphisms in the homology. From the definitions it is clear that the inclusions

$$U^* \cap W^s(q) \to U^* \cap A^s, \qquad U^{**} \cap W^s(q) \to U^{**} \cap A^s,$$
$$U^* \cap (A^s \cup A^0) \to U^* - (W^u(q) \cap U^*),$$
$$U^{**} \cap (A^s \cup A^0) \to U^{**} - (W^u(q) \cap U^{**}),$$

are all homology equivalences. Now consider the morphism of long exact sequences:

$$\begin{array}{ccccc}
\to H_i(U^* \cap A^s) & \to & H_i(U^* \cap (A^s \cup A^0)) & \to & H_i(U^* \cap (A^s \cup A^0), U^* \cap A^s) \\
\downarrow h_1 & & \downarrow h_2 & & \downarrow h_3 \\
\to H_i(U^{**} \cap A^s) & \to & H_i(U^{**} \cap (A^s \cup A^0)) & \to & H_i(U^{**} \cap (A^s \cup A^0), U^{**} \cap A)
\end{array}$$

h_1, h_2 and h_3 are all induced by $\widetilde{h}$, so h_1, h_2 are isomorphisms. By the Five Lemma, h_3 is then also an isomorphism. But since $\widetilde{h}(U^* \cap (A^s \cup A^0)) \subset U^{**} \cap A^s$, h_3 is the zero morphism. Hence $H_i(U^* \cap (A^s \cup A^0), U^* \cap A^s) = 0$ for all i. This is in contradiction with the following result, which follows from standard arguments:

if
$$\xi(x_1, \ldots, x_{m-1}) = x_1^2 + \ldots + x_k^2 - x_{k+1}^2 - \ldots - x_{m-1}^2$$

then

$$H_i(U^* \cap (A^s \cup A^0), U^* \cap A^s) = \begin{cases} \mathbf{Z} \text{ if } i = k \text{ or } i = m-k-1 \text{ provided } k \neq m-k-1; \\ \mathbf{Z} \oplus \mathbf{Z} \text{ if } i = k = m-k-1; \\ 0 \text{ otherwise.} \end{cases}$$

C) *The general case*: α, β, α', β' *real*

We assume that f has fixed points p, q and invariant manifolds H_p and H_q as in the theorem and $\alpha, \beta, \alpha', \beta' > 0$. Then, $\dim H_p = \dim W^u(p) + 1$, and $\dim H_q = \dim W^s(q) + 1$. Let $H_{p'}$, $H_{q'}$ denote the analogous manifolds for f'. Of course, if there is a conjugacy h between f and f', it does not follow that $h(H_p) = H_{p'}$ or $h(H_q) = H_{q'}$. If we knew that $h(H_p) = H_{p'}$ and $h(H_q) = H_{q'}$, we could apply the argument of case B for sequences $r_i \to r$ in $H_p \cap H_q$. For in $H_p \cap H_q$, the manifolds $W^u(p) \cap H_p \cap H_q$ and $W^s(q) \cap H_p \cap H_q$ have codimension one, and intersect "nearly" quasi-transversely. We say nearly because, since H_p and H_q are only C^1 manifolds, a quasi-transverse intersection of $W^u(p) \cap H_p \cap H_q$ and $W^s(q) \cap H_p \cap H_q$ is not defined. However, we can choose C^2 submanifolds $\widetilde{H}_p$ and $\widetilde{H}_q$ (not necessarily invariant) which are C^1 close to H_p and H_q, respectively, such that

$$H_p \cap \widetilde{H}_p \supset W^u(p), \quad H_q \cap \widetilde{H}_q \supset W^s(q),$$
$$W^u(p) \cap \widetilde{H}_p \cap \widetilde{H}_q = W^u(p) \cap H_p \cap H_q,$$
$$W^s(q) \cap \widetilde{H}_p \cap \widetilde{H}_q = W^s(q) \cap H_p \cap H_q,$$

and $W^s(q) \cap \widetilde{H}_p \cap \widetilde{H}_q$ intersects $W^u(p) \cap \widetilde{H}_p \cap \widetilde{H}_q$ quasi-transversely. Hence, the metric properties of the intersection of $W^u(p) \cap H_p \cap H_q$ and $W^s(q) \cap H_p \cap H_q$ in $H_p \cap H_q$ are those of a quasi-transversal intersection.

Since h does not map H_p to $H_{p'}$ or H_q to $H_{q'}$, we must modify our arguments. We shall show that there is a map σ from $h(H_p \cap H_q)$ to $H_{p'} \cap H_{q'}$ such that for any

$x \in h(H_p \cap H_q) - (W^u(p') \cup W^s(q'))$, we have $\rho(x, W^u(p')) \sim \rho(\sigma(x), W^u(p'))$ and $\rho(x, W^s(q')) \sim \rho(\sigma(x), W^s(q'))$.

Let N be a small disk neighborhood of p where we can define a C^2 coordinate chart such that the components of $W^s(p) \cap N$ and $W^u(p) \cap N$ containing p are coordinate planes. In the sequel, we will restrict ourselves to such a neighborhood N and to the components of $W^s(p)$ and $W^{ss}(p)$ containing p. Let $\pi_s : N \to W^s(p)$ be the natural projection and assume that the negative orbit of r lies in N.

Since $W^s(q)$ is transverse to H_p at r, we can characterize $W^{ss}(p)$ as the subset of $W^s(p)$ where the critical points of $\pi_s \mid W^s(q)$ accumulate. This characterization is based in the following fact:

If $v \in W^s(p)$ then $v \notin W^{ss}(p)$ if and only if there are a neighborhood V of v and a neighborhood U of r such that for each $v' \in V \cap W^s(p)$ there is an arbitrarily small neighborhood V' of v' in V such that, for $n \in \mathbf{N}$ suficiently big,

$$\{f^{-n}(W^s(q) \cap U) \cap V'\} \subset \{f^{-n}(W^s(q) \cap U) \cap V\}$$

is a homotopy equivalence. In this formula we should omit from $f^{-n}(W^s(q) \cap U)$ those points whose orbits from $W^s(q) \cap U$ to $f^{-n}(W^s(q) \subset U)$ leave N. From this dynamical characterization it follows that $h(W^{ss}(p)) = W^{ss}(p')$ and similarly $h(W^{uu}(q)) = W^{uu}(q')$.

Thus $h(H_p) \cap W^s(p')$ is an invariant C^0 curve of f' which meets $W^{ss}(p')$ only at p'. Since f' is C^1 linearizable on $W^s(p')$ [8], we see that $h(H_p) \cap W^s(p')$ has to lie in a small sector about $H_{p'} \cap W^s(p')$ which has width zero at $H_{p'} \cap W^u(p')$ as in the following figure.

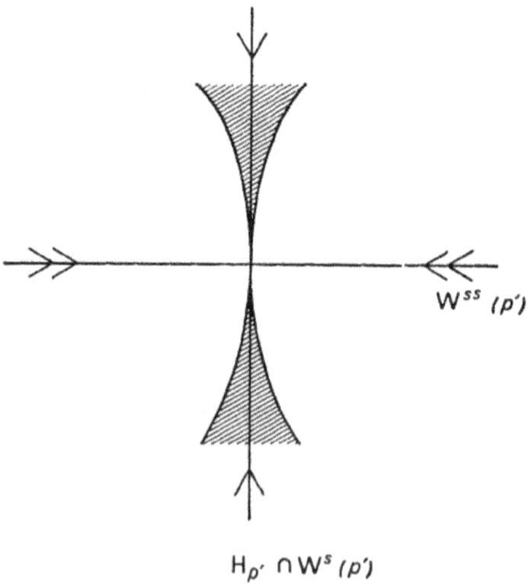

Fig. 5

34

From this it follows that for any smooth projection of a neighbourhood of $H_{p'}$ to $H_{p'}$, the distance of points in $h(H_{p'})$ to $W^u(p')$ is not changed essentially, *i.e.* the quotient of the new and old distances is uniformly bounded away from zero and infinity. It is not hard to choose such a projection σ_1 on $H_{p'}$ which does not essentially change (at least near r') the distances to $W^s(q')$.

Using the same type of arguments one can define a smooth local projection σ_2 on $H_{q'}$ in such a way that

— σ_2 maps $H_{p'}$ into itself
— σ_2 does not essentially change distances from $W^u(p')$.

As before, it follows that σ_2 does not essentially change the distances from points in $h(H_q)$ to $W^s(q')$. So $\sigma_2 \circ \sigma_1 \mid h(H_p \cap H_q) : h(H_p \cap H_q) \to H_{p'} \cap H_{q'}$ has the required properties.

To see that the whole reasoning of case B applies, we first choose metrics on H_p, H_q, $H_{p'}$, $H_{q'}$ induced by nearby C^2 manifolds $\tilde{H}_p$, $\tilde{H}_q$, $\tilde{H}_{p'}$, $\tilde{H}_{q'}$, respectively. Defining $\tilde{A}^s$, $\tilde{A}^0$, $\tilde{A}^u \subset \tilde{H}_p \cap \tilde{H}_q$ as before and projecting them into H_p and H_q to obtain A^s, A^0, A^u, we then have that the maps $W^u(p) \cap H_p \cap H_q \to A^u$, $W^s(q) \cap H_p \cap H_q \to A^s$, etc., are homotopy equivalences. Now the argument can be completed by observing that

(1) σ roughly preserves distances;
(2) σ induces, locally at r', homotopy equivalences from $W^s(q') \cap h(H_p \cap H_q)$ to $W^s(q') \cap (H_{p'} \cap H_{q'})$ and from $h(H_p \cap H_q) - W^u(p')$ to $(H_{p'} \cap H_{q'}) - W^u(p')$;
(3) if $r'_i \to r'$, $r'_i \in h(H_p \cap H_q)$ are such that $f^{-n_i}(r'_i)$, respectively $f^{m_i}(r'_i)$, has a limit in $W^s(p') - p'$, respectively $W^u(q') - q'$, then $\rho(r'_i, W^u(p')) \sim (\alpha')^{n_i}$ and $\rho(r'_i, W^s(q')) \sim (\beta')^{-m_i}$.

These facts follow easily from the constructions.

D) *The general case: α, α' are not real or β, β' are not real*

Let Q be the $df(p)$-invariant subspace of $T_pW^s(p)$ complementary to the eigenspace of α, and let $\tilde{Q}$ be the $df(q)$-invariant subspace of $T_qW^u(q)$ complementary to the eigenspace of β. Let $W^{ss}(p)$ be the invariant manifold in $W^s(p)$ tangent at p to Q, and let $W^{uu}(q)$ be the invariant manifold in $W^u(q)$ tangent at q to $\tilde{Q}$. If H_p and H_q are the invariant manifolds in the statement of Theorem (1.1), then $W^{ss}(p)$ is transverse to H_p at p and $W^{uu}(q)$ is transverse to H_q at q. The fact that $W^u(p)$ is transverse to H_q at r enables us to get a dynamical characterization of $W^{uu}(q)$. Similarly, we will obtain a dynamical characterization of $W^{ss}(p)$. Once these are obtained, it will follow that a conjugacy h from f to f' as in the statement of Theorem (1.1) will have the property that $h(W^{ss}(p)) = W^{ss}(p')$ and $h(W^{uu}(q)) = W^{uu}(q')$. From this, it will follow as in case C that $h(H_q \cap W^s(p))$ is in a small sector about $H_{p'} \cap W^s(p')$ which meets $W^{ss}(p')$ only at p'. Similarly, $h(H_q \cap W^u(q))$ is in a small sector about $H_{q'} \cap W^u(q')$ which meets $W^{uu}(q')$ only at q'. The arguments then proceed as in case C. In the present situation either or both of $W^u(p)$ and $W^s(q)$ will have codimension 2 in H_p and H_q, respectively. This changes the arguments slightly, but, since these changes are straightforward, they will be left to the reader. Thus to complete the proof of Theorem (1.1), we must show how $W^{uu}(q)$ is characterized dynamically.

Let U be a small disk neighborhood of q containing the forward orbit of r. We can take U as a coordinate chart with a C^2 coordinate system having the components of $W^u(q)$ and $W^s(q)$ containing q as coordinate planes. Let $\pi_u : U \to W^u(q)$ be the natural projection. For an integer $n > 0$ let f_U^n denote $f^n \mid \bigcap_{0 \leq j \leq n} f^{-j}(U)$. Define $W^{uu}_{loc}(q)$ to be the set of points $y \in W^u(q) \cap U$ such that

(*) for each sequence $0 < n_1 < n_2 < \ldots$, the critical points of $\pi_u \mid f_U^{n_i}(W^u(p))$ accumulate on y as $i \to \infty$; in case $\dim(W^u(p)) + \dim(W^s(q)) = \dim(M) - 1$), $f_U^{n_i}(W^u(p))$ accumulates on y.

We assert that $W^{uu}(q) = \bigcup_{i \geq 0} f^i(W^{uu}_{loc}(q))$. This implies a dynamical characterization of $W^{uu}(q)$.

To prove this assertion, first suppose that $f \mid U$ is linear and, near r, the set of critical points (or fold points) L of $\pi_u \mid W^u(p)$ is an affine subspace. Here we think of U as an open subset of Euclidean space via linearizing coordinates. Note that $\dim(L) = \dim(W^u(q)) - 1$. If the eigenvalue β is real, the assertion follows as in the previous case when the eigenvalues α, β were taken to be real. So assume β is not real. Assume also that H_q, $W^u(q)$ and $W^{uu}(q)$ are linear subspaces of U near q. Choose an affine subspace $L_1 \subset L$ which is complementary to H_q at r. Then $f^n(L_1)$ converges to $\tilde{Q}$ as $n \to \infty$. This implies that (*) holds for any $y \in W^{uu}(q)$ near q. Now let y be in $W^u(q) - W^{uu}(q)$ and near q. Since $f \mid H_q \cap W^u(q)$ near q is a rotation, there is a sequence of integers $n_1 < n_2 < \ldots$ such that $f_U^{n_i}(L \cap H_q)$ does not accumulate on πy as $\tau \to \infty$ where $\pi : W^u(q) \to W^u(q) \cap H_q$ is the natural projection. This implies that $f_U^{n_i}(L)$ does not accumulate on y as $\tau \to \infty$. This proves the assertion if $f \mid U$ is linear and L is affine. The extension to the case where f is not linear or L is not affine near r is straightforward. We leave the details to the reader. Theorem (1.1) is proved.

Lemma (1.2). — *Suppose f is an elementary diffeomorphism and x is a quasi-transversal intersection of $W^u(p,f)$ and $W^s(q,f)$ with p and q periodic points of f. Then f is not topologically conjugate to any Kupka-Smale diffeomorphism. Moreover, if f' is also elementary and h is a topological conjugacy between f and f', then $h(x)$ is a quasi-transversal intersection of $hW^u(p,f)$ and $hW^s(q,f)$.*

Proof. — This follows from the fact that a transversal intersection of two submanifolds W^s and W^u of a manifold M is topologically different from any quasi-transversal intersection of W^s and W^u. In a point of quasi-transversal intersection, either

$$\dim W^s + \dim W^u \neq \dim M$$

in which case the intersection is not a manifold of the right dimension or,

$$\dim W^s + \dim W^u = \dim M$$

in which case the intersection number is zero. See Chapter II, section 6.

Note that not all non-transversal intersections are topologically different from transversal ones.

Theorem **(1.3)**. — *Suppose f, p, q, α, β, and r are as in the hypotheses of Theorem* (1.1). *Then there is a residual set $\mathscr{P}_1 \subset \mathscr{P}$ such that if $\varphi \in \mathscr{P}_1$, then no φ_μ, $0 \leq \mu \leq 1$, is topologically conjugate to f.*

Proof. — There is a residual set $\mathscr{P}_0 \subset \mathscr{P}$ such that if $\varphi \in \mathscr{P}_0$, then

(1) each φ_μ is elementary or Kupka-Smale;
(2) if some φ_μ has a quasi-transversal orbit, then the hypotheses of Theorem (1.1) with $\varphi_\mu = f'$ are satisfied;
(3) there are at most countably many μ's for which φ_μ has a quasi-transversal orbit.

Now, since each φ_μ has at most one quasi-transversal orbit, it follows that the quotients of the logarithms of the moduli of quasi-transversal orbits occurring in $\{\varphi_\mu\}$ form a set which is at most countable. With standard arguments, one can show that there is a residual subset $\mathscr{P}_1$ of $\mathscr{P}_0$ such that if $\varphi \in \mathscr{P}_1$, then all the quotients occurring for φ are different from $\dfrac{\log|\alpha|}{\log|\beta|}$. Now Theorem (1.3) follows from Theorem (1.1) and Lemma (1.2).

Corollary **(1.4)**. — *If $\varphi \in \mathscr{B}$ is left stable, then all stable and unstable manifolds of periodic points of φ_b intersect transversally.*

Proof. — Since $\varphi \in \mathscr{B}$, we have that φ_b is elementary. Thus, if $\varphi_b = f$ has a quasi-transversal orbit, then we may assume the hypotheses of Theorem (1.1) and hence Theorem (1.3) are satisfied. By Theorem (1.3), φ is not left stable.

2. Necessary conditions for mild stability and stability

Suppose $\varphi \in \mathscr{B}$ and $\mathscr{O}(p)$ is a saddle-node for φ_b. We say $\mathscr{O}(p)$ is s-critical for φ_b if there is a periodic orbit $\mathscr{O}(q)$ such that $W^u(\mathscr{O}(q))$ has a non-transversal intersection with the strong stable foliation $\mathscr{F}^{ss}$ of $W^s(\mathscr{O}(p))$. Similarly, we say $\mathscr{O}(p)$ is u-critical if it is s-critical for φ_b^{-1}. If $\mathscr{O}(p)$ is either s-critical or u-critical but not both, we say that $\mathscr{O}(p)$ is semi-critical. If $\mathscr{O}(p)$ is both s-critical and u-critical, we say it is bicritical.

Proposition **(2.1)**. — *If φ is mildly stable, then it is left stable and the quasi-hyperbolic periodic orbit is not a Hopf orbit or bi-critical saddle-node.*

Proof. — The first statement is obvious and the second statement was proved in Section 5 of Chapter II. Now suppose $\mathcal{O}(p)$ is a bicritical saddle-node for φ_b. We assume p is a fixed point of φ_b. If q_1 and q_2 are periodic points of φ_b such that $W^u(q_1)$ has a non-transverse intersection with the strong stable foliation of $W^s(p)$ and $W^s(q_2)$ has a non-transverse intersection with the strong unstable foliation of $W^u(p)$, then, perturbing if necessary, we may assume that these intersections are quasi-transverse. For some μ near b, it then follows that $W^u(q_{1\mu}, \varphi_\mu)$ has a quasi-transverse intersection with $W^s(q_{2\mu}, \varphi_\mu)$. Here, of course, $q_{1\mu}$ and $q_{2\mu}$ denote the unique hyperbolic periodic points of φ_μ near q_1 and q_2, respectively. Using Theorem (1.3) we can perturb to φ' in $\mathcal{P}$ so that no φ'_ν is conjugate to φ_μ. This shows that φ is not mildly stable.

Proposition (**2.2**). — *If φ is mildly stable, then φ_b has no cycle of length bigger than one.*

Proof. — Suppose φ_b has a cycle of length bigger than one. We show φ is not mildly stable. Since $\varphi \in \mathcal{B}$, φ_b has a saddle-node orbit $\mathcal{O}(p_0)$ which is contained in every cycle. Indeed, the transversality of the stable and unstable manifolds implies that if φ_b had any cycle not containing a saddle nodes, then φ_b would have transversal homoclinic points. This would give φ_b infinitely many periodic points [34]. Let $\mathcal{O}(p_0), \ldots, \mathcal{O}(p_r)$ be the distinct orbits in the cycles of φ_b. Replacing $\{\varphi_\mu\}$ by some power $\{\varphi_\mu^n\}$, we assume that all the p_i's are fixed points. Let $W^{ss}(p_0)$, resp. $W^{uu}(p_0)$, denote the strong stable, resp. unstable, manifold of p_0. Since $L(\varphi_b)$ has finitely many orbits it follows that $W^u(p_i) \cap W^{ss}(p_0) = \emptyset$ and $W^s(p_i) \cap W^{uu}(p_0) = \emptyset$ for $0 < i \leq r$. Otherwise, φ_b would again have transversal homoclinic points.

By transversality, we have

$$\dim W^u(p_i) = \dim W^u(p_0) \quad \text{or} \quad \dim W^u(p_i) = \dim W^u(p_0) - 1$$

for each $0 \leq i \leq r$. Replacing $\{\varphi_\mu\}$ by $\{\varphi_\mu^{-1}\}$, if necessary, it is enough to consider the case in which there is a $1 \leq j \leq r$ such that $\dim W^u(p_j) = \dim W^u(p_0)$. This implies $\dim W^s(p_j) \cap W^u(p_0) = 0$ and hence p_0 is u-critical.

Observe that, having $\dim W^u(p_j) = \dim W^u(p_0)$ for some $1 \leq j \leq r$, we may indeed assume that for all $0 \leq i \leq r$

(a) $\dim W^u(p_i) = \dim W^u(p_0)$,
(b) $W^u(p_i)$ is transverse to the strong stable foliation $\mathcal{F}^{ss}$ of $W^s(p_0)$,
(c) $cl\, W^u(p_i) \cap W^{ss}(p_0) = \{p_0\}$.

For suppose *(a)* or *(b)* failed for some $1 \leq i \leq r$. Then p_0 is s-critical. Since $\dim W^u(p_j) = \dim W^u(p_0)$, we already know that p_0 is u-critical. Thus p_0 is bicritical and Proposition (2.1) has already ruled this out. Let us now suppose that *(b)* fails for $i = 0$; that is, $W^u(p_0)$ has a nontransverse intersection with $\mathcal{F}^{ss}$. We already know that $W^u(p_i)$ is transverse to $\mathcal{F}^{ss}$ for $1 \leq i \leq r$. Perturbing φ, if necessary, we may assume that $W^u(p_0)$ has a quasi-transversal intersection with some leaf F in $\mathcal{F}^{ss}$. Choose disks $D_1 \subset W^u(p_0)$ and $D_2 \subset F$ so that D_1 and D_2 have a quasi-transversal intersection. Let ξ

be an invariant center manifold in $W^s(p_0)$. Since $W^u(p_1, b)$ meets $W^s(p_0)$ transversally. and is transverse to $\mathscr{F}^{ss}$, it follows that each component of $W^u(p_1, b) \cap W^s(p_0)$ is an embedded curve whose closure near p_0 can be given as the graph of a smooth function from ξ to $W^{ss}(p_0)$ as in the next figure.

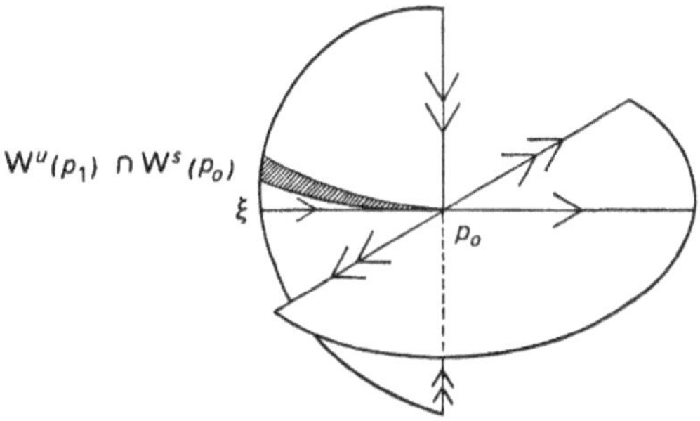

Fig. 6

As μ increases, pieces of $W^u(p_1, \mu)$ will sweep across the local part of $W^s(p_0)$ near p_0, so for each $\mu > b$ near b, $W^u(p_1, \mu)$ contains a disk $D_{1,\mu}$ C^2 near D_1.

On the other hand, $W^s(p_1, b)$ accumulates backward on $W^{ss}(p_0)$. For certain μ's near b and greater than b, $W^s(p_1, \mu)$ will contain a disk $D_{2,\mu}$ C^2 near D_2. Considering the continuous movements of $D_{1,\mu}$ and $D_{2,\mu}$ as μ varies, one sees that for certain μ's near b and greater than b, $D_{1,\mu}$ has a quasi-transverse intersection with $D_{2,\mu}$. Thus, if *(b)* fails for $i = 0$, there are μ's near b for which $W^u(p_{1,\mu})$ has a quasi-transverse intersection with $W^s(p_{1,\mu})$. In view of Theorem (1.3), we conclude that φ is not mildly stable. Finally if *(c)* failed for some i, then $cl\, W^u(p_i)$ would meet a fundamental domain for $W^{ss}(p_0)$. Since all intersections of $W^u(p_i)$ and $W^s(p_k)$ are transverse, this implies $W^u(p_i) \cap W^{ss}(p_0) \neq \emptyset$ as in the Morse-Smale case ([24]; Lemma (1.5)). As we have already mentioned, this gives transverse homoclinic points—an impossiblity.

Thus, we assume *(a)*, *(b)*, and *(c)* hold for all $0 \leq i \leq r$. We proceed to derive a contradiction from this, and, then, Proposition (2.2) will be proved.

From *(a)*, *(b)*, and *(c)* it follows that each component of $W^u(p_i) \cap W^s(p_0)$, $0 \leq i \leq r$, is an embedded curve whose closure near p_0 is the graph of a smooth function from the center manifold ξ to $W^{ss}(p_0)$.

Let p_{i_0} be a fixed point such that $W^u(p_{i_0}) \cap W^s(p_0) \neq \emptyset$, $W^u(p_0) \cap W^s(p_{i_0}) \neq \emptyset$, and for each $1 \leq i \leq r$ with $i \neq i_0$, $W^u(p_{i_0}) \cap W^s(p_i) = \emptyset$. Then, $W^u(p_{i_0}) \cap W^s(p_0)$ consists of finitely many curves which are permuted by φ_b. Also, the boundary of each such curve is $\{p_0, p_{i_0}\}$. By the boundary of a curve γ, we mean $cl\, \gamma - \gamma$, of course. Since $W^u(p_{i_0})$ is transverse to $W^s(p_0)$, we conclude that the closure of $W^s(p_0)$ near p_{i_0} is a finite union of closed half-spaces bounded by $W^s(p_{i_0})$. Also, since $W^u(p_0)$ has a transverse intersection with $W^s(p_{i_0})$ and $\dim W^u(p_0) = \dim W^u(p_{i_0})$, $W^u(p_0)$ accumulates on $W^u(p_{i_0})$ in the C^1 sense. Thus, we see that $W^u(p_0) \cap W^s(p_0)$ has a component γ which is an embedded curve whose boundary is $\{p_0, x\}$ where $x \in W^s(p_{i_0}) - \{p_{i_0}\} \cap W^u(p_0)$. This is depicted in the next figure.

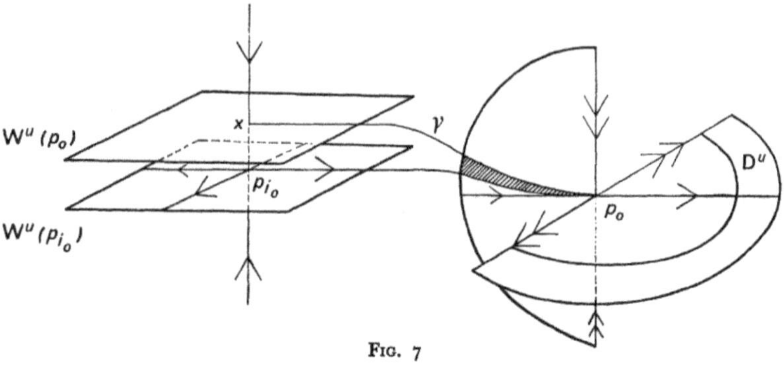

FIG. 7

We will show that the existence of γ leads to a contradiction. Let D be a small half-disk neighborhood about p_0 in $W^u(p_0)$ so that $\varphi_b(\partial D - W^{uu}(p_0)) \cap (\partial D - W^{uu}(p_0)) = \emptyset$. Let $D^u = cl(\varphi_b D - D)$ be a fundamental domain for $W^u(p_0)$. Choose $N > 0$ so that $n \geq N$ implies that $\varphi_b^{-n} x \in D$. It follows that, for $n \geq N$, $\varphi_b^{-n}\gamma$ is an arc whose intersection with $D^u - W^{uu}(p_0)$ has at least one component joining the two components of $\partial(D^u - W^{uu}(p_0))$. We may suppose that ∂D^u is transversal to each curve $\varphi_b^r \gamma$, $r \in \mathbf{Z}$. Hence, the set E of points $y \in D^u - W^{uu}(p_0)$, for which there is a sequence $y_i \in \varphi_b^{-n_i}(\gamma)$, $n_i \to \infty$, such that $y_i \to y$, is uncountable. But $E \subset cl\, W^s(p_0) \cap D^u$.

We assert that $E \cap W^s(p_0) = \emptyset$. For the moment, assume this assertion holds. Then $E \subset \bigcup_{1 \leq i \leq r} W^s(p_i)$. Since $\dim W^s(p_i) + \dim W^u(p_0) = \dim M$, this would put E in the countable set $D^u \cap \bigcup_{1 \leq i \leq r} W^s(p_i))$ which is a contradiction.

Thus, we must prove that $E \cap W^s(p_0) = \emptyset$. Suppose $E \cap W^s(p_0) \neq \emptyset$ and let $y \in E \cap W^s(p_0)$. Choose $y_i \in \varphi_b^{-n_i}(\gamma)$, $n_i \to \infty$, so that $y_i \to y$ as $i \to \infty$. We claim that y_i goes to infinity in $W^s(p_0)$; *i.e.* given any closed disc $F \subset W^s(p_0)$ transverse to D^u there is an $i_0 > 0$ so that $y_i \notin F$ for $i \geq i_0$. This follows from the facts that $\varphi_b^m \gamma \cap \varphi_b^r \gamma = \emptyset$ for $m \neq r$ and $F \cap D^u$ has only finitely many components. Now let D^s and D^{ss} be fundamental domains for $W^s(p_0)$ and $W^{ss}(p_0)$, respectively, and let D^{uu} be a fundamental domain for $W^{uu}(p_0)$. There is an $N > 0$ so that $\varphi_b^N(y) \in D^s$. Since $\varphi_b^N(y_i)$ converges to the point $\varphi_b^N(y)$ in D^s and $\varphi_b^N(y_i)$ goes to infinity in $W^s(p_0)$, there is an $m_i > 0$ such that $\varphi_b^{N+m_i}(y_i)$ accumulates on D^{uu}. Thus, $cl\, W^s(p_0) \cap D^{uu} \neq \emptyset$. But this implies that φ_b has transversal homoclinic points which is impossible. Thus $E \cap W^s(p_0) = \emptyset$ and Proposition (2.2) is proved.

Proposition **(2.3)**. — *If φ is mildly stable, then φ_b cannot have a non-critical 1-cycle.*

Proof. — As before, if φ_b has such a cycle, it must have a saddle-node $o(p)$ which is in the cycle. Since the cycle is non-critical, $W^u(o(p))$ is transverse to the strong stable foliation of $W^s(o(p))$ and $W^s(o(p))$ is transverse to the strong unstable foliation of $W^u(o(p))$. Let n_1 be the period of p. Using invariant manifold theory one sees that $W^u(o(p)) \cap W^s(o(p))$ is a finite union of C^∞ circles $S_{1b}, \ldots, S_{kb}$ which are permuted by $\varphi_b^{n_1}$. Taking n_1 larger, we may assume $\varphi_b^{n_1}(S_{ib}) = S_{ib}$ for $1 \leq i \leq k$. Replacing φ by a suitable C^∞ approximation, we may assume for μ near b that there is a C^∞ circle $S_{1\mu}$ near S_{1b} which is invariant by $\varphi_\mu^{n_1}$, and the rotation number of $\varphi_\mu^{n_1} \mid S_{1\mu}$ is not constant as a function of μ near b. From the recent work of M. Herman [9] we may find a μ near b such that $\varphi_\mu^{n_1} \mid S_{1\mu}$ is C^∞ conjugate to a geometric rotation through an angle α with $\alpha/2\pi$ irrational. By a small modification of φ to φ', we may make $\varphi_\mu'^{n_1} \mid S_{1\mu}$ C^∞ conjugate to a rotation through an angle α' with $\alpha'/2\pi$ rational. This means that φ'_μ has uncountably many periodic points of the same period. But, since each φ_μ is elementary, it has only finitely many periodic points of a given period. Thus φ is not mildly stable.

Remark. — The previous propositions leave open the possibility that a mildly stable arc might have a critical saddle-node in a 1-cycle at φ_b. We will prove in section 4 that this cannot occur if the stable or unstable manifold of the saddle-node is one dimensional. In particular, it cannot occur if $\dim M = 2$. In general, for $\dim M > 2$, we feel that this cannot occur, but we have no complete proof at this time. The next proposition shows that the situation regarding stability is better.

Proposition (**2.4**). — *If φ is stable, then φ_b has no cycles and no semi-critical saddle-nodes.*

Proof. — We have already taken care of cycles of length bigger than one and non-critical 1-cycles in Propositions (2.2) and (2.3). Suppose φ_b has a semi-critical 1-cycle containing the saddle-node orbit $o(p)$. We assume p is fixed by φ_b and $W^u(p)$ has a non-transverse intersection with the strong stable foliation $\mathscr{F}^{ss}$ of $W^s(p)$. The other cases are handled similarly. Perturbing, if necessary, we may assume that $\mathscr{F}^{ss}$ is a C^2 foliation, that all intersections of $W^u(p)$ with leaves of $\mathscr{F}^{ss}$ are transverse or quasi-transverse, and that all the eigenvalues of $d\varphi_b$ on $T_pW^{ss}(p)$ have multiplicity one with distinct norms. Let α be the weakest contracting eigenvalue of $d\varphi_b$ on $T_pW^{ss}(p)$, and let C_p be a C^1 invariant manifold in $W^{ss}(p)$ tangent at p to the eigenspace of α. Let H_p be a C^1 invariant manifold containing $W^u(p)$ and C_p as in the proof of Theorem (1.1). Let $x \in W^u(p) \cap W^s(p)$ be a quasi-transversal intersection of $W^u(p)$ and $\mathscr{F}_x^{ss}$ at x. If γ is a curve in $H_p \cap W^s(p)$ transverse to $W^u(p)$ at x, then γ is transverse to $\mathscr{F}_x^{ss}$ in $W^s(p)$ at x and, hence, projects diffeomorphically near x along the leaves of $\mathscr{F}^{ss}$ into an invariant center manifold ξ for φ_b. Let D be a fundamental domain for $W^{ss}(p)$. Thus D is a compact set in $W^{ss}(p)$ such that $\varphi_b^2 D \cap D = \emptyset$ and $W^{ss}(p) - \{p\} \subset \bigcup_{n \in \mathbb{Z}} \varphi_b^n(D)$. Let U be a small compact neighborhood of D. If $y \in \gamma$ is near x, then $\rho(y, x) \sim |\alpha|^{n(y)}$ where $n(y)$ is the least positive integer n such that $\varphi_b^{-n}(y) \in U$. Here, as in the proof of Theorem (1.1), we use $\rho(y, x) \sim |\alpha|^n$ to mean that $\dfrac{\rho(y, x)}{|\alpha|^n}$ is bounded and bounded away from zero independent of n.

Now let φ' be a perturbation of φ such that $b(\varphi') = b(\varphi)$, $|\alpha'| < |\alpha|$, and $\mathscr{F}^{ss}$ is still a C^2 foliation. Using Chapter II, we may assume that on the center manifold ξ through p, φ_b is the time-one map of a C^∞ vector field X_b which vanishes only at p. Also, there is a corresponding vector field X'_b for φ'_b. Suppose there is a continuous conjugacy $\{h_\mu\}$ between φ and φ'. By Theorems (3.2) and (3.4) of Chapter II, we know that

(1) h_b is C^∞ along $\xi - \{p\}$;

(2) h_b maps X_b to X'_b; and

(3) h_b maps the strong stable foliation $\mathscr{F}^{ss}$ to the corresponding strong stable foliation $\mathscr{F}'^{ss}$.

Since $\mathscr{F}'^{ss}$ is C^2 near $\mathscr{F}^{ss}$, all intersections of $W^u(p')$ and leaves of $\mathscr{F}'^{ss}$ will be transversal or quasi-transversal. As in the proof of Lemma (1.2), this implies that $h_b x$ is a quasi-transversal intersection of $W^u(p')$ and $\mathscr{F}'^{ss}_{h_b x}$. Applying the reasoning in section D of the proof of Theorem (1.1) to $W^{ss}(p)$, we see that, since H_p is transverse to $\mathscr{F}^{ss}_x$, the invariant manifold in $W^{ss}(p)$ tangent to the sum of the eigenspaces complementary to the eigenspace of α has a dynamical characterization. It is the set of points where the backward orbit of $\mathscr{F}^{ss}_x$ accumulates. Thus, $h_b(H_p \cap W^{ss}(p))$ is in a sector about $H'_p \cap W^{ss}(p')$ in $W^{ss}(p')$. This implies that if y is near x in γ and $\rho(x, y) \sim |\alpha|^{n(y)}$, then $\rho(h_b y, h_b x) \sim |\alpha'|^{n(y)}$. Hence,

$$\lim_{y \to x} \frac{\rho(h_b y, h_b x)}{\rho(y, x)} = 0. \tag{4}$$

If $\pi : W^s(p) \to \xi$ is the projection along the leaves of $\mathscr{F}^{ss}_b$, then (4) implies that h_b has derivative zero at $\pi(x)$. But this contradicts the fact that $h_b(X_b(\pi x)) = X'_b(h_b \pi(x))$.

The proof of the fact that stability of φ implies that φ_b has no saddle-node orbit $\mathscr{O}(p)$ which is semi-critical via some other periodic orbit $\mathscr{O}(q)$ is similar. If say $W^u(\mathscr{O}(q))$ has a non-transverse intersection with $\mathscr{F}^{ss}_b$, we repeat the preceding argument replacing $W^{ss}(\mathscr{O}(p))$ by $W^s(\mathscr{O}(q))$. The remaining case of $W^s(\mathscr{O}(q))$ having a non-transverse intersection with $\mathscr{F}^{uu}_b$ follows replacing $\{\varphi_\mu\}$ by $\{\varphi_\mu^{-1}\}$.

Remark. — We observe that stable arcs in $\mathscr{A}$ must lie in $\mathscr{B}$. This follows from a somewhat more general observation, namely that any stable arc $\{\varphi_\mu\}$, not necessarily starting in MS, with a bifurcation for $\mu = b$ such that the limit set of φ_b has finitely many orbits has the following properties:

— all stable, strong stable, unstable, and strong unstable manifolds of φ_b intersect transversally;

— φ_b has exactly one non-hyperbolic periodic orbit, which is either a flip or a non-critical saddle-node without a cycle; this non-hyperbolic periodic orbit unfolds generically; such a φ_b can have no cycles because it has finitely many orbits in its limit set;

— there is an $\varepsilon > 0$ such that if $\mu \in (b - \varepsilon, b + \varepsilon) - \{b\}$, then φ_μ is MS.

This statement follows from the previous arguments in the following way. If all the periodic orbits of φ_b are hyperbolic then, since the limit set has only finitely many orbits and φ_b is not structurally stable, there must be a non-transversal intersection of a stable and an unstable manifold. By the results of section one of this chapter the arc $\{\varphi_\mu\}$ would then be unstable. Also, the non-hyperbolic periodic points have to unfold generically otherwise we would not even have local stability. It is also clear that no more than one orbit of φ_b can be non-hyperbolic. If φ_b has a Hopf point or a non-critical 1-cycle, then there are nearby arcs with a smooth invariant circle with irrational rotation; as observed in the present section that also contradicts stability of the arc. Also in this section we saw that a saddle-node with criticallity and or a cycle of length bigger than one is impossible if the arc $\{\varphi_\mu\}$ is stable. If φ_b has a non-transverse intersection of the stable, strong stable, unstable, or strong unstable manifold of the non-hyperbolic periodic point with any of the other stable or unstable manifolds, then there is arbitrarily near $\{\varphi_\mu\}$ an arc $\{\varphi'_\mu\}$ and a $\bar{\mu} \in \mathbf{R}$ near b such that $\varphi_{\bar{\mu}}$ has a non-transversal intersection of a stable and an unstable manifold of hyperbolic periodic orbits. Again by the results of section 1 this implies $\{\varphi'_\mu\}$, and hence, $\{\varphi_\mu\}$ is unstable. Finally, we indicate why $\varphi_\mu \in \mathrm{MS}$ for μ near b and $\mu \neq b$. Since φ_b has no cycles and the limit set of φ_b has only finitely many orbits, the Birkhoff center of φ_b must be finite. Then the arguments in [15] show that the limit set of φ_b equals the Birkhoff center, and, hence, must also be finite. Because φ_b is elementary, it then follows that $\varphi_\mu \in \mathrm{MS}$ for μ near b and $\mu \neq b$.

With arguments similar to those in the preceding paragraph, one can show that a mildly stable arc in $\mathscr{A}$ must lie in $\mathscr{B}$. There are, however, left stable arcs in $\mathscr{A}$ which are not in $\mathscr{B}$: φ_b might, for example, have two Hopf points or flips. All such arcs are left conjugate to left stable arcs in $\mathscr{B}$.

3. Endomorphisms of the circle

In this section we first extend the notion of rotation number, defined by H. Poincaré for diffeomorphisms of the circle, to endomorphisms of degree one. Instead of a number we get in general a closed interval of the real line, which we call rotation set. These rotation sets are then used to analyze a class of 1-parameter families of endomorphisms with non-degenerate folds (see *(b)* below). We show that each such family must go through homoclinic trajectories with folds (see Theorem (3.7)). This result has a direct application to the bifurcation of diffeomorphisms exhibiting a saddle-node with one cycle described in the previous section. If the saddle-node has a one dimensional stable or unstable manifold, they must go through a non-transversal homoclinic orbit. An interesting question is if such families of endomorphisms, for which the rotation sets vary, have bifurcation sets of positive measure. For the diffeomorphism case, see [10].

A) *Rotation sets for endomorphisms of the circle*

We identify the circle S^1 with $\mathbf{R}/\mathbf{Z}$. By End(S) we denote the set of continuous maps $\Phi : S^1 \to S^1$ of degree 1. On End(S^1) we use the usual C^0 topology. For each

$\Phi \in \mathrm{End}(S^1)$ there is a $\bar{\Phi} : \mathbf{R} \to \mathbf{R}$ such that $\pi\bar{\Phi} = \Phi\pi$ where $\pi : \mathbf{R} \to S^1 \to \mathbf{R}/\mathbf{Z}$ is the canonical projection. $\bar{\Phi}$ is called a lift of Φ and it is unique up to the addition of an integer. Each such $\bar{\Phi}$ satisfies $\bar{\Phi}(x+1) = \bar{\Phi}(x) + 1$.

For $\Phi \in \mathrm{End}(S^1)$ with lift $\bar{\Phi}$ and $x \in \mathbf{R}$ we define the rotation number $\rho(\bar{\Phi}, x)$ to be $\limsup_{n\to\infty} \frac{1}{n}(\bar{\Phi}^n(x) - x)$ and the rotation set $\rho(\bar{\Phi})$ to be closure $\{\rho(\bar{\Phi}, x) \mid x \in \mathbf{R}\}$. Note that if we take a different lift, say $\bar{\bar{\Phi}}$, of Φ then $\rho(\bar{\Phi}, x)$ and $\rho(\bar{\bar{\Phi}}, x)$, and hence also $\rho(\bar{\Phi})$ and $\rho(\bar{\bar{\Phi}})$, are equal up to translation by some integer. If $x' \in \mathbf{R}$ with $\pi(x') = \pi(x)$ then $\rho(\bar{\Phi}, x') = \rho(\bar{\Phi}, x)$. Hence, up to translation by integers, $\rho(\bar{\Phi}, x)$ and $\rho(\bar{\Phi})$ are invariants of Φ, $\pi(x)$; if no confusion seems possible we may denote them by $\rho(\Phi, \pi(x))$, $\rho(\Phi)$. We note that $\rho(\Phi, p)$ and $\rho(\Phi)$, $p \in S^1$, are topological invariants: if $h : S^1 \to S^1$ is a homeomorphism, then

$$\rho(\Phi, p) = \rho(h\Phi h^{-1}, h(p)) \quad \text{and} \quad \rho(\Phi) = \rho(h\Phi h^{-1}).$$

Finally, if Φ is an orientation preserving homeomorphism, then $\rho(\Phi)$ is the usual rotation number [29].

Lemma (3.1). — *If* $\Phi \in \mathrm{End}(S^1)$ *and* $\bar{\Phi}$ *is a lift of* Φ *and if there is no periodic point of* Φ *with rotation number* $\frac{p}{q}$, $p \in \mathbf{Z}$, $q \in \mathbf{N}$, i.e. *if there is no* $x \in \mathbf{R}$ *with* $\Phi^q(\pi(x)) = \pi(x)$ *and* $\rho(\bar{\Phi}, x) = \frac{p}{q}$, *then* $\rho(\bar{\Phi})$ *is contained in* $\left\{x \in \mathbf{R} \mid x < \frac{p}{q}\right\}$ *or in* $\left\{x \in \mathbf{R} \mid x > \frac{p}{q}\right\}$.

Proof. — If, for some $x \in \mathbf{R}$, $\bar{\Phi}^q(x) - x = p$ then $\pi(x)$ is a periodic point with rotation number $\frac{p}{q}$; so this does not happen. Hence either $\bar{\Phi}^q(x) - x < p$ for all $x \in \mathbf{R}$ or $\bar{\Phi}^q(x) - x > p$ for all $x \in \mathbf{R}$. Since $\bar{\Phi}^q(x) - x$ is periodic in x (with period 1) there is some $\varepsilon > 0$ such that $\bar{\Phi}^q(x) - x < p - \varepsilon$ for all $x \in \mathbf{R}$ or $\bar{\Phi}^q(x) - x > p + \varepsilon$ for all $x \in \mathbf{R}$. But then $\rho(\bar{\Phi}) \subset \left\{x \in \mathbf{R} \mid x \leq \frac{p-\varepsilon}{q}\right\}$, respectively $\rho(\Phi) \subset \left\{x \in \mathbf{R} \mid x \geq \frac{p+\varepsilon}{q}\right\}$.

Corollary (3.2). — *If* $\Phi \in \mathrm{End}(S^1)$, $\alpha, \beta \in \rho(\Phi)$ *and* $\alpha \leq \frac{p}{q} \leq \beta$ *for some rational number* $\frac{p}{q}$, *then* Φ *has a periodic point with rotation number* $\frac{p}{q}$ *and hence* $\frac{p}{q} \in \rho(\Phi)$. *Since* $\rho(\Phi)$ *is closed* (by definition), $\rho(\Phi)$ *must be either a single point in* $\mathbf{R}$ *or a closed interval.*

Now we want to show that $\rho(\Phi)$ depends continuously on Φ. For this we introduce the following notation: if $\Phi \in \mathrm{End}(S^1)$, $\bar{\Phi}$ a lifting of Φ then $\rho_1(\bar{\Phi}), \rho_2(\bar{\Phi}) \in \mathbf{R}$ are such that $\rho(\bar{\Phi}) = [\rho_1(\bar{\Phi}), \rho_2(\bar{\Phi})]$.

Proposition (3.3). — *Let* $U \subset \mathrm{End}(S^1)$ *be some open set such that there is a continuous mapping* $\Phi \to \bar{\Phi}$ *which assigns to each* Φ *a lifting* $\bar{\Phi}$ (in order that $\bar{\Phi}$ depends continuously on Φ, U should not be too big). *Then the functions* $\Phi \mapsto \rho_1(\bar{\Phi})$ *and* $\Phi \mapsto \rho_2(\Phi)$ *on* U *are continuous.*

Proof. — We observe that for any rational number $\frac{p}{q}$, we have that $\frac{p}{q} < \rho_1(\bar{\Phi})$ is equivalent with $\bar{\Phi}^q(x) - x > p$ for all $x \in [0, 1]$ which is an "open condition", *i.e.* the set of $\Phi \in U$ with $\rho_1(\bar{\Phi}) > \frac{p}{q}$ is open. Analogously, the set of those $\Phi \in U$ with $\rho_2(\bar{\Phi}) < \frac{p}{q}$ is open.

Finally, $\frac{p}{q} \in (\rho_1(\bar{\Phi}), \rho_2(\bar{\Phi}))$ if and only if for some big $N \in \mathbf{N}$, there are $x, y \in [0, 1]$ with $\bar{\Phi}^{N.q}(x) - x > N.p + 1$ and $\bar{\Phi}^{N.q}(y) - y < N.p - 1$. Also this condition is open, hence $\rho_1(\bar{\Phi})$ and $\bar{\rho}_2(\bar{\Phi})$ depend continuously on Φ.

Proposition (3.4). — *Let* $\Phi \in \mathrm{End}(S^1)$ *with lifting* $\bar{\Phi}$ *and let* $\frac{p}{q} \in (\rho_1(\bar{\Phi}), \rho_2(\bar{\Phi}))$. *Then there is a periodic point* $s \in S^1$ *of* Φ *with rotation number* $\frac{p}{q}$ *such that* $W^u_\Phi(s) = S^1$, *where* $W^u_\Phi(s) = \bigcap_U [\bigcup_{i \in \mathbf{N}} \Phi^i(U)]$, *the intersection being taken over neighborhoods* U *of* s.

Proof. — We may assume that $\frac{p}{q} = 0$ (if not we replace Φ by Φ^q and choose an appropriate lifting). Let X denote the projection of the fixed point set of $\bar{\Phi}$ in S^1; since X is closed and non-empty, $S^1 - X$ consists of open intervals. Let U be such an interval and $\bar{U} = (\bar{s}_1, \bar{s}_2)$ a lifting of U. Then $\bar{\Phi}(x) - x$, for $x \in \bar{U}$, is always positive or always negative. In the first case, $U \subset W^u_\Phi(\pi(\bar{s}_1))$, in the second case, $U \subset W^u_\Phi(\pi(\bar{s}_2))$. If $\bigcup_{n \geq 0} \Phi^n(U) = S^1$ we may take $s = \pi(\bar{s}_1)$ or $s = \pi(s_2)$ and are done. If on the other hand for each component U of $S^1 - X$, $\bigcup_{n \geq 0} \Phi^n(U) \neq S^1$, then $\Phi^{-n}([0, 1]) \subset [-1, 2]$ for all $n \geq 0$, so we would clearly have $\rho(\bar{\Phi}) = \{0\}$. This would be against the assumptions.

Remark (3.5). — Let $\mathscr{B} \subset \mathrm{End}(S^1)$ denote the subset of those $\Phi \in \mathrm{End}(S^1)$ which are C^1, whose first derivative is of bounded variation and whose critical set consists of a finite and non-zero number of generic folds, *i.e.* points where Φ is locally C^0 conjugate with the map $y = x^2$. By a theorem of Block and Franke [2], every Φ in $\mathscr{B}$ has a periodic point. This is equivalent with saying that if $\Phi \in \mathscr{B}$, $\rho(\Phi)$ cannot consist of only one irrational real number. This together with the continuity of ρ_1 and ρ_2 implies that if $\mathbf{R} \ni \sigma \mapsto \Phi_\sigma$ is a continuous arc in $\mathrm{End}(S^1)$ with image in $\mathscr{B}$ then $\rho(\Phi_\sigma)$ is constant (as function of σ) or, for some $\bar{\sigma}$, $\rho(\Phi_{\bar{\sigma}})$ has a non-empty interior. This observation will be useful in the next section.

Finally, we pose the following conjecture which extends Herman's result for arcs of diffeomorphisms [10]. Let $\{\varphi_\mu\}$ be a C^1 arc of C^∞ endomorphisms of S^1 with rotation sets $\{[\rho_1(\mu), \rho_2(\mu)]\}$.

Conjecture **(3.6).** — *If the rotation set of φ_μ varies with μ, then the set of μ's for which either $\rho_1(\mu)$ or $\rho_2(\mu)$ is irrational has positive (Lebesgue) measure.*

B) *Some one-parameter families of endomorphisms of* S^1

In this sub-section we consider continuous (or smooth) families Φ_σ, $\sigma \in \mathbf{R}$, of endomorphisms of S^1 such that

— for each $\sigma \in \mathbf{R}$, $\Phi_\sigma \in \mathscr{B}$, see Remark (3.5);
— the rotation set $\rho(\Phi_\sigma)$ is non-constant, as a function of σ;
— for each rational number $\frac{p}{q}$, there is a locally finite set $\Sigma_{\frac{p}{q}} \subset R$ such that for $\sigma \notin \Sigma_{\frac{p}{q}}$, all periodic points of Φ_σ with rotation number $\frac{p}{q}$ are hyperbolic.

We shall prove for these families:

Theorem **(3.7).** — *For Φ_σ as above there are an interval $(\sigma_1, \sigma_2) \subset \mathbf{R}$, a hyperbolic periodic point s_σ, $\sigma_1 < \sigma < \sigma_2$ of Φ_σ such that $W^u_{\Phi_\sigma}(s_\sigma)$ is S^1 for all $\sigma \in (\sigma_1, \sigma_2)$ and a point t_σ, $\sigma_1 < \sigma < \sigma_2$ in the critical image of Φ^n_σ* (for some n) *such that:*

— *both s_σ and t_σ depend continuously on σ;*
— *the curves $\sigma \mapsto (s_\sigma, \sigma)$ and $\sigma \mapsto (t_\sigma, \sigma)$ in $S^1 \times (\sigma_1, \sigma_2)$ cross one another;*
— *if $\overline{\Phi}_\sigma$ is a lifting of Φ_σ, depending continuously on σ, and if $\overline{s}_\sigma$ is a lifting of s_σ, then t_σ is the projection of an end point of the interval $\overline{\Phi}^n_\sigma(\overline{s}_\sigma, \overline{s}_\sigma + 1)$.*

Remark **(3.8).** — For $\Phi \in \mathrm{End}(S^1)$, with lifting $\overline{\Phi}$ one can construct a 1-parameter family of endomorphisms Φ_σ by putting $\Phi_\sigma(\pi(x)) = \pi(\overline{\Phi}(x+\sigma))$. We note that in the set of C^r endomorphisms φ of S^1, $r \geq 2$, which are not diffeomorphisms, there is a residual subset for which the corresponding one parameter families of endomorphisms satisfy the assumptions in Theorem (3.7).

Proof. — Since $\rho(\Phi_\sigma)$ is non-constant, there is some σ_3 such that $\rho(\Phi_{\sigma_3})$ has interior points, see (3.5), and such that $\rho(\Phi_\sigma)$ is not locally constant on a neighborhood of σ_3. Choose a rational number $\frac{p}{q}$ in the interior of the rotation set. Since in any neighborhood of σ_3 there are infinitely many points where the function $\rho(\Phi_\sigma)$ is not locally constant, we can choose σ_4 such that Φ_{σ_4} has a hyperbolic periodic point s with rotation number $\frac{p}{q}$ such that $W^u_{\Phi_{\sigma_4}}(s) = S^1$ (see (3.4)), and such that $\rho(\Phi_\sigma)$ is not locally constant at σ_4. Now if follows that there is some neighborhood U of σ_4 in $\mathbf{R}$ and a continuous function $U \ni \sigma \mapsto s_\sigma \in S^1$ such that, for $\sigma \in U$, s_σ is a hyperbolic periodic point and such that $W^u_{\Phi_\sigma}(s_\sigma) = S^1$. Next, since $\rho(\Phi_\sigma)$ is not locally constant at σ_4 there is a $\overline{\sigma}$ in U for which $\rho(\varphi_{\sigma_4}) \neq \rho(\varphi_{\overline{\sigma}})$. Then one can find two rational numbers $\frac{p_1}{q_1}, \frac{p_2}{q_1}$ with $p_1 < p_2$ such that the closed interval $\left[\frac{p_1}{q_1}, \frac{p_2}{q_1}\right]$ is in $(\rho(\varphi_{\sigma_4}) - \rho(\varphi_{\bar\sigma})) \cup (\rho(\varphi_{\bar\sigma}) - \rho(\varphi_{\sigma_4}))$. We suppose $\left[\frac{p_1}{q_1}, \frac{p_2}{q_1}\right]$ is in $\rho(\varphi_{\sigma_4}) - \rho(\varphi_{\bar\sigma})$ since the other case is similar. Then, with $\bar\Phi_\sigma$, $\bar s_\sigma$ liftings of Φ_σ, s_σ,

$$\bar\Phi^{3q_1}_{\sigma_4}(\bar s_{\sigma_4}, \bar s_{\sigma_4} + 1) \supset [\bar s_{\sigma_4} + 1 + 3p_1, \bar s_{\sigma_4} + 3p_2]$$

and $$\bar\Phi^{3q_1}_{\bar\sigma}(\bar s_{\bar\sigma}, \bar s_{\bar\sigma} + 1) \subset \text{complement of } [\bar s_{\bar\sigma} + 1 + 3p_1, \bar s_{\bar\sigma} + 3p_2];$$

see also Lemma (3.1). This means that one of the endpoints of the interval $\Phi^{3q_1}_\sigma(\bar s_\sigma, \bar s_\sigma + 1)$, σ between σ_4 and $\bar\sigma$, has to cross over the interval $[\bar s_\sigma + 1 + 3p_1, \bar s_\sigma + 3p_2]$. Since this last interval has length at least 2, one of the endpoints of $\Phi^{3q_1}_\sigma(\bar s_\sigma, \bar s_\sigma + 1)$ crosses, after projection on S^1, over s_σ when σ goes from σ_4 to $\bar\sigma$.

Finally we have to show that if, for σ between σ_4 and $\bar\sigma$, one endpoint of $\bar\Phi^{3q_1}(\bar s_\sigma, \bar s_\sigma + 1)$ lies in $[\bar s_\sigma + 1 + 3p_1, \bar s_\sigma + 3p_2]$ then this endpoint is in the critical image of $\bar\Phi^{3q_1}_\sigma$. To show this, it is enough to show that neither $\bar\Phi^{3q_1}_\sigma(\bar s_\sigma)$ nor $\bar\Phi^{3q_1}_\sigma(\bar s_\sigma + 1)$ can be in $[\bar s_\sigma + 1 + 3p_1, s_\sigma + 3p_2]$. But this follows from the fact that the rotation number of s_σ is $\frac{p}{q}$ and that $\frac{p}{q}$ does not lie between $\frac{p_1}{q_1}$ and $\frac{p_2}{q_1}$ (because otherwise $\frac{p}{q} \notin \rho(\Phi_\sigma)$, see (3.2)).

Now it is clear that we can choose (σ_1, σ_2) to be an interval between σ_4 and $\bar\sigma$ such that if σ goes from σ_1 to σ_2, one endpoint of $\bar\Phi^{3q_1}_\sigma(\bar s_\sigma, \bar s_{\sigma+1})$ crosses over the interval $[\bar s_\sigma + 1 + 3p_1, \bar s_\sigma + 3p_2]$. We choose then $n = 3q_1$ and we choose t_σ to be the projection of that endpoint.

C) *Application to bifurcation theory*

The analysis of families of endomorphisms in the last section leads to results also when analyzing certain families of diffeomorphisms. An explicit formulation of what we need in this direction is:

Remark **(3.9)**. — Let $\varphi_\sigma : S^1 \to S^1$ be a one-parameter family $(\sigma \in \mathbf{R})$ of endomorphisms as in Theorem (3.7). Let $\Phi_{\sigma,\mu} : S^1 \times [0, 1] \to S^1 \times [0, 1]$ be a 2-parameter family of C^2 maps depending continuously on (σ, μ) such that:

— $\Phi_{\sigma,0}(x, t) = (\varphi_\sigma(x), T_\sigma(x))$ for some function T_σ;
— for μ positive, $\Phi_{\sigma,\mu}$ is a diffeomorphism into.

Then, for $\bar\mu$ sufficiently small, there is a $\sigma_{\bar\mu} \in (\sigma_1, \sigma_2)$ such that $\Phi_{\sigma_{\bar\mu},\bar\mu}$ has a non-transversal homoclinic point.

Even more holds: for $\bar\mu$ sufficiently small, and any curve $(\sigma, f(\sigma))$ with $0 < f(\sigma) < \bar\mu$, $\sigma \in (\sigma_1, \sigma_2)$, there is a σ_f such that $\Phi_{\sigma, f(\sigma_f)}$ has a non-transversal homoclinic point.

The proof of this statement follows from the following continuity considerations. The map $\Phi_{\sigma,\mu}$ has a hyperbolic periodic point $s_{\sigma,\mu} = (x_{\sigma,\mu}, t_{\sigma,\mu})$ such that $x_{\sigma,0} = s_\sigma$

as in Theorem (3.7). We observe that $W^s(s_{\sigma,\mu})$ depends continuously on (σ, μ) and $W^s(s_{\sigma,0}) = \{x_{\sigma,0}\} \times [0, 1]$ (note that $\varphi_\sigma : S^1 \to S^1$ was expanding at s_σ). One can choose a local unstable manifold $W^u_{loc}(s_{\sigma,\mu})$ as an embedding (or its image) which depends continuously on (σ, μ). For some k, $\varphi^k_{\sigma,0}(W^u_{loc}(x_{\sigma,0}))$ has a point T_σ in its critical image which moves over $s_{\sigma,0}$ by Theorem (3.7). From this and the above continuity we see that, for each small enough $\mu > 0$, there is a value of σ such that $\Phi^k_{\sigma,0}(W^u_{loc}(s_{\sigma,\mu}))$ is not transversal with respect to $W^s(s_{\sigma,\mu})$.

4. The saddle-node with 1-cycle in dimension 2

Let $\{\varphi_\mu\} : M \to M$ be an arc of diffeomorphisms of a compact 2-manifold M so that for $\mu < b$, φ_μ is Morse-Smale and such that for $\mu = b$, φ_b has a saddle-node at p (we assume p to be a fixed point of φ_b; the case where p is periodic can be handled analogously). Further we assume that φ_b has a 1-cycle containing p; *i.e.* there are non-trivial intersections of $W^u(p)$ with $W^s(p)$ and there is no periodic point q such that both $W^u(p) \cap W^s(q)$ and $W^u(q) \cap W^s(p)$ are non-empty.

Theorem **(4.1)**. — *Under the above hypotheses* $\{\varphi_\mu\}$ *is not mildly stable.*

Remark **(4.2)**. — It will be evident from the proof of Theorem (4.1) that the same result holds if $\dim M > 2$ and φ_b has a saddle-node in a 1-cycle whose stable or unstable manifold is one dimensional; *i.e.* a normally repelling or attracting saddle-node. An open dense set of these arcs will create homoclinic tangencies. Thus, in dimension two they will contain diffeomorphisms with infinitely many sinks or sources [22].

Proof of Theorem (4.1). — An arc $\{\varphi_\mu\}$ is already not mildly-stable if in any neighborhood of φ_μ there is a non mildly-stable arc. Hence we may, without loss of generality, impose generic conditions on $\{\varphi_\mu\}$. In particular, we assume that p unfolds generically. Also, we may assume that the eigenvalues of $(d\varphi_b)_p$ are 1 and α with $0 < \alpha < 1$; in case $\alpha > 1$ we take φ_μ^{-1}, in case $-1 < \alpha < 0$ we take φ_μ^2 and in case $\alpha < -1$ we take φ_μ^{-2} instead of φ.

From the fact that, for $\mu < b$, φ_μ is Morse-Smale and the fact that $\{\varphi_\mu\}$ unfolds generically at p, we may assume that, near p, φ_μ has two fixed points for $\mu < b$, one fixed point for $\mu = b$ and no fixed points for $\mu > b$; we also assume that

$$W^u(p, \varphi_b) \cap \partial(W^s(p, \varphi_b)) = \{p\}.$$

We first observe that $W^u(p, \varphi_b) \subset W^s(p, \varphi_b)$; this follows from the following two facts:

— $W^u(p, \varphi_b) - \{p\} \cap W^s(p, \varphi_b)$ is open in $W^u(p, \varphi_b) - \{p\}$ and non-empty;

— for any filtration $\{M_i\}$ of M for φ_b (see [20]), with $p \in M_{i_0} - M_{i_0-1}$, the set of points in $W^u(p, \varphi_b) - \{p\}$, which go eventually into M_{i_0-1} is open and equals $(W^u(p, \varphi_b) - \{p\}) - W^s(p, \varphi_b)$.

In the case that $W^u(p, \varphi_b)$ closes to a smooth circle, p has a non-critical 1-cycle. Then the instability here was proved in Proposition (2.3). Thus, we may suppose $W^u(p, \varphi_b)$ does not close to a smooth circle. This means that the projection of $W^u(p, \varphi_b)$ on a center manifold in $W^s(p, \varphi_b)$ along the strong stable foliation does not have maximal rank everywhere.

We shall approximate $\{\varphi_\mu\}$ by a family $\{\tilde{\varphi}_\mu\}$ such that non-transversal intersections of stable and unstable manifolds occur for certain μ's arbitrarily near the first bifurcation of $\{\tilde{\varphi}_\mu\}$. From this, we infer that $\{\tilde{\varphi}_\mu\}$, and, hence also $\{\varphi_\mu\}$, is not mildly stable. This will prove Theorem (4.1). For convenience of notation, let us assume that $\{\varphi_\mu\}$ is defined for μ near 0 and that $\mu = 0$ is the bifurcation point instead of $\mu = b$. For each diffeomorphism $\tilde{\varphi}_\mu$ we require that there is a smooth vector field X_μ, defined for μ close to 0 on a neighborhood of p, such that the time 1 map $X_{\mu,1}$ of X_μ, wherever defined, equals $\tilde{\varphi}_\mu$. Also we require that there are smooth coordinates y, z (which may depend on μ) such that X_μ locally has the form

$$X_\mu = Y_\mu(y)\,\frac{\partial}{\partial y} + Z_\mu(y)\cdot z\cdot\frac{\partial}{\partial z}$$

where Y and Z are smooth functions of (y, μ), $Z_0(0) < 0$, $\dfrac{\partial^2 Y_0(0)}{\partial y^2} > 0$, and $\left\{Y_\mu(y)\,\dfrac{\partial}{\partial y}\right\}$ is a saddle-node arc. The fact that such $\tilde{\varphi}_\mu$ exists arbitrarily close to φ_μ follows from [38], [39].

Choose, for $\mu \geq 0$ a fundamental domain $D_{2,\mu}$ for $\tilde{\varphi}_\mu$ in the positive y-axis, smoothly depending on μ. For some big m, $\tilde{\varphi}_0^m(D_{2,0})$ will again be in our coordinate

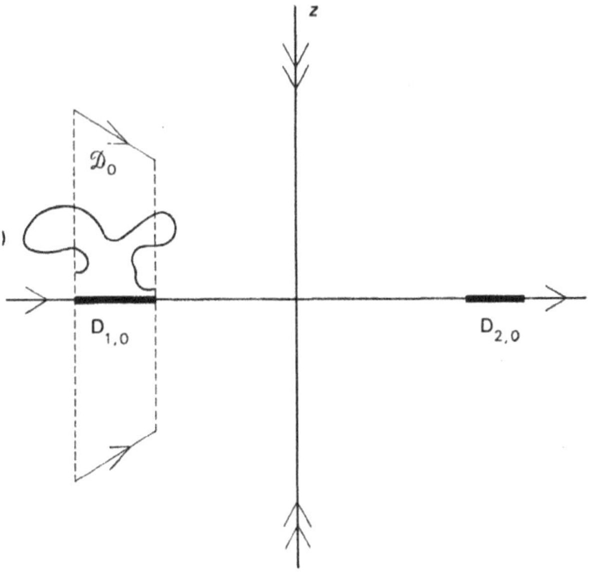

FIG. 8

49
7

neighborhood the same holds for $\widetilde{\varphi}_\mu^m(D_{2,\mu})$ if μ is near 0. Define $D_{1,\mu}$ to be the fundamental domain in the negative y-axis whose boundary is the projection, along the z-direction, of the boundary of $\widetilde{\varphi}_\mu^m(D_{2,\mu})$. We take a "rectangle" $\mathscr{D}_\mu$ so that its boundary consists of two pieces of X_μ-integral curves and two pieces of straight lines parallel to the z-axis, such that its projection on the y-axis is $D_{1,\mu}$, and such that for each $q \in \widetilde{\varphi}_\mu^m(D_{2,\mu})$, there is a positive m' such that $\widetilde{\varphi}_\mu^{m'}(q)$ is in the forward orbit of $\mathscr{D}_\mu$. See figure 8.

Let $A_\mu : \mathscr{D}_\mu \to M$, $\mu \geq 0$ be defined as follows:

$$A_0(\mathscr{D}_0) = D_{2,0} \quad \text{and} \quad A_0(y, z) = (a_0(y), 0)$$

where $a_0 : D_{1,0} \to D_{2,0}$ satisfies $(a_0)_* X_0 = X_0$ on $\{z = 0\}$. For $\mu > 0$ there is a positive T_μ such that the time T_μ map of X_μ, $(X_\mu)_{T_\mu}$, satisfies $(X_\mu)_{T_\mu} D_{1,\mu} = D_{2,\mu}$; we define A_μ to be $(X_\mu)_{T_\mu}$. For $\sigma \in [0, 1]$ we define $A_{\sigma,\mu} = (X_\mu)_\sigma \circ A_\mu$.

Next we define $B_{\sigma,\mu} = \widetilde{\varphi}_\mu^m \circ A_{\sigma,\mu}$. For μ close enough to 0, the image of $B_{\sigma,\mu}$ will be in our coordinate neighborhood in the part $\{y < 0\}$, and also there is some positive m' such that the image of $\widetilde{\varphi}_\mu^{m'} \circ B_{\sigma,\mu}$ is in the forward orbit of $\mathscr{D}_\mu$.

Consider now the quotient map

$$\{(y, z) \mid y < 0\} \to \{(y, z) \mid y < 0\}/(y, z) \sim \widetilde{\varphi}_\mu(y, z).$$

Under this quotient, $\mathscr{D}_\mu$ becomes an annulus $\widetilde{\mathscr{D}}_\mu$, $B_{\sigma,\mu}$ becomes a 2-parameter family of mappings $\widetilde{B}_{\sigma,\mu}$ of $\widetilde{\mathscr{D}}_\mu$ into itself and the projection $\pi(y, z) = (y, 0)$ goes over into $\widetilde{\pi}$, the annulus projection. We want to show that Remark (3.9) is applicable to $\widetilde{B}_{\sigma,\mu}$.

In order to prove this we need new coordinate functions $s_\mu : \widetilde{\mathscr{D}}_\mu \to \mathbf{R}/\mathbf{Z}$ and $w_\mu : \mathscr{D}_\mu \to \mathbf{R}$ such that $\widetilde{\pi}(s_\mu, w_\mu) = (s_\mu, 0)$ and such that X_μ has the form $\dfrac{\partial}{\partial s_\mu}$. We shall write s, w instead of s_0, w_0. For some circle endomorphism $\hat{\varphi} : R/Z \to R/Z$ and some map $W : \mathbf{R}/\mathbf{Z} \to \mathbf{R}$ we have $\widetilde{B}_{\sigma,0}(s, w) = (\hat{\varphi}(s + \sigma), W(s))$. The circle endomorphism $\hat{\varphi}$ is determined by $\widetilde{\varphi}_0$; $\widetilde{\varphi}_0$ was obtained from φ_0 by a small, but otherwise arbitrary, perturbation. Hence we may, and do, assume that the family $\{\hat{\varphi}_\sigma\}$ defined by $\hat{\varphi}_\sigma(s) = \hat{\varphi}(s + \sigma)$ satisfies the assumptions of Theorem (3.7); see also Remark (3.8). Now it is clear that $\widetilde{B}_{\sigma,\mu}$ satisfies the assumptions in Remark (3.9).

Now we observe that if $T_\mu + \sigma \in \mathbf{N}$, $\widetilde{B}_{\sigma,\mu}$ is an iterate of $\widetilde{\varphi}_\mu$ (up to the identifications). Hence if such a $\widetilde{B}_{\sigma,\mu}$ has a non-generic tangency of a stable and an unstable manifold then the same holds for $\widetilde{\varphi}_\mu$. Also, $\widetilde{B}_{\sigma,\mu}$ satisfies the conditions of $\Phi_{\sigma,\mu}$ in Remark (3.9). Let (σ_1, σ_2) and $\bar{\mu}$ be as in that remark. Now as $\mu \to 0$, we have $T_\mu \to \infty$. For each $\sigma \in (\sigma_1, \sigma_2)$ choose an $f(\sigma)$ such that $T_{f(\sigma)} + \sigma \in \mathbf{N}$, and $f(\sigma) < \bar{\mu}$. We may arrange for $\sigma \mapsto f(\sigma)$ to be continuous on (σ_1, σ_2). Then as in Remark (3.9), there is a $\sigma_f \in (\sigma_1, \sigma_2)$ so that $\widetilde{B}_{\sigma_f, f(\sigma_f)}$, and hence $\widetilde{\varphi}_{\sigma_f}$, has a non-transversal homoclinic point.

5. On the rigidity of the unfolding of the saddle-node

In Chapter II we have seen that a conjugacy between two C^∞ saddle-node bifurcations must satisfy very restrictive conditions. In particular, at the bifurcation parameter it must be C^∞ along central curves away from the fixed point, and it must preserve adapted saddle-node vector fields along these curves. In the present section, we shall present two more applications of the restrictive nature of such conjugacies. The first concerns certain arcs of diffeomorphisms of the circle, and the second concerns arcs between Anosov diffeomorphisms and so-called DA diffeomorphisms. These furnish more examples where mild conjugacies cannot be strengthened to conjugacies.

A) *One-parameter families of diffeomorphisms of* S^1

We consider one-parameter families $\varphi_\mu : S^1 \to S^1$ of C^∞-diffeomorphisms of S^1 with rotation number $\rho(\mu)$ increasing such that whenever $\rho(\bar{\mu})$ is rational, $\varphi_{\bar{\mu}}$ has two hyperbolic periodic points or one periodic point of saddle-node type which unfolds generically and such that whenever $\rho(\bar{\mu})$ is irrational, ρ is not locally constant in $\bar{\mu}$. Let $\varphi'_\mu : S^1 \to S^1$ be another such family of diffeomorphisms with rotation number $\rho'(\mu)$. If we assume that $\text{Image}(\rho) = \text{Image}(\rho')$, then there is a homeomorphism $h : \mathbf{R} \to \mathbf{R}$ such that $\rho'(h(\mu)) = \rho(\mu)$, see [5], [10]. For each μ, φ_μ and $\varphi'_{h(\mu)}$ are now conjugate: for $\rho(\mu)$ irrational this is Denjoy's theorem [6], for $\rho(\mu)$ rational it follows from the above description. So it is clear that the above two arcs are " mildly conjugate ".

A question, which was raised independently by R. Thom and S. Smale, is: are two arcs as above in general conjugate? We show here that the answer is in general negative.

Let $\{\varphi_\mu\}$, $\{\varphi'_\mu\}$ be a pair of arcs of diffeomorphisms of S^1 as above. Let $\bar{\mu}$ be a boundary point of $\rho^{-1}\left(\frac{p}{q}\right)$ for some rational $\frac{p}{q} \in \text{Image}(\rho)$. If there is a conjugacy (h, H_μ) between φ_μ and φ'_μ, *i.e.* $h : \mathbf{R} \to \mathbf{R}$ a homeomorphism and H_μ a conjugacy between φ_μ and $\varphi'_{h(\mu)}$ depending continuously on μ, then $h(\bar{\mu})$ is a boundary point of $(\rho')^{-1}\left(\frac{p}{q}\right)$. Now $\varphi_{\bar{\mu}}$ has a unique fixed point which is of saddle-node type. By Chapter II, Theorem (3.2), there is a unique smooth vector field X near the saddle-node orbit such that its time one map X_1 equals $(\varphi_{\bar{\mu}})^q$ (and such that $(\varphi_{\bar{\mu}})^q_* X = X$). Also for $\varphi'_{h(\mu)}$ there is such a vector field X'. Again by Chapter II, Theorem (3.2), $H_{\bar{\mu}}$ has to map X to X'. If we now extend X, and X', to all of S^1 so that $\varphi_{\bar{\mu}*} X = X$, $\varphi'_{h(\bar{\mu})*} X' = X'$, we obtain in general bivalued vector fields. This means that, on the complement of the periodic orbit, $H_{\bar{\mu}}$ has to respect two different vector fields. This is in general impossible.

B) *Arcs between Anosov and DA diffeomorphisms*

In [34], S. Smale showed that certain Anosov diffeomorphisms may be modified to give Axiom A diffeomorphisms with attractors having intricate topological properties.

Since the latter diffeomorphisms were obtained by modifying Anosov diffeomorphisms, he called them DA (for derived from Anosov) diffeomorphisms. Although Smale's construction was given for Anosov diffeomorphisms of the two torus, it works just as well with codimension one Anosov diffeomorphisms of the n-torus. A brief description of the construction is as follows.

Let L be an Anosov diffeomorphism of T^n and suppose that $\dim W^s(x) = 1$ for each $x \in T^n$. Since L is conjugate to a linear toral automorphism [16], it has a fixed point, say p. Assume that the contracting eigenvalue of p is positive. Locally, near p, one has the usual picture of a hyperbolic saddle fixed point as in Figure 9a.

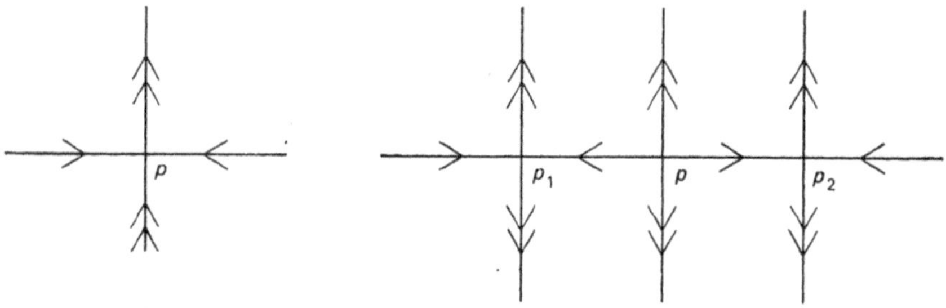

Fig. 9a

Fig. 9b

Fig. 9

Smale proposed to modify the diffeomorphism L in a disk neighborhood N of p to obtain a diffeomorphism g with two new saddle fixed points p_1 and p_2 on $W^s(p, L)$ and such that p is a fixed source of g (see Fig. 9b). This can be done so that g agrees with L off N, g satisfies Axiom A, and the non-wandering set of g consists of p and an $(n-1)$-dimensional hyperbolic attractor containing p_1 and p_2. In fact, as Williams pointed out in [42], one can choose g so that the foliation $\mathscr{F}^s = \{W^s(x, L) \mid x \in T^n\}$ is g-invariant. Of course, the unstable L-foliation is no longer g-invariant. Somewhat later in a private communication with us, Williams observed that a DA diffeomorphism g could be constructed from an arc in which a saddle-node occurs. The local picture is in Figure 9c.

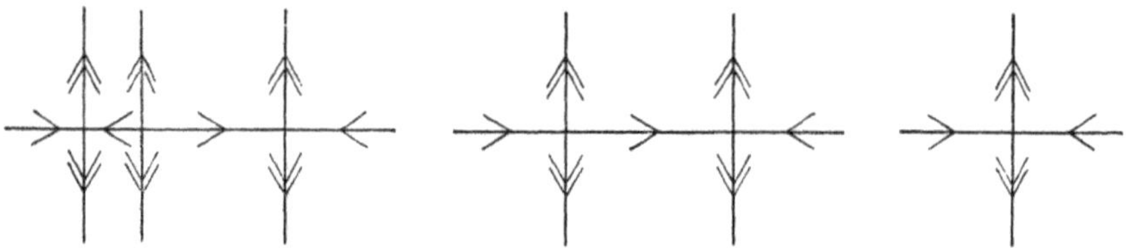

Fig. 9c

If one chooses such an arc carefully, then one can actually make the arc mildly stable with an isolated bifurcation point. This will be proved elsewhere. These mildly stable arcs are such that at the bifurcation point one has a saddle-node p whose stable manifold $W^s(p)$ is one dimensional and such that there is a hyperbolic periodic point q such that $W^u(q) \cap W^s(p) \neq \emptyset$. The next proposition shows that such an arc is never stable.

Proposition. — Suppose $\{\varphi_\mu\}$, $0 \leq \mu \leq 1$, *is an arc of diffeomorphisms of* T^n *so that* φ_0 *is Anosov with* $\dim W^s(x) = 1$ *for all* x. *Let* $b = b(\{\varphi_\mu\})$ *be the first bifurcation point of* $\{\varphi_\mu\}$ *and assume* $0 < b < 1$. *Suppose that* φ_b *has a saddle-node periodic point with* $\dim W^s(p) = 1$ *and* φ_b *has a hyperbolic periodic point* q *not in the orbit of* p *such that* $W^u(q) \cap W^s(p) \neq \emptyset$. *Then* $\{\varphi_\mu\}$ *is not stable.*

Proof. — We may assume, by perturbing $\{\varphi_\mu\}$ if necessary, that $W^u(q)$ is transverse to $W^s(p)$. Since q can be continued to a hyperbolic periodic point q_μ for $\mu < b$, and φ_μ is Anosov for $\mu < b$, it follows that $\dim W^u(q_\mu) = \dim M - 1$. As $W^u(q) \cap W^s(p) \neq \emptyset$, this intersection is zero dimensional, and, therefore p is s-critical. Now the method of proof of Proposition (2.4) may be applied to show that $\{\varphi_\mu\}$ is not stable.

C) *One-parameter families of vector fields*

As was pointed out in [41], the rigidity in the conjugacy of a saddle-node arc has consequences for the stability of one-parameter families of vector fields. Consider such a C^∞ one-parameter family $\{X_\mu\}$ on a 2-manifold M such that for $\mu = b$, X_b has a saddle-node closed orbit γ, *i.e.* a closed orbit which is attracting at one side and repelling at the other side and whose Poincaré map has first but not second order contact with the identity. We assume furthermore that γ unfolds generically; by this we mean that if S is a local cross section of X_b, intersecting γ, then the Poincaré map $P_\mu : S \to S$, μ near b, is a saddle-node arc, see section 3 of Chapter III. Let $\{X'_\mu\}$ be another one parameter family of vector fields on M (near $\{X_\mu\}$) so that for $\mu = b'$, $X'_{b'}$ has a saddle-node closed orbit γ' which unfolds generically. Let S' to be a local section of $X'_{b'}$ intersecting γ'. We assume for simplicity that neither $\{X_\mu\}$ nor $\{X'_\mu\}$ has any other saddle-node closed orbit. We say that these two families are topologically equivalent if there exist a homeomorphism $h : \mathbf{R} \to \mathbf{R}$ and a homeomorphism $H_\mu : M \to M$, depending continuously on μ, such that H_μ maps integral curves of X_μ to integral curves of $X'_{h(\mu)}$.

If such an equivalence (h, H_μ) exists, then $h(b) = b'$. Modifying H_μ along the orbits of X, we may assume that $H_\mu(S) = S'$ for μ near b. So $H_\mu \mid S$ has to conjugate the saddle-node arc $\{P_\mu\}$ with $\{P'_\mu\}$. By Chapter III, section 3 this implies that $H_b \mid S$ has to map X to X', where X, X' is the unique smooth vector field on S, S' such that its time 1 map X_1, X'_1 equals P_b, $P'_{b'}$. This means that $H_b \mid S$ is determined as soon as it is determined in two orbits of P_b in S, one on each side of $\gamma \cap S$.

Since H_b has to map separatrices, *i.e* stable or unstable manifolds of saddle points, to separatrices, the map $H_b \mid S$ is essentially fixed for each intersection of S with a separatrix; such an intersection is an orbit of P_b. So in general $\{X_\mu\}$ and $\{X'_\mu\}$ will not be topologically equivalent if at least two separatrices approach γ from the same side. This means that an arc $\{X_\mu\}$ which has a saddle-node closed orbit which is approached from the same side by at least two separatrices is not stable; it has a modulus of stability in the same way as arcs of diffeomorphisms with a tangency of stable and unstable manifolds. In [7] Guckenheimer incorrectly states that some of these arcs are stable.

IV — GLOBAL STABILITY

1. Introduction

In this chapter we complete the proofs of our main results concerning stability of one-parameter families (arcs) of diffeomorphisms, stated in the introduction to the paper. Necessary conditions for stability, mild and left stability at the first bifurcation point were provided in the previous chapter. We now show that these conditions are also sufficient. As a consequence we obtain a characterization for the stability of arcs containing several bifurcation values, under the basic assumption that the limit sets have finitely many orbits.

The main idea here consists of a suitable construction of tubular families or foliations, which will be used to define topological conjugacies. We begin by describing local tubular families for hyperbolic periodic orbits. This concept will then be extended to Hopf orbits, saddle-nodes and flips. The foliations are constructed for a family $\{\varphi_\mu\}$ in $\mathscr{B}$, as defined in the introduction, especially near its first bifurcation value b. In the case of left and mild stability, this interval is of the form $[\mu_1, b]$, for some $\mu_1 < b$. Otherwise, it is of the form $[\mu_1, \mu_2]$ with $b \in (\mu_1, \mu_2)$. For each μ in such intervals, we build up tubular families or foliations requiring them to be φ_μ-invariant. We shall usually assume that the periodic orbits of φ_μ are fixed points. In fact, if a periodic orbit of φ_μ has period k, we can consider φ_μ^k to define the foliation near one of its elements, and use φ_μ-iterates of the leaves to obtain the foliation near the others.

2. Local Tubular Families

We first recall, in a parametrized version, the notion of tubular family for a hyperbolic orbit [24], [25].

Let $\bar{x} \in M$ be a hyperbolic fixed point for $\varphi_{\bar{\mu}}$, $\bar{\mu} \in I$. Let $[\mu_1, \mu_2]$ be a neighborhood of $\bar{\mu}$ in I and U be a cell neighborhood of $\bar{x}$ in M. If these neighborhoods are small enough, there exists a continuous mapping $[\mu_1, \mu_2] \ni \mu \mapsto x_\mu \in U$, where x_μ is the unique (hyperbolic) fixed point for φ_μ in U and $\bar{x} = x_{\bar{\mu}}$. We denote by $W^u(x_\mu)$ and $W^s(x_\mu)$ the unstable and stable manifolds of φ_μ at x_μ. Given the family $\{\varphi_\mu\}$, $\mu \in I$, we define $\Phi : M \times I \to M$ by $\Phi(x, \mu) = (\varphi_\mu(x), \mu)$.

Definition (**2.1**). — *An unstable tubular family for* $\{\varphi_\mu\}$ *or* Φ *at* $(\bar{x}, \bar{\mu})$ *is a continuous foliation* $\mathscr{F}^u$ *of* $U \times [\mu_1, \mu_2]$, *such that for* $\mu \in [\mu_1, \mu_2]$

a) *the leaves are* C^k *discs varying continuously in the* C^k *topology*, $1 \leq k < \infty$, *and* $\mathscr{F}^u(x_\mu, \mu) = W^u(x_\mu) \times \{\mu\} \cap U \times \{\mu\}$,

b) *each leaf* $\mathscr{F}^u(y, \mu)$ *is contained in* $U \times \{\mu\}$ *for* $y \in U$,

c) *the foliation is* Φ*-invariant in the following sense*

$$\Phi(\mathscr{F}^u(y, \mu)) \supset \mathscr{F}^u(\varphi_\mu(y), \mu)$$

for $y, \varphi_\mu(y) \in U$.

Remark. — For our purposes it is enough to take the leaves of the foliation to be C^2. In fact, in their global setting, we can and will construct the foliation to be C^k $(k \geq 2)$ when restricted to any unstable manifold of a hyperbolic periodic orbit that intersects $W^s(x_\mu) \times \{\mu\}$. See Definition (3.1) of the next section, where even more differentiability is involved. The same remark applies to Definition (2.2) below.

Let us see how a similar foliation can be defined at a Hopf fixed point p of φ_b. We assume, for $\mu < b$ and μ near b, that φ_μ has a hyperbolic fixed point p_μ near p and $\dim W^s(p_\mu) = \dim W^s(p)$. As $\mu \to b$, $\mu < b$, the stable and unstable manifolds of p_μ in U converge to those of p in the C^k topology. So, we can use Definition (2.1) in this case, with the difference that we consider μ in some interval $[\mu_1, b]$.

Let now p be a saddle-node or a flip for φ_b. As a natural extension of Definition (2.1), we present below the concept of strong unstable foliation for $\{\varphi_\mu\}$ or Φ at (p, b).

First we need some basic facts about center manifolds as stated in Chapter II. Let U be a small cell neighborhood of p in M and $[\mu_1, \mu_2] \subset I$ a small interval with $b \in (\mu_1, \mu_2)$. Let W^c be a C^k center manifold for Φ at (p, b), $1 \leq k < \infty$. For each $\mu \in [\mu_1, \mu_2]$, $W^c_\mu = W^c \cap U \times \{\mu\}$ has dimension one and W^c_b is C^∞. We also consider the center stable manifold W^{cs} for Φ at (p, b), which is C^k and has dimension $s + 2$. Here s and u are the number of eigenvalues (with multiplicity) of $d\varphi_b(p)$ with norm less and bigger than one, respectively. Both W^c and W^{cs} are invariant by Φ, and W^c_μ and $W^{cs}_\mu = W^{cs} \cap U \times \{\mu\}$ are invariant by φ_μ. Frequently, here and in the sequel, we will identify a subset $V \times \{\mu\}$ of $M \times \{\mu\}$ with its projection V into M. We recall that W^{cs} is foliated by the strong stable foliation $\mathscr{F}^{ss}$, with leaves s-dimensional C^k discs transverse to W^c. Particular leaves are the strong stable manifolds through the fixed or periodic orbits of φ_μ near p. For each $\mu \in [\mu_1, \mu_2]$, W^{cs}_μ is a union of leaves of $\mathscr{F}^{ss}$. The foliation $\mathscr{F}^{ss}$ is invariant by Φ in the following sense: if $z, \Phi(z) \in U \times [\mu_1, \mu_2]$ and S is the leaf through z, then $\Phi(S)$ is contained in the leaf through $\Phi(z)$. Similarly, we can define the center unstable manifold W^{cu} for Φ at (p, b) and W^{cu} is foliated by the strong unstable foliation $\mathscr{F}^{uu}$. Particular leaves are the strong unstable manifolds of the fixed or periodic orbits of φ_μ near p. Our strong unstable tubular family (or foliation) is an extension of $\mathscr{F}^{uu}$ to a full neighborhood of (p, b) in $M \times I$. For later purposes we need this extended version of $\mathscr{F}^{uu}$, but no such version of $\mathscr{F}^{ss}$.

Suppose (p, b) is a saddle-node. As it can be seen from Chapter II, there are two possibilities for its unfolding. In the first one, for each $\mu < b$ and near b, there are two hyperbolic fixed points $p_{1,\mu}$ and $p_{2,\mu}$ of φ_μ near p, which collapse into p as $\mu \to b$ and then disappear for $\mu > b$. We also may assume that $\dim W^s(p_{2,\mu}) = s + 1$ and

$\dim W^s(p_{1,\mu}) = s$. The other possibility is similar, only the orbits $p_{1,\mu}$ and $p_{2,\mu}$ appear for $\mu > b$. Throughout this chapter we assume the first case.

If (p, b) is a flip, we also assume the following of four similar possibilities for its unfolding. For each $\mu < b$ and near b, there is a hyperbolic fixed point p_μ near p. For $\mu > b$, there is a hyperbolic fixed point $p_{1,\mu}$ and a hyperbolic period two point $p_{2,\mu}$ near p. We assume that $\dim W^s(p_{1,\mu}) = s$ and $\dim W^s(p_{2,\mu}) = s + 1$.

Definition (**2.2**). — *A strong unstable tubular family* $\mathscr{F}^{uu}$ *for* $\{\varphi_\mu\}$ *at the saddle-node or flip* (p, b) *is a continuous foliation of* $U \times [\mu_1, \mu_2]$ *such that*

a) *the leaves are* C^k *discs varying continuously in the* C^k *topology and*

$$\mathscr{F}^{uu}(p, b) = W^{uu}(p, b) \cap U \times \{b\},$$

b) *for each* μ, $U \times \{\mu\}$ *is a union of leaves transverse to* W^{cs}_μ,

c) *the foliation is* Φ*-invariant: if* $\mathscr{F}^{uu}(y, \mu)$ *is the leaf through* $(y, \mu) \in U \times \{\mu\}$, *then* $\Phi(\mathscr{F}^{uu}(y, \mu))$ *is the leaf through* $(\varphi_\mu(y), \mu) \in U \times \{\mu\}$.

These local unstable and strong unstable foliations have already been used in a similar context by several authors; see [26] for references. We provide a construction of them in Proposition (2.3) below, to give a clearer view of some of the main techniques of this chapter.

Proposition (**2.3**). — *There exists a strong unstable foliation for* $\{\varphi_\mu\}$ *at* (p, b), *where* p *is a saddle-node or a flip. Similarly, there exists an unstable foliation at a hyperbolic or a Hopf fixed point.*

Proof. — Let us first consider (p, b) to be a saddle-node or a flip. In $U \times [\mu_1, \mu_2]$ let W^{cs} be the center-stable manifold of Φ at (p, b). As before, Φ is defined by $\Phi(x, \mu) = (\varphi_\mu(x), \mu)$ and $U \ni p$, $[\mu_1, \mu_2] \ni b$ are small neighborhoods. In W^{cs}, we take a closed fundamental domain A. This is a set with the following property. $\varphi_b^2(A) \cap A = \emptyset$ and there is a neighborhood U' of (p, b) in W^{cs} such that if $x \in W^{cs}$ and its negative Φ-orbit leaves U', then the Φ-orbit of x has an element in A, which is unique except if it belongs to ∂A. Let us indicate the construction of such a fundamental domain when (p, b) is a saddle-node. Taking a C^2 coordinate system in the center-stable manifold of φ_b at p, we can write

$$\varphi_b(x,y) = (x + x^2 + x\langle c,y\rangle + o(|x|^3, \|x^2y\|, \|y\|^2), Ay + o(\|xy\|, \|y\|^2)),$$

where $c \in R^2$ and $\|A\| < 1$. Here $y = 0$ represents the one-dimensional center-manifold and $x = 0$ the s-dimensional strong stable manifold $W^{ss}(p)$. The stable manifold $W^s(p)$ of φ_b at p is the half-plane $x \leq 0$. We take in $W^s(p)$ a hemisphere H of small radius centered at p and slightly extended transversally across $W^{ss}(p)$. It is easy to check from the above expression that $\varphi_b(H) \cap H = \emptyset$. For $\mu_1 \leq \mu \leq b$, $H_\mu = H \times \{\mu\}$ also crosses $W^{ss}(p_1, \mu)$ and $W^{ss}(p_2, \mu)$ transversally in $U \times \{\mu\}$ and

$\varphi_\mu(H_\mu) \cap H_\mu = \emptyset$. Also, for $b < \mu < \mu_2$, $\varphi_\mu(H_\mu) \cap H_\mu = \emptyset$, where $H_\mu = H \times \{\mu\}$. We let A_μ be a closed region bounded by H_μ, $\varphi_\mu(H_\mu)$, and an annulus whose boundary lies in $H_\mu \cap \varphi_\mu(H_\mu)$. Then take $A = \bigcup_{\mu_1 \leq \mu \leq \mu_2} A_\mu$.

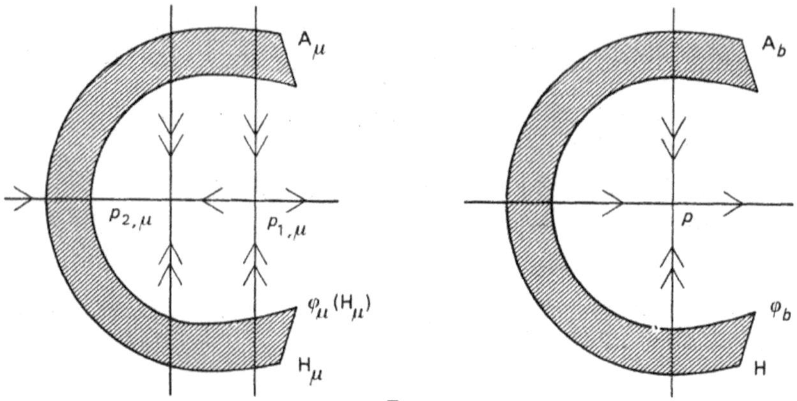

FIG. 10

In the case of a flip, A is diffeomorphic to $A_b \times [\mu_1, \mu_2]$, where A_b is an $s+1$-dimensional annulus. Similarly, if (p, b) is a Hopf point, A is diffeomorphic to $A_b \times [\mu_1, b]$, where A_b is an $(s+2)$-dimensional annulus. If $(\bar{x}, \bar{\mu})$ is a hyperbolic fixed point, A is diffeomorphic to $A_{\bar{\mu}} \times [\mu_1, \mu_2]$, where $A_{\bar{\mu}}$ is an s-dimensional annulus and $\bar{\mu} \in (\mu_1, \mu_2)$. Let us proceed with the construction of the strong unstable foliation for a saddle-node or a flip. Over a neighborhood U_1 of the exterior boundary of A we raise a fibration of class C^k, the fibers being u-dimensional discs transverse to W^{cs} and each of them contained in some $U \times \{\mu\}$. Over the neighborhood $\Phi(U_1)$ of the interior boundary of A, we have a similar fibration, the fibers being the Φ-images of those in U_1. Restricting to smaller neighborhoods of ∂A, we can extend this fibration to a full neighborhood of A. This is done as follows. Over a neighborhood of A, we raise another C^k fibration transverse to W^{cs} and the fibers contained in the sections $U \times \{\mu\}$, but not necessarily φ_μ-invariant. We now define a new fibration, which agrees with the first one in a neighborhood of ∂A and with the second one off a slightly bigger neighborhood of ∂A. Let π_1 and π_2 be the projections, into a neighborhood V of A, defined by the two fibrations. We define a C^k real function $\rho : V \to [0, 1]$, such that ρ is 1 near ∂A and 0 off a small neighborhood of ∂A. The required fibration is then given by the projection $\pi = \rho\pi_1 + (1-\rho)\pi_2$. Its fibers form the leaves of our foliation. We now simply define it over $W^{cs} - W^c$ through iterates Φ^n or φ_μ^n, all $n \geq 0$: if $(y, \mu) = \Phi^n(x, \mu)$ for $(x, \mu) \in A$ and the fiber through (x, μ) is $\mathcal{F}^{uu}(x, \mu)$, then the fiber through (y, μ) is $\Phi^n(\mathcal{F}^{uu}(x, \mu)) \cap U \times [\mu_1, \mu_2]$. By the generalized λ-lemma [24], [30], $\mathcal{F}^{uu}$ extends over W^c satisfying all the conditions of Definition (2.2). Notice that, on the center unstable manifold of Φ at (p, b), we get the usual strong unstable foliation. The construction at a hyperbolic fixed point is easier and can be done in a similar way. The same applies to a Hopf point (p, b), in which case we take the parameter μ in some interval $[\mu_1, b]$. This finishes the proof of the proposition.

We point out that these unstable and strong unstable foliations are not unique. There is a degree of freedom in their construction, as shown in Proposition (2.3). This is what enables us to globalize them in a compatible way with the tubular families of other periodic orbits. Such globalizations will be performed in the next section.

3. Global Tubular Families

In this section we construct compatible systems of tubular families or foliations for a family $\{\varphi_\mu\}$, μ near its first bifurcation point.

As before, we denote by $b \in I$ the first bifurcation point, so that φ_μ is Morse-Smale for $\mu < b$. From the previous chapter, $\{\varphi_\mu\}$ can be stable in one of our three senses only if one of the periodic orbits of φ_b is an elementary bifurcation (saddle-node, flip or Hopf). Moreover, the (strong) stable and unstable manifolds of all the periodic orbits of φ_b must have transversal intersections.

We recall that, since φ_μ is Morse-Smale for $\mu < b$, its periodic orbits can be partially ordered through the relation $p_{i,\mu} \geq p_{j,\mu}$ if $W^u(p_{i,\mu}) \cap W^s(p_{j,\mu}) \neq \emptyset$. See [24], [25]. We fix a total ordering for the periodic orbits compatible with this relation. If φ_b has a Hopf orbit or a flip, the same ordering applies as well to its periodic orbits: there can not be any cycle, since otherwise b would not be the first bifurcation point [18]. If φ_b has a saddle-node, we may have a cycle containing this orbit. If there is only a 1-cycle, the saddle-node will be counted as $p_{j,\mu}$ and $p_{j+1,\mu}$ for some positive integer j. However, if there is a cycle of larger length, then the saddle-node will be counted as $p_{j,\mu}$ $P_{j+j',\mu}$ where j and j' are positive integers and $j' > 1$. In all cases, this ordering of the periodic orbits will be used to build up global systems of foliations for $\{\varphi_\mu\}$, μ in a small interval in I. This interval is of the form $[\mu_1, b]$ when φ_b has a Hopf point or a saddle-node which is critical or has a cycle, and otherwise $[\mu_1, \mu_2]$ with $b \in (\mu_1, \mu_2)$.

Let $p_{1,\mu} > p_{2,\mu} > \ldots > p_{\ell,\mu}$ be the periodic orbits of φ_μ for some $\mu \in [\mu_1, \mu_2]$, all of them hyperbolic except at most one which is an elementary bifurcation. For convenience, we write p_i instead of $p_{i,\mu}$. Let $U_1, U_2, \ldots, U_\ell$ be neighborhoods of $p_1, p_2, \ldots, p_\ell$. We will consider local unstable or strong unstable foliations in $U_1 \times [\mu_1, \mu_2]$, $U_2 \times [\mu_1, \mu_2]$, $\ldots$, $U_\ell \times [\mu_1, \mu_2]$ as in Definitions (2.1) and (2.2). If the periodic point p_i is a saddle-node with no cycles or a flip, let $\mathscr{F}^u(p_i)$ denote its strong unstable foliation in $U_i \times [\mu_1, \mu_2]$. Otherwise, let $\mathscr{F}^u(p_i)$ denote the unstable foliation in $U_i \times [\mu_1, \mu_2]$. Since $\mathscr{F}^u(p_i)$ is a φ_μ-invariant foliation in $U_i \times [\mu_1, \mu_2]$, it naturally induces one in $\bigcup \varphi_\mu^n(U_i) \times \{\mu\}$ for $\mu \in [\mu_1, \mu_2]$ and $n \in N$ by simply taking iterates φ_μ^n of the leaves. In the sequel, we will be using this extended (or globalized) foliation, which will be still denoted by $\mathscr{F}^u(p_i)$.

Definition (3.1). — *The system of foliations* $\mathscr{F}^u(p_1), \mathscr{F}^u(p_2), \ldots, \mathscr{F}^u(p_\ell)$ *is compatible when*

a) *if a leaf* F *of* $\mathscr{F}^u(p_i)$ *intersects a leaf* S *of* $\mathscr{F}^u(p_j)$, $i \leq j$, *then* $F \supset S$,

b) *for all* $i \leq j$, *the restriction of* $\mathscr{F}^u(p_j)$ *to each leaf of* $\mathscr{F}^u(p_i)$ *is a foliation of class* C^k, $2 \leq k$.

Remark. — Note that the restriction of these foliations to unstable manifolds of hyperbolic periodic orbits are of class C^k. However, the unstable manifold of a non-critical saddle-node with no cycles or a flip is not a leaf of the strong unstable foliation. Thus, we do not demand differentiability of the foliations when restricted to these unstable manifolds.

Let us show the existence of such systems. First we consider the case where φ_b has a Hopf periodic orbit. This is the same as the parametrized version of the hyperbolic case. Although somewhat simpler than the saddle-node and the flip, it gives a pretty good idea of how to proceed in those cases as well.

Proposition (3.2). — *If* φ_b *has a Hopf periodic orbit, then there exists a compatible system of unstable foliations for* φ_μ, $\mu \in [\mu_1, b]$ *for some* μ_1 *near* b.

Proof. — Let $p_{1,\mu} > p_{2,\mu} > \ldots > p_{\ell,\mu}$ be an ordering of the periodic orbits of φ_μ for $\mu_1 \leq \mu \leq b$. We simply write p_i instead of $p_{i,b}$. By induction, we may assume that a compatible system of foliations $\mathscr{F}^u(p_1), \ldots, \mathscr{F}^u(p_n)$ has been constructed in neighborhoods $U_1 \times [\mu_1, b], \ldots, U_n \times [\mu_1, b]$, $p_i \in U_i$ for $1 \leq i \leq n$. Let us build $\mathscr{F}^u(p_{n+1})$. We will adapt the proof of Proposition (2.3) to guarantee the compatibility condition. Near p_{n+1}, we consider a closed annulus A_b as a fundamental domain for $W^s(p_{n+1})$ and denote by $\partial_{ex} A_b$ its exterior boundary. Then, the annulus A_μ with boundaries $\partial_{ex} A_b \times \{\mu\}$ and $\varphi_\mu(\partial_{ex} A_b \times \{\mu\})$ is also a fundamental domain for $W^s(p_{n+1}, \mu)$, $\mu \in [\mu_1, b]$ (we take μ_1 closer to b if necessary. If $W^u(p_n)$ intersects $W^s(p_{n+1})$ it must do so transversally. Thus, we can take a C^k fibration of $W^u(p_n)$ near the exterior boundary of A_b, whose fibers are discs transverse to $W^s(p_{n+1})$ with the same dimension as $W^u(p_{n+1})$. The image by φ_b of this fibration induces a similar one near the interior boundary of A_b. On the other hand, it is easy to get a second C^k fibration of $W^u(p_n)$ in a neighborhood of all of its intersection with A_b, if we do not require it to be φ_b-invariant. However, as in Proposition (2.3), this second fibration can be deformed to agree with the first one near the boundaries of A_b and, thus, it becomes φ_b-invariant. We now want to fiber in a similar way the leaves of $\mathscr{F}^u(p_n)$ near A_μ, $\mu_1 \leq \mu \leq b$ with μ_1 close to b. First, we observe that, in $U_n \times [\mu_1, b]$, the leaves of $\mathscr{F}^u(p_n)$ are C^k imbeddings of the disc $W^u(p_n) \cap U_n$, continuously parametrized by $(W^s(p_n) \cap U_n) \times [\mu_1, b]$. Thus, by restricting this parameter space, we have that the leaves of $\mathscr{F}^u(p_n)$ near A_μ are C^k close to $W^u(p_n)$. Using this parametrization and the fact that the foliation $\mathscr{F}^u(p_n)$ is φ_μ-invariant, we can fiber as above the leaves near $W^u(p_n) \cap A_b$. We get a φ_μ-invariant fibration with C^k fibers transverse to $W^s(p_{n+1}, \mu)$, $\mu_1 \leq \mu \leq b$, and varying continuously in the C^k topology. By construction, each fiber is contained in some leaf of

$\mathcal{F}^u(p_n)$. Also, the fibration is C^k when restricted to each leaf of $\mathcal{F}^u(p_n)$. Now we take $\mathcal{F}^u(p_{n-1})$. There are two cases to consider. If $W^u(p_{n-1})$ intersects $W^s(p_{n+1})$ but not $W^s(p_n)$, then $W^u(p_{n-1}) \cap A_b$ is compact and disjoint from $W^u(p_n) \cap A_b$. So, we can proceed as before, fibering the leaves of $\mathcal{F}^u(p_{n-1})$ near $W^u(p_{n-1}) \cap A$. Let us now suppose that $W^u(p_{n-1})$ intersects $W^s(p_n)$. Since $\mathcal{F}^u(p_{n-1})$ and $\mathcal{F}^u(p_n)$ are compatible by the induction hypothesis, the fibration of $\mathcal{F}^u(p_n)$ in a neighborhood V of $W^u(p_n) \cap A_b$ is also a fibration of $\mathcal{F}^u(p_{n-1})$ in V. On the other hand, $B = W^u(p_{n-1}) \cap (A_b - V)$ is compact. So, $\mathcal{F}^u(p_{n-1})$ can be fibered as above in a neighborhood of B. As in Proposition (2.3), we can average these two fibrations near ∂B to get a desired one in some neighborhood of $W^u(p_n) \cup W^u(p_{n-1})$ intersected with A_b. Its fibers form the leaves of our foliation. We repeat the argument to all $\mathcal{F}^u(p_i)$, $1 \leq i \leq n-2$. Once we have the φ_μ-invariant and compatible foliation $\mathcal{F}^u(p_{n+1})$ near $A_b \times [\mu_1, b]$, some μ_1 close to b, we just consider its positive iterates by φ_μ (or Φ). By the generalized λ-lemma [24], [30], it extends to the foliation $W^u(p_{n+1,\mu})$, $\mu \in [\mu_1, b]$, of the center unstable manifold of Φ at (p_{n+1}, b). This finishes the construction of the foliation $\mathcal{F}^u(p_{n+1})$. The proof of the theorem is complete.

Let us now consider the case where φ_b has a saddle-node. Let

$$p_{1,\mu} > p_{2,\mu} > \ldots > p_{l,\mu}$$

be an ordering for the periodic points of φ_μ, $\mu_1 \leq \mu < b$. We first consider the case where the saddle node is non-critical and φ_b has no cycles. We assume that $p_{j,\mu}$ and $p_{j+1,\mu}$ coalesce at $\mu = b$, giving rise to the saddle-node $p_j = p_{j+1}$.

We shall construct a compatible system of foliations for $\{\varphi_\mu\}$, with μ in some interval $[\mu_1, \mu_2]$ and $b \in (\mu_1, \mu_2)$. The foliation at the saddle-node $p_j = p_{j+1}$ will be a strong unstable foliation $\mathcal{F}^{uu}(p_j)$. Actually, the construction we just performed in Proposition (3.2) can be adapted to the present case as well as to the flip bifurcation. However, to prove the stability of the family, we also construct a one-dimensional center foliation in a neighborhood of the saddle-node $p_j = p_{j+1}$ in the center stable manifold of Φ at (p_j, b). As usual, $\Phi : M \times I \to M$ is defined by $\Phi(x, \mu) = (\varphi_\mu(x), \mu)$. This center foliation $\mathcal{F}^c$ should be φ_μ-invariant and compatible with the unstable foliations $\mathcal{F}^u(p_i)$, $i < j$, in the sense of Definition (3.1).

Proposition (3.3). — *If φ_b has a non critical saddle-node $p_j = p_{j+1}$ and no cycles, then*

(a) *there is a compatible center foliation $\mathcal{F}^c$ defined in a neighborhood of the center stable manifold* W^{cs} *of Φ at $(p_{j,b})$,*

(b) *there is a compatible system of unstable foliations $\mathcal{F}^u(p_i)$ for Φ or φ_μ, $\mu \in [\mu_1, \mu_2]$ for some $\mu_1 < b$ and $\mu_2 > b$. The foliation we consider at $p_j = p_{j+1}$ is a strong unstable one, $\mathcal{F}^{uu}$,*

(c) *the union of leaves of $\mathcal{F}^{uu}$ through a leaf of $\mathcal{F}^c$ forms a* C^1 *submanifold.*

Proof. — Since $p_1 > p_2 > \ldots > p_{j-1}$ are all hyperbolic, Proposition (3.2) provides a compatible system of unstable foliations $\mathcal{F}^u(p_1), \mathcal{F}^u(p_2), \ldots, \mathcal{F}^u(p_{j-1})$. The leaves of these foliations are taken to be C^k, $k \geq 2$. Let us construct a center foliation $\mathcal{F}^c$

at the saddle-node; its leaves will be C^{k-1}. In the center stable manifold of Φ at (p_j, b), we consider a fundamental domain as in Proposition (2.3). We define the center foliation $\mathcal{F}^c$ as the integral curves of a vector field X satisfying the following properties. X is tangent to the leaves of $\mathcal{F}^u(p_i)$ and it is C^{k-1} along these leaves for all $i<j$ such that $W^u(p_i)$ intersects $W^s(p_j)$. The vector field X is also transverse to the strong stable foliation of Φ at p_j and $d\Phi$-invariant. To get such a vector field, we proceed by induction taking $i=j-1, j-2, \ldots, 1$. If $W^u(p_{j-1})$ intersects W^{cs}, we take a C^{k-1} vector field X tangent to $W^u(p_{j-1})$ near its intersection with the exterior boundary of the fundamental domain A. Near the interior boundary, we just consider $d\Phi(X)$. It is easy to extend this vector field X to a full neighborhood of $W^u(p_{j-1}) \cap A$ in $W^u(p_{j-1})$. We now want to define X along the nearby leaves of $\mathcal{F}^u(p_{j-1})$. As we noticed in the proof of Proposition (3.2), these leaves are C^k imbeddings of a disc in $W^u(p_{j-1})$, continuously parametrized by their intersection with $W^s(p_{j-1})$. So we can project X into the leaves of $\mathcal{F}^u(p_{j-1})$ near the exterior boundary of A, consider its image by $d\Phi$ and extend it across A as above. Due to the fact that the saddle-node is noncritical, X is transverse to the strong stable foliation of Φ in the center stable manifold. By induction, let us suppose X defined along the leaves of $\mathcal{F}^u(p_{i+1}), \ldots, \mathcal{F}^u(p_{j-1})$ near the fundamental domain A. If $W^u(p_i) \cap A = \emptyset$, we proceed to the next foliation $\mathcal{F}^u(p_{i-1})$. If $W^u(p_i) \cap A \neq \emptyset$, but $W^u(p_i) \cap W^s(p_k) = \emptyset$ for $i<k<j$, we can proceed as before since $W^u(p_i) \cap A$ is compact and disjoint from $W^u(p_k) \cap A$. Finally, let $W^u(p_i) \cap W^s(p_n) \neq \emptyset$ for some $i<n<j$. The vector field X is already defined in a neighborhood V of $W^u(p_k)$ in A for $i<k<j$. Since $\mathcal{F}^u(p_i)$ is compatible with $\mathcal{F}^u(p_k)$, X is tangent to the leaves of $\mathcal{F}^u(p_i)$ in V. But $W^u(p_i) \cap (A - V)$ is compact, so we are again reduced to the previous case. Thus we can construct X as desired in a neighborhood of A in the center stable manifold. Now we take the integral curves of X and their positive iterates by Φ. By the λ-lemma [24], [30], this partial foliation extends to center manifolds and together they form a center foliation as asserted in (a). Parts (b) and (c) follow as in Proposition (3.2) for the hyperbolic periodic points and also as in Proposition (2.3) for the saddle-node. In this last case, we raise the strong unstable foliation from the center foliation we have just constructed. The proof of Proposition (3.3) is complete.

Remark. — The center foliation $\mathcal{F}^c$ is not a " classical " foliation in two ways

(1) as usual in the present work, we only required that the leaves of $\mathcal{F}^c$, which are C^{k-1} with $k \geq 2$, should vary continuously in the C^{k-1} topology. We did not demand the projection along the leaves into a transverse section (like a leaf of the strong stable foliation) to be differentiable,

(2) $\mathcal{F}^c$ has singularities, as shown in the picture

Fig. 11

Now let us consider the situation of a saddle-node $p_j = p_{j+j'}$ for φ_b which is critical or lies in a cycle and $j' \geq 1$. In the case the saddle-node is bicritical or it is part of a cycle, we wish to prove the stability of the family $\{\varphi_\mu\}$ for $\mu \leq b$ (left stability). When the saddle-node is critical but not bicritical and there is no cycle, we wish to prove that $\{\varphi_\mu\}$ is midly stable for $\mu < \mu_1$, some $\mu_1 > b$ and near b. Again, since φ_μ is Morse-Smale for $b < \mu \leq \mu_1$, some $\mu_1 > b$, $\{\varphi_\mu\}$ is stable in this range of the parameter. This follows from [25] or the next section. Thus, it is certainly enough to show that $\{\varphi_\mu\}$ is left stable at b. In conclusion, we can treat these three cases in the same way. As before, we need to construct a compatible system of unstable foliations for $\{\varphi_\mu\}$, $\mu \leq b$.

Proposition (3.4). — *If φ_b has a saddle-node $p_j = p_{j+j'}$, $j' \geq 1$, which is critical or, has a cycle, then there exists a compatible system of unstable foliations for $\mu \leq b$. The leaves of $\mathscr{F}^u(p_{j+j',\mu})$ are one dimension lower than those of $p_{j,\mu}$.*

Proof. — Similar to that of Proposition (3.2), using the unfolding of the saddle-node as in Proposition (2.3). In this case the leaves of $\mathscr{F}^u(p_j, \mu)$ cover a neighborhood of $W^{ss}(p_j, \mu) - \{p_{j,\mu}\}$ near $p_{j,\mu}$ where $W^{ss}(p_{j,\mu})$ is the strong stable manifold of $p_{j,\mu}$.
The leaves of $\mathscr{F}^u(p_{j,\mu})$ intersected with the center stable manifold W^{cs}_μ are represented by horizontal lines in the pictures below. The intersections of the leaves of $\mathscr{F}^u(p_{j+j',\mu})$ with W^{cs}_μ are represented by points.

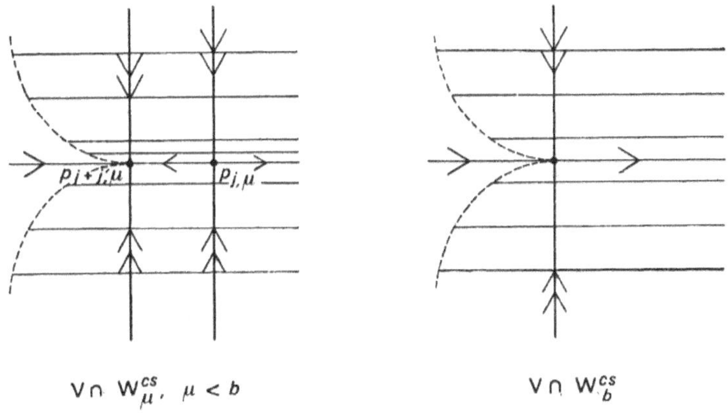

FIG. 12

Notice that the foliations $\mathscr{F}^u(p_{j,\mu})$ and $\mathscr{F}^u(p_{j+j',\mu})$ can be constructed to be compatible in a neighborhood V of $W^{ss}(p_j) - p_j$ in $M \times [\mu_1, b]$ for some $\mu_1 < b$.

Proposition (3.5). — *If φ_b has a flip point $p_{j'}$, then*

(a) *there is a compatible center foliation $\mathscr{F}^c$ defined in a neighborhood of the center stable manifold of Φ at (p_j, b);*

(b) *there is a compatible system of unstable foliations* $\mathcal{F}^u(p_i)$ *for* φ_μ, $\mu \in [\mu_1, \mu_2]$ *for some* $\mu_1 < b$ *and* $\mu_2 > b$; *the foliation we consider at the flip* p_j *is a strong unstable one*, $\mathcal{F}^{uu}$;

(c) *the leaves of* $\mathcal{F}^{uu}$ *through a leaf of* $\mathcal{F}^c$ *form a* C^1 *submanifold.*

Proof. — The proof is entirely similar to that of Proposition (3.3).

We finish this section by summing up in the next theorem the results we have obtained on compatible systems of unstable (strong unstable) foliations.

Theorem **(3.6)**. — *Let* b *be the bifurcation point of the family* $\{\varphi_\mu\}$ *in* $\mathcal{B}$. *Let* $\Phi : M \times I \to M$ *be defined by* $\Phi(x, \mu) = (\varphi_\mu(x), \mu)$. *Then,*

(1) *if* φ_b *has either a noncritical saddle-node and no cycles or a flip, there exists a compatible system of unstable foliations for* Φ *or* $\{\varphi_\mu\}$, $\mu < \mu_1$ *for some* μ_1 *near* b *and* $\mu_1 > b$. *The unstable foliation at the saddle-node or the flip is a strong unstable one,*

(2) *if* φ_b *has a saddle-node or a Hopf periodic orbit, there exists a compatible system of unstable foliations for* Φ *or* $\{\varphi_\mu\}$, $\mu \leq b$.

4. Stability

We culminate this chapter by showing the stability of the families of diffeomorphisms $\{\varphi_\mu\}$ in $\mathcal{B}$. Such families were introduced in Chapter III, studied there and in the previous sections of the present chapter. A family $\{\varphi_\mu\}$ in $\mathcal{B}$ is stable, mildly or left stable according to the structure of the diffeomorphism φ_b, where b is the first bifurcation point of $\{\varphi_\mu\}$. More specifically, the stability depends on the structure of the non-hyperbolic orbit of φ_b. We stress again that the transversality of the stable and unstable manifolds of the periodic orbits is necessary even for left stability, as was proved in Chapter III.

Our results on stability will follow rather naturally and in a unified way from the existence of compatible systems of (strong) unstable tubular families or foliations. Fitting together these foliations provides a kind of global coordinate system from which the topological conjugacies can be constructed. We define the conjugacies inductively on the stable manifolds of the periodic orbits, these orbits being totally ordered as in section 3.

We will make use of the following two known results. The first one states that $\mathcal{B}$ is open in the set $\mathcal{P}$ of C^∞ one-parameter families of C^∞ diffeomorphisms of M with the C^∞ topology. It corresponds to Theorem (2.5) of [19] and Theorem (3.1) of [20]. The second result is an easy extension (a parametrized version) of the Isotopy Extension Theorem [23].

Theorem **(4.1)**. — *Let* $\{\varphi_\mu\} \in \mathcal{B}$ *have the first bifurcation point* b. *There exists a neighborhood* U *of* $\{\varphi_\mu\}$ *such that if* $\{\tilde{\varphi}_\mu\} \in U$ *has first bifurcation point* $\tilde{b}$, *then* $\{\tilde{\varphi}_\mu\} \in \mathcal{B}$ *and* φ_b *and* $\tilde{\varphi}_{\tilde{b}}$ *have the same elementary bifurcation. Moreover, there is an order preserving one to one correspondence between the periodic orbits of* φ_b *and* $\tilde{\varphi}_{\tilde{b}}$.

Let N be a C^r compact manifold, $r \geq 1$, and A an open subset of $\mathbf{R}^s$. Let M be a C^∞ manifold with $\dim M > \dim N$. We indicate by $C^k_A(N \times A, M \times A)$ the set of C^k mappings $f : N \times A \to M \times A$ such that $\pi = \pi' f$, endowed with the C^k topology, $1 \leq k \leq r$. Here, π and π' denote the natural projections $\pi : N \times A \to A$, $\pi' : M \times A \to A$. Let $\mathrm{Diff}^k_A(M \times A)$ be the set of C^k diffeomorphisms φ of $M \times A$ such that $\pi' = \pi' \varphi$, again with the C^k topology.

Theorem (**4.2**). — *Let* $i \in C^k_A(N \times A, M \times A)$ *be an imbedding and* A' *a compact subset of* A. *Given neighborhoods* U *of* $i(N \times A)$ *in* $M \times A$ *and* V *of the identity in* $\mathrm{Diff}^k_A(M \times A)$, *there exists a neighborhood* W *of* i *in* $C^k_A(N \times A, M \times A)$ *such that for each* $j \in W$ *there exists* $\varphi \in V$ *satisfying* $\varphi i = j$ *restricted to* $N \times A'$ *and* $\varphi(x) = x$ *for all* $x \notin U$.

Now we prove the main theorem of the present chapter.

Theorem (**4.3**). — *Let* $\{\varphi_\mu\} \in \mathscr{B}$ *with first bifurcation point* b. *Then*

1) *if* φ_b *has a flip or a noncritical saddle-node with no cycles,* $\{\varphi_\mu\}$ *is stable,*
2) *if* φ_b *has a saddle-node which is not bicritical and has no cycles,* $\{\varphi_\mu\}$ *is mildly stable,*
3) *if* φ_b *has a Hopf periodic orbit or a saddle-node which is bicritical or has a cycle then* $\{\varphi_\mu\}$ *is left stable.*

Proof. — First, we observe that the statement in part (2) can be proved as in part (3). In fact, in case (2) we have that $\{\varphi_\mu\}$ is Morse-Smale for $b < \mu \leq \mu_1$, some $\mu_1 > b$, and so it is stable in this range of the parameter. This last fact is an easy consequence of the proof that Morse-Smale diffeomorphisms are stable. By the same reason, in all cases it is enough to consider the stability of $\{\varphi_\mu\}$ in $\mathscr{B}$ for μ near the first bifurcation point.

Let $p_{1,\mu} > p_{2,\mu} > \ldots > p_{\ell,\mu}$ be a total ordering of the periodic orbits of φ_μ for $\mu < b$. As in section 3, this ordering can be naturally extended for $\mu = b$ and even for $\mu > b$ when φ_b is as in parts (1) and (2) of the statement. By Theorem (3.6), we can take a compatible system of (strong) unstable foliations $\mathscr{F}^u(p_1), \mathscr{F}^u(p_2), \ldots, \mathscr{F}^u(p_\ell)$ defined in $M \times [\mu_1, \mu_2]$ or $M \times [\mu_1, b]$, for some $\mu_1 < b$ and $\mu_2 > b$ according to case (1) or (3). In case (1), the foliation is a strong unstable one at the flip or the saddle-node. Recall that this strong unstable foliation is constructed from a center foliation $\mathscr{F}^c$ in the center stable manifold of Φ at the flip or the saddle-node, where $\Phi : M \times I \to M \times I$ is defined by $\Phi(x, \mu) = (\varphi_\mu(x), \mu)$. By Theorem (4.1), for a nearby family $\{\widetilde{\varphi}_\mu\}$ with first bifurcation point b, we can consider a corresponding compatible system of (strong) unstable foliations and a compatible center foliation in case (1). Also, given a reparametrization (a homeomorphism) $\rho : I \to I$ near the identity and $\rho(b) = \widetilde{b}$, there is an order preserving continuous correspondence between the periodic orbits $p_{i,\mu}$ and $\widetilde{p}_{i,\rho(\mu)}$ of φ_μ and $\widetilde{\varphi}_{\rho(\mu)}$, for $\mu \in [\mu_1, \mu_2]$ or $\mu \in [\mu_1, b]$. This defines the conjugacy on the periodic orbits. In the construction of a global conjugacy we can take this reparametrization quite arbitrarily, except in the case of a noncritical saddle-node with no cycles. In this situation we must choose ρ as in Chapter II, so that we have a (continuous) conjugacy between Φ and $\widetilde{\Phi}$ restricted to center manifolds at the saddle-nodes.

Our global conjugacy H will be constructed on $M \times [\mu_1, \mu_2]$ or $M \times [\mu_1, b]$ inductively on the stable manifolds of the periodic orbits and, in case (1), at the center stable manifold of Φ at the saddle-node or the flip. Since these stable manifolds cover all of $M \times \{\mu\}$ for each $\mu \in [\mu_1, \mu_2]$ or $\mu \in [\mu_1, b]$, H will be defined on all of $M \times [\mu_1, \mu_2]$ or $M \times [\mu_1, b]$ and maps onto $M \times [\rho(\mu_1), \rho(\mu_2)]$ or $M \times [\rho(\mu_1), \widetilde{b}]$. We fix fundamental domains A_k for $W^s(p_k)$ or $W^{cs}(p_k)$ with exterior boundaries transverse to all $W^u(p_i)$, $i < k \leq \ell$. The conjugacy H will be constructed with the following properties:

a) $H(M \times \{\mu\}) = M \times \{\rho(\mu)\}$, μ in $M \times [\mu_1, \mu_2]$ or in $M \times [\mu_1, b]$,
b) it sends leaves of $\mathscr{F}^u(p_i)$ into leaves of $\mathscr{F}^u(\widetilde{p}_i)$, $1 \leq i \leq \ell$,
c) it is differentiable along each leaf of $\mathscr{F}^u(p_i)$ off the stable manifold of Φ at p_i, $1 \leq i \leq \ell$,
d) if $W^u(p_i) \cap W^s(p_k) \neq \emptyset$, $i < k$, then for each leaf F of $\mathscr{F}^u(p_i)$, in a neighborhood of $F \cap A_k$, H is C^1 near the inclusion map.

As the first step H takes the sources of $\{\varphi_\mu\}$ onto the sources of $\{\widetilde{\varphi}_{\rho(\mu)}\}$. Suppose H has been constructed on the stable manifolds of Φ at $p_1 > p_2 > \ldots > p_n$. Notice that the space of leaves of $\mathscr{F}^u(p_i)$ is parametrized by the intersection of the leaves with $W^s(p_i)$ or $W^{cs}(p_i)$. Thus, in particular, H defines a map from the space of leaves of $\mathscr{F}^u(p_i)$ onto the space of leaves of $\mathscr{F}^u(\widetilde{p}_i)$ for $1 \leq i \leq n$: $F \in \mathscr{F}^u(p_i)$ is associated to $\widetilde{F} \in \mathscr{F}^u(\widetilde{p}_i)$ if $H(F \cap W^s(p_i)) = \widetilde{F} \cap W^s(\widetilde{p}_i)$. Let us now consider $W^s(p_{n+1})$, $W^s(\widetilde{p}_{n+1})$ if p_{n+1} is either hyperbolic, a Hopf orbit or a saddle-node which is critical or has a cycle. Let $A_b = A_{n+1} \cap (M \times \{b\})$ and $\widetilde{A}_{\widetilde{b}} = \widetilde{A}_{n+1} \cap (M \times \{\widetilde{b}\})$. If $W^u(p_n) \cap A_b \neq \emptyset$, we choose H as a diffeomorphism near the inclusion of a neighborhood of $W^u(p_n) \cap \partial_{ex} A_b$ in $W^u(p_n)$ onto a neighborhood of $W^u(\widetilde{p}_n) \cap \partial_{ex} \widetilde{A}_{\widetilde{b}}$ in $W^u(\widetilde{p}_n)$. Such a diffeomorphism exists because $W^u(p_n) \cap A_b$ is close to $W^u(\widetilde{p}_n) \cap \widetilde{A}_{\widetilde{b}}$ for $\{\varphi_\mu\}$ near $\{\widetilde{\varphi}_\mu\}$. Near the interior boundary of A_b, H is defined by $H\varphi_b = \widetilde{\varphi}_{\widetilde{b}} H$. From Theorem (4.2) this partial diffeomorphism can be extended to all of $W^u(p_n) \cap A_b$. To extend H to the leaves of $\mathscr{F}^u(p_n)$ near A_{n+1}, $\mu_1 \leq \mu \leq b$, we proceed in an entirely similar way using Theorem (4.2), since we already know which leaf of $\mathscr{F}^u(\widetilde{p}_n)$ is associated to a given leaf of $\mathscr{F}^u(p_n)$. This completes the construction of the conjugacy on the intersection of the leaves of $\mathscr{F}^u(p_n)$ with the fundamental domain A_{n+1}. Next, we take $\mathscr{F}^u(p_{n-1})$ and suppose that $W^u(p_{n-1}) \cap A_b \neq \emptyset$. There are two cases to consider. If $W^u(p_{n-1}) \cap W^s(p_n) = \emptyset$, then $W^u(p_{n-1}) \cap A_b$ is compact and disjoint from $W^u(p_n) \cap A_b$. Thus, the construction of the conjugacy on the leaves of $\mathscr{F}^u(p_{n-1})$ restricted to A_{n+1} is the same as on the leaves of $\mathscr{F}^u(p_n)$. So, assume that $W^u(p_{n-1}) \cap W^s(p_n) \neq \emptyset$. Since $\mathscr{F}^u(p_{n-1})$ and $\mathscr{F}^u(p_n)$ are compatible, H is already defined on the leaves of $\mathscr{F}^u(p_{n-1})$ in a neighborhood V of $W^u(p_n) \cap A_b$ in A_{n+1}. But $B = W^u(p_{n-1}) \cap (A_b - V)$ is compact. So, we can extend H to B using Theorem (4.2) and also to the leaves of $\mathscr{F}^u(p_{n-1})$ in a neighborhood of B in A_{n+1}. We repeat the argument to all $\mathscr{F}^u(p_i)$, $1 \leq i \leq n-2$. If p_{n+1} is a

Hopf orbit or a critical saddle-node with no cycle, then H is defined on A_{n+1} satisfying $H\varphi_\mu = \widetilde{\varphi}_{\rho(\mu)}H$ or $H\varphi = \widetilde{\varphi}H$, $\mu_1 \leq \mu \leq b$ or $\mu_1 \leq \mu \leq \mu_2$ with $\mu_1 < b$ and $b < \mu_2$. The same equation allows us to define H on all of the stable manifold of φ at p_{n+1}. If the saddle-node occurs in a cycle so that $p_{n+1} = p_{n+1+j'}$ for some $j' \geq 1$, then we have only defined H on a neighborhood V of $W^{ss}(p_{n+1,b}) - \{p_{n+1,b}\}$ in $W^s(p_{n+1})$. We continue as before for p_{n+1+j} with $j < j'$. When we come to $p_{n+1+j'}$, the conjugacy H as already been defined on the part V of $W^s(p_{n+1+j'})$. We continue the process as above obtaining a conjugacy H_1 defined on $(\bigcup_{k \leq n+1+j'} W^u(p_k)) \cap W^s(p_{n+1+j'})$ such that $H_1 = H$ off a neighborhood of $W^s(p_{n+1+j'}) - V$ in $W^s(p_{n+1+j'})$. Thus, H can be defined on all of $W^s(p_{n+1})$.

Now we consider the case where p_{n+1} is a flip or a non-critical saddle-node with no cycles. Here, H will be defined in the center stable manifold of Φ at p_{n+1} as the "product" of two partial conjugacies. One of them, which we call H^c, is defined at the center manifold of Φ at p_{n+1}, as constructed in section 3 of Chapter II. It corresponds to a conjugacy on the space of leaves of the strong stable foliations $\mathcal{F}^{ss}(p_{n+1})$, $\mathcal{F}^{ss}(\widetilde{p}_{n+1})$. The other, which we call H^s, is to be defined on the space of leaves of the center foliations $\mathcal{F}^c(p_{n+1})$, $\mathcal{F}^c(\widetilde{p}_{n+1})$. However, since the center foliations may have several leaves going through the periodic orbits p_{n+1}, $\widetilde{p}_{n+1}$, we cannot express the stable manifolds of such orbits exactly as the product of these two foliations. That is the reason we will make a slight modification of the center foliation $\mathcal{F}^c(\widetilde{p}_{n+1})$ already constructed. The second partial conjugacy H^s will then send leaves of $\mathcal{F}^c(p_{n+1})$ onto leaves of the modified center foliation for $\widetilde{p}_{n+1}$. To do this, we first consider a continuous family of leaves $\{F_\mu\}$, $\mu_1 \leq \mu \leq \mu_2$ with $F_\mu \in \mathcal{F}^{ss}(p_{n+1}, \mu)$, such that F_μ meets all the leaves of the center foliation $\mathcal{F}^c(p_{n+1}, \mu)$. We also choose a family of disks $\{D_\mu\}$, $D_\mu \subset F_\mu$, such that the exterior boundary $\partial_{ex} D_\mu$ of D_μ is transverse to the unstable manifolds $W^u(p_i, \mu)$ for all $1 \leq i \leq n$ and each $\mu \in [\mu_1, \mu_2]$. We now form a fundamental domain $A = \bigcup A_\mu$ for the center stable manifold of p_{n+1} as follows. For each μ, the exterior boundary $\partial_{ex} A_\mu$ is made of the disk D_μ and a "cylinder" formed with the leaves of $\mathcal{F}^c(p_{n+1}, \mu)$ through the points of $\partial_{ex} D_\mu$. Notice that H^c determines a corresponding family $\{\widetilde{F}_{\rho(\mu)}\}$ of leaves of $\mathcal{F}^{ss}(\widetilde{p}_{n+1})$. We then take a family of disks $\{\widetilde{D}_{\rho(\mu)}\}$, $\widetilde{D}_{\rho(\mu)} \subset \widetilde{F}_{\rho(\mu)}$ for each μ, and construct a similar fundamental domain $\widetilde{A} = \bigcup \widetilde{A}_{\rho(\mu)}$ for the center stable manifold of $\widetilde{p}_{n+1}$. We now define $H^s : \bigcup D_\mu \to \bigcup \widetilde{D}_{\rho(\mu)}$ as follows. Of course, we want it to be compatible with the unstable foliations $\mathcal{F}^u(p_i)$ and the conjugacies already defined on the stable manifolds $W^s(p_i)$, $1 \leq i \leq n$. Near the exterior boundaries of D_μ, $\widetilde{D}_{\rho(\mu)}$ we proceed by induction on the indices $i = n, n-1, \ldots, 1$. We start with a diffeomorphism $H^s : V \to \widetilde{V}$ near the inclusion map, where V is a neighborhood of $W^u(p_n) \cap \partial_{ex} D_\mu$ in $W^u(p_n) \cap F_\mu$ and $\widetilde{V}$ is a neighborhood of $W^u(\widetilde{p}_n) \cap \partial_{ex} \widetilde{D}_{\rho(\mu)}$ in $W^u(\widetilde{p}_n) \cap \widetilde{F}_{\rho(\mu)}$. Similarly for the intersections of the leaves of $\mathcal{F}^u(p_n)$ with $\partial_{ex} D_\mu$ which are mapped onto the intersections of the corresponding leaves of $\mathcal{F}^u(\widetilde{p}_n)$ with $\partial_{ex} \widetilde{D}_{\rho(\mu)}$. Notice that the correspondence between the leaves of $\mathcal{F}^u(p_n)$ and those of $\mathcal{F}^u(\widetilde{p}_n)$ is determined by $H: W^s(p_n) \to W^s(\widetilde{p}_n)$ already defined. The extension to the leaves of $\mathscr{F}^u(p_{n-1}), \ldots, \mathscr{F}^u(p_1)$ is done in the same way as in the previous case. In order to define $H^s: \bigcup D_\mu \to \bigcup \widetilde{D}_{\rho(\mu)}$, we consider the annuli $A^*_\mu \subset D_\mu$ and $\widetilde{A}^*_{\rho(\mu)} \subset \widetilde{D}_{\rho(\mu)}$, each $\mu \in [\mu_1, \mu_2]$, obtained as follows. The exterior boundary of A^*_μ is $\partial_{ex} D_\mu$ and its interior boundary is the projection of $\varphi_\mu(\partial_{ex} D_\mu)$ into D_μ via the leaves of $\mathscr{F}^c(p_{n+1}, \mu)$. Similarly for $\widetilde{A}^*_{\rho(\mu)}$. Notice that $H^s: \partial_{ex} A^*_\mu \to \partial_{ex} \widetilde{A}^*_{\rho(\mu)}$ constructed above induces a map $H^s: \partial_{in} A^*_\mu \to \partial_{in} \widetilde{A}^*_{\rho(\mu)}$. In fact, for $x \in \partial_{in} A^*_\mu$ let ℓ_x be the leaf of $\mathscr{F}^c(p_{n+1}, \mu)$ through x and let y be the intersection of $\varphi_\mu^{-1}(\ell_x)$ with $\partial_{ex} \widetilde{A}^*_\mu$. If $\widetilde{\ell}$ is the leaf of $\mathscr{F}^c(\widetilde{p}_{n+1}, \rho(\mu))$ through $H^s(y)$, we set $H^s(x) = z$, where z is the intersection of $\widetilde{\varphi}_{\rho(\mu)}(\widetilde{\ell})$ with $\partial_{in} \widetilde{A}^*_{\rho(\mu)}$. We now extend H^s first to all of A^*_μ using the Isotopy Extension Theorem as before. Finally, we can extend H^s to all of D_μ sending leaves of $\mathscr{F}^u(p_i, \mu)$ to leaves of $\mathscr{F}^u(\widetilde{p}_i, \rho(\mu))$, $1 \leq i \leq n$. Again, this can be done as before since $W^u(p_i, \mu)$ and $W^u(\widetilde{p}_i, \rho(\mu))$ are transverse to D_μ and $\widetilde{D}_{\rho(\mu)}$, respectively. At this point, we would like to define the conjugacy $H: W^{cs}(p_{n+1}) \to W^{cs}(\widetilde{p}_{n+1})$ as the "product" of H^c and H^s using the strong stable and the center foliations. To do this we have to modify the center foliation $\mathscr{F}^c(\widetilde{p}_{n+1})$ in $W^{cs}(\widetilde{p}_{n+1})$. For each $\mu \in [\mu_1, \mu_2]$, it is enough to do so in the region bounded by $\widetilde{D}_{\rho(\mu)} - \widetilde{A}^*_{\rho(\mu)}$, $\widetilde{\varphi}_{\rho(\mu)}(\widetilde{D}_{\rho(\mu)})$ and the cylinder formed with the leaves of $\mathscr{F}^c(\widetilde{p}_{n+1})$ through points of $\partial_{in} \widetilde{A}^*_{\rho(\mu)}$. Let $\eta: \widetilde{D}_{\rho(\mu)} - \widetilde{A}^*_{\rho(\mu)} \to \widetilde{\varphi}_{\rho(\mu)}(\widetilde{D}_{\rho(\mu)})$ be the homeomorphism defined by $\eta(x) = y$, where $(H^s)^{-1}(x)$ and $\varphi_\mu (H^s)^{-1} \widetilde{\varphi}^{-1}_{\rho(\mu)}(y)$ belong to the same leaf of $\mathscr{F}^c(p_{n+1}, \mu)$. Let $\lambda: \widetilde{D}_{\rho(\mu)} - \widetilde{A}^*_{\rho(\mu)} \to \widetilde{\varphi}_{\rho(\mu)}(\widetilde{D}_{\rho(\mu)})$ be defined by $\lambda(x) = z$, where x and z belong to the same leaf of $\mathscr{F}^c(\widetilde{p}_{n+1}, \rho(\mu))$. If $\lambda = \eta$ then no modification of $\mathscr{F}^c(\widetilde{p}_{n+1}, \rho(\mu))$ is needed. So let us change $\mathscr{F}^c(\widetilde{p}_{n+1}, \rho(\mu))$ to get the second map to be equal to the first. Notice that $\eta(x)$ and $\lambda(x)$ belong to the same leaf of $\mathscr{F}^u(\widetilde{p}_i, \rho(\mu))$ for some $1 \leq i \leq n$. Moreover, $\lambda\eta^{-1}$ along such a leaf is C^1 near the identity map. Let $\widetilde{X}$ be the vector field whose integral curves are the leaves of $\mathscr{F}^c(\widetilde{p}_{n+1}, \rho(\mu))$. We modify $\widetilde{X}$ near but off $\widetilde{\varphi}_{\rho(\mu)}(\widetilde{D}_{\rho(\mu)})$ so that the corresponding map $\bar{\lambda}$ satisfies $\bar{\lambda}\eta^{-1} = 1$ on $\widetilde{\varphi}_{\rho(\mu)}(\widetilde{D}_{\rho(\mu)})$. Observe that the modification required for $\widetilde{X}$ along $W^u(\widetilde{p}_n, \rho(\mu))$ is well known and it can be performed in a parametrized way along the leaves of $\mathscr{F}^u(\widetilde{p}_n, \rho(\mu))$. Using the fact that the foliations $\mathscr{F}^u(\widetilde{p}_i, \rho(\mu))$ are compatible, we proceed by downward induction on the indices $1 \leq i \leq n$. It is clear that the new center foliation coincides with the previous one on the boundary $\partial\widetilde{A} = \bigcup \partial\widetilde{A}_{\rho(\mu)}$ of the fundamental domain $\widetilde{A}$. Once it is defined on $\widetilde{A}$, it can be defined on all of $W^{cs}(\widetilde{p}_{n+1})$ simply through iterations by $\widetilde{\Phi}$, where $\widetilde{\Phi}(x, \mu) = (\widetilde{\varphi}_\mu(x), \mu)$ as before. Now we can define H from $W^{cs}(p_{n+1})$ onto $W^{cs}(\widetilde{p}_{n+1})$ using the center foliation, the strong stable foliation and the conjugacies H^s and H^c on their spaces of leaves. H is clearly one to one, onto and also continuous since these foliations and the maps H^s, H^c are continuous. Moreover, H is differentiable along the leaves of $\mathscr{F}^u(p_i)$, $1 \leq i \leq n$, in $W^{cs}(p_{n+1})$. In fact, $\mathscr{F}^c$ was constructed in section 3 to be differentiable along these leaves; the same is true with the modified $\mathscr{F}^c$ and the map H^s constructed above. Being codimension one, the foliation $\mathscr{F}^{ss}$ is C^1 (see section 2, Chapter II). Finally, the differentiability of H^c off the periodic points is in section 3 and 4 of Chapter II. Together these facts imply our statement. Thus, the proof of the induction step is finished. Our map H is defined on $M \times [\mu_1, \mu_2]$ or $M \times [\mu_1, b]$ onto $M \times [\rho(\mu_1), \rho(\mu_2)]$ or $M \times [\rho(\mu_1), \tilde{b}]$, it is clearly one to one and satisfies the conjugacy equation $H\varphi_\mu = \tilde{\varphi}_{\rho(\mu)}H$. Let us prove that H is continuous. We have to show that H is continuous at the stable (center stable) manifolds of the periodic orbits of $\{\varphi_\mu\}$. Here we indicate by W^s both the stable and center stable manifolds; also, let I_1 indicate either $[\mu_1, \mu_2]$ or $[\mu_1, b]$. By construction, H is continuous along these manifolds and, in particular, along the stable manifolds of the sinks. By induction we may assume that $H: W \to \tilde{W}$ is a homeomorphism, where W and $\tilde{W}$ are the union of $W^s(p_{k,\mu})$ and $W^s(p_{k,\rho(\mu)})$ for $i < k \leq \ell$ and $\mu \in I_1$. Let us now show that H is also continuous at $W^s(p_{i,\mu})$. Consider a sequence $(x_n, \mu_n) \to (x, \mu) \in W^s(p_{i,\mu})$, $x_n \in M$ and $\mu_n \in I_1$. Since H restricted to $W^s(p_{i,\mu})$ is continuous, we may assume, via a subsequence, that $(x_n, \mu_n) \in W$ for all n. Let F_n, F be the leaves of $\mathscr{F}^u(p_i)$ containing (x_n, μ_n), (x, μ) and $\tilde{F}_n$, $\tilde{F}$ the leaves of $\mathscr{F}^u(\tilde{p}_i)$ containing $H(x_n, \mu_n)$ and $H(x, \mu)$, respectively. We have that $F_n \to F$ and since H is continuous restricted to the stable manifolds, we have $\tilde{F}_n \to \tilde{F}$. Thus, it is enough to show that the sequence $H(x_n, \mu_n)$ accumulates on $W^s(p_{i,\rho(\mu)})$. In fact, $H(x_n, \mu_n)$ cannot accumulate on W because $H: W \to W$ is a homeomorphism. Also, $H(x_n, \mu_n)$ cannot accumulate on the union $\tilde{Z}$ of $W^s(\tilde{p}_{j,\rho(\mu)})$, $1 \leq j < i$, because $\tilde{F}_n$ and $\tilde{Z}$ are far apart. This proves our assertion and so H is continuous on all of $M \times I_1$. This finishes the proof of the theorem.

We now complete the proof of our second main theorem stated in the introduction, the first part of which was done in section 2 of Chapter III.

Let us denote by $\mathscr{C} \subset \mathscr{P}$ the set of arcs $\{\varphi_\mu\}$ such that the limit set of each φ_μ has finitely many orbits, $\mu \in I = [0, 1]$. We also denote by $\mathscr{S} \subset \mathscr{C}$ the set of arcs $\{\varphi_\mu\}$ such that there are only finitely many bifurcation values for $\{\varphi_\mu\}$ say $b_1, \ldots, b_s$ in $(0, 1)$ and for each $1 \leq i \leq s$, φ_{b_i} has the following properties:

— all stable, strong stable, unstable and strong unstable manifolds intersect transversally
— φ_{b_i} has no cycles and has exactly one non-hyperbolic periodic orbit, which is either a flip or a non-critical saddle-node; this non-hyperbolic orbit unfolds generically.

As we mentioned before, it turns out that for arcs $\{\varphi_\mu\} \in \mathscr{S}$, φ_μ is a Morse-Smale diffeomorphism if μ is not a bifurcation value.

Theorem (4.4). — *The arcs in* $\mathscr{S}$ *are stable.*

Proof. — As in Theorem (4.1), it follows from [19], [20] that $\mathscr{S}$ is an open subset of $\mathscr{P}$. Thus, if $\{\varphi_\mu\} \in \mathscr{S}$ and has bifurcation values $b_1, \ldots, b_s$ in $(0, 1)$, then a nearby arc $\{\varphi'_\mu\}$ also belongs to $\mathscr{S}$ and has nearby bifurcation values $b'_1, \ldots, b'_s$ in $(0, 1)$. Moreover, the non-hyperbolic periodic orbits of φ_{b_i} and $\varphi'_{b'_i}$ are both saddle-nodes or flips. To produce a conjugacy between $\{\varphi_\mu\} \in \mathscr{S}$ and a nearby arc $\{\varphi'_\mu\}$, we first assume that

$\varphi_\mu = \varphi'_\mu$ for $\mu \in I - U$, where U is a small open subinterval of I containing at most one bifurcation value. In this case, by Theorem (3.6) we can construct unstable tubular families or foliations for $\{\varphi_\mu\}$ and $\{\varphi'_\mu\}$ and $\mu \in V$, where V is an open subinterval of I containing the closure $\overline{U}$ of U. Moreover, using the Isotopy Extension Theorem (4.2), we may construct these unstable foliations to be the same for $\{\varphi_\mu\}$ and $\{\varphi'_\mu\}$ if $\mu \in V - W$, where W is an open subinterval of I such that $\overline{U} \subset W$ and $\overline{W} \subset V$. Again using the Isotopy Extension Theorem, the construction of the conjugacy $\{h_\mu\}$ between $\{\varphi_\mu\}$ and $\{\varphi'_\mu\}$ for $\mu \in V$, as performed in Theorem (4.3), can be done so that h_μ is the identity map on M for $\mu \in V - W$. Since $\varphi_\mu = \varphi'_\mu$ for $\mu \in I - U$, we can extend this conjugacy to all of I by defining it to be the identity for $\mu \in I - V$. Finally, let $\{U_i\}$, $1 \leq i \leq n$, be a covering of I by small subintervals, each containing at most one bifurcation value. It is immediate that we can decompose any small perturbation of $\{\varphi_\mu\}$ into perturbations, each with support in one of the subintervals U_i. So the construction of the conjugacy between $\{\varphi_\mu\}$ and a nearby arc is reduced to the previous case. The proof of theorem is complete.

REFERENCES

[1] A. A. Andronov, E. A. Leontovich, I. I. Gordon, A. G. Maier, Theory of bifurcations of dynamic systems on a plane, *Isr. Program Scientific Transl.*, Jerusalem, 1971.

[2] L. Block, J. Franke, Existence of periodic points of maps of S^1, *Invent. Math.*, **22** (1973), 69-73.

[3] Th. Bröcker, L. C. Lander, Differentiable germs and catastrophes, *London Math. Soc. Lecture Notes*, **17**, Cambridge Univ. Press, 1975.

[4] P. Brunovski, On one-parameter families of diffeomorphisms, I and II, *Comment. Mat. Univ. Carolinae*, **11** (3) (1970), 559-582; **12** (4) (1971), 765-784.

[5] P. Brunovski, Generic properties of the rotation number of one-parameter diffeomorphisms of the circle, *Czech. Math. J.*, **24** (1974), 74-90.

[6] A. Denjoy, Les trajectoires à la surface du tore, *C. r. Acad. Sci.*, **223** (1946), 5-8.

[7] J. Guckenheimer, One-parameter families of vector fields on two-manifolds: another nondensity theorem, *Dynamical Systems*, Academic Press, 1973, 111-128.

[8] P. Hartman, On local homeomorphisms of Euclidean spaces, *Bol. Soc. Mat. Mexicana*, **5** (2) (1960), 220-241.

[9] M. Herman, Sur la conjugaison différentiable des difféomorphismes du cercle à des rotations, *Publ. Math. I.H.E.S.*, **49** (1979), 5-234.

[10] M. Herman, Mesure de Lebesgue et nombre de rotation, *Geometry and Topology, Springer Lecture Notes in Math.*, **597** (1977), 271-293.

[11] M. W. Hirsch, C. C. Pugh, Stable manifolds and hyperbolic sets, *Proc. Symp. Pure Math. A.M.S.*, **14** (1970), 133-164.

[12] M. W. Hirsch, C. C. Pugh, M. Shub, Invariant manifolds, *Springer Lecture Notes in Math.*, **583** (1977).

[13] I. Kupka, Contributions à la théorie des champs génériques, *Contributions to differential equations*, vol. 2 (1963), 475-484 and vol. 3 (1964), 411-420.

[14] D. C. Lewis, Formal power series transformations, *Duke Math. J.*, **5** (1939), 794-805.

[15] I. P. Malta, Hyperbolic Birkhoff centers, *Trans. A.M.S.*, **262** (1980), 181-193; announced in *An. Acad. Brasil. Ciências*, **51** (1979), (1), 27-29.

[16] A. Manning, There are no new Anosov diffeomorphisms on tori, *Amer. J. Math.*, **96** (1974), 422-429.

[17] J. E. Marsden, M. McCracken, The Hopf bifurcation and its applications, *Appl. Math. Sci.*, **19**, Springer-Verlag, 1976.

[18] W. C. de Melo, Moduli and stability of two-dimensional diffeomorphisms, *Topology*, **19** (1980), 9-21.

[18a] W. de MELO, J. PALIS and S. VAN STRIEN, Characterizing diffeomorphisms with modulus of stability one, *Dynamical Systems and Turbulence*, Warwick 1980, *Springer Lecture Notes in Math.*, **898** (1981), 266-286.
[19] S. NEWHOUSE, J. PALIS, Bifurcations of Morse-Smale dynamical systems, *Dynamical Systems*, Academic Press, 1973, 303-366.
[20] S. NEWHOUSE, J. PALIS, Cycles and bifurcation theory, *Astérisque*, **31** (1976), 43-140.
[21] S. NEWHOUSE, J. PALIS, F. TAKENS, Stable arcs of diffeomorphisms, *Bull. A.M.S.*, **82** (1976), 499-502.
[22] S. NEWHOUSE, The abundance of wild hyperbolic sets and non-smooth stable sets, *Publ. Math. I.H.E.S.*, **50** (1979), 101-152.
[23] R. PALAIS, Local triviality of the restriction map for embeddings, *Comm. Math. Helvet.*, **34** (1960), 305-312.
[24] J. PALIS, On Morse-Smale dynamical systems, *Topology*, **8** (1969), 385-405.
[25] J. PALIS, S. SMALE, Structural stability theorems, *Proc. Symp. Pure Math. A.M.S.*, **14** (1970), 223-232.
[26] J. PALIS, F. TAKENS, Topological equivalence of normally hyperbolic dynamical systems, *Topology*, **16** (1977), 335-345.
[26a] J. PALIS and F. TAKENS, *Stability of parametrized families of gradient vector fields*, Preprint IMPA, to appear.
[27] J. PALIS, A differentiable invariant of topological conjugacies and moduli of stability, *Astérisque*, **51** (1978), 335-346.
[28] J. PALIS, Moduli of stability and bifurcation theory, *Proc. Int. Congres of Math. Helsinki* (1978), 835-839.
[29] H. POINCARÉ, *Œuvres complètes*, t. 1, Gauthier-Villars, 1952, 137-158.
[30] C. C. PUGH, M. SHUB, Linearization of normally hyperbolic diffeomorphisms and flows, *Invent. Math.*, **10** (1970), 187-198.
[31] C. ROBINSON, Global structural stability of a saddle-node bifurcation, *Trans. A.M.S.*, **236** (1978), 155-172.
[32] D. RUELLE, F. TAKENS, On the nature of turbulence, *Comm. Math. Phys.*, **20** (1971), 167-192; A note concerning our paper on the nature of turbulence, *Comm. Math. Phys.*, **23** (1971), 343-344.
[33] S. SMALE, Stable manifolds for differential equations and diffeomorphisms, *Ann. Scuola Norm. Sup. Pisa*, **18** (1963), 97-116.
[34] S. SMALE, Differentiable dynamical systems, *Bull. A.M.S.*, **73** (1967), 747-817.
[35] J. SOTOMAYOR, Generic one parameter families of vector fields, *Publ. Math. I.H.E.S.*, **43** (1974), 4-56.
[36] J. SOTOMAYOR, Generic bifurcations of dynamical systems, *Dynamical Systems*, Academic Press 1973, 549-560.
[37] S. J. VAN STRIEN, Center manifolds are not C^∞, *Math. Z.*, **166** (1979), 143-145.
[38] F. TAKENS, Partially hyperbolic fixed points, *Topology*, **10** (1971), 133-147.
[39] F. TAKENS, Normal forms for certain singularities of vector fields, *Ann. Inst. Fourier*, **23** (1973), (2), 163-195.
[40] F. TAKENS, Forced oscillations and bifurcations, *Applications of Global Analysis* I, *Comm. of the Math. Inst.*, R.U. Utrecht (Holland), 1974.
[41] F. TAKENS, Global phenomena in bifurcations of dynamical systems with simple recurrence, *Jber. d. Dt. Math.-Verein*, **81** (1979), 87-96.
[42] R. WILLIAMS, The D.A. map of Smale and structural stability, *Proc. Symp. Pure Math. A.M.S.*, **14** (1970), 329-334.

Department of Mathematics,
University of North Carolina,
Chapel Hill, NC 27514, U.S.A.

I.M.P.A., Rua L. de Camões 68,
Rio de Janeiro, Brazil.

Department of Mathematics,
University of Groningen, Holland.

Manuscrit reçu le 22 avril 1981.

Annals of Mathematics, 118 (1983), 383–421

Stability of parametrized families of gradient vector fields

By J. Palis and F. Takens

Dedicated to René Thom on his sixtieth birthday.

Abstract

It is shown that the stable one-parameter families of gradient vector fields are open and dense.

Introduction

Gradient vector fields are a class of dynamical systems that appear naturally in several mathematical contexts. From the dynamics point of view, they are in general among the vector fields that exhibit the simplest orbit structures. Yet, and perhaps due to that same reason, they played a crucial role in the work of Morse and Smale relating some simple dynamical properties with the topology of the ambient manifold. Morse imposed a non-degeneracy condition at the singularities and Smale a transversality condition between their stable and unstable manifolds. Such a gradient vector field, nowadays called Morse-Smale, provides a decomposition of the manifold into a (disjoint) union of the varying-dimensional stable manifolds. Later, the Morse-Smale gradient fields were shown to be structurally stable [12]. That is, modulo a flow conjugacy, the orbit structure is unchanged under small perturbations of the original field. The key difficulty in proving this result is precisely the fact that the stable manifolds vary in dimension. For this reason, the concept of stable and unstable tubular families or foliations was introduced. Fitted together, they provide globally compatible coordinate systems, from which flow conjugacies are naturally constructed.

Some of these ideas will be extended here to analyse parametrized families of gradients. Following a common point of view in bifurcation theory, we want to describe how the orbit structure of the vector field can change and, especially, how stable such a vector field is. Our main objective is to show the existence of a dense open set of stable one-parameter families of gradients. This set, which we exhibit explicitly, plays in our context a role similar to Morse functions for real-valued maps or, more properly, to Morse-Smale fields for gradient flows.

Originally published in: *Ann. of Math. (2) 118 (1983), no. 3, 383–421,*

For a family of gradient vector fields, the values of the parameter for which the orbit structure changes are called bifurcation values. Generically, they are isolated and correspond to saddle-node singularities or to orbits of tangency between the stable and unstable manifolds of a pair of (hyperbolic) singularities. A substantial part of our work consists in describing the structure of the dynamics near an orbit of tangency.

We now state our main result in a more precise way.

Let M be a closed C^∞ manifold. As discussed in Section I.1, we can consider gradients of real functions either with respect to a fixed Riemannian metric or with respect to all possible Riemannian metrics. Although we restrict ourselves to the second case, the results hold as well for the first case. Let, then, $\mathfrak{X}_1^g(M)$ be the set of C^∞ one-parameter families of gradients endowed with the C^∞ (strong) Whitney topology. In $M \times \mathbf{R}$, let $\pi_2\colon M \times \mathbf{R} \to \mathbf{R}$ be the natural projection. We say that the families $\{X_\mu\}$ and $\{X'_\mu\}$ in $\mathfrak{X}_1^g(M)$, $\mu \in \mathbf{R}$, are equivalent if there exists a pair of homeomorphisms $H\colon M \times \mathbf{R} \to M \times \mathbf{R}$ and $\eta\colon \mathbf{R} \to \mathbf{R}$ such that $\pi_2 H = \eta\pi_2$ and, for each $\mu \in \mathbf{R}$, H_μ is a topological equivalence between X_μ and $X'_{\eta(u)}$, where H_μ is defined by $H(x, \mu) = (H_\mu(x), \eta(\mu))$. That is, H_μ sends orbits of X_μ into orbits of $X'_{\eta(\mu)}$. When H_μ is a flow conjugacy for each $\mu \in \mathbf{R}$, we say that $\{X_\mu\}$ and $\{X'_\mu\}$ are conjugate. A family $\{X_\mu\}$ is called (structurally) stable if it is equivalent to all nearby families. Finally, in order to pose the problems below, we say that $\{X_\mu\}$ has finite modulus of stability if we can describe all equivalence classes of families in some neighbourhood of $\{X_\mu\}$ with finitely many (real) parameters. In particular, stable families have modulus of stability zero.

MAIN THEOREM (The Stability Theorem). *There exists an open and dense $\mathcal{G} \subset \mathfrak{X}_1^g(M)$ such that if $\{X_\mu\}$ is in $\mathcal{G}$ then $\{X_\mu\}$ is stable.*

We present here a rough description of $\mathcal{G}$ and refer to the proper sections of Chapter I where one can find the details. $\{X_\mu\}$ belongs to $\mathcal{G}$ if it satisfies the following conditions:

(i) For each value of μ, where X_μ is not structurally stable, there is either one non-hyperbolic singularity, or there is an orbit of tangency, i.e., an orbit along which a stable and an unstable manifold intersect non-transversally (see Section 1);

(ii) If X_μ has a non-hyperbolic singularity, this is a generically unfolding saddle-node; in other words, the corresponding one-parameter family of functions $\{f_\mu\}$ has all its critical points unfolding generically (see Section 2, c, condition S.N.);

(iii) If X_μ has an orbit of tangency between stable and unstable manifolds W^s and W^u, then along this orbit the contact is parabolic, the non-transversality unfolds generically (with the parameter μ) and $\dim(W^u) + \dim(W^s) \geq \dim(M)$ (see Section 2, a, condition Q.T. 1, 2, 3);

(iv) In both of the above cases (saddle-node or tangency) a number of "extra" invariant manifolds appear (e.g. in the case of a saddle-node both the stable and the unstable manifold have a boundary; this boundary is considered as an "extra" invariant manifold); these "extra" invariant manifolds intersect each other and the usual stable and unstable manifolds transversally (see Section 2, d, condition T, and the comments below);

(v) Finally, in case X_μ has a saddle-node in s or an orbit of tangency between the unstable manifold of s and the stable manifold of $\tilde{s}$, the eigenvalues of the linear part of X_μ at s, respectively at s and $\tilde{s}$, have to satisfy certain non-resonance conditions (see Section 2, b, condition L.1, 2).

It seems to us that condition (v) is not needed for our conclusion; the others are much more natural and we believe that they are on the whole necessary.

To give an idea of the geometric arguments involved in our work, we describe the canonical splittings of the tangent spaces at singularities connected by an orbit of tangency. They suggested to us corresponding foliations that we use to construct topological equivalences (conjugacies) in our stability theorem.

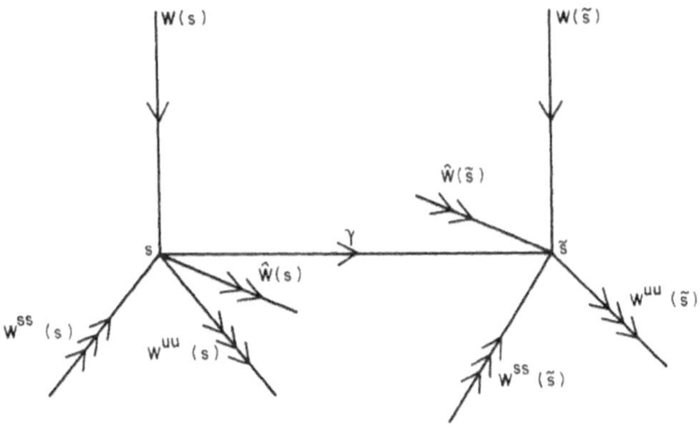

$W(s)$ is a one dimensional submanifold in $W^s(s)$ tangent to the direction of the weakest contraction.

$W(\tilde{s})$ is a one dimensional submanifold in $W^u(\tilde{s})$ tangent to the direction of weakest expansion.

$W^{ss}(s)$ has codimension one in $W^s(s)$ and is complementary to $W(s)$.

$W^{uu}(\tilde{s})$ has codimension one in $W^u(\tilde{s})$ and is complementary to $W(\tilde{s})$.

$W^{ss}(\tilde{s})$ has the same dimension as $W^{ss}(s)$, and is tangent to the directions of strongest contraction in $W^s(\tilde{s})$.

$W^{uu}(s)$ has the same dimension as $W^{uu}(\tilde{s})$, and is tangent to the directions of strongest expansion in $W^u(s)$.

$\hat{W}(s)$ is complementary to $W^{uu}(s)$ and $\dot{\gamma}(s)$ in the unstable manifold of s.

$\hat{W}(\tilde{s})$ is complementary to $W^{ss}(\tilde{s})$ and $\dot{\gamma}(\tilde{s})$ in the stable manifold of $\tilde{s}$. It has the same dimension as $\hat{W}(s)$.

In the direction of our main result, very recently G. Vegter (forthcoming thesis) proved a similar statement for 2- and 3-parameter families of gradients on 2-manifolds and for 2-parameter families on 3-manifolds. Even more recently, the second author of the present paper exhibited open sets of 8-parameter families with a modulus of stability and 16-parameter families with infinite moduli. Still, the following problems on k-parameter families are very relevant for small k.

Problem 1. Are the stable k-parameter families of gradients dense?

Problem 2. Are the k-parameter families of gradients with finite modulus of stability dense?

In general we pose the following question. For which configurations of bifurcating orbits (like degenerate singularities and orbits of tangency) does one have structural stability for generic k-parameter families of gradient vector fields? And for which configurations does one have finite moduli?

We are grateful to S. Newhouse for his cooperation at an early stage of the present work.

Contents

Introduction

Chapter I. Generic properties of families of gradient fields

Section 1. General observations

Section 2. Generic conditions for saddle-node and tangency

a) Quasi-transverse intersections

b) Linearizability of singularities of vector fields

c) Saddle-nodes

d) Invariant manifolds

e) Orbits of tangency

f) Conclusion

Section 3. Openness of the generic conditions and formulation of the main result

Chapter II. The dynamics near an orbit of tangency
Section 1. Smoothing structures
Section 2. Saddle connection in dimension two
Section 3. Invariant manifolds
a) Construction of the centre unstable manifold
b) Corollaries
Section 4. Construction of the conjugacy in the centre manifold

Chapter III. The main result: Global stability

I. Generic properties of families of gradient fields

1. *General observations*

In this chapter M is an m-dimensional compact C^∞-manifold. We investigate generic properties of one-parameter families of gradient vector fields on M. We define a *one-parameter family of gradients* as a family of vector fields $\{X_\mu\}$ on M, $\mu \in \mathbf{R}$, for which there are a family of functions $\{f_\mu\}$ and a family of Riemannian metrics $\{g_\mu\}$ such that $X_\mu = \operatorname{grad}_{g_\mu} f_\mu$; i.e., X_μ is the gradient of f_μ with respect to g_μ. We assume f_μ and g_μ, and hence X_μ, to depend C^∞ on $(x, \mu) \in M \times \mathbf{R}$. On the set of pairs, consisting of such a one-parameter family of functions and a one-parameter family of metrics, we take the strong C^∞-topology. We say that some property is *generic* for one-parameter families of gradient vector fields if there is a residual set of pairs $(\{f_\mu\}, \{g_\mu\})$ as above such that each $\{X_\mu = \operatorname{grad}_{g_\mu} f_\mu\}$ with $(\{f_\mu\}, \{g_\mu\})$ in the residual set has the property in question. In passing we note that there is also a notion of genericity of families of gradient vector fields on a *Riemannian* manifold: in that case the Riemannian metric is fixed and a generic gradient is just the gradient of a generic function. In this paper we restrict ourselves to the first type of genericity; in the case of generic gradients on a Riemannian manifold, some proofs are more complicated but the final results are analogous.

The genericity results are obtained by combining two types of bifurcations. First, we approximate the one-parameter family of functions $\{f_\mu\}$ by a generic one in the sense of [21]: in this way we get the singularities of grad f_μ generic. Then we consider generic conditions on intersections of stable and unstable manifolds. For this we need local perturbations of our gradient vector field off the set of singularities.

The lemma below shows that we can obtain any perturbation X' of $X = \operatorname{grad}_g f$ by perturbing the metric g (if $\operatorname{supp}(X - X')$ contains no singularities). This, combined with the method of proof of the Kupka-Smale theorem [5],

[15] and corresponding results on one-parameter families of vector fields [16], [17], yields the genericity results below.

LEMMA. *Let* $X = \operatorname{grad}_g f$ *be a gradient vector field on* M. *For any compact set* $K \subset M$, *so that for all* $x \in K$, $X(x) \neq 0$, *and for any neighbourhood* U *of* g *(in the* C^∞ *topology) there is a neighbourhood* V *of* X *in* $\mathfrak{X}(M)$, *the space of* C^∞ *vector fields on* M, *such that for any* $X' \in V$ *with* $\operatorname{supp}(X - X') \subset K$, *there is a* g' *in* U *with* $\operatorname{supp}(g - g') \subset K$ *and* $\operatorname{grad}_{g'} f = X'$. *Moreover, one may choose* g' *so that it depends differentiably on* X'.

Proof. For given X' we choose g' so that $\operatorname{grad}_{g'} f = X'$ and so that for each vector v, such that $v(f) = 0$, $g'(v, v) = g(v, v)$. It is clear that in this way g' is well defined if X' is sufficiently C^0 near X. Also, g' depends differentiably on X' and for $X' = X$, this construction yields $g' = g$. From this the lemma follows.

Remark. The same lemma, and proof, holds for the case where X, f, g and X' depend on a parameter μ in $\mathbf{R}$, or in $\mathbf{R}^p$.

Remark. The above perturbations can also be obtained by changing only the function instead of the metric. The proof is more complicated and will appear elsewhere.

We recall [10], [12] that a gradient vector field is structurally stable if and only if it is Morse-Smale, i.e., if all singularities are hyperbolic, stable and unstable manifolds intersect transversally, and hence the non-wandering set consists of the singularities only (for the definitions of these notions see [11], [14]). For this reason we call $\bar{\mu}$ a *regular parameter value* for a one-parameter family $\{X_\mu\}$ of gradients if $X_{\bar{\mu}}$ is Morse-Smale; otherwise we call $\bar{\mu}$ a *bifurcation value*. In view of the above considerations, we obtain:

THEOREM. *For one-parameter families of gradients* $\{X_\mu\}$ *the following properties are generic*:

a) *The set of regular parameter values is open and dense in* $\mathbf{R}$;

b) *For each bifurcation value* $\bar{\mu}$ *one has either that* $X_{\bar{\mu}}$ *has exactly one non-hyperbolic singularity, namely a saddle-node, while all the other singularities are hyperbolic and while all the stable and unstable manifolds intersect transversally, or that* $X_{\bar{\mu}}$ *has exactly one orbit along which a stable and an unstable manifold intersect non-transversally, while all the singularities are hyperbolic.*

In the next section we give generic properties of saddle-nodes and orbits of tangency (i.e. non-transversal intersections of stable and unstable manifolds) in the sense that for generic one-parameter families of gradients all saddle-nodes and tangencies have these properties. In Section 3, we show that there is even an open and dense set of one-parameter families of gradients satisfying all these generic conditions and that for these the bifurcation values are isolated.

2. *Generic conditions for saddle-node and tangency*

a. *Quasi-transverse intersections.* We first consider the case where $\{X_\mu\}$ is a one-parameter family of gradients which is generic in the sense of the last theorem and for which some $X_{\bar{\mu}}$ has an orbit γ of non-transversal intersection of a stable and an unstable manifold. Since all the singularities of $X_{\bar{\mu}}$ are hyperbolic, these stable and unstable manifolds depend (on compact parts) smoothly on μ for μ near $\bar{\mu}$; we denote them by W_μ^s and W_μ^u. So $\gamma \in W_{\bar{\mu}}^s \cap W_{\bar{\mu}}^u$ and for $r \in \gamma$, $T_r(W_{\bar{\mu}}^s) + T_r(W_{\bar{\mu}}^u)$ has dimension smaller than m. The first condition we consider is

(Q.T.1) $$\dim\left(T_r(W_{\bar{\mu}}^s) + T_r(W_{\bar{\mu}}^u)\right) = m - 1 \quad \text{for } r \in \gamma.$$

Again for $r \in \gamma$ we consider a smooth projection π of a neighbourhood of r in the normal space of $W_{\bar{\mu}}^s$ at r, or more formally, in $T_r(M)/T_r(W_{\bar{\mu}}^s)$, such that $\pi(W_{\bar{\mu}}^s) = 0$ and $(d\pi)_r$ is the canonical projection from $T_r(M)$ to $T_r(M)/T_r(W_{\bar{\mu}}^s)$. Note that π^u denotes the restriction of π to $W_{\bar{\mu}}^u$, while $D^2\pi^u$ denotes the intrinsic second order derivative of π^u in r; see [1]. This intrinsic second order derivative is a quadratic map from $\operatorname{Ker}(d\pi^u)_r = T_r(W_{\bar{\mu}}^u) \cap T_r(W_{\bar{\mu}}^s)$ to

$$\left(T_r(M)/T_r(W_{\bar{\mu}}^s)\right)/\operatorname{Im}(d\pi^u)_r \cong T_r(M)/\left(T_r(W_{\bar{\mu}}^s) + T_r(W_{\bar{\mu}}^u)\right).$$

This last vector space is one-dimensional if (Q.T.1) is satisfied; so in this case we have a quadratic form on $T_r(W^u) \cap T_r(W^s)$, which we also denote by $D^2\pi$ (it is only determined up to a multiplicative constant). This quadratic form cannot have maximal rank, because it is degenerate in the direction of $X_{\bar{\mu}}(r)$. The following condition we consider makes sense only in case (Q.T.1) is satisfied:

(Q.T.2) $D^2\pi$ is a quadratic form with co-rank one on $T_r(W_{\bar{\mu}}^s) \cap T_r(W_{\bar{\mu}}^u)$.

Finally we need a condition on the relative "speed" of W_μ^s and W_μ^u in γ (where μ is interpreted as time). For this we choose curves σ^u, σ^s in M with $\sigma^u(\mu) \in W_\mu^u$, $\sigma^s(\mu) \in W_\mu^s$ and $\sigma^u(\bar{\mu}) = \sigma^s(\bar{\mu}) = r \in \gamma$. Notice that $\dot{\sigma}^s(\bar{\mu}) - \dot{\sigma}^u(\bar{\mu}) \in T_r(M)$, after projection in $T_r(M)/(T_r(W_{\bar{\mu}}^u) + T_r(W_{\bar{\mu}}^s))$, is independent of the choice of σ^s and σ^u and describes this relative speed.

(Q.T.3) $\dot{\sigma}^s(\bar{\mu}) - \dot{\sigma}^u(\bar{\mu})$ is not contained in $T_r(W_{\bar{\mu}}^u) + T_r(W_{\bar{\mu}}^s)$.

THEOREM. *It is generic for one-parameter families of gradients $\{X_\mu\}$ to have the properties stated in the last theorem and also to have the property that for* *each $\bar{\mu}$, such that $X_{\bar{\mu}}$ has an orbit γ of non-transverse intersection of a stable and an unstable manifold,* (Q.T.1), (Q.T.2) *and* (Q.T.3) *hold.*

Moreover, if (Q.T. 1, 2, 3) *hold, one can construct local coordinates $x_1,\ldots,x_n$ in a neighbourhood of a point of γ, depending smoothly on μ such that in these coordinates one has for μ near $\bar{\mu}$:*

$$X_\mu = \frac{\partial}{\partial x_m};$$

$$W^s_\mu = \{x_1 = \cdots = x_{m-s} = 0\};$$

$$W^u_\mu = \{x_{u+1} = \cdots = x_{m-1} = 0, x_1 = f(x_{m-s+1},\ldots,x_u) + (\mu - \bar{\mu})\};$$

where $s = \dim(W^s_\mu)$, $u = \dim(W^u_\mu)$, $s + u \geq m$, and f is a homogeneous non-degenerate quadratic function (if $u + s = m$, one should read $f = 0$).

Proof. Let $r \in \gamma$. Near r we can take μ dependent coordinates $x'_1,\ldots,x'_m$ such that $X_\mu = \partial/\partial x'_m$ for μ near $\bar{\mu}$, and such that $r = (0,\ldots,0)$. Since W^u_μ and W^s_μ are invariant manifolds, i.e., are made of integral curves of X_μ, it is enough to study $W^u_\mu \cap \{x'_m = 0\}$ and $W^s_\mu \cap \{x'_m = 0\}$. By taking this intersection with $\{x'_m = 0\}$, the situation is completely analogous to that of quasi-transverse intersections of invariant manifolds of diffeomorphisms; see [7]. That (Q.T. 1, 2, 3) hold generically follows easily by the perturbation lemma: these conditions, when interpreted for $W^u_\mu \cap \{x'_m = 0\}$ and $W^s_\mu \cap \{x'_m = 0\}$ are just generic conditions on the intersection of two μ-dependent manifolds.

For the construction of the coordinates, at least for $\mu = \bar{\mu}$, we could refer to [7]. Since the μ-dependent case is somewhat more complicated, we give the whole construction here.

First of all we may, and do, assume that the coordinates $x'_1,\ldots,x'_m$ are such that $W^s_\mu = \{x'_1 = \cdots = x'_{m-s} = 0\}$ for μ near $\bar{\mu}$. Next we choose a μ-dependent manifold $\tilde{W}^u_\mu$ such that $W^u_\mu \subset \tilde{W}^u_\mu$, $\dim(W^u_\mu) + 1 = \dim(\tilde{W}^u_\mu)$, and such that the expression for $\tilde{W}^u_\mu$, with respect to the $x'_1,\ldots,x'_m$ coordinates, is independent of μ. The existence of such $\tilde{W}^u_\mu$ follows from (Q.T.3), while from (Q.T. 1, 3) it follows that $\tilde{W}^u_\mu$ is transverse to W^s_μ. This means that there is a coordinate change $x''_1 = x''_1(x'_1,\ldots,x'_m),\ldots,x''_m = x''_m(x'_1,\ldots,x'_m)$ such that

$$X_\mu = \frac{\partial}{\partial x''_m};$$

$$W^s_\mu = \{x''_1 = \cdots = x''_{m-s} = 0\};$$

$$\tilde{W}^u_\mu = \{x''_{u+1} = \cdots = x''_{m-1} = 0\};$$

and such that $T_r(W_\mu^u)$ is spanned by $\left(\frac{\partial}{\partial x_2''},\ldots,\frac{\partial}{\partial x_u''},\frac{\partial}{\partial x_m''}\right)_r$. This means that

$$W_\mu^u = \left\{x_{u+1}'' = \cdots = x_{m-1}'' = 0, \mu = \bar{\mu} + f'(x_1'',\ldots,x_u'')\right\}$$

with $f'(0) = 0$, $\frac{\partial f'}{\partial x_1''}(0) \neq 0$, and $\frac{\partial f'}{\partial x_i''}(0) = 0$ *for* $i = 2,\ldots,u$.

Next we modify the x_1'' coordinate,

$$x_1'''(x_1'',\ldots,x_m'') = f'(x_1'',\ldots,x_u'') - f'(0,\ldots,0,x_{m-s+1}'',\ldots,x_u'');$$

to complete the coordinate transformation we write $x_i''' = x_i''$. So now we have

$$X_\mu = \frac{\partial}{\partial x_m'''};$$

$$W_\mu^s = \left\{x_1''' = \cdots = x_{m-s}''' = 0\right\};$$

$$W_\mu^u = \left\{x_1''' = \cdots = x_{m-1}''' = 0, \mu = \bar{\mu} + x_1''' + f''(x_{m-s+1}''',\ldots,x_u''')\right\};$$

where $f''(0) = 0$, $df''(0) = 0$, and because of (Q.T.2), $d^2f''(0)$ has maximal rank.

The final coordinates are obtained by applying the Morse Lemma to f'': this gives $x_{m-s+1},\ldots,x_u$ as a function of $x_{m-s+1}''',\ldots,x_u'''$ so that f'' changes to a quadratic form; for the other coordinates we just take $x_i = x_i'''$.

b. *Linearizability of singularities of vector fields.* We recall some facts about differentiable linearizations of vector fields at singularities. If X is a C^∞ vector field on M with a singularity in s, then the linear part of X in s, $L_s(X)$, is an endomorphism of $T_s(M)$. Recall that if X is a gradient, all the eigenvalues of $L_s(X)$ are real. We say that X has a hyperbolic singularity in s if none of the eigenvalues of $L_s(X)$ are purely imaginary. In this situation there is a result of Sternberg [18], [19] according to which there is for each positive integer k an open and dense subset $\mathcal{O}_k$ in the space of linear endomorphisms of $T_s(M)$ such that if $L_s(X)$ belongs to $\mathcal{O}_k$, there is a C^k-coordinate system near s, with respect to which X is linear. Here we want somewhat more: we would like the linearization to depend smoothly on μ, when X depends smoothly on μ. Also, in the case of a saddle-node, where we have no hyperbolicity and cannot apply Sternberg's result, we still want some partial linearization. For this we need a linearization theorem for partially hyperbolic singularities [20].

For positive integers k and l, with $l \leq m$, there is an open and dense subset $\mathcal{O}_{k,l}$ in the set of l-tuples of real numbers such that, if a C^∞ vector field X on M has a singularity in s such that $L_s(X)$ has $(m-l)$ purely imaginary (or zero) eigenvalues, while for the non-purely imaginary eigenvalues $\lambda_1,\ldots,\lambda_l$, $(\mathrm{Re}\,\lambda_1,\ldots,\mathrm{Re}\,\lambda_l) \in \mathcal{O}_{k,l}$, then there are C^k-coordinates $x_1,\ldots,x_m$ on a neighbourhood of s, with respect to which X has the form

$$X = \sum_{i,j=1}^{l} A_{ij}(x_{l+1},\ldots,x_m)x_i\frac{\partial}{\partial x_j} + \sum_{n=l+1}^{m} X_n(x_{l+1},\ldots,x_m)\frac{\partial}{\partial x_n}$$

with all eigenvalues of $\left(\dfrac{\partial X_i}{\partial x_j}(0)\right)_{i,\,j\geq l+1}$ purely imaginary.

This result provides linearizations of $\{X_\mu\}$ which depend smoothly on μ: if s_μ is a continuous family of hyperbolic singularities of X_μ then each (s_μ, μ) is a singularity of X, the vector field on $M \times \mathbf{R}$ defined by $X(x, \mu) = X_\mu(x)$, to which we may apply the above result. One has to observe that it follows from the proof in [11], that one may assume that μ is one of the (partially) linearizing C^k-coordinates. In this way we obtain

Corollary. *For each positive integer k there is an open and dense subset $\mathcal{O}_{k,m}$ in the set of m-tuples of real numbers such that if $X_{\bar\mu}$, belonging to a one-parameter family $\{X_\mu\}$ of vector fields on M, has a hyperbolic singularity in s such that the m-tuple of real parts of eigenvalues of $L_s(X_{\bar\mu})$ belongs to $\mathcal{O}_{k,m}$, then there is a C^k μ-dependent linearization of $\{X_\mu\}$ near $(s, \bar\mu)$.*

In case $X_{\bar\mu}$, belonging to $\{X_\mu\}$, has a non-hyperbolic singularity in s, we have the more general

Corollary. *For positive integers k and l, $l \leq m$, there is an open and dense subset $\mathcal{O}_{k,l}$ in the set of all l-tuples of real numbers such that if $X_{\bar\mu}$, belonging to a one-parameter family $\{X_\mu\}$ of vector fields on M, has a singularity in s such that exactly $(m-l)$ eigenvalues of $L_s(X_{\bar\mu})$ are purely imaginary (or zero) and such that the real parts of the other eigenvalues form an l-tuple in $\mathcal{O}_{k,l}$, then there are μ-dependent C^k-coordinates $x_1,\ldots,x_m$ in a neighbourhood of s in which X_μ, for μ near $\bar\mu$, has the form:*

$$X_\mu = \sum_{i,j=1}^{l} x_i \cdot A_{ij}(x_{l+1},\ldots,x_m,\mu)\frac{\partial}{\partial x_j} + \sum_{n=l+1}^{m} X_n(x_{l+1},\ldots,x_m,\mu)\frac{\partial}{\partial x_n}$$

with all eigenvalues of $\left(\dfrac{\partial X_i}{\partial x_j}(0,\bar\mu)\right)_{i,\,j\geq l+1}$ purely imaginary.

Since the set of bifurcation values of a generic one-parameter family $\{X_\mu\}$ of gradients is discrete and since the above sets $\mathcal{O}_{k,l}$ are open and dense, the following properties are generic for such one-parameter families.

(L.1) For each $\bar\mu$ such that $X_{\bar\mu}$ has a saddle-node, say in s, $L_s(X_{\bar\mu})$ has one eigenvalue zero while the other (real) eigenvalues form an $(m-1)$-tuple in $\mathcal{O}_{2,m-1}$.

(L.2) For each $\bar{\mu}$ such that $X_{\bar{\mu}}$ has an orbit of tangency, say of $W^u(s)$ and $W^s(\tilde{s})$, the eigenvalues of $L_s(X_{\bar{\mu}})$, respectively $L_{\tilde{s}}(X_{\bar{\mu}})$, form an m-tuple in $\mathcal{O}_{k,m}$; moreover, all the eigenvalues have multiplicity one. We want $X_{\bar{\mu}}$ to be uniquely linearizable (up to linear transformations) on $W^s(s)$ and $W^u(\tilde{s})$ and to be C^{10} linearizable in neighbourhoods of both s and $\tilde{s}$. For this, k is taken to be the smallest integer greater than or equal to the maximum of: (i) the quotient of the strongest and weakest contracting eigenvalue of $L_s(X_{\bar{\mu}})$; (ii) the quotient of the strongest and weakest expanding eigenvalue of $L_{\tilde{s}}(X_{\bar{\mu}})$; (iii) 10.

c. *Saddle-nodes.* We say that a smooth vector field X on M has a saddle-node in s if $L_s(X)$ has one eigenvalue zero and no other eigenvalues on the imaginary axes and if, moreover, X, restricted to a center manifold through s, has the form $X = ax^2\frac{\partial}{\partial x} + O(|x|^3)$ with $a \neq 0$; for the definition and existence of centre manifolds see [2], [6]. If $X_{\bar{\mu}}$, belonging to a one-parameter family $\{X_\mu\}$ of vector fields, has a saddle-node in s, we say that it unfolds generically if there is a μ-dependent centre manifold passing through s (for $\mu = \bar{\mu}$) so that X_μ, restricted to this centre manifold, has the form

$$X_\mu = \left(ax^2 + b(\mu - \bar{\mu})\right)\frac{\partial}{\partial x} + O\left(|x|^3 + |x\cdot(\mu - \bar{\mu})| + |(\mu - \bar{\mu})|^2\right)$$

with $a, b \neq 0$.

It is well-known that the following property is generic for general one-parameter families $\{X_\mu\}$ of vector fields; e.g., see [16]. The fact that it is generic for one-parameter families of gradients follows from [21].

(S.N.) For each $\bar{\mu} \in \mathbf{R}$ and $s \in M$ such that $X_{\bar{\mu}}$ has a non-hyperbolic singularity in s, it has a generically unfolding saddle-node in s.

We observe, in relation to the linearizations just discussed, that if X_μ for $\mu = 0$ has a generically unfolding saddle-node in s and if the real parts of the non-zero eigenvalues of $L_s(X)$ form an $(m-1)$-tuple in $\mathcal{O}_{2,m-1}$ then there are μ-dependent C^2 coordinates $x_1, \ldots, x_m$ near s in which X_μ has the form

$$X_\mu = \left(ax_1^2 + b\mu x_1 + O\left(|x_1|^3 + |x_1\mu| + |\mu|^2\right)\right)\frac{\partial}{\partial x_1} + \sum_{i,j=2}^{m} A_{ij}(x_1, \mu)x_i\frac{\partial}{\partial x_j}$$

with no eigenvalues of $A_{ij}(0, 0)$ on the imaginary axis.

d. *Invariant manifolds.* We recall that if s is a singularity of a vector field X on M, the stable set of s, $W^s(s)$, consists of those points x in M for which $\varphi_t(x) \to s$ for $t \to \infty$ where φ_t is the flow of X. If s is a hyperbolic singularity, $W^s(s)$ is an injectively immersed submanifold; in this case we speak of the stable manifold. If s is a saddle-node, $W^s(s)$ is an injectively immersed submanifold with boundary. This boundary is called the *strong stable manifold* of s and is denoted by $W^{ss}(s)$. With $t \to -\infty$, we get the corresponding definitions for unstable sets and manifolds.

Let X be a vector field on M which has an orbit γ of non-transverse intersection of a stable and an unstable manifold, say $W^u(s)$ and $W^s(\tilde{s})$. We assume that X satisfies the conditions imposed on $X_{\bar{\mu}}$ in (Q.T.1), (Q.T.2), and (L.2). In this situation we define some invariant submanifolds (see the picture in the introduction).

Inside the stable manifold of s, we define two smaller invariant submanifolds: $W(s)$ and $W^{ss}(s)$. For the definition we use the fact that there are C^k coordinates linearizing X near s. According to [3] these coordinates, restricted to the stable manifold, are essentially unique (i.e. modulo a linear transformation). In fact, for this we need the complicated dependence of k on $X_{\bar{\mu}}$ in condition (L.2). As a consequence of the conditions on the eigenvalues, there is a unique curve $W(s)$ in $W^s(s)$ such that

a) In the linearizing coordinates, $W(s)$ is linear;

b) $T_s(W(s))$ is the eigenspace of $L_s(X)$ corresponding to the negative eigenvalue with smallest norm.

Also there is a unique invariant manifold $W^{ss}(s) \subset W^s(s)$ such that:

a) In the linearizing coordinates, $W^{ss}(s)$ is linear;

b) $T_s(W^{ss}(s))$ is an invariant subspace for $L_s(X)$ such that

$$T_s(W^s(s)) = T_s(W(s)) \oplus T_s(W^{ss}(s)).$$

Next we observe that, since $W^u(s)$ and $W^s(\tilde{s})$ have an orbit of quasi-transverse intersection, $\dim(W^s(\tilde{s})) \geq \dim(W^{ss}(s)) + 1$. Now we define a submanifold $W^{ss}(\tilde{s}) \subset W^s(\tilde{s})$ by requiring:

a) In linearizing coordinates near $\tilde{s}$, $W^{ss}(\tilde{s})$ is linear;

b) $\dim(W^{ss}(\tilde{s})) = \dim(W^{ss}(s))$;

c) $T_s(W^{ss}(\tilde{s}))$ is an invariant subspace of $L_{\tilde{s}}(X)$ such that the eigenvalues of $L_{\tilde{s}}(X) | T_{\tilde{s}}(W^{ss}(\tilde{s}))$ are smaller, i.e. more negative, than any of the eigenvalues of the map $L_{\tilde{s}}(X)$ induces on $T_{\tilde{s}}(M)/T_{\tilde{s}}(W^{ss}(\tilde{s}))$.

Replacing X by $-X$, interchanging s and $\tilde{s}$ and also stable and unstable, we obtain the definitions of $W(\tilde{s}) \subset W^u(\tilde{s})$, $W^{uu}(\tilde{s}) \subset W^u(\tilde{s})$ and $W^{uu}(s) \subset W^u(s)$.

The generic nature of the following condition (M) results from the same methods as the transversality of stable and unstable manifolds in the Kupka-Smale theorem [5], [15].

(M) We say that $\{X_\mu\}$ satisfies condition (M) if it satisfies all the previously stated generic conditions, and if for each $\bar{\mu}$, such that $X_{\bar{\mu}}$ has a saddle-node, say in s, all stable and unstable manifolds are transverse to $W^{ss}(s)$ and $W^{uu}(s)$, and if for each $\bar{\mu}$ such that $X_{\bar{\mu}}$ has an orbit of tangency of $W^u(s)$ and $W^s(\tilde{s})$, all stable and unstable manifolds are transverse to $W(s), W(\tilde{s})$, $W^{ss}(s)$, $W^{uu}(s)$, $W^{ss}(\tilde{s})$ and $W^{uu}(\tilde{s})$ (with the obvious exceptions, e.g.: $W(s)$ is not transverse to $W^s(s)$).

e. *Orbits of tangency.* Let X be a vector field on M with an orbit γ of tangency of $W^u(s)$ and $W^s(\tilde{s})$ as before. We assume X satisfies all the generic conditions imposed on $X_{\bar{\mu}}$ if $X_{\bar{\mu}}$ has an orbit of tangency; see above. For almost any point $r \in W^u(s)$, the integral curve $\varphi_t(r)$, $t \to -\infty$ approaches s in such a way that $\overline{(\varphi_t(-\infty, 0]))}$ is a C^1-curve which is tangent in s to the eigenspace of $L_s(X)$ of the smallest positive eigenvalue. Hence for a "generic orbit of tangency" $\gamma \subset W^u(s) \cap W^s(\tilde{s})$, the tangent vector of $\bar{\gamma}$ in s (respectively $\tilde{s}$) is in the eigenspace of $L_s(X)$ ($L_{\tilde{s}}(X)$) of the positive (negative) eigenvalue with smallest norm.

Next we observe that for $r \in W^u(s)$, there is a unique linear subspace $\tilde{W}^{uu}(r) \subset T_r(M)$ such that $(d\varphi_t)\tilde{W}^{uu}(r)$ has a limit for $t \to -\infty$ and such that this limit is $T_s(W^{uu}(s))$. We also observe that for almost any linear subspace S of $T_r(M)$ of the same dimension as $T_s(W^{ss}(s))$, $(d\varphi_t)S$ converges for $t \to -\infty$ to $T_s(W^{ss}(s))$. Corresponding statements hold for points in $W^s(\tilde{s})$ and $t \to +\infty$; in particular, there are linear subspaces $\tilde{W}^{ss}(r')$ for $r' \in W^s(\tilde{s})$. As above, it is clear that for a generic orbit of tangency γ, and $r \in \gamma$, one has $d\varphi_t(\tilde{W}^{uu}(r)) \to T_{\tilde{s}}(W^{uu}(\tilde{s}))$ for $t \to +\infty$ and $d\varphi_t(\tilde{W}^{ss}(r)) \to T_s(W^{ss}(s))$ for $t \to -\infty$. So we have the following generic condition:

(T) We say that $\{X_\mu\}$ satisfies condition (T) if it satisfies all the previously stated generic conditions, and if for each $\bar{\mu}$, such that $X_{\bar{\mu}}$ has an orbit of tangency γ of $W^u(s)$ and $W^s(\tilde{s})$, the tangent vector of $\bar{\gamma}$ at s (resp. $\tilde{s}$) exists and is in the eigenspace of smallest expansion (resp. contraction); moreover, for $r \in \gamma$, there are linear subspaces $\tilde{W}^{uu}(r)$ and $\tilde{W}^{ss}(r)$ in $T_r(M)$ such that

$$d\varphi_t\big(\tilde{W}^{uu}(r)\big) \to \begin{matrix} T_s(W^{uu}(s)) \\ T_{\tilde{s}}(W^{uu}(\tilde{s})) \end{matrix} \quad \text{for} \quad \begin{matrix} t \to -\infty \\ t \to +\infty \end{matrix}$$

and

$$d\varphi_t\big(\tilde{W}^{ss}(r)\big) \to \begin{matrix} T_s(W^{ss}(s)) \\ T_{\tilde{s}}(W^{ss}(\tilde{s})) \end{matrix} \quad \text{for} \quad \begin{matrix} t \to -\infty \\ t \to +\infty; \end{matrix}$$

φ_t denotes the flow of $X_{\bar{\mu}}$.

c. *Conclusion.* We now summarize the results of this section.

Definition. $\mathfrak{X}_1^g(M)$ denotes the set of one-parameter families of gradient vector fields on M with the strong Whitney topology; $\mathcal{G}(M)$, or $\mathcal{G}$, $\subset \mathfrak{X}_1^g(M)$ denotes the set of those arcs which satisfy all the conditions summed up so far, i.e., (Q.T., 1–3), (L., 1, 2), (S.N.), (M.), and (T.) plus of course the conclusion of the theorem in Section 1.

So the main result of this section is

THEOREM. *$\mathcal{G}(M)$ is a residual subset of $\mathfrak{X}_1^g(M)$.*

In the next section we prove that even $\mathcal{G}$ is open (and dense).

3. *Openness of the generic conditions and formulation of the main result*

THEOREM. *$\mathcal{G}(M)$ is an open subset of $\mathfrak{X}_1^g(M)$, and for $\{X_\mu\} \in \mathcal{G}(M)$ the bifurcation values are isolated.*

Proof. It is clear from the theory of generic one-parameter families of functions, that as far as the saddle-nodes are concerned the bifurcation values are isolated and the generic one-parameter families are open.

As in the case of Morse-Smale dynamical systems, e.g. see [10] and [12], one can prove that the transversality conditions are open, at least away from an orbit of tangency.

Now we consider a one-parameter family of gradients $\{X_\mu\} \in \mathcal{G}(M)$, such that $X_{\bar{\mu}}$ has an orbit of tangency, say of $W^u(s)$ and $W^s(\tilde{s})$. Since s and $\tilde{s}$ are hyperbolic singularities of $X_{\bar{\mu}}$, there is a neighbourhood $\mathcal{U}$ of $X_{\bar{\mu}}$ in $\mathfrak{X}(M)$, the space of C^∞ vector fields on M, and two continuous maps from $\mathcal{U}$ to M, $\mathcal{U} \ni X$, $s_X, \tilde{s}_X \in M$ such that $s_{X_{\bar{\mu}}} = s$, $\tilde{s}_{X_{\bar{\mu}}} = \tilde{s}$ and such that s_X and $\tilde{s}_X$ are hyperbolic singularities of X. It is not hard to verify that if $\mathcal{U}$ is a sufficiently small neighbourhood of $X_{\bar{\mu}}$ and if $X \in \mathcal{U}$ has an orbit of tangency of $W^u(s_X)$ and $W^s(\tilde{s}_X)$ near γ, then at this orbit of tangency all but one of the generic conditions of the previous sections are satisfied (except (Q.T.3) which concerns the μ-dependence). Also if $\{X'_\mu\}$ is a one-parameter family of gradients near $\{X_\mu\}$, then for any μ' such that $X'_{\bar{\mu}'}$ has an orbit of tangency of $W^u(s_{X'_{\bar{\mu}}})$ and $W^s(\tilde{s}_{X'_{\bar{\mu}'}})$ then at this tangency also (Q.T.3) is satisfied.

So the theorem follows if we can prove the following.

PROPOSITION. *For $X_{\bar{\mu}}$ as above there is a neighbourhood $\mathcal{U}' \ni X_{\bar{\mu}}$ in $\mathfrak{X}(M)$ such that if $X \in \mathcal{U}'$ has an orbit of tangency of a stable and unstable manifold, this is a tangency of $W^u(s_X)$ and $W^s(\tilde{s}_X)$ near γ.*

Proof. Let $r \in \gamma$ and let V be an arbitrarily small neighbourhood of r. From the methods of proof of the openness of Morse-Smale systems [4], it follows that there is a neighbourhood $\mathcal{U}''$ of $X_{\bar{\mu}}$ so that any orbit of tangency of $X \in \mathcal{U}''$ passes through V. This means that for $\mathcal{U}''$ small enough, we can only have tangencies of $W^u(s')$ and $W^s(\tilde{s}')$ if $W^u(s') \cap W^s(s) \neq \varnothing$ and $W^s(\tilde{s}') \cap W^s(\tilde{s}) \neq \varnothing$ ($s' = s$ and/or $\tilde{s}' = \tilde{s}$ is not excluded). We have to show that for such a tangency $s = s'$ and $\tilde{s} = \tilde{s}'$. We only prove that $s \neq s'$ and $\tilde{s} = \tilde{s}'$ lead to a contradiction; the other cases can be handled in the same way.

Suppose s'_X, for $X \in \mathcal{U}''$, is a hyperbolic singularity of X, depending continuously on X and different from s_X, and suppose r_i is a point of nontransversal intersection of $W^u(s'_{X_i})$ and $W^s(\tilde{s}_{X_i})$ with $X_i \to X$ and $r_i \to r$. We want to deduce a contradiction.

In this case $T_{r_i}(W^s(\tilde{s}_{X_i})) \to T_r(W^s(\tilde{s}))$. Also, by taking a subsequence, we may assume that $T_{r_i}(W^u(s'_{X_i}))$ has a limit. Clearly $T_r(W^u(s)) \subset \lim T_{r_i}(W^u(s'_{X_i}))$. Since $W^u(s')$ is transverse to $W^{ss}(s)$, $(\lim T_{r_i}(W^u(s'_{X_i}))) + \tilde{W}^{ss}(r) = T_r(M)$; for $\tilde{W}^{ss}(r)$ see Section 2 d. Hence $(\lim T_{r_i}(W^u(s'_{X_i}))) + T_r(W^s(\tilde{s})) = T_r(M)$ so $T_{r_i}(W^u(s'_{X_i})) + T_{r_i}(W^s(\tilde{s}_{X_i})) = T_r(M)$ which contradicts the fact that r_i is a point of tangency of $W^u(s'_{X_i})$ and $W^s(\tilde{s}_{X_i})$.

This concludes the proof of the proposition and hence of the theorem.

In the following chapters we prove

THEOREM. *One-parameter families of gradients $\{X_\mu\}$ which belong to $\mathcal{G}(M)$ are stable.*

Since $\mathcal{G}$ is dense in $\mathfrak{X}_1^g(M)$, this implies the stability theorem. We want to indicate how the proof of stability can be reduced to a local case in the sense of the parameter μ.

First we observe that we may assume that the neighbourhood of $\{X_\mu\}$ in $\mathcal{G}(M)$ is such that for $\{X'_\mu\} \in \mathcal{U}$, also $\{(1-\nu)X_\mu + \nu X'_\mu\} \in \mathcal{U}$ for $\nu \in [0, 1]$. Next we observe that it is enough to show that for any $(\mu_0, \nu_0) \in \mathbf{R} \times [0, 1]$ there is a neighbourhood V such that any two one-parameter families $\{(1 - N_i(\mu))X_\mu + N_i(\mu)X'_\mu\}$, $i = 1, 2$, such that $0 \leq N_1 \leq N_2 \leq 1$ and such that $\overline{\{(\mu, \nu) | N_1(\mu) < \nu < N_2(\mu)\}} \subset V$ are equivalent by an equivalence η: $\mathbf{R} \to \mathbf{R}$, H_μ: $M \to M$ with $\eta(\mu) = \mu$ and $H_\mu = \text{id}$ whenever $N_1(\mu) = N_2(\mu)$.

Using the fact that the bifurcation values are isolated we see that it is enough to construct η and H in the case that $\{X_\mu\}$ and $\{X'_\mu\}$ are equal, except for μ in some bounded neighbourhood $U \subset \mathbf{R}$, provided we also take care that for $\mu \notin U$, $\eta(\mu) = \mu$ and $H_\mu = \text{id}$. Due to this localization and the Openness Theorem, we can restrict ourselves to the case where the above neighbourhood U contains only one bifurcation value.

II. The dynamics near an orbit of tangency

In this chapter we investigate the vector field near an orbit of tangency. We do this by constructing various invariant manifolds to reduce the question to low dimensions. The invariant manifolds are in general only C^1 and we need more differentiability; for this we introduce in Section 1 the smoothing structures. In Section 2 we investigate the 2-dimensional case, and prove that two one-parameter families of vector fields in dimension 2, both having a generic orbit of tangency, are equivalent near the closure of that orbit. In Section 3 we construct invariant manifolds. The most important is W^c which is normally hyperbolic. In Section 4 another invariant manifold $W_\mu^{cc} \subset W_\mu^c$ is constructed. W_μ^{cc} is 2-dimensional, and so we can apply the results of Section 2 and then we extend the equivalence (or conjugacy) to W_μ^c and, thus, to $W^c = \bigcup_\mu W_\mu^c$. An extension of the equivalence to a full neighbourhood is now a simple application of [13]. However, we have to be much more careful here since the equivalence must be further extended to all of M. We leave this point to the next chapter dealing with global problems.

1. *Smoothing structures*

Let M be an m-dimensional C^s-manifold, $s \leq \infty$; here M does not have to be compact. Let X be a C^k vector field on M, $k < s$, and $N \subset M$ an invariant submanifold of class C^l, $l \leq k$; i.e., for each $x \in N$, $X(x) \in T_x(N)$. We say that a C^h-structure $\mathcal{S}$ on N, $l < h \leq k+1$, is a *smoothing structure* if:

1) The C^l-structure of N as a submanifold of M is the same as the C^l-structure induced by $\mathcal{S}$;

2) $X|N$ is C^{h-1} with respect to $\mathcal{S}$.

In this case we say that $q \in N$ is *regular* if there is a neighbourhood U of q in N such that U is a C^h-submanifold and such that $\mathcal{S}|U$ coincides with the submanifold structure. These notions were first explicitly introduced in [13], as far as we know; they are especially useful when working with normally hyperbolic invariant manifolds: in general they are only C^1 (hence vector fields on them can only be C^0!) but they always admit a C^2 smoothing structure; see below and [13].

Basic example 1. We take $M = \mathbf{R}^m$, $X = \sum_{i=1}^m X_i(x_1,\dots,x_m)\dfrac{\partial}{\partial x_i}$ and assume N is an invariant manifold and has the form

$$N = \{(x_1,\dots,x_n, f_{n+1}(x_1,\dots,x_n),\dots,f_m(x_1,\dots,x_n))\};$$

the functions X_i are C^k and the functions f_j are C^l, $l \leq k$. In this case there is a smoothing C^{l+1}-structure $\mathscr{S}$ on N. Namely: let π: $\mathbf{R}^m \to \mathbf{R}^n$ be the projection $\pi(x_1,\ldots,x_m) = (x_1,\ldots,x_n)$; we say that a function F: $N \to \mathbf{R}$ is C^{l+1} with respect to $\mathscr{S}$ if and only if $F \circ (\pi|N)^{-1}$: $\mathbf{R}^n \to \mathbf{R}$ is C^{l+1}. It is clear that the C^l-structure induced by $\mathscr{S}$ on N equals the C^l-structure as a submanifold. Now we have to show that X is C^l with respect to $\mathscr{S}$. The functions $\{x_i \circ \pi\}_{i=1}^n$ form a coordinate system on N. With respect to these coordinates, $X|N$ has the form

$$\sum_{i=1}^{n} X_i(x_1,\ldots,x_n, f_{n+1}(x_1,\ldots,x_n),\ldots,f_m(x_1,\ldots,x_n))\frac{\partial}{\partial x_i}.$$

This is clearly C^l.

Note finally that each point of N, where N is locally C^{l+1}, is regular.

Example 2, [13]. Let M be any C^s manifold with a C^k vector field X, $k < s$, and an invariant C^l submanifold N, $1 \le l \le k$. Then there is a C^{l+1} smoothing structure $\mathscr{S}$ on N where $\mathscr{S}$ is obtained by a globalization of the previous construction: let $\nu(M)$ be a small neighbourhood of the zero-section of the normal bundle of N; $\nu(N)$ admits a C^l and hence a C^{l+1} structure such that the canonical projection π: $\nu(N) \to N$ is C^l. The C^{l+1}-structure on N, for which this projection is C^{l+1} (we assume to have fixed a C^{l+1}-structure as above on $\nu(N)$), is denoted by $\tilde{\mathscr{S}}$. Also there is a C^l embedding α: $\nu(N) \to M$ such that for each $x \in N$, the zero normal vector 0_x is mapped to x. Since $l \ge 1$, we can approximate α by a C^{l+1} embedding $\tilde{\alpha}$ such that

(i) $N \subset \tilde{\alpha}(\nu(N))$;

(ii) $\pi \circ (\tilde{\alpha}^{-1}|N)$ is a C^l diffeomorphism.

We can now, as in the previous example, define the smoothing structure $\mathscr{S}$ on N by requiring that $\pi \circ (\alpha^{-1}|N)$ be a C^{l+1}-diffeomorphism from $(N, \mathscr{S})$ to $(N, \tilde{\mathscr{S}})$.

Example 3. Let again $M = \mathbf{R}^m$ and let X be a linear vector field with no multiple eigenvalues. Let $N \subset M$ be an invariant C^1 submanifold containing 0. Then N, at least in a neighbourhood of 0 in N, admits a smoothing C^h structure $\mathscr{S}$, $2 \le h \le \infty$, such that every point $x \in N$, where N is locally C^h, is regular.

This is another generalization of the first example. Since N is invariant, $T_0(N)$ is also invariant. Since X is linear and has no multiple eigenvalues, there is a complementary invariant subspace, say V, so that $T_0(\mathbf{R}^m) = V \oplus T_0(N)$.

If π: $\mathbf{R}^m \to T_0(N)$ is the composition of the identification $\mathbf{R}^m \simeq T_0(\mathbf{R}^m)$ with the projection on the second factor of T, $T_0(\mathbf{R}^m) = V \oplus T_0(N)$, $\pi|N$: $N \to T_0(N)$ is locally a diffeomorphism. Note that $(\pi|N)_*(X|N)$ is linear and hence C^{h-1}. So, if we take $\mathscr{S}$ to be the C^h structure on N, which makes $\pi|N$ a C^h diffeomorphism, we have the required smoothing structure.

This result will be applied also in situations where X is non-linear but where it admits a C^k-linearization. In that case one obtains a smoothing C^k structure for which every point of N, which was already C^k, is regular.

2. *Saddle connections in dimension two*

In this section we consider a 1-parameter family $\{X_\mu\}$ of gradient vector fields on $\mathbf{R}^2$ having a saddle connection or an orbit of tangency for $\mu = \bar{\mu}$ as in Chapter I. If we use invariant manifolds, the general case leads to this 2-dimensional case also. Since invariant manifolds are in general not C^∞ there are a few differences between the generic assumptions in Chapter I (in particular in property (L.2)) and the assumptions in the present section: here we assume X_μ to be only C^l but we still assume that X_μ admits C^l linearizations in a neighbourhood of s_μ, $\tilde{s}_\mu$, $l \geq 7$ (s_μ and $\tilde{s}_\mu$ are saddle points of X_μ such that $W^u(s_{\bar{\mu}})$ and $W^s(\tilde{s}_{\bar{\mu}})$ have an orbit γ of non-transverse intersection); we also assume that the contracting eigenvalue at s_μ is -1 and the expanding eigenvalue of $\tilde{s}_\mu$ is $+1$. We choose fixed C^l linearizing coordinates z_1^s, z_2^s near s_μ and C^l linearizing coordinates $z_1^{\tilde{s}}$, $z_2^{\tilde{s}}$ near $\tilde{s}_\mu$ (these coordinates depend of course on μ but this μ-dependence is not expressed in the notation; in the same way we sometimes write s instead of s_μ). We assume that in these coordinates

$$\{z_1^s = 0\} = W^s(s), \{z_2^s = 0\} = W^u(s), \{z_1^{\tilde{s}} = 0\} = W^u(\tilde{s})$$

and

$$\{z_2^{\tilde{s}} = 0\} = W^s(\tilde{s});$$

also we assume that the domain of definition of z_1^s, z_2^s, respectively $z_1^{\tilde{s}}, z_2^{\tilde{s}}$, contains a "linearizing box" $D^s = \{|z_1^s|, |z_2^s| \leq 1\}$, respectively $D^{\tilde{s}} = \{|z_1^{\tilde{s}}|, |z_2^{\tilde{s}}| \leq 1\}$.

S is a one-dimensional transverse section, intersecting $W^u(s_\mu)$ in r_μ^s and $W^s(\tilde{s}_\mu)$ in $r_\mu^{\tilde{s}}$; $r_{\bar{\mu}}^s = r_{\bar{\mu}}^{\tilde{s}}$. $B_1, \ldots, B_4$ are sides of D^s, $D^{\tilde{s}}$ as indicated below

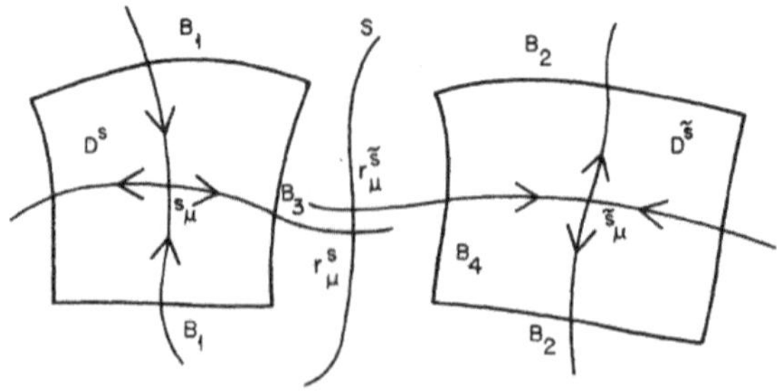

For $w \in S$, we define $\Lambda_s(w, \mu)$, for w near r_μ^s and μ near $\bar{\mu}$, as follows:

$$\Lambda_s(r_\mu^s, \mu) = 0$$

$$\Lambda_s(w, \mu) = e^{-T_s(w, \mu)} \quad \text{for } w \text{ above } r_\mu^s,$$

$$\Lambda_s(w, \mu) = -e^{-T_s(w, \mu)} \quad \text{for } w \text{ below } r_\mu^s;$$

$T_s(w, \mu)$ is the time needed by a $-X_\mu$ integral curve to go from w to B_1, so that $(X_\mu)_{-T_s(w,\mu)}(w) \in B_1$.

LEMMA 1. *As defined above, Λ_s is C^l and $\frac{\partial}{\partial w}(\Lambda_s(w, \mu)) \neq 0$ for w near r_μ^s.*

Proof. Let $t_\mu^s(w)$ be the time needed for an $-X_\mu$ integral curve to go from w to B_3 and let $\alpha_\mu^s(w)$ be the z_2^s coordinate of $(X_\mu)_{-t_\mu^s(w)}(w)$, $w \in S$. Both $t_\mu^s(w)$ and $\alpha_\mu^s(w)$ are clearly C^l in (μ, w); α_μ^s is even a diffeomorphism. Because the contracting eigenvalue at s is -1, $T_s(w, \mu) = t_\mu^s(w) - \ln|\alpha_\mu^s(w)|$. So, assuming that z_2^s, and hence α_μ^s, increases with height, we have

$$\Lambda_s(w, \mu) = \alpha_\mu^s(w) \cdot e^{-t_\mu^s(w)}.$$

From this the lemma follows.

We define $\Lambda_{\tilde{s}}(w, \mu)$ analogously: $\Lambda_{\tilde{s}}(r_\mu^{\tilde{s}}, \mu) = 0$ and if it takes an X_μ integral curve a time $T_{\tilde{s}}(w, \mu)$ to go from $w \in S$ to B_2, then $\Lambda_{\tilde{s}}(w, \mu) = \pm e^{-T_{\tilde{s}}(s,\mu)}$, $\Lambda_{\tilde{s}}(w, \mu)$ is positive for w above $r_\mu^{\tilde{s}}$. Finally, $\Lambda(w, \mu) = \Lambda_s(w, \mu) \cdot \Lambda_{\tilde{s}}(w, \mu)$. Note that $-\ln|\Lambda(w, \mu)|$ is the time it takes an X_μ integral curve to go from B_1 to B_2 through $w \in S$.

LEMMA 2. *$\Sigma = \{(w, \mu) | (\partial\Lambda/\partial w)(w, \mu) = 0\}$ is a C^{l-1} curve; i.e., $\Sigma = \{w = \sigma(\mu)\}$ for some C^{l-1} function σ, and $(d/d\mu)(\Lambda(\sigma(\mu), \mu))_{\mu=\bar{\mu}} = 0$, $(d^2/d\mu^2)(\Lambda(\sigma(\mu), \mu))|_{\mu=\bar{\mu}} < 0$.*

Proof. Λ is C^l by construction. Since $(\partial^2\Lambda/\partial w^2)(r_{\bar{\mu}}^s, \bar{\mu}) > 0$, Σ is, by the Implicit Function Theorem, C^{l-1}. Since by assumption our saddle connection unfolds generically, the distance between r_μ^s and $r_\mu^{\tilde{s}}$ is of the order of $|\mu - \bar{\mu}|$. From this it easily follows that $\Lambda(\sigma(\mu), \mu) \sim -(\mu - \bar{\mu})^2$.

We define $M(\mu) = \Lambda(\sigma(\mu), \mu)$; M is C^{l-1}.

In the rest of this section we are concerned with the problem of finding a conjugacy between two arcs $\{X_\mu\}, \{X'_\mu\}$ on $\mathbf{R}^2$ both satisfying all the conditions imposed so far on $\{X_\mu\}$. We assume that the bifurcation of $\{X'_\mu\}$ occurs for

$\mu = \bar{\mu}'$ and that γ' is the orbit of non-transversal intersection of $W^u(s_{\bar{\mu}'})$ and $W^s(\tilde{s}_{\bar{\mu}'})$. Λ' and M' are defined analogous to Λ and M.

LEMMA 3. *There is a local C^{l-3} diffeomorphism η of* **R** *with $\eta(\bar{\mu}) = \bar{\mu}'$ such that $M(\mu) = M'(\eta(\mu))$ and there is a local C^{l-5} diffeomorphism G from $S \times \mathbf{R}$ to $S' \times \mathbf{R}$ of the form $G(w, \mu) = (G_\mu(w), \eta(\mu))$ such that $\Lambda(w, \mu) = \Lambda'(G(w, \mu))$.*

Proof. By the previous lemma, the existence of η follows from the Morse Lemma. After this, and by Lemma 1, the existence of G follows by applying the "Morse Lemma with parameter"; see [9] or the appendix of [4]. The loss in differentiability follows from the fact that, by application of the Morse Lemma, the conjugating diffeomorphism has two degrees of differentiability less than the original function.

Now we come to the construction of a conjugacy H_μ between X_μ and $X'_{\eta(\mu)}$, near γ and for μ near $\bar{\mu}$, depending continuously on μ. We require H_μ to map B_1 to B'_1 (at least near $W^s(s)$, respectively $W^s(s')$) and the same for B_2 and B'_2.

H_μ has to preserve the parametrization (or integration time) along orbits. The time it takes the X_μ-orbit through $w \in S$ to go from B_1 to B_2 equals $-\ln|\Lambda(w, \mu)|$; in the same way Λ' is related to the integration time of X' orbits from B'_1 to B'_2. From this we see that H_μ has to map the X_μ-orbit through $w \in S$ to the $X'_{\eta(\mu)}$-orbit through $G_\mu(w) \in S'$.

Note that the conditions we imposed on H_μ determine H_μ in most points. In fact only $H_{\bar{\mu}}|\gamma$ is undetermined. The fact that H_μ can be extended and is smooth near γ for μ near $\bar{\mu}$ follows from the next lemma.

LEMMA 4. *Let $T_s(w, \mu)$ be the integration time needed by a $-X_\mu$ integral curve to go from $w \in S$ to B_1, and let $T'_{s'}(G_\mu(w), \eta(\mu))$ be the integration time needed by a $-X'_{\eta(\mu)}$ integral curve to go from $G_\mu(w) \in S'$ to B'_1; these functions are only defined for $w \notin S \cap W^u(s_\mu)$. Then $\Gamma(w, \mu) = T'_{s'}(G_\mu(w), \eta(\mu)) - T_s(w, \mu)$ can be extended to a C^{l-6} function of (w, μ).*

Proof. In the notation of the proof of Lemma 1, we have

$$T_s(w, \mu) = t^s_\mu(w) - \ln|\alpha^s_\mu(w)|$$

and

$$T'_{s'}(G_\mu(w), \eta(\mu)) = t'_{\eta(\mu)}(G_\mu(w)) - \ln\left|\left(\alpha'_{\eta(\mu)}(G_\mu(w))\right)\right|.$$

From the previous constructions we know that $(\alpha^s_\mu(w), \mu) \mapsto (\alpha'^{s'}_{\eta(\mu)}(G_\mu(w)), \eta(\mu))$ is a C^{l-5} diffeomorphism sending $(0, \mu)$ to $(0, \eta(\mu))$. Hence there is locally a positive C^{l-6} function F of (w, μ) such that $|\alpha'^{s'}_{\eta(\mu)}(G_\mu(w))| = F(w, \mu) \cdot |\alpha^s_\mu(w)|$.

From this we see that we can extend Γ as

$$t'^{s'}_{\eta(\mu)}(G_\mu(w)) - t^s_\mu(w) - F(w, \mu)$$

which is clearly C^{l-6}.

From the above lemma we conclude that for $w \in S$, $H_\mu(w) = (X'_{\eta(\mu)})_{\Gamma(w,\mu)} G_\mu(w)$. Hence $H_\mu | S$ is C^{l-6} and, because H_μ is a conjugacy between X_μ and $X'_{\eta(\mu)}$, H_μ is C^{l-6} in a neighbourhood of S. The extension of H_μ to a continuous conjugacy in a full neighbourhood of $\bar{\gamma}$ is now straightforward. In general H_μ will not be differentiable at $W^s(s_\mu)$ and $W^u(s_\mu)$. So we have:

THEOREM. *There is in the above situation a unique continuous conjugacy H between $\{X_\mu\}$ and $\{X'_\mu\}$ of the form*

$$H(x_1, x_2, \mu) = (G_\mu(x_1, x_2), \eta(\mu))$$

for μ near $\bar{\mu}$ and defined in a neighbourhood of $D^s \cup (\gamma \times \{\bar{\mu}\}) \cup D^{\tilde{s}}$, such that H maps B_1 to B'_1 and B_2 to B'_2.

This conjugacy H is C^{l-6} in a neighbourhood of $\gamma \times \{\bar{\mu}\}$ as a function of (x_1, x_2, μ).

3. *Invariant manifolds*

Let $\{X_\mu\}$ be a smooth one-parameter family of gradients on M belonging to $\mathcal{G}(M)$ (see Chapter I). Let γ be an orbit of tangency of stable and unstable manifolds of $X_{\bar{\mu}}$, say of $W^u(s)$ and $W^s(\tilde{s})$. We again refer the reader to the basic figure in the introduction.

a. *Construction of the centre unstable manifold.* This will be an invariant manifold W^{cu}, containing $\bar{\gamma}$ and such that in s, resp. $\tilde{s}, W^{ss}(s), W^{ss}(\tilde{s})$ are complementary to $T_s(W^{cu})$, resp. $T_{\tilde{s}}(W^{cu})$. The construction is as follows. We first choose fixed C^k-linearizations of $X_{\bar{\mu}}$ near s and near $\tilde{s}$. Near s we just take W^{cu} to be linear in the linearizing coordinates; thus W^{cu} is here C^k. Next we add to this part of W^{cu} the positive integral curves of its points (as long as they do not leave some fixed neighbourhood of $\bar{\gamma}$). By Chapter I, Section 2e, $\tilde{W}^{ss}(r)$ is complementary to $T_r(W^{cu})$ for $r \in \gamma$; the extended W^{cu} does not accumulate on $W^{ss}(\tilde{s})$. This means that $W^u(\tilde{s})$ is contained in the closure of the extended W^{cu} and that the tangent space of W^{cu} in points approaching $\tilde{s}$, have as limit the invariant subspace of $T_{\tilde{s}}(M)$, complementary to $W^{ss}(\tilde{s})$. This means that W^{cu} can be extended as a C^1 invariant manifold around $\tilde{s}$, using the linearization. It is clear that W^{cu} obtained this way has the required properties. Also it is C^k in points of γ and hence admits a smoothing structure as in Example 3 of Section 1.

In fact we want more: we want an invariant manifold W^{cu}_μ of X_μ, μ near $\bar{\mu}$, depending differentiably on μ (and of course such that $W^{cu}_{\bar{\mu}} = W^{cu}$) and such that W^{cu} contains the hyperbolic singularities of X_μ near s and near $\tilde{s}$. The construction of this μ dependent invariant manifold goes in the same way. Near s we use the μ dependent C^k-linearization and choose again W_μ^{cu} linear in these coordinates. Since W^{cu} intersects $W^s(\tilde{s})$ transversally, W_μ^{cu} will still intersect the stable manifold of X_μ near $W^s(\tilde{s})$. This means also that the rest of the construction can be carried out. So we have:

THEOREM. *In the above situation there is a C^1 submanifold $W^{cu} \subset M \times \mathbf{R}$, $W^{cu} = \bigcup_\mu W_\mu^{cu} \times \{\mu\}$ such that $W_{\bar{\mu}}^{cu}$ contains γ, s and $\tilde{s}$ and has a tangent space complementary to $W^{ss}(s)$ and $W^{ss}(\tilde{s})$. Now W^{cu} is C^k in a neighbourhood of every point of $\gamma \times \{\bar{\mu}\}$ and it is an invariant submanifold containing the singularities of X_μ near s and $\tilde{s}$.*

b. *Corollaries*. 1. W^{cu} and W_μ^{cu} admit smoothing C^k structures which can be made unique by fixing the C^k-linearizing coordinates at $\tilde{s}$.

2. Interchanging stable, unstable and s, $\tilde{s}$, we get the analogous construction of W^{cs} and W_μ^{cs} with smoothing structure.

3. We define W^c as $W^{cu} \cap W^{cs}$, and $W_\mu^c = W_\mu^{cu} \cap W_\mu^{cs}$; they are C^1, because W^{cu} and W^{cs} are transverse. W^c and, hence, W_μ^c, admits a smoothing C^k structure because W^c is smooth in a neighbourhood of every point of $\gamma \times \{\mu\}$.

Remark. Restricting to W_μ^c we have a one-parameter family of vector fields $\{X_\mu | W_\mu^c\}$ which have an orbit of tangency for $\mu = \bar{\mu}$. The situation is as in Chapter I, Section 2, except $\{X_\mu | W_\mu^c\}$ is only C^{k-1}, and has fixed C^k linearizations at s and $\tilde{s}$. Also this restricted vector field has the property that the stable manifold of s and the unstable manifold of $\tilde{s}$ both have dimension one.

4. *Construction of the conjugacy in the centre manifold*

We consider in this section the restriction of $\{X_\mu\}$ to a centre manifold W^c (depending on μ) as constructed in the previous section. Since we assumed to have C^k linearizations at s and $\tilde{s}$, we may assume $X_\mu | W_\mu^c$ to be C^{k-1} (using a smoothing structure) and to have μ dependent C^k-linearizations at s_μ and $\tilde{s}_\mu$ for μ near $\bar{\mu}$.

From now on we write X_μ instead of $X_\mu | W_\mu^c$; also we assume fixed linearizing coordinates $\tilde{z}_1^s, \ldots, \tilde{z}_n^s$ and $\tilde{z}_1^{\tilde{s}}, \ldots, \tilde{z}_n^{\tilde{s}}$ at s and $\tilde{s}$ are given (here we do not express the μ-dependence in the notation). From the fact that X_μ is obtained by restricting to a centre manifold, it follows that $W^s(s)$ and $W^u(\tilde{s})$ are one-dimensional. We assume that $W^u(s) = \{\tilde{z}_n^s = 0\}$ and that $W^s(s)$ is the $\tilde{z}_n^s$ axis; for $W^s(\tilde{s})$, $W^u(\tilde{s})$ and $\tilde{z}_n^{\tilde{s}}$ we make the corresponding assumptions. We also assume that the limit of the tangent vector of γ, when approaching s, resp. $\tilde{s}$, is in the direction of the $\tilde{z}_1^s$-axis, resp. $\tilde{z}_1^{\tilde{s}}$-axis. We define the invariant foliations $\mathscr{F}^u$ and $\mathscr{F}^s$ by requiring that locally near s, resp. $\tilde{s}$, the leaves are of the form $\{\tilde{z}_n^s = \text{constant}\}$, resp. $\{\tilde{z}_n^{\tilde{s}} = \text{constant}\}$, and by the requirement that they are invariant; i.e. if φ_t is the flow of X, φ_t maps leaves to leaves. Also, of course, $\mathscr{F}^u$ and $\mathscr{F}^s$ depend on μ; they are of class C^k.

$W^u(s)$ and $W^s(\tilde{s})$ are leaves of $\mathscr{F}^u$, $\mathscr{F}^s$, which have along γ a quadratic tangency. Hence there is a C^{k-1} surface of tangencies of $\mathscr{F}^u$ and $\mathscr{F}^s$, which is transverse to both $W^u(s)$ and $W^s(\tilde{s})$, contains γ and is invariant, in the sense that in each of its points, X is in its tangent space. As in Section 1 of this chapter, we can extend this surface to a C^1 invariant 2-dimensional manifold containing $\bar{\gamma}$ admitting a smoothing C^{k-1} structure. This invariant manifold we denote by W^{cc} or W_μ^{cc} if we want to stress the μ-dependence.

Finally we choose a co-dimension-one section S intersecting γ transversally. In S, the foliations $\mathscr{F}^u \cap S$, $\mathscr{F}^s \cap S$ and $W^{cc} \cap S$ are situated as follows:

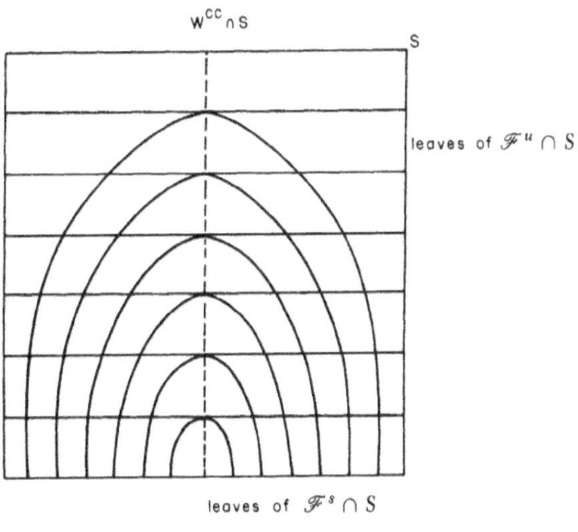

The positions of $W^u(s_\mu) \cap S$ and $W^s(s_\mu) \cap S$ are

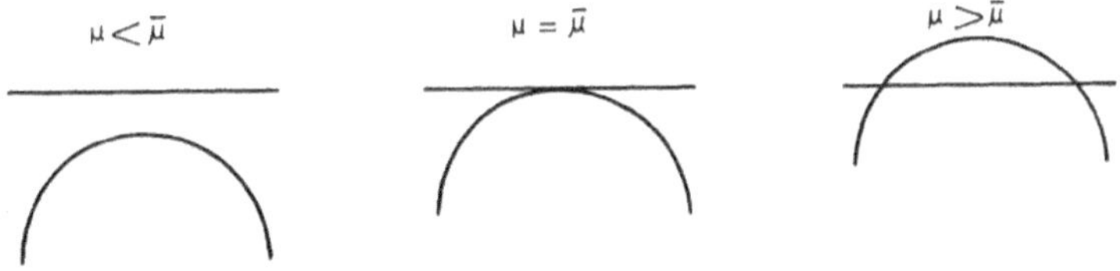

Let now $\{X'_\mu\}$ be another such one-parameter family of vector fields, with an orbit γ' of tangency between stable and unstable manifolds of s' and $\tilde{s}'$ of $X'_{\bar{\mu}'}$. After multiplying $\{X_\mu\}$ and $\{X'_\mu\}$ by smooth positive functions we may assume that the contracting eigenvalues of $dX_\mu(s)$ and $dX'_\mu(s')$ are -1 and the expanding eigenvalues of $dX_\mu(\tilde{s})$ and $dX'_\mu(\tilde{s}')$ are $+1$ for μ near $\bar{\mu}$, resp. $\bar{\mu}'$. The C^k linearizations at s', $\tilde{s}'$ and hence the foliations $\mathscr{F}'^s$, $\mathscr{F}'^u$, the manifold W'^{cc} and the section S' are defined for $\{X'_\mu\}$ in the same way as the corresponding objects for $\{X_\mu\}$.

We notice that we can apply the results of Section 2 of this chapter to $X|W^{cc}$ and $X'|W'^{cc}$. This yields a homeomorphism H_μ: $W^{cc}_\mu \to W'^{cc}_{\eta(\mu)}$ conjugating $X_\mu|W^{cc}$ with $X'_{\eta(\mu)}|W'^{cc}_{\eta(\mu)}$, where η is a one-dimensional homeomorphism with $\eta(\bar{\mu}) = \bar{\mu}'$. We now want to extend H_μ to a "neighbourhood" of W^{cc}, as a conjugacy between X_μ and $X'_{\eta(\mu)}$. We want to do this in such a way that H_μ maps $\mathscr{F}^u$ to $\mathscr{F}'^u$ and $\mathscr{F}^s$ to $\mathscr{F}'^s$, and hence W^{cc} to W'^{cc}.

Since H_μ is already fixed on W^{cc}_μ, $S \cap W^{cc}_\mu$ will not be mapped to $S' \cap W^{cc}_{\eta(\mu)}$. However, H_μ, when defined, induces a map H^S_μ from S to S' so that $H_\mu(x)$ and $H^S_\mu(x)$, for $x \in S$, are on the same orbit of $X'_{\eta(\mu)}$. $H^S_\mu|S \cap W^{cc}$ is already fixed and is C^{k-6} on $S \times \mathbf{R}$: namely we saw that $\{X_\mu|W^{cc}_\mu\}$ is C^{k-1} (using a smoothing structure which is regular along γ) so we can apply Section 2, Lemma 3 with $l = k - 1$.

Next we extend H^S_μ to all of S. Now H^S_μ has to respect the foliations, obtained by intersecting the stable and unstable foliations with S and S'. We choose μ-dependent coordinate functions κ_μ, κ'_μ on $S \cap W^{cc}_\mu$, respectively $S' \cap W'^{cc}_\mu$, so that $\kappa_\mu(x) = \kappa'_{\eta(\mu)}(H^S_\mu(x))$. Now H^S_μ has to map $F^{u,S}_\mu(x)$, the intersection of the $\mathscr{F}^u_\mu$-leaf through x with S, to $F^{u,S'}_{\eta(\mu)}(H_\mu(x))$. Since H^S_μ also has to preserve the stable foliation, $H^S_\mu|F^{u,S}_\mu(x)$ has to conjugate $g_{\mu,x}$ with $g'_{\eta(\mu), H^S_\mu(x)}$; $g_{\mu,x}$ is the function of $F^{u,S}_\mu(x)$ obtained by composing κ_μ with projection along leaves of $\mathscr{F}^s_\mu|S$ (see picture). The function $g'_{u,x}$ is defined analogously.

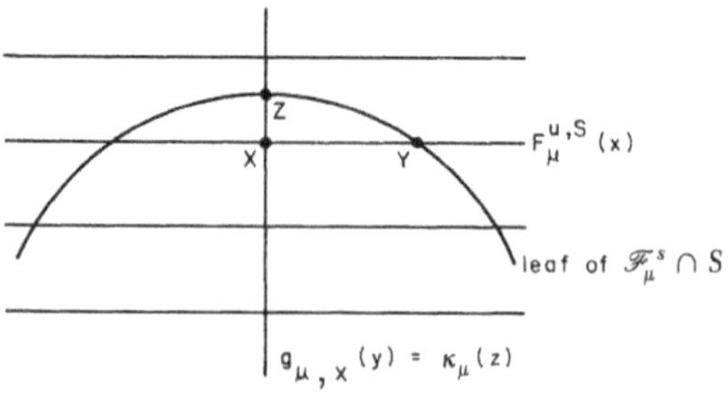

Now $g_{\mu,x}$ is a C^{k-6} two-parameter family of Morse functions; hence there is a C^{k-8} conjugacy H^S_μ of these two families; see [9] or the appendix of [4]. Since H^S_μ

respects the unstable foliation, and conjugates $\{g_{\mu,x}\}$ with $\{g_{\eta(\mu),H_\mu(x)}\}$, H_μ^S also respects the stable foliation.

Finally, $H_\mu | S \cap W^{cc}$ is C^{k-7}: see Section 2 and note that $l = k - 1$. We extend $H_\mu(S \cap W_\mu^{cc})$ to a C^{k-7} manifold S'_μ of co-dimension one transverse to W'^{cc}. We require now that $H_\mu(S) = S'_\mu$. This, together with the above H_μ^S, fixes H_μ in a neighbourhood of γ. Notice that H_μ is *not* defined on a neighbourhood of $\bar{\gamma}$. Near s and $\tilde{s}$ the domain of definition of H_μ is as follows:

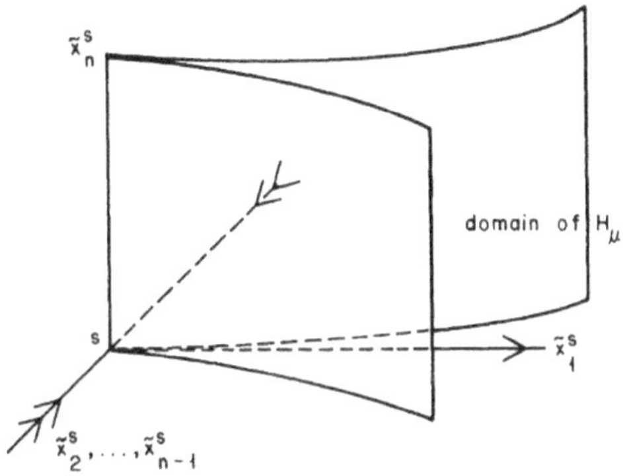

We call this a wedge shaped neighbourhood of γ. So we have

THEOREM. *For $\{X_\mu\}$ and $\{X'_\mu\}$ as above, there is a local conjugacy H_μ, defined in a wedge-shaped neighbourhood of γ, for μ near $\bar{\mu}$; $H_\mu(x)$ is C^{k-8} in (μ, x) and respects the stable and unstable foliations defined by the linearizations.*

Since we assumed $k \geq 10$ (Chapter I, *Section* 2 b, *property* (L.2)), $H_\mu(x)$ *is C^2.*

III. The main result: Global stability

We now prove that the families $\{X_\mu\} \in \mathcal{G}$, considered in the previous chapters, are globally stable. Thus, we will conclude our main result:

THE STABILITY THEOREM. *The (globally) stable one-parameter families of gradient vector fields are open and dense.*

Let us consider a family $\{X_\mu\} \in \mathcal{G}$ with a unique bifurcation value $\bar{\mu}$. Let $\{X'_\mu\}$ be a nearby family such that $X'_\mu = X_\mu$ for μ outside some neighbourhood of $\bar{\mu}$. We will construct a topological equivalence between $\{X_\mu\}$ and $\{X'_\mu\}$ that yields the identity map on $M \times \{\mu\}$ for all μ outside some neighbourhood of $\bar{\mu}$. Therefore, in view of the observation at the end of Section I.3, to prove our result we can and will restrict ourselves to such families $\{X_\mu\} \in \mathcal{G}$ with a unique bifurcation value.

So, let $\{X_\mu\} \in \mathcal{G}$ with a unique bifurcation value $\bar{\mu}$. We have that $X_{\bar{\mu}}$ exhibits either a saddle-node singularity or an orbit of tangency between some stable and some unstable manifolds of a pair of hyperbolic singularities. The same will happen for a nearby family $\{X'_\mu\}$, which will have a unique bifurcation value $\bar{\mu}'$. In either of the two cases above, we have to construct a local equivalence (conjugacy) between the families near the bifurcating orbit (the saddle-node or the orbit of tangency), in a way that it can be extended to all of M. To achieve such a goal, we make use of stable and unstable foliations (tubular families) associated to the singularities. Notice that we have already dealt with a particular kind of such foliations in the previous chapter, in order to define a local conjugacy in a wedge-shaped neighbourhood of an orbit of tangency in the centre manifold.

Let us first recall the notions of unstable and stable foliations for hyperbolic singularities. Let s be a hyperbolic singularity for $X_{\bar{\mu}}$. If U and $[\mu_1, \mu_2]$ are small neighbourhoods of s in M and of $\bar{\mu}$ in $\mathbf{R}$, respectively, then for each $\mu \in [\mu_1, \mu_2]$ there exists a unique singularity $s_\mu \in U$ for X_μ, $s = s_{\bar{\mu}}$. We indicate by $X_{\mu, t}$, $t \in \mathbf{R}$, the flow generated by X_μ.

Definition. An unstable foliation for $\{X_\mu\}$ at $(s, \bar{\mu})$ is a continuous foliation $\mathcal{F}^u$ of $U \times [\mu_1, \mu_2]$ such that

a) The leaves are C^k discs, $k \geq 2$, varying continuously in the C^k topology and $\mathcal{F}^u(s_\mu, \mu) = W^u(s_\mu) \cap (U \times \{\mu\})$;

b) Each leaf $\mathcal{F}^u(x, \mu)$, containing (x, μ), is contained in $U \times \{\mu\}$,

c) $\mathcal{F}^u$ is X_μ-invariant; that is,

$$X_{\mu, t}(\mathcal{F}^u(x, \mu)) \supset \mathcal{F}^u\big(X_{\mu, t}(x), \mu\big)$$

for $t \geq 0$, x and $X_{\mu, t}(x)$ in U.
Similarly, one defines a stable foliation for $\{X_\mu\}$ at $(s, \bar{\mu})$. They can be easily globalized through saturation by the flows $X_{\mu, t}$.

Let us now suppose that $X_{\bar{\mu}}$ has a saddle-node singularity $\bar{s}$. It follows from Chapter I, Section 2.c, that there are two possibilities for its unfolding. Either X_μ has two singularities s_1, s_2 near $\bar{s}$ for $\mu < \bar{\mu}$ or vice versa. Throughout we will assume that the first possibility occurs. Furthermore, in a small neighbourhood $U \times [\mu_1, \mu_2]$ of $(\bar{s}, \bar{\mu})$, we have an X_μ-invariant foliation $\mathcal{F}^{su}$ of codimension two, each leaf being transversal to the centre manifold $W^c = \bigcup_\alpha W^c_\mu$. The centre manifold is obtained as the intersection of the centre-stable manifold $W^{cs} = \bigcup_\mu W^{cs}_\mu$ and the centre-unstable manifold $W^{cu} = \bigcup_\mu W^{cu}_\mu$. The intersections of $\mathcal{F}^{su}$ with W^{cs} and W^{cu} yield strong stable and strong unstable foliations $\mathcal{F}^{ss}$ and

$\mathscr{F}^{uu}$, respectively. Special leaves are the strong stable and strong unstable manifolds of $(\bar{s}, \bar{\mu})$, $W^{ss}(\bar{s})$ and $W^{uu}(\bar{s})$. Thus, for each $\mu \in [\mu_1, \mu_2]$, W_μ^{cs} is a union of leaves of $\mathscr{F}^{ss}$ and W_μ^{cu} is a union of leaves of $\mathscr{F}^{uu}$. All these foliations are X_μ-invariant. The following definition corresponds to an extension of $\mathscr{F}^{uu}$ to a full neighbourhood $U \times [\mu_1, \mu_2]$ of $(\bar{s}, \bar{\mu})$.

Definition. A strong unstable foliation $\mathscr{F}^{uu}$ for $\{X_\mu\}$ at $(\bar{s}, \bar{\mu})$ is a continuous foliation of $U \times [\mu_1, \mu_2]$ such that

a) The leaves are C^1 discs varying continuously in the C^1 topology and $\mathscr{F}^{uu}(\bar{s}, \bar{\mu}) = W^{uu}(\bar{s})$,

b) Each leaf of $\mathscr{F}^{su}$ is foliated by $\mathscr{F}^{uu}$ and, for each μ, $U \times \{\mu\}$ is a union of leaves transverse to W_μ^{cs},

c) $\mathscr{F}^{uu}$ is X_μ-invariant; that is, if $\mathscr{F}^{uu}(x, \mu)$ is the leaf through (x, μ), then $\mathscr{F}^{uu}(X_{\mu,t}(x), \mu)$ is contained in $X_{\mu,t}(\mathscr{F}^{uu}(x, \mu))$ for $t \geq 0$, x and $X_{\mu,t}(x)$ in U. Similarly, one defines a strong stable foliation $\mathscr{F}^{ss}$ for $\{X_\mu\}$ at $(\bar{s}, \bar{\mu})$.

Let us now relate the global stable and unstable foliations associated to hyperbolic singularities with the local foliations at the saddle-node. Let $\{X_\mu\}$ belong to $\mathscr{G}$ with a unique bifurcation value $\bar{\mu}$ corresponding to a saddle-node $\bar{s}$. As mentioned before, X_μ has two singularities for $\mu < \bar{\mu}$. In what follows, we often indicate the singularities $s_{i,\bar{\mu}}$ of $X_{\bar{\mu}}$ simply by s_i and the saddle-node $\bar{s}$ by s. We can partially order the singularities of $X_{\bar{\mu}}$ by the relation: $s_n \leq s_m$ if $W^u(s_n) \cap W^s(s_m) \neq \varnothing$. Since $\bar{\mu}$ is the only bifurcation value for $\{X_\mu\}$, we can take a total ordering of the singularities $\{s_{i,\mu}\}$, $1 \leq i \leq l$, of X_μ for $\mu < \bar{\mu}$, compatible with the above partial ordering: if $i \neq j$ then $i < j$ whenever $s_{i,\mu} \leq s_{j,\mu}$. Let $s_{k,\mu}$ and $s_{k+1,\mu}$ be the two collapsing singularities; that is, $s_{k,\bar{\mu}} = s_{k+1,\bar{\mu}} = s$. Notice that the same ordering holds for $\mu > \bar{\mu}$ if we omit the indices k and $k+1$.

Definition. Let $\{X_\mu\}$ be as above. We say that the system of unstable and strong unstable foliations $\mathscr{F}^u(s_1), \ldots, \mathscr{F}^u(s_{k-1})$, $\mathscr{F}^{uu}(s)$ is compatible when:

a) If a leaf F of $\mathscr{F}^u(s_n)$ intersects a leaf $\tilde{F}$ of $\mathscr{F}^u(s_i)$, $n < i < k$, then $F \supset \tilde{F}$. Moreover, the restriction of $\mathscr{F}^u(s_i)$ to a leaf of $\mathscr{F}^u(s_n)$ is a foliation of class C^2;

b) If a leaf F of $\mathscr{F}^u(s_i)$, $i < k$, intersects a leaf $\tilde{F}$ of $\mathscr{F}^{uu}(s)$, then $F \supset \tilde{F}$. Moreover, the restriction of $\mathscr{F}^{uu}(s)$ to a leaf of $\mathscr{F}^u(s_i)$ is a foliation of class C^1. A similar definition holds for the foliations $\mathscr{F}^{ss}(s)$, $\mathscr{F}^s(s_{k+2}), \ldots, \mathscr{F}^s(s_l)$.

We now construct such a compatible system of foliations. We remark that the C^2 differentiability condition in (a) above will only be used in the construction of $\mathscr{F}^{uu}(s)$. In the other cases, a C^1 condition is sufficient.

PROPOSITION 3.1. *There exists a compatible system of foliations* $\mathscr{F}^u(s_1),\ldots,\mathscr{F}^u(s_{k-1}), \mathscr{F}^{uu}(s)$.

Proof. By induction, let us assume we have constructed $\mathscr{F}^u(s_1),\ldots,\mathscr{F}^u(s_i)$ as required, for some $i < k-1$. To construct $\mathscr{F}^u(s_{i+1})$, we first consider the union $A = \bigcup_\mu A_\mu$ of small codimension-one fences A_μ transverse to $W^s(s_{i+1,\mu})$ and to all $W^u(s_{n,\mu})$ for $n \leq i$ and μ in some small interval $[\mu_1, \mu_2]$ containing $\bar{\mu}$. We have that A is diffeomorphic to $S \times D \times [\mu_1, \mu_2]$. Here, S is a codimension-one small sphere in $W^s(s_{i+1})$ transverse to $X_{\bar{\mu}}$ and D^u is a small disc neighbourhood of s_{i+1} in $W^u(s_{i+1})$. Suppose that $W^u(s_i)$ intersects $W^s(s_{i+1})$ and, consequently, $A_{\bar{\mu}}$. Since this intersection is transversal, we can take a C^2 fibration of $W^u(s_i) \cap A_{\bar{\mu}}$ over $W^s(s_{i+1}) \cap A_{\bar{\mu}}$, the fibers being diffeomorphic to D^u. On the other hand, the leaves of $\mathscr{F}^u(s_i)$ are C^2 continuously parametrized by $D_{s_i} \times [\mu_1, \mu_2]$, where D_{s_i} is a small disc neighbourhood of s_i in $W^s(s_i)$. Thus, by shrinking D_{s_i} and $[\mu_1, \mu_2]$ if necessary, we can take a similar fibration along the intersection of each leaf $\mathscr{F}^u(s_i)$ with A. The fibers are diffeomorphic to D^u and, as imbeddings in A, they vary continuously in the C^2 topology. We take the fibers as leaves of $\mathscr{F}^u(s_{i+1})$. Let us now consider $\mathscr{F}^u(s_{i-1})$ and suppose that $W^u(s_{i-1})$ intersects $W^s(s_{i+1})$. There are two possibilities to consider: either $W^u(s_{i-1}) \cap W^s(s_i)$ is empty or not. In the first case, we just repeat the above construction. In the second case, $W^u(s_{i-1}) \cap A_{\bar{\mu}}$ is not compact since it must accumulate on $W^u(s_i) \cap A_{\bar{\mu}}$. But, in some neighbourhood of $W^u(s_i) \cap A_{\bar{\mu}}$, $W^u(s_{i-1}) \cap A_{\bar{\mu}}$ is already C^2 fibered because $\mathscr{F}^u(s_i)$ is compatible with $\mathscr{F}^u(s_{i-1})$.

Thus, we are again reduced to the first case and we can extend the fibration to all of $W^u(s_{i-1}) \cap A_{\bar{\mu}}$. Again, the leaves of $\mathscr{F}^u(s_{i-1})$ are C^2 continuously parametrized by $D_{s_{i-1}} \times [\mu_1, \mu_2]$, where $D_{s_{i-1}}$ is a disc neighbourhood of s_{i-1} in $W^s(s_{i-1})$. Therefore, by shrinking $D_{s_{i-1}}$ and $[\mu_1, \mu_2]$ if necessary, we can as before extend the fibration to all leaves of $\mathscr{F}^u(s_{i-1})$. By induction we construct a fibration of $A = \bigcup_\mu A_\mu$ over $\bigcup_\mu A_\mu \cap W^s(s_{i+1,\mu})$, the fibers being leaves of $\mathscr{F}^u(s_{i+1})$. To get $\mathscr{F}^u(s_{i+1})$, we positively saturate these leaves by the $X_{\mu,t}$, and add $W^u(s_{i+1,\mu})$ as leaves. By the λ-Lemma [10], these two sets of leaves fit together, forming $\mathscr{F}^u(s_{i+1})$. Clearly, $\mathscr{F}^u(s_{i+1})$ is compatible with $\mathscr{F}^u(s_1),\ldots,\mathscr{F}^u(s_i)$. It remains to construct $\mathscr{F}^{uu}(s)$.

First of all, in $W^{cs}(s)$ we take a cylinder C transverse to the strong stable foliation $\mathscr{F}^{ss}(s)$ and disjoint from all the connected components of $W^u(s_i)$ that do not intersect $W^{ss}(s)$ for $i \leq k-1$. Equivalently, C is taken so that $W^u(s_i) \cap C$ is transverse to $\mathscr{F}^{ss}(s)$ in C. Together with a disc D in a leaf of $\mathscr{F}^{ss}(s)$, C forms a fundamental domain of $W^s(s)$ slightly extended over $W^{ss}(s)$. Notice that D belongs to a leaf $\tilde{F}$ of $\mathscr{F}^{su}(s)$. For $\mu \in [\mu_1, \mu_2]$, we choose a continuous family of

C^2 cylinders C_μ and a continuous family of C^1 discs D_μ in $\tilde{F}$ so that $\cup_\mu(C_\mu \cup F_\mu)$ is diffeomorphic to $(C \cup D) \times [\mu_1, \mu_2]$. Once more, we take the interval $[\mu_1, \mu_2]$ sufficiently small. Before raising the foliation $\mathscr{F}^{uu}$ through points of $C_\mu \cup D_\mu$, we want to construct yet another foliation on the cylinders C_μ. Such a foliation, which we denote by $\mathscr{F}^c$, will play an important role in proving the stability of $\{X_\mu\}$ near the saddle-node. The leaves of $\mathscr{F}^c$ have dimension one and are taken along the leaves of $\mathscr{F}^u(s_i)$, $1 \leq i \leq k-1$. That is, the leaves of $\mathscr{F}^c$ subfoliate the leaves of $\mathscr{F}^u(s_i)$. Again, we proceed by induction on the indices i. We first construct a C^1 vector field Z along $W^u(s_{k-1}) \cap C$ transverse to $\mathscr{F}^{ss}(s)$ or, equivalently, to $\mathscr{F}^{su}(s)$. This vector field is then projected into the leaves of $\mathscr{F}^u(s_{k-1})$ intersected with the cylinders C_μ, for each $\mu \in [\mu_1, \mu_2]$. The integral curves of Z are leaves of $\mathscr{F}^c$. Now suppose that $W^u(s_{k-2})$ intersects C. Either $W^u(s_{k-2})$ intersects $W^s(s_{k-1})$ or not. If not we proceed as before to construct Z along the leaves of $\mathscr{F}^u(s_{k-2})$. In the other case, by the compatibility condition, some leaves of $\mathscr{F}^u(s_{k-2})$ intersect, and thus contain, leaves of $\mathscr{F}^u(s_{k-1})$. Moreover, $\mathscr{F}^u(s_{k-1})$ restricted to each of those leaves of $\mathscr{F}^u(s_{k-2})$ is a C^2 foliation. Thus, restricted to each leaf of $\mathscr{F}^u(s_{k-2})$ near $W^u(s_{k-1})$, Z is already defined and it is of class C^1. In particular, its integral curves are well defined. We can then extend Z to the leaves of $\mathscr{F}^u(s_{k-2})$ by observing that, outside a neighbourhood of $W^u(s_{k-1})$, we are reduced to the previous case. In this way, we construct the vector field Z whose integral curves yield the foliation $\mathscr{F}^c$.

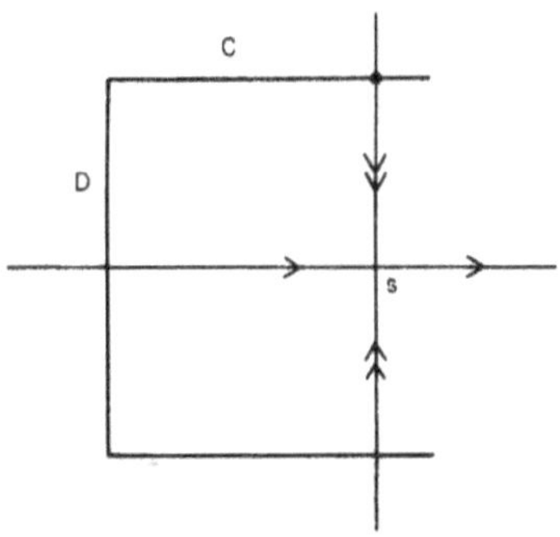

We now raise the foliation $\mathscr{F}^{uu}(s)$ over each $C_\mu \cup D_\mu$. Let $u = \dim W^{uu}(s)$. Using $\mathscr{F}^c$, we raise over the cylinders C_μ a $u+1$-dimensional foliation compatible with $\mathscr{F}^u(s_i)$, $i < k$. This can be done as before, proceeding by induction on the indices i. The intersections of these leaves with those of $\mathscr{F}^{su}(s)$ are taken as leaves of $\mathscr{F}^{uu}(s)$. In particular, the leaves of $\mathscr{F}^{uu}(s)$ are defined on the boundary of each D_μ. We then extend $\mathscr{F}^{uu}$ over all of D_μ with the desired properties. To have $\mathscr{F}^{uu}(s)$ defined in a full neighbourhood of $(s, \bar{\mu})$, we saturate its leaves through points of $C_\mu \cup D_\mu$ by the flows $X_{\mu,t}$, $t \geq 0$. The foliation we obtain converges to the uniquely defined strong unstable foliation in $W^u(s_{k,\mu})$ for $\mu \leq \bar{\mu}$. One can check directly this simple but relevant fact or see [2]. This completes the construction of $\mathscr{F}^{uu}(s)$ and, thus, of a compatible system of foliations $\mathscr{F}^u(s_1),\ldots,\mathscr{F}^u(s_{k-1})$, $\mathscr{F}^{uu}(s)$.

Dually, we can construct a compatible system of foliations $\mathscr{F}^{ss}(s)$, $\mathscr{F}^s(s_{k+2}),\ldots,\mathscr{F}^s(s_l)$.

Notice that $\mathscr{F}^{uu}(s)$ and $\mathscr{F}^{ss}(s)$ are integrable foliations in the following sense: there exists a foliation, namely $\mathscr{F}^{su}(s)$, such that for each point ω, $T_\omega F^{su}(\omega) = T_\omega F^{uu}(\omega) + T_\omega F^{ss}(\omega)$.

Before proving the stability of a family $\{X_\mu\}$ as above, we recall the well known Isotopy Extension Theorem [8].

Let N be a C^r compact manifold, $r \geq 1$, and A an open subset of $\mathbf{R}^s$. Let M be a C^∞ manifold with $\dim M > \dim N$. We indicate by $C^k_A(N \times A, M \times A)$ the set of C^k mappings f: $N \times A \to M \times A$ such that $\pi = \pi' f$, endowed with the C^k topology, $1 \leq k \leq r$. Here, π and π' denote the natural projections π: $N \times A \to A$, π': $M \times A \to A$. Let $\mathrm{Diff}^k_A(M \times A)$ be the set of C^k diffeomorphisms φ of $M \times A$ such that $\pi' = \pi'\varphi$, again with the C^k topology.

Isotopy Extension Theorem. *Let $i \in C^k_A(N \times A, M \times A)$ be an embedding and A' a compact subset of A. Given neighbourhoods U of $i(N \times A)$ in $M \times A$ and V of the identity in $\mathrm{Diff}^k_A(M \times A)$, there exists a neighbourhood W of i in $C^k_A(N \times A, M \times A)$ such that for each $j \in W$ there exists $\varphi \in V$ satisfying $\varphi i = j$ restricted to $N \times A'$ and $\varphi(x) = x$ for all $x \notin U$.*

We can now prove the part of our main result corresponding to a saddle-node bifurcating orbit. Let $\{X_\mu\} \in \mathscr{G}$ be as above, with a unique bifurcation value $\bar{\mu}$ associated to a saddle-node s for $X_{\bar{\mu}}$.

Theorem A. *The family $\{X_\mu\}$ is stable. Moreover, if $\{X'_\mu\}$ is close to $\{X_\mu\}$ and $X'_\mu = X_\mu$ except for μ near $\bar{\mu}$, then the conjugacy $H = \{H_\mu\}$ which we construct between $\{X_\mu\}$ and $\{X'_\mu\}$ is such that H_μ is the identity map on $M \times \{\mu\}$ for μ outside a small neighbourhood of $\bar{\mu}$.*

Proof. By the Openness Theorem, if $\{X'_\mu\}$ is close to $\{X_\mu\}$ then $\{X'_\mu\}$ also has a unique bifurcation value $\bar{\mu}'$ near $\bar{\mu}$ corresponding to a saddle-node s'. The singularities $s'_{1,\mu},\ldots,s'_{k,\mu}, s'_{k+1,\mu},\ldots,s'_{l,\mu}$ of $\{X'_\mu\}$ are close to the corresponding singularities of $\{X_\mu\}$ and $s'_{k,\bar{\mu}'} = s'_{k+1,\bar{\mu}'} = s'$. In view of the observation at the end of Section I.3, we may assume that $X'_\mu = X_\mu$ for μ outside a small interval (μ_1, μ_2) containing $\bar{\mu}$. The equivalence (in this case, a conjugacy) $H = \{H_\mu\}$ will be constructed on $M \times (\mu_3, \mu_4)$, so that H_μ is the identity on $M \times \{\mu\}$ for $\mu \notin (\mu_1, \mu_2)$, and (μ_3, μ_4) is a small interval containing (μ_1, μ_2). We first set a

C^0 reparametrization $\eta\colon (\mu_3, \mu_4) \to (\mu_3, \mu_4)$, close to the identity and equal to it outside (μ_1, μ_2), such that $\eta(\bar{\mu}) = \bar{\mu}'$. Later on, when constructing the conjugacy near the saddle-node, we will see that we need to impose a further condition on η. Next, we construct compatible systems of foliations $\mathscr{F}^u(s_1), \ldots, \mathscr{F}^u(s_{k-1})$, $\mathscr{F}^{uu}(s)$ and $\mathscr{F}^{ss}(s)$, $\mathscr{F}^s(s_{k+2}), \ldots, \mathscr{F}^s(s_l)$. Similarly for $\{X'_\mu\}$, we construct compatible systems $\mathscr{F}^u(s'_1), \ldots, \mathscr{F}^u(s'_{k-1})$, $\mathscr{F}^{uu}(s')$ and $\mathscr{F}^{ss}(s')$, $\mathscr{F}^s(s'_{k+2}), \ldots, \mathscr{F}^s(s'_l)$. Notice that these foliations can be taken to be the same for $\{X_\mu\}$ and $\{X'_\mu\}$ whenever $\mu \notin (\mu_1, \mu_2)$. This fact follows immediately from the construction of the foliations performed in Proposition 3.1, since $X_\mu = X'_\mu$ for $\mu \notin (\mu_1, \mu_2)$. The conjugacy will preserve the above foliations, so that it will be constructed by induction in their spaces of leaves. As we saw in Proposition 3.1, the space of leaves of $\mathscr{F}^u(s_i)$, $1 \le i \le k-1$, is a union of discs in $W^s(s_{i,\mu})$ and the space of leaves of $\mathscr{F}^s(s_j)$, $k+2 \le j \le l$, is a union of discs in $W^u(s_{j,\mu})$, $\mu_3 < \mu < \mu_4$. In the arguments we present in the sequel, these discs can be shrunk, if necessary, so that the leaves of $\mathscr{F}^u(s_i)$ are sufficiently close in the C^2 topology to the main leaf $W^u(s_{i,\bar{\mu}})$. This is similar for $\mathscr{F}^s(s_j)$. For each $i < k$, let A_i, A'_i be the union of small fences $A_{i,\mu}$, $A'_{i,\eta(\mu)}$ transverse to $W^s(s_{i,\mu})$, $W^s(s'_{i,\eta(\mu)})$ and S_i, S'_i be the union of the corresponding fundamental domains $S_{i,\mu}$, $S'_{i,\eta(\mu)}$ for these stable manifolds. We take $A_{i,\mu} = A'_{i,\mu}$ and $S_{i,\mu} = S'_{i,\mu}$ whenever $\mu \notin (\mu_1, \mu_2)$. The conjugacy will satisfy the following conditions:

a) It sends leaves of $\mathscr{F}^u(s_i)$, $\mathscr{F}^s(s_j)$ into leaves of $\mathscr{F}^u(s'_i)$, $\mathscr{F}^s(s'_j)$, respectively, for $i < k$ and $j > k+1$. It sends leaves of $\mathscr{F}^{su}(s)$, $\mathscr{F}^{uu}(s)$, $\mathscr{F}^{ss}(s)$ into those of $\mathscr{F}^{su}(s')$, $\mathscr{F}^{uu}(s')$, $\mathscr{F}^{ss}(s')$, respectively.

b) If $W^u(s_n) \cap W^s(s_i) \neq \varnothing$ for $n < i < k$, then for each leaf F of $\mathscr{F}^u(s_n)$, H restricted to $F \cap S_i$ is C^1 near the inclusion map, and similarly for the leaves of $\mathscr{F}^s(s_j)$, $j > k+1$.

c) For each leaf F of $\mathscr{F}^u(s_i)$ and each leaf E of $\mathscr{F}^s(s_j)$, $i < k$ and $j > k+1$, if $E \cap F \neq \varnothing$ then H restricted to $E \cap F$ is differentiable.

We begin to define H by sending the sources of $\{X_\mu\}$ into the corresponding ones for $\{X'_{\eta(\mu)}\}$. Let us suppose that H is defined on the union of the stable manifolds of $s_{n,\mu}$ for all $n < i < k$. That is, H is defined on the spaces of leaves of $\mathscr{F}^u(s_1), \ldots, \mathscr{F}^u(s_{i-1})$. Let us construct H on the stable manifolds $W^s(s_{i,\mu})$. Assume that $W^u(s_{i-1,\bar{\mu}})$ intersects $W^s(s_{i,\bar{\mu}})$. Let $C_{i-1,\bar{\mu}}$ be the intersection of $W^u(s_{i-1,\bar{\mu}})$ with the fundamental domain $S_{i,\bar{\mu}}$. Since $C_{i-1,\bar{\mu}}$ is compact, we can define a diffeomorphism $H\colon C_{i-1,\bar{\mu}} \to C'_{i-1,\bar{\mu}'}$, C^1 close to the inclusion map. This can be done because $S_{i,\bar{\mu}}$ and $S'_{i,\bar{\mu}}$ are C^1 close as well as $W^u(s_{i-1,\bar{\mu}})$ and $W^u(s'_{i-1,\bar{\mu}'})$ on compact parts. In a continuous way, the same can be done with the other leaves. Indeed, the leaves of $\mathscr{F}^u(s_{i-1})$ are continuously parameterized by their intersections with $W^s(s_{i-1,\mu})$, $\mu \in (\mu_3, \mu_4)$, where H is already defined. We send the intersections of the leaves with $S_{i,\mu}$ to the intersections of the corresponding leaves with $S'_{i,\eta(\mu)}$. This is achieved through a continuous family of diffeomorphisms, which are uniformly C^1 close to the inclusion maps.

Next, suppose that $C_{i-2,\bar{\mu}} = W^u(s_{i-2,\bar{\mu}}) \cap S_{i,\bar{\mu}}$ is nonempty. If $C_{i-2,\bar{\mu}}$ is compact, we repeat the argument. If not, then necessarily $W^u(s_{i-2,\bar{\mu}})$ intersects $W^s(s_{i-1,\bar{\mu}})$. Since $\mathscr{F}^u(s_{i-1})$ is compatible with $\mathscr{F}^u(s_{i-2})$, H is already defined on the leaves of $\mathscr{F}^u(s_{i-2})$ restricted to a neighbourhood $V \times (\mu_3, \mu_4)$ of $C_{i-1,\bar{\mu}}$ in $M \times (\mu_3, \mu_4)$. By the Isotopy Extension Theorem, we can extend H along the leaves of $\mathscr{F}^u(s_{i-2})$ satisfying all the desired properties. By induction, we have H defined on the union of $S_{i,\mu}$. Using the conjugacy equations $H_\mu X_\mu = X'_{\eta(\mu)} H_\mu$, we define H on the union of $W^s(s_{i,\mu})$, which is the space of leaves of $\mathscr{F}^u(s_i)$. Thus, we can construct H on the space of leaves of $\mathscr{F}^u(s_1), \ldots, \mathscr{F}^u(s_{k-1})$. Similarly, we can construct H on the spaces of leaves of $\mathscr{F}^s(s_{k+2}), \ldots, \mathscr{F}^s(s_l)$.

Notice that for $\mu \notin (\mu_1, \mu_2)$, H is the identity map whenever it is defined. We now extend the conjugacy to $U \times (\mu_3, \mu_4)$, where U is a neighbourhood of the saddle-node s in M. This is done in three steps. In the first step, we define H on the centre manifold W^c independently of what we have done so far. The last two steps consist in defining H on the centre-stable and centre-unstable manifolds W^{cs} and W^{cu}. This is done taking into account the already given definition of H on the spaces of leaves of $\mathscr{F}^u(s_1), \ldots, \mathscr{F}^u(s_{k-1})$ and $\mathscr{F}^s(s_{k+2}), \ldots, \mathscr{F}^s(s_l)$, respectively. The extension of H to a neighbourhood $U \times (\mu_3, \mu_4)$ is then immediate using the foliations $\mathscr{F}^{uu}(s)$ and $\mathscr{F}^{ss}(s)$. These foliations are integrable, since they are complementary subfoliations of $\mathscr{F}^{su}(s)$.

Let us proceed with the three steps above. The construction of H on W^c is done in [7], where it is proved that there is a unique model, up to conjugacy, for the unfolding of one-dimensional saddle-nodes. Such a construction requires a new reparametrization $\eta^*: \mathbf{R} \to \mathbf{R}$, $\eta^*(\bar{\mu}) = \bar{\mu}'$, which is again a homeomorphism C^0 close to the identity. As before, we can take $\eta^*(\mu) = \mu$ for $\mu \notin (\mu_1, \mu_2)$. Thus, we may assume that we have dealt with the reparametrization η^* since the beginning. To simplify the notation, we still indicate the reparametrization by η. Let us construct H on W^{cs}. It is enough to do so on the subsets $C_\mu \cup D_\mu$ of W^{cs}_μ defined in Proposition 3.1. Recall that on C_μ we constructed a centre foliation $\mathscr{F}^c$ transverse to $\mathscr{F}^{ss}$ and along the leaves of $\mathscr{F}^u(s_1), \ldots, \mathscr{F}^u(s_{k-1})$. We construct similar objects for $\{X'_\mu\}$ near s', observing that they can be taken to be the same for $\{X_\mu\}$ and $\{X'_\mu\}$ for $\mu \notin (\mu_1, \mu_2)$. Since H is known on W^c, we also demand that $H_\mu(D_\mu \cap W^c_\mu) = D'_{\eta(\mu)} \cap W^c_{\eta(\mu)}$ for all μ. We want H to send leaves of $\mathscr{F}^{ss}(s)$ and $\mathscr{F}^c(s)$ to leaves of $\mathscr{F}^{ss}(s')$ and $\mathscr{F}^c(s')$. Thus, H is completely defined on C_μ, once it is defined on ∂D_μ, and we only have to construct H on $\bigcup_\mu D_\mu$ and, in particular, on $\bigcup_\mu \partial D_\mu$.

As mentioned before, we want H to be compatible with its previous definition on the spaces of leaves of $\mathscr{F}^u(s_i)$, $i < k$. The procedure to do so is entirely similar to the one used to construct H on the stable manifolds $W^s(s_{i,\mu})$.

We first construct H on $\bigcup_\mu \partial D_\mu$ using induction on the indices i, $i < k$, and then extend it to all of $\bigcup_\mu D_\mu$. In this manner, we have H defined on the union of $C_\mu \cup D_\mu$, $\mu \in (\mu_3, \mu_4)$. Using the conjugacy equations $H_\mu X_\mu = X'_{\eta(\mu)} H_\mu$, we get H defined on all of W^{cs}. Similarly, it can be defined on W^{cu}. Since we want the conjugacy to send leaves of $\mathcal{F}^{uu}(s)$ and $\mathcal{F}^{ss}(s)$ into leaves of $\mathcal{F}^{uu}(s')$ and $\mathcal{F}^{ss}(s')$, respectively, H becomes well-defined in a neighbourhood $U \times (\mu_3, \mu_4)$ of $(s, \bar{\mu})$. Again using the conjugacy equations, we can define H on $V \times (\mu_3, \mu_4)$, where $V = \bigcup_\mu V_\mu$ and $V_\mu = \bigcup_{t \geq 0} X_{\mu, t}(U)$. So far, we have defined H on the stable manifolds of $s_{i,\mu}$ for $i < k + 2$, on the unstable manifolds of $s_{j,\mu}$ for $j > k - 1$ and on $V \times (\mu_3, \mu_4)$.

We now extend H to a neighbourhood of $(s_{k+2}, \bar{\mu})$ and, similarly, to neighbourhoods of $(s_j, \bar{\mu})$ for $j \geq k + 3$. These neighbourhoods are always of the form $U_j \times (\mu_3, \mu_4)$, where U_j is a neighbourhood of s_j in M. First suppose that $W^u(s)$ does not intersect $W^s(s_{k+2})$. We construct, as in Proposition 3.1, foliations $\mathcal{F}^u(s_{k+2})$ and $\mathcal{F}^u(s'_{k+2})$ compatible with $\mathcal{F}^u(s_i)$ and $\mathcal{F}^u(s'_i)$ for $i \leq k - 1$. We then construct H on $\bigcup_\mu W^s(s_{k+2,\mu})$ as in the first part of the proof. Using the complementary foliations $\mathcal{F}^u(s_{k+2})$, $\mathcal{F}^s(s_{k+2})$ for $\{X_\mu\}$ and $\mathcal{F}^u(s'_{k+2})$, $\mathcal{F}^s(s'_{k+2})$ for $\{X'_\mu\}$, we have H defined in a neighbourhood of $(s_{k+2}, \bar{\mu})$. So, let us now suppose that $W^u(s)$ intersects $W^s(s_{k+2})$. We have already constructed H on V and, thus, on $V_1 = W^u(s) \cap W^s(s_{k+2}) \cap V$. Inside a neighbourhood of $(s_{k+2}, \bar{\mu})$ and outside a small neighbourhood $N \times (\mu_3, \mu_4)$ of V_1, we perform the same constructions as above of $\mathcal{F}^u(s_{k+2})$, $\mathcal{F}^u(s'_{k+2})$ and of the conjugacy H. The crucial observation now is that, both in $V \times (\mu_3, \mu_4)$ and outside $N \times (\mu_3, \mu_4)$, H restricted to $F \cap E$ is C^1 close to the inclusion map for each leaf F of $\mathcal{F}^u(s_i)$, $i \leq k - 1$, and each leaf E of $\mathcal{F}^s(s_{k+2})$. Thus, proceeding by induction on the indices $i = k - 1, \ldots, 1$, we can extend H to a full neighbourhood of $(s_{k+2}, \bar{\mu})$ using the Isotopy Extension Theorem.

In a similar way, we construct H on full neighbourhoods of $(s_j, \bar{\mu})$ for $j \geq k + 2$ and, using the conjugacy equations as before, on all of the corresponding stable manifolds. Therefore, we have H defined on the stable manifolds of all singularities and thus on all of $M \times (\mu_3, \mu_4)$. By construction, H is continuous at the stable manifolds of $s_{j,\mu}$ for $j \geq k$. That H is continuous on the stable manifolds of $s_{i,\mu}$ for $i \leq k - 1$, follows from the fact that H sends leaves of $\mathcal{F}^u(s_i)$ to leaves of $\mathcal{F}^u(s'_i)$. Clearly, we also have that H is the identity map on $M \times (\mu_3 - \mu_1, \mu_4 - \mu_2)$. The proof of Theorem A is complete.

We now come to the main part of our result: the stability of a generic family of gradients going through an orbit of tangency.

Let $\{X_\mu\} \in \mathcal{G}$ with a unique bifurcation value $\bar{\mu}$ corresponding to an orbit of tangency γ between $W^u(s)$ and $W^s(\tilde{s})$, for a pair of singularities s and $\tilde{s}$ of $X_{\bar{\mu}}$. We take an ordering for the singularities $s_1, \ldots, s_k, s_{k+1}, \ldots, s_l$ of $X_{\bar{\mu}}$ so that if

$W^u(s_i) \cap W^s(s_j) \neq \varnothing$ then $i < j$. It follows from the Openness Theorem of Chapter I, Section 3 that the same ordering is prevalent for the singularities $\{s_{i,\mu}\}$ of X_μ for all $\mu \in R$, the correspondences $\mu \to s_{i,\mu}$ being continuous and $s_i = s_{i,\bar{\mu}}$. We may assume that the ordering above is such that $s = s_k = s_{k,\bar{\mu}}$, $\tilde{s} = s_{k+1} = s_{k+1,\bar{\mu}}$ and $\gamma \subset W^u(s) \cap W^s(\tilde{s})$. The arguments for the case $\tilde{s} = s_{k+j}$ for some $j > 0$ are similar. For μ close to $\bar{\mu}$, the stable manifolds $W^s(s_{k,\mu})$ are (uniquely) foliated by the strong stable foliations $\mathscr{F}^{ss}(s_{k,\mu})$, whose main leaves are the strong stable manifolds $W^{ss}(s_{k,\mu})$. In our case, this is particularly easy to see using the local coordinates that linearize X_μ as in Section I.2. We simply denote by $\mathscr{F}^{ss}(s_k)$ the union of $\mathscr{F}^{ss}(s_{k,\mu})$ for μ in some small interval (μ_3, μ_4) containing $\bar{\mu}$.

We want to construct a (singular) foliation $\mathscr{F}^{cu}(s_k)$ in a neighbourhood $U \times (\mu_3, \mu_4)$ of s_k in $M \times (\mu_3, \mu_4)$, whose main leaves are the centre-unstable manifolds $W^{cu}(s_{k,\mu})$ of $s_{k,\mu}$. As usual, we wish the foliation to be invariant by the flows $X_{\mu,t}$ for $t \geq 0$ and leaves to be C^1 imbedded discs varying continuously in the C^1 topology. Moreover, a leaf of $\mathscr{F}^{cu}(s_k)$ through a point, say (x, μ), transversely intersects $W^s(s_{k,\mu})$ in a C^1 curve transverse to $\mathscr{F}^{ss}(s_{k,\mu})$. The foliation is singular in the sense that two leaves may intersect each other; if so, they are tangent along their intersection, which is a C^1 submanifold, C^1 close to $W^u(s_k)$. In fact, many leaves intersect each other along $W^u(s_{k,\mu})$ for each $\mu \in (\mu_3, \mu_4)$. The picture indicates the intersections of some of the leaves of $\mathscr{F}^{cu}(s_k)$ with one of the stable manifolds $W^s(s_{k,\mu})$. Dually, we consider a centre-stable foliation $\mathscr{F}^{cs}(s_{k+1})$ in a neighbourhood of s_{k+1} in $M \times (\mu_3, \mu_4)$.

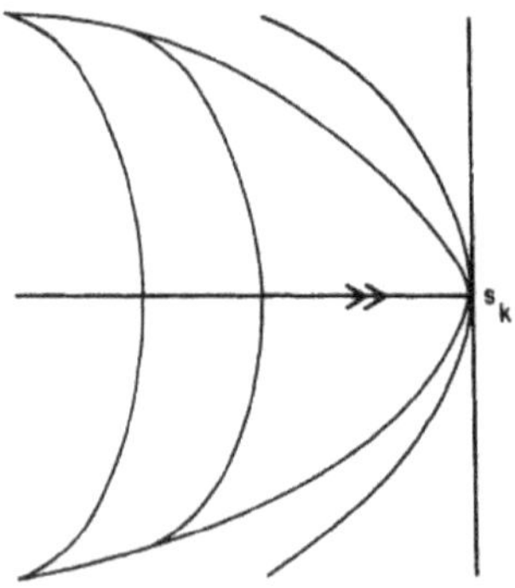

One of the key facts in proving the stability of $\{X_\mu\} \in \mathscr{G}$ as above is that we can construct compatible systems of foliations

$$\mathscr{F}^u(s_1), \ldots, \mathscr{F}^u(s_{k-1}), \mathscr{F}^{cu}(s_k) \quad \text{and} \quad \mathscr{F}^{cs}(s_{k+1}), \mathscr{F}^s(s_{k+2}), \ldots, \mathscr{F}^s(s_l).$$

PROPOSITION 3.2. *There exist compatible systems of foliations* $\mathcal{F}^u(s_1)$ $\ldots,\mathcal{F}^u(s_{k-1})$, $\mathcal{F}^{cu}(s_k)$ *and* $\mathcal{F}^{cs}(s_{k+1})$, $\mathcal{F}^s(s_{k+2}),\ldots,\mathcal{F}^s(s_l)$.

Proof. Let us show the existence of a compatible system of foliations $\mathcal{F}^u(s_1),\ldots,\mathcal{F}^u(s_{k-1})$, $\mathcal{F}^{cu}(s_k)$. By Proposition 3.1, there is a compatible system $\mathcal{F}^u(s_1),\ldots,\mathcal{F}^u(s_{k-1})$. To construct $\mathcal{F}^{cu}(s_k)$, we first observe that $W^u(s_i)$ is transverse to $W^{ss}(s_k)$ for each $i < k$ as stated in Property M of Section I.2. Thus, we can choose a continuous family of C^2 cylinders C_μ and continuous families of C^1 discs $D_\mu \subset F_\mu$ and $D_\mu^* \subset F_\mu^*$, F_μ and F_μ^* being leaves of $\mathcal{F}^{ss}(s_{k,\mu})$, so that $C_\mu \cup F_\mu \cup F_\mu^*$ is a fundamental domain for $W^s(s_{k,\mu})$ for each $\mu \in (\mu_3, \mu_4)$. The cylinders C_μ are taken so that they are transverse to $\mathcal{F}^{ss}(s_{k,\mu})$ and disjoint from all the components of $W^u(s_{i,\mu}) \cap W^s(s_{k,\mu})$, $i < k$, that do not intersect $W^{ss}(s_{k,\mu})$. We also require that X_μ be tangent to C_μ along its intersections with D_μ and D_μ^*.

As in Proposition 3.1, we construct a one-dimensional foliation $\mathcal{F}^c$ of C_μ, $\mu \in (\mu_3, \mu_4)$, along the leaves of $\mathcal{F}^u(s_i)$ for all $i < k$. That is, the leaves of $\mathcal{F}^c$ subfoliate the leaves of $\mathcal{F}^u(s_i)$. Let $u = \dim W^u(s_k)$. Proceeding by induction on the indices i, as in Proposition 3.1, we raise through the leaves of $\mathcal{F}^c$ a $(u + 1)$-dimensional foliation $\tilde{\mathcal{F}}^{cu}$ compatible with $\mathcal{F}^u(s_1),\ldots,\mathcal{F}^u(s_{k-1})$ and transverse to the stable manifolds $W^s(s_{k,\mu})$. This foliation yields a u-dimensional foliation $\mathcal{F}^u$ through the points of intersection of C_μ with D_μ and D_μ^*. We extend $\mathcal{F}^u$ over all of D_μ and D_μ^*, so that each leaf through a point of D_μ or D_μ^* is transverse to $W^s(s_{k,\mu})$ and $\mathcal{F}^u$ is compatible with $\mathcal{F}^u(s_1),\ldots,\mathcal{F}^u(s_{k-1})$. As usual, we demand that the leaves of both $\tilde{\mathcal{F}}^{cu}$ and $\mathcal{F}^u$ be C^1 imbedded discs varying continuously in the C^1 topology.

We can now describe the full set of leaves of $\mathcal{F}^{cu}(s_k)$. The first subset consists of leaves of $\tilde{\mathcal{F}}^{cu}$ and their positive iterates by the flows $X_{\mu,t}$. The remaining leaves are obtained as follows: for each leaf F of $\mathcal{F}^u$ contained say in $M \times \{\mu\}$, we take $\{X_{\mu,t}(F);\ t \geq 0\}$ as a leaf of $\mathcal{F}^{cu}(s_k)$. Clearly, $\mathcal{F}^{cu}(s_k)$ has all the required properties. The construction of $\mathcal{F}^{cs}(s_{k+1})$ is entirely similar.

THEOREM B. *Let* $\{X_\mu\}$ *be in* $\mathcal{G}$ *with an isolated bifurcation value* $\bar{\mu}$ *corresponding to an orbit of tangency. Then* $\{X_\mu\}$ *is stable. Furthermore, if* $\{X'_\mu\}$ *is close to* $\{X_\mu\}$ *and* $X_\mu = X'_\mu$ *except for* μ *near* $\bar{\mu}$, *then the equivalence* $H = \{H_\mu\}$ *which we construct between* $\{X_\mu\}$ *and* $\{X'_\mu\}$ *is such that* H_μ *is the identity map on* $M \times \{\mu\}$ *for* μ *outside some neighbourhood of* $\bar{\mu}$.

Proof. As we pointed out in Section 3 of Chapter I, it is enough to consider perturbations $\{X'_\mu\}$ of $\{X_\mu\}$ such that $X'_\mu = X_\mu$ for μ outside a small neighbourhood (μ_1, μ_2) of $\bar{\mu}$. By the Openness Theorem in the same section, if $\{X_\mu\}$ is close to $\{X'_\mu\}$ then $\{X'_\mu\}$ has a unique bifurcation value $\bar{\mu}'$ close to $\bar{\mu}$ corresponding to an orbit of tangency. Also, the singularities $s'_{1,\mu}, \ldots, s'_{k,\mu}, s'_{k+1,\mu}, \ldots, s'_{l,\mu}$ of $\{X'_\mu\}$ are close to the corresponding ones for $\{X_\mu\}$. As before, suppose that the orbit of tangency γ for $X_{\bar{\mu}}$ occurs between the unstable manifold of $s_{k,\bar{\mu}} = s_k$ and the stable manifold of $s_{k+1,\bar{\mu}} = s_{k+1}$. Similarly, let γ' be the orbit of tangency between $W^u(s'_k)$ and $W^s(s'_{k+1})$ for $X'_{\bar{\mu}'}$. As in Section 4 of Chapter II, we multiply the families by smooth positive functions so that the weakest contracting eigenvalues at $s_{k,\mu}$ and $s'_{k,\mu}$ are -1 and the weakest expanding eigenvalues at $s_{k+1,\mu}$ and $s'_{k+1,\mu}$ are $+1$ for all μ close to $\bar{\mu}$. Clearly the modified families, which we still denote by $\{X_\mu\}$ and $\{X'_\mu\}$, are respectively equivalent to the original ones since they have the same spaces of orbits. Thus, it is enough to construct an equivalence (actually a conjugacy) between the modified families $\{X_\mu\}$ and $\{X'_\mu\}$. We have partially done so in Sections 2 and 4 of Chapter II. There, we constructed a conjugacy $H = \{H_\mu\}$ between $\{X_\mu\}$ and $\{X'_\mu\}$ sending $W^{cc}(\gamma)$ to $W'^{cc}(\gamma')$.

If S and S' are cross sections through $r \in \gamma$ and $r' \in \gamma'$ as in the last chapter, then the conjugacy H is also defined from a neighbourhood of r in $W^c \cap S$ to a neighbourhood of r' in $W'^c \cap S'$. Notice that, by construction, H sends leaves of $\mathscr{F}^u(s_k)$ to leaves of $\mathscr{F}^u(s'_k)$. These unstable foliations are naturally given in coordinates linearizing X_μ and X'_μ near the singularities $s_{k,\mu}$ and $s'_{k,\mu}$, respectively, for μ close to $\bar{\mu}$. Similarly, H sends leaves of $\mathscr{F}^s(s_{k+1})$ to leaves of $\mathscr{F}^s(s'_{k+1})$. It also follows, by construction, that H_μ is the identity map on $M \times \{\mu\}$ for $\mu \notin (\mu_1, \mu_2)$, since $X_\mu = X'_\mu$ outside (μ_1, μ_2). We take the reparametrization of Section II.4, η: $\mathbf{R} \to \mathbf{R}$ close to the identity map, $\eta(\bar{\mu}) = \bar{\mu}'$ and $\eta(\mu) = \mu$ for $\mu \notin (\mu_1, \mu_2)$. Our goal is to globalize $H = \{H_\mu\}$ to all of $M \times (\mu_3, \mu_4)$, for some interval $(\mu_3, \mu_4) \supset [\mu_1, \mu_2]$. We want H_μ to conjugate X_μ and $X'_{\eta(\mu)}$ for all $\mu \in (\mu_3, \mu_4)$, and H_μ to be the identity map on M for μ outside (μ_1, μ_2). To do this, we construct compatible systems of foliations $\mathscr{F}^u(s_1), \ldots, \mathscr{F}^{cu}(s_k)$ and $\mathscr{F}^{cs}(s_{k+1}), \ldots, \mathscr{F}^s(s_l)$ for $\{X_\mu\}$ and $\mathscr{F}^u(s'_1), \ldots, \mathscr{F}^{cu}(s'_k)$ and $\mathscr{F}^{cs}(s'_{k+1}), \ldots, \mathscr{F}^s(s'_l)$ for $\{X'_\mu\}$. These systems of foliations are constructed in Proposition 3.2 and they are the same for $\{X_\mu\}$ and $\{X'_\mu\}$ on $M \times \{\mu\}$ for each $\mu \notin (\mu_1, \mu_2)$.

As before, we want to construct a conjugacy H that sends leaves of $\mathscr{F}^u(s_1), \ldots, \mathscr{F}^{cu}(s_k)$ to leaves of $\mathscr{F}^u(s'_1), \ldots, \mathscr{F}^{cu}(s'_k)$ and leaves of $\mathscr{F}^{cs}(s_{k+1}), \ldots, \mathscr{F}^s(s_l)$ to leaves of $\mathscr{F}^{cs}(s'_{k+1}), \ldots, \mathscr{F}^s(s'_l)$. As in Theorem A, we proceed by induction, defining H on the spaces of leaves of $\mathscr{F}^u(s_1), \ldots, \mathscr{F}^u(s_{k-1})$, that is, on the union on μ of the stable manifolds $W^s(s_1, \mu), \ldots, W^s(s_{k-1}, \mu)$. Using the centre foliation $\mathscr{F}^c(s_k)$ and the strong stable foliation $\mathscr{F}^{ss}(s_k)$ as in Theorem A for the saddle-node, we define H on the union of the stable manifolds $W^s(s_{k,\mu})$. In doing so, we use the previously defined conjugacy H on the space of leaves of $\mathscr{F}^{ss}(s_k)$. In fact, this space of leaves is the union on μ of $W(s_{k,\mu})$ and we already know H on $W^{cc}(\gamma)$. Similarly, we define H on the spaces of leaves of $\mathscr{F}^s(s_l),\ldots,\mathscr{F}^s(s_{k+2})$ and on $\bigcup_\mu W^u(s_{k+1,\mu})$.

We now extend the conjugacy defined in the last chapter from a neighbourhood of r in $W^c(\gamma)\cap S$ to a full neighbourhood of r in S. The (transversal) intersections of leaves of $\mathscr{F}^{cu}(s_k)$ with leaves of $\mathscr{F}^{cs}(s_{k+1})$ in S define a (singular) foliation whose main leaves are $W^{cu}(s_{k,\mu})\cap W^{cs}(s_{k+1,\mu})\cap S$. That is, the main leaves of this foliation are precisely $W^c_\mu\cap S$, where H is already defined and is, in fact, differentiable. Notice that, in $W^{cu}(s_k)$, we have the foliation $\mathscr{F}^u(s_k)$, and the intersection of the leaves with S yields a differentiable foliation $\tilde{\mathscr{F}}^u$. We subfoliate $\tilde{\mathscr{F}}^u$ with a differentiable foliation $\tilde{\mathscr{F}}^{uu}$ whose leaves are transversal to $W^{cs}(s_{k+1})$ and have the same dimension as $W^{uu}(s_{k+1})$. We perform a dual construction in $W^{cs}(s_{k+1})$, obtaining a differentiable foliation $\tilde{\mathscr{F}}^{ss}$ whose leaves are transversal to $W^{cu}(s_k)$ and have the same dimension as $W^{ss}(s_k)$. We can then define a differentiable foliation $\mathscr{F}^{su}$ on a full neighbourhood of r in S with the following properties. The leaves are transversal and complementary in dimension to W^c, their intersections with $W^{cu}(s_k)$ are the leaves of $\tilde{\mathscr{F}}^{uu}$ and their intersections with $W^{cs}(s_k)$ are the leaves of $\tilde{\mathscr{F}}^{ss}$.

We now define a foliation $\mathscr{F}^{ss}$ on a full neighbourhood of r in S as the intersections of the leaves of $\mathscr{F}^{su}$ with the leaves of $\mathscr{F}^{cs}(s_{k+1})$. Similarly, we define $\mathscr{F}^{uu}$ by taking the intersections of the leaves of $\mathscr{F}^{su}$ with the leaves of $\mathscr{F}^{cu}(s_k)$. The reason we called these foliations $\mathscr{F}^{ss}$ and $\mathscr{F}^{uu}$ is the crucial fact that their saturation by the flows $X_{\mu,t}$ for $t<0$ and $X_{\mu,t}$ for $t>0$, respectively, converge to $\mathscr{F}^{ss}(s_k)$ and $\mathscr{F}^{uu}(s_{k+1})$ (see [2]). We perform similar constructions for $\{X'_\mu\}$ obtaining foliations $\mathscr{F}'^{su}$, $\mathscr{F}'^{ss}$, $\mathscr{F}'^{uu}$.

Now it is clear how to extend H to a full neighbourhood N of r in S. Since H is already defined on W^c and on the spaces of leaves of $\mathscr{F}^{cu}(s_k)$ and $\mathscr{F}^{cs}(s_{k+1})$, it is enough to require that H send leaves of $\mathscr{F}^{su}$ to leaves of $\mathscr{F}'^{su}$. In particular, we remark that H sends leaves of $\mathscr{F}^{ss}$ to leaves of $\mathscr{F}'^{ss}$ and leaves of $\mathscr{F}^{uu}$ to leaves of $\mathscr{F}'^{uu}$. Using the conjugacy equations $H_\mu X_{\mu,t}=X'_{\eta(\mu),t}H_\mu$, we see that H becomes well-defined on the saturation $N_1=\{X_{\mu,t}(N);\, t\in\mathbf{R}\text{ and }\mu\in(\mu_3,\mu_4)\}$ of N. From the remark above, it continuously agrees with the previously defined H on the stable manifolds of $s_{k,\mu}$ and on the unstable manifolds of $s_{k+1,\mu}$. However, as observed in the last section of the previous chapter, N_1 only contains a wedge shaped neighbourhood of γ in $W^c(\gamma)$. So, to define H in a full neighbourhood of $(s_k,\bar{\mu})$ we just have to extend the foliations $\mathscr{F}^{ss}$ and $\mathscr{F}'^{ss}$ as well as to extend H to all of the space of leaves of $\mathscr{F}^{ss}$, that is, W^{cu}. Since H is already defined on the space of leaves of the complementary foliation $\mathscr{F}^{cu}(s_k)$, we simply see that H sends leaves of $\mathscr{F}^{ss}$ to leaves of $\mathscr{F}'^{ss}$ and leaves of $\mathscr{F}^{cu}(s_k)$ to leaves of $\mathscr{F}^{cu}(s'_k)$. Since we want $\mathscr{F}^{ss}$ to be invariant by the flows $X_{\mu,t}$ for $t\le 0$, it is enough to perform the above extension over a fundamental domain of

$W^u(s_k)$. Of course, we want $\mathscr{F}^{ss}$ to be compatible with $\mathscr{F}^s(s_{k+2}),\ldots,\mathscr{F}^s(s_l)$. But, by construction $\mathscr{F}^{ss}$ is differentiable when restricted to each leaf of these foliations. Thus, we can do this extension exactly as in Proposition 3.1 (or (3.2), proceeding by induction on the indices $j \geq k+2$, and similarly for the foliation $\mathscr{F}'^{ss}$.

The extension of H to all of $W^{cu}(s_k)$ is done as follows. First of all, we recall that H sends leaves of $\mathscr{F}^u(s_k)$ to leaves of $\mathscr{F}^u(s'_k)$. We extend H to a fundamental domain J of $W^u(s, k)$, so that it takes leaves of $\mathscr{F}^s(s_{k+2}),\ldots,\mathscr{F}^s(s_l)$ to leaves of $\mathscr{F}^s(s'_{k+2}),\ldots,\mathscr{F}^s(s'_l)$, respectively. As before, this is possible because H restricted to each leaf of these foliations is C^1 close to the inclusion map. Let $J' = H(J)$ be the corresponding fundamental domain of $W^u(s'_k)$. We raise over J a one-dimensional foliation $\mathscr{F}_1$ transversal to $W^u(s_k)$ in $W^{cu}(s_k)$ and compatible with $\mathscr{F}^s(s_{k+2}),\ldots,\mathscr{F}^s(s_l)$. We want to construct a similar foliation $\mathscr{F}'_1$ over J' in $W^{cu}(s'_k)$ starting, wherever defined, with the images by H of the leaves of $\mathscr{F}_1$. Again, this is possible because, wherever defined, the restriction of H to each leaf of one of the foliations $\mathscr{F}^s(s_{k+2}),\ldots,\mathscr{F}^s(s_l)$ is differentiable. Of course, $\mathscr{F}_1$ is defined on all of $W^{cu}(s_k)$ via saturation by the flows $X_{\mu,t}$ for $t \leq 0$ and similarly for $\mathscr{F}'_1$ on $W^{cu}(s'_k)$. Finally, H is defined on $W^{cu}(s_k)$ by simply requiring that it take leaves of $\mathscr{F}^u(s_k)$ and $\mathscr{F}_1$ to leaves of $\mathscr{F}^u(s'_k)$ and $\mathscr{F}'_1$, respectively. In this way, we have the conjugacy defined in a neighbourhood of $(s_k, \bar{\mu})$. In a dual way, we construct H in a neighbourhood of $(s_{k+1}, \bar{\mu})$.

Thus, we have $H = \{H_\mu\}$ defined on the stable manifolds of $s_{i,\mu}$ for $i \leq k+1$, on the unstable manifolds of $s_{j,\mu}$ for $j \geq k$ and also in a neighbourhood $U \times (\mu_3, \mu_4)$ of the closure of the orbit of tangency. Moreover, H_μ is the identity map on $M \times \{\mu\}$ for μ outside (μ_1, μ_2). Using the conjugacy equations $H_\mu X_{\mu,t} = H'_{\eta(\mu),t} H_\mu$, we can define H on the set V of positive X_μ-orbits of U. Exactly as in the last part of Theorem A, we can now extend H to a full neighbourhood of $s_{j,\mu}$ for $j \geq k+2$ and, thus, to the stable manifolds of $s_{j,\mu}$. In this way, we have constructed the conjugacy $H = \{H_\mu\}$ on all of $M \times (\mu_3, \mu_4)$ with the property that H_μ is the identity map on $M \times \{\mu\}$ for μ outside (μ_1, μ_2). This completes the proof of Theorem B and, consequently, of our main result.

Instituto de Matemática Pura e Aplicada, Rio de Janeiro, Brasil
Rijksuniversiteit, Gronigen, The Netherlands

References

[1] Th. Bröcker and L. Lander, *Differentiable Germs and Catastrophes*, Camb. Univ. Press, 1975, Cambridge.
[2] M. W. Hirsch, C. C. Pugh, M. Shub, *Invariant Manifolds*, Lecture Notes in Math. 583

(1977), Springer-Verlag.
[3] N. Kopell, Commuting diffeomorphisms, in *Global Analysis*, Proc. Symp. Pure Math. **14** A.M.S., 1970, Providence, Rhode Island.
[4] N. H. Kuiper, C^1-equivalence of functions near isolated critical points, in *Symp. on Infinite Dim. Topology*, ed. R. D. Anderson, Ann. of Math. Studies **69**, Princeton Univ. Press, 1972, Princeton, N.J.
[5] I. Kupka, Contributions à la théorie des champs génériques, Contr. Diff. Eq. **2** (1963), 457–484.
[6] O. E. Lanford III, Bifurcation of periodic solutions into invariant tori: the work of Ruelle and Takens, in *Nonlinear Problems in the Physical Sciences and Biology*, Lecture Notes in Math. **322** (1973), Springer-Verlag.
[7] S. Newhouse, J. Palis and F. Takens, Stable families of diffeomorphism, Publ. I.H.E.S. **57** (1983), 5–71.
[8] R. Palais, Local triviality of the restriction map for embeddings, Comm. Math. Helvet. **34** (1960), 305–312.
[9] ______, Morse theory on Hilbert manifolds, Topology **2** (1963), 299–340.
[10] ______, On Morse-Smale dynamical systems, Topology **8** (1969), 385–405.
[11] J. Palis and W. de Melo, *Geometric Theory of Dynamical Systems, an Introduction*, Springer Verlag, Heildelberg, 1982.
[12] J. Palis and S. Smale, Structural stability theorems, in *Global Analysis*, Proc. Symp. Pure Math. **14** A.M.S., 1970, Providence, Rhode Island.
[13] J. Palis and F. Takens, Topological equivalence of normally hyperbolic dynamical systems, Topology **16** (1977), 335–345.
[14] M. Shub, *Stabilité Globale des Systèmes Dynamiques*, Astérisque **56** (1978).
[15] S. Smale, Stable manifolds for differential equations and diffeomorphisms, Ann. Scuola Norm. Sup. Pisa (3) **17** (1963), 97–116.
[16] J. Sotomayor, Generic one parameter families of vector fields, Publ. I.H.E.S. **43** (1973), 5–46.
[17] ______, Generic bifurcations of dynamical systems, in *Dynamical Systems*, ed. M. M. Peixoto, Acad. Press 1973, New York.
[18] S. Sternberg, Local contractions and a theorem of Poincaré, Amer. J. Math. **79** (1957), 809–824.
[19] ______, On the structure of local homeomorphisms of euclidean space—II, Amer. J. Math. **80** (1958), 623–631.
[20] F. Takens, Partially hyperbolic fixed points, Topology **10** (1971), 133–147.
[21] E. C. Zeeman and D. J. A. Trotman, The classification of elementary catastrophes of codimension ≤ 5, in *Catastrophe Theory*, E. C. Zeeman, Addison-Wesley, 1977, London.

(Received January 22, 1982)
(Revised July 22, 1983)

A Note on the Inclination Lemma (λ-Lemma) and Feigenbaum's Rate of Approach

by

J. Palis

In recent years, striking results has been obtained concerning "universal constants" for cascades of bifurcations of dynamical systems such as maps of the interval and area preserving transformations (see [1], [3]). If $\mathfrak{F}_\mu$, $\mu \in \mathbb{R}$, is such a family and $\{\mu_n\}$, $\mu_1 < \ldots < \mu_n < \mu_{n+1} < \ldots$ with $\mu_n \to \mu_\infty$, are successive bifucation values of certain pattern, Feigenbaum suggested that the ratio $\mu_\infty - \mu_n/\mu_\infty - \mu_{n+1}$ should converge to a "universal" number, for a generic family in a (Banach) space E of transformations. Such a constant should be the expanding eigenvalue of the linear part of an operator F in the Banach space at a hyperbolic fixed point with codimension one stable manifold. The one-parameter family is a differentiable curve $L: I \subset \mathbb{R} \to E$ in E crossing the stable manifold transversally at $L(\mu_\infty)$ while Σ should also cross the unstable manifold transversally, where Σ is the codimension one submanifold of transformations exhibiting a (unique) bifurcating orbit of the required pattern.

The purpose of this note is to present an elementary discussion of the dynamical behavior of a transformation in Banach space, near a hyperbolic fixed point having its stable manifold of finite codimension. To relate it to part of Feigenbaum's program, we will be especially concerned with (local) iterates of submanifolds that intersect transversally the stable and unstable manifolds, respectivelly. Roughly speaking, the transformation being of class C^1, their iterates approach the stable and unstable manifolds, respectively, in

Originally published in: *Geometric dynamics (Rio de Janeiro, 1981), 630–635, Lecture Notes in Math. 1007*, © Springer, Berlin Heidelberg, 1983
Reprint by Springer-Verlag Berlin Heidelberg 2014

the C^1 sense. This is the content of the Inclination Lemma (λ-Lemma) in [6,7], and it was used in several contexts in Dynamics. The key novelty introduced in [1] was to show, in the codimension one case for a transformation of class C^2, that the (negative) iterates of a codimension one submanifold cutting the unstable manifold transversally approach the stable manifold as before but now at a geometric rate (the inverse of the expanding eigenvalue). With a different proof, we present this result in Theorem A for C^1 transformations with (one of the) partial derivatives being Lipschitz. The Lipschitz condition is necessary as shown in the counter-example we provide. Finally, in Theorem B we observe that a similar result is also valid in codimension k, $k \geq 1$ (corresponding to k-parameter families), under mild genericity conditions.

Two final remarks are in order. In the applications to compute Feigenbaum's constants ([1], [3]), the Banach space transformation is C^∞ (not only C^2 or C^1 plus Lipschitz). We think, however, that it makes sense to announce the precise theorem in this setting and, perhaps, to open up some perspective for k-parameter families of systems with such geometric view. We also observe that similar approach appears in the study of saddle-connections (crbits of tangency) generating moduli of stability (see [4], [5], [8]).

We now present Theorems A and B and the counter-example mentioned above.

Let $F: V(O) \subset E \to E$ be of class C^r, $r \geq 1$, with O being a hyperbolic fixed point of codimension k; i.e., its unstable manifold is of dimension k. Here, $V(O)$ is a neighborhood of the origin O in the (real) Banach space E. If $\rho_1, \rho_2, \ldots, \rho_k$ are the expanding eigenvalues of $DF(O)$, we assume that $|\rho_1| < |\rho_j|$ for $2 \leq j \leq k$ or $|\rho_1| = |\rho_2| < |\rho_j|$ for $2 \leq j \leq k$ in case ρ_1 is not real and ρ_2 is its conjugate. From the Stable Manifold Theorem (see [2]), there are C^r stable manifold W^s, unstable manifold W^u and strong unstable manifold $W^{uu} \subset W^u$, which is of dimension $k-1$ or $k-2$ depending on ρ_1 being real or not. These F-invariant submanifolds are C^r plus Lipschitz if F is C^r plus Lipschitz.

Theorem A. Let $F: V(O) \to E$ as above be of class C^1 and suppose O is a hyperbolic fixed point for F of codimension one. Let Σ be a codimension one submanifold intersecting W^u transversally and let L be a C^1 curve intersecting W^s transversally at z_∞. Then, there exist $n_o \in \mathbb{N}$ and a sequence of points $z_n \in L$, $n \geq n_o$, such that $F^n(z_n) \in \Sigma$. Moreover, if F is C^1 plus Lipschitz then $d(z_n, z_\infty)/d(z_{n+1}, z_\infty)$ converges to $|\rho|$ (ρ being the expanding eigenvalue).

Proof. The first part is an immediate consequence of the Inclination Lemma: given $\epsilon > 0$, a bounded disk D^u containing O in W^u and a small ("product") neighborhood $\mathcal{U}^u$ of D^u in $V(O)$, then there exist $n_o \in \mathbb{N}$ such that the connected components of $F^n(L) \cap \mathcal{U}$ through $F^n(z_\infty)$ is ϵ-C^1 close to D^u for all $n \geq n_o$.

To prove the second part, let us analyse somewhat more the inclination of the tangent vectors to the local inverse images by F of the submanifold Σ. Since F is not necessarily injective, we consider the components of $F^{-n}(\Sigma)$, $n \geq 0$, which are diffeomorphic to Σ. They exist, by the Implicit Function Theorem, and they will be denoted simply by Σ^{-n}. By the same Inclination Lemma, we have that given $\epsilon > 0$ and a (bounded) disk $D^s \subset W^s$ containing O and z_∞, there exist $n_o \in \mathbb{N}$ such that $\Sigma^{-n} \cap \mathcal{U}^s$ is ϵ-C^1 close to D^s for all $n \geq n_o$, where $\mathcal{U}^s$ is a neighborhood of D^s. We can express F as $F(x,y) = (Ax + \Phi^s(x,y), \rho y + \Phi^u(x,y))$, $x \in W^s$ and $y \in W^u$, where $\|A\| \leq a < 1$, $D\Phi^s(0,0) = 0$, $D\Phi^u(0,0) = 0$, $\Phi^s(0,y) = 0$, $\Phi^u(x,0) = 0$. Furthermore, $|\frac{\partial \Phi^u}{\partial x}(x,y)| \leq K|y|$ for some $K > 0$. By taking a finite number of iterates of the curve L by F, we may assume that in $\mathcal{U}^s$ all partial derivatives of Φ^s, Φ^u have norm bounded by a small number $\delta > 0$.

We now claim that, for ϵ and δ small enough and $n \geq n_o$, in every point $(x,y) \in \Sigma^{-n}$ and for every nonzero vector $v = (v^s, v^u)$ tangent to Σ^{-n} at (x,y), we have that $\lambda(v)/|y|$ is uniformly bounded, say, by a constant $C > 0$. Here, $\lambda(v)$ means the inclination of v; i.e., $\lambda(v) = |v^u|/\|v^s\|$. This readily implies the result. In fact, for $\tilde{y}_n = \Sigma^{-n} \cap W^u$ it is clear that $\lim |\tilde{y}_n|/|\tilde{y}_{n+1}| = |\rho|$, since $\frac{\partial \Phi^u}{\partial y}(0,0) = 0$. For $z_n = \Sigma^{-n} \cap L$, $z_n = (x_n, y_n)$, we conclude from our claim that $e^{-C\ell} \le |y_n|/|\tilde{y}_n| \le e^{C\ell}$, where ℓ is the diameter of a disk in W^s centered at O and containing z_∞ in its interior. From this we get

$$e^{-2C\ell}\cdot|\tilde{y}_n|/|\tilde{y}_{n+1}| \le |y_n|/|y_{n+1}| \le e^{2C\ell}\cdot|\tilde{y}_n|/|\tilde{y}_{n+1}|.$$

But, for any convergent subsequence $|y_k|/|y_{k+1}|$ its limit is the same as that of $d(z_k,z_\infty)/d(z_{k+1},z_\infty)$, which in turn is the same as that of $d(t_k,t_\infty)/d(t_{k+1},t_\infty)$, where $t_k = \Sigma^{-k} \cap F^j(L)$ and $t_\infty = F^j(t_\infty)$, for any (fixed) $j \in \mathbb{N}$. Thus, by making ℓ small, we conclude that $d(z_n,z_\infty)/d(z_{n+1},z_o)$ converges to $|\rho|$.

Let us prove the above claim. Formally we will be using the Inclination Lemma, but the proof we present can be easily adapted to also prove this Lemma. Let $(x,y) \in \Sigma^{-n}$ and $v = (v^s,v^u)$ be a nonzero tangent vector to Σ^{-n} at (x,y), and let $(w^s,w^u) = w = DF(v)$ and $(X,Y) = F(x,y)$. From the expressions for F and DF, we have that

$$\lambda(w) \ge [(\rho-\delta)\lambda(v) - K|y|]/[a + \delta + \delta\varepsilon].$$

Let $\rho-\delta = \tilde{\rho} > 1$, $a + \delta + \delta\varepsilon = \tilde{a} < 1$. We then get $1/\tilde{\rho}\,[\tilde{a}/|y|\cdot\lambda(w) + K] \ge \lambda(v)/|y|$. Thus, $1/\tilde{\rho}\,[\tilde{a}\cdot|Y|/|y|\cdot\lambda(W)/|Y| + K] \ge \lambda(v)/|y|$. Again, for δ small, we have that $\tilde{a}/\tilde{\rho}\cdot|Y|/|y| \le a^* < 1$ for some a^*. And so, $\lambda(v)/|y| \le a^*[\lambda(w)/|Y| + K]$. Proceeding by induction, using the fact that $a^* < 1$, we easily conclude our claim and Theorem A is proved.

The following result is a generalization of Theorem A and its proof is quite similar to the previous one.

<u>Theorem B</u>. Let $F\colon V(O) \to E$ as before be of class C^1. Suppose O is a hyperbolic fixed point of codimension $k \ge 1$. Let Σ be a co-

dimension k submanifold intersecting W^u transversally at w and let P be a k-dimensional surface intersecting W^s transversally at z_∞. Then, there exist $n_o \in \mathbb{N}$ and a sequence $z_n \in L$, $n \geq n_o$, such that $z_n \to z_\infty$, $F^n(z_n) \in \Sigma$ and $\lim F^n(z_n) = w$. Moreover, if F is C^1 plus Lipschitz and w is off W^{uu}, then $\lim d(z_n,z_\infty)/d(z_{n+1},z_\infty) = |\rho_1|$.

The Counter-Example. Let $V = \{(x,y) \in \mathbb{R}^2,\ y < 1\}$ and consider the diffeomorphism $F\colon V \to F(V) \subset \mathbb{R}^2$, defined as follows:

$$F(x,y) = (ax, a^{-1}y + \Phi^u(x,y)),$$

where $0 < a < 1$, and

i) $\Phi^u(x,y) = 0$ for $x \leq y$;

ii) $\Phi^u(x,y) = 0$ for $y \leq 0$;

iii) $\Phi^u(x,y) = 0$ for $a^{2k+1/2} \leq y/x \leq a^{2k}$, $k \in \mathbb{N}$;

iv) $\Phi^u(x,y) = -y/\log y \cdot \sin^2(2\pi \cdot \log(y/x)/\log a)$ otherwise.

One can check that F is of class C^1 (but not C^1 plus Lipschitz), O is a hyperbolic fixed point, a and a^{-1} are the contracting and expanding eigenvalues, the x-axis is the stable manifold and the y-axis is the unstable manifold. Consider Σ as the horizontal line $y = a^d$, $0 < d < a^{1/4}$, and L the vertical line $x = a$. As before, there exist a sequence $z_n \to z_\infty = (a,0)$ such that $F^n(z_n) \in \Sigma$. We will show that $d(z_n,z_\infty)/d(z_{n+1},z_\infty)$ does not converge to a^{-1}. To see this, observe that for n odd we have $z_n = a^{n+d}$. Let now n to be even. If we write $z_n = a^{n+d+\ell_n}$, then ℓ_n cannot converge zero. This is clear because $\Sigma_{j=1}^{n/2} \cdot 1/n-j > 1/2$ and so

$$-1/\log a^2 \cdot \sin^2(2\pi d) \cdot \Sigma_{j=1}^{n/2} 1/n-j > \delta > 0.$$

Thus, $d(z_n,z_\infty)/d(z_{n+1},z_\infty)$ does not converge to a^{-1}. In fact, this sequence does not converge at all. Also, by varying the line Σ, we can get subsequences of the above sequence converging to uncountably many different numbers.

References

[1] Collet, P. and Eckmann, J.-P., Iterated Maps of the Interval as Dynamical Systems, Progress in Physics, Birkhäuser (1980).

[2] Hirsch, M. and Pugh, C. and Shub, M., Invariant Manifolds, Lecture Notes in Math. 583, Springer-Verlag (1977).

[3] Lanford, O., A computer-assisted proof of the Feigenbaum conjectures, Bull. Am. Math. Soc. 6 (1982), 427-434.

[4] Newhouse, S. and Palis, J. and Takens, F., Bifurcations and stability of families of diffeomorphisms, Publ. Math. IHES, to appear.

[5] Palis, J., A differentiable invariant of topological conjugacies and moduli of stability, Astérisque 51 (1978), 335-346.

[6] Palis, J., On Morse-Smale dynamical systems, Topology 8 (1969), 385-405.

[7] Palis, J. and de Melo, W., Geometric Theory of Dynamical Systems, An Introduction, Springer-Verlag (1982).

[8] Palis, J. and Takens, F., Stability of parametrized families of gradient vector fields, Annals of Math., to appear.

Instituto de Matemática Pura e Aplicada (IMPA)
Estrada Dona Castorina, 110
22.460 - Rio de Janeiro, RJ - Brasil

Invent. math. 82, 397–422 (1985)

Inventiones mathematicae

Cycles and measure of bifurcation sets for two-dimensional diffeomorphisms

J. Palis[1] and F. Takens★[2]

[1] IMPA, Estr. Dona Castorina 110, Horto, Rio de Janeiro, Brazil
[2] Rijksuniversiteit Groningen, Mathematisch Instituut, Postbus 800, NL-Groningen, The Netherlands

Summary. We describe a large class of one-parameter families $\{\varphi_\mu\}$, $\mu \in \mathbb{R}$, of two-dimensional diffeomorphisms which are *stable* for $\mu < 0$, exhibit a *cycle* for $\mu = 0$, and thereafter have a bifurcation set of positive but arbitrarily small *relative measure* for μ in small intervals $[0, \delta]$. A main assumption is that the basic sets involved in the cycle have *limit capacities* that are not too large.

1. Introduction

It is well known that the creation of a cycle in a parametrized family of dynamical systems may lead to infinitely many bifurcations, and systems with a simple dynamical behavior may go through much more complicated ones for small variations of the parameter.

A point in the parameter space is called a bifurcation point if the corresponding system is structurally unstable: the system is not topologically conjugate to all nearby ones. If the same holds even when we restrict the system to its nonwandering set, Ω, the point is called an Ω-bifurcation point. The corresponding sets of such points are called the bifurcation and Ω-bifurcation set, respectively.

The present paper provides information about the (not yet well understood) structure of the bifurcation and Ω-bifurcation sets for certain one-parameter families of 2-dimensional diffeomorphisms. This is done near a value of the parameter where a cycle is created. Roughly speaking, we show that the Ω-bifurcation set has small relative measure if $d_c^u + d_c^s < 1$, where d_c^u is the maximum of the unstable limit capacities of the basic sets involved in the cycle and d_c^s is the maximum of the stable limit capacities. A similar theorem holds for the bifurcation set itself if we also take into account the unstable (stable) limit

★ The second author acknowledges hospitality and financial support from IMPA/CNPq during the period this paper was prepared capacities of the basic sets that are only positively (negatively) involved in the cycle. The exact content of these results (Theorems A and B) is given in Sect. 2.

We point out that in recent years a number of papers have discussed the relative Lebesgue measure (in the parameter space) of the bifurcation sets of parametrized families of diffeomorphisms and transformations. They motivated us to take up again the work initiated in [20].

In the case of families of quadratic maps of the interval or rational maps of the Riemann sphere, [9, 26, 3] exhibited bifurcation sets of positive measure in the parameter space with the corresponding maps being ergodic (see also [31, 13]). The expression "bifurcation sets" is used here in the sense that one expects such ergodic maps to be structurally unstable: this is indeed the case for rational maps of the Riemann sphere (see [36] for a discussion of their conjugacy classes), but in general it seems to be an interesting open problem for maps of the interval.

For diffeomorphisms of a compact manifold there was for some time a development with a somewhat different focus: a key question was whether the stable diffeomorphisms were dense in the space of all diffeomorphisms. Open sets of unstable diffeomorphisms were then provided in [33, 1, 38, 14, 32] among others, implying a negative answer to this question for all manifolds except the circle where the fact was already known to be true [24, 25]. The same negative response holds even when we consider the stability of a diffeomorphism restricted to its non-wandering set (Ω-stability). A common feature in these works is the persistence of non-transversal intersections (orbits of tangency) between stable and unstable manifolds associated to non trivial hyperbolic sets. Another interesting fact is that any of these examples can be obtained immediately after the first bifurcation in a generic one-parameter family of diffeomorphisms $\{\varphi_\mu\}$ with φ_μ, for $\mu<0$, being structurally stable (Axiom A and strong transversality). We observe that only in [14] the hyperbolic sets giving rise to persistent orbits of tangency are zero-dimensional Cantor sets.

On the other hand, the parametrized families of diffeomorphisms we describe have bifurcation sets that are neither full measure nor measure zero, when restricted to any small interval in the parameter space beginning with a first bifurcation point. Such a phenomenon already appears on the circle for families of diffeomorphisms whose corresponding rotation number varies with the parameter: this happens when a saddle-node bifurcation occurs in the absence of other periodic orbits. This fact was proved in [6], after the analytical example presented in [2]. The case we discuss here is of a different nature, being typical of families of diffeomorphisms having orbits of tangency; this requires the ambient space to have at least two dimensions. It concerns the creation of cycles when the parameter passes the first bifurcation point and thereafter homoclinic orbits of tangency and new horse-shoes. In this direction, [19, 20] presents an extensive analysis of the generic families $\{\varphi_\mu\}$ such that each φ_μ for $\mu<0$ is Morse-Smale (finite hyperbolic limit set and transversality condition) and φ_0 has finite α or ω-limit set which exhibits a cycle (see also [21]). As the predecessor to the present paper, [20] proves, say in dimension two, that for μ small the bifurcation set has small relative measure: given $\varepsilon>0$, there exists $\delta > 0$ such that the measure of the bifurcation set in $[0, \delta]$ is smaller than $\varepsilon\,\delta$. Our main purpose here is to extend considerably this result, at least in two dimensions, to the case where instead of only periodic orbits general basic sets occur in the cycle, provided that their limit capacities are not too big. Also the proof we present is somewhat simpler than the one in [20].

Notice that in the case of cycles with the so called "fat" basic sets it was already known [17] that the first tangency is part of a whole interval $0 \leqq \mu \leqq \delta$ of bifurcations.

We also point out, on the other hand, that the bifurcation set of $\{\varphi_\mu\}$ always has positive measure in $[0, \delta]$ no matter how small we take $\delta > 0$: there are subintervals of $[0, \delta]$ such that φ_μ exhibits a homoclinic tangency for a dense set of values of μ in the subintervals [17]. A remarkable consequence is that densily in these subintervals the corresponding diffeomorphisms have infinitely many sinks [16, 17, 30].

In conclusion, our parametrized families have, after a first bifurcation point, a bifurcation set of positive but small relative measure.

We illustrate the situation dealt with in [19, 20] and in the present paper with the figures below. In the first one we have Morse-Smale diffeomorphisms, their limit sets consisting of fixed saddles p and q, fixed sources r_1 and r_2 and fixed sinks s_1 and s_2. In Fig. 2, we indicate the resulting diffeomorphism we obtain by moving the stable manifold of q as to obtain a tangency with the unstable manifold of p. Our analysis concerns the measure of the bifurcation set for $\mu > 0$. In the present work, we include the possibility that p and q are part of basic sets whose location is indicated by Λ_1 and Λ_2.

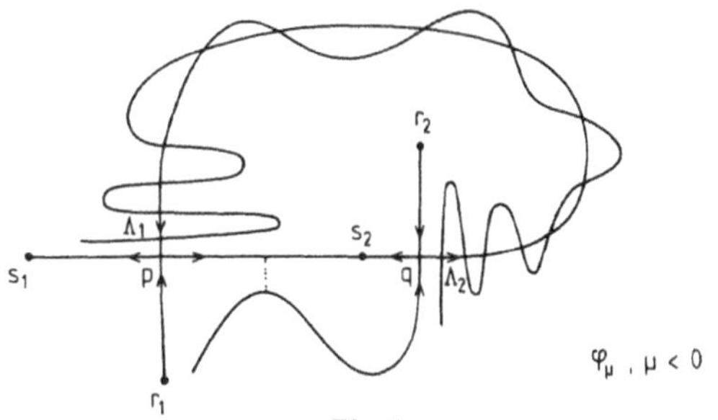

Fig. 1

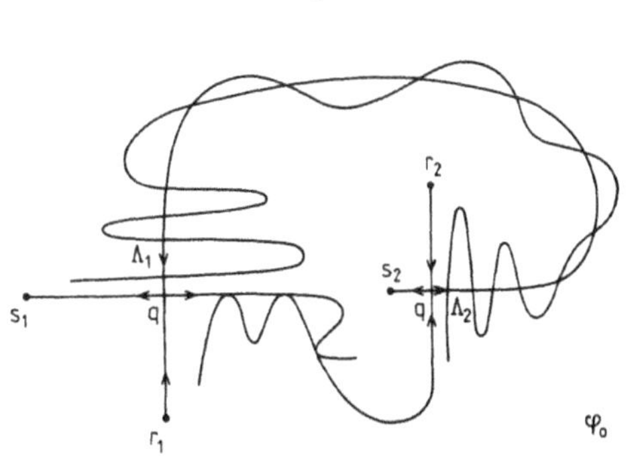

Fig. 2

The structure of the paper is as follows. Definitions, known facts and the precise statements of our results are presented in the next section. The proofs of these results are given in Sect. 3, which however depends on Sects. 4, 5 and 6. In particular, in Sects. 4 and 5 we discuss the possibility of tangencies between stable and unstable manifolds in relation to limit capacities related to the basic sets involved. In Sect. 6 we show that a controled loss of hyperbolicity near the tangency gets compensated by the overall hyperbolicity of the cycle; this basic fact is used in Sect. 3 to prove that $\Omega(\varphi_\mu)$ is hyperbolic for many values of $\mu>0$. There seems to be some resemblance between this hyperbolic property in our context and the expanding properties of the transformations exhibited in [9, 26, 3]. Finally, we believe that the results of Sect. 4 to 6 are of independent interest.

2. Statement of results

We begin by recalling some basic concepts and known results on hyperbolicity and stability.

Let M be a compact smooth manifold without boundary and let Diff(M) denote the set of C^2 diffeomorphisms of M endowed with the uniform C^2 Whitney topology. For $\varphi\in\mathrm{Diff}(M)$, we indicate by Per($\varphi$) the set of periodic points, by $L(\varphi)$ the limit set and by $\Omega(\varphi)$ the nonwandering set. Notice that $L(\varphi)=L^-(\varphi)\cup L^+(\varphi)$, where $L^-(\varphi)$ is the closure of the set of α-limit points of the orbits of φ and $L^+(\varphi)$ is the closure of the set of ω-limit points; they are called the negative and positive limit set, respectively. Clearly, $\Omega(\varphi)$ and $L(\varphi)$ are compact φ-invariant sets and $L(\varphi)\subset\Omega(\varphi)$.

A compact φ-invariant set Λ is called hyperbolic if there are a continuous splitting $T_\Lambda M=E^s\oplus E^u$, a constant $0<\lambda<1$, and a Riemannian norm $|\cdot|$ on TM such that, for $x\in\Lambda$,

i) $T_x\varphi(E^s_x)=E^s_{\varphi(x)}$, $T_x\varphi(E^u_x)=E^u_{\varphi(x)}$,

ii) $|T_x\varphi(v)|\leqq\lambda|v|$, $v\in E^x_x$ and $|T_x\varphi^{-1}(w)|\leqq\lambda|w|$, $w\in E^u_x$,

where $T_x\varphi$ denotes the derivative of φ at x. Let d be the induced distance function on M and let $A\subset M$ be a subset. We define the stable set of A by

$$W^s(A)=\{y\in M;\, d(\varphi^n(y),\varphi^n(A))\to 0 \text{ as } n\to\infty\}$$

and the unstable set of A by

$$W^u(A)=\{z\in M;\, d(\varphi^n(z),\varphi^n(A))\to 0 \text{ as } n\to-\infty\}.$$

If Λ is a hyperbolic set as above then, for any $x\in\Lambda$, $W^s(x)$ and $W^u(x)$ are C^2 injectively immersed copies of Euclidean spaces of dimensions $s=\dim E^s_x$ and $u=\dim E^u_x$ [7].

We say that φ satisfies *Axiom A* if $\Omega(\varphi)$ is hyperbolic and Per(φ) is dense in $\Omega(\varphi)$. In this case, $\Omega(\varphi)$ can be written as $\Omega(\varphi)=\Omega_1\cup\ldots\cup\Omega_k$, each Ω_i being closed, invariant and transitive [34, 35]. Each Ω_i is also isolated; i.e., there is a neighborhood U_i of Ω_i such that $\Omega_i=\bigcap_{n\in Z}\varphi^n(U_i)$; such U_i is called an isolating neighborhood for Ω_i. A hyperbolic set Λ with these properties is called a *basic set*. For a basic set Λ, we have $W^s(\Lambda)=\bigcup_{x\in\Lambda} W^s(x)$ and $W^u(\Lambda)=\bigcup_{x\in\Lambda} W^u(x)$ (see [8]). In [15], it was shown that if $L^-(\varphi)$ is hyperbolic then $\mathrm{Per}(f)$ is dense in $L^-(\varphi)$ and one can also write $L^-(\varphi)$ as a finite union of basic sets, $L^-(\varphi)=\Lambda_1 \cup\ldots\cup\Lambda_k$. Of course, the same is true for $L^+(\varphi)$ if this set is hyperbolic.

Hyperbolicity conditions already imply a kind of local stability. In fact, if Λ is a basic set with isolating neighborhood U, then for ψ close to φ one has that $\Lambda_\psi=\bigcap_{n\in\mathbb{Z}}\psi^n(U)$ is hyperbolic and there is a homeomorphism $h\colon \Lambda\to\Lambda_\psi$ such that $h\varphi(x)=\psi h(x)$ for all $x\in\Lambda$ [7, §7]. We say that Λ_ψ is canonically associated to Λ. To globalize stability of basic sets to *Ω-stability*, we need the notion of *cycle*. If $\Omega(\varphi)=\Omega_1\cup\ldots\cup\Omega_k$, we say that there is a j-cycle in $\Omega(\varphi)$, $j\geqq 1$, if for some sequence of distinct indices $i_1,\ldots,i_j$ we have $W^u(\Omega_{i_l})\cap W^s(\Omega_{i_{l+1}})\neq\emptyset$ for $1\leqq l\leqq j$ (here we take $i_{j+1}=i_1$). Similarly, if $L^-(\varphi)$ is hyperbolic and $L^-(\varphi)=\Lambda_1 \cup\ldots\cup\Lambda_k$ is its decomposition into basic sets, one can define the notion of a j-cycle on $L^-(\varphi)$ for $j\geqq 1$. Notice that if Ω is hyperbolic then it can not have 1-cycles. However, there are important cases where L^- is hyperbolic and yet for some basic set Λ_i of L^- we have a 1-*cycle* (see the last example of Sect. 7 of [19] or [20]). That is, there is a point $x\in W^u(y)\cap W^s(z)$ for some $y,z\in\Lambda_i$ and yet $x\notin L^-$; notice that in such a case x must be a point of tangency between $W^u(y)$ and $W^s(z)$. If there is a 1-cycle on $L^-(\varphi)$ (and if $L^-(\varphi)$ is hyperbolic) then $L^+(\varphi)$ may or may not be hyperbolic. In this paper we shall not deal with 1-cycles; they will be studied in a forthcoming paper.

We now state Smale's Ω-stability theorem [35]: if φ satisfies Axiom A and there are no cycles on $\Omega(\varphi)$, then φ is Ω-stable. An extension of it, due to Newhouse [15], says that if $L^-(\varphi)$ is hyperbolic and it has no cycles (including 1-cycles), then $\Omega(\varphi)=L^-(\varphi)$ and φ is Ω-stable. A similar statement holds for the positive limit set. On the other hand, if $L^-(\varphi)$ is hyperbolic and there is a cycle on it, then φ is not Ω-stable: we may perturb φ so to "enlarge" or "explode" the nonwandering and limit sets by creating new horse-shoes [22].

Even when the limit set is hyperbolic and it has no cycles, orbits of tangency between stable and unstable manifolds are an obstruction to global (structural) stability (that is, stability on all of M) as shown in [28]. If φ satisfies Axiom A, we say that it also satisfies the strong transversality condition if $W^u(x)$ and $W^s(y)$ are transversal for any pair of points $x\in\Omega(\varphi)$, $y\in\Omega(\varphi)$. We denote by AS the set of φ's satisfying these conditions and by $MS\subset AS$ the set of Morse-Smale ones; recall that $\Omega(\varphi)$ is finite for $\varphi\in MS$. The stability of elements in MS was shown in [23], and in general for elements in AS by Robbin [27] (see also de Melo [11] and Robinson [29]).

From this point on, we will assume that $\dim M=2$. One of the main reasons is that the stable and unstable foliations of a basic set Λ of a C^2 diffeomorphism can be extended to C^1 invariant foliations on a full neighborhood of Λ. This allows us, through the concept of limit capacity, to have a better control of how often the foliations associated to basic sets meet each other non-transversally. So, let $\dim M=2$ and let Λ be a basic set for φ such that $T_\lambda M=E^s\oplus E^u$ and $\dim E^s=\dim E^u=1$. Then, E^s and E^u can be extended to C^1 $T\varphi$-invariant line fields E^s_* and E^u_* on a neighborhood of Λ (Theorems 6.3

and 6.4 of [7]). Integrating E_*^s and E_*^u locally, we get C^1 foliations $\mathscr{F}^s$ and $\mathscr{F}^u$ on a neighborhood U of Λ such that $\varphi(\mathscr{F}_x^s)\subset\mathscr{F}_{\varphi(x)}^s$ and $\varphi(\mathscr{F}_x^u)\supset\mathscr{F}_{\varphi(x)}^u$ for $x\in U\cap\varphi^{-1}U\cap\varphi U$. The leaves are C^2 curves which vary continuously in the C^2 topology (on compact parts). A basic set Λ is called an attractor if $W^s(\Lambda)$ contains a neighborhood of Λ; similarly for a repellor with respect to its unstable set. We will be mostly interested in basic sets which are of saddle-type: neither attractors nor repellors. They are zero-dimensional and their stable and unstable sets have measure zero [4].

For a compact metric space C we define the *limit capacity* as follows. For $\varepsilon>0$ let $n(\varepsilon)$ denote the least number of ε-discs needed to cover C. The limit capacity of C is $d(C)=\limsup_{\varepsilon\to 0}(\operatorname{Log} n(\varepsilon)/-\operatorname{Log}\varepsilon)$. The local limit capacity at a point $c\in C$ is defined as the lim sup of the limit capacity of $C\cap D_\rho(c)$ as $\rho\to 0$; $D_\rho(c)$ denotes the closed disc of radius ρ around c. Clearly the limit capacity is invariant under Lipschitz homeomorphisms. Also, it satisfies the following property with respect to products: $d_A+d_B\geqq d_{A\times B}$ (see [37] and also [10]). For a basic set Λ of saddle-type we define the *unstable limit capacity* d^u as the limit capacity of the intersection of the unstable set of Λ with a C^1 line transverse to the unstable foliation. Since in this case the foliation is C^1, such a notion is well defined; i.e., it is independent of the transversal line. We can also define as above the *local unstable limit capacity* at some point of the unstable set. As in [14] for the notion of thickness, one can prove that the local unstable limit capacity is independent of the point in the unstable set and it is equal to the unstable limit capacity of Λ. Similarly, for the *stable limit capacity* $d^s(\Lambda)$. It is then clear that $d(\Lambda)\leqq d^u(\Lambda)+d^s(\Lambda)$ and that $d(\Lambda)$ is equal to the local limit capacity of Λ in any of its points. These statements hold in dimension two; in higher dimensions, the stable and unstable foliations are in general not differentiable and it is unclear to us how to define $d^s(\Lambda)$ and $d^u(\Lambda)$. However, when there is some C^2 normally hyperbolic 2-dimensional invariant submanifold, our analysis is still valid.

Now we come to our setting of bifurcating stable diffeomorphisms through the creation of cycles. We consider C^1 arcs $\{\varphi_\mu\}$, $\mu\in R$, in Diff(M) with the following properties:

i) $\varphi_\mu\in AS$ for $\mu<0$
ii) $L(\varphi_0)$ is hyperbolic and has a j-cycle, $j>1$.

We then ask how small is the measure of the bifurcation set for $\mu\in[0,\delta]$, whith $\delta>0$ being small. Notice that in any cycle for $L(\varphi_0)$ we must have a pair of basic sets say Λ_1 and Λ_2 of saddle-type and a pair of points $p\in\Lambda_1$ and $q\in\Lambda_2$ such that $W^u(p)$ and $W^s(q)$ have a point of tangency. Moreover, since $\varphi_\mu\in AS$ for $\mu<0$, the orbits of p and q must be periodic since otherwise we would have some bifurcation for $\mu<0$. The reason is that only periodic points in Λ_1 can have (local) unstable manifolds that are not accumulated from both sides by (local) unstable manifolds of points in Λ_1. Or, in another words, only periodic orbits can have their (local) unstable manifolds in the boundary of an (arcwise) connected component of the complement of Λ_1 (see Proposition 1 in [18]).

We assume that this *orbit of tangency* is *unique* and that it is of *quadratic contact* (quasi-transversal intersection): under the previous assumptions, these properties hold for an open and dense subset of one-parameter families. See [21] for the notion of quasi-transversal intersections of submanifolds in higher dimensions. Finally, we also assume this tangency to *unfold generically*, which again holds for an open and dense subset of families. This condition means the following. Let $l_{1,\mu}$ and $l_{2,\mu}$ be compact pieces of $W^u(p)$ and $W^s(q)$ so that they do not meet for $\mu<0$ and for $\mu=0$ they have a point of tangency in their interior. We want them to meet at two points say $x_{1,\mu}$ and $x_{2,\mu}$ for μ positive and small. Moreover, let $l(\mu)$ be the maximal distance to $l_{1,\mu}$ of points in $l_{2,\mu}$ between $x_{1,\mu}$ and $x_{2,\mu}$. We demand the (right) derivative of $l(\mu)$ to be positive at $\mu=0$.

If $\{\Lambda_i\}_{i=1}^k$ are the basic sets for $L(\varphi_0)$, we say that Λ_i is *involved in the cycle* if $W^u(\Lambda_i)\cap W^s(\Lambda_1)\neq\emptyset$ and $W^s(\Lambda_i)\cap W^u(\Lambda_2)\neq\emptyset$. As before, we assume here that the orbit of tangency occurs between a leaf of the unstable set $W^u(\Lambda_1)$ and a leaf of the stable set $W^s(\Lambda_2)$. Notice that only saddle-like basic sets can be involved in the cycle. We define the *unstable limit capacity* d_c^u *of the cycle* as the maximum of the unstable limit capacities of the basic sets involved. Similarly, one can define the *stable limit capacity of the cycle* d_c^s.

As mentioned before, any basic set of φ_0 gives rise in a canonical way to a basic set of φ_μ for small μ. We indicate by $\Lambda_{i,\mu}$ the basic set of φ_μ that corresponds to $\Lambda_i=\Lambda_{i,0}$, $1\leqq i\leqq k$. Notice that if Λ_i and Λ_l are involved in the cycle and if for some $\mu>0$ there is an orbit of tangency connecting $\Lambda_{i,\mu}$ and $\Lambda_{l,\mu}$, then φ_μ is not Ω-stable. In this case we say that μ is an Ω-bifurcation point. In general, we denote by $\Omega-B$ the set of values of μ for which φ_μ is not Ω-stable and call it the Ω-bifurcation set. Of course, $\Omega-B\subset B$ where B is the bifurcation set; that is, B is the set of values of μ for which φ_μ is not (structurally) stable.

To get global structural stability for φ_μ, $\mu>0$, one has also to consider the unstable limit capacities of the basic sets Λ_i such that $W^u(\Lambda_i)$ intersects $W^s(\Lambda_1)$. Each such Λ_i is said to be *positively involved with the tangency* since $W^u(\Lambda_i)$ contains the orbit of tangency in its closure. Similarly, Λ_l is *negatively involved with the tangency* if $W^s(\Lambda_l)$ intersects $W^u(\Lambda_2)$ or, equivalently, if $W^s(\Lambda_l)$ contains the orbit of tangency in its closure. We define the *unstable limit capacity of the tangency* d_t^u as the maximum of the unstable limit capacities of the basic sets positively involved with the tangency; the definition of *stable limit capacity of the tangency* d_t^s is similar. For the periodic basic sets we define the limit capacity as zero, even if they are sources or sinks. On the other hand, the unstable limit capacity of a nonperiodic repellor is one as well as the stable limit capacity of a nonperiodic attractor. Thus, for our second theorem below on global stability, they are automatically excluded of being involved with the tangency since we impose the condition that $d_t^u+d_t^s<1$.

We now formulate our results.

Theorem A. *Let* $\{\varphi_\mu\}$, $\mu\in\mathbb{R}$, *be defined in a closed surface* M, *satisfying*

i) $\varphi_\mu\in AS$ *for* $\mu<0$,
ii) $L(\varphi_0)$ *is hyperbolic and it has a j-cycle*, $j>1$.

Assume that there is a unique orbit of tangency between stable and unstable manifolds of orbits in $L(\varphi_0)$, *and that this tangency is of quadratic contact and* *unfolds generically. Then, if $d_c^u + d_c^s < 1$, for μ small the Ω-bifurcation set has small relative measure: given $\varepsilon > 0$ there is $\delta > 0$ such that the measure of the Ω-bifurcation set in $[0, \delta]$ is less than $\varepsilon\delta$.*

Theorem B. *Assume the same hypothesis of Theorem A. If, in addition, $d_t^u + d_t^s < 1$, then for μ small the bifurcation set has small relative measure.*

We remark that, in Theorems A and B, it is equivalent to say that $L(\varphi_0)$ is hyperbolic or $L^-(\varphi_0)$ is hyperbolic. In the cases of 1-cycles, that we will treat in a forthcoming paper, this difference is essential as already pointed out before. We also remark that Theorem B is a generalization, in two dimensions, of one of the main results in [20].

3. Proof of the main results

In this section we provide the proofs of Theorem A and B. We shall make use of results which are proved in the next three sections. We present these results in separate sections both because of their own independent interest and in order not to obscure the structure of the proofs of our main theorems.

Proof of Theorem A. Since $L(\varphi_0)$ is hyperbolic we can take a filtration $M_1 \subset M_2 \subset M$ such that

i) M_i is closed;
ii) $M_1 \subset \text{Int}(M_2)$;
iii) $\varphi_0(M_i) \subset \text{Int}(M_i)$;
iv) $M_2 - M_1$ contains exactly those basic sets which are involved in the cycle.

See [20] for details. As far as Ω-stability is concerned, we only have to deal with $M_2 - M_1$; more precisely, with the maximal invariant subsets in $M_2 - M_1$. $L(\varphi_0)$ is hyperbolic and $\mu = 0$ is the first bifurcation point; this implies (see also Sect. 2) that for $\mu \leqq 0$, $L(\varphi_\mu)$ depends continuously on μ. Let $L_c(\varphi_0)$ denote the union of the basic sets of φ_0 in $M_2 - M_1$ and define Σ by

$$\Sigma = (W^u(L_c(\varphi_0)) \cap W^s(L_c(\varphi_0))).$$

If U is a (small) compact neighborhood of Σ, then, for some integer n, depending on U, we have for each point $x \in (M_2 - M_1) - U$ that $\varphi_0^n(x)$ or $\varphi_0^{-n}(x) \notin M_2 - M_1$. This then also holds for φ_μ, μ small. Hence for such μ, all invariant sets, contained in $M_2 - M_1$, are actually contained in U. In fact we can choose U_μ, $\mu > 0$, depending continuously on μ, containing all invariant subsets of φ_μ in $M_2 - M_1$ and such that $\bigcap_{\mu > 0} U_\mu = \Sigma$. Our main purpose is to show that for many positive values of the parameter the new part of Ω in $\bigcap_{i \in \mathbb{Z}} \varphi_\mu^i(U_\mu)$ is hyperbolic.

Let x be a point in the orbit of tangency and let Σ' be $\Sigma - \{x\}$. Let $U_{\mu,x}$ be the component of U_μ containing x and $U'_\mu = U_\mu - U_{\mu,x}$.

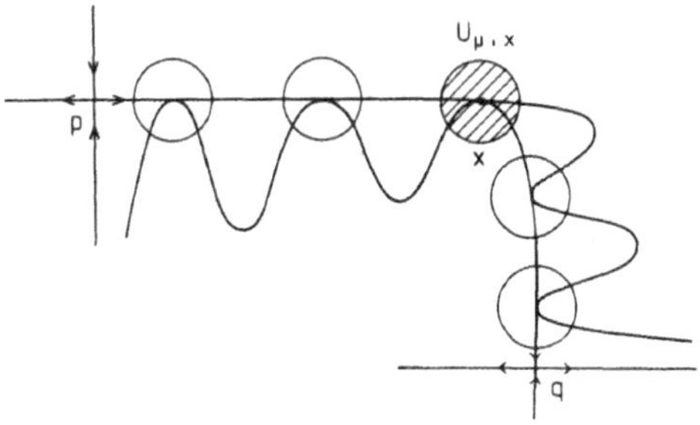

Fig. 3

Now we need to construct μ-dependent stable and unstable foliations in U'_μ: the tangencies between these foliations, when extended to a neighborhood of x, control the Ω-bifurcations. We only do the construction of the unstable foliation. We want this foliation to satisfy the following properties:

i) the foliation is C^0, but the leaves are differentiable and the tangent directions vary continuously;

ii) the foliation is invariant in the sense that if $F_{\mu,y}$ is a leaf through the point y, then $\varphi_\mu(F_{\mu,y}) \supset F_{\mu,\varphi_\mu(y)}$;

iii) in a neighborhood of the basic sets in U_μ, which are canonically associated with the basic sets in $L_c(\varphi_0)$, the foliation is an extension of the unstable manifolds.

As mentioned in Sect. 2 such a foliation, even C^1, can be constructed in a neighborhood of *one* basic set. In order to get the foliation in the whole of U_μ the only problem is to make the local constructions globally compatible with the invariance condition. This was done by de Melo in [11]. From now on these foliations will be denoted by $\mathscr{F}^u_\mu$ and $\mathscr{F}^s_\mu$.

Recall from Sect. 2 that the orbit of tangency must occur between the unstable and stable manifolds of periodic orbits. We assume these orbits to be fixed, for otherwise the arguments are the same. We denote them by p and q and the basic sets they belong to by Λ_1 and Λ_2; for small μ their continuous extensions are denoted by p_μ, q_μ, $\Lambda_{1,\mu}$, and $\Lambda_{2,\mu}$. Notice that p and q are different: otherwise we would have a 1-cycle. If there are, for $\mu=0$, intersections between 1-dimensional unstable manifolds and the stable manifold of p, they all have to be on one side of p (see the previous picture and the beginning of Sect. 5). The same remark holds for intersections with the unstable manifold of q. Since there is a cycle, there are always such intersections, in particular between $W^u(q)$ and $W^s(p)$; these intersections are transverse. As a consequence, the contracting eigenvalue at p and the expanding eigenvalue at q are positive.

As mentioned before we shall consider tangencies between leaves of the stable and unstable foliations; for this we extend $\mathscr{F}^u_\mu$ (applying φ_μ) and $\mathscr{F}^s_\mu$ (applying φ_μ^{-1}) to a neighborhood of the point of tangency x. See the figure below, which makes clear that such tangencies can only occur in a small disc near x bounded by segments of $W^u(p_\mu)$ and $W^s(q_\mu)$.

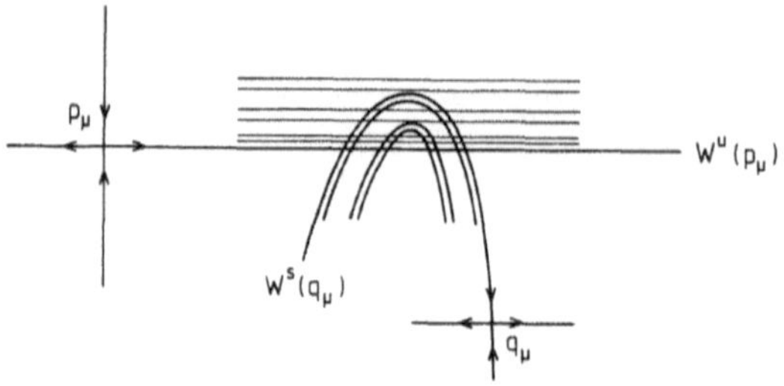

Fig. 4

In fact we don't need to consider all tangencies between leaves of $\mathscr{F}_\mu^u$ and $\mathscr{F}_\mu^s$, we only have to take into account the tangencies in those points r, whose entire orbit $\{\varphi_\mu^i(r)\}_{i\in\mathbb{Z}}$ stays inside U_μ: those points whose orbit leave U_μ cannot be part of $\Omega(\varphi_\mu)$.

Let D_μ^s be a fundamental domain in the branch of $W^s(p_\mu)$ containing the transverse intersections with one-dimensional unstable manifolds. For any point r near x, in $\bigcap_{i\in\mathbb{Z}} \varphi_\mu^i(U_\mu)$, μ being small, there is a negative interate $\varphi_\mu^{-j}(r)$ which is on a leaf of $\mathscr{F}_\mu^u$ through a point in D_μ^s. Now we define the set $A_\mu \subset W^s(p_\mu)$ as follows: $A_\mu \cap D_\mu^s$ is the set of intersection points of D_μ^s with leaves of $\mathscr{F}_\mu^u$ through points $\varphi_\mu^{-j}(r)$ as above. So $A_\mu \cap D_\mu^s$ is contained in a small neighborhood of D_μ^s intersected with the union of those $W^u(\Lambda_i)$ for which Λ_i is involved in the cycle. Moreover we require that A_μ is invariant under φ_μ. We point out that A_μ, as defined here agrees with A_μ as used in Sect. 5. The unstable foliation used there, is obtained from the present one by (i) taking the leaves of $\mathscr{F}_\mu^u$ intersecting $A_\mu \cap D_\mu^s$ and (ii) taking forward iterates of these leaves (as long as they stay in $V(p_\mu)$, see Sect. 5). The main difference between the unstable foliation in Sect. 5 and the one introduced here is the domain of definition. Notice that the domain of the new unstable foliation contains still the points where tangencies have to be controled. From now one we use $\mathscr{F}_\mu^u$ to denote this unstable foliation with its new domain. For the stable foliation $\mathscr{F}_\mu^s$ we perform the same construction; the analogous set in $W^u(q_\mu)$ is denoted by B_μ.

We observe that A_0 is a countable union of closed sets each with limit capacity smaller than or equal to d_c^u, namely A_0 is the intersection of $W^s(p_\mu)$ with unstable foliations of basic sets involved in the cycle. The analogous observation holds for B_0.

Thus we can apply the results of Sect. 4 and Sect. 5. In particular we conclude that for $\varepsilon > 0$, there are $\mu(\varepsilon) > 0$ and a subset $S(\varepsilon) \subset [0, \mu(\varepsilon)]$, such that

i) the measure of $S(\varepsilon)$ is bigger than $(1-\varepsilon)\cdot\mu(\varepsilon)$;

ii) for each $\mu \in S(\varepsilon)$, all intersections of leaves of the unstable and the stable foliations are transverse, in fact for any Riemannian metric there is a constant c such that in all these intersections near x, the angle between the stable and the unstable leaves is greater than $c\cdot\sqrt{\mu}$.

This result, together with the hyperbolicity result in Sect. 6, implies that for $\mu \in S(\varepsilon)$, ε small, the maximal invariant subset $\bigcap_{i\in\mathbb{Z}} \varphi_\mu^i(U_\mu)$ is hyperbolic. In fact this maximal invariant subset is a basic set containing the continuations of the basic sets involved in the cycle. To prove this fact we only have to show that this set contains a dense orbit, since by construction this set is clearly isolated. We observe that the unstable manifold of p_μ, $\mu > 0$, is dense in the continuation of the basic sets involved in the cycle, because it accumulates on the unstable manifold of q_μ. The other orbits of the new hyperbolic set pass through the disc near x bounded by small arcs of $W^u(p_\mu)$ and $W^s(q_\mu)$, so their stable manifolds intersect $W^u(p_\mu)$. Hence $W^u(p_\mu)$ is dense in the hyperbolic set. The same holds for $W^s(q_\mu)$. Since p_μ abd q_μ have heteroclinic points, this implies that the hyperbolic set has a dense orbit. Hence $\bigcap_{i\in\mathbb{Z}} \varphi_\mu^i(U_\mu)$ is indeed a basic set. Concluding, for these μ, $\Omega(\varphi_\mu)$ is hyperbolic and has no cycles; hence φ_μ is Ω-stable.

Proof of Theorem B. The structure of this proof is the same as that of the proof of Theorem A. The only difference is that we have to take into account all basic sets which are involved, positively involved, or negatively involved in the tangency.

4. Intersections of Cantor sets and scaled neighborhoods

A first question in which we are interested is the following. Let $C_1, C_2 \subset \mathbb{R}$ be two (compact) Cantor sets. What can we say about the measure of the set $\sigma_{C_1,C_2} = \{t \mid C_1 \cap (C_2 + t) \neq \emptyset\}$, in particular, when is this measure zero? The answer turns out to depend on the *limit capacities* of C_1 and C_2 which we defined in Sect. 2.

Proposition. *Let C_1 and C_2 as above be Cantor sets in $\mathbb{R}$ with limit capacity d_1, d_2 respectively. If $d_1 + d_2 < 1$, then the measure of σ_{C_1,C_2}, as defined above, is zero.*

Proof. We choose $\delta > 0$ so that $d_1 + d_2 + 2\delta < 1$. For ε sufficiently small, C_i can be covered by $\varepsilon^{-d_i-\delta}$ ε-balls (in this case intervals of length 2ε). Hence σ_{C_1,C_2} can be covered by $\varepsilon^{-d_1-\delta} \cdot \varepsilon^{-d_2-\delta}$ intervals of length $4 \cdot \varepsilon$. The total length of these intervals is $4 \cdot \varepsilon^{1-d_1-d_2-2\delta}$. This goes to zero when ε goes to zero; hence the proposition follows.

Remark. We need this proposition to prove that for certain one-parameter families of diffeomorphisms, the set of bifurcation values in certain regions of the parameter space has small measure. A criterion for the measure of σ_{C_1,C_2} to be positive, and even for σ_{C_1,C_2} to contain intervals, was obtained by Newhouse, e.g. see [14, 16, 17], as part of an analysis where the measure of the set of bifurcation values is big. We compare both criteria in the case of a "middle α-set". A *middle α-set* $F(\alpha)$ is an infinite intersection $\bigcap_{i=0}^{\infty} F_i(\alpha)$, where

$F_i(\alpha)$ is defined as follows: $F_0(\alpha)$ is the unit interval, and $F_{i+1}(\alpha)$ is obtained from $F_i(\alpha)$ by deleting from each interval of $F_i(\alpha)$ with length l an open interval of length $\alpha \cdot l$ situated in the middle; so $F_i(\alpha)$ consists of 2^i intervals, each of length $\left(\frac{1-\alpha}{2}\right)^i$. For $\alpha \in (0,1)$, $F(\alpha)$ is clearly a Cantor set. It follows easily from the definition of limit capacity that

$$d(F(\alpha)) = \frac{\operatorname{Log} 2}{\operatorname{Log} 2 - \operatorname{Log}(1-\alpha)}.$$

So, for $\alpha \in (\frac{1}{2}, 1)$, $\sigma_{F(\alpha), F(\alpha)}$ has measure zero. From Newhouse's criterion it follows that for $\alpha \in (0, \frac{1}{3})$, $\sigma_{F(\alpha), F(\alpha)}$ has positive measure and even contains intervals. F. Thayer has shown us that $\sigma_{F(\alpha), F(\alpha)}$ has still measure zero for $\alpha \in (\frac{1}{3}, \frac{1}{2}]$.

Remark. We shall use later that the conclusion of the proposition remains true if we replace both C_1 and C_2 by a countable union of Cantor sets each with limit capacity not greater than d_1, d_2 respectively.

Next we deal with a situation where $A, B \subset \mathbb{R}$ are closed subsets (not necessarily Cantor sets), both containing zero, but $A \subset \{t \geqq 0\} = \mathbb{R}_+$ and $B \subset \{t \leqq 0\} = \mathbb{R}_-$. Again we are interested in the measure of the set of those t for which $A \cap (B+t) \neq \emptyset$. In this case we assume A and B to have *scaling factors* λ and σ respectively, in the sense that $\lambda \cdot A = A$ and $\sigma \cdot B = B$; λ and σ are supposed to be positive and smaller than one. We define a *scalled* ε*-neighborhood*, $0 < \varepsilon < 1$, of A as

$$A_\varepsilon = \{x \in \mathbb{R} \mid \exists y \in A, |x-y| \leqq \varepsilon \cdot y\};$$

A_ε also has a scaling factor λ. Scaled neighborhoods of B are defined in the same way. In order to have a measure, in the scaled sense, of the distance between A and $(B+t)$, we define

$$\rho(t) = \inf\{\varepsilon \mid A_\varepsilon \cap (B_\varepsilon + t) \neq \emptyset\}.$$

Proposition. *There is a continuous, non-negative function* R, *defined on the torus* $T^2 = \mathbb{R}^2/\mathbb{Z}^2$, *such that*

$$\rho(t) = R\left(\frac{\operatorname{Log} t}{\operatorname{Log} \lambda} \bmod 1, \frac{\operatorname{Log} t}{\operatorname{Log} \sigma} \bmod 1\right).$$

Proof. All intersections of A_ε and $B_\varepsilon + t$ will lie in $[0,t]$. Because of the scaling, $([0,t],[0,t] \cap A_\varepsilon)$ is proportional to

$$([0,1],[0,1] \cap e^{\alpha \cdot \operatorname{Log} \lambda} \cdot A_\varepsilon) \quad \text{if } \alpha = \frac{-\operatorname{Log} t}{\operatorname{Log} \lambda} \bmod 1.$$

The same is true for $([0,t],[0,t] \cap (B_\varepsilon + t))$ and

$$([0,1],[0,1] \cap (e^{\beta \cdot \operatorname{Log} \sigma} \cdot B_\varepsilon + 1)) \quad \text{if } \beta = \frac{-\operatorname{Log} t}{\operatorname{Log} \sigma} \bmod 1.$$

From this it is clear that we can define a function with the required properties by

$$R(\alpha,\beta)=\inf\{\varepsilon \mid e^{\alpha\cdot \mathrm{Log}\,\lambda}\cdot A_\varepsilon \cap (e^{\beta\cdot \mathrm{Log}\,\sigma}\cdot B_\varepsilon+1)\neq\emptyset\}.$$

Definition. We say that closed sets A, B as above with scaling factors λ, σ, such that $\dfrac{\mathrm{Log}\,\lambda}{\mathrm{Log}\,\sigma}$ is irrational, *rarely intersect* if the measure of $R^{-1}(0), R$ as defined above, is zero. In case $\dfrac{\mathrm{Log}\,\lambda}{\mathrm{Log}\,\sigma}=\dfrac{p}{q}$ (p and q relatively prime) then we say that A and B *rarely intersect* if the (1-dimensional) measure of the intersection of $R^{-1}(0)$ with any circle in T^2 of the form

$$\{(q\cdot s \bmod 1, p\cdot s+c \bmod 1)\mid s\in\mathbb{R}\}, \qquad \text{where } c\in[0,1],$$

has measure zero.

Remark. Note that this definition is equivalent to the following one. A and B rarely intersect if for each $\alpha>0$, the set

$$\{t \mid \alpha A\cap(B+t)\neq\emptyset\}$$

has measure zero. So from the first proposition it is clear that if $d(A)+d(B)<1$, then A and B rarely intersect. Here, we use $d(A)$, or $d(B)$, to indicate the limit capacity of a compact neighborhood of 0 in A, B respectively; because of the scaling, the size of the neighborhood is immaterial.

The fact that we require this measure to be zero for all $\alpha>0$, or all $c\in[0,1]$, is because of the applications in the next section.

Definition. We say that a closed set $A(t)$ with scaling factor $\lambda(t)$ *depends continuously on* t if

(i) $A(t)\cap[0,1]$ is a continuous arc of closed sets (with respect to the Housdorff metric);

(ii) $\lambda(t)$ is a differentiable function of t with values in $(0,1)$.

We have to require the scaling factor to depend differentiably on the parameter t for technical reasons (see the proof of the next theorem). In our case the scaling factors are eigenvalues (or their inverses) of the diffeomorphism at the saddle points p_μ and q_μ. Clearly these eigenvalues depend differentiably on the parameter.

In case we have two closed sets $A(t)$, $B(t)$ as above, with scaling factors $\lambda(t)$, $\sigma(t)$, depending continuously on t, the corresponding function

$$R(t)\colon T^2\to\mathbb{R}$$

is defined as

$$(R(t))(\alpha,\beta)=\inf\{\varepsilon \mid e^{\alpha\cdot \mathrm{Log}\,\lambda(t)}\cdot A(t)_\varepsilon\cap(e^{\beta\cdot \mathrm{Log}\,\delta(t)}\cdot B(t)_\varepsilon+1)\neq\emptyset\}.$$

Observe that $R(t)$ depends continuously on t and that hence $(R(t))^{-1}(0)$ is an upper semi-continuous function of t; the same holds for $(R(t))^{-1}[0,\varepsilon]$.

Theorem. *Let $A(t)$ and $B(t)$ be closed sets in $\mathbb{R}_+, \mathbb{R}_-$ respectively with scaling factors $\lambda(t), \sigma(t)$ depending continuously on t. We assume that $A(0)$ and $B(0)$ rarely intersect.*

For every $\varepsilon > 0$, there are $t_0 > 0$ and $\varepsilon_1 > 0$ such that in $(0, t_0]$, the relative measure of those t, for which

$$(A(t)_{\varepsilon_1} \cap ((B(t))_{\varepsilon_1} + t) \neq \emptyset$$

is smaller than ε.

Proof. From the assumptions it follows that $(R(0))^{-1}(0)$ has measure zero. From the previous remarks we see that there is a closed subset $S \subset T^2$ such that

i) $R^{-1}(t)[0, \varepsilon_1] \subset S$ for ε_1 small and for all $t \in [0, t_1]$ for t_1 small;
ii) the measure of S is smaller than $\frac{1}{2}\varepsilon$

$\left(\text{in case } \frac{\operatorname{Log}\lambda(0)}{\operatorname{Log}\sigma(0)} = \frac{p}{q}\right.$, p and q relative prime, we require that the intersection of S with the circle $\{(q \cdot s \bmod 1, p \cdot s \bmod 1) | s \in \mathbb{R}\}$ has measure smaller than $\left.\frac{1}{2}\varepsilon\right)$.

Next we choose a C^3-function $f: T^2 \to \mathbb{R}$ such that:

i) $f \geqq 0$;
ii) $f \geqq 1$ on S;
iii) $\varepsilon_0 = \int_{T^2} f < \varepsilon$.

$\left(\text{In case } \frac{\operatorname{Log}\lambda(0)}{\operatorname{Log}\sigma(0)} = \frac{p}{q}\right.$, p and q relative prime we require the integral of f over the circle $\{(q \cdot s \bmod 1, p \cdot s \bmod 1) | s \in \mathbb{R}\}$ to be smaller than $\left.\varepsilon\right)$.

We shall show that there are (small) values of $t_0 \in (0, t_1)$, such that

$$I(t_0) = \frac{1}{t_0} \int_0^{t_0} f\left(\frac{\operatorname{Log} t}{\operatorname{Log}\lambda(t)}, \frac{\operatorname{Log} t}{\operatorname{Log}\sigma(t)}\right) dt < \varepsilon.$$

From this the theorem follows immediately. As a first step, we prove that there are arbitrarily small positive values of t_0 such that

$$\tilde{I}(t_0) = \frac{1}{t_0} \int_0^{t_0} f\left(\frac{\operatorname{Log} t}{\operatorname{Log}\lambda(0)}, \frac{\operatorname{Log} t}{\operatorname{Log}\sigma(0)}\right) dt < \varepsilon.$$

We prove this only in the case $\frac{\operatorname{Log}\lambda(0)}{\operatorname{Log}\sigma(0)}$ is irrational; the other case is similar but simpler because one works on a circle instead of on a torus.

For the Fourier's expansion

$$f(x_1, x_2) = \Sigma\, a_{m_1, m_2} \cdot e^{2\pi i(x_1 m_1 + x_2 m_2)}$$

we have $a_{0,0} = \varepsilon_0$ (since $\int_{T^2} f = \varepsilon_0$), and $|a_{m_1, m_2}| \leqq C \cdot (m_1^2 + m_2^2)^{-3/2}$ for some constant C (because the Fourier coefficients of the partial derivatives up to order three of f are bounded). With this we obtain for $\tilde{I}(t_0)$:

$$\tilde{I}(t_0)=\frac{1}{t_0}\int_0^{t_0}\Sigma a_{m_1,m_2} e^{2\pi i\left(m_1\frac{\operatorname{Log} t}{\operatorname{Log}\lambda(0)}+m_2\frac{\operatorname{Log} t}{\operatorname{Log}\sigma(0)}\right)} dt.$$

In this expression we substitute $s=\operatorname{Log} t$, so $dt=e^s\,ds$, $s_0=\operatorname{Log} t_0$, and use v, m as a shorthand for the vectors $\left(\frac{1}{\operatorname{Log}\lambda(0)}, \frac{1}{\operatorname{Log}\sigma(0)}\right)$ and (m_1, m_2) respectively; then:

$$\begin{aligned}\tilde{I}(t_0)&=e^{-s_0}\int_{-\infty}^{s_0}\Sigma a_{m_1,m_2} e^{2\pi i\langle m,v\rangle s}\cdot e^s\,ds\\ &=\Sigma a_{m_1,m_2}\cdot\frac{1}{1+2\pi i\langle m,v\rangle}\cdot e^{2\pi i\langle m,v\rangle s_0}.\end{aligned}$$

Define $F\colon T^2\to\mathbb{R}$ by

$$F(x_1,x_2)=\Sigma a_{m_1,m_2}\frac{1}{1+2\pi i\langle m,v\rangle} e^{2\pi i(m_1x_1+m_2x_2)}.$$

Since $|a_{m_1,m_2}|\leqq C\cdot(m_1^2+m_2^2)^{-3/2}$, this sum converges uniformly and hence F is continuous.

Since (i) $\tilde{I}(t_0)=F\left(\frac{\operatorname{Log}(t_0)}{\operatorname{Log}\lambda(0)}, \frac{\operatorname{Log}(t_0)}{\operatorname{Log}\sigma(0)}\right)$, (ii) $\frac{\operatorname{Log}\lambda(0)}{\operatorname{Log}\sigma(0)}$ is irrational, and (iii) $\int_{T^2} F=\varepsilon_0<\varepsilon$, there are arbitrarily small values t_0 such that $\tilde{I}(t_0)<\varepsilon$.

Finally we have to analyse the difference $|I(t_0)-\tilde{I}(t_0)|$. For this we observe that for constants $\bar{C}$ and $\bar{\bar{C}}$ we have

$$\begin{aligned}&\left|f\left(\frac{\operatorname{Log} t}{\operatorname{Log}\lambda(t)}, \frac{\operatorname{Log} t}{\operatorname{Log}\sigma(t)}\right)-f\left(\frac{\operatorname{Log} t}{\operatorname{Log}\lambda(0)}, \frac{\operatorname{Log} t}{\operatorname{Log}\sigma(0)}\right)\right|\\ &\leqq\bar{C}\cdot\left(\left|\frac{\operatorname{Log} t}{\operatorname{Log}\lambda(t)}-\frac{\operatorname{Log} t}{\operatorname{Log}\lambda(0)}\right|+\left|\frac{\operatorname{Log} t}{\operatorname{Log}\sigma(t)}-\frac{\operatorname{Log} t}{\operatorname{Log}\sigma(0)}\right|\right)\\ &=\bar{C}\cdot|\operatorname{Log} t|\cdot\left(\left|\frac{\operatorname{Log}\lambda(0)-\operatorname{Log}\lambda(t)}{\operatorname{Log}\lambda(t)\cdot\operatorname{Log}\lambda(0)}\right|+\left|\frac{\operatorname{Log}\sigma(0)-\operatorname{Log}\sigma(t)}{\operatorname{Log}\sigma(t)\cdot\operatorname{Log}\sigma(0)}\right|\right)\\ &\leqq\bar{\bar{C}}\cdot t\cdot|\operatorname{Log} t|.\end{aligned}$$

The fact that this last function goes to zero when t goes to zero implies that $\lim_{t_0\to 0}|I(t_0)-\tilde{I}(t_0)|=0$. This completes the proof of the theorem.

Remark. Note that we did not state nor prove that for any $t'\in(0,t_0]$, the relative measure of those $t\in(0,t']$, for which

$$(A(t))_{\varepsilon_1}\cap((B(t))_{\varepsilon_1}+t)\neq\emptyset$$

is smaller than ε. In the case where $d(A(0))+d(B(0))<1$, however, we expect this stronger statement to be true.

5. Intersections of stable and unstable foliations

We discuss here the possibilities of tangencies of stable and unstable foliations, depending on a parameter, and their angles of intersection as the parameter evolves.

More specifically we deal with the following situation. $\{\varphi_\mu\}$ is a smooth one-parameter family of diffeomorphisms on a closed 2-manifold. For μ near 0, φ_μ has two saddle points p_μ and q_μ. For $\mu=0$ there is a tangency in x of the unstable separatrix of $p_0=p$ and the stable separatrix of $q_0=q$. This tangency is quasi transversal and unfolds generically in the sense of [21], see also Sect. 2; for $\mu>0$ there are two transverse intersections of $W^u(p_\mu)$ and $W^s(q_\mu)$ near x, for $\mu<0$ there are none.

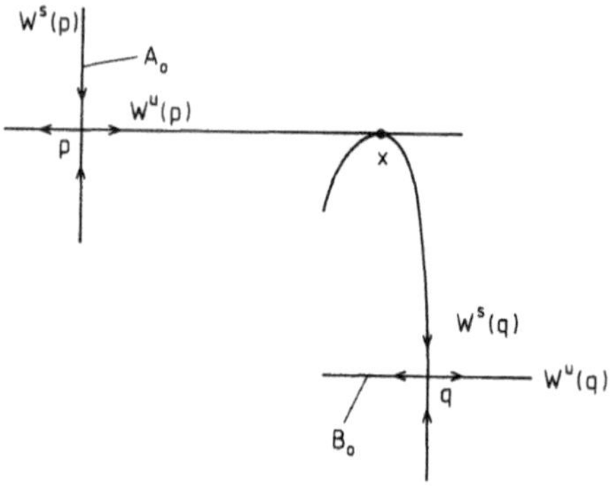

Fig. 5

Furthermore there are closed subsets A_μ, B_μ in $W^s(p_\mu)$, $W^u(q_\mu)$ respectively, with the following properties:

- both A_μ and B_μ are contained in only one branch of $W^s(p_\mu)$, $W^u(q_\mu)$ (including the saddle point p_μ, q_μ respectively) as indicated in the figure (in more detail: if l is a small arc in $W^s(q_0)$ containing x, the point of tangency, then A_0 is contained in the branch of $W^s(p_0)$ on which $\varphi^{-i}(l)$ does not accumulate; the same for B_0);
- A_μ and B_μ are invariant under the diffeomorphism φ_μ; this means that, for a coordinate of $W^s(p_\mu)$ which linearizes $\varphi_\mu|W^s(p_\mu)$, A_μ has a scaling factor $\lambda(\mu)$ which equals the contracting eigenvalue of $d\varphi_\mu$ at p_μ, analoguously for B_μ where the scaling factor $\sigma(\mu)$ is the inverse of the expanding eigenvalue of $d\varphi_\mu$ at q_μ;
- A_μ and B_μ depend continuously on μ in the sense of the Hausdorff metric applied to a compact neighborhood of p_μ, q_μ in $W^s(p_\mu)$, $W^u(q_\mu)$ respectively.

Finally there is for each point $a\in A_\mu$, and $b\in B_\mu$, a leaf $\mathscr{F}^u_{\mu,a}$, $\mathscr{F}^s_{\mu,b}$, through that point. These leaves form the *unstable* and the *stable foliations*, denoted by $\mathscr{F}^u$ and $\mathscr{F}^s$ or $\mathscr{F}^u_\mu$ and $\mathscr{F}^s_\mu$. The foliations $\mathscr{F}^u_\mu$ and $\mathscr{F}^s_\mu$ are continuous. On the other hand the leaves are differentiable curves and the foliations are invariant in the sense that $\varphi(\mathscr{F}^u_{\mu,a})\supset\mathscr{F}^u_{\mu,\varphi(a)}$ and $\varphi^{-1}(\mathscr{F}^s_{\mu,b})\supset\mathscr{F}^s_{\mu,\varphi^{-1}(b)}$; also the tangent directions of the leaves of the unstable (or stable) foliation are uniformly bounded away from the tangent directions of $W^s(p_\mu)$ (or $W^u(q_\mu)$). In Sect. 3 we first considered stable and unstable foliations with bigger domains of definition. In the present section we assume that $\mathscr{F}_\mu^u$ and $\mathscr{F}_\mu^s$ only consist of leaves intersecting A_μ and B_μ.

In order to formulate our theorem, we (may) need to further restrict the domain of definition of the unstable and stable foliation. We only describe this for the unstable foliation; for the stable foliation the construction is the same. Choose a μ-dependent fundamental domain D_μ^s in the branch of $W^s(p_\mu)$ which contains A_μ and a small neighborhood U_μ^s of D_μ^s; we assume that D_μ^s depends continuously on μ and that U_μ^s is a "uniform" family of neighborhoods. Also let W^s be a neighborhood of

- D_0^s and
- an arc in $W^s(p_0)$ from D_0^s to p_0 and
- an arc in $W^u(p_0)$ from p_0 to x as in the figure.

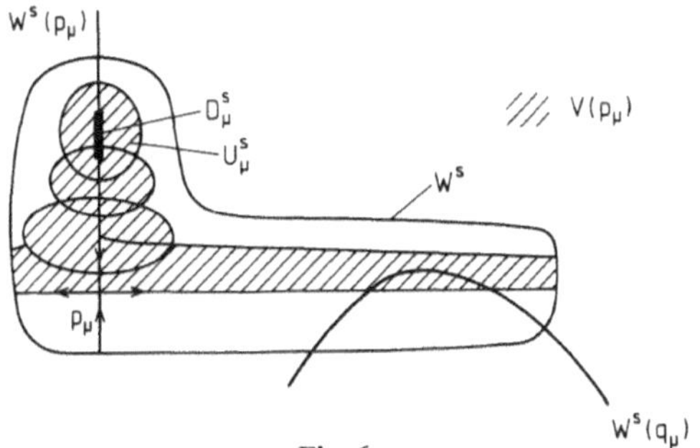

Fig. 6

Then $V(p_\mu)=\bigcup_{i=0}^{\infty}\{q\in W^s|\varphi^{-i}(q)\in U_\mu^s,\ \ \varphi^{-j}(q)\in W^s\, j=0,\dots,i\}$; in other words, $V(p_\mu)$ consists of the forward orbits starting in U_μ^s as long as they do not leave W^s. In the following we shall restrict the unstable foliation to $V(p_\mu)$ (and the stable foliation to a corresponding set $V(q_\mu)$).

We are interested in intersections and tangencies of leaves of the unstable and stable foliations $\mathscr{F}^u$ and $\mathscr{F}^s$ as they occur near x for $\mu>0$. Observe that if μ is positive, but sufficiently small, the disc enclosed by $W^u(p_\mu)$ and $W^s(q_\mu)$ (near x) is completely contained in $V_\mu^s\cap V_\mu^u$.

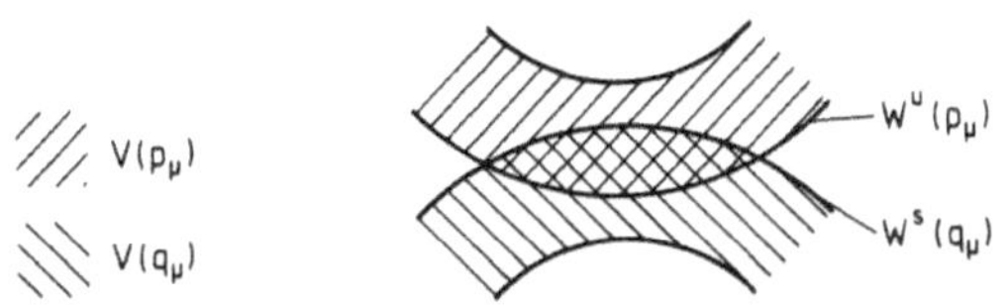

Fig. 7

All intersections of the unstable and the stable foliations near x are in this disc. Hence, for μ small, we don't lose anything when restricting the foliations to $V(p_\mu)$, $V(q_\mu)$.

Theorem. *If, in the above situation, the sum of the limit capacities $d(A_0)+d(B_0)$ is smaller than* 1, *(or, if A_0, B_0, are the countable union of Cantor sets each with limit capacity smaller than or equal to $d(A_0), d(B_0)$ with $d(A_0)+d(B_0)<1$), then for each $\varepsilon>0$ there are $\mu(\varepsilon)>0$ and a subset $S(\varepsilon)\subset[0,\mu(\varepsilon)]$ such that*

i) the measure of $S(\varepsilon)$ is bigger than $(1-\varepsilon)\cdot\mu(\varepsilon)$;

ii) for each $\mu\in S(\varepsilon)$, all intersections of leaves of the unstable and the stable foliations are transverse; in fact, for any given Riemannian metric there is a constant c (independent of μ) such that in all these intersections near x, the angle between the stable and unstable leaves is greater than $c\cdot\sqrt{\mu}$.

Proof. We first prove the theorem under simplifying assumptions and then show how these simplifying assumptions can be removed. So for the moment we assume

i) for μ near zero there are smooth (at least C^2) invariant foliations $\mathcal{G}^u_\mu$ and $\mathcal{G}^s_\mu$ defined on $V(p_\mu)$, $V(q_\mu)$ respectively containing $W^u(p_\mu)$, $W^s(q_\mu)$ as leaves;

ii) each leaf of $\mathcal{F}^u_\mu$ is contained in a leaf of $\mathcal{G}^u_\mu$ and each leaf of $\mathcal{F}^s_\mu$ is contained in a leaf of $\mathcal{G}^s_\mu$.

Let l_μ be the line of tangencies between the leaves of the foliations $\mathcal{G}^u_\mu$ and $\mathcal{G}^s_\mu$; since the tangency is quasi-transverse l_μ is at least C^1, also the projections along leaves from l_μ to $W^s(p_\mu)$ and to $W^u(q_\mu)$ are C^1. $\tilde{A}_\mu$ and $\tilde{B}_\mu\subset l_\mu$ are defined as the sets of intersection points of l_μ with leaves of $\mathcal{G}^u_\mu$, $\mathcal{G}^s_\mu$ containing points of A_μ, B_μ respectively. Let $\tilde{p}_\mu$ and $\tilde{q}_\mu$ denote the points of intersection of $W^u(p_\mu), W^s(q_\mu)$ respectively with l_μ. We choose a C^1 (μ-dependent) coordinate $\xi_\mu: l_\mu\to\mathbb{R}$ such that $\xi_\mu(\tilde{p}_\mu)=0$ and $\xi_\mu(\tilde{q}_\mu)=\mu$ (this is possible because the tangency unfolds generically). Observe that, although A_μ and B_μ have scaling factors $\lambda(\mu), \sigma(\mu)$, this may not be the case for $\tilde{A}_\mu$ and $\tilde{B}_\mu$ (for $\tilde{B}_\mu$ the scaling should of course be centered at μ); this is due to the fact that we can not make both projections of l_μ (namely on $W^s(p_\mu)$ and $W^u(q_\mu)$) linear with respect to ξ_μ (and linearizing coordinates on $W^s(p_\mu)$ and $W^u(q_\mu)$). For this reason we define a scaled approximation $\bar{A}_\mu$ of $\xi_\mu(\tilde{A}_\mu)$:

$$\bar{A}_\mu=\{t\in\mathbb{R}\,|\,\exists s_i\in\xi_\mu(\tilde{A}_\mu) \text{ such that } \lim_{i\to\infty} s_i\,\lambda(\mu)^{-i}=t\}.$$

Observe that if the projection of l_μ on $W^s(p_\mu)$ is linear in the above sense, then $\xi_\mu(\tilde{A}_\mu)=\bar{A}_\mu$. We also define such a scaled approximation $\bar{B}_\mu$ for $\xi_\mu(\tilde{B}_\mu)$, but in order to get a better agreement with the notation in the previous section we make 0 (instead of μ) the center of the scaling of $\bar{B}_\mu$:

$$\bar{B}_\mu=\{t\in\mathbb{R}\,|\,\exists s_i\in\xi_\mu(\tilde{B}_\mu) \text{ such that } \lim_{i\to\infty}(s_i-\mu)\cdot(\sigma(\mu))^{-i}=t\}.$$

We claim that for any $\varepsilon>0$, there are neighborhoods U, V of 0 such that for $\mu\in V$,

(1) $\xi_\mu(\tilde{A}_\mu)\cap U \subset (\bar{A}_\mu)_\varepsilon$ and
(2) $\xi_\mu(\tilde{B}_\mu)\cap U \subset ((\bar{B}_\mu)_\varepsilon + \mu)$;

$(\bar{A}_\mu)_\varepsilon$ and $(\bar{B}_\mu)_\varepsilon$ are scaled neighborhoods.

This claim is a consequence of the following

Lemma. *Let $A \subset \mathbb{R}$ be a closed set with scalling factor λ and let $\Psi: (\mathbb{R},0)\to(\mathbb{R},0)$ be a C^1 diffeomorphism. Then, for every $\varepsilon>0$ there is a neighborhood U of zero such that*

$$\Psi(A)\cap U \subset ((\Psi'(0))\cdot A)_\varepsilon.$$

Proof of the Lemma. Without loss of generality we may assume that $\Psi'(0)=1$. Then $\lim_{x\to 0}\frac{\Psi(x)}{x}=1$. Hence there is a neighborhood U of zero such that $\left|\frac{\Psi(x)-x}{x}\right|<\varepsilon$ for all $x\in U$, so $\Psi(A)\cap U\subset A_\varepsilon$.

Returning to the proof of the theorem (still under the simplifying assumptions), let $\varepsilon>0$ be as in the statement of the theorem. By the theorem of the previous section and the fact that $\bar{A}_0$ and $\bar{B}_0$ intersect rarely we conclude that there are $\mu_0>0$ and $\varepsilon_1>0$ such that the relative measure of those $\mu\in[0,\mu_0]$ for which

$$(\bar{A}_\mu)_{\varepsilon_1}\cap((\bar{B}_\mu)_{\varepsilon_1}+\mu)\neq\emptyset$$

is smaller than ε. Since, in the conclusion of this theorem, we may choose μ_0 arbitrarily small, we may assume that for $\mu\in[0,\mu_0]$:

i) $(\xi_\mu(\tilde{A}_\mu)\cap[0,\mu_0])\subset(\bar{A}_\mu)_{\frac{1}{2}\varepsilon_1}$;
ii) $(\xi_\mu(\tilde{B}_\mu)\cap[0,\mu_0])\subset((\bar{B}_\mu)_{\frac{1}{2}\varepsilon_1}+\mu)$;
iii) the disc, bounded by $W^u(p_\mu)$ and $W^s(q_\mu)$, near x is contained in $V(p_\mu)\cap V(q_\mu)$.

For μ_0, satisfying the above conditions, we define $\mu(\varepsilon)=\mu_0$. $S(\varepsilon)\subset[0,\mu(\varepsilon)]$ consists of those μ for which

$$(\bar{A}_\mu)_{\varepsilon_1}\cap((\bar{B}_\mu)_{\varepsilon_1}+\mu)=\emptyset.$$

Then (i), in the conclusion of the theorem, is true by construction. The first part of (ii) is also true, since the non-transverse intersections of leaves of $\mathscr{F}^u_\mu$ and $\mathscr{F}^s_\mu$ (passing through A_μ, B_μ respectively), must be in $\tilde{A}_\mu\cap\tilde{B}_\mu$, but for $\mu\in S(\varepsilon)$ this is empty. Now we come to the second part of (ii), the estimate for the angle of intersection.

Near x we take (μ-dependent) C^2 coordinates (v,w) such that

- leaves of $\mathscr{G}^u_\mu$ are horizontal lines, i.e., lines of the form $\{w=\text{constant}\}$;
- $l_\mu=\{v=0\}$;
- $\xi_\mu(0,w)=w$, i.e., on l_μ, ξ_μ and w agree.

We shall estimate angles with respect to these coordinates.

Let $\mu\in S(\varepsilon)$ $a\in\tilde{A}_\mu$ and $b\in\tilde{B}_\mu$. Then the distance from a to b is at least $\frac{1}{2}\cdot\varepsilon_1\cdot\mu$. In order to give a lower bound for the horizontal distance from l_μ to the

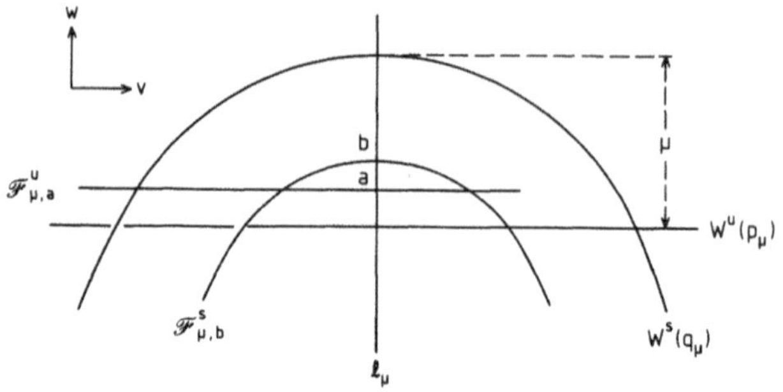

Fig. 8

intersection of $\mathscr{F}^u_{\mu,a}$ and $\mathscr{F}^s_{\mu,b}$, we use that $\mathscr{F}^u_{\mu,a}$ is a horizontal line and that $\mathscr{F}^s_{\mu,b}$ is the graph of a function $w=W(v)$ with $W'(0)=0$ and $W''>-K$ for some positive constant K. From this we deduce that this horizontal distance is at least $\sqrt{\frac{\varepsilon_1\cdot\mu}{K}}$. Due to the fact that, for $\mu=0$, the tangency in x is quasi-transverse, we also have $-K<W''<-K'$, for some positive K'. Then it follows that the angle of intersection between $\mathscr{F}^u_{\mu,a}$ and $\mathscr{F}^s_{\mu,b}$ is (modulo a bounded factor) proportional with the horizontal distance from l_μ to the point of intersection, so the angle is at least

$$c\cdot\sqrt{\varepsilon_1\cdot\mu}=c'\cdot\sqrt{\mu}.$$

Note that ε_1 depends on μ_0, but not on μ. This completes the proof of the theorem under the simplifying assumptions.

Elimination of the symplifying assumptions

(i) C^2 foliations like $\mathscr{G}^u_\mu$ and $\mathscr{G}^s_\mu$ need not exist. However, according to [12] there are such foliations, still denoted by $\mathscr{G}^u_\mu$ and $\mathscr{G}^s_\mu$, which are C^1 and whose leaves form a continuous family of C^2 curves with tangent directions varying C^1. This is enough for our purpose: because the tangent directions vary C^1, l_μ is a C^1 curve; because the foliation is C^1, the projection along leaves is C^1; finally, since the leaves form a continuous family of C^2 curves, the curvature of the leaves varies continuously. Thus we can still estimate the distance between l_μ and the intersections of leaves of $\mathscr{G}^u_\mu$ and $\mathscr{G}^s_\mu$.

(ii) We consider the consequences of $\mathscr{F}^u_\mu$ having leaves not contained in leaves of $\mathscr{G}^u_\mu$; these consideration also apply to $\mathscr{F}^s_\mu$ and $\mathscr{G}^s_\mu$. We have to extend the proof of our theorem in three points. First, for any $\varepsilon>0$ we can make the neighborhood U^s_μ of D^s_μ so small that for each leaf $\mathscr{F}^u_{\mu,a}$, containing a point $a\in D^s_\mu\cap A_\mu$, $\mathscr{F}^u_{\mu,a}\cap U^s_\mu$ is contained in the union of the leaves of $\mathscr{G}^u_\mu$ which pass through $(A_\mu)_\varepsilon$. Due to the positive invariance, this "engulfing" extends to all of $V(p_\mu)$. As a second point, after ε_1 appeared in the proof for the first time, we now replace the conditions (1) and (2) by

$$\xi_\mu((\tilde{A}_\mu)_{\frac{1}{4}\varepsilon_1}) \cap U \subset (\bar{A}_\mu)_{\frac{1}{2}\varepsilon_1} \quad \text{and} \tag{1'}$$

$$\xi_\mu((\tilde{B}_\mu)_{\frac{1}{4}\varepsilon_1}) \cap U \subset (\bar{B}_\mu)_{\frac{1}{2}\varepsilon_1} + \mu, \tag{2'}$$

where $(\tilde{A}_\mu)_{\frac{1}{4}\varepsilon_1}$ is defined to be the set of intersections of $\mathscr{G}^u_\mu$ leaves through $(A_\mu)_{\frac{1}{4}\varepsilon_1}$ with l_μ (and analogously for $(\tilde{B}_\mu)_{\frac{1}{4}\varepsilon_1}$). Then we take U^s_μ so small that the engulfing works inside U^s_μ with $(A_\mu)_{\frac{1}{4}\varepsilon_1}$. A third point is that we have to control the angle between leaves of $\mathscr{G}^u_\mu$ and $\mathscr{F}^u_\mu$. In fact we have to prove that the ratio between this angle and the distance to $W^u(p_\mu)$ goes to zero when approaching $W^u(p_\mu)$. This is the content of the following

Lemma. *Let p be a hyperbolic saddle point of a C^2-diffeomorphism φ on a 2-manifold and let $\mathscr{F}$ and $\mathscr{G}$ be invariant foliations containing $W^u(p)$ as a leaf. We assume that $\mathscr{G}$ is C^1 and that the tangent directions of $\mathscr{F}$ are bounded away from the tangent directions of $W^s(p_\mu)$. Then the ratio between the angle at which $\mathscr{F}$ and $\mathscr{G}$ intersect and the distance to $W^u(p)$ goes to zero when approaching $W^u(p)$.*

Proof. Since we are in a 2-dimensional situation we can take C^1 linearizing coordinates (v, w) so that the leaves of $\mathscr{G}$ are in the v-direction (and so that $W^s(p)$ is the w-axis) [5]. In these coordinates, $\varphi(v,w) = (\rho v, \lambda w)$ with $0 < |\lambda| < 1 < |\rho|$.

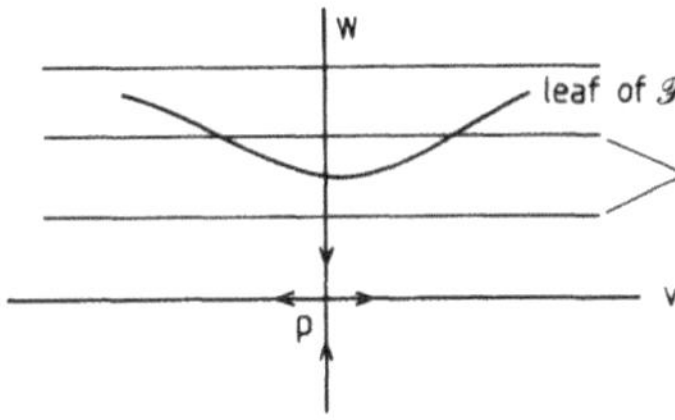

Fig. 9

In a neighborhood of p, the tangent directions of $\mathscr{F}$ can be given as $(1, W(v,w))$ (because these directions are bounded away from the vertical tangent direction of $W^s(p)$). Because $\mathscr{F}$ is invariant, $W(\rho \cdot v, \lambda \cdot w) = \lambda \cdot \rho^{-1} \cdot W(v,w)$. From this the lemma follows immediately.

With this last result, we then also get in the present case the $\sqrt{\mu}$-estimate for the angle between leaves of $\mathscr{F}^u_\mu$ and $\mathscr{F}^s_\mu$. This concludes the proof of our theorem.

6. Hyperbolicity

We consider a sequence of linear maps between 2-dimensional Euclidean spaces

$$\dots W_{i-1} \xrightarrow{A_{i-1}} W_i \xrightarrow{A_i} W_{i+1} \xrightarrow{A_{i+1}} \dots,$$

or $\{A_i\colon W_i \to W_{i+1}\}_{i\in\mathbf{Z}}$ which are hyperbolic except at some "exceptional values" of the index i. We shall prove, under certain conditions, that the non-hyperbolicity at the exceptional indices gets compensated by the overall hyperbolicity.

In detail, these Euclidean spaces are equipped with an orthogonal splitting $W_i = E_i^u \oplus E_i^s$ as a sum of 1-dimensional subspaces such that (at least in the non-exceptional cases),

$$A_i(E_i^u) = E_{i+1}^u,$$
$$A_i^{-1}(E_{i+1}^s) = E_i^s,$$

and such that for $v^u \in E_i^u$, $v^s \in E_i^s$, $\|Av^u\| \geqq \alpha\|v^u\|$ and $\|A^{-1}v^s\| \geqq \alpha\|v^s\|$ for some $\alpha > 1$; we use A instead of A_i, and even A^j for $A_{i+j-1} \circ \ldots \circ A_i$ if no confusion is possible.

For $i \in \text{Exc}$, the set of exceptional indices, the situation is different. In order to describe this we need a number of "constants": λ, λ', σ, σ', c and K, and a "variable" $\mu > 0$. It is important to note that these constants are independent of μ: the main result in this section is that if the conditions below are satisfied then, for μ sufficiently small, the family $\{A_i\}$ is uniformly hyperbolic (in spite of what happens at the exceptional index values). So for $i \in \text{Exc}$ we assume:

i) the angle between $A_i(E_i^u)$ and E_{i+1}^s is at least $c \cdot \mu^{\frac{1}{2}}$;

ii) the angle between $A_i^{-1}(E_{i+1}^s)$ and E_i^u is at least $c \cdot \mu^{\frac{1}{2}}$;

iii) the distortions of A_i and A_{i-1} are at most K, i.e.

$$\max(\|A_i\|, \|A_i^{-1}\|, \|A_{i-1}\|, \|A_{i-1}^{-1}\|) \leqq K;$$

iv) for $j \leqq \dfrac{\operatorname{Log}\mu}{\operatorname{Log}\sigma}$, $v^u \in E_{i+1}^u$ and

$$v^s \in E_i^s, \|A^{-j}v^u\| \leqq (\lambda')^j\|v^u\| \quad \text{and} \quad \|A^{-j}v^s\| \geqq \lambda^{-j}\|v^s\|$$

$\left(\dfrac{\operatorname{Log}\mu}{\operatorname{Log}\lambda} \text{ is also denoted by } T^{\text{entr}}\right)$;

v) for $j \leqq \dfrac{\operatorname{Log}\mu}{\operatorname{Log}\lambda}$, $v^u \in E_i^u$ and

$$v^s \in E_i^s, \|A^{-j}v^u\| \leqq (\lambda')^j\|v^u\| \quad \text{and } \|A^{-j}v^s\| \geqq \lambda^{-j}\|v^s\|$$

$\left(\dfrac{\operatorname{Log}\mu}{\operatorname{Log}\lambda} \text{ is also denoted by } T^{\text{entr}}\right)$;

vi) if $i' \neq i$ and $i' \in (i - T^{\text{entr}}, i + 1 + T^{\text{ex}})$, then $i' \notin \text{Exc}$.

This situation is abstracted from the case where we have a diffeomorphism φ_μ as described before:

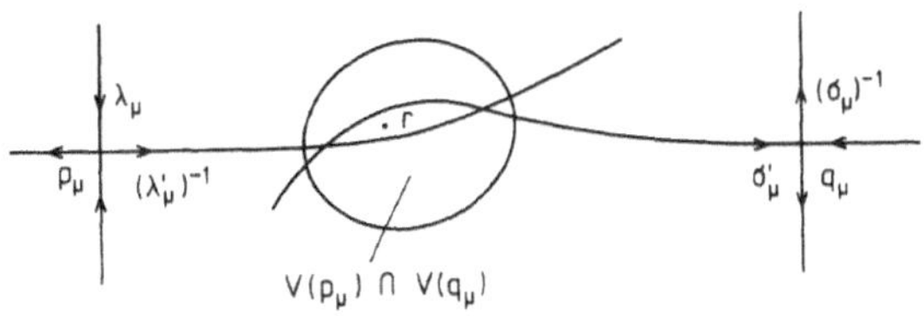

Fig. 10

Let $r \in V(p_\mu) \cap V(q_\mu)$ be a point such that $\varphi^i_\mu(r)$ returns to $V(p_\mu) \cap V(q_\mu)$ for arbitrarily big positive and negative values of i. If $\varphi^i_\mu(r)$ is not in $V(p_\mu) \cap V(q_\mu)$, then it is in the hyperbolic region, i.e., in the region where the stable and unstable foliations $\mathscr{F}^s_\mu$ and $\mathscr{F}^u_\mu$ are defined. In order to see the relation with the situation in the beginning of this section, one has to make the following identifications:

$$W_i = T_{\varphi^i(r)}(M);$$

$$A_i = d\varphi_\mu | T_{\varphi^i_\mu(r)};$$

$$\mathrm{Exc} = \{i \in \mathbb{Z} \,|\, \varphi^i_\mu(r) \in V(p_\mu) \cap V(q_\mu)\};$$

$$E^u_i = \begin{cases} T_{\varphi^i_\mu(r)}(\mathscr{F}^u_\mu) & \text{for } i \notin \mathrm{Exc}, \\ A_{i-1}(E^u_{i-1}) & \text{for } i \in \mathrm{Exc}; \end{cases}$$

$$E^s_i = \begin{cases} T_{\varphi^i_\mu(r)}(\mathscr{F}^s_\mu) & \text{for } i \notin \mathrm{Exc}, \\ A^{-1}_i(E^s_{i+1}) & \text{for } i \in \mathrm{Exc}. \end{cases}$$

We assume here that $\varphi_\mu(V(p_\mu) \cap V(q_\mu)) \cap (V(p_\mu) \cap V(q_\mu)) = \emptyset$. Also, λ, λ', σ, σ' are the contracting eigenvalues or their inverses, as indicated in the figure, for $\mu = 0$ (the dependence of these eigenvalues on μ is very inessential in this section). Finally, T^{ex} and T^{entr} correspond to the time which an orbit like $\varphi^i_\mu(r)$ spends in a neighborhood of q_μ (going forward) or in a neighborhood of p_μ (going backward) after (or before) being in $V(p_\mu) \cap V(q_\mu)$: this is a consequence of the fact that such an orbit, when passing through $V(p_\mu) \cap V(q_\mu)$, is at most at distance μ from $W^u(p_\mu)$ and $W^s(q_\mu)$. This explains the conditions (iv) and (v). Also condition (vi) is related with this: as long as an orbit stays near p_μ or q_μ it does not reenter $V(p_\mu) \cap V(q_\mu)$.

The above identifications are somewhat optimistic: due to nonlinearity we may be forced to choose λ, λ', σ and σ' somewhat closer to 1 and to choose T^{ex} and T^{entr} somewhat smaller. Since, however, in the proof of the following theorem the estimates allow such adjustments of the "constants", even a replacement of T^{ex} and T^{entr} by $\dfrac{\operatorname{Log}\mu}{\operatorname{Log}\sigma} - \bar{c}$, $\dfrac{\operatorname{Log}\mu}{\operatorname{Log}\sigma} - \bar{c}$ for a μ-independent constant $\bar{c}$, such modifications are immaterial for what follows (at least for μ small).

We also point out that the validity of the conditions (i) and (ii) in the present interpretation are a consequence of the fact that we could estimate the angle of intersection of $\mathscr{F}_\mu^u$ and $\mathscr{F}_\mu^s$, at least for $\mu \in S(\varepsilon)$ (see the theorem in the previous sections), as being at least of order $\mu^{\frac{1}{2}}$.

Finally, the Euclidean structures in W_i can be introduced in such a way that the conditions (i), ..., (vi) are satisfied and so that they are compatible with a Riemannian structure on the 2-manifold (in the sense that there are uniform estimates for the ratio of the lengths of vectors according to such a Riemannian metric and according to these Euclidean structures). In particular, the fact that we can satisfy condition (iii) is due to the compactness of $V(p_\mu) \cap V(q_\mu)$: there is a maximal distortion when an orbit goes through this set.

The consequence of the next theorem is that the orbits which stay in the hyperbolic region except when they pass through $V(p_\mu) \cap V(q_\mu)$ are *uniformly hyperbolic*. A detailed exposition of a proof of such hyperbolicity can also be found in [19, 20].

Theorem. *In the above situation, when $\mu > 0$ is sufficiently small $\{A_i\}_{i \in Z}$ is uniformly hyperbolic in the sense that there are splittings $\{W_i = F_i^u \oplus F_i^s\}_{i \in Z}$ and constants $\beta > 1$, $n \in N$, so that*

(a) $A_i(F_i^u) = F_{i+1}^u$;
(b) $A_i(F_i^s) = F_{i+1}^s$;
(c) *for* $v \in F_i^u$, $j \geqq n$, $\|A^j(v)\| \geqq \beta^j \cdot \|v\|$;
(d) *for* $v \in F_i^s$, $j \geqq n$, $\|A^j(v)\| \leqq \beta^{-j} \cdot \|v\|$.

β and n depend only on λ, λ', σ, σ', α, c, K and μ (see the comments above).

Proof. We shall only prove the expanding part, i.e. the existence of $F_i^u \subset W_i$ satisfying (a) and (c); the contracting part can be obtained by applying the same arguments to the inverse mappings. We shall prove it by constructing a sequence of cones $C_i^+ \subset W_i$ such that $A_i(C_i^+) \subset C_{i+1}^+$ and such that the length of each vector in C_i^+ grows exponentially under repeated application of A. The existence of $F_i^u \subset W_i$ then follows: just take a vector $0 \neq w \in W_0$ such that $A^i(w) \in C_i^+$ for all $i \in \mathbb{Z}$, positive and negative, and F_i^u the linear subspace generated by $A^i(w)$.

For the construction of the cones, we use orthogonal coordinates ξ_i, η_i on W_i such that E_i^u, E_i^s is the ξ_i-axis and η_i-axis, respectively.

We begin the construction of the cones in any W_i with $i \in \mathrm{Exc}$. We take $C_i^+ = \{|\eta_i| \leqq d \cdot \mu^{\frac{1}{2}} \cdot |\xi_i|\}$ and

$$C_{i+1}^+ = \{|\eta_{i+1}| \leqq D \cdot \mu^{-\frac{1}{2}} \cdot |\xi_{i+1}|\},$$

where d is so small and D so big (both independent of μ) that for μ sufficiently small, $A_i(C_i^+) \subset C_{i+1}^+$; this is possible because of condition (i). For $j \leqq T^{\mathrm{ex}} = \dfrac{\operatorname{Log} \mu}{\operatorname{Log} \sigma}$, we take $A^j(C_{i+1}^+) = C_{i+j+1}^+$. Then, for j equal to the integer part of T^{ex}, we have

$$C_{i+j+1}^+ = \{|\eta_{i+j+1}| \leqq \gamma \cdot |\xi_{i+j+1}|\},$$

where

$$\gamma \leqq D \cdot \mu^{-\frac{1}{2}} \cdot (\sigma \cdot \sigma')^{\frac{\operatorname{Log}\mu}{\operatorname{Log}\sigma}} = D \cdot \mu^{-\frac{1}{2}} \cdot \mu^{\frac{\operatorname{Log}\sigma + \operatorname{Log}\sigma'}{\operatorname{Log}\sigma}}.$$

Since $\frac{\operatorname{Log}\sigma+\operatorname{Log}\sigma'}{\operatorname{Log}\sigma}>1$, we have, for μ sufficiently small, $\gamma<d\cdot\mu^{\frac{1}{2}}$. This enables us to define C_l^+ for $l>i+T^{\mathrm{ex}}$, but l smaller than the next exceptional index:

$$C_l^+=\{|\eta_l|\leqq d\cdot\mu^{\frac{1}{2}}\cdot|\xi_l|\}.$$

In this way all the cones are defined so that $A_i(C_i^+)\subset C_{i+1}^+$ for μ sufficiently small.

In order to show that the vectors in C_i^+ grow exponentially, we only consider their ξ_i components. These grow exponentially except at exceptional indices. At exceptional indices there is a contraction which is at most of order $\mu^{\frac{1}{2}}$; this follows again from condition (i). This contraction is then followed by $T^{\mathrm{ex}}=\frac{\operatorname{Log}\mu}{\operatorname{Log}\sigma}$ iterates with expansion σ^{-1}. So the total expansion is at least

$$\mu^{\frac{1}{2}}\cdot\sigma^{-\frac{\operatorname{Log}\mu}{\operatorname{Log}\sigma}}=\mu^{\frac{1}{2}}\cdot\mu^{-1}=\mu^{-\frac{1}{2}}.$$

Hence the contraction at an exceptional index is well compensated in the next T^{ex} iterates in which no exceptional index occurs. This implies the exponential growth; the proof of the theorem is therefore complete.

References

G.A. Global analysis. Proc. Symp. Pure Math., vol. 14; Rhode Island: Am. Math. Soc. Providence, 1970

D.S. Dynamical systems. New York: Academic Press 1973

1. Abraham, R., Smale, S.: Nongenericity of Ω-stability. (In [G.A.])
2. Arnold, V.: Small denominators, I. Trans. Am. Math. Soc. (2nd series) **46**, 213–284 (1965)
3. Benedicks, M., Carleson, L.: On iterations of $1-ax^2$ on $(-1,+1)$. Institut Mittag-Leffler 1983 (Preprint)
4. Bowen, R., Ruelle, D.: The ergodic theory of Axiom A flows. Invent. Math. **29**, 181–202 (1975)
5. Hartman, P.: Ordinary differential equations. New York: Wiley 1964
6. Herman, M.: Mesure de Lebesgue et nombre de rotation. In: Geometry and Topology, Proceedings, Lect. Notes Math., vol. **597**. Berlin Heidelberg New York: Springer 1977
7. Hirsch, M., Pugh, C.: Stable manifolds and hyperbolic sets. (In [G.A.])
8. Hirsch, M., Palis, J., Pugh, C., Shub, M.: Neighbourhoods of hyperbolic sets. Invent. Math. **9**, 121–134 (1969/70)
9. Jacobson, M.V.: Absolutely continuous invariant measures for one-parameter families of one-dimensional maps. Commun. Math. Phys. **81**, 39–88 (1981)
10. Mañé, R.: On the dimension of the compact invariant sets of certain non-linear maps. In: Dynamical Systems and Turbulence, Proceedings. Lect. Notes Math., vol. **898**. Berlin Heidelberg New York: Springer 1980
11. De Melo, W.: Structural stability of diffeomorphisms on two-manifolds. Invent. Math. **21**, 233–246 (1973)
12. De Melo, W., Van Strien, S.: Diffeomorphisms with a finite number of moduli on surfaces. IMPA, Rio de Janeiro 1983 (Preprint)
13. Misiurewicz, M.: Absolutely continuous measures for certain maps of an interval. Publ. I.H.E.S. **53**, 17–52 (1981)
14. Newhouse, S.: Nondensity of axiom $A(a)$ on S^2. (In [G.A.])
15. Newhouse, S.: Hyperbolic limit sets. Trans. Am. Math. Soc. **167**, 125–150 (1972)
16. Newhouse, S.: Diffeomorphisms with infinitely many sinks. Topology **13**, 9–18 (1974)

17. Newhouse, S.: The abundance of wild hyperbolic sets and non-smooth stable sets for diffeomorphisms. Publ. I.H.E.S. **50**, 101–151 (1979)
18. Newhouse, S., Palis, J.: Hyperbolic nonwandering sets on two-dimensional manifolds. (In [D.S.])
19. Newhouse S., Palis, J.: Bifurcations of Morse-Smale dynamical systems. (In [D.S.])
20. Newhouse, S., Palis, J.: Cycles and bifurcation theory. Astérisque **31**, 44–140 (1976)
21. Newhouse, S., Palis, J., Takens, F.: Stable families of diffeomorphisms. Publ. I.H.E.S. **57**, 5–72 (1983)
22. Palis, J.: A note on Ω-stability. (In [G.A.])
23. Palis, J., Smale, S.: Structural stability theorems. (In [G.A.])
24. Peixoto, M.: Structural stability on two-manifolds. Topology **1**, 101–120 (1962)
25. Pliss, V.: On the structural stability of differential equations on the torus (in Russian). Vestn. Leningr. Univ. (M.R. 23, A 3884) **15**, 15–23 (1960)
26. Rees, M.: Positive measure sets of ergodic rational maps. Univ. of Minnesota 1982 (Preprint)
27. Robbin, J.: A structural stability theorem. Ann. Math. **94**, 447–493 (1971)
28. Robinson, C.: C^r structural stability implies Kupka-Smale. (In [D.S.])
29. Robinson, C.: Structural stability of C^1 diffeomorphisms. J. Differ. Equations **22**, 28–73 (1976)
30. Robinson, C.: Bifurcation to infinitely many sinks. Northwestern University 1982–1983 (Preprint)
31. Ruelle, D.: Applications conservant une mesure absolument continue par rapport à dx sur $[0,1]$. Commun. Math. Phys. **55**, 47–51 (1977)
32. Simon, C.: Instability in $\mathrm{Diff}^r(T^3)$ and the nongenericity of rational zeta functions. Trans. Am. Math. Soc. **174**, 217–242 (1972)
33. Smale, S.: Structurally stable systems are not dense. Am. J. Math. **88**, 491–496 (1966)
34. Smale, S.: Differentiable dynamical systems. Bull. Am. Math. Soc. **73**, 747–817 (1967)
35. Smale, S.: The Ω-stability theorem. (In: [G.A.])
36. Sullivan, D.: Quasi conformal homeomorphisms and dynamics III – Topological conjugacy classes of analytic endomorphisms. Ann. Math. (Submitted)
37. Takens, F.: Distinguishing deterministic and random systems. In: Barenblatt, G.I., Ioos, G., Joseph, D.D. (eds.): Non-linear Analysis and Mechanics. London: Pitman 1983
38. Williams, R.: The "*DA*" maps of Smale and structural stability. (In [G.A.])

Oblatum 9-XI-1984 & 8-VII-1985

Annals of Mathematics, **125** (1987), 337–374

Hyperbolicity and the creation of homoclinic orbits

By J. Palis and F. Takens*

Abstract

We consider one-parameter families $\{\varphi_\mu;\ \mu \in \mathbf{R}\}$ of diffeomorphisms on surfaces which display a *homoclinic tangency* for $\mu = 0$ and are *hyperbolic* for $\mu < 0$ (i.e., φ_μ has a hyperbolic nonwandering set); the tangency unfolds into *transversal homoclinic orbits* for μ positive. For many of these families, we prove that φ_μ is also hyperbolic for most small positive values of μ (which implies much regularity of the dynamical structure). A main assumption concerns the *limit capacities* of the basic set corresponding to the homoclinic tangency.

1. Introduction

In 1899, Poincaré first detected the existence of what he called *homoclinic orbits* (doubly asymptotic solutions) when studying the restricted three-body problem [12]. He expressed amazement about the dynamical complexity implied by the presence of what seemed to be a rather simple phenomenon. To recall this concept, let M be a 2-manifold, φ: $M \to M$ a diffeomorphism and p a fixed point for φ. We say that x is a *homoclinic point* associated with p if $\varphi^n(x) \to p$ for both $n \to +\infty$ and $n \to -\infty$; that is, when x has p as its ω and α-limit sets. When p is a periodic point of period k, we can define homoclinic points or orbits in a similar way by considering φ^k. We will be particularly interested in the case where p is a fixed (hyperbolic) *saddle*, i.e., when one eigenvalue of $d\varphi(p)$ has norm bigger than one and the other smaller than one. A homoclinic point means, in this situation, a point of intersection of the *stable* and *unstable manifolds* of p (see [3]). Depending on this intersection, we have a *transversal homoclinic point* or a *homoclinic point of tangency*. In both cases it is easy to see that a homoclinic point, say x, is *nonwandering*: for any given neighborhood U of x, there is an arbitrarily large integer n such that $\varphi^n U \cap U \neq \varnothing$. The

*The first author acknowledges hospitality and financial help from Rijksuniversiteit Groningen/Z.W.O./K.N.A.W. during the preparation of this paper. Similarly, the second author is thankful to IMPA/CNPq.

Originally published in: *Ann. of Math. (2) 125 (1987), no. 2, 337–374,*

union of the nonwandering points is called the nonwandering set, which is closed and denoted by $\Omega(\varphi)$.

Around 1935 Birkhoff proved the following relevant result in this context: every transversal homoclinic orbit of a 2-dimensional diffeomorphism is accumulated by periodic orbits [1]. In 1965, Smale generalized this result both in two and higher dimensions by showing that a transversal homoclinic orbit is contained in a hyperbolic set (in this case a "horseshoe") in which the periodic orbits are dense [15]. By a *hyperbolic set* we mean a closed, φ-invariant subset Λ of M such that the tangent bundle $T_\Lambda M$ splits into $d\varphi$-invariant sub-bundles E^s and E^u on which $d\varphi$ is uniformly contracting and uniformly expanding, respectively. A key fact is that through each point $x \in \Lambda$ we can construct a pair of "leaves", which are called stable and unstable manifolds of x and denoted by $W^s(x)$ and $W^u(x)$. The union of these leaves forms (partially defined) *foliations* denoted by $\mathscr{F}^s(\Lambda)$ and $\mathscr{F}^u(\Lambda)$ (see [3]). They have smooth leaves (C^r if φ is C^r), they are φ-invariant (φ sends leaves into leaves) and they are characterized as follows: given $x \in \Lambda$, the leaf $W^s(x)$ or $\mathscr{F}^s(x)$ consists of points y such that $d(f^n, y, f^n x) \to 0$ as $n \to \infty$, where d is a distance function of M; dually for $\mathscr{F}^u(\Lambda)$.

The results of Birkhoff and Smale confirm Poincaré's observation about how rich the dynamics of a transformation seemed to be because of the mere existence of a (transversal) homoclinic orbit. In the present paper we want to analyse, for a family of C^2 diffeomorphisms $\{\varphi_\mu;\ \mu \in \mathbf{R}\}$ of a closed surface, the global dynamical implications of creating a homoclinic orbit (of tangency) say, for $\mu = 0$ and letting it unfold to become transversal. It is well-known that such a process implies infinitely many changes (bifurcations) of the dynamical structure no matter how small we take the interval $(-\delta, \delta)$ of a parameter range. The set of values of μ for which these changes occur (*bifurcation set*) even contains intervals! This important fact is due to Newhouse [8]. However, as we show here, in many relevant cases the *relative measure* of the bifurcation set is very small, asymptotically going to zero with the diameter of the parameter range. Moreover φ_μ has no "new" attractors when μ is not in the bifurcation set, so not much "chaos" occurs in this 1-parameter family for μ near zero. We now turn to a description of the one-parameter families we intend to study. First, we assume that $\Omega(\varphi_\mu)$ is *persistently hyperbolic* for $\mu < 0$. (We will see in Chapter 4 that this implies Ω-stability.) Also, we suppose that $\Omega(\varphi_0)$ is a (closed) hyperbolic set together with a newly created homoclinic orbit of tangency. A second main and perhaps crucial ingredient concerns the *limit capacities* (capacity dimensions) of the stable and unstable foliations of the (basic) hyperbolic set that gives rise to the homoclinic tangency; their sum should be smaller than one.

Let us recall a few relevant concepts including the ones we have just mentioned, and then present the precise statement of our main result. For a diffeomorphism $\varphi: M \to M$, a closed subset $\Lambda \subset \Omega(\varphi)$ is called a *basic set* if it is hyperbolic, transitive (it has a dense orbit) and its subset of periodic orbits is dense. The basic sets that appear in our context are open in $\Omega(\varphi_\mu)$ for $\mu \leq 0$, except that for $\mu = 0$ there is *one* which is a limit as $\mu \nearrow 0$ of such a basic set for $\mu < 0$ and which has a saddle point with a homoclinic tangency. See the first section of Chapter 4 for more details. We define φ to be *hyperbolic* if $\Omega(\varphi)$ is hyperbolic, and *persistently hyperbolic* if every $\tilde{\varphi}$ which is C^r near φ ($r \geq 1$) is also hyperbolic. Here we will consider C^2 diffeomorphisms of a closed surface M endowed with the C^2 topology. We will show in Chapter 4 that if φ is persistently hyperbolic then φ is Ω-stable. That is, for every $\tilde{\varphi}$ near φ there is a homeomorphism $h: \Omega(\varphi) \to \Omega(\tilde{\varphi})$ such that $h\varphi = \tilde{\varphi}h$. A relevant fact about a C^2 diffeomorphism on a surface concerns the stable and unstable foliations of a basic set Λ: $\mathscr{F}^s(\Lambda)$ and $\mathscr{F}^u(\Lambda)$ can be extended to C^1 invariant foliations defined on a full neighborhood of Λ ([3]). This fact allows us to define intrinsically the following concept of "transverse dimensions" for $\mathscr{F}^s(\Lambda)$ and for $\mathscr{F}^u(\Lambda)$, which we call stable and unstable limit capacities of Λ. First, for a compact metric space C we define the *limit capacity* $d(C)$ as $\limsup \operatorname{Log} n(C, \varepsilon)/\operatorname{Log} 1/\varepsilon$ as $\varepsilon \to 0$, where $n(C, \varepsilon)$ is the minimum number of balls of radius ε needed to cover C. Notice that $d(C) = d(f(C))$ if f is a Lipschitz homeomorphism with Lipschitz inverse. For $\mathscr{F}^s(\Lambda)$ we consider, through a point of a leaf, a small transversal segment L and let C_L be the set of intersection points of the leaves with L. We then define the *stable limit capacity* $d^s(\Lambda)$ as $d(C_L)$. The above remark, on extending $\mathscr{F}^s(\Lambda)$ to a C^1 foliation on a neighborhood of Λ, shows as in [8], [11] that this definition does not depend on the transversal segment L. A similar procedure for $\mathscr{F}^u(\Lambda)$ gives the *unstable limit capacity* $d^u(\Lambda)$. If Λ is of *saddle type* (i.e., neither the leaves of $\mathscr{F}^s(\Lambda)$ nor those of $\mathscr{F}^u(\Lambda)$ are contained in Λ) and does not consist of a single periodic orbit, then C_L is a Cantor set and thus the stable limit capacity $d^s(\Lambda)$ can take any value between zero and one. Similarly for $d^u(\Lambda)$.

We now describe the parametrized families of diffeomorphisms which we consider. We say that a 1-parameter family $\varphi_\mu: M \to M$ of C^2 diffeomorphisms of a closed surface M has a *homoclinic Ω-explosion* at $\mu = 0$ if:

i) For $\mu < 0$, φ_μ is persistently hyperbolic;

ii) For $\mu = 0$, the nonwandering set $\Omega(\varphi_0)$ consists of a (closed) hyperbolic set $\tilde{\Omega}(\varphi_0) = \lim_{\mu \nearrow 0} \Omega(\varphi_\mu)$ together with a homoclinic orbit of tangency $\mathscr{O}$ associated with a fixed saddle point p, so that $\Omega(\varphi_0) = \tilde{\Omega}(\varphi_0) \cup \mathscr{O}$; the product of the eigenvalues of $d\varphi_0$ at p is different from one in norm;

iii) The separatrices have quadratic tangency along $\mathcal{O}$ unfolding generically (see Chapter 2); $\mathcal{O}$ is the only orbit of tangency between stable and unstable separatrices of periodic orbits of φ_0.

Since $\Omega_\mu = \Omega(\varphi_\mu)$ is hyperbolic for $\mu < 0$, we can write Ω_μ as a finite disjoint union of basic sets $\Lambda_1, \ldots, \Lambda_r$ (see Section 1, Chapter 4). From (ii) above, we have $p = \lim_{\mu \nearrow 0} p(\mu)$ where $p(\mu) \in \Lambda_i(\mu)$ for some basic set $\Lambda_i(\mu)$. Let us denote $\Lambda_i(0) = \lim_{\mu \nearrow 0} \Lambda_i(\mu)$ simply by Λ and call it the basic set associated with the homoclinic tangency $\mathcal{O}$. Finally, let us denote by B the set of values of $\mu > 0$ for which φ_μ is not persistently hyperbolic. As mentioned before, we will see in Chapter 4 that this set contains the set of μ's for which φ_μ is not Ω-stable (Ω-bifurcation set).

Main Theorem. *Let $\{\varphi_\mu;\ \mu \in \mathbf{R}\}$ be a family of diffeomorphisms of M with a homoclinic Ω-explosion at $\mu = 0$. Suppose that $d^s(\Lambda) + d^u(\Lambda) < 1$, where Λ is the basic set of φ_0 associated with the homoclinic tangency. Then*

$$\lim_{\delta \to 0} \frac{m(B \cap [0, \delta])}{\delta} = 0$$

where m denotes Lebesgue measure.

The theorem above considerably extends a result in [7] in two ways. First, in that paper the families $\{\varphi_\mu\}$ were such that $\Omega(\varphi_\mu)$ was finite (and hyperbolic), which we do not assume here. To deal with a non-trivial basic set (one that is not reduced to a periodic orbit) giving rise to a homoclinic tangency, we introduce the (transverse) limit capacities of the stable and unstable foliations. We then draw the conclusion that, if these capacities are not too large, we get mainly hyperbolic diffeomorphisms right after the unfolding of the tangency. Second, the result in [7] stated that $\liminf_{\delta \to 0} m(B \cap [0, \delta]/\delta = 0$, instead of the full limit as in the present paper. Another related, but more recent, work is [11], where we dealt with similar questions in the context of heteroclinic tangencies and larger cycles, i.e., when more than one basic set is involved (see Chapter 4). Also in that last paper we only proved that $\liminf_{\delta \to 0} m(B \cap [0, \delta])/\delta = 0$.

The content of the next chapters of this paper are as follows. In Chapter 2 we present a catalog of all possible types of 1-parameter families with homoclinic Ω-explosion that can exist on surfaces. The proof of the main result is in Chapter 4. It involves determining the locus of the nonwandering set and then establishing its hyperbolicity when there is a certain (positive) minimal angle between the leaves of the stable and unstable foliations of the basic set involved. The fact that this condition is frequently satisfied for μ positive and small depends heavily on Chapter 3. There, the assumption on the capacities is used to analyse intersection of (Cantor) sets when shifting one across the other. A special difficulty in our context is that the sets depend (although differentiably) on the parameter μ.

We finish by conjecturing a converse to our main result implying a strong correlation between hyperbolicity and limit capacity.

Conjecture. Suppose $\{\varphi_\mu\}$ has a homoclinic Ω-explosion at $\mu = 0$ and let Λ be the basic set associated with the homoclinic tangency of φ_0. Let $d^s(\Lambda)$ and $d^u(\Lambda)$ be the stable and unstable limit capacities of Λ. If $d^s(\Lambda) + d^u(\Lambda) > 1$, then

$$\liminf_{\delta \to 0} \frac{m(B \cap [0, \delta])}{\delta} > 0.$$

As usual, one may need some "generic" conditions on the family $\{\varphi_\mu\}$ to be able to verify the conjecture.

We can construct examples that corroborate this conjecture. For this we assume the sum of the stable and unstable *Hausdorff dimensions* to be greater than one; however, in a personal communication, F. Ledrapier and L. S. Young told us that in our context, limit capacity and Hausdorff dimension are equal. To get the examples we made much use of results in [4] about sets of fractional dimensions.

2. Classification of homoclinic Ω-explosions in dimension 2

1. *Introduction*. We consider one parameter families $\varphi_\mu \colon M \to M$ of diffeomorphisms with homoclinic Ω-explosion at $\mu = 0$ as defined in the introduction. We also denote the saddle point, corresponding to the homoclinic tangency for φ_0, by p; the continuation of this saddle point for φ_μ is denoted by p_μ.

We recall that the homoclinic tangency is supposed to satisfy some generic properties. In the first place the *tangency* should be of *first order* not of *second order*. This means that if r is in the orbit of tangency, then $W^u(p)$ and $W^s(p)$ have different curvature in r. We also assume that this *tangency unfolds generically*; e.g. see [9]. This means that near a point of tangency of φ, the separation between $W^u(p_\mu)$ and $W^s(p_\mu)$ is as indicated below:

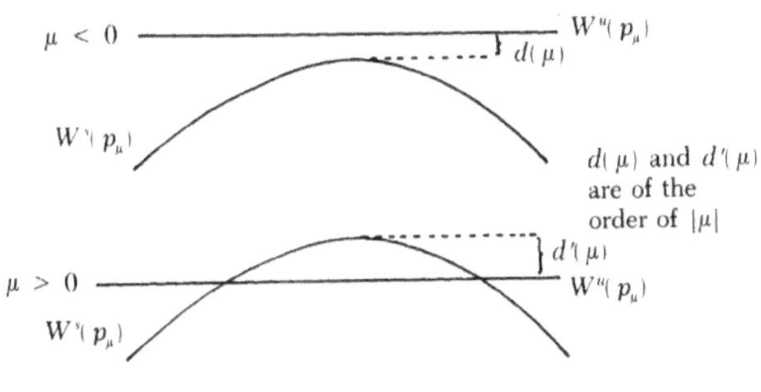

Finally, we assume that the product of the eigenvalues of $(d\varphi_0)_p$ has norm different from 1.

In the present chapter we shall distinguish these homoclinic Ω-explosions according to the following criteria, all of which refer to φ_0 (from now on we also denote φ_0 by φ):

a) *The signs of the eigenvalues* of $(d\varphi)_p$: here we have three possibilities, $++$, $+-$, and $--$ (in the $+-$ case one can of course distinguish between the cases where the positive eigenvalue is expanding or contracting but one can change from one case to the other by replacing φ by φ^{-1});

b) *The sides of tangencies*: by this we mean the following: if $q \in W^s_{\text{loc}}(p)$ and $r \in W^u_{\text{loc}}(p)$ are in the orbit of tangency (loc indicates that we take only a small interval around p in the corresponding separatrix), then the position of $W^u(p)$ near q and $W^s(p)$ near r can be as in the figures below:

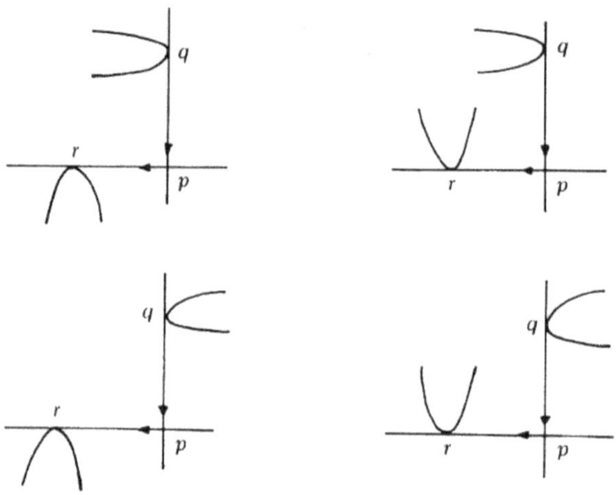

We refer to these differences as the different sides of tangency (if at least one of the eigenvalues of $(d\varphi)_p$ is negative we should take two successive points in the orbit of tangency in both $W^u_{\text{loc}}(p)$ and $W^s_{\text{loc}}(p)$);

c) *The connections*: these concern the distinction that still can be made if the sides of tangencies are known; for example if they are as in the first of the above cases, then the (global) unstable separatrix of p can still connect in two different ways with the piece of stable separatrix near q:

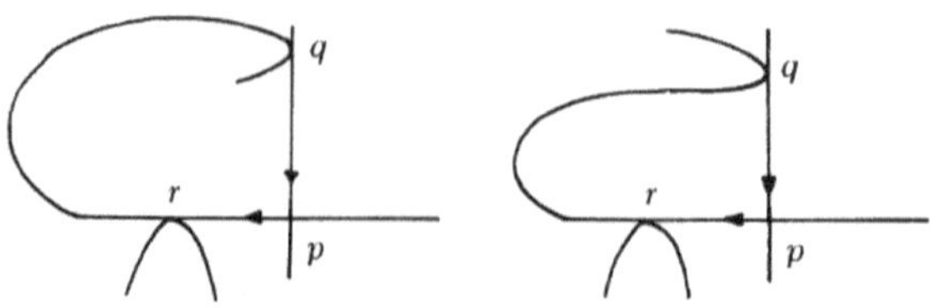

In these cases we speak of different connections (each branch of a stable or unstable separatrix has an orientation, given by the dynamics; points move in the positive direction; the different connections can be distinguished according to whether these two orientations agree or disagree in points of the orbit of tangency).

In all cases which may exist according to the above subdivision, we want to determine whether they can occur as a homoclinic Ω-explosion, and if so:

– if the ambient manifold can be orientable;

– if the saddle point p can be part of a non-trivial basic set or not.

In some cases the answer to these questions depends also on the so-called eigenvalue conditions. If λ, σ are the eigenvalues of $(d\varphi)_p$, with $0 < |\lambda| < 1 < |\sigma|$, we say that the *contracting*, resp. *expanding*, *eigenvalue* is *dominating* if $|\Lambda| \cdot |\sigma| < 1$, resp. $|\Lambda| \cdot |\sigma| > 1$. As we mentioned before we exclude the case $|\lambda| \cdot |\sigma| = 1$ (although this is not always necessary).

The rest of this chapter is organized as follows. In Section 2 we consider a number of typical cases and establish the answers to the above questions. These typical cases were selected in such a way that all the different arguments are displayed. Then, in Section 3 we treat all cases (according to the above subdivision), referring back to Section 2 for the main arguments. In Section 4 we construct examples of all the remaining cases which can occur as homoclinic Ω-explosions.

2. *Typical cases.*

A. *Premature tangencies.* We shall show that the case with eigenvalues $++$, and sides as in the figure below

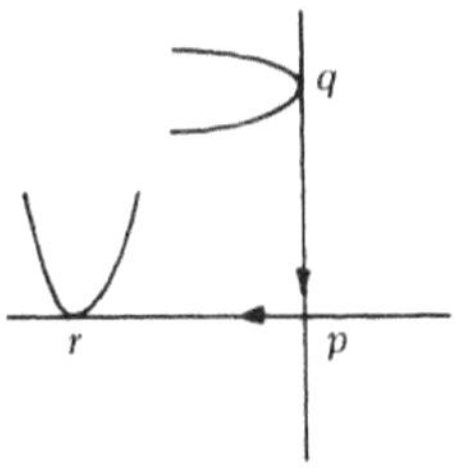

cannot occur as homoclinic Ω-explosion because then φ_μ would have orbits of tangency of $W^u(p_\mu)$ and $W^s(p_\mu)$ for negative values of μ arbitrarily close to zero.

In order to prove this we take as before $q \in W^s(p)$ and $r \in W^u(p)$ both in the orbit of tangency of φ. We also take $\tilde{q} = \varphi^n(q)$ for some (big) $n > 0$. We consider, for some neighborhood U of p the components of $W^u(p) \cap U$ and $W^s(p) \cap U$ containing r, q, and $\tilde{q}$; see the figure below on the left.

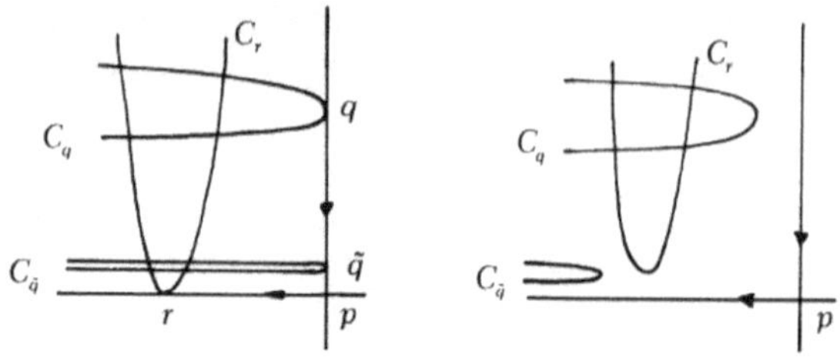

The components of $W^u(p) \cap U$ and $W^s(p) \cap U$, of r, q, and $\tilde{q}$, not containing p, are denoted by C_r, C_q, and $C_{\tilde{q}}$. For small negative values of μ one can follow these components and one gets, for example, a situation as in the above figure on the right: C_r and C_q moved slightly but $C_{\tilde{q}}$ almost disappeared. In fact if one lets μ go down from zero to negative values, $C_{\tilde{q}}$ moves to the left; the bigger n is in $\tilde{q} = \varphi^n(q)$, the faster $C_{\tilde{q}}$ moves. For $\mu = 0$, C_r and $C_{\tilde{q}}$ have four points of intersection. When they disappear there is a value of μ where they are tangent. Since $C_r \subset W^s(p_\mu)$ and $C_{\tilde{q}} \subset W^u(p_\mu)$, there is a sequence of negative values μ_i converging to zero for which $W^u(p_{\mu_i})$ and $W^s(p_{\mu_i})$ have an orbit of tangency and this is incompatible with φ_{μ_i} being hyperbolic. This argument was provided to us by J. Gheiner.

B. *Premature creation of horseshoes*. We consider the case where both eigenvalues of $(d\varphi)_p$ are positive and where the sides of tangencies and the connections are as indicated below (λ and σ are the contracting and expanding eigenvalues of $(d\varphi)_p$).

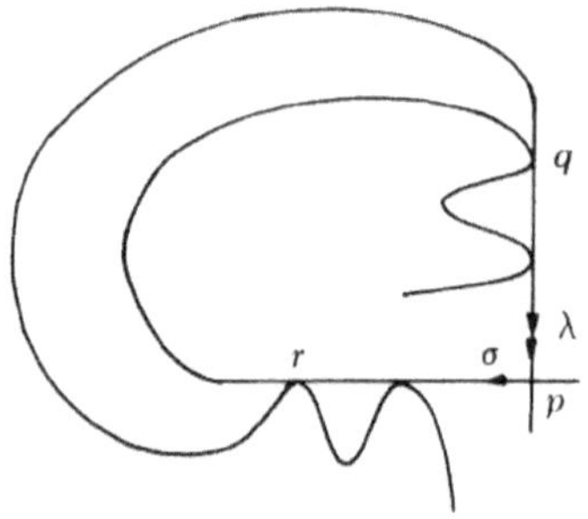

We claim that this case cannot occur as a homoclinic Ω-explosion if $\lambda^{-1} > \sigma$, i.e. if the contracting eigenvalue is dominating. In order to prove this we show that, if $\lambda^{-1} > \sigma$, $\varphi = \varphi_0$ has horseshoes which disappear when moving μ to small negative values. Of course this implies that there are also premature tangencies (as, also, premature tangencies imply premature creations or destructions of horseshoes). We only distinguish these cases by names, indicating the most easily visible reason why they cannot occur as homoclinic Ω-explosion.

We take C^1-linearizing coordinates in a neighborhood of p; for the existence of such coordinates see [2]. Let $r \in W^u_{\text{loc}}(p)$ be a point in the orbit of tangency; without loss of generality we may assume that r has coordinates $(-1, 0)$ (in the linearizing coordinates). Let R be a rectangle as indicated below: one side is part of $W^u_{\text{loc}}(p)$ and contains r in its interior; the length may be chosen for example equal to $1/2$; the height d is small. We consider negative iteratates of R. For $i < \operatorname{Log} d / \operatorname{Log} \lambda$, the height of $\varphi^{-1}(R)$ is smaller than one; let $i_0(d)$ be the

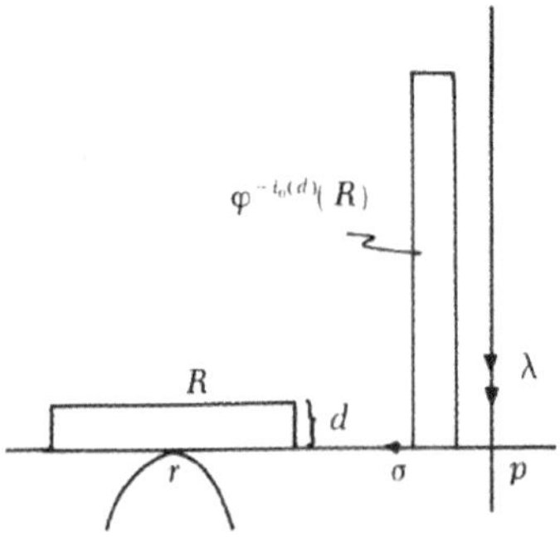

smallest integer for which this inequality is not satisfied. Then the distance from $\varphi^{-i_0(d)}(R)$ to $W^s_{\text{loc}}(p)$ is

$$\approx \sigma^{-i_0(d)} \approx d^{-\log \sigma / \log \lambda}$$

($\approx$ means: equal up to a multiplicative factor which is bounded and bounded away from zero, uniformly in d). Since $0 < \lambda < 1 < \sigma$ and $\lambda^{-1} > \sigma$, the distance from $\varphi^{-i_0(d)}(R)$ to $W^s_{\text{loc}}(p)$ is much bigger than d (for d small). For some integer j_0, independent of d, $r \in \varphi^{-j_0}(W^s_{\text{loc}}(p))$; here $W^s_{\text{loc}}(p)$ is the interval in $W^s(p)$ consisting of points at distance less than one from p. Then the distance from $\varphi^{-i(d_0)-j_0}(R)$ to $\varphi^{-j_0}(W^s_{\text{loc}}(p))$ is also much bigger than d (for d small). So we have the following situation

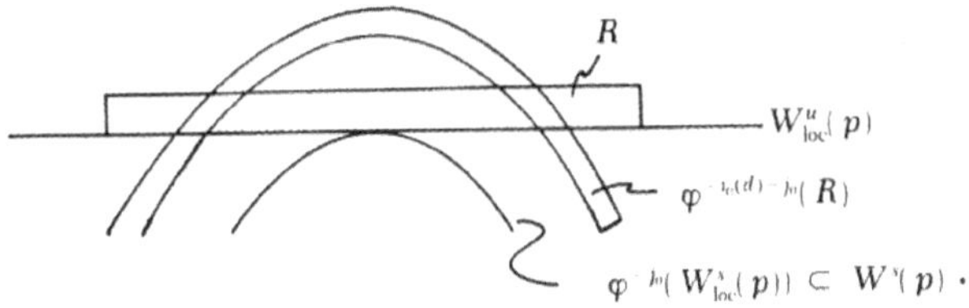

This means that in R we have a horseshoe, or at least a nonempty set (nonwandering points $\Lambda = \bigcap_{n \in \mathbf{Z}} \varphi^{n \cdot (i_0(d) + j_0)}(R)$. For $\mu < 0$ we define Λ_μ $\bigcap_{n \in \mathbf{Z}} \varphi^{n \cdot (i_0(d) + j_0)}(R)$ (we assume that $W^u_{\text{loc}}(p_\mu)$ is independent of μ, so that v can choose R also independent of μ). We have $\Lambda_0 = \Lambda \neq \varnothing$; for some $\mu_0 <$ $\Lambda_{\mu_0} = \varnothing$. Also, for $\mu \in (\mu_0, 0)$, $\Lambda_\mu \cap \partial R = \varnothing$. From this it is clear that the are Ω-bifurcations between μ_0 and 0. Since $|\mu_0|$ is of the order $d^{-\operatorname{Log} \sigma / \operatorname{Lo}}$

$|\mu_0|$ can be made arbitrarily small by taking d small. This proves that φ_μ has Ω-bifurcations for negative values of μ arbitrarily close to zero (when we apply the above argument for values of d going to zero).

C. *Infinitely many circles.* We consider the case where both eigenvalues of $(d\varphi)_p$ are positive and where the sides of tangencies and connections are as indicated below.

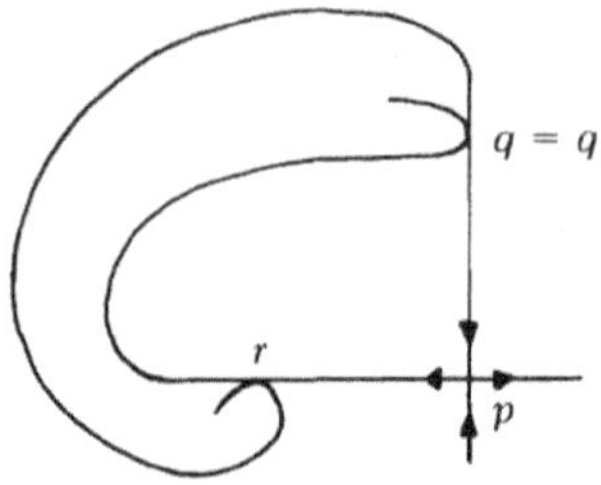

We assume that for $\mu \leq 0$, p_μ is not part of a non-trivial basic set. Then this situation cannot occur in a homoclinic Ω-explosion on a closed surface.

We prove this by contradiction. Let γ^s, γ^u be the branches of $W^s(p)$, $W^u(p)$ containing the orbit of tangency. Since for $\mu \leq 0$, p_μ is not part of a non-trivial basic set, γ^s and γ^u only meet in p and in the orbit of tangency. As indicated, q_1 is a point of the orbit of tangency and $q_i = \varphi^{i-1}(q_1)$. We define the closed curve C_i as the union of the segments in γ^u and γ^s joining q_{2i-1} and q_{2i}. Since γ^u and γ^s only meet in p and in the orbit of tangency, the closed curves C_i are double-point free and pairwise disjoint. We then prove that the complement of $\bigcup_{i=1}^k C_i$ is connected for any k (assuming that φ is defined on a connected surface). If this is proven we are done: for each closed surface M there is a number $n(M)$ such that any collection of more than $n(M)$ closed curves in M separates M in at least two components; e.g. see [13].

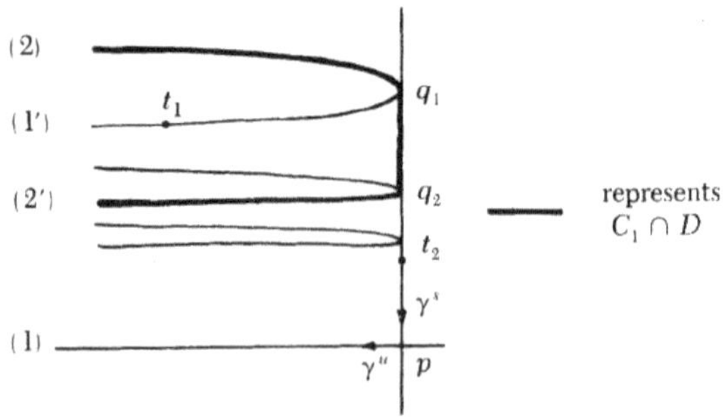

In the above figure we show part of $W^u(p)$ and $W^s(p)$ in a neighborhood D of p which is homeomorphic to a disc. Outside D the components of $\gamma^u \cap D$ are connected as indicated: (1) → (1′), (2) → (2′), etc. If the complement of C_1

were not connected, then the points t_1 and t_2 (see the above figure) would be in different components. But the segments in γ^u and γ^s joining t_1 with p and p with t_2 are disjoint from C_1; so the complement of C_1 is connected.

The arc in γ^u from t_1 to p is disjoint from $C_2, \ldots, C_k$. The arc in γ^s from p to t_2 becomes disjoint from $C_2, \ldots, C_k$ if we shift it slightly to the right. This means that the complement of $\bigcup_{i=1}^k C_i$ has just as many components as the complement of $\bigcup_{i=2}^k C_i$. Now the connectedness of the complement of $\bigcup_{i=1}^k C_i$ follows by induction. Note that we only used the fact that $\gamma^u \cap \gamma^s$ is the union of p and the orbit of tangency $\mathcal{O}$. This is weaker than the assumption that for $\mu \leq 0$, p_μ is not part of a non-trivial basic set.

D. *Basic sets and negative eigenvalues*. If a saddle point p is part of a non-trivial basic set Λ, then we indicate the (local) position of Λ with respect to p by the quadrant(s) (or sector(s)), separated by $W^u_{\text{loc}}(p)$ and $W^s_{\text{loc}}(p)$, in which there are points of $\Lambda \cap U$ for any small neighborhood U of p. In a situation where all homoclinic points are transverse, as for φ_μ, $\mu < 0$, we have the following. If γ^u and γ^s are branches of $W^u(p)$ and $W^s(p)$, then $\gamma^u \cap \gamma^p \supsetneqq \{p\}$ if and only if there is (part of) a basic set in the quadrant bounded by γ^u and γ^s. Also, if p is part of a non-trivial basic set Λ then $W^u(p)$ and $W^s(p)$ are accumulating on themselves at least from one side as indicated in the following figure.

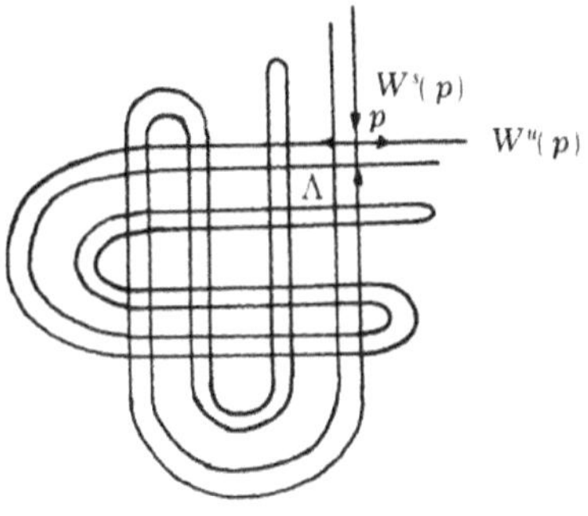

Returning to our 1-parameter families φ_μ with homoclinic Ω-explosion for $\mu = 0$ and φ_μ hyperbolic for $\mu < 0$, we see that if the sides of tangencies are as indicated below

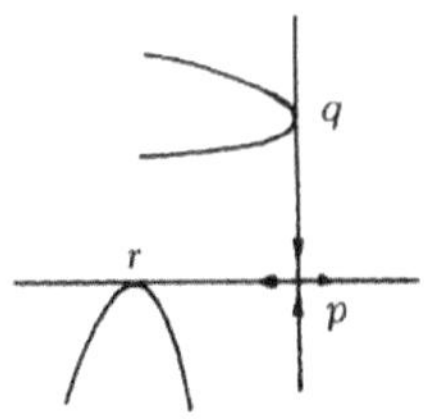

so that for $\mu < 0$ we have

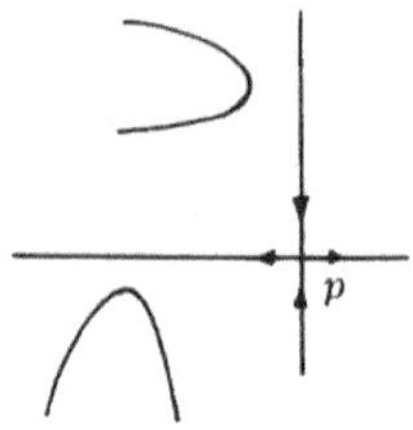

then the only quadrant where there might be a basic set is the upper right quadrant; for any other quadrant there would be tangencies of $W^u(p_\mu)$ and $W^s(p_\mu)$ for negative values of μ arbitrarily close to zero. For this reason, once the sides of tangencies are given, there is at most one quadrant which could be occupied by a basic set.

If at least one of the eigenvalues of $(d\varphi)_p$ is negative, then p cannot be part of a non-trivial basic set, because such a basic set would occupy at least two quadrants.

E. *Orientability*. We consider, as an example, a 1-parameter family φ_μ with homoclinic Ω-explosions, such that both eigenvalues of $(d\varphi)_p$ are positive and such that the sides of tangencies are as indicated below.

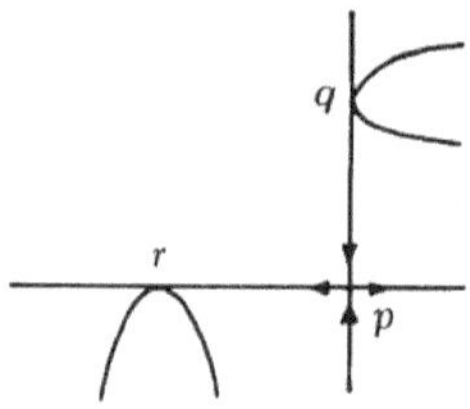

In this case we can still make the connections in two ways:

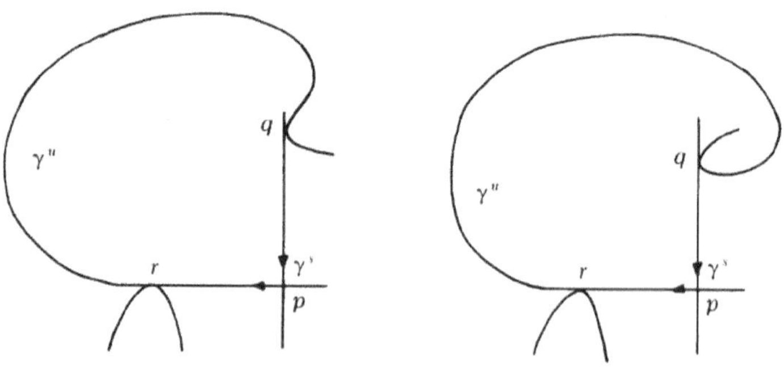

As observed in the introduction of this chapter, one can distinguish these cases by orienting the branches γ^u and γ^s in such a way that φ moves the points in the positive direction. These orientations are intrinsically defined in the sense that they are preserved by conjugacies. With these orientations of γ^u and γ^s one easily sees the differences between the above two cases: at a point of tangency, say q, the two orientations of γ^u and γ^s are in the same direction in the first case and in opposite directions in the second case. Clearly the same must hold at other points of tangency like r. In this way we know the orientation of the piece of γ^s near r.

There is another way of seeing the difference between the two cases: if we take a diffeomorphism from a neighborhood of r to a neighborhood of q, respecting γ^u and γ^s, with orientations, then this diffeomorphism is orientation reversing in the first case and orientation preserving in the second case. From this it follows that the first case cannot occur if the ambient surface is orientable: φ is orientation preserving since $(d\varphi)_p$ has positive eigenvalues and for some positive i, φ^i maps r to q preserving γ^u and γ^s with orientations.

In case $(d\varphi)_p$ has one negative eigenvalue (and hence φ cannot be orientation preserving) one can apply a similar argument using φ^2 instead of φ.

3. *General classification.* In this section we go systematically through all the cases as distinguished by: signs of eigenvalues, sides of tangencies, connections, and where necessary eigenvalue condition.

I. *Eigenvalues* $++$. In this case there are four possibilities of sides of tangencies:

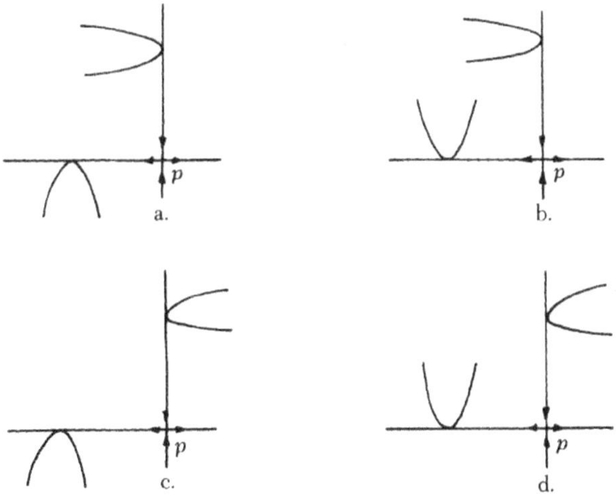

Before distinguishing according to the different connections, we observe that (b) cannot occur due to premature tangencies and that the cases (a) and (d) can be interchanged by replacing φ by φ^{-1}. So we only have to consider the cases (a) and (c). Consideration of the different connections again gives four cases.

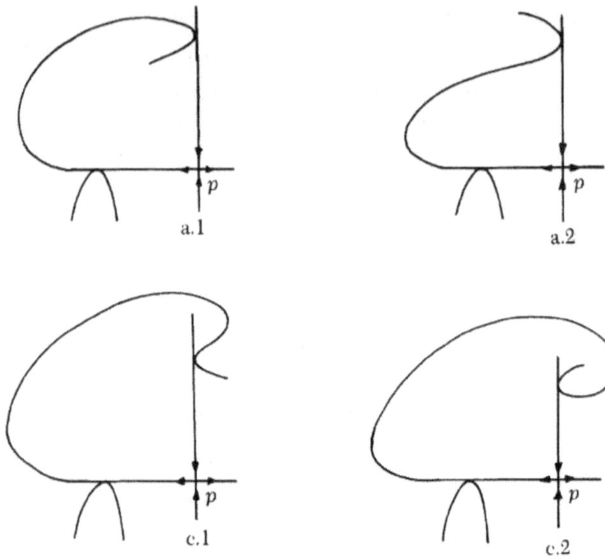

Case (a.1) can only occur if the expanding eigenvalue dominates the contracting eigenvalue; see (B) on the "premature creation of horseshoes"; also, it was given as an example in [5]. This case can occur both when p is part of a non-trivial basic set and when p is an isolated point in $\tilde{\Omega}(\varphi)$ (the hyperbolic part of the non-wandering set of φ). The ambient surface can be orientable or non-orientable in this case.

Case (a.2) Here p is part of a non-trivial basic set; see (C) *infinitely many circles*. There must be a basic set in the sector between γ^u and γ^s, the branches of $W^u(p)$ and $W^s(p)$ containing the orbit of tangency (because these branches must also have transverse intersections). This sector is however not allowed according to *basic sets and negative eigenvalues*. So this case cannot occur.

Case (c.1) was already considered in [5] as an example in the projective plane. It cannot occur in an orientable surface.* The saddle point p may be both isolated or not in $\tilde{\Omega}(\varphi)$.

Case (c.2) Again by *infinitely many circles* this case can only occur when p is part of a non-trivial basic set. As such it occurs in both orientable and non-orientable surfaces.

II. *Eigenvalues* $+-$. In this case there are two possibilities of sides of tangencies. We take the expanding eigenvalue as negative; the other case can then be obtained by replacement of φ by φ^{-1}.

---

*See *orientability*.

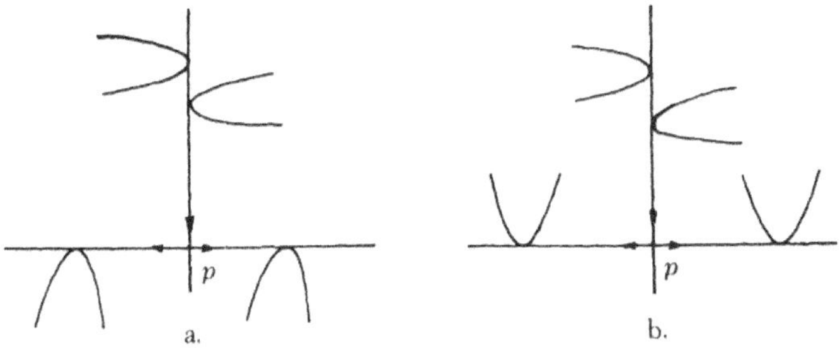

a. b.

Case (b) cannot occur because of premature tangencies. In case (a) we can make the connections in four different ways.

a.1 a.2

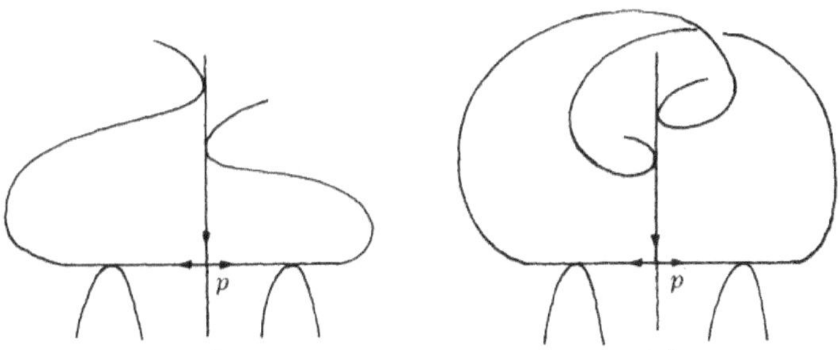

a.3 a.4

Case (a.1) can only occur if the expanding eigenvalue is dominating (see *premature creation of horseshoes*) and if p is isolated in $\tilde{\Omega}(\varphi)$ (see *basic sets and negative eigenvalues*). This case can occur in both orientable and non-orientable surfaces.

Case (a.2) can only occur on non-orientable surfaces when p is isolated in $\tilde{\Omega}(\varphi)$.

Case (a.3) *and* (a.4) cannot occur as homoclinic Ω-explosion. This follows from the argument in "infinitely many circles" together with the fact that p cannot be part of a basic set.

III. *Eigenvalues* $- -$. In this case there are two possibilities of sides of tangencies.

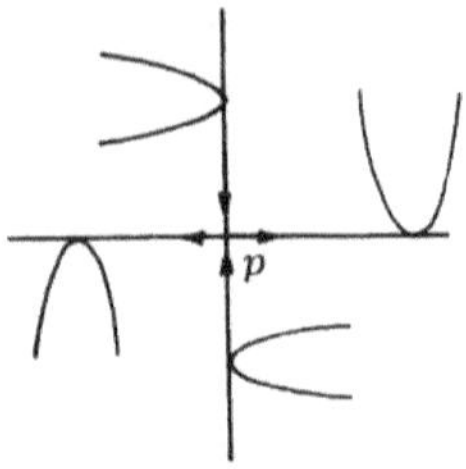

a.

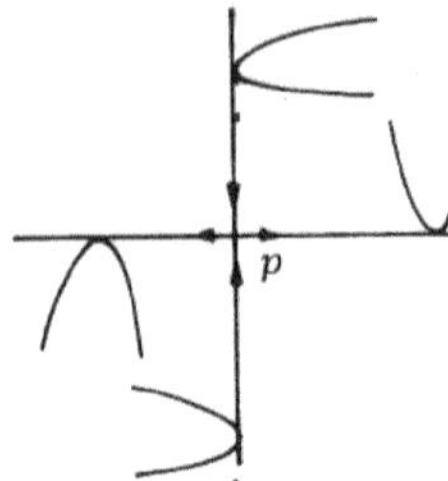

b.

Case (b) cannot occur because of premature tangencies. In case (a) there are four different ways to make connections.

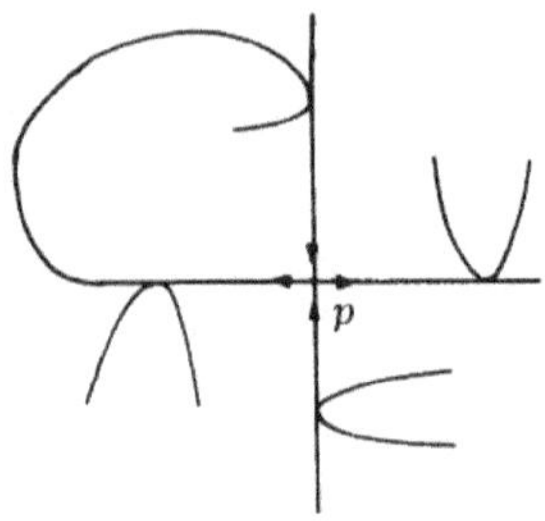

a.1

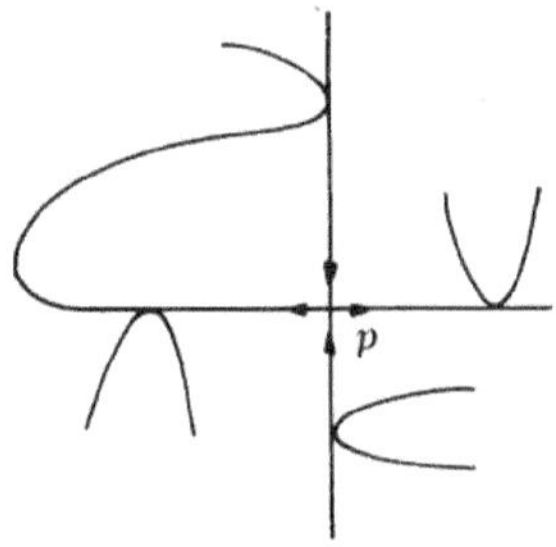

a.2

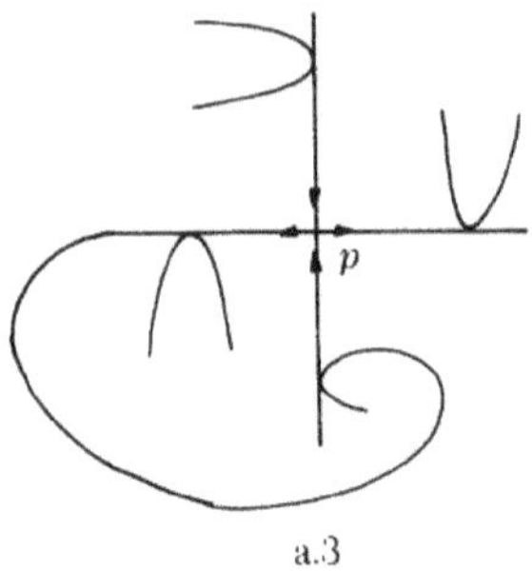

a.3

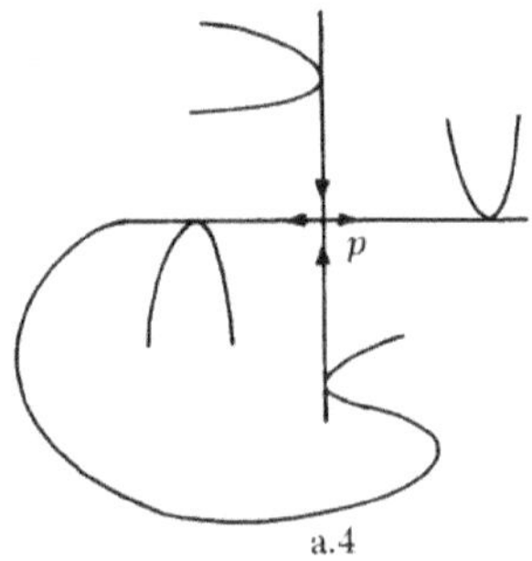

a.4

When φ is replaced by φ^{-1}, the cases (a.1) and (a.2) change to the cases (a.4) and (a.3) respectively. Both cases (a.1) and (a.2) (and hence also (a.3) and (a.4)) can be excluded because of premature creation of horseshoes. The argument goes as follows.

If the contracting eigenvalue dominates, we can immediately apply the arguments from the previous section. So we assume that the expanding eigenvalue dominates. We take a rectangle R, as indicated below, with small thickness d.

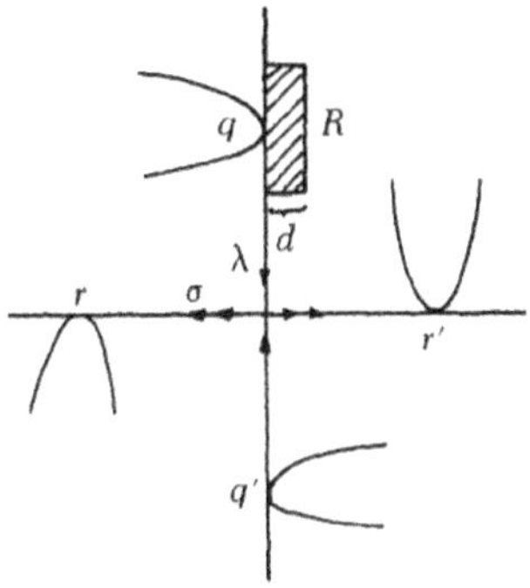

We assume that the point q of tangency is at distance one from p in C^1-linearizing coordinates. For i, such $\varphi^{-2i}(q) = r$, $R' = \varphi^{-2i}(R)$ has the following shape:

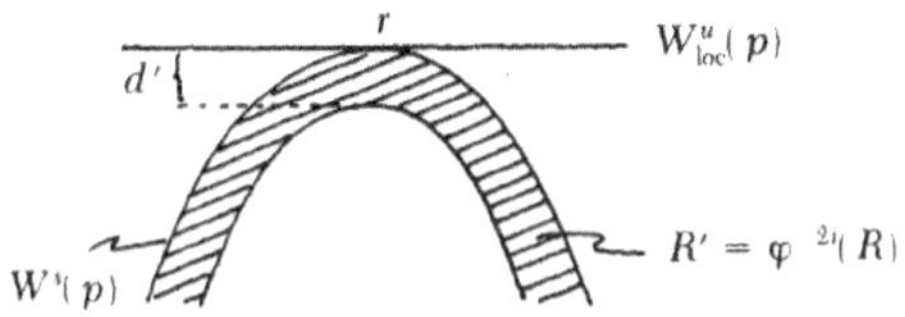

with d' of the same order as d. We denote the contracting and expanding eigenvalues of $(d\varphi)_p$ by λ and σ. Applying φ^{-2j} to R', where $j \approx \text{Log}(2d')/2\,\text{Log}\,|\lambda|$ we obtain $R'' = \varphi^{-2j}(R')$ as in the figure below.

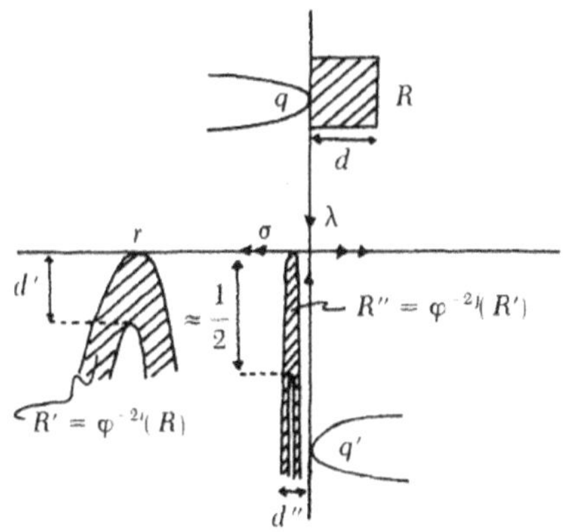

The maximal distance from R'' to $W^s_{\text{loc}}(p)$ is d''. Due to the eigenvalue condition, d'' is much smaller than d for d sufficiently small. From this it follows that $\varphi^{-1}(R'') = \varphi^{-2j-2i-1}(R)$ intersects R in such a way that there has to be a horseshoe, which disappears if μ moves to negative values; see "premature creation of horseshoes".

4. *List of possible homoclinic Ω-explosions.* Here we indicate examples of 1-parameter families of diffeomorphisms φ_μ with homoclinic Ω-explosion for all cases which were not excluded in Section 3. We shall do this by making in each case a sketch of the stable and unstable separatrices of the saddle point p of the diffeomorphism $\varphi = \varphi_0$. Since each case which can be realized on an orientable surface can also be realized on a non-orientable surface, we only give examples on non-orientable surfaces when necessary, i.e. when there is no "orientable example".

I. *Eigenvalues* $++$.

Case (a.1): p isolated in $\tilde{\Omega}(\varphi)$.

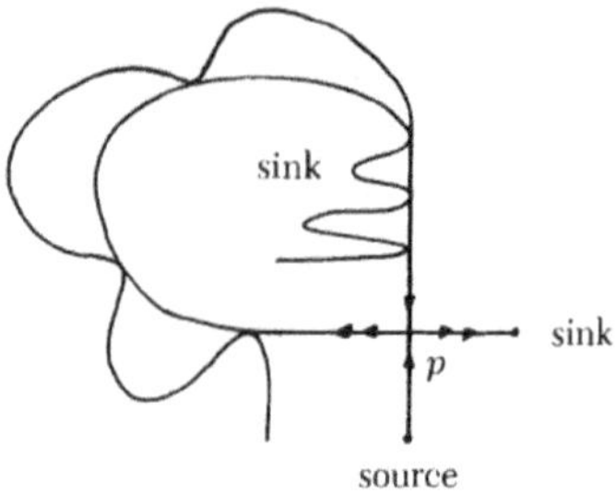

In order to see that this tangency is really a first bifurcation, we give a more careful description of the 1-parameter family φ_μ. We start with a diffeomorphism Φ as indicated below, with $\Omega(\Phi)$ consisting of one source and two sinks besides

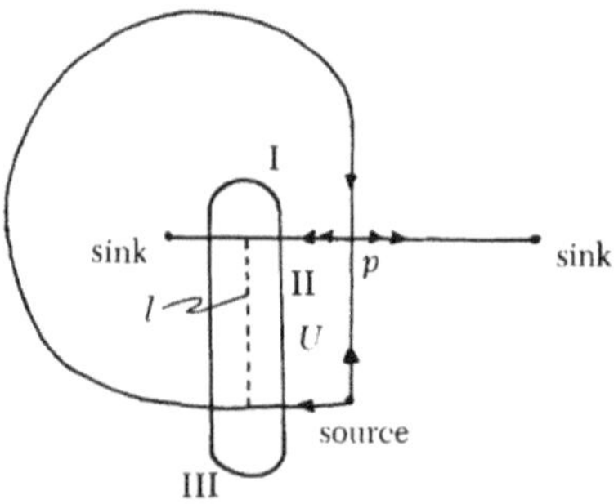

the saddle p. Let l be a line segment from $W^u(p)$ to $W^s(p)$, and U be a small neighbourhood of l as indicated. By $W^u(p)$ and $W^s(p)$, U is subdivided into three regions which are denoted by I, II, and III. We obtain our 1-parameter family by composing Φ with σ_μ, a 1-parameter family of diffeomorphisms with support in U which we describe below; $\varphi_\mu = \sigma_\mu \circ \Phi$.

For $\mu \leq -1$, σ_μ is the identity. For $\mu > -1$, σ_μ pushes the unstable separatrix down so that for $\mu = 0$ we have a generically unfolding tangency of $\sigma_\mu(W^u(p))$ and $W^s(p)$. Also we require that for any point x in region I, and

any $\mu \leq 0$

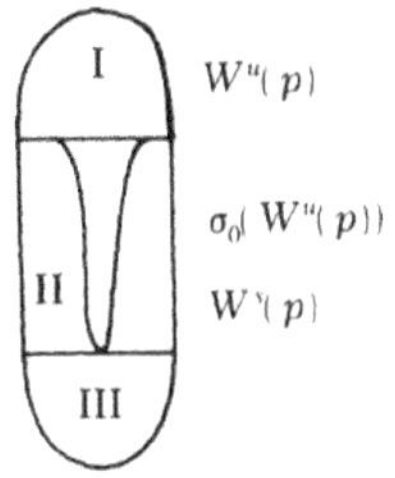

the distance of $\sigma_\mu(x)$ above $W^s(p)$ is at least equal to the distance of x above $W^u(p)$. We have to show that in this case φ_μ, $\mu < 0$, has no new non-wandering orbits. For this we only have to consider the orbits which continue to visit U: any orbit which does not pass through U, or which, after passing through U a finite number of times, never returns to U cannot be a new nonwandering orbit.

Observe that under forward iterations of Φ, the points of the regions I and III never return to U, while points of region II may return to region I. For $\mu < 0$, σ_μ maps region I inside the regions I and II, maps region II inside the regions II and III, and maps region III inside itself. From this it follows that the only possibility of having new nonwandering orbits is with points in region II which are mapped by Φ^n $(n > 0)$ back into region I and then by σ_μ again to region II. It is however the eigenvalue condition which assures that such orbits are wandering, at least if U is small enough. So this proves that we have a first bifurcation at $\mu = 0$.

In the other six cases below the arguments are the same. In some cases, exactly in those cases where there is no eigenvalue condition, the argument simplifies somewhat. Then, just from the analysis of which transitions are possible to realize through positive iterations of Φ, and through σ_μ, it follows that no orbit of φ_μ, $\mu < 0$, continues to come back to U.

In the cases below we indicate the line l along which to carry out the deformation as well as the modified $W^u(p)$ near l; a main point is that l is disjoint from the positive iterates $\Phi^n(l)$.

Case (a.2); p part of a non-trivial basic set.

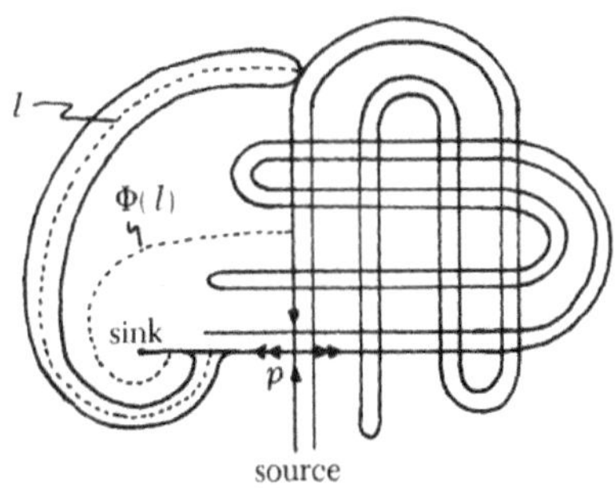

This is obtained from the usual horseshoe by a modification of $W^u(p)$ as indicated.

Case (c.1); p isolated in $\tilde{\Omega}(\varphi)$.

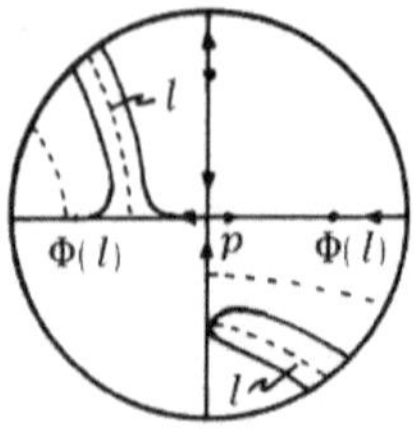

This is an example in the projective plane; in the boundary of this disc, antipodal points are identified.

Case (c.2); p part of a non-trivial basic set.

This case is more complicated in the sense that we have to make use of a "type-3 horseshoe", i.e., a horseshoe obtained by mapping a square D over itself as indicated below.

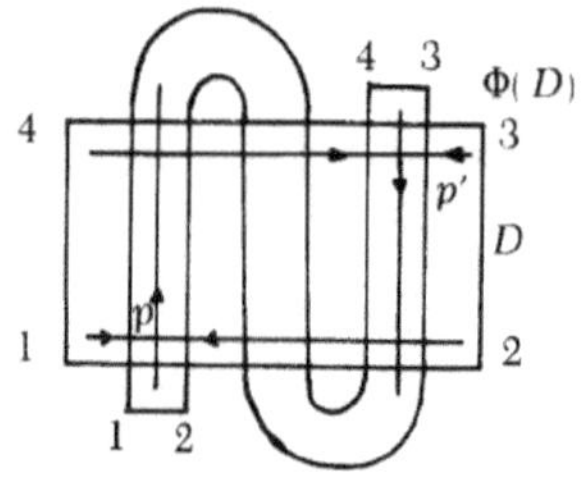

In this case there are two saddle points p and p' "fencing off" the hyperbolic set (see the beginning of Chapter 4). Below we indicate D, $\Phi(D) \setminus D$ and $\Phi^2(D) \setminus \Phi(D)$.

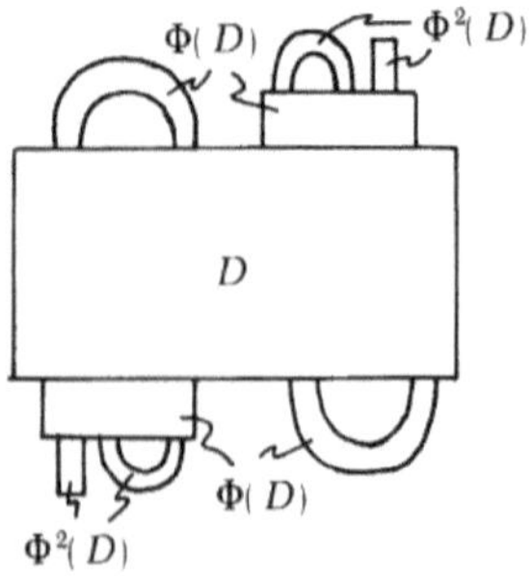

From this it is clear that the part of the plane, fenced off by the separatrices of p

and p' has the form:

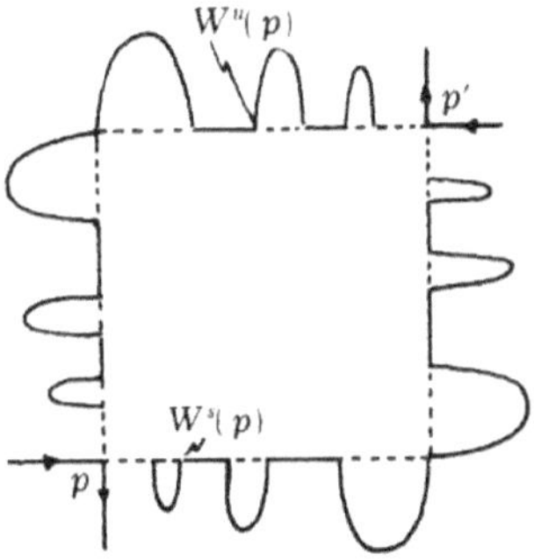

Finally we put this horseshoe in the projective plane and indicate the line l along which to push the unstable separatrix of p to its stable separatrix.

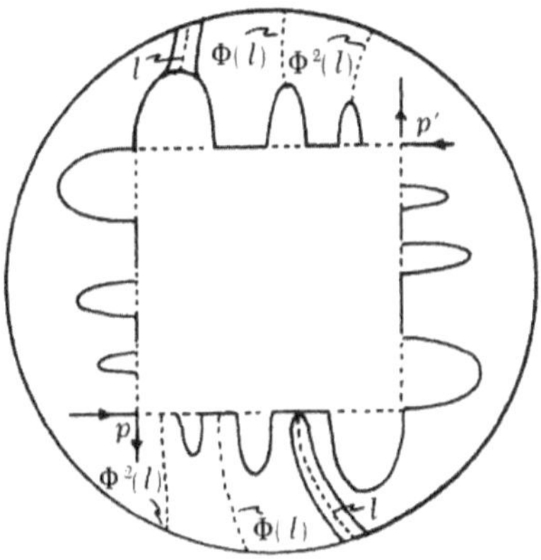

Case (c.2); p part of a non-trivial basic set.

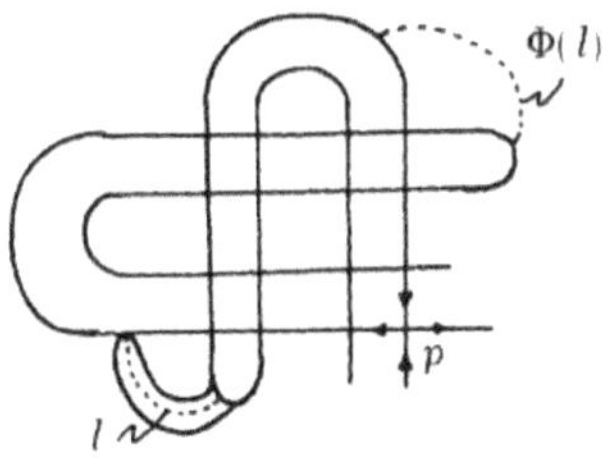

II. *Eigenvalues* $+-$.

Case (a.1); p isolated in $\tilde{\Omega}(\varphi)$.

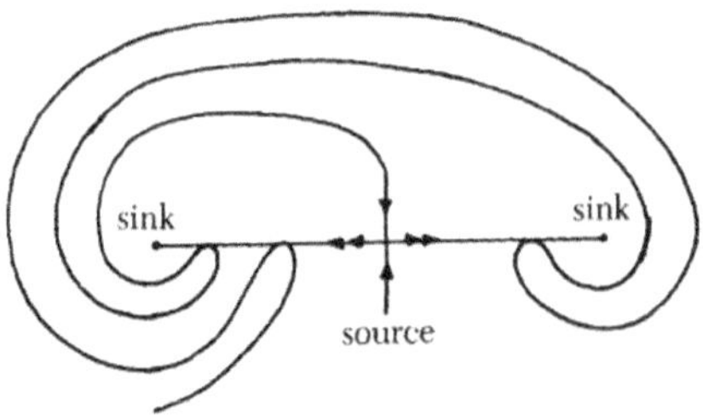

Case (a.2); p isolated in $\tilde{\Omega}(\varphi)$.

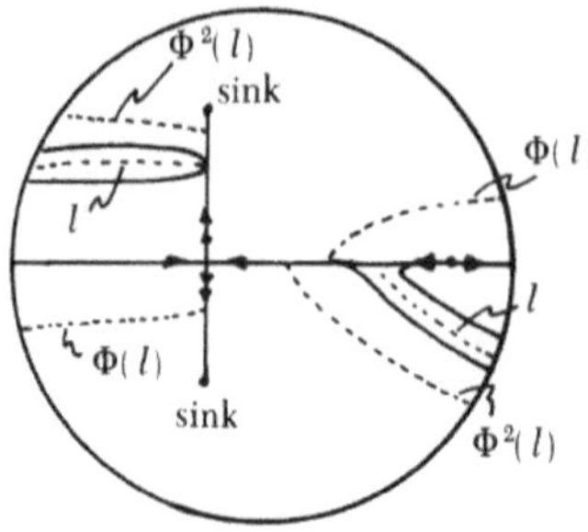

Also this is an example in the projective plane.

3. Intersections and almost-intersections of scaled sets

In this chapter we consider (almost) intersections of scaled sets (basically of Cantor type) which depend on a parameter. In the next chapter these results will be used to prove that, under certain hypotheses, the measure of the Ω-bifurcation set, after a homoclinic Ω-explosion, is very small. One can also compare the present exposition with [11]; there are several differences, mainly due to the fact that we show here that the "bifurcation set" B, as defined in the introduction, is relatively small in the sense that

$$\lim_{\mu\to 0} \frac{m(B\cap[0,\mu])}{\mu} = 0$$

(m denotes the Lebesgue measure on $\mathbf{R}$), whereas in [11] we only proved that the lim inf of the above expression is zero.

1. *Almost intersections of rigid scaled sets.* We say that a Cantor set, or more generally a closed subset $A \subset \mathbf{R}$, is *scaled* with *scaling factor* $\lambda \in (0,1)$ if A is invariant under scalar multiplication by λ. We recall the definition of limit capacity from the introduction. For a compact subset $C \subset \mathbf{R}$ we define the *limit capacity* $d(C)$ of C as

$$d(C) = \limsup_{\varepsilon\to 0} \frac{\operatorname{Log} n(C,\varepsilon)}{-\operatorname{Log}\varepsilon},$$

where $n(C,\varepsilon)$ is the minimal number of intervals of length ε needed to cover C. For a scaled subset $A \subset \mathbf{R}$, we define the limit capacity $d(A)$ as the limit capacity, in the above sense, of $A \cap [-r,+r]$ for some $r > 0$; due to the scaling, this definition is independent of r. We shall also use the following notation: If $A \subset \mathbf{R}$ is a subset, then $A^r = A \cap [-r,+r]$, A_ε is an ε-neighborhood of A and A^r_ε is the intersection of an ε-neighborhood of A with $[-r,+r]$, so that $A^r_\varepsilon = A_\varepsilon \cap [-r, +r]$. In this section we prove:

PROPOSITION. *Let $A, B \subset \mathbf{R}$ be scaled subsets of $\mathbf{R}$ with limit capacities $d(A)$ and $d(B)$ and let K be a fixed constant. For r, $\varepsilon > 0$,*

$$M_{r,\varepsilon} = \{t \in [0, r] | A_{\varepsilon \cdot r} \cap (B_{\varepsilon \cdot r} + t) \cap [-K \cdot r, K \cdot r] \neq \varnothing\}.$$

If $d(A) + d(B) < 1$, then for each $\delta > 0$ there is an $\varepsilon(\delta)$ such that $1/r \cdot m(M_{r,\varepsilon}) < \delta$ whenever $\varepsilon < \varepsilon(\delta)$; m denotes the Lebesgue measure.

Remark. One can consider values $t \in M_{r,\varepsilon}$ as the values for which the translate by t of B intersects, or *almost intersects*, A in a $(K \cdot r)$-neighborhood of 0. The constant K which may seem here somewhat artificial is forced upon us by the applications (see Chapter 4) which we have in mind.

Before we start the proof of this proposition we first state and prove the following lemma.

LEMMA. *Let A be a scaled set with limit capacity smaller than $d'(A)$. Then there is a constant a so that A^r can be covered with at most $a \cdot (\varepsilon/r)^{-d'(A)}$ intervals of length ε whenever $\varepsilon < r$.*

Proof. Since the limit capacity of A is smaller than $d'(A)$, there is a function $a(r) > 0$ of r such that A^r can be covered with at most $a(r) \cdot \varepsilon^{-d'(A)}$ intervals of length ε whenever $\varepsilon < r$. Let λ be the scaling factor of A. Because of the scaling, $A^{\lambda \cdot r}$ can be covered by $a(r) \cdot \varepsilon^{-d'(A)}$ intervals of length $\lambda \cdot \varepsilon$; on the other hand it can be covered by $a(\lambda \cdot r) \cdot (\lambda \cdot \varepsilon)^{-d'(A)}$ such intervals. So we may assume that $a(\lambda \cdot r) = \lambda^{d'(A)} \cdot a(r)$. From this, and the fact that we may assume that $a(r) = a(1)$ for $r \in (\lambda, 1]$, we conclude that there is a constant a such that $a(r) \leq a \cdot r^{d'(A)}$. This implies the lemma.

Proof of the proposition. According to the lemma there is a constant a such that $A^{(K+1) \cdot r}$ can be covered by at most

$$a \cdot \left(\frac{\varepsilon \cdot r}{(K+1) \cdot r}\right)^{-d'(A)}$$

intervals of length $\varepsilon \cdot r$ ($d'(A)$ is some number greater than $d(A)$). Hence, $A^{(K+1) \cdot r}_{\varepsilon \cdot r}$ can be covered by the same number of intervals of length $3 \cdot \varepsilon \cdot r$. In the same way, $B^{(K+1) \cdot r}_{\varepsilon \cdot r}$ can be covered by at most

$$b \cdot \left(\frac{\varepsilon \cdot r}{(K+1) \cdot r}\right)^{-d'(B)}$$

intervals of length $3 \cdot \varepsilon \cdot r$ (also $d'(B) > d(B)$, but we take $d'(A)$ and $d'(B)$ so that $d'(A) + d'(B) < 1$). This means that $M_{r,\varepsilon}$ can be covered by at most

$$a \cdot b \cdot \left(\frac{\varepsilon \cdot r}{(K+1) \cdot r} \right)^{-d'(A)-d'(B)}$$

intervals of length $6 \cdot \varepsilon \cdot r$. From this it follows that

$$m(M_{r,\varepsilon}) \leq 6 \cdot r \cdot a \cdot b \cdot \left(\frac{1}{K+1} \right)^{-d'(A)-d'(B)} \cdot \varepsilon^{1-d'(A)-d'(B)}.$$

Since $d'(A) + d'(B) < 1$, $1/r \cdot m(M_{r,\varepsilon})$ goes to zero for $\varepsilon \to 0$. This proves the proposition.

2. *Continuous* 1*-parameter families of scaled sets*. We say that a 1-parameter family $A(\mu) \subset \mathbf{R}$, $\mu \in \mathbf{R}$, of scaled sets is *continuous* if:

(i) For each $r > 0$, the map $\mu \mapsto A(\mu) \cap [-r, +r]$ is continuous with respect to the Hausdorff metric for compact subsets of $\mathbf{R}$;

(ii) The scaling factor $\lambda(\mu)$ of $A(\mu)$ is a Lipschitz function of μ.

For such 1-parameter families $A(\mu)$ we often denote $A(0)$ simply by A.

PROPOSITION. *Let $A(\mu)$ be a continuous 1-parameter family of scaled sets and let $K > 0$ be a fixed constant. Then, for each $\varepsilon > 0$ there is an $r(\varepsilon) > 0$ such that for each $0 < r < r(\varepsilon)$,*

$$A_{r\cdot\varepsilon} \supset (A(\mu) \cap [-(K+1) \cdot r, (K+1) \cdot r])$$

for all $0 < \mu < r$.

Proof. The proof of this proposition is complicated, mainly because of the fact that the scaling factor may not be constant. If $\lambda(\mu) \equiv \lambda(0)$, then the proof is quite simple: we just take $r(\varepsilon)$ so small that the distance, in the Hausdorff metric, between $A \cap [-1, +1]$ and $A(\mu) \cap [-1, +1]$ is less than $\varepsilon/K + 1$ for all $0 < \mu < r(\varepsilon)$.

From now on we concentrate on the consequences of the fact that the scaling factor may not be constant. To be more specific, we prove the proposition for the case where $A(\mu) = \{\lambda(\mu))^i | i \geq 0\} \cup \{0\}$, this being the simplest case with non-constant scaling factor. The general case can then be done in the same way.

LEMMA. *Let $p(\mu) = (\lambda(\mu))^i$ and $h = p(0)$. If $\Delta\lambda$ is so small that $1/2 \leq \mathrm{Log}(\lambda(0) + \Delta\lambda)/\mathrm{Log}\,\lambda(0)$, then for $|\lambda(\mu) - \lambda(0)| < \Delta\lambda$,*

$$|p(\mu) - p(0)| \leq \frac{\mathrm{Log}\, h}{\mathrm{Log}\,\lambda(0)} \cdot \left(\frac{h}{\lambda(0)} \right)^{1/2} \cdot \Delta\lambda.$$

Proof. We define $f(\lambda) = \lambda^i$; then $f'(\lambda) = i \cdot \lambda^{i-1}$. The maximum of $f'(\lambda)$, for $\lambda \in [\lambda(0) - \Delta\lambda, \lambda(0) + \Delta\lambda]$ is attained at $\lambda(0) + \Delta\lambda$ and this maximum is $i \cdot (\lambda(0) + \Delta\lambda)^{i-1}$. From this we get $|p(\mu) - p(0)| \leq i \cdot (\lambda(0) + \Delta\lambda)^{i-1} \cdot \Delta\lambda$. Since $i = \operatorname{Log} h / \operatorname{Log} \lambda(0)$, we find

$$|p(\mu) - p(0)| \leq \frac{\operatorname{Log} h}{\operatorname{Log} \lambda(0)} \cdot e^{((\operatorname{Log} h - \operatorname{Log} \lambda(0))/\operatorname{Log} \lambda(0)) \cdot \operatorname{Log}(\lambda(0) + \Delta\lambda)} \cdot \Delta\lambda$$

$$\leq \frac{\operatorname{Log} h}{\operatorname{Log} \lambda(0)} \cdot \left(\frac{h}{\lambda(0)} \right)^{1/2} \cdot \Delta\lambda$$

(for this last inequality we needed that $1/2 \leq \operatorname{Log}(\lambda(0) + \Delta\lambda)/\operatorname{Log} \lambda(0)$). For further use, let

$$M(h, \lambda(0)) = \frac{\operatorname{Log} h}{\operatorname{Log} \lambda(0)} \cdot \left(\frac{h}{\lambda(0)} \right)^{1/2}.$$

LEMMA. *For fixed $\lambda(0)$ and $h \in (0,1)$ variable, M attains its maximum for $h = e^{-2}$; this maximum value is $M_{\max}(\lambda_0) < -1/\operatorname{Log} \lambda(0) \cdot (\lambda(0))^{1/2}$.*

Proof.

$$\frac{dM}{dh} = \frac{1}{h \cdot \operatorname{Log} \lambda(0)} \cdot \left(\frac{h}{\lambda(0)} \right)^{1/2} + \frac{1}{2} \cdot \frac{\operatorname{Log} h}{\operatorname{Log} \lambda(0)} \cdot \left(\frac{1}{h \cdot \lambda(0)} \right)^{1/2}$$

$$= \frac{1}{\operatorname{Log} \lambda(0)} \cdot \left(\frac{1}{h \cdot \lambda(0)} \right)^{1/2} \cdot \left(1 + \tfrac{1}{2} \operatorname{Log} h\right).$$

This is zero when $h = e^{-2}$. Since $M(1, \lambda(0)) = 0$, $\lim_{h \searrow 0} M(h, \lambda(0)) = 0$, and $M(h, \lambda(0)) > 0$ for $h \in (0,1)$, $M(h, \lambda(0))$ attains its maximum for $h = e^{-2}$. Then

$$M\left(e^{-2}, \lambda(0)\right) = \frac{\operatorname{Log} e^{-2}}{\operatorname{Log} \lambda(0)} \cdot \left(\frac{e^{-2}}{\lambda(0)} \right)^{1/2}$$

$$= \frac{-2 \cdot e^{-1}}{(\lambda(0))^{1/2} \cdot \operatorname{Log} \lambda(0)}$$

$$< \frac{-1}{\operatorname{Log} \lambda(0) \cdot (\lambda(0))^{1/2}}.$$

Remark. From the above calculations we see that for $r \leq e^{-2}$, the maximum of $M(h, \lambda(0))$, for fixed $\lambda(0)$ and variable $h \in (0, r]$, is attained for $h = r$

and this maximum is

$$M(r, \lambda(0)) = \frac{\operatorname{Log} r}{\operatorname{Log} \lambda(0)} \cdot \frac{r^{1/2}}{(\lambda(0))^{1/2}} \leq r^{1/3};$$

for this last inequality we have to assume that r is sufficiently small, say $r \leq r_0$.

Now we go back to the proof of the proposition for

$$A(\mu) = \{(\lambda(\mu))^i | i \geq 0\} \cup \{0\}.$$

For some constant L, the variation in the scaling factor $\Delta\lambda$ is smaller or equal $L \cdot r(\varepsilon)$ for $0 < \mu < r(\varepsilon)$ (since $\lambda(\mu)$ is Lipschitz). Furthermore, we assume that $r(\varepsilon) \cdot (K+1) \leq r_0$, r_0 as in the above remark. Then, from the previous results it follows that $A(\mu) \cap [-(K+1) \cdot r, (K+1) \cdot r]$ is contained in an $L \cdot r \cdot ((K+1) \cdot r)^{1/3}$-neighborhood of $A(0)$. This means that, for given $\varepsilon > 0$, we have to choose $r(\varepsilon)$ so that $0 < r(\varepsilon) < r_0$ and so that $\varepsilon > L \cdot ((K+1) \cdot r(\varepsilon))^{1/3}$. For such a value of $r(\varepsilon)$ the conclusion of the proposition holds.

3. *Diffeomorphic images of scaled sets and the final result.*

PROPOSITION. *Let $A(\mu)$ be a continuous 1-parameter family of scaled sets and let Ψ_μ: $\mathbf{R} \to \mathbf{R}$ be a continuous 1-parameter family of C^1-diffeomorphisms with $\Psi_\mu(0) = 0$ and $\Psi'_\mu(0) = 1$. Let $K > 0$ be a constant. Then, for every $\varepsilon > 0$ there is an $r(\varepsilon) > 0$ such that*

$$A_{r \cdot \varepsilon}(\mu) \supset \left(\Psi_\mu(A(\mu)) \cap [-(K+1) \cdot r, (K+1) \cdot r]\right)$$

for all r, $\mu \in [0, r(\varepsilon)]$.

This is a simple consequence of the fact that for $\mu \in [0, r(\varepsilon)]$,

$$\frac{\Psi_\mu(x) - x}{|x|} \to 0$$

for $|x| \to 0$, uniformly in μ; see also [11].

In the case that $\Psi'_\mu(0) \neq 1$, one has to replace, in the conclusion, $A_{r \cdot \varepsilon}(\mu)$ by the $(r \cdot \varepsilon)$-neighborhood of $\Psi'_\mu(0) \cdot A(\mu)$.

In Section 1, we investigated the set of values of t such that A has an almost intersection with B, shifted over a distance t. In Section 2 we analyzed 1-parameter families of scaled sets depending continuously on a parameter μ. In the present section we gave an estimate on the deformation of a scaled set by a C^1-diffeomorphism depending on μ. Combining all this we obtain the following theorem.

THEOREM. *Let $A(\mu)$ and $B(\mu)$ be two continuous 1-parameter families of scaled sets so that $d(A) + d(B) < 1$. Let Ψ_μ, Φ_μ: $\mathbf{R} \to \mathbf{R}$ be two continuous*

1-parameter families of C^1-diffeomorphisms such that $\Psi_\mu(0) = 0$ and $\Phi_\mu(0) = \mu$. Let $K > 0$ be a constant and let $M_{r,\varepsilon}$ be the set of those $\mu \in [0, r]$ such that the distance between $\Psi_\mu(A(\mu)) \cap [-K \cdot r, K \cdot r]$ and $\Phi_\mu(B(\mu)) \cap [-K \cdot r, K \cdot r]$ is smaller than $\varepsilon \cdot r$. Then, for each $\delta > 0$, there is an $\varepsilon(\delta)$ such that

$$\frac{m(M_{r,\varepsilon(\delta)})}{r} < \delta$$

for r sufficiently small, where m denotes Lebesgue measure.

4. Proof of the main theorem

This chapter culminates with a proposition on the hyperbolicity of the nonwandering set (Section 4), which together with a key estimate in Section 3 yield our main result as stated in the introduction.

1. *Filtrations*. The 1-parameter families φ_μ: $M \to M$ with homoclinic Ω-explosion which we consider are persistently hyperbolic for $\mu < 0$. We first want to derive some consequences from this last assumption. For a diffeomorphism φ: $M \to M$, M a closed surface, we know that if $\Omega = \Omega(\varphi)$ is hyperbolic then the periodic points are dense in Ω, where Ω denotes the nonwandering set. Then it follows, e.g. see [14], that Ω is the disjoint union of a finite number of basic sets Λ_i. Recall that Λ is a *basic set* for a diffeomorphism φ if Λ is a compact φ-invariant hyperbolic set, in which the periodic orbits are dense, and which has a dense orbit (see the introduction).

For a basic set Λ we define its *stable set* $W^s(\Lambda)$ as $\{x \in M | \rho(\varphi^n(x), \Lambda) \to 0$ for $n \to \infty\}$, where ρ is some metric on M. $W^s(\Lambda)$ is the union of the stable sets of the points of Λ: $W^s(\Lambda) = \bigcup_{y \in \Lambda} W^s(y)$, where

$$W^s(y) = \{x \in M | \rho(\varphi^n(x), \varphi^n(y)) \to 0 \text{ for } n \to \infty\}.$$

Recall that these stable sets of points of Λ are (locally) leaves of the stable foliation of Λ; for y a periodic saddle point, $W^s(y)$ is its stable separatrix. Unstable sets are defined in the same way except that in that case one takes limits with $n \to \infty$. In the case where M is a compact surface and Λ is a basic set, of a diffeomorphism φ: $M \to M$, of saddle type (in particular, the stable and unstable sets of points in Λ are 1-dimensional), then we know from [6] the following fact. For each point $y \in \Lambda$ such that $W^s(y)$ is at most accumulated from one side by stable sets (or leaves of the stable foliation) of other points of Λ, then y is in the stable separatrix of a periodic point $p \in \Lambda$, and hence $W^s(y) = W^s(p)$. Intuitively this means that such a stable set $W^s(\Lambda)$ is fenced by stable separatrices of periodic points. Combining this fact with the corresponding fact about the unstable sets we see that Λ is fenced by stable and unstable separatrices of periodic points.

The next point we want to mention about persistently hyperbolic diffeomorphisms $\varphi: M \to M$, M a closed 2-manifold, is that φ cannot have cycles and hence is Ω-stable [16]. To prove this, assume that φ has a *cycle*, i.e. that there are basic sets $\Lambda_1, \Lambda_2, \ldots, \Lambda_k = \Lambda_1$, $k \geq 2$, such that $W^u(\Lambda_i) \cap W^s(\Lambda_{i+1}) \neq \varnothing$ for $i = 1, \ldots, k-1$. From [10] we know that, with an arbitrarily small perturbation of φ we obtain a new diffeomorphism $\tilde{\varphi}$ having one basic set Λ which contains basic (sub) sets $\tilde{\Lambda}_i \subsetneq \Lambda$ with $\tilde{\Lambda}_i$ close to Λ_i. This means that in a 1-parameter family of diffeomorphisms, connecting φ and $\tilde{\varphi}$, new parts of basic sets and hence new periodic points are created. This contradicts the assumption that φ is persistently hyperbolic and hence that we may assume that $\tilde{\varphi}$ as well as the diffeomorphisms between φ and $\tilde{\varphi}$ are hyperbolic.

Returning now to the 1-parameter family $\varphi_\mu: M \to M$, we recall that for $\mu = 0$ we have:

- $\varphi = \varphi_0$ is hyperbolic on $\lim_{\mu \nearrow 0} \Omega(\varphi_\mu) = \tilde{\Omega}(\varphi_0)$;
- φ has exactly one orbit of tangency between a stable and an unstable separatrix of periodic points, and this orbit of tangency is in fact an orbit of homoclinic tangency, i.e., of a tangency of the stable and unstable separatrix of the same saddle point p of $\varphi = \varphi_0$;
- the Ω-set of φ is the union of $\tilde{\Omega}(\varphi_0)$ and the orbit $\mathcal{O}$ of tangency.

Since, for $\mu < 0$, φ_μ is persistently hyperbolic, there are basic sets $\Lambda_i(\mu)$ depending continuously on μ such that $\Omega(\varphi_\mu)$ is the disjoint union of these basic sets. Since there are no cycles we may assume that

$$(*) \qquad W^u(\Lambda_i(\mu)) \cap W^s(\Lambda_l(\mu)) = \varnothing \quad \text{for all} \quad i < l \quad \text{and} \quad \mu < 0.$$

We define $\Lambda_i(0) = \lim_{\mu \nearrow 0} \Lambda_i(\mu)$, so that $\tilde{\Omega}(\varphi) = \bigcup \Lambda_i(0)$. We want to prove that also for $\mu = 0$, $(*)$ holds. Suppose not. Then there are $i < l$ so that

$$W^u(\Lambda_i(0)) \cap W^s(\Lambda_l(0)) \neq \varnothing .$$

We may assume that Λ_i and Λ_l are of saddle type. Since this intersection is empty for $\mu < 0$ and since the stable and unstable sets of these basic sets are fenced by stable and unstable separatrices of periodic points, there must be a tangency of the unstable separatrix of a periodic point in $\Lambda_i(0)$ with the stable separatrix of a periodic point in $\Lambda_l(0)$. But this extra tangency for $\mu = 0$ is against our assumptions.

In particular, all basic sets $\Lambda_i(\mu)$ are open in $\Omega(\varphi_\mu)$ for $\mu \leq 0$, except for $\Lambda_{i_0}(0)$. But, it is clear from the above that even $\Lambda_{i_0}(0)$ is the maximal φ_0-invariant set in some neighborhood of it. That is, there exists $V \supset \Lambda_{i_0}(0)$ such that $\bigcap_{n \in \mathbb{Z}} \varphi_0^n(V) = \Lambda_{i_0}(0)$. Then, from [3, §7], we conclude that $\varphi_0 | \Lambda_i(0)$ is conjugate to $\varphi_\mu | \Lambda_i(\mu)$ for all i and $\mu < 0$. Therefore, as we mentioned in the introduction, the basic sets we consider here are open in the nonwandering set or they are homeomorphic to the ones having this property. Moreover, we can even consider basic sets $\Lambda_i(\mu)$ for μ positive and small as (canonical) *continuation* of $\Lambda_i(0)$ in the following way. If we choose neighborhoods $V_i \supset \Lambda_i(0)$ in which $\Lambda_i(0)$ is the maximal φ_0-invariant set, then we take $\Lambda_i(\mu)$, μ positive and small, as the maximal invariant set for φ_μ in V_i. Again by [3, §7], we have that $\Lambda_i(\mu)$ is hyperbolic and $\varphi_\mu|\Lambda_i(\mu)$ is conjugate to $\varphi_0|\Lambda_i(0)$.

Let i_0 be the index of the basic set containing the saddle point p_μ of φ_μ whose stable and unstable separatrices make a tangency for $\mu = 0$ (we do not exclude that $\Lambda_{i_0}(\mu) = p_\mu$; in that case we call $\Lambda_{i_0}(\mu)$ a trivial basic set; otherwise we call it a non-trivial basic set). Observe that

$$W^s(\Lambda_{i_0}(0)) = W^s(\Lambda_{i_0}(0) \cup \mathcal{O})$$

and

$$W^u(\Lambda_{i_0}(0)) = W^u(\Lambda_{i_0}(0) \cup \mathcal{O}).$$

So if we replace $\Lambda_{i_0}(0)$ by $\Lambda_{i_0}(0) \cup \mathcal{O}$, then (*) still holds.

By [14] there is a filtration $M_1 \subset M_2 \subset M$ such that

(i) M_i is closed;

(ii) $M_1 \subset \operatorname{int}(M_2)$;

(iii) $\varphi(M_i) \subset \operatorname{int}(M_i)$;

(iv) $\bigcap_{n \in \mathbf{Z}} \varphi^n(M_2 - M_1) = \Lambda_{i_0}(0) \cup \mathcal{O}$.

This means that for any (small) neighborhood U of $\Lambda_{i_0}(0) \cup \mathcal{O}$ there is an $n(U)$ such that for each point $x \notin U$, $\varphi^{n(U)}(x)$ or $\varphi^{-n(U)}(x)$ is in the complement of $M_2 - \operatorname{Int}(M_1)$. Hence, for μ near zero one still has this same property for φ_μ instead of φ. This means that for any neighborhood U of $\Lambda_{i_0}(0) \cup \mathcal{O}$, there is a $\mu_0(U) > 0$ such that for any $\mu \in [0, \mu_0(U)]$, the nonwandering set $\Omega(\varphi_\mu)$ of φ_μ, intersected with $M_2 - M_1$, is contained in U. So for the rest of this chapter we may restrict our attention to such a neighborhood U, restricting μ to $[0, \mu_0(U)]$.

2. *Bounds on the region where new nonwandering points can appear.* Let again $\varphi_\mu: M \to M$ be a 1-parameter family of diffeomorphisms with a homoclinic Ω-explosion. As before, $\Omega(\varphi_\mu)$ denotes the set of nonwandering points of φ_μ. For $\mu < 0$, $\Omega(\varphi_\mu)$ is a hyperbolic set; the continuation of these hyperbolic sets, for $\mu \geq 0$, are denoted by $\tilde{\Omega}(\varphi_\mu)$. So $\Omega(\varphi_0) = \tilde{\Omega}(\varphi_0) \cup \mathcal{O}$, where $\mathcal{O}$ is the orbit of tangency; $\Omega(\varphi_0)$ and $\tilde{\Omega}(\varphi_0)$ will also be denoted by $\Omega(\varphi)$ and $\tilde{\Omega}(\varphi)$.

PROPOSITION. *Let r be a point in $\mathcal{O}$, the orbit of tangency of φ. Then for some constant K and $\mu > 0$, μ small, every orbit of $\Omega(\varphi_\mu) \setminus \tilde{\Omega}(\varphi_\mu)$ has a point in a $(K \cdot \mu)$-neighborhood of $W^u(p_\mu)$ near r. Again, p_μ denotes the saddle point of φ_μ with $\mathcal{O} \subset W^u(p_0) \cap W^s(p_0)$.*

Proof. We shall only give the proof in the case where the saddle point p has two positive eigenvalues, where the sides of tangencies and connections are as in

the figure below

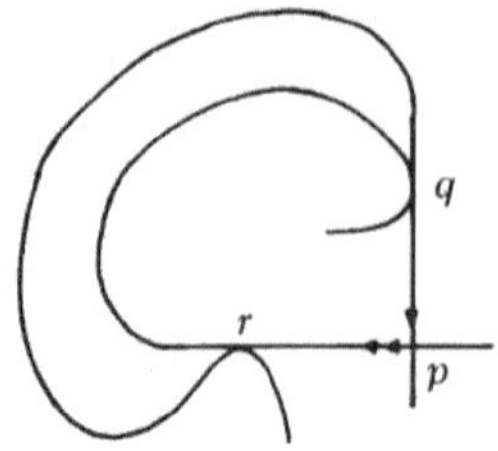

and where the expanding eigenvalue is dominating; see Chapter 2. We also assume that p is not part of a non-trivial basic set. The other cases can be treated similarly; see also the remark at the end of this section. As we have seen in the last section we can take a small neighborhood U_r of r, U_r being the component of r of a neighborhood U of $\Omega(\varphi)$, and then the "new Ω-orbits", i.e., the orbits of $\Omega(\varphi_\mu) \setminus \tilde{\Omega}(\varphi_\mu)$, will all pass through U_r and stay in U for $0 \leq \mu \leq \mu_0(U)$.

In the following considerations we assume that we have linearizing coordinates for φ_μ in a neighborhood of p_μ. We assume that in these coordinates $W^u_{\text{loc}}(p_\mu)$ and $W^s_{\text{loc}}(p_\mu)$ are independent of μ and that the coordinates of r and q are $(-1, 0)$ and $(0, 1)$; as usual r and q are points in the orbit of tangency in $W^u_{\text{loc}}(p_\mu)$, $W^s_{\text{loc}}(p_\mu)$. We assume that the tangency unfolds generically; to be more precise we assume the following situation near r for $\mu > 0$:

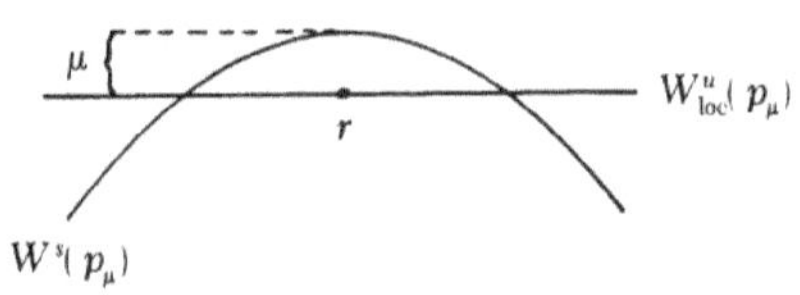

the maximum distance of $W^s(p_\mu)$ above $W^u_{\text{loc}}(p_\mu)$ is μ (here one only considers the part of $W^s(p_\mu)$ containing r). The proposition is a consequence of the following claim. For U_r and $\mu_0(U)$ as above and sufficiently small, and for any $x \in U_r$, $0 \leq \mu \leq \mu_0(U)$, there is a positive $n_0(x, \mu)$ such that, whenever $n' > n_0(x, \mu)$ and $\varphi_\mu^{-n'}(x) \in U_r$, then $\varphi_\mu^{-n'}(x)$ has height less than 2μ above $W^u(p_\mu)$.

To prove this claim, take a point $x \in U_r$ at distance d above $W^u_{\text{loc}}(p_\mu)$. If $\varphi_\mu^{-n}(x)$ is near q, then the distance from $\varphi_\mu^{-n}(x)$ to $W^s_{\text{loc}}(p_\mu)$ is of the order $d^{-\text{Log}\,\sigma_\mu/\text{Log}\,\lambda_\mu}$ where λ_μ and σ_μ are the contracting and expanding eigenvalues of $(d\varphi_\mu)_{p_\mu}$. Due to the dominance of σ, $d^{-\text{Log}\,\sigma_\mu/\text{Log}\,\lambda_\mu}$ is much smaller than d. We may assume that $\varphi_\mu^{-n}(x)$ is in a small neighborhood of q; otherwise it would never return to U_r under negative iterations of φ_μ (in which case the claim would be trivially true for this (x, μ)). Let m_0 be such that $\varphi^{m_0}(r) = q$. Then $\varphi_\mu^{-n-m_0}(x) \in U_r$, and the distance from $\varphi_\mu^{-n-m_0}(x)$ to $W^s(p_\mu)$ is of the order of

$d^{-\mathrm{Log}\,\sigma_\mu/\mathrm{Log}\,\lambda_\mu}$; i.e., there is a constant C, independent of d, such that the distance from $\varphi_\mu^{-n-m_0}(x)$ to $W^s(p_\mu)$ is smaller or equal to $C \cdot d^{-\mathrm{Log}\,\sigma_\mu/\mathrm{Log}\,\lambda_\mu}$. For d sufficiently small and μ near zero we have $C \cdot d^{-\mathrm{Log}\,\sigma_\mu/\mathrm{Log}\,\lambda_\mu} < \frac{1}{2}d$. We force d and μ to be so small that this last inequality holds by choosing U_r small. So if a point $x \in U_r$ has a negative iterate x' in U_r then $\rho(x', W^s(p_\mu)) < 1/2 \cdot \rho(x, W^u_{\mathrm{loc}}(p_\mu))$, where ρ is the distance function corresponding to the linearizing coordinates. If x' is below $W^u_{\mathrm{loc}}(p_\mu)$ then no further negative iterate of x will be in U_r. The maximum distance of $W^s(p_\mu)$ above $W^u_{\mathrm{loc}}(p_\mu)$ is μ so that $\rho(x', W^u_{\mathrm{loc}}(p_\mu)) < \mu + 1/2 \cdot \rho(x, W^u_{\mathrm{loc}}(p_\mu))$. This implies that after returning sufficiently often to U_r under negative iteration we will finally be at a distance of less than 2μ from $W^u_{\mathrm{loc}}(p_\mu)$. This completes the proof of the proposition in this case.

Remark. We observe that the shape of the region where the orbits of $\Omega(\varphi_\mu) \setminus \tilde{\Omega}(\varphi_\mu)$ pass near r, for $\mu > 0$, is as indicated in the figure below:

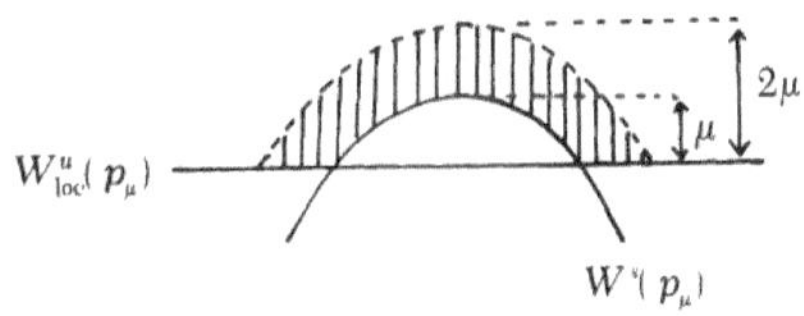

It is not hard to see that if p_μ is part of a non-trivial basic set for $\mu \le 0$, then the points of U_r can also return (under negative iteration) *between* $W^u_{\mathrm{loc}}(p_\mu)$ and $W^s_{\mathrm{loc}}(p_\mu)$, $\mu > 0$, as in the case of the formation of a cycle by heteroclinic tangency; see [11].

3. *Invariant foliations and further restrictions on the locus of the non-wandering set.* We continue with the same 1-parameter family $\varphi_\mu\colon M \to M$ with homoclinic Ω-explosion. As before, p_μ denotes the saddle point whose stable and unstable separatrices have a tangency for $\mu = 0$. For $\mu \le 0$, $\Lambda(\mu)$ denotes the basic set of $\Omega(\varphi_\mu)$ to which p_μ belongs. For $\mu \ge 0$, $\Lambda(\mu)$ is the basic set which is the continuation of the corresponding set for $\mu \le 0$.

As mentioned in the introduction, there are C^1-stable and unstable foliations $\mathscr{F}_\mu^s$ and $\mathscr{F}_\mu^u$, defined on a small neighborhood of $\Lambda(\mu)$. They are C^1 in the sense that the tangent directions depend C^1 on $x \in M$, and $\mu \in R$. For x in the domain of definition of these foliations, $E^s_\mu(x)$ and $E^u_\mu(x)$ denote these tangent spaces of the leaves of $\mathscr{F}_\mu^s$ and $\mathscr{F}_\mu^u$ through x. Since $p_\mu \in \Lambda(\mu)$, $\mathscr{F}_\mu^s$ and $\mathscr{F}_\mu^u$ are both defined in a neighborhood of p_μ. As usual r and q are points in the orbit of tangency in $W^u_{\mathrm{loc}}(p)$ and $W^s_{\mathrm{loc}}(p)$. We extend the domain of $\mathscr{F}_\mu^u$ along $W^u(p_\mu)$ until it includes a neighborhood of r; this extension is obtained by applying some positive iterates of φ_μ to $\mathscr{F}_\mu^u$. We also extend the domain of $\mathscr{F}_\mu^s$

until it includes r, but here we use negative iterates of φ_μ; the extension of the domain proceeds along $W^s(p_\mu)$. Near r, and for small μ, there is a C^1-curve l_μ where the leaves of $\mathscr{F}_\mu^u$ and $\mathscr{F}_\mu^s$ are tangent.

In $W^s(p_\mu)$ we define the set $A(\mu)$ as $(W^s(p_\mu) \cap \Lambda(\mu)) \cup O_\mu^s$, where O_μ^s is the φ_μ-orbit of a point $q_\mu \in W^s_{\text{loc}}(p_\mu)$ depending continuously on μ and such that $q_0 = q$. In the same way $B(\mu) \subset W^u(p_\mu)$ is defined as $(W^u(p_\mu) \cap \Lambda(\mu)) \cap O_\mu^u$ where O_μ^u is the φ_μ-orbit of a point $r_\mu \in W^u(p_\mu)$ depending continuously on μ and such that $r_0 = r$. We observe that $A(\mu)$ and $B(\mu)$ are scaled sets if we identify $W^s(p_\mu)$ and $W^u(p_\mu)$ with $\mathbf{R}$ in such a way that the induced φ_μ-action becomes linear; the scaling factors are $|\lambda_\mu|$ and $|\sigma_\mu|^{-1}$. We obtain the sets $\tilde{A}(\mu)$ and $\tilde{B}(\mu)$ after projection along the leaves of $\mathscr{F}_\mu^u$ and $\mathscr{F}_\mu^s$ into l_μ.

Notice that these projections along leaves are C^1, so that $\tilde{A}(\mu)$ and $\tilde{B}(\mu)$ are like the sets $\Psi_\mu(A(\mu))$ and $\Phi_\mu(B(\mu))$ in the final result of Chapter 3. It is this situation in which the results of Chapter 3 will be applied. Before this, we still need further preparations.

Let $F_\mu \subset W^s(p_\mu)$ be a fundamental domain, i.e., a pair of two intervals in the two branches of $W^s(p_\mu)$ such that φ_μ maps one endpoint of each interval into the other endpoint, and let $\varepsilon > 0$ be a small positive number. Then there are a (small) neighborhood U of $\Lambda(0) \cup \mathcal{O}$ and a $\mu_1 > 0$, such that for each $0 < \mu < \mu_1$, whenever for some $x \in F_\mu$ the leaf of $\mathscr{F}_\mu^u$ through x contains a point of U, then there is a point x' in $A(\mu)$ such that

$$\rho(x, x') < \varepsilon \cdot \rho(p_\mu, x'),$$

where ρ measures the distance with respect to a coordinate in $W^s(p_\mu)$ which linearizes the induced φ_μ action. In the notation of [11] this means that x' is in a scaled ε-neighborhood of $A(\mu)$. We define a *scaled ε-neighborhood* of a scaled set $A \subset \mathbf{R}$ as

$$\{x \in \mathbf{R} \mid \exists\, x' \in A \text{ such that } |x - x'| < \varepsilon \cdot |x'|\};$$

such a scaled ε-neighborhood is denoted by ${}_\varepsilon A$ (note the difference from the notation in [11]); this should be distinguished from A_ε (see Chapter 3) which denotes an ε-neighborhood of A.

We also assume that μ_1 is so small that a neighborhood U as above exists so that for $0 < \mu < \mu_1$, $\Omega(\varphi_\mu) \setminus \tilde{\Omega}(\varphi_\mu) \subset U$ (see the last section for the notation). The component of U containing r is denoted by U_r; $\pi_{s,\mu}$ and $\pi_{u,\mu}$ denote the projections of U_r on $W^s_{\text{loc}}(p_\mu)$, $W^u_{\text{loc}}(p_\mu)$ along the leaves of $\mathscr{F}_\mu^u$ and $\mathscr{F}_\mu^s$. Then, from the fact that for $0 < \mu < \mu_1$, $\Omega(\varphi_\mu) \setminus \tilde{\Omega}(\varphi_\mu) \subset U$, it follows that all nonwandering points of φ_μ, $0 < \mu < \mu_1$ near r, are contained in

$$U_r \cap \pi_{s,\mu}^{-1}({}_\varepsilon A(\mu)) \cap \pi_{u,\mu}^{-1}({}_\varepsilon B(\mu))$$

(the scaled ε-neighborhoods of $A(\mu)$ and $B(\mu)$ are obtained using identifications of $W^s(p_\mu)$ and $W^u(p_\mu)$ with $\mathbf{R}$, linearizing the induced φ_μ action). In the figure below we sketch such a set.

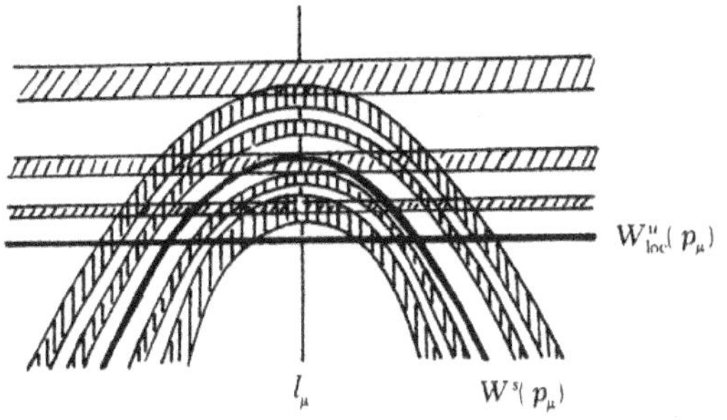

$\pi_{s,\mu}^{-1}({}_\varepsilon A(\mu))$

$\pi_{u,\mu}^{-1}({}_\varepsilon B(\mu))$

By use of fixed identifications of $W^u(p_\mu)$ and $W^s(p_\mu)$ with $\mathbf{R}$, it follows from the result in the previous section that we even have, for some constant K and $0 < \mu < \mu_1$,

$$\big(\Omega(\varphi_\mu) \cap U_r\big) \subset U_r \cap \pi_{s,\mu}^{-1}({}_\varepsilon A(\mu) \cap [-K\cdot\mu, K\cdot\mu]) \\ \cap \pi_{u,\mu}^{-1}({}_\varepsilon B(\mu) \cap [-K\cdot\mu, K\cdot\mu]).$$

We conclude this section with an estimate on the distances between the intersections of l_μ with $\pi_{s,\mu}^{-1}({}_\varepsilon A(\mu) \cap [-K\cdot\mu, K\cdot\mu])$ and with $\pi_{u,\mu}^{-1}({}_\varepsilon B(\mu) \cap [-K\cdot\mu, K\cdot\mu])$, for $0 < \mu < \mu_1$; these sets are denoted by ${}_\varepsilon\tilde{A}^{K\cdot\mu}(\mu)$ and ${}_\varepsilon\tilde{B}^{K\cdot\mu}(\mu)$. It was for this estimate that the results in Chapter 3 were derived and they will be used in Section 4 to prove hyperbolicity for many values of the parameter μ.

For these estimates we need, in the above considerations, ε and μ_1 variable. The possible choices of μ_1 depend on ε and so making one choice for each ε, we have $\mu_1(\varepsilon)$; we may, and do assume that $\mu_1(\varepsilon) \le \varepsilon$. For $0 < \mu < \mu_1(\varepsilon)$ we define

$$B_{\mu,\varepsilon} = \big\{\mu' \in (0,\mu) \mid \text{distance between } {}_\varepsilon\tilde{A}^{K\cdot\mu}(\mu') \text{ and} \\ {}_\varepsilon\tilde{B}^{K\cdot\mu}(\mu') \text{ is smaller than } \varepsilon\cdot\mu\big\}.$$

We observe that the definition of this set is closely related to the definition of the set $M_{r,\varepsilon}$ in the final result of Chapter 3. The reason why we use "B" here is that the above defined set is strongly related to the bifurcation set of $\{\varphi_\mu\}$.

PROPOSITION. *If the limit capacities $d(A(0))$ and $d(B(0))$ of $A(0)$ and $B(0)$ have sum smaller than one, then there is for each $\delta > 0$ an $\varepsilon(\delta) > 0$ such that*

$$\frac{1}{\mu}\cdot m\big(B_{\mu,\varepsilon(\delta)}\big) < \delta$$

for μ sufficiently small; m denotes Lebesgue measure.

Proof. This follows from the final result in Chapter 3 and the following observations.

1. For scaled and non-scaled neighborhoods we have the relation

$$_{\varepsilon}A^r \subset A^r_{\varepsilon \cdot r} \text{ or } {}_{\varepsilon}A \cap [-r, +r] \subset A_{\varepsilon \cdot r} \cap [-r, +r].$$

2. Because $\pi_{s,\mu}$ and $\pi_{u,\mu}$ are C^1, the ratio, between distances in $W^s_{\text{loc}}(p_\mu)$, $W^u_{\text{loc}}(p_\mu)$ and the corresponding distances in l_μ, is bounded and bounded away from zero.

4. *Hyperbolicity.* Keeping the notation of the last section we can state the result of this section as:

PROPOSITION. *For ε sufficiently small, and $\mu' \notin B_{\mu,\varepsilon}$ for some $0 < \mu' < \mu \leq \mu_1(\varepsilon)$, the nonwandering set of $\varphi_{\mu'}$ is hyperbolic and even $\varphi_{\mu'}$ is persistently hyperbolic.*

Remark. As observed in the beginning of the chapter, the main theorem of this paper is a direct consequence of the proposition above together with the estimate on $m(B_{\mu,\varepsilon(\delta)})$ in the last section.

Proof. This proof goes in the same way as the corresponding proof in [11] to which we refer for the details. Here we only want to give the geometric ideas behind the proof.

As in the last section, r is a point in the orbit of tangency, U is a small neighborhood of $\Lambda(0) \cap \mathcal{O}$, whose r-component is denoted by U_r, such that each point in $\Omega(\varphi_\mu) \setminus \tilde{\Omega}(\varphi_\mu)$ has its φ_μ-orbit passing through U_r and staying in U, for μ sufficiently small. So in order to prove hyperbolicity it is enough to show that the nonwandering points in U_r are uniformly hyperbolic.

Let now $0 < \mu' < \mu \leq \mu_1(\varepsilon)$, $\mu' \in B_{\mu,\varepsilon}$, be as in the theorem and let $x \in U_r \cap \Omega(\varphi_{\mu'})$. We want to show that at x the angle between $E^u_{\mu'}(x)$ and $E^s_{\mu'}(x)$ (the tangent directions to the corresponding invariant foliations; see Section 3) is at least of the order $\sqrt{\varepsilon \cdot \mu}$; in fact this angle is greater than or equal to $L \cdot \sqrt{\varepsilon \cdot \mu}$ where L is independent of x, μ, μ', and ε. This is based on the following three facts.

- If $F^u_{\mu'}(x)$ and $F^s_{\mu'}(x)$ are the leaves of $\mathscr{F}^u_{\mu'}$ and $\mathscr{F}^s_{\mu'}$ through x, then the intersections $l_{\mu'} \cap F^u_{\mu'}(x)$ and $l_{\mu'} \cap F^s_{\mu'}(x)$ have at least distance $\mu \cdot \varepsilon$; see the previous section.
- The curvature of a leaf of $\mathscr{F}^u_{\mu'}$ at x depends continuously on x and μ'; so, since we may assume that μ' is near zero and that x is near r, the curvature of leaves of $\mathscr{F}^u_{\mu'}$ in U_r is close to the curvature of $W^u(p)$ at r. A corresponding statement holds for the leaves of $\mathscr{F}^s_{\mu'}$.

Since the tangency of $W^u(p)$ and $W^s(p)$ at r is non-degenerate, $W^u(p)$ and $W^s(p)$ have different curvatures at r.

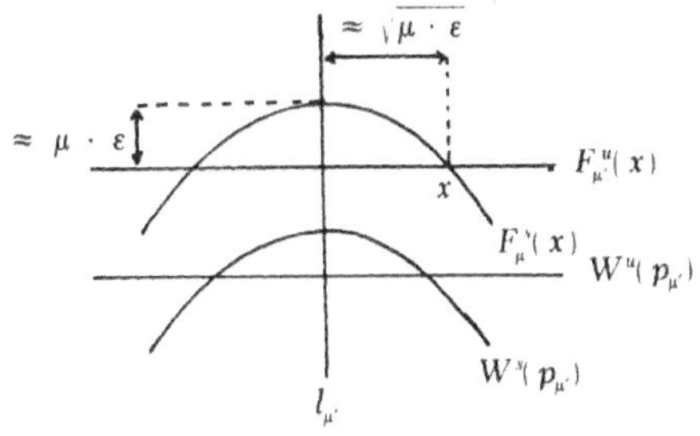

From this (see also the above figure) it follows that the distance from x to $l_{\mu'}$ is at least of the order $\sqrt{\mu \cdot \varepsilon}$. This in turn implies that the angle between $F^u_{\mu'}(x)$ and $F^s_{\mu'}(x)$ (or between $E^u_{\mu'}(x)$ and $E^s_{\mu'}(x)$) is at least of the order $\sqrt{\mu \cdot \varepsilon}$.

Next, in order to prove hyperbolicity, we construct a cone field over $\Omega(\varphi_{\mu'}) \cap U_r$; i.e., for each $x \in \Omega(\varphi_{\mu'}) \cap U_r$ we take a cone $C_{\mu'}(x) \subset T_x(M)$. The relation between such cone fields and hyperbolicity is as follows.

If, for each $x \in \Omega(\varphi_{\mu'}) \cap U_r$ and $n > 0$, such that $\varphi^n_{\mu'}(x) \in U_r$, $(d\varphi^n_{\mu'})_x C_{\mu'}(x) \subset C_{\mu'}(\varphi^n_{\mu'}(x))$ and if $(d\varphi^n_{\mu'})_x$ expands vectors in $C_{\mu'}(x)$ and contracts vectors in the complement of $(d\varphi^{-n}_{\mu'})_{\varphi^n_{\mu'}(x)} C_{\mu'}(\varphi^n_{\mu'}(x))$ by a factor which is exponential in n, then $\Omega(\varphi_{\mu'})$ is hyperbolic. This fact was already used in [11]. In the present situation we define $C_{\mu'}(x)$, for $x \in \Omega(\varphi_{\mu'}) \cap U_r$ to consist of those vectors in $T_x(M)$ which make an angle with $E^u_{\mu'}(x)$ smaller than half the angle between $E^u_{\mu'}(x)$ and $E^s_{\mu'}(x)$; see the following figure.

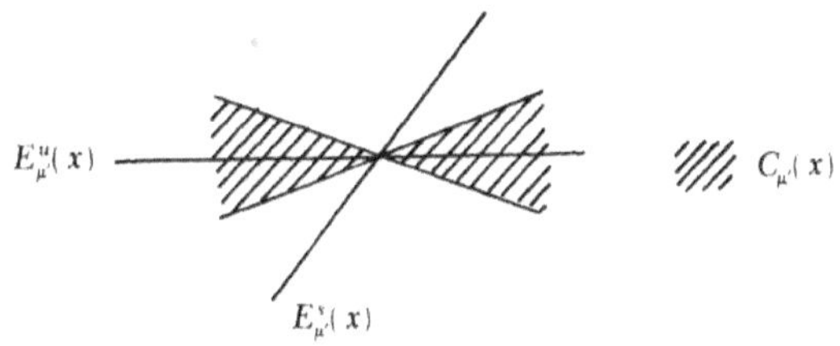

In order to see that whenever $x \in \Omega(\varphi_{\mu'}) \cap U_r$ and $\varphi^n_{\mu'}(x) \in U_r$, $n > 0$, then $(d\varphi^n_{\mu'})_x C_{\mu'}(x) \subset C_{\mu'}(\varphi^n_{\mu'}(x))$, one has to use the fact that the orbit $\varphi^i_{\mu'}(x)$ spends a long time, namely from $i = 1$ to $i = n - 1$, in the hyperbolic region, i.e., near $p_{\mu'}$ or near the basic set containing $p_{\mu'}$. Due to this fact the successive images of $C_{\mu'}(x)$ under $(d\varphi^i_{\mu})_x$, $i = 0, \ldots, n$, will have the following shapes (omitting indices μ').

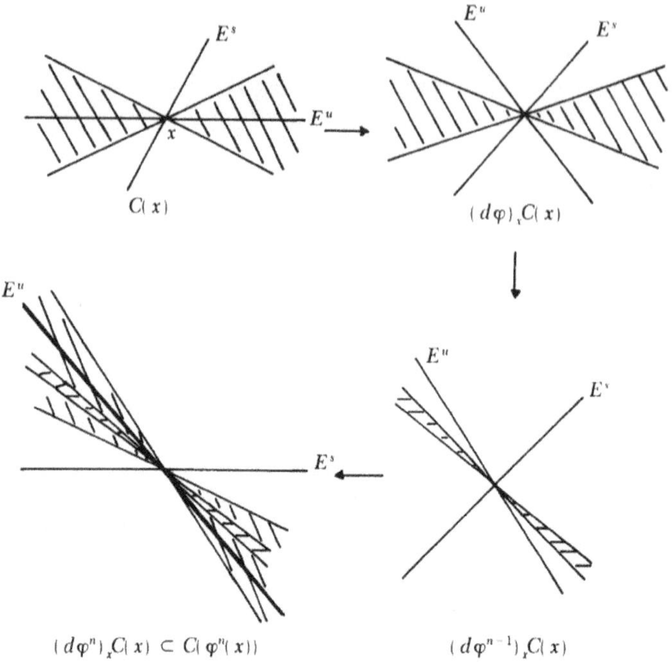

In order to justify these figures we observe that $d\varphi$ respects E^u and E^s *except* that

– $(d\varphi)_x$ does not respect the E^u direction and

– $(d\varphi)_{\varphi^{n-1}(x)}$ does not respect the E^s direction. This is a consequence of the construction of $\mathscr{F}^u$ and $\mathscr{F}^s$ as indicated in the figure below.

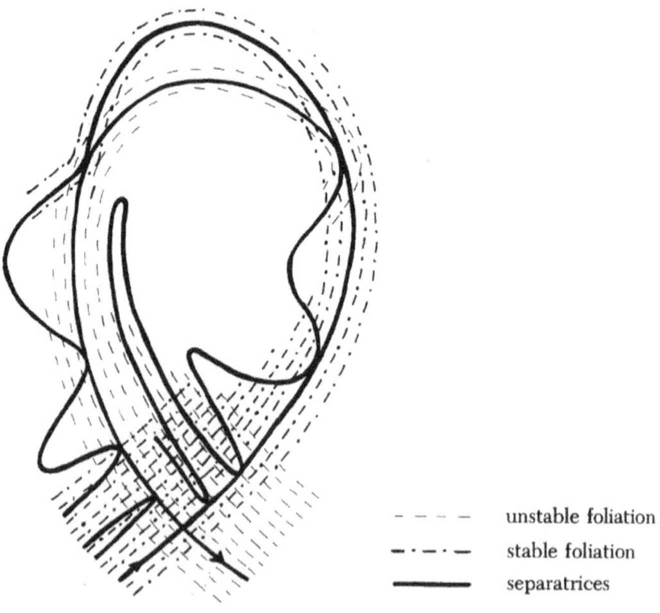

(To be formally correct it may be necessary to replace, in the above description, $\varphi(x)$ by $\varphi^{i_0}(x)$ and $\varphi^{n-1}(x)$ by $\varphi^{n-i_0}(x)$, i_0 independent of x, μ, and n. See the extension of the foliation in Section 3.)

The further estimates, see [11], are based on the fact that the smallness of the angles between $E^s_{\mu'}(x)$ and $E^u_{\mu'}(x)$, for small μ, is compensated by the long time which the orbit spends in the hyperbolic region.

To be more precise, these estimates are based on the following facts.

- The number of iterates which such an orbit spends in the hyperbolic region (even near p_μ) is at least of the order $\operatorname{Log}\mu/\operatorname{Log}|\lambda|$.
- In the hyperbolic region (near p_μ), a cone, symmetric around E^u, with opening coefficient α, see the figure below, gets mapped into a cone with opening coefficient approximately $|\lambda/\sigma|\cdot\alpha$, where λ and σ are the contracting and expanding eigenvalues of $(d\varphi)_p$:

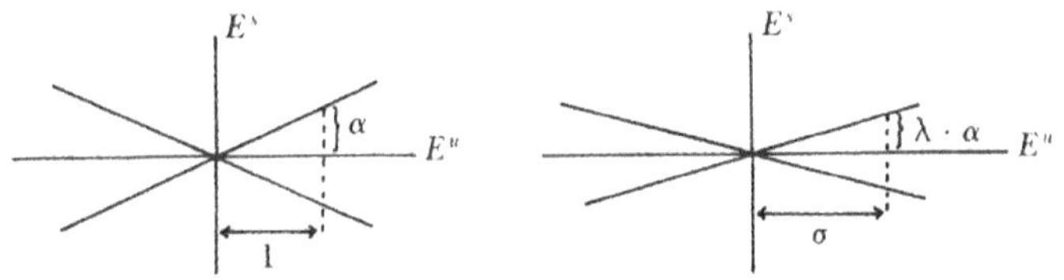

- The cone $(d\varphi_{\mu'})_x C_{\mu'}(x)$ is contained in a cone around $E^u_{\mu'}(\varphi_{\mu'}(x))$ with opening coefficient of the order $1/\sqrt{\mu\cdot\varepsilon}$ and $\mu\le\varepsilon$; see the last section.

This concludes our description of the proof of the hyperbolicity of $\varphi_{\mu'}$. From the proof it is also clear that $\varphi_{\mu'}$ is persistently hyperbolic.

Instituto Matemática Pura e Aplicada, Rio de Janiero, Brasil
University of Groningen, The Netherlands

References

[1] G. D. Birkhoff, Nouvelles recherches sur les systèmes dynamiques, Mem. Pont. Acad. Sci. Novi Lyncaei, 1 (1935), 85–216.
[2] P. Hartman, *Ordinary Differential Equations*, Wiley, New York, 1964.
[3] M. Hirsch and C. Pugh, Stable manifolds and hyperbolic sets, in *Global Analysis*, Proc. Symp. Pure Math., vol. 14, A. M. S., 1970, Providence, Rhode Island.
[4] J. M. Marstrand, Some fundamental geometrical properties of plane sets of fractional dimensions, Proc. London Math. Soc. (3) 4 (1954), 257–302.
[5] S. Newhouse and J. Palis, Bifurcations of Morse-Smale dynamical systems, in *Dynamical Systems*, Acad. Press, New York, 1973.
[6] ______, Hyperbolic nonwandering sets on two-dimensional manifolds, in *Dynamical Systems*, Acad. Press, New York, 1973.
[7] ______, Cycles and bifurcation theory, Astérisque, 31 (1976), 44–140.

[8] S. Newhouse, The abundance of wild hyperbolic sets and non-smooth stable sets for diffeomorphisms, Publ. Math. I.H.E.S. **50** (1979), 101–151.

[9] S. Newhouse, J. Palis and F. Takens, Bifurcations and stability of families of diffeomorphisms, Publ. Math. I.H.E.S. **57** (1983), 5–71.

[10] J. Palis, A note on Ω-stability, in *Global Analysis*, Proc. Symp. Pure Math., vol. 14, A. M. S., 1970, Providence, Rhode Island.

[11] J. Palis and F. Takens, Cycles and measure of bifurcation sets for two-dimensional diffeomorphism, Inventiones Math. **82** (1985), 397–422.

[12] H. Poincaré, *Les Méthodes Nouvelles de la Mécanique Céleste*, III, Gauthiers-Villars, 1899.

[13] H. Seifert and W. Threfall, *Lehrbuch der Topologie*, Chelsea, New York, 1947.

[14] M. Shub, Stabilité globale des systèmes dynamiques, Astérisque **56** (1978), 1–211.

[15] S. Smale, Diffeomorphisms with many periodic points, in *Differential and Combinatorial Topology*, Princeton University Press, 1965.

[16] ______, The Ω-stability theorem, in *Global Analysis*, Proc. Symp. Pure Math., vol. 14, A. M. S., 1970, Providence, Rhode Island.

(Received July 15, 1985)

ON THE C^1 Ω-STABILITY CONJECTURE

by J. PALIS*

In his remarkable proof of the C^1 *Stability Conjecture*, Mañé stated that from his result it became realistic to expect a proof of the similar conjecture for the nonwandering set, the C^1 *Ω-Stability Conjecture*. We show in this paper that this is indeed the case; actually, we just change the last argument in Section 1 of Mañé's paper. Besides some combinatorial reasoning, perhaps the novelty here is the following strategy: instead of showing directly that if f is Ω-stable then it satisfies Axiom A, we first show that it can be C^1 approximated by an Axiom A diffeomorphism. We then prove that f itself satisfies Axiom A. A key simple fact is that if f is Ω-stable then no cycle can exist among hyperbolic basic sets of f [7]. Using repeatedly one of Mañé's result we succeed in reaching cycles as above unless f satisfies Axiom A, thus proving that Ω-stability of a diffeomorphism implies that it satisfies Axiom A (see discussion below on the complete statement and converse).

We begin by briefly recalling some basic definitions and known facts concerning this conjecture.

Let M be a C^∞ compact manifold without boundary and let $\mathrm{Diff}^r(M)$ denote the set of C^r diffeomorphisms of M with the C^r topology for $r \geqslant 1$. For $f \in \mathrm{Diff}^r(M)$, we denote by $\Omega(f)$ its nonwandering set and by $P(f)$ its set of periodic points. We say that f is C^r structurally stable, resp. Ω-stable, if there exists a neighborhood $\mathcal{U}$ of f in $\mathrm{Diff}^r(M)$ such that for each $g \in \mathcal{U}$ there is a homeomorphism h of M, resp. $h : \Omega(f) \to \Omega(g)$, satisfying $hf(x) = gh(x)$ for all $x \in M$, resp. $x \in \Omega(f)$.

The C^r Stability Conjecture formulated in [6] states that $f \in \mathrm{Diff}^r(M)$ is structurally stable if and only if $\Omega(f)$ is hyperbolic and $\Omega(f) = \overline{P(f)}$ (Axiom A) and for every $x \in M$ its stable and unstable manifolds $W^s(x)$ and $W^u(x)$ are in general position (transversality condition).

We can formulate a similar conjecture for the Ω-stability as follows. If f satisfies Axiom A then by [8] we can write $\Omega(f)$ as a finite union of closed, transitive sets in which the periodic points are dense: $\Omega(f) = \Omega_1 \cup \ldots \cup \Omega_s$. Each Ω_i is called a basic

* I wish to thank the I.H.E.S. for its hospitality during part of the period of preparation of this paper.

Originally published in: *Publ. Math. Inst. Hautes Études Sci. no. 66 (1988), 211–215,*

set. A cycle on $\Omega(f)$ is a sequence $\Omega_{i_1}, \ldots, \Omega_{i_k}$ with points $x_1, y_1 \in \Omega_{i_1}, \ldots, x_k, y_k \in \Omega_{i_k}$ such that $W^s(x_1) \cap W^u(y_2) \neq \emptyset, \ldots, W^s(x_k) \cap W^u(y_1) \neq \emptyset$. The C^r Ω-Stability Conjecture states: f is Ω-stable if and only if it satisfies Axiom A and there are no cycles on $\Omega(f)$. The conjecture arises quite naturally from the following facts. On one hand, Smale's Ω-stability theorem [8] states that if f satisfies Axiom A and there are no cycles on $\Omega(f)$ then f is Ω-stable. On the other hand, if f satisfies Axiom A and there is a cycle on $\Omega(f)$ then by [7] f is not Ω-stable: it can be C^r approximated by g such that $\overline{P(g)} \not\approx \overline{P(f)}$.

Before showing the results we recall that Λ is a basic set for f if it is closed, hyperbolic and transitive, with a dense subset of periodic orbits. We also require it to be isolated, i.e. maximal invariant set for f in some neighborhood of it. If Λ is a basic set for f, and the maximal invariant set in some neighborhood $U(\Lambda)$, then Λ is persistent: if g is C^r close to f then there is a (unique) basic set $\Lambda(g)$ near Λ and f/Λ is conjugate to $g/\Lambda(g)$; $\Lambda(g)$ is called the continuation of Λ [2]. The stable manifold $W^s(\Lambda)$ consists of points whose ω-limit sets are in Λ; similarly for the unstable manifold $W^u(\Lambda)$. Also, as observed in [4], if f is Ω-stable then all periodic orbits of f are persistently hyperbolic: if g is near f then its periodic orbits are hyperbolic. That is, $f \in \mathscr{F}^r(M)$ where $\mathscr{F}^r(M)$ denotes the interior of the set of C^r diffeomorphisms whose periodic orbits are all hyperbolic. Finally, if $f \in \mathscr{F}^r(M)$ and Λ_i, $1 \leqslant i \leqslant s$, are basic sets for f then by [7] there can be no cycle on $\bigcup_i \Lambda_i$.

We can now present the main fact toward the proof of the Ω-Stability Conjecture.

Theorem A. — *If $f \in \mathscr{F}^1(M)$ then f can be C^1 approximated by an Axiom A diffeomorphism.*

This and our previous discussions have as immediate consequence the following result.

Corollary. — *The elements of an* open *and* dense *subset of $\mathscr{F}^1(M)$ satisfy Axiom* A *and the no-cycle condition.*

Proof of Theorem A. — Suppose first we want to show that f itself satisfies Axiom A. Since $f \in \mathscr{F}^1(M)$, following [4] we may suppose by induction that $\overline{P_k(f)}$ is hyperbolic for all $0 \leqslant k \leqslant j$, where $P_k(f)$ is the set of periodic points of index k (dimension of the stable manifold). We then have to prove that $\overline{P_{j+1}(f)}$ is hyperbolic. So let $\Lambda_1, \ldots, \Lambda_s$ be the decomposition of the union of the $\overline{P_k(f)}$ for $0 \leqslant k \leqslant j$ into (hyperbolic) basic sets. According to theorems I.4 through I.7 in [4], it is enough to show that $P_{j+1}(f)$ does not accumulate on $\bigcup_k \Lambda_k$. This is clearly the case for $P_1(f)$ since $P_0(f)$ is either empty or consists of sources. Thus we may consider only the basic sets corresponding to $\overline{P_k(f)}$ for $1 \leqslant k \leqslant j$. That f satisfies Axiom A is therefore reduced to the proof of the following statement:

(a) if $P_{j+1}(f)$ accumulates on $\bigcup_k \Lambda_k$ then we can create a cycle on $\bigcup_k \Lambda_k$ for some f^* C^1 close to f, $f^* = f$ on $\bigcup_k \Lambda_k$.

In fact, from the existence of a cycle, we can, as in [7], create new periodic points of the same index by small C^1 perturbations. We conclude that $f^* \notin \mathscr{F}^1(M)$ and so $f \notin \mathscr{F}^1(M)$ since $\mathscr{F}^1(M)$ is an open subset of $\mathrm{Diff}^1(M)$.

We now approximate f by g so that if $W^s(\Lambda_{j_1}(g))$ intersects $W^u(\Lambda_{j_2}(g))$ it does so persistently: the corresponding intersection for any $\tilde{g} \in C^1$ near g also occurs. After this we will prove that (a) above occurs for g; later in Theorem B we will show that f itself satisfies Axiom A. To see that we can take such a g close to f we consider a small neighborhood $\mathscr{U}$ of f on which the continuation of the basic sets Λ_k are defined. We claim that the set $g \in \mathscr{U}$ such that $W^s(\Lambda_{j_1}(g))$ intersects $W^u(\Lambda_{j_2}(g))$ persistently or not at all for $1 \leqslant j_1 \leqslant j_2 \leqslant s$ is of second category. This follows from the stable manifold theorems for basic sets [2], [3] including its continuous dependence on the map on compact parts and the existence of fundamental domains. For this last point, there is an easier proof in [1], [5] using a shadowing property.

So we consider such a g and introduce a partial ordering on the $\Lambda_k = \Lambda_k(g)$, $1 \leqslant k \leqslant s$, as follows: $\Lambda_{k_1} \leqslant \Lambda_{k_2}$ if $W^s(\Lambda_{k_1}) \cap W^u(\Lambda_{k_2}) \neq \emptyset$. The advantage of working with g is that such an intersection (and thus the ordering relation) is persistent. We now show how to choose a (finite) sequence of small perturbations of g so that, at each stage, either we have an Axiom A diffeomorphism or else, through the subsequent perturbation, a new pair of basic sets becomes directly ordered by the relation above and they do so in a persistent way. Clearly the statement implies the theorem: the number of basic sets is finite and we cannot reach a cycle among them since the diffeomorphisms are in $\mathscr{F}^1(M)$. To prove it, let us suppose $\overline{P_{j+1}(g)}$ to be nonhyperbolic and let Λ_{j_1} be such that $P_{j+1}(g)$ accumulates on it and, among the Λ_k with this property, Λ_{j_1} has maximum index. Then by theorem I.7 of [4], we can take Λ_{j_2} on which $P_{j+1}(g)$ accumulates and g_1 near g such that $\Lambda_{j_1}(g_1) < \Lambda_{j_2}(g_1)$ and we can make this relation persistent because of the index condition. We claim we may assume that $\Lambda_{j_1} \nless \Lambda_{j_2}$ which shows that we are indeed adding a new persistent relation $\Lambda_{j_1}(g_1) < \Lambda_{j_2}(g_1)$. In fact if $W^s(\Lambda_{j_1}) \cap W^u(\Lambda_{j_2}) \neq \emptyset$ we then choose $j_3 = j_2$ such that $P_{j+1}(g)$ accumulates on Λ_{j_3} and g_2 near g so that $\Lambda_{j_2}(g_2) < \Lambda_{j_3}(g_2)$. By persistence, we also have $\Lambda_{j_1}(g_2) < \Lambda_{j_2}(g_2)$. As in [7], using that $\Lambda_{j_2}(g_2)$ is a basic set, we can perturb g_2 to $\tilde{g}_2$ so that $\Lambda_{j_1}(\tilde{g}_2) < \Lambda_{j_3}(\tilde{g}_2)$ and, by the index condition, we may assume this relation to hold persistently. Notice that $j_3 \neq j_1$ for otherwise we reach a 2-cycle. Again, it may happen that already for g we have $\Lambda_{j_1} < \Lambda_{j_3}$. If so, as before, we do not perform any of the three previous perturbations; instead we take Λ_{j_4} in which $P_{j+1}(g)$ accumulates and g_3 near g so that $\Lambda_{j_3}(g_3) < \Lambda_{j_4}(g_3)$ and from that we get $\tilde{g}_3$ near g_3 such that $\Lambda_{j_1}(\tilde{g}_3) < \Lambda_{j_4}(\tilde{g}_3)$. In this way either we get an Axiom A diffeomorphism or else we achieve a new persistent relation $\Lambda_{j_1}(g^*) < \Lambda_{j_i}(g^*)$ for some g^* near g and some index j_i *except* in one of the following possibilities:

(b) $W^s(\Lambda_{j_1}(g))$ already intersects $W^u(\Lambda_i(g))$ for all Λ_i in which $P_{j+1}(g)$ accumulates,
(c) we reach a sequence $\Lambda_{i_1}, \ldots, \Lambda_{i_\ell}$ such that $P_{j+1}(g)$ accumulates on all of these basic sets and $\Lambda_{j_1} < \Lambda_{i_k}$ for all $1 \leqslant k \leqslant \ell$ and also there are maps $g_1, \ldots, g_\ell$ C^1 near g such that $\Lambda_{i_1}(g_1) < \Lambda_{i_2}(g_1), \ldots, \Lambda_{i_\ell}(g_\ell) < \Lambda_{i_1}(g_\ell)$.

In the second case we repeat our previous procedure choosing an element in this sequence with maximum index. Then, either we achieve an Axiom A diffeomorphism or a new persistent relation with this element as claimed, or else this sequence yields a 2-cycle for some g_k above, which is a contradiction. On the other hand, if (b) above holds, that is $\Lambda_{j_1} < \Lambda_i$ for all Λ_i on which $P_{j+1}(g)$ accumulates, then this property cannot be shared by any other basic set besides Λ_{j_1}, for otherwise we have a 2-cycle for g. So, in this last case we just put aside Λ_{j_1} and start anew: we choose some Λ_{j_2} on which $P_{j+1}(g)$ accumulates and having maximum index (except for that of Λ_{j_1}) among the basic sets which are accumulated upon by $P_{j+1}(g)$. Then, in this turn, we definitely either reach an Axiom A diffeomorphism or else we add a new persistent relation for a pair of basic sets. This follows from the fact that $\Lambda_{j_2} \not< \Lambda_{j_1}$ for, otherwise, we have a 2-cycle for g, which is not possible.

It is clear that we can now proceed inductively and the induction step is exactly the same as above. This proves statement (a) above for g which implies that $\overline{P_{j+1}(g)}$ is hyperbolic. Thus by induction $\overline{P_k(g)}$ is hyperbolic for all k and so g satisfies Axiom A. This finishes the proof of Theorem A.

We now reach the main goal of this paper proving the following theorem.

Theorem B. — *If* $f \in \mathrm{Diff}^1(M)$ *is* Ω*-stable then it satisfies Axiom* A.

Proof. — Using Theorem A we obtain an Axiom A diffeomorphism g C^1 close to f, and thus Ω-conjugate to f. As in Theorem A, by [4] we may assume that $\overline{P_k(f)}$ is hyperbolic for $0 \leqslant k \leqslant j$ and to show that $\Omega(f)$ is hyperbolic it is enough to show that $P_{j+1}(f)$ does not accumulate on $\overline{\bigcup_k P_k(f)}$. This means $\overline{P_{j+1}(f)}$ to be homogeneous in the sense that all periodic points have the same index. But the basic sets in $\Omega(g)$ are homogeneous and any pair of (periodic) points have stable and unstable manifolds that mutually intersect each other at points that belong to $\Omega(g)$. Thus if $h : \Omega(g) \to \Omega(f)$ is a conjugacy between $g/\Omega(g)$ and $f/\Omega(f)$ then it is enough to show that the image by h of each basic set in $\Omega(g)$ is homogeneous.

Suppose that this is not so. We now make use of Thom's transversality theorem and the continuous dependence of stable and unstable manifolds on the map on compact parts. By slightly perturbing f if necessary, we may assume that only one pair of periodic points of f with different indices have their stable and unstable manifolds mutually intersecting each other. Moreover, we can suppose that one of these points of intersection is in $\Omega(f)$: for hyperbolic $p, q \in P(f)$, if $W^s(p)$ and $W^u(q)$ have a point of transversal intersection then any point in $W^u(p) \cap W^s(q)$ is in $\Omega(f)$. But this is clearly absurd because f is Ω-conjugate to g, and so at least one of these two periodic orbits must be accumulated by others with a different index and having with them the mutual intersection property of stable and unstable manifolds. This concludes the proof of the theorem.

REFERENCES

[1] R. Bowen, On Axiom A diffeomorphisms, *Conference Board Math. Sciences*, **35**, *Amer. Math. Soc.*, 1977.
[2] M. Hirsch and C. Pugh, Stable manifolds and hyperbolic sets, Global Analysis, *Proc. AMS Symp. in Pure Math.*, **14** (1970), 133-163.
[3] M. Hirsch, J. Palis, C. Pugh and M. Shub, Neighborhoods of hyperbolic sets, *Invent. Math.*, **9** (1970), 121-134.
[4] R. Mañé, A proof of the C^1 Stability Conjecture, *Publ. Math. I.H.E.S.*, **66** (1987), 161-210.
[5] S. Newhouse, Lectures on dynamical systems, *Progress in Math.*, **8**, Birkhäuser, 1980.
[6] J. Palis and S. Smale, Structural stability theorems, *Proc. AMS Symp. in Pure Math.*, **14** (1970), 223-231.
[7] J. Palis, A note on Ω-stability, same *Proceedings*, **14** (1970), 221-222.
[8] S. Smale, The Ω-stability theorem, same *Proceedings*, **14** (1970), 289-297.

Instituto de Matemática Pura e Aplicada (IMPA)
Estrada Dona Castorina 110
CEP 22460 — Rio de Janeiro, RJ — Brazil.

Manuscrit reçu le 31 mars 1987.

Ann. scient. Éc. Norm. Sup.,
4e série, t. 22, 1989, p. 99 à 108.

CENTRALIZERS OF ANOSOV DIFFEOMORPHISMS ON TORI

BY J. PALIS AND J. C. YOCCOZ

ABSTRACT. — We prove here that the elements of an open and dense subset of Anosov diffeomorphisms on tori have trivial centralizers: they only commute with their own powers.

1. Introduction

Let M be a smooth connected compact manifold, and Diff(M) the group of C^∞ diffeomorphisms of M endowed with the C^∞ topology. The diffeomorphisms which satisfy Axiom A and the (strong) transversality condition—every stable manifold intersects transversely every unstable manifold—form an open subset $\mathfrak{A}(M)$ of Diff(M) and are, by Robbin [4] and a recent result of Mañé [2], exactly the C^1-structurally stable diffeomorphisms.

We continue here the study, initiated in [3], of centralizers of diffeomorphisms in $\mathcal{G}(M)$; the concepts we just mentioned are detailed there. We now treat the relevant case where M is the torus $\mathbb{T}^n$ and the diffeomorp'ˑms are such that all of $\mathbb{T}^n$ is a hyperbolic set (Anosov diffeomorphisms). Recall that for $f \in \mathrm{Diff}(M)$, its centralizer $Z(f)$ in Diff(M) is defined as the set of elements that commute with f. We say that f has trivial centralizer if $Z(f)$ is reduced to the iterates $\{f^n, n \in \mathbb{Z}\}$ of f.

CONJECTURE. — *There is an open and dense subset of $f \in \mathfrak{A}(M)$ which have trivial centralizer.*

N. Kopell [1] proved this conjecture for $M = S^1$. In [3], we investigated the conjecture in higher dimensions. We proved it for dim $M = 2$, and that (in any dimension) the centralizer is trivial for a residual set of $f \in \mathfrak{A}(M)$. Actually, one can state the question for any connected component of $\mathfrak{A}(M)$, where the topological dynamics stays the same. We then also proved the conjecture when the topological dynamics exhibit one periodic attractor or repellor [3].

The purpose of this paper is to settle the case of Anosov diffeomorphisms on tori.

THEOREM. — *For an open and dense subset of Anosov diffeomorphisms of the n-dimensional torus $\mathbb{T}^n$, the centralizer is trivial.*

Some basic facts about Anosov diffeomorphisms that we shall use in the sequel are collected in Section 2 and in Section 3 we recall a previous but recent result [3] that is

Originally published in: *Ann. Sci. École Norm. Sup. (4) 22 (1989), no. 1, 99–108,* also needed here. A basic proposition (Proposition 2) is stated in Section 4 and, from it and the result recalled in Section 3, we prove the main theorem. Sections 5 through 7 are dedicated to the proof of this basic proposition. First, in Section 5, we present a statement about the induced linear automorphism (Proposition 3) and show in Section 6 that it implies Proposition 2. Finally, in Section 7, we provide the proof of Proposition 3.

It is worthwhile noting that, while our techniques most probably can be generalized to the case of Anosov diffeomorphisms on infranilmanifolds (which are, conjecturally, the only ones to exist) they do not apply directly to an "abstract" Anosov diffeomorphism: the algebraic structure of the manifold is strongly used. We also point out that there is a version of the above conjecture for flows, which was solved by Sad [5].

2. Anosov diffeomorphisms on tori

We recall some basic material on Anosov diffeomorphisms on tori. For $A \in GL_n(\mathbb{Z}) = \mathrm{Aut}(H_1(\mathbb{T}^n, \mathbb{Z}))$, we denote by $\mathrm{Homeo}_A(\mathbb{T}^n)$ [resp. $\mathrm{Diff}_A(\mathbb{T}^n)$, resp. $\mathscr{D}_A(\mathbb{T}^n)$] the set of homeomorphisms (resp. diffeomorphisms, resp. Anosov diffeomorphisms) of M which induce A in homology. Then, $\mathscr{D}_A(\mathbb{T}^n)$ is non empty if and only if A is hyperbolic. We fix from now on some hyperbolic $A \in GL_n(\mathbb{Z})$. The main result about $\mathscr{D}_A(\mathbb{T}^n)$ is best expressed by considering the finite covering $\tilde{\mathscr{D}}_A(\mathbb{T}^n)$ of $\mathscr{D}_A(\mathbb{T}^n)$ formed by pairs $(f, p) \in \mathscr{D}_A(\mathbb{T}^n) \times \mathbb{T}^n$ such that $f(p) = p$. We recall that any $f \in \mathscr{D}_A(\mathbb{T}^n)$ has at least one fixed point.

PROPOSITION 0. — *For $(f, p) \in \tilde{\mathscr{D}}_A(\mathbb{T}^n)$, there is a unique $h = h_f$ in* $\mathrm{Homeo}_1(\mathbb{T}^n)$ *such that $hfh^{-1} = A$ and $h(p) = 0$. The map $f \mapsto h_f$ is continuous.*

Proof. — We only recall briefly the proof of unicity; more precisely, any homeomorphism h of $\mathbb{T}^n$ which commutes with A and fixes 0 belongs to $GL_n(\mathbb{Z})$. Indeed, replacing h by hB^{-1}, where B is induced by h in homology and therefore commutes with A, we can assume that $h \in \mathrm{Homeo}_1(\mathbb{T}^n)$. Then, for the lift $\tilde{h}$ of h to $\mathbb{R}^n$ which fixes the origin 0, we have

$$\sup_{\mathbb{R}^n} \| \tilde{h}(y) - y \| < +\infty.$$

From this we get, for any $y \in \mathbb{R}^n$,

$$\sup_{n \in \mathbb{Z}} \| A^n(\tilde{h}(y) - y) \| = \sup_{n \in \mathbb{Z}} \| \tilde{h}(A^n(y)) - A^n(y) \| < +\infty.$$

As A is hyperbolic, this implies that $\tilde{h}$ is the identity. ■

3. A previous basic result

The proof of the theorem is based on two results, one of which was proved in [3] and that we recall now.

For $f \in \mathfrak{A}(M)$, consider the following properties:

(i) The spectra of Df at the different fixed points of f are distinct;

(ii) At each fixed point of f, the eigenvalues of Df are simple, and there is no resonance neither between the stable eigenvalues nor between the unstable eigenvalues.

These properties are clearly satisfied by the elements of an open and dense subset $\mathfrak{A}^*(M)$ of $\mathfrak{A}(M)$. Let $\mathscr{U}$ be an open, connected, simply connected subset of $\mathfrak{A}^*(M)$ and $p: f \mapsto p(f)$ a continuous map from $\mathscr{U}$ to M such that $f(p(f)) = p(f)$. There exist finite sets I_j, $1 \leqq j \leqq 4$, and, with $I = \bigcup_{j=1}^{4} I_j$, a continuous map $\lambda = (\lambda_i)_{i \in I}$ from $\mathscr{U} \times I$ to $\mathbb{C}$ such that, for $f \in \mathscr{U}$:

– the $\lambda_i(f)$ with $i \in I_1$ (resp. $i \in I_3$) are the real stable (resp. unstable) eigenvalues of Df at $p = p(f)$;

– the $\lambda_i(f)$ with $i \in I_2$ (resp. $i \in I_4$) are the stable (resp. unstable) eigenvalues with strictly positive imaginary part of Df at p.

Let $\mathbb{K}_i$ be equal to $\mathbb{R}$ for $i \in I_1 \cup I_3$ and to $\mathbb{C}$ for $i \in I_2 \cup I_4$. By Sternberg's linearization theorem (*see* [3]), there exists, for $f \in \mathscr{U}$, a C^∞-diffeomorphism $k_s = k_s(f)$ from $\prod_{i \in I_1 \cup I_2} \mathbb{K}_i$ onto $W^s(f, p(f))$ such that $k_s^{-1} f k_s$ is the product of the homotheties of ratio $\lambda_i(f)$ on $\mathbb{K}_i$, for $i \in I_1 \cup I_2$. Furthermore, k_s is unique up to a product of homotheties in $\prod_{i \in I_1 \cup I_2} \mathbb{K}_i$, and we can choose the map $f \mapsto k_s(f)$ to be continuous. Similarly with $I_3 \cup I_4$, $W^u(f, p(f))$ and k_u. Then, for $f \in \mathscr{U}$, a diffeomorphism $g \in Z(f)$ must fix p [because of property (i)] and preserve the stable manifold $W^s(f, p(f))$ and the unstable manifold $W^u(f, p(f))$. Moreover $k_s^{-1} g k_s$ and $k_u^{-1} g k_u$ are products of homotheties on their factors $\mathbb{K}_i$, of respective ratios μ_i. We define the *compact part* of $Z(f)$ at $p(f)$ by the condition that all ratios $\operatorname{Log}|\mu_i| / \operatorname{Log}|\lambda_i|$, $i \in I$, are equal; this is a subgroup of $Z(f)$ which contains the iterates of f. We then proved in [3]:

PROPOSITION 1. — *There is an open and dense subset* $\mathscr{U}_1$ *of* $\mathscr{U}$ *such that for any* $f \in \mathscr{U}_1$ *the centralizer* $Z(f)$ *has trivial compact part at* $p(f)$.

Remark. – A similar result holds if we consider periodic orbits of a fixed period; here, we only need the case of fixed points.

4. A basic proposition and proof of the theorem

The other result which is basic to the proof of the theorem relates, for $(f, p) \in \tilde{\mathscr{D}}_A(\mathbb{T}^n)$ and $f \in \mathfrak{A}^*(\mathbb{T}^n)$, the local linearizing conjugacies k_s, k_u at p, introduced above, and the global conjugacy h_f.

Let $\mathscr{D}_1$ be the set of $(f, p) \in \tilde{\mathscr{D}}_A(\mathbb{T}^n)$ such that $f \in \mathfrak{A}^*(\mathbb{T}^n)$, and let $\mathscr{V}$ be a connected component of $\mathscr{D}_1$. We can then define I_j, $1 \leqq j \leqq 4$ as above. For $f \in \mathscr{V}$, let J be a non trivial subset of $I_1 \cup I_2$ (resp. $I_3 \cup I_4$); that is, J is neither the empty set nor the whole set. We then denote by $W_J(f)$ the image of $\prod_{i \in J} \mathbb{K}_i = \mathbb{K}_J$ under $k_s(f)$ [resp. $k_u(f)$].

PROPOSITION 2. — *There is an open and dense subset $\mathscr{V}_1$ of $\mathscr{V}$ such that for any $f \in \mathscr{V}_1$ and any non trivial subset* J *of either* $I_1 \cup I_2$ *or* $I_3 \cup I_4$, *the dimension of the linear subspace of* $\mathbb{R}^n$ *generated by* $h_f(W_J(f))$ *is strictly greater than the dimension of* $\mathbb{K}_J$.

Observe that the openness of $\mathscr{V}_1$ follows from the continuity of the maps $f \to h_f$, $f \to k_s(f)$, $f \to k_u(f)$. Before proving density, we show how to deduce the theorem from Propositions 1 and 2.

Proof of the theorem. — By Proposition 1 there is an open and dense set $\mathscr{V}_2$ in $\mathscr{V}_1$ of diffeomorphisms having centralizer with trivial compact part at p. When $\mathscr{V}$ varies among the components of $\mathscr{D}_1$, the union of these $\mathscr{V}_2$ is an open and dense subset of $\tilde{\mathscr{D}}_A(\mathbb{T}^n)$.

Let $f \in \mathscr{V}_2$, $g \in Z(f)$, and assume that g is not an iterate of f. Denote by $(\mu_i)_{i \in I}$ the ratios of the homotheties associated to g under the linearizing maps k_s, k_u. Then, as g does not belong to the compact part of $Z(f)$ at p, the ratios $\operatorname{Log}|\mu_i|/\operatorname{Log}|\lambda_i(f)|$ are not all equal. Replacing, if necessary, g by $g^k f^l$, with $k, l \in \mathbb{Z}$, we can assume that $Dg(p)$ is hyperbolic, but the stable manifold $W^s(g,p)$ of g at p is distinct from the stable manifold $W^s(f,p)$ and the unstable manifold $W^u(f,p)$ of f at p. Indeed this is true when none of these ratios is zero and they take both signs; and this last condition is satisfied by $g^k f^l$ (for some $k \geqq 1$, $l \in \mathbb{Z}$) if l/k is such that the ratios for g take values on both sides of l/k, none being equal to l/k. Replacing, if necessary, f by f^{-1}, we can assume that $W^s(f,p) \cap W^s(g,p) = W_J(f)$ for some non trivial subset J of $I_1 \cup I_2$.

The map $h_f g h_f^{-1} = B$ commutes with A and fixes O, hence it belongs to $GL_n(\mathbb{Z})$ (*see* § 2). Moreover, as $Dg(p)$ is hyperbolic, no sufficiently small neighbourhood of p contains a full orbit of g; the same must be true for B at O, which means that B is hyperbolic. Then $h_f(W_J(f))$ is equal to the intersection of the stable subspaces of A and B, a linear subspace with the same dimension as $\mathbb{K}_J$. This contradicts the definition of $\mathscr{V}_1$ and shows that any $f \in \mathscr{V}_2$ has trivial centralizer. ■

5. A statement about the induced linear automorphism

To prove that $\mathscr{V}_1$ is dense in $\mathscr{V}$, we consider some $(f,p) \in \mathscr{V}$, and a non trivial invariant subset J of $I_3 \cup I_4$, such that $h_f(W_J(f))$ is equal to a non trivial A-invariant linear subspace of the unstable subspace of A. We will show that there exist arbitrarily small perturbations (in the C^∞ topology) f' of f such that the linear subspace of $\mathbb{R}^n$ generated by $h_{f'}(W_J(f'))$ is equal to the unstable subspace of A.

We choose a minimal non zero A-invariant subspace E of $h_f(W_J(f))$; the dimension of E is 1 or 2, and the restriction of A to E is a similitude of ratio $\lambda \in \mathbb{C}$, with $|\lambda| > 1$. We denote by E^s (resp. E^u) the stable (resp. unstable) subspace of A in $\mathbb{R}^n$, and also the stable (resp. unstable) manifold $W^s(A, 0)$ in $\mathbb{T}^n$ [resp. $W^u(A, 0)$]. We choose an inner product

$\langle\ ,\ \rangle$ in $\mathbb{R}^n$, with associated norm $\|\ \|$, and a constant $\alpha \in (0, 1)$ such that the following properties hold:

(1) E^s and E^u are orthogonal for $\langle\ ,\ \rangle$;

(2) $\|A v\| \leqq \alpha \|v\|$, $\|A^{-1} w\| \leqq \alpha \|w\|$ for $v \in E^s$, $w \in E^u$;

(3) $\|A v\| = |\lambda| \|v\|$ for $v \in E$;

(4) $\|v\| \geqq 1$ for $v \in \mathbb{Z}^n - \{0\}$.

Denote by d the distance in $\mathbb{T}^n$ induced by $\|\ \|$. The restrictions of $\|\ \|$ to E^s, $E^u \subset \mathbb{R}^n$ define distances d_s, d_u on E^s, E^u and we denote by the same letters the images of these distances by the injections of E^s, E^u in $\mathbb{T}^n$. The closed ball in $\mathbb{T}^n$ (resp. $E^s \subset \mathbb{T}^n$, resp. $E^u \subset \mathbb{T}^n$) with center x and radius r for the distance d (resp. d_s, resp. d_u) will be denoted by $B(x, r)$ [resp. $B_s(x, r)$, resp. $B_u(x, r)$].

A main point in the proof of Proposition 2 is to find a certain number of homoclinic points for A near E in E^u, with "non interacting" orbits; this is made precise in the following statement.

PROPOSITION 3. — *Let* K *be a neighbourhood of* O *in* E^u, *and* m *be the dimension of* E^u. *There exist* $\delta_0 > 0$, *and, for any* $\varepsilon_0 > 0$, *points* $z_1, \ldots, z_m$ *in* E, $y_1, \ldots, y_m \in E^u$ *such that the following properties hold:*

(i) $z_i \in K$, *for* $1 \leqq i \leqq m$;

(ii) $d_u(z_i, y_i) < \varepsilon_0$, *for* $1 \leqq i \leqq m$;

(iii) $y_1, \ldots, y_m$ *is a basis of* E^u;

(iv) y_i *is homoclinic for* A *in* $\mathbb{T}^n$, *for* $1 \leqq i \leqq m$;

(v) $A^n(y_i) \notin B(z_j, \delta_0) = B_j$ *for* $n > 0$, $1 \leqq i, j \leqq m$;

(vi) $B_i \cap A^n(B_u(z_j, \delta_0)) = \varnothing$ *for* $n < 0$, $1 \leqq i, j \leqq m$;

(vii) $B_i \cap B_j = \varnothing$ *for* $i \neq j$.

We first assume Proposition 3 to be true and based on it we provide a proof of Proposition 2. We then will be left with proving Proposition 3, where only the hyperbolic linear automorphism A is involved.

6. Proof of Proposition 2

6.1. We first define the kind of perturbations of f we will consider.

We define the following subsets of $\mathbb{R}^n$:

$$K_1 = \{ x \in \mathbb{R}^n, |x_i| \leqq 1 \text{ for } 1 \leqq i \leqq n \},$$

$$K_2 = \{ x \in \mathbb{R}^n, |x_i| \leqq 1 \text{ for } 1 \leqq i \leqq m \text{ and } x_i = 0 \text{ for } m < i \leqq n \},$$

$$K_3 = \left\{ x \in \mathbb{R}^n, |x_i| \leqq \frac{1}{2} \text{ for } 1 \leqq i \leqq m \text{ and } x_i = 0 \text{ for } m < i \leqq n \right\},$$

where $x=(x_1, \ldots, x_n)$. Let $|x|=\left(\sum_{i=1}^{n} x_i^2\right)^{1/2}$. We fix some C^∞ function $\eta:[0,+\infty)\to[0,1]$, satisfying $\eta(t)=0$ for $t\geqq 1$ and $\eta(t)=1$ for $t\leqq 1/2$.

Let $\delta\in(0,1/2)$, $q_1, \ldots, q_l\in K_3$, $q'_1, \ldots, q'_l\in K_2$ $(l\geqq 1)$ be such that:

$$(1) \qquad |q_i-q_j|\geqq 2\delta \quad \text{for} \quad i\neq j;$$

$$(2) \qquad |q_i-q'_i|\leqq \frac{\delta}{2} \quad \text{for all } 1\leqq i\leqq l.$$

Then, the C^∞ vector field X in $\mathbb{R}^n$ defined by

$$X(y)=\sum_{j=1}^{l} \eta(\delta^{-1}|y-q_j|)(q'_j-q_j)$$

has support contained in the interior of K_1; it is tangent to K_2 along K_2, and for a given $y\in K_1$, by (1) at most one term in the sum defining $X(y)$ is non zero. If $(F_t)_{t\in\mathbb{R}}$ is the flow generated by X, one has, for $1\leqq i\leqq l$,

$$F_1(q_i)=q'_i.$$

Finally, given a neighbourhood $\tilde{\mathscr{W}}$ of the identity in $\operatorname{Diff}(\mathbb{R}^n)$, there exists $\varepsilon=\varepsilon(\delta,\tilde{\mathscr{W}})<\delta/2$ such that if we have the stronger condition

$$(2') \qquad |q_i-q'_i|\leqq\varepsilon \quad \text{for} \quad 1\leqq i\leqq l,$$

then F_1 belongs to $\tilde{\mathscr{W}}$.

6.2. We choose a diffeomorphism H of K_1 onto a compact neighborhood of p in $\mathbb{T}^n$, such that $H(0)=p$ and $H(K_2)\subset W^u(f,p)$. Let $\delta_0>0$ be as in Proposition 3 with $K=H(K_3)$. Let $\delta\in(0,1/2)$ be such that $|y-z|\leqq\delta$, for $y, z\in K_1$, implies that $d(h_f H(y), h_f H(z))<\delta_0$.

Let $\mathscr{W}$ be a neighbourhood of f in the C^∞ topology. We choose $\varepsilon=\varepsilon(\delta,\mathscr{W})<\delta/2$ such that if the data in (6.1) satisfy (1) and (2′), then $HF_1H^{-1}f$ belongs to $\mathscr{W}$. Then, by uniform continuity, there exists $\varepsilon_0>0$ such that if $y, z\in E^u$ satisfy $h_f^{-1}(z)\in K$, $d_u(y,z)<\varepsilon_0$, then $h_f^{-1}(y)\in H(K_2)$ and $|H^{-1}h_f^{-1}(y)-H^{-1}h_f^{-1}(z)|<\varepsilon$. For this $\varepsilon_0>0$ we apply Proposition 3 and get points $y_1, \ldots, y_m\in E^u$, $z_1, \ldots, z_m\in E$ satisfying properties (i)-(vii). We now construct as in (6.1) a flow $(F_t)_{t\in\mathbb{R}}$ with data $\delta, l=m$, $q_i=H^{-1}h_f^{-1}(z_i)$ and $q'_i=H^{-1}h_f^{-1}(y_i)$. For $0\leqq t\leqq 1$, let $f_t=HF_tH^{-1}f\in\mathscr{W}$; we may assume that $\mathscr{W}\subset\mathscr{D}_A(\mathbb{T}^n)$ and denote by h_t the global conjugacy of f_t and A (so that $h_0=h_f$).

6.3. We finally prove that f_1 belongs to $\mathscr{V}_1$, and more precisely that the linear subspace of $\mathbb{R}^n$ generated by $h_1(W_J(f_1))$ is E^u. As $\mathscr{W}$ is arbitrary and $f_1\in\mathscr{W}$, this will show that $\mathscr{V}_1$ is dense in $\mathscr{V}$.

With the notation of Proposition 3, the image under h_f of the support of HF_tH^{-1} $(0\leqq t\leqq 1)$ is contained in $\bigcup_{i=1}^{l} B_i$ (by the definition of δ); by property (vi), p does not belong to the support of HF_tH^{-1} and the manifolds $W_J(f)$, $W_J(f_1)$ coincide in a neighbourhood of p in $W^u(f, p)$. By property (v), we also have, for $1 \leqq i \leqq m$, $0 \leqq t \leqq 1$ and $n \geqq 0$:

$$f_t^n(h_f^{-1}(y_i)) = f^n(h_f^{-1}(y_i)) = h_f^{-1}(A^n(y_i)),$$

hence $h_f^{-1}(y_i)$ belongs to $W^s(f_t, p)$. On the other hand, by relation (vi), we have, for $1 \leqq i \leqq m$, $0 \leqq t \leqq 1$, and $n < 0$:

$$(3) \qquad f_t^n(h_f^{-1}(y_i)) = f^n HF_t^{-1}H^{-1}h_f^{-1}(y_i).$$

But $H^{-1}h_f^{-1}(y_i)$ belongs to K_2, hence $HF_t^{-1}H^{-1}h_f^{-1}(y_i)$ belongs to $W^u(f, p)$, and we conclude that $h_f^{-1}(y_i)$ belongs to $W^u(f_t, p)$, and indeed it is homoclinic for f_t.

As the set of homoclinic points of A is totally discontinuous, and h_t depends continuously on t, the map $t \to h_t(h_f^{-1}(y_i))$ is constant and equal to $h_0 h_f^{-1}(y_i) = y_i$. By construction of F_1, we have $HF_1H^{-1}(h_f^{-1}(z_i)) = h_f^{-1}(y_i)$ for $1 \leqq i \leqq m$, hence by relation (3) we have that $f_1^n(h_f^{-1}(y_i)) = f^n(h_f^{-1}(z_i))$ for $1 \leqq i \leqq m$, $n < 0$. But $h_f^{-1}(z_i)$ belongs to $W_J(f) \supset h_f^{-1}(E)$, and $W_J(f)$ and $W_J(f_1)$ coincide near p in $W^u(f, p)$, hence $h_f^{-1}(y_i)$ belongs to $W_J(f_1)$ for $1 \leqq i \leqq m$. As $h_1 h_f^{-1}(y_i) = y_i$, we get the conclusion we were after by property (iii) of Proposition 3. This concludes the proof of Proposition 2. ■

It remains to show Proposition 3, which will be done in the next section.

7. Proof of Proposition 3

7.1. The lemma we now state will be used at a crucial step in the proof of Proposition 3. Let $\theta > 1$, $N_0 \in \mathbb{N}^* = \mathbb{N} - \{0\}$ be such that $\theta^{N_0} > 6(N_0 + 1)$. Define $l_0 = 2(1/(\theta - 1) + N_0(\theta - 1)/3\theta)$, and denote by T the homothety of ratio θ in $\mathbb{R}$. Let $k \in \mathbb{N}^*$, and suppose we are given, for each $1 \leqq i \leqq k N_0$, a family $(L_{i,j})$ of intervals of length 1.

LEMMA. — *Let $l \geqq l_0$; assume that for each $1 \leqq i \leqq k N_0$ the centers of the intervals $L_{i,j}$ are mutually distant apart by at least $3\theta l/(\theta - 1)$. Then, the set:*

$$L = T^{k N_0}([0, l]) - \bigcup_{i=1}^{k N_0} \bigcup_j T^{k N_0 - i}(L_{i,j})$$

contains an interval of length l.

Proof. — We prove the assertion for $k = 1$; the assertion for $k \geqq 2$ follows by iterating k-times the process. The number of intervals $L_{i,j}$ (with fixed i, $1 \leqq i \leqq N_0$) contained in the interior of $[0, \theta^i l]$ is at most $[\theta^i(\theta - 1)/3\theta] + 1$, hence the number of components of L is less than:

$$1 + \sum_{i=1}^{N_0} \left(\left[\frac{\theta - 1}{3\theta} \theta^i \right] + 1 \right) \leqq 1 + N_0 + \frac{1}{3}\theta^{N_0} < \frac{1}{2}\theta^{N_0}.$$

On the other hand, the Lebesgue measure of the intersection $[0, \theta^i l] \cap (\bigcup_j L_{i,j})$ is less than $[\theta^i(\theta-1)/3\theta]+1$, hence the measure of $[0, \theta^{N_0} l]-L$ is less than:

$$\sum_{i=1}^{N_0} \theta^{N_0-i}\left(\left[\frac{\theta-1}{3\theta}\theta^i\right]+1\right) \leqq \theta^{N_0}\left(\frac{1}{\theta-1}+N_0\frac{\theta-1}{3\theta}\right)=\frac{1}{2}l_0\,\theta^{N_0},$$

For $l \geqq l_0$, the measure of L is therefore bigger than or equal to $(1/2)\,l\,\theta^{N_0}$ and the result follows.

7.2. We choose a unit vector w in E and denote by D the orthogonal subspace (with respect to $\langle\ ,\ \rangle$) of $\mathbb{R}w$ in E^u, and, for $\varepsilon \geqq 0$, by P_ε the endomorphism of E^u equal to the identity on $\mathbb{R}w$ and to the multiplication by ε on D. We may assume that $K \cap \mathbb{R}w = \{tw, |t| \leqq a\}$ for some $a>0$. We pick m distinct real numbers $u_1, \ldots, u_m$ in $((1/2)\,a\,|\lambda|^{-1}, (1/2)\,a)$ (recall that A/E is a similitude of ratio $\lambda \in \mathbb{C}$) and choose a basis $t_1, \ldots, t_m$ of E^u such that $P_0(t_i)=u_i w$ for $1 \leqq i \leqq m$.

For $\delta_1>0$, $1 \leqq i \leqq m$, let J_i (resp. J'_i) the line segment $\{u_i w+tw, 0 \leqq t \leqq \delta_1\}$ (resp. $\{t_i+tw, 0 \leqq t \leqq \delta_1\}$). Denote by V_i (resp. $V_{i,u}$) the δ_1-neighbourhood of J_i in $\mathbb{T}^n$ (resp. E^u) for the distance d (resp. d_u). We choose $\delta_1 \in (0, (1/4)\,a)$ sufficiently small so that the following properties hold:

(4) $$J_i \subset K, \qquad \text{for} \quad 1 \leqq i \leqq m;$$

(5) $$\bar V_i \cap \bar V_j = \varnothing, \qquad 0 \notin \bar V_i \quad \text{for} \ 1 \leqq i \leqq m;$$

(6) $$\bar V_i \cap A^n(\bar V_{j,u}) = \varnothing \quad \text{for all} \ 1 \leqq i, j \leqq m, \quad n<0;$$

(7) if $y_i \in J'_i$ for $1 \leqq i \leqq m$, then $y_1, \ldots, y_m$ is a basis of E^u.

Next we choose an open neighbourhood $\mathcal{O}$ of O in E^s such that $A(\mathcal{O}) \subset \mathcal{O}$ and $\mathcal{O} \cap \bar V_i = \varnothing$ for $1 \leqq i \leqq m$; let $\beta>0$ be such that for any $x \in E^u$, the ball $B_u(x, \beta)$ intersects $\mathcal{O}$. Define $r=2a+4\beta$.

7.3. In the context of the Lemma in (7.1), let $\theta=|\lambda|$, and $N_0 \geqq 1$ satisfying $\theta^{N_0}>6(N_0+1)$. This defines l_0; let N_1 be an integer such that $\theta^{N_1}>l_0 r \delta_1^{-1}$. For $\delta_0>0$, consider the union $\tilde B(\delta_0)$ of the balls $B_u(q, a)$ for $q \in B_s(0, \delta_0) \cap E^u$. We choose $\delta_0 \in (0, \delta_1/2)$ small enough to have:

(8) $A^n V'_i \cap V'_j = \varnothing$ for $0<n \leqq N_1$, $1 \leqq i, j \leqq m$, where V'_i is the δ_0-neighbourhood of J_i in $\mathbb{T}^n$;

(9) $d_u(q, q') \geqq (3\theta/(\theta-1)\,l_0+1)\,r$ for distinct q, q' in $B_s(0, \delta_0) \cap E^u$.

7.4. Let $\varepsilon_0 \in (0, \delta_0)$ be given (*see* Proposition 3). We choose $\varepsilon_1>0$ small enough so that we have:

(10) $$\| P_{\varepsilon_1}(y)-P_0(y) \| < \frac{1}{2}\varepsilon_0 \qquad \text{for} \quad y \in \bigcup_{i=1}^{m} J'_i.$$

As P_{ε_1} is a linear automorphism of E^u, any family $y_1, \ldots, y_m$ with $y_i \in P_{\varepsilon_1}(J_i')$ is a basis of E^u [*see* property (7) above]. Therefore we can choose $0<\varepsilon_2<(1/2)\,\varepsilon_0$ such that:

(11) any family $y_1, \ldots, y_m$, with $d_u(y_i, P_{\varepsilon_1}(J_i'))<\varepsilon_2$ for $1 \leqq i \leqq m$, is a basis of E^u.

We choose $k \in \mathbb{N}^*$ and define $N_2=N_1+k\,N_0$ such that $\alpha^{N_2}\beta<\varepsilon_2$ [where α is defined in property (2) of Section 5].

7.5. Fix $1 \leqq i \leqq m$; let D_1 be the linear subspace of E^u orthogonal to $A^{N_2} w$ and let $B=B_{D_1}(0, \beta)$ be the closed ball in D_1 with center O and radius β.

For $N_1<j \leqq N_2$, let L_j be the set of points x in $A^j(P_{\varepsilon_1}(J_i'))$ such that $x+A^{j-N_2}(B)$ intersects $\tilde{B}(\delta_0)$. The diameter of $A^{-N_2+j}(B)$ in E^u is at most 2β. If two points $x, x' \in L_j$ are such that $x+A^{j-N_2}(B)$ and $x'+A^{j-N_2}(B)$ intersect the same component of $\tilde{B}(\delta_0)$, we have:

$$d_u(x, x') \leqq 2a+4\beta=r. \tag{12}$$

Otherwise, by (9), we have:

$$d_u(x, x') \geqq \frac{3\theta}{\theta-1} l_0 r. \tag{13}$$

By the definition of N_1, the length of $A^{N_1}(P_{\varepsilon_1}(J_i'))$ is at least $l_0 r$; relations (12) and (13) show that we can apply the Lemma in (7.1) and conclude that there exists J_i'' of length bigger than or equal to $l_0 r$ contained in $A^{N_2}(P_{\varepsilon_1}(J_i')) - \bigcup_{j=N_1+1}^{N_2} A^{N_2-j}(L_j)$. Observe that $l_0 \geqq 1$, hence $l_0 r \geqq 2\beta$. By definition of β, there exists some $y_i' \in J_i''$ such that $y_i'+B$ intersects $\mathcal{O}$ in some point y_i''. Let $y_i=A^{-N_2}(y_i'')$, $y_i'''=A^{-N_2}(y_i')$ and $z_i=P_0(y_i''')$.

7.6. We finally check for y_i, z_i $(1 \leqq i \leqq m)$ the conclusions of Proposition 3. Clearly $y_i \in E^u$ and $z_i \in E$. Recall that $\mathcal{O} \subset E^s$.

Property (i). — We have $y_i''' \in P_{\varepsilon_1}(J_i')$, $P_0(P_{\varepsilon_1}(J_i'))=P_0(J_i')=J_i$, hence (i) is a consequence of (4).

Property (ii). — We have that

$$d_u(y_i, y_i''')<\alpha^{N_2}\beta<\varepsilon_2<\frac{1}{2}\varepsilon_0,$$

$$d_u(y_i''', z_i)<\frac{\varepsilon_0}{2} \quad [\textit{see} \text{ property (10) above}].$$

Property (iii). — This results from $d_u(y_i, y_i''')<\varepsilon_2$ and (11).

Property (iv). — Indeed $y_i'' \in E^u \cap \mathcal{O}$, and $y_i=A^{-N_2}(y_i'')$.

Property (v). — One has that $y_i \in B_i \subset V_i'$ for $1 \leqq i \leqq m$.

Let $1 \leqq i, j \leqq m$; for $1 \leqq n \leqq N_1$, $A^n(y_i)$ does not belong to V'_j according to (8) above. The same holds for $n \geqq N_2$ because then $A^n(y_i) = A^{n-N_2}(y''_i) \in \mathcal{O}$ and $\mathcal{O} \cap V'_j = \varnothing$. For $N_1 < n < N_2$, $A^n(y_i)$ does not belong to $\tilde{B}(\delta_0)$ by the construction of J''_i. As $\tilde{B}(\delta_0)$ contains $B_j \cap E^u$, we have again $A^n(y_i) \notin B_j$.

PROPERTIES (vi), (vii). – These follow immediately from (5), (6) and $\delta_0 < \delta_1$. □

REFERENCES

[1] N. KOPELL, *Commuting Diffeomorphisms, Global Analysis* (*Am. Math. Soc. Proc. Symp. Pure Math.*, Vol. 14, 1970, pp. 165-184).

[2] R. MAÑÉ, *A Proof of the C^1-Stability Conjecture* (*Publ. Math. I.H.E.S.*, Vol. 66, 1988).

[3] J. PALIS and J. C. YOCCOZ, *Rigidity of Centralizers of Diffeomorphisms* (*Ann. Sci. de l'E.N.S.*, this volume).

[4] J. ROBBIN, *A Structural Stability Theorem* (*Annals of Math.*, Vol. 94, 1971, pp. 447-493).

[5] P. SAD, *Centralizers of Vector Fields* (*Topology*, Vol. 18, 1979, pp. 97-104).

(Manuscrit reçu le 12 avril 1988,
révisé le 6 septembre 1988).

J. PALIS,
Instituto de Matemática Pura
e Aplicada (IMPA),
Estrada Dona Castorina 110,
22460 Jardim Botânico,
Rio de Janeiro, Brasil;

J. C. YOCCOZ,
Centre de Mathématiques,
École Polytechnique,
91128 Palaiseau Cedex
France.

Acta Math., 172 (1994), 91–136

Homoclinic tangencies for hyperbolic sets of large Hausdorff dimension

by

JACOB PALIS and JEAN-CHRISTOPHE YOCCOZ

Instituto de Matemática Pura e Aplicada
Rio de Janeiro, Brazil

Université de Paris-Sud
Orsay, France

Introduction

A fundamental concept in dynamics of a nongradient character is that of a homoclinic orbit, introduced by Poincaré in 1890 [P]: an orbit of intersection at large of the stable and unstable manifolds of a periodic saddle point. It is well known that when such an orbit is *transversal*, it must be accumulated by periodic saddles of the same index (dimension of the stable manifold) as the original saddle with respect to which the homoclinic orbit is doubly asymptotic, as shown by Birkhoff in two dimensions and Smale in general [Bi], [S]. In fact, in this last reference it was proved that transversal homoclinic orbits are always part of a hyperbolic Cantor set, a *horseshoe*, in which the periodic points are dense.

More recently, it has been realized that the creation and unfolding of a *homoclinic tangency*, say for a locally dissipative surface diffeomorphism, gives rise to a striking number of intricate and highly relevant dynamic phenomena: cascades of period doubling bifurcations [YA], infinitely many sinks [N], [R], [PT3], strange attractors of Hénon type [BC], [MV], and hyperbolic Cantor sets combined or not with the previous elements [NP], [PT1], [PT2]. Also, surface diffeomorphisms exhibiting a homoclinic tangency are certainly quite common among nonhyperbolic maps, i.e. maps whose limit set is not hyperbolic. Conjecturally, these homoclinically bifurcating diffeomorphisms may even be dense in the interior of the nonhyperbolic ones, which has turned out to be the case for C^∞ surface diffeomorphisms but in C^1 topology [AM].

Therefore, it seems to us that an important task in dynamics is to unfold the diffeomorphisms exhibiting a homoclinic tangency through k-parameter families and to inquire which of the above or other phenomena are more *common* or *prevalent* in terms of the Lebesgue measure in the parameter space. The main result in the present paper represents a contribution to such a program. Let us first explain it in a more informal

Originally published in: *Acta Math. 172 (1994), no. 1, 91–136,* way.

A homoclinic tangency may be associated to a single periodic orbit or more generally to a (hyperbolic) basic set. Recall that a basic set for a diffeomorphism is a compact, invariant, hyperbolic and transitive subset of the ambient manifold, which is the maximal invariant subset in some neighbourhood of it; moreover, periodic points are dense in it. We say that a basic set is nontrivial if it does not consist of a periodic orbit. It was proved in previous papers [NP], [PT1], [PT2] that *for a generic one-parameter unfolding if the Hausdorff dimension of the associated basic set is smaller than one, then the initial map exhibiting a homoclinic tangency is a Lebesgue density point of hyperbolic dynamics.* Here we prove a converse to the above statement: *if the Hausdorff dimension of the basic set is bigger than one, then for almost all one-parameter families of diffeomorphims the initial map is not a density point of hyperbolicity.*

To be somewhat more precise, let f be a surface diffeomorphism exhibiting a quadratic homoclinic tangency q between the stable and unstable manifolds of a periodic saddle point p, p being part of a basic set K with Hausdorff dimension $\mathrm{HD}(K)$ bigger than one. Amongst the germs of smooth families $(f_{s,t})$, $|s|<\eta$ and $|t|<\eta$, such that $f_{0,0}=f$, we consider those which unfold the homoclinic tangency at q with positive speed. After a local diffeomorphism in parameter space, we may assume that the homoclinic tangency happens along $t=0$. We then require that the relative variations with respect to s of the logarithms of the stable and unstable eigenvalues at $p_{s,t}$ on one hand, and of the stable and unstable Hausdorff dimensions of $K_{s,t}$ on the other, should not vanish at $s=t=0$. (Here, $p_{s,t}$ and $K_{s,t}$ indicate the continuations of p and K for $|s|$ and $|t|$ small.) These three transversality conditions define an open and dense subset $\mathcal{V}$ in the space of germs of smooth families $(f_{s,t})$, $f_{0,0}=f$; see §1. Let $\mathcal{F}^s(K_{s,t})$ and $\mathcal{F}^u(K_{s,t})$ be the stable and unstable foliations of $K_{s,t}$ and define for $\varepsilon>0$ small, $T_{s,\varepsilon}=\{t\in(-\varepsilon,\varepsilon) : \text{some leaf of } \mathcal{F}^u(K_{s,t}) \text{ is tangent near } q \text{ to some leaf of } \mathcal{F}^s(K_{s,t})\}$. We observe that often such orbits of tangency are still called homoclinic, and in fact in our case we even call them primary homoclinic tangencies, since they occur between pieces of leaves of $\mathcal{F}^u(K_{s,t})$ and $\mathcal{F}^s(K_{s,t})$ near the curves in $W^u(p)$ and $W^s(p)$ whose extreme points are p and q.

With the above notations and assumptions our result can be stated as follows.

THEOREM. *For each $f_{s,t}\in\mathcal{V}$, there is $c>0$ such that, for almost all $s\in(-\eta,\eta)$, we have*

$$\limsup_{\varepsilon\to 0}\frac{m(T_{s,\varepsilon})}{\varepsilon}>c$$

where $m(\cdot)$ indicates the Lebesgue measure of the set.

That such a statement could be true as well as its proof was much inspired by the remarkable result of Marstrand [Mar] concerning the positiveness of the Lebesgue measure of almost all linear projections of plane sets of Hausdorff dimension bigger than one. In our case, however, the situation is considerably more delicate due to the *lack of linearity and even smoothness* of the "projections" that we have to consider.

This paper is divided into four sections. The first one contains the precise setting of the problem and a more detailed statement of our result than the one presented above. It contains, moreover an indication of how the proof proceeds in the next three sections, each of them having a different character: *analytic, combinatorial* and *geometric*, respectively. The analytical part of the proof is inspired by Marstrand's theorem (§2). The most important objective in §3 is to establish a combinatorial lemma in the context of symbolic dynamics, which is one of the main new ingredients with respect to Kaufman's proof of Marstrand's theorem in [F]. Finally, in §4, we present geometric estimations on the first and second order variations with respect to parameters of the distance between stable manifolds of nearby points in a basic set.

1. The setting of the problem and statement of the result

1.1. Let M be a smooth surface and f a smooth diffeomorphism of M. Let Λ_1, Λ_2 be two (not necessarily distinct) basic sets of f, nontrivial, topologically mixing and of saddle type.

For $i=1,2$, let $p_i \in \Lambda_i$ be a periodic point. We assume that $W^s(p_1)$ and $W^u(p_2)$ have, at a point $q \in M$, a non-degenerate (i.e. quadratic) tangency.

1.2. We embed f in a smooth 2-parameter family $(f_{s,t})$ of smooth diffeomorphisms of M, with $f_{0,0}=f$. We will only be interested in the family for small s, t, i.e. $|s|, |t|<\eta$ with $\eta>0$ small enough. For small η, Λ_1 and Λ_2 have hyperbolic continuations $\Lambda_1(s,t), \Lambda_2(s,t)$ in $(-\eta,\eta)^2$. For $i=1,2$, $z \in \Lambda_i$, we denote by $z(s,t)$ the point in $\Lambda_i(s,t)$ associated to z, and write $W^s(z,s,t), W^u(z,s,t)$ for the stable and unstable manifolds of $z(s,t)$ relative to $f_{s,t}$.

Let us fix a small number $\varepsilon>0$ and local coordinates $(x,y) \in [-\varepsilon,\varepsilon]^2$ in a neighbourhood V of q such that:

— q has coordinates (0,0);

— the equation of the connected component of q in $W^u(p_2,0,0) \cap V$ is $\{y=0\}$;

— the equation of the connected component of q in $W^s(p_1,0,0) \cap V$ is $\{y=g_1(x)\}$, where $g_1 \in C^\infty([-\varepsilon,\varepsilon],[-\varepsilon,\varepsilon])$ satisfies $g_1(0)=g_1'(0)=0$, $1 \leqslant g_1''(x) \leqslant 2$ for $x \in [-\varepsilon,\varepsilon]$.

For small η, we can follow these connected components through $(-\eta,\eta)^2$. They will respectively have for equation

$$y = g_2(x,s,t) \quad (\text{with } g_2(x,0,0) \equiv 0),$$
$$y = g_1(x,s,t) \quad (\text{with } g_1(x,0,0) \equiv g_1(x)),$$

for smooth maps $g_1, g_2 \in C^\infty([-\varepsilon,\varepsilon]\times(-\eta,\eta)^2, [-\varepsilon,\varepsilon])$.

1.3. We will make three transversality hypotheses on the family $(f_{s,t})$. The first one is that the quadratic tangency of $W^s(p_1,f)$ and $W^u(p_2,f)$ unfolds generically. Using the implicit function theorem, this means that (with η small enough) we may assume that the coordinates s,t in parameter space are such that:

— for $t<0$ and all $s\in(-\eta,\eta)$, the function $x\mapsto g_1(x,s,t)-g_2(x,s,t)$ is strictly positive in $[-\varepsilon,\varepsilon]$;

— for $t=0$, the function $x\to g_1(x,s,t)-g_2(x,s,t)$ is positive and has a single zero in $[-\varepsilon,\varepsilon]$;

— for all $(x,s,t)\in[-\varepsilon,\varepsilon]\times(-\eta,\eta)^2$, we have

$$\partial_t(g_2-g_1)(x,s,t) \geqslant c > 0,$$

for some constant c (we take ε smaller if necessary).

On the other hand, there exists a neighbourhood U_1 of p_1 in $W^u_{\text{loc}}(p_1)\cap\Lambda_1$, a neighbourhood U_2 of p_2 in $W^s_{\text{loc}}(p_2)\cap\Lambda_2$, and, for $i=1,2$, a continuous map:

$$G_i\colon U_i \to C^\infty([-\varepsilon,\varepsilon]\times(-\eta,\eta)^2, [-\varepsilon,\varepsilon])$$

with the following properties:

— $G_i(p_i)=g_i$;

— for $(s,t)\in(-\eta,\eta)^2$ and $z\in U_1$ (resp. U_2), $\{y=G_i(z)(x,s,t)\}$ is the equation of the connected component of $W^s(z,s,t)\cap V$ (resp. $W^u(z,s,t)\cap V$) which corresponds (in an obvious meaning) to the component of $W^s(p_1,s,t)\cap V$ (resp. $W^u(p_2,s,t)\cap V$) considered above.

The continuity of G_1, G_2 guarantees that (restricting U_1, U_2 if necessary) we have a continuous map:

$$U_1\times U_2 \xrightarrow{T} C^\infty((-\eta,\eta),(-\eta,\eta))$$

with the following properties, for all $(z_1,z_2)\in U_1\times U_2$:

— if $t<T(z_1,z_2)(s)$, the function $x\to G_1(z_1)(x,s,t)-G_2(z_2)(x,s,t)$ is strictly positive in $[-\varepsilon,\varepsilon]$;

— if $t=T(z_1,z_2)(s)$, the same function is positive with a single zero in $[-\varepsilon,\varepsilon]$.

We can also assume that, for any $z_1 \in U_1$, $z_2 \in U_2$, $x \in [-\varepsilon, \varepsilon]$, $s, t \in (-\eta, \eta)$ we have:

$$\partial_t(G_2(z_2) - G_1(z_1))(x, s, t) \geqslant c > 0.$$

1.4. We now come to our two other transversality hypotheses.

For $s, t \in (-\eta, \eta)$, we define the unstable dimension $\Delta_1(s,t)$ of the basic set $\Lambda_1(s,t)$ of $f_{s,t}$ to be the Hausdorff dimension of $W^u_{\text{loc}}(p_1, s, t) \cap \Lambda_1(s,t)$. Similarly, the stable dimension $\Delta_2(s,t)$ of $\Lambda_2(s,t)$ is the Hausdorff dimension of $W^s_{\text{loc}}(p_2, s, t) \cap \Lambda_2(s,t)$.

We assume that

$$\Delta_1(0,0) + \Delta_2(0,0) > 1.$$

If the two basic sets Λ_1, Λ_2 coincide, $\Delta_1(0,0) + \Delta_2(0,0)$ is just the Hausdorff dimension of $\Lambda = \Lambda_1 = \Lambda_2$.

We also assume that

$$\partial_s \frac{\Delta_1(s,t)}{\Delta_2(s,t)}\bigg|_{(s,t)=(0,0)} \neq 0.$$

(It is known [Man] that Δ_1, Δ_2 are smooth functions of s, t.)

Let n_i, for $i=1,2$, be the period of the periodic point p_i of f. For $(s,t) \in (-\eta, \eta)^2$, let $\bar{\lambda}_1(s,t)$ be the logarithm of the modulus of the unstable eigenvalue of the fixed point $p_1(s,t)$ of $f^{n_1}_{s,t}$. Similarly, let $\bar{\lambda}_2(s,t)$ be the logarithm of the modulus of the stable eigenvalue of the fixed point $p_2(s,t)$ of $f^{n_2}_{s,t}$. We assume that

$$\partial_s \frac{\bar{\lambda}_1(s,t)}{\bar{\lambda}_2(s,t)}\bigg|_{(s,t)=(0,0)} \neq 0.$$

1.5. Before stating our main result, we introduce the following notations. Fix some Riemannian metric on M and denote by d_0 the associated distance.

For $i=1,2$ and $|s| < \eta$, let d_s be the distance on U_i defined by

$$d_s(z, z') = d_0(z(s,0), z'(s,0)).$$

For r small enough, let $B^i_s(r)$ be the d_s-ball in U_i of center p_i and radius r. Let $T_s(r)$ be the image of $B^1_s(r) \times B^2_s(r)$ by the map $(z_1, z_2) \mapsto T(z_1, z_2)(s)$.

THEOREM. *Under the hypotheses above, there are constants $r_1 > 0$, $c_1 > 0$ such that, if $0 < r < r_1$ and $s_0 \in (-\eta, \eta)$ the set*

$$\{(s,t) : |s - s_0| < |\log r|^{-1},\ t \in T_s(r)\}$$

has 2-dimensional Lebesgue measure bigger than $c_1 r |\log r|^{-1}$.

Remark. It is easy to see, and we will prove it later, that we have $T_s(r) \subset [-cr, cr]$, for some fixed $c > 0$. *Therefore, the conclusion of the theorem means that* $\{(s,t) : |s - s_0| <$

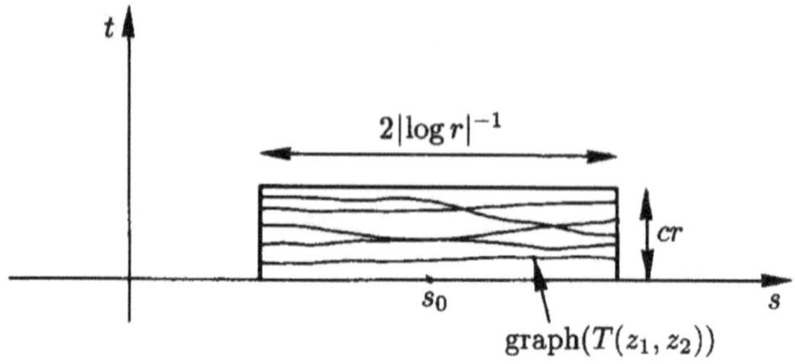

Fig. 1

$|\log r|^{-1}$, $t \in T_s(r)\}$ *has in the rectangle* $[s_0 - |\log r|^{-1}, s_0 + |\log r|^{-1}] \times [-cr, cr]$ *a relative Lebesgue measure bounded from below (independent of* s_0, r *if they are small enough).* See Figure 1.

COROLLARY. *There is a constant* $c_2 > 0$ *such that* for almost all $s \in (-\eta, \eta)$ *we have:*

$$\limsup_{r \to 0} \frac{m(T_s(r))}{r} > c_2$$

(where m is 1-dimensional Lebesgue measure).

Proof of the corollary. Take c_2 small enough. If the corollary is false, there exist $r_2 < r_1$ and a set $A \subset (-\eta, \eta)$ of positive measure such that, for $s \in A$ and $r < r_2$:

$$m(T_s(r)) < 2c_2 r.$$

Let $s_0 \in A$ and $r_3 < r_2$ be such that

$$m([s_0 - |\log r_3|^{-1}, s_0 + |\log r_3|^{-1}] \cap A) > (1 - c_2) 2 |\log r_3|^{-1}.$$

Using the remark which follows the theorem (that $m(T_s(r)) \leqslant cr$ for all $r < r_1$, $s \in (-\eta, \eta)$), we contradict the theorem if c_2 is small enough. □

1.6. The remaining part of the paper is devoted to the proof of the theorem. We give here a short account of the ideas underlying this proof.

In order to estimate the Lebesgue measure of the image of the map $T_s : (z_1, z_2) \mapsto T(z_1, z_2)(s)$, we equip $B^1_{s_0}(r) \times B^2_{s_0}(r)$ with a Radon measure μ and consider, for each s, its image ν_s under T_s. We study ν_s via its Fourier transform $\hat{\nu}_s$ (§2), and want to show that

$$\int \|\hat{\nu}_s\|^2_{L^2} \, ds < c < +\infty.$$

This would indeed easily imply that the support of ν_s (contained in the image of T_s) has positive Lebesgue measure for almost every s, and give an estimate from below for the mean value (with respect to s) of this measure.

This idea was used by Kaufman to give an elegant proof of Marstrand's theorem. The parameter s plays here the role of the angle in Marstrand's theorem, and the map T_s the role of the projection.

For this idea to work, the measure ν_s has to be absolutely continuous with respect to Lebesgue measure for almost all s. In Marstrand's theorem, this is a consequence of an energy estimate on the measure μ and of the variation of the angle. In our case, the energy estimate is essentially the same, but the map T_s depends in a much more complicated way on the parameter.

More precisely, there are various "angles" involved: at a not too small scale, the variation of the "angle" is assured by the relative variation of the logarithms of the eigenvalues (second transversality hypothesis, Proposition 1 in §3.6); at a very small scale, the variation of the angle is assured by the relative variation of the Hausdorff dimensions of Λ_1, Λ_2 (first transversality hypothesis).

The main problem arises from intermediate scales, which might create a singular part of the measures ν_s. In order to avoid this phenomenon, we have to delete, using a stopping time argument, part of the set $B^1_{s_0}(r)\times B^2_{s_0}(r)$, keeping a subset L supporting a positive proportion of μ but for which these intermediate scales do not occur (Proposition 2 in §3.6).

All these considerations rely on a quite good control of the map T_s (Proposition 3.10): approximate formulas for T_s and its first derivative with respect to the parameter s, and bounds for the second derivative. These estimates are themselves consequences of approximate formulas and bounds for the distance between stable manifolds of nearby points in a basic set, measured along these manifolds, and the variation of this distance with parameters (§4).

2. Proof of the theorem: the analysis

2.1. In this section we will give the analytical part of the proof of the theorem. As mentioned before, it is inspired by Kaufman's proof of a theorem of Marstrand presented in [F]. It requires geometrical estimates and a selection lemma that will be proved in later sections. We only state them here.

In the sequel, $c>0$, $r_1>0$, $\beta>1$ are constant independent of later choices.

Let $s_0\in(-\eta,\eta)$, $0<r<r_1$ and I be the interval $[s_0-|\log r|^{-1}, s_0+|\log r|^{-1}]$. Let

7–945201 *Acta Mathematica* 172. Imprimé le 29 mars 1994

$z_1, z'_1 \in B^1_{s_0}(r)$, $z_2, z'_2 \in B^2_{s_0}(r)$ and $\Delta_i = \Delta_i(s_0, 0)$ for $i = 1, 2$. We write

$$d = \sup(d_{s_0}(z_1, z'_1), d_{s_0}(z_2, z'_2)),$$
$$T(s) = T(z_1, z_2)(s) - T(z'_1, z'_2)(s), \quad s \in I.$$

It is assumed that there exist a measure μ_i on $B^i_{s_0}(r)$ (for $i=1,2$) and a compact set L in $B^1_{s_0}(r) \times B^2_{s_0}(r)$ such that the following properties hold:

(i) For any ball B_i of d_{s_0}-radius $\varrho \in (0, r)$ contained in $B^i_{s_0}(r)$, we have

$$c^{-1}\varrho^{\Delta_i} < \mu_i(B_i) < c\varrho^{\Delta_i}, \quad i = 1, 2;$$

(ii) For any $s \in I$, $L \subset B^1_s(r) \times B^2_s(r)$;

(iii) $\mu_1 \times \mu_2(L) > c^{-1} r^{\Delta_1 + \Delta_2}$;

(iv) Suppose that $d > r^\beta$; then the set

$$J = \{s \in I : |T(s)| < c^{-1} d\}$$

is empty or is an interval; in the last case we have, for $s \in J$:

$$|T'(s)| \geqslant c^{-1} d |\log r|,$$
$$|T''(s)| \leqslant cd |\log r|^2;$$

(v) Suppose that $d \leqslant r^\beta$, and that $(z_1, z_2) \in L$, $(z'_1, z'_2) \in L$; then the set

$$J = \{s \in I : |T(s)| < c^{-1} d^{1 + c|\log r|^{-1}}\}$$

is empty or is an interval; in the last case, we have, for $s \in J$:

$$|T'(s)| \geqslant c^{-1} d^{1 + c|\log r|^{-1}},$$
$$|T''(s)| \leqslant cd^{1 - c|\log r|^{-1}} |\log d|^2.$$

2.2. Under these assumptions we will now prove the theorem. Let μ be the restriction of $\mu_1 \times \mu_2$ to L. Let $s_0 \in (-\eta, \eta)$ and $0 < r < r_1$. For $s \in I$, let ν_s be the image of μ by the map $(z_1, z_2) \to T(z_1, z_2)(s)$. We will prove that for almost all $s \in I$, ν_s is absolutely continuous with respect to Lebesgue measure, and will obtain a bound from below for the Lebesgue measure of the support of ν_s. Because this support is contained in $T_s(r)$ (see (ii)), it will prove the theorem.

Let $\hat{\nu}_s$ be the Fourier transform of ν_s. For $p \in \mathbf{R}$, we have

$$|\hat{\nu}_s(p)|^2 = \iint_{L \times L} e^{2\pi i p T_{z,z'}(s)}\, d\mu(z)\, d\mu(z'),$$

with $z=(z_1,z_2)$, $z'=(z'_1,z'_2)$ and

$$T_{z,z'}(s)=T(z_1,z_2)(s)-T(z'_1,z'_2)(s).$$

For $p_0>0$, let

$$\mathcal{I}(p_0)=\int_I\int_{-p_0}^{+p_0}|\hat{\nu}_s(p)|^2\,dp\,ds=\iint_{L\times L}\mathcal{I}_{z,z'}(p_0)\,d\mu(z)\,d\mu(z'),$$

with

$$\mathcal{I}_{z,z'}(p_0)=\int_I\int_{-p_0}^{p_0}e^{2\pi ipT_{z,z'}(s)}\,dp\,ds=\frac{1}{\pi}\int_I\frac{\sin 2\pi p_0T_{z,z'}(s)}{T_{z,z'}(s)}\,ds.$$

2.3. Fix $z,z'\in L$ and just write T for $T_{z,z'}$. We estimate $\mathcal{I}_{z,z'}(p_0)=\mathcal{I}(p_0)$ in various cases. The letter d has the same meaning as in §2.1.

Case 1: $d>r^\beta$. With J as in §2.1 (iv), we have

$$\left|\int_{I-J}\frac{\sin 2\pi p_0T(s)}{T(s)}\,ds\right|\leqslant cd^{-1}|\log r|^{-1}.$$

On I, we use the following (classical) lemma.

LEMMA. *Let T be a C^2 monotonous function on an interval J. For any $p_0>0$:*

$$\left|\int_J\frac{\sin 2\pi pT(s)}{T(s)}\,ds\right|\leqslant c\left(\left(\inf_J|T'|\right)^{-1}+\frac{\sup_J|T''|(\sup_J T-\inf_J T)}{(\inf_J|T'|)^3}\right).$$

Proof. Let $u=T(s)$, $u_1=\sup_J T$, $u_0=\inf_J T$, $\varphi(u)=(T'\circ T^{-1}(u))^{-1}$.
One has, for $u\in[u_0,u_1]$,

$$|\varphi(u)|\leqslant\left(\inf_J|T'|\right)^{-1},$$
$$|\varphi'(u)|\leqslant\sup_J|T''|\left(\inf_J|T'|\right)^{-3}.$$

Also

$$\left|\int_J\frac{\sin 2\pi p_0T(s)}{T(s)}\,ds\right|=\left|\int_{u_0}^{u_1}\frac{\sin 2\pi p_0u}{u}\varphi(u)\,du\right|.$$

Let $\varphi_0=\varphi(0)$ if $0\in[u_0,u_1]$, $\varphi_0=\varphi(u_0)$ if $u_0>0$, $\varphi_0=\varphi(u_1)$ if $u_1<0$. We have

$$|\varphi(u)-\varphi_0|\leqslant\sup_J|T''|\left(\inf_J|T'|\right)^{-3}|u|,$$

hence

$$\int_{u_0}^{u_1}\frac{|\sin 2\pi p_0u|}{|u|}|\varphi(u)-\varphi_0|\,du\leqslant\sup_J|T''|\left(\inf_J|T'|\right)^{-3}(u_1-u_0);$$

On the other hand

$$\left|\int_{u_0}^{u_1} \frac{\sin 2\pi p_0 u}{u}\varphi_0\,du\right| = |\varphi_0|\left|\int_{2\pi p_0 u_0}^{2\pi p_1 u_1} \frac{\sin u}{u}\,du\right| \leqslant c|\varphi_0|,$$

which gives the lemma.

Using the lemma, the definition of J and §2.1 (iv), we get in Case 1:

$$|I_{z,z'}(p_0)| \leqslant cd^{-1}|\log r|^{-1}.$$

Case 2: $d<r^\beta$. Again we have, with the definition of J in §2.1 (v):

$$\left|\int_{I-J} \frac{\sin 2\pi p_0 T(s)}{T(s)}\,ds\right| \leqslant cd^{-1-c|\log r|^{-1}}|\log r|^{-1}.$$

On J, we use the lemma with the estimates of §2.1 (v):

$$\left|\int_{J} \frac{\sin 2\pi p_0 T(s)}{T(s)}\,ds\right| \leqslant cd^{-1-3c|\log r|^{-1}}|\log d|^2,$$

and we conclude that

$$|\mathcal{I}_{z,z'}(p_0)| \leqslant cd^{-1-3c|\log r|^{-1}}|\log d|^2.$$

2.4. Let $0<\varrho<2r$. By §2.1 (i), the set of (z,z') in $L\times L$ for which

$$d = \sup(d_{s_0}(z_1,z_1'), d_{s_0}(z_2,z_2')) < \varrho$$

has $\mu\times\mu$-measure at most $c(r\varrho)^{\Delta_1+\Delta_2}$.

Consequently:

$$\iint_{2^{-n}r\leqslant d<2^{1-n}r} \mathcal{I}_{z,z'}(p_0)\,d\mu(z)\,d\mu(z') \leqslant cr^{2(\Delta_1+\Delta_2)}2^{-n(\Delta_1+\Delta_2)}A_n,$$

with

$$A_n = \begin{cases} r^{-1}2^n|\log r|^{-1} & \text{if } 2^{-n}r > r^\beta, \\ (r/2^n)^{-1-3c|\log r|^{-1}}|\log(r/2^n)|^2 & \text{if } 2^{-n}r \leqslant r^\beta. \end{cases}$$

We take η small enough to have $\Delta_1+\Delta_2>c>1$ and r small enough to have

$$3c|\log r|^{-1} < \tfrac{1}{2}(\Delta_1+\Delta_2)\cdot\frac{\beta_{-1}}{\beta}.$$

Then we get

$$\sum_{n\geqslant 0} 2^{-n(\Delta_1+\Delta_2)}A_n \leqslant cr^{-1}|\log r|^{-1},$$

$$\mathcal{I}(p_0) \leqslant cr^{2\Delta_1+2\Delta_2-1}|\log r|^{-1}.$$

Letting p_0 go to ∞, we conclude that for almost all $s \in I$, ν_s has an L^2-density χ_s with respect to Lebesgue measure and that

$$\int_I \|\chi_s\|_{L^2}^2 \, ds \leqslant c r^{2\Delta_1+2\Delta_2-1} |\log r|^{-1}.$$

Therefore there is a set $A \subset I$ of Lebesgue measure $\geqslant c^{-1}|\log r|^{-1}$ such that, for $s \in A$:

$$\|\chi_s\|_{L^2}^2 \leqslant c r^{2\Delta_1+2\Delta_2-1}.$$

On the other hand, the total mass of ν_s satisfies:

$$\|\chi_s\|_{L^1} = \nu_s(\mathbf{R}) = \mu(L) > c^{-1} r^{\Delta_1+\Delta_2}.$$

By the Cauchy–Schwarz inequality, for $s \in A$, the support of ν_s has Lebesgue measure at least $c^{-1}r$. The theorem is proved. □

3. The selection lemma

Our goal in this section is twofold: after recalling some basic material on subshifts of finite type, we express the transversality hypothesis on Hausdorff dimensions in the theorem in a convenient form; we then proceed to construct a set L satisfying the assumptions of §2.1.

More precisely, the contents of this section are as follows. In §§3.1–3.5, we recall the basic facts that we need concerning subshifts of finite type and Gibbs measures. In §3.6, we translate our transversality hypotheses in the symbolic dynamics setting. The end of the section is then devoted to check, from the geometrical estimates given in Proposition 3 in §3.13, the conditions (i)–(v) of §2.1, for an appropriate set L. An outline is first given in §3.7.

3.1. Consider an integer $r \geqslant 2$, and a subshift of finite type Σ^+ of the unilateral full shift on r symbols $\{1, \ldots, r\}$. Let σ be the shift map, and $A = (a_{ij})_{1 \leqslant i,j \leqslant r}$ be the transition matrix determining Σ^+. We assume in the sequel that (Σ^+, σ) is topologically mixing.

For $\underline{x} = (x(l))_{l \geqslant 0}$ and $\underline{y} = (y(l))_{l \geqslant 0}$ in Σ^+, define:

$$v(\underline{x}, \underline{y}) = \inf\{l \geqslant 0 : x(l) \neq y(l)\},$$
$$d(\underline{x}, \underline{y}) = \exp(-v(\underline{x}, \underline{y})).$$

Then (Σ^+, d) is a compact ultrametric space, whose balls of positive radius are called cylinders. For a cylinder C, we denote by $v(C)$ the integer such that the diameter of C is $\exp(-v(C))$.

For any continuous function φ on Σ^+, we write $S_n\varphi=\sum_{l=0}^{n-1}\varphi\circ\sigma^l$ for $n\geqslant 0$. For $k,n\geqslant 0$ and $m\geqslant k+n$, $x\in\Sigma^+$, we have:

$$\min_{v(x,y)\geqslant m} S_{k+n}\varphi(y)\geqslant \min_{v(x,y)\geqslant m} S_n\varphi(y)+\min_{v(\sigma^n x,z)\geqslant m-n} S_k\varphi(z),$$
$$\min_{v(x,y)\geqslant m} S_{k+n}\varphi(y)\leqslant \min_{v(x,y)\geqslant m} S_n\varphi(y)+\max_{v(\sigma^n x,z)\geqslant m-n} S_k\varphi(z),$$
$$\min_{v(x,y)\geqslant m} S_{k+n}\varphi(y)\leqslant \max_{v(x,y)\geqslant m} S_n\varphi(y)+\min_{v(\sigma^n x,z)\geqslant m-n} S_k\varphi(z).$$

3.2. Let φ be a strictly positive continuous function on Σ^+. For distinct $x,y\in\Sigma^+$, define, with $m=v(x,y)$:

$$d_\varphi(x,y)=\exp\big(-\min_{v(x,z)\geqslant m} S_m\varphi(z)\big).$$

Putting also $d_\varphi(x,x)=0$, it follows from the inequalities above that d_φ is an ultrametric distance on Σ^+. Moreover, for distinct x,y and $0\leqslant n\leqslant v(x,y)=m$, we have:

$$\min_{v(x,z)\geqslant m} S_n\varphi(z)\leqslant \log\frac{d_\varphi(\sigma^n x,\sigma^n y)}{d_\varphi(x,y)}\leqslant \max_{v(z,x)\geqslant m} S_n\varphi(z).$$

We set $d=d_1$. The identity maps: $(\Sigma^+,d)\to(\Sigma^+,d_\varphi)$, $(\Sigma^+,d_\varphi)\to(\Sigma^+,d)$ are Hölder continuous. The balls of positive radius for d_φ are the cylinders; we write $|B|_\varphi$ for the d_φ-diameter of a subset B of Σ^+.

3.3. Let δ_φ be the Hausdorff dimension of (Σ^+,d_φ). For $1\leqslant j\leqslant r$, let

$$\Sigma_j^+=\{\underline{x}=(x(l))_{l\geqslant 0}\in\Sigma^+ : x(0)=j\}.$$

For $n\geqslant 0$, denote by Σ_j^n the set of cylinders C satisfying $C\subset\Sigma_j^+$, $v(C)\geqslant n+1$ which are maximal with these properties. They form a finite partition of Σ_j^+.

PROPOSITION. *Let $n\geqslant 0$. We have:*

$$\max_{1\leqslant j\leqslant r}\sum_{C\in\Sigma_j^n}\exp\big(-\delta_\varphi\min_C S_n\varphi\big)\geqslant 1,$$
$$\min_{1\leqslant j\leqslant r}\sum_{C\in\Sigma_j^n}\exp\big(-\delta_\varphi\max_C S_n\varphi\big)\leqslant 1.$$

Proof. For $\delta>0$ and a finite family $\mathcal{B}=(B_1,\dots,B_s)$ of cylinders, let

$$H_\delta(\mathcal{B})=\sum|B_i|_\varphi^\delta.$$

First, let $\delta<\delta_\varphi$; consider, for each $1\leqslant j\leqslant r$, a finite covering $\mathcal{B}_j$ of Σ_j^+ by cylinders. The proposition is trivial for $n=0$, so let $n\geqslant 1$. Let $1\leqslant j\leqslant r$, $C\in\Sigma_j^n$; let $1\leqslant k\leqslant r$ be such that $\sigma^n(C)=\Sigma_k^+$, and $\mathfrak{X}\mathcal{B}_C$ be the covering of C whose image under σ^n is $\mathcal{B}_k$; by §3.2, we have:

$$H_\delta(\mathcal{B}_C)\leqslant \exp\bigl(-\delta\min_C S_n\varphi\bigr)H_\delta(\mathcal{B}_k).$$

Therefore, if $\mathcal{C}_j$ is the covering of Σ_j^+ given by the various $\mathcal{B}_C$, $C\in\Sigma_j^n$, we have:

$$\max_{1\leqslant j\leqslant r} H_\delta(\mathcal{C}_j)\leqslant D\max_{1\leqslant k\leqslant r} H_\delta(\mathcal{B}_k),$$
$$D=\max_{1\leqslant j\leqslant r}\sum_{C\in\Sigma_j^n}\exp\bigl(-\delta\min_C S_n\varphi\bigr).$$

If we had $D\leqslant 1$, we would get arbitrarily fine coverings with bounded H_δ, contradicting $\delta<\delta_\varphi$. This gives the first inequality in the proposition.

Let now $\delta>\delta_\varphi$, $n\geqslant 1$. Let, for some $1\leqslant j\leqslant r$, $\mathcal{B}$ be a finite covering of Σ_j^+ by cylinders B with $v(B)\geqslant n+1$. For $C\in\Sigma_j^n$, let $\mathfrak{X}\mathcal{B}_C$ be the covering of C by those elements of $\mathcal{B}$ which meet C; if $\sigma^n(C)=\Sigma_k^+$, let $\mathcal{C}_C$ be the covering of Σ_k^+ image of $\mathcal{B}_C$ under σ^n. We have:

$$H_\delta(\mathcal{B})=\sum_{C\in\Sigma_j^n} H_\delta(\mathcal{B}_C),$$

and, by §3.2:

$$H_\delta(\mathcal{B}_C)\geqslant \exp\bigl(-\delta\max_C S_n\varphi\bigr)H_\delta(\mathcal{C}_C),$$

hence

$$H_\delta(\mathcal{B})\geqslant D'\inf_{C\in\Sigma_j^n} H_\delta(\mathcal{C}_C),$$
$$D'=\min_{1\leqslant j\leqslant r}\sum_{C\in\Sigma_j^n}\exp\bigl(-\delta\max_C S_n\varphi\bigr).$$

As $\mathcal{C}_c$ has fewer elements than $\mathcal{B}$ and $\delta>\delta_\varphi$, we must have $D'\leqslant 1$, proving the second inequality of the proposition. □

Corollary ([MM], [PV]). *The map $\varphi\mapsto\delta_\varphi$, defined on the strictly positive continuous functions on Σ^+, is continuous.*

Proof. Let $C_+(\Sigma^+)$ be the space of strictly positive continuous functions on Σ^+. For $\varphi\in C_+(\Sigma^+)$ and $n\geqslant 1$, define $\delta_n^\pm(\varphi)$ by:

$$\max_{1\leqslant j\leqslant r}\sum_{C\in\Sigma_j^n}\exp\bigl(-\delta_n^+(\varphi)\min_C S_n\varphi\bigr)=1,$$
$$\min_{1\leqslant j\leqslant r}\sum_{C\in\Sigma_j^n}\exp\bigl(-\delta_n^-(\varphi)\max_C S_n\varphi\bigr)=1;$$

we have

$$\delta_n^-(\varphi) \leqslant \delta_\varphi \leqslant \delta_n^+(\varphi)$$

by the proposition. On the other hand, the maps δ_n^+, δ_n^-, for $n \geqslant 1$, form an equicontinuous family on $C_+(\Sigma^+)$, and the sequence $(\delta_n^+ - \delta_n^-)_{n\geqslant 1}$ converge uniformly to 0 on compact subsets of $C_+(\Sigma^+)$. The corollary follows. □

3.4. Let $\gamma > 0$, and $C^\gamma(\Sigma^+)$ be the Banach algebra of Hölder continuous functions of exponent γ on (Σ^+, d). Here we simplify the notation, indicating an element $\underline{x} \in \Sigma^+$ by $x \in \Sigma^+$. For $\varphi \in C^\gamma(\Sigma^+)$ and $x, y \in \Sigma^+$, we have

$$|S_n\varphi(x) - S_n\varphi(y)| \leqslant C(\varphi),$$

for $n \leqslant v(x,y)$ and some constant $C(\varphi)$, v being defined as in (3.1) above. This is called *the bounded oscillation property of Birkhoff sums.*

For $\psi \in C^\gamma(\Sigma^+)$, the Perron–Frobenius operator $L_\psi: C^\gamma(\Sigma^+) \to C^\gamma(\Sigma^+)$ is defined by:

$$L_\psi(\chi)(x) = \sum_{\sigma y = x} \chi(y) \exp(-\psi(y)).$$

We recall Ruelle's theorem, and the relation to Hausdorff dimension ([Bo1], [Bo2], [Man]).

The spectrum of L_ψ is formed by a simple eigenvalue $\varrho_\psi > 0$ and a compact set contained in $\{|z| < \varrho_\psi\}$.

The eigenfunction h_ψ associated to ϱ_ψ is strictly positive; the complementary invariant hyperplane is the kernel of a probability measure ν_ψ on Σ^+, satisfying $L_\psi^*(\nu_\psi) = \varrho_\psi \nu_\psi$.

Normalizing h_ψ by $\nu_\psi(h_\psi) = 1$, the probability measure $\mu_\psi = h_\psi \nu_\psi$ is invariant under σ and ergodic.

3.5. The map $L: \psi \mapsto L_\psi$ from $C^\gamma(\Sigma^+)$ to $\mathcal{L}(C^\gamma(\Sigma^+))$ is analytic, with differential given by:

$$D_\psi L(\Delta\psi)(\chi) = L_\psi(\chi \Delta\psi).$$

The map $\varrho: \psi \mapsto \varrho_\psi$ from $C^\gamma(\Sigma^+)$ to $\mathbf{R}$ is analytic, with:

$$D_\psi \varrho(\Delta\psi) = \varrho_\psi \int \Delta\psi \, d\mu_\psi.$$

Let $\varphi \in C^\gamma(\Sigma^+)$, $\varphi > 0$. The Hausdorff dimension δ_φ of (Σ_A^+, d_φ) is the unique $\delta > 0$ such that $\varrho(\delta\varphi) = 1$. The map $\varphi \mapsto \delta_\varphi$ is analytic on $C_+^\gamma(\Sigma^+)$, with:

$$D_\varphi \delta(\Delta\varphi) = -\frac{\int \Delta\varphi \, d\mu_\psi}{\int \varphi \, d\mu_\psi} \delta_\varphi,$$

where $\psi = \delta_\varphi \varphi$.

Finally, let B be a d_φ-ball of radius r, $n = v(B)$, and A a measurable subset of B; we have:

$$c^{-1} r^\delta \leqslant \mu(B) \leqslant c r^\delta,$$
$$c^{-1} \frac{\mu(A)}{\mu(B)} \leqslant \mu(\sigma^n(A)) \leqslant c \frac{\mu(A)}{\mu(B)},$$

where $\mu = \mu_\psi$, $\delta = \delta_\varphi$ and c depend only on γ, $\|\varphi\|_\gamma$ and $\|\varphi^{-1}\|_0$.

3.6. Let us now translate in the setting of symbolic dynamics our geometrical transversality hypotheses.

Using a Markov partition for Λ_1, we choose a subshift of finite type Σ_1 of the full bilateral left-shift on symbols $\{1, \dots, r_1\}$, and a homeomorphism $h_1 : \Sigma_1 \to \Lambda_1$ such that:

$$h_1 \circ \sigma = f \circ h_1.$$

Similarly, we choose a subshift of finite type Σ_2 of the full bilateral left-shift on symbols $\{1, \dots, r_2\}$, and a homeomorphism $h_2 : \Sigma_2 \to \Lambda_2$ such that

$$h_2 \circ \sigma = f^{-1} \circ h_2.$$

Replacing if necessary f by some iterate, we assume that both subshifts are topologically mixing.

For $i \in \{1, 2\}$, let Σ_i^+ be the one-sided shifts on symbols $\{1, \dots, r_i\}$ and $\pi_i : \Sigma_i \to \Sigma_i^+$ be the canonical projections.

We recall that there are continuous linear operators $\Pi_i : C^\gamma(\Sigma_i) \to C^\gamma(\Sigma_i^+)$ and $\Theta_i : C^\gamma(\Sigma_i) \to C^\gamma(\Sigma_i)$ such that, for $\psi \in C^\gamma(\Sigma_i)$ $(i = 1, 2)$:

$$\Pi_i(\psi) \circ \pi_i = \psi + \Theta_i(\psi) - \Theta_i(\psi) \circ \sigma,$$

where $\pi_i : \Sigma_i \to \Sigma_i^+$ is the canonical projection.

We have fixed some Riemannian metric on M. For $z \in \Sigma_1$ and $s, t \in (-\eta, \eta)$, let:

$$\lambda_1(z, s, t) = \log \|T_{h_1(z)(s,t)} f_{s,t}|_{E^u}\|,$$

where E^u is the unstable subspace of $f_{s,t}$ at the point $h_1(z)(s,t)$ of the basic set. Similarly, for $z \in \Sigma_2$, $s, t \in (-\eta, \eta)$ let

$$\lambda_2(z, s, t) = \log \|T_{h_2(z)(s,t)} f_{s,t}^{-1}|_{E^s}\|.$$

We may assume that the Riemannian metrics is such that

$$\lambda_i(z,s,t) \geqslant c > 0, \quad i=1,2,$$

for $z\in\Sigma_i$, $s,t\in(-\eta,\eta)$.

For $s,t\in(-\eta,\eta)$, $i=1,2$, let $\lambda_i(s,t)$ be the map $z\mapsto\lambda_i(z,s,t)$ from Σ_i to $\mathbf{R}$. There exists $\gamma>0$ such that $\lambda_i\colon(s,t)\to\lambda_i(s,t)$ is a smooth map from $(-\eta,\eta)^2$ to $C^\gamma(\Sigma_i)$.

Let $\varphi_i(s,t)=\Pi_i(\lambda_i(s,t))$; then $(s,t)\mapsto\varphi_i(s,t)$ is a smooth map from $(-\eta,\eta)^2$ to $C^\gamma(\Sigma_i^+)$.

Let $(s,t)\in(-\eta,\eta)$. It is well-known (and we will prove in §4) that the composition

$$W^u_{\mathrm{loc}}(p_1,s,t)\cap\Lambda_1(s,t) \xrightarrow{h^{-1}_{1,s,t}} \Sigma_1 \xrightarrow{\pi_1} (\Sigma_1^+, d_{\varphi_1(s,t)})$$

is a biLipschitz homeomorphism on a neighbourhood of a_1 in Σ_1^+. Therefore the Hausdorff dimension $\Delta_1(s,t)$ of $W^u_{\mathrm{loc}}(p_1,s,t)\cap\Lambda_1(s,t)$ is the same as the Hausdorff dimension $\delta_1(s,t)$ of $(\Sigma_1^+, d_{\varphi_1(s,t)})$.

Similarly, the Hausdorff dimension $\Delta_2(s,t)$ of $W^s_{\mathrm{loc}}(p_2,s,t)\cap\Lambda_2(s,t)$ is the same as the Hausdorff dimension $\delta_2(s,t)$ of $(\Sigma_2^+, d_{\varphi_2(s,t)})$.

For $(s,t)\in(-\eta,\eta)^2$, let

$$\Delta\varphi_i(s,t)=\frac{\partial}{\partial s}(\varphi_i(s,t))=\Pi_i\left(\frac{\partial}{\partial s}\lambda_i(s,t)\right).$$

Let $\mu_{i,s,t}=\mu_{\psi_i(s,t)}$, with $\psi_i(s,t)=\delta_i(s,t)\varphi_i(s,t)$.

From §3.5 we have that

$$\frac{\partial}{\partial s}\log\delta_i(s,t)=-\frac{\mu_{i,s,t}(\Delta\varphi_i(s,t))}{\mu_{i,s,t}(\varphi_i(s,t))}$$

(and a similar formula holds with $\partial/\partial t$).

Taking η small enough, the transversality hypothesis on the Hausdorff dimensions is therefore equivalent to:

$$|\mu_{1,s,t}(\varphi_1(s,t))\mu_{2,s,t}(\Delta\varphi_2(s,t))-\mu_{2,s,t}(\varphi_2(s,t))\mu_{1,s,t}(\Delta\varphi_1(s,t))|\geqslant c>0.$$

For the eigenvalues of the periodic orbit, we have, for $i=1,2$:

$$\bar{\lambda}_i(s,t)=\sum_{j=0}^{n_i-1}\lambda_i(\sigma^j(a_i),s,t)=\sum_{j=0}^{n_i-1}\varphi_i(s,t)(\sigma^j a_i).$$

Hence the transversality hypothesis on the eigenvalues in §1.5 means that (taking η small enough) we have:

$$|S_{n_1}\varphi_1(s,t)(a_1)S_{n_2}\Delta\varphi_2(s,t)(a_2)-S_{n_2}\varphi_2(s,t)(a_2)S_{n_1}\Delta\varphi_1(s,t)a_1)|\geqslant c>0.$$

3.7. Let us go back to the setting and conditions of §2.1. We identify $W^u_{\text{loc}}(p_1, s, t) \cap \Lambda_1(s,t)$ with a neighbourhood of a_1 in Σ_1^+ and $W^s_{\text{loc}}(p_2, s, t) \cap \Lambda_2(s,t)$ with a neighbourhood of a_2 in Σ_2^+.

Let us fix $s_0 \in (-\eta, \eta)$ and set, for $i=1,2$:

$$\begin{aligned} &\varphi_i = \varphi_i(s_0, 0), \quad \Delta\varphi_i = \Delta\varphi_i(s_0, 0), \\ &\mu_i = \mu_{i, s_0, 0}, \\ &S_i = S_{n_i}\varphi_i(a_i), \quad \Delta S_i = S_{n_i}\Delta\varphi_i(a_i), \\ &J_i = \int \varphi_i \, d\mu_i, \quad \Delta J_i = \int \Delta\varphi_i \, d\mu_i. \end{aligned}$$

We use d_{φ_i} as distance on Σ_i^+ and denote by $B_i(r)$ the d_{φ_i}-ball of center a_i, radius r. Our transversality hypotheses mean that:

$$S_1 \Delta S_2 \neq S_2 \Delta S_1, \quad J_1 \Delta J_2 \neq J_2 \Delta J_1.$$

Let $z_1, z_1' \in B_1(r)$, $z_2, z_2' \in B_2(r)$ as in §2.1, and

$$T(s) = T(z_1, z_2)(s) - T(z_1', z_2')(s).$$

The main point of properties (iv), (v) in §2.1 is that $|T(s)|$ and $|T'(s)|$ should not *both* be too small at the same time.

Let $\nu_1 = v(z_1, z_1')$, $\nu_2 = v(z_2, z_2')$.

It will be a consequence of the geometrical estimates of Proposition 3 below that if both $|T(s)|$ and $|T'(s)|$ are small, then

$$\begin{aligned} D_1 &= |S_{\nu_1}\varphi_1(z_1) - S_{\nu_2}\varphi_2(z_2)|, \\ D_2 &= |S_{\nu_1}\Delta\varphi_1(z_1) - S_{\nu_2}\Delta\varphi_2(z_2)|, \end{aligned}$$

are both bounded.

That this cannot happen when ν_1, ν_2 are not too large follow from the hypothesis $S_1 \Delta S_2 \neq S_2 \Delta S_1$: this is the content of Proposition 1 below and will imply conditions (iv) of §2.1.

For large ν_1, ν_2, the Birkhoff sums above are related at most points, through Birkhoff's ergodic theorem, to the mean values of the functions considered.

Roughly speaking, we would like to have, for most z_1, z_2:

$$D_1 \text{ bounded} \implies \nu_1/\nu_2 \approx J_2/J_1;$$

hence

$$D_2 \approx \nu_2 |J_2/J_1 \Delta J_1 - \Delta J_2| \geqslant c^{-1} \nu_2$$

as $J_2 \Delta J_1 \neq J_1 \Delta J_2$.

The set L in condition (v) of §2.1 is thus constructed by first deleting exceptional points for Birkhoff's ergodic theorem.

But we have also to take the intermediate values of ν_1, ν_2 into account, which are covered neither by Proposition 1 nor by Birkhoff's theorem. This is done using a stopping-time argument. The precise construction of L is done in Proposition 2 below.

The rest of this section is as follows:

— in §3.8, we state and prove Proposition 1;

— in §3.9, we state Proposition 2, which is then proved in §§3.10–3.12;

— in §3.13, we state Proposition 3, to be proven in §4;

— in §§3.14 and 3.15 we finally deduce conditions (i)–(v) of §2.1 from Propositions 1, 2, 3.

3.8. PROPOSITION 1. *Assume that* $S_1 \Delta S_2 \neq S_2 \Delta S_1$. *Then there exist constants* $c_0 > 0$, $r_1 > 0$, $\beta_0 > 1$ *such that the following property holds: let* $0 < r < r_1$, $m_i = v(B_i(r))$, $\nu_i \in \mathbf{N}$ (*for* $i = 1, 2$); *if*

$$1 \leqslant \inf\left(\frac{\nu_1}{m_1}, \frac{\nu_2}{m_2}\right) \leqslant \beta_0,$$

then, for all $z_1 \in B_1(r)$, $z_2 \in B_2(r)$, *we have:*

$$\max(|S_{\nu_1}\varphi_1(z_1) - S_{\nu_2}\varphi_2(z_2)|, |S_{\nu_1}\Delta\varphi_1(z_1) - S_{\nu_2}\Delta\varphi_2(z_2)|) \geqslant c_0^{-1} |\log r|.$$

Proof. In the following, we write c for various positive constants depending only on z_i, r_i, φ_i, $\Delta\varphi_i$. Moreover, the dependence on φ_i, $\Delta\varphi_i$ is only through $\|\varphi_i\|_\gamma$, $\|\Delta\varphi_i\|_\gamma$, $\|\varphi_i^{-1}\|_0$, S_i, ΔS_i, J_i, ΔJ_i. We use repeatedly the bounded oscillation property for the Birkhoff sums of φ_i, $\Delta\varphi_i$, which follows from the Hölder continuity.

With $\varepsilon > 0$ small enough, assume that

$$|S_{\nu_1}\varphi_1(z_1) - S_{\nu_2}\varphi_2(z_2)| < \varepsilon |\log r|,$$

and for instance $m_1 \leqslant \nu_1 \leqslant m_1(1+\varepsilon)$, $m_2 \leqslant \nu_2$.

We have, for $i = 1, 2$:

$$\left| S_{m_i}\varphi_i(z_i) - \frac{m_i}{n_i} S_i \right| < c,$$

$$\left| \frac{m_i}{n_i} S_i - |\log r| \right| < c;$$

as φ_i is positive, we get:

$$c^{-1}(\nu_i-m_i)-c<S_{\nu_i}\varphi_i(z_i)-\frac{m_i}{n_i}S_i<c(\nu_i-m_i)+c,$$
$$-\varepsilon|\log r|<S_{\nu_1}\varphi_1(z_1)-S_{\nu_2}\varphi_2(z_2)<c+c(\nu_1-m_1)-c^{-1}(\nu_2-m_2).$$

For r_1 small enough, this implies:

$$\nu_2\leqslant m_2(1+c\varepsilon),$$
$$\left|S_{\nu_i}\varphi_i(z_i)-\frac{m_i}{n_i}S_i\right|<c\varepsilon|\log r|.$$

We obtain also in a similar way:

$$\left|S_{\nu_i}\Delta\varphi_i(z_i)-\frac{m_i}{n_i}\Delta S_i\right|<c\varepsilon|\log r|.$$

If we had also:

$$|S_{\nu_1}\Delta\varphi_1(z_1)-S_{\nu_2}\Delta\varphi_2(z_2)|<\varepsilon|\log r|,$$

we would get, as $c|\log r|\geqslant m_i\geqslant c^{-1}|\log r|$:

$$|S_1\Delta S_2-S_2\Delta S_1|\leqslant\frac{n_2}{m_2}\left(\left|S_1\left(\frac{m_2}{n_2}\Delta S_2-\frac{m_1}{n_1}\Delta S_1\right)\right|+\left|\Delta S_1\left(\frac{m_1}{n_1}S_1-\frac{m_2}{n_2}S_2\right)\right|\right)\leqslant c\varepsilon,$$

a contradiction for ε small enough. We take $\beta_0=1+\varepsilon$ and $c_0=\varepsilon^{-1}$. □

3.9. Proposition 2. *Assume that* $J_1\Delta J_2\neq J_2\Delta J_1$.

There exist constants $c_1,c_2>0$ *and, for any* $M>0$, *constants* $r(M)>0$, $\varepsilon(M)>0$ *such that, for any* $0<r<r(M)$, *we can find a compact subset* L *of* $B_1(r)\times B_2(r)$ *with the following properties*:

(i) $\mu_1\times\mu_2(L)>\varepsilon(M)\mu_1(B_1(r))\mu_2(B_2(r))$;

(ii) *for any distinct* $y_1,z_1\in B_1(r)$, $y_2,z_2\in B_2(r)$ *such that* $(y_1,y_2),(z_1,z_2)\in L$, *we have*:

$$\sup(|S_{\nu_1}\varphi_1(z_1)-S_{\nu_2}\varphi_2(z_2)|,|S_{\nu_1}\Delta\varphi_1(z_1)-S_{\nu_2}\Delta\varphi_2(z_2)|)\geqslant\sup(M,c_1(\nu_1+\nu_2-c_2|\log r|))$$

where $\nu_i=v(y_i,z_i)$ *for* $i=1,2$.

3.10. *Proof of Proposition* 2. *We may for instance assume that* $J_1\Delta J_2>J_2\Delta J_1$. We will write μ for $\mu_1\times\mu_2$.

Let $\eta>0$ be a small positive constant, to be chosen later, independent of M. As μ_i is ergodic for $i=1,2$, we can find a compact subset $K_i\subset\Sigma_i^+$ with $\mu_i(K_i)>1-\eta$ and an integer n_0 such that, for $n\geqslant n_0$ and $z_i\in K_i$, we have:

$$|S_n\varphi_i(z_i)-nJ_i|<\eta n,$$
$$|S_n\Delta\varphi_i(z_i)-n\Delta J_i|<\eta n.$$

For a cylinder $C\subset\Sigma_i^+$, we define

$$K_i(C)=\sigma^{-m}(K_i)\cap C,\quad m=v(C);$$

if η is small enough (independently of C), we have:

$$\mu_i(K_i(C))\geqslant \tfrac{1}{2}\mu_i(C)\quad \text{(see §3.5).}$$

LEMMA 1. *Assume that* $\eta<c^{-1}$. *For* $i=1,2$, *let* $z_i\in\Sigma_i^+$ *and* q_i,ν_i *be integers such that*

$$\bar\nu_i=\nu_i-q_i\geqslant n_0,\quad \sigma^{q_i}(z_i)\in K_i.$$

Then we have

$$\begin{aligned}(S_{\nu_2}\Delta\varphi_2(z_2)-S_{\nu_1}\Delta\varphi_1(z_1))+|S_{\nu_2}\varphi_2(z_2)-S_{\nu_1}\varphi_1(z_1)|\\ \geqslant (S_{q_2}\Delta\varphi_2(z_2)-S_{q_1}\Delta\varphi_1(z_1))-|S_{q_2}\varphi_2(z_2)-S_{q_1}\varphi_1(z_1)|+c(\bar\nu_1+\bar\nu_2).\end{aligned}$$

Proof. Writing

$$S_{\nu_i}\varphi_i=S_{q_i}\varphi_i+S_{\bar\nu_i}\varphi_i\circ\sigma^{q_i},$$

we get, as $\bar\nu_i\geqslant n_0$ and $\sigma^{q_i}(z_i)\in K_i$:

$$|S_{\nu_i}\varphi_i(z_i)-S_{q_i}\varphi_i(z_i)-\bar\nu_iJ_i|<\eta\bar\nu_i,$$
$$|S_{\nu_i}\Delta\varphi_i(z_i)-S_{q_i}\Delta\varphi_i(z_i)-\bar\nu_i\Delta J_i|<\eta\bar\nu_i.$$

It then follows that:

$$|S_{\nu_2}\varphi_2(z_2)-S_{\nu_1}\varphi_1(z_1)|\geqslant|\bar\nu_2J_2-\bar\nu_1J_1|-|S_{q_2}\varphi_2(z_2)-S_{q_1}\varphi_1(z_1)|-\eta(\bar\nu_1+\bar\nu_2);$$
$$S_{\nu_2}\Delta\varphi_2(z_2)-S_{\nu_1}\Delta\varphi_1(z_1)\geqslant\bar\nu_2\Delta J_2-\bar\nu_1\Delta J_1+S_{q_2}\Delta\varphi_2(z_2)-S_{q_1}\Delta\varphi_1(z_1)-\eta(\bar\nu_1+\bar\nu_2).$$

It remains to see that:

$$|\bar\nu_2J_2-\bar\nu_1J_1|+\bar\nu_2\Delta J_2-\bar\nu_1\Delta J_1\geqslant c(\bar\nu_1+\bar\nu_2).$$

But we have:

$$\bar{\nu}_2(J_1\Delta J_2-J_2\Delta J_1)=J_1(\bar{\nu}_2\Delta J_2-\bar{\nu}_1\Delta J_1)+\Delta J_1(\bar{\nu}_1 J_1-\bar{\nu}_2 J_2),$$
$$\bar{\nu}_1(J_1\Delta J_2-J_2\Delta J_1)=J_2(\bar{\nu}_2\Delta J_2-\bar{\nu}_1\Delta J_1)+\Delta J_2(\bar{\nu}_1 J_1-\bar{\nu}_2 J_2),$$

which implies the last inequality since $J_1>0$, $J_2>0$, $J_1\Delta J_2>J_2\Delta J_1$. □

3.11. We now proceed to the construction of L. Fix $M>0$, which we may assume big, and $0<r<r(M)$, with $r(M)$ small, to be determined later. For $i=1,2$, let $m_i=v(B_i(r))$. For $z_i\in B_i(r)$, we have:

$$|S_{m_i}\varphi_i(z_i)-|\log r||\leqslant c,$$
$$c^{-1}|\log r|\leqslant m_i\leqslant c|\log r|.$$

We distinguish two cases. (Recall that we are assuming that $J_1\Delta J_2>J_2\Delta J_1$.)

Case 1: $S_{m_2}\Delta\varphi_2(a_2)\geqslant S_{m_1}\Delta\varphi_1(a_1)$.

We have

$$\mu_i(K_i(B_i(r)))\geqslant \tfrac{1}{2}\mu_i(B_i(r)).$$

With an integer $m_0=m_0(M)\geqslant n_0$ to be chosen later, pick cylinders $C_i\subset B_i(r)$ satisfying:

$$m_i+m_0+c\geqslant v(C_i)\geqslant m_i+m_0,$$
$$\mu_i(K_i(B_i(r))\cap C_i)\geqslant \tfrac{1}{2}\mu_i(C_i),$$

and define

$$L=(C_1\cap K_1(B_1(r)))\times(C_2\cap K_2(B_2(r))).$$

We have

$$\mu(L)\geqslant \tfrac{1}{4}\mu_1(C_1)\mu_2(C_2)\geqslant \tfrac{1}{4}e^{-c(m_0+c)}\mu_1(B_1(r))\mu_2(B_2(r)),$$

hence condition (i) in Proposition 2 is satisfied if $\varepsilon(M)<\frac{1}{4}e^{-c(m_0+c)}$. Let $y_1,y_2,z_1,z_2,\nu_1,\nu_2$ be as in condition (ii) of Proposition 2.

We have

$$|S_{m_1}\varphi_1(z_1)-S_{m_2}\varphi_2(z_2)|\leqslant c$$

and, by the hypothesis of Case 1:

$$S_{m_2}\Delta\varphi_2(z_2)-S_{m_1}\Delta\varphi_1(z_1)\geqslant -c.$$

Also, $\sigma^{m_i}(z_i)\in K_i$ and $\nu_i-m_i=\bar{\nu}_i\geqslant m_0\geqslant n_0$; with $m_i=a_i$, we then get from Lemma 1:

$$\begin{aligned}(S_{\nu_2}\Delta\varphi(z_2)-S_{\nu_1}\Delta\varphi_1(z_1))+|S_{\nu_2}\varphi_2(z_2)-S_{\nu_1}\varphi_1(z_1)|&\geqslant c((\nu_1+\nu_2)-(m_1+m_2))-c'\\&\geqslant cm_0-c'.\end{aligned}$$

With $m_0 > cM$ and c_1, c_2^{-1} small enough, this implies the inequality in condition (ii) of Proposition 2.

3.12. *Case* 2: $S_{m_2}\Delta\varphi_2(a_2) < S_{m_1}\Delta\varphi_1(a_1)$.

The construction of L is more intricate, involving a stopping time argument.

Consider the family $\mathcal{F}$ of products $C = C_1 \times C_2$, where C_i is a cylinder contained in $B_i(r)$, such that, with $q_i = v(C_i)$, there exist points $u_1 \in C_1$, $u_2 \in C_2$ with:

$$|S_{q_1}\varphi_1(u_1) - S_{q_2}\varphi_2(u_2)| < M,$$
$$|S_{q_1}\Delta\varphi_1(u_1) - S_{q_2}\Delta\varphi_2(u_2)| < M.$$

We order $\mathcal{F}$ by inclusion, and denote by $\mathcal{F}_0$ the subfamily of maximal elements of $\mathcal{F}$. For $B > 0$, let $\mathcal{F}_B$ be the subfamily of $\mathcal{F}_0$ formed by the products $C = C_1 \times C_2 \in \mathcal{F}_0$ with:

$$v(C_i) \leqslant B|\log r|, \quad i = 1, 2.$$

LEMMA 2. *For* $M, B > c$, *we have:*

$$K_1(B_1(r)) \times K_2(B_2(r)) \subset \bigcup_{\mathcal{F}_B} C.$$

Proof. Let $u_i \in K_i(B_i(r))$, for $i = 1, 2$. For all $m \geqslant m_1$, select an integer $\tau(m) \geqslant m_2$ such that

$$|S_m\varphi_1(u_1) - S_{\tau(m)}\varphi_2(u_2)| \leqslant c,$$
$$\tau(m_1) = m_2, \quad \tau(m+1) \geqslant \tau(m).$$

(this is possible because $\varphi_1, \varphi_2 > 0$).

Let $\Delta_m = S_{\tau(m)}\Delta\varphi_2(u_2) - S_m\Delta\varphi_1(u_1)$ for $m \geqslant m_1$. We have $\tau(m+1) \leqslant \tau(m) + c$, hence:

$$|\Delta_m - \Delta_{m+1}| \leqslant c,$$

and, by the hypothesis of Lemma 2:

$$-c|\log r| < \Delta_{m_1} < c.$$

We also clearly have

$$c^{-1}m \leqslant \tau(m) \leqslant cm.$$

With $r(M)$ small enough, apply Lemma 1, taking $q_i = m_i$ and $B'|\log r| < \nu_i < B|\log r|$, $\nu_2 = \tau(\nu_1)$, $B' > c$; we get:

$$\Delta_{\nu_1} > \Delta_{m_1} - c' + c(\nu_1 + \nu_2 - m_1 - m_2) > 0,$$

hence there exists $m_1 \leqslant m \leqslant \nu_1$ such that

$$|S_m\varphi_1(u_1)-S_{\tau(m)}\varphi_2(u_2)| \leqslant c,$$
$$|\Delta_m| = |S_m\Delta\varphi_1(u_1)-S_{\tau(m)}\Delta\varphi_2(u_2)| \leqslant c.$$

Let C_1 be the smallest cylinder with $v(C_1)<m$ containing u_1, and C_2 be the smallest cylinder with $v(C_2)<\tau(m)$ containing u_2. We have

$$m-c \leqslant v(C_1) < m \leqslant \nu_1 < B|\log r|,$$
$$\tau(m)-c \leqslant v(C_2) < \tau(m) \leqslant \nu_2 < B|\log r|,$$

hence $C=C_1\times C_2$ belongs to $\mathcal{F}$; this proves Lemma 2. □

For $C\in\mathcal{F}_0$, define:

$$W_1(C) = \bigcup_{\substack{C'\in\mathcal{F}_0\\ C\cap C'\neq\varnothing}} C',$$
$$W_2(C) = \bigcup_{\substack{C'\in\mathcal{F}_0\\ C\cap C'\neq\varnothing}} W_1(C').$$

Let also $U=\bigcup_{C\in\mathcal{F}_B} C$.

Choose elements $C^1, ..., C^N$ of $\mathcal{F}_B$, with N maximal ($\mathcal{F}_B$ is finite), such that:

$$C^{i+1} \not\subset \bigcup_{j=0}^{i} W_2(C^j), \quad 1\leqslant i<N.$$

We then have

$$U\subset \bigcup_{j=1}^{N} W_2(C^j).$$

With an integer $m_0=m_0(M)\geqslant n_0$ to be chosen later, select as in Case 1, for $1\leqslant j\leqslant N$, a product of cylinders $\widehat{C}^j=\widehat{C}_1^j\times\widehat{C}_2^j\subset C^j=C_1^j\times C_2^j$ such that, for $i=1,2$:

$$m_0+v(C_i^j) \leqslant v(\widehat{C}_i^j) \leqslant m_0+c+v(C_i^j),$$
$$\mu_i(\widehat{C}_i^j\cap K_i(C_i^j)) \geqslant \tfrac{1}{2}\mu_i(\widehat{C}_i^j).$$

See Figure 2.

Finally, define

$$L = \bigcup_{j=1}^{N} (\widehat{C}_1^j\cap K_1(C_1^j))\times(\widehat{C}_2^j\cap K_2(C_2^j)).$$

8 – 945201 *Acta Mathematica* 172. Imprimé le 29 mars 1994

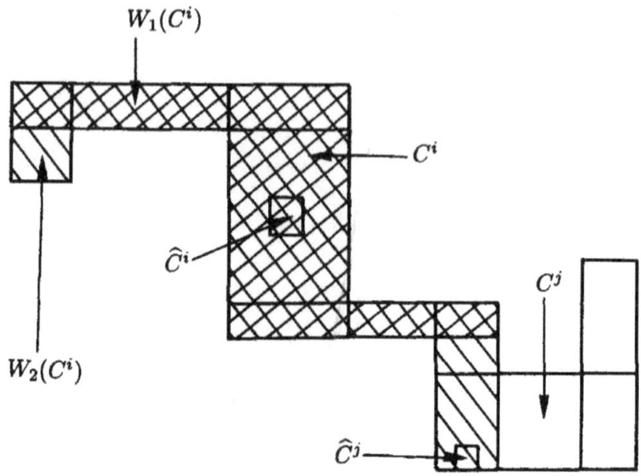

Fig. 2

We first check condition (i) of Proposition 2.

The C^j being disjoint by construction, we have:

$$\mu(L) \geqslant \tfrac{1}{4}\Sigma\mu(\widehat{C}^j) \geqslant \tfrac{1}{4}e^{-c(m_0+c)}\Sigma\mu(C^j).$$

On the other hand, if $C, C' \in \mathcal{F}_0$ satisfy $C \cap C' \neq \varnothing$, we must have for instance $C_1 \subset C_1'$, $C_2 \supset C_2'$; because φ_1, φ_2 are positive, this implies:

$$|v(C_i) - v(C_i')| \leqslant cM.$$

From this, it follows, for $C \in \mathcal{F}_0$:

$$\mu(W_1(C)) \leqslant ce^{cM}\mu(C),$$
$$\mu(W_2(C)) \leqslant ce^{cM}\mu(C).$$

By Lemma 2 we have:

$$U \supset K_1(B_1(r)) \times K_2(B_2(r)).$$

Therefore, we obtain

$$\tfrac{1}{4}\mu(B_1(r) \times B_2(r)) \leqslant \mu(U) \leqslant \Sigma\mu(W_2(C^j)) \leqslant ce^{cM}\Sigma\mu(C^j),$$
$$\mu(L) \geqslant \tfrac{1}{16}c^{-1}e^{-c(M+m_0+c)}\mu(B_1(r) \times B_2(r)),$$

and condition (i) is satisfied provided

$$\varepsilon(M) < \tfrac{1}{16}c^{-1}e^{-c(M+m_0+c)}.$$

Let $y_1, y_2, z_1, z_2, \nu_1, \nu_2$ be as in condition (ii) of Proposition 2. Let $1 \leqslant j, k \leqslant N$ be such that $(y_1, y_2) \in \widehat{C}^j$, $(z_1, z_2) \in \widehat{C}^k$.

First assume that $j<k$. Then, for $i=1,2$:

$$\nu_i \leqslant v(C_i^j) \leqslant B|\log r|,$$

hence, with c_2^{-1} small enough

$$\max(M, c_1(\nu_1+\nu_2-c_2|\log r|)) = M;$$

if the inequality in Proposition 2 was not valid, the minimal product $\widetilde{C}=\widetilde{C}_1 \times \widetilde{C}_2$ containing (y_1, z_1) and (y_2, z_2) (with $v(\widetilde{C}_i)=\nu_i$) would belong to $\mathcal{F}$. But then $\widetilde{C} \subset W_1(C^j)$ and $C^k \subset W_2(C^j)$, contradicting the choice of the C^l.

Assume now that $j=k$. With $q_i = v(C_i^j)$, we apply Lemma 1 to get:

$$\begin{aligned}&|S_{\nu_2}\Delta\varphi_2(z_2) - S_{\nu_1}\Delta\varphi_1(z_1)| + |S_{\nu_2}\varphi_2(z_2) - S_{\nu_1}\varphi_1(z_1)| \\ &\quad \geqslant (S_{q_2}\Delta\varphi_2(z_2) - S_{q_1}\Delta\varphi_1(z_1)) - |S_{q_2}\varphi_2(z_2) - S_{q_1}\varphi_1(z_1)| + c(\nu_1+\nu_2-q_1-q_2).\end{aligned}$$

From the definition of $\mathcal{F}, \mathcal{F}_B$, we have:

$$\begin{aligned}|q_i| &\leqslant B|\log r|,\\ |S_{q_2}\varphi_2(z_2) - S_{q_1}\varphi_1(z_1)| &\leqslant c+M,\\ |S_{q_2}\Delta\varphi_2(z_2) - S_{q_1}\Delta\varphi_1(z_1)| &\leqslant c+M,\end{aligned}$$

and the inequality of the proposition follows, provided c_1, c_2^{-1} are small enough and $m_0 > cM$ (recall that $\nu_i - q_i \geqslant m_0$).

The proof of Proposition 2 is complete. □

3.13. We now state the geometrical estimates proved in §4, and deduce from them the assumptions in §2.1.

PROPOSITION 3. *Let η be small enough. There exist constants $r_2>0$, $c>0$ such that, if $z_1, z_1' \in \Sigma_1^+$, $z_2, z_2' \in \Sigma_2^+$ satisfy:*

$$\begin{aligned}d_0(z_i, a_i) &< r_2 \quad \textit{for } i=1,2,\\ d_0(z_i', a_i) &< r_2 \quad \textit{for } i=1,2,\end{aligned}$$

then, writing

$$\begin{aligned}\tau_1(s) &= \log|T(z_1, z_2)(s) - T(z_1', z_2)(s)|,\\ \tau_2(s) &= \log|T(z_1', z_2)(s) - T(z_1', z_2')(s)|,\\ \nu_i &= v(z_i, z_i') \quad \textit{for } i=1,2,\end{aligned}$$

the following estimates, for $i=1,2$, hold:

$$\left|\tau_i(s)+\sum_{j=0}^{\nu_i-1}\lambda_i(\sigma^j z_i, s, T(z_1,z_2)(s))\right|\leqslant c,$$

$$\left|\frac{d}{ds}\left(\tau_i(s)+\sum_{j=0}^{\nu_i-1}\lambda_i(\sigma^j z_i, s, T(z_1,z_2)(s))\right)\right|\leqslant c,$$

$$\left|\frac{d^2}{ds^2}\tau_i(s)\right|\leqslant c\nu_i^2,$$

$$|T(z_1,z_2)(s)|\leqslant c\sup[d_s(z_1,z_1),d_s(z_2,a_2)]=cd_s,$$

$$\left|\frac{d}{ds}T(z_1,z_2)(s)\right|\leqslant cd_s|\log d_s|.$$

3.14. In the context of §2.1, let $r_1>0$, $\beta>1$ to be determined later.

Let $s_0\in(-\eta,\eta)$, $0<r<r_1$, $I=[s_0-|\log r|^{-1}, s_0+|\log r|^{-1}]$. Let $z_1,z_1'\in B^1_{s_0}(r)$, $z_2,z_2'\in B^2_{s_0}(r)$. Let $\Delta_i=\Delta_i(s_0,0)$,

$$\begin{aligned}
d&=\sup(d_{s_0}(z_1,z_1'),d_{s_0}(z_2,z_2')),\\
T(s)&=T(z_1,z_2)(s)-T(z_1',z_2')(s)\\
&=[T(z_1,z_2)(s)-T(z_1',z_2)(s)]+[T(z_1',z_2)(s)-T(z_1',z_2')(s)]\\
&=T_1(s)-T_2(s),\quad s\in I,\\
&\qquad |T_i(s)|=\exp\tau_i(s).
\end{aligned}$$

We assume r_1 small enough to have, for any $s\in(-\eta,\eta)$,

$$B^i_s(r_1)\subset B^i_0(r_2)$$

which means that we are in the domain of validity of Proposition 3.

Let $\nu_i=v(z_i,z_i')$ for $i=1,2$.

Let $\varphi_i=\varphi_i(s_0,0)$, $\Delta\varphi_i=\Delta\varphi_i(s_0,0)$, $\mu_i=\mu_{i,s_0,0}$ (cf. §3.9).

Property (i) of §2.1 follows from §3.5 and Proposition 3 (the metrics d_{s_0} and d_{φ_i} being equivalent).

We now check property (iv), using Proposition 1. We have seen in §3.7 that the hypothesis of Proposition 1 is satisfied ($S_1\Delta S_2\neq S_2\Delta S_1$).

According to Proposition 3 above, we have:

$$|T(z_1,z_2)(s)|=c\sup[d_s(z_1,a_1),d_s(z_2,a_2)]\leqslant cr^u,\quad 0<u<1.$$

Therefore:

$$\left|\sum_{j=0}^{\nu_i-1}\lambda_i(\sigma^j z_i,s,T(z_1,z_2)(s))-\sum_{j=0}^{\nu_i-1}\lambda_i(\sigma^j z_i,s_0,0)\right|<c\nu_i|\log r|^{-1},\quad s\in I.$$

Also

$$\left|\sum_{j=0}^{\nu_i-1}\lambda_i(\sigma^j z_i,s_0,0)-S_{\nu_i}\varphi_i(z_i)\right|<c,$$

hence $|\tau_i(s)+S_{\nu_i}\varphi_i(z_i)|<c\nu_i|\log r|^{-1}$ (clearly $\nu_i\geqslant c^{-1}|\log r|$).

We have:

$$\frac{d}{ds}\sum_{j=0}^{\nu_i-1}\lambda_i(\sigma^j z_i,s,T(z_1,z_2)(s))=\sum_{j=0}^{\nu_i-1}\frac{\partial}{\partial s}\lambda_i(\sigma^j z_i,s,T(z_1,z_2)(s))$$
$$+\frac{d}{ds}T(z_1,z_2)(s)\sum_{j=0}^{\nu_i-1}\frac{\partial}{\partial t}\lambda_i(\sigma^j z_i,s,T(z_1,z_2)(s),$$

with, as above:

$$\left|\sum_{j=0}^{\nu_i-1}\frac{\partial}{\partial s}\lambda_i(\sigma^j z_i,s,T(z_1,z_2)(s))-S_{\nu_i}\Delta\varphi_i(z_i)\right|<c\nu_i|\log r|^{-1}$$

and $|dT(z_1,z_2)(s)/ds|\leqslant r^u$. Therefore

$$\left|\frac{d}{ds}\tau_i(s)+S_{\nu_i}\Delta\varphi_i(z_i)\right|<c\nu_i|\log r|^{-1}.$$

Let $1<\beta<\beta_0$, with β_0 as in Proposition 1. We apply Proposition 1 with balls B_1,B_2 of radius cr containing $B^1_{s_0}(r),B^2_{s_0}(r)$ (the respective distances are equivalent). We assume for instance that

$$d=d_{s_0}(z_1,z_1')>r^\beta,$$

which implies, with the notations of Proposition 1:

$$m_1\leqslant\nu_1\leqslant\beta_0 m_1,$$

if r_1 is small enough.

If the set J of §2.1, property (iv), is empty, there is nothing to prove. Assume that J contains a point $s_1\in I$. We have

$$|\log d+S_{\nu_1}\varphi_1(z_1)|\leqslant c,$$

hence

$$|\tau_1(s)-\log d| \leqslant c, \quad s\in I,$$

and also

$$|\,|T_1(s_1)|-|T_2(s_1)|\,| < c^{-1}d.$$

Therefore

$$|\tau_1(s_1)-\tau_2(s_1)| \leqslant c.$$

But

$$|\tau_2(s_1)+S_{\nu_2}\varphi_2(z_2)| \leqslant c\nu_2|\log r|^{-1}$$

and

$$|\tau_2(s_1)-\log d| \leqslant c$$

imply $\nu_2 \leqslant c|\log r|$. We then get:

$$|S_{\nu_1}\varphi_1(z_1)-S_{\nu_2}\varphi_2(z_2)| \leqslant c.$$

Then, by Proposition 1 and the estimates above for $d\tau_i(s)/ds$, we get

$$\left|\frac{d}{ds}(\tau_1(s)-\tau_2(s))\right| > c^{-1}|\log r|, \quad s\in I.$$

Also

$$\left|\frac{d^2}{ds^2}(\tau_1(s)-\tau_2(s))\right| < c|\log r|^2.$$

This shows that J is an interval and gives the estimates of property (iv) in §2.1.

3.15. For $s\in I$, there is a constant c such that

$$B_{s_0}(c^{-1}r) \subset B_s(r) \subset B_{s_0}(cr).$$

Indeed, the distance d_s is equivalent to $d_{i,s,0}$, the distance d_{s_0} to the distance $d_{i,s_0,0}$, and the property is clearly true for the balls relative to these distances.

The hypothesis $J_1\Delta J_2 \neq J_2\Delta J_1$ of Proposition 2 is satisfied, as we have seen in §3.10. With M to be determined later, we apply Proposition 2 in balls $B_i(c^{-1}r)$, in order to satisfy property (ii) of §2.1. Property (iii) of §2.1 (with L as in Proposition 2) follows from the conclusion (i) of Proposition 2 and §3.5.

We now check property (v). We therefore assume that $(z_1,z_2)\in L$, $(z_1',z_2')\in L$ and $d\leqslant r^\beta$ (with $\beta>1$ as above). Again there is nothing to prove when J is empty, hence we assume that there exists $s_1\in I$ with

$$|\,|T_1(s_1)|-|T_2(s_1)|\,| < c^{-1}d^{1+c|\log r|^{-1}}.$$

Assume for instance that $d=d_{s_0}(z_1,z_1')$. We have:

$$|\log d+S_{\nu_1}\varphi_1(z_1)|\leqslant c,$$
$$|\tau_1(s)+S_{\nu_1}\varphi_1(z_1)|\leqslant c\nu_1|\log r|^{-1},$$

and therefore

$$|\tau_2(s_1)-\tau_1(s_1)|<c|\log r|^{-1}|\log d|,$$
$$|S_{\nu_1}\varphi_1(z_1)-S_{\nu_2}\varphi_2(z_2)|<c(\nu_1+\nu_2+|\log d|)|\log r|^{-1}.$$

But, $|\log d+S_{\nu_1}\varphi_1(z_1)|\leqslant c$ implies

$$c^{-1}|\log d|<\nu_1<c|\log d|,$$

and for the last inequality to hold we must then also have $c^{-1}|\log d|<\nu_2<c|\log d|$. Therefore

$$|S_{\nu_1}\varphi_1(z_1)-S_{\nu_2}\varphi_2(z_2)|<c|\log d|\,|\log r|^{-1}.$$

For M big enough (and r_1 small enough), we have

$$\sup(M,c_1(\nu_1+\nu_2-c_2|\log r|))>c|\log d|\,|\log r|^{-1},$$

hence, by Proposition 2:

$$|S_{\nu_1}\Delta\varphi_1(z_1)-S_{\nu_2}\Delta\varphi_2(z_2)|>\sup(M,c_1[(\nu_1+\nu_2)-c_2|\log r|]).$$

Then we will have, for $s\in I$, r_1 small enough, M big enough:

$$\left|\frac{d}{ds}(\tau_1(s)-\tau_2(s))\right|>\tfrac{1}{2}\sup(M,c_1(\nu_1+\nu_2-c_2|\log r|)),$$
$$\left|\frac{d^2}{ds^2}(\tau_1(s)-\tau_2(s))\right|<c|\log d|^2,$$

from which we deduce easily that J is an interval and that the estimates of property (v) of §2.1 hold in J.

We have thus reduced the proof of the theorem to the proof of Proposition 3.

4. The geometrical estimates

The aim of this section is to prove Proposition 3 and thus finish the proof of our main result.

In §4.1, using the smooth dependence on parameters of the stable and unstable foliations of the basic set Λ_1, we introduce appropriate local charts around each point of Λ_1.

These charts are then used in §§4.2–4.4 to obtain estimates of the distance between two nearby stable manifolds of Λ_1 along these manifolds and its variation with parameters. The main part of the calculation actually takes place in some group of jets; we therefore adopt a slightly more abstract setting to make this apparent.

In §4.5, we do some preparatory work in order to finally obtain in §4.6 the estimates in Proposition 3. The calculation is quite long but straightforward; it consists essentially in translating the estimates on the distances between stable manifolds (§4.4), via the implicit function theorem, to estimates on the parameter intervals corresponding to various tangencies as in Proposition 3.

4.1. We start from constants $\varepsilon>0$, $c_0>1$, $c_1>0$ and a continuous map L:

$$\Sigma_1 \to C^\infty([-c_0, c_0]^2 \times (-\eta, \eta)^2, M)$$

with the following properties.

(i) Let $z_1 \in \Sigma_1$, $s, t \in (-\eta, \eta)$; the map $L_{s,t}(z_1)$:

$$(x, y) \to L(z_1)(x, y, s, t)$$

is an embedding of $[-c_0, c_0]^2$ into M whose image $U_{s,t}(z_1)$ contains an ε-neighbourhood of $h_{1,s,t}(z_1)$; we have

$$\begin{aligned} &L_{s,t}(z_1)(0,0) = h_{1,s,t}(z_1), \\ &L_{s,t}(z_1)(x,0) \subset W^s(z_1, s, t), \quad x \in [-c_0, c_0], \\ &L_{s,t}(z_1)(0,y) \subset W^u(z_1, s, t), \quad y \in [-c_0, c_0]. \end{aligned}$$

(ii) Let $z_1 \in \Sigma_1$, $s, t \in (-\eta, \eta)$; the map $F_{s,t}(z_1)$:

$$F_{s,t}(z_1) = (L_{s,t}(\sigma z_1))^{-1} \circ f_{s,t} \circ L_{s,t}(z_1)$$

is defined on $[-1,1]^2$; it may be written in this square under the form

$$F_{s,t}(z_1)(x,y) = (\pm x \exp H_{s,t}(z_1)(x,y), \pm y \exp K_{s,t}(z_1)(x,y)),$$

where the + or − signs depend (continuously) only on z_1. The maps $F_{s,t}(z_1)$ together define a continuous map F:

$$\Sigma_1 \to C^\infty([-1,1]^2 \times (-\eta,\eta)^2, [-c_0, c_0]^2),$$

and similarly for $H_{s,t}(z_1)$, $K_{s,t}(z_1)$.

We have $K_{s,t}(z_1)(0,0)=\lambda_1(z_1,s,t)$ (cf. §3.6) and define $\varrho_1(z_1,s,t)=-H_{s,t}(z_1)(0,0)$; for $z_1\in\Sigma_1$, $s,t\in(-\eta,\eta)^2$, we have:

$$\lambda_1(z_1,s,t)\geqslant c_1>0,$$
$$\varrho_1(z_1,s,t)\geqslant c_1>0.$$

(iii) Let $x_1\in\Sigma_1(j)$, $z_1'\in\Sigma_1(k)$, with $j\neq k$ (i.e. they belong to distinct elements of the Markov partition). Then

$$U_{s,t}(z_1)\cap U_{s,t}(z_1')=\varnothing,\quad \text{for all } (s,t)\in(-\eta,\eta)^2.$$

4.2. We consider now the following slightly more general situation.

Let P be an open set in a parameter space $\mathbf{R}^d$. For $i\geqslant 0$, let $F_i\colon[-1,1]^2\times P\to\mathbf{R}^2$ be a map which may be written in the form:

$$F_i(x,y,p)=(x\exp H_i(x,y,p),y\exp K_i(x,y,p))$$

with smooth maps $H_i,K_i\colon[-1,1]^2\times P\to\mathbf{R}$.

Define

$$\varrho_i(p)=-H_i(0,0,p),$$
$$\lambda_i(p)=K_i(0,0,p),$$

and assume that, for some constant $c_1>0$, all $p\in P$, $i\geqslant 0$:

$$\varrho_i(p)\geqslant c_1,$$
$$\lambda_i(p)\geqslant c_1.$$

For $n\geqslant 0$, define

$$\varrho^{(n)}(p)=\sum_{i=0}^{n-1}\varrho_i(p),$$
$$\lambda^{(n)}(p)=\sum_{i=0}^{n-1}\lambda_i(p).$$

Assume that we have, for some constant $c_2>0$:

$$|\partial_x H_i(x,y,p)|\leqslant c_2,$$
$$|\partial_y H_i(x,y,p)|\leqslant c_2,$$
$$|\partial_x K_i(x,y,p)|\leqslant c_2,$$
$$|\partial_y K_i(x,y,p)|\leqslant c_2.$$

Let $\theta \in (0,1]$. Consider a smooth map

$$p \to (x_0(p), y_0(p))$$

with values in $(0,\theta]^2$. As long as $x_i(p) \in (0,\theta]$, $y_i(p) \in (0,\theta]$, we define:

$$(x_{i+1}(p), y_{i+1}(p)) = F_i(x_i(p), y_i(p), p).$$

Define

$$X_i(p) = \log x_i(p), \quad Y_i(p) = \log y_i(p);$$

then we have

$$|X_{i+1}(p) - X_i(p) + \varrho_i(p)| \leqslant c_2(x_i(p) + y_i(p)),$$
$$|Y_{i+1}(p) - Y_i(p) - \lambda_i(p)| \leqslant c_2(x_i(p) + y_i(p)).$$

From now on, we adopt the following convention: We denote by $c_1, c_2, c_3, \ldots$ constants which depend only on:

— bounds on F_i, H_i, K_i, uniform in i;
— bounds on the map g introduced in §4.4;
— bounds on the map $p \mapsto X_0(p)$.

We also use the letter c for such (unspecified) constants.

We assume that

$$\theta < \tfrac{1}{4} c_1 c_2^{-1} e^{-c_1/2} (1 - e^{-c_1})^{-1}.$$

LEMMA 1. *Assume that for some $n \geqslant 0$, we have*

$$0 < x_0(p) < \theta,$$
$$0 < y_0(p) < \theta \exp(-\lambda^{(n)}(p)).$$

Then $(x_i(p), y_i(p))$ is defined for $0 \leqslant i \leqslant n$ and we have:

$$|X_i(p) - X_0(p) + \varrho^{(i)}(p)| < \tfrac{1}{2} c_1,$$
$$|Y_i(p) - Y_0(p) - \lambda^{(i)}(p)| < \tfrac{1}{2} c_1.$$

Proof. This is clear if $i=0$; assume that it is true for $0 \leqslant i \leqslant j < n$. Then, for $0 \leqslant i \leqslant j$, we have:

$$0 < x_i(p) < \theta e^{c_1/2} e^{-ic_1} \quad (< \theta \text{ if } i > 0),$$
$$0 < y_i(p) < \theta e^{c_1/2} e^{(i-n)c_1} \quad (< \theta),$$

and therefore

$$c_2 \sum_{i=0}^{j} (x_i(p)+y_i(p)) < \frac{2\theta c_2 e^{c_1/2}}{1-e^{-c_1}} < \tfrac{1}{2} c_1$$

which proves the statement of the lemma for $j+1$. □

We recall, for further purposes the estimates

$$0 < x_i(p) < \theta e^{c_1/2} e^{-ic_1},$$
$$0 < y_i(p) < \theta e^{c_1/2} e^{(i-n)c_1},$$

under the hypotheses of the lemma.

Let us now study partial derivatives with respect to the parameters. We write $p=(p_1,...,p_d)$ and ∂_j for $\partial/\partial p_j$.

Let $1 \leqslant j \leqslant d$. For $0 \leqslant i \leqslant n$, define

$$J_i^{(j)} = \begin{pmatrix} \partial_j X_i \\ \partial_j Y_i \end{pmatrix}, \quad V_i^{(j)} = \begin{pmatrix} \partial_j H_i(x_i, y_i, p) \\ \partial_j K_i(x_i, y_i, p) \end{pmatrix},$$
$$M_i = \begin{pmatrix} 1 + x_i \partial_x H_i(x_i, y_i, p) & y_i \partial_y H_i(x_i, y_i, p) \\ x_i \partial_x K_i(x_i, y_i, p) & 1 + y_i \partial_y K_i(x_i, y_i, p) \end{pmatrix}.$$

Then, for $0 \leqslant i < n$, we have

$$J_{i+1}^{(j)} = M_i J_i^{(j)} + V_i^{(j)}.$$

For $0 \leqslant i \leqslant n$, define

$$M^{(i)} = M_{i-1} \dots M_0 = \begin{pmatrix} A^{(i)} & B^{(i)} \\ C^{(i)} & D^{(i)} \end{pmatrix},$$
$$M_{(i)} = M_{n-1} \dots M_i = \begin{pmatrix} A_{(i)} & B_{(i)} \\ C_{(i)} & D_{(i)} \end{pmatrix},$$
$$M_i = \begin{pmatrix} a_i & b_i \\ c_i & d_i \end{pmatrix},$$

with estimates (under the hypothesis of Lemma 1):

$$|a_i - 1| < c_2 e^{c_1/2} \theta e^{-ic_1},$$
$$|c_i| < c_2 e^{c_1/2} \theta e^{-ic_i},$$
$$|b_i| < c_2 e^{c_1/2} \theta e^{(i-n)c_i},$$
$$|d_i - 1| < c_2 e^{c_1/2} \theta e^{(i-n)c_i}.$$

Using the recursion formulas for the $A^{(i)}, A_{(i)}, \dots$ it is easy to prove that there exist constants $c_3, c_4 > 0$ depending only on c_1, c_2 such that if we assume that

$$\theta < c_3$$

then we will have, for $0\leqslant i\leqslant n$:

$$\begin{aligned}|A^{(i)}-1| &< c_4\theta,\\ |B^{(i)}| &< c_4\theta e^{(i-n)c_1},\\ |C^{(i)}| &< c_4\theta,\\ |D^{(i)}-1| &< c_4\theta e^{(i-n)c_1},\\ |A_{(i)}-1| &< c_4\theta e^{-ic_1},\\ |B_{(i)}| &< c_4\theta,\\ |C_{(i)}-1| &< c_4\theta e^{-ic_1},\\ |D_{(i)}-1| &< c_4\theta.\end{aligned}$$

We also have

$$M_{(0)}=M_{(i)}M^{(i)},$$

and therefore

$$\begin{aligned}B_{(0)} &= A_{(i)}B^{(i)}+B_{(i)}D^{(i)},\\ D_{(0)} &= C_{(i)}B^{(i)}+D_{(i)}D^{(i)},\\ |B_{(0)}-B_{(i)}| &< c\theta e^{(i-n)c_1},\\ |D_{(0)}-D_{(i)}| &< c\theta e^{(i-n)c_1}.\end{aligned}$$

On the other hand, we have:

$$\begin{aligned}J_n^{(j)} &= M_{(0)}J_0^{(j)}+\sum_{i=0}^{n-1}M_{(i+1)}V_i^{(j)},\\ \partial_jX_n &= A_{(0)}\partial_jX_0+B_{(0)}\partial_jY_0+\sum_{i=0}^{n-1}(A_{(i+1)}\partial_jH_i+B_{(i+1)}\partial_jK_i),\\ \partial_jY_n &= C_{(0)}\partial_jX_0+D_{(0)}\partial_jY_0+\sum_{i=0}^{n-1}(C_{(i+1)}\partial_jH_i+D_{(i+1)}\partial_jK_i).\end{aligned}$$

Suppose that on our domain we have

$$\begin{aligned}|\partial_j\partial_xH_i| &\leqslant c,\\ |\partial_j\partial_yH_i| &\leqslant c,\\ |\partial_j\partial_xK_i| &\leqslant c,\\ |\partial_j\partial_yK_i| &\leqslant c.\end{aligned}$$

The following lemma is then immediate.

LEMMA 2. *There are constants c_3, c_4 depending only on c_1, c_2 such that, if*

$$0 < x_0(p) < \theta,$$
$$0 < y_0(p) < \theta \exp(-\lambda^{(n)}(p)),$$
$$\theta < c_3 < \tfrac{1}{4} c_1 c_2^{-1} e^{-c_1/2}(1-e^{-c_2})^{-1},$$

then, one has:

$$\left|\left(\partial_j X_n + \sum_{i=0}^{n-1} \partial_j \varrho_i\right) - A_{(0)}\partial_j X_0 - B_{(0)}\left(\partial_j Y_0 + \sum_{i=0}^{n-1} \partial_j \lambda_i\right)\right| \leqslant c\theta,$$
$$\left|\partial_j Y_n - C_{(0)}\partial_j X_0 - D_{(0)}\left(\partial_j Y_0 + \sum_{i=0}^{n-1} \partial_j \lambda_i\right)\right| \leqslant c\theta,$$

where $|A_{(0)}-1| < c_4\theta$, $|B_{(0)}| < c_4\theta$, $|C_{(0)}| < c_4\theta$, $|D_{(0)}-1| < c_4\theta$.

4.3. We will also need estimate for second partial derivatives. Let $1 \leqslant j, k \leqslant d$. Define

$$J_i^{(j,k)} = \begin{pmatrix} \partial_j \partial_k X_i \\ \partial_j \partial_k Y_i \end{pmatrix}.$$

Then we have

$$J_{i+1}^{(j,k)} = M_i J_i^{(j,k)} + S_i(J_i^{(j)} \overset{s}{\otimes} J_j^{(k)}) + S_i^{(j)} J_i^{(k)} + S_i^{(k)} J_i^{(j)} + V_i^{(j,k)},$$

with

$$S_i^{(j)} = \begin{pmatrix} x_i \partial_x \partial_j H_i & y_i \partial_y \partial_j H_i \\ x_i \partial_x \partial_j K_i & y_i \partial_y \partial_j K_i \end{pmatrix}, \quad V_i^{(j,k)} = \begin{pmatrix} \partial_j \partial_k H_i \\ \partial_j \partial_k K_i \end{pmatrix},$$
$$S_i = \begin{pmatrix} x_i \partial_x H_i + x_i^2 \partial_x^2 H_i & x_i y_i \partial_x \partial_y H_i & y_i \partial_y H_i + y_i^2 \partial_y^2 H_i \\ x_i \partial_x K_i + x_i^2 \partial_x^2 K_i & x_i y_i \partial_x \partial_y K_i & y_i \partial_y K_i + y_i^2 \partial_y^2 K_i \end{pmatrix},$$

and we write the symmetric tensor product $J_i^{(j)} \overset{s}{\otimes} J_i^{(k)}$ with coordinates

$$\begin{pmatrix} \partial_j X_i \partial_k X_i \\ \partial_j X_i \partial_k Y_i + \partial_k X_i \partial_j Y_i \\ \partial_j Y_i \partial_k Y_i \end{pmatrix}.$$

We have

$$J_n^{(j,k)} = M_{(0)} J_0^{(j,k)} + \sum_{i=0}^{n-1} M_{(i+1)} R_i^{(j,k)},$$

with $R_i^{(j,k)} = S_i(J_i^{(j)} \overset{s}{\otimes} J_i^{(k)}) + S_i^{(j)} J_i^{(k)} + S_i^{(k)} J_i^{(j)} + V_i^{(j,k)}$.

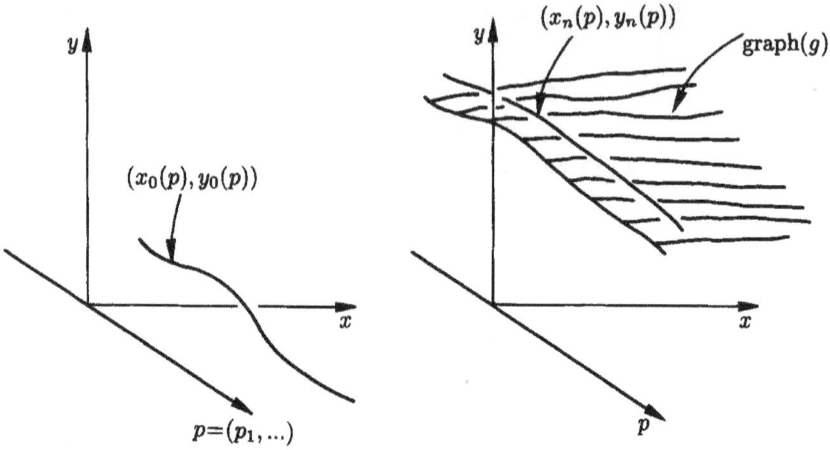

Fig. 3

Assume that there are constants c, C_0 such that all second partial derivatives of H_i, K_i are bounded in our domain by c, and that:

$$\|J_i^{(j)}\| < C_0 n,$$
$$\|J_i^{(k)}\| < C_0 n.$$

Then, for some constant $C_1 = C_1(C_0)$ we have:

$$\sup(\|S_i\|, \|S_i^{(j)}\|, \|S_i^{(k)}\|) \leqslant c \exp(-\inf(i, n-i)c_1),$$
$$\|J_n^{(j,k)} - M_{(0)} J_0^{(j,k)}\| < C_1 n^2.$$

4.4. We now fix θ such that

$$\theta < c_3, \quad \theta < \tfrac{1}{10} c_4^{-1} \quad \text{(cf. Lemma 2).}$$

We add the following new feature: there is a smooth map $g\colon [0,\theta] \times P \to [0,\theta]$, which satisfies

$$g(x,p) > c > 0, \quad x \in [0,\theta], \ p \in P,$$

and such that:

$$y_n(p) = g(x_n(p), p), \quad p \in P.$$

See Figure 3. We also assume that $p \to (x_0(p), y_0(p))$ satisfies the hypothesis of Lemma 2.

Write

$$\widetilde{\partial_j X_n} = \partial_j X_n + \sum_{i=0}^{n-1} \partial_j \varrho_i,$$

$$\widetilde{\partial_j Y_0} = \partial_j Y_0 + \sum_{i=0}^{n-1} \partial_j \lambda_i.$$

For $1 \leqslant j \leqslant d$, we have:

$$y_n \partial_j Y_n = \partial_j g + x_n \partial_x g \partial_j X_n,$$

with $|x_n \partial_x g| < ce^{-c_1 n}$ and $|\widetilde{\partial_j X_n} - \partial_j X_n| < cn$.

Joining this to the conclusion of Lemma 2, we obtain (as $y_n > c > 0$)

$$|\tilde{\partial}_j X_n - A_{(0)} \partial_j X_0 - B_{(0)} \widetilde{\partial_j Y_0}| < c,$$
$$|C_{(0)} \partial_j X_0 + D_{(0)} \widetilde{\partial_j Y_0}| < c + ce^{-c_1 n} |\widetilde{\partial_j X_n}|,$$

from which we deduce, as $|\partial_j X_0| \leqslant c$, for $n \geqslant c_5$:

$$|\widetilde{\partial_j X_n}| \leqslant c,$$
$$|\widetilde{\partial_j Y_0}| \leqslant c.$$

For second derivatives, we have, for $1 \leqslant j \leqslant k \leqslant d$:

$$\begin{aligned} y_n(\partial_j Y_n \partial_k Y_n + \partial_j \partial_k Y_n) = {} & \partial_j \partial_k g + x_n \partial_j \partial_x g \partial_k X_n \\ & + x_n \partial_k \partial_x g \partial_j X_n + (x_n \partial_x g + x_n^2 \partial_x^2 g) \partial_j X_n \partial_k X_n \\ & + x_n \partial_x g \partial_j \partial_k X_n. \end{aligned}$$

Then, we see from §4.3 and above that we have for $n \geqslant c_5$:

$$\|J_i^{(j)}\| < cn,$$
$$\|J_n^{(j,k)} - M_{(0)} J_0^{(j,k)}\| < cn^2.$$

We have

$$|\partial_j Y_n| < c, \quad |\partial_k Y_n| < c,$$

and therefore

$$|\partial_j \partial_k Y_n| < c + ce^{-c_1 n} |\partial_j \partial_k X_n|.$$

In the same way as for first derivatives, that allows to conclude that, for $n \geqslant c_6 \geqslant c_5$,

$$|\partial_j \partial_k Y_0| < cn^2.$$

4.5. We now proceed to prove Proposition 3. We use the notations of §4.1. Let $P=(-\eta,\eta)^2$, η small enough.

The maps

$$\Sigma_1 \xrightarrow{H} C^\infty([-1,1]^2\times P, \mathbf{R}),$$
$$\Sigma_1 \xrightarrow{K} C^\infty([-1,1]^2\times P, \mathbf{R})$$

are continuous, hence there is a constant $c_2>0$ such that, for all $z_1\in\Sigma_1$, $s,t\in(-\eta,\eta)^2$, $x,y\in[-1,1]$:

$$|\partial_x H_{s,t}(z_1)(x,y)|\leqslant c_2,$$
$$|\partial_y H_{s,t}(z_1)(x,y)|\leqslant c_2,$$
$$|\partial_x K_{s,t}(z_1)(x,y)|\leqslant c_2,$$
$$|\partial_y K_{s,t}(z_1)(x,y)|\leqslant c_2.$$

We determine then, from c_1 (in §4.1) and c_2 above, constants c_3, c_4 as in Lemma 2, and choose θ with

$$\theta<c_3, \quad \theta<\tfrac{1}{10}c_4^{-1}.$$

There exists $n_0>0$ such that, for all $s,t\in(-\eta,\eta)$, $z_1,z_1'\in\Sigma_1^+$ with $v(z_1,z_1')\geqslant n_0$, the point $h_{1,s,t}(z_1)$ lies in $U_{s,t}(z_1')$ and the equation of $W^s_{\mathrm{loc}}(z_1,s,t)\cap U_{s,t}(z_1')$ is

$$y=g_{z_1/z_1'}(x,s,t)$$

in the coordinate system given by $L_{s,t}(z_1')$.

For each z_1, z_1', the map $g_{z_1/z_1'}$ is smooth, and these maps together give a continuous map:

$$V_i \xrightarrow{G} C^\infty([-c_0,c_0]\times(-\eta,\eta)^2, [-c_0,c_0])$$

where $V_1=\{(z_1,z_1')\in\Sigma_1^+\times\Sigma_1^+ : v(z_1,z_1')\geqslant n_0\}$.

We choose $n_0>0$ big enough such that, for $(z_1,z_1')\in V_1$, the image of $g_{z_1/z_1'}$ is actually contained in $\left[-e^{-c_1/2}\frac{1}{2}\theta, e^{-c_1/2}\frac{1}{2}\theta\right]$.

We choose, once and for all, an integer m_0, multiple of the period of a_1, such that the point $q'=f^{m_0}(q)$ belongs to $U_{0,0}(a_1)$, with coordinates $(\theta_1,0)$, where $0<|\theta_1|\leqslant\frac{1}{2}\theta$.

We will consider tangencies near q' instead of q. The map T' related to q' and the map T related to q satisfy $T'(\sigma^{m_0}z_1,\sigma^{-m_0}z_2)=T(z_1,z_2)$. Once m_0 is fixed, it is clearly equivalent to prove Proposition 3 for T or T'. We will actually prove it for T'. But to keep notations simple, we assume that $m_0=0$, $T=T'$.

Let $0<\varepsilon_1<\frac{1}{2}|\theta_1|$ be a small number such that the component of q' in $W^u(a_2,0,0)\cap [\theta_1-\varepsilon_1,\theta_1+\varepsilon_1]\times\left[-\frac{1}{2}\theta,\frac{1}{2}\theta\right]$ in the coordinate system $L_{0,0}(a_1)$ has equation:

$$y=\varphi(x),\quad |x-\theta_1|\leqslant\varepsilon_1,$$

with $\varphi(\theta_1)=\varphi'(\theta_1)=0$, $|\varphi''(x)|>c>0$.

Let n_1 be an integer $\geqslant n_0$, and let $z_1,z_1'\in\Sigma_1^+$, $z_2\in\Sigma_2^+$ with $v(z_1,a_1)\geqslant n_1$, $v(z_1',a_1)\geqslant n_1$, $v(z_2,a_2)\geqslant n_1$. We claim that if ε_1 is small enough and n_1 is big enough the following properties hold, with constants independent of s, t, z_1, z_1', z_2 (provided η is also small enough).

(i) The connected component of $W^u(z_2,s,t)\cap[\theta_1-\varepsilon_1,\theta_1+\varepsilon_1]\times\left[-\frac{1}{2}\theta,\frac{1}{2}\theta\right]$ we are interested in has equation

$$y=\varphi_{z_2/z_1'}(x,s,t)$$

in the coordinate system $L_{s,t}(z_1')$, for a smooth function $\varphi_{z_2/z_1'}$; all partial derivatives of $\varphi_{z_2/z_1'}$ and $g_{z_1/z_1'}$ of order up to 3 are bounded by a constant c.

(ii) For $s,t\in(-\eta,\eta)$, $|x-\theta_1|<\varepsilon_1$, we have

$$\begin{aligned}|\partial_x^2\varphi_{z_2/z_1'}(x,s,t)|&>c>0,\\ |\partial_t\varphi_{z_2/z_1'}(x,s,t)|&>c>0,\\ |\partial_t(\varphi_{z_2/z_1'}-g_{z_1/z_1'}(x,s,t))|&>c>0,\\ |\partial_x^2(\varphi_{z_2/z_1'}-g_{z_1/z_1'}(x,s,t))|&>c>0.\end{aligned}$$

(iii) For $s,t\in(-\eta,\eta)$, $x,x'\in[\theta_1-\varepsilon_1,\theta_1+\varepsilon_1]$, we have

$$\begin{aligned}|\partial_x g_{z_1/z_1'}(x,s,t)|&\leqslant c|g_{z_1/z_1'}(x',s,t)|,\\ |\partial_x^2 g_{z_1/z_1'}(x,s,t)|&\leqslant c|g_{z_1/z_1'}(x',s,t)|.\end{aligned}$$

(iv) For $s,t\in(-\eta,\eta)$ the function $\partial_x\varphi_{z_1/z_1'}$ (resp. $\partial_x(\varphi_{z_2/z_1'}-g_{z_1/z_1'})$) has a (unique) zero $\hat{c}(s,t)$ (resp. $c(s,t)$) in $[\theta_1-\varepsilon_1,\theta_1+\varepsilon_1]$. We have

$$\begin{aligned}|\partial_s c|\leqslant c_7,&\quad |\partial_t c|\leqslant c_7,\\ |\partial_s\hat{c}|\leqslant c_7,&\quad |\partial_t\hat{c}|\leqslant c_7.\end{aligned}$$

and second partial derivatives are bounded by c.

(v) For $s\in(-\eta,\eta)$, the points $T(z_1,z_2)(s)$ and $T(z_1',z_2)(s)$ belong to $(-\eta,\eta)$; we write $\bar{t}=t-T(z_1',z_2)(s)$. In the parameter coordinates $(s,\bar{t})$, all estimates above are still valid.

9–945201 *Acta Mathematica* 172. Imprimé le 29 mars 1994

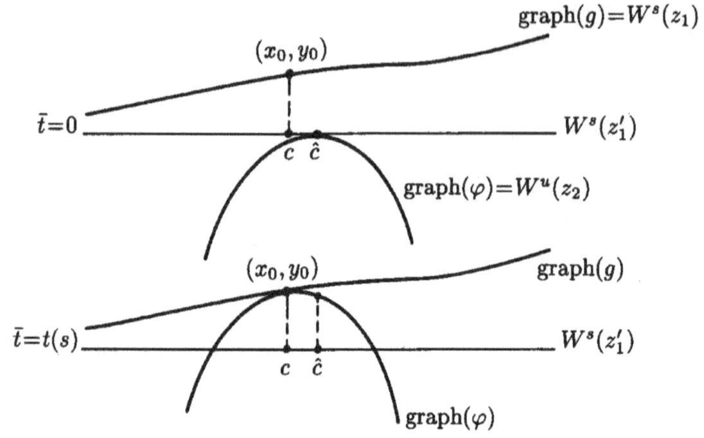

$|y_0(s,\bar t)| \sim |g(x,s,\bar t)| \sim \pi(s,\bar t) \sim t(s)$, for $|\bar t| \leqslant t(s)$

Fig. 4

All estimates and claims are straightforward, taking first ε_1 very small and then n_1 very big; estimate (iii) holds because the stable foliation of the basic set is, uniformly in s,t, of class $C^{1+\alpha}$ for some $\alpha>0$.

4.6. We fix points z_1, z'_1, z_2 as above, but we will assume that

$$v(z_1,a_1)\geqslant n_2, \quad v(z'_1,a_1)\geqslant n_2, \quad v(z_2,a_2)\geqslant n_2,$$

for an integer $n_2\geqslant n_1$, still to be chosen.

Let $n=v(z_1,z'_1)-n_0$; for $0\leqslant i<n$, let

$$F_i\colon [-1,1]^2\times(-\eta,\eta)^2\to\mathbf{R}^2$$

be the map $(x,y,s,t)\mapsto F_{s,t}(\sigma^i z'_1)(x,y)$.

By changing the signs of the coordinates in the coordinate systems $L_{s,t}(\sigma^i z'_1)$, $0\leqslant i\leqslant n$, we may assume that the F_i are exactly of the form considered in §4.2.

We will work with the parameter coordinates $(s,\bar t)$ considered in §4.5 (v). To keep notations simple, we just write (s,t) again for these new coordinates. We therefore have $T(z'_1,z_2)(s)=0$, which means:

$$\varphi(\hat c(s,0),s,0)\equiv 0.$$

We are interested in the function $T(z_1,z_2)(s)=t(s)$ which is defined by:

$$\varphi(c(s,t(s)),s,t(s))=g(c(s,t(s)),s,t(s)).$$

We have written φ for φ_{z_2/z'_1} and g for g_{z_1/z'_1}. See Figure 4.

Write $\tilde{g}=g_{\sigma^n z_1/\sigma^n z_1'}$. We have, as $v(\sigma^n z_1,\sigma^n z_1')$ is a fixed integer n_0:

$$\tfrac{1}{2}e^{-c_1/2}\theta > |\tilde{g}(x,s,t)| > c > 0.$$

By Lemma 1 in §4.2, we have therefore, for $x\in[\theta_1-\varepsilon_1,\theta_1+\varepsilon_1]$, $s,t\in(-\eta,\eta)$:

$$0<c<|g(x,s,t)|\exp\lambda^{(n)}(s,t)<\theta.$$

Let us just write $\pi(s,t)$ for $\exp-\lambda^{(n)}(s,t)$. We estimate $|c(s,t)-\hat{c}(s,0)|$. We have

$$|\hat{c}(s,0)-\hat{c}(s,t)|<c_7|t|,$$
$$\partial_x\varphi(\hat{c}(s,t),s,t)=0,$$
$$|\partial_x\varphi(c(s,t),s,t)|=|\partial_x g(c(s,t),s,t)|<c\pi(s,t),$$

hence (as $|\partial_u^2\varphi|>c>0$)

$$|\hat{c}(s,0)-c(s,t)|<c[\pi(s,t)+|t|].$$

But $\varphi(\hat{c}(s,0),s,0)=\partial_x\varphi(\hat{c}(s,0),s,0)=0$. Therefore

$$|\varphi(c(s,t),s,0)|<c[(\pi(s,t))^2+t^2].$$

On the other hand, writing $a(s)=\partial_t\varphi(c(s,t(s)),s,t(s))$, we have:

$$|a(s)|\geqslant c>0,$$

and, for $|t|\leqslant|t(s)|$,

$$|\partial_t\varphi(c(s,t(s)),s,t)-a(s)|\leqslant c|t(s)|.$$

Summarizing, the function $\chi\colon u\to\varphi(c(s,t(s)),s,u)$ for $|u|\leqslant|t(s)|$ satisfies

$$|\chi(0)|<c[\pi(s,t(s))^2+t(s)^2],$$
$$\chi(t(s))=g(c(s,t(s)),s,t(s)),$$
$$|a(s)|=|\chi'(t(s))|\geqslant c>0,$$
$$|\chi'(u)-\chi'(t(s))|<c|t(s)|.$$

Let $y_0(s,t)=g(c(s,t),s,t)$. We have

$$\left|\log\frac{|y_0(s,t)|}{\pi(s,t)}\right|<c,$$

hence $|t(s)-y_0(s,t(s))/a(s)|<c(\pi(s,t(s)))^2$.

In particular, writing $\pi(s)=\pi(s,t(s))$:

$$\left|\log\frac{|t(s)|}{\pi(s)}\right|<c.$$

Let us now estimate $t'(s)$. From the defining relation

$$\psi(s,t(s))=0,$$

with $\psi(s,t)=\varphi(c(s,t),s,t)-g(c(s,t),s,t)$, we get

$$t'(s)=-[\partial_t\psi(s,t(s))]^{-1}\partial_s\psi(s,t(s)),$$

where

$$\partial_s\psi=\partial_s\varphi+\partial_s c\partial_x\varphi-\partial_s y_0,$$
$$\partial_t\psi=\partial_t\varphi+\partial_t c\partial_x\varphi-\partial_t y_0.$$

We have

$$\partial_t\varphi(c(s,t(s)),s,t(s))=a(s),$$
$$|\partial_x\varphi(c(s,t(s)),s,t(s))|=|\partial_x g(c(s,t(s)),s,t(s))|<c\pi(s)$$

(by §4.5 (iii)),

$$|\partial_s\varphi(c(s,t(s)),s,t(s))-\partial_s\varphi(\hat{c}(s,0),s,0)|<c\pi(s)$$

(because $|t(s)|<c\pi(s)$, $|c(s,t)-\hat{c}(s,0)|<c\pi(s)$),

$$\partial_s\varphi(\hat{c}(s,0),s,0)=0$$

(because $\partial_x\varphi(\hat{c}(s,0),s,0)\equiv\varphi(\hat{c}(s,0),s,0)\equiv 0$).

On the other hand, with $x_0(s,t)=c(s,t)$, we have, in the notation of §4.2:

$$y_n(s,t)=\tilde{g}(x_n(s,t),s,t)$$

with estimates:

$$|\partial_s\log|x_0||<\frac{c_7}{\theta_1-\varepsilon_1},\quad |\partial_t\log|x_0||<\frac{c_7}{\theta_1-\varepsilon_1}.$$

Taking n_2 big enough, the calculations in §4.4 are valid and we get:

$$\left|\frac{\partial_s y_0}{y_0}+\sum_{i=0}^{n-1}\partial_s\lambda_i\right|\leqslant c,$$
$$|\partial_t y_0|<c\pi(s,t)n<c\pi(s,t)|\log\pi(s,t)|.$$

Therefore

$$|\partial_s\psi(s,t(s))-y_0(s,t(s))\partial_s\lambda^{(n)}(s,t(s))|<c\pi(s),$$
$$|\partial_t\psi(s,t(s))-a(s)|<c\pi(s)|\log\pi(s)|,$$

with $|y_0(s,t(s))-a(s)t(s)|<c(\pi(s))^2$.

We conclude that:

$$\left|\frac{t'(s)}{t(s)}+\sum_{i=0}^{n-1}\partial_s\lambda_i(s,t(s))\right|\leqslant c.$$

Let us now estimate from above the second derivative $t''(s)$. We have:

$$\partial_t\psi t''+\partial_t^2\psi(t')^2+2\partial_t\partial_s\psi t'+\partial_s^2\psi=0,$$

hence

$$|t''(s)|<c(|\partial_s^2\psi|+\pi(s)|\log\pi(s)|\,|\partial_s\partial_t\psi|+\pi^2(s)|\log\pi(s)|\,|\partial_t^2\psi|),$$

with

$$\partial_s^2\psi=\partial_s^2c\partial_x\varphi+(\partial_sc)^2\partial_x^2\varphi+2\partial_sc\partial_s\partial_x\varphi+\partial_s^2\varphi-\partial_s^2y_0,$$
$$\partial_s\partial_t\psi=\partial_s\partial_tc\partial_x\varphi+\partial_sc\partial_tc\partial_x^2\varphi+\partial_sc\partial_t\partial_x\varphi+\partial_tc\partial_s\partial_x\varphi+\partial_s\partial_t\varphi-\partial_s\partial_ty_0,$$
$$\partial_t^2\psi=\partial_t^2c\partial_x\varphi+(\partial_tc)^2\partial_x^2\varphi+2\partial_tc\partial_t\partial_x\varphi+\partial_t^2\varphi-\partial_t^2y_0.$$

According to §4.4, we have, for n_2 big enough:

$$\partial_s^2\log|y_0|<cn^2,$$
$$\partial_s\partial_t\log|y_0|<cn^2,$$
$$\partial_t^2\log|y_0|<cn^2,$$
$$\partial_s\log|y_0|<cn,$$
$$\partial_t\log|y_0|<cn,$$

therefore

$$|\partial_s^2y_0|<cn^2\pi(s),$$
$$|\partial_s\partial_ty_0|<cn^2\pi(s),$$
$$|\partial_t^2y_0|<cn^2\pi(s).$$

We already obtain:

$$|\partial_t^2\psi(s,t(s))|<c,$$
$$|\partial_s\partial_t\psi(s,t(s))|<c.$$

In the formula for $\partial_s^2\psi(s,t(s))$ we have:

$$|\partial_x g(c(s,t(s)),s,t(s))| = |\partial_x\varphi(c(s,t(s)),s,t(s))| < c\pi(s).$$

We compare $\partial_s c(s,t(s))$ with $\partial_s\hat{c}(s,0)$. We have

$$\begin{aligned}\partial_x\partial_s\varphi(\hat{c}(s,t),s,t)+\partial_x^2\varphi(\hat{c}(s,t),s,t)\partial_s\hat{c}(s,t)&\equiv 0,\\ \partial_x\partial_s(\varphi-g)(c(s,t),s,t)+\partial_x^2(\varphi-g)(c(s,t),s,t)\partial_s c(s,t)&\equiv 0,\end{aligned}$$

with

$$\begin{aligned}|\partial_x\partial_s\varphi(\hat{c}(s,0),s,0)-\partial_x\partial_s\varphi(c(s,t(s)),s,t(s))| &< c\pi(s),\\ |\partial_x^2\varphi(\hat{c}(s,0),s,0)-\partial_x^2(\varphi-g)(c(s,t(s)),s,t(s))| &< c\pi(s)\end{aligned}$$

(recall §4.5 (iii)), hence

$$|\partial_s c(s,t(s))-\partial_s\hat{c}(s,0)| < c(\pi(s)+|\partial_x\partial_s g(c(s,t(s)),s,t(s))|.$$

Let

$$\begin{cases}\tilde{x}_0(u,s,t)=u, & u\in[\theta_1-\varepsilon_1,\theta_1+\varepsilon_1],\\ \tilde{y}_0(u,s,t)=g(u,s,t), & u\in[\theta_1-\varepsilon_1,\theta_1+\varepsilon_1].\end{cases}$$

The discussion in §§4.2–4.4 applies to $(\tilde{x}_0,\tilde{y}_0)$ (depending now on three parameters u,s,t) and we get

$$|\partial_x\partial_s g(c(s,t(s)),s,t(s))| = |\partial_u\partial_s\tilde{y}_0| < cn^2\pi(s).$$

Therefore

$$|\partial_s c(s,t(s))-\partial_s\hat{c}(s,0)| < cn^2\pi(s).$$

But we observe that

$$\begin{aligned}\partial_s^2\varphi(\hat{c}(s,0),s,0)+\partial_s\hat{c}(s,0)\partial_s\partial_x\varphi(\hat{c}(s,0),s,0)&=0,\\ \partial_s\partial_x\varphi(\hat{c}(s,0),s,0)+\partial_s\hat{c}(s,0)\partial_x^2\varphi(\hat{c}(s,0),s,0)&=0,\end{aligned}$$

therefore

$$\begin{aligned}|\partial_s^2\varphi(c(s,t(s)),s,t(s))+\partial_s c(s,t(s))\partial_s\partial_x\varphi(c(s,t(s)),s,t(s))| &\leqslant cn^2\pi(s),\\ |\partial_s\partial_x\varphi(c(s,t(s)),s,t(s))+\partial_s c(s,t(s))\partial_x^2\varphi(c(s,t(s)),s,t(s))| &\leqslant cn^2\pi(s),\end{aligned}$$

and we conclude that

$$|\partial_s^2\psi(s,t(s))| < cn^2\pi(s),$$
$$|t''(s)| < cn^2\pi(s),$$
$$\left|\frac{d^2}{ds^2}\log|t(s)|\right| < cn^2.$$

Recapitulating, we have proved, as $\log\pi(s) = -\sum_{i=0}^{n-1}\lambda_i(s,t(s))$:

$$\left|\log|t(s)| + \sum_{i=0}^{n-1}\lambda_i(s,t(s))\right| < c,$$
$$\left|\frac{d}{ds}\log|t(s)| + \sum_{i=0}^{n-1}\partial_s\lambda_i(s,t(s))\right| < c,$$
$$\left|\frac{d^2}{ds^2}\log|t(s)|\right| < cn^2.$$

Let us see that this indeed gives the estimations for τ_1 as in Proposition 3. We have

$$\left|\sum_{i=0}^{n-1}\lambda_i(s,t(s)) - \sum_{i=0}^{n-1}\lambda_i(s,0)\right| < cn\pi(s) < c,$$
$$\left|\sum_{i=0}^{n-1}\partial_s\lambda_i(s,t(s)) - \sum_{i=0}^{n-1}\partial_s\lambda_i(s,0)\right| < cn\pi(s) < c.$$

As our t-variable here is really $\bar{t} = t - T(z_1', z_2)(s)$, we have in fact

$$\partial_s\lambda_i(s,0) = \frac{d}{ds}\lambda_1(\sigma^i z_1', s, T(z_1', z_2)(s))$$

where the notations of §3.10 are used in the right hand term. We have $\tau_1(s) = \log|t(s)|$. The integer ν_1 in Proposition 3 is here $n+n_0+m_0$: more precisely, in the sum $\sum_{j=0}^{n-1}\lambda_1(\sigma^j z_1', s, T(z_1', z_2)(s))$, the sum $\sum_{j=0}^{n-1}\lambda_i(s,0)$ missed the first m_0 terms (when we replaced q by q') and the last n_0 terms. But m_0, n_0 are fixed integers. Therefore we have proved the required estimates for $\tau_1(s)$. The estimates for τ_2 are true for the same reasons. The last two estimates in Proposition 3 are immediate, writing:

$$T(z_1,z_2) = T(z_1,z_2) - T(z_1,a_2) + T(z_1,a_2) - T(a_1,a_2).$$

Proposition 3 is, therefore, proved. □

References

[AM] ARAUJO, A. & MAÑÉ, R., On the existence of hyperbolic attractors and homoclinic tangencies for surface diffeomorphisms. To appear.

[BC] BENEDICKS, M. & CARLESON, L., The dynamics of the Hénon map. *Ann. of Math.*, 133 (1991), 73–169.
[Bi] BIRKHOFF, G. D., Nouvelles recherches sur les systèmes dynamiques. *Mem. Pont. Accad. Sci. Nuovi Lyncei*, 1 (1935), 85–216.
[Bo1] BOWEN, R., *Equilibrium States and the Ergodic Theory of Anosov Diffeomorphisms*. Lecture Notes in Math., 470. Springer-Verlag, 1975.
[Bo2] — Hausdorff dimensions of quasi-circles. *Inst. Hautes Études Sci. Publ. Math.*, 50 (1979), 11–25.
[F] FALCONER, K. J., *The Geometry of Fractal Sets*. Cambridge Univ. Press, 1985.
[Man] MAÑÉ, R., The Hausdorff dimension of horseshoes of diffeomorphisms of surfaces. *Bol. Soc. Brasil. Mat.* (*N.S.*), 20 (1990), 1–24.
[Mar] MARSTRAND, J. M., Some fundamental properties of plane sets of fractional dimensions. *Proc. London Math. Soc.*, 4 (1954), 257–302.
[MM] MANNING, A. & MCCLUSKEY, H., Hausdorff dimension for horseshoes. *Ergodic Theory Dynamical Systems*, 3 (1983), 251–261.
[MV] MORA, L. & VIANA, M., Abundance of strange attractors. *Acta Math.*, 171 (1993), 1–71.
[N] NEWHOUSE, S., The abundance of wild hyperbolic sets and nonsmooth stable sets for diffeomorphisms. *Inst. Hautes Études Sci. Publ. Math.*, 50 (1979), 101–151.
[NP] NEWHOUSE, S. & PALIS, J., Cycles and bifurcation theory. *Astérisque*, 31 (1976), 44–140.
[P] POINCARÉ, H., Sur le problème de trois corps et les equations de la dynamique (Mémoire couronné du prix de S. M. le roi Oscar II de Suède). *Acta Math.*, 13 (1890), 1–270.
[PT1] PALIS, J. & TAKENS, F., Cycles and measure of bifurcation sets for two-dimensional diffeomorphisms. *Invent. Math.*, 82 (1985), 397–422.
[PT2] — Hyperbolicity and creation of homoclinic orbits. *Ann. of Math.*, 125 (1987), 337–374.
[PT3] — *Hyperbolicity and Sensitive-Chaotic Dynamics at Homoclinic Bifurcations, Fractal Dimensions and Infinitely Many Attractors*. Cambridge Univ. Press, 1993.
[PV] PALIS, J. & VIANA, M., On the continuity of Hausdorff dimension and limit capacity for horseshoes, in *Dynamical Systems, Valparaiso, 1986*. Lecture Notes in Math., 1331. Springer-Verlag, 1988.
[R] ROBINSON, C., Bifurcation to infinitely many sinks. *Comm. Math. Phys.*, 90 (1983), 433–459.
[S] SMALE, S., Diffeomorphisms with many periodic points, in *Differential and Combinatorial Topology*, pp. 63–80. Princeton Univ. Press, 1965.
[YA] YORKE, J. A. & ALLIGOOD, K. T., Cascades of period doubling bifurcations: a prerequisite for horseshoes. *Bull. Amer. Math. Soc.*, 9 (1983), 319–322.

JACOB PALIS
Instituto de Matemática Pura e Aplicada
Estrada Dona Castorina, 110
Caixa Postal 34021
22460 Rio de Janeiro, RJ
Brazil
E-mail address: jpalis@impa.br

JEAN-CHRISTOPHE YOCCOZ
Département de Mathématiques
Université de Paris-Sud
Bâtiment 425
91405 Orsay Cedex
France

Received August 28, 1991
Received in revised form January 20, 1993

Annals of Mathematics, **140** (1994), 207–250

High dimension diffeomorphisms displaying infinitely many periodic attractors

By J. Palis and M. Viana

Introduction

We extend to higher dimensions a remarkable two-dimensional result of Newhouse proved in the seventies: many (residual subsets of an open set) smooth diffeomorphisms near one exhibiting a homoclinic tangency have infinitely many coexisting sinks. The saddle associated with the homoclinic tangency is taken to be sectionally dissipative. In this general setting we have to circumvent a major difficulty: the usual lack of (transversal) differentiability of invariant dynamic foliations of codimension higher than one. Also, in general there is no global smooth center manifold and so the problem cannot be reduced in this way to the two-dimensional case. Instead the question is solved by producing hyperbolic sets whose foliations are "essentially" differentiable (à la Whitney) with a large fractal dimension (thickness).

The recent development of (dissipative) dynamics has been much influenced by the discovery of some striking bifurcating phenomena such as the Lorenz-like attractors, the Hénon-like attractors, the Feigenbaum and Coullet-Tresser cascades of period doubling bifurcations and Newhouse's infinitely many coexisting sinks (attracting periodic orbits).

This last phenomenon, although more than twenty years old, has remained essentially a topic in the study of surface diffeomorphisms. That is, the known examples in higher dimensions were obtained by taking a two-dimensional prototype and "multiplying" it by a strongly contracting diffeomorphism. In the present paper we are able to extend the original result to other dimensions (even to infinite dimension) in its full strength, showing in general the abundance (from the topological point of view) of the diffeomorphisms displaying infinitely many coexisting sinks. Indeed, we prove the following result:

Main Theorem. *Near any smooth diffeomorphism exhibiting a homoclinic tangency associated to a sectionally dissipative saddle, there is a residual subset of an open set of diffeomorphisms such that each of its elements displays infinitely many coexisting sinks.*

Originally published in: *Ann. of Math. (2) 140 (1994), no. 1, 207–250,*

We point out that our methods also imply a one-parameter version of this theorem: A generic unfolding of such a (quadratic) homoclinic tangency yields residual subsets of intervals in the parameter line whose corresponding diffeomorphisms exhibit infinitely many sinks. For surface diffeomorphisms this version has been obtained in [Rob].

"Smooth" in the statement above means of class C^2 and "near" means closeness in the C^2 sense. Recall that a homoclinic tangency is just a tangency between the stable and unstable manifolds of a saddle periodic point. The saddle is called (*codimension-one*) *sectionally* or *strongly dissipative* if it has just one expanding eigenvalue (positive Lyapunov exponent) and the product of any two eigenvalues has norm less than one; i.e., any contracting eigenvalue is stronger than the expanding one. We observe that in order to get abundance of sinks when unfolding a homoclinic tangency, we assume that the associated saddle is sectionally dissipative for otherwise we are bound to obtain periodic orbits of smaller index (saddles, sources); see [Rom]. On the other hand, when we generically unfold such a quadratic homoclinic tangency in any dimension, we obtain a rather striking list of dynamical phenomena:

- Hénon-like strange attractors [MV], [V], based on the remarkable work of Benedicks-Carleson [BC];
- cascades of period doubling bifurcations of sinks [YA];
- residual subsets of open sets of intervals in the parameter line whose elements exhibit infinitely many coexisting sinks [N2], [N3], [Rob], and the present paper.

Thus, since homoclinic tangencies are a common bifurcating dynamic feature, evidence of the existence of diffeomorphisms with infinitely many coexisting sinks is rather abundant. It is quite possible, however, that in a parametrized form in terms of Lebesgue measure in the parameter space this is a rare phenomenon (measure zero). This is a subject of much interest and we refer the reader to [PT] for a discussion on this and other related questions.

As pointed out above, a basic difficulty in extending Newhouse's original result from two to higher dimensions is the usual lack of differentiability of invariant foliations (in our case, unstable foliations) of codimension greater than one, which are in general just Hölder continuous. This has been a major obstacle in making sense of (transversal) fractal dimensions of invariant Cantor sets. In particular, this is the case of *thickness*, a concept used by Newhouse to show that *many pairs of Cantor sets in the line necessarily intersect each other unless one lies in a gap of the other; see Section* 1. These intersections correspond to homoclinic tangencies that in turn generate sinks when unfolded. Also, the question treated in this paper has a semi-global character, involving the whole orbit of tangency. In particular, it cannot be solved by just assuming that we can locally linearize the map near the associated saddle, e.g., by supposing that it has nonresonant eigenvalues. Even more, we cannot apply the standard procedure of "reducing dimensions" through projection along the strong contracting directions onto a central manifold, which in fact does not exist in general. The key idea introduced here to circumvent these difficulties is to obtain a hyperbolic set whose higher-codimension unstable foliation is indeed "differentiable" (or "intrinsically differentiable") if we avoid the strong contracting directions. This idea may be useful in other similar dynamic situations involving fractal dimensions of hyperbolic sets or, more directly, the differentiability of their invariant foliations.

The formal definitions of intrinsic differentiability and thickness for the unstable foliation are given in Section 1 and developed in Sections 2 through 4. Specially, in Section 2 we briefly describe some basic properties of intrinsic differentiability. A new relevant condition on the homoclinic tangency is made explicit in Section 3: Assumption (3.2) in Proposition 3.2 yields the construction of hyperbolic sets whose unstable foliation is intrinsically differentiable, since we can then "avoid" the strong contracting directions as suggested above. In Section 1 we also present a sketch of the whole proof of our main result. It follows more closely a new proof of Newhouse's two-dimensional result presented in [PT] than the original papers. For the sake of simplifying the argument, we show in Section 5 that we may assume that the saddle associated to the homoclinic tangency has a unique least contracting eigenvalue. In Section 6 we construct thick invariant Cantor sets that appear when we unfold the homoclinic tangency and whose elements have stable and unstable manifolds that transversally intersect those of the associated saddle. This is done through a renormalization technique. Finally, in Section 7 we finish the proof of the theorem by just assembling together the facts established in the previous sections. We also briefly indicate how the parametrized version we stated above follows from these arguments.

1. Main ingredients and sketch of the proof

We begin by recalling a few notions and facts, mostly from [N1], [N3], which play a central role in the arguments below.

Let $K \subseteq \mathbb{R}$ be a Cantor set and $\hat{K}$ be its convex hull. A *presentation* of K is an ordering $\mathcal{U} = (U_n)_n$ of its gaps, i.e., of the connected components of $\hat{K} \setminus K$. For each n and $u \in \partial U_n$ let $\tau(K, \mathcal{U}, u) = \text{length}(C)/\text{length}(U_n)$, where C is the connected component of $\hat{K} \setminus (U_1 \cup \cdots \cup U_n)$ that contains u. The *thickness* of K is defined as

$$\tau(K) = \sup_{\mathcal{U}} \inf_{u} \tau(K, \mathcal{U}, u)$$

where the supremum is taken over all presentations of K and the infimum over all points $u \in \partial U_n$ as above. The *local thickness* of K *at* $x \in K$ is

$$\tau(K,x) = \lim_{\varepsilon \to 0}(\sup\{\tau(L): L \subset K \cap [x-\varepsilon, x+\varepsilon] \text{ a Cantor set}\}).$$

Recall that a basic set for a diffeomorphism is a compact, invariant, transitive, hyperbolic set, with a dense subset of periodic orbits, which is the maximal invariant set in a neighbourhood of it. A basic set is persistent: Any C^k-small perturbation of the map, $k \geq 1$, yields a unique basic set near the initial one, called the "analytic" or "smooth" continuation of it. Let now Λ be a (nontrivial) basic set of a C^2 diffeomorphism $\varphi: M \to M$, whose stable foliation is of codimension one, i.e., such that $\dim W^s(x) = m - 1$, $m = \dim M$, for all $x \in \Lambda$. Let $z \in W^s(\Lambda)$ and $\phi: [-a, a] \to M$ be a C^1 embedding transverse to $W^s(\Lambda)$ at $z = \phi(0)$. The *local stable thickness of* Λ *at* z is $\tau^s(\Lambda, z) = \tau(\phi^{-1}(W^s(\Lambda)), 0)$. This is independent of the choice of ϕ, as a consequence of the fact that (under the codimension-one assumption) the holonomy maps (i.e., the projections along the leaves) of the stable foliation of Λ can be extended to C^1 maps. Actually, this smoothness of the holonomy of $W^s(\Lambda)$, together with the transitivity of $\varphi \mid \Lambda$, also implies that $\tau^s(\Lambda, z)$ has the same value for every $z \in W^s(\Lambda)$. We denote by $\tau^s(\Lambda)$ this constant value and call it the (*local*) *stable thickness of* Λ. This is a strictly positive finite number and depends continuously on the diffeomorphism, in the sense that if Λ_ψ denotes the analytic continuation of Λ for a diffeomorphism ψ which is C^2-close to φ, then $\tau^s(\Lambda_\psi)$ is close to $\tau^s(\Lambda)$. *Local unstable thicknesses* $\tau^u(\Lambda, z)$ and $\tau^u(\Lambda)$ are defined in a similar way, when $W^u(\Lambda)$ has codimension one. In particular, both the stable thickness and the unstable thickness are well-defined if M is a surface.

Now we outline the proof of our main result. We start by recalling the main ideas in the proof of Newhouse's theorem ([N1], [N2], [N3]), as presented in [PT], and then describe the key ingredients involved in extending these arguments from two to higher dimensions. This extension will be carried out in the forthcoming sections.

Let us consider first the case of a homoclinic tangency (on a surface) involving a *thick horseshoe*. By this we mean that the homoclinic tangency is associated to a periodic point p belonging to some basic set Λ such that $\tau^u(\Lambda) \cdot \tau^s(\Lambda) > 1$. A crucial fact here is the *gap lemma*: if K_1, K_2 are Cantor sets in the real line such that $\tau(K_1) \cdot \tau(K_2) > 1$ and K_1 is not entirely contained in a gap of K_2 nor vice-versa, then $K_1 \cap K_2 \neq \emptyset$. By considering the line ℓ of tangencies between the stable and the unstable foliations of the basic set Λ and applying the lemma to $\ell \cap W^u(\Lambda)$ and $\ell \cap W^s(\Lambda)$, one concludes that there exists a C^2-open set of diffeomorphisms $\mathcal{N}$ (whose closure contains the diffeomorphism φ exhibiting the tangency) such that every $\psi \in \mathcal{N}$ has tangencies between leaves of the stable and unstable foliations of the analytic continuation Λ_ψ of Λ. In particular, a dense subset of diffeomorphisms $\psi \in \mathcal{N}$ exhibits homoclinic tangencies associated to periodic points in Λ_ψ (e.g., to the analytic continuation of p). We assume φ to be *dissipative* at the saddle-point involved in the tangency, i.e., $|\det D\varphi^k(p)| < 1$, k =period of p, and then the same holds for $\psi \in \mathcal{N}$. Recall that, under this dissipativeness assumption, a small perturbation of a diffeomorphism exhibiting a homoclinic tangency yields a periodic sink and that, moreover, a sink is persistent through small perturbations of the map. Thus, for each positive integer n, the subset of $\mathcal{N}$ consisting of elements with n sinks is open and dense. Then, by a standard Baire category argument, the elements of a residual subset of $\mathcal{N}$ display infinitely many coexisting sinks.

In order to prove Newhouse's theorem in the general 2-dimensional case we just assume the existence of a homoclinic tangency, not necessarily associated to a thick Cantor set. Then we proceed as follows to construct an open set $\mathcal{N}$ as above. We continue to take the diffeomorphism φ to be dissipative at the saddle-point p. Also, we may suppose that p belongs to a nontrivial hyperbolic basic set Λ_1; if this is not the case then we just perturb φ so as to create transverse homoclinic orbits together with a new orbit of tangency associated to p. We then consider $\tau^u(\Lambda_1) = \tau^u(\Lambda_1, p) > 0$ and further perturbations of the diffeomorphism are to be taken small enough so that Λ_1 and $\tau^u(\Lambda_1)$ persist essentially unchanged. The crucial step in the argument consists in showing that any small unfolding of the tangency produces new horseshoes Λ_2 with arbitrarily large stable thickness, in particular satisfying $\tau^s(\Lambda_2) \cdot \tau^u(\Lambda_1) > 1$. In [PT] such Λ_2 are constructed via a renormalization procedure as follows. Let $(\varphi_\mu)_{\mu\in(-\varepsilon,\varepsilon)}$, $\varphi_0 = \varphi$, be an arc of diffeomorphisms generically unfolding the tangency. Then one finds parameter values $\mu_n \to 0$ and small domains $Q_n \subset M$ converging to the tangency (see Figure 1), such that $\varphi^n_{\mu_n}(Q_n) \cap Q_n \neq \emptyset$ and, up to appropriate n-dependent rescaling of Q_n, the return maps $\varphi^n_{\mu_n} \mid Q_n$ converge in the C^2 topology to the endomorphism $\phi(x,y) = (1-2x^2, x)$ as $n \to \infty$ (convergence holds in the C^k topology, any $k \geq 1$, if the arc is taken to be C^∞). From the fact that ϕ is conjugate to $\psi(x,y) = (1 - 2|x|, x)$ one concludes that it has invariant hyperbolic sets with arbitrarily large thickness. Then we take Λ_2 to be the analytic continuation for $\varphi^n_\mu \mid Q_n$ of some of these ϕ-hyperbolic sets. One checks that leaves of $W^u(\Lambda_1)$ and $W^s(\Lambda_1)$ have some transverse intersections with, respectively, leaves of $W^s(\Lambda_2)$ and $W^u(\Lambda_2)$. Finally, a heteroclinic tangency associated to periodic points $p_1 \in \Lambda_1$, $p_2 \in \Lambda_2$, may be created, again by a small perturbation (thus with a negligible effect on Λ_2 and $\tau^s(\Lambda_2)$). This is done through an auxiliary saddle P that also originates from the 1-dimensional endomorphism:

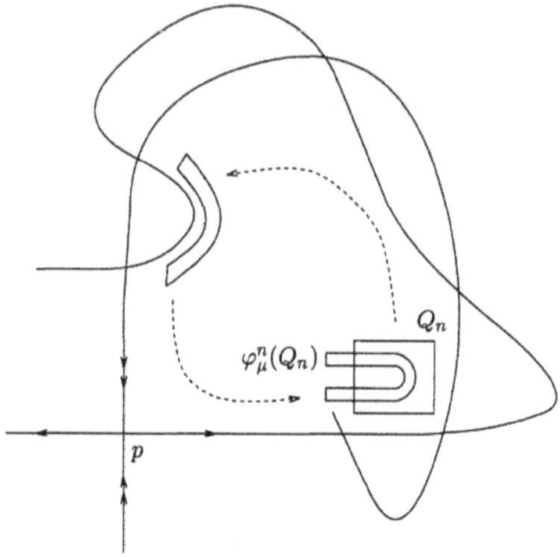

FIGURE 1

It exhibits a homoclinic tangency and its stable and unstable manifolds have points of transverse intersection with the dual invariant manifolds of points both in Λ_1 and Λ_2. Thus, arguing as in the particular case above, there exists an open set in $\mathrm{Diff}^2(M)$ with persistent heteroclinic tangencies involving Λ_1 and Λ_2 and so also with persistent homoclinic tangencies associated to Λ_1 or Λ_2.

Now we discuss in more detail than in the introduction the main difficulties in extending this result to higher dimensions. We also present in a more formal way the ideas to overcome them. As we said before, some of these ideas (like the intrinsic differentiability of invariant foliations or the definition and invariance of local thickness) are quite general and do not require the assumptions of codimension 1 or of sectional dissipativeness.

It is by now classic that the unfolding of a homoclinic tangency of an m-dimensional diffeomorphism leads to the creation of nontrivial basic sets ([Sm]). It is easy to check that we can get these basic sets together with new homoclinic tangencies outside them. Therefore, we may again assume that p is part of a nontrivial basic set Λ_1. However, it is not clear what $\tau^u(\Lambda_1)$ should be taken to mean now, since transverse sections to $W^u(\Lambda_1)$ are no longer lines. Another, perhaps even more serious difficulty arises from the fact that, $W^u(\Lambda_1)$ having codimension bigger than 1, the projections along its leaves may have a bad metric behaviour: In general they are not Lipschitz but just Hölder continuous.

In order to bypass these difficulties we proceed as follows. The first step is to show that the unfolding of a homoclinic tangency yields the formation, for arbitrarily small values $\tilde{\mu}$ of μ, of other homoclinic tangencies associated to periodic points $\tilde{p}_{\tilde{\mu}}$ of $\varphi_{\tilde{\mu}}$, such that:

(I) $D\varphi^{\ell}_{\tilde{\mu}}(\tilde{p}_{\tilde{\mu}})$, ℓ =period of $\tilde{p}_{\tilde{\mu}}$, has a unique weakest contracting eigenvalue which, as a consequence, is a real number.

Hence, we may assume right from the start that $D\varphi^k_0(p)$, k =period of p, satisfies (I). What we do then is to construct Λ_1 embedded in the manifold M in an infinitesimally 2-dimensional way, transversely to the strongest contracting directions of $D\varphi^k(p)$. By this we mean that Λ_1 is taken such that at every $x \in \Lambda_1$, the *intrinsic tangent space*

$$\mathrm{IT}_x \Lambda_1 = \mathrm{span}\left\{ v\colon \text{there is } (x_n)_n \in \Lambda_1^{\mathbb{N}} \text{ so that } x_n \to x \text{ and } \frac{x_n - x}{\|x_n - x\|} \to v \right\}$$

(for simplicity we consider here $M = \mathbb{R}^m$) is 2-dimensional and, like the plane generated by the unstable and the weakest stable directions, is (uniformly) transverse to the codimension-two plane generated by the strongest contracting directions. By *abus de langage*, we say that the angle (meaning its trigonometric tangent) between the above 2-planes is bounded from above. For such a Λ_1 we prove that the projection maps $\pi\colon \Sigma_0 \cap W^u(\Lambda_1) \to \Sigma_1 \cap W^u(\Lambda_1)$ along leaves of $W^u(\Lambda_1)$ are *intrinsically differentiable*, Σ_0, Σ_1 being transversal sections to leaves of $W^u(\Lambda_1)$. That is, there exists a continuous map $\Sigma_0 \cap W^u(\Lambda_1) \ni x \mapsto \partial\pi(x) \in \mathcal{L}(\mathbb{R}^{m-1}, \mathbb{R}^{m-1})$ such that for every $x \in \Sigma_0 \cap W^u(\Lambda_1)$, we have

$$\frac{\pi(x) - \pi(z) - \partial\pi(x) \cdot (x - z)}{\|x - z\|} \to 0 \quad \text{as } z \to x \text{ with } z \in \Sigma_0 \cap W^u(\Lambda_1),$$

where Σ_1 is identified with a disk in $\mathbb{R}^{m-1}$. Here and in what follows $\mathcal{L}(\mathbb{R}^p, \mathbb{R}^q)$ denotes the space of linear maps from $\mathbb{R}^p$ to $\mathbb{R}^q$. Then we define the *local unstable thickness* of Λ_1 at $x \in \Lambda_1$ by $\tau^u(\Lambda_1, x) = \tau(\tilde{\pi}(W^u(\Lambda_1) \cap \Sigma_0), \tilde{\pi}(x))$, where Σ_0 is a transversal section to $W^u(\Lambda_1)$ at x and $\tilde{\pi}\colon W^u(\Lambda_1) \cap \Sigma_0 \to \mathbb{R}$ is any intrinsically differentiable map such that $\partial\tilde{\pi} \mid \mathrm{IT}_x(W^u(\Lambda_1) \cap \Sigma_0)$ is a bijection. *Intrinsic differentiability of the unstable foliation allows us to check that this definition does not depend on the choices of Σ_0 and $\tilde{\pi}$ and to prove that $\tau^u(\Lambda_1, x)$ is strictly positive and independent of $x \in \Lambda_1$. Also it varies continuously with the diffeomorphism $\varphi \in \mathrm{Diff}^2(M)$.* We then show that by arbitrarily small perturbations (essentially not affecting Λ_1 and $\tau^u(\Lambda_1)$) one obtains a hyperbolic basic set Λ_2 with codimension-one stable foliation and large stable thickness, namely $\tau^u(\Lambda_1)\tau^s(\Lambda_2) > 1$. This is done by a natural extension to higher dimensions of the renormalization scheme mentioned above. We also check that Λ_1 and Λ_2 may be taken to be heteroclinically related (existence of mutual transverse intersections between leaves of their stable and unstable foliations) and, moreover, to exhibit a tangency q between

$W^u(p_1)$ and $W^s(p_2)$, $p_1 \in \Lambda_1$, $p_2 \in \Lambda_2$ periodic points. We recall that the stable foliation of Λ_2, being of codimension one, admits an extension to a C^1 foliation $\mathcal{F}^s(\Lambda_2)$ defined in a neighbourhood of Λ_2 (which we may assume to contain q). An implicit function argument shows that there exists an intrinsically differentiable map $\pi_1 : W^u(\Lambda_1) \cap W^s_{loc}(p_1) \to M$ such that $\pi_1(p_1) = q$ and each $\pi_1(x)$, $x \in W^u(\Lambda_1) \cap W^s_{loc}(p_1)$, is a point of tangency between leaves of $W^u(\Lambda_1)$ and $\mathcal{F}^s(\Lambda_2)$ (the image of π_1 is the set of tangencies, see Figure 2).

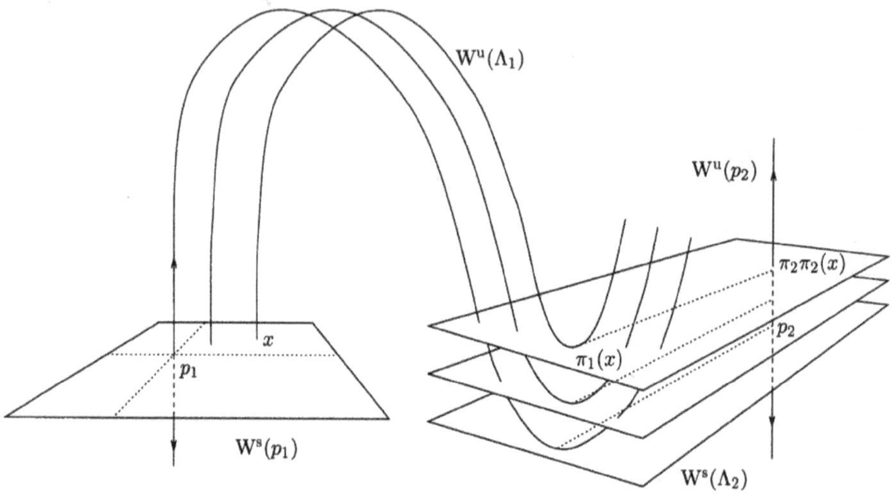

FIGURE 2

We let π_2 be the projection along the leaves of $\mathcal{F}^s(\Lambda_2)$ onto $W^u_{loc}(p_2)$, which we identify with an interval in $\mathbb{R}$. Then the theorem follows by an application of the gap lemma to the Cantor sets $\pi_2 \circ \pi_1(W^u(\Lambda_1) \cap W^s_{loc}(p_1))$ and $W^s(\Lambda_2) \cap W^u_{loc}(p_2)$.

2. Intrinsically smooth maps

In this section we describe a notion of "intrinsic" differentiability of functions on compact subsets of $\mathbb{R}^m$, closely related to Whitney [W], and we list some of its basic properties. The main result is Proposition 2.10, on intrinsic differentiability of invariant sections of (contracting) bundle morphisms.

Let $X \subset \mathbb{R}^m$ be a compact set and $\varphi : X \to \mathbb{R}^n$ be continuous. We say that φ is *intrinsically* C^1 *on* X if there exists a continuous map $\Delta\varphi : X \times X \to \mathcal{L}(\mathbb{R}^m, \mathbb{R}^n)$ such that

$$\varphi(x) - \varphi(z) = \Delta\varphi(x, z) \cdot (x - z) \quad \text{for all } x, z \in X.$$

Such a $\Delta\varphi$ (which is, in general, far from unique) is called an *intrinsic derivative of* φ. We say that φ is *intrinsically* $C^{1+\gamma}$ *on* X if it admits some γ-Hölder continuous intrinsic derivative.

Example 2.1. Let $\varphi\colon U \to \mathbb{R}^n$, U an open rectangle in $\mathbb{R}^m$, be a C^1 map. Then $\varphi \mid X$ is intrinsically C^1 on X, for every compact $X \subset U$: One may take $\Delta\varphi(x,z) = (\Delta_i\varphi_j(x,z))_{i,j}$ where, denoting $y^{(i)} = (x_1,\dots,x_i,z_{i+1},\dots,z_m)$,

$$\Delta_i\varphi_j(x,z) = \begin{cases} \frac{\varphi_j(y^{(i)})-\varphi_j(y^{(i-1)})}{x_i-z_i}, & \text{if } x_i \neq z_i \\ \frac{\partial\varphi_j}{\partial x_i}(y^{(i)}), & \text{if } x_i = z_i. \end{cases}$$

If, moreover, φ is C^2 then this $\Delta\varphi$ is Lipschitz continuous and so φ is intrinsically $C^{1+\gamma}$ on X for every $0 < \gamma \le 1$.

Example 2.2. Let $\varphi\colon \mathbb{R}^2 \to \mathbb{R}^2$ be given by $\varphi(x,y) = (\sigma x, \rho y)$ with $1 < \sigma < \rho$ and let $X \subset \mathbb{R}^2$ be defined as follows. Take K to be the standard middle-third Cantor set and let $f\colon K \to K$ be the unique continuous map satisfying $f(0) = 0$, $f(1) = 1$ and $f\big(\frac{p}{3^{q-1}} \pm \frac{1}{3^q}\big) = f\big(\frac{p}{3^{q-1}}\big) \pm \frac{1}{3^{2q}}$ for all $p,q \in \mathbb{N}$. This is an intrinsically C^1 function on K with $\Delta f(k,k) = 0$ for all $k \in K$. We take $X = \operatorname{graph}(f)$ and then

$$\Delta\varphi(x,z) = \begin{pmatrix} \sigma & 0 \\ \rho\Delta f(k,\ell) & 0 \end{pmatrix}, \quad \text{denoting } x = (k, f(k)), z = (\ell, f(\ell)),$$

is an intrinsic derivative for φ on X. Observe that $\|\Delta\varphi(x,x)\| = \sigma$ for all $x \in X$; compare with the previous construction.

This second example illustrates the main point in the proof that Λ_1, as constructed in Section 3, has intrinsically C^1 unstable foliation (Proposition 3.4); one uses the geometry of the domain to obtain an intrinsic derivative with the smallest norm of the derivative of the map "restricted to factors". The proof of the following properties of intrinsically C^1 (or $C^{1+\gamma}$) maps is immediate.

LEMMA 2.3 (Chain Rule). *Let* $\varphi\colon X \to \mathbb{R}^n$ *and* $\psi\colon Y \to \mathbb{R}^p$ *be intrinsically* C^1 *(resp.* $C^{1+\gamma}$*) with* $\varphi(X) \subset Y$. *Then* $\psi \circ \varphi$ *is intrinsically* C^1 *(resp.* $C^{1+\gamma}$*) and one may take* $\Delta(\psi\circ\varphi)(x,z) = \Delta\psi(\varphi(x),\varphi(z)) \cdot \Delta\varphi(x,z)$.

LEMMA 2.4 (Uniform limits). *Let a sequence* $\varphi_k\colon X \to \mathbb{R}^n$ *of intrinsically* C^1 *maps converge uniformly to* $\varphi\colon X \to \mathbb{R}^n$ *and admit intrinsic derivatives* $\Delta\varphi_k$ *converging uniformly to* $\Phi\colon X \times X \to \mathcal{L}(\mathbb{R}^m,\mathbb{R}^n)$. *Then* φ *is intrinsically* C^1 *and one may take* $\Delta\varphi = \Phi$.

LEMMA 2.5 (Restrictions). *Let* $\varphi\colon X_1 \times X_2 \to \mathbb{R}^n$ *and for* $x_1 \in X_1$ *and* $x_2 \in X_2$ *define maps* $\varphi_{x_1}\colon X_2 \to \mathbb{R}^n$ *and* $\varphi^{x_2}\colon X_1 \to \mathbb{R}^n$ *by* $\varphi_{x_1}(x_2) = \varphi(x_1,x_2) = \varphi^{x_2}(x_1)$.

(a) *If* φ *is intrinsically* C^1 *then the same holds for every* φ_{x_1} *and* φ^{x_2}, *with* $\Delta\varphi_{x_1}(x_2,z_2)\cdot v_2 = \Delta\varphi((x_1,x_2),(x_1,z_2))\cdot(0,v_2)$ *and* $\Delta\varphi^{x_2}(x_1,z_1)\cdot v_1 = \Delta\varphi((x_1,x_2),(z_1,x_2))\cdot(v_1,0)$.

(b) *If all* φ_{x_1} *and* φ^{x_2} *admit intrinsic derivatives* $\Delta\varphi_{x_1}$ *and* $\Delta\varphi^{x_2}$ *which, moreover, vary continuously with* x_1 *and* x_2 *then* φ *is intrinsically* C^1 *with* $\Delta\varphi((x_1,x_2),(z_1,z_2))\cdot(v_1,v_2) = \Delta\varphi^{z_2}(x_1,z_1)\cdot v_1 + \Delta\varphi_{x_1}(x_2,z_2)\cdot v_2$.

Remark 2.6. The following simple remark will be used in the forthcoming sections. Let $\varphi\colon X \to \mathbb{R}^n$ be Lipschitz continuous and $U \subset X \times X$ be such that $\{\|x-z\|\colon (x,z)\in U\}$ is bounded away from zero. Then there is a Lipschitz continuous map $\Delta\colon U \to \mathcal{L}(\mathbb{R}^m,\mathbb{R}^n)$ such that $\varphi(x)-\varphi(z) = \Delta(x,z)(x-z)$ for every $(x,z)\in U$.

Let us also observe that intrinsic differentiability is a local property. We say that $\varphi\colon X \to \mathbb{R}^n$ is *locally intrinsically* C^1 (resp. $C^{1+\gamma}$) if every $x \in X$ has a neighbourhood $V_x \subset X$ such that $\varphi \mid V_x$ is intrinsically C^1 (resp. $C^{1+\gamma}$).

LEMMA 2.7. $\varphi\colon X \to \mathbb{R}^n$ *is intrinsically* C^1 *if and only if there exists a continuous map* $\partial\varphi\colon X \to \mathcal{L}(\mathbb{R}^m,\mathbb{R}^n)$ *such that*

$$\lim_{\substack{z\to x\\ z\in X}} \frac{\varphi(x)-\varphi(z)-\partial\varphi(x)\cdot(x-z)}{\|x-z\|} = 0 \quad \textit{for every } x\in X.$$

In particular, φ *is intrinsically* C^1 *if and only if it is locally intrinsically* C^1.

Proof. The "only if" affirmative is proved by fixing an intrinsic derivative $\Delta\varphi$ of φ and taking $\partial\varphi(x) = \Delta\varphi(x,x)$. For the proof of the "if" part we take $\partial\varphi$ as in the statement and define $\theta\colon X\times X \to \mathcal{L}(\mathbb{R}^m,\mathbb{R}^n)$ by

$$\theta(x,z)\cdot(x-z) = \varphi(x)-\varphi(z)-\partial\varphi(x)\cdot(x-z) \quad \text{and}$$
$$\theta(x,z)\cdot v = 0 \quad \text{whenever } v\cdot(x-z)=0.$$

Then $\theta(x,z)$ depends continuously on $(x,z)\in X\times X$ and $\Delta\varphi(x,z) = \partial\varphi(x) + \theta(x,z)$ is an intrinsic derivative for φ. The second part of the lemma is an immediate consequence of the first one. □

Remark 2.8. Notice that, for a sequence $(\varphi_k)_k$ converging uniformly to $\varphi\colon X\to\mathbb{R}^n$, having $\partial\varphi_k$ converging uniformly to some $\Phi\colon X \to \mathcal{L}(\mathbb{R}^m,\mathbb{R}^n)$ is not sufficient to assure that φ is intrinsically C^1.

Remark 2.9. The same construction permits us to check that φ is intrinsically $C^{1+\gamma}$ if and only if $\partial\varphi$ as above can be found which is γ-Hölder continuous and, moreover, satisfies

$$\|\varphi(x)-\varphi(z)-\partial\varphi(x)(x-z)\| \le C\,\|x-z\|^{1+\gamma}$$

for some $C > 0$ and every $x, z \in X$. In particular, by Whitney [W] (see also [St]), any intrinsically $\mathcal{C}^{1+\gamma}$ function, $\gamma > 0$, admits a $\mathcal{C}^{1+\gamma}$ extension to an open neighbourhood of its domain. However, in the present paper we make no use of such extension, except for the simple one-dimensional situation in Lemma 4.4 where an explicit construction is provided.

Proposition 3.4 will be proved by means of the following general result on contracting bundle morphisms, which is a version for intrinsic differentiability of Theorem 6.1 in [HP]. Let $X_0 \subset X$ be compact subsets of $\mathbb{R}^m$ and $B = \overline{\mathrm{B}}_R(0)$ be a compact ball in $\mathbb{R}^n$. Let $f\colon X_0 \to X$ be a homeomorphism and $\tilde{F}\colon X_0 \times B \to X \times B$ be a continuous map of the form $\tilde{F}(x, v) = (f(x), F_x(v))$. Assume that, for some $c < 1$ and every $x \in X_0$, F_x has Lipschitz constant $\le c$, so that there is a unique continuous section $\tilde{\sigma}\colon X \to B$ satisfying $\tilde{\sigma}(f(x)) = F_x(\tilde{\sigma}(x))$ for all $x \in X_0$ (*$\tilde{F}$-invariance*).

PROPOSITION 2.10.
(a) *Suppose that $\tilde{F}$ and f^{-1} are intrinsically $\mathcal{C}^1$ (on their domains) and there is $a < 1$ such that*

$$\|\Delta F_y(v, w)\| \cdot \|\Delta f^{-1}(x, z)\| \le a, \tag{2.1}$$

for any x,$z \in X$, $y \in X_0$ and v,$w \in B$. Then $\tilde{\sigma}$ is intrinsically $\mathcal{C}^1$ on X.
(b) *Suppose that $\tilde{F}$ and f^{-1} are intrinsically $\mathcal{C}^{1+\gamma}$ (on their domains) and there is $b < 1$ such that*

$$\|\Delta F_y(v, w)\| \cdot \|\Delta f^{-1}(x, z)\|^{1+\gamma} \le b \tag{2.2}$$

for any x,$z \in X$, $y \in X_0$ and v,$w \in B$. Then $\tilde{\sigma}$ is intrinsically $\mathcal{C}^{1+\gamma}$ on X.

Proof. Let S (resp. $\mathcal{S}$) be the space of continuous maps $\sigma : X \to B$ (resp. $\Sigma\colon X \times X \to \mathcal{L}(\mathbb{R}^m, \mathbb{R}^n)$), endowed with the sup-norm. Define $\mathcal{F}\colon S \times \mathcal{S} \to S \times \mathcal{S}$ by $\mathcal{F}(\sigma, \Sigma) = (\sigma', \Sigma')$ with

$$\sigma'(x) = F_{\overline{x}}(\sigma(\overline{x})) \tag{2.3}$$

$$\begin{aligned}\Sigma'(x, z) = {}& \Delta F^{\sigma(\overline{z})}(\overline{x}, \overline{z}) \circ \Delta f^{-1}(x, z)\\ &+ \Delta F_{\overline{x}}(\sigma(\overline{x}), \sigma(\overline{z})) \circ \Sigma(\overline{x}, \overline{z}) \circ \Delta f^{-1}(x, z)\end{aligned} \tag{2.4}$$

where, for simplicity, we denote $\overline{x} = f^{-1}(x)$, $\overline{z} = f^{-1}(z)$. By construction (recall also Lemma 2.5), if Σ is an intrinsic derivative for σ then Σ' is an intrinsic derivative for σ'. Observe now that $\mathcal{F}$ is a fiber contraction:

(i) Given $(\sigma_1, \Sigma_1), (\sigma_2, \Sigma_2) \in S \times \mathcal{S}$ and denoting $(\sigma_i', \Sigma_i') = \mathcal{F}(\sigma_i, \Sigma_i)$, $1 \le i \le 2$, and $\overline{x} = f^{-1}(x)$, we have

$$\begin{aligned}\|\sigma_1' - \sigma_2'\| &= \sup_x \|F_{\overline{x}}(\sigma_1(\overline{x})) - F_{\overline{x}}(\sigma_2(\overline{x}))\| \le c \sup_x \|\sigma_1(\overline{x}) - \sigma_2(\overline{x})\|\\ &\le c\|\sigma_1 - \sigma_2\| \quad (c < 1);\end{aligned}$$

(ii) Given $\sigma \in S$ and $\Sigma_1, \Sigma_2 \in \mathcal{S}$ and denoting $(\sigma', \Sigma_i') = \mathcal{F}(\sigma, \Sigma_i)$, $1 \le i \le 2$, and $\overline{x} = f^{-1}(x)$, $\overline{z} = f^{-1}(z)$, we have

$$\begin{aligned}\|\Sigma_1' - \Sigma_2'\| &\le \sup_x(\|\Delta F_{\overline{x}}(\sigma(\overline{x}), \sigma(\overline{z}))\| \cdot \|\Sigma_1(\overline{x},\overline{z}) - \Sigma_2(\overline{x},\overline{z})\| \cdot \|\Delta f^{-1}(x,z)\|) \\ &\le a\,\|\Sigma_1 - \Sigma_2\| \quad (a<1).\end{aligned}$$

It follows ([HP]) that $\mathcal{F}$ has a unique fixed point $(\overline{\sigma}, \tilde{\Sigma})$ and $\mathcal{F}^n(\sigma, \Sigma) \to (\overline{\sigma}, \tilde{\Sigma})$ as $n \to +\infty$, for every $(\sigma, \Sigma) \in S \times \mathcal{S}$. Clearly $\overline{\sigma} = \tilde{\sigma}$. Choose now $\sigma_0 \in S$ intrinsically $\mathcal{C}^1$ and Σ_0 an intrinsic derivative of σ_0. Writing $\mathcal{F}^n(\sigma_0, \Sigma_0) = (\sigma_n, \Sigma_n)$ we have that Σ_n is an intrinsic derivative of σ_n for every n. Since $(\sigma_n)_n \to \tilde{\sigma}$ and $(\Sigma_n)_n \to \tilde{\Sigma}$, it follows from Lemma 2.4 that $\tilde{\Sigma}$ is an intrinsic derivative for $\tilde{\sigma}$. This proves (a). Now we show that under the assumptions of (b) $\tilde{\Sigma}$ is γ-Hölder continuous. Fix $C > 0$ large enough so that ΔF_x, $x \in X_0$, ΔF^v, $v \in B$, and Δf^{-1} are all (C, γ)-Hölder continuous. Let $x_1, x_2, z \in X$. Subtracting the equalities

$$\tilde{\Sigma}(x_i, z) = \Delta F^{\tilde{\sigma}(\overline{z})}(\overline{x}_i, \overline{z}) \circ \Delta f^{-1}(x_i, z) + \Delta F_{\overline{x}_i}(\tilde{\sigma}(\overline{x}_i), \tilde{\sigma}(\overline{z})) \circ \tilde{\Sigma}(\overline{x}_i, \overline{z}) \circ \Delta f^{-1}(x_i, z)$$

(recall (2.4)) we obtain

$$\begin{aligned}\|\tilde{\Sigma}(x_1, z) - \tilde{\Sigma}(x_2, z)\| \le\ & C\|\overline{x}_1 - \overline{x}_2\|^\gamma \cdot \|\Delta f^{-1}\| + \|\Delta F^*\| \cdot C\|x_1 - x_2\|^\gamma \\ &+ C\|\tilde{\sigma}(\overline{x}_1) - \tilde{\sigma}(\overline{x}_2)\|^\gamma \|\tilde{\Sigma}\| \cdot \|\Delta f^{-1}\| \\ &+ \|\Delta F_*\| \|\tilde{\Sigma}(\overline{x}_1, \overline{z}) - \tilde{\Sigma}(\overline{x}_2, \overline{z})\| \cdot \|\Delta f^{-1}\| \\ &+ \|\Delta F_*\| \cdot \|\tilde{\Sigma}\| \cdot C\|x_1 - x_2\|^\gamma,\end{aligned}$$

where $\|\Delta F^*\| = \sup_{v,x,z} \|\Delta F^v(x,z)\|$ and $\|\Delta F_*\| = \sup_{x,v,w} \|\Delta F_x(v,w)\|$. Noting that $\tilde{\sigma}$ is $\|\tilde{\Sigma}\|$-Lipschitz continuous we get

$$\begin{aligned}\|\tilde{\Sigma}(x_1, z) - \tilde{\Sigma}(x_2, z)\| \le\ & C_1\|x_1 - x_2\|^\gamma \\ &+ (\|\Delta F_*\|\,\|\Delta f^{-1}\|)\|\tilde{\Sigma}(f^{-1}x_1, f^{-1}z) - \tilde{\Sigma}(f^{-1}x_2, f^{-1}z)\|\end{aligned}$$

with $C_1 = (C\|\Delta f^{-1}\|)^{1+\gamma} + C\|\Delta F^*\| + (C\|\Delta f^{-1}\|\,\|\tilde{\Sigma}\|)^{1+\gamma} + C\|\Delta F_*\|\,\|\tilde{\Sigma}\|$. By recurrence and using the fact that

$$\begin{aligned}\|\Delta F_*\|\,\|\Delta f^{-1}\|\|f^{-1}y_1 - f^{-1}y_2\|^\gamma &\le \|\Delta F_*\|\,\|\Delta f^{-1}\|^{1+\gamma}\|y_1 - y_2\|^\gamma \\ &\le b\|y_1 - y_2\|^\gamma\end{aligned}$$

we find, for every $k \ge 1$

$$\begin{aligned}\|\tilde{\Sigma}(x_1, z) - \tilde{\Sigma}(x_2, z)\| \le\ & C_1\Big(\sum_0^{k-1} b^i\Big)\|x_1 - x_2\|^\gamma \\ &+ (\|\Delta F_*\|\,\|\Delta f^{-1}\|)^k\|\tilde{\Sigma}(f^{-k}x_1, f^{-k}z) - \tilde{\Sigma}(f^{-k}x_2, f^{-k}z)\|.\end{aligned}$$

On the other hand

$$\|\Delta F_*\|\,\|\Delta f^{-1}\| \leq \max\{\|\Delta F_*\|, \|\Delta F_*\|\,\|\Delta f^{-1}\|^{1+\gamma}\} \leq \max\{c, b\} < 1$$

and so passing to the limit as $k \to +\infty$ leads to

$$\|\tilde{\Sigma}(x_1, z) - \tilde{\Sigma}(x_2, z)\| \leq C_1(\sum_0^\infty b^i)\|x_1 - x_2\|^\gamma.$$

In the same way one proves that $\tilde{\Sigma}$ is γ-Hölder continuous in the second variable. □

Remark 2.11. It is an interesting question whether the conclusion of the proposition still holds under the slightly more general (and more natural) assumptions

$$\|\Delta F_{f^{-1}(x)}(v, w)\| \cdot \|\Delta f^{-1}(x, x)\| \leq a < 1, \quad \text{for all } x \in X, v, w \in B$$
$$\|\Delta F_{f^{-1}(x)}(v, w)\| \cdot \|\Delta f^{-1}(x, x)\|^{1+\gamma} \leq b < 1, \quad \text{for all } x \in X, v, w \in B.$$

We close with another simple example which will be of future use.

Example 2.12. It is a well known fact that the map $\chi\colon\, u \to (I+u)^{-1}$ is defined and smooth on $\{u \in \mathcal{L}(\mathbb{R}^m, \mathbb{R}^m)\colon\, \|u\| < 1\}$. In what follows we let $\Delta\chi$ be some (fixed) intrinsic derivative for χ on, say, $\{u\colon\, \|u\| \leq 1/2\}$; observe that we must have $\|\Delta\chi(0,0)\| = \|\mathrm{D}\chi(0)\| = 1$. Let $R > 0$ be fixed and $B = \overline{\mathrm{B}}_R(0)$ denote the closed R-ball in $\mathcal{L}(\mathbb{R}^m, \mathbb{R}^n)$. Let also $H \in \mathcal{L}(\mathbb{R}^{m+n}, \mathbb{R}^{m+n})$ be fixed and write $H = \begin{pmatrix} a & b \\ c & d \end{pmatrix}$, $a \in \mathcal{L}(\mathbb{R}^m, \mathbb{R}^m)$, $b \in \mathcal{L}(\mathbb{R}^n, \mathbb{R}^m)$, $c \in \mathcal{L}(\mathbb{R}^m, \mathbb{R}^n)$, $d \in \mathcal{L}(\mathbb{R}^n, \mathbb{R}^n)$. Suppose $\|b\|\|a^{-1}\| \leq 1/(2R)$, so that the graph transform induced by H

$$\Gamma\colon h \mapsto (c + dh) \cdot (a + bh)^{-1} = (ca^{-1} + dha^{-1}) \cdot \chi(bha^{-1})$$

is well defined on B. For $h_1, h_2 \in B$ and $g \in \mathcal{L}(\mathbb{R}^m, \mathbb{R}^n)$, we let

$$\begin{aligned}\Delta\Gamma(h_1, h_2) \cdot g = {} & (dga^{-1}) \cdot \chi(bh_1a^{-1}) \\ & + (ca^{-1} + dh_2a^{-1}) \cdot \Delta\chi(bh_1a^{-1}, bh_2a^{-1}) \cdot (bga^{-1}).\end{aligned}$$

Then, clearly, $\Delta\Gamma$ is an intrinsic derivative for Γ on B. Observe that $\Delta\Gamma(h_1, h_2)$ is (uniformly) close to $g \mapsto dga^{-1}$ if $\|b\|\|a^{-1}\|$ is close to zero.

3. Intrinsically smooth foliations of basic sets

Let q_0 be a transverse homoclinic point of some hyperbolic fixed (or periodic) point p of a $\mathcal{C}^2$ diffeomorphism $\varphi\colon\, M \to M$. We assume $q_0 \notin \mathrm{W}^{\mathrm{ss}}(p)$ and another mild (open and dense) transversality condition to be stated in

(3.2) below. Then our goal is to prove that there exists a hyperbolic basic set Λ_1 containing p and q_0 and whose unstable foliation is intrinsically $\mathcal{C}^1$. For the sake of simplicity we restrict to the case when φ is $\mathcal{C}^2$-linearizable on a neighbourhood U of p. Apart from the corresponding nonresonance assumptions no other conditions on the eigenvalues of $\mathrm{D}\varphi(p)$ (or the dimensions of $\mathrm{W^u}(p)$, $\mathrm{W^s}(p)$) are required for this construction. Moreover, a dual result holds when $q_0 \notin \mathrm{W^{uu}}(p)$.

Example 3.1. Let $f\colon \mathbb{R}^3 \to \mathbb{R}^3$ be a $\mathcal{C}^\infty$ diffeomorphism such that for (x, y, z) close enough to the origin, $f(x, y, z) = (\sigma x, \lambda y, \theta z)$ with $\sigma > 1 > \lambda > \theta > 0$ and $\lambda > \sigma\theta$. Suppose moreover that $p = (0, 0, 0)$ has a transverse homoclinic point of the form $q_0 = (0, 0, \delta)$, δ small. Let us parametrize a segment γ of $\mathrm{W^u}(p)$ near q_0 by $[-\varepsilon, \varepsilon] \ni x \mapsto (x, y(x), z(x))$ and assume that $y'(0) \neq 0$. For $n \geq 1$ sufficiently large $f^n(\gamma)$ intersects $\{x = \alpha\}$, α small, in the point $\tilde{q}_n = (\alpha, \lambda^n y(\alpha\sigma^{-n}),\ \theta^n z(\alpha\sigma^{-n}))$. On the other hand $q_n = f^n(q_0) = (0, 0, \theta^n\delta)$ and so

$$\frac{\|\tilde{q}_n - (\alpha, 0, 0)\|}{\|q_n - (0, 0, 0)\|} \approx \operatorname{const} y'(0) \left(\frac{\lambda}{\sigma\theta}\right)^n \to \infty.$$

Therefore, the projection from $\{x = 0\}$ to $\{x = \alpha\}$ along $\mathrm{W^u}(p)$ is not even Lipschitz on $\{p\} \cup \{q_n\colon n \geq 0\}$. This illustrates the main obstruction for an invariant foliation of a basic set (with codimension bigger than 1) to be smooth.

Starting our construction of Λ_1, let us denote by $\sigma_1, \ldots, \sigma_u$, $\lambda_1, \ldots, \lambda_s$, $u + s = m$, the eigenvalues of $\mathrm{D}\varphi(p)$, with $|\sigma_u| \geq \cdots \geq |\sigma_1| > 1 > |\lambda_1| \geq \cdots \geq |\lambda_s|$. We define $1 \leq w \leq s$ by $|\lambda_1| = \cdots = |\lambda_w| > |\lambda_{w+1}| \geq \cdots \geq |\lambda_s|$ and let $\mathrm{E^s} = \mathrm{E^w} \oplus \mathrm{E^{ss}}$ be the invariant splitting such that $\mathrm{D}\varphi(p) \mid \mathrm{E^w}$ has eigenvalues $\lambda_1, \ldots, \lambda_w$ and $\mathrm{D}\varphi(p) \mid \mathrm{E^{ss}}$ has eigenvalues $\lambda_{w+1}, \ldots, \lambda_s$. We choose $\mathcal{C}^2$ linearizing coordinates $(\xi_1, \ldots, \xi_u, \zeta_1, \ldots, \zeta_s)$ on U and, clearly, we may assume that

(A1) $\mathrm{W^u_{loc}}(p) \subset \{\zeta_1 = \cdots = \zeta_s = 0\}$ and $\mathrm{W^s_{loc}}(p) \subset \{\xi_1 = \cdots = \xi_u = 0\}$;
(A2) $\mathrm{E^w} = \{0^u\} \times \mathbb{R}^w \times \{0^{s-w}\}$ and the strong stable manifold (tangent to $\mathrm{E^{ss}}$ at p) satisfies $\mathrm{W^{ss}_{loc}}(p) \subset \{\xi_1 = \cdots = \xi_u = \zeta_1 = \cdots = \zeta_w = 0\}$.

Up to a convenient choice of riemannian metric we have, for $\sigma = |\sigma_1|$, $\lambda = |\lambda_1| = |\lambda_w|$ and $\theta = |\lambda_{w+1}|$,

(B1) $(\sigma - \varepsilon)\,\|v\| \leq \|\mathrm{D}\varphi(p)v\|$, for all $v \in \mathrm{E^u}$

(B2) $(\lambda - \varepsilon)\,\|v\| \leq \|\mathrm{D}\varphi(p)v\| \leq (\lambda + \varepsilon)\,\|v\|$, for all $v \in \mathrm{E^w}$

(B3) $\|\mathrm{D}\varphi(p)v\| \leq (\theta + \varepsilon)\,\|v\|$, for all $v \in \mathrm{E^{ss}}$

where $\varepsilon > 0$ is fixed small enough so that $\theta + 2\varepsilon < \lambda - 2\varepsilon < \lambda + 2\varepsilon < \sigma - 2\varepsilon$. (In the case $w = s$, i.e., if all contracting eigenvalues have the same norm, $E^{ss} = \{0\}$, $W^{ss}(p) = \{p\}$ and we leave θ undefined). Fix points $q \in W^s_{loc}(p)$ and $r = \varphi^{-N}(q) \in W^u_{loc}(p)$ in the orbit of q_0. Take

$$V = V_\delta = \{\|(\xi_1, \ldots, \xi_u)\| \le \delta\} \times \{\|(\zeta_1, \ldots, \zeta_s)\| \le \rho\}$$

where $\delta > 0$ is small and $\rho > 0$ is fixed in such a way that $\{p, q\} \subset \operatorname{int}(V) \subset V \subset U$. Let $n = n(\delta)$ be minimum such that $r \in \operatorname{int}(\varphi^n(V))$. (We suppose δ conveniently adjusted so that $\varphi^{n+N}(V)$ cuts V in two cylinders as in Figure 3.)

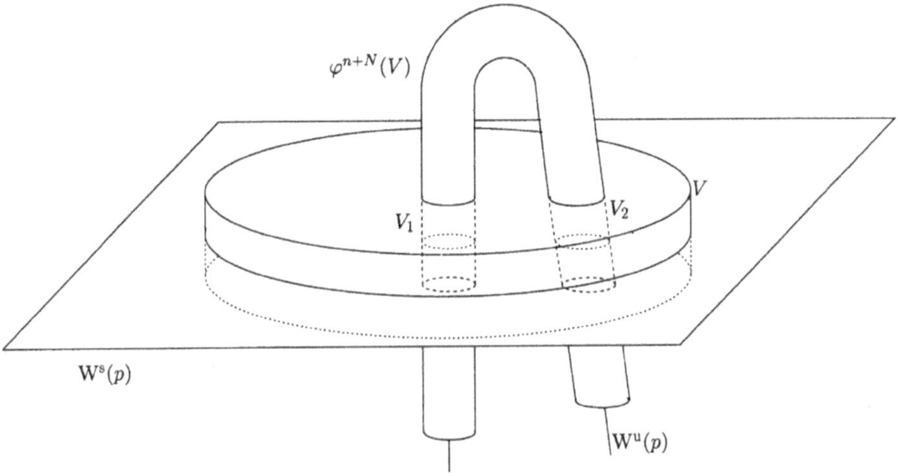

FIGURE 3

We let $\Lambda = \bigcap_{k\in\mathbb{Z}} \varphi^{(n+N)k}(V)$ and then take simply $\Lambda_1 = \bigcup_1^{n+N} \varphi^i(\Lambda)$. The next proposition expresses the crucial geometric property of Λ: the intrinsic tangent space to $W^u_{loc}(\Lambda)$ at every $x \in W^u_{loc}(\Lambda)$ is contained in a $(u+w)$-dimensional subspace whose angle to $\mathbb{R}^{u+w} \times \{0^{s-w}\}$ is uniformly bounded from above. We denote by (v_{uw}, v_{ss}) the components of a vector $v \in \mathbb{R}^m$ with respect to the splitting $\mathbb{R}^m = \mathbb{R}^{u+w} \times \mathbb{R}^{s-w}$.

PROPOSITION 3.2. *There are $C > 0$ and $\gamma > 0$, independent of the n chosen in the definition of Λ, and there exists some (C,γ)-Hölder continuous map $A\colon W^u_{loc}(\Lambda) \times W^u_{loc}(\Lambda) \to \mathcal{L}(\mathbb{R}^{u+w}, \mathbb{R}^{s-w})$ such that*

$$\|A(x,z)\| \le C \quad \textit{and} \quad (x-z)_{ss} = A(x,z)\cdot(x-z)_{uw} \quad \textit{for all } x, z \in W^u_{loc}(\Lambda).$$

Proof. Let $W = W^u(\Lambda) \cap V$, $\psi = \varphi^{n+N}$ and V_1, V_2 be the two connected components (cylinders) of $\psi(V) \cap V$. For $k \ge 0$ let $U_k \subset W \times W$ be the set of pairs (x, z) such that

(i) $\psi^{-i}(x)$, $\psi^{-i}(z)$ belong to the same V_{j_i}, $j_i = 1, 2$, for $0 \le i < k$;
(ii) $\psi^{-k}(x)$, $\psi^{-k}(z)$ belong to different V_j's.

Observe that $\widetilde{W} = \bigcup_{k \ge 0} U_k$ is dense in $W \times W$: in fact $(x, z) \in W \times W \setminus \widetilde{W}$ if and only if x and z belong to the same local unstable leaf. We construct A on each U_k, by recurrence on k, and then it extends from $\widetilde{W}$ to $W \times W$ by uniformity. The definition of $A \mid U_0$ is rather arbitrary. Clearly, there is $K > 1$ (independent of n) such that (see Figure 4)

$$\|(x-z)_{\mathrm{ss}}\| \le K \text{ and } \|(x-z)_{\mathrm{uw}}\| \ge K^{-1} \quad \text{for all } x \in V_i,\ z \in V_j,\ i \ne j.$$

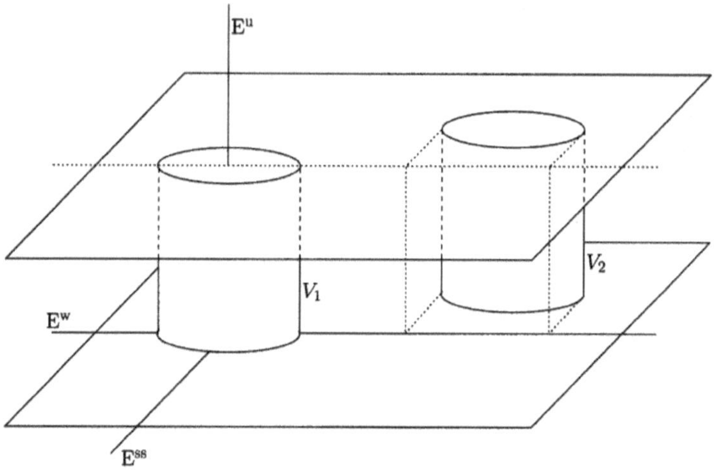

FIGURE 4

Therefore, by Remark 2.6, for $C_0 > 0$ sufficiently large (depending only on K) there are C_0-Lipschitz continuous maps $A_0\colon U_0 \to \mathcal{L}(\mathbb{R}^{u+w}, \mathbb{R}^{s-w})$ with norm $\le C_0$ and satisfying

$$(x-z)_{\mathrm{ss}} = A_0(x,z) \cdot (x-z)_{\mathrm{uw}} \quad \text{for all } (x,z) \in U_0.$$

We take $A \mid U_0$ to be any such map. Let now $(x, z) \in U_k$, $k \ge 1$, and denote $\overline{x} = \psi^{-1}(x)$, $\overline{z} = \psi^{-1}(z)$. Observe that $(\overline{x}, \overline{z}) \in U_{k-1}$. We write

$$\varphi^n = \begin{pmatrix} L^n & 0 \\ 0 & T^n \end{pmatrix}, \quad L \in \mathcal{L}(\mathbb{R}^{u+w}, \mathbb{R}^{u+w}), \quad T \in \mathcal{L}(\mathbb{R}^{s-w}, \mathbb{R}^{s-w})$$

with (recall above) $\|L^{-1}\| \le (\lambda - \varepsilon)^{-1}$ and $\|T\| \le (\theta + \varepsilon)$. Moreover, we choose as in Example 2.1 a Lipschitz continuous intrinsic derivative

$$\Delta\phi = \begin{pmatrix} \Delta_{\mathrm{uw}}\phi_{\mathrm{uw}} & \Delta_{\mathrm{ss}}\phi_{\mathrm{uw}} \\ \Delta_{\mathrm{uw}}\phi_{\mathrm{ss}} & \Delta_{\mathrm{ss}}\phi_{\mathrm{ss}} \end{pmatrix}$$

of $\phi = \varphi^N$ on a neighbourhood of $\{p, r\}$. Then $\Delta\psi(\eta,\zeta) = \Delta\phi(\varphi^n\eta, \varphi^n\zeta) \circ \varphi^n$ defines an intrinsic derivative for ψ and so, denoting $y = (\varphi^n\overline{x}, \varphi^n\overline{z})$; we have

$$\begin{aligned}(x-z)_{\mathrm{uw}} &= \Delta_{\mathrm{uw}}\phi_{\mathrm{uw}}(y)\cdot L^n\cdot(\overline{x}-\overline{z})_{\mathrm{uw}} + \Delta_{\mathrm{ss}}\phi_{\mathrm{uw}}(y)\cdot T^n\cdot(\overline{x}-\overline{z})_{\mathrm{ss}}\\ &= (\Delta_{\mathrm{uw}}\phi_{\mathrm{uw}}(y) + \Delta_{\mathrm{ss}}\phi_{\mathrm{uw}}(y)\cdot T^n\cdot A(\overline{x},\overline{z})\cdot L^{-n})\cdot L^n\cdot(\overline{x}-\overline{z})_{\mathrm{uw}}\end{aligned}$$

and analogously

$$(x-z)_{\mathrm{ss}} = (\Delta_{\mathrm{uw}}\phi_{\mathrm{ss}}(y) + \Delta_{\mathrm{ss}}\phi_{\mathrm{ss}}(y)\cdot T^n\cdot A(\overline{x},\overline{z})\cdot L^{-n})\cdot L^n\cdot(\overline{x}-\overline{z})_{\mathrm{uw}}.$$

We want to define

$$\begin{aligned}A(x,z) &= (\Delta_{\mathrm{uw}}\phi_{\mathrm{ss}}(y) + \Delta_{\mathrm{ss}}\phi_{\mathrm{ss}}(y)T^nA(\overline{x},\overline{z})L^{-n})\\ &\quad\circ(\Delta_{\mathrm{uw}}\phi_{\mathrm{uw}}(y) + \Delta_{\mathrm{ss}}\phi_{\mathrm{uw}}(y)T^nA(\overline{x},\overline{z})L^{-n})^{-1}\end{aligned} \tag{3.1}$$

so that, automatically, $(x-z)_{\mathrm{ss}} = A(x,z)\cdot(x-z)_{\mathrm{uw}}$. *In order to show that* (3.1) *makes sense we need the following generic (open and dense) hypothesis:*

$$\mathrm{D}_{\mathrm{uw}}\phi_{\mathrm{uw}}(r) \text{ is an isomorphism.} \tag{3.2}$$

Here D *denotes the usual derivative and* (3.2) *means that unstable/weak-stable directions are not sent to strong-stable directions by* φ^N. Notice now that if $\delta > 0$ is small (and then n is large) V_1 and V_2 are small and so $y = (\varphi^{-N}(x), \varphi^{-N}(z))$ must be close to either (r,r) or (p,p). This means that (up to reducing δ) we may assume $\Delta_{\mathrm{uw}}\phi_{\mathrm{uw}}(y)$ to be close to $\mathrm{D}_{\mathrm{uw}}\phi_{\mathrm{uw}}(r) = \Delta_{\mathrm{uw}}\phi_{\mathrm{uw}}(r,r)$ or $\mathrm{D}_{\mathrm{uw}}\phi_{\mathrm{uw}}(p) = \Delta_{\mathrm{uw}}\phi_{\mathrm{uw}}(p,p)$. On the other hand, if n is large we also have

$$\|\Delta_{\mathrm{ss}}\phi_{\mathrm{uw}}(y)T^nA(\overline{x},\overline{z})L^{-n}\| \le \mathrm{const}C_0\left(\frac{\theta+\varepsilon}{\lambda-\varepsilon}\right)^n$$

small. It follows from (3.2) that $(\Delta_{\mathrm{uw}}\phi_{\mathrm{uw}}(y) + \Delta_{\mathrm{ss}}\phi_{\mathrm{uw}}(y)T^nA(\overline{x},\overline{z})L^{-n})$ is invertible and the norm of its inverse is at most

$$a_1 = 1 + \max\{\|(\mathrm{D}_{\mathrm{uw}}\phi_{\mathrm{uw}}(p))^{-1}\|, \|(\mathrm{D}_{\mathrm{uw}}\phi_{\mathrm{uw}}(r))^{-1}\|\}.$$

Moreover, the same argument also shows that if n is taken large enough the norm of $(\Delta_{\mathrm{uw}}\phi_{\mathrm{ss}}(y) + \Delta_{\mathrm{ss}}\phi_{\mathrm{ss}}(y)T^nA(\overline{x},\overline{z})L^{-n})$ is at most

$$a_2 = 1 + \max\{\|\mathrm{D}_{\mathrm{uw}}\phi_{\mathrm{ss}}(p)\|, \|\mathrm{D}_{\mathrm{uw}}\phi_{\mathrm{ss}}(r)\|\}.$$

This proves that (3.1) is indeed defined and gives $\|A(x,z)\| \le C_0$, as long as we take from the beginning $C_0 \ge a_1a_2$. Successive repetition of this procedure extends the definition of A to $\widetilde{W} = \bigcup_{k\ge0} U_k$ (keeping $\|A\| \le C_0$) and now the proof will be complete if we show that such A is (C,γ)-Hölder continuous for some $\gamma > 0$ and $C > C_0$. Let $(x,z) \in U_k$ and $(\tilde{x},z) \in U_{\tilde{k}}$ with $k \le \tilde{k}$. Denote $x_i = \psi^{-i}(x)$, $\tilde{x}_i = \psi^{-i}(\tilde{x})$ and $z_i = \psi^{-i}(z)$, for $i \ge 0$. If $k < \tilde{k}$ then x_k and $\tilde{x}_k$ belong to different V_j's and so $\|A(x_k,z_k) - A(\tilde{x}_k,z_k)\| \le 2C_0 \le 2C_0K\|x_k - \tilde{x}_k\|$.

If, on the contrary, $k = \tilde{k}$ then both (x_k, z_k) and $(\tilde{x}_k, z_k)$ belong to U_0 and so, by construction, $\|A(x_k, z_k) - A(\tilde{x}_k, z_k)\| \le C_0\|x_k - \tilde{x}_k\|$. Now we proceed by induction: Let $0 \le i < k$ and assume that

$$\|A(x_{i+1}, z_{i+1}) - A(\tilde{x}_{i+1}, z_{i+1})\| \le C\|x_{i+1} - \tilde{x}_{i+1}\|^\gamma$$

for some $0 < \gamma < 1$ and some large $C > 0$. Clearly,

$$\|x_i - \tilde{x}_i\| \ge (\|\mathrm{D}\varphi^{-1}\|)^{-(n+N)}\|x_{i+1} - \tilde{x}_{i+1}\|.$$

On the other hand, expressing $A(x_i, z_i)$ and $A(\tilde{x}_i, z_i)$ in terms of $A(x_{i+1}, z_{i+1})$, $A(\tilde{x}_{i+1}, z_{i+1})$ and then subtracting we get

$$\begin{aligned}\|A(x_i, z_i) - A(\tilde{x}_i, z_i)\| \le a_3\,\|x_i - \tilde{x}_i\| \\ + a_4\left(\frac{\theta+\varepsilon}{\lambda-\varepsilon}\right)^n \|A(x_{i+1}, z_{i+1}) - A(\tilde{x}_{i+1}, z_{i+1})\|\end{aligned}$$

with a_3, a_4 depending only on $\phi = \varphi^N$. Taking $C > 0$ sufficiently large with respect to a_3 and $\operatorname{diam}(U)$ we obtain $a_3\|x_i - \tilde{x}_i\| \le (C/2)\|x_i - \tilde{x}_i\|^\gamma$. On the other hand

$$\begin{aligned} a_4 &\left(\frac{\theta+\varepsilon}{\lambda-\varepsilon}\right)^n \|A(x_{i+1}, z_{i+1}) - A(\tilde{x}_{i+1}, z_{i+1})\| \\ &\le C a_4 \|\mathrm{D}\varphi^{-1}\|^{\gamma N}\left(\frac{\theta+\varepsilon}{\lambda-\varepsilon}\cdot\|\mathrm{D}\varphi^{-1}\|^\gamma\right)^n \|x_i - \tilde{x}_i\|^\gamma \\ &\le \frac{C}{2}\|x_i - \tilde{x}_i\|^\gamma, \end{aligned}$$

as long as we fix $\gamma < \log((\lambda-\varepsilon)/(\theta+\varepsilon))/\log\|\mathrm{D}\varphi^{-1}\|$ (and take n large enough). Hence $\|A(x_i, z_i) - A(\tilde{x}_i, z_i)\| \le C\|x_i - \tilde{x}_i\|^\gamma$ and by induction this gives $\|A(x,z) - A(\tilde{x}, z)\| \le C\|x - \tilde{x}\|^\gamma$. By symmetry, Hölder continuity in the second variable also follows. □

Remark 3.3. It is also clear from the argument that $A(p,p) = 0$, i.e., $\mathrm{IT}_p\,\mathrm{W}^{\mathrm{u}}_{\mathrm{loc}}(\Lambda_1) \subset \mathrm{E}^{\mathrm{u}} \oplus \mathrm{E}^{\mathrm{w}}$.

PROPOSITION 3.4. *For Λ_1 as above the map $\mathcal{F}\colon \mathrm{W}^{\mathrm{u}}(\Lambda_1) \ni x \mapsto \mathrm{T}_x\mathrm{W}^{\mathrm{u}}(x)$ is intrinsically $\mathcal{C}^1$ on compact parts of $\mathrm{W}^{\mathrm{u}}(\Lambda_1)$.*

Proof. Given any compact $K \subset \mathrm{W}^{\mathrm{u}}(\Lambda_1)$ there is $k \ge 0$ such that $K \subset \bigcup_1^{n+N}\varphi^{k+i}(W)$ where, as before, $W = \mathrm{W}^{\mathrm{u}}(\Lambda)\cap V$. Note that $\mathcal{F}(\varphi^{k+i}(x)) = \mathrm{D}\varphi^{k+i}(x)\cdot\mathcal{F}(x)$ for all $x \in \mathrm{W}^{\mathrm{u}}(\Lambda_1)$. Therefore, the proposition will follow if we prove that the restriction of $\mathcal{F}$ to W is intrinsically $\mathcal{C}^1$ and we proceed to do this. The argument goes as follows. Let

$$\mathrm{D}\psi = \begin{pmatrix}\mathrm{D_u}\psi_{\mathrm{u}} & \mathrm{D_s}\psi_{\mathrm{u}}\\ \mathrm{D_u}\psi_{\mathrm{s}} & \mathrm{D_s}\psi_{\mathrm{s}}\end{pmatrix} \quad\text{and}\quad \mathrm{D}\phi = \begin{pmatrix}\mathrm{D_u}\phi_{\mathrm{u}} & \mathrm{D_s}\phi_{\mathrm{u}}\\ \mathrm{D_u}\phi_{\mathrm{s}} & \mathrm{D_s}\phi_{\mathrm{s}}\end{pmatrix}$$

denote the expression of the derivatives of $\psi = \varphi^{n+N}$ and $\phi = \varphi^N$ in the splitting $\mathbb{R}^m = \mathbb{R}^u \times \mathbb{R}^s$. Note that for $x \in W$

$$\mathrm{D}\psi(x) = \mathrm{D}\phi(\varphi^n x) \cdot \begin{pmatrix} U^n & 0 \\ 0 & S^n \end{pmatrix}$$

with $U \in \mathcal{L}(\mathbb{R}^u, \mathbb{R}^u)$, $S \in \mathcal{L}(\mathbb{R}^s, \mathbb{R}^s)$ satisfying $\|U^{-1}\| \leq (\sigma - \varepsilon)^{-1}$ and $\|S\| \leq (\lambda + \varepsilon)$. We use Proposition 2.10 with $f = \psi$, $X_0 = W$, $X = f(W)$, B a closed disk of radius $R = \|\mathrm{D_u}\phi_\mathrm{s}\|\|(\mathrm{D_u}\phi_\mathrm{u})^{-1}\| + 1$ in $\mathcal{L}(\mathbb{R}^u, \mathbb{R}^s) \simeq \mathbb{R}^{u \times s}$ and $\tilde{F}\colon X_0 \times B \to X \times B$ being the graph transform induced by $\mathrm{D}\psi$: given $x \in X_0$ and $h \in B$

$$F_x(h) = (\mathrm{D_u}\psi_\mathrm{s}(x) + \mathrm{D_s}\psi_\mathrm{s}(x) \cdot h) \cdot (\mathrm{D_u}\psi_\mathrm{u}(x) + \mathrm{D_s}\psi_\mathrm{u}(x) \cdot h)^{-1}. \tag{3.3}$$

Observe that, as long as n is taken sufficiently large (by choosing δ small),

$$\left\|\mathrm{D_s}\psi_\mathrm{u}(x) \cdot h \cdot (\mathrm{D_u}\psi_\mathrm{u}(x))^{-1}\right\| \leq R\,\|\mathrm{D_s}\phi_\mathrm{u}\|\,\left\|(\mathrm{D_u}\phi_\mathrm{u})^{-1}\right\|(\lambda + \varepsilon)^n(\sigma - \varepsilon)^{-n} \ll 1$$

and, analogously, $\|\mathrm{D_s}\psi_\mathrm{s}(x) \cdot h \cdot (\mathrm{D_u}\psi_\mathrm{u}(x))^{-1}\| \ll 1$. Therefore $F_x(h)$ is well defined by (3.3) and it belongs to B: $\|F_x(h)\| \leq \|\mathrm{D_u}\psi_\mathrm{s}(x) \cdot (\mathrm{D_u}\psi_\mathrm{u}(x))^{-1}\| + 1 \leq R$. Clearly, $\tilde{F}$ is intrinsically $\mathcal{C}^1$ (it extends to a $\mathcal{C}^1$ map on a neighbourhood of $X_0 \times B$) and the combination of the constructions in Lemma 2.5 and Example 2.12 yields an intrinsic derivative

$$\Delta\tilde{F}((x_1, h_1), (x_2, h_2)) = \begin{pmatrix} \Delta f(x_1, x_2) & 0 \\ \Delta F^{h_1}(x_1, x_2) & \Delta F_{x_2}(h_1, h_2) \end{pmatrix}$$

satisfying, for some $C_1 > 0$ (depending only on ϕ),

$$\|\Delta F_y(h_1, h_2)\| \leq C_1 \left(\frac{\lambda + \varepsilon}{\sigma - \varepsilon}\right)^n \quad \text{for all } y \in X_0 \text{ and all } h_1, h_2 \in B.$$

It is also clear that f^{-1} is intrinsically $\mathcal{C}^1$ but choosing Δf^{-1} in such a way that the hypothesis of Proposition 2.10 holds is somewhat delicate. Using Proposition 3.2 we may write for $x, z \in X$

$$\begin{aligned} &f^{-1}(x) - f^{-1}(z) \\ &\quad = ((f^{-1}(x) - f^{-1}(z))_\mathrm{uw}, A(f^{-1}(x), f^{-1}(z))(f^{-1}(x) - f^{-1}(z))_\mathrm{uw}) \\ &\quad = (L^{-n}(\phi^{-1}(x) - \phi^{-1}(z))_\mathrm{uw}, A(f^{-1}(x), f^{-1}(z))L^{-n}(\phi^{-1}(x) - \phi^{-1}(z))_\mathrm{uw}). \end{aligned}$$

Therefore, fixing $\Delta\phi^{-1}$ an intrinsic derivative for ϕ^{-1},

$$\Delta f^{-1}(x, z) = \begin{pmatrix} L^{-n} & 0 \\ A(f^{-1}(x), f^{-1}(z)) \cdot L^{-n} & 0 \end{pmatrix} \cdot \Delta\phi^{-1}(x, z)$$

is an intrinsic derivative for f^{-1}. Recall moreover that $\|L^{-1}\| \leq (\lambda - \varepsilon)^{-1}$. Thus, for some $C_2 > 0$ depending only on C_0 and ϕ, $\|\Delta f^{-1}(x, z)\| \leq C_2(\lambda -$

$\varepsilon)^{-n}$, for all $x, z \in X$. In this way, having fixed $\varepsilon > 0$ and $\delta > 0$ small enough, we have

$$\|\Delta F_y(h_1, h_2)\| \cdot \|\Delta f^{-1}(x, z)\| \le C_1 C_2 \left(\frac{(\lambda+\varepsilon)}{(\lambda-\varepsilon)(\sigma-\varepsilon)}\right)^n \le \frac{1}{2}, \tag{3.4}$$

for all $y \in X_0$, $x, z \in X$ and $h_1, h_2 \in B$. This proves that the invariant section $\tilde{\sigma}$ of $\tilde{F}$ is intrinsically $\mathcal{C}^1$ and the proposition follows since clearly $\mathrm{T}_x \mathrm{W}^{\mathrm{u}}(x) = \mathrm{graph}(\tilde{\sigma}(x))$ for every $x \in W$. □

PROPOSITION 3.5. *Let $x \in \mathrm{W}^{\mathrm{u}}(\Lambda_1)$, Σ_0 and Σ_1 be (small) $\mathcal{C}^1$ sections transverse to $\mathrm{W}^{\mathrm{u}}(x)$ and $\pi\colon \Sigma_0 \cap \mathrm{W}^{\mathrm{u}}(\Lambda_1) \to \Sigma_1 \cap \mathrm{W}^{\mathrm{u}}(\Lambda_1)$ denote the projection along the leaves of $\mathrm{W}^{\mathrm{u}}(\Lambda_1)$. Then π is intrinsically $\mathcal{C}^1$.*

Proof. Fix coordinates $(\eta, \zeta) = (\eta_1, \ldots, \eta_u, \zeta_1, \ldots, \zeta_s)$ such that $\Sigma_0 \subset \{\eta = (0, 0, \ldots, 0)\}$, $\Sigma_1 \subset \{\eta = (1, 0, \ldots, 0)\}$ and each leaf intersecting Σ_0 in a point $(0, z)$ can be written as the graph of a function $\zeta = G(\eta; z)$. Then

$$\mathrm{graph}\left(\frac{\partial G}{\partial \eta}(\eta; z)\right) = \mathrm{T}_{(\eta, G(\eta; z))} \mathrm{W}^{\mathrm{u}}(\eta, G(\eta; z))$$

and so, by the previous proposition, $(\eta, G(\eta; z)) \mapsto \frac{\partial G}{\partial \eta}(\eta; z)$ is an intrinsically $\mathcal{C}^1$ map. For $-\varepsilon < t < 1 + \varepsilon$ we define $g(t; z) = G(t\tilde{\eta}; z)$, where $\tilde{\eta} = (1, 0, \ldots, 0)$, and then

$$H\colon (t, g(t; z)) \mapsto \frac{dg}{dt}(t; z) = \frac{\partial G}{\partial \eta_1}(t\tilde{\eta}; z)$$

is intrinsically $\mathcal{C}^1$. Notice that $\pi(0, z) = (\tilde{\eta}, G(\tilde{\eta}; z)) = (\tilde{\eta}, g(1; z))$. We fix an intrinsic derivative ΔH of H and define, for each t, $\Delta H_t(\zeta_1, \zeta_2) \cdot v = \Delta H((t, \zeta_1), (t, \zeta_2)) \cdot (0, v)$ (recall Lemma 2.5). For $(0, z_1), (0, z_2) \in \Sigma_0 \cap \mathrm{W}^{\mathrm{u}}(\Lambda_1)$ we denote by $\Gamma(t; z_1, z_2) \in \mathcal{L}(\mathbb{R}^s, \mathbb{R}^s)$ the unique solution of the initial value problem:

$$\begin{cases} \frac{d\Gamma}{dt}(t; z_1, z_2) = \Delta H_t(g(t; z_1), g(t; z_2)) \cdot \Gamma(t; z_1, z_2) \\ \Gamma(0; z_1, z_2) = \mathrm{id}. \end{cases}$$

Then $(z_1, z_2) \mapsto \Gamma(t; z_1, z_2)$ is continuous and we claim that it is an intrinsic derivative for $z \mapsto g(t; z)$. In fact, putting $\theta(t) = g(t; z_1) - g(t; z_2) - \Gamma(t; z_1, z_2)(z_1 - z_2)$, we get

$$\begin{cases} \frac{d\theta}{dt}(t) = \Delta H_t(g(t; z_1), g(t; z_2)) \cdot \theta(t) \\ \theta(0) = 0. \end{cases}$$

Hence θ is identically zero, which proves the claim. In particular $(z_1, z_2) \mapsto \Gamma(1; z_1, z_2)$ is an intrinsic derivative for $z \mapsto g(1; z)$ and so π is intrinsically $\mathcal{C}^1$, as we stated. □

Remark 3.6. In particular π is bi-Lipschitz continuous and so it preserves metric invariants such as Hausdorff dimension or limit capacity.

Remark 3.7. If one assumes that φ is of class $\mathcal{C}^3$ then stronger conclusions follow from the arguments above: Using Proposition 2.10(b) one gets that $\mathcal{F}\colon \mathrm{W}^{\mathrm{u}}(\Lambda_1) \ni x \mapsto \mathrm{T}_x\mathrm{W}^{\mathrm{u}}(x)$ is intrinsically $\mathcal{C}^{1+\gamma}$ for some $\gamma > 0$ and so the same holds for the holonomy maps $\pi\colon \Sigma_0 \cap \mathrm{W}^{\mathrm{u}}(\Lambda_1) \to \Sigma_1 \cap \mathrm{W}^{\mathrm{u}}(\Lambda_1)$ of $\mathrm{W}^{\mathrm{u}}(\Lambda_1)$. On the other hand, this last conclusion will also be obtained (in the case $w = 1$) in Proposition 4.3, by a different approach which applies also if φ is just $\mathcal{C}^2$.

4. Thickness in higher dimensions

Let Λ_1 be as constructed in the previous section. Here we assume that $w = 1$; i.e., $\mathrm{D}\varphi(p)$ has a unique (necessarily real) weakest contracting eigenvalue $\lambda = \lambda_1$. Then we consider $\pi : \Lambda_1 \cap \mathrm{W}^{\mathrm{s}}_{\mathrm{loc}}(p) \to \mathbb{R}$ to be an arbitrary intrinsically $\mathcal{C}^1$ map such that $\ker(\Delta\pi(p,p))$ does not contain $\mathrm{IT}_p(\Lambda_1 \cap \mathrm{W}^{\mathrm{s}}_{\mathrm{loc}}(p)) = \mathrm{E}^{\mathrm{w}}$ (i.e., $\Delta\pi(p,p) \mid \mathrm{E}^{\mathrm{w}}$ is bijective) and we show that $\tau(\pi(\Lambda_1 \cap \mathrm{W}^{\mathrm{s}}_{\mathrm{loc}}(p)), \pi(p))$ is independent of the choice of π. We call

$$\tau^{\mathrm{u}}(\Lambda_1, p) = \tau(\pi(\Lambda_1 \cap \mathrm{W}^{\mathrm{s}}_{\mathrm{loc}}(p)), \pi(p)), \quad \text{any } \pi \text{ as above},$$

the *local unstable thickness* of Λ_1 at p and prove that it is strictly positive and varies continuously with the diffeomorphism: A $\mathcal{C}^2$-small perturbation of φ yields a small variation of $\tau^{\mathrm{u}}(\Lambda_1, p)$.

We keep the notation of Section 3. Let $\pi_{\mathrm{w}} : \Lambda_1 \cap \mathrm{W}^{\mathrm{s}}_{\mathrm{loc}}(p) \to \mathbb{R}$ be the restriction to $\Lambda_1 \cap \mathrm{W}^{\mathrm{s}}_{\mathrm{loc}}(p) \subset \{0^u\} \times \mathbb{R}^s$ of the projection $(\xi_1, \ldots, \xi_u, \zeta_1, \ldots, \zeta_s) \mapsto \zeta_1$. We claim that π_{w} is a homeomorphism onto its image K^{w} and moreover π_{w}^{-1} is intrinsically $\mathcal{C}^{1+\gamma}$ on K^{w}. In fact, by Proposition 3.2, for every $x, z \in \Lambda_1 \cap \mathrm{W}^{\mathrm{s}}_{\mathrm{loc}}(p)$ we have

$$(x - z)_{\mathrm{ss}} = A(x, z) \cdot (0^u, \pi_{\mathrm{w}}(x) - \pi_{\mathrm{w}}(z))$$

and so (for $C \geq \|A\|$ as in the proposition)

$$|\pi_{\mathrm{w}}(x) - \pi_{\mathrm{w}}(z)| \leq \|x - z\| \leq (C + 1)\, |\pi_{\mathrm{w}}(x) - \pi_{\mathrm{w}}(z)|\,.$$

This proves that π_{w} is invertible and π_{w}^{-1} is Lipschitz continuous. Moreover,

$$\Delta\pi_{\mathrm{w}}^{-1}(s, t)\colon \mathbb{R} \ni v \mapsto (v, A(\pi_{\mathrm{w}}^{-1}(s), \pi_{\mathrm{w}}^{-1}(t)) \cdot (0^u, v)) \in \mathbb{R}^s$$

defines a Hölder continuous intrinsic derivative for π_{w}^{-1}.

Let now $\pi\colon \Lambda_1 \cap \mathrm{W}^{\mathrm{s}}_{\mathrm{loc}}(p) \to \mathbb{R}$ be an intrinsically $\mathcal{C}^1$ map such that the kernel of $\Delta\pi(p,p)$ does not contain $\mathrm{IT}_p(\Lambda_1 \cap \mathrm{W}^{\mathrm{s}}_{\mathrm{loc}}(p)) = \mathrm{graph}(A(p,p) \mid \{0^u\} \times \mathbb{R})$. Then $\pi \circ \pi_{\mathrm{w}}^{-1}$ is intrinsically $\mathcal{C}^1$ with $\Delta(\pi \circ \pi_{\mathrm{w}}^{-1})(0,0) = \Delta\pi(p,p) \cdot$

$\Delta\pi_{\mathrm{w}}^{-1}(0,0) \neq 0$. Since $\pi(\Lambda_1 \cap W^s_{\mathrm{loc}}(p)) = (\pi \circ \pi_{\mathrm{w}}^{-1})(K^{\mathrm{w}})$, we conclude that $\tau(\pi(\Lambda_1 \cap W^s_{\mathrm{loc}}(p)), \pi(p)) = \tau(K^{\mathrm{w}}, 0)$, as a consequence of the following simple result.

LEMMA 4.1. *Let $K \subset \mathbb{R}$ be a Cantor set, $y \in K$ and $g \colon K \to \mathbb{R}$ be an intrinsically C^1 map with $\Delta g(y,y) \neq 0$. Then $\tau(g(K),g(y)) = \tau(K,y)$.*

Proof. Fix $\delta > 0$ small and take $\varepsilon > 0$ such that

$$(4.1) \quad |\Delta g(x_1, z_1)| \leq (1+\delta)\,|\Delta g(x_2, z_2)| \quad \text{for all } x_1, x_2, z_1, z_2 \in [y-\varepsilon, y+\varepsilon].$$

Clearly, g is a homeomorphism on $[y-\varepsilon, y+\varepsilon]$ if ε is small enough. Hence, given a Cantor set $L \subset [y-\varepsilon, y+\varepsilon]$, $\mathcal{U} = (U_n)_n$ is a presentation of L if and only if $g(\mathcal{U}) = (g(U_n))_n$ is a presentation of $g(L)$. From (4.1) we get that for every $u \in \partial U_n$, $n \geq 1$,

$$\tau(g(L), g(\mathcal{U}), g(u)) \leq (1+\delta)\tau(L,\mathcal{U}, u).$$

Therefore, $\tau(g(L)) \leq (1+\delta)\tau(L)$ for every such L and so $\tau(g(K), g(y)) \leq (1+\delta)\tau(K, y)$. Since δ is arbitrary we conclude that $\tau(g(K), g(y)) \leq \tau(K, y)$ and the reverse inequality is proved in the same way. □

For future use, let us state also the next result, which follows from the previous properties by well-known arguments: Part (a) is a direct consequence of Proposition 3.5; for part (b) one also uses the transitivity of $\varphi \mid \Lambda_1$.

PROPOSITION 4.2.

(a) *Let $q \in W^u(p)$, Σ be a C^1 section transverse to $W^u(p)$ at the point q and $\pi\colon W^u(\Lambda_1) \cap \Sigma \to \mathbb{R}$ be an intrinsically C^1 map such that $\mathrm{IT}_q(W^u(\Lambda_1) \cap \Sigma)$ is not contained in* $\ker(\Delta\pi(q,q))$. *Then $\tau(\pi(W^u(\Lambda_1) \cap \Sigma),\pi(q)) = \tau^u(\Lambda_1,p)$.*

(b) *More generally, given $z \in W^u(\Lambda_1)$, Σ a transverse section to $W^u(\Lambda_1)$ at z and $\pi\colon W^u(\Lambda_1) \cap \Sigma \to \mathbb{R}$ submersion with $\mathrm{IT}_z(W^u(\Lambda_1) \cap \Sigma) \not\subset$* $\ker(\Delta\pi(z,z))$, *then $\tau(\pi(W^u(\Lambda_1) \cap \Sigma),\pi(z))$ is equal to $\tau^u(\Lambda_1,p)$.*

Our strategy to prove that $\tau^u(\Lambda_1, p) = \tau(K^{\mathrm{w}}, 0)$ is positive is to write K^{w} as an invariant set of some $C^{1+\gamma}$ expanding map, so that the reasoning in Proposition 6 of [N3] (in the language of [PT, Ch. IV]) can be applied to it. For the construction of this map we proceed as follows. Let $\widetilde{W}$ be the connected component of $W^s(p) \cap V$ containing p. By construction, for every $x \in \Lambda_1 \cap V$ the connected component of $W^u(x) \cap V$ containing x intersects $\widetilde{W}$ in a unique point, which we denote by $\pi_s(x)$. Observe also that every $\widetilde{W}_\ell = \{z\colon \psi^j(z) \in V \text{ for } 0 \leq j < \ell \text{ and } \psi^\ell(z) \in \widetilde{W}\}$, $\ell \geq 1$, has exactly 2^ℓ components, denoted by $W_{j,\ell}$, $1 \leq j \leq 2^\ell$. For each $\ell \geq 1$ we define a map

$$g_\ell\colon K^{\mathrm{w}} \to K^{\mathrm{w}} \quad \text{by } g_\ell = \pi_{\mathrm{w}} \circ \pi_{\mathrm{s}} \circ \psi^{-\ell} \circ \pi_{\mathrm{w}}^{-1}.$$

Clearly, g_ℓ is intrinsically $\mathcal{C}^1$ and it follows from the next result that it is even intrinsically $\mathcal{C}^{1+\gamma}$.

PROPOSITION 4.3. *There are $C > 0$ and $\gamma \in (0,1)$ such that for every $\ell \geq 1$ and $1 \leq j \leq 2^\ell$ the map $\pi_s \mid (W_{j,\ell} \cap \Lambda_1)\colon W_{j,\ell} \cap \Lambda_1 \to \widetilde{W}$ admits a (C,γ)-Hölder continuous intrinsic derivative.*

For the sake of clearness we postpone the proof of this result to the end of the section. Recall also Remark 3.7.

Observe that g_ℓ maps K^w onto K^w in a 2^ℓ to 1 way: For each $1 \leq j \leq 2^\ell$ and $K_{j,\ell} = \pi_w(\psi^\ell(W_{j,\ell} \cap \Lambda_1))$ the map $g_\ell \mid K_{j,\ell}\colon K_{j,\ell} \to K^w$ is a homeomorphism. It is also easy to check that the $K_{j,\ell}$, $1 \leq j \leq 2^\ell$, have their convex hulls $\hat{K}_{j,\ell}$ two by two disjoint. Moreover, for $\ell \geq 1$ large enough every $(g_\ell \mid K_{j,\ell})$ is an expansion:

$$\|\Delta g_\ell(t,r)\| \geq 2 \quad \text{(say) for every } \; t,r \in K_{j,\ell}. \tag{4.2}$$

We now fix $\ell \geq 1$ so that (4.2) holds and denote $g = g_\ell$ and $K_j = K_{j,\ell}$. At this stage the fact that K^w has strictly positive local thickness can be proved just by translating, in a more or less straightforward way, the arguments in [N3], [PT, Ch. IV] into our language of intrinsic differentiability. Alternatively, one can show that g can in fact be extended to a $\mathcal{C}^{1+\gamma}$ map G defined on a neighbourhood of K^w, which is easy to prove in this one-dimensional setting.

LEMMA 4.4. *Let $K \subset \mathbb{R}$ be a Cantor set and $g\colon K \to \mathbb{R}$ be intrinsically $\mathcal{C}^1$ (resp. intrinsically $\mathcal{C}^{1+\gamma}$). Then there exists $G\colon \mathbb{R} \to \mathbb{R}$ a $\mathcal{C}^1$ (resp. $\mathcal{C}^{1+\gamma}$) extension of g.*

Proof. Fix $\theta\colon [0,1] \to [0,1]$ a $\mathcal{C}^\infty$ function such that $\theta(x) = 0$ and $\theta(1-x) = 1$ whenever $0 \leq x \leq 1/3$. Let also $a_0 = \inf K$, $b_0 = \sup K$. We set: $G(x) = g(x)$ if $x \in K$;

$$\begin{aligned} G(x) = (g(a) + \Delta g(a,a)(x-a)) \left(1 - \theta\left(\frac{x-a}{b-a}\right)\right) \\ + (g(b) + \Delta g(b,b)(x-b))\theta\left(\frac{x-a}{b-a}\right) \end{aligned}$$

if x belongs to a gap (a,b) of K; $G(x) = g(b_0) + \Delta g(b_0,b_0)(x-b_0)$ if $x > b_0$ and $G(x) = g(a_0) + \Delta g(a_0,a_0)(x-a_0)$ if $x < a_0$. We also define $H\colon \mathbb{R} \to \mathbb{R}$ by putting: $H(x) = \Delta g(x,x)$ if $x \in K$;

$$\begin{aligned} H(x) = \Delta g(a,a)\left(1 - \theta\left(\frac{x-a}{b-a}\right)\right) + \Delta g(b,b)\theta\left(\frac{x-a}{b-a}\right) \\ + \theta'\left(\frac{x-a}{b-a}\right)\left(\Delta g(b,a) - \Delta g(b,b) + \left(\frac{x-a}{b-a}\right)(\Delta g(b,b) - \Delta g(a,a))\right) \end{aligned}$$

if $x \in (a,b)$, a gap of K; $H(x) = \Delta g(b_0, b_0)$ if $x > b_0$ and $H(x) = \Delta g(a_0, a_0)$ if $x < a_0$. It is a simple exercise to check that H is continuous ($\mathcal{C}^\gamma$ if Δg is $\mathcal{C}^\gamma$) and $G'(x) = H(x)$ for every $x \in \mathbb{R}$. □

This means that K^{w} is a *dynamically defined Cantor set*, in the same sense as in [PT, Ch. IV] and our claim that $\tau(K^{\mathrm{w}}, 0) > 0$ follows.

Let us now observe that (K^{w}, G), as we constructed it, varies continuously with the diffeomorphism φ. In order to explain this affirmative we let $\tilde{\varphi}$ be a diffeomorphism $\mathcal{C}^2$-close to φ and denote by $K^{\mathrm{w}}(\tilde{\varphi}), K_j(\tilde{\varphi}), \hat{K}_j(\tilde{\varphi})$ and $G(\tilde{\varphi})$ the objects obtained by performing the above construction for $\tilde{\varphi}$ (note that we may, and do, take $\ell(\tilde{\varphi}) = \ell$ if $\tilde{\varphi}$ is close enough to φ). Then it is straightforward to check that $(K^{\mathrm{w}}(\tilde{\varphi}), G(\tilde{\varphi}))$ is close to (K^{w}, G) in the sense that

(a) corresponding endpoints of $\hat{K}_j(\tilde{\varphi})$ and $\hat{K}_j$ are close;

(b) $G(\tilde{\varphi})$ is $\mathcal{C}^1$-close to G and their derivatives have nearby Hölder constants (see also the proof of Proposition 4.3 below).

It follows, see again [PT, Ch. IV], that $\tau(K^{\mathrm{w}}(\tilde{\varphi}), 0)$ is close to $\tau(K^{\mathrm{w}}, 0)$ and this proves that the local unstable thickness $\tau^{\mathrm{u}}(\Lambda_1, p)$ is a continuous function of the diffeomorphism, as we claimed.

We close this section by presenting the proof of Proposition 4.3.

Proof. We keep the notation from the proof of Proposition 3.2. Moreover, we use C and γ as generic notation for constants ($C > 0$ large, $0 < \gamma < 1$) depending only on the diffeomorphism φ. Let $\mathcal{S} = \{S(x)\colon x \in \Lambda_1\}$ where $S(x)$ denotes the connected component of $\mathrm{W}^{\mathrm{s}}(x) \cap V$ containing x. Note that $W_{j,\ell} \in \mathcal{S}$ for every $1 \le j \le 2^\ell$, $\ell \ge 1$. For $S, \tilde{S} \in \mathcal{S}$ let $\pi(S, \tilde{S}; \cdot)\colon S \cap \Lambda_1 \to \tilde{S} \cap \Lambda_1$ be the projection along the leaves of $\mathrm{W}^{\mathrm{u}}(\Lambda_1)$, inside V. By Proposition 3.5 $\pi(S, \tilde{S}; \cdot)$ is C-Lipschitz continuous. On the other hand, as observed before, $\{\|x - z\|\colon (x,z) \in (S \times S) \cap U_0, S \in \mathcal{S}\}$ is bounded away from zero. Hence (see Remark 2.6) there is a C-Lipschitz continuous map $\tilde{\Delta}(S, \tilde{S}; \cdot, \cdot)\colon ((S \cap \Lambda_1) \times (S \cap \Lambda_1)) \cap U_0 \to \mathcal{L}(\mathbb{R}^s, \mathbb{R}^s)$ such that

$$\pi(S, \tilde{S}; x) - \pi(S, \tilde{S}; z) = \tilde{\Delta}(S, \tilde{S}; x, z) \cdot (x - z) \tag{4.3}$$

for every $(x,z) \in ((S \cap \Lambda_1) \times (S \cap \Lambda_1)) \cap U_0$. Here and in what follows we identify each $S \in \mathcal{S}$ with its image under the projection $(\xi_1, \dots, \xi_u, \zeta_{u+1}, \dots, \zeta_m) \mapsto (\zeta_{u+1}, \dots, \zeta_m)$. We write the expression of $\tilde{\Delta}(S, \tilde{S}; x, z)$ with respect to the splitting $\mathbb{R}^s = \mathbb{R}^1 \times \mathbb{R}^{s-1}$

$$\tilde{\Delta}(S, \tilde{S}; x, z) = \begin{pmatrix} \Delta_{\mathrm{ww}} & \Delta_{\mathrm{ws}} \\ \Delta_{\mathrm{sw}} & \Delta_{\mathrm{ss}} \end{pmatrix} (S, \tilde{S}; x, z)$$

and then define

$$\Delta(S,\tilde{S};x,z) = \begin{pmatrix} \Delta_{\mathrm{w}} = \Delta_{\mathrm{ww}} + \Delta_{\mathrm{ws}} \cdot A(x,z) & 0 \\ \\ \Delta_{\mathrm{s}} = \Delta_{\mathrm{sw}} + \Delta_{\mathrm{ss}} \cdot A(x,z) & 0 \end{pmatrix} (S,\tilde{S};x,z),$$

where $A(x,z)$ is as given by Proposition 3.2. Then $\Delta(S,\tilde{S};\cdot,\cdot)$ is (C,γ)-Hölder continuous and satisfies (4.3). Moreover, Δ_{w} is uniformly bounded away from zero and infinity, as a consequence of that same proposition. We also need some information on how $\Delta(S,\tilde{S};\cdot,\cdot)$ varies with $S,\tilde{S} \in \mathcal{S}$ and this is easy to get. Let us denote $\operatorname{dist}(S,\tilde{S}) = \sup_{x\in S\cap\Lambda_1} \|\pi(S,\tilde{S};x) - x\|$. It is easy to see that $\tilde{\Delta}$ may be taken such that $\|\tilde{\Delta}(S,\tilde{S};x,z) - \mathrm{id}\| \le C \operatorname{dist}(S,\tilde{S})$ and so

$$\left|\Delta_{\mathrm{w}}(S,\tilde{S};x,z) - 1\right| \le C \operatorname{dist}(S,\tilde{S}) \tag{4.4}$$

for every $x,z \in S\cap\Lambda_1$. Note that if $S \in \mathcal{S}$ then $\psi(S) \subset S'$ for some $S' \in \mathcal{S}$. We let $\tilde{\Delta}\psi(\cdot,\cdot)$ be a Lipschitz continuous intrinsic derivative for ψ and denote by $\tilde{\Delta}\psi(S;\cdot,\cdot)$ its restriction to $(S\cap\Lambda_1)\times(S\cap\Lambda_1)$. Through the above identification we may think of $\tilde{\Delta}\psi(S;\cdot,\cdot)$ as taking values in $\mathcal{L}(\mathbb{R}^s,\mathbb{R}^s)$. Proceeding as before for Δ we construct a new, (C,γ)-Hölder continuous, intrinsic derivative $\Delta\psi(S;\cdot,\cdot)$ for $\psi \mid (S\cap\Lambda_1)$ having the form

$$\Delta\psi(S;x,z) = \begin{pmatrix} \Delta\psi_{\mathrm{w}} & 0 \\ \Delta\psi_{\mathrm{s}} & 0 \end{pmatrix} (S;x,z)$$

with respect to the splitting $\mathbb{R}^s = \mathbb{R}^1 \times \mathbb{R}^{s-1}$. In the same way we obtain a (C,γ)-Hölder continuous intrinsic derivative for $\psi^{-1} \mid (\psi(S)\cap\Lambda_1)$:

$$\Delta\psi^{-1}(S;x',z') = \begin{pmatrix} \Delta\psi_{\mathrm{w}}^{-1} & 0 \\ \Delta\psi_{\mathrm{s}}^{-1} & 0 \end{pmatrix} (S;x',z').$$

Observe that Proposition 3.2 implies that $\Delta\psi_{\mathrm{w}}$ is uniformly bounded away from zero and infinity; moreover, independently of all the choices, we have

$$\Delta\psi_{\mathrm{w}}^{-1}(S;\psi(x),\psi(z)) \cdot \Delta\psi_{\mathrm{w}}(S;x,z) = 1 \quad \text{for every } x,z \in S\cap\Lambda_1.$$

We also point out that this construction yields

$$\left|\Delta\psi_{\mathrm{w}}(\tilde{S};\pi(S,\tilde{S};x),\pi(S,\tilde{S};z)) - \Delta\psi_{\mathrm{w}}(S;x,z)\right| \le C \operatorname{dist}(S,\tilde{S})^{\gamma} \tag{4.5}$$

for every $x,z \in S\cap\Lambda_1$. Now we fix $S_0,\tilde{S}_0 \in \mathcal{S}$ and construct an intrinsic derivative $\Delta\pi(S_0,\tilde{S}_0;\cdot,\cdot)$ for $\pi(S_0,\tilde{S}_0;\cdot)$ in the following way. For $(x,z) \in ((S_0\cap\Lambda_1)\times(S_0\cap\Lambda_1))\cap U_0$ we set simply $\Delta\pi(S_0,\tilde{S}_0;x,z) = \Delta(S_0,\tilde{S}_0;x,z)$. Let now $(x,z) \in ((S_0\cap\Lambda_1)\times(S_0\cap\Lambda_1))\cap U_k$, i.e., $k \ge 1$ is the minimum integer such that $(\psi^{-k}(x),\psi^{-k}(z)) \in U_0$. We denote $\tilde{x} = \pi(S_0,\tilde{S}_0;x)$, $x_j = \psi^{-j}(x), \tilde{x}_j = \psi^{-j}(\tilde{x})$ and analogously for $\tilde{z}, z_j$ and $\tilde{z}_j$, $j \ge 0$. We also let

$S_j = S(x_j) = S(z_j)$, $\tilde{S}_j = S(\tilde{x}_j) = S(\tilde{z}_j)$, $0 \le j \le k$. Then we define

$$\Delta\pi(S_0, \tilde{S}_0; x, z) = \Delta\psi^k(\tilde{S}_k; \tilde{x}_k, \tilde{z}_k) \cdot \Delta(S_k, \tilde{S}_k; x_k, z_k) \cdot \Delta\psi^{-k}(S_k; x, z).$$

We are left to check that with this definition $\Delta\pi(S_0, \tilde{S}_0; \cdot, \cdot)$ is (C, γ)-Hölder continuous: we consider $x, z, w \in S_0 \cap \Lambda_1$ and prove that

$$\left|\Delta\pi(S_0, \tilde{S}_0; x, z) - \Delta\pi(S_0, \tilde{S}_0; x, w)\right| \le C\,|z - w|^\gamma \tag{4.6}$$

with $(x, z) \in U_k$, $(x, w) \in U_\ell$ and $(z, w) \in U_m$, $k, \ell, m \ge 0$. Clearly, $j \mapsto \operatorname{dist}(S_j, \tilde{S}_j)$ is exponentially decreasing and $j \mapsto |z_j - w_j|$ is exponentially increasing, at least while $j \le m$. It follows that, if $|z - w|$ is small enough depending on $D = \sup\{\operatorname{dist}(S', S'') \colon S', S'' \in \mathcal{S}\}$, then there exists $s < m$ such that

$$|z_s - w_s| \ge \operatorname{dist}(S_s, \tilde{S}_s). \tag{4.7}$$

Note that for the purpose of proving (4.6) it is no restriction to assume that $|z-w|$ is small with respect to D and we do so from now on. We fix s minimum satisfying (4.7) and observe that there is $\lambda < 1$ (depending only on φ) such that

$$\operatorname{dist}(S_j, \tilde{S}_j) \le C\lambda^{j-s}\,|z - w|^\gamma \quad \text{for } j \ge s, \tag{4.8}$$

$$|z_j - w_j| \le C\lambda^{s-j}\,|z - w|^\gamma \quad \text{for } 0 \le j \le s. \tag{4.9}$$

This is an easy consequence of the definition of s and the exponential variation of $\operatorname{dist}(S_j, \tilde{S}_j)$ and $|z_j - w_j|$ mentioned above. Since the $\pi(S_i, \tilde{S}_i; \cdot)$ admit a uniform Lipschitz constant, we also get that

$$|\tilde{z}_j - \tilde{w}_j| \le C\lambda^{s-j}\,|z - w|^\gamma \quad \text{for } 0 \le j \le s. \tag{4.10}$$

Now we take $r = \min\{k, \ell, s\}$. By construction

$$\Delta\pi(S_0, \tilde{S}_0; x, z) = \begin{pmatrix} \Delta\pi_{\rm w} & 0 \\ \Delta\pi_{\rm s} & 0 \end{pmatrix} (S_0, \tilde{S}_0; x, z)$$

with $\Delta\pi_{\rm w}(S_0, \tilde{S}_0; x, z) = \Delta\psi^r_{\rm w}(\tilde{S}_r; \tilde{x}_r, \tilde{z}_r) \cdot \Delta\pi_{\rm w}(S_r, \tilde{S}_r; x_r, z_r) \cdot \Delta\psi^{-r}_{\rm w}(S_r; x, z)$, and analogously for (x, w). Then

$$\frac{\Delta\pi_{\rm w}(S_0, \tilde{S}_0; x, z)}{\Delta\pi_{\rm w}(S_0, \tilde{S}_0; x, w)} = \frac{\Delta\pi_{\rm w}(S_r, \tilde{S}_r; x_r, z_r)}{\Delta\pi_{\rm w}(S_r, \tilde{S}_r; x_r, w_r)} \cdot R(x, z, w), \quad \text{where}$$

$$R(x, z, w) = \prod_1^r \frac{\Delta\psi_{\rm w}(\tilde{S}_i; \tilde{x}_i, \tilde{z}_i) \cdot \Delta\psi_{\rm w}(S_i; x_i, w_i)}{\Delta\psi_{\rm w}(\tilde{S}_i; \tilde{x}_i, \tilde{w}_i) \cdot \Delta\psi_{\rm w}(S_i; x_i, z_i)}.$$

From (4.9), (4.10) and the uniform Hölder continuity and boundedness of $\Delta\psi_{\rm w}(S_i; \cdot, \cdot)$, $\Delta\psi_{\rm w}(\tilde{S}_i; \cdot, \cdot)$ it follows that

$$|R(x, z, w) - 1| \le C\,|z - w|^\gamma. \tag{4.11}$$

We claim that

$$\left|\frac{\Delta\pi_{\mathrm{w}}(S_r,\tilde{S}_r;x_r,z_r)}{\Delta\pi_{\mathrm{w}}(S_r,\tilde{S}_r;x_r,w_r)}-1\right|\le C\,|z-w|^{\gamma}. \tag{4.12}$$

Note that this implies (4.6) in an easy way. In fact, from (4.11), (4.12) we get $|\Delta\pi_{\mathrm{w}}(S_0,\tilde{S}_0;x,z)-\Delta\pi_{\mathrm{w}}(S_0,\tilde{S}_0;x,w)|\le C|z-w|^{\gamma}$ and then, since $\Delta\pi_{\mathrm{s}}(S_0,\tilde{S}_0;x,y)=A(\tilde{x},\tilde{y})\cdot\Delta\pi_{\mathrm{w}}(S_0,\tilde{S}_0;x,y)$, for $y=z$ or w, the same holds for $\Delta\pi_{\mathrm{s}}$. Therefore, the proposition will follow once we have proved this claim. Suppose first $r=k$; then, because $r\le s<m$, we also have $r=\ell$. It follows that

$$\Delta\pi_{\mathrm{w}}(S_r,\tilde{S}_r;x_r,y)=\Delta_{\mathrm{w}}(S_r,\tilde{S}_r;x_r,y),\quad y=z_r \text{ or } w_r,$$

and so (4.12) is a direct consequence of (4.9) and the fact that Δ_{w} is Hölder continuous and bounded away from zero and infinity. Let now $r=s$. Once more by definition,

$$\Delta\pi_{\mathrm{w}}(S_r,\tilde{S}_r;x_r,z_r)=\Delta_{\mathrm{w}}(S_k,\tilde{S}_k;x_k,z_k)\cdot\prod_{r+1}^{k}\frac{\Delta\psi_{\mathrm{w}}(\tilde{S}_i;\tilde{x}_i,\tilde{z}_i)}{\Delta\psi_{\mathrm{w}}(S_i;x_i,z_i)}.$$

Note that $\mathrm{dist}(S_j,\tilde{S}_j)\le\sigma^{r-j}\,\mathrm{dist}(S_r,\tilde{S}_r)$ for $r\le j\le m$ and so, using (4.5),

$$\left|\prod_{r+1}^{k}\frac{\Delta\psi_{\mathrm{w}}(\tilde{S}_i;\tilde{x}_i,\tilde{z}_i)}{\Delta\psi_{\mathrm{w}}(S_i;x_i,z_i)}-1\right|\le C\,|z-w|^{\gamma}.$$

On the other hand, by (4.4) and (4.8), $|\Delta_{\mathrm{w}}(S_k,\tilde{S}_k;x_k,z_k)-1|\le C|z-w|^{\gamma}$ and this proves that $|\Delta\pi_{\mathrm{w}}(S_r,\tilde{S}_r;x_rz_r)-1|\le C|z-w|^{\gamma}$. In this same way one shows that $|\Delta\pi_{\mathrm{w}}(S_r,\tilde{S}_r;x_r,w_r)-1|\le C|z-w|^{\gamma}$ and then (4.12) follows immediately. □

5. Unique least contracting eigenvalue

Let $(\varphi_\mu)_\mu$ be a C^2 one-parameter family of diffeomorphisms which at $\mu=0$ goes through a homoclinic tangency associated to a hyperbolic fixed (or periodic) point p of φ_0. We prove here that under a few generic assumptions to be stated below such a family exhibits, at $\tilde{\mu}_j\to 0$, homoclinic tangencies associated to periodic points $\tilde{p}_j\to p$ such that

(I) $\mathrm{D}\varphi_{\tilde{\mu}_j}^{\ell_j}(\tilde{p}_j)$, $\ell_j=$ period of $\tilde{p}_j$, has a unique weakest contracting eigenvalue which, therefore, is a real number.

A dual statement, for expanding eigenvalues, can be obtained in the same way. No assumptions on the dimensions of $\mathrm{W^u}(p)$ or $\mathrm{W^s}(p)$ are required in this section.

Observe first that, generically, either $D\varphi_0(p)$ satisfies the property in (I) or else

(II) $D\varphi_0(p)$ has exactly two weakest contracting eigenvalues and these are complex conjugate numbers.

In the first case there is nothing to prove, so we assume from now on that (II) holds. Notation analogous to that of Section 3 is used; this means that $w = 2$, $\lambda_1 = \lambda e^{i\phi}$, $\lambda_2 = \lambda e^{-i\phi}$ with $\lambda > |\lambda_3|$ and $\phi \in \mathbb{R} \setminus \{k\pi : k \in \mathbb{Z}\}$. For the sake of simplicity we assume that there are C^2 μ-dependent coordinates $(\xi_1, \ldots, \xi_u, \zeta_1, \ldots, \zeta_s)$ linearizing the φ_μ, for $|\mu|$ small, on a neighbourhood U of the analytic continuation p_μ of p. Moreover, we may take these coordinates to satisfy conditions (A1)–(B3) of Section 3. Here we may even assume that $D\varphi_\mu(p_\mu) \mid E^w$ is conformal with respect to the euclidean metric induced by the coordinates ζ_1, ζ_2. We also suppose that the tangency is quadratic and the family $(\varphi_\mu)_\mu$ unfolds it generically. Then we may take, say for $\mu \geq 0$, points $q_\mu \in W^s_{loc}(p_\mu)$, $r_\mu \in W^u_{loc}(p_\mu)$ depending continuously on μ, such that $\varphi_\mu^N(r_\mu) = q_\mu$ for some fixed $N \geq 1$, r_0, q_0 belong to the orbit of tangency and r_μ, q_μ are points of transverse intersection of $W^u(p_\mu), W^s(p_\mu)$ for every $\mu > 0$. Recall, moreover, that $W^u(p_\mu)$ and $W^s(p_\mu)$ also have, for a sequence of parameter values $\mu = \mu_j \to 0$, points of tangential intersection. For each fixed $\mu = \mu_j$ and every sufficiently large $n \geq 1$ we may take, as in Section 3, a neighbourhood $V = V(j,n)$ of $\{p_\mu, q_\mu\}$ such that $\Lambda(j,n) = \bigcap_{k\in\mathbb{Z}} \varphi_\mu^{k(n+N)}(V)$ is a φ_μ^{n+N}-invariant hyperbolic set and $\varphi_\mu^{n+N} \mid \Lambda(j,n)$ is conjugate to the 2-shift. Then, given any periodic point $\tilde{p} \in \Lambda(j,n)$, one may find parameter values $\tilde{\mu}$ arbitrarily close to μ_j for which $\varphi_{\tilde{\mu}}$ has homoclinic tangencies associated to (the analytic continuation of) $\tilde{p}$. We consider $\tilde{p} = \tilde{p}(j,n)$ to be the unique φ_μ^{n+N}-fixed point in $\Lambda(j,n) \setminus \{p_\mu\}$. Clearly, the orbit of $\tilde{p}$ passes arbitrarily close to p_μ if j and n are sufficiently large. Therefore, in order to conclude our argument it is now sufficient to show:

PROPOSITION 5.1. *Given $j \geq 1$ (large) there exist arbitrarily large values of $n = n_j$ for which $D\varphi_\mu^{n+N}(\tilde{p})$ has a unique weakest contracting eigenvalue.*

Proof. For the sake of notational simplicity we continue to denote $\mu = \mu_j$ and $\tilde{p} = \tilde{p}(j,n)$. We make use of the following elementary fact whose proof we omit.

LEMMA 5.2. *Let $P \in GL(\mathbb{R}^p)$, $Q \in GL(\mathbb{R}^q)$ satisfy $\|P^{-1}\|\,\|Q\| < 1$. Let* $L = \begin{pmatrix} A & B \\ C & D \end{pmatrix} \in GL(\mathbb{R}^p \times \mathbb{R}^q)$ *and denote*

$$L^{-1} = \begin{pmatrix} A^- & B^- \\ C^- & D^- \end{pmatrix} \quad \textit{and} \quad L_n = \begin{pmatrix} A & B \\ C & D \end{pmatrix} \begin{pmatrix} P^n & 0 \\ 0 & Q^n \end{pmatrix}, \quad n \geq 1.$$

Assume that $A \in \mathrm{GL}(\mathbb{R}^p)$ *(and so also* $D^- \in \mathrm{GL}(\mathbb{R}^q)$*). Then there are* $n_0 \geq 1$ *and* $C_0 > 0$*, depending only on* $(\|P^{-1}\|\,\|Q\|, \|L\|, \|L^{-1}\|, \|A^{-1}\|, \|(D^-)^{-1}\|)$*, such that for every* $n \geq n_0$ *there exist linear maps* $f\colon \mathbb{R}^p \to \mathbb{R}^q$*,* $g\colon \mathbb{R}^q \to \mathbb{R}^p$ *satisfying*

(a) $\|f\| \leq C_0$ *and* $\|g\| \leq C_0(\|P^{-1}\|\,\|Q\|)^n$;

(b) $\mathrm{graph}\,(f), \mathrm{graph}\,(g)$ *are* L_n*-invariant.*

Moreover, given $\delta > 0$ *there is* $n_1 = n_1(\delta) \geq n_0$ *such that for* $n \geq n_1$

(c) $\|(\pi_f L_n \pi_f^{-1})P^{-n} - A\| \leq \delta$ *and* $\|Q^n(\pi_g L_n^{-1} \pi_g^{-1}) - D^-\| \leq \delta$*, where* $\pi_f\colon \mathrm{graph}\,(f) \to \mathbb{R}^p$ *and* $\pi_g\colon \mathrm{graph}\,(g) \to \mathbb{R}^q$ *are the canonical projections.*

Let then

$$\mathrm{D}\varphi_\mu^N = \begin{pmatrix} A_{\mathrm{uu}} & A_{\mathrm{uw}} & A_{\mathrm{us}} \\ A_{\mathrm{wu}} & A_{\mathrm{ww}} & A_{\mathrm{ws}} \\ A_{\mathrm{su}} & A_{\mathrm{sw}} & A_{\mathrm{ss}} \end{pmatrix}, \quad \mathrm{D}\varphi_\mu(p) = \begin{pmatrix} \Lambda_{\mathrm{u}} & 0 & 0 \\ 0 & \Lambda_{\mathrm{w}} & 0 \\ 0 & 0 & \Lambda_{\mathrm{s}} \end{pmatrix}$$

be the expressions of $\mathrm{D}\varphi_\mu^N$ and $\mathrm{D}\varphi_\mu(p)$ with respect to the splitting $\mathbb{R}^m = \mathbb{R}^u \times \mathbb{R}^2 \times \mathbb{R}^{s-2} = \mathrm{E}^{\mathrm{u}} \times \mathrm{E}^{\mathrm{w}} \times \mathrm{E}^{\mathrm{ss}}$. We also denote

$$\Delta_\mu = \begin{pmatrix} A_{\mathrm{uu}} & A_{\mathrm{uw}} \\ A_{\mathrm{wu}} & A_{\mathrm{ww}} \end{pmatrix} \quad \text{and} \quad \mathrm{D}\varphi_\mu^{-N} = \begin{pmatrix} A^-_{\mathrm{uu}} & A^-_{\mathrm{uw}} & A^-_{\mathrm{us}} \\ A^-_{\mathrm{wu}} & A^-_{\mathrm{ww}} & A^-_{\mathrm{ws}} \\ A^-_{\mathrm{su}} & A^-_{\mathrm{sw}} & A^-_{\mathrm{ss}} \end{pmatrix}$$

and assume the generic property (cf. (3.2))

(5.1) $\Delta_{\mu=0}(r_0)$, and so also $A^-_{\mathrm{ss}}(\mu = 0, q_0)$, is an isomorphism.

Then a first application of the lemma yields, for j and n sufficiently large, linear maps $f_{\mathrm{uw}}\colon \mathbb{R}^u \times \mathbb{R}^2 \to \mathbb{R}^{s-2}$, $f_{\mathrm{ss}}\colon \mathbb{R}^{s-2} \to \mathbb{R}^u \times \mathbb{R}^2$, whose graphs are invariant under $\mathrm{D}\varphi_\mu^{n+N}(\tilde{p})$ and satisfy

(i) $\mathrm{D}\varphi_\mu^{n+N}(\tilde{p}) \mid \mathrm{graph}(f_{\mathrm{uw}})$ is conjugate to $K_1 \begin{pmatrix} \Lambda_{\mathrm{u}}^n & 0 \\ 0 & \Lambda_{\mathrm{w}}^n \end{pmatrix}$ for some isomorphism $K_1 \in \mathrm{GL}(\mathbb{R}^u \times \mathbb{R}^2)$ which is close to $\Delta_\mu(\varphi_\mu^{-N}(\tilde{p}))$ and so also to $\Delta_{\mu=0}(r_0)$;

(ii) $\mathrm{D}\varphi_\mu^{-(n+N)}(\tilde{p}) \mid \mathrm{graph}(f_{\mathrm{ss}})$ is conjugate to $\Lambda_{\mathrm{s}}^{-n} K_2$, some $K_2 \in \mathrm{GL}(\mathbb{R}^{s-2})$ close to $A^-_{\mathrm{ss}}(\tilde{p})$ and therefore also to $A^-_{\mathrm{ss}}(\mu = 0, q_0)$.

Observe, on the other hand, that $A_{\mathrm{uu}}(r_\mu)$ is an isomorphism, since $\mathrm{W}^{\mathrm{u}}(p_\mu)$ and $\mathrm{W}^{\mathrm{s}}(p_\mu)$ intersect transversely at r_μ. Then, by using the lemma again, we obtain linear maps $f_{\mathrm{u}}\colon \mathbb{R}^u \to \mathbb{R}^2$, $f_{\mathrm{w}}\colon \mathbb{R}^2 \to \mathbb{R}^u$ such that the pre-images $\pi_{\mathrm{uw}}^{-1}(\mathrm{graph}(f_{\mathrm{u}}))$, $\pi_{\mathrm{uw}}^{-1}(\mathrm{graph}(f_{\mathrm{w}}))$ of their graphs under the canonical projection $\pi_{\mathrm{uw}}\colon \mathrm{graph}(f_{\mathrm{uw}}) \to \mathbb{R}^u \times \mathbb{R}^2$ are $\mathrm{D}\varphi_\mu^{n+N}(\tilde{p})$-invariant subspaces satisfying

(iii) $\mathrm{D}\varphi_\mu^{n+N}(\tilde{p}) \mid \pi_{\mathrm{uw}}^{-1}(\mathrm{graph}(f_{\mathrm{u}}))$ is conjugate to $K_3 \Lambda_{\mathrm{u}}^n$ for some $K_3 \in \mathrm{GL}(\mathbb{R}^u)$ close to $A_{\mathrm{uu}}(r_\mu)$;

(iv) $D\varphi_\mu^{-(n+N)}(\tilde{p}) \mid \pi_{\mathrm{uw}}^{-1}(\mathrm{graph}(f_{\mathrm{w}}))$ is conjugate to $\Lambda_{\mathrm{w}}^{-n} K_0$ for some $K_0 \in \mathrm{GL}(\mathbb{R}^2)$ close to $\Delta_{\mathrm{ww}}^{-}(q_\mu)$;

($\Delta_{\mathrm{uu}}^{-}, \dots, \Delta_{\mathrm{ww}}^{-}$ denoting the entries of Δ_μ^{-1}). Altogether, and recalling properties (B1)–(B3) of Section 3, we see this implies that for $j \geq 1$ sufficiently large there is $c > 1$ such that for $n \gg j$ the isomorphism $D\varphi_\mu^{(n+N)}(\tilde{p})$ has exactly

(a) u eigenvalues with absolute value $\geq c^{-1}(\sigma - \varepsilon)^n$;
(b) 2 eigenvalues, say $\tilde{\lambda}_1, \tilde{\lambda}_2$, with $|\tilde{\lambda}_1|, |\tilde{\lambda}_2| \in [c^{-1}(\lambda - \varepsilon)^n, c(\lambda + \varepsilon)^n]$;
(c) $(s-2)$ eigenvalues with absolute value $\leq c(\theta + \varepsilon)^n$.

This reduces the completion of our argument to showing that there exist arbitrarily large values of n for which $|\tilde{\lambda}_1| \neq |\tilde{\lambda}_2|$. Note first that $A_{\mathrm{uu}}(\mu = 0, r_0)$ is not an isomorphism, since $W^{\mathrm{u}}(p), W^{\mathrm{s}}(p)$ are tangent at r_0. We assume this tangency to be quasi-transversal, meaning that $\dim \ker(A_{\mathrm{uu}}(\mu = 0, r_0)) = 1$. Since $\ker(\Delta_{\mathrm{ww}}^{-}) = A_{\mathrm{wu}}(\ker(A_{\mathrm{uu}}))$ and $\ker(A_{\mathrm{uu}}) = \Delta_{\mathrm{uw}}^{-}(\ker(\Delta_{\mathrm{ww}}^{-}))$, it follows that $\ker(\Delta_{\mathrm{ww}}^{-}(\mu = 0, q_0))$ is also 1-dimensional. We fix $\{v_1, v_2\}$ to be a basis of $\mathbb{R}^2 \simeq \pi_{\mathrm{uw}}^{-1}(\mathrm{graph}(f_{\mathrm{w}}))$, orthogonal with respect to the euclidean metric induced by the coordinates ζ_1, ζ_2 and such that $v_2 \in \ker(\Delta_{\mathrm{ww}}^{-}(\mu = 0, q_0))$. Referring always to this metric, we denote by $C(v, \alpha)$ the closed cone in $\mathbb{R}^2$ of amplitude $\alpha > 0$ around the direction of a (nonzero) vector v. By construction $K_0(C(v_1, 8\pi/18)) \subset C(\tilde{v}_0, \pi/18)$, with $\tilde{v}_0 = \Delta_{\mathrm{ww}}^{-}(\mu = 0, q_0) \cdot v_1$, if K_0 is sufficiently near $\Delta_{\mathrm{ww}}^{-}(\mu = 0, q_0)$, i.e., if j and n are large enough (recall (iv)). Also, since Λ_{w} is conformal, $\Lambda_{\mathrm{w}}^{-n}(C(\tilde{v}_0, \pi/18)) = C(\tilde{v}_n, \pi/18)$, $\tilde{v}_n = \Lambda_{\mathrm{w}}^{-n} \cdot \tilde{v}_0$. On the other hand, since the eigenvalues of Λ_{w} are not real, there exist infinitely many values of n for which $\tilde{v}_n = \Lambda_{\mathrm{w}}^{-n} \cdot \tilde{v}_0 \in C(v_1, 6\pi/18)$ and so $C(\tilde{v}_n, \pi/18) \subset C(v_1, 7\pi/18)$. This means that for such values of n, $\Lambda_{\mathrm{w}}^{-n} K_0$ has a strictly invariant cone $\Lambda_{\mathrm{w}}^{-n} K_0(C(v_1, 8\pi/18)) \subset C(v_1, 7\pi/18)$ and so its two eigenvalues $(\tilde{\lambda}_1)^{-1}$ and $(\tilde{\lambda}_2)^{-1}$ must have different norms. This completes our proof. □

Remark 5.3. It is also clear from (a)–(c) that $\dim W^*(\tilde{p}) = \dim W^*(p)$, $* = \mathrm{u}$ or s, and $D\varphi_\mu^{n+N}(\tilde{p})$ is sectionally dissipative if $D\varphi_0(p)$ is.

6. Renormalization and thick basic sets

Let φ_0 be a C^2 diffeomorphism with a (nondegenerate) homoclinic tangency associated to a fixed (or periodic) point p. We suppose here that $\dim W^{\mathrm{u}}(p) = 1$ and $D\varphi_0(p)$ is sectionally dissipative, i.e., the product of any two of its eigenvalues has norm less than one. The main goal is to show that the unfolding of such a tangency by a generic one-parameter family of diffeomorphisms $(\varphi_\mu)_\mu$ yields the formation, for arbitrarily small values of μ, of hyperbolic basic sets $\Lambda_2 = \Lambda_2(\mu)$ having (codimension one stable foliation and) arbitrarily large stable thickness $\tau^s(\Lambda_2)$. For future use we also check that

(α) Λ_1 and Λ_2 are heteroclinically related; i.e., there exist some mutual transverse intersections between their stable and unstable leaves;

(β) there exist periodic points $p_1 \in \Lambda_1$, $p_2 \in \Lambda_2$ such that $W^u(p_1)$ has a nontransverse intersection with $W^s(p_2)$.

Actually, for the proof of (α), (β) we consider only the case when $D\varphi_0(p)$ has a unique least contracting eigenvalue. We suppose in addition that, besides the homoclinic tangency, the point p also has transverse homoclinic orbits and, moreover, these involve the same separatrix of $W^u(p)$ (resp. the same connected component of $W^s(p) \setminus W^{ss}(p)$) as the tangency. Since any diffeomorphism with a homoclinic tangency may be approximated by another satisfying these conditions, they represent no restriction to the purpose of proving our main result.

For the construction of Λ_2 we first deduce a higher-dimensional version of the renormalization scheme in [TY], [PT, Ch. III]. The basic idea is to show that, under a few generic assumptions, the unfolding of $(\varphi_\mu)_\mu$ contains the unfolding of families of nearly quadratic diffeomorphisms. More precisely, one shows that restrictions of iterates of φ_μ to appropriate domains close to the tangency have the form

$$(x, Y) \mapsto (x^2 + \nu, xA) + \varepsilon(\nu, x, Y) \quad \text{for some } A \in \mathbb{R}^{m-1} \text{ and } \varepsilon\ C^2\text{-small},$$

when written in conveniently chosen coordinates $(x, Y) \in \mathbb{R} \times \mathbb{R}^{m-1}$ and parameter ν.

Let us describe this construction in more detail. We assume once more that the φ_μ, μ small, admit C^2 μ-dependent linearizing coordinates $(\xi, Z) \in \mathbb{R} \times \mathbb{R}^{m-1}$ on a neighbourhood of p. We fix these coordinates in such a way that $W^s_{loc}(p_\mu) \subset \{\xi = 0\}$ and $W^u_{loc}(p_\mu) \subset \{Z = 0\}$. The assumption on the eigenvalues of $D\varphi_0(p)$ means that we may choose the norm in $\mathbb{R}^m$ to be such that

$$|\sigma_\mu| \cdot \|S_\mu\| < 1 \quad \text{(for every small } \mu) \tag{6.1}$$

where σ_μ is the expanding eigenvalue of $D\varphi_\mu(p_\mu)$ and $S_\mu = D\varphi_\mu \mid E^s(p_\mu)$. Let $q_0 = (0, Q_0) \in W^s_{loc}(p)$ and $r_0 = (\rho_0, 0) \in W^u_{loc}(p)$ belong to the orbit of tangency, say $r_0 = \varphi_0^{-N}(q_0)$, $N \geq 1$. Then for (μ, ξ, Z) close to $(0, \rho_0, 0)$ we may write $\varphi_\mu^N(\xi, Z)$ as

$$\begin{aligned}(v\mu + a_2 Z + b(\xi - \rho_0)^2 + b_1(\xi - \rho_0)\mu + b_2\mu^2 + h(\mu, \xi - \rho_0, Z),\\ Q_0 + V\mu + A_1(\xi - \rho_0) + A_2 Z + H(\mu, \xi - \rho_0, Z))\end{aligned}$$

where we have $v, b, b_1, b_2 \in \mathbb{R}$, $a_2 \in \mathcal{L}(\mathbb{R}^{m-1}, \mathbb{R})$, $V, A_1 \in \mathcal{L}(\mathbb{R}, \mathbb{R}^{m-1})$, $A_2 \in \mathcal{L}(\mathbb{R}^{m-1}, \mathbb{R}^{m-1})$ and

$$\text{(6.2)} \qquad \mathrm{D}h = 0, \quad \mathrm{D}H = 0, \quad \partial_{\xi\xi}h = \partial_{\xi\mu}h = \partial_{\mu\mu}h = 0 \quad \text{at } (0,0,0).$$

We also assume that the homoclinic tangency is quadratic and that it is generically unfolded by the family $(\varphi_\mu)_\mu$, which corresponds to having $b \neq 0$ and $v \neq 0$. First we introduce n-dependent reparametrizations

$$\text{(6.3)} \qquad \tilde{\nu} = \tilde{\theta}_n(\mu) = v\sigma_\mu^{2n}\mu + (a_2\sigma_\mu^{2n}S_\mu^n Q_0 - \rho_0\sigma_\mu^n).$$

Let $K > 0$ (large) be fixed. It is not difficult to check that for $n \geq 1$ sufficiently large $\tilde{\theta}_n$ maps some small interval I_n, close to zero in the μ-space, diffeomorphically onto $[-K, K]$. Denoting $\theta_n = (\tilde{\theta}_n \mid I_n)^{-1}$, we now introduce (n, μ)-dependent coordinates $(\tilde{x}, \tilde{Y})$ defined by

$$\text{(6.4)} \qquad (\xi, Z) = \psi_{n,\mu}(\tilde{x}, \tilde{Y}) = (\sigma_\mu^{-2n}\tilde{x} + \rho_0\sigma_\mu^{-n}, \sigma_\mu^{-n}\tilde{Y} + Q_0 + V\mu)$$

and take $\Psi_n \colon [-K, K]^{m+1} \to \mathbb{R} \times M$ to be given by

$$\Psi_n(\tilde{\nu}, \tilde{x}, \tilde{Y}) = (\mu, \xi, Z), \quad \mu = \theta_n(\tilde{\nu}), \quad (\xi, Z) = \psi_{n,\mu}(\tilde{x}, \tilde{Y}).$$

Let us also denote $\Phi \colon \mathbb{R} \times M \to \mathbb{R} \times M$, $\Phi(\mu, \eta) = (\mu, \varphi_\mu(\eta))$. Now a direct calculation yields the expression of Φ^{n+N} in the coordinates $(\tilde{\nu}, \tilde{x}, \tilde{Y})$:

$$\begin{aligned}
&(\Psi_n^{-1}\Phi^{n+N}\Psi_n)(\tilde{\nu}, \tilde{x}, \tilde{Y}) \\
&\quad = (\tilde{\nu}, b\tilde{x}^2 + b_1\tilde{x}\sigma_\mu^n\mu + b_2\sigma_\mu^{2n}\mu^2 + \tilde{\nu} + a_2\sigma_\mu^n S_\mu^n(\tilde{Y} + V\sigma_\mu^n\mu) + \sigma_\mu^{2n}h(\mu, x_n, Y_n), \\
&\qquad A_1\tilde{x} + A_2(S_\mu^n\tilde{Y} + \sigma_\mu^n S_\mu^n(Q_0 + V\mu)) + \sigma_\mu^n H(\mu, x_n, Y_n))
\end{aligned}$$

where $\mu = \theta_n(\nu)$, $x_n = \sigma_\mu^{-n}\tilde{x}$, $Y_n = S_\mu^n(\sigma_\mu^{-n}\tilde{Y} + Q_0 + V\mu)$. Observe that $|\mu|, |x_n| \leq \text{const}|\sigma_\mu|^{-n}$ and $\|Y_n\| \leq \text{const}\|S_\mu\|^n$ and, in view of (6.1), (6.2), this implies that $\sigma_\mu^{2n}h(\mu, x_n, Y_n)$ and $\sigma_\mu^n H(\mu, x_n, Y_n) \to 0$ as $n \to \infty$, uniformly on $(\tilde{\nu}, \tilde{x}, \tilde{Y}) \in [-K, K]^{m+1}$. Using also (6.3) one concludes that the sequence $(\Psi_n^{-1}\Phi^{n+N}\Psi_n)(\tilde{\nu}, \tilde{x}, \tilde{Y})$ converges uniformly to

$$\tilde{\chi}(\tilde{\nu}, \tilde{x}, \tilde{Y}) = \left(\tilde{\nu}, b\tilde{x}^2 + b_1\frac{\rho_0}{v}\tilde{x} + b_2\frac{\rho_0^2}{v^2} + \tilde{\nu}, A_1\tilde{x}\right)$$

when $n \to \infty$. Moreover, essentially the same argument applies to the derivatives and we conclude in this way that $(\Psi_n^{-1}\Phi^{n+N}\Psi_n) \to \tilde{\chi}$ in the C^2 topology. Finally, we introduce

$$\nu = b\tilde{\nu} + \frac{b_1\rho_0}{2v} - \frac{b_1^2\rho_0^2}{4v^2} + \frac{bb_2\rho_0^2}{v^2}; \quad x = b\tilde{x} + \frac{b_1\rho_0}{2v}; \quad Y = \tilde{Y} + A_1\frac{b_1\rho_0}{2vb}$$

and then, immediately, the expression of Φ^{n+N} with respect to (ν, x, Y) converges (in the C^2 topology) to

$$\chi(\nu, x, Y) = (\nu, x^2 + \nu, Ax), \quad \text{where } A = \frac{A_1}{b} \in \mathbb{R}^{m-1}. \tag{6.5}$$

Remark 6.1. As a consequence, the generic unfolding of a homoclinic tangency associated to a sectionally dissipative saddle yields the formation of sinks close to the orbit of tangency. In fact, $\chi_\nu(x, y) = (x^2 + \nu, Ax)$ has attracting periodic orbits, say for ν close to zero. The presence of such orbits is a persistent phenomenon under small perturbations. It follows that for arbitrarily small values of μ ($\mu = \theta_n(\nu)$, ν close to zero) φ_μ has an attracting periodic orbit contained in a (const $|\mu|$)-neighbourhood of the orbit of tangency.

Now the construction of Λ_2 proceeds in the same way as in the two-dimensional case; we sketch the main points and refer the reader to [PT, Ch. VI] for more details. The crucial fact here is the existence for the map $x \mapsto x^2 - 2$, and so also for χ_{-2}: $(x, Y) \mapsto (x^2 - 2, Ax)$, of invariant expanding Cantor sets K_j with thickness $\tau(K_j) \to +\infty$ as $j \to +\infty$. Moreover, these K_j are transitive and have a dense subset of periodic orbits. It follows that each K_j has, for n large and $\mu = \theta_n(\nu)$, ν close to -2, an analytic continuation as a hyperbolic basic set $K_j(n, \mu)$ of $(\psi_{n,\mu} \circ \varphi_\mu^{n+N} \circ \psi_{n,\mu}^{-1})$. In particular, the $K_j(n, \mu)$ have codimension-1 stable foliation and stable thickness $\tau^s(K_j(n, \mu))$ close to $\tau(K_j) \gg 1$. Then, we just take $\Lambda_2 = \psi_{n,\mu}(K_j(n, \mu))$ with j and n large and $\mu = \theta_n(\nu)$, ν close to -2.

Let, moreover, $P = (2, 2A)$ and Q be any other (hyperbolic) periodic point of χ_{-2}. We denote by $P(\nu)$ and $P(n, \mu)$ the analytic continuations of P for, respectively, χ_ν, and $(\psi_{n,\mu} \circ \varphi_\mu^{n+N} \circ \psi_{n,\mu}^{-1})$, n large, $\mu = \theta_n(\nu)$, ν close to -2. We also introduce analogous notations for Q. Now, it is easy to check that

- $\mathrm{W}^s(Q(\nu))$ has transverse intersections with $\mathrm{W}^u(P(\nu))$ for all ν close enough to -2;
- $\mathrm{W}^u(Q(\nu))$ intersects $\mathrm{W}^s(P(\nu))$ if and only if $\nu \leq -2$ and this intersection is transverse if $\nu < -2$.

Combining this with the fact that (compact parts of) stable and unstable manifolds of hyperbolic periodic points vary continuously with the maps, one concludes that for each large n there are μ-values $\tilde{\mu}_n = \theta_n(\tilde{\nu}_n)$, $\tilde{\nu}_n$ close to -2, for which (see Figure 5)

- $P(n, \tilde{\mu}_n)$ and $Q(n, \tilde{\mu}_n)$ are heteroclinically related;
- $\mathrm{W}^u(Q(n, \tilde{\mu}_n))$ also has nontransverse intersections with $\mathrm{W}^s(P(n, \tilde{\mu}_n))$.

This applies, in particular, when Q is a periodic point in any of the K_j. Therefore, these comments reduce the proof of properties (α), (β) to checking

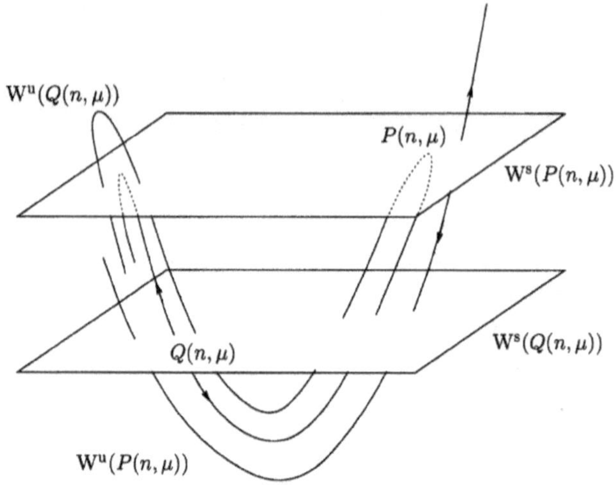

FIGURE 5

that

$$p_\mu \text{ is heteroclinically related to } \hat{P}(n,\mu) = \psi_{n,\mu}(P(n,\mu)) \tag{6.6}$$

for arbitrarily large n and $\mu = \theta_n(\nu)$, ν in a fixed neighbourhood J of -2.

The basic strategy to prove this is the same as in two dimensions. Let us first give a brief outline of it. We show that there exist compact domains $\sigma^u(n,\mu) \subset W^u(\hat{P}(n,\mu))$, $\Sigma^s(n,\mu) \subset W^s(\hat{P}(n,\mu))$, $\sigma^u \subset W^u(p)$, $\Sigma^s \subset W^s(p)$ such that (see Figure 6)

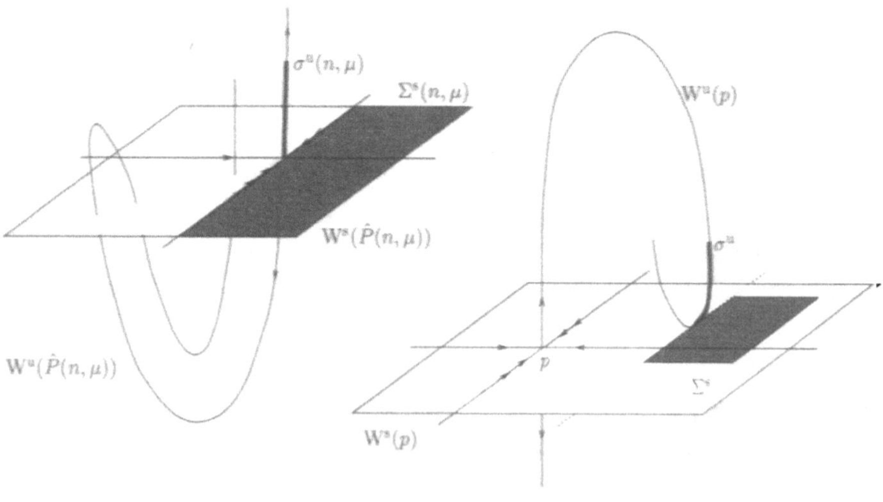

FIGURE 6

- $\hat{P}(n,\mu) \in \partial\sigma^{\mathrm{u}}(n,\mu) \cap \partial\Sigma^{\mathrm{s}}(n,\mu)$;
- the point of tangency q_0 belongs to $\partial\sigma^{\mathrm{u}} \cap \partial\Sigma^{\mathrm{s}}$;
- $\sigma^{\mathrm{u}}(n,\mu) \to \sigma^{\mathrm{u}}$ and $\Sigma^{\mathrm{s}}(n,\mu) \to \Sigma^{\mathrm{s}}$ as $n \to \infty$ (and so $\mu \to 0$).

Moreover, our assumption on the existence of transverse homoclinic orbits associated to p, together with the construction of Σ^{s}, σ^{u}, to be described below, assure that $\mathrm{W}^{\mathrm{u}}(p)$ (resp. $\mathrm{W}^{\mathrm{s}}(p)$) accumulates on σ^{u} (resp. Σ^{s}) "from the appropriate side" so that it eventually intersects Σ^{s} (resp. σ^{u}) transversely. Hence, for n large $\mathrm{W}^{\mathrm{u}}(p_\mu)$ (resp. $\mathrm{W}^{\mathrm{s}}(p_\mu)$) cuts $\Sigma^{\mathrm{s}}(n,\mu)$ (resp. $\sigma^{\mathrm{u}}(n,\mu)$) and the affirmative (6.6) above follows.

On the other hand, this strategy requires considerably more care in the present higher-dimensional setting, especially in the construction of $\Sigma^{\mathrm{s}}(n,\mu)$, Σ^{s}. We concentrate on this, the argument for $\sigma^{\mathrm{u}}(n,\mu)$, σ^{u} being analogous (and simpler). As mentioned before, we may suppose that p_μ has a unique weakest contracting eigenvalue, which we denote by λ_μ. We also assume once more the transversality condition (5.1):

$$\Delta_{\mu=0}(r_0) \text{ is an isomorphism.} \tag{6.7}$$

First we deduce the following result on existence of invariant splittings in a neighbourhood of the tangency. As before, we denote $\mathrm{E}^{\mathrm{ss}} = \{0\} \times \{0\} \times \mathbb{R}^{m-2}$, $\mathrm{E}^{\mathrm{uw}} = \mathbb{R} \times \mathbb{R} \times \{0^{m-2}\}$. We also let $\sigma = |\sigma_0|$, $\lambda = |\lambda_0|$, $\theta = \|\mathrm{D}\varphi_0(p) \mid \mathrm{E}^{\mathrm{ss}}\|$ and fix $\varepsilon > 0$ such that $\theta + 2\varepsilon < \lambda - 2\varepsilon$. Moreover, we continue to denote by U a neighbourhood of p where C^2 linearizing coordinates for the φ_μ are defined.

PROPOSITION 6.2. *There exist neighbourhoods B of q_0 and I of $0 \in \mathbb{R}$ and constants $C_0 > 0$ and $n_0 \geq 1$ such that for $n \geq n_0$ and $\mu \in I$ there is a Hölder continuous splitting $\mathrm{T}_B M = \mathrm{E}^{\mathrm{uw}}_{n,\mu} \oplus \mathrm{E}^{\mathrm{ss}}_{n,\mu}$ satisfying*

(a) $\dim \mathrm{E}^{\mathrm{uw}}_{n,\mu}(z) = 2$ *and* $\dim \mathrm{E}^{\mathrm{ss}}_{n,\mu}(z) = m-2$ *for* $z \in B$;

(b) $\mathrm{E}^{\mathrm{ss}}_{n,\mu}$ *admits an integral foliation* $\mathcal{F}^{\mathrm{ss}}_{n,\mu}$.

(c) $\mathrm{D}\varphi_\mu^{n+N}(z) \cdot \mathrm{E}^*_{n,\mu}(z) = \mathrm{E}^*_{n,\mu}(\varphi_\mu^{n+N}(z))$ *for* $* = \mathrm{uw}$ *or* ss *and for every* $z \in B \cap \varphi_\mu^{-(n+N)}(B)$ *such that* $\varphi_\mu^i(z) \in U$ *for* $0 \leq i \leq n$.

(d) *For every z as in (c) we have*

$$C_0^{-1}(\lambda-\varepsilon)^n \|v\| \leq \left\|\mathrm{D}\varphi_\mu^{n+N}(z) \cdot v\right\| \leq C_0(\sigma+\varepsilon)^n \|v\| \quad \textit{if } v \in \mathrm{E}^{\mathrm{uw}}_{n,\mu}(z)$$
$$\textit{and } \left\|\mathrm{D}\varphi_\mu^{n+N}(z) \cdot v\right\| \leq C_0(\theta+\varepsilon)^n \|v\| \quad \textit{if } v \in \mathrm{E}^{\mathrm{ss}}_{n,\mu}(z).$$

Proof. The construction of the subbundles $\mathrm{E}^{\mathrm{uw}}_{n,\mu}$, $\mathrm{E}^{\mathrm{ss}}_{n,\mu}$ is done by a standard fixed point argument analogous to that in [HP, Theorems 6.1, 6.2] or Section 2 in the present paper. Thus we just present an outline of it, leaving the details to the reader. We let γ^{u} be a segment in $\mathrm{W}^{\mathrm{u}}_{\mathrm{loc}}(p)$ containing p and r_0 in its interior and we take B and I to be small neighbourhoods of q_0 and $0 \in \mathbb{R}$, respectively, according to certain conditions to be stated in the sequel.

For simplicity we choose B to have the form $B_u \times B_s$, B_u a neighbourhood of $0 \in \mathbb{R} \simeq E^u$ and B_s a neighbourhood of Q_0 in $\mathbb{R}^{m-1} \simeq E^s$. Then we denote $\partial^u B = \partial B_u \times B_s$ and $\partial^s B = B_u \times \partial B_s$. Since we are assuming the tangency to be nondegenerate we may fix γ^u, I, B_u and B_s in such a way that, for $\mu \in I$, $\varphi_\mu^N(\gamma^u)$ does not intersect $\partial^s B$ (see Figure 7). Then, for n sufficiently large,

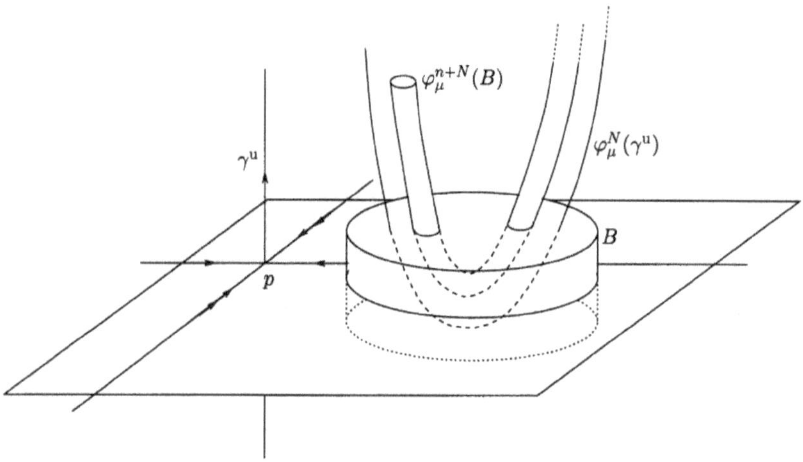

FIGURE 7

we have

$$\varphi_\mu^{n+N}(\partial^u B \cap X_{n,\mu}) \cap B = \emptyset \quad (\text{actually } \partial^u B \cap X_{n,\mu} = \emptyset) \tag{6.8}$$

$$\varphi_\mu^{n+N}(\partial^s B \cap X_{n,\mu}) \cap \partial^s B = \emptyset \tag{6.9}$$

where $X_{n,\mu}$ denotes the set of points z such that $\varphi_\mu^i(z) \in U$ for every $0 \le i \le n$. Let us describe first the construction of $E_{n,\mu}^{uw}$. We let E_1 be the constant (parallel) 2-dimensional vector bundle on $\partial^s B$ given by $E_1(z) = D\varphi_0^N(r_0) \cdot E^{uw}$. By (6.7) this is transversal to E^{ss} and so, for n large and any $z \in \partial^s B \cap X_{n,\mu}$, $D\varphi_\mu^n(z) \cdot E_1(z)$ is close to E^{uw}. It follows that $E_2(\eta) = D\varphi_\mu^{n+N}(z) \cdot E_1(z)$, $z = \varphi_\mu^{-(n+N)}(\eta)$, defines a two-dimensional bundle on $\varphi_\mu^{n+N}(\partial^s B \cap X_{n,\mu}) \cap B$, with $E_2(\eta)$ uniformly close to $D\varphi_0^N(r_0) \cdot E^{uw}$ if B and I are small and n is large. Then, recall also (6.9), we may take a two-dimensional bundle $\tilde{E}$ on B of class C^1 coinciding with E_1, E_2 on their domains and such that $\tilde{E}(z)$ is uniformly close to $D\varphi_0^N(r_0) \cdot E^{uw}$; given any $\delta > 0$ then, up to choosing B and I small and n large enough, we may assume that $\text{angle}(\tilde{E}(z), D\varphi_0^N(r_0) \cdot E^{uw}) \le \delta$ for every $z \in B$. In what follows we fix $0 < \delta \ll \text{angle}(D\varphi_0^N(r_0) \cdot E^{uw}, E^{ss})$. Then we define $\mathcal{X} = \mathcal{X}(\delta, \tilde{E})$ to be the space of continuous 2-dimensional bundles E such that

- $\text{angle}(E(\eta), D\varphi_0^N(r_0) \cdot E^{uw}) \le 2\delta$ for every $\eta \in B$ and

- $E(\eta) = \tilde{E}(\eta)$ for every $\eta \in B \setminus \varphi_\mu^{n+N}(B \cap X_{n,\mu})$.

We also introduce the graph-transform $\mathcal{T}: \mathcal{X} \to \mathcal{X}$ given by

$$\mathcal{T}(E)(\eta) = \mathrm{D}\varphi_\mu^{n+N}(z) \cdot E(z), \quad z = \varphi_\mu^{-(n+N)}(\eta)$$

if $\eta \in \varphi_\mu^{n+N}(B \cap X_{n,\mu})$ and $\mathcal{T}(E)(\eta) = \tilde{E}(\eta)$ otherwise. Observe that by construction every $E \in \mathcal{X}$ is transversal to E^{ss}. It follows in a fairly easy way that, at least for n large, $\mathcal{T}$ is well-defined ($\mathcal{T}(\mathcal{X}) \subset \mathcal{X}$) and a $\mathcal{C}^0$-contraction. Hence $\mathcal{T}$ has a unique fixed point, which we call $\mathrm{E}^{\mathrm{uw}}_{n,\mu}$. Moreover, the same type of hyperbolicity argument as in [HP] or Section 2 shows that $\mathrm{E}^{\mathrm{uw}}_{n,\mu}$ is Hölder continuous. Thus $\mathrm{E}^{\mathrm{uw}}_{n,\mu}$ satisfies (a), and (c). On the other hand, (d) is a direct consequence of the transversality of $\mathrm{E}^{\mathrm{uw}}_{n,\mu}$ to E^{ss} and the fact that most iterations are done inside the linearizing neighbourhood U. A similar argument allows us to obtain $\mathrm{E}^{\mathrm{ss}}_{n,\mu}$. We start by taking $E_1(\eta) = \mathrm{E}^{\mathrm{ss}}$, for $\eta \in \partial^{\mathrm{u}} B$, and then proceed as for $\mathrm{E}^{\mathrm{uw}}_{n,\mu}$, only this time taking negative iterates: we set $E_2(z) = \mathrm{D}\varphi_\mu^{-(n+N)}(\eta) \cdot E_1(\eta)$, $\eta = \varphi_\mu^{(n+N)}(z)$, for $z \in \varphi_\mu^{-(n+N)}(\partial^{\mathrm{u}} B) \cap X_{n,\mu} \cap B$. As before, we extend E_1, E_2 to a $\mathcal{C}^1$ $(m-2)$-dimensional bundle $\tilde{E}$ on B and then we define $\mathcal{X}$ to be the space of continuous $(m-2)$-dimensional bundles E on B such that

- $\mathrm{angle}(E(z), \mathrm{E}^{\mathrm{ss}}) \leq 2\delta$ for every $z \in B$ and
- $E(z) = \tilde{E}(z)$ for every $z \in B \setminus (\varphi_\mu^{-(n+N)}(B) \cap X_{n,\mu})$.

The graph-transform $\mathcal{T}: \mathcal{X} \to \mathcal{X}$ is defined by

$$\mathcal{T}(E)(z) = \mathrm{D}\varphi_\mu^{-(n+N)}(\eta) \cdot E(\eta), \quad \eta = \varphi_\mu^{n+N}(z)$$

if $z \in \varphi_\mu^{-(n+N)}(B) \cap X_{n,\mu}$ and $\mathcal{T}(E)(z) = \tilde{E}(z)$ otherwise. Then, as in the previous case, we conclude that $\mathcal{T}$ has a unique fixed point, which is a Hölder continuous invariant subbundle, and we take this to be $\mathrm{E}^{\mathrm{ss}}_{n,\mu}$. Moreover, $\mathrm{E}^{\mathrm{ss}}_{n,\mu}$ can be integrated to a φ_μ^{n+N}-invariant foliation $\mathcal{F}^{\mathrm{ss}}_{n,\mu}$. This can be seen as follows. First, when extending E_1, E_2 above we may take $\tilde{E}$ to coincide with the tangent bundle of a $\mathcal{C}^1$ foliation $\tilde{\mathcal{F}}$ of B by $(m-2)$-dimensional submanifolds. Now consider the space $\hat{\mathcal{X}}$ of foliations $\mathcal{F}$ of B by $(m-2)$-submanifolds of class $\mathcal{C}^1$ such that the tangent spaces to leaves of $\mathcal{F}$ vary continuously with the point and satisfy

- $\mathrm{angle}(\mathrm{T}_z\mathcal{F}(z), \mathrm{E}^{\mathrm{ss}}) \leq 2\delta$ for $z \in B$ and
- $\mathrm{T}_z\mathcal{F}(z) = \tilde{E}(z)$ if $z \in B \setminus (\varphi_\mu^{-(n+N)}(B) \cap X_{n,\mu})$.

We also introduce the graph-transform $\hat{\mathcal{T}}: \hat{\mathcal{X}} \to \hat{\mathcal{X}}$ given by

$$\hat{\mathcal{T}}(\mathcal{F})(z) = \text{connected component of } \varphi_\mu^{-(n+N)}(\mathcal{F}(\varphi_\mu^{n+N}(z))) \cap B \text{ containing } z,$$

if $z \in \varphi_\mu^{-(n+N)}(B) \cap X_{n,\mu}$ and $\hat{\mathcal{T}}(\mathcal{F})(z) = \tilde{\mathcal{F}}(z)$ otherwise. The fact that negative iterations expand the leaves of foliations $\mathcal{F} \in \hat{\mathcal{X}}$ assures that $\hat{\mathcal{T}}$ is well defined, at least if n is large enough. Moreover, the same calculations as before show that $\hat{\mathcal{T}}$ is a contraction, with respect to the C^0-distance between tangent bundles. It follows that $\hat{\mathcal{T}}$ has a unique fixed point $\mathcal{F}^{\mathrm{ss}}_{n,\mu}$ and then, clearly, we must have $\mathrm{T}_z\mathcal{F}^{\mathrm{ss}}_{n,\mu}(z) = \mathrm{E}^{\mathrm{ss}}_{n,\mu}(z)$ for every $z \in B$. □

Remark 6.3. For future use let us also state explicitly the following consequences of the renormalization techniques above:

(a) Given $\delta > 0$ we have $\mathrm{angle}(\mathrm{E}^{\mathrm{uw}}_{n,\mu}(z), \mathrm{D}\varphi_0^N(r_0)\cdot\mathrm{E}^{\mathrm{uw}}) \le \delta$ for every $(\mu, z) \in I \times B$, as long as B and I are small enough and n is sufficiently large.

(b) Given B and I small and $\delta > 0$ then for large n, $\mathrm{angle}(\mathrm{E}^{\mathrm{ss}}_{n,\mu}(z), \mathrm{E}^{\mathrm{ss}}) \le \delta$ for every $(\mu, z) \in I \times B$.

Keeping the same notation as before, we take B, I and δ to be fixed (small) and n to be large, depending on μ. In particular, we suppose $\theta_n(J) \subset I$. We observe that for $* =$ uw or ss and all the values of μ and n under consideration, we must have $\mathrm{E}^{*}_{n,\mu}(\hat{P}(n,\mu)) = \mathrm{graph}(f_*)$, recall (i) and (ii) in Section 5. Thus, $\mathrm{D}\varphi_\mu^{n+N} \mid \mathrm{E}^{\mathrm{uw}}_{n,\mu}(\hat{P}(n,\mu))$ is conjugate to $K\begin{pmatrix}\sigma_\mu^n & 0\\ 0 & \lambda_\mu^n\end{pmatrix}$ for some $K \in \mathrm{GL}(\mathbb{R}^2)$ close to $\Delta_{\mu=0}(r_0)$. We let $\sigma(n,\mu), \lambda(n,\mu)$ be its eigenvalues, $|\sigma(n,\mu)| \ge |\lambda(n,\mu)|$. The convergence to (6.5) implies that $\sigma(n,\mu)$ is close to 4 (if n is large and $\mu = \theta_n(\nu)$, ν close to -2). Hence, $\lambda(n,\mu)$ has the same sign as $(\det \Delta_{\mu=0}(r_0)(\lambda_0\sigma_0)^n)$. We claim that for the purpose of proving our main theorem it is no restriction to suppose that

$$\lambda(n,\mu) > 0 \quad \text{for } \mu \in \theta_n(J) \text{ and } n \text{ arbitrarily large.} \tag{6.10}$$

This can be justified as follows. If $(\lambda_0\sigma_0) > 0$ and $\det \Delta_{\mu=0}(r_0) > 0$ then there is nothing to prove. If $(\lambda_0\sigma_0) < 0$ we just restrict to the n-values having the appropriate parity so that $\det \Delta_{\mu=0}(r_0) \cdot (\lambda_0\sigma_0)^n > 0$. Suppose now $(\lambda_0\sigma_0) > 0$ and $\det \Delta_{\mu=0}(r_0) < 0$. We note first that, by the arguments developed previously in this section, there are μ-values $\hat{\mu}_n = \theta_n(\hat{\nu}_n)$, $\hat{\nu}_n$ close to -2, for which $\hat{P}(n,\hat{\mu}_n)$ has homoclinic tangencies. On the other hand, in the present situation we have $\lambda(n,\hat{\mu}_n) < 0 < \sigma(n,\hat{\mu}_n)$. This means that the $\varphi_{\hat{\mu}_n}$ have homoclinic tangencies falling in the second of the previous two cases. Moreover, it is clear that if the theorem holds for each $\varphi_{\hat{\mu}_n}$ then, since $\hat{\mu}_n \to 0$, it also holds for φ_0. Therefore, we may, without any loss of generality, assume that (6.10) holds.

Finally, we prove (6.6) (for the values of n as in (6.10)). For simplicity we use $\approx$ to represent equality up to a (multiplicative) constant not depending on n or δ and const to denote such a constant. Distances and angles refer always to the euclidean metric associated to the coordinates (ξ, Z) above. We begin by introducing the foliations $\mathcal{F}^{\mathrm{s}}_\mu = (\{\xi = \mathrm{const}\})$ and $\mathcal{F}^{\mathrm{u}}_\mu = (\varphi_\mu^N(\{Z =$

const})) defined on B. We suppose B small enough so that the leaves of these two foliations intersect transversely outside a hypersurface

$$L_\mu = \varphi_\mu^N(\tilde{L}_\mu), \quad \tilde{L}_\mu = \{(\xi, Z) \colon \partial_\xi \varphi_\mu^N(\xi, Z) \in \mathrm{E}^{\mathrm{s}}\}.$$

Due to the quadratic nature of the tangency we even have

$$\text{(6.11)} \qquad \operatorname{angle}(\mathrm{T}_z\mathcal{F}_\mu^{\mathrm{u}}(z), \mathrm{T}_z\mathcal{F}_\mu^{\mathrm{s}}(z)) \approx \operatorname{dist}(z, L_\mu) \quad \text{for } z \in B.$$

A fairly simple calculation shows that $\tilde{L}_\mu$ may be written as a graph $\xi = g(\mu, Z)$ and, moreover, $\varphi_\mu^{-N}(\hat{P}(n,\mu))$ is at a distance $\approx |\sigma_\mu|^{-n}$ of $\tilde{L}_\mu$, in $\{\xi > g(\mu, Z)\}$ or $\{\xi < g(\mu, Z)\}$ depending on whether b is positive or negative. It follows that $\hat{P}(n,\mu)$ is placed at a distance $d \approx |\sigma_\mu|^{-n}$ of L_μ, to the side determined by the vector $(0, A)$ (recall (6.5)). Now we let S be the $(d/2)$-neighbourhood of $\hat{P}(n,\mu)$ in its stable manifold. For n sufficiently large

$$\text{(6.12)} \qquad \operatorname{angle}(\mathrm{T}_z S, \mathrm{E}^{\mathrm{s}}) \le \delta\, |\sigma_\mu|^{-n} \quad \text{for } z \in S.$$

This is a direct consequence of the fact that the (local) stable manifold of $P(n,\mu) = \psi_{n,\mu}^{-1}(\hat{P}(n,\mu))$ converges to a hyperplane $\{x = \text{const}\}$ as $n \to \infty$, together with the form of the coordinate changes (6.4). Note also that, by construction,

$$\operatorname{dist}(z, L_\mu) \ge \text{const}\, |\sigma_\mu|^{-n} \quad \text{for } z \in S.$$

We take $\hat{\mathrm{E}}^{\mathrm{w}}$ to be the line field on S given by $\hat{\mathrm{E}}^{\mathrm{w}}(z) = \mathrm{E}^{\mathrm{uw}}_{n,\mu}(z) \cap \mathrm{T}_z S$. Remark 6.3(a) and (6.12) imply that $\hat{\mathrm{E}}^{\mathrm{w}}$ is almost colinear with $(0, A)$: if n is large then $\operatorname{angle}(\hat{\mathrm{E}}^{\mathrm{w}}(z), (0, A)) \le 2\delta$ for $z \in S$. We orient $\hat{\mathrm{E}}^{\mathrm{w}}$ in such a way that $\hat{\mathrm{E}}^{\mathrm{w}} \cdot (0, A) > 0$ and for each $\eta_0 \in \Gamma_0 = \mathrm{W}^{\mathrm{ss}}(\hat{P}(n,\mu)) \cap S$ we define $W_0(\eta_0)$ to be the positive trajectory of η_0 under $\hat{\mathrm{E}}^{\mathrm{w}}$. Now we consider the set $\Gamma \supset \Gamma_0$ of points $\eta = \varphi_\mu^{-(n+N)}(\eta_0)$ with $\eta_0 \in \Gamma_0$ and $\varphi_\mu^{-(i+N)}(\eta_0) \in U$ for $0 \le i \le n$. We point out that $\Gamma \cap B$ contains a neighbourhood of $\hat{P}(n,\mu)$ in $\mathrm{W}^{\mathrm{ss}}(\hat{P}(n,\mu))$ with fixed radius $\approx \operatorname{diam}(B)$. This may be seen from the following remarks: (i) $\varphi_\mu^{-N}(\Gamma_0)$ contains a neighbourhood of $\varphi_\mu^{-N}(\hat{P}(n,\mu))$ in $\mathrm{W}^{\mathrm{ss}}(\varphi^{-N}(\hat{P}(n,\mu)))$ with radius $\approx |\sigma_\mu|^{-n}$ and, by (6.7), it is transversal to E^{uw} at every point; (ii) the expansion during the n (negative) iterations inside U is $\ge \text{const}|\lambda_\mu|^{-n} \gg |\sigma_\mu|^n$. For each $i \ge 1$ and $\eta \in \Gamma$, we define $W_i(\eta) = \varphi_\mu^{-(n+N)}(W_{i-1}(\varphi_\mu^{(n+N)}(\eta)))$. Since $\hat{\mathrm{E}}^{\mathrm{w}}$ is $\varphi_\mu^{(n+N)}$-invariant and $\lambda(n,\mu)$ is positive, we have $W_i(\eta) \supset W_{i-1}(\eta)$ for every $i > 1$ and $\eta \in \Gamma$. We claim that, as long as they stay inside B, the $W_i(\eta)$ remain nearly colinear with $(0, A)$:

$$\text{(6.13)} \qquad \operatorname{angle}(\mathrm{T}_z W_i(\eta), (0, A)) \le 2\delta \quad \text{for } z \in W_i(\eta),\ i \ge 1,\ \eta \in \Gamma.$$

In order to prove this we note first that $\mathrm{T}_z W_i(\eta) \subset \mathrm{E}^{\mathrm{uw}}(z)$ and thus, by Remark 6.3(a), $\operatorname{angle}(\mathrm{T}_z W_i(\eta), \mathrm{D}\varphi_0^N(r_0) \cdot \mathrm{E}^{\mathrm{uw}}) \le \delta$. Therefore, (6.13) will follow if we prove

(6.14) $\quad \operatorname{angle}(\mathrm{T}_z W_i(\eta), \mathrm{E}^{\mathrm{s}}) \le \delta\,|\sigma_\mu|^{-n} \quad$ for $z \in W_i(\eta),\ i \ge 1,\ \eta \in \Gamma$.

We do this in an inductive way. Let $i \ge 1$ and suppose that (6.14) holds for every $\mathrm{T}_\zeta W_{i-1}(\eta_0)$, $\eta_0 \in \Gamma_0 \subset \Gamma$ and $\zeta \in W_{i-1}(\eta_0)$ (note that for $i = 1$ this is contained in (6.12)). Let us also point out that, due to the way we have oriented $\hat{\mathrm{E}}^{\mathrm{w}}$, the $W_j(\eta_0)$, $j = 0, 1, \ldots, i-1$, grow in the direction *opposite* to L_μ. Thus, we have $\operatorname{dist}(\zeta, L_\mu) \ge \operatorname{const}|\sigma_\mu|^{-n}$ for every $\zeta \in W_{i-1}(\eta_0)$, $\eta_0 \in \Gamma_0$. In view of (6.11) and the induction hypothesis this implies $\operatorname{angle}(\mathrm{T}_\zeta W_{i-1}(\eta_0), \mathrm{T}_\zeta \mathcal{F}^{\mathrm{u}}_\mu(\zeta)) \ge \operatorname{const}|\sigma_\mu|^{-n}$. Then, $\operatorname{angle}(\mathrm{D}\varphi_\mu^{-N} \cdot \mathrm{T}_\zeta W_{i-1}(\eta_0), \mathrm{E}^{\mathrm{u}}) \ge \operatorname{const}|\sigma_\mu|^{-n}$ and so $\operatorname{angle}(\mathrm{D}\varphi_\mu^{-(n+N)} \cdot \mathrm{T}_\zeta W_{i-1}(\eta_0), \mathrm{E}^{\mathrm{u}}) \ge \operatorname{const}|\lambda_\mu|^{-n}$. This means that $\operatorname{angle}(\mathrm{T}_z W_i(\eta), \mathrm{E}^{\mathrm{s}}) \le |\lambda_\mu|^n < \delta|\sigma_\mu|^{-n}$, for $z = \varphi_\mu^{-(n+N)}(\zeta)$, $\eta = \varphi_\mu^{-(n+N)}(\eta_0)$, and so the proof of (6.13), (6.14) is complete. In particular, for each $\eta \in \Gamma \cap B$ the curve $\bigcup_{i \ge 1} W_i(\eta)$ contains a segment $W(\eta)$ nearly colinear with $(0, A)$ and connecting η to the boundary of B. We just let $\Sigma^{\mathrm{s}}(n,\mu) = \bigcup_{\eta \in \Gamma \cap B} W(\eta) \subset \mathrm{W}^{\mathrm{s}}(\hat{P}(n,\mu))$ and then (6.14), Remark 6.3(b) and the fact that $\hat{P}(n,\mu) \to q_0$ imply that these $\Sigma^{\mathrm{s}}(n,\mu)$ converge, in the C^1 sense, to a domain $\Sigma^s \subset \mathrm{W}^{\mathrm{s}}(p)$ containing q_0 in its boundary. Finally, it is not difficult to check that, given any segment of $\mathrm{W}^{\mathrm{u}}(p)$ intersecting transversely the connected component of $\mathrm{W}^{\mathrm{s}}(p) \setminus \mathrm{W}^{\mathrm{ss}}(p)$ that contains q_0, its positive iterates eventually intersect Σ^{s} transversely.

7. Proof of the main result

Finally, we explain how the ideas and results in the previous sections fit together to prove our main theorem. We start with a general C^2 diffeomorphism φ with a homoclinic tangency associated to a sectionally dissipative saddle p. We show how to obtain, after a certain number of C^2-small perturbations of φ, a new diffeomorphism contained in the closure of an open set $\mathcal{N} \subset \operatorname{Diff}^2(M)$ exhibiting persistent homoclinic tangencies. For the sake of clearness we divide the proof into four steps. In order not to overload the notation we denote all perturbed diffeomorphisms by φ; on the other hand we state carefully the properties obtained after each perturbation.

Step 1. Up to a first perturbation, we may suppose that φ satisfies the (generic) assumptions of Section 5: The homoclinic tangency is quadratic (and quasi-transversal); the saddle point p is C^2-linearizable and has either 1 (real) or 2 (complex) weakest contracting eigenvalues; finally, the transversality condition (5.1) holds. Then, proceeding as in Section 5, we obtain a diffeomorphism $\tilde{\varphi}$ arbitrarily close to φ, exhibiting homoclinic tangencies associated to a periodic saddle $\tilde{p}$ which (besides being also sectionally dissipative, recall Remark 5.3), has a unique least contracting eigenvalue. In the sequel we still denote by φ and p such $\tilde{\varphi}$ and $\tilde{p}$, respectively.

Step 2. We may again suppose that the saddle p is $\mathcal{C}^2$-linearizable and the homoclinic tangency is quadratic and satisfies (5.1). Also, we may assume that the point of tangency is not contained in the strong stable manifold of p. Then for diffeomorphisms arbitrarily close to φ, p has transverse homoclinic intersections together with a new homoclinic tangency. Note that, by continuity, these transverse intersections are also outside $W^{ss}(p)$ and they satisfy (3.2). Hence, by Section 3, p belongs to a nontrivial basic set Λ_1 with intrinsically $\mathcal{C}^1$ unstable foliation. Moreover, the results of Section 4 apply to Λ_1 since, after Step 1, we have a unique weak-stable direction. In particular, $\tau^u(\Lambda_1)$ is strictly positive and remains bounded away from zero under further (small) perturbations.

Step 3. The new transverse and nontransverse homoclinic orbits are constructed (Step 2) in such a way that they involve the same connected component of $W^s(p) \setminus W^{ss}(p)$ and of $W^u(p) \setminus \{p\}$. It is also clear that we may continue to suppose that the saddle p is $\mathcal{C}^2$-linearizable. Moreover, we may assume that this new homoclinic tangency is again quadratic and satisfies the generic condition (6.7): Otherwise we just replace φ by some nearby diffeomorphism for which this holds. Then, by generically unfolding this tangency as in Section 6, we obtain arbitrarily small perturbations of φ exhibiting basic sets Λ_1 and Λ_2 such that: $\tau^u(\Lambda_1)\tau^s(\Lambda_2) > 1$ and

- the leaves of $W^u(\Lambda_1)$ (resp. $W^s(\Lambda_1)$) have transverse intersections with those of $W^s(\Lambda_2)$ (resp. $W^u(\Lambda_2)$);
- there are periodic points $p_1 \in \Lambda_1$, $p_2 \in \Lambda_2$ such that $W^u(p_1)$ and $W^s(p_2)$ also have a point q of nontransverse intersection (i.e., a tangency).

Step 4. Let U_2 be a neighbourhood of Λ_2 such that $W^s(\Lambda_2)$ admits an extension to a $\mathcal{C}^1$ foliation $\mathcal{F}_2^s$ defined on U_2. By $\mathcal{C}^1$ we mean here that the tangent spaces to the leaves $T_z\mathcal{F}_2^s(z)$ vary in a $\mathcal{C}^1$ fashion with the point z. Clearly, we may take q to belong to U_2 and then an implicit function argument allows us to define the "line" of tangencies between the leaves of $W^u(\Lambda_1)$ and $\mathcal{F}_2^s$ near q. This is based on the following:

LEMMA 7.1 (Implicit function). *Let $X \subset \mathbb{R}^m$ be compact and $I \subset \mathbb{R}$ be a compact interval. Let $F\colon X \times I \to \mathbb{R}$ be intrinsically $\mathcal{C}^1$ and $(x_0,t_0) \in X \times \operatorname{int}(I)$ be such that*

$$F(x_0,t_0) = 0 \quad \text{and} \quad \Delta F_{x_0}(t_0,t_0) \neq 0. \tag{7.1}$$

Then there exist $V \subset X$ a compact neighbourhood of x_0 and a unique intrinsically C^1 map $f\colon V \to I$ such that $f(x_0) = t_0$ and $F(x, f(x)) = 0$ for every $x \in V$.

Proof. Let $\alpha > 0$, V_0 a neighbourhood of x_0 and $\delta > 0$ (small) be such that $|\Delta F_x(s,t)| \geq \alpha$ for every $x \in V_0$ and $s,t \in [t_0-\delta, t_0+\delta]$. We take $V \subset V_0$ so that $|F(x,t_0)| \leq \alpha\delta/2$ for every $x \in V$. Then, for $x \in V$, $F(x,t_0-\delta) \cdot F(x,t_0+\delta) < 0$ and so there is $f(x) \in [t_0-\delta, t_0+\delta]$ such that $F(x, f(x)) = 0$. Moreover, $f(x)$ is unique since ΔF_x is never zero on $[t_0-\delta, t_0+\delta]$. The same kind of argument shows that the function $f\colon V \to I$ defined in this way is continuous. Finally, $\Delta f(x,z) = -\Delta F^{f(x)}(x,z)/\Delta F_z(f(x), f(z))$ (where we use the same notation as in Lemma 2.5) is an intrinsic derivative for f. □

In order to apply the lemma we first fix $U \subset U_2$ a small neighbourhood of q and ξ_s a C^1 vector field on U orthogonal to the leaves of $\mathcal{F}_2^s$. By Proposition 3.5, $W^u(\Lambda_1) \cap U$ contains an intrinsically C^1 diffeomorphic image Y of $X \times I$ where I is a compact interval and X is a small compact neighbourhood of p_1 in $\Lambda_1 \cap W^s_{loc}(p_1)$. We let ξ_u be some intrinsically C^1 vector field on Y tangent to the leaves of $W^u(\Lambda_1)$ and then we define $F(y) = \xi_u(y) \cdot \xi_s(y)$. Hypothesis (7.1) in the lemma corresponds to having a quadratic tangency at q and, up to considering an additional perturbation of φ, we may assume this to be the case. As a conclusion we get that there exist V_1 a compact neighbourhood of p_1 in $\Lambda_1 \cap W^s_{loc}(p_1)$ and $\pi_1\colon V_1 \to W^u(\Lambda_1) \cap U$ an intrinsically C^1 map such that each $\pi_1(x)$, $x \in V_1$, is a point of tangency between $W^u(x)$ and some leaf of $\mathcal{F}_2^s$. We also introduce $\pi_2\colon U \to W^u_{loc}(p_2)$, the projection along the leaves of $\mathcal{F}_2^s$ onto $W^u_{loc}(p_2)$ which we identify with an interval in $\mathbb{R}$ (via some C^1 diffeomorphism). If necessary, we perform a last perturbation of φ so that $\Delta\pi_1(p_1,p_1) \cdot IT_{p_1}(\Lambda_1 \cap W^s_{loc}(p_1))$ is not tangent to the stable leaf $\mathcal{F}_2^s(q)$. Then $(\pi_2 \circ \pi_1)$ is intrinsically C^1 and $\Delta(\pi_2 \circ \pi_1)(p_1,p_1)$ is bijective. We let $K_1 = (\pi_2 \circ \pi_1)(V_1)$ and K_2 be a small compact neighbourhood of p_2 in $\Lambda_2 \cap W^u_{loc}(p_2)$ and then

- K_1 and K_2 intersect each other at p_2;
- $\tau(K_1,p_2) \cdot \tau(K_2,p_2) = \tau^u(\Lambda_1) \cdot \tau^s(\Lambda_2) > 1$.

Now from the gap lemma and the continuous variation of the thickness, we get that for a whole C^2-open set $\mathcal{N}$ of perturbations of φ the corresponding sets K_1, K_2 intersect each other, which corresponds to heteroclinic tangencies involving Λ_1 and Λ_2. Since Λ_1 and Λ_2 are heteroclinically related, it follows in an easy way that given any periodic point $\tilde{p}$ in $\Lambda_1 \cup \Lambda_2$ (e.g., $\tilde{p} = p$) a dense subset of elements of $\mathcal{N}$ exhibits homoclinic tangencies associated to (the analytic continuation of) $\tilde{p}$. As explained in Section 1, this implies that residually in $\mathcal{N}$ the maps have infinitely many coexisting sinks. This completes our proof.

On the other hand, the parametrized version of the theorem now follows by the fact that, given a generic one-parameter family of diffeomorphisms passing through a homoclinic tangency, *the previous arguments can be carried out making use only of perturbations along the parameter*: At each step the new, perturbed, diffeomorphism is taken belonging to the initial family. We just observe a few points in this direction. First of all, the perturbation described in Section 5 is of that kind and so, as before, we may assume that the periodic point associated to the homoclinic tangency has a unique weakest contracting eigenvalue. Of course, in performing such a perturbation along the parameter line we also have to assure that all the conditions concerning the initial homoclinic tangency are still valid for the new one. Through another small perturbation we obtain a basic set Λ_1 as in Section 3, with a homoclinic tangency associated to a periodic point $p_1 \in \Lambda_1$ satisfying all the previous transversality conditions. Then, again as before, an additional small perturbation inside the parametrized family yields a basic set Λ_2 as in Section 6: It has large stable thickness (the product with the unstable thickness of Λ_1 is greater than 1) and, for some periodic point $p_2 \in \Lambda_2$, p_1 and p_2 are heteroclinically related (mutual transverse intersections between their invariant manifolds) and moreover $W^u(p_1)$, $W^s(p_2)$ also exhibit an orbit of tangency. The last part of the proof is then similar to that of the main theorem.

Instituto de Matemática Pura e Aplicada (IMPA), Rio de Janeiro, Brazil
Universidade do Porto, Portugal

References

[BC] M. Benedicks and L. Carleson, On iterations of $1-ax^2$ on $(-1,1)$, Ann. of Math. **122** (1985), 1–24.

[HP] M. Hirsch and C. Pugh, Stable manifolds and hyperbolic sets, Proc. A.M.S. Symp. Pure Math., **14**(1970), 133–163.

[MV] L. Mora and M. Viana, Abundance of strange attractors, Acta Math. **171**(1993), 1–71.

[N1] S. Newhouse, Non-density of Axiom A(a) on S^2, Proc. A.M.S., Symp. Pure Math. **14** (1970), 191–202.

[N2] ———, Diffeomorphisms with infinitely many sinks, Topology **13**(1974), 9–18.

[N3] ———, The abundance of wild hyperbolic sets and nonsmooth stable sets for diffeomorphisms, Publ. Math. I.H.E.S. **50**(1979), 101–151.

[PT] J. Palis and F. Takens, *Hyperbolicity and Sensitive-Chaotic Dynamics at Homoclinic Bifurcations*, Cambridge University Press, 1992.

[Rob] C. Robinson, Bifurcation to infinitely many sinks, Comm. Math. Phys. **90**(1983), 433–459.

[Rom] N. Romero, Persistence of homoclinic tangencies in higher dimensions, to appear Ergod. Th. and Dynam. Sys.

[Sm] S. SMALE, Diffeomorphisms with many periodic points, *Differential and Combinatorial Topology*, Princeton Math'l Series, Princeton Univ. Press (1965), 63–80.

[St] E. STEIN, *Singular Integrals and Differentiability Properties of Functions*, Princeton Math'l Series, Princeton University Press, 1970.

[TY] L. TEDESCHINI-LALLI and J.A. YORKE, How often do simple dynamical processes have infinitely many coexisting sinks?, Comm. Math. Phys. **106**(1986), 635–657.

[V] M. VIANA, Strange attractors in higher dimensions, Bull. Braz. Math. Soc. **24**(1993), 13–62.

[W] H. WHITNEY, Analytic extensions of differentiable functions defined in closed sets, Trans. A.M.S. **36**(1934), 63–89.

[YA] J.A. YORKE and K.T. ALLIGOOD, Cascades of period-doubling bifurcations: a prerequisite for horseshoes, Bull. A.M.S. **9**(1983), 319–322.

(Received November 10, 1992)

Ann. Inst. Henri Poincaré,

Vol. 14, n° 4, 1997, p. 439-456 *Analyse non linéaire*

On the arithmetic sum of regular Cantor sets

by

J. PALIS

IMPA, Rio de Janeiro, Brazil.
E-mail: jpalis@impa.br

and

J. C. YOCCOZ

Dep. Math., Université Paris-Sud, Orsay, France.

ABSTRACT. – We prove here that for any pair of regular Cantor sets in the line, either its arithmetic sum has Lebesgue measure zero or the pair can be approximated (through the image of a smooth diffeomorphism) by another one whose arithmetic sum contains an interval. The latter occurs when the Hausdorff dimension of the product of the sets is bigger than one.

RÉSUMÉ. – Nous démontrons que pour toute paire d'ensembles de Cantor réguliers sur la droite, ou bien leur somme arithmétique est de mesure de Lebesgue nulle, ou bien la paire peut être approchée (par l'image d'un difféomorphisme) par une autre dont la somme arithmétique contient un intervalle. Le dernier cas se produit quand la dimension de Hausdorff du produit des ensembles est plus grande que 1.

1. INTRODUCTION

Besides its perhaps more classical role in number theory and related topics, as it can be seen in the reference to Hall in [PT2], the arithmetic sum (difference) of Cantor sets plays a central role in dynamic bifurcations,

Annales de l'Institut Henri Poincaré - Analyse non linéaire - 0294-1449
Vol. 14/97/04/$ 7.00/© Gauthier-Villars

Originally published in: *Ann. Inst. H. Poincaré Anal. Non Linéaire* 14 (1997), no. 4, 439–456 , © Gauthier-Villars, Paris 1997
Reprint by Springer-Verlag Berlin Heidelberg 2014 with kind permission

especially homoclinic bifurcations. These bifurcations in turn, can be a key to better understand the dark realm of dynamics: we conjecture that bifurcating homoclinic systems are dense in the complement of the hyperbolic or stable ones [PT2, Ch. 7].

Roughly speaking, the arithmetic sum of Cantor sets appear naturally in dynamics, when we analyse the set of intersections (often the set of tangencies) of invariant foliations, say one-dimensional stable and unstable foliations, and how this set varies with a parameter. If such foliations are associated to the same invariant hyperbolic set and their leaves are tangent along a transversal line, we are lead to analyse the sum of the two Cantor sets thus obtained, to study the bifurcations that appear when varying the diffeomorphism. They include, in the dissipative case, infinitely many simultaneous sinks, strange attractors, and cascades of period doubling bifurcations (see [PT2]). The only known way to produce infinitely many sinks, due to Newhouse, is to show that the sum of the Cantor sets contains intervals. That is what we prove here for a dense subset of pairs of regular Cantor sets whose Cartesian product has Hausdorff dimension bigger than one. The precise statement is presented in Section 2. Notice that if the Hausdorff dimension of the product of the Cantor sets is smaller than one, then its sum has Lebesgue measure zero [PT2]. In this last situation, in the context of homoclinic bifurcations, we have a total prevalence of hyperbolicity for the nearby diffeomorphisms [PT1], whereas in the latter it is shown with techniques quite different from the ones exhibited here, that this is definitively not the case [PY].

Finally, we have been conjecturing for more than a decade that the result in the present paper should be true for generic, or even an open and dense subset of pairs of regular Cantor sets and also for affine Cantor sets in general [P]. The conjecture may be of central interest in bifurcations of homoclinic tangencies.

2. DEFINITIONS AND RESULTS

We consider the following setting:

– a compact non trivial interval I;

– $C^{1+\alpha}$ diffeomorphisms $\varphi_1, \ldots, \varphi_p$ of I onto disjoint subintervals of I, which satisfy $|D\varphi_i(x)| < 1$, $1 \leq i \leq p$, $x \in I$.

– a matrix $A = (a_{ij})$, $1 \leq i, j \leq p$, with $a_{ij} \in \{-1, 1\}$.

We assume that the subshift of finite type Σ_A of the full shift on p elements is *infinite* and *topologically mixing*. For $1 \leq i_o \leq p$, let:

$$K(i_o) = \bigcup \bigcap_{k \geq 1} \varphi_{i_1} \dots \varphi_{i_k}(I)$$

the union being taken over all sequences $(i_o, i_1 \dots)$ such that $a_{i_k i_{k+1}} = 1$ for $k \geq 0$. This is a Cantor subset of I; a Cantor subset of $\mathbf{R}$ obtained in this way is called *regular*.

The $K(i_o)$, for $1 \leq i_o \leq p$, are called *twin* regular Cantor sets.

The diffeomorphic image of a regular Cantor set is regular: conjugate the φ_i by the diffeomorphism.

For $K = K(i_o)$ as above, we define:

$$\Phi(K) = \{\varphi = \varphi_{i_1} \dots \varphi_{i_k} \mid k \geq 1,\ a_{i_\ell i_{\ell+1}} = 1 \text{ for } 0 \leq \ell < k\}.$$

For $\varphi = \varphi_{i_1} \dots \varphi_{i_k} \in \Phi(K)$, we have:

$$\varphi(I) \cap K(i_o) = \varphi(K(i_k)). \tag{1}$$

This shows that the *cylinder* $\varphi(I) \cap K(i_o)$ is *regular*; $\varphi(I)$ is the *box* of $\varphi(I) \cap K(i_o)$. Two cylinders (or boxes) are either disjoint, or one is contained in the other.

We recall that the arithmetic sum of two subsets A' and A'' of the line is defined as

$$A' + A'' = \{t \in \mathbf{R}; \exists\, a' \in A', a'' \in A'' \text{ such that } a' + a'' = t\}$$

We can now state our result.

THEOREM. – *Let K', K'' be two regular Cantor sets, of respective Hausdorff dimensions d', d''. Assume that $d' + d'' > 1$. Then, given $\delta > 0$, one can find C^∞ diffeomorphisms h of $\mathbf{R}$, arbitrarily near the identity, such that any point in $h(K') + K''$ is at a distance less than δ from some nontrivial interval in $h(K') + K''$.*

It is actually sufficient to find h such that $h(K') + K''$ contains *one* nontrivial interval. Indeed, let $J'_1, \dots, J'_k$ and $J''_1, \dots, J''_k$ be disjoint cylinders of K' and K'', respectively, such that any point in $K' + K''$ contains some $J'_i + J''_i$ in its $\delta/2$ neighbourhood. Gluing together diffeomorphisms $h_1 \dots h_k$ near the identity such that $h_i(J'_i) + J''_i$ contains one nontrivial interval, we get the conclusion of the theorem (observe that J'_i, resp. J''_i, is regular with the same Hausdorff dimension as K', respectively K'').

3. BASIC ESTIMATES

We give here the basic estimates for the regular set K', with constants $c'_1, c'_2 \ldots$. Similar estimates hold for K'', with constants $c''_1, c''_2 \ldots$. Let I', $\varphi'_1, \ldots, \varphi'_p$ be as in 2., defining K'.

There exist c'_1, c'_2, c'_3, c'_4 such that:

(2) $$0 < c_1'^{-1} < |D\varphi'_i(x)| < c'_2 < 1, \quad 1 \le i \le p, x \in I'$$

(3) $$|\varphi'_i(x) - \varphi'_j(y)| > c_3'^{-1} > 0, \quad 1 \le i, j \le p, i \ne j, x, y \in I'$$

(4) $$|\log |D\varphi'_i(x)| - \log |D\varphi'_i(y)|| \le c'_4 |x - y|^\alpha, \quad 1 \le i \le p, \ x, y \in I'.$$

From (2) and (4), we get:

Distortion estimate. – For $\varphi \in \Phi(K')$, $x, y \in I'$.

(5) $$|\log |D\varphi(x)| - \log |D\varphi(y)|| \le c'_5 = c'_4(1 - c_2'^{\alpha})^{-1} |I'|^\alpha.$$

Let $Q' = \varphi(I')$, $\varphi \in \Phi(K')$, be a box of K', with $Q' \ne I'$. Writing $\varphi = \varphi' \varphi'_i$, we have $|Q'| \le c'_2 |I'| \max |D\varphi'|$ and the distance from Q' to $K' - Q'$ is at least $c_3'^{-1} \min |D\varphi'|$. Hence we get from (5):

(6) $$d(Q', K' - Q') \ge c_6'^{-1} |Q'|, \quad \text{with} \quad c'_6 = c'_3 \, e^{c'_5} \, c'_2 |I'|.$$

Let m' be the d'-dimensional Hausdorff measure. Because of the topologically mixing property, each twin of K' contains a diffeomorphic image of any other, so they all have Hausdorff dimension d'. Moreover, regularity implies that there are constants c'_7, c'_8 such that:

(7) $$0 < c_7'^{-1} < m'(K'_1) < c'_8$$

for any K'_1 twin of K' (including itself). See [F], [PT2].

LEMMA 1. – *Let J' be a cylinder of K', with box Q'; one has:*

$$c_9'^{-1} |Q'|^{d'} \le m'(J') \le c'_{10} |Q'|^{d'},$$

with $c'_9 = c'_7 \exp(c'_5 d') |I'|^{d'}$, $c'_{10} = c'_8 \exp(c'_5 d') |I'|^{-d'}$.

Proof. – One has $J' = \varphi(K_1')$, $Q' = \varphi(I')$ for some $\varphi \in \Phi(K')$ and some twin K_1' of K'. From this, we get:

$$|Q'| = |D\varphi(x_o)|\,|I'|, \quad \text{for some} \quad x_o \in I'$$

$$m'(J') = \int_{K_1'} |D\varphi(x)|^{d'}\, dm'(x) = m'(K_1')|D\varphi(x_1)|^{d'}$$

for some $x_1 \in I'$. The lemma then follows from (5) and (7). □

Let $\beta > 0$. A *β-decomposition* of K' is a partition of K' in (disjoint) cylinders whose boxes have length between β and $c_1'^{-1}\beta$. Such a decomposition exists and is non trivial if $\beta < |I'|$ (see (2)). Let π' be a β-decomposition of K'.

LEMMA 2. – *We have:*

$$m'(J_1') \le c_{11}'\, m'(J_2'), \quad \textit{for} \quad J_1', J_2' \in \pi'$$

with $c_{11}' = c_9' c_{10}' c_1'^{d'}$.

Proof. – Trivial from Lemma 1 and the definition of a β-decomposition. □

LEMMA 3. – *Let Q_0' of length greater than β. The number N of elements of π' contained in Q_0' satisfies:*

$$c_{12}'^{-1}\left(\beta^{-1}|Q_0'|\right)^{d'} \le N \le c_{13}'\left(\beta^{-1}|Q_0'|\right)^{d'}$$

with $c_{12}' = c_9' c_{10}' \quad c_{13}' = c_9' c_{10}' c_1'^{d'}$.

Proof. – One has, if J_0' is the cylinder of Q_0':

$$c_9'^{-1}\, |Q_0'|^{d'} \le m'(J_0') \le c_{10}'|Q_0'|^{d'}, \quad \text{by Lemma 1;}$$

$$m'(J_0') = \sum_{\substack{J' \in \pi' \\ J' \subset Q_0'}} m'(J'),$$

and $c_9'^{-1} c_1'^{-d'} \beta^{d'} \le m'(J') \le c_{10}'\beta^{d'}$ for $J' \in \pi'$.

The lemma follows. □

4. THE ENERGY ESTIMATES

We now consider K', K'' as in the theorem. We let $K = K' \times K'' \subset \mathbf{R}^2$, $D = d' + d'' > 1$, $m_D = m' \times m''$ the D-dimensional Hausdorff measure on K. The Euclidean metrics on $\mathbf{R}^2$ is noted $\|\ \ \|$.

A *β-decomposition* of K is a partition of K which is product of two β-decompositions of K', K''. In a similar way, we define cylinders and boxes for K.

For simplicity, we assume that:

$$(8) \qquad |I'| \leq 1, \quad |I''| \leq 1, \quad |I'| + |I''| \geq 1,$$

which can be obtained by an homothety of the *same* ratio on K', K''. We also assume that c'_1, c''_1 have been taken so that:

$$(9) \qquad |I'| \geq c_1'^{-1}, \quad |I''| \geq c_1''^{-1},$$

in such that a way that $\pi(0) = \{K\}$ is a 1-decomposition of K.

Let π be a β-decomposition of K.

LEMMA 4. – *For $J_1, J_2 \in \pi$, one has:*

$$m_D(J_1) \leq c_{11}\, m_D(J_2), \quad \textit{with} \quad c_{11} = c'_{11} c''_{11}$$

Proof. – Direct consequence of Lemma 2.

LEMMA 5. – *Let $\beta_1 < (c'_1 + c''_1)^{-1} \beta$ and π_1 a β_1-decomposition of K. For $J \in \pi_0$, the number N of elements of π_1 contained in J satisfies:*

$$c_{14}^{-1} \left(\beta\beta_1^{-1}\right)^D \leq N \leq c_{14} \left(\beta\beta_1^{-1}\right)^D$$

with $c_{14} = c'_9 c''_9 c'_{10} c''_{10} c_1'^{d'} c_1''^{d''}$.

Proof. – Use Lemma 3, knowing that if $Q' \times Q''$ is the box of J, one has $c_1'^{-1} \beta \leq |Q'| \leq \beta$, $c_1''^{-1} \beta \leq |Q''| \leq \beta$. □

DEFINITION. – The *energy* of a cylinder J of K is the value of the integral:

$$\mathcal{I}(J) = \iint_{J \times J} \frac{dm_D(z)\, dm_D(z')}{\|z - z'\|}.$$

LEMMA 6. – *One has*

$$\mathcal{I}(K) \leq c_{15} = 2^D \left(1 - 2^{1-D}\right)^{-1} c_3'^{d'} c_3''^{d''} (c'_8 c''_8)^2 \exp(c'_5 d' + c''_5 d'').$$

Proof. – Let $z = (x', x'') \in K$, $r > 0$, and $B_r(z)$ the Euclidean ball of center z, radius r. Let $Q = Q' \times Q''$ be the smallest box of K containing $B_r(z) \cap K$; write $Q' = \varphi'(I')$, $Q'' = \varphi''(I'')$ with $\varphi' \in \Phi(K')$, $\varphi'' \in \phi(K'')$. Let i (resp. j) be such that $x' \in \varphi'\varphi'_i(I')$ (resp. $x'' \in \varphi''\varphi''_j(I'')$). By the minimality of Q, the interval of center x' (resp. x'') and radius r must intersect $\varphi'\varphi'_k(I')$ (resp. $\varphi''\varphi''_\ell(I'')$) for some $k \neq i$ (resp. $\ell \neq j$). Then, we get from (3) that:

$$r \geq (\inf |D\varphi'|) c_3'^{-1}$$
$$r \geq (\inf |D\varphi''|) c_3''^{-1}.$$

If J', J'' are the cylinders associated to Q', Q'', we obtain (using (5)):

$$m'(J') = \int_{K_1'} |D\varphi'(y)|^{d'} \, dm'(y) \leq (c_3' r)^{d'} \, c_8' \exp(c_5' d')$$
$$m''(J'') = \int_{K_1'} |D\varphi''(y)|^{d''} \, dm''(y) \leq (c_3'' r)^{d''} \, c_8'' \exp(c_5'' d''),$$

where K_1', K_1'' are twins of K', K''. This gives:

$$m_D(B_r(z) \cap K) \leq c_{16}\, r^D, \text{ with } c_{16} = c_8' c_8'' c_3'^{d'} c_3''^{d''} \exp(c_5' d' + c_5'' d'').$$

One has $K \subset B_2(z)$ by (8), so we can write

$$K - \{z\} = \bigcup_{k=0}^{\infty} ((B_{2^{1-k}}(z) - B_{2^{-k}}(z)) \cap K),$$

to get

$$\int_K \frac{dm_D(z')}{||z - z'||} \leq \sum_{k=0}^{\infty} 2^k c_{16} (2^{1-k})^D = 2^D c_{16} (1 - 2^{1-D})^{-1}.$$

Using $m_D(K) \leq c_8' c_8''$, we get the lemma. □

LEMMA 7. – *Let J be an element of a β-decomposition of K. One has:*

$$\mathcal{I}(J) \leq c_{17} \beta^{2D-1} \text{ with } c_{17} = c_{15} c_1'^{2d'} c_1''^{2d''} e^{2c_5' d' + 2c_5'' d''} \, Max\left(c_1' e^{c_5'}, c_1'' e^{c_5''}\right).$$

Proof. – Write $J = J' \times J''$, Q', Q'' for the boxes of J', J''; let $\varphi' \in \Phi(K)$, $\varphi'' \in \Phi(K'')$ be such that $Q' = \varphi'(I')$, $Q'' = \varphi''(I'')$, $J' = \varphi'(K_1')$, $J'' = \varphi''(K_1'')$, for some twins K_1', K_1'' of K', K''. The estimate of Lemma 6 holds with $K_1 = K_1' \times K_1''$ instead of K. One has:

$$\inf |D\varphi'| \geq e^{-c_5'} |Q'| \, |I'|^{-1} \geq e^{-c_5'} \, c_1'^{-1} \beta \text{ (see (8))}$$
$$\inf |D\varphi''| \geq e^{-c_5''} \, c_1''^{-1} \beta$$

and, for $z = (x', x'')$, $z' = (y', y'') \in K_1$:

$$||\varphi' \times \varphi''(z) - \varphi' \times \varphi''(z')|| \geq \min(\inf |D\varphi'|, \inf |D\varphi''|)||z - z'||.$$

We then get:

$$\iint_{J \times J} \frac{dm_D(z)dm_D(z')}{||z - z'||}$$

$$\leq \iint_{K_1 \times K_1} \frac{dm_D(z)dm_D(z')|D\varphi'(x')|^{d'}|D\varphi'(y')|^{d'}|D\varphi''(x'')|^{d''}|D\varphi''(y'')|^{d''}}{||z - z'|| \min(\inf |D\varphi'|, \inf |D\varphi''|)}$$

$$\leq \beta^{-1} c_{15} \max(c_1' e^{c_5'}, c_1'' e^{c_5''}) \exp(2c_5' d' + 2c_5'' d'') \left(\frac{|Q'|}{|I'|}\right)^{2d'} \left(\frac{|Q''|}{|I''|}\right)^{2d''}$$

which gives the lemma, using (9). □

5. MEASURE OF PROJECTIONS

For $\lambda > 0$, let π_λ the linear map from $\mathbf{R}^2$ to $\mathbf{R}$ defined by:

$$\pi_\lambda(x, y) = \lambda x + y.$$

Let J be a cylinder of K. We write m_J for the restriction of m_D to J and $m_{J,\lambda}$ for the image of m_J by π_λ.

LEMMA. – *Let J be an element of a β-decomposition π of K. Assume that for some $\lambda \in [1/2, 1]$, $\gamma > 0$, we have:*

$$(10) \qquad \int_{-\infty}^{+\infty} |\hat{m}_{J,\lambda}(\zeta)|^2 \, d\zeta \leq \gamma^{-1} \beta^{2D-1}.$$

Let J_o be a measurable subset of J satisfying $m_D(J_o) \geq k\, m_D(J)$, with $0 < k \leq 1$. Then, one has:

$$m(\pi_\lambda(J_o)) \geq c_{18}^{-1} k^2 \gamma \beta, \text{ with } m \text{ the Lebesgue measure on } \mathbf{R}$$

and $c_{18} = \left(c_9' c_9'' c_1'^{d'} c_1''^{d''}\right)^2$.

Proof. – Let m_{J_o} be the restriction of m_J to J_o, and $m_{J_o,\lambda}$ its image by π_λ. The hypothesis implies that $m_{J,\lambda}$ (resp. $m_{J_o,\lambda}$) is absolutely continuous with respect to m and its density g (resp. g_o) satisfy:

$$(11) \qquad \int_{-\infty}^{+\infty} (g_o(u))^2 \, du \leq \int_{-\infty}^{+\infty} (g(u))^2 \, du < \gamma^{-1} \beta^{2D-1}.$$

By Cauchy-Schwarz's inequality, one has:

$$(12) \qquad m(\pi_\lambda(J_o)) \int_{-\infty}^{+\infty} (g_o(u))^2\,du \geq \left[\int_{-\infty}^{+\infty} g_o(u)\,du\right]^2.$$

On the other hand, one has, with

$$(13) \qquad \int_{-\infty}^{+\infty} g_o(u)\,du = m_{J_o,\lambda}(\mathbf{R}) = m_D(J_o)$$
$$\geq k\, m_D(J) \geq k\, c_9'^{-1} c_9''^{-1} c_1'^{-d'} c_1''^{-d''} \beta^D$$

by Lemma 1-Lemma 8 follows from (10), (12), (13). □

LEMMA 9. – *Let $\beta_1 < (c_1' + c_1'')^{-1}\beta$, π_1 a β_1-decomposition of K and $J \in \pi$. Let J_o be the union of at least one eight of all elements of π_1 which are contained in J. If λ, γ are as in Lemma 8 and (10) holds, then we have:*

$$m(\pi_\lambda(J_o)) \geq c_{19}^{-1}\gamma\beta,$$

with $c_{19} = 64 c_{18} c_{11}^2$.

Proof. – By Lemma 4, we can take $k = \frac{1}{8}c_{11}^{-1}$ in Lemma 8. □

LEMMA 10. – *There is a universal constant c such that, for any cylinder J of K, we have:*

$$\int_{1/2}^{1}\left(\int_{-\infty}^{+\infty} |\hat{m}_{J,\lambda}(\zeta)|^2\,d\zeta\right)d\lambda \leq c\mathcal{I}(J).$$

Proof. – This is a trivial consequence of a deep inequality of Kaufman; see the proof of Marstrand's theorem in [F]. □

6. TWO COUNTING LEMMAS

LEMMA 11. – *Let $E = E_1 \cup \ldots \cup E_N$ be a partition of a finite set E. Then at least half of the elements of E belong to a E_j such that $\#E_j \geq \frac{1}{2N}\#E$.*

Proof. – Immediate. □

Let $\beta_1 < (c_1' + c_1'')^{-1}\beta$, and π_1 (resp. π) a β_1 (resp. β) decomposition of K. We consider a subset $\tilde{\pi}_1$ of π_1 such that:

$$(14) \qquad \#\tilde{\pi}_1 \geq (1-\rho_1)\#\pi_1,$$

with some $\rho_1 \in [0; 1/2]$. We then define, for $J \in \pi$:

$$\pi_1(J) = \{J_1 \in \pi_1, J_1 \subset J\}$$
$$\tilde{\pi}_1(J) = \pi_1(J) \cap \tilde{\pi}_1,$$

and

$$\pi_2 = \left\{ J \in \pi \mid \#\tilde{\pi}_1(J) \geq \frac{1}{2}(1-\rho_1)\#\pi_1(J) \right\}$$

LEMMA. – *We have*

$$\#\pi_2 \geq (1-\rho)\#\pi, \textit{ with } \rho = 4c_{14}^2\rho_1.$$

Proof. – By definition of π_2, one has:

$$(1-\rho_1)\#\pi_1 \leq \#\tilde{\pi}_1 \leq \sum_{J\in\tilde{\pi}} \#\pi_1(J) + \frac{1}{2}(1-\rho_1) \sum_{J\in\pi-\pi_2}\#\pi_1(J)$$

which implies:

$$\frac{1}{2}(1-\rho_1) \sum_{J\in\pi-\pi_2} \#\pi_1(J) \leq \rho_1 \sum_{J\in\pi_2} \#\pi_1(J).$$

Using Lemma 5, we get (as $\rho_1 \leq 1/2$):

$$\#(\pi - \pi_2) \leq 4c_{14}^2\rho_1\#\pi_2. \qquad \square$$

7. STARTING THE CONSTRUCTION

The construction we are going to make depends on a certain number of parameters, which are:

– an integer $N \geq 1$;

– a positive number $\varepsilon \in (0, 1/2]$

– a decreasing sequence $1 = \beta_o > \beta_1 > \cdots > \beta_{2N} > 0$

These parameters will be specified later.

We define $\pi(0) = \{K\}$ and choose, for $1 \leq i \leq 2N$, a β_i-decomposition $\pi(i) = \pi'(i) \times \pi''(i)$ of K.

We assume that

$$\beta_i > (c_1' + c_1'')\beta_{i+1} \tag{15}$$

so that each element of $\pi(i+1)$ is contained in one element of $\pi(i)$. From Lemmas 7 and 10, we have:

$$\int_{1-\varepsilon}^{1}\left(\int_{-\infty}^{+\infty}|\hat{m}_{J,\lambda}(\zeta)|^2\,d\zeta\right)d\lambda \le cc_{17}\,\beta_i^{2D-1} \quad \text{if } J\in\pi(i). \tag{16}$$

From this we deduce:

(17)

$$\int_{1-\varepsilon}^{1}\left[\sum_{i=0}^{2N}2^{-i}\beta_i^{1-2D}[\#\pi(i)]^{-1}\sum_{J\in\pi(i)}\int_{-\infty}^{+\infty}|\hat{m}_{J,\lambda}(\zeta)|^2\,d\zeta\right]d\lambda \le 2cc_{17}.$$

We can therefore find $\lambda\in[1-\varepsilon,1]$, *fixed in the following* such that we have, for $0\le i\le 2N$:

$$[\#\pi(i)]^{-1}\sum_{J\in\pi(i)}\int_{-\infty}^{+\infty}|\hat{m}_{J,\lambda}(\zeta)|^2\,d\zeta \le 2^{i+1}\varepsilon^{-1}cc_{17}\beta_i^{2D-1}. \tag{18}$$

For $1\le i\le 2N$, let $\pi_1(i)$ be the subset of $\pi(i)$ formed by the J which satisfy:

$$\int_{-\infty}^{+\infty}|\hat{m}_{J,\lambda}(\zeta)|^2\,d\zeta \le c_{20}\left(16c_{14}^2\right)^{i-1}\varepsilon^{-1}\,\beta_i^{2D-1} \tag{19}$$

with $c_{20}=16cc_{17}$.

From (18), (19), we get:

$$\#\pi_1(i)\ge\left(1-\frac{1}{4}\left(8c_{14}^2\right)^{1-i}\right)\#\pi(i). \tag{20}$$

We now define inductively subsets $\tilde{\pi}(i)$ of $\pi(i)$. Define $\rho_i=\frac{1}{2}(8c_{14}^2)^{1-i}$, for $1\le i\le 2N$. Let:

$$\tilde{\pi}(2N)=\pi_1(2N).$$

For $i<2N$, if $\tilde{\pi}(i+1)$ has already been defined, let

$$\pi_2(i)=\left\{J\in\pi(i)\mid \#\tilde{\pi}(i+1,J)\ge\frac{1}{2}(1-\rho_{i+1})\#\pi(i+1,J)\right\} \tag{21}$$

where the notations $\pi(i+1,J)$, $\tilde{\pi}(i+1,J)$ is as before Lemma 12: the sets of elements of $\pi(i+1)$, $\tilde{\pi}(i+1)$ which are contained in J. Then let

$$\tilde{\pi}(i)=\pi_1(i)\cap\pi_2(i). \tag{22}$$

We show inductively that

$$\#\tilde{\pi}(i) \geq (1-\rho_i)\#\pi(i). \tag{23}$$

This is true for $i = 2N$ by (20). Assuming it true for $i+1$, one can apply Lemma 12 to get:

$$\#\pi_2(i) \geq \left(1 - \frac{1}{2}\rho_i\right)\#\pi(i),$$

which, taking (20), (22) into account, gives (23). Observe that $\rho_1 = \frac{1}{2}$, and $\rho_i \in [0, 1/2]$ for $1 \leq i \leq 2N$. Finally, we define:

$$\tilde{\pi}(0) = \pi(0) = \{K\}.$$

8. THE MAIN ESTIMATES

8.1. For $1 \leq i \leq 2N$, $k \in \mathbf{Z}$, let:

$$M(i,k) = [(2k-2)\beta_i, (2k+2)\beta_i].$$

For $J \in \pi(i)$, the diameter of $\pi_\lambda(J)$ is less than $2\beta_i$, so we have $\pi_\lambda(J) \subset M(i,k)$ for at least one (and at most 3) $k \in \mathbf{Z}$.

We choose a *partition* of $\pi(i)$ in subsets $\pi(i,k)$ such that $J \in \pi(i,k)$ implies $\pi_\lambda(J) \subset M(i,k)$. We use the notation:

$$\pi(i+1, J, k) = \pi(i+1, k) \cap \pi(i+1, J)$$
$$\tilde{\pi}(i+1, J, k) = \pi(i+1, k) \cap \tilde{\pi}(i+1, J), \quad J \in \pi(i).$$

8.2. Let $0 \leq i \leq 2N-1$, $J \in \tilde{\pi}(i)$.

By (21), (22) and Lemma 5, one has:

$$\#\tilde{\pi}(i+1, J) \geq \frac{1}{4}\#\pi(i+1, J) \geq \frac{1}{4}c_{14}^{-1}\left(\beta_i\beta_{i+1}^{-1}\right)^D. \tag{24}$$

On the other hand, the number of $k \in \mathbf{Z}$ such that $M(i+1,k)$ intersects $\pi_\lambda(J)$ (whose diameter is $\leq 2\beta_i$) is at most $3 + \beta_i\beta_{i+1}^{-1}$, therefore assuming

$$\beta_i > 3\beta_{i+1} \tag{25}$$

we get

$$\#\{k \in \mathbf{Z}, \tilde{\pi}(i+1, J, k) \neq \phi\} \leq 2\beta_i\beta_{i+1}^{-1}. \tag{26}$$

Let $A(i+1, J)$ be the set of $k \in \mathbf{Z}$ such that:

(27) $$\#\tilde{\pi}(i+1, J, k) \geq (16c_{14})^{-1}\left(\beta_i\beta_{i+1}^{-1}\right)^{D-1}.$$

From Lemma 11, we have:

(28) $$\sum_{k\in A(i+1,J)} \#\tilde{\pi}(i+1, J, k) \geq \frac{1}{2}\#\tilde{\pi}(i+1, J) \geq \frac{1}{8}\#\pi(i+1, J).$$

If we define:

$$J_o = \bigcup_{\substack{k\in A(i+1,J)\\ J_1\in\tilde{\pi}(i+1,J,k)}} J_1,$$

we can apply Lemma 9, using (19) and (28) to conclude that:

(29) $$m(\pi_\lambda(J_o)) \geq c_{21}^{-1}c_{22}^{1-i}\,\varepsilon\beta_i, \text{ with } c_{21} = c_{19}c_{20}, c_{22} = 16c_{14}^2.$$

As $\pi_\lambda(J_o)$ is included in the union of $M(i+1, k)$ for $k \in A(i+1, J)$, one deduces from (29) that:

(30) $$\#A(i+1, J) \geq (4c_{21})^{-1}\,c_{22}^{1-i}\,\varepsilon\beta_i\beta_{i+1}^{-1}.$$

8.3. Keeping the same notations as above, we assume $i \leq 2N-2$. For $\ell \in \mathbf{Z}$, $J_1 \in \tilde{\pi}(i+1, J)$, we define:

$$\begin{aligned}\chi(\ell, J_1) &= 1 \quad \text{if} \quad \tilde{\pi}(i+2, J_1, \ell) \neq \phi\\ &= 0 \quad \text{if} \quad \tilde{\pi}(i+2, J_1, \ell) = \phi.\end{aligned}$$

From (30), we have

(31) $$\sum_{\ell}\chi(\ell, J_1) \geq (4c_{21})^{-1}\,c_{22}^{1-i}\,\varepsilon\beta_{i+1}\beta_{i+2}^{-1}.$$

Let $k \in A(i+1, J)$, and let $\chi(\ell, k) = \sum_{J_1\in\tilde{\pi}(i+1,J,k)} \chi(\ell, J_1)$. If $\chi(\ell, k) \neq 0$, $M(i+2, \ell)$ must intersect $M(i+1, k)$, so there are at most $2\beta_{i+1}\beta_{i+2}^{-1}$ such ℓ. On the other hand, by (31):

(32) $$\sum_{\ell}\chi(\ell, k) \geq (4c_{21})^{-1}\,c_{22}^{-i}\,\varepsilon\beta_{i+1}\beta_{i+2}^{-1}\#\tilde{\pi}(i+1, J, k).$$

Let $B(i+2, J, k)$ be the set of $\ell \in \mathbf{Z}$ such that:

(33) $$\chi(\ell, k) \geq (16c_{21})^{-1}c_{22}^{1-i}\,\varepsilon\,\#\tilde{\pi}(i+1, J, k).$$

Then we have by (32) and the remark preceding it:

(34) $$\sum_{\ell\in B(i+2,J,k)} \chi(\ell, k) \geq (8c_{21})^{-1}\,c_{22}^{1-i}\,\varepsilon\,\beta_{i+1}\beta_{i+2}^{-1}\#\tilde{\pi}(i+1, J, k).$$

As trivially $\chi(\ell, k) \leq \#\tilde{\pi}(i+1, J, k)$, we conclude that:

(35) $$\#B(i+2, J, k) \geq (8c_{21})^{-1}\,c_{22}^{-1}\,\varepsilon\,\beta_{i+1}\beta_{i+2}^{-1}.$$

9. PERTURBATIONS

We choose once and for all a C^∞ function η on $\mathbf{R}$ which satisfies:

$$\eta(x) = 0, \text{ for } x \leq -2, \ \eta(x) = 1 \text{ for } x \geq -1, \ 0 \leq \eta(x) \leq 1, \ \forall x.$$

Let Q' be a box of $\pi'(i)$, $1 \leq i \leq 2N$; by (6) we have:

$$d(Q', K' - Q') > c_{23}^{-1} \beta_i, \text{ with } c_{23} = c'_6 c'_1. \tag{36}$$

Define $\eta_{Q'}$ by (if $Q' = [a, b]$):

$$\eta_{Q'}(x) = \eta(6c_{23}\beta_i^{-1}(x - a)) + \eta(6c_{23}\beta_i^{-1}(b - x)) - 1. \tag{37}$$

The following properties hold:

$$\eta_{Q'}(x) = 1 \quad \text{if} \quad d(x, Q') \leq (6c_{23})^{-1} \beta_i; \tag{38}$$

$$\eta_{Q'}(x) = 0 \quad \text{if} \quad d(x, Q') \geq (3c_{23})^{-1} \beta_i \tag{39}$$

$$0 \leq \eta_{Q'}(x) \leq 1, \quad \forall x \tag{40}$$

$$||D^k \eta_{Q'}||_{C^0} \leq (6c_{23}\beta_i^{-1})^k ||D^k \eta||_{C^0}, \text{ for } k \geq 0. \tag{41}$$

For $t \in [-1/2, +1/2]$, define a map $h_{Q',t}$ by:

$$h_{Q',t}(x) = x + 16\beta_{i+1} t \, \eta_{Q'}(x). \tag{42}$$

We require that

$$48c_{23}||D\eta||_{C^0} \beta_{i+1} < \beta_i \tag{43}$$

so that, by (41), $h_{Q',t}$ is a C^∞ diffeomorphism.

Let $T = (t(Q'))_{Q' \in \pi'(i)}$ be a family with $t(Q') \in [-1/2, 1/2]$ for all Q'. Then $h_{Q'_1, t(Q'_1)} -$ Id and $h_{Q'_2, t(Q'_2)} -$ Id, for distinct $Q'_1, Q'_2 \in \pi'(i)$, have disjoint support by (39), (36); hence the various $h_{Q', t(Q')}$ glue in a diffeomorphism h_T which satisfy, by (41):

$$||D^k(h_T - \mathrm{id}_{\mathbf{R}})||_{C^0} \leq 8\beta_{i+1}(6c_{23}\beta_i^{-1})^k \, ||D^k \eta||_{C^0}. \tag{44}$$

We denote by $C(i)$ the cube $\left[-\frac{1}{2}, \frac{1}{2}\right]^{\pi'(i)}$, equipped the Lebesgue measure m.

If h is a diffeomorphism of the form $h = h_{T_i} \circ h_{T_{i+1}} \circ \ldots \circ h_{T_{2N-1}}$, with $T_j \in C(j)$ for $i \leq j < 2N$, we get from (40), (42) that:

$$|h(x) - x| \leq 8 \sum_{j>i} \beta_j \leq 16\beta_{i+1} \quad \text{by (25).} \tag{45}$$

10. INDUCTION

10.1. We begin with some easy consequences of the estimates in 8. For $0 \le i \le 2N-2$, $J \in \tilde{\pi}(i)$, let:

$$B(i+2,J) = \bigcup_{k \in A(i+1,J)} B(i+2,J,k).$$

If $\ell \in B(i+2,J,k)$, $M(i+2,\ell)$ must intersect $M(i+1,k)$, so ℓ can belong to at most 3 of the $B(i+2,J,k)$.

Then using (30) and (35), we obtain:

$$\#B(i+2,J) \ge c_{24}^{-1} c_{25}^{-i}\, \varepsilon^2 \beta_i \beta_{i+2}^{-1}, \tag{46}$$

with $c_{24} = 96 c_{21}^2 c_{22}^{-2}$, $c_{25} = c_{22}^2$.

Let $\ell \in B(i+2,J)$, and $N(\ell,J)$ be the number of $J_1 \in \tilde{\pi}(i+1,J)$ such that $\tilde{\pi}(i+2,J_1,\ell) \neq \phi$. By (33) and (27), one has:

$$N(\ell,J) \ge c_{26}^{-1} c_{22}^{-i}\, \varepsilon \left(\beta_i \beta_{i+1}^{-1}\right)^{D-1}, \quad c_{26} = 256 c_{14} c_{21} c_{22}^{-1}. \tag{47}$$

10.2. Let $J_1, J_2 \in \tilde{\pi}(i+1,J)$, such that $\tilde{\pi}(i+2,J_p,\ell) \neq \phi$ for $p = 1,2$, and let a_1, a_2 be the smallest points of $\pi_\lambda(J_1)$, $\pi_\lambda(J_2)$. If we assume that J_1, J_2 have the same component in $\pi'(i+1)$, one must have, like in (36), $|a_1 - a_2| \ge \frac{1}{2}\tilde{c}_{23}^{-1}\,\beta_{i+1}$, where $\tilde{c}_{23} = c_6'' c_9''$. On the other hand, $\pi_\lambda(J_p)$ intersects $M(i+2,\ell)$ and has diameter less than $2\beta_{i+1}$ for $p = 1,2,\ldots$, hence $|a_1 - a_2| \le 2\beta_{i+1} + 4\beta_{i+2} \le 4\beta_{i+1}$. We conclude that at most $(8\tilde{c}_{23}+1)$ elements $\tilde{J}$ of $\tilde{\pi}(i+1,J)$ with the *same* component in $\pi'(i+1)$ can verify $\tilde{\pi}(i+2,\tilde{J},\ell) \neq \phi$.

Using (47), we see that for $\ell \in B(i+2,J)$, we can find w_i elements $J_1 \in \tilde{\pi}(i+1,J)$ with *distinct* components in $\pi'(i+1)$ such that $\tilde{\pi}(i+2,J_1,\ell) \neq \phi$, where:

$$w_i = c_{27}^{-1} c_{22}^{-i}\, \varepsilon \left(\beta_i \beta_{i+1}^{-1}\right)^{D-1}, \quad c_{27} = c_{26}(8\tilde{c}_{23}+1). \tag{48}$$

10.3. We now construct a diffeomorphism of the form:

$$h = h_{T_1} \circ h_{T_3} \circ \ldots \circ h_{T_{2N-1}}, \quad \text{with} \quad T_{2j+1} \in C(2j+1).$$

Denote $h_j = h_{T_{2j+1}} \circ \ldots \circ h_{T_{2N-1}}$ so that $h_N = \mathrm{id}$, $h_j = h_{T_{2j+1}} \circ h_{j+1}$, $h = h_o$. We will determine the T_{2j+1} inductively, starting from $T_{2N-1} = 0$, in order to have the following property:

> $P(i)$: For any $J \in \tilde{\pi}(2i)$, and any $\ell \in B(2i+2,J)$, any point in $M(2i+2,\ell)$ is at a distance less than $4\beta_{2N}$ from $\pi_\lambda(h_i(J') \times J'')$, where $J = J' \times J''$.

We take $T_{2N-1} = 0$, so that $h_{N-1} = \text{id}$. Each interval $M(2N, \ell)$ has diameter $4\beta_{2N}$, so $P(N-1)$ is a consequence of the definition of $B(2N, J)$.

We now assume that $T_{2i+1}, \dots, T_{2N-3}$ have been determined in order to have $P(i)$, and we will choose T_{2i-1} in order to satisfy $P(i-1)$.

Let $J \in \tilde{\pi}(2i-2)$, $\ell \in B(2i, J)$, and $x \in M(2i, \ell)$; by 10.2 one can find $J_1, \dots, J_{w_{2i-2}} \in \tilde{\pi}(2i-1, J)$, having disjoint projections $J'_1, \dots, J'_{w_{2i-2}}$ on K', such that:

$$\tilde{\pi}(2i, J_k, \ell) \neq \phi, \qquad 1 \leq k \leq w_{2i-2}.$$

Let $1 \leq k \leq w_{2i-2}$, and $\widetilde{J}_k \in \tilde{\pi}(2i, J_k, \ell)$. Let

$$M_k = \bigcup_{n \in B(2i+2, \widetilde{J}_k)} \widetilde{M}(2i+2, n),$$

where $\widetilde{M}(2i+2, n)$ is the interval with same centre and half length than $M(2i+2, n)$. The $\widetilde{M}(2i+2, n)$ have disjoint interiors so, by (46):

$$(49) \qquad m(M_k) \geq 2c_{24}^{-1} c_{25}^{-2i} \varepsilon^2 \beta_{2i}.$$

As each $\widetilde{M}(2i+2, n)$ must intersect $M(2i, \ell)$, M_k is contained in the interval of center x, radius $5\beta_{2i}$. On the other hand, let

$$E(J, x, k) = \left\{ t \in \left[-\frac{1}{2}, \frac{1}{2} \right], x - 16\beta_{2i}\lambda t \notin M_k \right\}.$$

We assume $\varepsilon < \frac{1}{4}$, hence $\lambda \in \left[\frac{3}{4}, 1\right]$, and we get, from (49):

$$(50) \qquad m(E(J, x, k)) \leq 1 - (8c_{24})^{-1} c_{25}^{-2i} \varepsilon^2.$$

Consider now the subset $E(J, x)$ of $C(2i-1)$ formed by the T_{2i-1} such that, for $1 \leq k \leq w_{2i-2}$, the coordinate t_k of T_{2i-1} corresponding to J'_k belongs to $E(J, x, k)$. As these coordinates are distinct, we obtain:

$$(51) \qquad m(E(J, x)) \leq \left(1 - (8c_{24})^{-1} c_{25}^{-2i} \varepsilon^2\right)^{w_{2i-2}}.$$

We now take the union E of the sets $E(J, x)$, where J runs over $\tilde{\pi}(2i-2)$, and x is any point of the form $p\beta_{2i+2}$, $p \in \mathbf{Z}$ in $M(2i, \ell)$ for some $\ell \in B(2i, J)$, some $J \in \tilde{\pi}(2i-2)$. The number of these sets is at most $2\beta_{2i+2}^{-1} c_{14} \beta_{2i-2}^{-D}$ (Lemma 5, and (8)). We assume:

$$(52) \qquad 2\beta_{2i+2}^{-1} c_{14} \beta_{2i-2}^{-D} \left(1 - (8c_{24})^{-1} c_{25}^{-2i} \varepsilon^2\right)^{w_{2i-2}} < 1,$$

which implies $m(E) < 1$, and we choose $T_{2i-1} \in C(2i-1) - E$.

10.4. We now check that $P(i-1)$ is satisfied by this choice of T_{2i-1}. Let $J, \ell, x \ldots$ be as above. Let $x_o = p\beta_{2i+2}$, $p \in \mathbf{Z}$, be such that $x_o \in M(2i, \ell)$ and $|x - x_o| < \beta_{2i+2}$. As $T_{2i-1} \notin E(J, x_o)$, there exists $1 \leq k \leq w_{2i-2}$ such that $x_o - 16\beta_{2i}\lambda t_k \in M_k$ (where t_k is the coordinate of T_{2i-1} corresponding to J'_k). This means that $x_o - 16\beta_{2i}\lambda t_k \in \widetilde{M}(2i+2, n)$ for some $n \in B(2i+2, \widetilde{J}_k)$; as $|x - x_o| < \beta_{2i+2}$, $x - 16\beta_{2i}\lambda t_k \in \widetilde{M}(2i+2, n)$. By the induction hypothesis $(P(i))$, there exists $z = (z', z'') \in \widetilde{J}_k$ such that:

$$|x - 16\beta_{2i}\lambda t_k - \pi_\lambda(h_i(z'), z'')| \leq 4\beta_{2N},$$

which can be rewritten as:

$$|x - \pi_\lambda(h_i(z') + 16\beta_{2i}t_k, z'')| \leq 4\beta_{2N}.$$

Finally, from (38), (42), (45) we have $h_i(z') + 16\beta_{2i}t_k = h_{T_{2i-1}} \circ h_i(z')$ if we assume that

$$96 c_{23}\,\beta_{2i+1} < \beta_{2i}. \tag{53}$$

This proves $P(i-1)$.

11. CONCLUSION

11.1. We have made on the β_i's the assumptions (15), (25), (43), (53) and (52) the first four reduce to a single assumption:

$$\beta_i > c_{28}\beta_{i+1} \quad \text{for} \quad 0 \leq i < 2N. \tag{54}$$

On the other hand, as $e^x < (1-x)^{-1}$ for $0 < x < 1$, and using formula (48) for w_{2i-2}, (52) is implied by:

$$\beta_{2i+2}^{-1} < (2c_{14})^{-1}\,\beta_{2i-2}^{D} \exp\left[c_{29}^{-1} c_{30}^{-i}\, \varepsilon^3 \left(\beta_{2i-2}\beta_{2i-1}^{-1}\right)^{D-1}\right] \tag{55}$$

with $c_{29} = 8c_{24}c_{27}c_{22}^{-2}$, $c_{30} = c_{25}^2 c_{22}^2$, for $1 \leq i \leq N-1$.

11.2. We now choose the *infinite* sequence $(\beta_i)_{i\geq 0}$ as:

$$\beta_i = \varepsilon^{n_i}, \qquad n_i = a^{2^i} - a,$$

for some $a > 1$. It is then easy to check that $\beta_o = 1$, and (54), (55) are satisfied if we have $a > c_{31}$ (where c_{31} depends only on c_{28}, c_{29}, c_{30}, c_{14} and D, but *not* on ε).

We now fix a closed neighbourhood $\mathcal{U}$ of the identity in $\mathrm{Diff}^\infty(\mathbf{R})$. It is not difficult (but fastidious) to check that one can choose ε and $a > c_{31}$ such that the following hold:

If we fix $N \geq 1$ and construct from the finite sequence $\beta_o = 1, \ldots, \beta_{2N}$ a number $\lambda \in [1 - \varepsilon, 1]$ as in 7. and a diffeomorphism h as in 10., then, for the diffeomorphism $H_N = \lambda h$, we have:

(a) $H_N \in \mathcal{U}$, for all $N \geq 1$,

(b) the sequence $H_N -$ Id is bounded in the C^r topology, for all $r \geq 1$.

Then we can extract from H_N a subsequence which converges in the C^∞-topology to a C^∞ diffeomorphism H in $\mathcal{U}$. By $P(0)$ for H_{N_k}, each diffeomorphism H_{N_k} in this subsequence has at least one interval I_k of length $4\beta_2$ such that any point in I_k is at distance less than $4\beta_{2N_k}$ from $H_{N_k}(K') + K''$. There exists an interval I length $2\beta_2$ which is contained in infinitely many I_k. Passing to the limit, I must be contained in $H(K') + K''$.

This proves the theorem. $\square$

REFERENCES

[F] K. J. FALCONER, The geometry of fractal sets, *Cambridge Univ. Press*, 1985.

[PT1] J. PALIS and F. TAKENS, Cycles and measure of bifurcation sets for two-dimensional diffeomorphisms, *Inventiones Math.*, Vol. **82**, 1985, pp. 397-422.

[P] J. PALIS, Homoclinic orbits, hyperbolic dynamics and dimension of Cantor sets, *Contemporary Math.*, Vol. **58**, 1987, pp. 203-215.

[PT2] J. PALIS and F. TAKENS, Hyperbolicity and sensitive-chaotic dynamics at homoclinic bifurcations, *Cambridge Univ. Press*, 1993, 2nd edition 1994.

[PY] J. PALIS and J-C. YOCCOZ, Homoclinic tangencies for hyperbolic sets of large Hausdorff dimension, *Acta Math.*, Vol. **172**, 1994, pp. 91-136.

Astérisque
261, 2000, p. 335–347

A GLOBAL VIEW OF DYNAMICS AND A CONJECTURE ON THE DENSENESS OF FINITUDE OF ATTRACTORS

by

Jacob Palis

To Adrien Douady for his lasting contribution to mathematics (July 1995)

Abstract. — A view on dissipative dynamics, i.e. flows, diffeomorphisms, and transformations in general of a compact boundaryless manifold or the interval is presented here, including several recent results, open problems and conjectures. It culminates with a conjecture on the denseness of systems having only finitely many attractors, the attractors being sensitive to initial conditions (chaotic) or just periodic sinks and the union of their basins of attraction having total probability. Moreover, the attractors should be stochastically stable in their basins of attraction. This formulation, dating from early 1995, sets the scenario for the understanding of most nearby systems in parametrized form. It can be considered as a probabilistic version of the once considered possible existence of an open and dense subset of systems with dynamically stable structures, a dream of the sixties that evaporated by the end of that decade. The collapse of such a previous conjecture excluded the case of one dimensional dynamics: it is true at least for real quadratic maps of the interval as shown independently by Swiatek, with the help of Graczyk [GS], and Lyubich [Ly1] a few years ago. Recently, Kozlovski [Ko] announced the same result for C^3 unimodal mappings, in a meeting at IMPA. Actually, for one-dimensional real or complex dynamics, our main conjecture goes even further: for most values of parameters, the corresponding dynamical system displays finitely many attractors which are periodic sinks or carry an absolutely continuous invariant probability measure. Remarkably, Lyubich [Ly2] has just proved this for the family of real quadratic maps of the interval, with the help of Martens and Nowicki [MN].

1. Introduction and Main Conjecture

In the sixties two main theories in dynamics were developed, one of which was designed for conservative systems and called KAM for Kolmogorov-Arnold-Moser. A later important development in this area, in the eighties, was the Aubry-Mather theory

1991 ***Mathematics Subject Classification.*** — 58F10, 58F15.
Key words and phrases. — Attractor, physical measure, homoclinic orbit, stability.

This work has been partially supported by Pronex-Dynamical Systems, Brazil.

Originally published in: *Astérisque no. 261 (2000), 335–347,*
© Astérisque 261, Société Mathématique de France, Paris 2000
Reprint by Springer-Verlag Berlin Heidelberg 2014 with kind permission

for periodic (and Cantori) motions, which has been more recently further improved by Mather. Other outstanding results have been obtained even more recently by Eliasson, Herman, Mañé and others.

The focus of this paper, however, will be the surprising unfolding of the other theory that has been constructed for general systems (nonconservative, dissipative) and called hyperbolic: it deals with systems with hyperbolic limit sets. This means for a diffeomorphism f on a manifold M, that the tangent bundle to M at L, the limit set of f, splits up into df-invariant continuous subbundles $T_L M = E^s \oplus E^u$ such that $df \mid E^s$, $df^{-1} \mid E^u$ are contractions, with respect to some Riemmanian metric. As for most of the concepts in the sequel, a similar definition holds for a non-invertible map g, requiring $dg \mid E^u$ to be invertible. And for a flow X_t, $t \in \mathbb{R}$, we add to the splitting a subbundle in the direction of the vector field that generates the flow

$$T_L M = E^s \oplus E^u \oplus E^0$$

and require for some Riemannian metric and some constants C, $0 < \lambda < 1$, that

$$||dX_t \mid E^s||, \quad ||dX_{-t} \mid E^u|| \leq C\, e^{\lambda t}, \quad t \in \mathbb{R}.$$

See [**PT**], specially chapter seven, for details and many of the notions presented here.

The concept of hyperbolicity was introduced by Smale, with important contributions to its development as a theory being also given by some of his students at the time, as well as Anosov, Arnold, Sinai and others. Initially, it was created to help pursue the "lost dream" referred to in the abstract: to find an open and dense subset of dynamically (structurally) stable systems; i.e., systems that when slightly perturbed in the C^r-topology, $r \geq 1$, remain with the same dynamics, modulo homeomorphisms of the ambiente space that preserve orbit structures, in the case of flows, or are conjugacies, in the case of transformations. It has actually transcended this objective, loosing through a series of counter-examples its projected character of much universality, i.e. its validity for an open dense subset of systems. But it became the ground basis for a notable evolution that dynamics experienced in the last twenty five years or so. Still, based on previous results, specially by Anosov and by ourselves, Smale and I were able to formulate in 1967 what was called the Stability Conjecture, that would fundamentally tie together hyperbolicity and dynamical stability: a system is C^r-stable if its limit set is hyperbolic and, moreover, stable and unstable manifolds meet transversally at all points. For stability restricted to the limit set, the transversality condition is substituted by the nonexistence of cycles among the transitive (dense orbit), hyperbolic subsets of the limit set.

The theory of hyperbolic systems, i.e. systems with hyperbolic limit sets, was quite developed especially for flows and diffeomorphisms, and it was perhaps even near completion (an exageration!), by the end of the sixties and beginning of the seventies. That included some partial classifications, and an increasing knowledge of their ergodic properties, due to Sinai, Bowen, Ruelle, Anosov, Katok, Pesin, Franks, Williams, Shub, Manning, among several others.

More or less at the same time, the proof of one side of the Stability Conjecture was completed through the work of Robbin, Robinson and de Melo. However, the outstanding part of this basic question from the 60's was proved to be true only in the middle 80's, in a remarkable work of Mañé [**M2**] for diffeomorphisms, followed ten years later by an again remarkable paper of Hayashi [**Ha**] for flows: C^1 dynamically stable systems must be hyperbolic. Before, by 1980, Mañé had proved the two-dimensional version of the result, but independent and simultaneous proofs were also provided by Liao and Sannami. Other partial contributions should be credited to Pliss, Doering, Hu and Wen. A high point in Hayashi's work is his connecting lemma creating homoclinic orbits by C^1 small perturbations of a flow or diffeomorphism: an unstable manifold accumulating on some stable one can be C^1 perturbed to make it intersect one another (the creation of homoclinic or heteroclinic orbits). This fact has been at this very moment sparkling some advance of dynamics in the lines proposed here, as it will be pointed out later.

While the ergodic theory of dynamical systems, as suggested by Kolmogorov and more concretely by Sinai, was being successfully developed, the hope of proposing some global structure for dynamics in general grew dimmer and dimmer in the seventies. This was due to new intricate dynamical phenomena that were presented or suggested all along the decade. First, Newhouse [**N**] extended considerably his previous results, showing that infinitely many sinks occur for a residual subset of an open set of C^2 surface diffeomorphisms near one exhibiting a homoclinic tangency. Perhaps equally or even more striking at the time, was the appearance of attractors having sensitivity with respect to initial conditions in their basin.

Although there are several possible definitions of an attractor, here we will just require it to be invariant, transitive (dense orbit) and attracting all nearby future orbits or at least a Lebesgue positive measure set in the ambient manifold. If A is an attractor for f with basin $B(A)$, we say that it is sensitive to initial conditions, or chaotic if there is $\varepsilon > 0$ such that with total probability on $B(A) \times B(A)$, for each pair of points (x, y) there is an integer $n > 0$ so that $f^n(x)$ and $f^n(y)$ are more than ε apart, where the distances are considered with respect to some Riemannian metric. The definition for flows is entirely similar. Chaotic attractors became also known as strange. The first one, beyond the hyperbolic attractors which are not just sinks, is due to Lorenz [**L**]. Proposed numerically by Lorenz in 1963, it's a rather striking fact that only in the middle seventies most of us became acquainted with Lorenz-like attractors through the examples of Guckenheimer and Williams, which we now call geometric ones. It is still an open and interesting question if the original Lorenz's equations

$$\dot{x} = -10x + 10y$$
$$\dot{y} = 28x - y - xz$$
$$\dot{z} = -(8/3)z + xy$$

in fact correspond to a flow displaying a strange attractor.

Subsequently, again based on numerical experiments, Hénon [**He**] asked about the possible existence of a strange attractor, but now in two dimensions, for certain quadratic diffeomorphisms of the plane

$$f_{a,b}(x,y) = (1 - ax^2 + y, bx)$$

for $a \approx 1.4$ and $b \approx 0.3$. Finally, by the end of the decade, Feigenbaum [**F**] and independently Coullet-Tresser [**CT**] suggested another kind of attractor, now for quadratic maps of the interval and related to a limiting map of a sequence of transformations exhibiting period-doubling bifurcations of periodic orbits. Almost immediately after that, Jacobson [**J**] exhibited strange attractors in the same setting. All this, together with the unsuccessful atempts of the sixties, led to a common belief that perhaps no such a global scenario for dynamics was possible.

However, a series of important results on strange attractors for maps, concerning their persistence, i.e. their existence for a positive Lebesgue measure set in parameter space, and the fact that they carry physical or SRB (Sinai-Ruelle-Bowen) invariant measures, provided by Jacobson [**J**] for the interval, Benedicks-Carleson [**BC**], Mora-Viana [**MV**], Benedicks-Young [**BY1**], [**BY2**] and Diaz-Rocha-Viana [**DRV**] for Hénon-like maps (small C^r perturbations of Hénon maps, $r \geq 1$), were about to take place in the next fifteeen years or so. Perhaps even more striking is the recent proof that they are stochastically stable, a recent remarkable result of Benedicks-Viana [**BV1**], [**BV2**]. In proving this fact, Benedicks and Viana first showed for Hénon-like attractors, that there are "no holes" in the basin of attraction with respect to the SRB measure, a question I heard from Ruelle and Sinai more than a decade ago: a.e. in the attractor with respect to the SRB measure, there are stable manifolds and their union covers Lebesgue a.a. points in the basin of attraction (and, thus, the union of the stable manifolds is dense in the basin of attraction) [**BV1**]. The concepts of SRB measure and stochastic stability will be presented below. Almost simultaneously, after previous pioneering work of Arnold and Herman (see [**Ar**], [**H**]), the theory of one-dimensional dynamics experienced a great advance, due to Yoccoz, Sullivan, McMullen, Lyubich, Douady, Hubbard, Swiatek and an impressive number of other mathematicians (see [**dMS**] and [**dM**]).

Such developments, as well as my own work with Takens and Yoccoz, [**PT1**], [**PT2**], [**PY1**], and many inspiring conversations with colleagues, former and present students, among them de Melo, Pujals, Takens, Yoccoz and above all Viana, made me progressively acquire a new global view of dynamics, emphasizing a much more probabilistic approach to the question. I was then able to formulate, just prior to the meeting at the Université Paris-Sud, Orsay in honour of A. Douady, in 1995, the following conjecture:

Global conjecture on the finitude of attractors and their metric stability

(I) *Denseness of finitude of attractors – there is a C^r $(r \geq 1)$ dense set D of dynamics such that any element of D has finitely many attractors whose union of basins of attraction has total probability;*

(II) *Existence of physical (SRB) measures – the attractors of the elements in D support a physical measure;*

(III) *Metric stability of basins of attraction – for any element in D and any of its attractors, for almost all small C^r perturbations in generic k-parameter families of dynamics, $k \in \mathbb{N}$, there are finitely many attractors whose union of basins is nearly (Lebesgue) equal to the basin of the initial attractor; such perturbed attractors support a physical measure;*

(IV) *Stochastic stability of attractors – the attractors of elements in D are stochastically stable in their basins of attraction;*

(V) *For generic families of one-dimensional dynamics, with total probability in parameter space, the attractors are either periodic sinks or carry an absolutely continuous invariant measure.*

As mentioned in the abstract, Lyubich [**Ly2**] solved the last item of the conjecture for families of quadratic maps of the interval, setting the stage for its full solution for one-dimensional real dynamics.

We close this section by briefly recalling the notions of physical or SRB measures and sthocastic stability. Let A be an attractor for a C^r, $r \geq 1$, map and let $B(A)$ be its basin of attraction. We call an invariant probability measure μ with support in A, a physical or SRB measure if

$$\lim_{n\to\infty} \frac{1}{n} \sum_{i=0}^{n-1} g(f^i(x)) = \int g\, d\mu$$

for every continuous function g and for a positive Lebesgue measure set of x in $B(A)$. Notice that if the attractor is just a periodic sink, then we take the Dirac measure equally distributed at its orbit as SRB measure. To define stochastic stability, we first assume that μ is an SRB measure as above, with the property holding for a.e. $x \in B(A)$ (like is the case of Lorenz and Hénon-like attractors as commented above). Let now $\{f_n\}$ be independent, identically distributed random variables in the space of C^r maps, $r \geq 1$, and $f_n \in B_\varepsilon(f)$ with (a convenient) probability distribution θ_ε, where $B_\varepsilon(f)$ denotes the ε-neighbourhood of f. For a point $z_0 \in B(A)$, let $z_j = f_j \circ \cdots \circ f_1(z_0)$, $j \geq 1$. We say that (f, A, μ) is stochastically stable on the basin of attraction $B(A)$ if for every neighbourhood V of μ in the weak* topology, the weak* limit of $\frac{1}{n}\sum_{j=0}^{n-1}\delta_{z_j}$ is in V, for every small enough $\varepsilon > 0$ and $(m \times \theta_\varepsilon^{\mathbb{N}})$-almost all $(z_0, f_1, f_2, \dots)$. Here, δ_{z_j} stands for the Dirac measure at z_j and m for Lebesgue measure. We observe that when we consider parametrized families of maps with finitely many parameters, the distribution θ_ε is to be interpreted as random (Lebesgue) choice of parameters. And this is precisely the meaning we want to give to stochastic stability in the conjecture above.

There is another, formally not equivalent, definition of stochastic stability: instead of different maps $f_1, \dots, f_j$ above, one considers only the initial one f but lets the images of an initial point have small random fluctuations. That is, trajectories $(x_0, x_1, x_2, \dots)$ of the perturbed system are obtained by letting each x_{i+1} be chosen at random in a small neighbourhood of $f(x_i)$. However, in all known stable cases, like the hyperbolic systems treated by Kifer and Young (see [**K**]), as well as the Hénon-like diffeomorphisms, dealt with by Benedicks-Viana, both definitions can be applied.

In the case of flows, the definition is basically the same, the random fluctuations taking place at "infinitesimal" intervals of time dt. More formally, one considers a stochastic differential equation like (see [**K**])

$$dx = X_0(x)\, dt + dY(x),$$

where X_0 is the initial (unperturbed) vector field and dY is a stochastic differential, for instance $dY = \varepsilon dW$ where dW corresponds to the standard Brownian motion. Then, for ε small, the weak* limit of $\frac{1}{T}\int_0^T \delta_{\xi_t}$ should be close to the SRB measure of the vector field X_0 for almost all trajectories of the stochastic flow ξ_t associated to this equation.

At this point, the following remark is in order concerning the definition of attractor. Essentially in all known cases, the basin of attraction is a neighborhood of the set. Still we don't know that such a property is common to a dense subset of all dynamics. For possible relevant situations where the basin contains a set of positive probability but not a full neighborhood of the attractor, random perturbations might cause orbits to escape from the basin of the attractor. In fact this is the case for the random perturbations of flows through Brownian motion, even when the basin is a neighborhood. A related possible situation corresponds to having more than one SRB measure with supports contained in a same (transitive) attractor, so that random orbits cross several of the basins of these measures. In these cases, stochastic stability should mean that the weak* limit of $\frac{1}{n}\sum_{j=0}^{n-1}\delta_{z_j}$ is close to the convex hull of those SRB measures if ε is small (see above and [**V2**]).

2. Other conjectures and some recent results

In his famous essay on the stability of the solar system, written around 1890, Poincaré introduced the notion of homoclinic orbits: in the past and future they converge to the same periodic orbit or are in the intersection of the stable and unstable manifolds (sets) of such a periodic orbit. Subsequently, not only he expressed amazement with the inherent dynamical complexity, but stressed its importance:

"Rien n'est plus propre à nous donner une idée de la complication du problème de trois corps et en général de tous les problèmes de dynamique..."

Indeed, on one hand we have that a transversal homoclinic orbit for a diffeomorphism implies the presence of the dynamics of the horseshoe, as proved by Smale in the early sixties (see [**PT**]). On the other, a generic unfolding of a homoclinic tangency of a dissipative (determinant of the Jacobian smaller than one) surface diffeomorphism yields a rather striking number of rich, intricate dynamical phenomena:

(a) residual subsets of intervals in the parameter line whose corresponding diffeomorphisms display infinitely many coexisting sinks
(b) cascade of period-doubling bifurcations of periodic points (sinks)
(c) positive Lebesgue measure sets in the parameter line whose corresponding diffeomorphism display Hénon-like attractors.

These facts were proved in the late seventies and the eighties by Newhouse, Yorke-Alligood and Mora-Viana extending the fundamental work of Benedicks-Carleson (see [**PT**]). They are also valid in higher dimensions, when the dimension of the unstable manifold of the periodic point is one (codimension-one case) and the product of any two eigenvalues of the derivative of the map at this point has norm less than one (sectionally dissipative). See [**PV**] for the existence of infinitely many coexisting sinks in such a case. It is to be noticed that there are also plenty of values of the parameter such that the corresponding diffeomorphisms have infinitely many coexisting Hénon-like attractors [**C**].

The results above show how Poincaré once and again had a great mathematical insight. In accordance with his view, I have proposed some time ago the following conjectures:

Conjecture I. — *C^r near any surface diffeomorphism exhibiting one of the above bifurcating phenomena (a), (b), (c), there exists a diffeomorphism displaying a homoclinic tangency, $r \geq 1$. Thus, C^r near any of such bifurcating phenomena we may find all the others. The same in higher dimensions, the homoclinic tangency being now of codimension-one and sectionally dissipative.*

Conjecture II. — *In any dimension, the diffeomorphisms exhibiting either a homoclinic tangency or a (finite) cycle of hyperbolic periodic orbits with different stable dimensions (heterodimensional cycles) are C^r dense in the complement of the closure of the hyperbolic ones, $r \geq 1$.*

Notice that for surface diffeomorphisms, it is not possible to have an heterodimensional cycle. So, we have conjectured that diffeomorphisms exhibiting Hénon-like attractors or repellers (or infinitely many sinks or sources) are dense in the complement of the hyperbolic ones. The same question may be posed for non-invertible maps. Also for flows, but in this case one has to add, to homoclinic tangencies and heterodimensional cycles, flows exhibiting singular attractors, like the Lorenz-like ones, and singular cycles, i.e. cycles involving periodic orbits and singularities, studied in [**LP**]. This conjecture for flows will be mentioned again at the end of the paper, when discussing new Lorenz-like attractors. In general, we have presented in [**PT**], page 134, a notion of a homoclinic bifurcating dynamical system and formulate a less explicit but similar conjecture in terms of a such a notion.

Progress has been made on conjecture I, by Ures [**U**], who has showed that for all known cases of diffeomorphisms exhibiting Hénon-like attractors, they may be C^r approximated by one with homoclinic tangencies. Similarly for the limiting map of cascades of period-doubling bifurcations in many interesting cases [**CE1, CE2**]. On the other hand, Pujals and Sambarino seem to be making good progress in showing in any dimension, that a diffeomorphism with infinitely many sinks can be C^1 approximated by one displaying a homoclinic tangency, which is codimension one and sectionally dissipative.

More concretely, Pujals and Sambarino [**PS**] have provided a positive solution to conjecture II for surface diffeomorphisms and $r = 1$. From the comments above, this implies that conjecture I would also be true in the same setting. Remarkably, they also obtain from their method a similar result for surface diffeomorphisms whose topological entropy changes under small C^1 perturbations.

Related to conjecture II, we wish to pose the following weaker version of it:

Conjecture III. — *The subset of dynamical systems that either have their limit set consisting of finitely many hyperbolic periodic orbits or else they have transversal homoclinic orbits, are C^r dense in the set of all dynamical systems, $r \geq 1$.*

Again, Pujals and Sambarino, made some progress towards a positive answer to this question in dimensions higher than two and $r = 1$.

Somewhat related to our main conjecture, stated in the previous section, and initially motivated by [**TY**], we formulate yet another conjecture.

Conjecture IV. — *When unfolding a homoclinic tangency for a codimension-one sectionally dissipative diffeomorphism, the set of parameter values corresponding to diffeomorphisms with infinitely many sinks or infinitely many Hénon-like attractors has (Lebesgue) measure zero. The same for generic k-parameter families of dynamical systems, $k \in \mathbb{N}$.*

In all the previous assertions concerning homoclinic tangencies for surface diffeomorphisms, the corresponding periodic point may be part of a larger hyperbolic (basic) set. The problem is then translated to the understanding of the arithmetic difference of Cantor sets in the line, obtained from the intersections of the stable and unstable foliations of the hyperbolic set with a line transverse to the leaves at the homoclinic tangency. Such Cantor sets are regular (bounded geometry) if the diffeomorphism is of class C^2 [**PT**]. One can also consider a topology for such Cantor sets. Trying to go beyond the notion of thickness of Cantor sets used by Newhouse (see [**PT**]), I asked if for a residual subset of pairs C_1, C_2 of regular Cantor sets, either $C_1 - C_2$ has measure zero or else it contains an interval. The first case would correspond to $HD(C_1 \times C_2) < 1$ and the second to $HD(C_1 \times C_2) > 1$, where HD stands for Hausdorff dimension. Recently, Moreira and Yoccoz [**Mor**], [**MorY1**], [**MorY2**] have proved this fact, even more strongly, for an open and dense subset of pairs of Cantor sets, extending partial previous results in [**PT1**], [**PT2**] and [**PY1**]. The question for affine Cantor sets is still open:

Conjecture V. — *If C_1, C_2 are affine Cantor sets and $HD(C_1 \times C_2) > 1$, then $C_1 - C_2$ contains an interval. In particular, the question can be posed for most or an open and dense subset of pairs of affine Cantor sets.*

I believe that the result of Moreira-Yoccoz is a major contribution to a more profound understanding of the unfolding of homoclinic bifurcations, which we consider to be central to dynamics. In the previous works, it has been shown that if $HD(C_1 \times C_2) < 1$ then the set of parameter values corresponding to hyperbolicity has full (Lebesgue) density at the value corresponding to the homoclinic tangency [**PT1**], [**PT2**]. Conversely, this is not so if $HD(C_1 \times C_2) > 1$ as proved in [**PY1**]. The results of Moreira-Yoccoz enriches the picture: the set of parameter values corresponding to hyperbolicity or generalized homoclinic tangencies (tangencies between stable and unstable leaves of the foliations) has full density at that same parameter value. In another development, it is shown in [**PY2**] that at the homoclinic bifurcation parameter value, attractors have Lebesgue density equal to zero. This is another indication that attractors perhaps are not so abundant, as Newhouse's result may at first glance suggest.

Concerning attractors in particular carrying an SRB measure, a focusing point of the main conjecture in this article, there has been an explosion of new relevant results, adding to the important development on strange attractors mentioned before. This raises the hope of more progress in the questions posed here as well as in some partial classification of attractors with the properties mentioned in that conjecture.

First of all, Alves [**A**] has shown the existence of SRB measures for Viana's attractors with multidimensional positive Lyapunov exponents [**V1**]. Together with Bonatti, they are announcing a series of results aiming at the following one: a positive Lyapunov exponent on (Lebesgue) many points suffices for the existence of SRB measures [**ABV**].

These results have to do with an important class of diffeomorphisms called partially hyperbolic. This means that there is a continuous invariant decomposition

$$TM = E^1 \oplus E^2$$

where E^1 is uniformly expanding or uniformly contracting and it dominates E^2 [**PT**, page 161]. Shub [**S**], Mañé [**M1**] and Bonatti-Diaz [**BD**] have constructed partially hyperbolic but not hyperbolic diffeomorphisms which are robustly transitive: every C^1 small perturbation is also transitive. Very recently, Diaz-Pujals-Ures [**DPU**] have shown that partial hyperbolicity is in fact a necessary condition for C^1 robust transitiveness in three dimensions. Some time earlier, Pugh and Shub had shown that for volume preserving diffeomorphisms, partial hyperbolicity is a main ingredient for robust ergodicity; see [**PSh**] and references therein.

Concerning flows, Rovella [**R**] has constructed a few years ago an important variation of Lorenz geometric attractor, which is not robust but only persists for perturbations of the flow corresponding to a positive Lebesgue measure set of parameter values. The global spiralling attractors constructed in [**PRV1**], [**PRV2**], whose reduced one-dimensional map exhibits infinitely many critical points, and the critical geometric Lorenz attractors in [**LV1**], [**LV2**] also display measure theoretical persistence. In this setting, de Melo and Martens [**dMM**] have built up families of Lorenz-like maps that are full in the sense that exhibit all combinatorial types. An oustanding question that remained open since the seventies is whether there exist robust transitive attractors for flows containing a singularity with more than one expanding eigenvalue. This has been positively answered very recently by Bonatti-Pumariño-Viana [**BPV**].

The theory of singular or Lorenz-like attractors, specially in dimension three, attained sharp progress in recent years through a series of papers by Morales, Pacifico and Pujals. They have constructed new relevant types of singular attractors in [**MPa**], [**MPu1**], [**MPu2**], [**MPP1**], [**MPP2**] and [**MPP3**]. In some cases, these attractors are across the boundary of hyperbolic flows. They also proved the following partial characterization [**MPP4**]: robustly transitive singular sets, i.e. invariant sets containing singularities, of three dimensional flows are necessarily attractors or repellers and they are singularly hyperbolic, i.e. there exists a continuous invariant decomposition $TM = E^s \oplus E^{cu}$ such that E^s is uniformly contracting, E^{cu} is volume expanding and E^s dominates E^{cu}. Moreover, the derivatives of the flow at the singularities contained in the attractor have eigenvalues with the same relative distribution as in the geometric Lorenz attractors: the unstable (positive) eigenvalue has norm bigger than that of the weak stable (least negative) one. Here again, as well as in [**DPU**] above, Hayashi's connecting lemma is used. Such developments, I believe, should be very inspiring in tackling Conjecture II for flows, at least in three dimensions.

Let me finish by briefly remarking that the scenario we have proposed for most dynamical systems in finite dimension, may apply to some infinite dimensional systems whose solutions are often only defined for positive time (semiflow), like dissipative evolution equations (Euler, Navier-Stokes). Putting the view presented here together with several conjectures/results in this setting, we would have the following conjecture: For most such systems in parametrized form (finitely many parameters) and most initial conditions, the solutions are global and converge to finitely many finite-dimensional attractors with nice ergodic properties, as above.

As a sign of much activity in the area, many of the references below are to appear. They are, however, available at the Internet, for instance at the authors web page at IMPA.

References

[ABV] J.F. Alves, C. Bonatti and M. Viana, *SRB measures for partially hyperbolic systems: the expanding case*, to appear in Invent. Math.

[A] J.F. Alves, *SRB measures for nonhyperbolic systems with multidimensional expansion*, thesis IMPA, 1997, to appear in Ann. Sci. ENS.

[Ar] V.I. Arnold, *On the mapping of the circle into itself*, Izvestia Akad. Nauk., USSR Math. Series bf 25 (1961), 21–86.

[BC] M. Benedicks and L. Carleson, *The dynamics of the Hénon maps*, Ann. of Math. 133 (1991), 73–169.

[BV1] M. Benedicks and M. Viana, *Solution of the basin problem for Hénon-like attractors*, preprint 1999.

[BV2] M. Benedicks and M. Viana, *Stochastic stability and other statistical properties for Hénon-like maps*, in preparation.

[BY1] M. Benedicks and L.-S. Young, *Absolutely continuous invariant measures and random perturbations for certain one-dimensional maps*, Ergod. Th. & Dynam. Sys. 12 (1992), 13–37.

[BY2] M. Benedicks and L.-S. Young, *SBR-measures for certain Hénon maps*, Invent. Math. 112 (1993), 541–576.

[BD] C. Bonatti and L.J. Díaz, *Nonhyperbolic transitive diffeomorphisms*, Ann. of Math. 143 (1996), 357–396.

[BPV] C. Bonatti, A. Pumariño, and M. Viana, *Lorenz attractors with arbitrary expanding dimension*, C.R. Acad. Sci. Paris 325 (1997), 883-888.

[C] E. Colli, *Infinitely many coexisting strange attractors*, to appear, Annales de l'Inst. Henri Poincaré, Analyse Non linéaire 15 (1998), 539-579.

[CE1] E. Catsigeras and H. Enrich, *Homoclinic tangencies near cascades of period doubling bifurcations*, to appear, Ann. de l'Inst. Henri Poincaré Analyse Nonlinéaire.

[CE2] E. Catsigeras and H. Enrich, *Persistence of the Feigenbaum attractor in one-parameter families*, preprint 1997.

[CT] P. Coullet and C. Tresser, *Iterations d'endomorphismes et group de renormalisation*, CRAS Paris 287 (1978), 577–580.

[dMM] W. de Melo and M. Martens, *Universal models for Lorenz maps*, preprint 1997, to appear, Ergodic Th. and Din. Syst.

[dM] W. de Melo, *Rigidity and renormalization in one-dimensional dynamics*, Procs. International Congress of Mathematicians, ICM98-Berlin.

[dMS] W. de Melo and S. van Strien, *One-dimensional dynamics*, Springer-Verlag, 1993.

[DRV] L.J. Díaz, J. Rocha, and M. Viana, *Strange attractors in saddle-node cycles: prevalence and globality*, Invent. Math. 125 (1996), 37– 74.

[DPU] L.J. Díaz, E. Pujals, and R. Ures, *Normal hyperbolicity and robust transitivity*, preprint 1997, to appear in Acta Math.

[F] M. Feigenbaum, *Qualitative universality for a class of nonlinear transformations*, J. Stat. Phys. 19 (1978), 25–52.

[GS] J. Graczyk and G. Swiatek, *Generic hyperbolicity in the logistic family*, Annals of Math. 146 (1997), 1–52.

[Ha] S. Hayashi, *Connecting invariant manifolds and the solution of the C^1 stability and Ω-stability conjectures for flows*, Annals of Math. 145 (1997), 81–137.

[He] M. Hénon, *A two dimensional mapping with a strange attractor*, Comm. Math. Phys. 50 (1976), 69–77.

[H] M. Herman. *Sur la conjugaison différentiable des difféomorphims du cercle à des rotations*, Publ. Math. 49 (1979), 5–233.

[J] M. Jacobson, *Absolutely continuous invariant measures for one-parameter families of one-dimensional maps*, Comm. Math. Phys. 81 (1981), 39–88.

[K] Y. Kifer, *Random perturbations of dynamical systems*, Birkhäuser, 1988.

[Ko] O. Kozlovski, *Structural stability in one-dimensional dynamics*, PhD thesis, Univ. Amsterdam, 1997.

[L] E.N. Lorenz, *Deterministic nonperiodic flow*, J. Atmosph. Sci. 20 (1963), 130–141.

[LP] R. Labarca and M.J. Pacifico, *Stability of singular horseshoes*, Topology 25 (1986), 337–352.

[LV1] S. Luzzatto and M. Viana, *Positive Lyapunov exponents for Lorenz-like maps with criticalities*, these proceedings.

[LV2] S. Luzzatto and M. Viana, *Lorenz-like attractors without invariant foliations*, in preparation.

[Ly1] M. Lyubich, *Dynamics of quadratic polynomials I-II*, Acta Math., 178 (1997), 185–297.

[Ly2] M. Lyubich, *Almost every real quadratic map is either regular or stochastic*, preprint 1997.

[M1] R. Mañé, *Contribution to the stability conjecture*, Topology 17 (1978), 383–396.

[M2] R. Mañé, *A proof of the C^1 stability conjecture*, Publ. Math. I.H.E.S. 66 (1988), 161–210.

[MN] M. Martens and T. Nowicki, *Invariant measures for Lebesgue typical quadratic maps*, these proceedings.

[Mor] C. Moreira, *Stable intersections of Cantor sets and homoclinic bifurcations*, Annales de l'Institut Henri Poincaré, Analyse Non Linéaire, vol. 13, nº 6, (1996), 741–781.

[MorY1] C. Moreira and J.-C. Yoccoz, *Stable intersections of regular Cantor sets with large Hausdorff dimensions*, preprint 1998.

[MorY2] C. Moreira and J.-C. Yoccoz, *Stable homoclinic tangencies for hyperbolic sets of large Hausdorff dimension*, in preparation.

[MV] L. Mora and M. Viana, *Abundance of strange attractors*, Acta Math. 171 (1993), 1–71.

[MPa] C. Morales and M.J. Pacifico, *New singular strange attractors arising from hyperbolic systems*, preprint 1997.

[MPu1] C. Morales and E. Pujals, *Singular strange attractors on the boundary of Morse-Smale systems*, Ann. Sci. E.N.S. 30 (1997), 693–718.

[MPu2] C. Morales and E. Pujals, *Strange attractors containing a singularity with two positive multipliers*, Comm. Math. Phys. 196 (1998), 671-679.

[MPP1] C. Morales, M.J. Pacifico, and E. Pujals, *New strange attractors on the boundary of hyperbolic systems*, preprint 1997.

[MPP2] C. Morales, M.J. Pacifico, and E. Pujals, *Global attractors from the explosion of singular cycles*, C.R. Acad. Sci. Paris 325 (1997), 1317-1322.

[MPP3] C. Morales, M.J. Pacifico, and E. Pujals, *Singular hyperbolic sets*, Proc. Amer. Math. Soc. 127 (1999), 3343-3400.

[MPP4] C. Morales, M.J. Pacifico, and E. Pujals, *On C^1 robust singular transitive sets for three-dimensional flows*, C.R. Acad. Sci. Paris 326 (1998), 81-86.

[N] S. Newhouse, *The abundance of wild hyperbolic sets and nonsmooth stable sets for diffeomorphisms*, Publ. Math. I.H.E.S. 50 (1979), 101–151.

[PRV1] M.J. Pacifico, A. Rovella, and M. Viana, *Infinite-modal maps with global chaotic behaviour*, Annals of Math. 148 (1998), 1-44.

[PRV2] M.J. Pacifico, A. Rovella, and M. Viana, *Persistence of global spiraling attractors*, in preparation.

[PS] E. Pujals and M. Sambarino, *Homoclinic tangencies and hyperbolicity for surface diffeomorphisms: A conjecture of Palis*, preprint 1998, to appear in Annals of Math.

[PSh] C. Pugh and M. Shub, *Stable ergodicity and julienne quasi-conformality*, preprint 1997.

[PT] J. Palis and F. Takens, *Hyperbolicity and sensitive-chaotic dynamics at homoclinic bifurcations*, Cambridge University Press, 1993.

[PT1] J. Palis and F. Takens, *Cycles and measure of bifurcation sets for two-dimensional diffeomorphisms*, Inventiones Math. 82 (1985), 397–422.

[PT2] J. Palis and F. Takens, *Hyperbolicity and the creation of homoclinic orbits*, Annals of Math. 125 (1987), 337–374.

[PV] J. Palis and M. Viana, *High dimension diffeomorphisms displaying infinitely many periodic attractors*, Annals of Math. 140 (1994), 207–250.

[PY1] J. Palis and J.-C. Yoccoz, *Homoclinic tangencies for hyperbolic sets of large Hausdorff dimension*, Acta Mathematica, vol. 172 (1994), 91–136.

[PY2] J. Palis and J.-C. Yoccoz, *Non-uniform hyperbolic horseshoes arising from homoclinic bifurcations*, to appear in C.R. Acad. Sci. Paris.

[R] A. Rovella, *The dynamics of perturbations of the contracting Lorenz attractor*, Bull. Braz. Math. Soc. vol. 24, nº 2 (1993), 233–259.

[S] M. Shub, *Topological transitive diffeomorphisms on T^4*, Lecture Notes in Math. 206 (1971), 39, Springer-Verlag.

[TY] L. Tedeschini-Lalli and J.A. Yorke, *How often do simple dynamical processes have infinitely many coexisting sinks?* Comm. Math. Phys. 106 (1986), 635–657.

[U] R. Ures, *On the approximation of Hénon-like attractors by homoclinic tangencies*, Ergodic Theory and Dynamical Systems, vol. 15 (1995), 1223–1230.

[V1] M. Viana, *Multidimensional nonhyperbolic attractors*, Publ. Math. IHES (1998).

[V2] M. Viana, *Dynamics: a probabilistic and geometric perspective*, Procs. International Congress of Mathematicians, ICM98-Berlin.

J. Palis, Instituto de Matemática Pura e Aplicada - IMPA, Est. D. Castorina 110, 22460-320 Rio de Janeiro, Brazil • *E-mail* : jpalis@impa.br

C. R. Acad. Sci. Paris, t. 333, Série I, p. 867–871, 2001
Systèmes dynamiques/*Dynamical Systems*

Fers à cheval non uniformément hyperboliques engendrés par une bifurcation homocline et densité nulle des attracteurs

Jacob PALIS [a], Jean-Christophe YOCCOZ [b]

[a] IMPA, Estrada D. Castorina 110, 22460-320 Rio de Janeiro, Brésil
[b] Collège de France, 3, rue d'Ulm, 75005 Paris, France
Courriel : jpalis@impa.br; jean-c.yoccoz@college-de-france.fr

(Reçu le 11 juillet 2001, accepté le 3 août 2001)

Résumé. Soit f_0 un difféomorphisme d'une surface tel que l'ensemble maximal invariant dans un ouvert V soit constitué d'un fer à cheval et d'une orbite de tangence quadratique entre feuilletages stable et instable de ce fer à cheval. Supposons que la dimension du fer à cheval soit supérieure à mais assez voisine de 1. Nous annonçons que, pour la plupart des difféomorphismes f proches de f_0, l'ensemble maximal invariant par f dans V est un fer à cheval non uniformément hyperbolique, dont la dynamique s'apparente à celle des attracteurs de Hénon. © 2001 Académie des sciences/Éditions scientifiques et médicales Elsevier SAS

Non-uniformly hyperbolic horseshoes unleashed by homoclinic bifurcations and density zero of attractors

Abstract. *Let f_0 be a surface diffeomorphism such that the maximal invariant set in an open set V is the union of a horseshoe and a quadratic tangency between the stable and unstable foliations of this horseshoe. We assume that the dimension of the horseshoe is larger than but close to one. We announce that, for most diffeomorphisms f close to f_0, the maximal f-invariant set in V is a non-uniformly hyperbolic horseshoe, with dynamics of the same type as met in Hénon attractors.* © 2001 Académie des sciences/Éditions scientifiques et médicales Elsevier SAS

1. Soit f_0 un difféomorphisme de classe C^∞ d'une surface M^2 possédant un ensemble basique K_0, non trivial et de type selle. On rappelle qu'un ensemble basique est une partie compacte invariante hyperbolique localement maximale dans laquelle f_0 a une orbite dense ; il est non trivial s'il n'est pas réduit à une orbite périodique, de type selle si ses ensembles stables et instables sont d'intérieur vide.

On suppose qu'il existe des points périodiques p_s, p_u dans K_0 et un point $q \in M^2 - K_0$ tels que les variétés stable $W^s(p_s)$ et instable $W^u(p_u)$ aient en q un contact quadratique.

On impose de plus les conditions suivantes, destinées à garantir que l'ensemble compact invariant Λ_0, union de K_0 et de l'orbite de q, soit localement maximal :

Note présentée par Étienne GHYS.

S0764-4442(01)02139-5/FLA

- pour tout entier N assez grand, q est un point d'intersection isolé de $f_0^{-N}(W^{\mathrm{s}}_{\mathrm{loc}}(K_0))$ et $f_0^N(W^{\mathrm{u}}_{\mathrm{loc}}(K_0))$;
- lorsque $p_{\mathrm{s}}, p_{\mathrm{u}}$ appartiennent à la même orbite périodique, les branches de $W^{\mathrm{s}}(p_{\mathrm{s}}) - \{p_{\mathrm{s}}\}$, $W^{\mathrm{u}}(p_{\mathrm{u}}) - \{p_{\mathrm{u}}\}$ qui contiennent q rencontrent K_0.

2. Il est bien connu que l'ensemble basique K_0, et en particulier ses point périodiques p_{s}, p_{u}, possèdent une continuation hyperbolique dans un voisinage $\mathcal{U}$ du difféomorphisme f_0 dans $\mathrm{Diff}^\infty(M^2)$. La condition de tangence de $W^{\mathrm{s}}(p_{\mathrm{s}})$, $W^{\mathrm{u}}(p_{\mathrm{u}})$ au voisinage de q définit dans $\mathcal{U}$ une hypersurface lisse $\mathcal{U}_0$, séparant $\mathcal{U}$ en deux régions $\mathcal{U}_+$, $\mathcal{U}_-$ ($\mathcal{U}_-$ étant celle où $W^{\mathrm{s}}(p_{\mathrm{s}})$, $W^{\mathrm{u}}(p_{\mathrm{u}})$ ne s'intersectent pas au voisinage de q).

Soit V un voisinage ouvert de Λ_0, assez petit pour que

$$\Lambda_0 = \bigcap_{n\in\mathbb{Z}} f_0^{-n}(V).$$

On souhaite alors, pour $\mathcal{U}$ assez petit et $g \in \mathcal{U}$, étudier l'ensemble invariant

$$\Lambda(g) = \bigcap_{n\in\mathbb{Z}} g^{-n}(V).$$

Il est facile de voir que :
- $\Lambda(g)$ est compact, donc localement maximal ;
- pour $g \in \mathcal{U}_-$, $\Lambda(g)$ est la continuation hyperbolique $K(g)$ de K_0 ;
- pour $g \in \mathcal{U}_0$, $\Lambda(g)$ est l'union de $K(g)$ et de l'orbite de tangence prescrite par l'appartenance à $\mathcal{U}_0$.

C'est donc dans $\mathcal{U}_+$ que la situation est intéressante.

3. Il est commode de considérer des familles lisses à un paramètre $(f_t)_{t\in(-t_0,t_0)}$, transverses en $t=0$ à $\mathcal{U}_0$ (et telles que $f_t \in \mathcal{U}_-$ pour $t<0$, $f_t \in \mathcal{U}_+$ pour $t>0$) ; on posera alors $\Lambda_t = \Lambda(f_t)$. Rappelons alors un théorème de Takens et un des auteurs.

THÉORÈME ([8]). – *Lorsque la dimension de Hausdorff de K_0 est <1, $\Lambda(f_t)$ est hyperbolique (c'est même un ensemble basique) pour un ensemble de paramètres E de densité totale en* 0 :

$$\lim_{\varepsilon\to 0} \varepsilon^{-1}\,\mathrm{Leb}\big(E \cap [0,\varepsilon]\big) = 1.$$

Inversement, nous avons montré :

THÉORÈME ([9]). – *Lorsque la dimension de Hausdorff de K_0 est >1, pour presque toute famille (f_t), l'ensemble E des valeurs du paramètre pour lesquelles $\Lambda(f_t)$ est hyperbolique vérifie*

$$\lim_{\varepsilon\to 0} \inf \varepsilon^{-1}\mathrm{Leb}\big(E \cap [0,\varepsilon]\big) < 1.$$

On consultera [9] pour un énoncé plus précis, et [6] pour un énoncé plus fort.

On notera d_{s} (resp. d_{u}) la dimension de Hausdorff transverse de $W^{\mathrm{s}}_{\mathrm{loc}}(K_0)$ (resp. $W^{\mathrm{u}}_{\mathrm{loc}}(K_0)$).

La dimension de Hausdorff de K_0 est alors $d_{\mathrm{s}} + d_{\mathrm{u}}$.

Notre résultat est le suivant.

THÉORÈME. – *Il existe un voisinage D de $\{d_{\mathrm{s}} + d_{\mathrm{u}} \leqslant 1\}$ dans $(0,1)^2$, indépendant de f_0, tel que, si $(d_{\mathrm{s}}, d_{\mathrm{u}}) \in D$, alors Λ_t est un fer à cheval non uniformément hyperbolique pour un ensemble E de paramètres de densité totale en* 0.

Nous expliquons ci-après ce que nous entendons par fer à cheval non uniformément hyperbolique. Parmi les nombreux travaux traitant de dynamique non uniformément hyperbolique, mentionnons [4,1,5,3,2]. On pourra aussi consulter [7] où un scénario global est proposé.

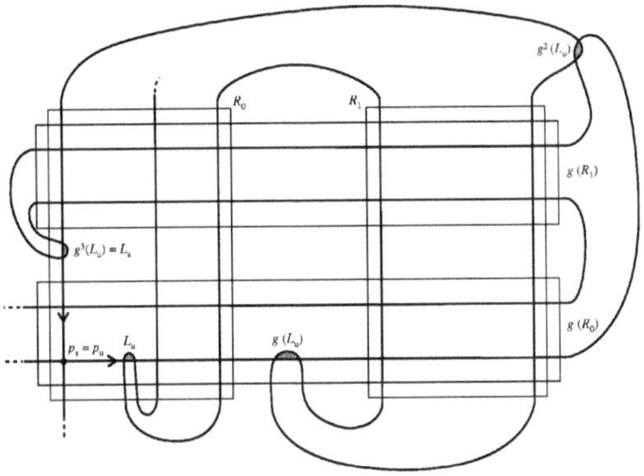

Figure 1.

Notre démonstration fournit pour D la forme explicite :

$$D = \{(d_s + d_u)^2 + [\max(d_s, d_u)]^2 \leqslant d_s + d_u + \max(d_s, d_u)\},$$

mais il ressort aussi de la démonstration que ceci n'est certainement pas optimal. En fait, il ne parait pas exclu de généraliser nos arguments de façon à éliminer toute restriction sur d_s, d_u, c'est-à-dire sur f_0.

4. On choisit, et fixe dans la suite, un système fini $(R_a)_{a \in \mathcal{A}}$ de rectangles disjoints induisant une partition de Markov de la continuation hyperbolique K_t de K_0. On pose $R = \bigcup R_a$.

Pour $t > 0$, les courbes $W^s(\mathcal{O}(p_s))$ et $W^u(\mathcal{O}(p_u))$ délimitent une région lenticulaire L_u telle que Λ_t est l'ensemble maximal invariant dans $R \cup \bigcup_{0<i<N_0} f_t^i(L_u)$. L'entier N_0 ne dépend que de f_0 ; on peut s'arranger pour que L_u et $f_t^{N_0}(L_u) := L_s$ soient contenus dans R, mais que $f_t^i(L_u)$ ne rencontre pas R pour $0 < i < N_0$.

La dynamique est ainsi constituée

– des applications de transition « hyperboliques »

$$R_a \cap f_t^{-1}(R_{a'}) \longrightarrow f_t(R_a) \cap R_{a'}$$

associées à la partition de Markov ;

– du « pli » $f_t^{N_0} \colon L_u \to L_s$.

Voir figure ci-dessus, où $p_s = p_u$ et $N_0 = 3$.

5. Considérons des triplets (P, Q, n) tels que

– P (resp. Q) est un sous-rectangle essentiellement vertical (resp. horizontal) de l'un des rectangles R_a, d'épaisseur $|P|$ (resp. $|Q|$) approximativement constante ;

– la restriction de f_t^n à P est un difféomorphisme de P sur Q, satisfaisant à une condition de cône uniforme et de petite distortion.

Pour deux tels triplets (P, Q, n), (P', Q', n'), on peut envisager deux types de composition.

La composition simple (ou hyperbolique) est licite dès que P' et Q sont contenus dans le même rectangle R_a ; on pose $P'' = P \cap f_t^{-n}(P')$, $Q'' = Q' \cap f_t^{n'}(Q)$, $n'' = n + n'$.

La composition parabolique est plus délicate. Il est nécessaire que Q traverse L_u, P' traverse L_s et $f_t^{N_0}(Q \cap L_u)$ rencontre P' suivant deux composantes connexes $S^{\pm}$; on exige de plus que la distance

$\mathrm{d}(P', Q)$ entre le sommet de la bande parabolique $f_t^{N_0}(Q \cap L_\mathrm{u})$ et la bande verticale P' vérifie

$$\mathrm{d}(P', Q) \geqslant \max\big(|P'|, |Q|\big)^{1-\eta},$$

où $0 < \eta \ll 1$ ne dépend que de f_0. Sous ces conditions, la composition parabolique est définie et fournit deux triplets :

$$\widehat{P}^{\pm} = f_t^{-(n+N_0)}(S^{\pm}),$$
$$\widehat{Q}^{\pm} = f_t^{n'}(S^{\pm}),$$
$$\widehat{n}^{+} = \widehat{n}^{-} = n + n' + N_0.$$

On notera $\mathcal{R}$ l'ensemble des triplets (P, Q, n) obtenus à partir des applications de transition initiales (avec $n = 1$) par saturation pour la composition simple ou parabolique.

6. Soit $(P, Q, n) \in \mathcal{R}$. Pour $\delta \geqslant |Q|$, on dira que Q est δ-critique s'il existe $(P', Q', n') \in \mathcal{R}$ tel que

$$|P'| \leqslant \delta, \quad \mathrm{d}(P', Q) < \delta^{1-\eta};$$

de même, pour $\delta \geqslant |P|$, on dira que P est δ-critique s'il existe $(P'', Q'', n'') \in \mathcal{R}$ tel que

$$|Q''| \leqslant \delta, \quad \mathrm{d}(P, Q'') < \delta^{1-\eta}.$$

Ces définitions nous permettent d'introduire l'ensemble E des bons paramètres : ce sont ceux tels que, pour tout $(P, Q, n) \in \mathcal{R}$, si l'on pose $\delta = \max(|P|, |Q|)$, alors soit P, soit Q n'est pas δ-critique.

7. Pour un bon paramètre $t \in E$, la dynamique est analysée de la façon suivante.

Appelons primitif un triplet $(P, Q, n) \in \mathcal{R}$ qui ne peut être obtenu à partir de triplets plus simples par composition simple. Appelons courbe stable une intersection décroissante $\omega = \bigcap_{k \geqslant 0} P_k$, pour une suite $(P_k, Q_k, n_k) \in \mathcal{R}$ avec $n_k \to +\infty$. Notons $\mathcal{R}_\infty$ l'ensemble des courbes stables.

Soit (P, Q, n) un triplet primitif; notons $\mathcal{R}_\infty(P)$ l'ensemble des courbes stables $\omega \in \mathcal{R}_\infty$ telles que P est le plus petit rectangle primitif contenant ω ; lorsque $\omega \in \mathcal{R}_\infty(P)$, l'image $f_t^n(\omega)$ est contenue dans une courbe stable $T(\omega)$; de plus, on définit ainsi une bijection T de $\mathcal{R}_\infty(P)$ sur l'ensemble des courbes stables contenues dans le rectangle $\mathcal{R}_a$ contenant Q.

Les ensembles $\mathcal{R}_\infty(P)$ (pour (P, Q, n) primitif) sont disjoints, et leur union $\mathcal{W}$ est formée des courbes stables qui ne sont contenues que dans un nombre fini de rectangles primitifs. Considérons alors

$$T\colon \mathcal{W} \longrightarrow \mathcal{R}_\infty.$$

On procède de façon classique : on associe à T un opérateur de Ruelle–Perron–Frobenius, dépendant d'un paramètre de dimension d ; cet opérateur a une valeur propre dominante, et on ajuste d de sorte que cette valeur propre soit égale à 1. La valeur $d = \widetilde{d}_\mathrm{s}$ obtenue est la dimension de Hausdorff transverse de l'union $\widetilde{\mathcal{R}}_\infty$ des courbes stables $\omega \in \mathcal{R}_\infty$. L'opérateur fournit aussi une mesure de probabilité T-invariante μ_T sur $\mathcal{R}_\infty$, équivalente à la mesure de Hausdorff en dimension $\widetilde{d}_\mathrm{s}$. La dimension de Hausdorff du complémentaire de $\mathcal{W}$ dans $\mathcal{R}_\infty$ est $< \widetilde{d}_\mathrm{s}$.

L'application T se relève en une application $\widetilde{T} : \widetilde{\mathcal{W}} = \bigsqcup_{\mathcal{W}} \omega \to \widetilde{\mathcal{R}}_\infty$; la mesure μ_T se relève en une unique mesure de probabilité $\widetilde{T}$-invariante $\mu_{\widetilde{T}}$. Finalement, on récupère une mesure de probabilité f_t-invariante μ_t à partir de $\mu_{\widetilde{T}}$ car la condition d'intégrabilité requise :

$$\sum_{(P,Q,n) \text{ primitif}} n|P|^{\widetilde{d}_\mathrm{s}} < +\infty$$

est vérifiée. Les exposants de Lyapunov de μ_t sont non nuls.

8. Le numéro précédent donne une description assez satisfaisante de la dynamique pour les points qui rencontrent $\widetilde{\mathcal{R}}_\infty$ sous itération positive. L'assertion suivante signifie dans une certaine mesure que cette propriété est presque sûrement vérifiée (pour les bons paramètres).

Soit $\widetilde{\omega}$ une courbe instable (intersection d'une suite décroissante de rectangles Q_ℓ). Considérons l'ensemble $E_{\mathrm{s}}(\widetilde{\omega})$ des points de $\widetilde{\omega}$ qui restent dans $R \cup \bigcup_{0<i<N_0} f^i(L_{\mathrm{u}})$ sous itération positive, et la partie $\mathcal{E}(\widetilde{\omega}) \subset E_{\mathrm{s}}(\widetilde{\omega})$ formée des points qui ne rentrent jamais dans $\widetilde{\mathcal{R}}_\infty$ sous itération positive. Nous montrons que

$$\mathrm{HD}(E_{\mathrm{s}}(\widetilde{\omega})) = \widetilde{d}_{\mathrm{s}},$$
$$\mathrm{HD}(\mathcal{E}(\widetilde{\omega})) < \widetilde{d}_{\mathrm{s}}.$$

9. Dans la description de l'ensemble E des bons paramètres évoquée au § 6, futur et passé jouent le même rôle. Par conséquent, les propriétés de f_t sous itération positive décrites dans les deux numéros précédents sont également valables pour l'itération négative de f_t.

10. Le point crucial de la démonstration est bien sûr de prouver que l'ensemble E des bons paramètres défini en § 6 est de densité totale en 0, dès que les dimensions d_{s}, d_{u} vérifient $(d_{\mathrm{s}}, d_{\mathrm{u}}) \in D$. Dans cette optique, on définit pour $t \sim 2^{-M}$, une suite d'échelles $(\varepsilon_k)_{k \geqslant 0}$ par

$$\varepsilon_0 = 2^{-M}, \quad \varepsilon_{k+1} = \varepsilon_k^{1+\Delta},$$

où $0 < \Delta \ll 1$

À la $k^{\text{ième}}$ étape, on dispose d'intervalles $I_{k,\ell}$ de longueur ε_k formés de paramètres qui satisfont à une série d'inégalités, entre autres :

- le nombre de triplets $(P, Q, N) \in \mathcal{R}$ tels que $|P| \geqslant \varepsilon_k$ et P est $|P|$-critique est au plus $\varepsilon_k^{1-d_{\mathrm{s}}-d_{\mathrm{u}}-C\eta}$;
- de même pour le nombre de triplets tels que $|Q| \geqslant \varepsilon_k$ et Q est $|Q|$-critique.

On subdivise chaque $I_{k,l}$ en $\varepsilon_k^{-\Delta}$ intervalles de longueur ε_{k+1}, ne conservant à l'étape suivante que ceux qui vérifient les inégalités requises à l'échelle ε_{k+1}. L'hypothèse $(d_{\mathrm{s}}, d_{\mathrm{u}}) \in D$ permet de montrer qu'on en garde suffisamment pour que E soit de densité totale en 0.

Références bibliographiques

[1] Benedicks M., Carleson L., The dynamics of the Hénon map, Ann. of Math. 133 (1991) 73–169.

[2] Benedicks M., Viana M., Solution of the basin problem for Hénon-like attractors, Invent. Math. 143 (2001) 375–434.

[3] Benedicks M., Young L.-S., SRB-measures for certain Hénon maps, Invent. Math. 112 (1993) 541–576.

[4] Jakobson M., Absolutely continuous invariant measures for one-parameter families of one-dimensional maps, Commun. Math. Phys. 81 (1981) 39–88.

[5] Mora L., Viana M., Abundance of strange attractors, Acta Math. 171 (1993) 1–71.

[6] Moreira C.G., Yoccoz J.-C., Stable intersections of regular Cantor sets with large Hausdorff dimensions, Ann. of Math. 154 (2001) 45–96.

[7] Palis J., A global view of dynamics and a conjecture on the denseness of finitude of attractors, Astérisque 261 (1999) 339–351.

[8] Palis J., Takens F., Hyperbolicity and the creation of homoclinic orbits, Ann. of Math. 125 (1987) 337–374.

[9] Palis J., Yoccoz J.-C., Homoclinic tangencies for hyperbolic sets of large Hausdorff dimension, Acta Math. 172 (1994) 91–136.

C. R. Acad. Sci. Paris, t. 333, Série I, p. 475–480, 2001
Systèmes dynamiques/*Dynamical Systems*

Homoclinic tangencies and fractal invariants in arbitrary dimension *

Carlos Gustavo MOREIRA, Jacob PALIS, Marcelo VIANA

IMPA, Estrada D. Castorina 110, 22460-320 Rio de Janeiro, Brazil
E-mail: gugu@impa.br; jpalis@impa.br; viana@impa.br

(Reçu le 28 juin 2001, accepté le 2 juillet 2001)

Abstract. For generic families of diffeomorphisms in arbitrary dimension unfolding a homoclinic tangency associated to a hyperbolic basic set, we prove that uniform hyperbolicity prevails in parameter space if and only if the Hausdorff dimension of the basic set is less than 1. © 2001 Académie des sciences/Éditions scientifiques et médicales Elsevier SAS

Tangences homoclines et invariants fractaux en dimension arbitraire

Résumé. *Pour des familles génériques de difféomorphismes en dimension quelconque, qui développent une tangence homocline associée à un ensemble basique hyperbolique, nous montrons que les dynamiques uniformément hyperboliques sont prévalentes dans l'espace des paramètres si et seulement si la dimension de Hausdorff de l'ensemble basique est plus petite que* 1. © 2001 Académie des sciences/Éditions scientifiques et médicales Elsevier SAS

Version française abrégée

Les résultats sur les tangences homoclines des difféomorphismes des surfaces, obtenus dans le courant des trois dernières décennies par Newhouse, Palis, Takens, Yoccoz et Moreira dévoilèrent une relation profonde entre des invariants fractaux d'ensembles hyperboliques et les comportements dynamiques typiques, dans une famille paramétrée, après la bifurcation homocline. (*Voir* [5,6,9,3] et [7], chapitre 7, pour un panorama de la théorie.)

Voici, en quelques mots, comment se manifeste cette relation : si la dimension de Hausdorff de l'ensemble hyperbolique associé à la tangence est plus petite que 1, les dynamiques hyperboliques sont prédominantes (densité de Lebesgue totale) dans l'espace des paramètres près de la tangence; si la dimension de Hausdorff est plus grande que 1, l'hyperbolicité n'est pas prédominante dans l'espace des paramètres; en revanche, l'union de l'hyperbolicité et des tangences persistantes l'est.

L'extension de ces résultats aux dimensions supérieures présente des difficultés importantes, liées au manque de régularité des feuilletages invariants des ensembles hyperboliques. Par exemple, on ne sait même pas si la dimension de Hausdorff locale est bien définie pour les ensembles hyperboliques dans les variétés de dimension plus grande que 2. Une partie de ces difficultés fut résolue par Palis et Viana [8],

Note présentée par Jean-Christophe YOCCOZ.

S0764-4442(01)02085-7/FLA

à l'aide d'une notion de différentiabilité à la Whitney, dans leur généralisation en dimension arbitraire du théorème de Newhouse [4] sur la coexistence d'une infinité de puits.

Le but de ce travail est de démontrer que, pourtant, le paradigme

l'hyperbolicité est prédominante $\Leftrightarrow$ *la dimension de Hausdorff est plus petite que* 1

peut être étendu aux bifurcations homoclines en dimension arbitraire.

Plus précisément, nous considérons des familles à un paramètre de difféomorphismes $\varphi_\mu : M^n \to M^n$, $\mu \in (-1,1)$, de classe C^2 qui développent une tangence homocline associée à un point périodique p_0 inclus dans un fer à cheval Λ_0 de φ_0. Nous supposons que φ_μ est hyperbolique (axiome A) pour $\mu < 0$, ce qui implique que les valeurs propres les plus faibles, contractante et dilatante, de p_0 sont réelles.

Soient d_s et d_u les dimensions de Hausdorff de $W^s(p_0) \cap \Lambda_0$ et $W^u(p_0) \cap \Lambda_0$, respectivement. Nous montrons que, pour un sous-ensemble résiduel de difféomorphismes φ_0, les dimensions d_s et d_u dépendent continûment de φ_0, et on a :

(i) si $d_s + d_u < 1$, alors $H = \{\mu > 0 \mid \varphi_\mu$ est hyperbolique$\}$ a densité de Lebesgue totale en $\mu = 0$;

(ii) si $d_s + d_u > 1$, alors $T_s = \{\mu > 0 \mid \varphi_\mu$ présente des tangences homoclines persistantes associées à la continuation hyperbolique Λ_μ de $\Lambda_0\}$ est un ensemble ouvert avec densité de Lebesgue positive en $\mu = 0$.

La démonstration est basée sur une extension des techniques introduites par Newhouse–Palis [5], Palis–Takens [6] et Moreira–Yoccoz [2,3] pour montrer les résultats correspondants en dimension 2, et par Palis–Viana [8] pour étudier les bifurcations homoclines de codimension 1 en dimension arbitraire.

Un des lemmes les plus importants dans ce travail est la construction (peut-être après une perturbation de φ_0) de feuilletages stable-fort et instable-fort de codimension 1 pour des sous-ensembles hyperboliques de Λ_0 avec presque les mêmes dimensions d_s et d_u. Ces feuilletages sont utilisés pour réduire l'étude des géométries des feuilletages stable et instable près de la tangence homocline initiale à une situation proche du cas bidimensionel.

Il est intéressant de remarquer que la dimension critique 1 est indépendante de la dimension n de l'espace ambiant. Donc, dans un certain sens, le deuxième cas devient relativement plus fréquent quand n croît.

---

1. Introduction

Over the last three decades, a number of results have been proved by Newhouse, Palis, Takens, Yoccoz and Moreira, unveiling a deep connection between fractal invariants of hyperbolic sets and the typical dynamics observed in the unfolding of a homoclinic tangency by a parametrized family of surface diffeomorphisms. (*See* [5,6,9,3] and [7], chapter 7, for an overview.)

In brief terms, if the Hausdorff dimension of the hyperbolic set involved in the tangency is less than 1, then hyperbolicity prevails in parameter space (full Lebesgue density at the tangency parameter); if the Hausdorff dimension is larger than 1, then hyperbolicity alone is not prevalent, but the union of hyperbolicity and persistent tangencies is.

The extension of these results to higher dimension involves considerable difficulties, related to the lack of smoothness of invariant foliations of hyperbolic sets of codimension larger than 1. For instance, it is not even known whether the local Hausdorff dimension is well defined for hyperbolic sets in high dimension manifolds. Part of these difficulties could be handled by Palis and Viana [8], by means of a notion of differentiability "à la Whitney", in their generalization to arbitrary dimension of Newhouse's theorem [4] on coexistence of infinitely many sinks.

Our goal in this work is to prove that, nevertheless, the paradigm

hyperbolicity prevails $\Leftrightarrow$ *the Hausdoff dimension is smaller than* 1 does extend to homoclinic bifurcations in any dimension.

More precisely, we consider 1-parameter families of diffeomorphisms $\varphi_\mu : M^n \to M^n$, $\mu \in (-1,1)$ of class C^2, unfolding a generic homoclinic tangency associated to a periodic point p_0 contained in a horseshoe Λ_0 of φ_0. We suppose that φ_μ is hyperbolic (Axiom A) for $\mu < 0$, which implies that the weakest contracting and weakest expanding eigenvalues of p_0 are real numbers.

Let d_s and d_u be the Hausdorff dimension of $W^s(p_0) \cap \Lambda_0$ and $W^u(p_0) \cap \Lambda_0$, respectively. We show that, for a residual subset of diffeomorphisms, d_s and d_u are continuous with respect to φ_0, and we have:

(i) if $d_s + d_u < 1$, then $H = \{\mu > 0 \mid \varphi_\mu \text{ is hyperbolic}\}$ has full Lebesgue density at $\mu = 0$;

(ii) if $d_s + d_u > 1$, then $T_s = \{\mu > 0 \mid \varphi_\mu$ presents persistent homoclinic tangencies associated to the hyperbolic continuation Λ_μ of $\Lambda_0\}$ is an open set with positive lower density at $\mu = 0$.

The proof is based on an extension of the techniques introduced by Newhouse–Palis [5], Palis–Takens [6], and Moreira–Yoccoz [2,3] to prove the corresponding results in dimension 2, and by Palis–Viana [8] to investigate homoclinic bifurcations of codimension one in arbitrary dimension.

One of our main lemmas is the construction (perhaps after perturbation of φ_0) of strong-stable and strong-unstable foliations of codimension 1 for hyperbolic subsets of Λ_0 with almost the same dimensions d_s and d_u. These foliations are used to (essentially) reduce the study of the geometries of the stable and unstable foliations near the initial homoclinic tangency to the bidimensional case.

It is interesting to note that the separating dimension 1 is independent of the dimension n of the ambient space. Thus, in some sense, the second case above becomes more frequent when n increases.

2. Geometry of horseshoes: the upper dimension

Let Λ be a horseshoe for a diffeomorphism $\varphi : M \to M$, and P be some Markov partition of Λ. We call *vertical n-cylinder* any subset of Λ defined by prescribing the first n symbols in the backward itinerary with respect to P. Let $\mathcal{V}_n$ be the set of vertical n-cylinders. There is a dual notion of *horizontal n-cylinder*, where one considers forward itinerary.

Fix $\varepsilon > 0$ small. For each $V \in \mathcal{V}_n$ let $D_s(V) = \sup\{\operatorname{diam}(W^s_\varepsilon(x) \cap V) \mid x \in \Lambda \cap V\}$. We define $\widetilde{\lambda}_n$ by the relation

$$\sum_{V \in \mathcal{V}_n} D_s(V)^{\tilde{\lambda}_n} = 1$$

and we let the *upper stable dimension* be given by $\overline{d}_s(\Lambda) = \lim_{n\to\infty} \widetilde{\lambda}_n$. There is a dual notion of *upper unstable dimension* $\overline{d}_u(\Lambda)$, dealing with W^u_ε instead of W^s_ε.

It is not difficult to show that the limit always exists, and $\overline{d}_s$ is an upper semicontinuous function of φ. Moreover,

$$\operatorname{HD}\big(W^s_\varepsilon(x) \cap \Lambda\big) \leqslant \overline{d}_s(\Lambda) \quad \text{for each } x \in \Lambda.$$

The following result shows that the equality holds in most cases where $\overline{d}_s < 1$:

THEOREM 1. – *Suppose $\overline{d}_s(\Lambda) < 1$. Then there is a residual set of diffeomorphisms $\mathcal{R}$ in a neighborhood of φ such that, for all $\psi \in \mathcal{R}$ and all $x \in \Lambda_\psi$,*

$$\operatorname{HD}\big(\Lambda_\psi \cap W^s_\varepsilon(x)\big) = \overline{d}_s(\Lambda_\psi).$$

In addition, $\mathcal{R}$ may be taken so that its elements are points of continuity of $\overline{d}_s$.

The idea of the proof is to consider k-parameter families of diffeomorphisms ψ, for some large k, and to show that for each $V \in \mathcal{V}_n$ we have $\operatorname{diam}(W^s_\varepsilon(x) \cap V) \geqslant D_s(V)^{1-\delta}$ for every $x \in \Lambda_\psi$, with large probability in parameter space. This gives that, for most parameters, most of the vertical cylinders have subexponential variation of their horizontal diameters. Using this fact, we obtain lower estimates for $\mathrm{HD}(W^s_\varepsilon(x) \cap \Lambda_\psi)$ that imply the conclusion of the theorem.

Now we can state our:

MAIN RESULT. – *There are open sets $\mathcal{U}$ and $\mathcal{V}$ of families of diffeomorphisms unfolding a homoclinic tangency as in the Introduction, such that $\mathcal{U} \cup \mathcal{V}$ is dense and*

(i) *for families in $\mathcal{U}$ we have*

$$\overline{d}_s(\Lambda) + \overline{d}_u(\Lambda) < 1 \quad \textit{and} \quad \lim_{\delta \to 0} m(H \cap [0, \delta])/\delta = 1,$$

where $H = \{\mu \mid \varphi_\mu$ is hyperbolic$\}$;

(ii) *for families in $\mathcal{V}$ we have*

$$\overline{d}_s(\Lambda) + \overline{d}_v(\Lambda) > 1 \quad \textit{and} \quad \liminf_{\delta \to 0} m(T_s \cap [0, \delta])/\delta > 0,$$

where $T_s = \{\mu \mid \varphi_\mu$ presents a persistent homoclinic tangency associated to $\Lambda_\mu\}$.

In the next sections we indicate main ideas of the proof of this theorem.

3. Prevalence of hyperbolicity

In this section we outline the proof of the first part of the main theorem: hyperbolicity prevails in parameter space when the sum of the stable and unstable upper dimensions is less than 1. The first main step is a reduction to the case when the horseshoe Λ does not intersect the strong-stable manifold, nor the strong-unstable manifold of the point p.

THEOREM 2. – *Assume $\overline{d}_s(\Lambda) < 1$ (resp. $\overline{d}_u(\Lambda) < 1$). For an open and dense subset of maps in some neighborhood of φ we have*

$$\Lambda \cap W^{ss}(p) = \{p\} \quad (\textit{resp.}\ \Lambda \cap W^{uu}(p) = \{p\}).$$

By continuity, $\overline{d}_s(\Lambda_\psi) < 1$ for any ψ in a whole neighborhood of φ. The density statement in the theorem is proved by deformation. We embed ψ in a one-parameter family of diffeomorphisms ψ_ν such that $\Lambda_\nu \cap W^s(p_{\psi_\nu})$ moves with respect to $W^{ss}(p_{\psi_\nu})$ when ν varies. The fact that the upper Hausdorff dimension is less than 1 ensures that $\Lambda_{\psi_\nu} \cap W^s(p_{\psi_\nu})$ intersects $W^{ss}(p_{\psi_\nu})$ only for a zero Lebesgue measure set of parameters ν. This implies density, and openness is clear from compactness of Λ_{ψ_ν} and of a fundamental domain of $W^{ss}(p_{\psi_\nu})$.

Theorem 2 has the important geometric consequence that $\Lambda \cap W^s(p)$ is contained in a cuspidal region around the weak-stable direction of p. Let K^s be the projection of $\Lambda \cap W^s(p)$ to this weak-stable direction along the strong-stable foliation of p. Then, for any $\delta > 0$ there exists $\varepsilon > 0$ such that each $x \in \Lambda \cap W^s(p)$ in the ε-neighborhood of p is within distance $\varepsilon\delta$ from some $y \in K_s$. We say that $\Lambda \cap W^s(p)$ is *well represented* by K_s near p.

Then a similar fact is true replacing $W^s(p)$ by any other cross-section Σ to the unstable direction at any $q \in W^u(p)$: the intersection $W^u(p) \cap \Sigma$ is well represented by K_s near q. Using also smoothness of the unstable leaves, together with Hölder continuity of their tangent bundle, we conclude that the intersection of $W^u(\Lambda)$ with a neighborhood of q is well-represented by the product of K_s by $T_q W^u(p)$ near q.

We consider q to be the point of homoclinic tangency. The previous paragraph, together with a dual statement for $W^s(\Lambda)$, mean that the stable and unstable foliations of Λ are well represented by affine models

$$K_s \times T_q W^u(p) \quad \text{and} \quad K_u \times T_q W^s(p)$$

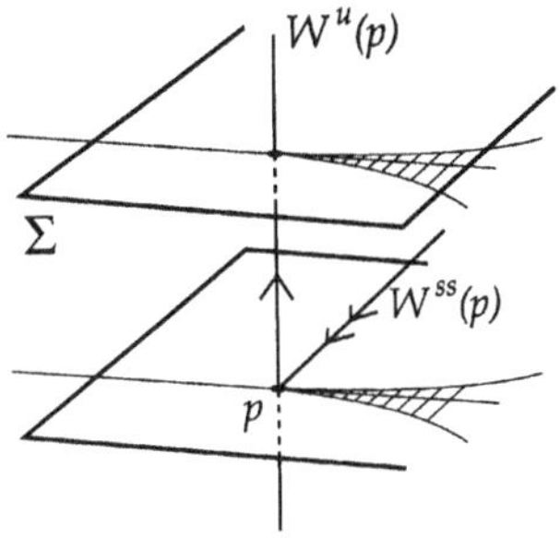

Figure 1.

near q. By continuity of the invariant foliations this remains true when the tangency is unfolded along a one-parameter family φ_μ: if μ is close to zero the $W^{\mathrm{s}}(\Lambda_\mu)$ and $W^{\mathrm{u}}(\Lambda_\mu)$ are well represented by translates

$$\left[K_{\mathrm{s}} \times \mathrm{T}_q W^{\mathrm{u}}(p)\right] + \mu v_1 \quad \text{and} \quad \left[K_{\mathrm{u}} \times \mathrm{T}_q W^{\mathrm{s}}(p)\right] + \mu v_2$$

for some vectors v_1, v_2. So, along the lines of [6] and [10], to ensure that the two foliations are transverse and, in fact, the angles at their intersection are bounded away from zero for most parameters, it suffices to prove a corresponding fact for these affine models:

LEMMA 3. – *Let K_1, K_2 be subsets of $\mathbb{R}^+$ with sum of limit capacities $c(K_1) + c(K_2) < 1$. Let V_1 and V_2 be subspaces of $\mathbb{R}^n$ with $\dim V_1 = u - 1$, $\dim V_2 = s - 1$, and $u + s = n$, and $w_1 \notin V_1$ and $w_2 \notin V_2$ be vectors such that the closed half-subspaces $\mathbb{R}^+ w_1 \oplus V_1$ and $\mathbb{R}^+ w_2 \oplus V_2$ intersect at the origin. Then most translates of $K_1 w_1 \oplus V_1$ and $K_2 w_2 \oplus V_2$ do not intersect: for any vectors v_1, and v_2 of $\mathbb{R}^n$ in general position, and for any $\delta > 0$,*

$$\mathrm{d}(K_1 w_1 \oplus V_1 + \mu v_1, K_2 w_2 \oplus V_2 + \mu v_2) > 3\varepsilon\delta$$

for a fraction of parameters in $[-\varepsilon, \varepsilon]$ that goes to 1 when $\varepsilon \to 0$.

4. Global strong-stable and strong-unstable foliations

In order to prove the second part of our main result, we consider compact subsets of our basic set that are invariant by some iterate of the dynamics at the first bifurcation parameter and have good geometric properties (perhaps after a small perturbation of the family). More precisely, in case (ii) of the theorem we prove that, generically, there are disjoint subsets Λ_1 and Λ_2 of Λ, invariant by some iterate $\varphi_0^{n_0}$, with the following properties:

(a) Λ_1 has a globally defined strong-stable foliation, invariant under $\varphi_0^{n_0}$;
(b) Λ_2 has a globally defined strong-unstable foliation, invariant under $\varphi_0^{n_0}$;
(c) $\overline{d}_{\mathrm{s}}(\Lambda_1) < 1$, $\overline{d}_{\mathrm{u}}(\Lambda_2) < 1$, and $\overline{d}_{\mathrm{s}}(\Lambda_1) + \overline{d}_{\mathrm{u}}(\Lambda_2) > 1$;
(d) for each $p_1 \in \Lambda_1$ and $p_2 \in \Lambda_2$, the projections $\pi_{\mathrm{ss}} \mid (\Lambda_1 \cap W^{\mathrm{s}}(p_1))$ along strong-stable leaves and $\pi_{\mathrm{uu}} \mid (\Lambda_2 \cap W^{\mathrm{u}}(p_2))$ are injective and have *intrinsically* $\mathrm{C}^{1+\varepsilon}$ inverses, in the sense of [8];
(e) either Λ_1 contracts area in the weak (bidimensional) direction or Λ_2 expands area in that weak direction.

The main and most delicate points of the above statement are items (a) and (b). Let us give some ideas of their proof.

First, we show that, perhaps after a small perturbation, Λ has a periodic point whose weakest, contracting and expanding, eigenvalues are real. Next, we use the existence of this periodic point to show that a sizable proportion of the vertical (and of the horizontal) n-cylinders in an advanced stage $n \gg 1$ of the construction of Λ are exponentially squeezed along the corresponding direction.

Finally, we consider families of perturbations with a large number of parameters, whose effect is to rotate the rectangles of a Markov partition of Λ with enough independence, in order to show that, with large probability, there are invariant subsets of Λ with almost the same stable and unstable upper dimensions as Λ, and admitting invariant strong-stable (respectively, strong-unstable) cone fields.

5. Positive density of persistent tangencies

Conditions (a) to (d) of the previous section imply that the projection of $\Lambda_1 \cap W^{\mathrm{s}}(p_1)$ onto a weak-stable (one-dimensional) manifold, along the strong-stable foliation, is a regular $C^{1+\varepsilon}$ Cantor set K^{s}, whose conjugation class does not depend on $p_1 \in \Lambda_1$ nor on the choice of the weak-stable leaf. The same holds for $\pi_{\mathrm{uu}}(\Lambda_2 \cap W^{\mathrm{u}}(p_2)) = K^{\mathrm{u}}$.

We adapt arguments of [1] to show the next lemma. It is assumed that the eigenvalues of φ at the point p satisfy a generic non-resonance condition.

LEMMA 4. – *If there are $\lambda > 0$ and $t \in \mathbb{R}$ such that K^{s} intersects $\lambda K^{\mathrm{u}} + t$ stably, then $(\varphi_\mu)_\mu$ persistently exhibits positive density of stable tangencies at $\mu = 0$.*

In order to get the stable intersection between K^{s} and $\lambda K^{\mathrm{u}} + t$, we first prove that condition (e) implies that either K^{s} or K^{u} is a C^2-regular Cantor set on the central leaf. This is important to show that, generically, K^{s} and K^{u} satisfy the Scale recurrence lemma, the main technical tool of [2,3]. Afterwards, we consider again a family with a large number of parameters of local perturbations of the diffeomorphism φ_0 and show that for most parameters there are λ and t such that K^{s} stably intersects $\lambda K^{\mathrm{u}} + t$. This final step is adapted from the perturbation arguments of [2,3]. This allows us to to use the previous lemma and conclude the proof of part (ii) of the main theorem.

* Work partially supported by Pronex-Dynamical Systems, CNPq 001/2000, and Faperj.

References

[1] Moreira C.G., Stable intersections of Cantor sets and homoclinic bifurcations, Ann. Inst. Henri-Poincaré, Anal. Non Linéaire 13 (1996) 741–781.
[2] Moreira C.G., Yoccoz J.-C., Stable intersections of regular Cantor sets with large Hausdorff dimensions, Ann. Math. (to appear).
[3] Moreira C.G., Yoccoz J.-C., Tangences homoclines stables pour des ensembles hyperboliques de grande dimension fractale (in preparation).
[4] Newhouse S., The abundance of wild hyperbolic sets and nonsmooth stable sets for diffeomorphisms, Publ. Math. Inst. Hautes Études Sci. 50 (1979) 101–151.
[5] Newhouse S., Palis J., Cycles and bifurcation theory, Astérisque 31 (1976) 44–140.
[6] Palis J., Takens F., Hyperbolicity and the creation of homoclinic orbits, Ann. Math. 125 (1987) 337–374.
[7] Palis J., Takens F., Hyperbolicity and Sensitive Chaotic Dynamics at Homolinic Bifurcations: Fractal Dimensions and Infinitely Many Attractors, Cambridge Univ. Press, 1992.
[8] Palis J., Viana M., High dimension diffeomorphisms displaying infinitely many periodic attractors, Ann. Math. 140 (1994) 207–250.
[9] Palis J., Yoccoz J.-C., Homoclinic tangencies for hyperbolic sets of large Hausdorff dimension, Acta Math. 172 (1994) 91–136.
[10] Takens F., Homoclinic bifurcations, in: Proc. Intern. Congress of Math., Berkeley, 1986, pp. 1229–1236.

Available online at www.sciencedirect.com
SCIENCE DIRECT
ELSEVIER
Ann. I. H. Poincaré – AN 22 (2005) 485–507
ANNALES DE L'INSTITUT HENRI POINCARÉ
ANALYSE NON LINÉAIRE
www.elsevier.com/locate/anihpc

A global perspective for non-conservative dynamics

Une perspective globale pour la dynamique non-conservative

J. Palis [1]

Instituto Nacional de Matemática Pura e Aplicada, Est. Dona Castorina 110, 22460-320, Jardim Botânico, RJ, Brazil

Received 19 January 2005; accepted 28 January 2005

Available online 27 April 2005

Abstract

Since Poincaré's fundamental work on the qualitative study of differential equations in 1881, the viewpoint of pursuing the description of the long range behavior of trajectories for "most" systems has been somewhat present in dynamics. About eight decades later, Smale was aiming exactly at that purpose when he introduced in the early sixties the concept of hyperbolic systems, which turns out to correspond to a large class of robust systems that gave rise to a basic theory of modern dynamics. Yet, hyperbolic systems are not typical since not every system can be approximated by a hyperbolic one, as can be seen by the famous example of the butterfly attractor provided by Lorenz, which is not hyperbolic and still robust or totally persistent under small perturbations of the initial flow. A string of other counter-examples were constructed in the late sixties and early seventies, and is to be noticed that all of them are related to cycles in dynamics, introduced by Poincaré, which will be focused here.

In the present paper, we wish to provide some perspective about this major question, proposing in qualitative terms what could be the main characteristics of a typical system. Indeed, we shall discuss partial successes and a possible strategy for proving a global conjecture on the finitude of large basin attractors and their stochastic stability for non-conservative dynamics; i.e. C^r flows, diffeomorphisms, and transformations of compact, boundaryless manifolds or the interval, $r \geqslant 1$. We shall also impose the union of the attracting basins to have total probability in the ambient space (phase space). Thus, the aim of our conjecture is a description in a rather simple conceptual way of the long range behavior of a typical (positive) trajectory of a typical dynamical system: each trajectory has only finitely many choices (of attractors) where to accumulate upon in the future. We discuss some recent related results, including homoclinic bifurcations, dynamical robustness and persistence, as well as systems with a dominated or partially hyperbolic decomposition.

My perspective is that we should expect, no so long in the future, substantial progress about the main and some of the other conjectures in this paper in the C^1 topology context and for one-dimensional dynamics, as conveyed in the introduction. A key difficulty in higher dimensions and the C^r topology, $r > 1$, arises from the question whether densely, in the space of non-conservative dynamics, the systems display a dense subset of periodic orbits in their limit or nonwandering set. The question

E-mail address: jpalis@impa.br (J. Palis).

[1] Partially supported by CNPq-Brazil and FAPERJ.

doi:10.1016/j.anihpc.2005.01.001

has been positively settled only in the C^1 topology, a result due to Pugh in the sixties, and it remains open otherwise except for dimension one or flows on orientable surfaces. Still, in my view, the C^1 case is already iluminating of the darker realm of dynamics. Above all, the idea of having most systems with only finitely many attractors toward which almost all orbits are attracted to, seems to me a tempting one to be considered even in special and yet relevant settings.

Résumé

Depuis le travail fondamental de Poincaré sur l'étude qualitative des équations différentielles en 1881, l'idée de chercher la description du comportement à long terme des trajectoires pour « la plupart» des systèmes a été présente d'une certaine façon en dynamique. Environ huit décennies plus tard, Smale s'attelait exactement à cette tâche quand il introduisit, au début des années soixante, le concept de systèmes hyperboliques, qui correspondent finalement à une grande classe de systèmes robustes qui donnèrent naissance à une théorie de base de la dynamique moderne. Cependant, les systèmes hyperboliques ne sont pas typiques puisque n'importe quel système ne peut pas être approché par un système hyperbolique, comme le montre le célèbre attracteur « papillon» découvert par Lorenz, qui n'est pas hyperbolique mais toujours robuste ou totalement persistant par des petites perturbations du flot initial. Une série d'autres contre-exemples furent construits à la fin des années soixante et au début des années soixante-dix, et l'on doit remarquer que tous sont liés à des cycles en dynamique, introduits par Poincaré, et que nous allons considérer ici.

Dans cet article, nous souhaitons donner une vision de cette question majeure, proposer en des termes qualitatifs ce que pourraient être les principales caractéristiques d'un système typique. En effet, nous discuterons des succès partiels et d'une possible stratégie pour prouver une conjecture globale sur la finitude des attracteurs des grands bassins et leur stabilité stochastique pour les dynamiques non-conservatives, à savoir, les flots, les difféomorphismes et les transformations des variétés compactes sans bords ou de l'intervalle de classe C^r pour $r \geqslant 1$. Nous imposerons aussi que l'union des bassins attracteurs soit de probabilité un dans l'espace ambiant (l'espace des phases). Ainsi, le but de notre conjecture est une description, d'une manière relativement simple conceptuellement, du comportement à long terme d'une trajectoire (positive) typique d'un système dynamique typique : chaque trajectoire n'a qu'un nombre fini de choix (d'attracteurs) où s'accumuler dans le futur.

Nous discuterons de quelques résultats connexes récents, notamment des bifurcations homocliniques, de la robustesse dynamique et de la persistance, de même que des systèmes avec une décomposition dominée ou partiellement hyperbolique.

Mon point de vue est que nous devrions connaître à moyen terme des progrès substanciels à propos de la conjecture principale et d'autres conjectures de cet article dans le contexte de la topologie C^1 et pour la dynamique de dimension un, comme cela a été expliqué dans l'introduction. Une difficulté majeure en dimension supérieure et en topologie C^r pour $r > 1$ provient de la question de savoir si, de façon dense dans l'espace des dynamiques non-conservatives, les systèmes possèdent un sous-ensemble dense d'orbites périodiques dans leur ensemble limite ou leur ensemble « non-récurrent». La question a été résolue seulement dans le cas C^1 : c'est un résultat de Pugh dans les années soixante ; elle reste ouverte dans les autres cas, sauf en dimension un ou pour le flot sur des surfaces orientables. De mon point de vue, même le cas C^1 est déjà éclairant dans le domaine inconnu de la dynamique. Avant tout, l'idée d'avoir souvent des systèmes avec un nombre fini d'attracteurs vers lesquels presque toutes les orbites sont également attirées me semble être une idée attrayante à considérer même dans des contextes particuliers mais pourtant pertinents.

1. Introduction

We shall present in this paper a global scenario for dissipative or, more precisely, non-conservative dynamics, i.e. C^r flows, diffeomorphisms and transformations, $r \geqslant 1$, on compact boundaryless manifolds or intervals on the real line. The main focus concerns a conjecture on the denseness, in the C^r topology, of systems having only finitely many attractors, the attractors being sensitive to initial conditions (chaotic) or just periodic sinks and the union of their basins of attraction having total Lebesgue probability. The attractors should also be stochastically stable in their basins of attraction with respect to random perturbations in a finite dimensional space of parameters. The conjecture is formally presented and discussed in Section 2. In Sections 2 and 3, other related conjectures are presented as well as a number of recent partial results. From them we have inferred a possible strategy to prove that our global scenario may indeed be valid. Basic definitions and some known or motivating results are presented in Section 2.

Our conjecture corresponds to a probabilistic and subtle version of another main one proposed by Smale in the sixties, but by the end of that decade many counter-examples to it had been provided, including by Smale himself. This fact certainly led for quite a while to a vacuum with respect to the possibility of formulating a global scenario for dynamics, through which we could provide key properties of a typical dynamical system. Smale's proposal stated that systems with a hyperbolic structure, form an open and dense subset of all systems in the C^r topology, $r \geqslant 1$ (see Section 2).

Most remarkably, towards the end of the sixties Kolmogorov when visiting IMPA, Rio de Janeiro, stated to colleagues there that the global study of dynamical systems could not go very far without the use of new additional mathematical tools, like probabilistic ones. Of course, the work of Sinai, Arnold and Anosov, among others, had already such a flavor at that point. But, I interpreted Kolmogorov's view, as expressed to me by Elon Lima sometime afterwards, as referring to Smale's proposal of a global scenario for dynamics, since counter-examples to the denseness of hyperbolic systems were already known. For years to come, I kept such a remarkable insight in my mind.

So, I take the occasion to pay Kolmogorov and Smale my tribute.

The collapse of Smale's conjecture excluded the case of flows on disks and surfaces, as well as for one-dimensional dynamics. Indeed, new excellent results have just been announced proving it in the real one-dimensional case, as shall be discussed in Section 2.5.

On the other hand, for one-dimensional real or complex dynamics, the conjecture presented here goes even further than indicated in the Abstract:

- For almost all parameter values, the corresponding dynamical systems display finitely many attractors which are periodic sinks or carry an absolutely continuous invariant probability measure.

Concerning the main conjecture, there is the following string of outstanding recent results in the real one-dimensional case, that culminates with its full solution for unimodal maps, i.e., maps with only one critical point.

A first breakthrough is due to Lyubich [71], using results by Martens and Nowicki [74] and previously Sullivan [115] and McMullen [76], among others. To state his result, we consider the quadratic family

$$f_x(x) = ax(1-x), \quad f_a : [0,1] \to [0,1], \quad 0 < a \leqslant 4,$$

that plays a key role in the theory of dynamics of interval maps. We then have

Lyubich [71]: For almost every a in $(0,4]$, f_a either has a hyperbolic attractor (sink) attracting almost all orbits or else it is chaotic, i.e., it has an absolute continuous invariant probability measure, which is unique and ergodic (and so a SRB measure).

Following that, there was an intense activity with other authors joining in the efforts to successfully achieve a definite picture of the typical dynamics for unimodal maps.

Avila, de Melo and Lyubich [17]: Same conclusion as above in the analytic case, assuming negative Schwarz derivative and making use of a key structure (laminations corresponding to the topological conjugacy classes) in the space of infinitely renormalizable maps.

Avila and Moreira: The main conjecture is true for non-degenerate analytic families, in particular the above quadratic one [20], and even just generic C^k families, $k \geqslant 2$, of unimodal maps [18,21]. Moreover, surprisingly, there exists an explicit formula that gives the eigenvalues of periodic orbits as a function of corresponding kneading sequences and of the kneading sequence of the critical orbit, valid for almost all parameters in any generic analytic family of unimodal maps [19]. They also announced that the main conjecture holds in the complement of a set of positive codimension.

In my perspective these last results shall soon be fully extended to the multimodal case.

As mentioned above, we shall discuss in Section 3 other relevant conjectures and questions, particularly concerning homoclinic bifurcations and robustness of transitive sets. Motivated by such questions and some of the recent results presented here, we have set up a possible strategy to prove the validity of the global scenario for non-conservative dynamics as proposed in 1995 in a meeting in honor of A. Douady [90].

I wish to thank W. de Melo, C.G. Moreira and specially to M. Viana for their precious comments and information while I was preparing this text. I also want to mention that Viana's book with Bonatti and Diaz [28] has a broad account of present day dynamics, being more comprehensive in the non-conservative case.

2. Basic definitions and results: the global conjecture on the finitude of attractors

In this section we shall discuss the basic ingredients leading to our main conjecture. We also discuss Smale's global conjecture in the sixties proposing the class of hyperbolic (or structurally stable) systems as paradigm for typical dynamics.

2.1. Sinai–Ruelle–Bowen probability measure

This remarkable notion was introduced by Sinai [112] and proven by him to exist for Anosov or globally hyperbolic diffeomorphisms, as defined in Section 2.4 below. Subsequently, it was shown to exist by Ruelle [107] and Ruelle and Bowen [32] for hyperbolic diffeomorphisms and flows.

The definition is expressed in rather simple terms, but its content is quite meaningful in describing the dynamics of attractors when it's possible to show that they carry SRB-measures, which is usually a hard task.

Let A be an attractor for f, i.e. there is a set of points in the phase space with positive Lebesgue probability whose future orbits tend to A, as the number of iterates tends to infinite. The set of orbits attracted to A in the future is called its basin. The definition for flows is entirely similar. Let μ be an f-invariant probability measure on A. Then μ is called a SRB (Sinai–Ruelle–Bowen) measure for (f, A) if we have for any continuous map g, that

$$\lim_{n\to\infty} \frac{1}{n} \sum g\big(f^i(x)\big) = \int g \,\mathrm{d}\mu$$

$x \in E \subset B(A)$ basin of attraction for A, with $m(E) > 0$, where m denotes Lebesgue measure. It's common to have E with total probability in $B(A)$ and so, in such a case, the above convergence holds for a typical trajectory attracted to A in the future.

2.2. Stochastic stability

For attractors carrying SRB-measures, we can investigate the question of their stability or persistence when we slightly perturb the dynamics, more particularly from a probabilistic viewpoint. Thus, we shall follow Kolmogorov's perspective. Indeed, the more classical concept of structural stability, introduced by Andronov and Pontryagin in the 30's, remarkably turned out to be too close or even equivalent to hyperbolicity (see Section 2.5). However, there are open sets of structurally unstable systems, and so we consider instead the concept of stochastic stability.

Let A be an attractor for f, μ be a SRB invariant probability for (f, A) and f be an element of a finite dimensional parameter-space of C^r maps, $r \geqslant 1$. Random Lebesgue choice of parameters gives rise to maps f_j, $j = 0, 1, \ldots$, that are ε-near f in parameter space, some $\varepsilon > 0$. Let $z_j = f_j \circ \cdots \circ f_1(z_0)$, $z_0 \in B(A)$.

We call (f, A, μ) stochastically stable if given a neighborhood V of μ in the weak topology, the weak limit of $\frac{1}{n}\sum_{j=0}^{n-1} \delta_{z_j}$ is in V for a.a. $(z_0, f_1, f_2, \ldots)$ if ε is small, where δ_{z_j} stands for the Dirac measure at z_j.

Variants of this definition, including the case of flows, are briefly presented in [90] and references provided; see also [28] for a comprehensive discussion.

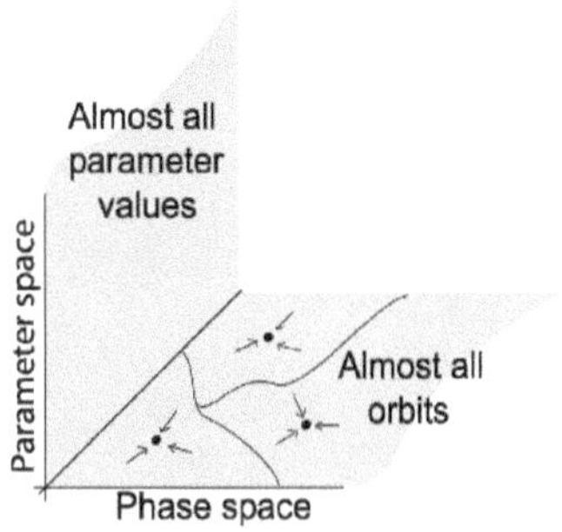

Fig. 1.

2.3. Attractors

As defined in Subsection 2.1 on SRB-measures above, attractors should be the positive limit set for the orbits starting at a subset of the phase space (space of events) with positive Lebesgue probability. Often, they attract the orbits of a full measure set in a neighborhood of them. We add to the definition of attractor that it must be transitive, i.e., it must display a dense orbit, so it cannot be decomposed into strictly smaller attractors. The simplest example for a diffeomorphism f is that of an attracting periodic orbit: the eigenvalues of $\mathrm{d}f^n(p)$ have norm less than one, where p is the periodic orbit and n its period. Another well known case is that of the whole torus T^2 for the map induced by the linear one L on $\mathbb{R}^2$ defined by $L(1,0) = (2,1)$, $L(0,1) = (1,1)$.

Attractors are of paramount importance in dynamics since they mold the future behavior of many orbits, i.e. the ones in their basin of attraction. What we want mainly to discuss in the present paper is our conjecture that a finite number of attractors should in fact attract Lebesgue almost all future orbits, at least for a dense subset of dynamical systems in the C^r topology, $r \geqslant 1$. We can go even further when considering finite dimensional parameter families of dynamics, by requiring that such a property holds for dynamical systems corresponding to Lebesgue almost all parameter values. The precise statement of the conjecture is at the end of the present section.

We want to emphasize that we shall play, in probabilistic terms, with *almost all orbits in the phase space* (space of events) of dynamical systems corresponding to *almost all parameter values*, as suggested by Fig. 1.

So, figuratively what we wish to propose is that "small accidents" both in terms of Lebesgue measure in the parameter space of systems as well as in the phase space are not "very relevant" in order to provide such a global scenario for dynamics.

We now give the definition of a hyperbolic system and, in particular, of a hyperbolic attractor, and then briefly present important examples of both hyperbolic and non-hyperbolic attractors. They all shall have at least some degree of persistence. Formally, they shall exist for a positive probability set of parameter values. Or, sometimes even more sharply, for an open set in the space of parameters, in which case the attractors are called robust. The hyperbolic attractors are robust and, historically, the only ones known to be robust until the appearance of Lorenz-like attractors, as discussed in Section 2.6.

2.4. Hyperbolicity

A diffeomorphism $f: M \to M$ is called hyperbolic if on its limit set L, that is the closure of the sets where the orbits accumulate in the future or in the past, we have a decomposition of the tangent bundle of M at L, $T_L M = E^s \oplus E^u$, such that $\mathrm{d}f|E^s$, $\mathrm{d}f^{-1}|E^u$ are uniform contractions.

In the case of a flow X_t, $t \in \mathbb{R}$, generated by a vector field X, we require that

$$T_L M = E^s \oplus E^0 \oplus E^u.$$

Here, E^0 is one-dimensional and tangent to X, outside its singularities (a finite number) and

$$\|dX_t \mid E^s\|, \|dX_{-t} \mid E^u\| \leqslant Ce^{\lambda t}, \quad t \in \mathbb{R},$$
$$C > 0, \qquad 0 < \lambda < 1.$$

In the cases where such bundle decompositions occur in the whole ambient manifold, we refer to such globally hyperbolic diffeomorphisms or flows as Anosov ones. Recall that associated to the orbits of a hyperbolic set there are stable and unstable manifolds, corresponding to the contracting and expanding sub-bundles by the action of the derivative of the dynamical system. Often we refer to a hyperbolic system as satisfying Axiom A or more precisely when its non-wandering set, which may be larger than the limit set, is hyperbolic. For our discussion here, such a difference is not very relevant. Sometimes, by *abus de langage*, we refer to hyperbolic systems as having hyperbolic limit sets and all stable and unstable manifolds transversal. They form a robust set of dynamics, i.e., an open set in every C^r topology, $r \geqslant 1$. Also, sometimes we refer to an Axiom A system as being hyperbolic and without cycles among the different closed, invariant and transitive subsets, in which the (hyperbolic) limit set can be decomposed. They, again, form a robust set of dynamics [95,114].

A very notable result in dynamics states in a simplified form (see additional condition below concerning invariant manifolds) that

$$\text{hyperbolicity} \overset{C^1}{\sim} \text{structural stability}$$

We recall that a diffeomorphism f is C^r structurally stable, or just dynamically stable or just stable, if for any C^r small perturbation g of f, there is a homeomorphism h of the phase space such that $hf(x) = gh(x)$, for all points x in the phase space. For flows, we require the existence of a homeomorphism h sending trajectories of the initial flow to the trajectories of any small C^r perturbation.

A beautiful page of dynamics is the solution, in the C^1 topology, of the so called *Stability Conjecture*, stating that *a system is C^r structurally stable if and only if it is hyperbolic and all the stable and unstable manifolds associated to the orbits in the limit set are in general position*, i.e., *transversal*. In the equivalence indicated above, we didn't mention explicitly the transversality condition on stable and unstable manifolds because it corresponds to the easiest part of the conjecture.

That stability holds for hyperbolic systems has been proved, in various settings, by Anosov [10], Palis [89], Palis and Smale [91], Robbin [103], de Melo [39], Robinson [104] in the sixties and early seventies. The converse was completed in the eighties by Mañé [73] for diffeomorphisms and in the nineties by Hayashi [51] for flows, after earlier contributions by Liao [66,67], Sannami [109] and Mañé [72].

Predecessors to such works are the pioneering ones by Andronov and Pontryagin for flows on disks and later on Peixoto for flows on orientable surfaces, as cited in the next subsection.

2.5. Conjecture (Smale)

Smale's conjecture concerning a typical dynamical system formulated in the sixties, can simply be stated as follows: Every system can be approximated by a hyperbolic one. If the conjecture was true, the hyperbolic systems by denseness and robustness would be *typical* of dynamics.

A few years afterwards counter examples abounded, due to Smale [114], Abraham and Smale [1], Simon [111] and others. More strikingly, Newhouse [86] provided counter-examples for C^2 surface diffeomorphisms using arithmetic difference of "thick" Cantor sets (thickness is a kind of fractal dimension). He did so by unfolding homoclinic tangencies, as in Section 3.2, to generate open sets of nonhyperbolic diffeomorphisms containing dense subsets of maps exhibiting homoclinic tangencies. Newhourse's idea does not hold in the C^1 category, by Ures [119]. There were also the beautiful Lorenz butterfly attractor that came to our attention only in the early seventies and the Hénon attractor, both of which we shall much comment in the sequel.

In fact, Smale's conjecture was known for flows on disks [9] and orientable surfaces [99], and consequently for diffeomorphisms of the circle. Smale was initially inspired by these results that correspond to pioneering works by Andronov and Pontryagin in the thirties and Peixoto in late fifties.

More recently, it has been independently proved by Swiatek and Graczyk [49] and Lyubich [71] that the hyperbolic maps are dense among real quadratic ones. Subsequently, Kozlovski [59,60] extended this result to C^3 unimodal mappings. A remarkable recent theorem has been announced in two parts by Kozlovski, Shen and Van Strien [61,62], namely the density of the hyperbolic transformations among C^r maps of the interval, for $r \geqslant 2$, the case of $r = 1$ being also included by a much previous theorem of Jakobson [54]. The result of Kozlovski, Shen and Van Strien not only answers Smale's conjecture in one real dimension, but opens the way for proving the main conjecture in the present paper in full generality for maps of the interval or the circle.

It is important to observe that for the one-dimensional complex case, i.e., rational maps of the Riemman sphere, the question remains wide open.

2.6. Beyond hyperbolic attractors

For dynamics, the most interesting examples of attractors are the ones that are robust under small perturbations of the dynamics, as in the hyperbolic and Lorenz-like cases, or which are at least probability persistent (i.e., the attractor exists for a positive probability in parameter space), as in the Hénon-like case.

As mentioned before, the simplest attractors are the ones consisting of a simple fixed or periodic hyperbolic orbit. In the hyperbolic case, the other extreme corresponds to attractors consisting of the whole ambient manifold, like the one induced by hyperbolic toral automorphisms. This example turns out to be a particular case of Anosov diffeomorphisms for which we assume (uniform) hyperbolicity at all points of the manifold [10]. In this vein, it is still an open question whether it is true for all Anosov diffeomorphisms that the whole manifold consists of a (transitive) attractor. This is generally not true for Anosov flows already in dimension three [46].

The remarkable Lorenz "butterfly" attractor, whose original equations are in Fig. 2, was the first known case of a *fully persistent or robust (it exists for all sufficiently small variations of the initial parameter values) and (transitive) attractor which is not hyperbolic* [70]. Notice that its singularity $(0, 0, 0)$ is accumulated by hyperbolic

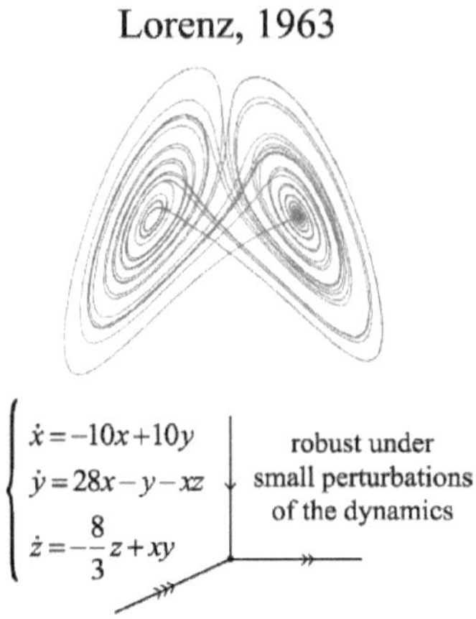

Fig. 2.

periodic orbits. Actually, Lorenz focused on the property that the long range behaviour of the orbits in the basin of attraction is sensitive to initial conditions: with total probability, orbits with different initial conditions spread apart in the future as much as the diameter of the attractor. And he did so not with a rigorous mathematical proof, but with rather convincing computational arguments. Attractors with such characteristics are called Lorenz-like. Only, recently, Tucker [118] provided a computer aided proof that the original Lorenz's equations indeed corresponds to a sensitive (or chaotic) robustly transitive non-hyperbolic attractor containing the singularity $(0, 0, 0)$. A charming account of the Lorenz attractor's beautiful history is in Viana [123].

Only about ten years after Lorenz's work, a number of concrete Lorenz-like attractors were exhibited by Afraimovich, Bykov and Shil'nikov [2] and Guckenheimer and Williams [50], for which the authors provided mathematical proofs that they are sensitive to initial conditions and, moreover, robustly transitive and yet not hyperbolic. In the eighties Robinson [105] and Rychlik [108] showed that such attractors may be obtained, again in three dimensions, through bifurcations taking place in certain families of cubic differential equations – not quadratic, like above.

A new kind of attractor in three dimensions, the contracting Lorenz attractor or Lorenz–Rovella attractor, which is probability persistent but not robust, was obtained by Rovella [106] upon a previous work by Arneodo, Coullet and Tresser [11]: It contains a hyperbolic singularity with real eigenvalues (but now the sum of any two eigenvalues is negative), it is probability persistent in terms of Lebesgue probability or just *probability persistent*, but not robust (Fig. 3).

Another remarkable example is the one named after Hénon [52]. Like Lorenz, he provided ingenious computational arguments suggesting the existence of an attractor for a two-parameter family of quadratic diffeomorphisms of the plane. It would display a long range sensitivity of orbits near it with respect to their initial conditions. Hénon pointed to the existence of folds and expansion along lines in the attractor and a fractal structure in a transversal direction. The challenge was then to provide a formal proof of the existence of such a chaotic attractor with some degree of persistence.

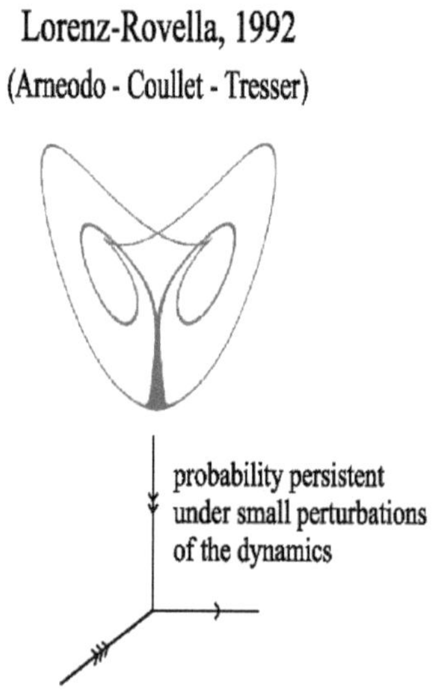

Fig. 3.

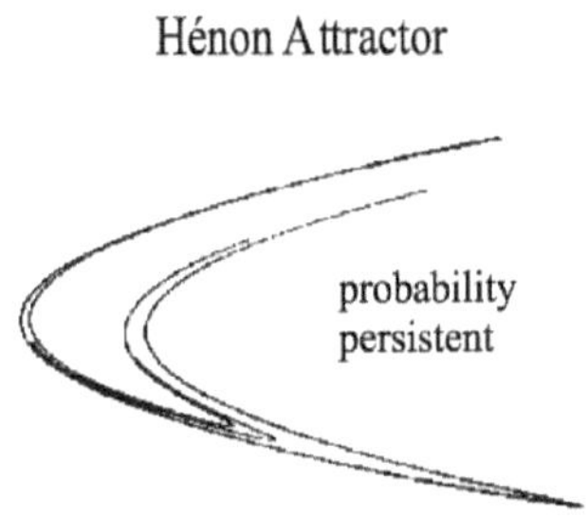

Fig. 4.

The equations and figure (Fig. 4) provided by Hénon were as follows:

$$f_{a,b}(x, y) = (1 - ax^2 + y, bx), \quad \text{for } a \cong 1.4 \text{ and } b \cong 0.3.$$

I also wish to mention Kolmogorov's pioneering and classic work on hydrodynamical attractors as presented by Arnold [14], as well as the charming paper of May on nonlinear dynamics and some applications [75].

By the last part of the seventies and going into the eighties, there appeared a string of very creative results giving rise to new insights on how dynamics could develop in the future. First, there was the work by Hénon mentioned above, proposing a new kind of attractor now named after him. Following that, there was the work by Feigenbaum [45] and in parallel Coullet and Tresser [37], concerning period doubling bifurcations for quadratic families of interval maps and the notable fact that the limiting ratio of approach of bifurcating parameter values converging to a limit point is a universal constant (independently of the specific quadratic family).

These works contributed to sparkle a much more robust development than before particularly of one real dimensional dynamics.

Then, there was the result of Jakobson [55], exhibiting new attractors of parametrized families of unimodal maps of the interval, including the quadratic family. They are *probability persistent* (but not robust) under small perturbations of the parameter. It was a key predecessor to the remarkable work by Benedicks and Carleson [23] showing the existence of a probability persistent Hénon-like attractor, which was obtained for the above coefficients (parameters) a near 2 and b small. Subsequently, Mora and Viana [79] showed that Hénon-like attractors occur in the unfolding of quadratic homoclinic tangencies associated to dissipative fixed or periodic hyperbolic points. As a consequence, the attractor appears near the values originally suggested by Hénon whose figure is displayed above, i.e. $a \approx 1.4$ and $b \approx 0.3$. This result was extended to higher dimensions, when the unstable manifold of the associated fixed or periodic point has dimension one [121], in the sectionally dissipative case as in Section 3.2. Also, Viana [122] exhibited new non-hyperbolic examples in higher dimensions which are, in fact, robust.

2.7. *Global conjecture on the finitude of attractors and their metric stability*

With the basic definitions, facts and examples presented above, we are now ready to formally state our main global conjecture, introduced in 1995 in a meeting in Paris in honor of A. Douady [90]. As mentioned in the abstract, we are considering the set of all C^r ($r \geqslant 1$) dynamics, i.e. flows diffeomorphisms or transformations of a compact, smooth, boundaryless manifold or closed interval on the real line.

Global Conjecture.

- There is a dense set D of dynamics such that any element of D has finitely many attractors whose union of basins of attraction has total probability;

- The attractors of the elements in D support a physical (SRB) measure;
- For any element in D and any of its attractors, for almost all small perturbations in generic k-parameter families of dynamics, $k \in \mathbb{N}$, there are finitely many attractors whose union of basins is nearly (Lebesgue) equal to the basin of the initial attractor; each such perturbed attractor supports a physical measure;
- Stochastic stability of attractors – the attractors of elements in D are stochastically stable in their basins of attraction;
- For generic families of one-dimensional dynamics, with total probability in parameter space, the attractors are either periodic sinks or carry an absolutely continuous invariant measure.

One can also ask whether the first two items above are valid with total probability in parameter space.

2.8. More on the existence of SRB probabilities – stochastic stability

All the attractors that have been presented so far carry physical or SRB probability measures. Before we discuss some new results concerning other classes of dynamical systems whose attractors enjoy the same property, we introduce the notion of dominated decomposition.

Dominated Decomposition – Definition

Let f be a C^r map on M and Λ an invariant set for f. We say that Λ has a dominated decomposition, see Fig. 5, if there exist $C > 0$ and $0 < \lambda < 1$ such that $T_\Lambda M = E^1 \oplus E^2$, E^1 and E^2 being $\mathrm{d}f$-invariant and

$$\frac{\|\mathrm{d}f^n(x)v^1\|}{\|\mathrm{d}f^n(x)v^2\|} \leqslant C\lambda^n \frac{\|v^1\|}{\|v^2\|}, \quad \forall v = (v_1, v_2),\ v_2 \neq 0,\ \forall n \geqslant 1.$$

In the figure, domination means that the angle between $\mathrm{d}f^n(x)v$ and E^2 decreases exponentially fast when $n \to \infty$. We say that E^2 dominates E^1 and that $f|\Lambda$ has i-dominated decomposition, where $1 \leqslant i < \dim M$, if $i = \dim E^2$.

In particular, we call f partially hyperbolic when we have a decomposition $T_\Lambda M = E^s \oplus E^c \oplus E^u$, where E^s is uniformly contracting, E^u is uniformly expanding, E^u dominates $E^{cs} = E^s \oplus E^c$ and $E^{cu} = E^c \oplus E^u$ dominates E^s. The definition can be adapted when $E^u = 0$, $E^s \neq 0$ or $E^s = 0$, $E^u \neq 0$.

Concerning SRB measures, recall that they were introduced by Sinai, Ruelle, and Bowen for hyperbolic systems, see Section 2.1. Such measures were shown to exist for probability persistent non-hyperbolic attractors: first, for interval maps by Jakobson in his pioneering work [55] and much more recently for Hénon-like attractors by Benedicks and Young [26]. In the latter case, subsequently, Benedicks and Viana [25] showed the remarkable fact that the SRB measure has full Lebesgue probability in a neighborhood of the attractor and it is unique.

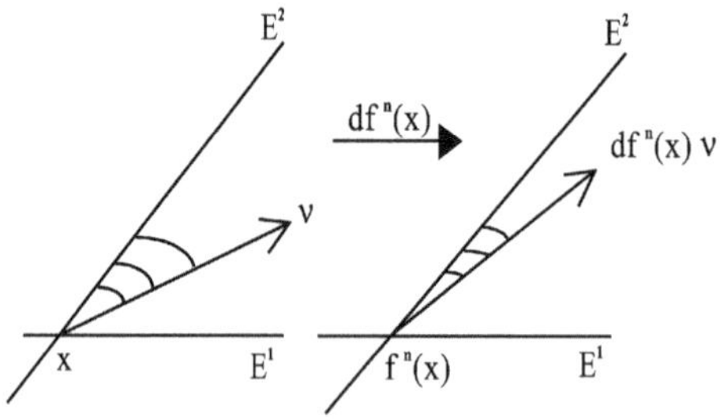

Fig. 5.

On the other hand, SRB measures were constructed for Lorenz-like attractors by Bunimovich and Sinai [33] and Kifer [58], for Lorenz–Rovella attractors by Metzger [77], who also announced the same result for higher dimensional Lorenz attractors such as the ones constructed by Bonatti, Pumariño and Viana [31].

More recently, there were important developments on SRB measures for partially hyperbolic attractors, notably the following theorem due to Alves, Bonatti and Viana [5]:

Theorem. *Let $K \subset M$ be a compact forward invariant set for a C^2 diffeomorphism on M. Suppose the subbundle E^{cs} is uniformly contracting on K, and E^{cu} is non-uniformly expanding on a positive Lebesgue measure set $H \subset K$. Then H is covered* mod 0, *i.e., with total Lebesgue probability, by the basins of finitely many SRB measures.*

The authors also introduced the notion of cu-Gibbs states for an attractor Λ with dominated decomposition $T_\Lambda M = E^{cs} \oplus E^{cu}$, after the pioneering work of Pesin and Sinai [100] on the existence of u-Gibbs states for an attractor Λ with dominated decomposition $T_\Lambda M = E^{cs} \oplus E^{u}$. Notice that SRB measures are u-Gibbs states and, as a partial converse, ergodic u-Gibbs states with $k = \dim E^{cs}$ negative Lyapunov exponents are in fact SRB measures. Vasquez [120] extended considerably the theory of cu-Gibbs states and obtained results on their continuity with respect to the diffeomorphism, which he called statistical stability, as well as on the existence and finitude of SRB measures for diffeomorphisms with dominated decomposition. Before that, for certain parametrized families of diffeomorphisms, Dolgopyat [44] has even proved the differentiability of SRB measures, from which he inferred not only uniqueness but also good statistical properties for them.

It is also much worthwhile mentioning a recent important result by Tsujii [117]: *partially hyperbolic surface endomorphisms of class C^r with one-dimensional uniformly expanding subbundle and r large enough, say $r \geqslant 19$, generically (residually) carry finitely many ergodic SRB measures whose union of basins of attraction has total Lebesgue probability.* So, in this setting the first part of the main conjecture in the present paper is confirmed.

Actually, we believe that Tsujii's results can be generalized to higher dimensions for partially hyperbolic diffeomorphisms (maps) with one-dimensional central direction.

Problem. Show that generic partially hyperbolic maps as in Tsujii's work are stochastically stable in their basins of attraction.

Finally, there is the following relevant conjecture on SRB measures formulated by Viana in 1997:

Conjecture 1. *Generically, non uniformly hyperbolic diffeomorphisms with a uniformly expanding subbundle carry finitely many SRB measures whose union of basins of attraction has total Lebesgue probability.*

Concerning stability, let us emphasize, although not so surprisingly in view of Section 2.4, that *none* of the previous attractors are structurally stable, except for the hyperbolic ones. However, all of them are stochastically stable. For the hyperbolic case we refer to Kifer [57] and also Young [126]. For Hénon-like attractors the result is due to Benedicks and Viana [24]. For Lorenz and Lorenz–Rovella attractors it is due to Kifer [58] and Metzger [78], respectively.

3. A strategy for proving the global conjecture on the finitude of attractors: to focus on homoclinic tangencies

I believe that it might be possible to prove the main conjecture in this paper when there are no homoclinic tangencies in a robust way. At this point, a famous sentence by Poincaré in his classic treatise "*Méthods Nouvelles de la Mécanique Céleste*" comes to mind: "Rien n'est plus propre à nous donner une idée de la complication du problème des trois corps et en général de tous les problèmes de Dynamique ...", referring to homoclinic orbits and the geometry of the corresponding stable and unstable manifolds. More concretely, we shall mention in the sequel that a robust absence of homoclinic tangencies implies, at least in the C^1 topology, a certain amount of hyperbolicity, namely a dominated decomposition. Since hyperbolic systems satisfy the main conjecture, this may also be the case here. And, if so, we could concentrate our efforts in the fundamental and very difficult problem of understanding the dynamics arising from unfolding homoclinic or heteroclinic cycles, as much as the attractors and their basins are concerned.

This will be discussed in the sequel. Let us first see what is known in terms of a weak form of hyperbolicity, in the present case dominated decomposition, when robustly we have no homoclinic tangencies.

Theorem. *Let f be a diffeomorphism and let Λ be a compact, maximal invariant set with a dense subset of hyperbolic periodic orbits of unstable index $i > 0$. Suppose that there are no homoclinic tangencies associated to periodic hyperbolic orbits with unstable index i in any C^1 small continuation of Λ. Then, there is an i-dominated decomposition of $f|\Lambda$.*

Let us recall the meaning of continuation of a maximal invariant set in the statement of the theorem. Let $\Lambda = \bigcap_{n} \in \mathbb{Z} f^n(U)$, where U is a neighborhood of Λ and let $\mathcal{U}$ be a C^1 neighborhood of f. For $g \in \mathcal{U}$, we call $\Lambda_g = \bigcap_{n\in\mathbb{Z}} g^n(U)$ the g-continuation of Λ.

This theorem is due to Pujals and Sambarino [102] in dimension two and to Wen [124] in higher dimensions. It is certainly relevant to pursue a proof of the main conjecture in such a setting:

Conjecture 2. *There is a residual (or at least dense) subset R of C^r diffeomorphisms, $r \geqslant 1$, such that: If $f \in R$ and Λ is a maximal invariant set as in the theorem above, with no homoclinic tangencies associated to periodic points in a robust way, then f exhibits only finitely many attractors inside Λ. Such attractors are stochastically stable.*

A particularly interesting case in this direction is when f, Λ are as before, but now it is assumed that the periodic points with biggest unstable index in Λ are dense in it. These conjectures can in particular also be posed for partially hyperbolic pairs f, Λ; see Conjecture 1 in Section 2.8.

3.1. Bifurcation theory

We now turn to bifurcation theory, a classical and central area in dynamics, as set forth by Poincaré. The list of contributions is very notable and we shall highlight some of them:

- Birkhoff [27] in the 30's has shown that transversal homoclinic orbits for surface diffeomorphisms are accumulated by periodic orbits (so, infinitely many of them). About three decades later, Smale [113,114] went much further by showing that they are part of what he called a *horseshoe* in any dimension. Before, in the 20's, there was the remarkable work by Van der Pol on relaxation oscillations [40] to model radio transmissions and electric circuits, among other applications. Following up this pioneer work, there were a series of papers by Cartwright and Littlewood [34,35], Littlewood [68,69] and Levinson [65], in the 40's and 50's in which Van der Pol's equations with parameters were considered and new kind of solutions were found, going through dynamic bifurcations when varying parameters. In his autobiography, Littlewood called his work in this topic "the monster", since he seemed to continuously work hard but without a clue of *at which port he would arrive if any* ... Then, in the late 70's and early 80's, Levi revisited the problem disposing of new developments that were not at the time available to Littlewood, such as the horseshoe, sometimes with small and sometimes large Hausdorff dimensions, Newhouse's infinitely many simultaneous sinks and other intricate phenomena that are created when unfolding (bifurcating) quadratic homoclinic tangencies; see Section 2.5. He showed [64] that all

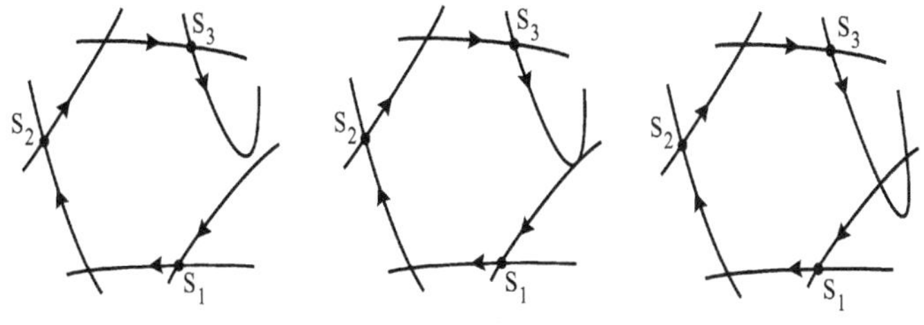

Fig. 6.

such phenomena were present in cross-sections to 3-dimensional flows generated by Van der Pol's equations. Such equations which were motivated by applications, clearly led to the development of deep mathematics.

- Closely following Poincaré's fundamental work, Andronov and school [6–8] set the ground for bifurcation theory in dynamics, influencing generations to come, noteworthily his work with Pontrjagin on structural stability that has been already mentioned.
- In the 60's and 70's, Arnold and school were a source of fine questions and results, in particular much extending the previous work on bifurcations, as can be seen in [12] and [13]. I wish to add here the work of Il'Yashenko and co-authors, as in [53].
- Much relevant has also been the work of Shil'nikov and school, mostly devoted to bifurcation theory and more particularly to the unfolding of homoclinic tangencies and creation of horseshoes [3,4] and [110].
- An extensive study of bifurcations and (structural) stability of parametrized families of dynamics were performed in the seventies and going through the next two decades by Newhouse, Palis, Takens [87,88,92,93] and Dias Carneiro [38] and Viana, Yoccoz, Moreira [83,84,96,97], whom are also referred to in the sequel on topics discussing homoclinic bifurcations.

See also [15,28,53,95] for more results and further references.

We have generally been baffled by the creation of cycles in dynamics, in particular 1-cycles, which correspond to homoclinic tangencies; a 3-cycle is in Fig. 6. Actually, the creation of cycles is often unavoidable when we consider parametrized families of systems and vary the parameters. In most cases, the unfolding of cycles gives rise to a very complicated, chaotic evolution of the dynamics in terms of parameters. In particular, they are a source of counter-examples to the denseness of hyperbolicity in the space of all dynamical systems endowed with any C^r topology, $r \geqslant 1$ (the only unknown case being that of C^1 diffeomorphisms on surfaces, see Ures [119]).

Still, we expect for a dense set of parameters to have only finitely many attractors for the corresponding dynamical systems. In fact, the same may be true with total Lebesgue probability in the space of parameters.

3.2. *Unfolding homoclinic tangencies*

In the sentence at the beginning of Section 3, Poincaré was referring to transversal homoclinic orbits of surface diffeomorphisms. One can imagine that he would possibly be as enthusiastic with respect to Birkhoff and Smale's results, mentioned in Section 3.1. They highlight even more so the richness of the dynamics implied, in all dimensions, by the presence of a transversal homoclinic orbit. Indeed, Smale's horseshoe corresponded to a new prototype dynamical model: its maximal invariant set is like the product of Cantor sets in which the periodic orbits are dense.

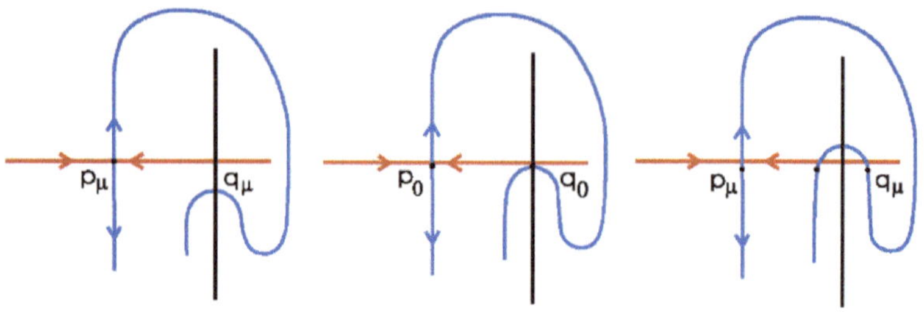

Fig. 7.

Similarly striking is the following list of complex phenomena discovered in the last three and half decades. It concerns the unfolding of a homoclinic tangency q_0 associated to a fixed (periodic) orbit p_0 with positive speed at $\mu = 0$, for a C^2 family f_μ, μ small, of *surface diffeomorphisms*:

- There are infinitely many simultaneous sinks residually in some open subsets of parameter values, as shown by Newhouse [85,86]. This renowned result was extended to higher dimensions by Palis and Viana [96], when $W^u(p_0) = 1$ and p_0 is sectionally dissipative, i.e., the product of any two different eigenvalues of $\mathrm{d}f_0$ has norm smaller than one.
- There are Hénon-like attractors for some parameter values, due to Mora and Viana [79], extending Benedicks and Carleson [23], and to Viana [121] in general in the codimension-one and sectionally dissipative case.
- There are infinitely many simultaneous Hénon-like attractors for dense subsets of open sets of parameter values, as shown by Colli [36].
- A fast growth of the number of periodic points occurs residually in an open Newhouse set as above, due to Kaloshin and Hunt [56], based on Shil'nikov, Gonchenko and Turaev [47].

In view of these dynamical phenomena, one is lead to consider the question of whether the coexistence of infinitely many sinks or Hénon-like attractors or just attractors are probability persistent, i.e., they can occur with positive Lebesgue probability in parameter space:

Conjecture 3. *In the unfolding of a homoclinic tangency, the coexistence of infinitely many sinks or, more generally, attractors is not a probability persistent phenomenon. In other words, with total probability in the parameter line, the corresponding maps exhibit only finitely many attractors, in particular sinks, in a neighborhood of the orbit of* q_0 *and* p_0 *(see Fig. 7).*

A very nice recent result concerning this conjecture for surface diffeomorphisms is due to Gorodetski and Kaloshin [48], much extending Tedeschini and Lalli and Yorke [116]. It says that, *for any positive integer N, infinitely many coexisting localized sinks with cyclicity bounded by N or with period relatively large with respect to N is not probability persistent.* To understand this result, we need to introduce the definition of a localized sink and its cyclicity.

Consider Fig. 8, where q_0 is a homoclinic tangency, say quadratic, associated to the fixed (periodic) hyperbolic point p_0 and let V and U be fixed small disjoint neighborhoods of the segments $[f_0(q_0), p_0]$, $[p_0, f_0^{-1}(q_0)]$ in the stable and unstable manifolds and of p_0, respectively.

Let $\Lambda_\mu = \bigcap_{n\in\mathbb{Z}} f_\mu^n(V \cup U)$.

We can now define a periodic sink for f_μ as being localized if it belongs to Λ_μ. In such a case, it's called (Λ_μ, s)-localized if it visits U exactly s times during its minimal period. Then, we call s its cyclicity.

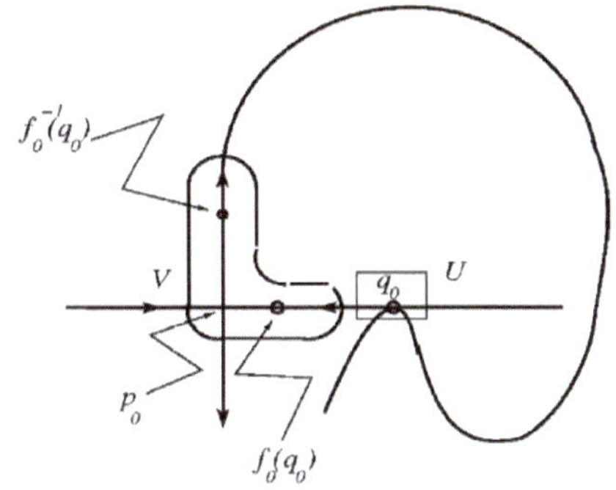

Fig. 8.

We close for the moment the discussion on the unfolding of homoclinic tangencies. We shall return to it in Section 3.4, but from a somewhat different point of view. We shall consider homoclinic tangencies associated to fixed or periodic orbits, that may or not be part of a larger hyperbolic set, say K, at the initial parameter value for the unfolding. We shall see that the Hausdorff dimension of K is a crucial ingredient in establishing some prevalence of hyperbolicity or non-uniform hyperbolicity for the dynamics, in terms of the parameter near its initial value.

3.3. More on the dichotomy hyperbolicity – homoclinic bifurcations

Along the lines discussed at the introduction to the present section, we have set up the following conjecture, which we consider by itself relevant:

Conjecture 4. *In any dimension, every diffeomorphism can be C^r approximated, $r \geqslant 1$, by a hyperbolic one or one exhibiting either a homoclinic tangency or a (finite) cycle of hyperbolic periodic orbits with different stable dimensions – called a heterodimensional cycle* [28,87].

As mentioned, hyperbolic diffeomorphisms are not dense (except for the circle and unknown for $r = 1$ on two-dimensional surfaces). What the conjecture states is that it is enough to complement them with diffeomorphisms exhibiting just two different dynamical structures, which are relatively simple to formulate. Yet, such diffeomor-

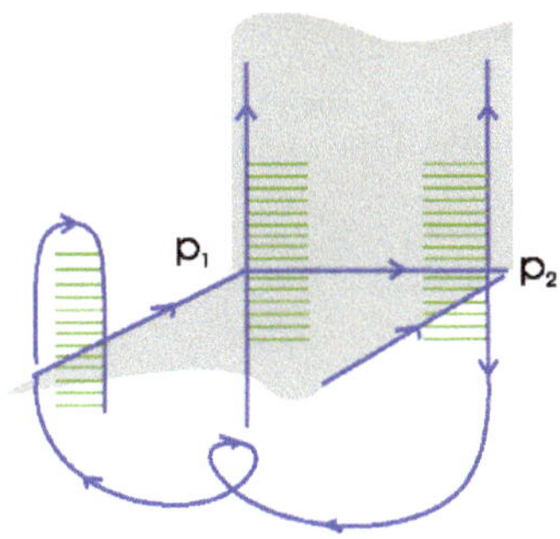

Fig. 9.

phisms display an array of varying intricate dynamics when subject to bifurcations, say in parametrized form. They may be in fact at a door of the difficult realm of dynamics needed to be better understood. Let me point out that there is a number of impressive results concerning heterodimensional cycles due to Bonatti, Diaz, Pujals and Rocha, like the ones in [30,41,43], also presented in [28].

We have a version of Conjecture 4 for flows. To state it, we have to consider the notion of a singular cycle, introduced in [63] and [22]. Here, a singular cycle for a flow just means a cycle involving singularities, at least one of them, and periodic orbits. Similar to the diffeomorphism case, a heterodimensional cycle for a flow is a cycle involving finitely many hyperbolic periodic orbits with different stable dimensions (indices).

Conjecture 5. *In any dimension, every flow (vector field) can be C^r approximated by a hyperbolic one or by one displaying a homoclinic tangency or a singular cycle or a heterodimensional cycle.*

The latter can only occur in dimensions greater than three. In dimension three, we can demand even more:

Conjecture 6. *Every three-dimensional flow can be C^r approximated by a hyperbolic one or by one displaying a singular cycle or a Lorenz-like attractor or repellor.*

In this spirit, Conjecture 5 may be formulated in a somewhat stronger form, substituting singular cycle by singular hyperbolic set in the sense of [80,81] and [82].

These conjectures have been the subject of much interest and some remarkable results have been obtained in the C^1 context. First of all, a breakthrough was due to Pujals and Sambarino [102] when proving Conjecture 4 for C^1 surface diffeomorphisms. Just now, Pujals is presenting in a long pre-publication [101] a similar fact for three-dimensional diffeomorphisms restricted to contracting regions. Partial advancements were provided by Wen [124,125] and an announcement has been made by Hayashi.

Concerning the conjectures for flows, Arroyo and Rodriguez and Hertz [16] provided a proof of Conjecture 5 in dimension three. In particular, they made an ingenious use of the work of Pujals–Sambarino mentioned above. Conjecture 6 relative to flows in dimension three remains open.

We conclude this part by very briefly mentioning some outstanding results relative to an important topic, certainly most relevant to our discussion at the beginning of this section. We shall do that first in the context of diffeomorphisms and then flows.

Let Λ be a maximal (in some neighborhood of it) invariant, transitive set for a diffeomorphism f on a closed (as usual here) manifold M. We say that Λ is robustly transitive if Λ_g is transitive, where Λ_g is the continuation of Λ and g is any map C^1 close to f. Similarly for flows.

Assume Λ is robustly transitive. Then:

- For C^1 diffeomorphisms, first, Mañé [72] has shown that if $\dim M = 2$, then Λ is a hyperbolic set. Then, in the three-dimensional case, Λ has to be partially hyperbolic, a result due to Diaz, Pujals and Ures [42], and in higher dimensions, Λ has at least to display a dominated decomposition, as proved by Bonatti, Diaz and Pujals [29].
- For C^1 flows, Morales, Pacifico and Pujals [82] proved the striking fact that on 3-manifolds a robustly transitive set is a Lorenz-like attractor or reppeler.

3.4. Homoclinic bifurcations and Hausdorff dimension

We have now a somewhat better understanding of the (complicated) dynamics that arises when we unfold homoclinic tangencies, say associated to a hyperbolic set Λ. Still, we have to go much further to be able, for instance, to respond to the challenge represented by Conjecture 3 in Section 3.2. Already a major question is

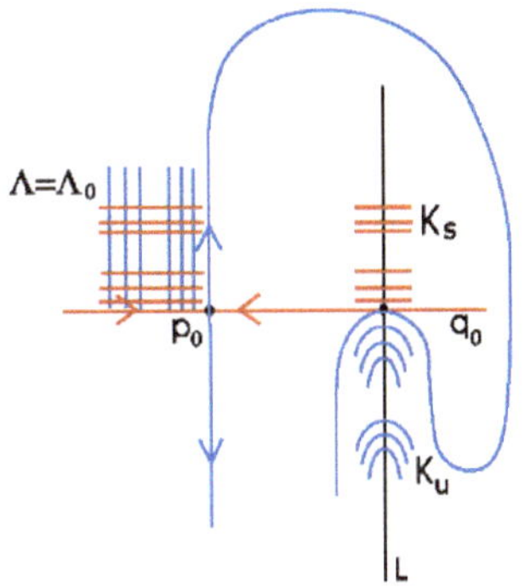

Fig. 10.

whether densely in the parameter line the corresponding maps have only finitely many localized attractors, i.e., in a neighborhood of p_0 and of the orbit of q_0, as in figure in Section 3.2.

So, let us again consider the unfolding with positive speed at $\mu = 0$ of a quadratic homoclinic tangency q_0 for a C^2 family f_μ of diffeomorphisms, associated to a fixed (periodic) point p_0 of a hyperbolic set $\Lambda = \Lambda_0$. We take μ small and let Λ_μ be the continuation of Λ. Also, let K_s, K_u be Cantor sets obtained by intersecting (for the first time from points in Λ) the leaves of the stable and unstable foliations of Λ with L, where L is a line transversal to the stable manifold of p_0. One may consider L as the unfolding parameter line. See Fig. 10. Finally, let d_s and d_u be the Hausdorff dimensions of K_s and K_u, respectively. Notice that $HD(\Lambda) = d_s + d_u$, where $HD(\Lambda)$ is the Hausdorff dimension of Λ.

We have:

- If $HD(\Lambda) < 1$, then hyperbolicity is fully prevalent at $\mu = 0$. That is, the set of parameter values corresponding to which f_μ is hyperbolic has Lebesgue density one at $\mu = 0$, Newhouse and Palis [87] and Palis and Takens [93,94].
- If $HD(\Lambda) > 1$, then hyperbolicity is not fully prevalent at $\mu = 0$, Palis and Yoccoz [97].
- If $HD(\Lambda) > 1$, then the arithmetic difference $K_s - K_u$ robustly contains non-trivial intervals. From that, one infers that homoclinic tangencies for f_μ have positive density at $\mu = 0$, implying the previous result in a strong form, Moreira and Yoccoz [84].

Notice that when $HD(\Lambda) < 1$, then $K_s - K_u$ has Lebesgue measure equals to zero.

From this string of facts, one can state the following dichotomy principle concerning the unfolding on surfaces of a homoclinic tangency associated to a hyperbolic set Λ:

The hyperbolic maps in the parametric family are prevalent if and only if $HD(\Lambda) < 1$.

Such a dichotomy was suggested by the author about two decades ago.

The most recent result in this direction is the following one, going quite deeply into the dynamical structure of f_μ after the unfolding of the homoclinic tangency:

- If $(d_s + d_u)^2 + (\max(d_s, d_u))^2 < d_s + d_u + \max(d_s, d_u)$, then non-uniform hyperbolicity is fully prevalent at $\mu = 0$. In particular, the set of parameter values corresponding to which the maps f_μ have localized attractors is of density zero at $\mu = 0$, Palis and Yoccoz [98]. Localized here has the same meaning as in Gorodetski and Kaloshin's result in Section 3.2, but now V is a neighborhood the whole hyperbolic set Λ and U is as before.

In our present proof of the last result, outlined in [98] and the full proof soon to appear, we treat the unfolding of a quadratic tangency between stable and unstable manifolds of two different periodic orbits in the hyperbolic set Λ. But, certainly the same should be true in the homoclinic case. We observe, from such a proof, that for many parameter values (total density at $\mu = 0$), the continuation of the forward maximal invariant set in a neighborhood $U \cup V$ as above has Hausdorff dimension smaller than two. Similarly, for the continuation of the backward maximal invariant set. Notice that, for such parameter values, there may be tangencies between stable and unstable manifolds for the corresponding maps, but no attractors (nor repellors).

Let us add a few words about the proof. It goes by showing that, for most parameters, the limit set is hyperbolic, in a delicate, non-uniform sense. Essentially, although the limit set may contain tangencies, they correspond to very special points: at "most" points there are transversal directions which are (asymptotically) contracted by forward and backward iterates, respectively. The proof of this fact requires a very careful analysis of how trajectories return close to the tangencies and, even, the very definition of what a "tangency" is. To ensure hyperbolic behavior, such returns should not be too frequent nor too close. This is achieved by parameter exclusions, which turns out to be less and less significant near the original tangency parameter (Lebesgue density zero). The rate of formation of tangencies is a crucial ingredient, and it is closely related to the Hausdorff dimension of the original horseshoe. The assumption, that this dimension is not far from one, ensures that the number of tangencies that must be considered at each stage grows fairly slowly. This implies that a fairly small amount (in measure) of parameter exclusions is needed. Returns close to the tangencies yield quadratic type folds. The condition on the frequency and depth of returns is used to ensure that folds always are "ironed-out" before a new return occurs. In this way, one never has to deal with contacts of order bigger than two.

One can certainly expect to have a similar general result without imposing any restriction on the Hausdorff dimension of the original horseshoe, at the price of having to deal with higher order contacts.

On the horizon lies the case of area preserving maps. An example is the famous family of standard maps on the torus, for which the limit set is the whole ambient space and, thus, has Hausdorff dimension two:

$$f_\mu(x, y) = \bigl(-y + 2x + \mu \sin(2\pi x), x\bigr).$$

In this case, one has to study contacts of all order simultaneously. The main open question concerning this family is whether there exists a positive Lebesgue measure set of parameter values μ, for which the corresponding maps are ergodic and non-uniformly hyperbolic.

3.5. *Unfolding homoclinic tangencies in higher dimensions*

To have a grasp of typical dynamics in all dimensions, we certainly must pursue a deep study of the unfolding of homoclinic tangencies in such a generality.

In doing so, we shall face some new and serious difficulties in comparison with the two-dimensional case. One of them is that in general the stable and unstable foliations of hyperbolic sets, when having codimension bigger than one, are not differentiable or even Lipshitz, independenty of the smoothness of the dynamics. As a consequence, the invariance along leaves of such foliations of some key instruments of nowadays dynamics, like Hausdorff dimension, are simply no longer available in higher dimensions. This is because at least one of the foliations must have codimension bigger than one.

Still, it's a promising fact that the obstacles, such as the one just mentioned, have been in some cases successfully surpassed. It was so in several of the results in this paper, particularly in the extension of Newhouse's infinitely many sinks phenomenon to higher dimensions in Section 3.2. Here, we comment on a recent result that goes much in the direction we are conveying.

As before, we consider one-parameter families of diffeomorphisms f_μ of class C^2, $\mu \in (-1,1)$, that generically (positive speed in μ at $\mu = 0$) unfold a quadratic homoclinic tangency q_0 associated to a hyperbolic horseshoe $\Lambda = \Lambda_0$.

A reasonable hypothesis, as done on surfaces, see Section 3.4, is to assume that f_μ is hyperbolic (Axiom A) for $\mu < 0$. So, we are creating a cycle, a homoclinic tangency, at $\mu = 0$. This hypothesis implies that the weak stable and unstable eigenvalues of $\mathrm{d}f_0$ at p_0 are well defined and that they are real numbers. Moreira, Viana and myself then showed (see a résumé in [83]) that the dichotomy

$$\textit{hyperbolicity is prevalent} \quad \Leftrightarrow \quad \textit{Hausdorff dimension} < 1$$

is true in all dimensions.

Prevalence of hyperbolicity means: the set of μ's for which hyperbolicity holds for the maximal invariant set of f_μ in a neighborhood of Λ and q_0 as in Section 3.4, has total density at $\mu = 0$. And by Hausdorff dimension smaller than one, we mean $HD_s(\Lambda) + HD_u(\Lambda) < 1$ where

$$HD_s(\Lambda) = HD\big(\Lambda \cap W^s(p_0)\big), \qquad HD_u(\Lambda) = HD\big(\Lambda \cap W^u(p_0)\big).$$

Formally, we state the result as follows:

There are open sets R_1 and R_2 of one-parameter families of C^k diffeomorphisms, $k \geqslant 2$, unfolding a homoclinic tangency as set above, such that

(a) $R_1 \cup R_2$ is dense;
(b) for families in R_1, we have

$$HD_s(\Lambda) + HD_u(\Lambda) < 1 \quad \text{and} \quad \lim_{\delta \to 0} \frac{m(H \cap [0,\delta])}{\delta} = 1$$

where $H = \{\mu \mid f_\mu \text{ is hyperbolic}\}$ and m stands for Lebesgue measure;
(c) for families in R_2, we have for $\delta > 0$

$$HD_s(\Lambda) + HD_u(\Lambda) > 1 \quad \text{and} \quad \liminf_{\delta \to 0} \frac{m(T_s \cap [0,\delta])}{\delta} > 0$$

where $T_s = \{\mu \mid f_\mu \text{ presents a persistent homoclinic tangency associated to } \Lambda_\mu\}$ and Λ_μ is the continuation of Λ, μ small.

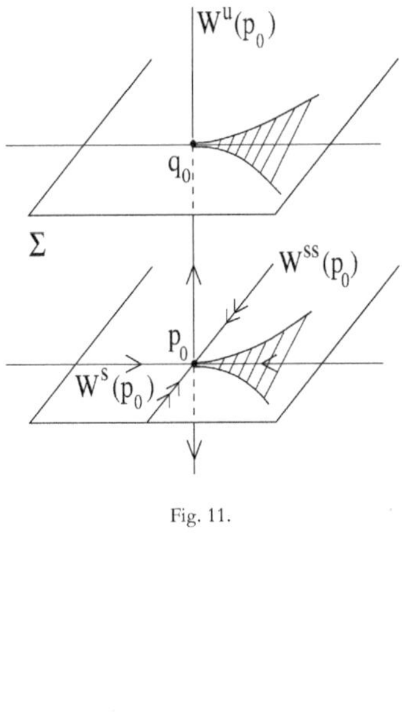

Fig. 11.

We briefly comment the proof of this result:

Part (b) is reminiscent of the Palis–Takens' result on surfaces [94]. The proof here goes by ensuring that, up to small perturbation, the horseshoe avoids the strong stable and unstable directions. This is possible because its Hausdorff dimension is small. It implies that $\Lambda \cap W^s(p)$ is contained in a cuspidal region around the weak stable direction of p_0. Similarly for $\Lambda \cap W^u(p)$ with respect to the weak unstable direction of p_0. See Fig. 11. It has the geometric consequence that the horseshoe is roughly two-dimensional. This permits to mimic the two-dimensional arguments to obtain uniform transversality of stable and unstable foliations of the continuation of the original horseshoe for most parameters.

Part (c) is considerably more delicate. The main ingredient is the construction, after perturbation of f, of strong-stable and strong-unstable foliations of codimension-one for hyperbolic subsets of Λ with almost the same Hausdorff dimensions.

These foliations are used to reduce the study of the geometries of the stable and unstable foliations near the initial homoclinic tangency to the two-dimensional case. Another key step is to obtain stable tangencies, in terms of parameters, of stable and unstable manifolds of periodic orbits as in Moreira and Yoccoz [84].

References

[1] R. Abraham, S. Smale, Nongenericity of Ω-stability, in: Global Analysis, Berkeley 1968, in: Proc. Sympos. Pure Math., vol. XIV, Amer. Math. Soc., 1970.

[2] V.S. Afraimovich, V.V. Bykov, L.P. Shil'nikov, On the appearance and structure of the Lorenz attractor, Dokl. Acad. Sci. USSR 234 (1977) 336–339.

[3] V.S. Afraimovich, L.P. Shil'nikov, The accessible transitions from Morse–Smale systems to systems with several periodic motions, Izv. Akad. Nauk SSSR 38 (1974) 1248–1288;
English translation: V.S. Afraimovich, L.P. Shil'nikov, Math. USSR-Izv. 8 (1976) 1325–1370.

[4] V.S. Afraimovich, L.P. Shil'nikov, On bifurcations of codimension 1, leading to the appearance of fixed points of a countable set of tori, Dokl. Akad. Nauk SSSR 262 (1982) 777–780;
English translation: V.S. Afraimovich, L.P. Shil'nikov, Soviet Math. Dokl. 25 (1982) 101–105.

[5] J.F. Alves, C. Bonatti, M. Viana, SRB measures for partially hyperbolic systems whose central direction is mostly expanding, Invent. Math. 140 (2000) 351–398.

[6] A. Andronov, E.A. Leontovich, Some cases of dependence of limit cycles on parameters, Uchen. Zap. Gor'kov. Univ. 6 (1939) 3–24.

[7] A. Andronov, E.A. Leontovich, I.I. Gordon, A.G. Majer, Qualitative Theory of Dynamical Systems of Second Order, Nauka, Moscow, 1966, p. 569.

[8] A. Andronov, E.A. Leontovich, I.I. Gordon, A.G. Majer, The Theory of Bifurcations of Dynamical Systems on the Plane, Nauka, Moscow, 1967, p. 487.

[9] A. Andronov, L. Pontryagin, Systèmes grossiers, Dokl. Akad. Nauk USSR 14 (1937) 247–251.

[10] D.V. Anosov, Geodesic flows on closed Riemannian manifolds of negative curvature, Proc. Steklov Math. Inst. 90 (1967) 1–235.

[11] A. Arnéodo, P. Coullet, C. Tresser, Possible new strange attractors with spiral structure, Comm. Math. Phys. 79 (1981) 673–679.

[12] V.I. Arnold, Small denominators I: On the mapping of a circle to itself, Izv. Akad. Nauk Math. Ser. 25 (1961) 21–86;
English translation: V.I. Arnold, Transl. Amer. Math. Soc. 46 (1965) 213–284.

[13] V.I. Arnold, Lectures on bifurcations and versal families, Uspekhi Math. Nauk 27 (1972) 119–184;
English translation: V.I. Arnold, Russian Math. Surveys 27 (1973) 54–123.

[14] V.I. Arnold, Kolmogorov's hydrodynamics attractors. turbulence and stochastic processes: Kolmogorov's ideas 50 years on, Proc. Roy. Soc. London Ser. A 434 (1991) 19–22;
English translation: V.I. Arnold, Russian Math. Surveys 27 (1973) 54–123.

[15] V.I. Arnold, V.S. Afraimovich, Yu. Ilyasheko, L.P. Shilnikov, Bifurcation Theory and Catastrophe Theory, Springer, 1999.

[16] A. Arroyo, F. Rodriguez Hertz, Homoclinic bifurcations and uniform hyperbolicity for three-dimensional flows, Ann. Inst. H. Poincaré 20 (2003) 805–841.

[17] A. Avila, M. Lyubich, W. de Melo, Regular or stochastic dynamics in real analytic families of unimodal maps, Invent. Math. 154 (2003) 451–550.

[18] A. Avila, C.G. Moreira, Phase-parameter relation sharp statistical properties of unimodal maps, Contemp. Math., in press.

[19] A. Avila, C.G. Moreira, Statistical properties of unimodal maps: physical measures, periodic orbits and pathological laminations, Publ. Math. IHES, in press.

[20] A. Avila, C.G. Moreira, Statistical properties of unimodal maps: the quadratic family, Ann. of Math., in press.

[21] A. Avila, C.G. Moreira, Statistical properties of unimodal maps: smooth families with negative Schwarzian derivative, Astérisque 286 (2003) 81–118.
[22] R. Bamón, R. Labarca, R. Mañé, M.J. Pacifico, The explosion of singular cycles, Publ. Math. IHES 78 (1993) 207–232.
[23] M. Benedicks, L. Carleson, The dynamics of the Hénon map, Ann. of Math. 133 (1991) 73–169.
[24] M. Benedicks, M. Viana, Random perturbations and statistical properties of Hénon-like maps, Ann. Inst. H. Poincaré Anal. Non Linéaire.
[25] M. Benedicks, M. Viana, Solution of the basin problem for Hénon-like attractors, Invent. Math. 143 (2001) 375–434.
[26] M. Benedicks, L.-S. Young, SBR-measures for certain Hénon maps, Invent. Math. 112 (1993) 541–576.
[27] G.D. Birkhoff, Nouvelles recherches sur les systèmes dynamiques, Mem. Pont. Acad. Sci. Novi. Lyncaei 1 (1935) 85–216.
[28] C. Bonatti, L. Diaz, M. Viana, Dynamics Beyond Uniform Hyperbolicity, Encyclopaedia Math. Sci., vol. 102, Springer, 2004.
[29] C. Bonatti, L.J. Díaz, E. Pujals, A C^1-generic dichotomy for diffeomorphisms: weak forms of hyperbolicity or infinitely many sinks or sources, Ann. of Math. 158 (2003) 355–418.
[30] C. Bonatti, L.J. Díaz, E. Pujals, J. Rocha, Robust transitivity and heterodimensional cycles, Astérisque 286 (2003) 187–222.
[31] C. Bonatti, A. Pumariño, M. Viana, Lorenz attractors with arbitrary expanding dimension, C. R. Acad. Sci. Paris, Sér. I Math. 325 (1997) 883–888.
[32] R. Bowen, D. Ruelle, The ergodic theory of Axiom A flows, Invent. Math. 29 (1975) 181–202.
[33] L.A. Bunimovich, Ya.G. Sinai, Stochasticity of the attractor in the Lorenz model, in: Nonlinear Waves, in: Proc. Winter School, Nauka, Moscow, 1980, pp. 212–226.
[34] M. Cartwright, J. Littlewood, On non-linear differential equations of the second order, J. London Math. Soc. 20 (1945) 127–153.
[35] M. Cartwright, J. Littlewood, On non-linear differential equations of the second order, Ann. of Math. 48 (1947) 472–494.
[36] E. Colli, Infinitely many coexisting strange attractors, Ann. Inst. H. Poincaré Anal. Non Linéaire 15 (1998) 539–579.
[37] P. Coullet, C. Tresser, Itérations d'endomorphims et groupe de renormalization, C. R. Acad. Sci. Paris, Sér. I 287 (1978) 577–580.
[38] M.J. Dias Carneiro, J. Palis, Bifurcations and global stability of families of gradients, Publ. Math. IHES 70 (1989) 70–163.
[39] W. de Melo, Structural stability of diffeomorphisms on two-manifolds, Invent. Math. 21 (1973) 233–246.
[40] B. Van der Pol, On relaxation oscillations, Philos. Mag. Ser. 7 2 (1926) 978–992.
[41] L.J. Díaz, Robust nonhyperbolic dynamics and heterodimensional cycles, Ergodic Theory Dynam. Systems 15 (1995) 291–315.
[42] L.J. Díaz, E. Pujals, R. Ures, Partial hyperbolicity and robust transitivity, Acta Math. 183 (1999) 1–43.
[43] L.J. Díaz, J. Rocha, Large measure of hyperbolic dynamics when unfolding heterodimensional cycles, Nonlinearity 10 (1997) 857–884.
[44] D. Dolgopyat, On differentiability of SRB states for partially hyperbolic systems, Invent. Math. 155 (2004) 359–449.
[45] M. Feigenbaum, Qualitative universality for a class of nonlinear transformations, J. Statist. Phys. 19 (1978) 25–52.
[46] J. Franks, R. Williams, Anomalous Anosov Flows. Global Theory of Dynamical Systems, Lecture Notes in Math., vol. 819, Springer, 1980.
[47] S.V. Gonchenko, L.P. Shil'nikov, D.V. Turaev, Dynamical phenomena in systems with structurally unstable Poincaré homoclinic orbits, Chaos 6 (1996) 15–31.
[48] A. Gorodetski, V. Kaloshin, How often surface diffeomorphisms have infinitely many sinks and hyperbolicity of periodic points near a homoclinic tangency, in press.
[49] J. Graczyk, G. Swiatek, Generic hyperbolicity in the logistic family, Ann. of Math. 146 (1997) 1–52.
[50] J. Guckenheimer, R.F. Williams, Structural stability of Lorenz attractors, Publ. Math. IHES 50 (1979) 59–72.
[51] S. Hayashi, Connecting invariant manifolds and the solution of the C^1 stability and Ω-stability conjectures for flows, Ann. of Math. 145 (1997) 81–137.
[52] M. Hénon, A two dimensional mapping with a strange attractor, Comm. Math. Phys. 50 (1976) 69–77.
[53] Yu. Ilyashenko, W. Li, Nonlocal Bifurcations, Math. Surveys and Monographs, vol. 66, Amer. Math. Soc., 1999.
[54] M. Jakobson, On smooth mappings of the circle into itself, Math. USSR-Sb. 14 (1971) 161–185.
[55] M. Jakobson, Absolutely continuous invariant measures for one-parameter families of one-dimensional maps, Comm. Math. Phys. 81 (1981) 39–88.
[56] V.Yu. Kaloshin, B.R. Hunt, Stretched exponential estimate on growth of the number of periodic points for prevalent diffeomorphisms, in press.
[57] Yu. Kifer, Ergodic Theory of Random Perturbations, Birkhäuser, 1986.
[58] Yu. Kifer, Random Perturbations of Dynamical Systems, Birkhäuser, 1988.
[59] O. Kozlovski, Getting rid of the negative Schwarzian derivative condition, Ann. of Math. 152 (2000) 743–762.
[60] O. Kozlovski, Axiom A maps are dense in the space of unimodal maps in the C^k topology, Ann. of Math. 157 (2003) 1–43.
[61] O. Kozlovski, W. Shen, S. van Strien, Density of hyperbolicity in dimension one, Preprint, Warwick, 2004.
[62] O. Kozlovski, S. van Strien, W. Shen, Rigidity for real polynomials, Preprint, Warwick, 2003.
[63] R. Labarca, M.J. Pacifico, Stability of singular horseshoes, Topology 25 (1986) 337–352.
[64] M. Levi, Qualitative analysis of the periodically forced relaxation oscillations, Mem. Amer. Math. Soc. 32 (244) (1981).
[65] N. Levinson, A second order differential equations with singular solutions, Ann. of Math. 50 (1949) 127–153.
[66] S.-T. Liao, Hyperbolicity properties of the non-wandering sets of certain 3-dimensional systems, Acta Math. Sci. 3 (1983) 361–368.
[67] S.T. Liao, On the stability conjecture, Chinese Ann. of Math. 1 (1980) 9–30.

[68] J. Littlewood, On non-linear differential equations of the second order, III, Acta Math. 97 (1957) 267–308.
[69] J. Littlewood, On non-linear differential equations of the second order, IV, Acta Math. 98 (1957) 1–110.
[70] E.N. Lorenz, Deterministic nonperiodic flow, J. Atmosph. Sci. 20 (1963) 130–141.
[71] M. Lyubich, Almost every real quadratic map is either regular or stochastic, Ann. of Math. 156 (2002) 1–78.
[72] R. Mañé, An ergodic closing lemma, Ann. of Math. 116 (1982) 503–540.
[73] R. Mañé, A proof of the C^1 stability conjecture, Publ. Math. IHES 66 (1988) 161–210.
[74] M. Martens, T. Nowicki, Invariant measures for Lebesgue typical quadratic maps. Géométrie complexe et systèmes dynamiques (Orsay, 1995), Astérisque 261 (2001) 239–252.
[75] R.M. May, Simple mathematical models with very complicated dynamics, Nature 261 (1976) 459–467.
[76] C. McMullen, Complex Dynamics and Renormalization, Ann. of Math. Stud., vol. 142, Princeton University Press, 1994.
[77] R. Metzger, Sinai–Ruelle–Bowen measures for contracting Lorenz maps and flows, Ann. Inst. H. Poincaré Anal. Non Linéaire 17 (2000) 247–276.
[78] R. Metzger, Stochastic stability for contracting Lorenz maps and flows, Comm. Math. Phys. 212 (2000) 277–296.
[79] L. Mora, M. Viana, Abundance of strange attractors, Acta Math. 171 (1993) 1–71.
[80] C. Morales, M.J. Pacifico, A dichotomy for three-dimensional vector fields, Ergodic Theory Dynam. Systems 23 (2003) 1575–1600.
[81] C. Morales, M.J. Pacifico, E. Pujals, Singular hyperbolic systems, Proc. Amer. Math. Soc. 127 (1999) 3393–3401.
[82] C. Morales, M.J. Pacifico, E. Pujals, Robust transitive singular sets for 3-flows are partially hyperbolic attractors and repellers, Ann. of Math. 160 (2004) 1–58.
[83] C.G. Moreira, J. Palis, M. Viana, Homoclinic tangencies and fractal invariants in arbitrary dimension, C. R. Acad. Sci. Paris, Sér. I Math. 333 (2001) 475–480.
[84] C.G. Moreira, J.-C. Yoccoz, Stable intersections of regular Cantor sets with large Hausdorff dimensions, Ann. of Math. 154 (2001) 45–96; Corrigendum in: C.G. Moreira, J.-C. Yoccoz, Ann. of Math. 154 (2001) 527.
[85] S. Newhouse, Diffeomorphisms with infinitely many sinks, Topology 13 (1974) 9–18.
[86] S. Newhouse, The abundance of wild hyperbolic sets and nonsmooth stable sets for diffeomorphisms, Publ. Math. IHES 50 (1979) 101–151.
[87] S. Newhouse, J. Palis, Cycles and bifurcation theory, Astérisque 31 (1976) 44–140.
[88] S. Newhouse, J. Palis, F. Takens, Bifurcations and stability of families of diffeomorphisms, Publ. Math. IHES 57 (1983) 5–71.
[89] J. Palis, On Morse–Smale dynamical systems, Topology 8 (1969) 385–405.
[90] J. Palis, A global view of Dynamics and a conjecture on the denseness of finitude of attractors. Géométrie complexe et systèmes dynamiques (Orsay, 1995), Astérisque 261 (1995) 335–347.
[91] J. Palis, S. Smale, Structural stability theorems, in: Global Analysis, Berkeley 1968, in: Proc. Sympos. Pure Math., vol. XIV, Amer. Math. Soc., 1970, pp. 223–232.
[92] J. Palis, F. Takens, Stability of parametrized families of gradient vector fields, Ann. of Math. 118 (1983) 383–421.
[93] J. Palis, F. Takens, Cycles and measure of bifurcation sets for two-dimensional diffeomorphisms, Invent. Math. 82 (1985) 397–422.
[94] J. Palis, F. Takens, Hyperbolic and the creation of homoclinic orbits, Ann. of Math. 1987 (1987) 337–374.
[95] J. Palis, F. Takens, Hyperbolicity and Sensitive-Chaotic Dynamics at Homoclinic Bifurcations, Cambridge University Press, 1993.
[96] J. Palis, M. Viana, High dimension diffeomorphisms displaying infinitely many periodic attractors, Ann. of Math. 140 (1994) 207–250.
[97] J. Palis, J.-C. Yoccoz, Homoclinic tangencies for hyperbolic sets of large Hausdorff dimension, Acta Math. 172 (1994) 91–136.
[98] J. Palis, J.-C. Yoccoz, Fers à cheval non uniformément hyperboliques engendrés par une bifurcation homocline et densité nulle des attracteurs, C. R. Acad. Sci. Paris, Sér. I Math. 333 (2001) 867–871.
[99] M. Peixoto, Structural stability on two-dimensional manifolds, Topology 1 (1962) 101–120.
[100] Ya. Pesin, Ya. Sinai, Gibbs measures for partially hyperbolic attractors, Ergodic Theory Dynam. Systems 2 (1982) 417–438.
[101] E. Pujals, Density of hyperbolicity and homoclinic bifurcation for 3d-diffeomorphisms in attracting regions, IMPA's preprint server, http://www.preprint.impa.br.
[102] E. Pujals, M. Sambarino, Homoclinic tangencies and hyperbolicity for surface diffeomorphisms, Ann. of Math. 151 (2000) 961–1023.
[103] J. Robbin, A structural stability theorem, Ann. of Math. 94 (1971) 447–493.
[104] C. Robinson, Structural stability of vector fields, Ann. of Math. 99 (1974) 154–175;
Errata in: C. Robinson, Ann. of Math. 101 (1975) 368.
[105] C. Robinson, Homoclinic bifurcation to a transitive attractor of Lorenz type, Nonlinearity 2 (1989) 495–518.
[106] A. Rovella, The dynamics of perturbations of the contracting Lorenz attractor, Bull. Braz. Math. Soc. 24 (1993) 233–259.
[107] D. Ruelle, A measure associated with Axiom A attractors, Amer. J. Math. 98 (1976) 619–654.
[108] M. Rychlik, Lorenz attractors through Shil'nikov-type bifurcation. Part 1, Ergodic Theory Dynam. Systems 10 (1990) 793–821.
[109] A. Sannami, The stability theorems for discrete dynamical systems on two-dimensional manifolds, Nagoya Math. J. 90 (1983) 1–55.
[110] L.P. Shil'nikov, On the generation of periodic motion from a trajectory doubly asymptotic to an equilibrium state of saddle type, Math. USSR-Sb. 6 (1968) 428–438.
[111] R. Simon, A 3-dimensional Abraham–Smale example, Proc. Amer. Math. Soc. 34 (1972) 629–630.
[112] Ya. Sinai, Gibbs measures in ergodic theory, Russian Math. Surveys 27 (1972) 21–69.

[113] S. Smale, Diffeomorphisms with many periodic points, in: Differential and Combinatorial Topology, Princeton University Press, 1965.
[114] S. Smale, Differentiable dynamical systems, Bull. Amer. Math. Soc. 73 (1967) 747–817.
[115] D. Sullivan, Bounds, quadratic differentials and renormalization conjectures, Ann. Math. Soc. Centennial Publ. 2 (1992) 417–466.
[116] L. Tedeschini-Lalli, J. Yorke, How often do simple dynamical systems have infinitely many coexisting sinks? Comm. Math. Phys. 106 (1986) 635–657.
[117] M. Tsujii, Physical measures for partially hyperbolic surface endomorphisms, Acta Math., in press.
[118] W. Tucker, A rigorous ODE solver and Smale's 14th problem, Found. Comput. Math. 2 (2002) 53–117.
[119] R. Ures, Abundance of hyperbolicity in C^1 topology, Ann. Sci. École Norm. Sup. 28 (1995) 747–760.
[120] C. Vasquez, Statistical stability for diffeomorphisms with dominated splitting, in press.
[121] M. Viana, Strange attractors in higher dimensions, Bull. Braz. Math. Soc. 24 (1993) 13–62.
[122] M. Viana, Multidimensional nonhyperbolic attractors, Publ. Math. IHES 85 (1997) 63–96.
[123] M. Viana, What's new on Lorenz strange attractors? Math. Intelligencer 22 (2000) 6–19.
[124] L. Wen, Homoclinic tangencies and dominated splittings, Nonlinearity 15 (2002) 1445–1469.
[125] L. Wen, Generic diffeomorphisms away from homoclinic tangencies and heterodimensional cycles, Bull. Braz. Math. Soc. 35 (2004) 419–452.
[126] L.-S. Young, Stochastic stability of hyperbolic attractors, Ergodic Theory Dynam. Systems 6 (1986) 311–319.

NON-UNIFORMLY HYPERBOLIC HORSESHOES ARISING FROM BIFURCATIONS OF POINCARÉ HETEROCLINIC CYCLES

by JACOB PALIS *and* JEAN-CHRISTOPHE YOCCOZ*

ABSTRACT

In the present paper, we advance considerably the current knowledge on the topic of bifurcations of heteroclinic cycles for smooth, meaning C^∞, parametrized families $\{g_t \mid t \in \mathbf{R}\}$ of surface diffeomorphisms. We assume that a quadratic tangency q is formed at $t = 0$ between the stable and unstable lines of two periodic points, not belonging to the same orbit, of a (uniformly hyperbolic) horseshoe K (see an example at the Introduction) and that such lines cross each other with positive relative speed as the parameter evolves, starting at $t = 0$ and the point q. We also assume that, in some neighborhood W of K and of the orbit of tangency $o(q)$, the maximal invariant set for $g_0 = g_{t=0}$ is $K \cup o(q)$, where $o(q)$ denotes the orbit of q for g_0. We then prove that, when the Hausdorff dimension HD(K) is bigger than one, but not much bigger (see (H.4) in Section 1.2 for a precise statement), then for most t, $|t|$ small, g_t is a non-uniformly hyperbolic horseshoe in W, and so g_t has no attractors in W. Most t, and thus most g_t, here means that t is taken in a set of parameter values with Lebesgue density one at $t = 0$.

CONTENTS

Abstract . . . 1
1. Introduction . . . 1
2. Markov partition and folding map . . . 14
3. Affine-like maps . . . 18
4. Structure of parameter space . . . 27
5. Classes of affine-like iterates and the transversality relation . . . 29
6. Some properties of the classes $\mathcal{R}(I)$ . . . 40
7. Estimates for the classes $\mathcal{R}(I)$ . . . 55
8. Number of children and dimension estimates . . . 67
9. Strong regularity and parameter selection . . . 87
10. The well-behaved part of the dynamics for strongly regular parameters . . . 139
11. Some further geometric properties of the invariant set . . . 174
Appendix A: Composition formulas for affine-like maps . . . 194
Appendix B: On the Lipschitz regularity of $\widetilde{\mathbf{R}}_+^\infty$ . . . 207
Appendix C: A toy model for the transversality relation . . . 212
References . . . 216

1. Introduction

1.1. *The context.* — One of the most challenging problems in the theory of dynamical systems is to understand some of the main features, like creation of attractors, of the orbit structure of the dynamics arising from bifurcations of homoclinic or heteroclinic cycles. Typically, the cycles we consider display an orbit of non-transversal intersection between some stable and unstable manifolds of fixed or periodic orbits, whose unfolding leads to dynamics with a rich orbit structure. For surface diffeomorphisms, such orbit of non-transversal intersection of stable and unstable lines correspond to a homoclinic or

* Partially support by CNPq and FAPERJ, Brazil.

Originally published in *Publ. Math. Inst. Hautes Études Sci. no. 110 (2009), 1-217,* heteroclinic tangency. The existence of cycles was much emphasized by Poincaré in late 19th century, in his classic Nouvelles Méthodes de la Mécanique Céleste [Po], where he stated that "rien n'est plus propre à nous donner une idée de la complication de tous les problèmes de dynamique". The question of bifurcations of cycles often arises when we consider parametrized families of dynamics, like in the present work: they are created and then bifurcate, as the parameter evolves.

In the case of surface diffeomorphisms, the richness of the dynamics obtained from unfolding a homoclinic or heteroclinic tangency is implicit in the work of Cartright-Littlewood [CL] more than sixty years ago and thereafter in several articles by the authors, especially the latter; see Levi [L] for an account of some explicit consequences of these works. In between, in the early sixties, and along the same line, a new and fundamental dynamical structure was exhibited by Smale [S]: the horseshoe map associated to a transversal homoclinic orbit. Besides that, Levi also made use of the following remarkable result of Newhouse [N]: under some mild conditions, the unfolding of a homoclinic tangency for C^r, $r \geq 2$, surface diffeomorphisms leads, in the C^r topology, to open sets of diffeomorphisms such that none of its elements is (uniformly) hyperbolic. That is, hyperbolic diffeomorphisms are not dense in the set of all such maps, which had been an important conjecture by Smale. He also showed that such an unfolding leads to the existence of open sets of diffeomorphisms with a dense (actually, Baire second category) subset of elements displaying each of them infinitely many simultaneous periodic attractors (sinks) or repellors (sources). We refer especially to [BDV], for a comprehensive presentation of the concepts and results that we have just mentioned.

Abundance of other more intricate kind of attractors, the so called Hénon-like ones, was proved to be also present in the unfolding of such cycles. This was another striking fact. It resulted from the pioneering work of Benedicks-Carleson [BC], and those of Mora-Viana [MV] and Colli [C]. Attractors here mean invariant sets that attract future orbits of points of a positive Lebesgue measure set in the phase space (space of events).

In view of all these intricacies inherent to homoclinic and heteroclinic bifurcations, a new global conjecture has been proposed in [P1] (see also [P2] and [P3]) concerning the orbit structure of a typical dynamical system: in particular, systems with finitely many attractors should be dense in the universe of dynamics, i.e. C^r flows, diffeomorphisms and maps, with $r \geq 1$. Also, their basins of attraction should cover the whole phase space, except for a Lebesgue zero measure set. Several other conjectures were formulated in the above works that together compose a global scenario for dynamics.

As mentioned in the Abstract, the present paper represents a contribution to the understanding of the dynamics arising from bifurcating a cycle of a C^∞ surface diffeomorphism. We consider one-parameter families of diffeomorphisms g_t containing the initial bifurcating diffeomorphism, say g_0 at parameter value $t = 0$. We assume that g_t is hyperbolic for $t < 0$ and $|t|$ small. We suppose that the cycle is formed by a (hyperbolic) horseshoe K and an orbit of tangency $o(q)$ between stable and unstable manifolds of different periodic orbits of K. We assume the maximal invariant set in a small neighbor-

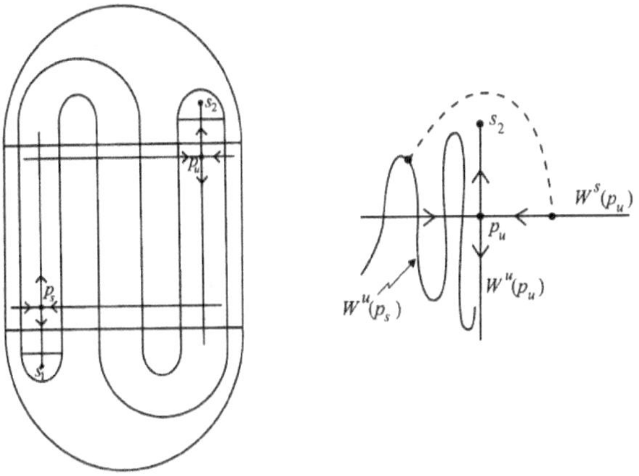

FIG. 1. — Horseshoe and homoclinic tangency

hood W of $K \cup o(q)$ to consist precisely of $K \cup o(q)$. A main novelty is that we allow the Hausdorff dimension of K to be larger than one, but not too far from one. We show that right after the bifurcation, i.e. for $t > 0$ small, most diffeomorphisms are non-uniformly hyperbolic in W and so they display no attractors nor repellors in the neighborhood W of $K \cup o(q)$. This means that the parameter values corresponding to diffeomorphisms displaying no attractors nor repellors should have total density, that is density one, at $t = 0$. The concept is again discussed in the next subsection.

An example of the creation of a heteroclinic cycle associated to a (hyperbolic) horseshoe is indicated in Figure 1. Initially we have the classic Smale's horseshoe map (diffeomorphism) on the two-sphere S^2 with two saddle fixed points p_s, p_u with positive eigenvalues, a fixed point repellor outside the figure and two fixed point attractors s_1 and s_2. The rectangle inside the figure is sent by the map to the snake-shaped piece, while the bigger top half-disk is sent to the small one around s_2 and the lower bigger half disk is sent to the small one around s_1. At the right hand side of the figure, we show how to move a small neighborhood of a point in the stable line of p_u so as to create a tangency with the unstable manifold of p_s. This is done through a one-parameter family of diffeomorphisms; until we create such a tangency the corresponding map remains hyperbolic, i.e. having a hyperbolic limit set with no cycles among its basic sets.

Our results considerably extend those in [PT], [NP] obtained for the case when the Hausdorff dimension HD(K) is smaller than one. They were announced in [PY3].

Of course, we expect the same results to be true for all cases $0 < HD(K) < 2$. To achieve that, it seems to us that our methods need to be considerably sharpened: we have to study deeper the dynamical recurrence of points near tangencies of higher order (cubic, quartic ...) between stable and unstable curves. We also expect our results to be true in higher dimensions (see [MPV]). Finally, we hope that the ideas introduced in the present paper might be useful in broader contexts. In the horizon lies a famous question concerning the standard family of area preserving maps (see [BDV]): we ask whether we can find sets of positive Lebesgue probability in parameter space such that the corresponding maps display non-zero Lyapunov exponents in sets of positive Lebesgue probability in phase space.

1.2. *The setting and a first formulation of the main result.* — Let f be a smooth, i.e. C^∞ diffeomorphism of a smooth surface M.

Recall that a *basic set* is a compact hyperbolic transitive locally maximal invariant set. A basic set is a *horseshoe* if it is infinite and is neither an attractor nor a repellor. A horseshoe is topologically a Cantor set.

We assume that there exists a basic set K for f, points $p_s, p_u \in K$, $q \in M - K$ such that the following properties hold:

(H1) p_s and p_u are periodic points and belong to distinct periodic orbits;
(H2) $W^s(p_s)$ and $W^u(p_u)$ have a quadratic tangency at q;
(H3) there exists a neighbourhood U of K, a neighbourhood V of the orbit $\mathcal{O}(q)$ of q, such that $K \cup \mathcal{O}(q)$ is the maximal invariant set in $U \cup V$.

We would like to understand, when U, V are appropriately small and g is C^∞ close to f, the maximal invariant set

$$\Lambda_g = \bigcap_{\mathbf{Z}} g^{-n}(U \cup V). \tag{1.1}$$

Observe that the smaller set

$$K_g = \bigcap_{\mathbf{Z}} g^{-n}(U) \tag{1.2}$$

is a horseshoe which is the hyperbolic continuation of K.

Let $\mathcal{U}$ be an appropriately small neighbourhood of f in $\mathrm{Diff}^\infty(M)$. We still denote by p_s, p_u the continuation of these hyperbolic periodic points in $\mathcal{U}$. The condition that $W^s(p_s)$, $W^u(p_u)$ have a quadratic tangency near q defines a codimension 1 hypersurface $\mathcal{U}_0$ through f in $\mathcal{U}$. It divides $\mathcal{U}$ into regions $\mathcal{U}_+, \mathcal{U}_-$ such that, for $g \in \mathcal{U}_-$, $W^s(p_s)$ and $W^u(p_u)$ do not intersect near q while, for $g \in \mathcal{U}_+$, $W^s(p_s)$ and $W^u(p_u)$ have two transverse intersection points near q (for obvious dynamical reasons, the intersection is actually infinite in this case; we are really considering here the intersection derived from the continuation of large *compact* curves contained in $W^s(p_s)$ and $W^u(p_u)$).

When $g \in \mathcal{U}_-$, we clearly have

$$\Lambda_g = K_g. \tag{1.3}$$

When $g \in \mathcal{U}_0$, we have

$$\Lambda_g = K_g \cup \mathcal{O}(q_g), \tag{1.4}$$

where q_g is the tangency point close to q given by the definition of $\mathcal{U}_0$. In the subsequent sections, we will omit the dependence on g in the notation and just write q for q_g when $g \in \mathcal{U}_0$.

The interesting case is therefore $g \in \mathcal{U}_+$.

It is actually not realistic to try to understand Λ_g for all $g \in \mathcal{U}_+$. One of the reasons is the so-called Newhouse's phenomenon [N]: there exists an open set $\mathcal{N} \subset \mathcal{U}_+$, with $\mathcal{U}_0 \subset \overline{\mathcal{N}}$, such that, residually in $\mathcal{N}$, Λ_g has infinitely many periodic sinks or sources and so its full dynamical description appears to be beyond reach. See also [BC], [MV], [C] for similar results involving Hénon-like attractors.

Still, we can and shall consider most $g \in \mathcal{U}_+$ in the following sense.

We will say that a subset $\mathcal{P} \subset \mathcal{U}_+$ contains most $g \in \mathcal{U}_+$ if, for any smooth 1-parameter family $(g_t)_{t\in(-t_0,t_0)}$ which is transverse to $\mathcal{U}_0$ at $t = 0$ (with $g_t \in \mathcal{U}_+$ for $t > 0$), we have

$$\lim_{t\to 0} \frac{1}{t} \operatorname{Leb}(s \in (0, t], g_s \in \mathcal{P}) = 1. \tag{1.5}$$

Denote by $W^s(K)$ (resp. $W^u(K)$) the stable set (resp. unstable set) of K for f. This is a partial foliation with a $C^{1+\alpha}$ Cantor transverse structure; denote by d_s^0 (resp. d_u^0) the transverse Hausdorff dimension of $W^s(K)$ (resp. $W^u(K)$). The Hausdorff dimension of K is equal to $d_s^0 + d_u^0$. We then have, in some contrast to Newhouse's phenomenon:

Theorem. — [PT], [NP] *Assume that* $d_s^0 + d_u^0 < 1$. *Then, for most* $g \in \mathcal{U}_+$, Λ_g *is a horseshoe.*

On the other hand, by [PY1], the same conclusion does not hold when $d_s^0 + d_u^0 > 1$. The paper [MY] gives substantially more geometric information in this case, specially concerning tangencies between stable and unstable manifolds (lines) in the hyperbolic continuation K_g of K. These results have been extended to higher dimensions, as announced in [MPV] and complete proofs to appear in the near future.

In the present work, we investigate the maximal invariant set Λ_g, for most $g \in \mathcal{U}_+$, provided that the dimensions d_s^0, d_u^0 satisfy (see Figure 2)

(H4) $(d_s^0 + d_u^0)^2 + (\max(d_s^0, d_u^0))^2 < d_s^0 + d_u^0 + \max(d_s^0, d_u^0)$.

Our results can essentially be summarized as:

Main Theorem. — *Assume that* (H1), (H2), (H3), (H4) *hold. Then, for most* $g \in \mathcal{U}_+$, Λ_g *is a non-uniformly hyperbolic horseshoe.*

The meaning of a non-uniformly hyperbolic horseshoe in the present context will be explained somewhat in the next section and more completely in the rest of the paper. We can, however, comment that, for most $g \in \mathcal{U}_+$, Λ_g will be a saddle-like object in the sense that both the stable set $W^s(\Lambda_g)$ and the unstable set $W^u(\Lambda_g)$ have Lebesgue measure zero and, so, it carries no attractors nor repellors. It will be (non-uniformly) hyperbolic

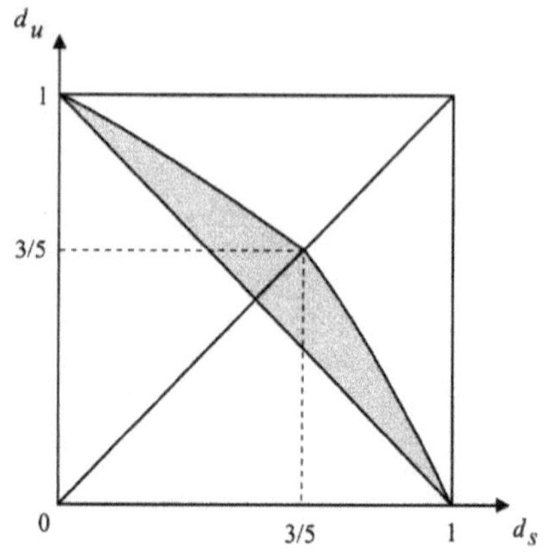

FIG. 2. — The dimension condition

in the sense that we will construct geometric invariant measures, à la Sinai-Ruelle-Bowen [Si, Ru, BR], on $\Lambda_g \subset W^s(\Lambda_g)$ and $\Lambda_g \subset W^u(\Lambda_g)$ with non-zero Lyapunov exponents. Such properties of the invariant set Λ_g are made especially precise in Sections 10 and 11, the last ones in the paper. They yield some rephrasing of the main result in these terms, which is presented at the end of Section 11.

Remark **1.** — 1. In the case when $d_s^0 + d_u^0 < 1$, mentioned above and studied in [PT], [NP], it is not necessary to assume that p_s, p_u belong to *distinct* periodic orbits. It is probably not necessary in our case either, at least as far as the qualitative statements are concerned. But, this assumption seems to make the technicalities significantly easier in what is already a very long construction.

2. The properties that we are proving for non-uniformly hyperbolic horseshoes are also true for uniformly hyperbolic ones, and indeed it is possible for Λ_g to be uniformly hyperbolic for a positive proportion of parameters even when $d_s^0 + d_u^0 \geqslant 1$. Therefore we will assume in the following that $d_s^0 + d_u^0 \geqslant 1$.

3. The tools that we develop probably allow to give, after some more work, further geometric and dynamical information on Λ_g (for most $g \in \mathcal{U}_+$) beyond that given in Sections 10 and 11. This could be the subject of subsequent investigations.

1.3. *A summary of the next sections of the paper.* — Sections 2–4 consist mainly of preparatory work.

In Section 2, we introduce a Markov partition by smooth disjoint rectangles $(R_a)_{a\in\mathcal{A}}$ for the horseshoe K. The dynamics in the neighbourhood U of K is given by the transition maps from one rectangle to another, which enjoy a nice hyperbolic behaviour. To understand the dynamics in the larger set $U \cup V$, we need to control the dynamics along a finite part of the orbit of q, stretching from the moment this orbit goes out of $R := \bigsqcup R_a$ until it comes back to R. The region of exit of R and the region of entry into R are two parabolic tongues L_u and L_s, and the transition map

$$G = g^{N_0} : L_u \to L_s$$

is a folding map which share many features with the Henon quadratic polynomial diffeomorphisms of the plane.

Section 3 is essentially a summary of our previous work [PY2] (which was written having the present paper in mind). The important concept of affine-like map is introduced. The basic idea, which goes back to the early stages of the hyperbolic theory, is to describe maps that present hyperbolic features in an implicit way exhibiting preference for coordinates with a macroscopic range. Concretely, if a two-dimensional diffeomorphism contracts the vertical coordinate y and expands the horizontal coordinate x, we use y_0 and x_1 as independent variables associated with a point (x_0, y_0) and its image (x_1, y_1), writing x_0 and y_1 as functions of y_0 and x_1.

Cone conditions are easy to formulate in this setting. A nice feature of this implicit representation of the dynamics is that it is time-symmetric: the map and its inverse satisfy symmetric formulas. Another even more important feature is that this formalism is well-adapted to the right concept of distortion (for 2-dimensional maps), yielding appropriate control on the partial derivatives of order two.

Composition of two affine-like maps which satisfy the same cone condition is also affine-like, and the distortion is only slightly bigger than the distortion of the two maps. Besides this "simple" composition, we study "parabolic" compositions of the form $F_1 \circ G \circ F_0$, where F_0, F_1 are affine-like and G is the folding map of Section 2. When the relative positions of the parabolic strip $G(Q_0)$ (where Q_0 is the image of F_0) and P_1 (the domain of F_1) are appropriate, the domain of $F_1 \circ G \circ F_0$ has two connected components and the restrictions $F^{\pm}$ of $F_1 \circ G \circ F_0$ to each component is affine-like. A control of the distortion of F^+ and F^- is also obtained.

In Section 4, the general structure of the parameter space is introduced. The parameter coordinate is normalized by the relative speed at the quadratic tangency of the tips of the stable and unstable manifolds. Then, with ε_0 very small, the starting interval $I_0 := [\varepsilon_0, 2\varepsilon_0]$ for the parameter selection process is introduced. A small parameter τ (with $\tau \ll 1$ but still $\varepsilon_0^{\tau} \ll 1$) determines a sequence of scales $(\varepsilon_k)_{k\geq 0}$ in parameter space through the formula $\varepsilon_{k+1} = \varepsilon_k^{1+\tau}$. At level k, we have disjoint parameter intervals of length ε_k (starting from level 0 with I_0). Each parameter interval of level k that has been selected is divided into $[\varepsilon_k^{-\tau}]$ disjoint candidates of length ε_{k+1}. These candidates will pass a test to decide whether they are selected at level $k+1$.

The test takes two forms. First, in Section 5, a property of the parameter interval called regularity (see below) will be introduced; candidates which do not possess this property are discarded. Such a property is sufficient to develop in Sections 5–8 some basic combinatorial and quantitative properties, but it is not well-adapted to an inductive scheme. Hence, in Section 9, a stronger property called strong regularity is introduced, and candidates have to satisfy this property in order to be selected.

Sections 5–7 constitute in some sense a single logical step: in Section 5, certain classes of restrictions of iterates of g_t are inductively defined, and the definition is only possible because of properties that are inductively proved in Sections 6 and 7.

In Section 5, the goal is to define, for each parameter interval I which is a candidate (i.e. its parent interval at the immediately upper level has been tested as regular), a class $\mathcal{R}(\mathrm{I})$ of I-persistent affine-like iterates. An I-persistent affine-like iterate is a triple $(\mathrm{P}, \mathrm{Q}, n)$ where P is a vertical-like strip in some rectangle R_a depending on $t \in \mathrm{I}$, Q is a horizontal-like strip in some rectangle $\mathrm{R}_{a'}$, depending on $t \in \mathrm{I}$, and the restriction of g_t^n to P is a diffeomorphism onto Q which is affine-like.

However, we do not want to have in $\mathcal{R}(\mathrm{I})$ all I-persistent affine-like iterates: we will argue about them by induction (on n, for instance) and in order to do this, we want to obtain them in some explicit constructive way. Therefore, a number of Axioms, (R1)–(R7), are introduced and together they completely determine the class $\mathcal{R}(\mathrm{I})$. The most important feature of these Axioms is the following: every element of $\mathcal{R}(\mathrm{I})$ consisting of more than one iteration of g_t can be obtained from simpler elements of $\mathcal{R}(\mathrm{I})$ by simple or parabolic composition; in this context, the notions of parent and simple or non-simple child introduced here, play a relevant role; simple composition is allowed in $\mathcal{R}(\mathrm{I})$ whenever it makes sense; and parabolic compositions of elements of $\mathcal{R}(\mathrm{I})$ is allowed if and only if a certain *transversality relation* is satisfied.

Thus, the definition of $\mathcal{R}(\mathrm{I})$ is reduced to the definition of this transversality relation, which is presented in Section 5.4. The intuitive notion behind the formal definition is the following: an element $(\mathrm{P}, \mathrm{Q}, n)$ (with Q crossing the domain L_u of G) should be I-transverse to an element $(\mathrm{P}', \mathrm{Q}', n')$ (with P′ crossing the image L_s of G) if the distance $\delta(\mathrm{Q}, \mathrm{P}')$ between the tip of the parabolic-like strip $\mathrm{G}(\mathrm{Q} \cap \mathrm{L}_u)$ and P′ satisfies

$$\delta(\mathrm{Q}, \mathrm{P}') \geq \max_{\mathrm{I}}(|\mathrm{Q}|^{1-\eta}, |\mathrm{P}'|^{1-\eta})$$

for all $t \in \mathrm{I}$, where $|\mathrm{Q}|$, $|\mathrm{P}'|$ are the widths of the strips Q and P′. Actually, as the distance $\delta(\mathrm{Q}, \mathrm{P}')$ is expected to vary with the parameter with derivative close to one, and we want a uniform control in I, it is more natural to ask that

$$\delta(\mathrm{Q}, \mathrm{P}') \geq \max_{\mathrm{I}}(|\mathrm{Q}|^{1-\eta}, |\mathrm{P}'|^{1-\eta}, |\mathrm{I}|)$$

for all $t \in \mathrm{I}$. Here η is a small positive constant, fixed once and for all. However, a number of properties, presented in Section 6, are very helpful, and they require a formal definition of the transversality relation that is more complicated than this. In Appendix C, we explain why this seemed complication is rather necessary. We use the notation $\mathrm{Q} \pitchfork_{\mathrm{I}} \mathrm{P}'$ to say that Q is I-transverse to P′.

For the starting interval I_0, it follows from the formal definition that the transversality relation is never satisfied; therefore, parabolic composition is not allowed and the class $\mathcal{R}(I_0)$ is exactly the one associated with the symbolic dynamics given by the Markov partition. Thus, the construction of $\mathcal{R}(I_0)$ is easy. For smaller parameter intervals, what is clear from the defining axioms (R1)–(R7) is that there can be at most one class $\mathcal{R}(I)$ satisfying them. That such a class exists (for candidate intervals whose parent is regular) is stated in Section 5.4 but will only be completely proven at the end of Section 7.

We conclude Section 5 with the introduction of some concepts that play an important role in the rest of the paper. We say that a strip P (from an element $(P, Q, n) \in \mathcal{R}(I)$, with P crossing L_s) is I-*transverse* if one can find finitely many elements $(P_\alpha, Q_\alpha, n_\alpha) \in \mathcal{R}(I)$ such that

- every point in L_u which stays in $U \cup V$ under negative iteration is contained in one of the Q_α;
- every Q_α is I-transverse to P.

Strips P crossing L_s which are not I-transverse are said to be I-*critical*. The I-critical strips can be viewed as representing some "critical region" at the $|I|$-scale. One defines symmetrically I-transversality and I-criticality for horizontal strips Q crossing L_u. Elements $(P, Q, n) \in \mathcal{R}(I)$ such that both P and Q are I-critical are said to be I-*bicritical* and correspond to returns from the critical region to itself (in time n). Given a constant $\beta > 1$, one says that the parameter interval is β-*regular* if any I-bicritical (P, Q, n) satisfies $|P| < |I|^\beta$, $|Q| < |I|^\beta$ for all $t \in I$. Intuitively, this means that no short return to the critical set is allowed. The value that we choose for $\beta > 1$ is announced in Section 5.6 and explained in Section 9.3. It depends only on the transverse Hausdorff dimensions d_s^0, d_u^0 and the eigenvalues of the periodic points p_s, p_u. It is easy to see that the starting interval I_0 is β-regular.

In Section 6, we prove a number of properties of the transversality relation and the classes $\mathcal{R}(I)$.

The first one is natural (but already requires some non trivial induction): children are born from their parents. Let us explain what it means. Let $(P, Q, n) \in \mathcal{R}(I)$, and let $(\widetilde{P}, \widetilde{Q}, \widetilde{n})$ be the element of $\mathcal{R}(I)$ such that $P \subset \widetilde{P}$, $P \neq \widetilde{P}$ and $\widetilde{P}$ is the thinnest rectangle with this property; one says that P is a child of $\widetilde{P}$ and that $\widetilde{P}$ is the parent of P. There are two cases; either $n = \widetilde{n} + 1$ and one says that P is a simple child; (P, Q, n) is obtained by simple composition of $(\widetilde{P}, \widetilde{Q}, \widetilde{n})$ with an element of length 1; or $n > \widetilde{n} + 1$ and one says that P is a non-simple child; one then proves that (P, Q, n) is obtained by parabolic composition of $(\widetilde{P}, \widetilde{Q}, \widetilde{n})$ with some element (P_1, Q_1, n_1).

An easy and very natural property of the transversality relation, proved in Section 5.6, is that transversality is hereditary: let $P_1' \subset P_1$, $Q_0' \subset Q_0$, $I' \subset I$; if $Q_0 \pitchfork_I P_1$ holds, then $Q_0' \pitchfork_{I'} P_1'$ also holds. A non-intuitive property of the transversality relation, but one which is useful at many places, is a partial converse called concavity (Section 6.3): with P_1, P_1', Q_0, Q_0', I, I′ as above, if both $Q_0 \pitchfork_I P_1'$ and $Q_0' \pitchfork_{I'} P_1$ hold, then $Q_0 \pitchfork_I P_1$ also holds; if both $Q_0 \pitchfork_{I'} P_1$ and $Q_0' \pitchfork_I P_1'$ hold, then $Q_0 \pitchfork_I P_1$ also holds; if both $Q_0' \pitchfork_I P_1$ and $Q_0 \pitchfork_{I'} P_1'$ hold, then $Q_0 \pitchfork_I P_1$ also holds.

The most important result in Section 6 is a structure theorem for new rectangles in Section 6.5. One considers an element (P, Q, n) which belongs to $\mathcal{R}(I)$ but not to $\mathcal{R}(\tilde{I})$, where $\tilde{I}$ is the parameter interval containing I of the level immediately inferior (one says that $\tilde{I}$ is the parent of I). Then there is a unique way to write (P, Q, n) as the result of a sequence of $k \geqslant 1$ parabolic compositions, possible in $\mathcal{R}(I)$ but not in $\mathcal{R}(\tilde{I})$, of elements $(P_0, Q_0, n_0), \ldots, (P_k, Q_k, n_k) \in \mathcal{R}(\tilde{I})$. This fundamental result is used with decisive effect at several places in Sections 7 and 9.

In Section 6.6 are proven several estimates on the width of strips, based on the structure theorem. As elements of $\mathcal{R}(I)$ are constructed using parabolic composition, one cannot hope for a uniform exponential estimate for the width of the strips in terms of the number n of iterations. However, we are able to prove a stretched exponential uniform estimate in Section 6.6.2. Another important result of Section 6.6 is that "thick" I-critical rectangles actually belong to some class $\mathcal{R}(I^*)$ for a parameter interval much bigger than I. Actually, because the definition of I-criticality supposes the knowledge of the full class $\mathcal{R}(I)$, which has not been constructed at this point, we introduce a substitute for this: a strip P (with $(P, Q, n) \in \mathcal{R}(I)$) is I-special if the thinnest $\tilde{I}$-defined strip $\tilde{P}$ (with $\tilde{I}$ the parent of I) is $\tilde{I}$-critical.

We complete in Section 7 the proof of the existence of the class $\mathcal{R}(I)$, for a candidate interval whose parent is β-regular. Actually, the long calculations are performed in Appendix A and we use in Section 7 the estimates derived from these calculations. We check that the uniform cone condition of the affine-like iterates in $\mathcal{R}(I)$ is satisfied, and that their distortion is bounded. After obtaining in Section 7.4 estimates for the class $\mathcal{R}(I_0)$ of the starting interval (which are simpler and better because no parabolic composition is involved), we prove in Section 7.5 a technical estimate related to parabolic composition. In Section 7.6, we deal with the relative speed of the strips when the parameter varies (the derivative of the quantity $\delta(Q, P')$ which comes up in the transversality relation); this is clearly of capital importance if we are to succeed. A point which is worth mentioning is that we are not able to obtain speed estimates for all pairs of strips (actually, it is easy to see that such estimates do not exist); we have to restrict ourselves to I-special strips, a case which, fortunately, is sufficient for our purposes. In Section 7.7, we investigate the oscillation of the widths of the strips with the parameter. While it is just not true that the relative oscillation is bounded (in the sense that the maximum over a parameter interval is bounded by a constant times the minimum), the result that we get (again only for I-special intervals) will allow us to argue as if it was.

At the end of Section 7, the construction of the classes $\mathcal{R}(I)$ is completed, for every parameter interval I whose parent $\tilde{I}$ is regular. But we still don't know whether a single interval I is regular.

Section 8 is a transition between the construction of the classes $\mathcal{R}(I)$ in Sections 5–7 and the heavy work of parameter selection in Section 9. We collect a number of properties of the transversality relation and the classes $\mathcal{R}(I)$ that were not needed before, but will be useful later. We also develop several quantitative estimates that will turn out to be crucial both in the parameter selection process of Section 9 and in the analysis of the dynamics for strongly regular parameters in Sections 10 and 11. We investigate in particular (Section 8.2), for a given element $(\widetilde{P}, \widetilde{Q}, \widetilde{n})$, the number of elements (P, Q, n) such that P is a non-simple child of $\widetilde{P}$; we show that, for every $\varepsilon > 0$, there are at most $\varepsilon^{-c'\eta}$ such non-simple children with width $|P|$ larger than $\varepsilon|\widetilde{P}|$. The constant η here is small and is the one occurring in the definition of the transversality relation. The meaning of this estimate is that the presence of non-simple children is not too significant from the point of view of Hausdorff (or box) dimension, as it is made clear in Section 8.3. Using Poincaré (or Dirichlet)-like series, we show that, for every $(P^*, Q^*, n^*) \in \mathcal{R}(I)$, every $\varepsilon > 0$, the number of elements $(P, Q, n) \in \mathcal{R}(I)$ with $P \subset P^*$, $|P| \geqslant \varepsilon|P^*|$ is at most $\varepsilon^{-d_s^*}$, with d_s^* very close to d_s^0. In Section 8.4, we transfer this information to parameter space, combining it with the result on relative speed of strips in Section 7.6: we show that, for any thin enough $\widetilde{I}$-critical strip Q^*, the proportion of candidates $I \subset \widetilde{I}$ such that Q^* is I-critical is at most $|\widetilde{I}|^{1-d_s^+}$, where again d_s^+ is close to d_s^0.

Section 9 is the longest one in the paper and deals with the parameter selection process. The concept of regularity is very useful to develop a number of properties of the classes R(I), as we did in Sections 5–8. Unfortunately, we are not able to prove (and it is probably false) that, given a β-regular interval $\widetilde{I}$, most candidates $I \subset \widetilde{I}$ at the next level are β-regular. [It is a consequence of the structure theorem of Section 6 that all candidates are $\overline{\beta}$-regular, where $\overline{\beta} = \beta(1+\tau)^{-1}$ is very close to β; this allows us to obtain all qualitative consequences of regularity for all candidates; but obviously we cannot repeat this at many successive levels of parameter intervals, because we need to keep $\beta > 1$.] The problem with the concept of regularity is that it is dealing with only one scale $|\widetilde{I}|^{\beta}$; it could happen a priori that for a regular parameter interval $\widetilde{I}$ we have many $\widetilde{I}$-bicritical $(P, Q, n) \in \mathcal{R}(I)$ with $|P|$ or $|Q|$ only slightly below the threshold $|\widetilde{I}|^{\beta}$ (and therefore above the next threshold $|I|^{\beta}$ for candidates $I \subset \widetilde{I}$); for each such (P, Q, n), we have to eliminate candidates I such that (P, Q, n) is I-bicritical, and no candidate will survive this selection process if there are too many (P, Q, n).

The solution to this difficulty is to introduce the condition of strong regularity, which implies regularity and gives a quantitative control at all scales. Actually, the strong regularity condition involves two parts, and three sets of inequalities (SR1), (SR2), (SR3).

In the first part (Section 9.1), one controls the size of the critical locus, in two slightly different ways expressed by (SR1), (SR2); both amount to say that the "dimension" of the critical locus is not much larger than $d_s^0 + d_u^0 - 1$, but (SR1) is a direct "box-counting" estimate, while (SR2) is more subtle.

The second part of the strong regularity condition, by far the most subtle one, is a quantitative estimate for the number of bicritical elements at all scales. Because of the inductive nature of the argument, which relies in an essential way on the structure theorem of Section 6, we need to control the number of elements $(P, Q, n) \in \mathcal{R}(I)$ such that P is I_α-critical, Q is I_ω-critical and $|P| \geqslant x$ for some $t \in I$. Here, I_α and I_ω are parameter intervals containing I, and the control will depend on I_α, I_ω and x. The formulas in Section 9.2 present a phase transition with respect to the width parameter x. Discussing this phase transition in Section 9.3 leads naturally to the hypothesis (H4) on the transverse dimensions d_s^0, d_u^0: a small calculation shows that (H4) is exactly what one needs to obtain β-regularity with $\beta > 1$. We also show in Section 9.3 that a strongly regular parameter interval is regular.

In Section 9.4, we show that the starting interval is strongly regular. Then, in Section 9.5, we show that most candidates in a strongly regular parameter interval satisfy (SR1); The argument for parameter selection is based on the result of Section 8.4 mentioned above. The same is done in Section 9.6 for condition (SR2), but the proof is more complicated and involves (SR3).

In the rest of Section 9 we prove that, given a strongly regular parameter interval $\widetilde{I}$, most candidates $I \subset \widetilde{I}$ at the next level satisfy (SR3) (the proportion of failed candidates turns out to be not larger than $C|I|^{\tau^2}$; the same is true for (SR1) and (SR2)). The easy cases are dealt with in Section 9.7. In Section 9.8, we estimate the number of bicritical elements in $\mathcal{R}(I)$ that are not $\widetilde{I}$-defined ($\widetilde{I}$ is the strongly regular parent of I). This is a rather long but straightforward calculation based on the structure theorem of Section 6.5. Then, we are left with the most difficult case $I = I_\alpha = I_\omega$, x small (below the threshold of phase transition), but not too small (the estimate is trivial in this last case). In this case, and in this case only, we are forced to take into account that for an $\widetilde{I}$-bicritical element (P, Q, n) to be I-bicritical, the two events (of the "random" variable I) "P is I-critical" and "Q is I-critical" must occur simultaneously. The strategy (explained at the end of Section 9.7) to deal with this case is to divide the class of I-bicritical rectangles in a (large, but not too large) number of adequately defined subclasses. These subclasses are defined in Section 9.9, their number is estimated in Section 9.10 and the calculation which validates the strategy is performed in Sections 9.11 and 9.12. In Section 9.13, we explain which (very few) modifications have to be made when dealing with the condition $|Q| \geqslant x$ instead of $|P| \geqslant x$ (the setting is not symmetric at this level because the formulations of (SR3) when $d_s^0 \geqslant d_u^0$ and $d_s^0 \leqslant d_u^0$ are different). Finally, in Section 9.14 we conclude by defining precisely the exponents appearing in (SR1), (SR2), (SR3) which were so far only approximately defined. At this point, we know that all parameters in the starting interval $I_0 = [\varepsilon_0, 2\varepsilon_0]$, except for a subset of relative measure $\leqslant C\varepsilon_0^{\tau^2}$, are strongly regular.

It is worth mentioning that up to the end of Section 9, we never consider the dynamics for a single parameter, only for parameter intervals. In the last two sections, we study the dynamics for a strongly regular parameter value, i.e. the intersection of a decreasing sequence (I_m) of strongly regular parameter intervals.

In Section 10, we study the dynamics on the set of stable curves. A *stable curve* ω is the decreasing intersection of a sequence of vertical-like strips P_k, where $(P_k, Q_k, n_k) \in \mathcal{R} = \bigcup_m \mathcal{R}(I_m)$. The set of stable curves is denoted by $\mathcal{R}_+^\infty$, their union by $\widetilde{\mathcal{R}}_+^\infty$. We show in Section 10.5 that $\widetilde{\mathcal{R}}_+^\infty$ is a lamination by C^{1+Lip} curves with Lipschitz holonomy. In order to define a map on $\widetilde{\mathcal{R}}_+^\infty$ (which is not invariant under g), we introduce in Section 10.1 the concept of prime element in $\mathcal{R}$, i.e. one which cannot be written as the simple composition of shorter elements. The number of prime factors in a decomposition is controlled in Section 10.2. Let then ω be a stable curve which is not contained in infinitely many prime elements P_k, and let (P, Q, n) be such that P is the thinnest prime element containing ω. The image $g^n(\omega)$ is contained in a stable curve ω' and we set $T^+(\omega) = \omega'$, $\widetilde{T}^+/\omega = g^n/\omega$. This defines a map T^+ from a subset $\mathcal{D}_+$ of $\mathcal{R}_+^\infty$ onto $\mathcal{R}_+^\infty$ which lifts to a map $\widetilde{T}^+$ from the union $\widetilde{\mathcal{D}}_+$ of curves in $\mathcal{D}_+$ to $\widetilde{\mathcal{R}}_+^\infty$.

The map T^+ is Bernoulli in the following sense: its domain $\mathcal{D}_+$ splits into countably many pieces $\mathcal{R}_+^\infty(P)$ indexed by prime elements, and each piece is sent homeomorphically by T^+ onto the intersection of $\mathcal{R}_+^\infty$ with some rectangle R_a of the Markov partition.

The map T^+ is uniformly expanding (with countably many branches) and we introduce a one parameter family of weighted transfer operators in the spirit of classical uniformly hyperbolic maps. One has only to be careful because the presence of countably many branches is the source of some problems, which are dealt with in Section 10.3 using the estimates of Section 8 on the number of children.

As expected, the transfer operators L_d, considered in the appropriate function space, turn out to have a positive eigenfunction h_d associated with a dominant eigenvalue $\lambda_d > 0$. There is a unique value d_s such that $\lambda_{d_s} = 1$. This value turns out to be, unsurprisingly, the transverse Hausdorff dimension of the lamination $\widetilde{\mathcal{R}}_+^\infty$. This is proved in Section 10.10 (Theorem 4), but the proof is more difficult than in the classical case, because the transverse geometry of $\widetilde{\mathcal{R}}_+^\infty$ is complicated and requires some delicate handling.

The transfer operator also allows us to identify, as usual, a measure μ_d with prescribed Jacobian and an invariant measure $\nu_d = h_d \mu_d$. For $d = d_s$, the μ_d-measure (or ν_d-measure) of the set of stable curves contained in any vertical-like strip P is proportional to $|P|^{d_s}$.

The set $\widetilde{\mathcal{R}}_+^\infty - \widetilde{\mathcal{D}}_+$ where $\widetilde{T}^+$ is not defined, has transverse dimension smaller than d_s, hence is negligible in a geometrical sense. One can lift the T^+-invariant measure $\nu = \nu_{d_s}$ to a $\widetilde{T}^+$-invariant measure $\widetilde{\nu}$ which is ergodic and then spread it to a g-invariant measure on Λ.

In Section 11, the last in the paper, we pursue the study of the dynamics of g_t on $\Lambda = \Lambda_{g_t}$ for a strongly regular parameter t, looking now beyond the well-behaved set $\widetilde{\mathcal{R}}_+^\infty$ which was studied in Section 10. In the first part (Sections 11.1–11.5), we study the intersection of the invariant set Λ with an unstable curve ω^* (defined as a stable curve, exchanging P's and Q's). The main part of this intersection is a countable disjoint union of dynamical copies of the set $\mathcal{R}_+^\infty$ studied in Section 10. There are also at most countably many critical points, corresponding to quadratic tangencies between stable curves and images under G of unstable curves. And, finally, there is an exceptional set (formed by points which come very close to the critical locus infinitely many times); but this exceptional set is small; its Hausdorff dimension is explicitly controlled by a value smaller by a definite amount than the dimension d_s of $\omega^* \cap \Lambda$.

In the second part of Section 11, we prove that the invariant set Λ is a saddle-like object in the metric sense: both its stable set $W^s(\Lambda)$ and its unstable set $W^u(\Lambda)$ have

Lebesgue measure 0. So, no attractors nor repellors are present on Λ. One actually expects more: certainly the Hausdorff dimension of $W^s(\Lambda)$ should be strictly less than 2, probably it is close to $1 + d_s$, and perhaps even equal to $1 + d_s$. However, we stick to the simpler, but still very meaningful result: it implies that Λ_g carries no attractor nor a repellor for most g.

One has a nice combinatorial decomposition of the restricted stable set $W^s(\Lambda, R)$, but to compute Lebesgue measure (or Hausdorff dimension), one has to transport the pieces of this decomposition by affine-like iterates of g of high order. This is easy to do as far as Lebesgue measure is concerned, because bounded distortion of affine-like maps mean also bounded distortion of measure (bounded relative oscillation of Jacobians). This is much more delicate with respect to Hausdorff dimension: the geometry of the pieces after iteration can get very distorted.

In Appendix A, we recall all formulas related to the implicit representation of affine-like maps; many of them can already be found in [PY2], but we have also to consider the derivatives with respect to parameter, a setting which was not considered in [PY2]. We also perform a number of estimates, both for simple and parabolic composition, which are used in Section 7.

In Appendix B, we prove a result related to Proposition 51 in Section 10.5, which states the transversally Lipschitz regularity of the lamination $\widetilde{\mathcal{R}}_+^\infty$.

In Appendix C, we give some justification for what seems to be a convoluted definition of the transversality relation in Section 5.4.

We wish to thank Sylvain Crovisier for carefully reading a first version of this text and making many valuable suggestions to improve several conceptual parts of the paper, hopefully making it more amenable to read. We also wish to thank W. de Melo and M. Viana for fruitful conversations.

2. Markov partition and folding map

2.1. *Markov partition and related charts.* — We will choose once and for all a finite system of smooth charts

$$I_a^s \times I_a^u \xrightarrow{\approx} R_a \subset M, \qquad a \in \mathcal{a}$$

indexed by a finite alphabet $\mathcal{a}$. Each chart depends smoothly on $g \in \mathcal{U}$; the intervals I_a^s, I_a^u are compact; the rectangles R_a are disjoint.

Let $R = \bigcup_{\mathcal{a}} R_a$. We choose the charts in order to have:

(MP1) for each $g \in \mathcal{U}$, K_g is the maximal invariant set in int R; for each $g \in \mathcal{U}$, $a \in \mathcal{a}$, one has

$$g(\partial I_a^s \times I_a^u) \cap R = \emptyset, \tag{2.1}$$

$$g^{-1}(I_a^s \times \partial I_a^u) \cap R = \emptyset; \tag{2.2}$$

(MP2) for each $g \in \mathcal{U}$, the family $(R_a \cap K_g)_{a \in \mathcal{a}}$ induces a Markov partition for the horseshoe K_g. Moreover, no rectangle R_a meets both the orbit of p_s and the orbit of p_u.

Let

$$\mathcal{B} = \{(a, a') \in \mathcal{a}^2, f(R_a) \cap R_{a'} \cap K_f \neq \emptyset\}. \tag{2.3}$$

The Markov partition provides a coding which is a topological conjugacy between the horseshoe K_g and the subshift of finite type of $\mathcal{a}^{\mathbf{Z}}$ defined by $\mathcal{B}$.

2.2. *The parabolic tongues* L_u, L_s. — Denote by a_s, $a_u \in \mathcal{a}$ the letters such that $p_s \in R_{a_s}$, $p_u \in R_{a_u}$. We choose the corresponding charts in order to have:

(MP3) for each $g \in \mathcal{U}$, the equation of the local stable manifold $W^s_{\text{loc}}(p_s)$ is $\{x_{a_s} = 0\}$, the equation of the local unstable manifold $W^u_{\text{loc}}(p_u)$ is $\{y_{a_u} = 0\}$.

We have written x_a (resp. y_a) for the coordinate in I^s_a (resp. I^u_a). We also choose the rectangles R_a in order to have, for some integer $N_0 \geqslant 2$:

(MP4) for each $g \in \mathcal{U}_0$, there are points q_s, q_u in the orbit of q such that

- for $n \geq 0$, $g^n(q_s)$ and $g^n(p_s)$ belong to the interior of the same rectangle;
- for $n \leq 0$, $g^n(q_u)$ and $g^n(p_u)$ belong to the interior of the same rectangle;
- $q_s = g^{N_0}(q_u)$ and $g^i(q_u)$ does not belong to R for $0 < i < N_0$.

Consider small pieces of $W^s(p_s)$, $W^u(p_u)$ which are tangent at q_u for $g \in \mathcal{U}_0$. When $g \in \mathcal{U}_+$, these pieces will meet in two points and bound a compact lenticular region $L_u \subset \text{int}\, R_{a_u}$. Taking the image under g^{N_0}, we get another lenticular region $L_s \subset \text{int}\, R_{a_s}$. These regions are called *parabolic tongues*. See Figure 3.

Define then, for $g \in \mathcal{U}_+$

$$\widehat{R} = R \cup \bigcup_{0<i<N_0} g^i(L_u). \tag{2.4}$$

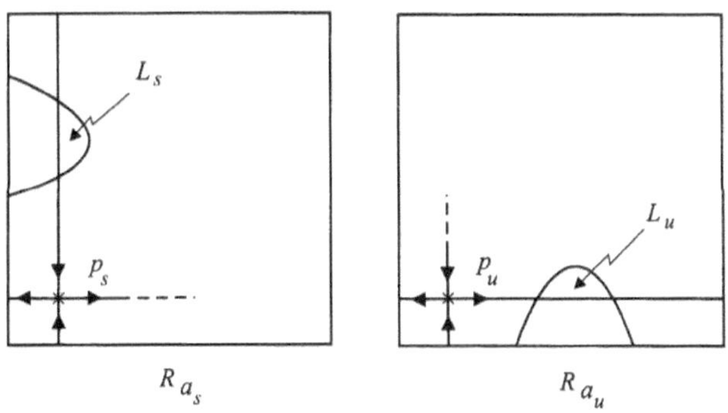

FIG. 3. — The parabolic tongues L_s, L_u

The maximal invariant set we are interested in is

$$\Lambda_g = \bigcap_{\mathbf{Z}} g^{-n}(\widehat{\mathrm{R}}). \tag{2.5}$$

Observe that $\mathrm{W}^s_{\mathrm{loc}}(p_s)$ bounds the set K_g in R_{a_s} and $\mathrm{W}^u_{\mathrm{loc}}(p_u)$ bounds the set K_g in R_{a_u}. The maximal invariant set Λ_g meets the boundary of $\widehat{\mathrm{R}}$ (in fact, the boundary of L_s, L_u, and their images), but it is still locally maximal because it would also be the maximal invariant set in a slight enlargement of $\widehat{\mathrm{R}}$.

We also define

$$\mathrm{W}^s(\Lambda_g, \widehat{\mathrm{R}}) = \bigcap_{n\geqslant 0} g^{-n}(\widehat{\mathrm{R}}), \tag{2.6}$$

$$\mathrm{W}^u(\Lambda_g, \widehat{\mathrm{R}}) = \bigcap_{n\leqslant 0} g^{-n}(\widehat{\mathrm{R}}). \tag{2.7}$$

The dynamics in $\widehat{\mathrm{R}}$ are generated by

- the transition maps related to the Markov partition:

$$g : \mathrm{R}_a \cap g^{-1}(\mathrm{R}_{a'}) \to g(\mathrm{R}_a) \cap \mathrm{R}_{a'}, \quad \text{for } (a, a') \in \mathcal{B};$$

- the folding map $\mathrm{G} := g^{\mathrm{N}_0}$ from L_u onto L_s.

2.3. *The folding map* G. — For simplicity, we write (x_s, y_s), (x_u, y_u) for the coordinates in $\mathrm{R}_{a_s} \supset \mathrm{L}_s$, $\mathrm{R}_{a_u} \supset \mathrm{L}_u$.

The folding map G is *parabolic* in the sense of [PY2]; let us recall this definition.

Consider the graph Γ_{G} of the restriction G of g^{N_0} to the component of $\mathrm{R}_{a_u} \cap g^{-\mathrm{N}_0}(\mathrm{R}_{a_s})$ which contains L_u (for $g \in \mathcal{U}_+$; we then follow this component in the rest of $\mathcal{U}$). Using the corresponding charts, we can view Γ_{G} as a surface in $\mathrm{I}^s_{a_u} \times \mathrm{I}^u_{a_u} \times \mathrm{I}^s_{a_s} \times \mathrm{I}^u_{a_s}$. Denote by π the projection from $\mathrm{I}^s_{a_u} \times \mathrm{I}^u_{a_u} \times \mathrm{I}^s_{a_s} \times \mathrm{I}^u_{a_s}$ onto $\mathrm{I}^u_{a_u} \times \mathrm{I}^s_{a_s}$. For $\mathcal{U}$ small enough, from the quadratic tangency at q and (MP3) we deduce that:

(P1) the restriction of π to Γ_{G} is a fold map (in the sense of singulary theory).

Denote by $\Gamma_0 \subset \mathrm{I}^u_{a_u} \times \mathrm{I}^s_{a_s}$ the smooth curve which is the image of the critical locus of this fold map. It divides $\mathrm{I}^u_{a_u} \times \mathrm{I}^s_{a_s}$ into two regions Γ_+, Γ_- such that $\Gamma_+ \cup \Gamma_0$ is the image of the fold map. We can reformulate (P1) as:

(P′1)

(i) for $(y^0, x^0) \in \Gamma_0$, the image $\mathrm{G}(\{y_u = y^0\})$ meets $\{x_s = x^0\}$ in a single point, interior to both curves, at which the curves have a quadratic tangency;
(ii) for $(y^0, x^0) \in \Gamma_-$, the curves $\mathrm{G}(\{y_u = y^0\})$ and $\{x_s = x^0\}$ do not intersect;
(iii) for $(y^0, x^0) \in \Gamma_+$, the curves $\mathrm{G}(\{y_u = y^0\})$ and $\{x_s = x^0\}$ intersect transversally in two points.

As G is a diffeomorphism, the tangents to Γ_0 are never vertical or horizontal. Therefore, we can and will choose a smooth function θ on $I^u_{a_u} \times I^s_{a_s}$ such that

(P2) $\theta \equiv 0$ on Γ_0, $\theta > 0$ on Γ_+, $\theta < 0$ on Γ_-;

(P3) the partial derivatives θ_y, θ_x of θ do not vanish on $I^u_{a_u} \times I^s_{a_s}$.

Remark **2.** — The choice of θ is far from unique. One could for instance choose θ of the form

$$\theta(y_u, x_s) = \varepsilon_u y_u + \varepsilon_s \chi(x_s), \tag{2.8}$$

with $\varepsilon_s, \varepsilon_u \in \{-1, +1\}$ and χ monotone increasing. We prefer not to specify a particular choice in order to keep a time-symmetric setting between positive and negative iterations.

From θ, we define a smooth function w on Γ_G by

$$w^2 = \theta \circ \pi \tag{P4}$$

(there are two choices for w; the other is $-w$).

Then, from (P3) we obtain smooth maps Y_u, X_s implicitly defined by

$$\begin{aligned} w^2 &= \theta(Y_u(w, x_s), x_s) \\ &= \theta(y_u, X_s(w, y_u)). \end{aligned} \tag{P5}$$

On the graph Γ_G, we can use either (x_u, y_u) or (x_s, y_s) or (w, y_u) or (x_s, w) as coordinates; therefore we can factorize G as $G_+ \circ G_0 \circ G_-$:

$$(x_u, y_u) \xrightarrow{G_-} (w, y_u) \xrightarrow{G_0} (x_s, w) \xrightarrow{G_+} (x_s, y_s) \tag{2.9}$$

with

$$\begin{aligned} G_0(w, y_u) &= (X_s(w, y_u), w), \\ G_0^{-1}(x_s, w) &= (w, Y_u(w, x_s)), \\ G_+(x_s, w) &= (x_s, Y_s(w, x_s)), \\ G_-^{-1}(w, y_u) &= (X_u(w, y_u), y_u). \end{aligned} \tag{P6}$$

The last two formulas define smooth maps Y_s, X_u and the partial derivatives $Y_{s,w}$, $X_{u,w}$ do not vanish as G_+, G_- are diffeomorphisms. Observe that the map G_0 is very similar to a quadratic Hénon-like map.

A simple explicit example for G, G_+, G_-, G_0, θ is given in [PY2].

3. Affine-like maps

This section is essentially a summary of [PY2].

3.1. *Definition and implicit representation.* — Let I_0^s, I_0^u, I_1^s, I_1^u be non trivial compact intervals, x_0, y_0, x_1, y_1 the corresponding coordinates. Consider a smooth diffeomorphisms F whose domain is a *vertical strip*

$$P = \{\varphi^-(y_0) \leqslant x_0 \leqslant \varphi^+(y_0)\} \subset I_0^s \times I_0^u$$

and whose image is a *horizontal strip*

$$Q = \{\psi^-(x_1) \leqslant y_1 \leqslant \psi^+(x_1)\} \subset I_1^s \times I_1^u.$$

We say that F is *affine-like* if

(AL1) the restriction to the graph of F of the projection onto $I_0^u \times I_1^s$ is a diffeomorphism onto $I_0^u \times I_1^s$.

This allows us to define smooth maps A, B on $I_0^u \times I_1^s$ such that

$$(3.1) \qquad F(x_0, y_0) = (x_1, y_1) \quad \Longleftrightarrow \quad \begin{cases} x_0 = A(y_0, x_1), \\ y_1 = B(y_0, x_1). \end{cases}$$

The pair (A, B) is the *implicit representation* (or definition) of the affine-like map F. See Figure 4. In the formulas below, we shall most of the time omit the arguments of the functions considered, which should be obvious from the context. We will write A_x, A_y, A_{xx}, B_x, B_y, ... for partial derivatives.

On the graph of F, we have

$$(3.2) \qquad \begin{aligned} dx_0 &= A_y dy_0 + A_x dx_1, \\ dy_1 &= B_y dy_0 + B_x dx_1, \end{aligned}$$

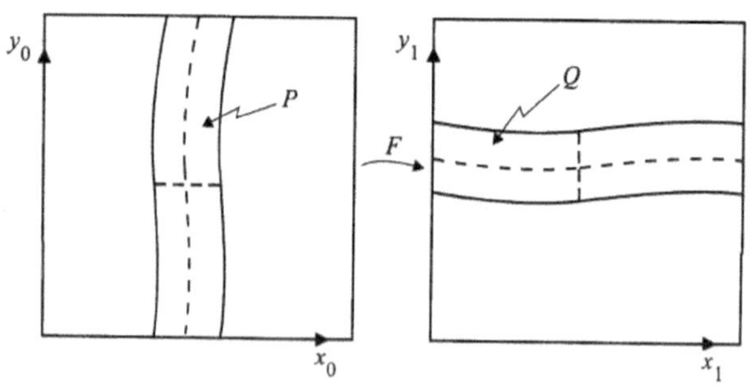

FIG. 4. — An affine-like map

which leads to

$$\mathrm{DF} = \mathrm{A}_x^{-1}\begin{pmatrix} 1 & -\mathrm{A}_y \\ \mathrm{B}_x & \mathrm{A}_x\mathrm{B}_y - \mathrm{A}_y\mathrm{B}_x \end{pmatrix}, \tag{3.3}$$

$$\mathrm{DF}^{-1} = \mathrm{B}_y^{-1}\begin{pmatrix} \mathrm{A}_x\mathrm{B}_y - \mathrm{A}_y\mathrm{B}_x & \mathrm{A}_y \\ -\mathrm{B}_x & 1 \end{pmatrix}, \tag{3.4}$$

$$\det \mathrm{DF} = \mathrm{A}_x^{-1}\mathrm{B}_y. \tag{3.5}$$

The main advantage of the implicit representation is the symmetry between positive and negative iteration.

3.2. *Cone condition and distortion.* — Let $\lambda, u, v > 0$ satisfy

$$1 < uv \leqslant \lambda^2. \tag{3.6}$$

Let $(\mathrm{X}_0, \mathrm{Y}_0)$ be a tangent vector at some point in the domain of F, and let $(\mathrm{X}_1, \mathrm{Y}_1)$ be its image under TF. The usual cone condition with parameters (λ, u, v) is:

(AL2)

(i) if $|\mathrm{Y}_0| \leqslant u|\mathrm{X}_0|$, then $|\mathrm{Y}_1| \leqslant v^{-1}|\mathrm{X}_1|$ and $|\mathrm{X}_1| \geqslant \lambda|\mathrm{X}_0|$;
(ii) if $|\mathrm{X}_1| \leqslant v|\mathrm{Y}_1|$, then $|\mathrm{X}_0| \leqslant u^{-1}|\mathrm{Y}_0|$ and $|\mathrm{Y}_0| \geqslant \lambda|\mathrm{Y}_1|$.

This is readily seen to be equivalent to

(AL′2)
$$\lambda|\mathrm{A}_x| + u|\mathrm{A}_y| \leqslant 1,$$
$$\lambda|\mathrm{B}_y| + v|\mathrm{B}_x| \leqslant 1,$$

everywhere on $\mathrm{I}_0^u \times \mathrm{I}_1^s$.

We will also need to control partial derivatives of second order of A, B. By (3.5), the partial derivatives A_x, B_y do not vanish on $\mathrm{I}_0^u \times \mathrm{I}_1^s$. It turns out that the right way to look at partial derivatives of second order is to consider the six functions

$$\partial_x \log|\mathrm{A}_x|,\ \partial_y \log|\mathrm{A}_x|,\ \mathrm{A}_{yy},$$
$$\partial_y \log|\mathrm{B}_y|,\ \partial_x \log|\mathrm{B}_y|,\ \mathrm{B}_{xx}.$$

We define the *distortion* of an affine-like map F, and denote by D(F), the maximal absolute value attained by any one of these six functions on $\mathrm{I}_0^u \times \mathrm{I}_1^s$.

We also define the *width* of the domain P of F by

$$|\mathrm{P}| := \max|\mathrm{A}_x|, \tag{3.7}$$

and the width of the image Q by

$$|\mathrm{Q}| := \max|\mathrm{B}_y|. \tag{3.8}$$

Observe that we have $\max|\mathrm{A}_x| \leqslant \mathrm{C}\min|\mathrm{A}_x|$, where the constant C only depends on D(F) and the lengths of the intervals $\mathrm{I}_0^u, \mathrm{I}_1^s$. The same estimate holds for B_y.

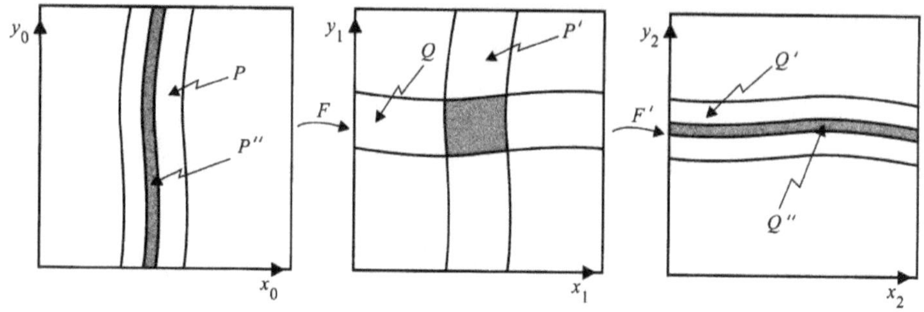

FIG. 5. — Simple composition

3.3. *Simple composition.* — The composition of two affine-like maps is not always affine-like. However, the composition of two affine-like maps which also satisfy the same cone condition (AL2) will again be affine-like and satisfy the same cone condition (actually a better one).

More precisely, let I_0^s, I_0^u, I_1^s, I_1^u, I_2^s, I_2^u be compact intervals. Let $F : P \to Q$ and $F' : P' \to Q'$ be affine-like maps with domains $P \subset I_0^s \times I_0^u$, $P' \subset I_1^s \times I_1^u$ and images $Q \subset I_1^s \times I_1^u$, $Q' \subset I_2^s \times I_2^u$. We assume that both F and F′ satisfy (AL2) (or (AL′2)) with parameters λ, u, v. The composition $F'' = F' \circ F$ has domain $P'' = P \cap F^{-1}(P')$ and image $Q'' = Q' \cap F'(Q)$. It satisfies (AL1) and (AL2) with parameters λ^2, u, v (cf. [PY2]). See Figure 5.

Let (A, B), (A′, B′), (A″, B″) be the implicit representations of F, F′, F″ respectively. Define

$$\Delta := 1 - A'_y B_x > 1 - u^{-1} v^{-1} > 0. \tag{3.9}$$

The partial derivatives of first order of A″, B″ are given by

$$\begin{aligned} A''_x &= A_x A'_x \Delta^{-1}, \\ B''_y &= B_y B'_y \Delta^{-1}, \end{aligned} \tag{3.10}$$

$$\begin{aligned} A''_y &= A_y + A'_y A_x B_y \Delta^{-1}, \\ B''_x &= B'_x + B_x A'_x B'_y \Delta^{-1}. \end{aligned} \tag{3.11}$$

From (3.10), we get

$$\begin{aligned} C^{-1} &\leqslant \frac{|P''|}{|P||P'|} \leqslant C, \\ C^{-1} &\leqslant \frac{|Q''|}{|Q||Q'|} \leqslant C, \end{aligned} \tag{3.12}$$

where the constants are uniform once u, v are fixed and the distortions are uniformly bounded.

The formulas for the partial derivatives of second order are derived in [PY2] and recalled in Appendix A. From formulas (A.12) to (A.21), one obtains the following estimate for the distortion:

$$D(F'') \leqslant \max\Big\{D(F) + C|Q|(D(F)+D(F')), D(F') + C|P'|(D(F)+D(F'))\Big\}, \quad (\mathbf{3.13})$$

where C depends only on u, v.

3.4. *Properties of the Markov partition.* — We choose charts for the Markov partition discussed in Section 2.1 in order to have the following property, for some λ, u, v satisfying (3.6):

(MP5) for any $(a, a') \in \mathcal{B}$, any $g \in \mathcal{U}$, the transition map $g_{a,a'}$ from $P_{aa'} = R_a \cap g^{-1}(R_{a'})$ onto $Q_{aa'} = R_{a'} \cap g(R_a)$ is affine-like and also satisfies the cone condition (AL2).

These values of (λ, u, v) will be fixed in what follows.

To any finite word $\underline{a} = (a_0, \ldots, a_n)$ with transitions in $\mathcal{B}$, we have a composition

$$g_{\underline{a}} = g_{a_{n-1}a_n} \circ \ldots \circ g_{a_0 a_1}$$

which satisfies also (AL1) and (AL2).

Moreover, as the widths decrease exponentially with the number of iterations, it follows from (3.13) that there exists $D_0 > 0$ such that all $g_{\underline{a}}$ satisfy

(MP6) $D(g_{\underline{a}}) \leq D_0$.

Finally, the following simple property will be useful in Section 8.

(MP7) for any $(a, a') \in \mathcal{B}$, any $(x, y) \in P_{aa'}$, any $(x', y') \in Q_{aa'}$, we have

$$x_a^- + C^{-1} < x < x_a^+ - C^{-1},$$
$$y_{a'}^- + C^{-1} < y' < y_{a'}^+ - C^{-1},$$

where $I_a^s = [x_a^-, x_a^+]$, $I_{a'}^u = [y_{a'}^-, y_{a'}^+]$ and C is an appropriate large constant.

3.5. *Parabolic composition.* — Let G be the folding map of Section 2.3, satisfying properties (P1)–(P6).

Let also I_0^s, I_0^u, I_1^s, I_1^u be compact intervals; let F_0 be an affine-like map from a vertical strip $P_0 \subset I_0^s \times I_0^u$ to a horizontal strip $Q_0 \subset I_{a_u}^s \times I_{a_u}^u$; let F_1 be an affine-like map from a vertical strip $P_1 \subset I_{a_s}^s \times I_{a_s}^u$ to a horizontal strip $Q_1 \subset I_1^s \times I_1^u$.

We recall from [PY2] how, under appropriate hypotheses, the composition $F_1 \circ G \circ F_0$ defines two affine-like maps $F^\pm$ with domains $P^\pm \subset P_0$ and image $Q^\pm \subset Q_1$. See Figure 6.

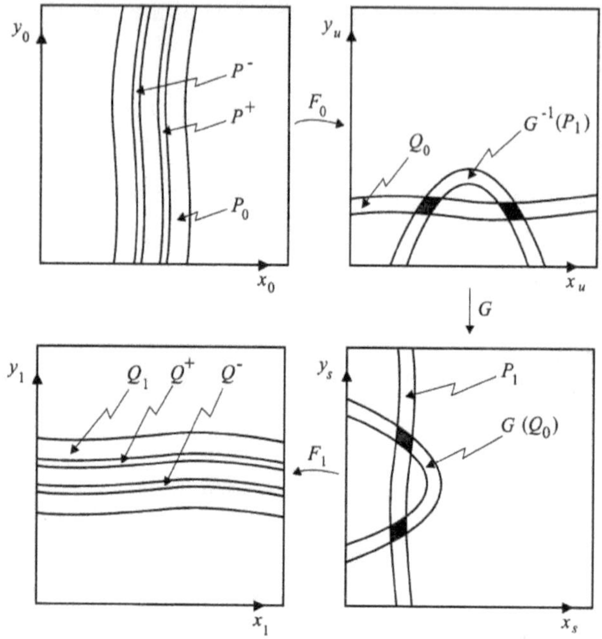

FIG. 6. — Parabolic composition

Let (A_0, B_0), (A_1, B_1) be implicit representations of F_0, F_1, respectively. We assume that

(PC1) $$|A_{1,y}| < b, \qquad |A_{1,yy}| < b,$$
$$|B_{0,x}| < b, \qquad |B_{0,xx}| < b,$$

with $b \ll 1$. In the system

(3.14) $$x_u = X_u(w, y_u), \qquad y_u = B_0(y_0, x_u),$$

we can, as $|B_{0,x}| \ll 1$, eliminate y_u and solve for x_u to define

(3.15) $$x_u = X(w, y_0).$$

Similarly, in the system

(3.16) $$y_s = Y_s(w, x_s), \qquad x_s = A_1(y_s, x_1),$$

we eliminate x_s, solve for y_s to get

$$y_s = Y(w, x_1). \tag{3.17}$$

The next step is to define

$$C(w, y_0, x_1) := w^2 - \theta\Big(B_0(y_0, X(w, y_0)), A_1(Y(w, x_1), x_1)\Big). \tag{3.18}$$

This quantity has the following geometrical interpretation. Fix values y_0^*, x_1^* for y_0, x_1. The image $G_- \circ F_0(\{y_0 = y_0^*\})$ is the graph

$$\gamma_0 = \Big\{y_u = B_0(y_0^*, X(w, y_0^*))\Big\}; \tag{3.19}$$

symmetrically, $G_+^{-1} \circ F_1^{-1}(\{x_1 = x_1^*\})$ is the graph

$$\gamma_1 = \Big\{x_s = A_1(Y(w, x_1^*), x_1^*)\Big\}. \tag{3.20}$$

Then, $C(w, y_0^*, x_1^*)$ gives the relative position of the two curves γ_0 and $G_0^{-1}(\gamma_1)$ (or equivalently $G_0(\gamma_0)$ and γ_1). More precisely, it is positive for all w if the two curves do not intersect; it vanishes at the intersection points and is negative between the intersection points.

It follows from (PC1) just above that

$$|C_w - 2w| \ll 1, \tag{3.21}$$

$$|C_{ww} - 2| \ll 1. \tag{3.22}$$

Therefore, for fixed values of y_0 and x_1, C has a unique minimum as a function of w; we denote by $\overline{C}(y_0, x_1)$ the corresponding minimum value. We have $\overline{C}(y_0^*, x_1^*) > 0$ (resp. $= 0$, resp. < 0) if and only if the curves γ_0 and $G_0^{-1}(\gamma_1)$ do not intersect (resp. are tangent, resp. have two transverse intersection points).

In order to consider parabolic compositions, we shall require that $\overline{C}(y_0, x_1) < 0$ everywhere on $I_0^u \times I_1^s$. Setting

$$\delta = \delta(Q_0, P_1) = \min_{y_0, x_1} -\overline{C}(y_0, x_1) \tag{3.23}$$

we actually want to have

$$\delta > b^{-1}(|P_1| + |Q_0|). \tag{\textbf{PC2}}$$

The geometric interpretation of this requirement is clear: the displacement of one of the rectangles and the image of the other should be much bigger than the sum of their widths. In other words, the distance between the tip of the parabolic strip $G_0^{-1}(P_1)$ and the horizontal strip Q_0 should be much bigger than the widths of these strips.

Assume now that (PC1) and (PC2) are satisfied; the equation $C(w, y_0, x_1) = 0$ defines two smooth functions

$$w = W^{\pm}(y_0, x_1) \tag{3.24}$$

with $W^+ > W^-$. One then defines

$$A^{\pm}(y_0, x_1) := A_0\Big(y_0, X(W^{\pm}(y_0, x_1), y_0)\Big), \tag{3.25}$$

$$B^{\pm}(y_0, x_1) := B_1\Big(Y(W^{\pm}(y_0, x_1), x_1), x_1\Big). \tag{3.26}$$

As shown in [PY2], the pair (A^+, B^+) (resp. (A^-, B^-)) implicitly defines an affine-like map F^+ (resp. F^-).

Denote by P^+ (resp. P^-) the domain of F^+ (resp. F^-) and by Q^+ (resp. Q^-) the image of F^+ (resp. F^-). Then P^+ and P^- are the two components of $P_0 \cap (G \circ F_0)^{-1}(P_1)$, Q^+ and Q^- are the two components of $Q_1 \cap (F_1 \circ G)(Q_0)$; F^+ (resp. F^-) is the restriction of $F_1 \circ G \circ F_0$ to P^+ (resp. P^-).

The formulas for the partial derivatives of $A^{\pm}$, $B^{\pm}$ are derived in [PY2] and recalled in Appendix A (see in particular (A.42)). The partial derivative C_w in these formulas is of order $\delta^{\frac{1}{2}}$ at the points under consideration (see (A.84) in Appendix A). From this one obtains the following estimate for the widths:

$$C^{-1} \leqslant \frac{|P^{\pm}|}{|P_0||P_1|\delta^{-\frac{1}{2}}} \leqslant C, \tag{3.27}$$

$$C^{-1} \leqslant \frac{|Q^{\pm}|}{|Q_0||Q_1|\delta^{-\frac{1}{2}}} \leqslant C, \tag{3.28}$$

where the constants are uniform once b is fixed and the distortions are uniformly bounded.

From [PY2, Theorem 3.7], we also have the following estimate for the distortion of $F^{\pm}$ (see also Appendix A.4): assuming that b is small enough (in terms of the partial derivatives of first order of X_u, Y_s, θ), we have

$$D(F^{\pm}) \leq \max\Big\{D(F_0) + C|Q_0|\delta^{-1}, D(F_1) + C|P_1|\delta^{-1}\Big\}, \tag{3.29}$$

provided that $D(F_0) + D(F_1) \leq \delta^{-\frac{1}{2}}$. The constant C in (3.29) depends only on the partial derivatives of first order of X_s, Y_u, θ.

Let us observe that, while conditions (PC1), (PC2) are *necessary* in order to consider parabolic composition, they will not be *sufficient* for our construction: in Section 5, the requirement for parabolic composition will be much more restrictive than (PC2).

3.6. *Simple properties of $\delta(Q_0, P_1)$ and related quantities.*

3.6.1. The setting in the same than in the last Subsection, with maps F_0, F_1 satisfying the hypotheses as in that Subsection. We have defined

$$\delta(Q_0, P_1) = \min_{y_0} \min_{x_1} -\overline{C}(y_0, x_1). \tag{3.30}$$

In the following sections, we also need to consider

$$\delta_L(Q_0, P_1) := \max_{y_0} \min_{x_1} -\overline{C}(y_0, x_1), \tag{3.31}$$

$$\delta_R(Q_0, P_1) := \min_{y_0} \max_{x_1} -\overline{C}(y_0, x_1), \tag{3.32}$$

$$\delta_{LR}(Q_0, P_1) := \max_{y_0} \max_{x_1} -\overline{C}(y_0, x_1). \tag{3.33}$$

All together, δ, δ_L, δ_R, δ_{LR} are the values of $-\overline{C}$ at the four corners of the rectangle $I_0^u \times I_1^s$ of definition of $\overline{C}$. From the formula (A.78) of Appendix A, we have, for any t in I

$$C^{-1}|P_1| \leqslant |\overline{C}_x| \leqslant C|P_1|, \tag{3.34}$$

$$C^{-1}|Q_0| \leqslant |\overline{C}_y| \leqslant C|Q_0|. \tag{3.35}$$

This gives

$$C^{-1}|Q_0| \leqslant \delta_L(Q_0, P_1) - \delta(Q_0, P_1) \leqslant C|Q_0|, \tag{3.36}$$

$$C^{-1}|P_1| \leqslant \delta_R(Q_0, P_1) - \delta(Q_0, P_1) \leqslant C|P_1|, \tag{3.37}$$

$$C^{-1}|Q_0| \leqslant \delta_{LR}(Q_0, P_1) - \delta_R(Q_0, P_1) \leqslant C|Q_0|, \tag{3.38}$$

$$C^{-1}|P_1| \leqslant \delta_{LR}(Q_0, P_1) - \delta_L(Q_0, P_1) \leqslant C|P_1|. \tag{3.39}$$

Let F_0' be another affine-like map, implicitly represented by (A_0, B_0), satisfying the same hypotheses than F_0. Let C' be the function constructed from F_0' and F_1 as C was from F_0 and F_1.

3.6.2. We first assume that the image Q_0' of F_0' is *disjoint* from the image Q_0 of F_0. Assume for instance that Q_0' is above Q_0. For any $y_0^* \in I_0^u, y_0'^* \in I_0'^u$, the curve $G_- \circ F_0'(\{y_0' = y_0'^*\})$ is above the curve $G_- \circ F_0(\{y_0 = y_0^*\})$. This means that we have, for any w

$$B_0(y_0^*, X(w, y_0^*)) \leqslant B_0'(y_0'^*, X'(w, y_0'^*)). \tag{3.40}$$

As θ is monotone in the first variable, we will have, for all $w, x_1, y_0^*, y_0'^*$

$$C(w, y_0^*, x_1) < C'(w, y_0'^*, x_1), \tag{3.41}$$

if $\theta_y > 0$, and the opposite inequality if $\theta_y < 0$. In the first case, we will have

$$\delta(Q_0, P_1) > \delta_L(Q'_0, P_1), \tag{3.42}$$

$$\delta_R(Q_0, P_1) > \delta_{LR}(Q'_0, P_1). \tag{3.43}$$

In the second case, the same is true with Q_0, Q'_0 exchanged.

3.6.3. Assume now that the image Q'_0 of F'_0 is *contained* in Q_0.

Let $y_0^{\pm}$ be the endpoints of I_0^u. For any $y_0'^*$, the curve $G_- \circ F'_0(\{y'_0 = y_0'^*\})$ is between the curves $G_- \circ F_0(\{y_0 = y_0^{\pm}\})$. This means that we have, for any w

$$B_0(y_0^-, X(w, y_0^-)) \leqslant B'_0(y_0'^*, X'(w, y_0'^*)) \leqslant B_0(y_0^+, X(w, y_0^+)) \tag{3.44}$$

(assuming $B_{0,y} > 0$; otherwise, exchange y_0^+ and y_0^-).

As θ is monotone in the first variable, we have, for all w, $x_1, y_0'^*$

$$\min_{y_0} C(w, y_0, x_1) \leqslant C'(w, y_0'^*, x_1) \leqslant \max_{y_0} C(w, y_0, x_1). \tag{3.45}$$

This now gives

$$\delta(Q_0, P_1) \leqslant \delta(Q'_0, P_1), \tag{3.46}$$

$$\delta_R(Q_0, P_1) \leqslant \delta_R(Q'_0, P_1), \tag{3.47}$$

$$\delta_L(Q_0, P_1) \geqslant \delta_L(Q'_0, P_1), \tag{3.48}$$

$$\delta_{LR}(Q_0, P_1) \geqslant \delta_{LR}(Q'_0, P_1). \tag{3.49}$$

3.6.4. The setting is the same than in Section 3.6.2 and we will obtain slightly stronger estimates (which will be useful in Section 8) under a slightly stronger hypothesis. We not only assume that the image Q'_0 of F'_0 is contained in Q_0, but that one has, for any $(x_0, y_0) \in F_0^{-1}(Q'_0)$

$$y_0^- + C^{-1} < y_0 < y_0^+ - C^{-1}, \tag{3.50}$$

for some fixed large constant C. We will now obtain, instead of (3.44)

$$\begin{aligned} B_0(y_0^- + C^{-1}, X(w, y_0^- + C^{-1})) &\leqslant B'_0(y_0'^*, X'(w, y_0'^*)) \\ &\leqslant B_0(y_0^+ - C^{-1}, X(w, y_0^+ - C^{-1})) \end{aligned} \tag{3.51}$$

or the reverse inequalities. This gives

$$\begin{aligned} B_0(y_0^-, X(w, y_0^-)) + C^{-1}|Q_0| &\leqslant B'_0(y_0'^*, X'(w, y_0'^*)) \\ &\leqslant B_0(y_0^+, X(w, y_0^+)) - C^{-1}|Q_0| \end{aligned} \tag{3.52}$$

(if $B_{0,y} > 0$; otherwise, exchange y_0^+ and y_0^-). As the partial derivative θ_y is bounded away from 0, we get also, for all w, $x_1, y_0'^*$

$$\min_{y_0} C(w, y_0, x_1) + C^{-1}|Q_0| \leqslant C'(w, y_0', x_1) \leqslant \max_{y_0} C(w, y_0, x_1) - C^{-1}|Q_0| \tag{3.53}$$

which finally gives

$$\delta(Q_0, P_1) + C^{-1}|Q_0| \leqslant \delta(Q_0', P_1), \tag{3.54}$$

$$\delta_R(Q_0, P_1) + C^{-1}|Q_0| \leqslant \delta_R(Q_0', P_1), \tag{3.55}$$

$$\delta_L(Q_0, P_1) - C^{-1}|Q_0| \geqslant \delta_L(Q_0', P_1), \tag{3.56}$$

$$\delta_{LR}(Q_0, P_1) - C^{-1}|Q_0| \geqslant \delta_{LR}(Q_0', P_1). \tag{3.57}$$

3.6.5. All the discussion in Sections 3.6.2, 3.6.3, 3.6.4 have an obvious symmetric counterpart exchanging Q's and P's.

4. Structure of parameter space

4.1. *One-parameter families.* — From now on, we fix a one-parameter family $(g_t)_{t\in(-t_0,t_0)}$ in $\mathcal{U}$. We assume that the family is transverse to $\mathcal{U}_0$ at $t = 0$, with $g_t \in \mathcal{U}_+$ for $t > 0$ and $g_t \in \mathcal{U}_-$ for $t < 0$.

Observe that g_0 satisfies exactly the same assumptions as f, provided $\mathcal{U}$ is small enough. Therefore, we may and shall, assume that $g_0 = f$.

We will first reparametrize the family in order to make some computations simpler. Consider the folding map $G_t = g_t^{N_0}$ of Section 2.3. If t_0 is small enough, G_t is a fold map for all values of $t \in (-t_0, t_0)$. Moreover, we can in properties (P2), (P3) of Section 2.3 choose a function θ which depends smoothly on t.

From (MP3), Section 2.2, the values $y_u = 0$, $x_s = 0$ of the arguments of θ correspond to $W^u_{\text{loc}}(p_u)$ and $W^s_{\text{loc}}(p_s)$ respectively. Therefore, the transversality of the family to $\mathcal{U}_0$ is equivalent to

$$\frac{\partial}{\partial t}\theta(0, 0, t)\,|_{t=0} > 0. \tag{4.1}$$

Taking t_0 small enough, we can therefore reparametrize our family in order to have

$$\theta(0, 0, t) \equiv t, \quad t \in (-t_0, t_0). \tag{4.2}$$

4.2. *Some important constants.* — The constants λ, u, v, D_0, b of Sections 3.4 and 3.5 depend only on the initial diffeomorphism f provided $\mathcal{U}$ is small enough and are now fixed.

Throughout the rest of the paper, we will use three main constants ε_0, η, τ which satisfy

$$0 < \varepsilon_0 \ll \eta \ll \tau \ll 1. \tag{4.3}$$

We roughly explain the meaning of each constant:

- ε_0 is the maximal width of the parabolic tongues L_u, L_s. It is also the size of the parameter interval we start with.
- η is involved in the transversality relation (defined in Section 5) which allows parabolic composition: instead of the condition (PC2) of Section 3.5, roughly speaking we will ask that

$$\delta \geqslant (|P_1| + |Q_0|)^{1-\eta}. \tag{4.4}$$

  As the widths $|P_1|$ and $|Q_0|$ are not larger than ε_0 which is much smaller than η, this requirement is much more restrictive than (PC2) and will allow to control distortion.
- τ relates the successive scales of the parameter intervals we will consider through the formula $\varepsilon_{k+1} = \varepsilon_k^{1+\tau}$.

Another important number β appears in the definition of regularity in Section 5, which controls the recurrence of the "critical locus". This number will be chosen explicitly in Section 9.3 (see also Section 5.6) in terms of d_s^0, d_u^0 and the eigenvalues of f at the periodic points p_s, p_u; the condition (H4) involving d_s^0, d_u^0 in Section 1 is required because we must have $\beta > 1$.

Finally, we will use the generic letter C (with indices or other decorations) for various constants which depend on the initial diffeomorphism f, but not on ε_0, η, τ.

4.3. *Parameter intervals.* — The starting parameter interval will be

$$I_0 := [\varepsilon_0, 2\varepsilon_0], \tag{4.5}$$

where, as explained above, ε_0 will be taken very small. This is the only parameter interval at level 0.

At level k, we will deal with parameter intervals of length ε_k, where the sequence of scales ε_k is defined inductively by

$$\varepsilon_{k+1} = \varepsilon_k^{1+\tau}. \tag{4.6}$$

The constant τ is small, but ε_0 is much smaller and in particular we will have $\varepsilon_0^{\tau^2} \ll 1$. Every parameter interval of level k is divided into $[\varepsilon_k^{-\tau}]$ parameter intervals with disjoint interiors of level $k+1$.

The remaining part, if any, is discarded; it is of length $< \varepsilon_{k+1}$; the total length discarded in this way is smaller than $\varepsilon_1 \ll \varepsilon_0$.

Let $\tilde{I}$ be a parameter interval of level k and I be a parameter interval of level $k+1$ contained in $\tilde{I}$. We say that $\tilde{I}$ is the *parent* of I and that I is a *child* of $\tilde{I}$.

4.4. *The selection process.* — In Section 5, we will define what it means for a parameter interval to be *regular*. The starting interval I_0 will be regular.

Given a regular parameter interval $\widetilde{I}$ of level k, we divide it into its children: these parameter intervals of level $k+1$ are the *candidates*. We then test each candidate for regularity and discard those which are not regular. We then proceed to level $k+1$ with each surviving candidate.

The *regular* parameters are those which are the intersection of a decreasing sequence of regular parameter intervals. For such parameters, we are able to carry out some analysis of the maximal invariant set Λ_{g_t}.

4.5. *Strongly regular parameters.* — The regularity property is, in some sense, the minimal requirement that is needed to keep control on the geometry and dynamics of the maximal invariant set. However, this requirement is of an essentially qualitative character and this leads in particular to the following difficulty: we are not able to estimate which proportion of the children of a regular parameter interval are also regular.

To circumvent this problem, we define in Section 9 a stronger property for parameter intervals, called *strong regularity*. It implies regularity, and is better adapted to the inductive selection process. It also gives additional geometric information on the maximal invariant set.

When $\widetilde{I}$ is a strongly regular parameter interval of level k, we will show in Section 9 that most candidates of level $k+1$ contained in $\widetilde{I}$ are also strongly regular. The proportion of discarded candidates is less than α_k, with

$$\sum_{k\geqslant 0} \alpha_k \ll 1, \tag{4.7}$$

the $\ll$ sign means that the sum gets arbitrarily small as ε_0 goes to zero. Then we can conclude that most parameters are strongly regular in the sense that they are equal to the intersection of decreasing sequences of strongly regular parameter intervals.

The non-uniformly hyperbolic horseshoes that are the subject of our study are exactly the maximal invariant set Λ_g for strongly regular $g \in \mathcal{U}_+$.

5. Classes of affine-like iterates and the transversality relation

5.1. *Affine-like iterates.* — Let I be a parameter interval of some level.

Definition **1.** — *An* I*-persistent affine-like iterate is a triple* (P, Q, n) *such that*

- P *is a vertical strip in some* R_a, *depending smoothly on* $t \in I$;
- Q *is a horizontal strip in some* $R_{a'}$, *depending smoothly on* $t \in I$;
- n *is a nonnegative integer;*

- *for each* $t \in \mathrm{I}$, *the restriction of* g_t^n *to* P_t *is an affine-like map onto* Q_t, *i.e. property* (AL1) *of Section* 3.1 *holds;*
- *for each* $t \in \mathrm{I}$, *each* $m \in [0, n]$, *we have* $g_t^m(\mathrm{P}_t) \subset \widehat{\mathrm{R}}$.

Example **1.**

1. For $n = 0$, the I-persistent affine-like iterates are the $(\mathrm{R}_a, \mathrm{R}_a, 0)$, $a \in \mathcal{a}$.
2. For $n = 1$, the I-persistent affine-like iterates are the $(\mathrm{P}_{aa'}, \mathrm{Q}_{aa'}, 1)$, $(a, a') \in \mathcal{B}$.
3. More generally, for any finite word $\underline{a} = (a_0, \ldots, a_n)$ with transitions in $\mathcal{B}$, the map $g_{\underline{a}}$ of Section 3.4 defined an I-persistent affine-like iterate $(\mathrm{P}_{\underline{a}}, \mathrm{Q}_{\underline{a}}, n)$.

Notation. — If P is a vertical strip $\{\varphi_-(y) \leqslant x \leqslant \varphi_+(y)\}$ we denote by $\partial\mathrm{P}$ the vertical part of the boundary, i.e. the two graphs $\{x = \varphi^{\pm}(y)\}$. Similarly for horizontal strips.

If $(\mathrm{P}, \mathrm{Q}, n)$ is an I-persistent affine-like iterate and I′ is a parameter interval contained in I, $(\mathrm{P}, \mathrm{Q}, n)$ also defines by restriction an I′-persistent affine-like iterate. A slightly less trivial property is given by

Proposition **1.** — *Let* $(\mathrm{P}, \mathrm{Q}, n)$, $(\mathrm{P}', \mathrm{Q}', n')$ *be* I*-persistent affine-like iterates. We have*

(a) *if* $n = n'$, *then either* $\mathrm{P} = \mathrm{P}'$ *and* $\mathrm{Q} = \mathrm{Q}'$ *for all* $t \in \mathrm{I}$ *or* $\mathrm{P} \cap \mathrm{P}' = \emptyset$, $\mathrm{Q} \cap \mathrm{Q}' = \emptyset$ *for all* $t \in \mathrm{I}$.

(b) *if* $n < n'$, *then either* $\mathrm{P} \supset \mathrm{P}'$, $\partial\mathrm{P} \cap \mathrm{P}' = \emptyset$ *for all* $t \in \mathrm{I}$ *or* $\mathrm{P} \cap \mathrm{P}' = \emptyset$, *for all* $t \in \mathrm{I}$.

Remark **3.** — Throughout the paper, except in Section 9 (where we break the symmetry assuming $d_s^0 \geqslant d_u^0$), we will keep a time-symmetric setting. Thus every property stated for the domains P's is also valid for the images Q's. This apply for instance to part b) of the proposition.

Proof. — By the definition of an I-persistent affine-like iterate, for all $t \in \mathrm{I}$, P is a connected component of $\mathrm{R} \cap g_t^{-n}(\mathrm{R})$ and also of $\bigcap_{0 \leqslant m \leqslant n} g_t^{-m}(\widehat{\mathrm{R}})$.

(a) If $n = n'$ and $\mathrm{P} \cap \mathrm{P}' \neq \emptyset$ for some $t_0 \in \mathrm{I}$, we must have $\mathrm{P} = \mathrm{P}'$ at t_0 and hence, $\mathrm{P} \cap \mathrm{P}' \neq \emptyset$ for t close to t_0. It follows that $\mathrm{P} = \mathrm{P}'$ for all $t \in \mathrm{I}$, and also $\mathrm{Q} = \mathrm{Q}'$ for all $t \in \mathrm{I}$.

(b) Assume that $n < n'$ and $\mathrm{P} \cap \mathrm{P}' \neq \emptyset$ for some $t_0 \in \mathrm{I}$, then $\mathrm{P}' \subset \mathrm{P}$ at t_0 (since $\bigcap_{0 \leqslant m \leqslant n'} g_{t_0}^{-m}(\widehat{\mathrm{R}})$ is contained in $\bigcap_{0 \leqslant m \leqslant n} g_{t_0}^{-m}(\widehat{\mathrm{R}})$), hence $\mathrm{P}' \cap \mathrm{P} \neq \emptyset$ for t close to t_0 and $\mathrm{P}' \subset \mathrm{P}$ for all $t \in \mathrm{I}$.

Let $t \in \mathrm{I}$, $z \in \partial\mathrm{P}$; then, $g_t^n(z)$ belongs to the vertical boundary of R and $g_t^{n+1}(z) \notin \widehat{\mathrm{R}}$; therefore, $z \notin \mathrm{P}'$. This proves that $\partial\mathrm{P} \cap \mathrm{P}'$ is empty for all $t \in \mathrm{I}$. □

5.2. *The classes* $\mathcal{R}(\mathrm{I})$*: general overview.* — It would be nice to work with the class of all I-persistent affine-like iterates, but with this approach one faces two problems:

- I-persistent affine-like iterates do not satisfy a uniform cone condition, and they do not have uniformly bounded distortion;

– even if we force such uniformity in the definition, a major problem is that we lack some control on the way in which long I-persistent affine-like iterates are constructed from shorter ones by simple or parabolic composition.

To overcome these problems, we will construct, for every candidate parameter interval I (see Section 4.4: it means that $I = I_0$ or that the parent interval of I is regular, a notion that will be defined in Section 5.6) a subset $\mathcal{R}(I)$ of the set of all I-persistent affine-like iterates. All elements of $\mathcal{R}(I)$ with $n > 1$ will be obtained, from the very definition of $\mathcal{R}(I)$, from shorter ones by simple or parabolic composition. The elements of $\mathcal{R}(I)$ will turn out to satisfy a uniform cone condition and have uniformly bounded distortion.

The main ingredient in the definition of $\mathcal{R}(I)$ is a *transversality relation* which is an appropriate strengthening of condition (PC2) in Section 3.5. Simple composition is allowed whenever it makes sense, but parabolic composition is only allowed when this transversality relation holds.

The definition of the transversality relation, given later in this section, is quite involved; this is because we want some combinatorial properties proved in Section 6 to be satisfied. Such properties make our later work much easier.

The precise requirements on $\mathcal{R}(I)$ are the properties (R1)–(R6) formulated in the next subsection and property (R7) formulated in Section 5.4.

While it is clear that at most one class $\mathcal{R}(I)$ can satisfy (R1)–(R7), it is by no means obvious that such a class exists. Actually, the proof of this fact, stated at the end of Section 5.4, will only be completed at the end of Section 7.

The proof of the existence of $\mathcal{R}(I)$ is based on a double induction:

- an induction on the level of the parameter interval I, starting with $I_0 = [\varepsilon_0, 2\varepsilon_0]$ at level 0 (see Section 5.5 for this first step).
- for a given candidate parameter interval I, an induction on the length n of the I-persistent affine-like iterates under consideration.

In this induction scheme, all properties required for $\mathcal{R}(I)$ are proved simultaneously. Actually, several other properties (coherence, concavity, ... see Section 6) of the class $\mathcal{R}(I)$ are needed in the induction process; these properties are defined and proved (inductively!) in Section 6. Of particular importance is a structure theorem in Section 6.5.

5.3. *Defining properties for the special class of affine-like iterates* $\mathcal{R}(I)$. — Let I be a candidate parameter interval of some level.

The class $\mathcal{R}(I)$ of I-persistent affine-like iterates that we want to construct should satisfy the following properties (R1)–(R7).

(R1) For any word $\underline{a} = (a_0, \dots, a_n)$ with transitions in $\mathcal{B}$, the element $(P_{\underline{a}}, Q_{\underline{a}}, n)$ (see Example 1 above) belongs to $\mathcal{R}(I)$.

For the starting interval $I_0 = [\varepsilon_0, 2\varepsilon_0]$, it will turn out that one obtains in this way all elements of $\mathcal{R}(I_0)$.

Recall from (MP5), (MP6) in Section 3.4 that all $(P_{\underline{a}}, Q_{\underline{a}}, n)$ with $n > 0$ satisfy for all $t \in I_0$ a uniform cone condition (AL2) with parameters λ, u, v (satisfying $1 < uv \leqslant \lambda^2$), and have distortion bounded by D_0. Let $u_0 = \frac{u}{(uv)^{1/4}}$, $v_0 = \frac{v}{(uv)^{1/4}}$.

(R2) All $(P, Q, n) \in \mathcal{R}(I)$ satisfy for all $t \in I$ the cone condition (AL2) with parameters λ, u_0, v_0 and have distortion bounded by $2D_0$ for all $t \in I$.

Let (P, Q, n), (P', Q', n') be elements of $\mathcal{R}(I)$ such that $Q \subset R_a$, $P' \subset R_a$ for some $a \in \mathcal{a}$. As both iterates satisfy the cone condition (AL2) with parameters λ, u_0, v_0, we know from Section 3.3 that the simple composition defined by

$$P'' = P \cap g^{-n}(P'), \quad Q'' = Q' \cap g^{n'}(Q), \quad n'' = n + n', \tag{5.1}$$

is an (I-persistent) affine-like iterate.

The next condition states that it should also belong to $\mathcal{R}(I)$.

(R3) The class $\mathcal{R}(I)$ is stable under simple composition.

We now turn to parabolic composition.

We first define two special elements which belong to $\mathcal{R}(I)$ according to (R1): define (P_s, Q_s, n_s) (resp. (P_u, Q_u, n_u)) to be the element $(P_{\underline{a}}, Q_{\underline{a}}, n)$ with maximal length n such that $L_s \subset P_s$ for all $t \in I_0$ (resp. $L_u \subset Q_u$ for all $t \in I_0$). We have that $p_s \in P_s$ and $p_u \in Q_u$. See Figure 7.

We obviously have, for all $t \in I_0$

$$\begin{aligned} C^{-1}\varepsilon_0 \leqslant |P_s| \leqslant C\varepsilon_0, \\ C^{-1}\varepsilon_0 \leqslant |Q_u| \leqslant C\varepsilon_0. \end{aligned} \tag{5.2}$$

The next condition guarantees that property (PC1) in Section 3.5 is satisfied.

(R4) Let (A, B) be the implicit representation of an affine-like iterate $(P, Q, n) \in \mathcal{R}(I)$.

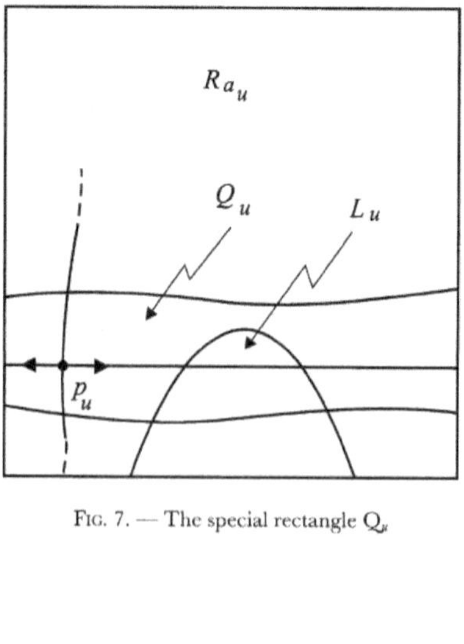

FIG. 7. — The special rectangle Q_u

(a) If $P \subset P_s$, then for all $t \in I$ we have

$$|A_y| \leqslant C\varepsilon_0, \qquad |A_{yy}| \leqslant C\varepsilon_0.$$

(b) If $Q \subset Q_u$, then for all $t \in I$ we have

$$|B_x| \leqslant C\varepsilon_0, \qquad |B_{xx}| \leqslant C\varepsilon_0.$$

Let (P_0, Q_0, n_0), (P_1, Q_1, n_1) be elements in $\mathcal{R}(I)$ with $Q_0 \subset Q_u$, $P_1 \subset P_s$. In these circumstances, we will define in Section 5.4 a *transversality relation* denoted by $Q_0 \pitchfork_I P_1$ which may or may not hold. When it holds, it implies condition (PC2) of Section 3.5 for all $t \in I$ (see (R7) below).

(R5) If (P_0, Q_0, n_0), (P_1, Q_1, n_1) as above satisfy $Q_0 \pitchfork_I P_1$, then both I-persistent affine-like iterates obtained from the parabolic composition $g_t^{n_1} \circ G_t \circ g_t^{n_0}$ belong to $\mathcal{R}(I)$.

Writing (P^+, Q^+, n) and (P^-, Q^-, n) for these two iterates, we have $n = n_0 + n_1 + N_0$. The domains P^+ and P^- are the two components of $g^{-n_0}(Q_0 \cap G_t^{-1}(P_1))$; the images Q^+ and Q^- are the two components of $g^{n_1}(P_1 \cap G_t(Q_0))$. See Figure 6, Section 3.5.

When (P_0, Q_0, n_0), (P_1, Q_1, n_1) satisfy $Q_0 \subset Q_u$, $P_1 \subset P_s$, $Q_0 \pitchfork_I P_1$, we say that their parabolic composition is allowed in $\mathcal{R}(I)$.

(R6) Any $(P, Q, n) \in \mathcal{R}(I)$ with $n > 1$ can be obtained from shorter elements by simple composition or (allowed) parabolic composition.

Typically, an element of $\mathcal{R}(I)$ can be obtained in many ways by composition of shorter ones. We say that an element of $\mathcal{R}(I)$ is *prime* if it cannot be obtained by *simple* composition of shorter ones. Prime elements play a key role in the description of the dynamics for regular parameters in Section 10.

It is pretty clear from conditions (R1), (R3), (R5), (R6) alone that there is at most one class $\mathcal{R}(I)$ satisfying these conditions. The existence of $\mathcal{R}(I)$, i.e. the proof of the consistency of conditions (R1)–(R6), is much more delicate. There is actually a seventh property (R7) formulated in the next subsection and related to the condition (PC2) for parabolic composition.

Parent-child terminology and notations for compositions. — Let (P, Q, n), $(\widetilde{P}, \widetilde{Q}, \widetilde{n}) \in \mathcal{R}(I)$ with $P \subset \widetilde{P}$, $n > \widetilde{n}$. When there is no $(\widehat{P}, \widehat{Q}, \widehat{n}) \in \mathcal{R}(I)$ with $P \subset \widehat{P} \subset \widetilde{P}$ and $n > \widehat{n} > \widetilde{n}$, we say that P is a *child* of $\widetilde{P}$ and that $\widetilde{P}$ is the *parent* of P.

When $n = \widetilde{n} + 1$, we say that P is a *simple* child of $\widetilde{P}$. The image $g^{\widetilde{n}}(P)$ is equal to the intersection of $\widetilde{Q}$ with some $P_{a,a'}$ and (P, Q, n) is the simple composition of $(\widetilde{P}, \widetilde{Q}, \widetilde{n})$ with $(P_{a,a'}, Q_{a,a'}, 1)$.

When $n > \widetilde{n} + 1$, we say that P is a *non-simple* child of $\widetilde{P}$. The image $g^{\widetilde{n}}(P)$ is contained in $L_u \cap \widetilde{Q}$. We will prove in Section 6.2 that (P, Q, n) is obtained by parabolic composition of $(\widetilde{P}, \widetilde{Q}, \widetilde{n})$ with some (P_1, Q_1, n_1)

Let $(P_0, Q_0, n_0), (P_1, Q_1, n_1) \in \mathcal{R}(I)$. If Q_0, P_1 are contained in a same rectangle R_a, the simple composition $(P, Q, n) \in \mathcal{R}(I)$ of these elements will be written as

$$(P, Q, n) = (P_0, Q_0, n_0) * (P_1, Q_1, n_1). \tag{5.3}$$

If $Q_0 \subset Q_u$, $P_1 \subset P_s$ and $Q_0 \pitchfork_I P_1$, any $(\widehat{P}, \widehat{Q}, \widehat{n})$ of the two elements obtained by the corresponding allowed parabolic composition will be written as

$$(\widehat{P}, \widehat{Q}, \widehat{n}) \in (P_0, Q_0, n_0) \,\square\, (P_1, Q_1, n_1). \tag{5.4}$$

5.4. *Definition of the transversality relation.* — Let I be a parameter interval of some level, and let (P_0, Q_0, n_0), (P_1, Q_1, n_1) be elements of $\mathcal{R}(I)$ which satisfy $Q_0 \subset Q_u$, $P_1 \subset P_s$.

From (R4) the condition (PC1) of Section 3.5 is satisfied provided ε_0 small enough. Denote by (x_0, y_0) (resp. (x_1, y_1)) the coordinates in the rectangle containing P_0 (resp. Q_1). A function $\overline{C}(y_0, x_1)$ was defined in Section 3.5, together with

$$\delta(Q_0, P_1) = \min_{y_0} \min_{x_1} -\overline{C}(y_0, x_1). \tag{5.5}$$

In Section 3.5, we were asking for δ to be much larger than $|P_0|$ and $|Q_1|$. Recall from Section 3.6 the definitions

$$\delta_L(Q_0, P_1) := \max_{y_0} \min_{x_1} -\overline{C}(y_0, x_1), \tag{5.6}$$

$$\delta_R(Q_0, P_1) := \min_{y_0} \max_{x_1} -\overline{C}(y_0, x_1), \tag{5.7}$$

$$\delta_{LR}(Q_0, P_1) := \max_{y_0} \max_{x_1} -\overline{C}(y_0, x_1). \tag{5.8}$$

Preliminary Definition. — *We write* $Q_0 \overline{\pitchfork}_I P_1$ *if the following holds*

(T1) *for all* $t \in I$,

$$\delta_{LR}(Q_0, P_1) \geqslant 2|I|,$$

(T2) *for some* $t_0 \in I$,

$$\delta_R(Q_0, P_1) \geqslant 2|Q_0|^{1-\eta},$$

(T3) *for some* $t_1 \in I$,

$$\delta_L(Q_0, P_1) \geqslant 2|P_1|^{1-\eta}.$$

Definition. — *We say that* Q_0, P_1 *are* I*-transverse and write* $Q_0 \pitchfork_I P_1$ *if there exist a parameter interval* $\widetilde{I} \supset I$, *elements* $(\widetilde{P}_0, \widetilde{Q}_0, \widetilde{n}_0)$, $(\widetilde{P}_1, \widetilde{Q}_1, \widetilde{n}_1) \in \mathcal{R}(\widetilde{I})$ *with* $\widetilde{P}_1 \supset P_1$, $\widetilde{Q}_0 \supset Q_0$ *such that* $\widetilde{Q}_0 \overline{\pitchfork}_{\widetilde{I}} \widetilde{P}_1$.

Remark **4.**

1. Taking $\widetilde{\mathrm{I}} = \mathrm{I}$, $\widetilde{\mathrm{P}}_0 = \mathrm{P}_0$, $\widetilde{\mathrm{Q}}_1 = \mathrm{Q}_1$, it is obvious that if $\mathrm{Q}_0 \,\overline{\pitchfork}_{\mathrm{I}}\, \mathrm{P}_1$, then $\mathrm{Q}_0 \pitchfork_{\mathrm{I}} \mathrm{P}_1$.
2. In view of our inductive procedure, all $(\widetilde{\mathrm{P}}_0, \widetilde{\mathrm{Q}}_0, \widetilde{n}_0)$, $(\widetilde{\mathrm{P}}_1, \widetilde{\mathrm{Q}}_1, \widetilde{n}_1)$ which have to be considered have been constructed before $(\mathrm{P}_0, \mathrm{Q}_0, n_0)$, $(\mathrm{P}_1, \mathrm{Q}_1, n_1)$.
3. As mentioned before, the definition of the transversality relation is quite involved. One reason for this is that we wish to obtain several properties (heredity, concavity, . . .) that will be proven in the next subsections. Some justification for the choice of quantifiers in (T1), (T2), (T3) can be found in Appendix C.

At first sight, it appears that properties (T2), (T3) above are not quite sufficient to guarantee condition (PC2) of parabolic composition (Section 3.5), because they involve only one value of the parameter. The next property takes care of this problem.

(R7) If $(\mathrm{P}_0, \mathrm{Q}_0, n_0)$, $(\mathrm{P}_1, \mathrm{Q}_1, n_1) \in \mathcal{R}(\mathrm{I})$ satisfy $\mathrm{Q}_0 \subset \mathrm{Q}_u$, $\mathrm{P}_1 \subset \mathrm{P}_s$ and $\mathrm{Q}_0 \pitchfork_{\mathrm{I}} \mathrm{P}_1$ holds, then, for all $t \in \mathrm{I}$, we have

$$\delta(\mathrm{Q}_0, \mathrm{P}_1) \geqslant \mathrm{C}^{-1}\left(|\mathrm{P}_1|^{1-\eta} + |\mathrm{Q}_0|^{1-\eta}\right).$$

Now that properties (R1)–(R7) have been introduced, we can state the

Theorem **1.** — *For each candidate parameter interval* I, *there exists exactly one class* $\mathcal{R}(\mathrm{I})$ *of I-persistent affine-like iterates which satisfies* (R1)–(R7).

As mentioned before, uniqueness is clear. The proof of existence will only be completed at the end of Section 7.

5.5. *The class* $\mathcal{R}(\mathrm{I}_0)$. — We claim that the class $\mathcal{R}(\mathrm{I}_0)$ is exactly formed by the iterates associated to the horseshoe K considered in Section 3.4, i.e. the elements $(\mathrm{P}_{\underline{a}}, \mathrm{Q}_{\underline{a}}, n)$ with $\underline{a} = (a_0, \ldots, a_n)$ a word with transitions in $\mathcal{B}$.

Indeed, such elements must belong to $\mathcal{R}(\mathrm{I}_0)$ by (R1). They satisfy (R2), (R3) and (R6). We show below that, for $(\mathrm{P}_0, \mathrm{Q}_0, n_0)$,$(\mathrm{P}_1, \mathrm{Q}_1, n_1)$ of this form with $\mathrm{Q}_0 \subset \mathrm{Q}_u$,$\mathrm{P}_1 \subset \mathrm{P}_s$, the transversality relation $\mathrm{Q}_0 \pitchfork_{\mathrm{I}} \mathrm{P}_1$ is never satisfied. Then it follows that (R5), (R7) are vacuously satisfied. Finally, property (R4) will be proved in Proposition 16 of Section 7.4.

Let $(\mathrm{P}_0, \mathrm{Q}_0, n_0)$,$(\mathrm{P}_1, \mathrm{Q}_1, n_1)$ as above (with $\mathrm{Q}_0 \subset \mathrm{Q}_u$,$\mathrm{P}_1 \subset \mathrm{P}_s$). The notations are those of Sections 3.5 and 5.4. Assume for instance that in R_{a_u} the foliation $\mathrm{W}^u_{\mathrm{loc}}(\mathrm{K})$ is contained in $\{y_{a_u} \geqslant 0\}$, and that in R_{a_s} the foliation $\mathrm{W}^s_{\mathrm{loc}}(\mathrm{K})$ is contained in $\{x_{a_s} \geqslant 0\}$. This is equivalent to say that θ is monotone increasing in both variables.

There exists y_0^* such that the image under $\mathrm{G}_- \circ g^{n_0}$ of the horizontal segment $\{y = y_0^*\} \cap \mathrm{P}_0$ lies in $\{y_u > 0\}$. Similarly, there exists x_1^* such that the image under $\mathrm{G}_+^{-1} \circ g^{-n_1}$ of the vertical segment $\{x = x_1^*\} \cap \mathrm{Q}_1$ lies in $\{x_s > 0\}$. Then, from formulas (3.18)–(3.20) of Section 3.5 and the monotonicity of θ, we have, for all $t \in \mathrm{I}_0$ and w

$$\mathrm{C}(w, y_0^*, x_1^*) \geqslant -\theta(0, 0, t). \tag{5.9}$$

As we have $\theta(0,0,t)=t$ by the normalization of Section 4.1, we obtain, for $t=\varepsilon_0$:

$$(\mathbf{5.10}) \qquad -\overline{\mathrm{C}}(y_0^*, x_1^*) \leqslant |\mathrm{I}_0|,$$

and therefore, for the same value of t:

$$(\mathbf{5.11}) \qquad \delta(\mathrm{Q}_0, \mathrm{P}_1) \leqslant |\mathrm{I}_0|.$$

We have also, for all $t \in \mathrm{I}_0$

$$(\mathbf{5.12}) \qquad \delta(\mathrm{Q}_0, \mathrm{P}_1) \leqslant 2|\mathrm{I}_0|.$$

If condition (T1) in Section 5.4 is satisfied, we have from (3.36)–(3.39) that at $t=\varepsilon_0$ either $|\mathrm{Q}_0|$ or $|\mathrm{P}_1|$ (or both) is at least of the order of ε_0. But then Proposition 16 in Section 7.4 guarantees that the same width stays at least of the order of ε_0 for all $t \in \mathrm{I}_0$. Then either (T2) or (T3) will not be satisfied.

We have thus shown that the transversality relation is never satisfied in $\mathcal{R}(\mathrm{I}_0)$. This completes the initial step in the proof of the theorem in Section 5.4 (except for the proof of Proposition 16 in Section 7.4).

5.6. *Criticality, bicriticality, and the regularity property.* — We introduce some terminology and some concepts related to the transversality relation. This includes the regularity property mentioned in the selection process of Section 4.4.

5.6.1. The following obvious but fundamental property, which we may sum up by saying that transversality is hereditary, was forced into the definition of the transversality relation.

Proposition **2.** — *Let* $\widetilde{\mathrm{I}} \supset \mathrm{I}$ *be parameter intervals. Let* $(\mathrm{P}_0, \mathrm{Q}_0, n_0), (\mathrm{P}_1, \mathrm{Q}_1, n_1) \in \mathcal{R}(\mathrm{I})$ *and* $(\widetilde{\mathrm{P}}_0, \widetilde{\mathrm{Q}}_0, \widetilde{n}_0), (\widetilde{\mathrm{P}}_1, \widetilde{\mathrm{Q}}_1, \widetilde{n}_1) \in \mathcal{R}(\widetilde{\mathrm{I}})$. *Assume that* $\mathrm{Q}_0 \subset \widetilde{\mathrm{Q}}_0 \subset \mathrm{Q}_u$ *and* $\mathrm{P}_1 \subset \widetilde{\mathrm{P}}_1 \subset \mathrm{P}_s$. *If* $\widetilde{\mathrm{Q}}_0$ *and* $\widetilde{\mathrm{P}}_1$ *are* $\widetilde{\mathrm{I}}$*-transverse, then* Q_0 *and* P_1 *are* I*-transverse.*

Corollary **1.** — *Let* $\widetilde{\mathrm{I}} \supset \mathrm{I}$ *be parameter intervals, and let* $(\mathrm{P}_0, \mathrm{Q}_0, n_0), (\mathrm{P}_1, \mathrm{Q}_1, n_1) \in \mathcal{R}(\widetilde{\mathrm{I}}) \cap \mathcal{R}(\mathrm{I})$ *be such that* $\mathrm{Q}_0 \subset \mathrm{Q}_u$, $\mathrm{P}_1 \subset \mathrm{P}_s$. *If their parabolic composition is allowed in* $\mathcal{R}(\widetilde{\mathrm{I}})$, *it is also allowed in* $\mathcal{R}(\mathrm{I})$.

Proof. — This is the case $\widetilde{\mathrm{Q}}_0 = \mathrm{Q}_0$, $\widetilde{\mathrm{P}}_1 = \mathrm{P}_1$ of the proposition. □

Corollary **2.** — *Let* $\widetilde{\mathrm{I}} \supset \mathrm{I}$ *be parameter intervals. Then* $\mathcal{R}(\widetilde{\mathrm{I}})$ *is contained in* $\mathcal{R}(\mathrm{I})$.

Remark **5.** — This is a slight abus de langage of no consequence: properly speaking, we mean that the restriction to I of any $(\mathrm{P}, \mathrm{Q}, n) \in \mathcal{R}(\widetilde{\mathrm{I}})$ belongs to $\mathcal{R}(\mathrm{I})$.

Proof. — This is a consequence, by induction of the length n, of Axiom (R6) and Corollary 1. □

The following result is also an easy consequence of the definition of the transversality relation.

Proposition **3.** — *Let* $I \subset \widehat{I}$ *be parameter intervals, and let* (P_0, Q_0, n_0), (P_1, Q_1, n_1) *be elements of* $\mathcal{R}(I)$ *such that* $Q_0 \subset Q_u$ *and* $P_1 \subset P_s$. *Assume that* $Q_0 \pitchfork_I P_1$ *holds but* $Q_0 \pitchfork_{\widehat{I}} P_1$ *does not hold. Then there exists* $t \in \widehat{I}$ *such that* $\delta_{LR}(Q_0, P_1) < 2|\widehat{I}|$.

Proof. — Let $\widetilde{I} \supset I$, and let elements $(\widetilde{P}_0, \widetilde{Q}_0, \widetilde{n}_0)$, $(\widetilde{P}_1, \widetilde{Q}_1, \widetilde{n}_1) \in \mathcal{R}(\widetilde{I})$ with $\widetilde{P}_1 \supset P_1$, $\widetilde{Q}_0 \supset Q_0$ such that $\widetilde{Q}_0 \overline{\pitchfork}_{\widetilde{I}} \widetilde{P}_1$ holds. As $Q_0 \pitchfork_{\widehat{I}} P_1$ does not hold, $\widetilde{I}$ is strictly contained in $\widehat{I}$ and $\widetilde{Q}_0 \overline{\pitchfork}_{\widehat{I}} \widetilde{P}_1$ does not hold. But conditions (T2), (T3) for $\widetilde{Q}_0 \overline{\pitchfork}_{\widetilde{I}} \widetilde{P}_1$ imply the same for $\widetilde{Q}_0 \overline{\pitchfork}_{\widehat{I}} \widetilde{P}_1$. Hence (T1) for $\widetilde{Q}_0 \overline{\pitchfork}_{\widehat{I}} \widetilde{P}_1$ does not hold and there exists $t \in \widehat{I}$ such that $2|\widehat{I}| > \delta_{LR}(\widetilde{Q}_0, \widetilde{P}_1) \geqslant \delta_{LR}(Q_0, P_1)$. □

5.6.2. Let I be a (candidate) parameter interval, and let (P_0, Q_0, n_0), (P_1, Q_1, n_1) be elements of $\mathcal{R}(I)$ such that $Q_0 \subset Q_u$ and $P_1 \subset P_s$. When Q_0 and P_1 are not I-transverse, we say that:

- Q_0 and P_1 are I-*separated* if $G_t(Q_0) \cap P_1 \cap \Lambda = \emptyset$ for all $t \in I$; this happens in particular when $\delta_{LR}(Q_0, P_1) < 0$ for all $t \in I$;
- Q_0 and P_1 are I-*critically related* otherwise.

5.6.3. We assume in this subsection that I is a parameter interval for which the class $\mathcal{R}(I)$ has been fully constructed with properties (R1)–(R7).

Let $(P, Q, n) \in \mathcal{R}(I)$. An I-*decomposition* of P is a finite family $(P_\alpha, Q_\alpha, n_\alpha)$ of elements of $\mathcal{R}(I)$ such that the P'_αs are disjoint, strictly contained in P and satisfy, for all $t \in I$

$$W^s(\Lambda, \widehat{R}) \cap P = \bigsqcup_\alpha (W^s(\Lambda, \widehat{R}) \cap P_\alpha), \tag{5.13}$$

where $W^s(\Lambda, \widehat{R})$ was defined in Section 2.2. We say that P is I-*decomposable* if it admits an I-*decomposition*. Then, there is a coarsest one, namely by the children of P.

Remark **6.** — We will see in Section 8 that any P has only finitely many children.

Let $(P, Q, n) \in \mathcal{R}(I)$. We say that Q is I-*transverse* if either $Q \cap Q_u = \emptyset$ or $Q \subset Q_u$ and there exists an I-*decomposition* $(P_\alpha, Q_\alpha, n_\alpha)$ of P_s such that, for any α, Q and P_α are either I-transverse or I-separated.

We say that Q is I-*critical* when it is not I-transverse. This is always the case if $Q \supset Q_u$.

Let $(P, Q, n), (P', Q', n') \in \mathcal{R}(I)$ satisfy $Q' \subset Q$. As transversality is hereditary (Proposition 2), if Q is I-transverse, then Q′ is also I-transverse.

All these notions are also symmetrically defined exchanging P's and Q's,future and past.

We say that $(P, Q, n) \in \mathcal{R}(I)$ is I-*bicritical* if both P and Q are I-critical. The corresponding iterate should be thought as a piece of the dynamics corresponding to a return of the "critical region" to itself.

Definition **2.** — *Let $\beta > 1$. We say that the candidate parameter interval* I *is β-regular (or just regular when the value of β is fixed) if any* I*-bicritical element* $(P, Q, n) \in \mathcal{R}(I)$ *satisfies, for all $t \in I$:*

$$|P| < |I|^{\beta}, \qquad |Q| < |I|^{\beta}. \tag{5.14}$$

5.6.4. For $i = s, u$, let $\lambda(p_i)$ (resp. $\mu(p_i)$) be the stable (resp. unstable) eigenvalue of the periodic point p_i. Define

$$\omega_s = -\frac{\log|\lambda(p_s)|}{\log|\mu(p_s)|}, \qquad \omega_u = -\frac{\log|\mu(p_u)|}{\log|\lambda(p_u)|}. \tag{5.15}$$

As we have

$$C^{-1}\varepsilon_0 \leqslant |P_s| \leqslant C\varepsilon_0, \qquad C^{-1}\varepsilon_0 \leqslant |Q_u| \leqslant C\varepsilon_0, \tag{5.16}$$

we will have

$$C^{-1}\varepsilon_0^{\omega_s} \leqslant |Q_s| \leqslant C\varepsilon_0^{\omega_s}, \qquad C^{-1}\varepsilon_0^{\omega_u} \leqslant |P_u| \leqslant C\varepsilon_0^{\omega_u}. \tag{5.17}$$

Remark **7.** — The ratios ω_s, ω_u are smooth functions of the parameter, therefore the relative variation of the quantities $\varepsilon_0^{\omega_s}$, $\varepsilon_0^{\omega_u}$ in I_0 is negligible.

Proposition **4.** — *Assume that $1 < \beta < 1 + \min(\omega_s, \omega_u)$. Then the starting interval I_0 is β-regular.*

Proof. — Let $\underline{a}^s$ (resp. $\underline{a}^u$) the admissible word in the alphabet $\mathcal{A}$ such that $P_s = P_{\underline{a}^s}$ (resp. $Q_u = Q_{\underline{a}^u}$). By the property (MP2) of Section 2.1, these two words do not have a letter in common.

Let $(P_{\underline{a}}, Q_{\underline{a}}, n)$ an I_0-bicritical rectangle. As $P_{\underline{a}} \cap P_s$ is non-empty, either $\underline{a}$ starts with $\underline{a}^s$ or $\underline{a}^s$ starts with $\underline{a}$. Similarly, either $\underline{a}$ ends with $\underline{a}^u$ or $\underline{a}^u$ ends with $\underline{a}$. As $\underline{a}^s$, $\underline{a}^u$ do not have a common letter, $\underline{a}$ starts with $\underline{a}^s$, ends with $\underline{a}^u$ and $n > n_s + n_u$. In other words, we have

$$(P_{\underline{a}}, Q_{\underline{a}}, n) = (P_s, Q_s, n_s) * (P', Q', n') * (P_u, Q_u, n_u), \tag{5.18}$$

for some $(P', Q', n') \in \mathcal{R}(I_0)$. We conclude that $|P_{\underline{a}}| < |I_0|^{\beta}$, $|Q_{\underline{a}}| < |I_0|^{\beta}$ as required. □

Remark **8.** — The final choice of β is made in Section 9.3. Beside the requirement $1 < \beta < 1 + \min(\omega_s, \omega_u)$ we will ask when $d_s^0 \geqslant d_u^0$ that

$$\beta < \frac{(1-d_u^0)(d_s^0+d_u^0)}{d_s^0(2d_s^0+d_u^0-1)} \tag{5.19}$$

(if $d_s^0 \leqslant d_u^0$, exchange d_s^0 and d_u^0). Condition (H4) in Section 1.2 is actually obtained by asking that the right-hand term in the last inequality is larger than 1.

5.6.5.

Proposition **5.** — *Let* $\widetilde{\mathrm{I}} \supset \mathrm{I}$ *be parameter intervals, and let* $(\mathrm{P}, \mathrm{Q}, n) \in \mathcal{R}(\widetilde{\mathrm{I}})$. *If* Q *is* $\widetilde{\mathrm{I}}$*-transverse, then it is also* I*-transverse.*

Proof. — If $\mathrm{Q} \cap \mathrm{Q}_u = \emptyset$, this is obvious. Assume therefore that $\mathrm{Q} \subset \mathrm{Q}_u$. Then there exists an $\widetilde{\mathrm{I}}$-decomposition $(\mathrm{P}_\alpha, \mathrm{Q}_\alpha, n_\alpha)_\alpha$ of P_s by elements of $\mathcal{R}(\widetilde{\mathrm{I}})$ such that for all α, Q and P_α are either $\widetilde{\mathrm{I}}$-transverse or $\widetilde{\mathrm{I}}$-separated.

First observe that $(\mathrm{P}_\alpha, \mathrm{Q}_\alpha, n_\alpha) \in \mathcal{R}(\mathrm{I})$ and therefore this is also an I-decomposition of P_s. By Corollary 1, if Q and P_α are $\widetilde{\mathrm{I}}$-transverse, they are also I-transverse. On the other hand, it is obvious from the definition that if Q and P_α are $\widetilde{\mathrm{I}}$-separated they are also I-separated. The result follows. □

Proposition **6.** — *Let* I *be a parameter interval, and let* $(\mathrm{P}, \mathrm{Q}, n) \in \mathcal{R}(\mathrm{I})$. *If* Q *is* I*-transverse, then* P *is* I*-decomposable.*

Proof. — Let us first assume that $\mathrm{Q} \cap \mathrm{Q}_u = \emptyset$. Let $a \in \mathcal{A}$ be such that $\mathrm{Q} \subset \mathrm{R}_a$. We have

$$\mathrm{R}_a \cap \mathrm{W}^s(\Lambda, \widehat{\mathrm{R}}) \subset \bigcup_{(a,a')\in\mathcal{B}} \left(\mathrm{P}_{a,a'} \cap \mathrm{W}^s(\Lambda, \widehat{\mathrm{R}})\right) \cup \mathrm{L}_u; \tag{5.20}$$

for each $a' \in \mathcal{A}$ such that $(a, a') \in \mathcal{B}$, we have the simple child of P:

$$\left(\mathrm{P}(a'), \mathrm{Q}(a'), n+1\right) = (\mathrm{P}, \mathrm{Q}, n) * (\mathrm{P}_{aa'}, \mathrm{Q}_{aa'}, 1), \tag{5.21}$$

and together they form by (5.20) an I-decomposition of P (the canonical one).

Let us now assume that $\mathrm{Q} \subset \mathrm{Q}_u$. As Q is I-transverse, there is an I-decomposition $(\mathrm{P}_\alpha, \mathrm{Q}_\alpha, n_\alpha)_\alpha$ of P_s such that, for each α, Q and P_α are not I-critically related. For each α such that Q and P_α are I-transverse, let $(\mathrm{P}_\alpha^\pm, \mathrm{Q}_\alpha^\pm, n_\alpha + n + \mathrm{N}_0)$ be the two elements produced by the allowed parabolic composition. Together with the simple children defined by (5.21), they form an I-decomposition of P. □

Corollary **3.** — *Let* I *be a* β*-regular parameter interval and let* $(\mathrm{P}, \mathrm{Q}, n) \in \mathcal{R}(\mathrm{I})$. *If* P *is* I*-critical and* $|\mathrm{P}| > |\mathrm{I}|^\beta$ *or* $|\mathrm{Q}| > |\mathrm{I}|^\beta$ *for some* $t \in \mathrm{I}$, *then* P *is* I*-decomposable.*

Proof. — Indeed, by the very definition of regularity, Q cannot be I-critical. □

The decomposability of "fat" critical rectangles is crucial to our analysis.

6. Some properties of the classes $\mathcal{R}(I)$

6.1. *Preliminaries.* — In this section,we start the proof of the existence of the class $\mathcal{R}(I)$. For $I = I_0$, this was done essentially in Section 5.5. Therefore we will, unless specified otherwise, assume that the level of the interval I is > 0. Moreover, the induction hypothesis will guarantee that the classes $\mathcal{R}(\widetilde{I})$, with $\widetilde{I}$ strictly larger than I, have already been constructed with the required properties (R1)–(R7).

Once I is fixed, the construction will be by induction on the length n of an element (P, Q, n). We will denote by $\mathcal{R}_N(I)$ the set of elements (P, Q, n) in $\mathcal{R}(I)$ with $n \leqslant N$. For small N, parabolic composition does not come into play, and therefore the elements of $\mathcal{R}_N(I)$ are those of $\mathcal{R}_N(I_0)$.

Assuming that $\mathcal{R}_{N-1}(I)$ has already been constructed with the required properties, we will consider elements of length N that can be obtained from shorter elements by simple or (allowed) parabolic composition. For these new elements, we will not only prove (R2), (R4), (R7) (in Section 7) as required, but also many other properties that are detailed in the next subsections. The structure theorem of Section 6.5 is particularly important. All these properties are assumed for $\mathcal{R}_{N-1}(I)$, and then they will all be proved for $\mathcal{R}_N(I)$.

6.2. *Children are born from their parent.* — Let $(P, Q, n) \in \mathcal{R}_N(I)$, and let $(\widetilde{P}, \widetilde{Q}, \widetilde{n}) \in \mathcal{R}_{N-1}(I)$ be such that $\widetilde{P}$ is the parent of P. Recall that P is a simple child if $n = \widetilde{n} + 1$, non-simple otherwise. When P is simple, there exists $(a, a') \in \mathcal{B}$ such that (P, Q, n) is the simple composition of $(\widetilde{P}, \widetilde{Q}, \widetilde{n})$ and $(P_{a,a'}, Q_{a,a'}, 1)$.

Proposition **7.** — *Assume that* P *is a non-simple child of* $\widetilde{P}$. *Then there exists* $(P_1, Q_1, n_1) \in \mathcal{R}_{N-1}(I)$ *such that* $\widetilde{Q} \pitchfork_I P_1$ *holds and*

$$(P, Q, n) \in (\widetilde{P}, \widetilde{Q}, \widetilde{n}) \,\square\, (P_1, Q_1, n_1).$$

Proof. — 1. We first prove that (P, Q, n) can be written as a *parabolic* composition of shorter elements. Otherwise, by (R6), we have

$$(P, Q, n) = (P_0, Q_0, n_0) * (P_1, Q_1, n_1), \tag{6.1}$$

with $n_0, n_1 > 0$. As P is a non-simple child, we must have $n_1 > 1$. Let then $(\widetilde{P}_1, \widetilde{Q}_1, \widetilde{n}_1)$ be the element of $\mathcal{R}(I)$ such that $\widetilde{P}_1$ is the parent of P_1. If P_1 was a simple child, we would have, for some $(a, a') \in \mathcal{B}$

$$(P_1, Q_1, n_1) = (\widetilde{P}_1, \widetilde{Q}_1, \widetilde{n}_1) * (P_{aa'}, Q_{aa'}, 1), \tag{6.2}$$

(**6.3**) $$(P, Q, n) = ((P_0, Q_0, n_0) * (\widetilde{P}_1, \widetilde{Q}_1, \widetilde{n}_1)) * (P_{ad'}, Q_{ad'}, 1),$$

in contradiction with the hypothesis that P is a simple child. Therefore, P_1 is a non-simple child; by induction on the length, we can write

(**6.4**) $$(P_1, Q_1, n_1) \in (\widetilde{P}_1, \widetilde{Q}_1, \widetilde{n}_1) \,\square\, (P_2, Q_2, n_2)$$

for some $(P_2, Q_2, n_2) \in \mathcal{R}_{N-2}(I)$ with $\widetilde{Q}_1 \pitchfork_I P_2$.

Let $(\widetilde{P}_0, \widetilde{Q}_0, \widetilde{n}_0) := (P_0, Q_0, n_0) * (\widetilde{P}_1, \widetilde{Q}_1, \widetilde{n}_1)$. We then have $\widetilde{Q}_0 \subset \widetilde{Q}_1$ and hence $\widetilde{Q}_0 \pitchfork_I P_2$ from Proposition 2 in Section 5.6.

Thus, the parabolic composition of $(\widetilde{P}_0, \widetilde{Q}_0, \widetilde{n}_0)$ and (P_2, Q_2, n_2) is allowed; we obviously have:

(**6.5**) $$(P, Q, n) \in (\widetilde{P}_0, \widetilde{Q}_0, \widetilde{n}_0) \,\square\, (P_2, Q_2, n_2)$$

which proves our claim.

2. We next write

(**6.6**) $$(P, Q, n) \in (P_0, Q_0, n_0) \,\square\, (P_1, Q_1, n_1)$$

with n_0 maximal, and want to show that $P_0 = \widetilde{P}$.

Assume that this does not hold, which means $n_0 < \widetilde{n}$. Let (P'_0, Q'_0, n'_0) be the element of $\mathcal{R}_{N-1}(I)$ such that P'_0 is the child of P_0 containing P. As $n_0 < \widetilde{n}$, we have $n'_0 < n$. As $g_t^{n_0}(P) \subset L_u$ by (6.6), P'_0 must be a non-simple child. Then, from the induction hypothesis, we can write

(**6.7**) $$(P'_0, Q'_0, n'_0) \in (P_0, Q_0, n_0) \,\square\, (P'_1, Q'_1, n'_1)$$

for some $(P'_1, Q'_1, n'_1) \in \mathcal{R}(I)$ with $Q_0 \pitchfork_I P'_1$, $P'_1 \supset P_1$. Let $(\widetilde{P}_1, \widetilde{Q}_1, \widetilde{n}_1)$ be the element of $\mathcal{R}_{N-2}(I)$ such that $\widetilde{P}_1$ is the parent of P_1 We have thus $P'_1 \supset \widetilde{P}_1$. As Q_0 is I-transverse to P'_1, it is also I-transverse to $\widetilde{P}_1$ (Proposition 2), and we can define $(\widetilde{P}_0, \widetilde{Q}_0, \widetilde{n}_0) \in \mathcal{R}_{N-1}(I)$ by

(**6.8**) $$(\widetilde{P}_0, \widetilde{Q}_0, \widetilde{n}_0) \in (P_0, Q_0, n_0) \,\square\, (\widetilde{P}_1, \widetilde{Q}_1, \widetilde{n}_1),$$

and $\widetilde{P}_0 \supset P$. If P_1 was a simple child of $\widetilde{P}_1$, P would be a simple child of $\widetilde{P}_0$. Therefore, by the induction hypothesis, we can write

(**6.9**) $$(P_1, Q_1, n_1) \in (\widetilde{P}_1, \widetilde{Q}_1, \widetilde{n}_1) \,\square\, (P_2, Q_2, n_2)$$

for some (P_2, Q_2, n_2) with $\widetilde{Q}_1 \pitchfork_I P_2$. But then from Proposition 2, we have $\widetilde{Q}_0 \pitchfork_I P_2$ and

(**6.10**) $$(P, Q, n) \in (\widetilde{P}_0, \widetilde{Q}_0, \widetilde{n}_0) \,\square\, (P_2, Q_2, n_2)$$

which contradicts the maximality of n_0. The proof is complete. □

6.3. *Coherence and concavity.*

6.3.1. We present together the following two properties of $\mathcal{R}(\mathrm{I})$ because the proofs are interwoven.

The coherence property, asserted in the next proposition, means that larger rectangles are constructed before thinner ones.

Proposition **8** *(Coherence). — Let* $(\mathrm{P}, \mathrm{Q}, n) \in \mathcal{R}_{\mathrm{N}}(\mathrm{I})$, $\tilde{\mathrm{I}}$ *be the parent of* I, *and* $(\mathrm{P}', \mathrm{Q}', n') \in \mathcal{R}(\tilde{\mathrm{I}})$. *If* $\mathrm{P}' \subset \mathrm{P}$, *then* $(\mathrm{P}, \mathrm{Q}, n)$ *belongs to* $\mathcal{R}(\tilde{\mathrm{I}})$.

The coherence property implies that the parent-child relationship is independent of the parameter interval under consideration, provided the child is already constructed.

Proposition **9** *(Concavity). — Let* $\tilde{\mathrm{I}}$ *be the parent of* I.

1. *Let* $(\mathrm{P}'_0, \mathrm{Q}'_0, n'_0)$, $(\mathrm{P}'_1, \mathrm{Q}'_1, n'_1)$ *be elements of* $\mathcal{R}_{\mathrm{N}}(\mathrm{I})$, $(\mathrm{P}_0, \mathrm{Q}_0, n_0)$, $(\mathrm{P}_1, \mathrm{Q}_1, n_1)$ *be elements of* $\mathcal{R}_{\mathrm{N}}(\mathrm{I}) \cup \mathcal{R}(\tilde{\mathrm{I}})$ *(with possibly* n_0 *or* $n_1 > \mathrm{N}$*) such that*
$$\mathrm{Q}_0 \subset \mathrm{Q}'_0 \subset \mathrm{Q}_u, \qquad \mathrm{P}_1 \subset \mathrm{P}'_1 \subset \mathrm{P}_s.$$
If both $\mathrm{Q}_0 \pitchfork_{\mathrm{I}} \mathrm{P}'_1$ *and* $\mathrm{Q}'_0 \pitchfork_{\mathrm{I}} \mathrm{P}_1$ *hold, then* $\mathrm{Q}'_0 \pitchfork_{\mathrm{I}} \mathrm{P}'_1$ *also holds.*
2. *Let* I' *be a parameter interval containing* $\tilde{\mathrm{I}}$, *let* $(\mathrm{P}'_0, \mathrm{Q}'_0, n'_0)$, $(\mathrm{P}_1, \mathrm{Q}_1, n_1)$ *be elements of* $\mathcal{R}_{\mathrm{N}}(\mathrm{I}')$, $(\mathrm{P}_0, \mathrm{Q}_0, n_0)$ *an element of* $\mathcal{R}(\mathrm{I}')$ *(with possibly* $n_0 > \mathrm{N}$*) with*
$$\mathrm{Q}_0 \subset \mathrm{Q}'_0 \subset \mathrm{Q}_u, \qquad \mathrm{P}_1 \subset \mathrm{P}_s.$$
If both $\mathrm{Q}_0 \pitchfork_{\mathrm{I}'} \mathrm{P}_1$ *and* $\mathrm{Q}'_0 \pitchfork_{\mathrm{I}} \mathrm{P}_1$ *hold, then* $\mathrm{Q}'_0 \pitchfork_{\mathrm{I}'} \mathrm{P}_1$ *also holds.*
3. *Let* I' *be a parameter interval containing* $\tilde{\mathrm{I}}$, *let* $(\mathrm{P}_0, \mathrm{Q}_0, n_0)$, $(\mathrm{P}'_1, \mathrm{Q}'_1, n'_1)$ *be elements of* $\mathcal{R}_{\mathrm{N}}(\mathrm{I}')$, $(\mathrm{P}_1, \mathrm{Q}_1, n_1)$ *an element of* $\mathcal{R}(\mathrm{I}')$ *(with possibly* $n_1 > \mathrm{N}$*) with*
$$\mathrm{P}_1 \subset \mathrm{P}'_1 \subset \mathrm{P}_s, \qquad \mathrm{Q}_0 \subset \mathrm{Q}_u.$$
If both $\mathrm{Q}_0 \pitchfork_{\mathrm{I}'} \mathrm{P}_1$ *and* $\mathrm{Q}_0 \pitchfork_{\mathrm{I}} \mathrm{P}'_1$ *hold, then* $\mathrm{Q}_0 \pitchfork_{\mathrm{I}'} \mathrm{P}'_1$ *also holds.*

The concavity property is very helpful in the sequel. The proof of the proposition will help to explain why the definition of the transversality relation had to be complicated.

In the following subsections, we will successively:

- prove Proposition 8 for $\mathcal{R}_{\mathrm{N}}(\mathrm{I})$, assuming Proposition 8 and Proposition 9 for $\mathcal{R}_{\mathrm{N}-1}(\mathrm{I})$;
- prove Proposition 9 for $\mathcal{R}_{\mathrm{N}}(\mathrm{I})$, assuming Proposition 8 for $\mathcal{R}_{\mathrm{N}}(\mathrm{I})$.

6.3.2. *Proof of Proposition 8. Proof.* — Let $(\widetilde{P}, \widetilde{Q}, \widetilde{n})$ be the element of $\mathcal{R}(\widetilde{I})$ such that $\widetilde{P}$ is the parent of P' in $\mathcal{R}(\widetilde{I})$. By enlarging P' if necessary, one can assume that $P' \subset P \subset \widetilde{P}$, $P \neq \widetilde{P}$.

If P' is a simple child, we have $P = P'$ and we are done. Assume that P' is a non-simple child of $\widetilde{P}$.

Let $(\widehat{P}, \widehat{Q}, \widehat{n})$ be the element of $\mathcal{R}_N(I)$ such that $\widehat{P}$ is the child of $\widetilde{P}$ in $\mathcal{R}(I)$ containing P; as $\widehat{P}$ also contains P' it has to be a non-simple child.

Applying Proposition 7 twice, we find $(P_1, Q_1, n_1) \in \mathcal{R}_{N-1}(I)$, $(P'_1, Q'_1, n'_1) \in \mathcal{R}(\widetilde{I})$ such that $\widetilde{Q} \pitchfork_I P_1$, $\widetilde{Q} \pitchfork_{\widetilde{I}} P'_1$ both hold and

$$(\widehat{P}, \widehat{Q}, \widehat{n}) \in (\widetilde{P}, \widetilde{Q}, \widetilde{n}) \,\square\, (P_1, Q_1, n_1), \tag{6.11}$$

$$(P', Q', n') \in (\widetilde{P}, \widetilde{Q}, \widetilde{n}) \,\square\, (P'_1, Q'_1, n'_1). \tag{6.12}$$

As $P' \subset P \subset \widehat{P}$, we have $P'_1 \subset P_1$. By Proposition 8 for $\mathcal{R}_{N-1}(I)$, we have $(P_1, Q_1, n_1) \in \mathcal{R}(\widetilde{I})$. By Proposition 9, part 3, for $\mathcal{R}_{N-1}(I)$, we have $\widetilde{Q} \pitchfork_{\widetilde{I}} P_1$ and $(\widehat{P}, \widehat{Q}, \widehat{n}) \in \mathcal{R}(\widetilde{I})$.

As $P' \subset P \subset \widehat{P}$ and P' is a child of $\widetilde{P}$ in $\mathcal{R}(\widetilde{I})$, we must have $P' = P = \widehat{P}$ and the proof is complete. □

6.3.3. *Proof of Proposition 9.*

1. With (P_0, Q_0, n_0), (P'_0, Q'_0, n'_0), (P_1, Q_1, n_1), (P'_1, Q'_1, n'_1) as in the first part of Proposition 9, we assume that both $Q_0 \pitchfork_I P'_1$ and $Q'_0 \pitchfork_I P_1$ hold.

   By definition of the transversality relation, there exist parameter intervals I_1, I_2 containing I, elements $(\widetilde{P}_0, \widetilde{Q}_0, \widetilde{n}_0) \in \mathcal{R}(I_1)$, $(\widetilde{P}'_1, \widetilde{Q}'_1, \widetilde{n}'_1) \in \mathcal{R}_N(I_1)$, $(\widetilde{P}'_0, \widetilde{Q}'_0, \widetilde{n}'_0) \in \mathcal{R}_N(I_2)$, $(\widetilde{P}_1, \widetilde{Q}_1, \widetilde{n}_1) \in \mathcal{R}(I_2)$ such that $Q_0 \subset \widetilde{Q}_0$, $P'_1 \subset \widetilde{P}'_1$, $Q'_0 \subset \widetilde{Q}'_0$, $P_1 \subset \widetilde{P}_1$ and

$$\widetilde{Q}_0 \,\overline{\pitchfork}_{I_1}\, \widetilde{P}'_1, \tag{6.13}$$

$$\widetilde{Q}'_0 \,\overline{\pitchfork}_{I_2}\, \widetilde{P}_1. \tag{6.14}$$

   If we have either $Q'_0 \subset \widetilde{Q}_0$ or $P'_1 \subset \widetilde{P}_1$, we can already conclude that $Q'_0 \pitchfork_I P'_1$. Assume therefore that $\widetilde{Q}_0 \subset Q'_0$, $\widetilde{P}_1 \subset P'_1$. Assume also for instance that $I_1 \subset I_2$. We have $(\widetilde{P}_1, \widetilde{Q}_1, \widetilde{n}_1) \in \mathcal{R}(I_2)$ and $(\widetilde{P}'_1, \widetilde{Q}'_1, \widetilde{n}'_1) \in \mathcal{R}_N(I_1)$. By the coherence property (Proposition 8) for $\mathcal{R}_N(I_1)$, the element $(\widetilde{P}'_1, \widetilde{Q}'_1, \widetilde{n}'_1)$ belongs to $\mathcal{R}(I_2)$. We will show that

$$\widetilde{Q}'_0 \,\overline{\pitchfork}_{I_2}\, \widetilde{P}'_1 \tag{6.15}$$

   which implies that Q'_0 and P'_1 are I-transverse.

   We check properties (T1)–(T3) of Section 5.4. For all $t \in I_2$, by the estimate (3.49) of Section 3.6 and (T1) for $\widetilde{Q}'_0 \,\overline{\pitchfork}_{I_2}\, \widetilde{P}_1$, we have

$$\delta_{LR}(\widetilde{Q}'_0, \widetilde{P}'_1) \geqslant \delta_{LR}(\widetilde{Q}'_0, \widetilde{P}_1) \geqslant 2|I_2|. \tag{6.16}$$

Next, by (3.47) and (T2) for $\widetilde{Q}'_0 \,\overline{\pitchfork}_{I_2}\, \widetilde{P}_1$, there exists $t_0 \in I_2$ such that

(6.17) $$\delta_R(\widetilde{Q}'_0, \widetilde{P}'_1) \geqslant \delta_R(\widetilde{Q}'_0, \widetilde{P}_1) \geqslant 2|\widetilde{Q}'_0|^{1-\eta}.$$

Finally, by (3.48) and (T3) for $\widetilde{Q}_0 \,\overline{\pitchfork}_{I_1}\, \widetilde{P}'_1$, there exists $t_1 \in I_1 \subset I_2$ such that

(6.18) $$\delta_L(\widetilde{Q}'_0, \widetilde{P}'_1) \geqslant \delta_L(\widetilde{Q}_0, \widetilde{P}'_1) \geqslant 2|\widetilde{P}'_1|^{1-\eta}.$$

We have proved (6.15) and this concludes the proof of the first statement in the proposition.

2. We assume now that, with I', (P_0, Q_0, n_0), (P'_0, Q'_0, n'_0), (P_1, Q_1, n_1) as in the second part of the proposition, both $Q_0 \pitchfork_{I'} P_1$ and $Q'_0 \pitchfork_I P_1$ hold.

By definition of the transversality relation, there exist parameter intervals $\widehat{I} \supset I$, $\widehat{I'} \supset I'$ and elements $(\widetilde{P}_0, \widetilde{Q}_0, \widetilde{n}_0) \in \mathcal{R}(\widehat{I'})$, $(\widetilde{P}_1, \widetilde{Q}_1, \widetilde{n}_1) \in \mathcal{R}_N(\widehat{I'})$, $(\widetilde{P}'_0, \widetilde{Q}'_0, \widetilde{n}'_0)$, $(\widetilde{P}'_1, \widetilde{Q}'_1, \widetilde{n}'_1) \in \mathcal{R}_N(\widehat{I})$ such that $Q_0 \subset \widetilde{Q}_0$, $Q'_0 \subset \widetilde{Q}'_0$, $P_1 \subset \widetilde{P}_1$, $P_1 \subset \widetilde{P}'_1$ and

(6.19) $$\widetilde{Q}_0 \,\overline{\pitchfork}_{\widehat{I'}}\, \widetilde{P}_1,$$

(6.20) $$\widetilde{Q}'_0 \,\overline{\pitchfork}_{\widehat{I}}\, \widetilde{P}'_1$$

both hold. If either $Q'_0 \subset \widetilde{Q}_0$ or $I' \subset \widehat{I}$, we conclude immediately that $Q'_0 \pitchfork_{I'} P_1$ holds. Assume therefore that $\widetilde{Q}_0 \subset Q'_0$ and $\widehat{I} \subset I'$. Let P_1^* be the largest of $\widetilde{P}_1, \widetilde{P}'_1$.

We claim that (P_1^*, Q_1^*, n_1^*) always belongs to $\mathcal{R}(\widehat{I'})$. This is clear if $\widetilde{P}'_1 \subset \widetilde{P}_1$; on the other hand, if $\widetilde{P}_1 \subset \widetilde{P}'_1$, it follows from coherence (Proposition 8) for $\mathcal{R}_N(\widehat{I})$. As $\widetilde{Q}_0 \subset \widetilde{Q}'_0$, it also follows from coherence for $\mathcal{R}_N(\widehat{I})$ that $(\widetilde{P}'_0, \widetilde{Q}'_0, \widetilde{n}'_0)$ belongs to $\mathcal{R}(\widehat{I'})$.

We will show that

(6.21) $$\widetilde{Q}'_0 \,\overline{\pitchfork}_{\widehat{I'}}\, P_1^*$$

holds, which implies the required conclusion $Q'_0 \pitchfork_{I'} P_1$.

We check properties (T1)–(T3) of Section 5.4. For all $t \in \widehat{I'}$, we have, from (3.49) and (T1) for $\widetilde{Q}_0 \,\overline{\pitchfork}_{\widehat{I'}}\, \widetilde{P}_1$

(6.22) $$\delta_{LR}(\widetilde{Q}'_0, P_1^*) \geqslant \delta_{LR}(\widetilde{Q}_0, \widetilde{P}_1) \geqslant 2|\widehat{I'}|.$$

From (T2) for $\widetilde{Q}'_0 \,\overline{\pitchfork}_{\widehat{I}}\, \widetilde{P}'_1$, there exists $t_0 \in \widehat{I} \subset I' \subset \widehat{I'}$ such that

(6.23) $$\delta_R(\widetilde{Q}'_0, \widetilde{P}'_1) \geqslant 2|\widetilde{Q}'_0|^{1-\eta}.$$

Then, by (3.47), for the same t_0, we have

(6.24) $$\delta_R(\widetilde{Q}'_0, P_1^*) \geqslant \delta_R(\widetilde{Q}'_0, \widetilde{P}'_1) \geqslant 2|\widetilde{Q}'_0|^{1-\eta}.$$

When $P_1^* = \widetilde{P}'_1$, it follows directly from (T3) for $\widetilde{Q}'_0 \,\overline{\pitchfork}_{\widehat{I}}\, \widetilde{P}'_1$ that we have

$$\delta_L(\widetilde{Q}'_0, P_1^*) \geqslant 2|P_1^*|^{1-\eta} \tag{6.25}$$

for some $t_1 \in \widehat{I} \subset \widehat{I}'$.

When $P_1^* = \widetilde{P}_1$, we use (T3) for $\widetilde{Q}_0 \,\overline{\pitchfork}_{\widehat{I}'}\, \widetilde{P}_1$ and (3.48) to find $t_1 \in \widehat{I}'$ such that

$$\delta_L(\widetilde{Q}'_0, P_1^*) \geqslant \delta_L(\widetilde{Q}_0, \widetilde{P}_1) \geqslant 2|P_1^*|^{1-\eta}. \tag{6.26}$$

We have thus proved (6.21). This proves the second part of the proposition. The third part is proven in a symmetric way, exchanging P's and Q's.

6.3.4. *Further forms of concavity.* — One obtains more general statements than in Proposition 9 by combining its different parts.

Corollary **4.** — *Let* I' *be a parameter interval containing the parent of* I, *and let* (P'_0, Q'_0, n'_0), (P'_1, Q'_1, n'_1), (P_0, Q_0, n_0), (P_1, Q_1, n_1) *be such that*

$$Q_0 \subset Q'_0 \subset Q_u, \qquad P_1 \subset P'_1 \subset P_s.$$

1. *Assume that* $(P'_0, Q'_0, n'_0), (P'_1, Q'_1, n'_1) \in \mathcal{R}_N(I)$, *and that* $(P_0, Q_0, n_0), (P_1, Q_1, n_1) \in \mathcal{R}(I')$. *If both* $Q_0 \pitchfork_{I'} P_1$ *and* $Q'_0 \pitchfork_I P'_1$ *hold, and* $\min(n_0, n_1) \leqslant N$, *then* $Q'_0 \pitchfork_{I'} P'_1$ *also holds.*
2. *Assume that* $(P_0, Q_0, n_0), (P'_1, Q'_1, n'_1) \in \mathcal{R}_N(I)$, *and that* $(P'_0, Q'_0, n'_0) \in \mathcal{R}_N(I')$, $(P_1, Q_1, n_1) \in \mathcal{R}(I')$ *(with possibly* $n_1 > N$*). If both* $Q'_0 \pitchfork_{I'} P_1$ *and* $Q_0 \pitchfork_I P'_1$ *hold, then* $Q'_0 \pitchfork_{I'} P'_1$ *also holds.*
3. *Assume that* $(P'_0, Q'_0, n'_0), (P_1, Q_1, n_1) \in \mathcal{R}_N(I)$, *and that* $(P'_1, Q'_1, n'_1) \in \mathcal{R}_N(I')$, $(P_0, Q_0, n_0) \in \mathcal{R}(I')$ *(with possibly* $n_0 > N$*). If both* $Q_0 \pitchfork_{I'} P'_1$ *and* $Q'_0 \pitchfork_I P_1$ *hold, then* $Q'_0 \pitchfork_{I'} P'_1$ *also holds.*

Proof. — 1. For the first statement of the corollary, we first observe that, by coherence for $\mathcal{R}_N(I)$, the elements (P'_0, Q'_0, n'_0), (P'_1, Q'_1, n'_1) belong to $\mathcal{R}(I')$.

Assume for instance that $n_0 \leqslant N$. From $Q'_0 \pitchfork_I P'_1$ and Proposition 2, we have $Q_0 \pitchfork_I P'_1$. Then, from $Q_0 \pitchfork_I P'_1$ and $Q_0 \pitchfork_{I'} P_1$, we have $Q_0 \pitchfork_{I'} P'_1$ by Proposition 9 for $\mathcal{R}_N(I)$. Finally, from $Q_0 \pitchfork_{I'} P'_1$ and $Q'_0 \pitchfork_I P'_1$, we have $Q'_0 \pitchfork_{I'} P'_1$ again by Proposition 9 for $\mathcal{R}_N(I)$.

2. For the second statement of the corollary, we first observe that, by coherence for $\mathcal{R}_N(I)$, the element (P'_1, Q'_1, n'_1) belong to $\mathcal{R}(I')$.

From $Q'_0 \pitchfork_{I'} P_1$ and Proposition 2, we have $Q_0 \pitchfork_{I'} P_1$. Then, from $Q_0 \pitchfork_I P'_1$ and $Q_0 \pitchfork_{I'} P_1$, we have $Q_0 \pitchfork_{I'} P'_1$ by Proposition 9 for $\mathcal{R}_N(I)$. Finally, from $Q_0 \pitchfork_{I'} P'_1$ and $Q'_0 \pitchfork_{I'} P_1$, we have $Q'_0 \pitchfork_{I'} P'_1$ again by Proposition 9 for $\mathcal{R}_N(I)$.

3. The proof of the third statement of the corollary is the same than for the second, exchanging P's and Q's. □

6.4. *Another transversality criterion.*

Proposition **10.** — *Let* (P_0, Q_0, n_0), (P_1, Q_1, n_1), (P'_1, Q'_1, n'_1) *be elements of* $\mathcal{R}_N(I)$ *such that* $Q_0 \subset Q_u$ *and* $P_1 \subset P'_1 \subset P_s$. *Assume that* $Q_0 \pitchfork_I P_1$ *holds and that* $2|P'_1|^{1-\eta} \leqslant |I|$ *for some* $t_1 \in I$. *Then* Q_0 *and* P'_1 *are also* I*-transverse.*

Proof. — By definition of the transversality relation, there exist $\tilde{I} \supset I$, $(\tilde{P}_0, \tilde{Q}_0, \tilde{n}_0)$, $(\tilde{P}_1, \tilde{Q}_1, \tilde{n}_1) \in \mathcal{R}(\tilde{I})$ such that $Q_0 \subset \tilde{Q}_0$, $P_1 \subset \tilde{P}_1$ and $\tilde{Q}_0 \overline{\pitchfork}_{\tilde{I}} \tilde{P}_1$.

If $P'_1 \subset \tilde{P}_1$ this already implies that $Q_0 \pitchfork_I P'_1$. Let us assume that $\tilde{P}_1 \subset P'_1$. We will show that $\tilde{Q}_0 \overline{\pitchfork}_{\tilde{I}} P'_1$ holds. By coherence (Proposition 8), we have $(P'_1, Q'_1, n'_1) \in \mathcal{R}(\tilde{I})$.

Let us check (T1)–(T3).

By (T1) for $Q_0 \pitchfork_I P_1$ and (3.49) in Section 3.6, we have, for all $t \in \tilde{I}$:

$$\delta_{LR}(\tilde{Q}_0, P'_1) \geqslant \delta_{LR}(\tilde{Q}_0, \tilde{P}_1) \geqslant 2|\tilde{I}|. \tag{6.27}$$

By (T2) for $Q_0 \pitchfork_I P_1$ and (3.47), there exists $t_0 \in \tilde{I}$ such that

$$\delta_R(\tilde{Q}_0, P'_1) \geqslant \delta_R(\tilde{Q}_0, \tilde{P}_1) \geqslant 2|\tilde{Q}_0|^{1-\eta}. \tag{6.28}$$

Finally, we have, for all $t \in I$, by (3.39)

$$\begin{aligned} \delta_L(\tilde{Q}_0, P'_1) &\geqslant \delta_{LR}(\tilde{Q}_0, P'_1) - C|P'_1| \\ &\geqslant 2|I| - C|P'_1|. \end{aligned} \tag{6.29}$$

But, for $t = t_1$, we have, if ε_0 is small enough

$$2|I| - C|P'_1| \geqslant 4|P'_1|^{1-\eta} - C|P'_1| \geqslant 2|P'_1|^{1-\eta}. \tag{6.30}$$

We have shown that $\tilde{Q}_0 \overline{\pitchfork}_{\tilde{I}} P'_1$ holds, and this implies that $Q_0 \pitchfork_I P'_1$ holds as required. □

6.5. *A structure theorem for new rectangles.*

6.5.1. *Associativity of parabolic composition.* — Let I be a parameter interval, and let (P_0, Q_0, n_0), (P_1, Q_1, n_1), (P_2, Q_2, n_2) be elements in $\mathcal{R}(I)$ such that $Q_0 \subset Q_u$, $Q_1 \subset Q_u$, $P_1 \subset P_s$, $P_2 \subset P_s$. We assume that both $Q_0 \pitchfork_I P_1$ and $Q_1 \pitchfork_I P_2$ hold.

Parabolic composition of (P_0, Q_0, n_0), (P_1, Q_1, n_1) produces two elements $(P^+_{01}, Q^+_{01}, n^+_{01})$, $(P^-_{01}, Q^-_{01}, n^-_{01})$.

As Q^+_{01} and Q^-_{01} are contained in Q_1, it follows from Proposition 2 that both $Q^+_{01} \pitchfork_I P_2$ and $Q^-_{01} \pitchfork_I P_2$ hold.

In the same way, parabolic composition of (P_1, Q_1, n_1), (P_2, Q_2, n_2) produces two elements $(P^+_{12}, Q^+_{12}, n^+_{12})$, $(P^-_{12}, Q^-_{12}, n^-_{12})$ such that both $Q_0 \pitchfork_I P^+_{12}$ and $Q_0 \pitchfork_I P^-_{12}$ hold.

It is clear that the four elements of $\mathcal{R}(I)$ obtained by parabolic composition of $(P^+_{01}, Q^+_{01}, n^+_{01})$ or $(P^-_{01}, Q^-_{01}, n^-_{01})$ with (P_2, Q_2, n_2) are the same as the four elements obtained by the parabolic composition of (P_0, Q_0, n_0) with $(P^+_{12}, Q^+_{12}, n^+_{12})$ or $(P^-_{12}, Q^-_{12}, n^-_{12})$. Their domains are the components of $P_0 \cap (G_t \circ g_t^{n_0})^{-1} P_1 \cap (G_t \circ g_t^{n_1} \circ G_t \circ g_t^{n_0})^{-1} P_2$. If (P, Q, n) is any of these four elements, we will write

$$(P, Q, n) \in (P_0, Q_0, n_0) \square (P_1, Q_1, n_1) \square (P_2, Q_2, n_2). \tag{6.31}$$

The same considerations extend immediately, by induction on k, to the case of elements $(P_0, Q_0, n_0), \ldots, (P_k, Q_k, n_k)$ such that $P_i \subset P_s$ for $0 < i \leq k$, $Q_i \subset Q_u$ for $0 \leq i < k$, and $Q_i \pitchfork_I P_{i+1}$ holds for $0 \leq i < k$. Then the successive parabolic compositions of $(P_0, Q_0, n_0), \ldots, (P_k, Q_k, n_k)$ produce 2^k elements and we will write for any such element (P, Q, n):

$$(P, Q, n) \in (P_0, Q_0, n_0) \square \cdots \square (P_k, Q_k, n_k). \tag{6.32}$$

6.5.2. *Statement of the structure theorem.* — We have seen in Section 5.5 that parabolic composition is never allowed in the class $\mathcal{R}(I_0)$ associated to the starting interval $I_0 = [\varepsilon_0, 2\varepsilon_0]$. This class consists exactly of the affine-like iterates associated to the Markov partition of the initial horseshoe K_{g_t}.

On the other hand, for elements (P, Q, n) belonging to some class $\mathcal{R}(I)$ but which are not (restrictions of) an element of $\mathcal{R}(I_0)$, parabolic composition must occur. The following theorem gives some rather precise information on this process.

Theorem **2.** — *Let* I *be a parameter interval of level* > 0, $\widetilde{I}$ *be the parent interval, and let* (P, Q, n) *be an element of* $\mathcal{R}_N(I)$ *which is not (the restriction of) an element of* $\mathcal{R}(\widetilde{I})$. *Then there exists* $k > 0$, *elements* $(P_0, Q_0, n_0), \ldots, (P_k, Q_k, n_k)$ *of* $\mathcal{R}(\widetilde{I})$ *such that* $Q_i \subset Q_u$ *for* $0 \leq i < k$, $P_i \subset P_s$ *for* $0 < i \leq k$, $Q_i \pitchfork_I P_{i+1}$ *holds for* $0 \leq i < k$, $Q_i \pitchfork_{\widetilde{I}} P_{i+1}$ *does not hold for* $0 \leq i < k$ *and*

$$(P, Q, n) \in (P_0, Q_0, n_0) \square \cdots \square (P_k, Q_k, n_k).$$

Moreover, these elements are uniquely determined by these conditions, P_i *is* $\widetilde{I}$*-critical for* $0 < i \leq k$ *and* Q_i *is* $\widetilde{I}$*-critical for* $0 \leq i < k$. *The rectangle* P_0 *is the thinnest* $\widetilde{I}$*-defined vertical rectangle containing* P, *the rectangle* Q_k *is the thinnest* $\widetilde{I}$*-defined horizontal rectangle containing* Q.

The rest of Section 6.5 is devoted to the proof of the theorem.

6.5.3. *Proof.* — We will first introduce a concept, relative to an element (P, Q, n) as in the theorem above, that leads to the determination of the (P_i, Q_i, n_i).

Let m, p be integers such that $0 \leq m \leq p \leq n$. We say that $[m, p]$ is an $\widetilde{I}$-interval if there exists $(\widetilde{P}, \widetilde{Q}, \widetilde{n}) \in \mathcal{R}(\widetilde{I})$ such that

$$g_t^m(P) \subset \widetilde{P} \quad \text{for all } t \in I \text{ and } \widetilde{n} = p - m.$$

Lemma **1.** — *The union of two* $\widetilde{I}$*-intervals with non empty intersection is an* $\widetilde{I}$*-interval.*

Proof. — Let $[m, p]$, $[m', p']$ be two $\widetilde{\mathrm{I}}$-intervals with non-empty intersection, and let $(\widetilde{\mathrm{P}}, \widetilde{\mathrm{Q}}, \widetilde{n})$, $(\widetilde{\mathrm{P}}', \widetilde{\mathrm{Q}}', \widetilde{n}')$ be the corresponding elements of $\mathcal{R}(\widetilde{\mathrm{I}})$. Without loss of generality, we may assume that $m < m' \leq p < p'$. Replacing if necessary $\widetilde{\mathrm{P}}$ by a larger rectangle, we also assume that the element $(\widehat{\mathrm{P}}, \widehat{\mathrm{Q}}, \widehat{n})$ of $\mathcal{R}(\widetilde{\mathrm{I}})$ such that $\widehat{\mathrm{P}}$ is the parent of $\widetilde{\mathrm{P}}$ satisfies $m + \widehat{n} < m'$. There are now two cases:

(a) $p = m'$.

Let R_a be the rectangle containing $\widetilde{\mathrm{Q}}$. Then $\mathrm{R}_a \supset \widetilde{\mathrm{Q}} = g^{\widetilde{n}}(\widetilde{\mathrm{P}}) \supset g^{p}(\mathrm{P}) = g^{m'}(\mathrm{P})$; thus $\widetilde{\mathrm{P}}'$ is also contained in R_a and the simple composition

$$(\widetilde{\mathrm{P}}'', \widetilde{\mathrm{Q}}'', \widetilde{n}'') := (\widetilde{\mathrm{P}}, \widetilde{\mathrm{Q}}, \widetilde{n}) * (\widetilde{\mathrm{P}}', \widetilde{\mathrm{Q}}', \widetilde{n}') \tag{6.33}$$

is defined. We have $m + \widetilde{n}'' = p'$ and $g_t^m(\mathrm{P}) \subset \mathrm{P}''$.

(b) $p > m'$.

Then, $\widetilde{\mathrm{P}}$ is not a simple child of $\widehat{\mathrm{P}}$, because otherwise we would have $\widehat{n} = \widetilde{n} - 1 \geq m' - m$. By Proposition 7, there exists $(\widetilde{\mathrm{P}}_0, \widetilde{\mathrm{Q}}_0, \widetilde{n}_0)$ in $\mathcal{R}(\widetilde{\mathrm{I}})$ such that

$$(\widetilde{\mathrm{P}}, \widetilde{\mathrm{Q}}, \widetilde{n}) \in (\widehat{\mathrm{P}}, \widehat{\mathrm{Q}}, \widehat{n}) \,\square\, (\widetilde{\mathrm{P}}_0, \widetilde{\mathrm{Q}}_0, \widetilde{n}_0). \tag{6.34}$$

The element $(\widetilde{\mathrm{P}}_0, \widetilde{\mathrm{Q}}_0, \widetilde{n}_0)$ of $\mathcal{R}(\widetilde{\mathrm{I}})$ is associated to the $\widetilde{\mathrm{I}}$-interval $[\widehat{m}, p]$, where $\widehat{m} = m + \widehat{n} + \mathrm{N}_0$. We have $m + \widehat{n} < m'$ and $g_t^{m+\widehat{n}}(\mathrm{P}) \subset \mathrm{L}_u$, hence also $m + \widehat{n} + \mathrm{N}_0 = \widehat{m} \leq m'$.

To conclude the proof, we argue by induction on the total length $p' - m$ of the interval considered. The case $p' - m = 0$ is trivial. In the other case, we have the $\widetilde{\mathrm{I}}$-intervals $[\widehat{m}, p]$ and $[m', p']$ with $m < \widehat{m} \leq m'$ and hence by induction $[\widehat{m}, p']$ is an $\widetilde{\mathrm{I}}$-interval. Let $(\widetilde{\mathrm{P}}_1, \widetilde{\mathrm{Q}}_1, \widetilde{n}_1)$ be the corresponding element of $\mathcal{R}(\widetilde{\mathrm{I}})$; we have $g_t^{\widehat{m}}(\mathrm{P}) \subset \widetilde{\mathrm{P}}_1 \subset \widetilde{\mathrm{P}}_0$. From (6.34), $\widehat{\mathrm{Q}} \pitchfork_{\widetilde{\mathrm{I}}} \widetilde{\mathrm{P}}_0$ holds, hence also does $\widehat{\mathrm{Q}} \pitchfork_{\widetilde{\mathrm{I}}} \widetilde{\mathrm{P}}_1$ by Proposition 2. Then, the parabolic composition of $(\widehat{\mathrm{P}}, \widehat{\mathrm{Q}}, \widehat{n})$ and $(\widetilde{\mathrm{P}}_1, \widetilde{\mathrm{Q}}_1, \widetilde{n}_1)$ is allowed and defines an element of $\mathcal{R}(\widetilde{\mathrm{I}})$ which guarantees that $[m, p']$ is an $\widetilde{\mathrm{I}}$-interval. □

6.5.4. We will now show that the $(\mathrm{P}_i, \mathrm{Q}_i, n_i)$ in the theorem are uniquely determined by their properties. Indeed, define $m_0 = 0$, $p_0 = n_0$ and for $i > 0$:

$$m_i = p_{i-1} + \mathrm{N}_0, \qquad p_i = m_i + n_i. \tag{6.35}$$

Lemma **2.** — *The maximal $\widetilde{\mathrm{I}}$-intervals are exactly the $[m_i, p_i]$, $0 \leq i \leq k$, with associated elements $(\mathrm{P}_i, \mathrm{Q}_i, n_i)$.*

Proof. — First, the $[m_i, p_i]$ are indeed $\widetilde{\mathrm{I}}$-intervals with associated elements $(\mathrm{P}_i, \mathrm{Q}_i, n_i)$. To complete the proof, it is sufficient to show that no $\widetilde{\mathrm{I}}$-interval $[m, p]$ can intersect a gap (p_i, m_{i+1}). Assume by contradiction that there exists such a $[m, p]$ with associated element $(\widetilde{\mathrm{P}}, \widetilde{\mathrm{Q}}, \widetilde{n})$ and minimal $\widetilde{n} = p - m$. As $g_t^{\ell}(\mathrm{P}) \subset g_t^{\ell - p_i}(\mathrm{L}_u)$ does not intersect R for $p_i < \ell < m_{i+1}$, we must have $m \leqslant p_i$ and $m_{i+1} \leqslant p$. By property (R6) of $\mathcal{R}(\widetilde{\mathrm{I}})$ (Section 5.3) and the minimality of $\widetilde{n}$, there exists $(\widetilde{\mathrm{P}}_0, \widetilde{\mathrm{Q}}_0, \widetilde{n}_0)$, $(\widetilde{\mathrm{P}}_1, \widetilde{\mathrm{Q}}_1, \widetilde{n}_1)$ in $\mathcal{R}(\widetilde{\mathrm{I}})$ such that

$$(\widetilde{\mathrm{P}}, \widetilde{\mathrm{Q}}, \widetilde{n}) \in (\widetilde{\mathrm{P}}_0, \widetilde{\mathrm{Q}}_0, \widetilde{n}_0) \,\square\, (\widetilde{\mathrm{P}}_1, \widetilde{\mathrm{Q}}_1, \widetilde{n}_1) \tag{6.36}$$

with $\widetilde{n}_0 = p_i - m \leqslant n_i$, $\widetilde{n}_1 = p - m_{i+1} \leqslant n_{i+1}$. But then, from $\widetilde{Q}_0 \supset Q_i$, $\widetilde{P}_1 \supset P_{i+1}$ and $\widetilde{Q}_0 \pitchfork_{\widetilde{I}} \widetilde{P}_1$, we deduce from Proposition 2 that $Q_i \pitchfork_{\widetilde{I}} P_{i+1}$ holds, a contradiction. □

6.5.5. Lemma 2 allows us to *define* k as being the number of maximal $\widetilde{I}$-intervals minus one, and to define the $(P_i, Q_i, n_i) \in \mathcal{R}(\widetilde{I})$ as the elements of $\mathcal{R}(\widetilde{I})$ associated to the successive maximal $\widetilde{I}$-intervals. Observe that the maximal $\widetilde{I}$-intervals $[m_i, p_i]$, $(0 \leqslant i \leqslant k)$ must indeed satisfy $m_0 = 0$, $m_{i+1} = p_i + N_0$ for $0 \leqslant i < k$: every $\ell \in [0, n]$ not contained in an $\widetilde{I}$-interval is such that $g_t^{\ell - N_1}(P) \subset L_u$ for some $0 < N_1 < N_0$ and then no $\widetilde{I}$-interval intersects with $(\ell - N_1, \ell - N_1 + N_0)$, while the degenerates intervals$\{\ell - N_1\}$, $\{\ell - N_1 + N_0\}$ are $\widetilde{I}$-intervals. We observe also that $Q_i \pitchfork_{\widetilde{I}} P_{i+1}$ does not hold because otherwise $[m_i, p_{i+1}]$ would be an $\widetilde{I}$-interval.

6.5.6. Let $0 \leqslant i < k$. Let us assume by induction over i that P_j is $\widetilde{I}$-critical for $0 < j \leqslant i$, Q_j is $\widetilde{I}$-critical for $0 \leqslant j < i$, $Q_j \pitchfork_I P_{j+1}$ holds for $0 \leqslant j < i$ and that we have an element of $\mathcal{R}_N(I)$:

$$(P^{(i)}, Q^{(i)}, p_i) \in (P_0, Q_0, n_0) \,\square \cdots \square\, (P_i, Q_i, n_i) \tag{6.37}$$

such that $P \subset P^{(i)}$. The assumption is vacuously true for $i = 0$. We will prove it at step $i + 1$. For $i = k$, it gives the properties stated in the theorem for the (P_i, Q_i, n_i).

6.5.7. We first prove that Q_i is $\widetilde{I}$-critical. Assume by contradiction that Q_i is $\widetilde{I}$-transverse. Then P_i is $\widetilde{I}$-decomposable. Let then $(\widehat{P}_i, \widehat{Q}_i, \widehat{n}_i)$ be an element of $\mathcal{R}(\widetilde{I})$ such that $\widehat{P}_i$ is a child of P_i intersecting $g_t^{m_i}(P \cap \Lambda)$. Let $(\widehat{P}^{(i)}, \widehat{Q}^{(i)}, \widehat{n}^{(i)})$ be the element of $\mathcal{R}_N(I)$ such that $\widehat{P}^{(i)}$ is the child of $P^{(i)}$ containing P. Both $\widehat{P}_i$ and $\widehat{P}^{(i)}$ are non-simple children by Section 6.5.5.

We apply Proposition 7 twice. We find $(\widetilde{P}_{i+1}, \widetilde{Q}_{i+1}, \widetilde{n}_{i+1})$ in $\mathcal{R}_{N-1}(I)$, $(\widetilde{P}'_{i+1}, \widetilde{Q}'_{i+1}, \widetilde{n}'_{i+1})$ in $\mathcal{R}(\widetilde{I})$ such that both $Q_i \pitchfork_{\widetilde{I}} \widetilde{P}'_{i+1}$ and $Q^{(i)} \pitchfork_I \widetilde{P}_{i+1}$ hold and

$$(\widehat{P}_i, \widehat{Q}_i, \widehat{n}_i) \in (P_i, Q_i, n_i) \,\square\, (\widetilde{P}'_{i+1}, \widetilde{Q}'_{i+1}, \widetilde{n}'_{i+1}), \tag{6.38}$$

$$(\widehat{P}^{(i)}, \widehat{Q}^{(i)}, \widehat{n}^{(i)}) \in (P^{(i)}, Q^{(i)}, p_i) \,\square\, (\widetilde{P}_{i+1}, \widetilde{Q}_{i+1}, \widetilde{n}_{i+1}). \tag{6.39}$$

We must have $m_i + \widehat{n}_i > n$, because otherwise $[m_i, m_i + \widehat{n}_i]$ would be an $\widetilde{I}$-interval strictly larger than $[m_i, p_i]$. Thus we have $\widetilde{n}'_{i+1} > \widetilde{n}_{i+1}$. But then, from $Q_i \pitchfork_{\widetilde{I}} \widetilde{P}'_{i+1}$ and $Q^{(i)} \pitchfork_I \widetilde{P}_{i+1}$, we deduce by Corollary 4 (part 2) that $Q_i \pitchfork_{\widetilde{I}} \widetilde{P}_{i+1}$. Parabolic composition yields an element $(\overline{P}_i, \overline{Q}_i, \overline{n}_i) \in \mathcal{R}(\widetilde{I})$ with $P_i \supsetneq \overline{P}_i \supsetneq \widehat{P}_i$, in contradiction with the definition of $\widehat{P}_i$.

6.5.8. The proof that P_{i+1} is $\widetilde{I}$-critical is rather similar. We assume by contradiction that it is $\widetilde{I}$-transverse. Then Q_{i+1} is $\widetilde{I}$-decomposable and we can find $(P^*_{i+1}, Q^*_{i+1}, n^*_{i+1}) \in \mathcal{R}(\widetilde{I})$ such that Q^*_{i+1} is a child of Q_{i+1} intersecting $g^{p_{i+1}}(P \cap \Lambda)$. It is a non-simple child as $g^{p_{i+1}}(\mathrm{P} \cap \Lambda)$ is contained in L_u. By Proposition 7, there exists $(\mathrm{P}_*^{(i)}, \mathrm{Q}_*^{(i)}, n_*^{(i)}) \in \mathcal{R}(\widetilde{\mathrm{I}})$ such that $\mathrm{Q}_*^{(i)} \pitchfork_{\widetilde{\mathrm{I}}} \mathrm{P}_{i+1}$ holds and

$$(\mathbf{6.40}) \qquad (\mathrm{P}_{i+1}^*, \mathrm{Q}_{i+1}^*, n_{i+1}^*) \in (\mathrm{P}_*^{(i)}, \mathrm{Q}_*^{(i)}, n_*^{(i)}) \,\square\, (\mathrm{P}_{i+1}, \mathrm{Q}_{i+1}, n_{i+1}).$$

We must have $n_*^{(i)} > p_i$, because otherwise $[p_i - n_*^{(i)}, p_{i+1}]$ would be an $\widetilde{\mathrm{I}}$-interval strictly larger than $[m_{i+1}, p_{i+1}]$. Then, by coherence for $\mathcal{R}_{\mathrm{N}}(\mathrm{I})$, we have $(\mathrm{P}^{(i)}, \mathrm{Q}^{(i)}, p_i) \in \mathcal{R}(\widetilde{\mathrm{I}})$ and thus $i = 0$.

Let $(\widehat{\mathrm{P}}^{(i)}, \widehat{\mathrm{Q}}^{(i)}, \widehat{n}^{(i)})$, $(\widetilde{\mathrm{P}}_{i+1}, \widetilde{\mathrm{Q}}_{i+1}, \widetilde{n}_{i+1})$ be as in Section 6.5.7. From $\mathrm{Q}_*^{(i)} \pitchfork_{\widetilde{\mathrm{I}}} \mathrm{P}_{i+1}$ and $\mathrm{Q}^{(i)} \pitchfork_{\mathrm{I}} \widetilde{\mathrm{P}}_{i+1}$, we deduce by Corollary 4 (part 2) (if $\widetilde{n}_{i+1} \geqslant n_{i+1}$) or Proposition 9 (if $\widetilde{n}_{i+1} \leqslant n_{i+1}$) that $\mathrm{Q}^{(i)} \pitchfork_{\widetilde{\mathrm{I}}} \mathrm{P}_{i+1}$. This means that $\mathrm{Q}_0 \pitchfork_{\widetilde{\mathrm{I}}} \mathrm{P}_1$ and we conclude that $[0, p_1]$ is an $\widetilde{\mathrm{I}}$-interval, a contradiction.

6.5.9. We now prove that $\mathrm{Q}^{(i)}$ and P_{i+1} are I-transverse. Let $(\widehat{\mathrm{P}}^{(i)}, \widehat{\mathrm{Q}}^{(i)}, \widehat{n}^{(i)})$, $(\widetilde{\mathrm{P}}_{i+1}, \widetilde{\mathrm{Q}}_{i+1}, \widetilde{n}_{i+1})$ be as in Section 6.5.7. If $\widetilde{n}_{i+1} \leq n_{i+1}$, we have $\mathrm{Q}^{(i)} \pitchfork_{\mathrm{I}} \widetilde{\mathrm{P}}_{i+1}$ by (6.39) and thus also $\mathrm{Q}^{(i)} \pitchfork_{\mathrm{I}} \mathrm{P}_{i+1}$ by Proposition 2. Let us assume that $\widetilde{n}_{i+1} > n_{i+1}$.

We claim that, under this hypothesis, Q_{i+1} is $\widetilde{\mathrm{I}}$-critical. Indeed, if it was $\widetilde{\mathrm{I}}$-transverse, P_{i+1} would be $\widetilde{\mathrm{I}}$-decomposable and we would find an element $(\widehat{\mathrm{P}}_{i+1}, \widehat{\mathrm{Q}}_{i+1}, \widehat{n}_{i+1})$ of $\mathcal{R}(\widetilde{\mathrm{I}})$ such that $\widehat{\mathrm{P}}_{i+1}$ is a child of P_{i+1} intersecting $g^{m_{i+1}}(\mathrm{P} \cap \Lambda)$. By coherence (Proposition 8), we should have $\widehat{\mathrm{P}}_{i+1} \supset \widetilde{\mathrm{P}}_{i+1}$ and $[m_{i+1}, m_{i+1} + \widehat{n}_{i+1}]$ would be an $\widetilde{\mathrm{I}}$-interval larger than $[m_{i+1}, p_{i+1}]$, a contradiction which proves our claim.

As $(\mathrm{P}_{i+1}, \mathrm{Q}_{i+1}, n_{i+1})$ is $\widetilde{\mathrm{I}}$-bicritical, and the parent interval $\widetilde{\mathrm{I}}$ is always assumed to be β-regular, we have, for all $t \in \widetilde{\mathrm{I}}$

$$(\mathbf{6.41}) \qquad |\mathrm{P}_{i+1}| < |\widetilde{\mathrm{I}}|^{\beta}$$

and thus also (with ε_0 small enough)

$$(\mathbf{6.42}) \qquad 2|\mathrm{P}_{i+1}|^{1-\eta} < |\mathrm{I}|.$$

It now follows from Proposition 10 and $\mathrm{Q}^{(i)} \pitchfork_{\mathrm{I}} \widetilde{\mathrm{P}}_{i+1}$ that $\mathrm{Q}^{(i)}$ and P_{i+1} are I-transverse.

6.5.10. We finally show that Q_i and P_{i+1} are I-transverse. When $i = 0$, we have $\mathrm{Q}^{(0)} = \mathrm{Q}_0$, so this has been done in Section 6.5.9.

When $i > 0$, $(\mathrm{P}_i, \mathrm{Q}_i, n_i)$ is $\widetilde{\mathrm{I}}$-bicritical and, therefore, we have for all $t \in \widetilde{\mathrm{I}}$:

$$(\mathbf{6.43}) \qquad |\mathrm{Q}_i| < |\widetilde{\mathrm{I}}|^{\beta},$$

and thus also

$$(\mathbf{6.44}) \qquad 2|\mathrm{Q}_i|^{1-\eta} < |\mathrm{I}|.$$

It follows from Proposition 10 and $\mathrm{Q}^{(i)} \pitchfork_{\mathrm{I}} \mathrm{P}_{i+1}$ that Q_i and P_{i+1} are I-transverse.

To conclude the induction step of Section 6.5.6, we simply observe that the parabolic composition of $(P^{(i)}, Q^{(i)}, n^{(i)})$ and $(P_{i+1}, Q_{i+1}, n_{i+1})$ is allowed in $\mathcal{R}(I)$; it produces an element $(P^{(i+1)}, Q^{(i+1)}, p_{i+1}) \in \mathcal{R}(I)$ such that $P^{(i+1)}$ intersects P and therefore contains P.

The last assertion in the theorem follows from Section 6.5.5. The proof of the theorem is now complete.

6.6. *Width estimates.*

6.6.1. I-*special rectangles.* — As long as $\mathcal{R}(I)$ has not been fully constructed, we cannot decide whether a rectangle P or Q is I-critical or not. On the other hand, we can decide whether $\widetilde{P}$ or $\widetilde{Q}$ is $\widetilde{I}$-critical or not when $\widetilde{I}$ strictly contains I and $(\widetilde{P}, \widetilde{Q}, \widetilde{n}) \in \mathcal{R}(\widetilde{I})$. The following definition is therefore useful.

Let $(P, Q, n) \in \mathcal{R}_N(I)$. We say that P is I-*special* if either $(P, Q, n) \in \mathcal{R}(I_0)$ (i.e. no parabolic composition is ever involved in the construction of P) or, denoting by $\widetilde{I}$ the parent of I and by $(\widetilde{P}, \widetilde{Q}, \widetilde{n})$ the element of $\mathcal{R}(\widetilde{I})$ such that $P \subset \widetilde{P}$ and $\widetilde{P}$ is smallest possible, we have that $\widetilde{P}$ is $\widetilde{I}$-critical. We define similarly the property for Q.

There are therefore three cases:

- $(P, Q, n) \in \mathcal{R}(I_0)$; in this case, P is always I-special.
- $(P, Q, n) \in \mathcal{R}(\widetilde{I}) - \mathcal{R}(I_0)$; in this case, P is I-special iff it is $\widetilde{I}$-critical.
- $(P, Q, n) \in \mathcal{R}(I) - \mathcal{R}(\widetilde{I})$; in this case, we apply the structure theorem in the last subsection and see that $\widetilde{P}$ is the rectangle P_0 in the statement of this theorem; thus P is I-special iff P_0 is $\widetilde{I}$-critical.

Let us also observe that if $(P, Q, n) \in \mathcal{R}(\widetilde{I})$ and P is I-special, i.e. $\widetilde{I}$-critical, then it is $\widetilde{I}$-special by Proposition 3 of Section 5.6.

The same discussion holds for Q, replacing P_0 by Q_k in the case where $(P, Q, n) \in \mathcal{R}(I) - \mathcal{R}(\widetilde{I})$.

The following result will be useful in Section 7.

Proposition **11.** — *Let* $(P_0, Q_0, n_0), (P_1, Q_1, n_1) \in \mathcal{R}_N(I)$ *such that* $Q_0 \subset Q_u$, $P_1 \subset P_s$. *Assume that* $Q_0 \pitchfork_I P_1$ *holds and that* I, Q_0, P_1 *are maximal with this property: any* $\widehat{I} \supset I, (\widehat{P}_0, \widehat{Q}_0, \widehat{n}_0), (\widehat{P}_1, \widehat{Q}_1, \widehat{n}_1) \in \mathcal{R}_N(\widehat{I})$ *with* $Q_0 \subset \widehat{Q}_0 \subset Q_u$, $P_1 \subset \widehat{P}_1 \subset P_s$ *such that* $\widehat{Q}_0 \pitchfork_{\widehat{I}} \widehat{P}_1$ *holds must verify* $\widehat{I} = I, \widehat{Q}_0 = Q_0, \widehat{P}_1 = P_1$. *Then* Q_0 *and* P_1 *are* I-*special.*

Proof. — We may assume that the level of I is > 0. Let $\widetilde{I}$ be the parent of I, and $(\widetilde{P}_0, \widetilde{Q}_0, \widetilde{n}_0), (\widetilde{P}_1, \widetilde{Q}_1, \widetilde{n}_1) \in \mathcal{R}_N(\widetilde{I})$ be such that $\widetilde{Q}_0$ (resp. $\widetilde{P}_1$) is the smallest $\widetilde{I}$-defined rectangle containing Q_0 (resp. P_1). We have to show that Q_0, P_1 are $\widetilde{I}$-critical.

Assume for instance by contradiction that $\widetilde{Q}_0$ is $\widetilde{I}$-transverse. Let $(P_\alpha, Q_\alpha, n_\alpha)$ an $\widetilde{I}$-decomposition of P_s such that, for each α, $\widetilde{Q}_0$ and P_α are either $\widetilde{I}$-separated or $\widetilde{I}$-transverse. Let α such that $P_\alpha \cap P_1 \neq \emptyset$. Then $\widetilde{Q}_0$ and P_α are $\widetilde{I}$-transverse. If

$P_1 \subset P_\alpha$, it follows from Proposition 2 that $\widetilde{Q}_0$ and P_1 are $\widetilde{I}$-transverse. If $P_1 \supset P_\alpha$, then $(P_1, Q_1, n_1) \in \mathcal{R}_N(\widetilde{I})$ by coherence for $\mathcal{R}_N(I)$, and it follows then by Corollary 4 from $\widetilde{Q}_0 \pitchfork_{\widetilde{I}} P_\alpha$ and $Q_0 \pitchfork_I P_1$ that $\widetilde{Q}_0 \pitchfork_{\widetilde{I}} P_1$ holds.

Thus $\widetilde{Q}_0 \pitchfork_{\widetilde{I}} P_1$ always holds, in contradiction with the assumption of the lemma. This shows that $\widetilde{Q}_0$ is $\widetilde{I}$-critical. The proof that $\widetilde{P}_1$ is $\widetilde{I}$-critical is similar. □

6.6.2. *Uniform stretched exponential estimates for widths.* — The next proposition is a substitute for the uniform exponential estimates for widths that are characteristic of the uniformly hyperbolic dynamics. We denote by γ the constant

$$\gamma := \frac{\log \frac{3}{2}}{\log 2} \in (0, 1). \tag{6.45}$$

Proposition **12.** — *Let* (P, Q, n) *be an element of* $\mathcal{R}_N(I)$. *For all* $t \in I$, *we have*

$$|P| \leqslant C_0 \exp(-n^\gamma)$$

with the stronger estimate

$$|P| \leqslant C_0 \exp(-2n^\gamma)$$

when P *is* I-*special.*

The constant C_0 depends only on the constants in the formulas (3.12) and (3.27) for the widths in simple and parabolic composition.

Proof. — For $(P, Q, n) \in \mathcal{R}(I_0)$, we have an exponential estimate for $|P|$ which implies the weaker estimate of the I-special case of the proposition.

We assume now that the level of I is > 0.

For $(P, Q, n) \in \mathcal{R}(\widetilde{I}) - \mathcal{R}(I_0)$ (with $\widetilde{I}$ the parent of I), the result is true by the induction hypothesis (as observed in Section 6.6.1, if P is I-special, it is also $\widetilde{I}$-special).

We now assume that $(P, Q, n) \in \mathcal{R}(I) - \mathcal{R}(\widetilde{I})$ and apply the structure theorem of Section 6.5. We write

$$(P, Q, n) \in (P_0, Q_0, n_0) \square \cdots \square (P_k, Q_k, n_k) \tag{6.46}$$

as in the statement of the theorem. Let us denote by (P'_1, Q'_1, n'_1) the rectangle in $\mathcal{R}_{N-1}(I)$ defined by $(P_1, Q_1, n_1) \square \cdots \square (P_k, Q_k, n_k)$ such that Q is contained in Q'_1. As P_1 is $\widetilde{I}$-critical, P'_1 is I-special. We have

$$(P, Q, n) \in (P_0, Q_0, n_0) \square (P'_1, Q'_1, n'_1). \tag{6.47}$$

We use (3.27) (Section 3.5), condition (R7) for $\mathcal{R}_{N-1}(I)$ (Section 5.4) and the induction hypothesis to write

$$|P| \ll |P_0||P'_1|^{1/2} \leqslant C_0^{3/2} \exp(-n_0^\gamma - n_1'^\gamma). \tag{6.48}$$

As $n = n_0 + n'_1 + N_0$, this gives the first statement of the proposition.

Assume now that P is I-special. Then P_0 is $\tilde{I}$-critical, hence $\tilde{I}$-special (cf. Section 6.6.1).

When $n_0 \geqslant n'_1$, we have from the choice of γ that

$$(\mathbf{6.49}) \qquad 2n_0^\gamma + n_1'^\gamma \geqslant 2(n_0 + n'_1)^\gamma,$$

and therefore the same estimate $|P| \ll |P_0||P'_1|^{1/2}$ now leads to $|P| \leqslant C_0 \exp(-2n^\gamma)$.

On the other hand, when $n_0 \leqslant n'_1$, we proceed as follows. We now have

$$(\mathbf{6.50}) \qquad n_0^\gamma + 2n_1'^\gamma \geqslant 2(n_0 + n'_1)^\gamma.$$

As (P_0, Q_0, n_0) is $\tilde{I}$-bicritical and $\tilde{I}$ is regular, we have $|P_0| \leqslant |\tilde{I}|^\beta$ for all $t \in \tilde{I}$. From (3.27), we have

$$(\mathbf{6.51}) \qquad |P_0| \leqslant C|P_0||P'_1|\delta(Q_0, P'_1)^{-1/2}.$$

From $Q_0 \pitchfork_I P'_1$, and (T1), (R7) (Section 5.4), we have $\delta(Q_0, P'_1) \geqslant |I|$ for all $t \in I$. Using also $|P_0| \leqslant |\tilde{I}|^\beta$, we obtain

$$(\mathbf{6.52}) \qquad |P| \ll |P_0|^{1/2}|P'_1| \leqslant C_0^{3/2} \exp(-n_0^\gamma - 2n_1'^\gamma)$$

and we conclude again that $|P| \leqslant C_0 \exp(-2n^\gamma)$ from (6.50). □

6.6.3. The following simple estimate says that the width of any rectangle not in $\mathcal{R}(I_0)$ is uniformly small with ε_0.

Proposition **13.** — *Let* $(P, Q, n) \in \mathcal{R}_N(I)$. *If* (P, Q, n) *does not belong to* $\mathcal{R}(I_0)$, *it satisfies, for all* $t \in I$

$$|P| < \varepsilon_0^{1/2}, \qquad |Q| < \varepsilon_0^{1/2}.$$

Proof. — We may assume that the level of I is > 0 and that (P, Q, n) does not belong to $\mathcal{R}(\tilde{I})$, where $\tilde{I}$ is the parent of I. Write

$$(\mathbf{6.53}) \qquad (P, Q, n) \in (P_0, Q_0, n_0) \boxdot (P'_1, Q'_1, n'_1)$$

as in the proof of Proposition 12. The estimate (6.51) is still valid. From condition (R7) for $\mathcal{R}_{N-1}(I)$ (Section 5.4), we obtain $|P'_1| \ll \delta(Q_0, P'_1)$ and thus $|P| \ll |P_0|(\delta(Q_0, P'_1))^{1/2}$. As $\delta(Q_0, P'_1)$ is at most of the order of ε_0 and $|P_0|$ is small, we get the required estimate for P. The proof for $|Q|$ is symmetric. □

Lemma **3.** — *Let* (P_0, Q_0, n_0), (P_1, Q_1, n_1) *be elements of* $\mathcal{R}_{N-1}(I)$ *such that* $Q_0 \subset Q_u$ *and* $P_1 \subset P_s$. *If* $Q_0 \pitchfork_I P_1$ *holds then we have* $\delta(Q_0, P_1) > |I|$ *for all* $t \in I$.

Proof. — Let $\widetilde{\mathrm{I}} \supset \mathrm{I}$, and let elements $(\widetilde{\mathrm{P}}_0, \widetilde{\mathrm{Q}}_0, \widetilde{n}_0), (\widetilde{\mathrm{P}}_1, \widetilde{\mathrm{Q}}_1, \widetilde{n}_1) \in \mathcal{R}(\widetilde{\mathrm{I}})$ with $\widetilde{\mathrm{P}}_1 \supset \mathrm{P}_1$, $\widetilde{\mathrm{Q}}_0 \supset \mathrm{Q}_0$ such that $\widetilde{\mathrm{Q}}_0 \pitchfork_{\widetilde{\mathrm{I}}} \widetilde{\mathrm{P}}_1$ holds. From condition (T1), we have $\delta_{\mathrm{LR}}(\widetilde{\mathrm{Q}}_0, \widetilde{\mathrm{P}}_1) \geqslant 2|\widetilde{\mathrm{I}}|$ for all $t \in \mathrm{I}$. Using (3.36)–(3.39) and property (R7) for $\mathcal{R}_{\mathrm{N}-1}(\mathrm{I})$ we obtain, for all $t \in \mathrm{I}$

$$\delta(\mathrm{Q}_0, \mathrm{P}_1) \geqslant \delta(\widetilde{\mathrm{Q}}_0, \widetilde{\mathrm{P}}_1)$$

$$\geqslant \delta_{\mathrm{LR}}(\widetilde{\mathrm{Q}}_0, \widetilde{\mathrm{P}}_1) - \mathrm{C}(|\mathrm{Q}_0| + |\mathrm{P}_1|) > \frac{1}{2}\delta_{\mathrm{LR}}(\widetilde{\mathrm{Q}}_0, \widetilde{\mathrm{P}}_1) \geqslant |\widetilde{\mathrm{I}}| \geqslant |\mathrm{I}|. \tag{6.54}$$ □

In the following proposition, the setting and the notations are those of the structure theorem in Section 6.5: $\widetilde{\mathrm{I}}$ is the parent interval of I, (P, Q, n) is an element in $\mathcal{R}_{\mathrm{N}}(\mathrm{I})$ but not in $\mathcal{R}(\widetilde{\mathrm{I}})$, $k > 0$ and $(\mathrm{P}_i, \mathrm{Q}_i, n_i) \in \mathcal{R}(\widetilde{\mathrm{I}})$, $0 \leqslant i \leqslant k$ are as given by the conclusion of the theorem.

Proposition **14.** — *For all $t \in \mathrm{I}$, we have*

$$|\mathrm{P}| \leqslant \mathrm{C}^k |\mathrm{P}_0||\mathrm{P}_1| \cdots |\mathrm{P}_k||\mathrm{I}|^{-\frac{k}{2}}.$$

Moreover, for all $t \in \mathrm{I}$, we have

$$|\mathrm{P}_i| < |\widetilde{\mathrm{I}}|^{\beta} \quad \textit{for } 0 < i < k, \qquad |\mathrm{P}_k| < |\widetilde{\mathrm{I}}|.$$

In particular, we have always $|\mathrm{P}| < |\mathrm{P}_0||\mathrm{I}|^{1/3}$.

Proof. — For $0 \leqslant i \leqslant k$, let $(\mathrm{P}^{(i)}, \mathrm{Q}^{(i)}, p_i)$ be the element in $\mathcal{R}_{\mathrm{N}}(\mathrm{I})$ such that

$$(\mathrm{P}^{(i)}, \mathrm{Q}^{(i)}, p_i) \in (\mathrm{P}_0, \mathrm{Q}_0, n_0) \,\square \cdots \square\, (\mathrm{P}_i, \mathrm{Q}_i, n_i)$$

and $\mathrm{P} \subset \mathrm{P}^{(i)}$. We show by induction on i that

$$|\mathrm{P}^{(i)}| \leqslant \mathrm{C}^i |\mathrm{P}_0| \cdots |\mathrm{P}_i||\mathrm{I}|^{-\frac{i}{2}}. \tag{6.55}$$

As $\mathrm{P}^{(0)} = \mathrm{P}_0$, this is true for $i = 0$. By Lemma 3 above, we have, for all $t \in \mathrm{I}$,

$$\delta(\mathrm{Q}^{(i)}, \mathrm{P}_{i+1}) > |\mathrm{I}|,$$

so the required estimate for P_{i+1} follows from (3.27) in Section 3.5. This proves the first statement of the proposition.

The estimate $|\mathrm{P}_i| < |\widetilde{\mathrm{I}}|^{\beta}$ for $0 < i < k$ is true because $(\mathrm{P}_i, \mathrm{Q}_i, n_i)$ is $\widetilde{\mathrm{I}}$-bicritical and $\widetilde{\mathrm{I}}$ is β-regular. To estimate $|\mathrm{P}_k|$, we first observe that, by Proposition 3 in Section 5.6.1, there exists $t \in \widetilde{\mathrm{I}}$ such that $\delta_{\mathrm{LR}}(\mathrm{Q}_{k-1}, \mathrm{P}_k) < 2|\widetilde{\mathrm{I}}|$. Then, from Corollary 8 for $\mathcal{R}(\widetilde{\mathrm{I}})$ (in Section 7.6), we obtain, as $\mathrm{Q}_{k-1}, \mathrm{P}_k$ are $\widetilde{\mathrm{I}}$-critical, hence $\widetilde{\mathrm{I}}$-special, that $\delta_{\mathrm{LR}}(\mathrm{Q}_{k-1}, \mathrm{P}_k) < 4|\widetilde{\mathrm{I}}|$ for all $t \in \widetilde{\mathrm{I}}$. Finally, by property (R7) for $\mathcal{R}_{\mathrm{N}-1}(\mathrm{I})$, $|\mathrm{P}_k|$ is much smaller than $\delta_{\mathrm{LR}}(\mathrm{Q}_{k-1}, \mathrm{P}_k)$ for all $t \in \mathrm{I}$, which gives the required estimate.

The last statement in the proposition is an easy consequence of the first two. □

In the following corollary, the setting is the same that in the proposition. For $0 \leqslant i \leqslant k$, $(P^{(i)}, Q^{(i)}, p_i)$ is the element defined in the proof of the proposition.

Corollary **5.** — *Assume that* P *is* I*-special. Then, for all* $t \in I$, *we have*

$$|P^{(i)}| < |\widetilde{I}|^{\beta(i+1)-\frac{(1+\tau)i}{2}} \quad \textit{for } 0 \leqslant i < k,$$

$$|P| < |\widetilde{I}|^{\beta k+1-\frac{(1+\tau)k}{2}}.$$

In particular, one has always $|P| < |I|^{\beta+1/3}$.

Proof. — As P is I-special, P_0 is $\widetilde{I}$-critical. Therefore, (P_0, Q_0, n_0) is $\widetilde{I}$-bicritical and $|P_0| < |\widetilde{I}|^\beta$ for all $t \in I$. The estimates of the corollary now follow easily from those of the proposition and (6.55) above. □

Corollary **6.** — *Let* $(P^*, Q^*, n^*) \in \mathcal{R}_N(I)$. *Assume that* P^* *is* I*-special and that* $|P^*| \geqslant |I|^{\beta+1/3}$ *for some* $t \in I$. *If* $|P^*| \geqslant \varepsilon_0^{\beta+1/3}$ *for some* $t \in I$, *let* $I^* = I_0$. *Otherwise, let* $I^* \supset I$ *be the smallest parameter interval such that* $|P^*| < |I^*|^{\beta+1/3}$ *for all* $t \in I$. *Then* $(P^*, Q^*, n^*) \in \mathcal{R}_N(I^*)$.

Proof. — If $I = I_0$, there is nothing to prove. Assume that the level of I is > 0, and let $\widetilde{I}$ be the parent interval. Corollary 5 above show that we must have $(P^*, Q^*, n^*) \in \mathcal{R}_N(\widetilde{I})$. Then P^* is $\widetilde{I}$-critical, hence $\widetilde{I}$-special. Iterating the argument gives the corollary. □

Observe that, once the existence of the class $\mathcal{R}(I)$ is established (for every candidate I in a β regular parent interval $\widetilde{I}$), the corollary implies that every candidate is $\overline{\beta}$-regular, with $\overline{\beta} = (1+\tau)^{-1}\beta$. Indeed, I-critical elements are either very thin ($\ll |I|^\beta$), or already $\widetilde{I}$-defined, and $|\widetilde{I}|^\beta = |I|^{\overline{\beta}}$.

7. Estimates for the classes $\mathcal{R}(I)$

7.1. *Uniform cone condition.* — In this subsection, we will check that all elements $(P, Q, n) \in \mathcal{R}_N(I)$ satisfy the cone condition (AL2) of Section 3.2 for the parameters λ, u_0, v_0 of Section 5.3: we have $u_0 = \frac{u}{(uv)^{1/4}}$, $v_0 = \frac{v}{(uv)^{1/4}}$, where all $(P, Q, n) \in \mathcal{R}(I_0)$ satisfy (AL2) with parameters λ, u, v.

Let (A, B) be the implicit representation of the affine-like iterate (P, Q, n); we have to prove that

(AL2)
$$\lambda|A_x| + u_0|A_y| \leqslant 1,$$
$$\lambda|B_y| + v_0|B_x| \leqslant 1.$$

Let $u_1 = \frac{u}{(uv)^{1/8}}$, $v_1 = \frac{v}{(uv)^{1/8}}$. We will prove that, for all $t \in I$, we have

(7.1)
$$|A_y| < u_1^{-1}, \qquad |B_x| < v_1^{-1}.$$

This is sufficient to obtain (AL2): we already know that if $(P, Q, n) \in \mathcal{R}(I_0)$ them (AL2) is satisfied; on the other hand, if $(P, Q, n) \notin \mathcal{R}(I_0)$, then, for all $t \in I$, we have from Proposition 13

$$(7.2) \qquad |A_x| < \varepsilon_0^{\frac{1}{2}}, \qquad |B_y| < \varepsilon_0^{\frac{1}{2}}.$$

With ε_0 small enough, (7.1) and (7.2) give (AL2).

Let us now proceed with the proof of (7.1). When $(P, Q, n) \in \mathcal{R}(I_0)$, we have the stronger estimate:

$$(7.3) \qquad |A_y| < u^{-1}, \qquad |B_x| < v^{-1}.$$

Assume that $(P, Q, n) \notin \mathcal{R}(I_0)$. By the structure theorem (Section 6.5), (P, Q, n) is obtained from the parabolic composition of shorter elements $(\widetilde{P}, \widetilde{Q}, \widetilde{n})$, $(\widehat{P}, \widehat{Q}, \widehat{n}) \in \mathcal{R}_{N-1}(I)$. Denote by $(\widetilde{A}, \widetilde{B})$ the implicit representation of the affine-like iterate $(\widetilde{P}, \widetilde{Q}, \widetilde{n})$. We use formula (A.86) of Appendix A to obtain

$$(7.4) \qquad |A_y - \widetilde{A}_y| \leqslant C|\widetilde{P}||\widetilde{Q}|(\delta(\widetilde{Q}, \widehat{P}))^{-\frac{1}{2}}.$$

From (R7), $\delta(\widetilde{Q}, \widehat{P})$ is much larger than $|\widetilde{Q}|$. We have therefore

$$(7.5) \qquad |A_y - \widetilde{A}_y| \leqslant |\widetilde{P}| \leqslant C_0 \exp(-\widetilde{n}^{\gamma}),$$

where we have used Proposition 12 in the last inequality. We only use (7.5) when $\widetilde{n}$ is large (because $\widetilde{Q} \subset Q_u$), and the series $\sum \exp(-m^{\gamma})$ is convergent. Therefore (7.1) is a consequence of (7.3) and (7.5).

The proof of (AL2), i.e., the first part of condition (R2) in Section 5.3, is now complete.

7.2. *Bounded distortion.* — We now check the second half of property (R2) in Section 5.3. We have to prove that, for all $(P, Q, n) \in \mathcal{R}_N(I)$, we have the following estimate on distortion:

$$(7.6) \qquad D(g_t^n/P) \leqslant 2D_0.$$

Here, the constant D_0 corresponds to the stronger estimate we obtain from (MP6) when $(P, Q, n) \in \mathcal{R}(I_0)$:

$$(7.7) \qquad D(g_t^n/P) \leqslant D_0.$$

Define

$$(7.8) \qquad D_I(N) = \sup_{(P,Q,n) \in \mathcal{R}_N(I)} \sup_{t \in I} D(g_t^n/P).$$

From (7.7), we have $D_I(N) \leqslant D_0$ for $N = o(\log \varepsilon_0^{-1})$, because no parabolic composition is involved in this case. We set

(**7.9**) $$D_I^s(N) := \max_{\substack{n>0, n'>0 \\ n+n' \leqslant N}} D_I^s(n, n')$$

with

(**7.10**) $$D_I^s(n, n') := D_I(n) + C \exp(-n^\gamma)(D_I(n) + D_I(n')).$$

We also set

(**7.11**) $$D_I^p(N) := \max_{\substack{n \gg 0, n' \gg 0 \\ n+n'+N_0 \leqslant N}} D_I^p(n, n')$$

with

(**7.12**) $$D_I^p(n, n') := D_I(n) + C \exp(-\eta n^\gamma).$$

We claim that, if $D_I(n)$ for $n < N$ is not too large so that the condition in (3.29) (Section 3.5) is satisfied, we have

(**7.13**) $$D_I(N) \leqslant \max(D_I^s(N), D_I^p(N)).$$

Indeed, this follows from (3.13) (Section 3.3) for simple composition and (3.29) (Section 3.5) for parabolic composition; the term $C|P_1|\delta^{-1}$ in (3.29) is smaller than $|P_1|^\eta$ by condition (R7) for $\mathcal{R}_{N-1}(I)$; then one uses Proposition 12.

It is now clear that (7.6) follows from $D_I(N) \leqslant D_0$ for $N = o(\log \varepsilon_0^{-1})$ and (7.9)–(7.13).

7.3. *Estimates for the special rectangles* P_s *and* Q_u. — In Section 7.5, we will check the estimates contained in condition (R4) of Section 5.3 concerning the class $\mathcal{R}(I)$.

These estimates, which are related to parabolic composition, are valid for an element (P, Q, n) of $\mathcal{R}(I)$ which satisfies $Q \subset Q_u$ (or $P \subset P_s$).

In the present section, we will be concerned with the affine-like iterates which are directly associated with the elements (P_s, Q_s, n_s) and (P_u, Q_u, n_u).

We will make the computations for (P_s, Q_s, n_s) the other case is obviously symmetric. We will assume that the periodic point p_s is *fixed*: the general case is completely similar, but the notations are more awkward.

In this subsection, we just write (x, y) for the coordinates in the rectangle R_{a_s} containing p_s; we denote by (A, B) the implicit representation of the affine-like iterate

(**7.14**) $$G_t : (R_{a_s}) \cap g_t^{-1}(R_{a_s}) \to g_t(R_{a_s}) \cap R_{a_s}.$$

For $n \geqslant 0$, we denote by $(A^{(n)}, B^{(n)})$ the implicit representation of the nth iterate of this restriction.

As the equation of $W^s_{\mathrm{loc}}(p_s)$ is $\{x = 0\}$ (cf. (MP3) in Section 2.2), we have

$$A(y, 0, t) \equiv 0, \tag{7.15}$$

from which we deduce

$$\begin{aligned} |A_y(y, x, t)| &\leqslant C|x|, \\ |A_t(y, x, t)| &\leqslant C|x|, \\ |A_{yy}(y, x, t)| &\leqslant C|x|, \\ |A_{yt}(y, x, t)| &\leqslant C|x|. \end{aligned} \tag{7.16}$$

Denote by $\mu = \mu(t)$ the unstable eigenvalue of Dg_t at p_s. For all t, x, y, n, we have

$$C^{-1}\mu^{-n} \leqslant |A_x^{(n)}(y, x, t)| \leqslant C\mu^{-n}. \tag{7.17}$$

Let $(x_i, y_i)_{0\leqslant i\leqslant n}$ be an orbit of g_t in R_{a_s}. For all $0 \leqslant \ell \leqslant m \leqslant n$, we have:

$$C^{-1}\mu^{m-\ell}|x_\ell| \leqslant |x_m| \leqslant C\mu^{m-\ell}|x_\ell|. \tag{7.18}$$

Proposition **15.** — *The following estimates hold:*

$$|A_y^{(n)}(y_0, x_n, t)| \leqslant C|x_0| \leqslant C\mu^{-n}|x_n|, \tag{7.19}$$

$$|A_t^{(n)}(y_0, x_n, t)| \leqslant Cn|x_0| \leqslant Cn\mu^{-n}|x_n|, \tag{7.20}$$

$$|A_{yy}^{(n)}(y_0, x_n, t)| \leqslant C|x_0| \leqslant C\mu^{-n}|x_n|. \tag{7.21}$$

Proof of (7.19). — From formula (3.11) in Section 3.3, we have:

$$A_y^{(n)}(y_0, x_n, t) = A_y^{(n-1)}(y_0, x_{n-1}, t) + A_y A_x^{(n-1)} B_y^{(n-1)} \Delta^{-1}, \tag{7.22}$$

with $B_y^{(n-1)}\Delta^{-1}$ exponentially small with n and, using (7.16)–(7.18):

$$|A_y(y_{n-1}, x_n, t) A_x^{(n-1)}(y_0, x_{n-1}, t)| \leqslant C|x_0|. \tag{7.23}$$

The inequality (7.19) is now clear. □

Proof of (7.20). — We use here formulas (A.6), (A.10) of Appendix A which give

$$\begin{aligned} &|A_t^{(n)}(y_0, x_n, t) - A_t^{(n-1)}(y_0, x_{n-1}, t)| \\ &\quad \leqslant C\mu^{-n}\Big(|A_t(y_{n-1}, x_n, t)| + |B_t^{(n-1)}(y_0, x_{n-1}, t)||A_y(y_{n-1}, x, t)|\Big), \end{aligned} \tag{7.24}$$

$$\text{(7.25)} \qquad |\mathrm{B}_t^{(n)}(y_0, x_n, t) - \mathrm{B}_t^{(n-1)}(y_1, x_n, t)| \leqslant \mathrm{C}|\mathrm{B}_y^{(n-1)}|_{\mathrm{C}^0}\left(1 + |\mathrm{A}_t^{(n-1)}|_{\mathrm{C}^0}\right).$$

As $|\mathrm{B}_y^{(n)}|_{\mathrm{C}_0}$ is exponentially small, we deduce from (7.25) that

$$\text{(7.26)} \qquad |\mathrm{B}_t^{(n)}(y_0, x_n, t)| \leqslant \mathrm{C},$$

and then, from (7.24), (7.16) that (7.20) holds. □

Proof of (7.21). — We use formulas (A.6), (A.11), (A.18), (A.20) of Appendix A to obtain

$$\text{(7.27)} \qquad \mathrm{A}_{yy}^{(n)} = \mathrm{A}_{yy}^{(n-1)} + 2\mathrm{A}_{xy}^{(n-1)}\mathrm{X}_y + \mathrm{A}_{xx}^{(n-1)}\mathrm{X}_y^2 + \mathrm{A}_x^{(n-1)}\mathrm{X}_{yy},$$

with

$$\text{(7.28)} \qquad \mathrm{X}_y = \mathrm{A}_y\mathrm{B}_y^{(n-1)}\Delta^{-1},$$

$$\text{(7.29)} \qquad \Delta = 1 - \mathrm{A}_y\mathrm{B}_x^{(n-1)},$$

$$\begin{aligned} \mathrm{X}_{yy} = \mathrm{B}_y^{(n-1)}\Delta^{-1}\Big(&\mathrm{A}_{yy}\mathrm{B}_y^{(n-1)}\Delta^{-1} + \mathrm{A}_y\partial_y \log|\mathrm{B}_y^{(n-1)}| \\ &+ \mathrm{A}_y\mathrm{X}_y\partial_x \log|\mathrm{B}_y^{(n-1)}| - \mathrm{A}_y\Delta_y\Delta^{-1}\Big), \qquad \text{(7.30)} \end{aligned}$$

$$\text{(7.31)} \qquad -\Delta_y = \mathrm{A}_{yy}\mathrm{B}_y^{(n-1)}\mathrm{B}_x^{(n-1)}\Delta^{-1} + \mathrm{A}_y\mathrm{B}_{xy}^{(n-1)} + \mathrm{A}_y\mathrm{B}_{xx}^{(n-1)}\mathrm{X}_y.$$

In these formulas, $\mathrm{A}^{(n-1)}$, $\mathrm{B}^{(n-1)}$ and their derivatives are taken at (y_0, x_{n-1}, t), A, B and their derivatives are taken at (y_{n-1}, x_n, t). The terms $\mathrm{B}_x^{(n-1)}$, $\mathrm{B}_{xx}^{(n-1)}$, $\partial_x \log|\mathrm{B}_y^{(n-1)}|$, $\partial_y \log|\mathrm{B}_y^{(n-1)}|$, Δ^{-1} are bounded by the uniform cone condition and the uniform distortion; the terms $\mathrm{B}_y^{(n-1)}$, $\mathrm{B}_{xy}^{(n-1)}$, $\mathrm{A}_x^{(n-1)}$, $\mathrm{A}_{xx}^{(n-1)}$, $\mathrm{A}_{xy}^{(n-1)}$ are exponentially small. Also, from (7.16) we have:

$$\text{(7.32)} \qquad \begin{aligned} &|\mathrm{A}_y(y_{n-1}, x_n, t)| \leqslant \mathrm{C}|x_n|, \\ &|\mathrm{A}_{yy}(y_{n-1}, x_n, t)| \leqslant \mathrm{C}|x_n|. \end{aligned}$$

We conclude that we can write

$$\text{(7.33)} \qquad \mathrm{A}_{yy}^{(n)}(y_0, x_n, t) = \mathrm{A}_{yy}^{(n)}(y_0, x_{n-1}, t) + \mu^{-n}x_n r_n$$

with r_n exponentially small; this leads to (7.21). □

Corollary **7.** — *For the special rectangle* $(\mathrm{P}_s, \mathrm{Q}_s, n_s)$, *we have:*

$$\begin{aligned} &|\mathrm{A}_y^{(n_s)}|_{\mathrm{C}^0} \leqslant \mathrm{C}\varepsilon_0, \\ &|\mathrm{A}_{yy}^{(n_s)}|_{\mathrm{C}^0} \leqslant \mathrm{C}\varepsilon_0, \\ &|\mathrm{A}_t^{(n_s)}|_{\mathrm{C}^0} \leqslant \mathrm{C}\varepsilon_0 \log \varepsilon_0^{-1}. \end{aligned}$$

Proof. — We have only to observe that μ^{-n_s} is of order ε_0. □

Obviously, the same estimates hold for the other special element (P_u, Q_u, n_u).

7.4. *Further estimates for the class* $\mathcal{R}(I_0)$. — In this subsection, we derive estimates for the class $\mathcal{R}(I_0)$ from the estimates for (P_s, Q_s, n_s), (P_u, Q_u, n_u) obtained in the last subsection and from the estimates for simple composition found in Appendix A.3. These estimates are typically better than the estimates for classes $\mathcal{R}(I)$ with smaller I, which will be obtained in the end of Section 7, because only simple composition is involved into the construction of elements of $\mathcal{R}(I_0)$.

Proposition **16.** — *Let* (P, Q, n) *be an element of* $\mathcal{R}(I_0)$, *and let* (A, B) *be the associated implicit representation. For all* t, y, x, *we have*

$$|A_t| \leqslant C, \qquad |B_t| \leqslant C,$$
$$|A_{yt}| \leqslant C, \qquad |B_{xt}| \leqslant C,$$
$$|\partial_t \log |A_x|| \leqslant Cn, \qquad |\partial_t \log |B_y|| \leqslant Cn.$$

If moreover $P \subset P_s$, *we have, for all* t, y, x

$$|A_y| \leqslant C\varepsilon_0, \qquad |A_{yy}| \leqslant C\varepsilon_0, \qquad |A_t| \leqslant C\varepsilon_0 |\log \varepsilon_0|.$$

Similarly, if $Q \subset Q_u$, *we have, for all* t, y, x

$$|B_x| \leqslant C\varepsilon_0, \qquad |B_{xx}| \leqslant C\varepsilon_0, \qquad |B_t| \leqslant C\varepsilon_0 |\log \varepsilon_0|.$$

Proof. — The widths of rectangles in $\mathcal{R}(I_0)$ are exponentially small with the number of iterations. Then, the estimate for A_t, B_t follow by a simple induction on n from formula (A.67) in Appendix A.3. Similarly, one derives the estimates for A_{yt}, B_{xt}, $\partial_t \log |A_x|$, $\partial_t \log |B_y|$ from formulas (A.72), (A.73).

When $P \subset P_s$, we write

$$(P, Q, n) = (P_s, Q_s, n_s) * (P', Q', n')$$

for some $(P', Q', n') \in \mathcal{R}(I_0)$. As $|P_s| \leqslant C\varepsilon_0$, the estimates for A_y, A_{yy}, A_t follow from Corollary 7 and formulas (A.66), (A.71), (A.67) in Appendix A.3. The case where $Q \subset Q_u$ is similar. □

7.5. *Proof of the property* (R4) *of Section 5.3.* — We have to show that, for an element (P, Q, n) in $\mathcal{R}_N(I)$, with associated implicit representation (A, B), we have

(**7.34**) $$|A_y| \leqslant C\varepsilon_0, \qquad |A_{yy}| \leqslant C\varepsilon_0$$

whenever $P \subset P_s$ and

$$(\mathbf{7.35}) \qquad |B_x| \leqslant C\varepsilon_0, \qquad |B_{xx}| \leqslant C\varepsilon_0$$

whenever $Q \subset Q_u$. We will deal only with (7.34), the other case being symmetric.

We have already proved (7.34), (7.35) when (P, Q, n) belongs to $\mathcal{R}(I_0)$. We may therefore assume that (P, Q, n) does not belong to $\mathcal{R}(I_0)$. We assume that $P \subset P_s$ and prove (7.34).

In this case, the structure theorem of Section 6.5 guarantees that (P, Q, n) can be obtained from the parabolic composition of shorter elements (P_0, Q_0, n_0), (P_1, Q_1, n_1). Let (A_0, B_0), (A_1, B_1) be the implicit representations associated to these iterates.

From formulas (A.86), (A.91) in Appendix A.4, we have

$$|A_y - A_{0,y}| \leqslant C|P_0||Q_0|\delta(Q_0, P_1)^{-1/2},$$
$$|A_{yy} - A_{0,yy}| \leqslant C|P_0||Q_0|\delta(Q_0, P_1)^{-1/2}.$$

As $P \subset P_s$, we have also $P_0 \subset P_s$, hence $|P_0| \leqslant C\varepsilon_0$. On the other hand, by property (R7) for $\mathcal{R}_{N-1}(I)$, $|Q_0|$ is for all $t \in I$ much smaller than $\delta(Q_0, P_1)$. We have then, from Proposition 12 in Section 6.6.2

$$(\mathbf{7.36}) \qquad |Q_0|\delta(Q_0, P_1)^{-1/2} \leqslant |Q_0|^{1/2} \leqslant C\exp\left(-\frac{1}{2}n_0^{\gamma}\right),$$

which gives

$$(\mathbf{7.37}) \qquad |A_y - A_{0,y}| \leqslant C\varepsilon_0, \exp\left(-\frac{1}{2}n_0^{\gamma}\right),$$

$$(\mathbf{7.38}) \qquad |A_{yy} - A_{0,yy}| \leqslant C\varepsilon_0 \exp\left(-\frac{1}{2}n_0^{\gamma}\right).$$

As $P_0 \subset P_s$, the estimate (7.34) follows immediately by induction on n (starting with Proposition 16) from (7.37), (7.38). The proof of (7.35) is similar.

7.6. *Relative speeds of special rectangles.* — Let (P_0, Q_0, n_0), (P_1, Q_1, n_1) be elements of $\mathcal{R}_N(I)$ such that $Q_0 \subset Q_u$, $P_1 \subset P_s$.

The displacements $\delta(Q_0, P_1)$, $\delta_L(Q_0, P_1)$, $\delta_R(Q_0, P_1)$, $\delta_{LR}(Q_0, P_1)$ were introduced in formulas (3.30)–(3.33) of Section 3.6 and are the values at the four corners of the rectangle of definition of the function $\overline{C}(y_0, x_1)$ introduced in Section 3.5 as

$$(\mathbf{7.39}) \qquad \overline{C}(y_0, x_1) = \min_w C(w, y_0, x_1).$$

All these quantities also depend on the parameter t, and we want in this section to estimate the variation with the parameter of the displacements, which amounts to estimate the partial derivative C_t.

Let (A_0, B_0), (A_1, B_1) be the implicit representations for (P_0, Q_0, n_0), (P_1, Q_1, n_1) respectively. As will be seen below, an estimate for C_t depends very much on estimates for the partial derivatives $A_{1,t}$, $B_{0,t}$. Good estimates for these two quantities are not available for all (P_0, Q_0, n_0), (P_1, Q_1, n_1). We will obtain good estimates when Q_0, P_1 are I-special, which is sufficient in the applications.

Proposition **17.** — *Let* (P, Q, n) *be an element of* $\mathcal{R}_N(I)$ *with* $P \subset P_s$. *Let* (A, B) *be the implicit representation of* (P, Q, n). *If* P *is* I-*special, we have for all* t, y, x

$$|A_t| \leqslant \varepsilon_0^{\frac{1}{2}}.$$

Proof. — If $(P, Q, n) \in \mathcal{R}(I_0)$, we have from Proposition 14 the stronger estimate:

$$|A_t| \leqslant C\varepsilon_0 \log \varepsilon_0^{-1}.$$

We now assume that the level of I is > 0. Let $\widetilde{I}$ be the parent of I. If $(P, Q, n) \in \mathcal{R}(\widetilde{I})$, P is also $\widetilde{I}$-special (cf. Section 6.6.1) and the estimate of the proposition is true by induction.

We now assume that $(P, Q, n) \notin \mathcal{R}(\widetilde{I})$. We apply the structure theorem in Section 6.5: let $k \geqslant 1$ and (P_i, Q_i, n_i), $0 \leqslant i \leqslant k$, be the elements of $\mathcal{R}(\widetilde{I})$ given by the statement of the theorem. We also denote, for $0 \leqslant i \leqslant k$, by $(P^{(i)}, Q^{(i)}, n^{(i)})$ the element of $\mathcal{R}(\widetilde{I})$ such that $P \subset P^{(i)}$ and

$$(P^{(i)}, Q^{(i)}, n^{(i)}) \in (P_0, Q_0, n_0) \,\square \cdots \square\, (P_i, Q_i, n_i). \tag{7.40}$$

We have $(P^{(k)}, Q^{(k)}, n^{(k)}) = (P, Q, n)$. As P is I-special, P_0 is $\widetilde{I}$-critical, hence $P^{(i)}$ is I-special for $0 \leqslant i \leqslant k$. Moreover, for $0 \leqslant i < k$, Q_i, P_{i+1} are $\widetilde{I}$-critical, hence $Q^{(i)}$, P_{i+1} are I-special.

Let $(A^{(i)}, B^{(i)})$, (A_i, B_i) be the implicit representations associated to $(P^{(i)}, Q^{(i)}, n^{(i)})$, (P_i, Q_i, n_i). The estimate (A.96) in Appendix A.4 gives

$$|A_t^{(i+1)} - A_t^{(i)}| \leqslant C|P^{(i)}|\delta(Q^{(i)}, P_{i+1})^{-1/2}(1 + |A_{i+1,t}| + |B_t^{(i)}|). \tag{7.41}$$

Here we have from the induction hypothesis $|B_t^{(i)}| \leqslant C\varepsilon_0$ since $Q^{(i)}$ is I-special and $|A_{i+1,t}| \leqslant C\varepsilon_0$ since P_{i+1} is I-special.

From Lemma 3 in Section 6.6.3 we have $\delta(Q^{(i)}, P_{i+1}) > |I|$ for all $t \in I$. From Corollary 5 in Section 6.6.3, we have

$$|P^{(i)}| \leqslant C^i |\widetilde{I}|^{\beta(i+1) - \frac{(1+\tau)i}{2}}. \tag{7.42}$$

Plugging these estimates into (7.41) above gives

$$|A_t^{(i+1)} - A_t^{(i)}| \leqslant C^{i+1} |\widetilde{I}|^{(i+1)(\beta - \frac{1+\tau}{2})}. \tag{7.43}$$

As $\beta - \frac{1+\tau}{2} > \frac{1}{2}$, we obtained the required estimate for $|A_t|$ by first summing over i and then on the successive levels of the parameter intervals under consideration.

The proof for B_t is similar. □

Corollary **8.** — *Let* (P_0, Q_0, n_0), (P_1, Q_1, n_1) *be elements of* $\mathcal{R}_N(I)$ *with* $Q_0 \subset Q_u$, $P_1 \subset P_s$. *Assume that* Q_0 *and* P_1 *are* I*-special. Then, the partial derivative with respect to the parameter of the function* C *(introduced in Section 3.5) satisfies*

$$|C_t + 1| \leqslant C\varepsilon_0^{\frac{1}{2}}.$$

Proof. — From formula (A.35) in Appendix A, using the notations there, we have

$$-C_t = \theta_x \overline{X}_t + \theta_y \overline{Y}_t + \theta_t, \tag{7.44}$$

$$\overline{X}_t = (A_{1,t} + A_{1,y} Y_{s,t}) \Delta_1^{-1}, \tag{7.45}$$

$$\overline{Y}_t = (B_{0,t} + B_{0,x} X_{u,t}) \Delta_0^{-1}, \tag{7.46}$$

with Δ_0^{-1}, Δ_1^{-1} uniformly bounded. The value of θ_t is taken at $(\overline{X}, \overline{Y}, t)$, with

$$|\overline{X}| \leqslant C\varepsilon_0, \qquad |\overline{Y}| \leqslant C\varepsilon_0. \tag{7.47}$$

On the other hand, we have, in Section 4.1, normalized the parameter in order to have

$$\theta_t(0, 0, t) \equiv 1. \tag{7.48}$$

We, therefore, have

$$|\theta_t(\overline{X}, \overline{Y}, t) - 1| \leqslant C\varepsilon_0. \tag{7.49}$$

In (7.45) and (7.46), we have $|A_{1,y}| < C\varepsilon_0$, $|B_{0,x}| < C\varepsilon_0$, by (R4) and $|A_{1,t}| < \varepsilon_0^{\frac{1}{2}}$, $|B_{0,t}| < \varepsilon_0^{\frac{1}{2}}$ by Proposition 17. The Corollary follows, as θ_x, θ_y, $Y_{s,t}$, $X_{u,t}$ are uniformly bounded. □

Corollary **9.** — *Let* (P_0, Q_0, n_0), (P_1, Q_1, n_1) *be elements of* $\mathcal{R}_N(I)$ *with* $Q_0 \subset Q_u$, $P_1 \subset P_s$. *Assume that* Q_0 *and* P_1 *are* I*-special and that* $Q_0 \overline{\pitchfork}_I P_1$ *holds. Then, we have*

$$\max_I \delta_{LR}(Q_0, P_1) \leqslant 2 \min_I \delta_{LR}(Q_0, P_1).$$

Proof. — Indeed, by condition (T1) of Section 5.4, we have $\min_I \delta_{LR}(Q_0, P_1) \geqslant 2|I|$ and by Corollary 8 above, we have $\max_I \delta_{LR}(Q_0, P_1) - \min_I \delta_{LR}(Q_0, P_1) < 2|I|$. □

We will have a more general version of this statement in the next Subsection (Proposition 19).

7.7. *Variation of width of special rectangles.* — Our main purpose now is to prove property (R7) of Section 5.4:

(R7) If (P_0, Q_0, n_0), $(P_1, Q_1, n_1) \in \mathcal{R}(I)$ satisfy $Q_0 \subset Q_u$, $P_1 \subset P_s$ and $Q_0 \pitchfork_I P_1$ holds, then, for all $t \in I$, we have

$$\delta(Q_0, P_1) \geqslant C^{-1}\Big(|P_1|^{1-\eta} + |Q_0|^{1-\eta}\Big).$$

A priori, the transversality condition gives some control through (T2), (T3) in Section 5.4 only for *some* values of the parameter. However, from Corollary 9 above, we know that the order of magnitude of $\delta(Q_0, P_1)$ is the same through I, at least when Q_0, P_1 are I-special and $Q_0 \overline{\pitchfork}_I P_1$ holds. Therefore, to obtain (R7), we do need to control how the widths $|P_1|$ and $|Q_0|$ vary through I. Good estimates will be obtained when Q_0, P_1 are I-special, and this will turn out to be sufficient due to Proposition 11 in Section 6.6.1.

Proposition **18.** — *Let* (P, Q, n) *be an element of* $\mathcal{R}_N(I)$ *with* $P \subset P_{ss}$, (A, B) *be the associated implicit representation. If* P *is* I*-special, we have, for all* t, y, x

$$|\partial_t \log |A_x|| \leqslant C\frac{\log|P|}{|I|\log|I|}, \tag{7.50}$$

$$|A_{yt}| \leqslant C. \tag{7.51}$$

Obviously, there is a similar statement exchanging P and Q, A and B, x and y.

Proof. — When $(P, Q, n) \in \mathcal{R}(I_0)$, n and $|\log|P||$ are of the same order; as $|I||\log|I||$ is always smaller than $\varepsilon_0 \log \varepsilon_0^{-1}$, the estimates in Proposition 16 of Section 7.4 imply the inequalities above.

We will now assume that the level of I is > 0. Let $\widetilde{I}$ be the parent of I. If $(P, Q, n) \in \mathcal{R}(\widetilde{I})$, P is also $\widetilde{I}$-special (cf. Section 6.6.1) and $|I||\log|I|| < |\widetilde{I}||\log|\widetilde{I}||$. Therefore the estimates of the proposition follow from the induction hypothesis.

We now assume that $(P, Q, n) \notin \mathcal{R}(\widetilde{I})$. We apply the structure theorem in Section 6.5: let $k \geqslant 1$ and (P_i, Q_i, n_i), $0 \leqslant i \leqslant k$, be the elements of $\mathcal{R}(\widetilde{I})$ given by the statement of the theorem. We also denote, for $0 \leqslant i \leqslant k$, by $(P^{(i)}, Q^{(i)}, n^{(i)})$ the element of $\mathcal{R}(\widetilde{I})$ such that $P \subset P^{(i)}$ and

$$(P^{(i)}, Q^{(i)}, n^{(i)}) \in (P_0, Q_0, n_0) \,\square \cdots \square\, (P_i, Q_i, n_i). \tag{7.52}$$

We have $(P^{(k)}, Q^{(k)}, n^{(k)}) = (P, Q, n)$. As P is I-special, P_0 is $\widetilde{I}$-critical, hence $P^{(i)}$ is I-special for $0 \leqslant i \leqslant k$. Moreover, for $0 \leqslant i < k$, Q_i, P_{i+1} are $\widetilde{I}$-critical, hence $Q^{(i)}$, P_{i+1} are I-special.

Let $(A^{(i)}, B^{(i)})$, (A_i, B_i) be the implicit representations associated to $(P^{(i)}, Q^{(i)}, n^{(i)})$, (P_i, Q_i, n_i). The estimates (A.103), (A.104) in Appendix A.4 give, with $\delta = \delta(Q^{(i)}, P_{i+1})$

$$|\partial_t \log |A_x^{(i+1)}| - \partial_t \log |A_x^{(i)}||$$

$$\leqslant C(\delta^{-1} + \delta^{-\frac{1}{2}}(|B_{xt}^{(i)}| + |A_{i+1,yt}|) + |\partial_t \log |A_{i+1,x}||), \tag{7.53}$$

(**7.54**) $$|A_{yt}^{(i+1)} - A_{yt}^{(i)}| \leqslant C\delta^{-\frac{1}{2}}|P^{(i)}|(1 + |Q^{(i)}|K),$$

with

(**7.55**) $$K = \delta^{-1} + \delta^{-\frac{1}{2}}(|B_{xt}^{(i)}| + |A_{i+1,yt}|) + |\partial_t \log |A_x^{(i)}|| + |\partial_t \log |B_y^{(i)}||.$$

Assume, by the induction hypothesis, that we have

(**7.56**) $$|B_{xt}^{(i)}| \leqslant C_0,$$

(**7.57**) $$|A_{i+1,yt}| \leqslant C_0,$$

(**7.58**) $$|\partial_t \log |A_{i+1,x}|| \leqslant C_0 \frac{\log |P_{i+1}|}{|\widetilde{I}| \log |\widetilde{I}|},$$

(**7.59**) $$|\partial_t \log |A_x^{(i)}|| \leqslant C_0 \frac{\log |P^{(i)}|}{|I| \log |I|},$$

(**7.60**) $$|\partial_t \log |B_y^{(i)}|| \leqslant C_0 \frac{\log |Q^{(i)}|}{|I| \log |I|}.$$

Here C_0 is large but independent of ε_0. This means that the term $\delta^{-\frac{1}{2}}(|B_{xt}^{(i)}| + |A_{i+1,yt}|)$ in (7.53) and (7.55) is dominated by δ^{-1}. As $|I| = |\widetilde{I}|^{1+\tau}$, in order to prove (7.50) by induction, we need to have, in view of (7.53):

(**7.61**) $$C|I||\log |I||\delta^{-1} + CC_0|\widetilde{I}|^{\tau}|\log |P_{i+1}|| + C_0|\log |P^{(i)}|| \leqslant C_0|\log |P^{(i+1)}||.$$

We have here $\delta \geqslant |I|$ from Lemma 3 in Section 6.6.3 and, by (3.27):

(**7.62**) $$|\log |P^{(i+1)}|| \geqslant |\log |P^{(i)}|| + |\log |P_{i+1}|| - \frac{1}{2}|\log |I|| - C.$$

Therefore, (7.61) will hold as far as

(**7.63**) $$\left(\frac{C_0}{2} + C\right)|\log |I|| + C_0 C \leqslant C_0(1 - C|\widetilde{I}|^{\tau})|\log |P_{i+1}||.$$

From (R7) for $\mathcal{R}_{N-1}(I)$, we know that $|P_{i+1}|$ is much smaller than δ. On the other hand, as Q_j and P_{i+1} are not $\widetilde{I}$-transverse, by Proposition 3 in Section 5.6.1, $\delta_{LR}(Q_j, P_{i+1})$ (which is larger than δ) is smaller than $2|\widetilde{I}|$ for some $t \in \widetilde{I}$; it then follows from Corollary 8 in the last subsection applied to Q_j, P_{i+1} (which are $\widetilde{I}$-critical, hence $\widetilde{I}$-special), that $\delta_{LR}(Q_j, P_{i+1})$, and thus also δ, stay smaller than $C|\widetilde{I}|$ for all $t \in \widetilde{I}$. We therefore have

(**7.64**) $$|\log |P_{i+1}|| \geqslant \log |\widetilde{I}| = (1 + \tau) \log |I|,$$

from which (7.63) follows if we take $C_0 \geqslant 3C$ (provided ε_0 is small enough). This completes the proof of the induction step for (7.50).

To prove (7.51), we estimate the right-hand side of (7.54). From Corollary 5 in Section 6.6.3, we have

$$|P^{(i)}| \leqslant C^i |\widetilde{I}|^{\beta(i+1)-\frac{(1+\tau)i}{2}}. \tag{7.65}$$

From (R7) for $\mathcal{R}_{N-1}(I)$, we know that $|Q^{(i)}|$ is much smaller than δ. We have seen above that $|I| < \delta < C|\widetilde{I}|$.

This gives, as $\beta > 1$

$$\delta^{-\frac{1}{2}} |P^{(i)}| \leqslant |\widetilde{I}|^{(i+1)\beta/2}, \tag{7.66}$$

$$\delta^{-\frac{3}{2}} |P^{(i)}||Q^{(i)}| \leqslant |\widetilde{I}|^{(i+1)\beta/2}, \tag{7.67}$$

$$\delta^{-\frac{1}{2}} |P^{(i)}||Q^{(i)}||\partial_t \log |A_x^{(i)}|| \leqslant |\widetilde{I}|^{(i+1)\beta/2}, \tag{7.68}$$

$$\delta^{-\frac{1}{2}} |P^{(i)}||Q^{(i)}||\partial_t \log |B_y^{(i)}|| \leqslant |\widetilde{I}|^{(i+1)\beta/2}. \tag{7.69}$$

This leads to:

$$|A_{yt}^{(i+1)} - A_{yt}^{(i)}| \leqslant C|\widetilde{I}|^{\beta/2(i+1)}. \tag{7.70}$$

We can now sum over i and then over the different levels of parameter intervals to obtain (7.51). The proof of Proposition 18 is complete. □

Proof of Property (R7). — Let (P_0, Q_0, n_0), (P_1, Q_1, n_1) be elements of $\mathcal{R}_N(I)$ such that $Q_0 \subseteq Q_u$, $P_1 \subset P_s$ and $Q_0 \pitchfork_I P_1$ holds.

If $\widehat{I} \supset I$, $(\widehat{P}_0, \widehat{Q}_0, \widehat{n}_0)$, $(\widehat{P}_1, \widehat{Q}_1, \widehat{n}_1) \in \mathcal{R}_N(\widehat{I})$ satisfy $Q_0 \subset \widehat{Q}_0 \subset Q_u$, $P_1 \subset \widehat{P}_1 \subset P_s$, and the inequality in (R7) is satisfied for $\widehat{I}, \widehat{Q}_0, \widehat{P}_1$ then it is also satisfied for I, Q_0, P_1.

We can therefore assume that I, Q_0, P_1 are maximal with the property $Q_0 \pitchfork_I P_1$. From Proposition 11 in Section 6.6.1, Q_0, P_1 are then I-special.

The maximality property implies that $Q_0 \overline{\pitchfork}_I P_1$ holds. Then, from Corollary 9 in Section 7.6, we have

$$\max_I \delta_{LR}(Q_0, P_1) \leqslant 2 \min_I \delta_{LR}(Q_0, P_1). \tag{7.71}$$

We have also $\min_I \delta_{LR}(Q_0, P_1) \geqslant 2|I|$ from (T1) for $Q_0 \overline{\pitchfork}_I P_1$.

From Proposition 18, we have either $\max_I |Q_0| < |I|^2$ or $\max_I |Q_0| \leqslant C \min_I |Q_0|$, and similarly for P_1. It then follows from (T2), (T3) for $Q_0 \overline{\pitchfork}_I P_1$ that, for all $t \in I$

$$\delta(Q_0, P_1) \geqslant C^{-1}(|Q_0|^{1-\eta} + |P_1|^{1-\eta})$$

for all $t \in I$.

The proof of property (R7) is complete. □

Once (R7) has been obtained, we have a stronger form of Corollary 9 in Section 7.6:

Proposition **19.** — *Let* (P_0, Q_0, n_0), (P_1, Q_1, n_1) *be elements of* $\mathcal{R}_N(I)$ *with* $Q_0 \subset Q_u$, $P_1 \subset P_s$. *Assume that* $Q_0 \pitchfork_I P_1$ *holds. Then, we have*

$$\max_I \delta_{LR}(Q_0, P_1) \leqslant 3 \min_I \delta(Q_0, P_1).$$

Proof. — If $\widehat{I} \supset I$, $(\widehat{P}_0, \widehat{Q}_0, \widehat{n}_0)$, $(\widehat{P}_1, \widehat{Q}_1, \widehat{n}_1) \in \mathcal{R}_N(\widehat{I})$ satisfy $Q_0 \subset \widehat{Q}_0 \subset Q_u$, $P_1 \subset \widehat{P}_1 \subset P_s$, and the inequality in the Proposition is satisfied for $\widehat{I}, \widehat{Q}_0, \widehat{P}_1$ then it is also satisfied for I, Q_0, P_1.

We can therefore assume that I, Q_0, P_1 are maximal with the property $Q_0 \pitchfork_I P_1$. From Proposition 11 in Section 6.6.1, Q_0, P_1 are then I-special. The maximality property also implies that $Q_0 \overline{\pitchfork}_I P_1$ holds. By Corollary 9 and (R7), we have then

$$\begin{aligned}\max_I \delta_{LR}(Q_0, P_1) &\leqslant 2 \min_I \delta_{LR}(Q_0, P_1)\\ &\leqslant 2 \min_I (\delta(Q_0, P_1) + C|Q_0| + C|P_1|)\\ &\leqslant 3 \min_I \delta(Q_0, P_1).\end{aligned}$$ □

The existence and properties of the classes $\mathcal{R}(I)$, for a candidate parameter interval I are now fully justified and the proof of the theorem in Section 5.4 is complete. What we do not know at this moment is whether there exists, besides I_0, any regular parameter interval at all! This will be the subject of Section 9. Before, we develop in the next section some results that will turn out to be essential in Sections 9, 10 and 11.

8. Number of children and dimension estimates

In this section (except in Section 8.4), we fix a candidate parameter interval I (in particular, the parent interval $\widetilde{I}$ is always assumed to be β-regular) and obtain some properties of the class $\mathcal{R}(I)$ which will be important in Sections 9 and 10.

In Section 8.1, we collect some results that could have been proved earlier, but whose proof was deferred because the results were not necessary for the construction of the class $\mathcal{R}(I)$.

In Section 8.2, we establish some bounds on the number of children of a given rectangle which will be useful both in Sections 8.3, 8.4 and in Sections 9, 10 and 11.

In Section 8.3, we prove that, in some appropriate sense, the transverse stable and unstable "dimensions" of the class $\mathcal{R}(I)$ are very close to those of $\mathcal{R}(I_0)$.

Finally, in Section 8.4, we establish an estimate in parameter space which will be essential in Section 9.

8.1. *Further criteria for transversality.* — The following results are useful variants of Proposition 10 in Section 6.4.

Proposition **20.** — *Let* (P_0, Q_0, n_0), (P_1, Q_1, n_1), (P'_1, Q'_1, n'_1) *be elements of* $\mathcal{R}(I)$ *such that* $Q_0 \subset Q_u$, $P_1 \subset P'_1 \subset P_s$. *If* P'_1 *is* I*-transverse and* $Q_0 \pitchfork_I P_1$ *holds, then* $Q_0 \pitchfork_I P'_1$ *also holds.*

There is a symmetric statement exchanging P's and Q's.

Proof. — There exists an I-decomposition $(P_\alpha, Q_\alpha, n_\alpha)$ of Q_u such that, for any α, Q_α and P'_1 are either I-transverse or I-separated. There exists α such that Q_α and Q_0 do intersect. As $Q_0 \pitchfork_I P_1$ holds, Q_α and P'_1 must be I-transverse. If $Q_\alpha \supset Q_0$, it follows from Proposition 2 that Q_0 and P'_1 are I-transverse. If $Q_\alpha \subset Q_0$, the same conclusion follows from concavity. □

Proposition **21.** — *Let* (P_0, Q_0, n_0), (P_1, Q_1, n_1), (P'_1, Q'_1, n'_1) *be elements of* $\mathcal{R}(I)$ *such that* $Q_0 \subset Q_u$, $P_1 \subset P'_1 \subset P_s$. *Assume that* $Q_0 \pitchfork_I P_1$ *holds and that* $|P'_1| \leqslant \frac{1}{2}|Q_0|$ *for all* $t \in I$. *Then* Q_0 *and* P'_1 *are also* I*-transverse.*

Again, here is a symmetric statement exchanging P's and Q's.

Proof. — This follows closely the proof of Proposition 10. By definition of the transversality relation, there exist $\widetilde{I} \supset I$, $(\widetilde{P}_0, \widetilde{Q}_0, \widetilde{n}_0)$, $(\widetilde{P}_1, \widetilde{Q}_1, \widetilde{n}_1) \in \mathcal{R}(\widetilde{I})$ such that $Q_0 \subset \widetilde{Q}_0$, $P_1 \subset \widetilde{P}_1$ and $\widetilde{Q}_0 \overline{\pitchfork}_{\widetilde{I}} \widetilde{P}_1$.

If $P'_1 \subset \widetilde{P}_1$ this already implies that $Q_0 \pitchfork_I P'_1$. Let us assume that $\widetilde{P}_1 \subset P'_1$. We will show that $\widetilde{Q}_0 \overline{\pitchfork}_{\widetilde{I}} P'_1$ holds. By coherence (Proposition 8), we have $(P'_1, Q'_1, n'_1) \in \mathcal{R}(\widetilde{I})$.

Let us check (T1)–(T3).

By (T1) for $Q_0 \pitchfork_I P_1$ and (3.49) in Section 3.5, we have, for all $t \in \widetilde{I}$:

$$\delta_{LR}(\widetilde{Q}_0, P'_1) \geqslant \delta_{LR}(\widetilde{Q}_0, \widetilde{P}_1) \geqslant 2|\widetilde{I}|. \tag{8.1}$$

By (T2) for $Q_0 \pitchfork_I P_1$ and (3.47), there exists $t_0 \in \widetilde{I}$ such that

$$\delta_R(\widetilde{Q}_0, P'_1) \geqslant \delta_R(\widetilde{Q}_0, \widetilde{P}_1) \geqslant 2|\widetilde{Q}_0|^{1-\eta}. \tag{8.2}$$

Finally, for this same value t_0, we have

$$\begin{aligned} \delta_L(\widetilde{Q}_0, P'_1) &\geqslant \delta_R(\widetilde{Q}_0, P'_1) - C|P'_1| \\ &\geqslant 2|Q_0|^{1-\eta} - C|P'_1| \\ &\geqslant 2(2|P'_1|)^{1-\eta} - C|P'_1| \\ &\geqslant 2|P'_1|^{1-\eta}, \end{aligned} \tag{8.3}$$

if ε_0 is small enough. This proves that $\widetilde{Q}_0 \overline{\pitchfork}_{\widetilde{I}} P'_1$ holds and thus also $Q_0 \pitchfork_I P'_1$ holds. □

Corollary **10.** — *Let* $(P_1, Q_1, n_1) \in \mathcal{R}(I)$. *Assume that* P_1 *is* I*-critical. Then there exists* $(P_0, Q_0, n_0) \in \mathcal{R}(I)$ *such that* Q_0 *and* P_1 *are* I*-critically related and* $|Q_0| < \max(|I|^\beta, 2|P_1|)$ *for some* $t \in I$.

Proof. — Let $(P, Q, n) \in \mathcal{R}(I)$ such that Q and P_1 are I-critically related and $|Q| \geqslant 2|P_1|$ for all $t \in I$. We claim that Q is I-critical. Indeed, assume on the contrary that Q is I-transverse. Let (P_i, Q_i, n_i) be an I-decomposition of P_s such that, for each i, Q is either I-separated from P_i or I-transverse to P_i. If P_1 is contained in some P_i, Q is I-separated from P_1 if it is I-separated from P_i, and I-transverse to P_1 if it is I-transverse to P_i. If P_1 contains some P_i such that Q is I-transverse to P_i, then Q is I-transverse to P_1 by the proposition above. The remaining case is where the P_i contained in P_1 form an I-decomposition of P_1 and Q is I-separated from each of them; then Q is I-separated from P_1. In all cases, we get a contradiction. This proves the claim.

Consider now the following inductively constructed sequence of I-decompositions of Q_u, starting with the canonical one by children of Q_u. We stop the process when one element $(P_\alpha, Q_\alpha, n_\alpha)$ at least in the decomposition is such that $Q_\alpha \pitchfork_I P_1$ does not hold and $|Q_\alpha| < \max(|I|^\beta, 2|P_1|)$ for some $t \in I$: it satisfies then the conclusion of the Corollary.

To go from one I-decomposition to the next one, we keep those $(P_\alpha, Q_\alpha, n_\alpha)$ such that Q_α is I-transverse to P_1, or I-separated from P_1. Because P_1 is I-critical, there are other elements $(P_\alpha, Q_\alpha, n_\alpha)$ in the decomposition, which are I-critically related to P_1 and satisfy $|Q_\alpha| \geqslant \max(|I|^\beta, 2|P_1|)$ for all $t \in I$ by assumption. The claim above shows that Q_α is I-critical and therefore I-decomposable (Corollary 3 in Section 5.6.5). We replace such a Q_α by its children.

It is clear that the process has to stop, and the proof of the Corollary is complete. □

Proposition **22.** — *Let* $I' \supset I$ *be a parameter interval and let* (P_0, Q_0, n_0), (P_1, Q_1, n_1) *be elements of* $\mathcal{R}(I')$ *such that* $Q_0 \subset Q_u$, $P_1 \subset P_s$. *Assume that* $Q_0 \pitchfork_I P_1$ *holds and that we have* $2|I'| < |P_1|^{1-\eta}$, *for all* $t \in I'$. *Then* Q_0 *and* P_1 *are also* I'*-transverse.*

Again, there is a symmetric statement exchanging P's and Q's.

Proof. — Let $(\widehat{P}_0, \widehat{Q}_0, \widehat{n}_0)$, $(\widehat{P}_1, \widehat{Q}_1, \widehat{n}_1) \in \mathcal{R}(I')$ satisfy $Q_0 \subset \widehat{Q}_0 \subset Q_u$, $P_1 \subset \widehat{P}_1 \subset P_s$; then $2|I'| < |\widehat{P}_1|^{1-\eta}$ for all $t \in I'$, and if $\widehat{Q}_0$ and $\widehat{P}_1$ are I'-transverse, then Q_0 and P_1 are I'-transverse.

It is therefore sufficient to consider the case where Q_0, P_1, I are maximal with the properties $Q_0 \pitchfork_I P_1, I \subset I'$.

If $I = I'$, we are done. We therefore assume that I is strictly smaller than I'. In this case, by Proposition 11 in Section 6.6.1, Q_0 and P_1 are I-special. Maximality also guarantees that $Q_0 \overline{\pitchfork}_I P_1$.

We show that $Q_0 \overline{\pitchfork}_{I'} P_1$. Properties (T2), (T3) for $Q_0 \overline{\pitchfork}_I P_1$ imply the same for $Q_0 \overline{\pitchfork}_{I'} P_1$.

For the value t_1 given by (T3), we have

$$\delta_{LR}(Q_0, P_1) \geqslant \delta_L(Q_0, P_1) \geqslant 2|P_1|^{1-\eta} \geqslant 4|I'|. \tag{8.4}$$

By Corollary 8 in Section 7.6, this implies

$$\delta_{LR}(Q_0, P_1) \geqslant 2|I'|, \quad \forall t \in I',$$

which is (T1) for $Q_0 \overline{\pitchfork}_{I'} P_1$. This contradicts the maximality of I and proves the proposition. □

8.2. *Estimates on the number of children.* — We start with some preliminary results.

Proposition **23.** — *Let* $I' \supset I$ *be a parameter interval, and let* $(\widetilde{P}, \widetilde{Q}, \widetilde{n})$ *be an element of* $\mathcal{R}(I')$. *We assume that* $\widetilde{Q}$ *is* I'*-transverse. Then, any element* (P, Q, n) *in* $\mathcal{R}(I)$ *such that* P *is a child of* $\widetilde{P}$ *is already an element of* $\mathcal{R}(I')$.

Proof. — We can assume that P is a non-simple child. Then (P, Q, n) is obtained by parabolic composition in $\mathcal{R}(I)$ of $(\widetilde{P}, \widetilde{Q}, \widetilde{n})$ with some $(P_1, Q_1, n_1) \in \mathcal{R}(I)$. As $\widetilde{Q}$ is I'-transverse, there exists an I'-decomposition $(P_\alpha, Q_\alpha, n_\alpha)$ of P_s such that each P_α is I'-separated or I'-transverse with $\widetilde{Q}$. Let α_0 be such that P_{α_0} and P_1 intersect. Then, $\widetilde{Q} \pitchfork_{I'} P_{\alpha_0}$ holds, and also $\widetilde{Q} \pitchfork_I P_1$; if we had $P_1 \subsetneqq P_{\alpha_0}$, this would imply that $\widetilde{Q}$ is I-transverse to the parent $\widetilde{P}_1$ of P_1 and P would not be a child of $\widetilde{P}$. Therefore, we must have $P_{\alpha_0} \subset P_1$. By coherence (Proposition 8), we have that $(P_1, Q_1, n_1) \in \mathcal{R}(I')$. By concavity (Proposition 9), from $\widetilde{Q} \pitchfork_I P_1$ and $\widetilde{Q} \pitchfork_{I'} P_{\alpha_0}$, we deduce that $\widetilde{Q} \pitchfork_{I'} P_1$ also holds and $(P, Q, n) \in \mathcal{R}(I')$. □

Proposition **24.** — *Let* $I_0 \supset I_1 \supset I$ *be the largest parameter interval such that*

$$|I_1|^\beta < \left(\frac{1}{2}|I|\right)^{\frac{1}{1-\eta}}. \tag{8.5}$$

Let $(\widetilde{P}, \widetilde{Q}, \widetilde{n})$, (P, Q, n) *be elements of* $\mathcal{R}(I)$ *such that* P *is a non-simple child of* $\widetilde{P}$. *Let* (P_1, Q_1, n_1), $(\widetilde{P}_1, \widetilde{Q}_1, \widetilde{n}_1)$ *be the elements of* $\mathcal{R}(I)$ *such that*

$$(P, Q, n) \in (\widetilde{P}, \widetilde{Q}, \widetilde{n}) \,\square\, (P_1, Q_1, n_1)$$

and $\widetilde{P}_1$ *is the parent of* P_1.

Then, (P_1, Q_1, n_1) *belongs to* $\mathcal{R}(I_1)$, $\widetilde{P}_1$ *is* I*-critical,* $\widetilde{Q}_1$ *is* I_1*-transverse and we have*

$$2|\widetilde{P}_1|^{1-\eta} > |I| \tag{8.6}$$

for all $t \in I$.

Remark **9.** — As parabolic composition is possible, we have $I \neq I_0$; then, as $\beta > (1-\eta)^{-1}$, we must have $I_1 \supsetneq I$ and I_1 is β-regular.

Proof. — As P is a child of $\widetilde{P}$, $\widetilde{Q} \pitchfork_I \widetilde{P}_1$ does not hold. Then, it follows from Proposition 19 above that $\widetilde{P}_1$ is I-critical, and from Proposition 11 that (8.6) holds. Then, by definition of I_1, we have:

$$(\mathbf{8.7}) \qquad |\widetilde{P}_1| > |I_1|^{\beta}$$

for all $t \in I$. Let us show that $(\widetilde{P}_1, \widetilde{Q}_1, \widetilde{n}_1)$ belongs to $\mathcal{R}(I_1)$. Otherwise, there would exist $I_2 \supset I$, with parent interval $\widetilde{I}_2 \subset I_1$, such that $(\widetilde{P}_1, \widetilde{Q}_1, \widetilde{n}_1)$ belongs to $\mathcal{R}(I_2)$ but not to $\mathcal{R}(\widetilde{I}_2)$. Then, we would have from Corollary 5 in Section 6.6.3 (as $\widetilde{P}_1$ is I-critical, hence I_2-special) that $|\widetilde{P}_1| < |I_2|^{\beta+1/3}$ for all $t \in I_2$, in contradiction to (8.7).

Therefore, $(\widetilde{P}_1, \widetilde{Q}_1, \widetilde{n}_1)$ belongs to $\mathcal{R}(I_1)$. As $I_1 \supsetneq I$, I_1 is β-regular; $(\widetilde{P}_1, \widetilde{Q}_1, \widetilde{n}_1)$ cannot be I_1-bicritical in view of (8.7); $\widetilde{P}_1$ is I_1-critical and hence $\widetilde{Q}_1$ is I_1-transverse. Proposition 23 then shows that $(P_1, Q_1, n_1) \in \mathcal{R}(I_1)$. □

Corollary **11.** — *Assume that the level of* I *is* > 0, *and let* $\widetilde{I}$ *be the parent interval of* I. *Let* $(\widetilde{P}, \widetilde{Q}, \widetilde{n})$ *be an element of* $\mathcal{R}(\widetilde{I})$, *such that* $|\widetilde{Q}|^{1-\eta} > 2|I|$ *for all* $t \in \widetilde{I}$. *Then all children of* $\widetilde{P}$ *in* $\mathcal{R}(I)$ *belong already to* $\mathcal{R}(\widetilde{I})$.

Proof. — Indeed, let $(P, Q, n) \in \mathcal{R}(I)$ such that P is a child of $\widetilde{P}$. We can assume that P is a non-simple child. Let I_1, (P_1, Q_1, n_1), $(\widetilde{P}_1, \widetilde{Q}_1, \widetilde{n}_1) \in \mathcal{R}(I_1)$ be as in Proposition 24. We have $I_1 \supset \widetilde{I}$. As $\widetilde{Q} \pitchfork_I P_1$, we can apply Proposition 21 in Section 8.1 to obtain that $\widetilde{Q} \pitchfork_{\widetilde{I}} P_1$ holds and $(P, Q, n) \in \mathcal{R}(\widetilde{I})$. □

Corollary **12.** — *Let* $(\widetilde{P}, \widetilde{Q}, \widetilde{n})$ *be an element of* $\mathcal{R}(I)$. *The number of* $(P, Q, n) \in \mathcal{R}(I)$ *such that* P *is a child of* $\widetilde{P}$ *is finite.*

Proof. — We argue by induction on the level of the parameter interval I.

If I is the starting interval I_0, $\widetilde{P}$ has only simple children and the assertion is obvious. Assume that $I \subsetneq I_0$. The number of simple children is finite, and we have to show that the same is true for the number of non-simple children. For every non-simple child P of $\widetilde{P}$, let I_1, (P_1, Q_1, n_1), $(\widetilde{P}_1, \widetilde{Q}_1, \widetilde{n}_1) \in \mathcal{R}(I_1)$ be as in Proposition 24. By the induction hypothesis, there is for each fixed $\widetilde{P}_1$ only a finite number of possibilities for P_1. On the other hand, in view of relation (8.6), there are obviously only a finite number of possibilities for $\widetilde{P}_1$. The induction step is complete, and this completes the proof. □

We want to make the finiteness assertion quantitative, and will do that in two distinct ways. In each case, we have to estimate in the proof of Corollary 12 the number of possibilities for $\widetilde{P}_1$, and the number of possibilities for P_1 once $\widetilde{P}_1$ is fixed.

Proposition **25.** — *Let* $(\widetilde{P}, \widetilde{Q}, \widetilde{n})$ *be an element of* $\mathcal{R}(I)$. *The number of* $(P, Q, n) \in \mathcal{R}(I)$ *such that* P *is a child of* $\widetilde{P}$ *is at most* $|I|^{-c\eta}$.

The constant c, as the other constants denoted by C, and the content c' in the next proposition, depends only on the initial diffeomorphism f, not on $\tau \gg \eta \gg \varepsilon_0$.

Proof. — We argue again by induction on the level of I, following the proof of Corollary 12. The notations are those of Proposition 24. When $I = I_0$, the number of (simple) children is at most the number of rectangles in the Markov partition, which is much smaller than $\varepsilon_0^{-c\eta}$ when ε_0 is small enough.

When $I \neq I_0$, the number of possibilities for P_1 when $\widetilde{P}_1$ is fixed in at most $|I_1|^{-c\eta}$ by the induction hypothesis. We have to estimate the number of possibilities for $\widetilde{P}_1$. We know that $\widetilde{Q} \pitchfork_I \widetilde{P}_1$ does not hold, but $\widetilde{Q} \pitchfork_I P_1$ holds.

As $\widetilde{P}_1$ is I-critical, it is I-special. We have from (8.6) and Proposition 18 in Section 7.7

$$\max_I |\widetilde{P}_1| \leqslant C \min_I |\widetilde{P}_1|. \tag{8.8}$$

From Proposition 19 in Section 7.7, we have also, using (8.8)

$$\max_I \delta(\widetilde{Q}, \widetilde{P}_1) < \max_I \delta(\widetilde{Q}, P_1) \leqslant C \min_I \delta(\widetilde{Q}, P_1) < C \min_I (\delta(\widetilde{Q}, \widetilde{P}_1) + |\widetilde{P}_1|). \tag{8.9}$$

Lemma **4.** — *We have for all* $t \in I$

$$\delta(\widetilde{Q}, \widetilde{P}_1) \leqslant C |\widetilde{P}_1|^{1-\eta}. \tag{8.10}$$

Proof. — As $\widetilde{Q} \overline{\pitchfork}_I \widetilde{P}_1$ does not hold, at least one of the following three properties must be true:

$$\delta_{L,R}(\widetilde{Q}, \widetilde{P}_1) < 2|I| \quad \text{for some } t_0 \in I; \tag{8.11}$$

$$\delta_R(\widetilde{Q}, \widetilde{P}_1) < 2|\widetilde{Q}|^{1-\eta} \quad \text{for all } t \in I; \tag{8.12}$$

$$\delta_L(\widetilde{Q}, \widetilde{P}_1) < 2|\widetilde{P}_1|^{1-\eta} \quad \text{for all } t \in I. \tag{8.13}$$

If (8.13) holds, we are done.

If (8.12) holds, we argue as follows. As $\widetilde{Q} \pitchfork_I P_1$ holds but $\widetilde{Q} \pitchfork_I \widetilde{P}_1$ does not hold, it follows from Proposition 20 in Section 8.1 that $|\widetilde{P}_1| > \frac{1}{2}|\widetilde{Q}|$ for some $t_0 \in I$. Then, from (8.8) and (8.9), we have

$$\begin{aligned} \max_I \delta(\widetilde{Q}, \widetilde{P}_1) &\leqslant C \min_I (\delta(\widetilde{Q}, \widetilde{P}_1) + |\widetilde{P}_1|) \leqslant C \min_I (|\widetilde{Q}|^{1-\eta} + |\widetilde{P}_1|) \\ &\leqslant C \max_I |\widetilde{P}_1|^{1-\eta} \leqslant C \min_I |\widetilde{P}_1|^{1-\eta}. \end{aligned}$$

Finally, assume that (8.11) holds. As $\widetilde{\mathrm{Q}} \pitchfork_{\mathrm{I}} \mathrm{P}_1$ holds but $\widetilde{\mathrm{Q}} \pitchfork_{\mathrm{I}} \widetilde{\mathrm{P}}_1$ does not hold, it follows from Proposition 9 in Section 6.4 that $2|\widetilde{\mathrm{P}}_1|^{1-\eta} > |\mathrm{I}|$ for all $t \in \mathrm{I}$. Then, from (8.8) and (8.9), we have

$$\begin{aligned}\max_{\mathrm{I}} \delta(\widetilde{\mathrm{Q}}, \widetilde{\mathrm{P}}_1) &\leqslant \mathrm{C} \min_{\mathrm{I}} (\delta(\widetilde{\mathrm{Q}}, \widetilde{\mathrm{P}}_1) + |\widetilde{\mathrm{P}}_1|) \leqslant \mathrm{C}(|\mathrm{I}| + \max_{\mathrm{I}} |\widetilde{\mathrm{P}}_1|) \\ &\leqslant \mathrm{C} \max_{\mathrm{I}} |\widetilde{\mathrm{P}}_1|^{1-\eta} \leqslant \mathrm{C} \min_{\mathrm{I}} |\widetilde{\mathrm{P}}_1|^{1-\eta}. \qquad \square\end{aligned}$$

We are now able to estimate the number of possibilities for $\widetilde{\mathrm{P}}_1$ and show that this number is at most

$$\textbf{(8.14)} \qquad \mathrm{C}|\mathrm{I}|^{-\frac{\eta}{1-\eta}}.$$

This indeed follows from (8.9), (8.6) and the fact that if two distinct $\widetilde{\mathrm{P}}_1$ are not disjoint, the ratio of their widths is bounded away from 1 (so the $\widetilde{\mathrm{P}}_1$ at a given scale are disjoint; one then sums over scales). The total number of children is thus bounded by

$$\textbf{(8.15)} \qquad \mathrm{C} + 2\mathrm{C}|\mathrm{I}|^{-\frac{\eta}{1-\eta}} |\mathrm{I}_1|^{-c\eta},$$

where I_1 was the largest parameter interval satisfying

$$\textbf{(8.16)} \qquad |\mathrm{I}_1|^{\beta} < \left(\frac{1}{2}|\mathrm{I}|\right)^{\frac{1}{1-\eta}}.$$

If $|\mathrm{I}| > 2\varepsilon_0^{\beta(1-\eta)}$, we have $\mathrm{I}_1 = \mathrm{I}_0$; in this case, the term $|\mathrm{I}_1|^{-c\eta}$ in (8.15) is unnecessary because $\widetilde{\mathrm{P}}_1$ has only simple children. If $|\mathrm{I}| \leqslant 2\varepsilon_0^{\beta(1-\eta)}$, we have

$$\textbf{(8.17)} \qquad |\mathrm{I}_1|^{\beta(1+\tau)^{-1}} \geqslant \left(\frac{1}{2}|\mathrm{I}|\right)^{\frac{1}{1-\eta}}$$

and the term in (8.15) is bounded by $|\mathrm{I}|^{-c\eta}$ provided that ε_0 is small enough and

$$\textbf{(8.18)} \qquad c\eta > \frac{\eta}{1-\eta} + c\eta \frac{1+\tau}{1-\eta} \beta^{-1}.$$

As η, τ are very small, any choice of $c > \frac{\beta}{\beta-1}$ yields (8.18). Then, as $c > 1$, such a choice is also convenient when $|\mathrm{I}| > 2\varepsilon_0^{\beta(1-\eta)}$, and this concludes the proof of Proposition 25. □

In Proposition 25, we have estimated the total number of children in terms of the level of the parameter interval.

In the next proposition, we are interested, not in the total number of children, but in the number of children of a given width. The estimate is independent on the level of the parameter interval.

Proposition **26.** — *Let* $(\widetilde{P}, \widetilde{Q}, \widetilde{n})$ *be an element of* $\mathcal{R}(I)$. *For any* $\varepsilon > 0$, *the number of elements* $(P, Q, n) \in \mathcal{R}(I)$ *such that* P *is a non-simple child of* $\widetilde{P}$ *satisfying* $|P| \geqslant \varepsilon|\widetilde{P}|$ *for some* $t \in I$, *is at most* $\varepsilon^{-c'\eta}$.

Proof. — Let $\varepsilon > 0$, and let (P, Q, n) be an element of $\mathcal{R}(I)$ such that P is a non-simple child of $\widetilde{P}$. We assume, for some $t_0 \in I$, that:

(8.19) $$|P| \geqslant \varepsilon|\widetilde{P}|.$$

Let (P_1, Q_1, n_1), $(\widetilde{P}_1, \widetilde{Q}_1, \widetilde{n}_1) \in \mathcal{R}(I)$ be as in Proposition 24. From (3.27), we have, for all $t \in I$:

(8.20) $$|P| \leqslant C|\widetilde{P}||P_1|\delta(\widetilde{Q}, P_1)^{-\frac{1}{2}}.$$

Property (R7) guarantees that, for all $t \in I$

(8.21) $$\delta(\widetilde{Q}, P_1) \geqslant C^{-1}|P_1|^{1-\eta}.$$

Combining (8.19), (8.20), (8.21), we have, for some $t_0 \in I$

(8.22) $$\delta(\widetilde{Q}, P_1) \geqslant C^{-1}\varepsilon^{2\frac{1-\eta}{1+\eta}}.$$

As we always have

(8.23) $$\delta(\widetilde{Q}, P_1) \leqslant C\varepsilon_0,$$

there is no non-simple child satisfying (8.19) unless $\varepsilon < \varepsilon_0^{\frac{1}{2}}$; we will assume that this holds in the sequel.

From Lemma 4 above, we have, for all $t \in I$:

(8.24) $$\delta(\widetilde{Q}, \widetilde{P}_1) \leqslant C|\widetilde{P}_1|^{1-\eta},$$

and thus, from (3.36), also

$$\begin{aligned} \delta(\widetilde{Q}, P_1) &\leqslant \delta_R(\widetilde{Q}, \widetilde{P}_1) \\ &\leqslant \delta(\widetilde{Q}, \widetilde{P}_1) + C|\widetilde{P}_1| \\ &\leqslant C|\widetilde{P}_1|^{1-\eta}. \end{aligned} \qquad \textbf{(8.25)}$$

Combining (8.22) and (8.25), we get, for some $t_0 \in I$

(8.26) $$|\widetilde{P}_1| \geqslant C^{-1}\varepsilon^{\frac{2}{1+\eta}},$$

an inequality which actually holds for all $t \in I$ in view of (8.8).

As in the proof of Proposition 25, the number of $(\widetilde{P}_1, \widetilde{Q}_1, \widetilde{n}_1)$ for which both (8.25), (8.26) hold is easily seen to be at most

$$\text{(8.27)} \qquad C\varepsilon^{-\frac{2\eta}{1+\eta}}.$$

To estimate the number of P_1 for a given $\widetilde{P}_1$, we will apply Proposition 25 in an appropriate way.

Define a parameter interval $\widehat{I} \supset I$ as follows. If $|I|^\beta \geqslant \varepsilon^2$, let $\widehat{I} = I$. If $|I|^\beta < \varepsilon^2$, let $\widehat{I} \supset I$ be the largest parameter interval such that $|\widehat{I}|^\beta < \varepsilon^2$. We have thus in any case

$$\text{(8.28)} \qquad |\widehat{I}| \geqslant \min(\varepsilon_0, \varepsilon^{\frac{2(1+\tau)}{\beta}}).$$

As $\widetilde{P}_1$ is I-critical, hence I-special, it follows from (8.26) and Corollary 6 in Section 6.6.3 that $(\widetilde{P}_1, \widetilde{Q}_1, \widetilde{n}_1)$ belongs to $\mathcal{R}(\widehat{I})$. Moreover, when $\widehat{I} \neq I$, we have $|\widetilde{P}_1| > |\widehat{I}|^\beta$ for all $t \in I$ from (8.26) and the definition of $\widehat{I}$. Therefore, $(\widetilde{P}_1, \widetilde{Q}_1, \widetilde{n}_1)$ cannot be $\widehat{I}$-bicritical. As $\widetilde{P}_1$ is $\widehat{I}$-critical, $\widetilde{Q}_1$ is $\widehat{I}$-transverse and we conclude from Proposition 23 that (P_1, Q_1, n_1) also belongs to $\mathcal{R}(\widehat{I})$. The same is also obviously true when $\widehat{I} = I$.

We apply Proposition 25: for each fixed $(\widetilde{P}_1, \widetilde{Q}_1, \widetilde{n}_1)$, the number of children P_1 is at most $|\widehat{I}|^{-c\eta}$.

From (8.28), we have, for some appropriate c_0

$$\text{(8.29)} \qquad |\widehat{I}|^{-c\eta} \leqslant \varepsilon^{-c_0\eta}.$$

Combining this with the previous estimate in (8.27) for the number of possibilities for $\widetilde{P}_1$ gives therefore the required estimate. □

8.3. *A dimension estimate.* — The goal of this subsection is to obtain a bound on the number of elements (P, Q, n) in $\mathcal{R}(I)$ with width $|P|$ bounded from below. This is a first step towards estimating the transverse dimension of the stable set $W^s(\Lambda)$, which is necessary in order to achieve our parameter selection in Section 9.

Let I be a parameter interval, and let (P^*, Q^*, n^*) be an element of $\mathcal{R}(I)$. We introduce, in the spirit of Laplace, Dirichlet and Poincaré, the series

$$\text{(8.30)} \qquad \Theta(P^*, I, s) = \sum \max_I |P|^s,$$

where the sum runs over elements $(P, Q, n) \in \mathcal{R}(I)$ such that $P \subset P^*$. Here s is a complex variable and the series is at first a formal object, but we will soon see that it is uniformly convergent in a half-plane $\{\operatorname{Re} s > \sigma_0\}$. The goal of this subsection is to obtain a nice estimate for σ_0 and for Θ in this half-plane.

The dependence of the estimate on P^* is also quite straightforward, through the simple scaling factor $\max_I |P^*|^s$.

Let us recall that we denote by d_s^0 the transverse Hausdorff dimension of the stable foliation $W^s(K)$ of the horseshoe K for the value 0 of the parameter. Let us also denote by d_s^t the transverse Hausdorff dimension of the stable foliation $W^s(K_{g_t})$ for the value t of the parameter. It is well-known that d_s^t depends smoothly on the parameter. The transverse Hausdorff dimension controls in a precise way the number of cylinders (for the Markov partition) of a given size; more precisely, as these cylinders correspond exactly to the elements of $\mathcal{R}(I_0)$, we know that, for any $t \in I_0$ and all $\varepsilon > 0$ the number of $(P, Q, n) \in \mathcal{R}(I_0)$ such that $|P| \geqslant \varepsilon$ is exactly of the order of $\varepsilon^{-d_s^t}$.

Let us say that P (with $(P, Q, n) \in \mathcal{R}(I)$) is a *simple* descendant of P^* if (P, Q, n) is the simple composition of (P^*, Q^*, n^*) by an element of $\mathcal{R}(I_0)$. Consider the series

(8.31) $$\Theta_0(P^*, s) = \sum \max_I |P|^s,$$

where the sum runs over simple descendants of P^*. By what we have just recalled, for any $t \in I_0$, the series converge in a half-plane $\{\operatorname{Re} s > d_s^0 + C\varepsilon_0\}$ and satisfies, for $s \geqslant d_s^0 + 2C\varepsilon_0$

(8.32) $$|\Theta_0(P^*, s)| < C \max_I |P^*|^s (\operatorname{Re} s - d_s^0)^{-1}.$$

For $\Theta(P^*, I, s)$ with a parameter interval $I \neq I_0$, we have to allow a slightly larger margin with relation to the initial value d_s^0.

Proposition **27.** — *Let* $(P^*, Q^*, n^*) \in \mathcal{R}(I)$. *The series* $\Theta(P^*, I, s)$ *is uniformly convergent in the half-plane* $\{\operatorname{Re} s \geqslant d_s^0 + \varepsilon_0^{\frac{1}{3}d_s^0}\}$ *and we have, for* $s \geqslant d_s^0 + \varepsilon_0^{\frac{1}{3}d_s^0}$

$$|\Theta(P^*, I, s) - \Theta_0(P^*, s)| \leqslant \max_I |P^*|^s \varepsilon_0^{\frac{1}{20}d_s^0}.$$

Proof. — Let (P, Q, n) be an element of $\mathcal{R}(I)$ with $P \subset P^*$. Consider the intermediary rectangles

$$P^* = P(0) \subset P(1) \subset \cdots \subset P(\ell) = P$$

with $P(i)$ the parent of $P(i+1)$. Let

(8.33) $$\ell_0 < \ell_1 < \cdots < \ell_{k-1}$$

be the indices such that $P(\ell_j + 1)$ is a non-simple child of $P(\ell_j)$. Obviously, P is a simple descendant of P^* iff $k = 0$.

We also define for $0 \leqslant j \leqslant k$ elements $(P^{(j)}, Q^{(j)}, n^{(j)}) \in \mathcal{R}(I_0)$ by the following properties

(8.34) $$(P(\ell_0), Q(\ell_0), n(\ell_0)) = (P(0), Q(0), n(0)) * (P^{(0)}, Q^{(0)}, n^{(0)}),$$

(8.35) $$(P(\ell_j), Q(\ell_j), n(\ell_j)) = (P(\ell_{j-1}+1), Q(\ell_{j-1}+1), n(\ell_{j-1}+1)) * (P^{(j)}, Q^{(j)}, n^{(j)}),$$

(8.35) $$(P(\ell_0), Q(\ell_0), n(\ell_0)) \quad , n^{(0)}),$$

$$(P(\ell_j), Q(\ell_j), n(\ell_j)), \,_{-1} + 1))$$

(8.35)

$$(\mathrm{P}, \mathrm{Q}, n) = (\mathrm{P}(\ell_{k-1}+1), \mathrm{Q}(\ell_{k-1}+1), n(\ell_{k-1}+1)) \tag{8.36}$$
$$* \, (\mathrm{P}^{(k)}, \mathrm{Q}^{(k)}, n^{(k)}).$$

We now estimate the widths from (3.12), for all $t \in \mathrm{I}$:

$$|\mathrm{P}(\ell_0)| \leqslant \mathrm{C}|\mathrm{P}^*||\mathrm{P}^{(0)}|, \tag{8.37}$$

$$|\mathrm{P}(\ell_j)| \leqslant \mathrm{C}|\mathrm{P}(\ell_{j-1}+1)||\mathrm{P}^{(j)}|, \tag{8.38}$$

$$|\mathrm{P}| \leqslant \mathrm{C}|\mathrm{P}(\ell_{k-1}+1)||\mathrm{P}^{(k)}|. \tag{8.39}$$

From (3.27) and property (R7), we also have:

$$|\mathrm{P}(\ell_j+1)| < \varepsilon_0^{\frac{1}{2}}|\mathrm{P}(\ell_j)|. \tag{8.40}$$

Define m_j for $0 \leqslant j < k$ to be the largest integer such that, for all $t \in \mathrm{I}$

$$|\mathrm{P}(\ell_j+1)| \leqslant 2^{-m_j}\varepsilon_0^{\frac{1}{2}}|\mathrm{P}(\ell_j)|. \tag{8.41}$$

From Proposition 26, for each fixed $\mathrm{P}(\ell_j)$, the number of non-simple children $\mathrm{P}(\ell_j+1)$ satisfying (8.41) is at most

$$\left(2^{m_j+1}\varepsilon_0^{-\frac{1}{2}}\right)^{c'\eta}. \tag{8.42}$$

Combining (8.37), (8.38), (8.39) and (8.41), we also have

$$\max_{\mathrm{I}}|\mathrm{P}| \leqslant \mathrm{C}^{k+1}\max_{\mathrm{I}}|\mathrm{P}^*|\left(\prod_0^k \max_{\mathrm{I}}|\mathrm{P}^{(j)}|\right)\varepsilon_0^{\frac{k}{2}}2^{-\Sigma_0^{k-1}m_j}, \tag{8.43}$$

with the usual convention that $\Sigma_0^{k-1}m_j = 0$ when $k=0$.

We will take this to the power s and sum over P. We introduce (corresponding to the term $|\mathrm{P}^{(j)}|^s$)

$$\Theta_0(s) := \sum_{(\widehat{\mathrm{P}},\widehat{\mathrm{Q}},\widehat{n})\in\mathcal{R}(\mathrm{I}_0)} \max_{\mathrm{I}}|\widehat{\mathrm{P}}|^s$$
$$= \sum_a \Theta(\mathrm{R}_a, \mathrm{I}, s), \tag{8.44}$$

and also

$$\theta(s) := \sum_{m\geqslant 0}(\mathrm{C}\varepsilon_0^{\frac{1}{2}}2^{-m})^s(2^{m+1}\varepsilon_0^{-\frac{1}{2}})^{c'\eta}. \tag{8.45}$$

The function Θ_0 is controlled by (8.32), while θ satisfies

$$\theta(s) = 2^{c'\eta}\mathrm{C}^s\varepsilon_0^{\frac{1}{2}(s-c'\eta)}\left(1-2^{-(s-c'\eta)}\right)^{-1}, \tag{8.46}$$

and therefore, for $C^{-1} < s < C$:

$$C^{-1}\varepsilon_0^{\frac{1}{2}(s-c'\eta)} \leqslant \theta(s) \leqslant C\varepsilon_0^{\frac{1}{2}(s-c'\eta)}. \tag{8.47}$$

From (8.32), we have, for $s > d_s^0 + 2C\varepsilon_0$:

$$\Theta_0(s) \leqslant C(s - d_s^0)^{-1}. \tag{8.48}$$

In particular, for $s \geqslant d_s^0 + \varepsilon_0^{1/3d_s^0}$, we have

$$\Theta_0(s) \leqslant C\varepsilon_0^{-\frac{1}{3}d_s^0}, \tag{8.49}$$

$$\Theta_0(s)\theta(s) \leqslant C\varepsilon_0^{\frac{1}{10}d_s^0}. \tag{8.50}$$

But, from (8.43), we have for real s

$$\Theta(P^*, I, s) - \Theta_0(P^*, s) \leqslant C^s \max_I |P^*|^s \sum_{k>0} \Theta_0^{k+1}(s)\theta^k(s), \tag{8.51}$$

and therefore we deduce from (8.50) that the series defining Θ is uniformly convergent in the half plane $\{\operatorname{Re} s \geqslant d_s^0 + \varepsilon_0^{\frac{1}{3}d_s^0}\}$.

For $s > d_s^0 + \varepsilon_0^{\frac{1}{5}d_s^0}$, we have, from (8.48), (8.47):

$$\Theta_0(s) \leqslant C\varepsilon_0^{-\frac{1}{5}d_s^0}, \tag{8.52}$$

$$\Theta_0^2\theta(s) \leqslant C\varepsilon_0^{\frac{1}{15}d_s^0}, \tag{8.53}$$

which gives the second part of the proposition. □

Corollary **13.** — *Let* $d_s^* = d_s^0 + \varepsilon_0^{\frac{1}{5}d_s^0}$, $(P^*, Q^*, n^*) \in \mathcal{R}(I)$, $\varepsilon > 0$. *The number of* $(P, Q, n) \in \mathcal{R}(I)$ *with* $P \subset P^*$, $|P| > \varepsilon|P^*|$ *is at most* $\varepsilon^{-d_s^*}$.

Proof. — Indeed, the number of simple descendants P with $\max_I |P| > \varepsilon \max_I |P^*|$ is of order $\leqslant \varepsilon^{-d_s^0 + C\varepsilon_0}$, and the number of non-simple ones with $\max_I |P| > \varepsilon \max_I |P^*|$ is $\ll \varepsilon^{-d_s^*}$ from the proposition. □

8.4. *Transfer to parameter space.*

8.4.1. Our goal in this subsection will be to prove the following result, which expresses a transfer of the dimension estimate of Section 8.2 to parameter space.

Proposition **28.** — *Let* $\tilde{I}$ *be a regular parameter interval. Let* (P^*, Q^*, n^*) *be an element of* $\mathcal{R}(\tilde{I})$ *such that* Q^* *is* $\tilde{I}$*-critical and*

$$(\mathbf{8.54}) \qquad |Q^*| \leqslant \frac{1}{2}|\tilde{I}|^{(1+\tau)(1-\eta)^{-1}}$$

for all $t \in \tilde{I}$. *Then, the number of candidates* $I \subset \tilde{I}$ *of the next level, such that* Q^* *is* I*-critical, is at most* $|\tilde{I}|^{-\tau d_s^+}$, *where* $d_s^+ = d_s^0 + C\eta\tau^{-1}$ *can be made arbitrarily close to* d_s^0.

Recall that the total number of candidates is $|\tilde{I}|^{-\tau}$. Proposition 28 is the key estimate that will allow us in Section 9 to proceed with the selection process for parameters. The rest of the section is devoted to the proof of Proposition 28.

8.4.2. We make some general observations, that could have been made much earlier, but are only useful now.

Let (P, Q, n), (P_0, Q_0, n_0), (P'_0, Q'_0, n'_0), be elements of $\mathcal{R}(I)$ such that $P \subset P_s$, $Q'_0 \subset Q_0 \subset Q_u$, and $P'_0 \neq P_0$. From property (MP7) of the Markov partition (Section 3.4), it is easy to see that the condition (3.50) of Section 3.6.4 is satisfied. As explained in this subsection, we have then, for any $t \in I$

$$(\mathbf{8.55}) \qquad \delta(Q_0, P) + C^{-1}|Q_0| \leqslant \delta(Q'_0, P_s),$$

$$(\mathbf{8.56}) \qquad \delta_R(Q_0, P) + C^{-1}|Q_0| \leqslant \delta_R(Q'_0, P),$$

$$(\mathbf{8.57}) \qquad \delta_L(Q_0, P) - C^{-1}|Q_0| \geqslant \delta_L(Q'_0, P),$$

$$(\mathbf{8.58}) \qquad \delta_{LR}(Q_0, P) - C^{-1}|Q_0| \geqslant \delta_{LR}(Q'_0, P).$$

Let now (P, Q, n), (P_0, Q_0, n_0), (P_1, Q_1, n_1), be elements of $\mathcal{R}(I)$ such that $Q \subset Q_u$, $P_0 \subset P_s$, $P_1 \subset P_s$ and $P_0 \cap P_1 = \emptyset$. From the discussion in Section 3.6.2, we have either

$$(\mathbf{8.59}) \qquad \begin{aligned} &\delta_R(Q, P_0) < \delta(Q, P_1),\\ &\delta_{LR}(Q, P_0) < \delta_L(Q, P_1), \end{aligned}$$

or the same inequalities after exchanging P_0 and P_1.

Proposition **29.** — *Assume that* (8.59) *holds (see Figure 8).*

1. *If* Q *and* P_1 *are* I*-separated, then* Q *and* P_0 *are* I*-separated.*
2. *If* Q *and* P_0 *are* I*-transverse, and* $|P_1|^{1-\eta} \leqslant |I|$ *for some* $t \in I$, *then* Q *and* P_1 *are* I*-transverse.*

Proof. — 1. Fix $t \in I$. We will assume that $G(Q \cap L_u) \cap P_1 \cap \Lambda$ is empty and show that $G(Q \cap L_u) \cap P_0 \cap \Lambda$ is also empty.

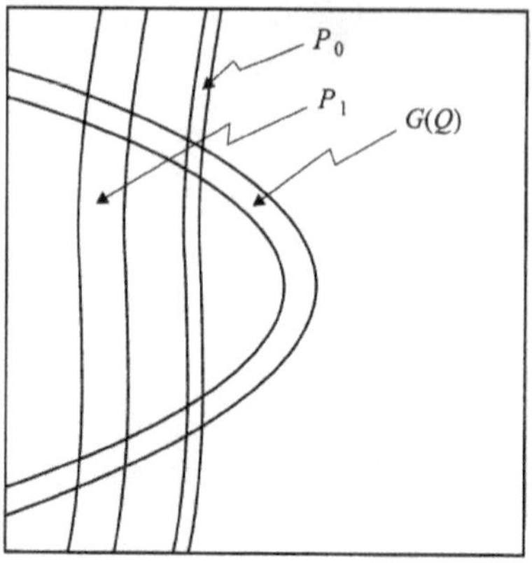

FIG. 8. — Proposition 29, part 2

Let γ a vertical-like curve which is the intersection of a decreasing sequence of simple descendants of P_1. The curve γ is contained in $W^s(\Lambda, \widehat{R}) \cap P_1$.

If γ does not intersect $G(Q \cap L_u)$, P_0 does also not intersect $G(Q \cap L_u)$ and the conclusion holds.

On the other hand, if the intersection of γ with $G(Q \cap L_u)$ is not connected, γ intersects the image $G(\gamma' \cap L_u)$, where γ' is any intersection of a decreasing sequence of simple descendants of Q; the intersection would be contained in $G(Q \cap L_u) \cap P_1 \cap \Lambda$, which is empty by hypothesis.

Therefore we can assume that the intersection γ_0 of γ with $G(Q \cap L_u)$ is non empty and connected. Let O_0 be the Jordan domain whose boundary is the union of γ_0 and the arc in the boundary of $G(Q \cap L_u)$ with the same endpoints than γ_0.

The intersection of P_0 with $G(Q \cap L_u)$ is contained in O_0. We will show that $O_0 \cap \Lambda$ is empty, which implies that $G(Q \cap L_u) \cap P_0 \cap \Lambda$ is empty.

Consider $(G \circ g^n)^{-1}(O_0) \subset P$. Part of the boundary of this Jordan domain is an horizontal segment contained in the boundary of P. The other part of the boundary is $(G \circ g^n)^{-1}(\gamma_0)$ which does not intersect $W^u(\Lambda, \widehat{R})$. In particular, it does not intersect L_s (in case $L_s \cap P$ is not empty), and it does not cross any $Q_{a,a'}$, $(a, a') \in \mathcal{B}$. Therefore, either it does not intersect any $Q_{a,a'}$ at all, in which case we can already conclude that $(G \circ g^n)^{-1}(O_0) \cap \Lambda$ is empty, or it intersects a single $Q_{a,a'}$. In this last case, there exists a simple child $Q^{(1)}$ of Q (defined by $(P_{a,a'}, Q_{a,a'}, 1) * (P, Q, n) = (P^{(1)}, Q^{(1)}, n+1)$) with the following property: let γ_1 be the subarc of γ_0 defined by $\gamma_1 = \gamma_0 \cap G(Q^{(1)} \cap L_u)$, and O_1 be the Jordan domain whose boundary is the union of γ_1 and the arc in the boundary of $G(Q^{(1)} \cap L_u)$ with the same endpoints than γ_1; then $O_0 \cap \Lambda = O_1 \cap \Lambda$.

We now apply to $Q^{(1)}, O_1, \gamma_1$ the same arguments that we used for Q, O_0, γ_0. Either we conclude that $O_1 \cap \Lambda$ is empty or we find a simple child $Q^{(2)}$ of $Q^{(1)}$, a subarc $\gamma_2 \subset \gamma_1$ and a Jordan domain O_2 whose boundary contains γ_2 such that $O_2 \cap \Lambda = O_1 \cap \Lambda$.

Iterating, either we conclude at some stage that $O_l \cap \Lambda$ is empty (and then we have $O_0 \cap \Lambda = O_l \cap \Lambda = \emptyset$), or we construct a decreasing sequence $Q^{(l)}$ of simple descendants of Q such that $\gamma \cap G(Q^{(l)} \cap L_u)$ is not empty for all l. But this is impossible, because the intersection γ' of the $Q^{(l)}$ is contained in $W^u(\Lambda, \widehat{R})$ and the intersection $\gamma \cap G(\gamma' \cap L_u)$ would be contained in $G(Q \cap L_u) \cap P_1 \cap \Lambda$.

2. Assume now that Q and P_0 are I-transverse, and that $|P_1|^{1-\eta} \leqslant |I|$ for some $t \in I$.

Let $\widetilde{I} \supset I$ and $(\widetilde{P}, \widetilde{Q}, \widetilde{n})$, $(\widetilde{P}_0, \widetilde{Q}_0, \widetilde{n}_0) \in \mathcal{R}(\widetilde{I})$ be such that $\widetilde{Q} \supset Q$, $\widetilde{P}_0 \supset P_0$ and $\widetilde{Q} \pitchfork_{\widetilde{I}} \widetilde{P}_0$ holds. If $P_1 \subset \widetilde{P}_0$, we immediately conclude that Q and P_1 are I-transverse. We assume, therefore, that $P_1 \cap \widetilde{P}_0 = \emptyset$; replacing (P_0, Q_0, n_0) by $(\widetilde{P}_0, \widetilde{Q}_0, \widetilde{n}_0)$, and (P, Q, n) by $(\widetilde{P}, \widetilde{Q}, \widetilde{n})$, we can also assume that (P, Q, n), $(P_0, Q_0, n_0) \in \mathcal{R}(\widetilde{I})$ and $Q \pitchfork_{\widetilde{I}} P_0$ holds.

Let $(\widetilde{P}_1, \widetilde{Q}_1, \widetilde{n}_1)$, be the element of $\mathcal{R}(\widetilde{I})$ with $P_1 \subset \widetilde{P}_1$ and smallest $\widetilde{P}_1$.

We will prove that Q and $\widetilde{P}_1$ are $\widetilde{I}$-transverse. We assume by contradiction that it is not the case. We have, for all $t \in \widetilde{I}$, that

$$\textbf{(8.60)} \qquad \delta_{LR}(Q, \widetilde{P}_1) > \delta_{LR}(Q, P_0) \geqslant 2|\widetilde{I}|,$$

and also for some $t_0 \in \widetilde{I}$,

$$\textbf{(8.61)} \qquad \delta_R(Q, \widetilde{P}_1) > \delta_R(Q, P_0) \geqslant 2|Q|^{1-\eta}.$$

Therefore, we must have, for all $t \in \widetilde{I}$, that

$$\textbf{(8.62)} \qquad \delta_L(Q, \widetilde{P}_1) < 2|\widetilde{P}_1|^{1-\eta}.$$

We cannot have in this case $\widetilde{P}_1 = P_1$, because, for all $t \in I$,

$$\textbf{(8.63)} \qquad \delta_L(Q, P_1) > \delta_{LR}(Q, P_0) \geqslant 2|\widetilde{I}|,$$

and (8.62), (8.63) together would contradict the hypothesis of the proposition. Therefore, $\widetilde{P}_1$ strictly contains P_1 and $\widetilde{I}$ strictly contains I. But, then, applying the structure theorem of Section 6.5 to the child of $\widetilde{P}_1$ which contains P_1, we obtain that $\widetilde{Q}_1$ is $\widetilde{I}$-critical. As $\widetilde{I}$ is β-regular, it then follows from (8.62), (8.63) and (3.39) that $\widetilde{P}_1$ is $\widetilde{I}$-transverse. This implies that there exists $(P', Q', n') \in \mathcal{R}(\widetilde{I})$ with $Q \cap Q' \neq \emptyset$ such that $Q' \pitchfork_{\widetilde{I}} \widetilde{P}_1$ holds.

When $Q \subset Q'$, it follows that $Q \pitchfork_{\widetilde{I}} \widetilde{P}_1$ holds.

When $Q' \subset Q$ and $P_0 \subset \widetilde{P}_1$, it follows by concavity from $Q' \pitchfork_{\widetilde{I}} \widetilde{P}_1$ and $Q \pitchfork_{\widetilde{I}} P_0$ that $Q \pitchfork_{\widetilde{I}} \widetilde{P}_1$ holds.

When $Q' \subset Q$ and $P_0 \cap \widetilde{P}_1 = \emptyset$, we consider a thin simple descendant P_1^* of $\widetilde{P}_1$ such that $|P_1^*|^{1-\eta} < |\widetilde{I}|$ for some $t_1 \in \widetilde{I}$. We have, as in (8.60), $\delta_{LR}(Q, P_1^*) \geqslant 2|\widetilde{I}|$ for all $t \in \widetilde{I}$, and, as in (8.61), $\delta_R(Q, P_1^*) \geqslant 2|Q|^{1-\eta}$ for some $t_0 \in \widetilde{I}$. But we have also, at $t_1 \in \widetilde{I}$, $\delta_L(Q, P_1^*) > \delta_{LR}(Q, P_0) \geqslant 2|\widetilde{I}| > 2|P_1^*|^{1-\eta}$. Therefore, $Q \pitchfork_{\widetilde{I}} P_1^*$ holds. We conclude by concavity from $Q' \pitchfork_{\widetilde{I}} \widetilde{P}_1$ and $Q \pitchfork_{\widetilde{I}} P_1^*$ that $Q \pitchfork_{\widetilde{I}} \widetilde{P}_1$ holds.

We thus obtain a contradiction in all cases. This proves that $Q \pitchfork_{\widetilde{I}} \widetilde{P}_1$ holds and thus also $Q \pitchfork_I P_1$. □

8.4.3. We now switch back to the setting of Proposition 28.

Let $(P, Q, n) \in \mathcal{R}(\tilde{I})$ with $P \subset P_s$. We say that P is *eventually* $\tilde{I}$*-transverse* to Q^* if there exists an $\tilde{I}$-decomposition $(P_\alpha, Q_\alpha, n_\alpha)$ of P such that Q^* and P_α are $\tilde{I}$-transverse for every α. We say that P is *eventually* $\tilde{I}$-Q^**-critical* if it is neither $\tilde{I}$-separated from Q^* nor eventually $\tilde{I}$-transverse to Q^*.

Lemma **5.** — *If* P *is eventually* $\tilde{I}$*-transverse to* Q^* *and* $2|P|^{1-\eta} \leqslant |\tilde{I}|$ *holds for some* $t \in \tilde{I}$, *then* $Q^* \pitchfork_{\tilde{I}} P$ *holds.*

Proof. — This is an immediate consequence of Proposition 10 in Section 6.4. □

Lemma **6.** — *If* P *is eventually* $\tilde{I}$-Q^**-critical, then* P *is* $\tilde{I}$*-critical.*

Proof. — Assume on the contrary that P is $\tilde{I}$-transverse. Let $(P_\alpha, Q_\alpha, n_\alpha)$ be an $\tilde{I}$-decomposition of Q_u such that, for each α, Q_α and P are either $\tilde{I}$-separated or $\tilde{I}$-transverse.

If $Q^* \subset Q_\alpha$ for some α, Q^* and P would be $\tilde{I}$-separated if Q_α and P are $\tilde{I}$-separated, and $\tilde{I}$-transverse if Q_α and P are $\tilde{I}$-transverse.

If there exists α such that $Q_\alpha \subset Q^*$ and $Q_\alpha \pitchfork_{\tilde{I}} P$ holds, then $Q^* \pitchfork_{\tilde{I}} P$ also holds by Proposition 10 in Section 6.4.

In the remaining case, the $Q_\alpha \subset Q^*$ form an $\tilde{I}$-decomposition of Q^* and they are all $\tilde{I}$-separated from P; this imply that Q^* itself is $\tilde{I}$-separated from P.

In all cases, we get a contradiction. The proof of the lemma is complete. □

Lemma **7.** — *If* P *is eventually* $\tilde{I}$-Q^**-critical and* $|P| > |\tilde{I}|^\beta$ *holds for some* $t \in \tilde{I}$, *then some child of* P *is also eventually* $\tilde{I}$-Q^**-critical.*

Proof. — We assume by contradiction that none of the children is eventually $\tilde{I}$-Q^*-critical.

By Lemma 6 and Corollary 3 (in Section 5.6.5), P is $\tilde{I}$-critical and thus $\tilde{I}$-decomposable. If all children of P were eventually $\tilde{I}$-transverse to Q^*, we would put together the corresponding $\tilde{I}$-decompositions and obtain that P is eventually $\tilde{I}$-transverse to Q^*. If all children of P were $\tilde{I}$-separated from Q^*, P would be $\tilde{I}$-separated from Q^* Therefore some child of P is eventually $\tilde{I}$-transverse to Q^*, and some other child of P is $\tilde{I}$-separated from Q^*.

We will show that Q^* is $\tilde{I}$-transverse. We will construct an $\tilde{I}$-decomposition $(P_\alpha, Q_\alpha, n_\alpha)$ of P_s such that every P_α is either $\tilde{I}$-separated from Q^* or $\tilde{I}$-transverse to Q^*.

Actually, it is sufficient to have an $\tilde{I}$-decomposition such that every P_α is either $\tilde{I}$-separated from Q^* or eventually $\tilde{I}$-transverse to Q^*.

Starting from the trivial decomposition of P_s, we have at step i an $\tilde{I}$-decomposition $(P_\alpha^{(i)}, Q_\alpha^{(i)}, n_\alpha^{(i)})$. As long as there is one $(P_\alpha^{(i)}, Q_\alpha^{(i)}, n_\alpha^{(i)})$ with $P \subset P_\alpha^{(i)}$, we observe that $P_\alpha^{(i)}$ is $\tilde{I}$-critical and therefore $\tilde{I}$-decomposable and break it into its children to go to step $i + 1$.

After a finite number of steps, each $P^{(i)}_\alpha$ is either a child of P or disjoint from P.

The $P^{(i)}_\alpha$ which are children of P are either eventually $\tilde{I}$-transverse to Q^* or $\tilde{I}$-separated from Q^* by assumption.

The $P^{(i)}_\alpha$ which are disjoint from P may sit on one or the other side of P. On one side, we apply the first part of Proposition 29 to Q^*, $P^{(i)}_\alpha$ and a child of P which is $\tilde{I}$-separated from Q^* to conclude that $P^{(i)}_\alpha$ is $\tilde{I}$-separated from Q^*.

We claim that those on the other side are eventually $\tilde{I}$-transverse to Q^*. Indeed, let $P^{(i)}_\alpha$ be such a rectangle. If it is $\tilde{I}$-transverse to Q^*, we are done. Assume this is not the case. Let $\overline{P}^{(i)}_\alpha$ be a simple descendant of $P^{(i)}_\alpha$ such that $|\overline{P}^{(i)}_\alpha|^{1-\eta} < |\tilde{I}|$ for some $t \in \tilde{I}$. By Proposition 29 (part 2) applied to $\tilde{I}, Q^*, P_0, \overline{P}^{(i)}_\alpha$, we have $Q^* \pitchfork_{\tilde{I}} \overline{P}^{(i)}_\alpha$. As $P^{(i)}_\alpha$ is not $\tilde{I}$-transverse to Q^*, it must be $\tilde{I}$-critical (by Proposition 20 in Section 8.1), hence $\tilde{I}$-decomposable. We replace $P^{(i)}_\alpha$ by its children and repeat the argument till the rectangles are thin enough to apply directly Proposition 29 (part 2) to $\tilde{I}, Q^*, P_0, P^{(i)}_\alpha$.

This proves our claim and the proof of the lemma is complete. □

Lemma **8.** — *If* P_0, P_1 *are eventually* $\tilde{I}$-Q^**-critical and disjoint, then we have* $|P_0| \leqslant C|\tilde{I}|$, $|P_1| \leqslant C|\tilde{I}|$ *for all* $t \in \tilde{I}$.

Proof. — Exchanging P_0, P_1 if necessary, we can assume that (8.59) holds for Q^*, P_0, P_1.

From Lemma 7, we can find $(\widehat{P}_0, \widehat{Q}_0, \widehat{n}_0)$, $(\widehat{P}_1, \widehat{Q}_1, \widehat{n}_1)$ in $\mathcal{R}(\tilde{I})$ with $\widehat{P}_0 \subset P_0$, $\widehat{P}_1 \subset P_1$, such that both $\widehat{P}_0, \widehat{P}_1$ are eventually $\tilde{I}$-Q^*-critical and we have

$$(\mathbf{8.64}) \qquad |\widehat{P}_0| < |\tilde{I}|^\beta, \quad |\widehat{P}_1| < |\tilde{I}|^\beta \quad \text{for all } t \in \tilde{I}.$$

As $\widehat{P}_0$ is not $\tilde{I}$-separated from Q^*, we must have

$$(\mathbf{8.65}) \qquad \delta_{LR}(Q^*, \widehat{P}_0) \geqslant 0$$

for some $t_0 \in \tilde{I}$.

From (8.54) and (8.64), as $Q^* \overline{\pitchfork}_{\tilde{I}} \widehat{P}_1$ does not hold, we must have

$$(\mathbf{8.66}) \qquad \delta_{LR}(Q^*, \widehat{P}_1) < 2|\tilde{I}|$$

for some $t_1 \in I$.

Observe that Q^* is $\tilde{I}$-critical by assumption and that P_0, P_1 are $\tilde{I}$-critical by Lemma 6. From Corollary 8 in Section 7.6, it follows that, for $i = 0, 1$

$$(\mathbf{8.67}) \qquad \max_{\tilde{I}} \delta_{LR}(Q^*, P_i) - \min_{\tilde{I}} \delta_{LR}(Q^*, P_i) \leqslant 2|\tilde{I}|,$$

$$(\mathbf{8.68}) \qquad \max_{\tilde{I}} \delta_{L}(Q^*, P_i) - \min_{\tilde{I}} \delta_{L}(Q^*, P_i) \leqslant 2|\tilde{I}|.$$

From Proposition 18 in Section 7.7, for $i = 0, 1$, we have either $|\mathrm{P}_i| < |\widetilde{\mathrm{I}}|$ for all $t \in \widetilde{\mathrm{I}}$ (the required conclusion), or

$$\max_{\widetilde{\mathrm{I}}} |\mathrm{P}_i| \leqslant \min_{\widetilde{\mathrm{I}}} |\mathrm{P}_i|. \tag{8.69}$$

Recall that, by (8.55)–(8.58), we have, for $i = 0, 1$, if $\mathrm{P}_i \neq \widehat{\mathrm{P}}_i$

$$\delta_{\mathrm{LR}}(\mathrm{Q}^*, \mathrm{P}_i) > \delta_{\mathrm{LR}}(\mathrm{Q}^*, \widehat{\mathrm{P}}_i) + \mathrm{C}^{-1}|\mathrm{P}_i|, \tag{8.70}$$

$$\delta_{\mathrm{L}}(\mathrm{Q}^*, \mathrm{P}_i) < \delta_{\mathrm{L}}(\mathrm{Q}^*, \widehat{\mathrm{P}}_i) - \mathrm{C}^{-1}|\mathrm{P}_i| \tag{8.71}$$

for any $t \in \widetilde{\mathrm{I}}$.

If $\mathrm{P}_0 = \widehat{\mathrm{P}}_0$, we have $|\mathrm{P}_0| < |\widetilde{\mathrm{I}}|$ by (8.64). If $\mathrm{P}_0 \neq \widehat{\mathrm{P}}_0$, we have

$$\begin{aligned}
\mathrm{C}^{-1}|\mathrm{P}_0| &\leqslant \mathrm{C}^{-1}|\mathrm{P}_0| + \delta_{\mathrm{LR}}(\mathrm{Q}^*, \widehat{\mathrm{P}}_0) \quad \text{at } t_0, \text{ from (8.63)}\\
&\leqslant \delta_{\mathrm{LR}}(\mathrm{Q}^*, \mathrm{P}_0) \quad \text{at } t_0, \text{ from (8.68)}\\
&\leqslant \delta_{\mathrm{L}}(\mathrm{Q}^*, \mathrm{P}_1) \quad \text{at } t_0\\
&\leqslant \delta_{\mathrm{L}}(\mathrm{Q}^*, \mathrm{P}_1) + \mathrm{C}|\widetilde{\mathrm{I}}| \quad \text{at } t_1, \text{ from (8.65)}\\
&\leqslant \delta_{\mathrm{LR}}(\mathrm{Q}^*, \widehat{\mathrm{P}}_1) + \mathrm{C}|\widetilde{\mathrm{I}}| \quad \text{at } t_1\\
&\leqslant \mathrm{C}|\widetilde{\mathrm{I}}| \quad \text{from (8.64).}
\end{aligned}$$

If $\mathrm{P}_1 = \widehat{\mathrm{P}}_1$, we have $|\mathrm{P}_1| < |\widetilde{\mathrm{I}}|$ by (8.64). If $\mathrm{P}_1 \neq \widehat{\mathrm{P}}_1$, we have

$$\begin{aligned}
2|\widetilde{\mathrm{I}}| - \mathrm{C}^{-1}|\mathrm{P}_1| &\geqslant \delta_{\mathrm{LR}}(\mathrm{Q}^*, \widehat{\mathrm{P}}_1) - \mathrm{C}^{-1}|\mathrm{P}_1| \quad \text{at } t_1, \text{ from (8.64)}\\
&\geqslant \delta_{\mathrm{L}}(\mathrm{Q}^*, \widehat{\mathrm{P}}_1) - \mathrm{C}^{-1}|\mathrm{P}_1| \quad \text{at } t_1\\
&\geqslant \delta_{\mathrm{L}}(\mathrm{Q}^*, \mathrm{P}_1) \quad \text{at } t_1, \text{ from (8.69)}\\
&\geqslant \delta_{\mathrm{L}}(\mathrm{Q}^*, \mathrm{P}_1) - \mathrm{C}|\widetilde{\mathrm{I}}| \quad \text{at } t_0, \text{ from (8.65)}\\
&\geqslant \delta_{\mathrm{LR}}(\mathrm{Q}^*, \widehat{\mathrm{P}}_0) - \mathrm{C}|\widetilde{\mathrm{I}}| \quad \text{at } t_0\\
&\geqslant -\mathrm{C}|\widetilde{\mathrm{I}}| \quad \text{from (8.63).}
\end{aligned}$$

We have proven the required estimates. □

8.4.4. Consider the set Π of elements $(\mathrm{P}, \mathrm{Q}, n) \in \mathcal{R}(\widetilde{\mathrm{I}})$ which are eventually $\widetilde{\mathrm{I}}$-Q^*-critical, satisfy

$$|\mathrm{P}| \leqslant |\widetilde{\mathrm{I}}|^{1+\tau} \tag{8.72}$$

for all $t \in \widetilde{\mathrm{I}}$ and are maximal (in P) with respect to these two properties.

Lemma **9.** — *We have*

$$\#\Pi \leqslant |\widetilde{I}|^{-\tau d_s^+},$$

where $d_s^+ = d_s^0 + C\eta\tau^{-1}$ *is as in the statement of Proposition* 28.

Proof. — Assume that Π is non-empty. From Lemma 8, there exists $C_0 > 0$ and a unique element $(P_0, Q_0, n_0) \in \mathcal{R}(\widetilde{I})$ with the following properties:

- $P \subset P_0$ for all $(P, Q, n) \in \Pi$
- $|P_0| > C_0|\widetilde{I}|$ for some $t \in \widetilde{I}$
- every child P_1 of P_0 which contains a rectangle P with $(P, Q, n) \in \Pi$ satisfies $|P_1| \leqslant C_0|\widetilde{I}|$ for all $t \in \widetilde{I}$.

As P_0 is eventually $\widetilde{I}$-Q^*-critical, P_0 is $\widetilde{I}$-critical by Lemma 6.

There are two kind of elements $(P, Q, n) \in \Pi$:

- those such that P is a child of P_0; the number of such elements is at most $|\widetilde{I}|^{-c\eta}$ by Proposition 25;
- those such that the parent $\widetilde{P}$ of P is contained in some child P_1 of P_0.

In this last case, from the definition of Π we have

$$(\mathbf{8.73}) \qquad |\widetilde{P}| > |\widetilde{I}|^{1+\tau} \quad \text{for some } t \in \widetilde{I}.$$

As P is $\widetilde{I}$-critical by Lemma 6, $\widetilde{P}$ is also $\widetilde{I}$-critical, hence $\widetilde{I}$-special and we have, from Proposition 18 in Section 7.7, that

$$(\mathbf{8.74}) \qquad |\widetilde{P}| > C^{-1}|\widetilde{I}|^{1+\tau} \quad \text{for all } t \in \widetilde{I}.$$

Let P_1 be a child of P_0; we have $|P_1| \leqslant C_0|\widetilde{I}|$ for all $t \in \widetilde{I}$ by the definition of P_0. For given P_1, the number of possible $\widetilde{P}$ is therefore bounded by $C|\widetilde{I}|^{-\tau d_s^*}$ from Corollary 13, with $d_s^* = d_s^0 + \varepsilon_0^{\frac{1}{3}d_s^0}$.

The number of P for given $\widetilde{P}$, and the number of children P_1 of P_0, are both bounded from Proposition 25 by $|\widetilde{I}|^{-c\eta}$. Summing up, the total number of elements of Π is bounded by

$$|\widetilde{I}|^{-c\eta} + |\widetilde{I}|^{-2c\eta}(C|\widetilde{I}|^{\tau})^{-d_s^*},$$

in accordance with the statement of Lemma 9, choosing appropriately the constant in the definition of d_s^+. □

8.4.5. *Proof of Proposition 28.* — By Lemma 6 and Corollary 3 (in Section 5.6.5), if $(P, Q, n) \in \mathcal{R}(\widetilde{I})$ is such that P is eventually $\widetilde{I}$-Q^*-critical, and $|P| \geqslant |\widetilde{I}|^\beta$ for some $t \in \widetilde{I}$, then P is $\widetilde{I}$-decomposable.

Therefore, there exists an $\widetilde{I}$-decomposition $(P_\alpha, Q_\alpha, n_\alpha)$ of P_s such that every $(P_\alpha, Q_\alpha, n_\alpha)$ is either eventually $\widetilde{I}$-separated from Q^* or eventually $\widetilde{I}$-transverse to Q^* or an element of Π.

Let $I \subset \widetilde{I}$ be a candidate interval of the next level, i.e. $|I| = |\widetilde{I}|^{1+\tau}$, such that Q^* is I-critical.

We claim that there exists $(P, Q, n) \in \Pi$ such that P is eventually I-Q^*-critical.

Indeed, every $(P_\alpha, Q_\alpha, n_\alpha)$ which is eventually $\widetilde{I}$-transverse to Q^* (resp. eventually $\widetilde{I}$-separated from Q^*) is a fortiori I-transverse to Q^* (resp. I-separated from Q^*). If every $(P_\alpha, Q_\alpha, n_\alpha) \in \Pi$ was also either eventually I-transverse to Q^* or eventually I-separated from Q^*, we would obtain a decomposition of P_s which expresses that Q^* is I-transverse. This proves the claim.

On the other hand, fix $(P, Q, n) \in \Pi$. We show that there are at most C_1 candidates $I \subset \widetilde{I}$ such that P is eventually I-Q^*-critical. Together with Lemma 9, this will imply the statement of Proposition 28 (after modifying the value of the constant in the definition of d_s^+).

Both P (by Lemma 6) and Q^* (by assumption) are $\widetilde{I}$-critical, hence $\widetilde{I}$-special. By Corollary 8 in Section 7.6, we have, for all $t \in \widetilde{I}$

$$\left| \frac{d}{dt} \delta_{LR}(Q^*, P) - 1 \right| \leqslant C\varepsilon_0^{\frac{1}{2}}. \tag{8.75}$$

If, for all $t \in I$, we have

$$\delta_{LR}(Q^*, P) < 0, \tag{8.76}$$

then P is I-separated from Q^*.

We claim that if, for all $t \in I$, we have

$$\delta(Q^*, P) > 2|I|, \tag{8.77}$$

then P is eventually I-transverse to Q^*. As we have $|P| \leqslant |I|$ from the definition of Π and $|Q| < |I|$ from the assumption of the proposition, we have, for all $t \in \widetilde{I}$

$$\delta_{LR}(Q^*, P) < \delta(Q^*, P) + C(|P| + |Q^*|) < \delta(Q^*, P) + C|I|, \tag{8.78}$$

which allows to conclude.

Finally, we prove the claim.

Observe first that, if $(P', Q', n') \in \mathcal{R}(I)$ satisfies $P' \subset P$, $|P'| \leqslant \frac{1}{2}|\widetilde{I}|^{(1+\tau)(1-\eta)^{-1}}$ for all $t \in \widetilde{I}$, then $Q^* \overline{\pitchfork}_I P'$ holds. Indeed, $\delta_{LR}(Q^*, P')$, $\delta_L(Q^*, P')$, $\delta_R(Q^*, P')$ are all larger than $\delta(Q^*, P)$; then (T3) follows from (8.75) above; as both $|Q^*|^{1-\eta}$, $|P|^{1-\eta}$ are smaller than $|I|$, (T1) and (T2) also follow from (8.75).

On the other hand, if $(P', Q', n') \in \mathcal{R}(I)$ satisfies $P' \subset P$ and P' is not I-transverse to Q^*, from the observation (applied to a thin simple descendant of P′) and Proposition 20, we deduce that P′ is I-critical hence I-decomposable, and we can replace it by its children. Starting with P, we iterate the process till getting rectangles thin enough for the observation to apply. This proves the claim and thus also Proposition 28.

9. Strong regularity and parameter selection

As it was mentioned in Section 4.5, regularity is a rather qualitative property which is not appropriate for the quantitative estimates needed for parameter selection. We will introduce in this section, a stronger quantitative property, that we call *strong regularity*.

Think of each bicritical element as a return from the "critical region" to itself. We want to control the number of bicritical elements of a given width (including, of course, that there are no "fat" bicritical elements). In order to do this, we also need a control on the size of the critical region itself.

In the whole of Section 9, we fix a parameter interval $\widetilde{I}$ which is assumed by induction to be strongly regular (the definition is given at the end of Section 9.2). In Section 9.3, we check that strong regularity implies β-regularity for an appropriate $\beta > 1$. Therefore the properties proven in Sections 5–8 will be applicable. The aim of Section 9 is to estimate how many candidates $I \subset \widetilde{I}$ fail to be strongly regular.

In Section 9.1, the estimates (SR1), (SR2) on the size of the critical region are presented. In Section 9.2, the estimates (SR3) on the number of bicritical elements are introduced. This leads to the definition of strong regularity at the end of this Subsection, namely that the full set of estimates (SR1), (SR2), (SR3) must be satisfied, and to the choice of β in Section 9.3, where condition (H4) of Section 1.2 is finally explained. It is also proven in Section 9.3 that strong regularity implies β-regularity. In Section 9.4, we check that the starting interval I_0 is strongly regular. Then, in Section 9.5, we estimate how many candidates $I \subset \widetilde{I}$ fail to satisfy (SR1). Condition (SR2) is more delicate and is dealt with in Section 9.6.

In the rest of Sections 9, 9.7–9.13, we estimate the number of candidates which fail to satisfy (SR3); a general overview of the strategy to do this is first presented in Section 9.7. The conclusions of the process of parameter selection are presented in Section 9.14.

9.1. *Partitions and size of the critical locus.*

Definition **3.** — *We denote by* $\mathcal{C}_+(I)$ *the set of* (P, Q, n) *in* $\mathcal{R}(I)$ *such that* P *is* I*-critical,* $|P| \leqslant |I|^{1+\tau}$ *for all* $t \in I$, *and* P *is maximal with this property: the parent* $\widetilde{P}$ *of* P *satisfies* $|\widetilde{P}| > |I|^{1+\tau}$ *for some* $t \in I$.

Obviously, if (P, Q, n), (P', Q', n') are distinct elements in $\mathcal{C}_+(I)$, P and P′ are disjoint. Moreover, if $(\widehat{P}, \widehat{Q}, \widehat{n})$ belongs to $\mathcal{R}(I)$, $\widehat{P}$ is I-critical, and $|\widehat{P}| \leqslant |I|^{1+\tau}$ for all $t \in I$, there is a unique $(P, Q, n) \in \mathcal{C}_+(I)$ such that $\widehat{P} \subset P$.

Exchanging P's and Q's, we define $\mathcal{C}_-(\mathrm{I})$ in a similar way. The sets $\mathcal{C}_+(\mathrm{I})$, $\mathcal{C}_-(\mathrm{I})$ correspond to the I-critical locus at the $|\mathrm{I}|^{1+\tau}$ scale.

We will need in the sequel to consider $\widehat{\mathrm{I}}$-criticality (for some parameter interval $\widehat{\mathrm{I}} \supset \mathrm{I}$) for rectangles in $\mathcal{R}(\mathrm{I})$ but not in $\mathcal{R}(\widehat{\mathrm{I}})$.

Definition **4.** — *Let* $\widehat{\mathrm{I}} \supset \mathrm{I}$ *be parameter intervals, and let* $(\mathrm{P}, \mathrm{Q}, n) \in \mathcal{R}(\mathrm{I})$. *We say that* P *is thin* $\widehat{\mathrm{I}}$*-critical if there exists* $(\widehat{\mathrm{P}}, \widehat{\mathrm{Q}}, \widehat{n}) \in \mathcal{R}(\widehat{\mathrm{I}})$ *with* $\mathrm{P} \subset \widehat{\mathrm{P}}$, $\widehat{\mathrm{P}}$ *is* $\widehat{\mathrm{I}}$*-critical and*

$$\min_{\widehat{\mathrm{I}}} |\widehat{\mathrm{P}}|^{1-\eta} \leqslant 2|\widehat{\mathrm{I}}|.$$

The notion is useful in connection in Proposition 22 of Section 8.1, as in the following

Lemma **10.** — *Let* I *be a candidate parameter interval (with regular parent* $\widetilde{\mathrm{I}}$*), and let* $(\mathrm{P}, \mathrm{Q}, n)$ *be an element of* $\mathcal{R}(\mathrm{I})$ *which is not the restriction of an element of* $\mathcal{R}(\widetilde{\mathrm{I}})$. *Let* $k > 0$ *and* $(\mathrm{P}_i, \mathrm{Q}_i, n_i)$, *for* $0 \leqslant i \leqslant k$, *be the elements of* $\mathcal{R}(\widetilde{\mathrm{I}})$ *given by the structure theorem (of Section* 6.5*). Then* Q_i *and* P_{i+1} *are thin* $\widetilde{\mathrm{I}}$*-critical for* $0 \leqslant i < k$.

Proof. — As Q_i is I-transverse but not $\widetilde{\mathrm{I}}$-transverse to P_{i+1}, this follows from Proposition 22 in Section 8.1. □

Definition **5.** — *Let* $\widetilde{\mathrm{I}}$ *be a regular interval,* $(\mathrm{P}, \mathrm{Q}, n) \in \mathcal{C}_+(\widetilde{\mathrm{I}})$. *We denote by* $\mathcal{C}r(\mathrm{P})$ *the set of candidates* $\mathrm{I} \subset \widetilde{\mathrm{I}}$ *such that* P *contains a thin* I*-critical rectangle. We define symmetrically* $\mathcal{C}r(\mathrm{Q})$ *for* $(\mathrm{P}, \mathrm{Q}, n) \in \mathcal{C}_-(\widetilde{\mathrm{I}})$.

Lemma **11.** — *Let* $\widehat{\mathrm{I}}$ *be the largest parameter interval containing* $\widetilde{\mathrm{I}}$ *with* $|\widehat{\mathrm{I}}|^{\beta} \leqslant |\widetilde{\mathrm{I}}|^{(1+\tau)^2}$. *Let* $\mathrm{I} \subset \widetilde{\mathrm{I}}$ *be a candidate interval and let* $(\mathrm{P}, \mathrm{Q}, n) \in \mathcal{C}_+(\mathrm{I})$. *Then* $(\mathrm{P}, \mathrm{Q}, n)$ *belongs to* $\mathcal{R}(\widehat{\mathrm{I}})$, *and there exists* $(\widetilde{\mathrm{P}}, \widetilde{\mathrm{Q}}, \widetilde{n}) \in \mathcal{C}_+(\widetilde{\mathrm{I}})$ *with* $\mathrm{P} \subset \widetilde{\mathrm{P}}$ *and* $\mathrm{I} \in \mathcal{C}r(\widetilde{\mathrm{P}})$.

Proof. — We first prove that $(\mathrm{P}, \mathrm{Q}, n)$ belongs to $\mathcal{R}(\widehat{\mathrm{I}})$. Let $\mathrm{I}' \supset \mathrm{I}$ be a parameter interval distinct from I_0 with parent $\widetilde{\mathrm{I}}'$; assume that $(\mathrm{P}, \mathrm{Q}, n)$ belongs to $\mathcal{R}(\mathrm{I}')$ but not to $\mathcal{R}(\widetilde{\mathrm{I}}')$. By the structure theorem of Section 6.5, the element $(\mathrm{P}_0, \mathrm{Q}_0, n_0) \in \mathcal{R}(\widetilde{\mathrm{I}}')$ with P_0 smallest containing P is $\widetilde{\mathrm{I}}'$-bicritical and satisfy $|\mathrm{P}_0| < |\widetilde{\mathrm{I}}'|^{\beta}$ for all $t \in \widetilde{\mathrm{I}}'$. This contradicts the maximality of P with respect to the property $\max_{\mathrm{I}} |\mathrm{P}| \leqslant |\mathrm{I}|^{1+\tau}$ if $|\widetilde{\mathrm{I}}'|^{\beta} \leqslant |\mathrm{I}|^{1+\tau}$ and shows that $(\mathrm{P}, \mathrm{Q}, n)$ belongs to $\mathcal{R}(\widehat{\mathrm{I}})$.

In particular, $(\mathrm{P}, \mathrm{Q}, n)$ belongs to $\mathcal{R}(\widetilde{\mathrm{I}})$ and P is $\widetilde{\mathrm{I}}$-special. From Proposition 18 in Section 7.7, we have $|\mathrm{P}| < \mathrm{C}|\mathrm{I}|^{1+\tau} < |\widetilde{\mathrm{I}}|^{1+\tau}$ for all $t \in \widetilde{\mathrm{I}}$. Therefore, there exists $(\widetilde{\mathrm{P}}, \widetilde{\mathrm{Q}}, \widetilde{n}) \in \mathcal{C}_+(\widetilde{\mathrm{I}})$ with $\mathrm{P} \subset \widetilde{\mathrm{P}}$. Finally, we observe that P is thin I-critical and therefore I belongs to $\mathcal{C}r(\widetilde{\mathrm{P}})$. □

We will state several inequalities related to the size of the sets $\mathcal{C}_+(\mathrm{I})$, $\mathcal{C}_-(\mathrm{I})$. All these inequalities are part of the definition of strong regularity: they have to be satisfied by a strongly regular parameter interval. In Sections 9.5, 9.6, we will see that they are satisfied by most candidates in a strongly regular parameter interval $\widetilde{\mathrm{I}}$. These estimates are used in Sections 9.12, 9.13 in the control of the number of bicritical elements.

We will control the cardinalities of $\mathcal{C}_+(\mathrm{I})$, $\mathcal{C}_-(\mathrm{I})$, through:

$$(\mathrm{SR1})_s \qquad \#\mathcal{C}_+(\mathrm{I}) \leqslant \mathrm{C}\Big(\frac{|\mathrm{I}|}{\varepsilon_0}\Big)^{\sigma} \varepsilon_0^{-\tau d_s^0},$$

$$(\mathrm{SR1})_u \qquad \#\mathcal{C}_-(\mathrm{I}) \leqslant \mathrm{C}\Big(\frac{|\mathrm{I}|}{\varepsilon_0}\Big)^{\sigma} \varepsilon_0^{-\tau d_u^0}.$$

The exponent σ will be completely specified in Section 9.5. It is very close to $1 - d_s^0 - d_u^0$ when $\tau, \eta, \varepsilon_0$ are small.

We need also to control $\mathcal{C}_+(\mathrm{I})$, $\mathcal{C}_-(\mathrm{I})$ in another, more complicated, way.

$$(\mathrm{SR2})_s \qquad \sum_{(\mathrm{P}_\alpha, \mathrm{Q}_\alpha, n_\alpha)\in\mathcal{C}_+(\mathrm{I})} \max_{\mathrm{I}} |\mathrm{Q}_\alpha|^{\rho_u} \leqslant \mathrm{C}|\mathrm{Q}_s|^{\rho_u}\Big(\frac{|\mathrm{I}|}{\varepsilon_0}\Big)^{\sigma_u},$$

$$(\mathrm{SR2})_u \qquad \sum_{(\mathrm{P}_\omega, \mathrm{Q}_\omega, n_\omega)\in\mathcal{C}_-(\mathrm{I})} \max_{\mathrm{I}} |\mathrm{P}_\omega|^{\rho_s} \leqslant \mathrm{C}|\mathrm{P}_u|^{\rho_s}\Big(\frac{|\mathrm{I}|}{\varepsilon_0}\Big)^{\sigma_s}.$$

The exponents $\rho_s, \rho_u, \sigma_s, \sigma_u$ will be specified in Sections 9.6 and 9.14. When $\tau, \eta, \varepsilon_0$ are small, they are respectively close to d_s^0, d_u^0, $1 - d_s^0$, $1 - d_u^0$. We write $|\mathrm{P}_u|$, $|\mathrm{Q}_s|$ for the maximum of these quantities over I_0.

We actually need a stronger version of $(\mathrm{SR2})_s$, better suited for induction purposes. Let $\mathrm{I} \subset \widetilde{\mathrm{I}}$ be a candidate interval.

Definition **6.** — *We denote by* $\widehat{\mathcal{C}}_+(\mathrm{I})$ *to be the set of* $(\mathrm{P}, \mathrm{Q}, n) \in \mathcal{C}_+(\widetilde{\mathrm{I}})$ *such that* $\mathrm{I} \in \mathcal{C}r(\mathrm{P})$. *We define symmetrically* $\widehat{\mathcal{C}}_-(\mathrm{I})$. *We also define* $\widehat{\mathcal{C}}_+(\mathrm{I}_0) := \{(\mathrm{P}_s, \mathrm{Q}_s, n_s)\}$, $\widehat{\mathcal{C}}_-(\mathrm{I}_0) := \{(\mathrm{P}_u, \mathrm{Q}_u, n_u)\}$.

Let $((\mathrm{P}_i, \mathrm{Q}_i, n_i))_i$ *be a finite family of elements of* $\mathcal{R}(\mathrm{I})$ *with the* P_i *disjoint and each* P_i *contained in some* P *with* $(\mathrm{P}, \mathrm{Q}, n) \in \widehat{\mathcal{C}}_+(\mathrm{I})$. *We ask that, for any such family*

$$(\mathrm{SR2})'_s \qquad \sum_i \max_{\mathrm{I}} |\mathrm{Q}_i|^{\rho_u} \leqslant \mathrm{C}|\mathrm{Q}_s|^{\rho_u}\Big(\frac{|\mathrm{I}|}{\varepsilon_0}\Big)^{\sigma_u}.$$

We define symmetrically $(\mathrm{SR2})'_u$.

Observe that, by Lemma 11, $(\mathrm{SR2})_s$ is a consequence of $(\mathrm{SR2})'_s$.

The heuristics behind $(\mathrm{SR2})_s$ is the following: in the mean, one expects that elements of $\mathcal{R}(\mathrm{I})$ more or less satisfy

$$\textbf{(9.1)} \qquad |\mathrm{P}|^{d_s^0} \sim |\mathrm{Q}|^{d_u^0}$$

and, for $(\mathrm{P}, \mathrm{Q}, n) \in \widehat{\mathcal{C}}_+(\mathrm{I})$, one should have

$$\textbf{(9.2)} \qquad |\mathrm{P}| \sim |\mathrm{I}|$$

which explains the relation between $(\mathrm{SR1})_s$ and $(\mathrm{SR2})_s$.

9.2. *Classes and number of bicritical rectangles.* — Once the size of the critical locus is under control, we must pay attention to the number of bicritical rectangles, which represent the returns of the critical locus to itself under the dynamics.

In order to have an appropriate induction scheme, we need to bound the number of bicritical rectangles according to all width scales and also according to the level of criticality (i.e., the distance to critical locus) of both P and Q. As we will see in the next subsection, the number of bicritical elements experiments a "phase transition" which is crucial for our argument but brings a lot of complications.

Let I be a candidate interval as above, and let I_α, I_ω be parameter intervals such that $I \subset I_\alpha \cap I_\omega$. Let also x be a positive number.

Definition **7.** — *We denote by* $Bi_+(I, I_\alpha, I_\omega; x)$ *the set of elements* $(P, Q, n) \in \mathcal{R}(I)$ *such that* P *is thin* I_α*-critical,* Q *is thin* I_ω*-critical and* $|P| \geqslant x$ *for some* $t \in I$.

Similarly, $Bi_-(I, I_\alpha, I_\omega; x)$ *is the set of elements* $(P, Q, n) \in \mathcal{R}(I)$ *such that* P *is thin* I_α*-critical,* Q *is thin* I_ω*-critical and* $|Q| \geqslant x$ *for some* $t \in I$.

At this point, we have to break the symmetry between past and future, P's and Q's, stable and unstable directions: the estimates are indeed not symmetric, except when $d_s^0 = d_u^0$, i.e., in the conservative case of area-preserving diffeomorphisms.

We will assume that $d_s^0 \geqslant d_u^0$ (and $d_s^0 + d_u^0 \geqslant 1$). The case $d_u^0 \geqslant d_s^0$ is obviously symmetric.

For I, I_α, I_ω, x as above we want to have, for a fixed large enough constant A (we will see in Section 9.14 that A = 3 is a convenient choice)

$$(SR3)_s \qquad \#Bi_+(I, I_\alpha, I_\omega; x) \leqslant \varepsilon_0^{-A\tau} B,$$

with

$$\textbf{(9.3)} \qquad B = \max(B_0, B_1),$$

$$\textbf{(9.4)} \qquad B_0 = \left(\frac{x}{\varepsilon_0 |P_u|}\right)^{-\rho_0} \left(\frac{|I_\alpha|}{\varepsilon_0}\right)^{\sigma_0+\sigma_1} \left(\frac{|I_\omega|}{\varepsilon_0}\right)^{\sigma_0},$$

$$\textbf{(9.5)} \qquad B_1 = \left(\frac{x}{\varepsilon_0 |P_u|}\right)^{-\rho_1} \left(\frac{|I_\alpha|}{\varepsilon_0}\right)^{\sigma_1} \left(\min\left(\frac{|I_\alpha|}{\varepsilon_0}, \frac{|I_\omega|}{\varepsilon_0}\right)\right)^{\sigma_0}.$$

Here $|P_u|$ denotes the supremum over I_0 of the width of P_u: the exponents ρ_0, ρ_1, σ_0, σ_1 will be specified more precisely later, but anyway they satisfy

$$\textbf{(9.6)} \qquad \rho_0 = d_s^0 + o(1),$$

$$\textbf{(9.7)} \qquad \rho_1 = \frac{d_s^0}{d_s^0 + d_u^0}(2d_s^0 + d_u^0 - 1) + o(1),$$

(9.8) $$\sigma_0 = 1 - d_s^0 + o(1),$$

(9.9) $$\sigma_1 = d_s^0 - d_u^0 + o(1).$$

The meaning of the $o(1)$ terms in these formulas is that they become arbitrarily small when $\tau \gg \eta \gg \varepsilon_0$ are small enough.

For the $\mathcal{B}i_-$ sets, we should have:

$(\text{SR3})_u$ $$\#\mathcal{B}i_-(\mathrm{I}, \mathrm{I}_\alpha, \mathrm{I}_\omega; x) \leqslant \varepsilon_0^{-\mathrm{A}\tau}\mathrm{B}',$$

with

(9.10) $$\mathrm{B}' = \max(\mathrm{B}'_0, \mathrm{B}'_1),$$

(9.11) $$\mathrm{B}'_0 = \Big(\frac{x}{\varepsilon_0|\mathrm{Q}_s|}\Big)^{-\rho'_0}\Big(\frac{|\mathrm{I}_\alpha|}{\varepsilon_0}\Big)^{\sigma_0+\sigma_1}\Big(\frac{|\mathrm{I}_\omega|}{\varepsilon_0}\Big)^{\sigma_0},$$

(9.12) $$\mathrm{B}'_1 = \Big(\frac{x}{\varepsilon_0|\mathrm{Q}_s|}\Big)^{-\rho'_1}\Big(\frac{|\mathrm{I}_\alpha|}{\varepsilon_0}\Big)^{\sigma_1}\Big(\min\Big(\frac{|\mathrm{I}_\alpha|}{\varepsilon_0}, \frac{|\mathrm{I}_\omega|}{\varepsilon_0}\Big)\Big)^{\sigma_0},$$

(9.13) $$\rho'_0 = \frac{d_u^0}{d_s^0}\rho_0 = d_u^0 + o(1),$$

(9.14) $$\rho'_1 = \frac{d_u^0}{d_s^0}\rho_1 = \frac{d_u^0}{d_s^0 + d_u^0}(2d_s^0 + d_u^0 - 1) + o(1).$$

Observe that the formulas (9.7), (9.14) for ρ_1, ρ'_1 are *not* symmetric.

Definition **8.** — *A parameter interval* I *is strongly regular if its parent is (when* $\mathrm{I} \neq \mathrm{I}_0$*) and if it satisfies the conditions* (SR1), (SR2)′ *(hence also* (SR2)*) of Section* 9.1 *and* $(\text{SR3})_s$, $(\text{SR3})_u$ *for all* $\mathrm{I}_\alpha \supset \mathrm{I}$, $\mathrm{I}_\omega \supset \mathrm{I}$, $0 < x < 1$.

Remark **10.**

1. At this point, the definition of strong regularity is not complete because the exponents ρ_0, ρ_1, ρ'_0, ρ'_1, σ_0, σ_1, A have not been completely specified. These exponents should be viewed for the present time as parameters constrained by (9.6)–(9.9) and (9.13), (9.14).
2. The inequalities $(\text{SR3})_s$, $(\text{SR3})_u$ form a family parametrized not only by I, but also by the parameter intervals $\mathrm{I}_\alpha \supset \mathrm{I}$ and $\mathrm{I}_\omega \supset \mathrm{I}$ and the real number $1 > x > 0$. Because each inequality, at least when $\mathrm{I} = \mathrm{I}_\alpha$ or $\mathrm{I} = \mathrm{I}_\omega$, is only obtained after parameter selection, we will discretize the continuous variable x by considering only the values $x = 2^{-l}$, $l \geqslant 0$. There is still an infinite number of inequalities, but we will see that they are trivially satisfied if l is large enough.

9.3. *Phase transition and the choice of β.* — We comment on the estimate $(SR3)_s$. First, observe that B does not depend on I. The reason will appear in Section 9.8 when we show that most elements in $Bi_+(I, I_\alpha, I_\omega; x)$ belong actually to $\mathcal{R}(\widetilde{I})$.

From the formulas (9.6), (9.7), we have

$$\rho_1 < \rho_0. \tag{9.15}$$

Set

$$x_{cr} := \varepsilon_0 |P_u| \Big(\max\Big(\frac{|I_\alpha|}{\varepsilon_0}, \frac{|I_\omega|}{\varepsilon_0} \Big) \Big)^{\frac{\sigma_0}{\rho_0 - \rho_1}}. \tag{9.16}$$

Then, we have $B = B_0$ for $x \leqslant x_{cr}$ and $B = B_1$ for $x \geqslant x_{cr}$: this is the "phase transition" mentioned earlier. Roughly speaking, the reason for this phase transition is that, when $I = I_\alpha = I_\omega$, we are able to eliminate, in the scale transition from $\widetilde{I}$ to I, more bicritical elements of small width ($x < x_{cr}$) than of large width ($x > x_{cr}$).

We have

$$\rho_0 - \rho_1 = \frac{d_s^0(1 - d_s^0)}{d_s^0 + d_u^0} + o(1), \tag{9.17}$$

$$\frac{\sigma_0}{\rho_0 - \rho_1} = \frac{d_s^0 + d_u^0}{d_s^0} + o(1) > 1. \tag{9.18}$$

For $x = x_{cr}$, we have

$$B = B_{cr} := \Big(\frac{|I_\alpha|}{\varepsilon_0}\Big)^{\sigma_0 + \sigma_1} \Big(\frac{|I_\omega|}{\varepsilon_0}\Big)^{\sigma_0} \Big(\max\Big(\frac{|I_\alpha|}{\varepsilon_0}, \frac{|I_\omega|}{\varepsilon_0} \Big) \Big)^{-\frac{\rho_0 \sigma_0}{\rho_0 - \rho_1}}. \tag{9.19}$$

Assume $I_\alpha = I_\omega$; we then have

$$B_{cr} = \Big(\frac{|I_\alpha|}{\varepsilon_0}\Big)^{\sigma_1 + \sigma_0 \frac{\rho_0 - 2\rho_1}{\rho_0 - \rho_1}}. \tag{9.20}$$

Here, the exponent satisfies

$$\sigma_1 + \sigma_0 \frac{\rho_0 - 2\rho_1}{\rho_0 - \rho_1} = 2 - 2d_s^0 - 2d_u^0 + o(1) < 0. \tag{9.21}$$

As $|I_\alpha| \leqslant \varepsilon_0$, we have $B_{cr} \geqslant 1$. As B is a decreasing function of x, we have $\varepsilon_0^{-A\tau} B < 1$ (in which case $(SR3)_s$ means that the Bi_+ set is empty!) iff $B_1 < \varepsilon_0^{A\tau}$ which corresponds to

$$x > \bar{x} := \varepsilon_0^{1 + A\tau/\rho_1} |P_u| \Big(\frac{|I_\alpha|}{\varepsilon_0}\Big)^{\frac{\sigma_0 + \sigma_1}{\rho_1}}. \tag{9.22}$$

The exponent here satisfies

$$\frac{\sigma_0+\sigma_1}{\rho_1} = \frac{1-d_u^0}{d_s^0}\frac{d_s^0+d_u^0}{2d_s^0+d_u^0-1} + o(1). \qquad \textbf{(9.23)}$$

We are finally able to justify the assumption (H4) of our Main Theorem stated in Section 1.2! Indeed, with $d_s^0 \geqslant d_u^0$, it means that

$$2(d_s^0)^2 + (d_u^0)^2 + 2d_s^0 d_u^0 < 2d_s^0 + d_u^0 \qquad \textbf{(H4)}$$

and this is exactly what is needed to guarantee that

$$\frac{\sigma_0+\sigma_1}{\rho_1} > 1. \qquad \textbf{(9.24)}$$

The discussion for $(\mathrm{SR3})_u$ is similar; the critical threshold is

$$x'_{cr} := \varepsilon_0|\mathrm{Q}_s|\Big(\max\Big(\frac{|\mathrm{I}_\alpha|}{\varepsilon_0}, \frac{|\mathrm{I}_\omega|}{\varepsilon_0}\Big)\Big)^{\frac{\sigma_0}{\rho'_0-\rho'_1}}, \qquad \textbf{(9.25)}$$

with

$$\rho'_0 - \rho'_1 = \frac{d_u^0(1-d_s^0)}{d_s^0+d_u^0} + o(1) = \frac{d_u^0}{d_s^0}(\rho_0-\rho_1), \qquad \textbf{(9.26)}$$

$$\frac{\sigma_0}{\rho'_0-\rho'_1} = \frac{\sigma_0}{\rho_0-\rho_1}\frac{d_s^0}{d_u^0} = \frac{d_s^0+d_u^0}{d_u^0} + o(1) > 1. \qquad \textbf{(9.27)}$$

When $\mathrm{I}_\alpha = \mathrm{I}_\omega$, we have

$$\mathrm{B}'_{cr} := \Big(\frac{|\mathrm{I}_\alpha|}{\varepsilon_0}\Big)^{\sigma_1+\sigma_0\frac{\rho'_0-2\rho'_1}{\rho'_0-\rho'_1}} = \mathrm{B}_{cr} \geqslant 1. \qquad \textbf{(9.28)}$$

Thus, we have $\varepsilon_0^{-\mathrm{A}\tau}\mathrm{B}' < 1$ iff

$$x > \overline{x}' := \varepsilon_0^{1+\mathrm{A}\tau/\rho'_1}|\mathrm{Q}_s|\Big(\frac{|\mathrm{I}_\alpha|}{\varepsilon_0}\Big)^{\frac{\sigma_0+\sigma_1}{\rho'_1}}. \qquad \textbf{(9.29)}$$

We have here

$$\frac{\sigma_0+\sigma_1}{\rho'_1} = \frac{\sigma_0+\sigma_1}{\rho_1}\frac{d_s^0}{d_u^0} \geqslant \frac{\sigma_0+\sigma_1}{\rho_1}. \qquad \textbf{(9.30)}$$

Choice of β. — We will choose the constant β (related to the regularity property) in order to have

$$(\mathbf{9.31}) \qquad 1 < \beta < \frac{1-d_u^0}{d_s^0}\,\frac{d_s^0+d_u^0}{2d_s^0+d_u^0-1}$$

and also, from Proposition 4 in Section 5.6.4

$$(\mathbf{9.32}) \qquad \beta < 1+\min(\omega_s,\omega_u).$$

From (9.31), we will have

$$(\mathbf{9.33}) \qquad \beta < \frac{\sigma_0+\sigma_1}{\rho_1} \leqslant \frac{\sigma_0+\sigma_1}{\rho_1'}.$$

Then, in (9.22), (9.29), we will have

$$(\mathbf{9.34}) \qquad \overline{x} < |I_\alpha|^\beta, \qquad \overline{x}' < |I_\alpha|^\beta.$$

Proposition **30.** — *If a candidate interval satisfies* $(SR3)_s$ *and* $(SR3)_u$, *then it is* β*-regular. In particular, strong regularity implies* β*-regularity.*

Proof. — We argue by induction on the level of the parameter interval. For the starting interval I_0, we already know from Proposition 4 that it is β-regular (independently of $(SR3)_s$, $(SR3)_u$). Assume that $I \neq I_0$ satisfies $(SR3)_s$, $(SR3)_u$ and that $(P, Q, n) \in \mathcal{R}(I)$ is I-bicritical. Assume also, for instance, that

$$(\mathbf{9.35}) \qquad \max_I |Q| \leqslant \max_I |P|$$

and, by contradiction that

$$(\mathbf{9.36}) \qquad \max_I |P| \geqslant |I|^\beta.$$

From Corollary 5 in Section 6.6, we know that $(P, Q, n) \in \mathcal{R}(\widetilde{I})$ ($\widetilde{I}$ being the parent of I). As (P, Q, n) is $\widetilde{I}$-bicritical, we must have, by the induction hypothesis

$$(\mathbf{9.37}) \qquad \max_{\widetilde{I}} |P| < |\widetilde{I}|^\beta.$$

Therefore, P would be thin I-critical; similarly Q would be thin I-critical. But in view of (9.34), the estimate $(SR3)_s$ says that such a (P, Q, n) satisfying (9.36) does not exist. □

9.4. *The starting interval.* — We establish in this subsection the starting point of an inductive construction of strongly regular parameters. We assume that the ρ exponents satisfy

(9.38) $$\rho_u > d_u^0 + C\varepsilon_0, \qquad \rho_s > d_s^0 + C\varepsilon_0, \qquad \rho_0 > d_s^0 + C\varepsilon_0, \qquad \rho_0' > d_u^0 + C\varepsilon_0,$$

with C large enough. The exponents σ do not come into play for the starting interval.

Proposition **31.** — *The starting interval is strongly regular.*

Proof. — We start by checking (SR1) for I_0.

The set $\mathcal{C}_+(I_0)$ consists of the elements $(P, Q, n) \in \mathcal{R}(I_0)$ such that $P \subset P_s$, $|P| \leqslant \varepsilon_0^{1+\tau}$ for all $t \in I_0$, and which are maximal with this property. Such a (P, Q, n) is therefore the simple composition of (P_s, Q_s, n_s) with an element $(P', Q', n') \in \mathcal{R}(I_0)$ satisfying, according to (3.12) in Section 3.3

$$C^{-1}\varepsilon_0^\tau < |P'| < C\varepsilon_0^\tau$$

for all $t \in I_0$. As recalled in the beginning of Section 8.3, for each $t \in I_0$, the number of P' with $|P'| > C^{-1}\varepsilon_0^\tau$ is of order $\varepsilon_0^{-\tau d_s^t}$ which is also the order of $\varepsilon_0^{-\tau d_s^0}$. This proves $(SR1)_s$. The estimate $(SR1)_u$ is obtained in the same way.

We turn to $(SR2)_s'$. Let $((P_i, Q_i, n_i))_i$ be a finite family of elements of $\mathcal{R}(I_0)$ with the P_i disjoint and contained in P_s. We have to prove that

$(SR2)_s'$ $$\sum_i \max_{I_0} |Q_i|^{\rho_u} \leqslant C|Q_s|^{\rho_s}.$$

We can write each (P_i, Q_i, n_i) as the simple composition of (P_s, Q_s, n_s) with an element $(P_i', Q_i', n_i') \in \mathcal{R}(I_0)$. The P_i' are disjoint. We need to have

(9.39) $$\sum_i \max_{I_0} |Q_i'|^{\rho_u} \leqslant C.$$

This will be a consequence from the existence of equilibrium measures for Hölder potentials on regular Cantor sets defined by expansive $C^{1+\alpha}$ maps: fix a parameter $t \in I_0$; choose an horizontal segment in each rectangle of the Markov partition, and let J be their union; the intersection of J with the local stable foliation $W^s(K, R)$ is a regular Cantor set K_s; there exists on K_s a probability measure (a Gibbs state for the appropriate potential) such that, for each $(P, Q, n) \in \mathcal{R}(I_0)$, the measure of the cylinder of K_s defined by P, divided by $|Q|^{d_u^t}$, is uniformly bounded away from 0 and ∞. As the P_i' are disjoint, this proves that, for each $t \in I_0$, we have

(9.40) $$\sum_i |Q_i'|^{d_u^t} \leqslant C.$$

But Proposition 15 in Section 7.4 shows that, for each $(P, Q, n) \in \mathcal{R}(I_0)$

$$\textbf{(9.41)} \qquad \min_{I_0} |Q'_t| \geqslant \left(\max_{I_0} |Q'_t|\right)^{1+C\varepsilon_0}.$$

Therefore (9.39) is a consequence of (9.40), (9.41) as

$$\textbf{(9.42)} \qquad \rho_u > d_u^0 + C\varepsilon_0.$$

The estimate $(SR2)_u$ is proved in the same way, under the requirement

$$\textbf{(9.43)} \qquad \rho_s > d_s^0 + C\varepsilon_0.$$

Let us now prove $(SR3)_s$ and $(SR3)_u$. The only case to consider is $I_\alpha = I_\omega = I_0$. Then $Bi_+(I_0, I_0, I_0; x)$ (resp. $Bi_-(I_0, I_0, I_0; x)$) is the set of $(P, Q, n) \in \mathcal{R}(I_0)$ such that $P \subset P_s$, $Q \subset Q_u$ and $|P| \geqslant x$ (resp. $|Q| \geqslant x$) for some $t \in I_0$. We write (P, Q, n) as a simple composition

$$\textbf{(9.44)} \qquad (P, Q, n) = (P_s, Q_s, n_s) * (P', Q', n') * (P_u, Q_u, n_u))$$

(cf. (5.18) in Section 5.6.4). The same argument than for $(SR1)_s$ above now gives

$$\textbf{(9.45)} \qquad \#Bi_+(I_0, I_0, I_0; x) \leqslant C\left(\frac{x}{\varepsilon_0|P_u|}\right)^{-(d_s^0+C\varepsilon_0)},$$

and similarly

$$\textbf{(9.46)} \qquad \#Bi_-(I_0, I_0, I_0; x) \leqslant C\left(\frac{x}{\varepsilon_0|Q_s|}\right)^{-(d_u^0+C\varepsilon_0)}.$$

Therefore, we obtain $(SR3)_s$ and $(SR3)_u$ as we have:

$$\textbf{(9.47)} \qquad \rho_0 > d_s^0 + C\varepsilon_0,$$

$$\textbf{(9.48)} \qquad \rho'_0 > d_u^0 + C\varepsilon_0. \qquad \square$$

9.5. *The induction step for* (SR1). — In Proposition 28 of Section 8.3, we introduced $d_s^+ = d_s^0 + C\eta\tau^{-1}$. Let also $d_u^+ = d_u^0 + C\eta\tau^{-1}$. The exponent σ in the estimates (SR1) will be defined as

$$\textbf{(9.49)} \qquad \sigma = \min(1 - d_u^+ - (1+\tau)d_s^* - \tau - c\eta\tau^{-1}, 1 - d_s^+ - (1+\tau)d_u^* - \tau - c\eta\tau^{-1}).$$

The aim of this Subsection is to prove the following result.

Proposition **32.** — *Assume that the parent interval* $\widetilde{I}$ *is* β*-regular and satisfies one of the two inequalities* (SR1). *Then all candidates* $I \subset \widetilde{I}$ *satisfy the same inequality except perhaps for a proportion not larger than* $C|\widetilde{I}|^{\tau^2}$.

Before proving the Proposition above, we first state and prove

Proposition **33.** — *For any* $(P, Q, n) \in \mathcal{C}_+(\widetilde{I})$, *we have*

$$\#Cr(P) \leqslant C|\widetilde{I}|^{-\tau d_u^+}.$$

Proof. — Let $(P^*, Q^*, n^*) \in \mathcal{R}(\widetilde{I})$ be an element such that $P^* \subset P$ and

$$|P^*| \leqslant \frac{1}{2}|\widetilde{I}|^{(1+\tau)(1-\eta)^{-1}} \tag{9.50}$$

for all $t \in \widetilde{I}$. By Proposition 28 in Section 8.3, there are at most $|\widetilde{I}|^{-\tau d_u^+}$ candidates $I \subset \widetilde{I}$ such that P^* is I-critical.

Let now $I \in Cr(P)$. We will prove that I is within distance $C|\widetilde{I}|^{1+\tau}$ either from the boundary of $\widetilde{I}$, or from a candidate $I' \subset \widetilde{I}$ such that P^* is I'-critical. This implies the estimate of the Proposition.

By definition, there exists $(P_1, Q_1, n_1) \in \mathcal{R}(I)$ such that $P_1 \subset P$, P_1 is I-critical and

$$|P_1|^{1-\eta} \leqslant 2|I| \quad \text{for some } t_1 \in I. \tag{9.51}$$

We take such a (P_1, Q_1, n_1) with P_1 maximal. Then, (P_1, Q_1, n_1) belongs in fact to $\mathcal{R}(\widetilde{I})$: indeed, otherwise, by the structure theorem, the element $(\widetilde{P}_1, \widetilde{Q}_1, \widetilde{n}_1) \in \mathcal{R}(\widetilde{I})$ with smallest $\widetilde{P}_1 \supset P_1$ would be $\widetilde{I}$-bicritical and we would have $|\widetilde{P}_1| < |\widetilde{I}|^{\beta} < 2|I|$ for all $t \in \widetilde{I}$.

As P_1 is I-critical, it is I-special. From Proposition 18 in Section 7.7, we obtain that

$$|P_1|^{1-\eta} \leqslant C|I| \quad \text{for all } t \in I. \tag{9.52}$$

By Corollary 10 in Section 8.1, there exists $(P_0, Q_0, n_0) \in \mathcal{R}(I)$ such that Q_0, P_1 are I-critically related and $|Q_0| < \max(|I|^{\beta}, 2|P_1|)$ for some $t_0 \in I$. In particular, we have

$$|Q_0|^{1-\eta} \leqslant C|I| \quad \text{for some } t_0 \in I. \tag{9.53}$$

We take such a (P_0, Q_0, n_0) with Q_0 maximal. We claim that (P_0, Q_0, n_0) belongs to $\mathcal{R}(\widetilde{I})$ and that Q_0 is $\widetilde{I}$-critical.

Indeed, let $(\widetilde{P}_0, \widetilde{Q}_0, \widetilde{n}_0)$ the element of $\mathcal{R}(\widetilde{I})$ such that $\widetilde{Q}_0$ contains Q_0 and is the smallest with this property.

Assume by contradiction that $\widetilde{Q}_0$ is $\widetilde{I}$-transverse. Let $(P_\alpha, Q_\alpha, n_\alpha)$ be an $\widetilde{I}$-decomposition of P_s such that, for each α, $\widetilde{Q}_0$ and P_α are either $\widetilde{I}$-separated or $\widetilde{I}$-transverse. If $P_1 \subset P_\alpha$ for some α, we would conclude that $\widetilde{Q}_0$ and P_1 are $\widetilde{I}$-separated if $\widetilde{Q}_0$ and P_α are $\widetilde{I}$-separated, and $\widetilde{Q}_0$ and P_1 are $\widetilde{I}$-transverse if $\widetilde{Q}_0$ and P_α are $\widetilde{I}$-transverse. If there exists α such that $P_\alpha \subset P_1$ and $\widetilde{Q}_0 \pitchfork_{\widetilde{I}} P_\alpha$ holds, then $\widetilde{Q}_0 \pitchfork_{\widetilde{I}} P_1$ holds also by Proposition 10 in Section 6.4. In the remaining case, the $P_\alpha \subset P_1$ form an $\widetilde{I}$-decomposition of P_1 and they are all $\widetilde{I}$-separated from $\widetilde{Q}_0$; this imply that P_1 itself is $\widetilde{I}$-separated from $\widetilde{Q}_0$. The contradiction obtained in all cases show that $\widetilde{Q}_0$ is $\widetilde{I}$-critical.

But then, if $Q_0 \neq \widetilde{Q}_0$, the element $(\widetilde{P}_0, \widetilde{Q}_0, \widetilde{n}_0)$ is $\widetilde{I}$-bicritical by the structure theorem and we would have $|\widetilde{Q}_0| < |\widetilde{I}|^{\beta}$ for all $t \in \widetilde{I}$, contradicting the maximality of Q_0.

Thus we know that both Q_0 and P_1 are $\widetilde{I}$-defined and $\widetilde{I}$-special.In particular, from Proposition 18 in Section 7.7 and (9.53), we have

$$|Q_0|^{1-\eta} \leqslant C|I| \quad \text{for all } t \in \widetilde{I}. \tag{9.54}$$

If we had $\delta_{LR}(Q_0, P_1) > C_0|I|$ for all $t \in I$ and C_0 large enough, we would have $Q_0 \,\overline{\pitchfork}_I\, P_1$ from the estimates on $|Q_0|, |P_1|$ above. On the other hand, if we had $\delta_{LR}(Q_0, P_1) < 0$ for all $t \in I$, Q_0 and P_1 would be I-separated. As they are I-critically related, we must have, for some $t \in I$ and $C_0 > 2$

$$0 \leqslant \delta_{LR}(Q_0, P_1) \leqslant C_0|I|. \tag{9.55}$$

Assume that I is not within distance $4C_0|I|$ from the boundary of $\widetilde{I}$. Let J be the $3C_0|I|$-neighborhood of I contained in $\widetilde{I}$. From Corollary 8 in Section 7.6, the quantity $\delta_{LR}(Q_0, P_1)$ will be larger than $2C_0|I|$ at the upper endpoint of J and less than $-C_0|I|$ at the lower endpoint. As P_1 and P^* are both contained in P and $|P| < |I|$ for all $t \in I$, we have, (taking C_0 larger if necessary) that, for all $t \in I$

$$|\delta_{LR}(Q_0, P^*) - \delta_{LR}(Q_0, P_1)| \leqslant C_0|I|. \tag{9.56}$$

Therefore, the value of $\delta_{LR}(Q_0, P^*)$ at the lower endpoint of J is less than 0 and the value at the upper endpoint is at least $C_0|I|$. Then, in view of the estimates on $|Q_0|, |P^*|$ and Lemma 3 in Section 6.6.3, there is a candidate interval I' intersecting J such that P^* is I'-critical. As explained above, this allows to obtain the conclusion of the proposition. □

Proof of Proposition 32. Let $(\widetilde{P}, \widetilde{Q}, \widetilde{n}) \in \mathcal{C}_+(\widetilde{I})$; we bound the number of $(P, Q, n) \in \mathcal{C}_+(I)$ (for some candidate $I \subset \widetilde{I}$) with $P \subset \widetilde{P}$. Recall that $(P, Q, n) \in \mathcal{R}(\widetilde{I})$ by Lemma 11 in Section 9.1. Let $\widehat{P}$ be the parent of P (if $P \neq \widetilde{P}$). It satisfies $|\widehat{P}| \geqslant |\widetilde{I}|^{(1+\tau)^2}$ for some $t \in \widetilde{I}$. By Corollary 13 in Section 8.3, the number of possible $\widehat{P}$ (for fixed $\widetilde{P}$) is at most $|\widetilde{I}|^{-d_s^*\tau(1+\tau)}$. From Proposition 25 in Section 8.2, each $\widehat{P}$ has at most $|\widetilde{I}|^{-c\eta}$ children. We conclude that the number of possible P for each $\widetilde{P}$ is at most $|\widetilde{I}|^{-d_s^*\tau(1+\tau)-c\eta}$.

We obtain therefore, in view of Lemma 11, Proposition 33 and the induction hypothesis

$$\begin{aligned}
\sum_{I \subset \widetilde{I}} \#\mathcal{C}_+(I) &\leqslant |\widetilde{I}|^{-c\eta-\tau(1+\tau)d_s^*} \sum_{\mathcal{C}_+(\widetilde{I})} \#Cr(\widetilde{P}) \\
&\leqslant \#\mathcal{C}_+(\widetilde{I})|\widetilde{I}|^{-c\eta-\tau(1+\tau)d_s^*} \max \#Cr(\widetilde{P}) \\
&\leqslant C\#\mathcal{C}_+(\widetilde{I})|\widetilde{I}|^{-c\eta-\tau d_u^+-\tau(1+\tau)d_s^*} \\
&\leqslant C\left(\frac{|\widetilde{I}|}{\varepsilon_0}\right)^{\sigma} \varepsilon_0^{-\tau d_s^0} |\widetilde{I}|^{-c\eta-\tau d_u^+-\tau(1+\tau)d_s^*}.
\end{aligned} \tag{9.57}$$

As the total number of candidates is $|\widetilde{\mathrm{I}}|^{-\tau}$, we will have, except for a proportion at most $\mathrm{C}|\widetilde{\mathrm{I}}|^{\tau^2}$ of candidates I

$$\#\mathcal{C}_+(\mathrm{I}) \leqslant \left(\frac{|\widetilde{\mathrm{I}}|}{\varepsilon_0}\right)^{\sigma} \varepsilon_0^{-\tau d_s^0} |\widetilde{\mathrm{I}}|^{\tau-\tau^2-c\eta-\tau d_u^+-\tau(1+\tau)d_s^*}$$

$$\leqslant \left(\frac{|\mathrm{I}|}{\varepsilon_0}\right)^{\sigma} \varepsilon_0^{-\tau d_s^0} |\widetilde{\mathrm{I}}|^{\tau-\tau^2-c\eta-\tau d_u^+-\tau(1+\tau)d_s^*-\tau\sigma}. \tag{9.58}$$

In order to obtain the required estimate, we want to have

$$\tau - \tau^2 - c\eta - \tau d_u^+ - \tau(1+\tau)d_s^* - \tau\sigma \geqslant 0, \tag{9.59}$$

which amounts to

$$\sigma \leqslant 1 - d_u^+ - (1+\tau)d_s^* - \tau - c\eta\tau^{-1}. \tag{9.60}$$

In the same way, to obtain $(\mathrm{SR1})_u$ for all but a proportion $\mathrm{C}|\widetilde{\mathrm{I}}|^{\tau^2}$ of candidates I from $(\mathrm{SR1})_u$ for $\widetilde{\mathrm{I}}$, we will ask that

$$\sigma \leqslant 1 - d_s^+ - (1+\tau)d_u^* - \tau - c\eta\tau^{-1}. \tag{9.61}$$

In conclusion, we take σ to be the largest number satisfying (9.60), (9.61) above and the proof of the proposition is complete.

9.6. *The induction step for* $(\mathrm{SR2})'$.

9.6.1. Recall that the exponents ρ_0' (in $(\mathrm{SR3})_u$), ρ_u (in $(\mathrm{SR2})_s'$) and d_u^* are all close to d_u^0 when $\tau \gg \eta \gg \varepsilon_0$ are small. Similarly, ρ_0 (in $(\mathrm{SR3})_s$), ρ_s (in $(\mathrm{SR2})_u'$) and d_s^* are all close to d_s^0. We now assume moreover that

$$\rho_0' - \varepsilon_0^{\tau} > \rho_u > d_u^* + \varepsilon_0^{\tau}, \qquad \rho_0 - \varepsilon_0^{\tau} > \rho_s > d_s^* + \varepsilon_0^{\tau}. \tag{9.62}$$

The exponents σ_s, σ_u in $(\mathrm{SR2})'$ are defined as

$$\sigma_s := 1 - 3\tau - d_s^+, \qquad \sigma_u := 1 - 3\tau - d_u^+. \tag{9.63}$$

The aim of this subsection is to prove the following result

Proposition **34.** — *Assume that the parent interval* $\widetilde{\mathrm{I}}$ *is strongly regular. Then all candidates* $\mathrm{I} \subset \widetilde{\mathrm{I}}$ *satisfy* $(\mathrm{SR2})_s'$ *and* $(\mathrm{SR2})_u'$ *except perhaps for a proportion not larger than* $\mathrm{C}|\widetilde{\mathrm{I}}|^{\tau^2}$.

Proof. — We first explain the general idea (for $(\mathrm{SR2})_s'$).

Let $(P, Q, n) \in \mathcal{C}_+(\widetilde{I})$. Let $((P_i, Q_i, n_i))_i$ be a finite family of elements of $\mathcal{R}(\widetilde{I})$ with the P_i disjoint and contained in P. As $(SR2)'_s$ is satisfied by $\widetilde{I}$, and P is contained in some P^* with $(P^*, Q^*, n^*) \in \widehat{\mathcal{C}}_+(\widetilde{I})$ by Lemma 11, the supremum over such families of the quantity

$$\sum_i \max_{\widetilde{I}} |Q_i|^{\rho_u}$$

is finite. Denote this supremum by $c(P)$. We have, still from $(SR2)'_s$, that

$$\sum_{\mathcal{C}_+(\widetilde{I})} c(P) \leqslant C|Q_s|^{\rho_u}\left(\frac{|\widetilde{I}|}{\varepsilon_0}\right)^{\sigma_u}. \tag{9.64}$$

Let now $I \subset \widetilde{I}$ be a candidate interval such that $I \in Cr(P)$. Let $((P_i, Q_i, n_i))_i$ be a finite family of elements of $\mathcal{R}(I)$ with the P_i disjoint and contained in P. Denote by $c(P, I)$ the supremum (in $\mathbf{R} \cup +\infty$) over all such families of the quantity

$$\sum_i \max_{I} |Q_i|^{\rho_u}.$$

We also consider two related quantities $c(P, I, old)$ and $c(P, I, new)$ obtained by taking the supremum of the same quantity $\sum_i \max_I |Q_i|^{\rho_u}$ over a more restricted set of families: we still ask that the P_i are disjoint and contained in P, but for $c(P, I, old)$ we ask moreover that all (P_i, Q_i, n_i) belong to $\mathcal{R}(\widetilde{I})$, and for $c(P, I, new)$ that none of them belong to $\mathcal{R}(\widetilde{I})$.

Clearly, we have

$$c(P, I) \leqslant c(P, I, old) + c(P, I, new),$$

and also

$$c(P, I, old) \leqslant c(P). \tag{9.65}$$

On the other hand, by Lemma 11, the candidate I will satisfy $(SR2)'_s$ iff

$$\sum_{\widehat{\mathcal{C}}_+(I)} c(P, I) \leqslant C|Q_s|^{\rho_u}\left(\frac{|I|}{\varepsilon_0}\right)^{\sigma_u}. \tag{9.66}$$

In each term of the sum, we separate the old and new part and will deal successively with these two terms.

9.6.2. The sum over the old parts is easily controlled for most candidates. We have

$$\sum_{I \subset \widetilde{I}} \sum_{\widehat{\mathcal{C}}_+(I)} c(P, I, old) \leqslant \sum_{I \subset \widetilde{I}} \sum_{\widehat{\mathcal{C}}_+(I)} c(P) \quad \text{by (9.72)}$$

$$\leqslant \sum_{\mathcal{C}_+(\widetilde{\mathrm{I}})} c(\mathrm{P})\#\mathrm{C}r(\mathrm{P})$$

$$\leqslant \mathrm{C}|\widetilde{\mathrm{I}}|^{-\tau d_u^+} \sum_{\mathcal{C}_+(\widetilde{\mathrm{I}})} c(\mathrm{P}) \quad \text{by Proposition 33}$$

(9.67)
$$\leqslant \mathrm{C}|\widetilde{\mathrm{I}}|^{-\tau d_u^+} |\mathrm{Q}_s|^{\rho_u} \left(\frac{|\widetilde{\mathrm{I}}|}{\varepsilon_0}\right)^{\sigma_u}, \quad \text{by (9.71).}$$

Consequently, all candidates I except for a proportion at most $|\widetilde{\mathrm{I}}|^{2\tau^2}$ will satisfy

(9.68)
$$\sum_{\widehat{\mathcal{C}}_+(\mathrm{I})} c(\mathrm{P}, \mathrm{I}, old) \leqslant \mathrm{C}|\widetilde{\mathrm{I}}|^{\tau - 2\tau^2 - \tau d_u^+} |\mathrm{Q}_s|^{\rho_u} \left(\frac{|\widetilde{\mathrm{I}}|}{\varepsilon_0}\right)^{\sigma_u}.$$

From the definition $\sigma_u := 1 - 3\tau - d_u^+$ we obtain, for all candidates I except for a proportion at most $|\widetilde{\mathrm{I}}|^{2\tau^2}$

(9.69)
$$\sum_{\widehat{\mathcal{C}}_+(\mathrm{I})} c(\mathrm{P}, \mathrm{I}, old) \leqslant |\mathrm{Q}_s|^{\rho_u} \left(\frac{|\mathrm{I}|}{\varepsilon_0}\right)^{\sigma_u}.$$

9.6.3. The sum over the new parts is more complicated.

Let $\mathrm{I} \subset \widetilde{\mathrm{I}}$ be a candidate and let $(\mathrm{P}', \mathrm{Q}', n')$ be an element in $\mathcal{R}(\mathrm{I})$ but not in $\mathcal{R}(\widetilde{\mathrm{I}})$ such that P′ is contained in some P with $(\mathrm{P}, \mathrm{Q}, n) \in \widehat{\mathcal{C}}_+(\mathrm{I})$. For a finite family of such elements with disjoint P′, we must bound the sum $\mathrm{S} := \sum \max_{\mathrm{I}} |\mathrm{Q}'|^{\rho_u}$.

We apply the structure theorem of Section 6.5. We obtain an integer $k > 0$ and elements $(\mathrm{P}'_l, \mathrm{Q}'_l, n'_l) \in \mathcal{R}(\widetilde{\mathrm{I}})$ for $0 \leqslant l \leqslant k$ such that

- $(\mathrm{P}', \mathrm{Q}', n')$ is obtained from the parabolic composition of the $(\mathrm{P}'_l, \mathrm{Q}'_l, n'_l)$;
- for $0 \leqslant l < k$, Q'_l and P'_{l+1} are $\widetilde{\mathrm{I}}$-critical and $\mathrm{Q}'_l \pitchfork_{\widetilde{\mathrm{I}}} \mathrm{P}'_{l+1}$ does not hold;
- for all $t \in \mathrm{I}$, $|\mathrm{Q}'| \leqslant \mathrm{C}^k |\mathrm{I}|^{-k/2} \prod |\mathrm{Q}'_l|$ (cf. Proposition 13 in Section 6.6.3).

Recall that, by Lemma 10 in Section 9.1, Q'_l and P'_{l+1} are thin $\widetilde{\mathrm{I}}$-critical for $0 \leqslant l < k$. Observe that P'_0 is also thin $\widetilde{\mathrm{I}}$-critical because it is contained in the rectangle P such that $(\mathrm{P}, \mathrm{Q}, n) \in \mathcal{C}_+(\widetilde{\mathrm{I}})$ and $\mathrm{P}' \subset \mathrm{P}$.

We claim that P'_k is contained in some $\widetilde{\mathrm{P}}$ with $(\widetilde{\mathrm{P}}, \widetilde{\mathrm{Q}}, \widetilde{n}) \in \widehat{\mathcal{C}}_+(\widetilde{\mathrm{I}})$. As P'_k is thin $\widetilde{\mathrm{I}}$-critical, it is sufficient to prove that a simple descendant $\overline{\mathrm{P}}$ of P'_k with $\max_{\widetilde{\mathrm{I}}} |\overline{\mathrm{P}}| < |\widetilde{\mathrm{I}}|^{1+\tau}$ is $\widetilde{\mathrm{I}}$-critical. If $\overline{\mathrm{P}}$ was $\widetilde{\mathrm{I}}$-transverse, there would exist $(\overline{\mathrm{P}}', \overline{\mathrm{Q}}', \overline{n}')$ with $\overline{\mathrm{Q}}' \cap \mathrm{Q}'_{k-1} \neq \emptyset$ and $\overline{\mathrm{Q}}' \pitchfork_{\widetilde{\mathrm{I}}} \overline{\mathrm{P}}$. As $\mathrm{Q}'_{k-1} \pitchfork_{\mathrm{I}} \mathrm{P}'_k$ also holds, we would deduce by concavity that $\mathrm{Q}'_{k-1} \pitchfork_{\widetilde{\mathrm{I}}} \mathrm{P}'_k$ holds, a contradiction which proves the claim.

Consider now a finite family of such $(\mathrm{P}', \mathrm{Q}', n')$ with disjoint P′. In the sum $\mathrm{S} = \sum \max_{\mathrm{I}} |\mathrm{Q}'|^{\rho_u}$, we first do the partial sum over those $(\mathrm{P}', \mathrm{Q}', n')$ which share the same fixed integer k and the same fixed $(\mathrm{P}'_l, \mathrm{Q}'_l, n'_l)$ for $0 \leqslant l < k$. Except for the fact that the parabolic composition of the (P'_l, Q'_l, n'_l) for $0 \leqslant l \leqslant k$ produces 2^k elements, the disjointness of the P' implies the disjointness of the P'_k. Therefore, from the estimate for $|Q'|$ above, we are able to use $(SR2)'_s$ for $\widetilde{I}$ to bound the partial sum by

$$S(Q'_0, \ldots, Q'_{k-1}) := 2^k\left(C|Q_s|^{\rho_u}\left(\frac{|\widetilde{I}|}{\varepsilon_0}\right)^{\sigma_u}\right)C^{k\rho_u}|I|^{-k\rho_u/2}\prod_{0\leqslant l<k}\max_{\widetilde{I}}|Q'_l|^{\rho_u}. \tag{9.70}$$

When we now sum over $Q'_0, \ldots, Q'_{k-1}$, we are led to introduce

$$S'_{bi} := \sum \max_{\widetilde{I}} |Q|^{\rho_u}, \tag{9.71}$$

where the sum is taken over all $(P, Q, n) \in \mathcal{R}(\widetilde{I})$ such that P and Q are thin $\widetilde{I}$-critical. This sum will be estimated below. In terms of S'_{bi} we have for the full some S

$$\begin{aligned} S &\leqslant \sum_{k\geqslant 1}\sum_{Q'_0,\ldots,Q'_{k-1}} S(Q'_0, \ldots, Q'_{k-1}) \\ &\leqslant C|Q_s|^{\rho_u}\left(\frac{|\widetilde{I}|}{\varepsilon_0}\right)^{\sigma_u}\sum_{k\geqslant 1}(2C^{\rho_u}|I|^{-\rho_u/2}S'_{bi})^k \\ &\leqslant C|Q_s|^{\rho_u}\left(\frac{|\widetilde{I}|}{\varepsilon_0}\right)^{\sigma_u}|I|^{-\rho_u/2}S'_{bi}, \end{aligned} \tag{9.72}$$

provided $|I|^{-\rho_u/2}S'_{bi} < C^{-1}$ with C large enough.

In the next subsection, we will prove the

Proposition **35.** — *Let θ_u be a number (independent of $\tau \gg \eta \gg \varepsilon_0$) such that $d^0_u/2 < \theta_u < 1 - d^0_s$. Under the hypotheses of the proposition, we have (if $\tau \gg \eta \gg \varepsilon_0$ are small enough)*

$$S'_{bi} \leqslant |\widetilde{I}|^{\theta_u}. \tag{9.73}$$

Remark **11.** — It is easy to check that hypothesis (H4) on d^0_u, d^0_s implies that $d^0_u/2 < 1 - d^0_s$.

The estimate of the proposition is sufficient to get $C|I|^{-\rho_u/2}S'_{bi} < |\widetilde{I}|^{\tau\sigma_u}$ and thus

$$S \leqslant |Q_s|^{\rho_u}\left(\frac{|I|}{\varepsilon_0}\right)^{\sigma_u}. \tag{9.74}$$

The proof of $(SR2)_s$ is therefore complete except for the estimate for S'_{bi}.

Proof of Proposition 35. It is easy to relate S'_{bi} to the cardinality of the sets $Bi_-(\widetilde{I}, \widetilde{I}, \widetilde{I}; x)$. Indeed, we have clearly

$$S'_{bi} \leqslant \sum_{l\geqslant 0} \#Bi_-(\widetilde{I}, \widetilde{I}, \widetilde{I}; 2^{-l})2^{(1-l)\rho_u}. \tag{9.75}$$

To estimate the cardinality of $Bi_-(\widetilde{I}, \widetilde{I}, \widetilde{I}; 2^{-l})$, we will use $(SR3)_u$ except when l is very large; in this case, the following result will provide a slightly better estimate.

Proposition **36.** — *Assume that* $(SR1)_u$ *holds for* $\widetilde{I}$. *Then, for any candidate interval* I, *any* $x > 0$, *the number of* $(P, Q, n) \in \mathcal{R}(I)$ *with* Q *thin* I*-critical and* $\max_I |Q| \geqslant x$ *is at most*

$$C\Big(\frac{|\widetilde{I}|}{\varepsilon_0}\Big)^{\sigma} \varepsilon_0^{-\tau d_u^0} \Big(\frac{x}{|I|}\Big)^{-d_u^*}.$$

Proof. — Let $(P, Q, n) \in \mathcal{R}(I)$ with Q thin I-critical. We claim that $Q \subset \widetilde{Q}$ for some $(\widetilde{P}, \widetilde{Q}, \widetilde{n}) \in \mathcal{C}_-(\widetilde{I})$. Indeed, let $(P', Q', n') \in \mathcal{R}(I)$ such that $Q \subset Q'$, Q' is I-critical, and $\min_I |Q'|^{1-\eta} \leqslant 2|I|$. Replacing if necessary Q' by the smallest $\widetilde{I}$-defined rectangle containing Q', we may assume that $(P', Q', n') \in \mathcal{R}(\widetilde{I})$ (see Corollary 5 in Section 6.6.3). Then, as Q' is $\widetilde{I}$-special, we have from Proposition 18 in Section 7.7 that $\max_{\widetilde{I}} |Q'|^{1-\eta} \leqslant C|I|$. Therefore Q', and also Q is contained in $\widetilde{Q}$ for some $(\widetilde{P}, \widetilde{Q}, \widetilde{n}) \in \mathcal{C}_-(\widetilde{I})$.

In the estimate of the Proposition, we can now assume that $x < |I|$, otherwise there is no (P, Q, n) with the required properties. Let $(\widetilde{P}, \widetilde{Q}, \widetilde{n}) \in \mathcal{C}_-(\widetilde{I})$; we have $\max_{\widetilde{I}} |\widetilde{Q}| \leqslant |I|$. From Corollary 13 in Section 8.3, the number of $(P, Q, n) \in \mathcal{R}(I)$ with Q thin I-critical, $Q \subset \widetilde{Q}$, and $\max_I |Q| \geqslant x$ is at most $(\frac{x}{|I|})^{-d_u^*}$.

As the cardinality of $\mathcal{C}_-(\widetilde{I})$ is at most $C(|\widetilde{I}|/\varepsilon_0)^{\sigma} \varepsilon_0^{-\tau d_u^0}$ by $(SR1)_u$, we obtain the required estimate. □

Corollary **14.** — *One has*

$$\#Bi_-(I, I, I; x) \leqslant B'^{\#} := C\Big(\frac{x}{\varepsilon_0|Q_s|}\Big)^{-d_u^*} \Big(\frac{|I|}{\varepsilon_0}\Big)^{\frac{\sigma}{1+\tau}+d_u^*} \varepsilon_0^{-\tau(d_u^0+\frac{\sigma}{1+\tau})} |Q_s|^{-d_u^*}.$$

The estimate of the Corollary does not take into account that, for the elements $(P, Q, n) \in Bi_-(I, I, I; x)$, P is I-critical. Therefore, it is not surprising that it is worse than $(SR3)_u$ for middle-sized x. Indeed, on one side, the constant $\varepsilon_0^{-\tau(d_u^0+\frac{\sigma}{1+\tau})} |Q_s|^{-d_u^*}$ is large (much larger than $\varepsilon_0^{-A\tau}$). On the other side, comparing the exponents of $(\frac{x}{\varepsilon_0|Q_s|})$ and $(\frac{|I|}{\varepsilon_0})$ in $B'^{\#}$, B'_1, B'_0 when $\tau \gg \eta \gg \varepsilon_0$ are small, one gets

- $-d_u^*$ (close to $-d_u^0$) and $\frac{\sigma}{1+\tau} + d_u^*$ (close to $1 - d_s^0$) for $B'^{\#}$;
- $-\rho'_1$ (close to $-\frac{d_u^0}{d_s^0+d_u^0}(2d_s^0 + d_u^0 - 1) > -d_u^0$) and $\sigma_0 + \sigma_1$ (close to $1 - d_u^0 \geqslant 1 - d_s^0$) for B'_1;
- $-\rho'_0$ (close to $-d_u^0$) and $2\sigma_0 + \sigma_1$ (close to $2 - d_u^0 - d_s^0 > 1 - d_u^0$) for B'_0.

The only range where $B'^{\#}$ is smaller than $B' = \max(B'_0, B'_1)$ is when x is very small because, although ρ'_0 and d_u^* are both close to d_u^0, we have $\rho'_0 > d_u^*$. Forgetting about powers of ε_0^τ, we define $\overline{x}'_{\min}$ (for $\widetilde{I}$) by

$$\left(\frac{\overline{x}'_{\min}}{\varepsilon_0|Q_s|}\right)^{-d_u^*}\left(\frac{|I|}{\varepsilon_0}\right)^{\frac{\sigma}{1+\tau}+d_u^*}|Q_s|^{-d_u^*}=\left(\frac{\overline{x}'_{\min}}{\varepsilon_0|Q_s|}\right)^{-\rho_0'}\left(\frac{|I|}{\varepsilon_0}\right)^{2\sigma_0+\sigma_1}, \tag{9.76}$$

which amounts to

$$\frac{\overline{x}'_{\min}}{\varepsilon_0|Q_s|}=\left[\left(\frac{|I|}{\varepsilon_0}\right)^{2\sigma_0+\sigma_1-\frac{\sigma}{1+\tau}-d_u^*}|Q_s|^{d_u^*}\right]^{\frac{1}{\rho_0'-d_u^*}}. \tag{9.77}$$

The quantity $\sigma_u^* := 2\sigma_0+\sigma_1-\frac{\sigma}{1+\tau}-d_u^*$ is close to $1-d_u^0$ and the exponent $\frac{1}{\rho_0'-d_u^*}$ is very large.

We now come back to the estimation of S'_{bi}. We divide the sum over l in the right-hand term of (9.82) in three parts:

– $\overline{x}' \geqslant 2^{-l} \geqslant x'_{cr}$.

Here $\overline{x}' = \varepsilon_0^{1+A\tau/\rho_1'}|Q_s|(\frac{|\widetilde{I}|}{\varepsilon_0})^{\frac{\sigma_0+\sigma_1}{\rho_1'}}$ is defined (in (9.29)) by the relation $\varepsilon_0^{-A\tau}B_1'=1$ and satisfies $\overline{x}' < |\widetilde{I}|^\beta$. The set $Bi_-(\widetilde{I},\widetilde{I},\widetilde{I};x)$ is empty for $x>\overline{x}'$. In the range under consideration, the cardinality of the set $Bi_-(\widetilde{I},\widetilde{I},\widetilde{I};2^{-l})$ is bounded by $\varepsilon_0^{-A\tau}B_1'$. We get, as $\rho_1'<\rho_u$

$$S_1' := \sum_{\overline{x}'\geqslant 2^{-l}\geqslant x'_{cr}} \#Bi_-(\widetilde{I},\widetilde{I},\widetilde{I};2^{-l})2^{(1-l)\rho_u} \leqslant C\overline{x}'^{\rho_u}. \tag{9.78}$$

As $\overline{x}' < |\widetilde{I}|^\beta$, we have

$$S_1' \ll |\widetilde{I}|^{d_u^0} \leqslant |\widetilde{I}|^{1-d_s^0}. \tag{9.79}$$

– $x'_{cr} \geqslant 2^{-l} \geqslant \overline{x}'_{\min}$.

In this range, the cardinality of the set $Bi_-(\widetilde{I},\widetilde{I},\widetilde{I};2^{-l})$ is bounded by $\varepsilon_0^{-A\tau}B_0'$. As we have $\rho_0'>\rho_u+\varepsilon_0^\tau$, we get

$$S_2' := \sum_{x'_{cr}\geqslant 2^{-l}\geqslant \overline{x}'_{\min}} \#Bi_-(\widetilde{I},\widetilde{I},\widetilde{I};2^{-l})2^{(1-l)\rho_u}$$

$$\leqslant C\varepsilon_0^{-A\tau}(\rho_0'-\rho_u)^{-1}\left(\frac{\overline{x}'_{\min}}{\varepsilon_0|Q_s|}\right)^{\rho_u-\rho_0'}(\varepsilon_0|Q_s|)^{\rho_u}\left(\frac{|\widetilde{I}|}{\varepsilon_0}\right)^{2\sigma_0+\sigma_1}$$

$$\leqslant C\varepsilon_0^{-\tau(A+1)}(\varepsilon_0|Q_s|)^{\rho_u}|Q_s|^{-d_u^*\frac{\rho_0'-\rho_u}{\rho_0'-d_u^*}}\left(\frac{|\widetilde{I}|}{\varepsilon_0}\right)^{2\sigma_0+\sigma_1-\sigma_u^*\frac{\rho_0'-\rho_u}{\rho_0'-d_u^*}}. \tag{9.80}$$

Here, $2\sigma_0+\sigma_1$ and σ_u^* are respectively close to $2-d_u^0-d_s^0$ and $1-d_u^0$, $\frac{\rho_0'-\rho_u}{\rho_0'-d_u^*}$ belongs to $(0,1)$, and ρ_u is close to d_u^0. This allows to obtain, when $\tau \gg \eta \gg \varepsilon_0$ are small enough

$$S_2' \ll |\widetilde{I}|^{\theta_u}. \tag{9.81}$$

– $\overline{x}'_{\min} \geqslant 2^{-l}$.

In this range, we use $B'^{\#}$ to bound the cardinality of the set $Bi_-(\widetilde{I}, \widetilde{I}, \widetilde{I}; 2^{-l})$. As we have $\rho_u > d_u^* + \varepsilon_0^\tau$, we get, using (9.83)

$$S'_3 := \sum_{\overline{x}'_{\min} \geqslant 2^{-l}} \#Bi_-(\widetilde{I}, \widetilde{I}, \widetilde{I}; 2^{-l}) 2^{(1-l)\rho_u}$$

$$\leqslant C\varepsilon_0^{-\tau(d_u^0 + \frac{\sigma}{1+\tau})} (\rho_u - d_u^*)^{-1} \left(\frac{\overline{x}'_{\min}}{\varepsilon_0 |Q_s|}\right)^{\rho_u - \rho'_0} (\varepsilon_0 |Q_s|)^{\rho_u} \left(\frac{|\widetilde{I}|}{\varepsilon_0}\right)^{2\sigma_0 + \sigma_1}$$

$$\textbf{(9.82)} \qquad \leqslant C\varepsilon_0^{-\tau(1+d_u^0 + \frac{\sigma}{1+\tau})} (\varepsilon_0 |Q_s|)^{\rho_u} |Q_s|^{-d_u^* \frac{\rho'_0 - \rho_u}{\rho'_0 - d_u^*}} \left(\frac{|\widetilde{I}|}{\varepsilon_0}\right)^{2\sigma_0 + \sigma_1 - \sigma_u^* \frac{\rho'_0 - \rho_u}{\rho'_0 - d_u^*}}.$$

Up to a meaningless power of ε_0^τ, this is the same bound as for S'_2.

Summing the three estimates for S'_1, S'_2, S'_3 give the required estimate for S'_{bi}. The proof of $(SR2)'_s$ is now complete.

Proof of $(SR2)'_u$. — The proof of $(SR2)'_u$ is essentially symmetric to the proof of $(SR2)'_s$, but we have to be somewhat careful because we used (SR3) which is not symmetric.

We divide the sums of the $\max_I |P_i|^{\rho_s}$ (over a finite family of elements of $\mathcal{R}(\widetilde{I})$ with the Q_i disjoint and contained in a Q with $(P, Q, n) \in \widehat{\mathcal{C}}_-(I)$) into an old an a new part as for $(SR2)_s$. The old part is dealt with exactly as above. For the new part, we introduce

$$\textbf{(9.83)} \qquad S_{bi} := \sum \max_{\widetilde{I}} |P|^{\rho_s},$$

where the sum is taken over all $(P, Q, n) \in \mathcal{R}(\widetilde{I})$ such that P and Q are thin $\widetilde{I}$-critical. We now claim that

$$\textbf{(9.84)} \qquad S_{bi} \leqslant |\widetilde{I}|^{\theta_s},$$

where θ_s is any fixed constant in $(d_s^0/2, 1 - d_u^0)$ independent of $\tau \gg \eta \gg \varepsilon_0$. The deduction of $(SR2)'_u$ from this estimate is the same as for $(SR2)'_s$. To prove the claim, we proceed as in the proof of Proposition 35. The estimate corresponding to Proposition 36 and its Corollary is now

$$\textbf{(9.85)} \qquad \#Bi_+(I, I, I; x) \leqslant B^{\#} := C\left(\frac{x}{\varepsilon_0 |P_u|}\right)^{-d_s^*} \left(\frac{|I|}{\varepsilon_0}\right)^{\frac{\sigma}{1+\tau} + d_s^*} \varepsilon_0^{-\tau(d_s^0 + \frac{\sigma}{1+\tau})} |P_u|^{-d_s^*}.$$

The threshold where $B^{\#}$ gets better than $B = \max(B_0, B_1)$ is $\overline{x}_{\min}$ with

$$\textbf{(9.86)} \qquad \frac{\overline{x}_{\min}}{\varepsilon_0 |P_u|} = \left[\left(\frac{|I|}{\varepsilon_0}\right)^{\sigma_s^*} |P_u|^{d_s^*}\right]^{\frac{1}{\rho_0 - d_s^*}}$$

with $\sigma_s^* := 2\sigma_0 + \sigma_1 - \frac{\sigma}{1+\tau} - d_u^*$ close to $1 - d_s^0$. We have

$$S_{bi} \leqslant \sum_{l \geqslant 0} \#Bi_+(\widetilde{I}, \widetilde{I}, \widetilde{I}; 2^{-l}) 2^{(1-l)\rho_s}, \tag{9.87}$$

and the right-hand sum is divided into three parts:

– $\overline{x} \geqslant 2^{-l} \geqslant x_{cr}$.

Here $\overline{x} = \varepsilon_0^{1+A\tau/\rho_1} |P_u| (\frac{|\widetilde{I}|}{\varepsilon_0})^{\frac{\sigma_0+\sigma_1}{\rho_1}}$ is defined by the relation $\varepsilon_0^{-A\tau} B_1 = 1$ and satisfies $\overline{x} < |\widetilde{I}|^{\beta}$. The set $Bi_+(\widetilde{I}, \widetilde{I}, \widetilde{I}; x)$ is empty for $x > \overline{x}$. In the range under consideration, the cardinality of the set $Bi_+(\widetilde{I}, \widetilde{I}, \widetilde{I}; 2^{-l})$ is bounded by $\varepsilon_0^{-A\tau} B_1$. We get, as $\rho_1 < \rho_s$

$$S_1 := \sum_{C\overline{x} \geqslant 2^{-l} \geqslant x_{cr}} \#Bi_+(\widetilde{I}, \widetilde{I}, \widetilde{I}; 2^{-l}) 2^{(1-l)\rho_s} \leqslant C\overline{x}^{\rho_s} \leqslant |\widetilde{I}|^{1-d_u^0}. \tag{9.88}$$

– $x_{cr} \geqslant 2^{-l} \geqslant \overline{x}_{\min}$.

In this range, the cardinality of the set $Bi_+(\widetilde{I}, \widetilde{I}, \widetilde{I}; 2^{-l})$ is bounded by $\varepsilon_0^{-A\tau} B_0$. As we have $\rho_0 > \rho_s + \varepsilon_0^{\tau}$, we get

$$\begin{aligned} S_2 &:= \sum_{x_{cr} \geqslant 2^{-l} \geqslant \overline{x}_{\min}} \#Bi_+(\widetilde{I}, \widetilde{I}, \widetilde{I}; 2^{-l}) 2^{(1-l)\rho_s} \\ &\leqslant C\varepsilon_0^{-A\tau} (\rho_0 - \rho_s)^{-1} \left(\frac{\overline{x}_{\min}}{\varepsilon_0 |P_u|} \right)^{\rho_s - \rho_0} (\varepsilon_0 |P_u|)^{\rho_s} \left(\frac{|\widetilde{I}|}{\varepsilon_0} \right)^{2\sigma_0 + \sigma_1} \\ &\leqslant C\varepsilon_0^{-\tau(A+1)} (\varepsilon_0 |P_u|)^{\rho_s} |P_u|^{-d_s^* \frac{\rho_0 - \rho_s}{\rho_0 - d_s^*}} \left(\frac{|\widetilde{I}|}{\varepsilon_0} \right)^{2\sigma_0 + \sigma_1 - \sigma_s^* \frac{\rho_0 - \rho_s}{\rho_0 - d_s^*}}. \end{aligned} \tag{9.89}$$

Here, $2\sigma_0 + \sigma_1$ and σ_s^* are respectively close to $2 - d_u^0 - d_s^0$ and $1 - d_s^0$, $\frac{\rho_0 - \rho_u}{\rho_0 - d_s^*}$ belongs to $(0, 1)$, and ρ_s is close to d_s^0. This allows to obtain

$$S_2 \ll |\widetilde{I}|^{\theta_s}. \tag{9.90}$$

– the sum S_3 over $\overline{x}_{\min} \geqslant 2^{-l}$ has the same bound as for S_2, up to a meaningless power of ε_0^{τ}.

Summing the three estimates for S_1, S_2, S_3 give the required estimate for S_{bi}. The proof of $(SR2)'_u$ is complete. □

9.7. *The induction step for* $(SR3)_s$*: general overview and easy cases.*

9.7.1. *Very small values of* x. — Let $I \subset \widetilde{I}$ be a candidate interval. When x is very small, one obtains directly a (trivial) estimate for the cardinality of $Bi_+(I, I_\alpha, I_\omega; x)$ which turns out to be better than $(SR3)_s$. Indeed, from Corollary 13 in Section 8.3, one certainly has

$$\#Bi_+(I, I_\alpha, I_\omega; x) \leqslant Cx^{-d_s^*}. \quad \textbf{(9.91)}$$

This very rough estimate is nevertheless better than $(SR3)_s$ when x is very small because $\rho_0 > d_s^*$ (see (9.69) above). More precisely, we define

$$x_{\min} := |\widetilde{I}|^{C(\rho_0 - d_s^*)^{-1}}, \quad \textbf{(9.92)}$$

with C large enough. For $x \leqslant x_{\min}$, we have $x^{-d_s^*} \ll B_0$ for all intervals I_α, I_ω containing I. Therefore we have proved the

Proposition **37.** — *The estimate* $(SR3)_s$ *is satisfied for all candidates* I, *all* $I_\alpha, I_\omega \supset I$, *as soon as* $x \leqslant x_{\min}$.

9.7.2. *Old and new elements.* — Let $I \subset \widetilde{I}$ be a candidate interval.

Definition **9.** — *We denote by* $Bi_+^{old}(I, I_\alpha, I_\omega; x)$ *the set of* $(P, Q, n) \in Bi_+(I, I_\alpha, I_\omega; x)$ *which belong to* $\mathcal{R}(\widetilde{I})$, *by* $Bi_+^{new}(I, I_\alpha, I_\omega; x)$ *the set of* $(P, Q, n) \in Bi_+(I, I_\alpha, I_\omega; x)$ *which do not belong to* $\mathcal{R}(\widetilde{I})$.

We will estimate separately the cardinalities of the two sets. The estimate for $Bi_+^{new}(I, I_\alpha, I_\omega; x)$ is based on the structure theorem of Section 6.5 and is valid for all candidates $I \subset \widetilde{I}$. In Section 9.8, we will prove the

Proposition **38.** — *Assume that* $\widetilde{I}$ *is strongly regular. There exists a constant* $\sigma_2 > 0$, *depending only on* d_u^0, d_s^0, ω_u, *such that one has, for all candidates* $I \subset \widetilde{I}$, *all* $I_\alpha, I_\omega \supset I$, *all* $x \geqslant x_{\min}$

$$\#Bi_+^{new}(I, I_\alpha, I_\omega; x) \leqslant |\widetilde{I}|^{\sigma_2} B,$$

where $\varepsilon_0^{-A\tau} B$ *is the bound for* $\#Bi_+(I, I_\alpha, I_\omega; x)$ *predicted by* $(SR3)_s$.

On the other hand, the estimate for $Bi_+^{old}(I, I_\alpha, I_\omega; x)$ is trivial (from the strong regularity of $\widetilde{I}$) when both I_α and I_ω contain $\widetilde{I}$. Indeed, one has in this case

$$Bi_+^{old}(I, I_\alpha, I_\omega; x) \subset Bi_+(\widetilde{I}, I_\alpha, I_\omega; x) \quad \textbf{(9.93)}$$

(we don't have always equality because the maximum value of |P| is taken over I in one case and over $\widetilde{I}$ in the other case). Combining this with Proposition 38, we get

Proposition **39.** — *Assume that* $\widetilde{I}$ *is strongly regular. Then* $(SR3)_s$ *is satisfied for all candidates* $I \subset \widetilde{I}$, *all* $I_\alpha, I_\omega \supset \widetilde{I}$, *all* $x \geqslant x_{\min}$.

Proof. — We have only to observe that the series $\sum_{k\geqslant 0} \varepsilon_k^{\sigma_2}$ (where $\varepsilon_k = \varepsilon_0^{(1+\tau)^k}$ is the length of intervals of generation k), related to the iterated application of Proposition 38 is convergent with sum $\leqslant C\varepsilon_0^{\sigma_2}$. □

9.7.3. *The case* $I = I_\alpha \neq I_\omega$. — Let $I \subset \tilde{I}$ be a candidate interval. We consider here a set $Bi_+^{old}(I, I, I_\omega; x)$ with $I_\omega \supset \tilde{I}$. We will estimate the size of this set for most candidates $I \subset \tilde{I}$, assuming that $\tilde{I}$ is strongly regular.

Let (P, Q, n) be an element of $Bi_+^{old}(I, I, I_\omega; x)$. As P in thin I-critical, there exists $(P^*, Q^*, n^*) \in \mathcal{R}(I)$ such that $P \subset P^*$, P^* is I-critical and $\min_I |P^*|^{1-\eta} \leqslant 2|I|$. By coherence, $(P^*, Q^*, n^*) \in \mathcal{R}(\tilde{I})$ and P^* is $\tilde{I}$-special; then, by Proposition 18 in Section 7.7, we have $\max_{\tilde{I}} |P^*|^{1-\eta} \leqslant C|I|$. It follows that there exists $(\tilde{P}, \tilde{Q}, \tilde{n}) \in \mathcal{C}_+(\tilde{I})$ such that $P \subset P^* \subset \tilde{P}$. Moreover, we have $I \in Cr(\tilde{P})$ by definition of $Cr(\tilde{P})$. Observe also that P^* is $\tilde{I}$-critical, hence P is thin $\tilde{I}$-critical, and that $\max_{\tilde{I}} |P| \geqslant \max_I |P| \geqslant x$.

Definition **10.** — *For* $(\tilde{P}, \tilde{Q}, \tilde{n}) \in \mathcal{C}_+(\tilde{I})$, *we denote by* $Bi_+(\tilde{P})$ *the set of* $(P, Q, n) \in Bi_+(\tilde{I}, \tilde{I}, I_\omega; x)$ *such that* $P \subset \tilde{P}$.

We have just seen that

$$Bi_+^{old}(I, I, I_\omega; x) \subset \bigsqcup_{\substack{(\tilde{P},\tilde{Q},\tilde{n}) \in \mathcal{C}_+(\tilde{I}) \\ I \in Cr(\tilde{P})}} Bi_+(\tilde{P}). \tag{9.94}$$

As the sets $Bi_+(\tilde{P})$ are disjoint, we have

$$\sum_{\mathcal{C}_+(\tilde{I})} \#Bi_+(\tilde{P}) \leqslant \#Bi_+(\tilde{I}, \tilde{I}, I_\omega; x). \tag{9.95}$$

We now sum over candidates $I \subset \tilde{I}$ the estimate for the cardinality of $Bi_+^{old}(I, I, I_\omega; x)$ deduced from the inclusion above. We obtain, using Proposition 33 in Section 9.5 and (9.102)

$$\begin{aligned}
\sum_{I \subset \tilde{I}} \#Bi_+^{old}(I, I, I_\omega; x) &\leqslant \sum_{\mathcal{C}_+(\tilde{I})} \#Bi_+(\tilde{P}) \#Cr(\tilde{P}) \\
&\leqslant C|\tilde{I}|^{-\tau d_u^+} \sum_{\mathcal{C}_+(\tilde{I})} \#Bi_+(\tilde{P}) \\
&\leqslant C|\tilde{I}|^{-\tau d_u^+} \#Bi_+(\tilde{I}, \tilde{I}, I_\omega; x).
\end{aligned} \tag{9.96}$$

From now on, we assume that the exponents σ_0 (close to $1 - d_s^0$) and σ_1 (close to $d_s^0 - d_u^0$) in (SR3) satisfy

$$\sigma_0 + \sigma_1 \leqslant 1 - 3\tau - d_u^+. \tag{9.97}$$

We also assume that

$$\rho_0 \geqslant d_s^* + \varepsilon_0^{\tau^2/2}. \tag{9.98}$$

Proposition **40.** — *Assume that* $\widetilde{\mathrm{I}}$ *is strongly regular. Then, except for a proportion of candidates not greater than* $|\widetilde{\mathrm{I}}|^{\tau^2}$, *the estimate* (SR3)$_s$ *is satisfied by* $\mathrm{B}i_+(\mathrm{I}, \mathrm{I}, \mathrm{I}_\omega; x)$ *for all* $x \geqslant x_{\min}$ *and all* $\mathrm{I}_\omega \supset \widetilde{\mathrm{I}}$.

Proof. — In view of Proposition 38, we have only to consider $\mathrm{B}i_+^{old}(\mathrm{I}, \mathrm{I}, \mathrm{I}_\omega; x)$. For fixed $x = 2^{-l} \geqslant x_{\min}$ and fixed $\mathrm{I}_\omega \supset \widetilde{\mathrm{I}}$, it follows from (9.96) that we have

$$\#\mathrm{B}i_+^{old}(\mathrm{I}, \mathrm{I}, \mathrm{I}_\omega; x) \leqslant \mathrm{C}|\widetilde{\mathrm{I}}|^{\tau - 2\tau^2} |\widetilde{\mathrm{I}}|^{-\tau d_u^+} \#\mathrm{B}i_+(\widetilde{\mathrm{I}}, \widetilde{\mathrm{I}}, \mathrm{I}_\omega; x), \tag{9.99}$$

except for a proportion of candidates no greater than $|\widetilde{\mathrm{I}}|^{2\tau^2}$.

Let $x_{cr} = \varepsilon_0 |\mathrm{P}_u| (\frac{|\mathrm{I}_\omega|}{\varepsilon_0})^{\frac{\sigma_0}{\rho_0 - \rho_1}}$. For $x \leqslant x_{cr}$, we have

$$\#\mathrm{B}i_+(\widetilde{\mathrm{I}}, \widetilde{\mathrm{I}}, \mathrm{I}_\omega; x) \leqslant \varepsilon_0^{-\mathrm{A}\tau} \Big(\frac{x}{\varepsilon_0 |\mathrm{P}_u|}\Big)^{-\rho_0} \Big(\frac{|\widetilde{\mathrm{I}}|}{\varepsilon_0}\Big)^{\sigma_0 + \sigma_1} \Big(\frac{|\mathrm{I}_\omega|}{\varepsilon_0}\Big)^{\sigma_0}, \tag{9.100}$$

while for $x \geqslant x_{cr}$, we have

$$\#\mathrm{B}i_+(\widetilde{\mathrm{I}}, \widetilde{\mathrm{I}}, \mathrm{I}_\omega; x) \leqslant \varepsilon_0^{-\mathrm{A}\tau} \Big(\frac{x}{\varepsilon_0 |\mathrm{P}_u|}\Big)^{-\rho_1} \Big(\frac{|\widetilde{\mathrm{I}}|}{\varepsilon_0}\Big)^{\sigma_0 + \sigma_1}. \tag{9.101}$$

As $\sigma_0 + \sigma_1 \leqslant 1 - 3\tau - d_u^+$, we have

$$\mathrm{C}|\widetilde{\mathrm{I}}|^{\tau - 2\tau^2 - \tau d_u^+} \ll |\widetilde{\mathrm{I}}|^{\tau(\sigma_0 + \sigma_1)}, \tag{9.102}$$

and therefore, when a candidate I satisfies (9.99), we have

$$\#\mathrm{B}i_+^{old}(\mathrm{I}, \mathrm{I}, \mathrm{I}_\omega; x) \ll \varepsilon_0^{-\mathrm{A}\tau}\mathrm{B} = \varepsilon_0^{-\mathrm{A}\tau} \max(\mathrm{B}_0, \mathrm{B}_1). \tag{9.103}$$

Thus, to get the required estimate, it is sufficient to remove a proportion $\leqslant |\widetilde{\mathrm{I}}|^{2\tau^2}$ of candidates for each $x = 2^{-l} \geqslant x_{\min}$ and each $\mathrm{I}_\omega \supset \widetilde{\mathrm{I}}$. The number of values of l is $\leqslant \mathrm{C}(\rho_0 - d_s^*)^{-1} \leqslant \mathrm{C}\varepsilon_0^{-\tau^2/2} \log_2 |\widetilde{\mathrm{I}}|$. The number of possibilities for I_ω is the level k of the interval I. From $|\mathrm{I}| = \varepsilon_0^{(1+\tau)^k}$, we have

$$k = \frac{\log(\frac{\log |\mathrm{I}|}{\log \varepsilon_0})}{\log(1 + \tau)}. \tag{9.104}$$

Now, we have, for ε_0 small enough

$$\mathrm{C}|\widetilde{\mathrm{I}}|^{2\tau^2} \varepsilon_0^{-\tau^2/2} \log_2 |\widetilde{\mathrm{I}}| \frac{\log(\frac{\log |\mathrm{I}|}{\log \varepsilon_0})}{\log(1 + \tau)} \ll |\widetilde{\mathrm{I}}|^{\tau^2}, \tag{9.105}$$

and the proof of the proposition is complete. □

9.7.4. *The case* $I = I_\omega \neq I_\alpha$. — The case is essentially symmetric to the case that we have just considered above. We will now assume that (9.98) holds and that

$$\textbf{(9.106)} \qquad \sigma_0 \leqslant 1 - 3\tau - d_s^+.$$

Proposition **41.** — *Assume that* $\widetilde{I}$ *is strongly regular. Then, except for a proportion of candidates not greater than* $|\widetilde{I}|^{\tau^2}$, *the estimate* $(SR3)_s$ *is satisfied by* $Bi_+(I, I_\alpha, I; x)$ *for all* $x \geqslant x_{\min}$ *and all* $I_\alpha \supset \widetilde{I}$.

Proof. — The only difference with the proof of Proposition 40 is that we now have, for each $x = 2^{-l} \geqslant x_{\min}$, each $I_\alpha \supset \widetilde{I}$

$$\textbf{(9.107)} \qquad \#Bi_+^{old}(I, I_\alpha, I; x) \leqslant C|\widetilde{I}|^{\tau - 2\tau^2}|\widetilde{I}|^{-\tau d_s^+}\#Bi_+(\widetilde{I}, I_\alpha, \widetilde{I}; x),$$

except from a proportion of candidates no greater than $|\widetilde{I}|^{2\tau^2}$, where $Bi_+(\widetilde{I}, I_\alpha, \widetilde{I}; x)$ is now controlled by

$$\textbf{(9.108)} \qquad \#Bi_+(\widetilde{I}, I_\alpha, \widetilde{I}; x) \leqslant \varepsilon_0^{-A\tau}\left(\frac{x}{\varepsilon_0|P_u|}\right)^{-\rho_0}\left(\frac{|\widetilde{I}|}{\varepsilon_0}\right)^{\sigma_0}\left(\frac{|I_\alpha|}{\varepsilon_0}\right)^{\sigma_0+\sigma_1},$$

for $x \leqslant x_{cr} = \varepsilon_0|P_u|(\frac{|I_\alpha|}{\varepsilon_0})^{\frac{\sigma_0}{\rho_0-\rho_1}}$, and by

$$\textbf{(9.109)} \qquad \#Bi_+(\widetilde{I}, I_\alpha, \widetilde{I}; x) \leqslant \varepsilon_0^{-A\tau}\left(\frac{x}{\varepsilon_0|P_u|}\right)^{-\rho_1}\left(\frac{|\widetilde{I}|}{\varepsilon_0}\right)^{\sigma_0}\left(\frac{|I_\alpha|}{\varepsilon_0}\right)^{\sigma_1},$$

for $x \geqslant x_{cr}$.

Using $\sigma_0 \leqslant 1 - 3\tau - d_s^+$ allows to conclude as in the proof of Proposition 40. □

9.7.5. *The case* $I = I_\alpha = I_\omega$, *x large.* — Except for the proof of Proposition 38, the only case left in the induction step for $(SR3)_s$ is $I = I_\alpha = I_\omega$, $x \geqslant x_{\min}$. In this case, we have

$$\textbf{(9.110)} \qquad B_0 = \left(\frac{x}{\varepsilon_0|P_u|}\right)^{-\rho_0}\left(\frac{|I|}{\varepsilon_0}\right)^{2\sigma_0+\sigma_1},$$

$$\textbf{(9.111)} \qquad B_1 = \left(\frac{x}{\varepsilon_0|P_u|}\right)^{-\rho_1}\left(\frac{|I|}{\varepsilon_0}\right)^{\sigma_0+\sigma_1}.$$

Observe that we have

$$\textbf{(9.112)} \qquad Bi_+^{old}(I, I, I; x) \subset Bi_+^{old}(I, I, \widetilde{I}; x).$$

In the proof of Proposition 40, we have shown that for $x = 2^{-l} \geqslant \widetilde{x}_{cr} := \varepsilon_0|P_u|(\frac{|\widetilde{I}|}{\varepsilon_0})^{\frac{\sigma_0}{\rho_0-\rho_1}}$, we have

$$\textbf{(9.113)} \qquad \#Bi_+^{old}(I, I, \widetilde{I}; x) \leqslant \varepsilon_0^{-A\tau}B_1,$$

except for a proportion $\leqslant |\widetilde{I}|^{2\tau^2}$ of candidates I. We get therefore the required estimate for $Bi_+^{old}(I, I, I; x)$ when $x \geqslant \widetilde{x}_{cr}$.

Proposition **42.** — *Assume that* $\widetilde{\mathrm{I}}$ *is strongly regular. Then, except for a proportion of candidates not greater than* $|\widetilde{\mathrm{I}}|^{\tau^2}$, *the estimate* (SR3)$_s$ *is satisfied by* $\mathrm{B}i_+(\mathrm{I}, \mathrm{I}, \mathrm{I}; x)$ *for all* $x \geqslant \widetilde{x}_{cr}$, *where*

$$(\mathbf{9.114}) \qquad \widetilde{x}_{cr} = \varepsilon_0 |\mathrm{P}_u| \left(\frac{|\widetilde{\mathrm{I}}|}{\varepsilon_0} \right)^{\frac{\sigma_0}{\rho_0 - \rho_1}}.$$

9.7.6. *The case* $\mathrm{I} = \mathrm{I}_\alpha = \mathrm{I}_\omega$, $x_{\min} \leqslant x \leqslant \widetilde{x}_{cr}$: *idea of the proof.* — This remaining case is by far the hardest! In this range of x, the required bound for the cardinality of $\mathrm{B}i_+(\mathrm{I}, \mathrm{I}, \mathrm{I}; x)$ is given by B_0 (except at the very top of the range); the exponent $2\sigma_0 + \sigma_1$ (close to $2 - d_s^0 - d_u^0$) in (9.110) means that we have to take into account the criticality of both P and Q in the selection process.

Let $\mathrm{I} \subset \widetilde{\mathrm{I}}$ be a candidate interval and let $(\mathrm{P}, \mathrm{Q}, n) \in \mathrm{B}i_+^{old}(\mathrm{I}, \mathrm{I}, \mathrm{I}; x)$. The argument at the beginning of Section 9.7.3 shows that there exists $(\mathrm{P}_\alpha, \mathrm{Q}_\alpha, n_\alpha) \in \mathcal{C}_+(\widetilde{\mathrm{I}})$ such that $\mathrm{P} \subset \mathrm{P}_\alpha$ and $\mathrm{I} \in Cr(\mathrm{P}_\alpha)$. Similarly, there exists $(\mathrm{P}_\omega, \mathrm{Q}_\omega, n_\omega) \in \mathcal{C}_-(\widetilde{\mathrm{I}})$ such that $\mathrm{Q} \subset \mathrm{Q}_\omega$ and $\mathrm{I} \in Cr(\mathrm{Q}_\omega)$. We also observe that $(\mathrm{P}, \mathrm{Q}, n) \in \mathrm{B}i_+(\widetilde{\mathrm{I}}, \widetilde{\mathrm{I}}, \widetilde{\mathrm{I}}; x)$.

Definition **11.** — *For* $(\mathrm{P}_\alpha, \mathrm{Q}_\alpha, n_\alpha) \in \mathcal{C}_+(\widetilde{\mathrm{I}})$, $(\mathrm{P}_\omega, \mathrm{Q}_\omega, n_\omega) \in \mathcal{C}_-(\widetilde{\mathrm{I}})$, *we denote by* $\mathrm{B}i_+(\mathrm{P}_\alpha, \mathrm{Q}_\omega)$ *the set of* $(\mathrm{P}, \mathrm{Q}, n) \in \mathrm{B}i_+(\widetilde{\mathrm{I}}, \widetilde{\mathrm{I}}, \widetilde{\mathrm{I}}; x)$ *such that* $\mathrm{P} \subset \mathrm{P}_\alpha$ *and* $\mathrm{Q} \subset \mathrm{Q}_\omega$.

We have just seen that, for any candidate I, we have

$$(\mathbf{9.115}) \qquad \mathrm{B}i_+^{old}(\mathrm{I}, \mathrm{I}, \mathrm{I}; x) \subset \bigsqcup_{\substack{(\mathrm{P}_\alpha, \mathrm{Q}_\alpha, n_\alpha) \in \mathcal{C}_+(\widetilde{\mathrm{I}}) \\ \mathrm{I} \in Cr(\mathrm{P}_\alpha)}} \; \bigsqcup_{\substack{(\mathrm{P}_\omega, \mathrm{Q}_\omega, n_\omega) \in \mathcal{C}_-(\widetilde{\mathrm{I}}) \\ \mathrm{I} \in Cr(\mathrm{Q}_\omega)}} \mathrm{B}i_+(\mathrm{P}_\alpha, \mathrm{Q}_\omega).$$

As the sets $\mathrm{B}i_+(\mathrm{P}_\alpha, \mathrm{Q}_\omega)$ are disjoint, we have

$$(\mathbf{9.116}) \qquad \sum_{\mathcal{C}_+(\widetilde{\mathrm{I}})} \sum_{\mathcal{C}_-(\widetilde{\mathrm{I}})} \#\mathrm{B}i_+(\mathrm{P}_\alpha, \mathrm{Q}_\omega) \leqslant \#\mathrm{B}i_+(\widetilde{\mathrm{I}}, \widetilde{\mathrm{I}}, \widetilde{\mathrm{I}}; x).$$

If we were able to prove (for a "random" candidate I) that the events $\mathrm{I} \in Cr(\mathrm{P}_\alpha)$ and $\mathrm{I} \in Cr(\mathrm{Q}_\omega)$ are roughly "independent", i.e. if the proportion of candidates in $Cr(\mathrm{P}_\alpha) \cap Cr(\mathrm{Q}_\omega)$ was roughly the product $|\widetilde{\mathrm{I}}|^{\tau(2 - d_s^+ - d_u^+)}$ of the proportions in $Cr(\mathrm{P}_\alpha)$ and $Cr(\mathrm{Q}_\omega)$ (cf. Proposition 33 in Section 9.5), we would be able to proceed as in the proof of Proposition 40. But this is unfortunately not the case.

Instead, we will use some degree of independence, in the range of x under consideration, of the variables P_α and Q_ω in $\mathrm{B}i_+(\mathrm{P}_\alpha, \mathrm{Q}_\omega)$. To explain the technique, consider first the unrealistic model case where we assume that

$$(\mathbf{9.117}) \qquad \#\mathrm{B}i_+(\mathrm{P}_\alpha, \mathrm{Q}_\omega) = b_+(\mathrm{P}_\alpha) b_-(\mathrm{Q}_\omega),$$

for some functions b_+, b_- on $\mathcal{C}_+(\widetilde{\mathrm{I}})$, $\mathcal{C}_-(\widetilde{\mathrm{I}})$, respectively.

From (9.115), we would get

$$(\mathbf{9.118}) \qquad \#\mathrm{B}i_+^{old}(\mathrm{I}, \mathrm{I}, \mathrm{I}; x) \leqslant \phi_+(\mathrm{I}) \phi_-(\mathrm{I}),$$

with

$$\phi_+(\mathrm{I}) = \sum_{\mathrm{I}\in Cr(\mathrm{P}_\alpha)} b_+(\mathrm{P}_\alpha), \tag{9.119}$$

$$\phi_-(\mathrm{I}) = \sum_{\mathrm{I}\in Cr(\mathrm{Q}_\omega)} b_-(\mathrm{Q}_\omega). \tag{9.120}$$

We now average *separately* ϕ_+ and ϕ_-. We obtain

$$\sum_{\mathrm{I}} \phi_+(\mathrm{I}) \leqslant \Big(\max_{\mathcal{C}_+(\widetilde{\mathrm{I}})} \#Cr(\mathrm{P}_\alpha)\Big) \sum_{\mathcal{C}_+(\widetilde{\mathrm{I}})} b_+(\mathrm{P}_\alpha), \tag{9.121}$$

$$\sum_{\mathrm{I}} \phi_-(\mathrm{I}) \leqslant \Big(\max_{\mathcal{C}_-(\widetilde{\mathrm{I}})} \#Cr(\mathrm{Q}_\omega)\Big) \sum_{\mathcal{C}_-(\widetilde{\mathrm{I}})} b_-(\mathrm{Q}_\omega), \tag{9.122}$$

where $Cr(\mathrm{P}_\alpha)$ and $Cr(\mathrm{Q}_\omega)$ are estimated by Proposition 33 in Section 9.5. From (9.116), we have

$$\sum_{\mathcal{C}_+(\widetilde{\mathrm{I}})} b_+(\mathrm{P}_\alpha) \sum_{\mathcal{C}_-(\widetilde{\mathrm{I}})} b_-(\mathrm{Q}_\omega) = \sum_{\mathcal{C}_+(\widetilde{\mathrm{I}})\times\mathcal{C}_-(\widetilde{\mathrm{I}})} \#Bi_+(\mathrm{P}_\alpha, \mathrm{Q}_\omega) \leqslant \#Bi_+(\widetilde{\mathrm{I}}, \widetilde{\mathrm{I}}, \widetilde{\mathrm{I}}; x). \tag{9.123}$$

It is then sufficient to eliminate candidates for which either ϕ_+ or ϕ_- is much above its average value to obtain $(\mathrm{SR3})_s$ for the remaining intervals.

As (9.117) does not hold, we will find an appropriate substitute as follows.

We will subdivide each class $Bi_+(\mathrm{P}_\alpha, \mathrm{Q}_\omega)$ into subclasses $Bi_+(\mathrm{P}_\alpha, \mathrm{Q}_\omega, \ell)$; the index ℓ runs through a finite large set L dependent on $\widetilde{\mathrm{I}}$ and x but independent on P_α and Q_ω. Moreover, we will have functions $b_+(\mathrm{P}_\alpha, \ell)$, $b_-(\mathrm{Q}_\omega, \ell)$ on $\mathcal{C}_+(\widetilde{\mathrm{I}}) \times \mathrm{L}$, $\mathcal{C}_-(\widetilde{\mathrm{I}}) \times \mathrm{L}$, respectively, such that,

$$\#Bi_+(\mathrm{P}_\alpha, \mathrm{Q}_\omega, \ell) \leqslant b_+(\mathrm{P}_\alpha, \ell), b_-(\mathrm{Q}_\omega, \ell). \tag{9.124}$$

We then set, for each $\ell \in \mathrm{L}$:

$$\phi_{+,\ell}(\mathrm{I}) = \sum_{\mathrm{I}\in Cr(\mathrm{P}_\alpha)} b_+(\mathrm{P}_\alpha, \ell), \tag{9.125}$$

$$\phi_{-,\ell}(\mathrm{I}) = \sum_{\mathrm{I}\in Cr(\mathrm{Q}_\omega)} b_-(\mathrm{Q}_\omega, \ell). \tag{9.126}$$

We average each of these functions to get, in view of Proposition 33 in Section 9.5,

$$\sum_{\mathrm{I}} \phi_{+,\ell}(\mathrm{I}) \leqslant \mathrm{C}|\widetilde{\mathrm{I}}|^{-\tau d_u^+} b_+(\ell), \tag{9.127}$$

$$\sum_{\mathrm{I}} \phi_{-,\ell}(\mathrm{I}) \leqslant \mathrm{C}|\widetilde{\mathrm{I}}|^{-\tau d_s^+} b_-(\ell), \tag{9.128}$$

with

$$b_+(\ell) = \sum_{C_+(\widetilde{I})} b_+(P_\alpha, \ell), \tag{9.129}$$

$$b_-(\ell) = \sum_{C_-(\widetilde{I})} b_-(Q_\omega, \ell). \tag{9.130}$$

For each ℓ, we will have

$$\phi_{+,\ell}(I) \leqslant |\widetilde{I}|^{\tau(1-d_u^+-3\tau)} b_+(\ell), \tag{9.131}$$

$$\phi_{-,\ell}(I) \leqslant |\widetilde{I}|^{\tau(1-d_s^+-3\tau)} b_-(\ell) \tag{9.132}$$

except for a proportion of candidates not greater than $C|\widetilde{I}|^{3\tau^2}$. Set

$$\widehat{B} = \sum_{L} b_+(\ell) b_-(\ell). \tag{9.133}$$

Because we need to eliminate candidates for each ℓ, L should not be too large. We will show in Section 9.10 that

$$\# L \leqslant |\widetilde{I}|^{-\tau^2}. \tag{9.134}$$

Taking into account that we must eliminate candidates for each $x = 2^{-l} \geqslant x_{\min}$, the total proportion of the failed candidates is at most $|\widetilde{I}|^{\tau^2}$ (cf. proof of Proposition 40 in Section 9.7.3). On the other hand, for the surviving candidates, the discussion above gives

$$\begin{aligned}
\# Bi_+^{old}(I, I, I; x) &\leqslant \sum_{I \in Cr(P_\alpha) \cap Cr(Q_\omega)} \# Bi_+(P_\alpha, Q_\omega) \\
&\leqslant \sum_{L} \sum_{I \in Cr(P_\alpha)} \sum_{I \in Cr(Q_\omega)} b_+(P_\alpha, \ell) b_-(Q_\omega, \ell) \\
&= \sum_{L} \phi_{+,\ell}(I) \phi_{-,\ell}(I) \\
&\leqslant |\widetilde{I}|^{\tau(2-d_s^+-d_u^+-6\tau)} \widehat{B}.
\end{aligned} \tag{9.135}$$

If we are able to prove that

$$|\widetilde{I}|^{\tau(2-d_s^+-d_u^+-6\tau)} \widehat{B} \leqslant \varepsilon_0^{-A\tau} B, \tag{9.136}$$

where B is the bound from $(SR3)_s$, we get the required conclusion.

In the next four subsections, we will

- prove Proposition 38 (estimate on the size of $Bi_+^{new}(I, I_\alpha, I_\omega; x)$);
- define precisely L and the subclasses $Bi_+(P_\alpha, Q_\omega, \ell)$;
- bound the cardinality of L (prove (9.134));
- obtain an appropriate estimate for $\widehat{B}$.

9.8. *Size of* $Bi_+^{new}(I, I_\alpha, I_\omega; x)$.

9.8.1. The goal of Section 9.8 is to prove Proposition 38 in Section 9.7.2. Let us recall the statement: there exists a constant $\sigma_2 > 0$, depending only on d_u^0, d_s^0, ω_u, such that, if $\widetilde{I}$ is strongly regular, one has, for all candidates $I \subset \widetilde{I}$, all $I_\alpha, I_\omega \supset I$, all $x \geqslant x_{\min}$

$$(\mathbf{9.137}) \qquad \#Bi_+^{new}(I, I_\alpha, I_\omega; x) \leqslant |\widetilde{I}|^{\sigma_2} B,$$

where $\varepsilon_0^{-A\tau} B$ is the bound for $\#Bi_+(I, I_\alpha, I_\omega; x)$ predicted by $(SR3)_s$.

It is sufficient to prove this estimate when I_α, I_ω contain $\widetilde{I}$. Indeed, let $\widetilde{I}_\alpha = \widetilde{I} \cup I_\alpha$, $\widetilde{I}_\omega = \widetilde{I} \cup I_\omega$. We have $Bi_+^{new}(I, I_\alpha, I_\omega; x) \subset Bi_+^{new}(I, \widetilde{I}_\alpha, \widetilde{I}_\omega; x)$. On the other hand, the bounds $\varepsilon_0^{-A\tau} B$, $\varepsilon_0^{-A\tau} \widetilde{B}$ predicted by $(SR3)_s$ for $\#Bi_+(I, I_\alpha, I_\omega; x)$, $\#Bi_+(I, \widetilde{I}_\alpha, \widetilde{I}_\omega; x)$ respectively satisfy $B \geqslant |\widetilde{I}|^{C\tau} \widetilde{B}$. Therefore the estimate (9.144) for $\#Bi_+^{new}(I, \widetilde{I}_\alpha, \widetilde{I}_\omega; x)$ imply the estimate for $\#Bi_+^{new}(I, I_\alpha, I_\omega; x)$ with a slightly smaller value of $\sigma_2 > 0$.

We will therefore now assume that both I_α, I_ω contain $\widetilde{I}$.

9.8.2. Let $I \subset \widetilde{I}$ be a candidate interval, let I_α, I_ω be parameter intervals containing $\widetilde{I}$, let $x = 2^{-l} \geqslant x_{\min}$ and let $(P, Q, n) \in \#Bi_+(I, I_\alpha, I_\omega; x)$.

We apply the structure theorem of Section 6.5. We obtain an integer $k > 0$ and elements $(P_i, Q_i, n_i) \in \mathcal{R}(\widetilde{I})$ for $0 \leqslant i \leqslant k$ such that

- $(P, Q, n) \in (P_0, Q_0, n_0) \boxdot \cdots \boxdot (P_k, Q_k, n_k)$;
- for $0 \leqslant i < k$, Q_i and P_{i+1} are $\widetilde{I}$-critical and $Q_i \pitchfork_{\widetilde{I}} P_{i+1}$ does not hold;
- for all $t \in I$, $|P| \leqslant C^k |I|^{-k/2} \prod_{0 \leqslant i \leqslant k} |P_i|$ (cf. Proposition 14 in Section 6.6.3).

For $0 \leqslant i \leqslant k$, denote by $x_i = 2^{-l_i}$ the largest integral negative power of 2 such that $|P_i| \geqslant x_i$ for some $t \in \widetilde{I}$.

Lemma **12.** — *We have* $(P_0, Q_0, n_0) \in Bi_+(\widetilde{I}, I_\alpha, \widetilde{I}; x_0)$, $(P_k, Q_k, n_k) \in Bi_+(\widetilde{I}, \widetilde{I}, I_\omega; x_k)$, *and* $(P_i, Q_i, n_i) \in Bi_+(\widetilde{I}, \widetilde{I}, \widetilde{I}; x_i)$ *for* $0 < i < k$.

Remark **12.** — This lemma is the reason why we need to consider different levels of criticality for P and Q.

Proof. — For $0 \leqslant i < k$, Q_i and P_{i+1} are thin $\widetilde{I}$-critical by Lemma 10 in Section 9.1. As P is thin I_α-critical, there exists $(P', Q', n') \in \mathcal{R}(I_\alpha)$ with $P \subset P'$ and $\min_{I_\alpha} |P'|^{1-\eta} \leqslant 2|I_\alpha|$. As $I_\alpha \supset \widetilde{I}$ and P_0 is the thinnest $\widetilde{I}$-defined rectangle containing P, one has $P_0 \subset P'$ and P_0 is thin I_α-critical. Similarly, Q_k is thin I_ω-critical. In view of the definition of the x_i, the proof of the lemma is complete. □

From the definition of the x_i and the estimate for $|P|$ above, we get

$$(\mathbf{9.138}) \qquad x \leqslant C^k |I|^{-\frac{k}{2}} \prod_0^k x_i.$$

Let us write

$$\begin{aligned}
\#(x_0) &:= \#\mathrm{B}i_+(\widetilde{\mathrm{I}}, \mathrm{I}_\alpha, \widetilde{\mathrm{I}}; x_0),\\
\#(x_k) &:= \#\mathrm{B}i_+(\widetilde{\mathrm{I}}, \widetilde{\mathrm{I}}, \mathrm{I}_\omega; x_k),\\
\#(x_i) &:= \#\mathrm{B}i_+(\widetilde{\mathrm{I}}, \widetilde{\mathrm{I}}, \widetilde{\mathrm{I}}; x_i) \quad \text{for } 0 < i < k.
\end{aligned}$$

Then, as each parabolic composition produces two elements, we have

$$\#\mathrm{B}i_+^{new}(\mathrm{I}, \mathrm{I}_\alpha, \mathrm{I}_\omega; x) \leqslant \sum_{k>0} 2^k \sum_{x_0,\dots,x_k} \prod_0^k \#(x_i) \tag{9.139}$$

where the x_i in the sum are of the form 2^{-l_i} and must satisfy (9.138).

The term $\#(x_i)$ is estimated by the induction hypothesis $(\mathrm{SR3})_s$ for $\widetilde{\mathrm{I}}$. The bound for $\#(x_i)$ has a phase transition at a threshold $x_{i,cr}$, with (cf. (9.16))

$$\begin{aligned}
x_{0,cr} &= \varepsilon_0 |\mathrm{P}_u| \left(\frac{|\mathrm{I}_\alpha|}{\varepsilon_0}\right)^{\frac{\sigma_0}{\rho_0-\rho_1}},\\
x_{k,cr} &= \varepsilon_0 |\mathrm{P}_u| \left(\frac{|\mathrm{I}_\omega|}{\varepsilon_0}\right)^{\frac{\sigma_0}{\rho_0-\rho_1}},\\
x_{i,cr} &= \varepsilon_0 |\mathrm{P}_u| \left(\frac{|\widetilde{\mathrm{I}}|}{\varepsilon_0}\right)^{\frac{\sigma_0}{\rho_0-\rho_1}}, \quad \text{for } 0 < i < k.
\end{aligned} \tag{9.140}$$

We divide the sum in the right-hand part of (9.139) into two parts. In the first sum, denoted by S_l, we put the terms for which every x_i is above the threshold $x_{i,cr}$. In the second sum, denoted by S_s, we put the terms for which at least one of the x_i is below $x_{i,cr}$.

9.8.3. *Terms with all x_i large.* — Let us consider a term in S_l. All $\#(x_i)$ are bounded by $\varepsilon_0^{-\mathrm{A}\tau}\mathrm{B}_1$ and we have

$$\prod_0^k \#(x_i) \leqslant \varepsilon_0^{-(k+1)\mathrm{A}\tau} \left(\prod_0^k x_i\right)^{-\rho_1} (\varepsilon_0|\mathrm{P}_u|)^{(k+1)\rho_1} \left(\frac{|\mathrm{I}_\alpha|}{\varepsilon_0}\right)^{\sigma_1} \left(\frac{|\widetilde{\mathrm{I}}|}{\varepsilon_0}\right)^{(k+1)\sigma_0+k\sigma_1}. \tag{9.141}$$

In view of (9.138), the right-hand side is bounded by

$$\mathrm{C}\left(\frac{x}{\varepsilon_0|\mathrm{P}_u|}\right)^{-\rho_1} \left(\frac{|\mathrm{I}_\alpha|}{\varepsilon_0}\right)^{\sigma_1} \left(\frac{|\widetilde{\mathrm{I}}|}{\varepsilon_0}\right)^{\sigma_0} \mathrm{Z}^k, \tag{9.142}$$

with

$$\mathrm{Z} := \mathrm{C}\varepsilon_0^{-2\mathrm{A}\tau} \left(\varepsilon_0|\mathrm{P}_u||\mathrm{I}|^{-\frac{1}{2}}\right)^{\rho_1} \left(\frac{|\widetilde{\mathrm{I}}|}{\varepsilon_0}\right)^{\sigma_0+\sigma_1}. \tag{9.143}$$

For all $0 \leqslant i \leqslant k$, from (9.140) and $C^{-1}\varepsilon_0^{\omega_u} \leqslant |P_u| \leqslant C^{-1}\varepsilon_0^{\omega_u}$ (cf. (9.17) in Section 5.6.4), we have $x_{i,\sigma} > |\widetilde{I}|^C$ for some large enough C. Therefore, for each $k > 0$, the number of terms in S_ℓ is at most $(C|\log_2|\widetilde{I}||)^{k+1} \leqslant (C|\log_2|\widetilde{I}||)^{2k}$.

We thus obtain for S_ℓ the bound

$$S_\ell \leqslant C\Big(\frac{x}{\varepsilon_0|P_u|}\Big)^{-\rho_1}\Big(\frac{|I_\alpha|}{\varepsilon_0}\Big)^{\sigma_1}\Big(\frac{|\widetilde{I}|}{\varepsilon_0}\Big)^{\sigma_0}\sum_{k>0} Z_1^k, \tag{9.144}$$

with

$$Z_1 := 2(C|\log_2|\widetilde{I}||)^2 Z. \tag{9.145}$$

In view of (9.143), we have

$$Z_1 \ll \Big(\frac{|\widetilde{I}|}{\varepsilon_0}\Big)^{\sigma_0+\sigma_1-\rho_1}|\widetilde{I}|^{\frac{1}{3}\rho_1}, \tag{9.146}$$

where the exponent $\sigma_0 + \sigma_1 - \rho_1$ is strictly positive from the hypothesis (H4), as we have seen in Section 9.3 (cf. (9.24)).

As $Z_1 < \frac{1}{2}$, we obtain from (9.144) that

$$\begin{aligned} S_\ell &\leqslant \frac{1}{2}\Big(\frac{x}{\varepsilon_0|P_u|}\Big)^{-\rho_1}\Big(\frac{|\widetilde{I}|}{\varepsilon_0}\Big)^{2\sigma_0+\sigma_1-\rho_1}\Big(\frac{|I_\alpha|}{\varepsilon_0}\Big)^{\sigma_1}|\widetilde{I}|^{\frac{1}{3}\rho_1} \\ &\leqslant \frac{1}{2}\Big(\frac{x}{\varepsilon_0|P_u|}\Big)^{-\rho_1}\Big(\min\frac{|I_\alpha|}{\varepsilon_0},\frac{|I_\omega|}{\varepsilon_0}\Big)^{\sigma_0}\Big(\frac{|I_\alpha|}{\varepsilon_0}\Big)^{\sigma_1}|\widetilde{I}|^{\frac{1}{3}\rho_1} \\ &= \frac{1}{2}B_1|\widetilde{I}|^{\frac{1}{3}\rho_1}, \end{aligned} \tag{9.147}$$

where we have used that both I_α, I_ω contain $\widetilde{I}$ and $\sigma_0 + \sigma_1 - \rho_1 > 0$.

9.8.4. *Terms with some x_i small.* — Consider a term in the sum S_s. Let J be the non-empty subset of indices $i \in \{0,\dots,k\}$ for which $x_i < x_{i,\sigma}$, and write $j = \#J$.

We first estimate the product $\prod_i \#(x_i)$. As $\rho_0 > \rho_1$, we have from (9.138)

$$\prod_J x_i^{-\rho_0}\prod_{J^c} x_i^{-\rho_1} \leqslant \big(C^{-k}|I|^{\frac{k}{2}}x\big)^{-\rho_0}. \tag{9.148}$$

As we also have $\widetilde{I} \subset I_\alpha$, $\widetilde{I} \subset I_\omega$, we obtain

$$\prod_i \#(x_i) \leqslant C\varepsilon_0^{-A\tau}\Big(\frac{x}{\varepsilon_0|P_u|}\Big)^{-\rho_0}\Big(\frac{|I_\alpha|}{\varepsilon_0}\Big)^{\sigma_0+\sigma_1}\Big(\frac{|I_\omega|}{\varepsilon_0}\Big)^{\sigma_0} Y_0^{j-1} Y_1^k, \tag{9.149}$$

with

$$\textbf{(9.150)} \qquad Y_0 = (\varepsilon_0|P_u|)^{\rho_0-\rho_1}\left(\frac{|\widetilde{I}|}{\varepsilon_0}\right)^{\sigma_0},$$

$$\textbf{(9.151)} \qquad Y_1 = C\varepsilon_0^{-A\tau}|I|^{-\rho_0/2}(\varepsilon_0|P_u|)^{\rho_1}\left(\frac{|\widetilde{I}|}{\varepsilon_0}\right)^{\sigma_0+\sigma_1}.$$

We have

$$\textbf{(9.152)} \qquad Y_0 < 1,$$

and, as $C^{-1}\varepsilon_0^{\omega_u} \leqslant |P_u| \leqslant C^{-1}\varepsilon_0^{\omega_u}$ (cf. (5.17)), we can rewrite Y_1 as

$$\textbf{(9.153)} \qquad Y_1 = C\left(\frac{\widetilde{I}}{\varepsilon_0}\right)^{\sigma_0+\sigma_1-\frac{1}{2}\rho_0(1+\tau)}\varepsilon_0^{\rho_1-\frac{1}{2}\rho_0(1+\tau)+\rho_1\omega_u-A\tau}.$$

The exponent $\sigma_0+\sigma_1-\frac{1}{2}\rho_0(1+\tau)$ is close to $1-d_u^0-1/2d_s^0$ when $\tau \gg \eta \gg \varepsilon_0$ are small. We have $1-d_u^0-1/2d_s^0 > 0$ from (H4) (see also the remark after Proposition 35). The exponent $\rho_1-\frac{1}{2}\rho_0(1+\tau)$ is close to $\frac{1}{2}d_s^0(3d_s^0+d_u^0-2)(d_s^0+d_u^0)^{-1}$ when $\tau \gg \eta \gg \varepsilon_0$ are small. From $d_s^0+d_u^0 \geqslant 1$, $d_s^0 \geqslant d_u^0$, we have $3d_s^0+d_u^0-2 \geqslant 0$ (with equality iff $d_u^0 = d_s^0 = 1/2$). On the other hand, the exponent $\rho_1\omega_u$ is close to $\omega_u d_s^0(2d_s^0+d_u^0-1)(d_s^0+d_u^0)^{-1} > 0$ when $\tau \gg \eta \gg \varepsilon_0$ are small. All this means that we can find a number $\sigma_2 > 0$, depending only on d_u^0, d_s^0, ω_u (and only on d_u^0, d_s^0 when $(d_u^0, d_s^0) \neq (1/2, 1/2)$) such that

$$\textbf{(9.154)} \qquad Y_1 \leqslant |\widetilde{I}|^{2\sigma_2}.$$

Consider now, for fixed k, the number of terms in S_s. From Proposition 14 in Section 6.6.3, we have $x_i \leqslant |\widetilde{I}|$ for each i; then (9.138) and $x \geqslant x_{\min}$ imply that $x_i > x_{\min}$ for each i. As $x_{\min} = |\widetilde{I}|^{C(\rho_0-d_s^*)^{-1}}$, the number of terms is bounded by

$$\textbf{(9.155)} \qquad \left(C(\rho_0-d_s^*)^{-1}\log_2|\widetilde{I}|^{-1}\right)^{k+1}.$$

Using $k+1 \leqslant 2k$ for $k>0$ and (9.149), (9.152), we obtain the following bound for S_s

$$\textbf{(9.156)} \qquad S_s \leqslant \left(\frac{x}{\varepsilon_0|P_u|}\right)^{-\rho_0}\left(\frac{|I_\alpha|}{\varepsilon_0}\right)^{\sigma_0+\sigma_1}\left(\frac{|I_\omega|}{\varepsilon_0}\right)^{\sigma_0}\sum_{k>0} Y^k$$

with

$$\textbf{(9.157)} \qquad Y = 2C(\rho_0-d_s^*)^{-2}(\log|\widetilde{I}|^{-1})^2 Y_1.$$

From (9.98) in Section 9.7.3, we have $(\rho_0-d_s^*)^{-2} \leqslant \varepsilon_0^{-\tau^2} \leqslant |\widetilde{I}|^{-\tau^2}$. We obtain therefore

$$\textbf{(9.158)} \qquad \sum_{k>0} Y^k < \frac{1}{2}|\widetilde{I}|^{\sigma_2}.$$

We have thus shown that

$$\text{(9.159)} \qquad S_s \leqslant \frac{1}{2} B_0 |\widetilde{I}|^{\sigma_2}.$$

Taking σ_2 smaller if necessary, we can assume that

$$\text{(9.160)} \qquad 0 < \sigma_2 \leqslant \frac{1}{3}\rho_1.$$

The estimates (9.147) and (9.159) complete the proof of Proposition 38.

9.9. *Subclasses of bicritical elements.*

9.9.1. *Complement to the structure theorem.* — The following result will be used several times in the construction of the classes of bicritical elements.

Proposition **43.** — *Let* $I \subset \widetilde{I}$ *be a candidate interval. Let* $(P, Q, n), (P', Q', n') \in \mathcal{R}(I)$ *with* $P \subset P'$. *Assume that* (P, Q, n) *does not belong to* $\mathcal{R}(\widetilde{I})$. *Let* $k > 0$ *be the integer and* $(P_i, Q_i, n_i) \in \mathcal{R}(\widetilde{I})$ *for* $0 \leqslant i \leqslant k$ *be the elements given by the structure theorem.*

1. *If* (P', Q', n') *belongs to* $\mathcal{R}(\widetilde{I})$, *we have* $P' \supset P_0$.
2. *If* (P', Q', n') *does not belong to* $\mathcal{R}(\widetilde{I})$, *there exists* $0 < j \leqslant k$ *and* $(P'_j, Q'_j, n'_j) \in \mathcal{R}(\widetilde{I})$ *with* $P'_j \supset P_j$ *such that*

$$(P', Q', n') \in (P_0, Q_0, n_0) \square \cdots \square (P_{j-1}, Q_{j-1}, n_{j-1}) \square (P'_j, Q'_j, n'_j)$$

is the decomposition associated to (P', Q', n') *by the structure theorem.*

Remark **13.** — Taking $j = 0$ and $(P'_0, Q'_0, n'_0) = (P', Q', n')$, the first case can be considered as a special case of the second case.

Proof. — The first case follows from the fact that P_0 is the thinnest $\widetilde{I}$-defined rectangle containing P.

We now assume that (P', Q', n') does not belong to $\mathcal{R}(\widetilde{I})$. By coherence, P_0 is also the thinnest $\widetilde{I}$-defined rectangle containing P'. Therefore, (P_0, Q_0, n_0) is also the first element in the decomposition associated to (P', Q', n') by the structure theorem.

Let $(\widehat{P}, \widehat{Q}, \widehat{n}) \in (P_1, Q_1, n_1) \square \cdots \square (P_k, Q_k, n_k)$ be the element such that

$$(P, Q, n) \in (P_0, Q_0, n_0) \square (\widehat{P}, \widehat{Q}, \widehat{n}).$$

We claim that, when $k > 1$, i.e. when $(\widehat{P}, \widehat{Q}, \widehat{n})$ does not belong to $\mathcal{R}(\widetilde{I})$,

$$(\widehat{P}, \widehat{Q}, \widehat{n}) \in (P_1, Q_1, n_1) \square \cdots \square (P_k, Q_k, n_k)$$

is the decomposition associated to $(\widehat{P}, \widehat{Q}, \widehat{n})$ by the structure theorem. Indeed, this follows easily from the characterization of the (P_i, Q_i, n_i) in terms of maximal $\widetilde{I}$-intervals (see Lemma 2 in Section 6.5.2).

Define similarly $(\widehat{P}', \widehat{Q}', \widehat{n}')$ such that

$$(P', Q', n') \in (P_0, Q_0, n_0) \square (\widehat{P}', \widehat{Q}', \widehat{n}').$$

As $P \subset P'$, we have $\widehat{P} \subset \widehat{P}'$. If $k = 1$, we have $(\widehat{P}, \widehat{Q}, \widehat{n}) = (P_1, Q_1, n_1)$, we take $j = 1$, $(P'_1, Q'_1, n'_1) = (\widehat{P}', \widehat{Q}', \widehat{n}')$ to satisfy the conclusions of the proposition. If $k > 1$, the claim above allows also to conclude by induction on k. □

9.9.2. *Bound elements.*

Definition **12.** — *Let* $(P_\alpha, Q_\alpha, n_\alpha) \in \mathcal{C}_+(\widetilde{I})$, $(P_\omega, Q_\omega, n_\omega) \in \mathcal{C}_-(\widetilde{I})$. *An element* $(P, Q, n) \in Bi_+(P_\alpha, Q_\omega)$ *is bound if* $n \leqslant n_\alpha + n_\omega$. *Otherwise, we say that* (P, Q, n) *is free. We will denote by* $Bi_+(P_\alpha, Q_\alpha, \lozenge)$ *the subset of bound elements of* $Bi_+(P_\alpha, Q_\omega)$.

Thus, $\lozenge$ *is an element of* L. *On the other hand, free elements will correspond to many elements of* L. *Recall that we have* $x_{\min} \leqslant x \leqslant \widetilde{x}_{cr}$. *When* $x \ll \widetilde{x}_{cr}$, *most elements are free. If we would allow* $x \gg \widetilde{x}_{cr}$, *on the opposite, most elements would be bound.*

Proposition **44.** — *For any* $(P_\alpha, Q_\alpha, n_\alpha), (P_\omega, Q_\omega, n_\omega) \in \mathcal{R}(\widetilde{I})$, *and any* $n \leqslant n_\alpha + n_\omega$, *there is at most one element* $(P, Q, n) \in \mathcal{R}(\widetilde{I})$ *of length n such that* $P \subset P_\alpha$, $Q \subset Q_\omega$.

Proof. — We argue by induction on the level of the parameter interval.

When $\widetilde{I}$ is the starting interval I_0, no parabolic composition is involved and the result follows from usual symbolic dynamics: as $n \leqslant n_\alpha + n_\omega$, the word associated to a bound element is determined by its initial and final parts.

Assume that the result holds for parameter intervals strictly larger than $\widetilde{I}$. Denote by $\widetilde{I}_1$ the parent interval of $\widetilde{I}$.

Assume first that both $(P_\alpha, Q_\alpha, n_\alpha)$ and $(P_\omega, Q_\omega, n_\omega)$ belong to $\mathcal{R}(\widetilde{I}_1)$. We claim that any bound element also belongs to $\mathcal{R}(\widetilde{I}_1)$, which allow us to conclude the proof by the induction hypothesis. Indeed, if (P, Q, n) satisfies $P \subset P_\alpha$, $Q \subset Q_\omega$ and does not belong to $\mathcal{R}(\widetilde{I}_1)$, we apply the structure theorem: it gives elements (P_0, Q_0, n_0), $(P_k, Q_k, n_k) \in \mathcal{R}(\widetilde{I}_1)$ such that P_0 is the thinnest rectangle containing P defined over $\widetilde{I}_1$, Q_k is the thinnest rectangle containing Q defined over $\widetilde{I}_1$, and $n \geqslant n_0 + n_k + N_0$. Therefore, $n_0 \geqslant n_\alpha$, $n_k \geqslant n_\omega$ and $n > n_\alpha + n_\omega$.

We now consider the case when, for instance, $(P_\alpha, Q_\alpha, n_\alpha)$ does not belong to $\mathcal{R}(\widetilde{I}_1)$. We now apply the structure theorem to $(P_\alpha, Q_\alpha, n_\alpha)$ and also to an element $(P, Q, n) \in \mathcal{R}(\widetilde{I})$ with $P \subset P_\alpha$, $Q \subset Q_\omega$, $n \leqslant n_\alpha + n_\omega$. From Proposition 43, we obtain integers $0 < j \leqslant k$, elements $(P_i, Q_i, n_i) \in \mathcal{R}(\widetilde{I}_1)$ for $0 \leqslant i \leqslant k$ such that

$$(P, Q, n) \in (P_0, Q_0, n_0) \square \cdots \square (P_k, Q_k, n_k) \tag{9.161}$$

and also $(\widetilde{P}_j, \widetilde{Q}_j, \widetilde{n}_j) \in \mathcal{R}(\widetilde{I}_1)$ such that $\widetilde{P}_j \supset P_j$ and

$$(P_\alpha, Q_\alpha, n_\alpha) \in (P_0, Q_0, n_0) \square \cdots \square (P_{j-1}, Q_{j-1}, n_{j-1}) \square (\widetilde{P}_j, \widetilde{Q}_j, \widetilde{n}_j). \tag{9.162}$$

Similarly, there exists m with $0 \leqslant m \leqslant k$ and $(\widetilde{P}'_m, \widetilde{Q}'_m, \widetilde{n}'_m)$ in $\mathcal{R}(\widetilde{I}_1)$ such that $\widetilde{Q}'_m \supset Q_m$ and

- either $m = k$, $(\widetilde{P}'_m, \widetilde{Q}'_m, \widetilde{n}'_m) = (P_\omega, Q_\omega, n_\omega)$;
- or $m < k$ and we have,

$$(P_\omega, Q_\omega, n_\omega) \in (\widetilde{P}'_m, \widetilde{Q}'_m, \widetilde{n}'_m) \square \cdots \square (P_k, Q_k, n_k). \tag{9.163}$$

The relations $n \leqslant n_\alpha + n_\omega$, $n_j \geqslant \widetilde{n}_j$, $n_m \geqslant \widetilde{n}'_m$ imply that $m \leqslant j$.

If $m < j$, the sequence (P_i, Q_i, n_i) and the choice of the result (out of two possibilities) in each parabolic composition are completely determined by P_α and Q_ω: the assertion of the proposition follows.

When $m = j$, the sequence (P_i, Q_i, n_i) for $i \neq m = j$ is determined by P_α, Q_ω; but we also have $P_j \subset \widetilde{P}_j$, $Q_j = Q_m \subset \widetilde{Q}_m$ and $n_j \leqslant \widetilde{n}_j + \widetilde{n}'_m$, so by the induction hypothesis (P_j, Q_j, n_j) is also determined by P_α, Q_ω and n. Again, the choices of the results in the parabolic compositions are also determined by P_α, Q_ω. The proof of the proposition is complete. □

Recall that, by Proposition 12 in Section 6.6.2, we have

$$\max_{\widetilde{I}} |P| \leqslant C_0 \exp(-n^\gamma) \tag{9.164}$$

for any $(P, Q, n) \in \mathcal{R}(\widetilde{I})$, with $\gamma = \log\frac{3}{2}/\log 2 > \frac{1}{2}$.

If $(P, Q, n) \in Bi_+(P_\alpha, Q_\omega, \Diamond)$, we have $\max_{\widetilde{I}} |P| \geqslant x \geqslant x_{\min} = |\widetilde{I}|^{C(\rho_0 - d_s^*)^{-1}}$, and therefore

$$n \leqslant (\rho_0 - d_s^*)^{-2} (\log |\widetilde{I}|)^2. \tag{9.165}$$

We thus shall define

$$b_+(P_\alpha, \Diamond) = b_-(Q_\omega, \Diamond) = (\rho_0 - d_s^*)^{-1} \log |\widetilde{I}|^{-1}, \tag{9.166}$$

and we will indeed have, from Proposition 44,

$$\# Bi_+(P_\alpha, Q_\omega, \Diamond) \leqslant b_+(P_\alpha, \Diamond) b_-(Q_\omega, \Diamond). \tag{9.167}$$

9.9.3. *Decomposition of a free element.* — Let $(P_\alpha, Q_\alpha, n_\alpha) \in \mathcal{C}_+(\widetilde{I})$, $(P_\omega, Q_\omega, n_\omega) \in \mathcal{C}_-(\widetilde{I})$ and let $(P, Q, n) \in Bi_+(P_\alpha, Q_\omega)$ be a *free* element. We will analyze with respect to the structure theorem the way in which (P, Q, n), $(P_\alpha, Q_\alpha, n_\alpha)$, $(P_\omega, Q_\omega, n_\omega)$ have been created. This will allow us in the sequel to define various subclasses of free elements.

Denote by $\widehat{I}_0$ the largest parameter interval such that $(P, Q, n) \in \mathcal{R}(\widehat{I}_0)$. Elements (P, Q, n) for which $\widehat{I}_0$ is the starting interval I_0 are said to have depth 0. They form a first free subclass of $Bi_+(P_\alpha, Q_\omega)$ denoted by $Bi_+(P_\alpha, Q_\omega, 0)$.

We now assume that $\widehat{I}_0 \neq I_0$ and denote by $\widetilde{I}_0$ the parent interval of $\widehat{I}_0$. We apply the structure theorem. We obtain an integer $k > 0$, elements $(P_0, Q_0, n_0), \dots, (P_k, Q_k, n_k)$ in $\mathcal{R}(\widetilde{I}_0)$ such that

$$(P, Q, n) \in (P_0, Q_0, n_0) \boxdot \cdots \boxdot (P_k, Q_k, n_k). \tag{9.168}$$

By Proposition 43, we find $0 \leqslant j \leqslant k$ and $(\widetilde{P}_j, \widetilde{Q}_j, \widetilde{n}_j) \in \mathcal{R}(\widetilde{I}_0)$ such that either $j = 0$, $(\widetilde{P}_j, \widetilde{Q}_j, \widetilde{n}_j) = (P_\alpha, Q_\alpha, n_\alpha)$ (if $(P_\alpha, Q_\alpha, n_\alpha) \in \mathcal{R}(\widetilde{I}_0)$) or $j > 0$ and

$$(P_\alpha, Q_\alpha, n_\alpha) \in (P_0, Q_0, n_0) \boxdot \cdots \boxdot (\widetilde{P}_j, \widetilde{Q}_j, \widetilde{n}_j). \tag{9.169}$$

Similarly, we find $0 \leqslant m \leqslant k$ and $(\widetilde{P}'_m, \widetilde{Q}'_m, \widetilde{n}'_m) \in \mathcal{R}(\widetilde{I}_0)$ such that either $m = k$, $(\widetilde{P}'_m, \widetilde{Q}'_m, \widetilde{n}'_m) = (P_\omega, Q_\omega, n_\omega)$ or $m < k$ and

$$(P_\omega, Q_\omega, n_\omega) \in (\widetilde{P}'_m, \widetilde{Q}'_m, \widetilde{n}'_m) \boxdot \cdots \boxdot (P_k, Q_k, n_k). \tag{9.170}$$

We also have $P_j \subset \widetilde{P}_j$, $Q_m \subset \widetilde{Q}'_m$. Moreover, as (P, Q, n) is free, we must have $j \leqslant m$ and, when $j = m$, we must also have $n_j = n_m > \widetilde{n}_j + \widetilde{n}'_m$.

We say that (P, Q, n) is *fully decomposed* if one has here $j < m$ or $j = m$ and $(P_j, Q_j, n_j) \in \mathcal{R}(I_0)$. Such elements are said to have depth one.

Assume that (P, Q, n) is not fully decomposed. Then, we have $j = m$, $P_j \subset \widetilde{P}_j$, $Q_j \subset \widetilde{Q}'_j$ and the largest parameter interval $\widehat{I}_1$ for which $(P_j, Q_j, n_j) \in \mathcal{R}(\widehat{I}_1)$ is not the starting interval I_0. We denote by $\widetilde{I}_1$ the parent interval. We rewrite

$$\begin{aligned}
(P^1, Q^1, n^1) &:= (P_j, Q_j, n_j),\\
(P^1_\alpha, Q^1_\alpha, n^1_\alpha) &:= (\widetilde{P}_j, \widetilde{Q}_j, \widetilde{n}_j),\\
(P^1_\omega, Q^1_\omega, n^1_\omega) &:= (\widetilde{P}'_j, \widetilde{Q}'_j, \widetilde{n}'_j),
\end{aligned} \tag{9.171}$$

and proceed with these elements as we did with (P, Q, n), $(P_\alpha, Q_\alpha, n_\alpha)$, $(P_\omega, Q_\omega, n_\omega)$: we will find integers $0 \leqslant j_1 \leqslant m_1 \leqslant k_1$ (with $k_1 > 0$), elements (P^1_i, Q^1_i, n^1_i) for $0 \leqslant i \leqslant k_1$ and also $(P^2_\alpha, Q^2_\alpha, n^2_\alpha)$, $(P^2_\omega, Q^2_\omega, n^2_\omega)$, all in $\mathcal{R}(\widetilde{I}_1)$, such that

$$\begin{aligned}
(P^1, Q^1, n^1) &\in (P^1_0, Q^1_0, n^1_0) \boxdot \cdots \boxdot (P^1_{k_1}, Q^1_{k_1}, n^1_{k_1}),\\
(P^1_\alpha, Q^1_\alpha, n^1_\alpha) &\in (P^1_0, Q^1_0, n^1_0) \boxdot \cdots \boxdot (P^1_{j_1-1}, Q^1_{j_1-1}, n^1_{j_1-1}) \boxdot (P^2_\alpha, Q^2_\alpha, n^2_\alpha),\\
(P^1_\omega, Q^1_\omega, n^1_\omega) &\in (P^2_\omega, Q^2_\omega, n^2_\omega) \boxdot (P^1_{m_1+1}, Q^1_{m_1+1}, n^1_{m_1+1}) \boxdot \cdots\\
&\qquad \boxdot (P^1_{k_1}, Q^1_{k_1}, n^1_{k_1}).
\end{aligned} \tag{9.172}$$

Again, we say that (P^1, Q^1, n^1) is fully decomposed if either $j_1 < m_1$ or $j_1 = m_1$ and $(P^1_{j_1}, Q^1_{j_1}, n^1_{j_1})$ is defined over the starting interval I_0; otherwise we set

$$(P^2, Q^2, n^2) := (P^1_{j_1}, Q^1_{j_1}, n^1_{j_1}), \tag{9.173}$$

and we go on. The sequence of parameter intervals $\widehat{I}_0 \subset \widehat{I}_1 \subset \cdots$ is strictly increasing and therefore the process will stop. We define inductively the *depth* of (P, Q, n) to be the depth of (P^1, Q^1, n^1) plus one.

9.9.4. *Size of the subclass of depth* 0. — We will define in this subsection $b_+(P_\alpha, 0)$, $b_-(Q_\omega, 0)$ in order to have

$$\#Bi_+(P_\alpha, Q_\omega, 0) \leqslant b_+(P_\alpha, 0) b_-(Q_\omega, 0). \tag{9.174}$$

Let $(P_\alpha, Q_\alpha, n_\alpha) \in \mathcal{C}_+(\widetilde{I})$, $(P_\omega, Q_\omega, n_\omega) \in \mathcal{C}_-(\widetilde{I})$ and let $(P, Q, n) \in Bi_+(P_\alpha, Q_\omega, 0)$. Then both $(P_\alpha, Q_\alpha, n_\alpha)$, $(P_\omega, Q_\omega, n_\omega) \in \mathcal{R}(I_0)$ and (P, Q, n) is obtained by a simple composition

$$(P, Q, n) = (P_\alpha, Q_\alpha, n_\alpha) * (P', Q', n') * (P_\omega, Q_\omega, n_\omega). \tag{9.175}$$

We have here, by definition of $Bi_+(P_\alpha, Q_\omega)$ and $\mathcal{C}_+(\widetilde{I})$

$$x \leqslant \max_{\widetilde{I}} |P| \leqslant C \max_{\widetilde{I}} |P_\alpha| \max_{\widetilde{I}} |P'| \max_{\widetilde{I}} |P_\omega| \leqslant C|\widetilde{I}|^{1+\tau} \max_{\widetilde{I}} |P'| \max_{\widetilde{I}} |P_\omega|. \tag{9.176}$$

This gives

$$\max_{\widetilde{I}} |P'| \geqslant C^{-1} |\widetilde{I}|^{-(1+\tau)} \left(\max_{\widetilde{I}} |P_\omega|\right)^{-1} x. \tag{9.177}$$

From the reminder at the beginning of Section 8.3, we can thus define, as $d_s^0 + C\varepsilon_0 < d_s^* < \rho_s$,

$$b_+(P_\alpha, 0) = \begin{cases} (C|\widetilde{I}|^{1+\tau} x^{-1})^{\rho_s} & \text{if } (P_\alpha, Q_\alpha, n_\alpha) \in \mathcal{R}(I_0), \\ 0 & \text{otherwise,} \end{cases} \tag{9.178}$$

$$b_-(Q_\omega, 0) = \begin{cases} (\max_{\widetilde{I}} |P_\omega|)^{\rho_s} & \text{if } (P_\omega, Q_\omega, n_\omega) \in \mathcal{R}(I_0), \\ 0 & \text{otherwise.} \end{cases} \tag{9.179}$$

Then, (9.174) is satisfied.

9.9.5. *Subclasses of higher depth.* — Let $(P_\alpha, Q_\alpha, n_\alpha) \in \mathcal{C}_+(\widetilde{I})$, $(P_\omega, Q_\omega, n_\omega) \in \mathcal{C}_-(\widetilde{I})$ and let $(P, Q, n) \in Bi_+(P_\alpha, Q_\omega)$ be an element of depth $s > 0$.

Let us first restate and extend somewhat the notations and the setting of Section 9.9.2. We set

$$\begin{aligned} (P^0, Q^0, n^0) &:= (P, Q, n), \\ (P^0_\alpha, Q^0_\alpha, n^0_\alpha) &:= (P_\alpha, Q_\alpha, n_\alpha), \\ (P^0_\omega, Q^0_\omega, n^0_\omega) &:= (P_\omega, Q_\omega, n_\omega). \end{aligned} \tag{9.180}$$

We have

- a strictly increasing sequence of parameter intervals

$$\widehat{\mathrm{I}}_0 \subset \widehat{\mathrm{I}}_1 \subset \cdots \subset \widehat{\mathrm{I}}_{s-1} \subset \widehat{\mathrm{I}}_s = \mathrm{I}_0 \tag{9.181}$$

  with $\widetilde{\mathrm{I}} \subset \widehat{\mathrm{I}}_0$; we denote by $\widetilde{\mathrm{I}}_r$ the parent interval of $\widehat{\mathrm{I}}_r$ for $0 \leqslant r < s$;
- a sequence $(\mathrm{P}^r, \mathrm{Q}^r, n^r)$, $0 \leqslant r \leqslant s$ such that $(\mathrm{P}^r, \mathrm{Q}^r, n^r)$ belongs to $\mathcal{R}(\widehat{\mathrm{I}}_r)$ but not to $\mathcal{R}(\widetilde{\mathrm{I}}_r)$ for $r < s$; also $(\mathrm{P}^s, \mathrm{Q}^s, n^s) \in \mathcal{R}(\widehat{\mathrm{I}}_{s-1})$;
- two sequences $(\mathrm{P}^r_\alpha, \mathrm{Q}^r_\alpha, n^r_\alpha)$, $(\mathrm{P}^r_\omega, \mathrm{Q}^r_\omega, n^r_\omega)$, $0 \leqslant r \leqslant s$; for each $r < s$, resp. $r = s$, the two elements belong to $\mathcal{R}(\widetilde{\mathrm{I}}_r)$, resp. $\mathcal{R}(\widetilde{\mathrm{I}}_{s-1})$;
- two sequences $(\mathrm{P}^r_+, \mathrm{Q}^r_+, n^r_+)$, $(\mathrm{P}^r_-, \mathrm{Q}^r_-, n^r_-)$, $0 < r \leqslant s$; for each r, the two elements belong to $\mathcal{R}(\widehat{\mathrm{I}}_{r-1})$.

These data are related by the following properties: for each $0 < r \leqslant s$, we have

$$(\mathrm{P}^{r-1}, \mathrm{Q}^{r-1}, n^{r-1}) \in (\mathrm{P}^r_-, \mathrm{Q}^r_-, n^r_-) \,\square\, (\mathrm{P}^r, \mathrm{Q}^r, n^r) \,\square\, (\mathrm{P}^r_+, \mathrm{Q}^r_+, n^r_+), \tag{9.182}$$

$$(\mathrm{P}^{r-1}_\alpha, \mathrm{Q}^{r-1}_\alpha, n^{r-1}_\alpha) \in (\mathrm{P}^r_-, \mathrm{Q}^r_-, n^r_-) \,\square\, (\mathrm{P}^r_\alpha, \mathrm{Q}^r_\alpha, n^r_\alpha), \tag{9.183}$$

$$(\mathrm{P}^{r-1}_\omega, \mathrm{Q}^{r-1}_\omega, n^{r-1}_\omega) \in (\mathrm{P}^r_\omega, \mathrm{Q}^r_\omega, n^r_\omega) \,\square\, (\mathrm{P}^r_+, \mathrm{Q}^r_+, n^r_+). \tag{9.184}$$

The process stops at step s because of one of the two following cases occur

(a) $(\mathrm{P}^s, \mathrm{Q}^s, n^s)$ does not belong to $\mathcal{R}(\widetilde{\mathrm{I}}_{s-1})$; then, by the structure theorem, there exists an integer $h > 0$, elements $(\mathrm{P}^s_0, \mathrm{Q}^s_0, n^s_0) \cdots (\mathrm{P}^s_h, \mathrm{Q}^s_h, n^s_h)$ in $\mathcal{R}(\widetilde{\mathrm{I}}_{s-1})$ with

$$(\mathrm{P}^s, \mathrm{Q}^s, n^s) \in (\mathrm{P}^s_0, \mathrm{Q}^s_0, n^s_0) \,\square \cdots \square\, (\mathrm{P}^s_h, \mathrm{Q}^s_h, n^s_h) \tag{9.185}$$

and also

$$\mathrm{P}^s_0 \subset \mathrm{P}^s_\alpha, \qquad \mathrm{Q}^s_h \subset \mathrm{Q}^s_\omega. \tag{9.186}$$

(b) $(\mathrm{P}^s, \mathrm{Q}^s, n^s)$ belongs to $\mathcal{R}(\mathrm{I}_0)$; in this case we set $h = 0$.

We also observe that the parabolic compositions in (9.182) through (9.184) take place in $\mathcal{R}(\widehat{\mathrm{I}}_{r-1})$ but not in $\mathcal{R}(\widetilde{\mathrm{I}}_{r-1})$; in (9.185), they take place in $\mathcal{R}(\widehat{\mathrm{I}}_{s-1})$ but not in $\mathcal{R}(\widetilde{\mathrm{I}}_{s-1})$.

A subclass $\mathrm{B}i_+(\mathrm{P}_\alpha, \mathrm{Q}_\omega, \ell)$, i.e. an element of L, distinct from the two $(\lozenge, 0)$ that we already know is determined by the following data

- the depth $s(> 0)$;
- the sequence $\widehat{\mathrm{I}}_0 \subset \cdots \subset \widehat{\mathrm{I}}_s = \mathrm{I}_0$;
- the integer $h \geqslant 0$;
- when $h > 1$, for each $0 < i < h$, the largest negative integral power of 2, denoted by x_i, such that $\max_{\widetilde{\mathrm{I}}} |\mathrm{P}^s_i| \geqslant x_i$;
- when $h > 0$, the largest negative integral powers of 2, denoted by x_0, x_h, such that $\max_{\widetilde{\mathrm{I}}} |\mathrm{P}_-| \geqslant x_0$, $\max_{\widetilde{\mathrm{I}}} |\mathrm{P}_+| \geqslant x_h$; here, the elements $(\mathrm{P}_-, \mathrm{Q}_-, n_-)$, $(\mathrm{P}_+, \mathrm{Q}_+, n_+)$ are determined by $\mathrm{P} \subset \mathrm{P}_-$, $\mathrm{Q} \subset \mathrm{Q}_+$ and

$$(\mathrm{P}_-, \mathrm{Q}_-, n_-) \in (\mathrm{P}^1_-, \mathrm{Q}^1_-, n^1_-) \square \cdots \square (\mathrm{P}^s_-, \mathrm{Q}^s_-, n^s_-) \tag{9.187}$$
$$\square (\mathrm{P}^s_0, \mathrm{Q}^s_0, n^s_0),$$

$$(\mathrm{P}_+, \mathrm{Q}_+, n_+) \in (\mathrm{P}^s_h, \mathrm{Q}^s_h, n^s_h) \square (\mathrm{P}^s_+, \mathrm{Q}^s_+, n^s_+) \square \cdots \tag{9.188}$$
$$\square (\mathrm{P}^1_+, \mathrm{Q}^1_+, n^1_+).$$

One has by construction $\mathrm{P}_- \subset \mathrm{P}_\alpha$, $\mathrm{Q}_+ \subset \mathrm{Q}_\omega$.

Thus, we group together in a subclass $Bi_+(\mathrm{P}_\alpha, \mathrm{Q}_\omega, \ell)$ the elements of $Bi_+(\mathrm{P}_\alpha, \mathrm{Q}_\omega)$ who share the same data; the elements of L, distinct from $\lozenge$, 0, are the sets of data for which at least one subclass $Bi_+(\mathrm{P}_\alpha, \mathrm{Q}_\omega, \ell)$ is non-empty, for some $(\mathrm{P}_\alpha, \mathrm{Q}_\alpha, n_\alpha)$ in $\mathcal{C}_+(\widetilde{\mathrm{I}})$, $(\mathrm{P}_\omega, \mathrm{Q}_\omega, n_\omega)$ in $\mathcal{C}_-(\widetilde{\mathrm{I}})$.

The definition of the set L is now complete.

9.9.6. *Sizes of subclasses of higher depth.* — The context and notations are the same as above. We want to define $b_+(\mathrm{P}_\alpha, \ell)$ and $b_-(\mathrm{Q}_\omega, \ell)$ in order to satisfy (9.124) in Section 9.8.

We first observe that $(\mathrm{P}_\alpha, \mathrm{Q}_\alpha, n_\alpha)$ determines $(\mathrm{P}^1_-, \mathrm{Q}^1_-, n^1_-), \ldots, (\mathrm{P}^s_-, \mathrm{Q}^s_-, n^s_-)$, $(\mathrm{P}^s_\alpha, \mathrm{Q}^s_\alpha, n^s_\alpha)$ and the result of parabolic compositions between these elements. Similarly, $(\mathrm{P}_\omega, \mathrm{Q}_\omega, n_\omega)$ determines $(\mathrm{P}^1_+, \mathrm{Q}^1_+, n^1_+), \ldots, (\mathrm{P}^s_+, \mathrm{Q}^s_+, n^s_+)$, $(\mathrm{P}^s_\omega, \mathrm{Q}^s_\omega, n^s_\omega)$ and the result of parabolic compositions between these elements. Therefore, the only "freedom" for the element $(\mathrm{P}, \mathrm{Q}, n)$ in the subclass $Bi_+(\mathrm{P}_\alpha, \mathrm{Q}_\omega, \ell)$ is through $(\mathrm{P}^s, \mathrm{Q}^s, n^s)$, and this freedom is constrained by the relations $\mathrm{P}^s \subset \mathrm{P}^s_\alpha$, $\mathrm{Q}^s \subset \mathrm{Q}^s_\omega$.

Consider first a subclass with $h = 0$, i.e., $(\mathrm{P}^s, \mathrm{Q}^s, n^s) \in \mathcal{R}(\mathrm{I}_0)$. The widths of the strips are related as follows: for every $t \in \widetilde{\mathrm{I}}$, we have

$$\mathrm{C}^{-1} \frac{|\mathrm{P}^s|}{|\mathrm{P}^s_\alpha||\mathrm{P}^s_\omega|} \leqslant \frac{|\mathrm{P}|}{|\mathrm{P}_\alpha||\mathrm{P}_\omega|} \leqslant \mathrm{C} \frac{|\mathrm{P}^s|}{|\mathrm{P}^s_\alpha||\mathrm{P}^s_\omega|}. \tag{9.189}$$

This allows us to take, as in the case of depth 0,

$$b_+(\mathrm{P}_\alpha, \ell) = \begin{cases} (\mathrm{C}|\widetilde{\mathrm{I}}|^{1+\tau} x^{-1})^{\rho_s} & \text{if } (\mathrm{P}^s_\alpha, \mathrm{Q}^s_\alpha, n^s_\alpha) \in \mathcal{R}(\mathrm{I}_0), \\ 0 & \text{otherwise}, \end{cases} \tag{9.190}$$

$$b_-(\mathrm{Q}_\omega, \ell) = \begin{cases} \max_{\widetilde{\mathrm{I}}} |\mathrm{P}_\omega|^{\rho_s} & \text{if } (\mathrm{P}^s_\omega, \mathrm{Q}^s_\omega, n^s_\omega) \in \mathcal{R}(\mathrm{I}_0), \\ 0 & \text{otherwise}. \end{cases} \tag{9.191}$$

Consider now a subclass with $h > 0$, i.e., case (a) in Section 9.9.4 above.

By Lemma 10 (Section 9.1), for $h > 1$ and $0 < i < h$, P^s_i and Q^s_i are thin $\widetilde{\mathrm{I}}_{s-1}$-critical, hence element $(\mathrm{P}^s_i, \mathrm{Q}^s_i, n^s_i)$ belongs to $Bi_+(\widetilde{\mathrm{I}}_{s-1}, \widetilde{\mathrm{I}}_{s-1}, \widetilde{\mathrm{I}}_{s-1}; x_i)$. In the parabolic composition

$$(\mathrm{P}, \mathrm{Q}, n) \in (\mathrm{P}_-, \mathrm{Q}_-, n_-) \square (\mathrm{P}^s_1, \mathrm{Q}^s_1, n^s_1) \square \cdots \square (\mathrm{P}^s_{h-1}, \mathrm{Q}^s_{h-1}, n^s_{h-1})$$
$$\square (\mathrm{P}_+, \mathrm{Q}_+, n_+),$$

we have $\delta(Q_-, P_1^s) > |\widehat{I}_{s-1}|, \ldots, \delta(Q_{h-1}^s, P_+) > |\widehat{I}_{s-1}|$ for all $t \in \widetilde{I}$ by Lemma 3 (Section 6.6.3), hence, using (3.27)

$$\max_{\widetilde{I}} |P| \leqslant C^h |\widehat{I}_{s-1}|^{-h/2} \max_{\widetilde{I}} |P_-| \max_{\widetilde{I}} |P_1^s| \cdots \max_{\widetilde{I}} |P_+|.$$

This gives, by definition of the x_i

$$x \leqslant \left(C |\widehat{I}_{s-1}|^{-\frac{1}{2}} \right)^h x_0 x_1 \cdots x_h. \tag{9.192}$$

Thus, the data of every subclass must satisfy (9.192). Assuming that (9.192) holds, we set $b_+(P_\alpha, \ell) = 0$ if $(P_\alpha, Q_\alpha, n_\alpha) \notin \mathcal{R}(\widehat{I}_0)$. When $(P_\alpha, Q_\alpha, n_\alpha) \in \mathcal{R}(\widehat{I}_0)$, we set

$$b_+(P_\alpha, \ell) = 2^h \Big(\prod_{0<i<h} \# Bi_+(\widetilde{I}_{s-1}, \widetilde{I}_{s-1}, \widetilde{I}_{s-1}; x_i) \Big) \# Bi_+(P_\alpha, \widetilde{I}_{s-1}; x_0). \tag{9.193}$$

Here, $Bi_+(P_\alpha, \widetilde{I}_{s-1}, x_0)$ is by definition the set of elements (P_-, Q_-, n_-) in $\mathcal{R}(\widetilde{I})$ such that $P_- \subset P_\alpha$, Q_- is thin $\widetilde{I}_{s-1}$-critical and $\max_{\widetilde{I}} |P_-| \geqslant x_0$.

Similarly, when (9.192) holds, we set $b_-(Q_\omega, \ell) = 0$ if $(P_\omega, Q_\omega, n_\omega) \notin \mathcal{R}(\widehat{I}_0)$. When $(P_\omega, Q_\omega, n_\omega) \in \mathcal{R}(\widehat{I}_0)$, we set

$$b_-(Q_\omega, \ell) = \# Bi_+(\widetilde{I}_{s-1}, Q_\omega; x_h), \tag{9.194}$$

where now $Bi_+(\widetilde{I}_{s-1}, Q_\omega; x_h)$ is the set of elements (P_+, Q_+, n_+) in $\mathcal{R}(\widetilde{I})$ such that $Q_+ \subset Q_\omega$, P_+ is thin $\widetilde{I}_{s-1}$-critical and $\max_{\widetilde{I}} |P_+| \geqslant x_h$.

The factor 2^h in (9.193) takes care of the possible results of the "free" parabolic compositions, i.e., those compositions which are not constrained by $(P_\alpha, Q_\alpha, n_\alpha)$ or $(P_\omega, Q_\omega, n_\omega)$.

The definition of L, b_+, b_- is now complete, and relation (9.124) is satisfied.

9.10. *The size of the index set* L. — It is not difficult from (9.192) to see that the index set L is finite, but we need an explicit bound on its cardinality (cf. (9.134)).

We assume that, with C large enough,

$$\rho_0 > d_s^* + C\tau^{-2} \frac{\log\log \varepsilon_0^{-1}}{\log \varepsilon_0^{-1}}, \tag{9.195}$$

where $\frac{\log\log \varepsilon_0^{-1}}{\log \varepsilon_0^{-1}} \ll \tau^2$ when ε_0 is small enough.

Proposition **45.** — *The index set* L *satisfies*

$$\# L \leqslant |\widetilde{I}|^{-\tau^2}.$$

Proof. — It is sufficient to count the subclasses of strictly positive depth (there are only two subclasses besides these ones).

In a first step, we fix the depth $s > 0$ and the sequence of intervals $\widehat{I}_0 \subset \cdots \subset \widehat{I}_{s-1} \subset \widehat{I}_s = I_0$. There is one subclass with $h = 0$ and we will estimate the number of subclasses with $h > 0$ (the integer h itself is *not* fixed). This number is bounded by the number of $(h+1)$-tuples $(x_0, \dots, x_h)$ (of negative integral powers of 2) which satisfy (9.192).

By Proposition 14 (Section 6.6.3), we have

$$x_i < |\widetilde{I}_{s-1}|^{\beta} \quad \text{for } 0 < i < h, \tag{9.196}$$

$$x_h < |\widetilde{I}_{s-1}|. \tag{9.197}$$

As $P_- \subset P_\alpha$, we also have, for a non-empty subclass

$$x_0 < |\widetilde{I}|^{1+\tau}. \tag{9.198}$$

We rewrite (9.192) as

$$\frac{x_0}{|\widetilde{I}|^{1+\tau}} \left(\prod_{0<i<h} \frac{x_i}{|\widetilde{I}_{s-1}|^{\beta}} \right) \frac{x_h}{|\widetilde{I}_{s-1}|} \geqslant \frac{x}{|\widetilde{I}|^{1+\tau} |\widetilde{I}_{s-1}|^{\beta(h-1)+1}} \left(C^{-1} |\widehat{I}_{s-1}|^{\frac{1}{2}} \right)^h. \tag{9.199}$$

Using $\beta > 1$, and taking base-two logarithms, it is sufficient to bound the number of non-negative integral solutions of

$$n_0 + \cdots + n_h \leqslant A_0 - A_1 h, \tag{9.200}$$

with

$$A_0 = \log_2(|\widetilde{I}| x^{-1}), \tag{9.201}$$

$$A_1 = \frac{1}{3} \log_2 |\widehat{I}_{s-1}|^{-1}. \tag{9.202}$$

We have $x \leqslant x^*_{cr} \leqslant \widetilde{x}_{cr} = \varepsilon_0 |P_u| (\frac{|\widetilde{I}|}{\varepsilon_0})^{\frac{\sigma_0}{\rho_0 - \rho_1}} \ll |\widetilde{I}|$, because $\sigma_0 > \rho_0 - \rho_1$ and $|P_u| \ll 1$. Therefore A_0 is large; it is obvious that A_1 is also large. By taking A_0 slightly larger and A_1 slightly smaller, we can assume that both A_0, A_1 are integers. The number of non-negative integral solutions of (9.200) (including those with $h = 0$) is then the coefficient of z^{A_0} in the power series for

$$\chi(z) := \sum_{h \geqslant 0} z^{A_1 h} (1 - z)^{-h-2} = (1 - z)^{-1} (1 - z - z^{A_1})^{-1}. \tag{9.203}$$

We estimate this coefficient by a Cauchy integral on the circle $|z| = 1 - 2A_1^{-1} \log A_1$. On this circle, we have

$$|z^{A_1}| < A_1^{-1}, \tag{9.204}$$

(**9.205**) $$|\chi(z)| < \frac{1}{2}A_1^2(\log A_1)^{-2}.$$

The number of solutions of (9.200) is therefore not greater than

(**9.206**) $$A_1^2(\log A_1)^{-2}(1-2A_1^{-1}\log A_1)^{-A_0}.$$

In view of (9.202), this quantity is smaller than

(**9.207**) $$(\log|\widehat{I}_{s-1}|)^2 \exp\left(CA_0\frac{\log|\log|\widehat{I}_{s-1}||}{|\log|\widehat{I}_{s-1}||}\right).$$

This is a bound for the number of subclasses with fixed depth s and fixed sequence $\widehat{I}_0 \subset \cdots \subset \widehat{I}_{s-1}$. We have now to sum over these remaining data. Observe that (9.207) depends on $|\widehat{I}_{s-1}|$, *not* on the depth s and the intervals $\widehat{I}_r$, $0 \leqslant r < s-1$.

Fix an interval $\widehat{I}$ with $\widetilde{I} \subset \widehat{I} \subset I_0$, $\widehat{I} \neq I_0$. Let $S(\widehat{I})$ be the number of parameter intervals I^* with $\widetilde{I} \subset I^* \subset \widehat{I}$, $I^* \neq \widehat{I}$. Every I^* in this range may or may not be one of the $\widehat{I}_r$, for a sequence $\widehat{I}_0 \subset \cdots \subset \widehat{I}_{s-1}$ terminating with $\widehat{I}_{s-1} = \widehat{I}$; in other terms, there are exactly $2^{S(\widehat{I})}$ such sequences (of various lengths). This means that the total number of subclasses (with $s > 0$) is bounded by

(**9.208**) $$\sum_{\widehat{I}} 2^{S(\widehat{I})}(\log|\widehat{I}|)^2 \exp\left(CA_0\frac{\log|\log|\widehat{I}||}{|\log|\widehat{I}||}\right).$$

We have here

(**9.209**) $$\frac{\log|\log|\widehat{I}||}{|\log|\widehat{I}||} \leqslant \frac{\log\log\varepsilon_0^{-1}}{\log\varepsilon_0^{-1}},$$

(**9.210**) $$|\log|\widetilde{I}|| = (1+\tau)^{S(\widehat{I})}|\log|\widehat{I}||,$$

(**9.211**) $$S(\widehat{I}) \leqslant 2\tau^{-1}\log_2\left(\frac{\log|\widetilde{I}|^{-1}}{\log\varepsilon_0^{-1}}\right) =: S_{\max}.$$

The sum (9.208) is thus bounded by

$$C2^{S_{\max}}(\log\varepsilon_0^{-1})^2\exp\left(CA_0\frac{\log\log\varepsilon_0^{-1}}{\log\varepsilon_0^{-1}}\right)$$

(**9.212**) $$\leqslant (\log|\widetilde{I}|^{-1})^{2\tau^{-1}}\exp\left(CA_0\frac{\log\log\varepsilon_0^{-1}}{\log\varepsilon_0^{-1}}\right).$$

As $x \geqslant x_{\min} := |\widetilde{I}|^{C(\rho_0-d_s^*)^{-1}}$, we have

(**9.213**) $$A_0 \leqslant C(\rho_0-d_s^*)^{-1}\log|\widetilde{I}|^{-1}.$$

If the constant in (9.195) is large enough, we obtain

$$\textbf{(9.214)} \qquad \mathrm{CA}_0 \frac{\log\log \varepsilon_0^{-1}}{\log \varepsilon_0^{-1}} < \frac{1}{2}\tau^2 \log |\widetilde{\mathrm{I}}|^{-1}.$$

Introducing this in (9.212) concludes the proof of the proposition. □

9.11. *The size of* $\widehat{\mathrm{B}}$.

9.11.1. According to the roadmap exposed in Section 9.7.6, we have now to estimate the quantity set in Section 9.7.6

$$\textbf{(9.140)} \qquad \widehat{\mathrm{B}} = \sum_{\mathrm{L}} b_+(\ell) b_-(\ell)$$

with

$$\textbf{(9.136)} \qquad b_+(\ell) = \sum_{\mathcal{C}_+(\widetilde{\mathrm{I}})} b_+(\mathrm{P}_\alpha, \ell),$$

$$\textbf{(9.137)} \qquad b_-(\ell) = \sum_{\mathcal{C}_-(\widetilde{\mathrm{I}})} b_-(\mathrm{Q}_\omega, \ell).$$

Consider first the bound elements. In view of (9.173), we have:

$$\textbf{(9.215)} \qquad \begin{aligned} b_+(\Diamond) &= \#\mathcal{C}_+(\widetilde{\mathrm{I}})(\rho_0 - d_s^*)^{-1} \log |\widetilde{\mathrm{I}}|^{-1}, \\ b_-(\Diamond) &= \#\mathcal{C}_-(\widetilde{\mathrm{I}})(\rho_0 - d_s^*)^{-1} \log |\widetilde{\mathrm{I}}|^{-1}. \end{aligned}$$

Consider next the class of depth 0, and also the classes of higher depth with $h = 0$: in view of (9.178)–(9.179) and (9.190)–(9.191), we have in these cases

$$\textbf{(9.216)} \qquad b_+(\ell) \leqslant (\mathrm{C}|\widetilde{\mathrm{I}}|^{1+\tau} x^{-1})^{\rho_s} \#\mathcal{C}_+(\widetilde{\mathrm{I}}),$$

$$\textbf{(9.217)} \qquad b_-(\ell) \leqslant \sum_{\mathcal{C}_-(\widetilde{\mathrm{I}})} \Big(\max_{\widetilde{\mathrm{I}}} |\mathrm{P}_\omega|\Big)^{\rho_s}.$$

Also, the number of such classes, according to the discussion in the proof of Proposition 45 is not larger than

$$\textbf{(9.218)} \qquad 2^{\mathrm{S}_{\max}} \leqslant \left(\frac{\log |\widetilde{\mathrm{I}}|^{-1}}{\log \varepsilon_0^{-1}}\right)^{2\tau^{-1}}.$$

9.11.2. The remaining subclasses are more complicated! Formulas (9.193), (9.194) suggest an induction. We thus assume that $(\mathrm{SR3})_s$ is satisfied for all parameter intervals containing $\widetilde{\mathrm{I}}$. We have, for a class of depth $s > 0$ with $h > 0$:

$$b_+(\ell) \leqslant 2^h \left(\prod_{0<i<h} (\#\mathrm{B}i_+(\widetilde{\mathrm{I}}_{s-1}, \widetilde{\mathrm{I}}_{s-1}, \widetilde{\mathrm{I}}_{s-1}; x_i)) \right) \#\mathrm{B}i_+(\widetilde{\mathrm{I}}, \widetilde{\mathrm{I}}, \widetilde{\mathrm{I}}_{s-1}; x_0), \tag{9.219}$$

$$b_-(\ell) \leqslant \#\mathrm{B}i_+(\widetilde{\mathrm{I}}, \widetilde{\mathrm{I}}_{s-1}, \widetilde{\mathrm{I}}; x_h). \tag{9.220}$$

We recall from (9.199) that

$$x \leqslant \left(\mathrm{C}|\widehat{\mathrm{I}}_{s-1}|^{-\frac{1}{2}} \right)^h x_0 x_1 \cdots x_h.$$

Observe that, from (9.16), the critical value $\widehat{x}_{cr}$ in each of the $\mathrm{B}i_+$ sets above is the same and equal to

$$\widehat{x}_{cr} := \varepsilon_0 |\mathrm{P}_u| \left(\frac{|\widetilde{\mathrm{I}}_{s-1}|}{\varepsilon_0} \right)^{\frac{\sigma_0}{\rho_0 - \rho_1}}. \tag{9.221}$$

As in Section 9.8, we separate the subclasses into two parts: those for which every x_i is above the critical value $\widehat{x}_{cr}$ and the others.

9.11.3. *Subclasses with all x_i large.* — In this case, we have from $(\mathrm{SR3})_s$

$$\#\mathrm{B}i_+(\widetilde{\mathrm{I}}_{s-1}, \widetilde{\mathrm{I}}_{s-1}, \widetilde{\mathrm{I}}_{s-1}; x_i) \leqslant \varepsilon_0^{-\mathrm{A}\tau} \left(\frac{x_i}{\varepsilon_0 |\mathrm{P}_u|} \right)^{-\rho_1} \left(\frac{|\widetilde{\mathrm{I}}_{s-1}|}{\varepsilon_0} \right)^{\sigma_0 + \sigma_1}, \quad \text{for } 0 < i < h, \tag{9.222}$$

$$\#\mathrm{B}i_+(\widetilde{\mathrm{I}}, \widetilde{\mathrm{I}}, \widetilde{\mathrm{I}}_{s-1}; x_0) \leqslant \varepsilon_0^{-\mathrm{A}\tau} \left(\frac{x_0}{\varepsilon_0 |\mathrm{P}_u|} \right)^{-\rho_1} \left(\frac{|\widetilde{\mathrm{I}}|}{\varepsilon_0} \right)^{\sigma_0 + \sigma_1}, \tag{9.223}$$

$$\#\mathrm{B}i_+(\widetilde{\mathrm{I}}, \widetilde{\mathrm{I}}_{s-1}, \widetilde{\mathrm{I}}; x_h) \leqslant \varepsilon_0^{-\mathrm{A}\tau} \left(\frac{x_h}{\varepsilon_0 |\mathrm{P}_u|} \right)^{-\rho_1} \left(\frac{|\widetilde{\mathrm{I}}_{s-1}|}{\varepsilon_0} \right)^{\sigma_1} \left(\frac{|\widetilde{\mathrm{I}}|}{\varepsilon_0} \right)^{\sigma_0}. \tag{9.224}$$

Multiplying these inequalities, we obtain, taking (9.192) into account

$$b_+(\ell) b_-(\ell) \leqslant \mathrm{A}_2^h \mathrm{A}_3, \tag{9.225}$$

with

$$\mathrm{A}_2 = 2\varepsilon_0^{-2\mathrm{A}\tau} \left(\frac{|\widetilde{\mathrm{I}}_{s-1}|}{\varepsilon_0} \right)^{\sigma_0 + \sigma_1} \left(\mathrm{C}\varepsilon_0 |\mathrm{P}_u| |\widehat{\mathrm{I}}_{s-1}|^{-\frac{1}{2}} \right)^{\rho_1}, \tag{9.226}$$

$$\mathrm{A}_3 = \mathrm{C}^{\rho_1} \left(\frac{x}{\varepsilon_0 |\mathrm{P}_u|} \right)^{-\rho_1} \left(\frac{|\widetilde{\mathrm{I}}|}{\varepsilon_0} \right)^{\sigma_0 + \sigma_1}. \tag{9.227}$$

9.11.4. *Subclasses with some x_i small.*

Lemma **13.** — *Let* $I \subset \widetilde{I}$ *be a candidate interval. Let* $(P, Q, n) \in \mathcal{R}(I)$. *If* P *and* Q *are thin* I_0*-critical, then we have* $\max_I |P| \leqslant C\varepsilon_0 |P_u|$ *and* $\max_I |Q| \leqslant C\varepsilon_0 |Q_s|$.

Proof. — We prove the first statement by induction on the level of the largest interval I_1 such that $(P, Q, n) \in \mathcal{R}(I_1)$. If $I_1 = I_0$, the statement follows from the fact that (P, Q, n) is the simple composition of (P_s, Q_s, n_s), some $(P', Q', n') \in \mathcal{R}(I_0)$ and (P_u, Q_u, n_u) (see Section 9.4). If $I_1 \neq I_0$ and $\widetilde{I}_1$ is the parent interval, consider the thinnest $\widetilde{I}_1$-defined rectangle P_0 containing P. As P is thin I_0-critical, P_0 is also thin I_0-critical. By Lemma 10 in Section 9.1, Q_0 is thin I_0-critical. By the induction hypothesis, we have $\max_{\widetilde{I}_1} |P_0| \leqslant C\varepsilon_0 |P_u|$. As $P \subset P_0$, we are able to conclude.

The proof of the other statement is symmetric. □

For $0 < i < h$, the cardinality of $\mathcal{B}i_+(\widetilde{I}_{s-1}, \widetilde{I}_{s-1}, \widetilde{I}_{s-1}; x_i)$ is controlled by (9.229) if $x_i \geqslant \widehat{x}_{cr}$ and by

$$\#\mathcal{B}i_+(\widetilde{I}_{s-1}, \widetilde{I}_{s-1}, \widetilde{I}_{s-1}; x_i) \leqslant \varepsilon_0^{-A\tau} \left(\frac{x_i}{\varepsilon_0 |P_u|}\right)^{-\rho_0} \left(\frac{|\widetilde{I}_{s-1}|}{\varepsilon_0}\right)^{2\sigma_0+\sigma_1}, \tag{9.228}$$

if $x_i \leqslant \widehat{x}_{cr}$. Similarly, we have (9.230) if $x_0 \geqslant \widehat{x}_{cr}$ and

$$\#\mathcal{B}i_+(\widetilde{I}, \widetilde{I}, \widetilde{I}_{s-1}; x_0) \leqslant \varepsilon_0^{-A\tau} \left(\frac{x_0}{\varepsilon_0 |P_u|}\right)^{-\rho_0} \left(\frac{|\widetilde{I}|}{\varepsilon_0}\right)^{\sigma_0+\sigma_1} \left(\frac{|\widetilde{I}_{s-1}|}{\varepsilon_0}\right)^{\sigma_0}, \tag{9.229}$$

if $x_0 \leqslant \widehat{x}_{cr}$; we have (9.231) if $x_h \geqslant \widehat{x}_{cr}$ and

$$\#\mathcal{B}i_+(\widetilde{I}, \widetilde{I}_{s-1}, \widetilde{I}; x_h) \leqslant \varepsilon_0^{-A\tau} \left(\frac{x_h}{\varepsilon_0 |P_u|}\right)^{-\rho_0} \left(\frac{|\widetilde{I}_{s-1}|}{\varepsilon_0}\right)^{\sigma_0+\sigma_1} \left(\frac{|\widetilde{I}|}{\varepsilon_0}\right)^{\sigma_0} \tag{9.230}$$

if $x_0 \leqslant \widehat{x}_{cr}$.

By Lemma 13, all the terms $\frac{x_j}{\varepsilon_0 |P_u|}$ are bounded by C. As $\rho_0 > \rho_1$, we can when necessary replace the exponent ρ_1 (for those x_j which are larger than $\widehat{x}_{cr}$) by ρ_0. On the other hand comparing the σ exponents, we see that we always have an additional factor $(\frac{|\widetilde{I}_{s-1}|}{\varepsilon_0})^{\sigma_0}$ when $x_j \leqslant \widehat{x}_{cr}$ compared with $x_j \geqslant \widehat{x}_{cr}$ (including the cases $j = 0, h$). As at least one x_j is $\leqslant \widehat{x}_{cr}$, multiplying the estimates together gives

$$b_+(\ell) b_-(\ell) \leqslant \widetilde{A}_2^h \widetilde{A}_3, \tag{9.231}$$

with

$$\widetilde{A}_2 = 2\varepsilon_0^{-2A\tau} \left(\frac{|\widetilde{I}_{s-1}|}{\varepsilon_0}\right)^{\sigma_0+\sigma_1} \left(C\varepsilon_0 |P_u| |\widehat{I}_{s-1}|^{-\frac{1}{2}}\right)^{\rho_0}, \tag{9.232}$$

$$\widetilde{A}_3 = C^{\rho_0} \left(\frac{x}{\varepsilon_0 |P_u|}\right)^{-\rho_0} \left(\frac{|\widetilde{I}|}{\varepsilon_0}\right)^{2\sigma_0+\sigma_1}. \tag{9.233}$$

9.11.5. *Partial sums of the $b_+(\ell)b_-(\ell)$.* — We observe that in both (9.225) and (9.231), our estimate for $b_+(\ell)b_-(\ell)$ depends on the class ℓ only through $\widehat{I}_{s-1}$ and h.

We first sum over subclasses with a fixed depth s and fixed sequence $\widehat{I}_0 \subset \cdots \subset \widehat{I}_{s-1}$, using the same method of generating series as in the proof of Proposition 45.

To deal with the two cases above at the same time, we observe that $\sigma_0 + \sigma_1 - \frac{1}{2}\rho_0(1+\tau)$ is close to $1 - d_u^0 - \frac{1}{2}d_s^0$ when $\tau \gg \eta \gg \varepsilon_0$ are small, with $1 - d_u^0 - \frac{1}{2}d_s^0 > 0$ from (H4). A fortiori $\sigma_0 + \sigma_1 - \frac{1}{2}\rho_1(1+\tau) > 0$ is positive and bounded away from 0. Thus, A_2 and $\widetilde{A}_2$ are larger when $\widetilde{I}_{s-1}$ is larger; the largest case is $\widetilde{I}_{s-1} = I_0$, which gives

$$\max(A_2, \widetilde{A}_2) \leqslant \widehat{A}_2 := 2\varepsilon_0^{-2A\tau}\left(C\varepsilon_0^{\frac{1}{2}(1-\tau)}|P_u|\right)^{\rho_1}. \tag{9.234}$$

We have $\max(A_3, \widetilde{A}_3) \leqslant C\widetilde{B}$, where $\varepsilon_0^{-A\tau}\widetilde{B}$ is the bound from $(SR3)_s$ for the cardinality of $Bi_+(\widetilde{I}, \widetilde{I}, \widetilde{I}; x)$.

We set

$$\begin{aligned}\chi_1(z) &= \sum_{h>0} \widehat{A}_2^h z^{A_1 h}(1-z)^{-h-2}\\ &= \widehat{A}_2 z^{A_1}(1-z)^{-2}(1-z-\widehat{A}_2 z^{A_1})^{-1}.\end{aligned} \tag{9.235}$$

The partial sum of $b_+(\ell)b_-(\ell)$ is thus not larger than $C\widetilde{B}$ times the coefficient of z^{A_0} in the power series for $\chi_1(z)$. Recall that A_0, A_1 were defined in (9.201), (9.202).

We estimate this coefficient by Cauchy integration on the circle $\{|z| = 1 - A_0^{-1} - \widehat{A}_2\}$, on which we have

$$|1-z|^{-2} \leqslant (\widehat{A}_2 + A_0^{-1})^{-2} \leqslant A_0^2, \tag{9.236}$$

$$|1 - z - \widehat{A}_2 z^{A_1}|^{-1} \leqslant A_0, \tag{9.237}$$

$$|\chi_1(z)| \leqslant \widehat{A}_2 A_0^3, \tag{9.238}$$

$$|z^{-A_0}| \leqslant C(1+\widehat{A}_2)^{A_0}. \tag{9.239}$$

The partial sum of $b_+(\ell)b_-(\ell)$ is therefore dominated by

$$C(1+\widehat{A}_2)^{A_0}\widehat{A}_2 A_0^3 \widetilde{B}. \tag{9.240}$$

9.11.6. *Sum of the $b_+(\ell)b_-(\ell)$ over all free subclasses with $h > 0$.* — We now have to sum over sequences $\widehat{I}_0 \subset \cdots \subset \widehat{I}_{s-1}$ and depth s; but (9.240) is independent of these data and the same remarks as in the proof of Proposition 45 apply. So, we finally obtain for the sum of $b_+(\ell)b_-(\ell)$ over subclasses with $s > 0$ and $h > 0$, a bound by

$$C(1+\widehat{A}_2)^{A_0}\widehat{A}_2 A_0^3 2^{S_{\max}}\widetilde{B} \tag{9.241}$$

with $S_{\max} := 2\tau^{-1}\log_2(\frac{\log|\widetilde{I}|^{-1}}{\log\varepsilon_0^{-1}})$ (cf. (9.211)).

Recall that (cf. (9.213))

$$\textbf{(9.242)} \qquad A_0 \leqslant C(\rho_0 - d_s^*)^{-1} \log |\widetilde{I}|^{-1}.$$

As $|P_u|$ is of the order of $\varepsilon_0^{\omega_u}$, we have

$$\textbf{(9.243)} \qquad \widehat{A}_2 \leqslant \varepsilon_0^{\rho_1/2}.$$

This gives

$$\textbf{(9.244)} \qquad (1+\widehat{A}_2)^{A_0} \leqslant |\widetilde{I}|^{-C\varepsilon_0^{\rho_1/2}(\rho_0 - d_s^*)^{-1}}.$$

From (9.195), we have

$$\textbf{(9.245)} \qquad (\rho_0 - d_s^*)^{-1} \ll \tau^2 \log \varepsilon_0^{-1}.$$

The bound that we finally obtain (from (9.241)) for the sum of the $b_+(\ell)b_-(\ell)$ over subclasses with $s>0$ and $h>0$ is therefore

$$\textbf{(9.246)} \qquad \varepsilon_0^{\rho_1/3} |\widetilde{I}|^{-\varepsilon_0^{\rho_1/3}} (\log |\widetilde{I}|^{-1})^{3\tau^{-1}} \widetilde{B}.$$

9.11.7. We summarize the calculations in this subsection in

Proposition **46.** — *The quantity* $\widehat{B} = \sum_L b_+(\ell)b_-(\ell)$ *is bounded by* $\widehat{B}_1 + \widehat{B}_2 + \widehat{B}_3$, *with*

$$\widehat{B}_1 = (\#\mathcal{C}_+(\widetilde{I}))(\#\mathcal{C}_-(\widetilde{I}))(\log |\widetilde{I}|^{-1})^2 (\rho_0 - d_s^*)^{-2},$$

$$\widehat{B}_2 = \left(\frac{\log |\widetilde{I}|^{-1}}{\log \varepsilon_0^{-1}}\right)^{2\tau^{-1}} (C|\widetilde{I}|^{1+\tau} x^{-1})^{\rho_s} (\#\mathcal{C}_+(\widetilde{I})) \sum_{\mathcal{C}_-(\widetilde{I})} \left(\max_{\widetilde{I}} |P_\omega|\right)^{\rho_s},$$

$$\widehat{B}_3 = \varepsilon_0^{\rho_1/3} |\widetilde{I}|^{-\varepsilon_0^{\rho_1/3}} (\log |\widetilde{I}|^{-1})^{3\tau^{-1}} \widetilde{B}.$$

9.12. *End of the induction step for* $(SR3)_s$.

9.12.1. In order to complete the induction step for $(SR3)_s$, it is sufficient, in view of (9.136) (Section 9.7.6), to show that

$$\textbf{(9.247)} \qquad \widehat{B} |\widetilde{I}|^{\tau(2 - d_s^+ - d_u^+ - 6\tau)} \leqslant \varepsilon_0^{-A\tau} B$$

where B is the bound in $(SR3)_s$.

We will bound each of the three expressions $\widehat{B}_i \varepsilon_0^{A\tau} |\widetilde{I}|^{\tau(2-d_s^+-d_u^+-6\tau)}$, $i = 1, 2, 3$, with $\widehat{B}_i$ defined in Proposition 46.

9.12.2. First consider $\widehat{B}_3$. For $x \leqslant \widetilde{x}_{cr}$, we have

$$\widetilde{B} = \widetilde{B}_0 = |\widetilde{I}|^{-\tau(2\sigma_0+\sigma_1)} B_0 \leqslant |\widetilde{I}|^{-\tau(2\sigma_0+\sigma_1)} B, \tag{9.248}$$

and, therefore,

$$\widehat{B}_3 |\widetilde{I}|^{\tau(2-d_s^+-d_u^+-6\tau)} \leqslant \varepsilon_0^{\rho_1/3} (\log|\widetilde{I}|^{-1})^{3\tau^{-1}} |\widetilde{I}|^{\omega} B, \tag{9.249}$$

with $\omega = \tau(2 - d_s^+ - d_u^+ - 6\tau - 2\sigma_0 - \sigma_1) - \varepsilon_0^{\rho_1/3}$.

We choose the exponents σ_0, σ_1 in order to have

$$\omega > \tau^2 \tag{9.250}$$

which is guaranteed by

$$2\sigma_0 + \sigma_1 \leqslant 2 - d_s^+ - d_u^+ - 8\tau. \tag{9.251}$$

Then, as $\varepsilon_0^{\rho_1/3} (\log|\widetilde{I}|^{-1})^{3\tau^{-1}} |\widetilde{I}|^{\tau^2} \ll 1$ the required estimate holds for $\widehat{B}_3$.

9.12.3. Next consider $\widehat{B}_1$. We use $(\mathrm{SR1})_s$, $(\mathrm{SR1})_u$ to control the sizes of $\mathcal{C}_+(\widetilde{I})$, $\mathcal{C}_-(\widetilde{I})$, and (9.245) to estimate $(\rho_0 - d_s^*)^{-1}$. This gives

$$\begin{aligned} &\widehat{B}_1 \varepsilon_0^{A\tau} |\widetilde{I}|^{\tau(2-d_s^+-d_u^+-6\tau)} \\ &\leqslant C \log^2 |\widetilde{I}|^{-1} (\rho_0 - d_s^*)^{-2} \left(\frac{|\widetilde{I}|}{\varepsilon_0}\right)^{2\sigma} \varepsilon_0^{\tau(A-d_s^0-d_u^0)} |\widetilde{I}|^{\tau(2-d_s^+-d_u^+-6\tau)} \\ &\leqslant C\tau^4 \log^2 \varepsilon_0^{-1} \log^2 |\widetilde{I}|^{-1} \left(\frac{|\widetilde{I}|}{\varepsilon_0}\right)^{2\sigma+\tau(2-d_s^+-d_u^+-6\tau)} \varepsilon_0^{\tau(A-d_s^0-d_u^0+2-d_s^+-d_u^+-6\tau)}. \end{aligned} \tag{9.252}$$

On the other hand, in the range $x \leqslant \widetilde{x}_{cr}$, B_0 is smaller at $\widetilde{x}_{cr}$ where it is equal to

$$\begin{aligned} \left(\frac{\widetilde{x}_{cr}}{\varepsilon_0 |P_u|}\right)^{-\rho_0} \left(\frac{|I|}{\varepsilon_0}\right)^{2\sigma_0+\sigma_1} &= \left(\frac{|\widetilde{I}|}{\varepsilon_0}\right)^{-\frac{\rho_0\sigma_0}{\rho_0-\rho_1}} \left(\frac{|I|}{\varepsilon_0}\right)^{2\sigma_0+\sigma_1} \\ &= \left(\frac{|\widetilde{I}|}{\varepsilon_0}\right)^{-\frac{\rho_0\sigma_0}{\rho_0-\rho_1}+(1+\tau)(2\sigma_0+\sigma_1)} \varepsilon_0^{\tau(2\sigma_0+\sigma_1)}. \end{aligned} \tag{9.253}$$

We ask that

$$-\frac{\rho_0\sigma_0}{\rho_0-\rho_1} + (1+\tau)(2\sigma_0+\sigma_1) \leqslant 2\sigma + \tau(2 - d_s^+ - d_u^+ - 6\tau) - \tau^2, \tag{9.254}$$

and

$$A - d_s^0 - d_u^0 + 2 - d_s^+ - d_u^+ - 6\tau \geqslant 2\sigma_0 + \sigma_1 + 1. \tag{9.255}$$

Then, we will have, for all $x_{\min} \leqslant x \leqslant \widetilde{x}_{cr}$

$$\begin{aligned} \widehat{B}_1 \varepsilon_0^{A\tau} |\widetilde{I}|^{\tau(2-d_s^+-d_u^+-6\tau)} &\leqslant C\tau^4 \log^2 \varepsilon_0^{-1} \log^2 |\widetilde{I}|^{-1} \left(\frac{|\widetilde{I}|}{\varepsilon_0}\right)^{\tau^2} \varepsilon_0^{\tau} B_0 \\ &\ll B_0. \end{aligned} \tag{9.256}$$

9.12.4. In the case of $\widehat{B}_2$, we use $(SR1)_s$ to control the size of $\mathcal{C}_+(\widetilde{I})$ and $(SR2)_u$ to control $\sum_{\mathcal{C}_-(\widetilde{I})}(\max_{\widetilde{I}}|P_\omega|)^{\rho_s}$. This gives

$$\widehat{B}_2\varepsilon_0^{A\tau}|\widetilde{I}|^{\tau(2-d_s^+-d_u^+-6\tau)}$$
$$\leqslant C\Big(\frac{\log|\widetilde{I}|^{-1}}{\log\varepsilon_0^{-1}}\Big)^{2\tau^{-1}} x^{-\rho_s}\Big(\frac{|\widetilde{I}|}{\varepsilon_0}\Big)^{\sigma+\sigma_s}|P_u|^{\rho_s}\varepsilon_0^{\tau(A-d_s^0)}|\widetilde{I}|^{(1+\tau)\rho_s+\tau(2-d_s^+-d_u^+-6\tau)}$$

(9.257)
$$\leqslant C\Big(\frac{\log|\widetilde{I}|^{-1}}{\log\varepsilon_0^{-1}}\Big)^{2\tau^{-1}}\Big(\frac{x}{\varepsilon_0|P_u|}\Big)^{-\rho_s}\Big(\frac{|\widetilde{I}|}{\varepsilon_0}\Big)^{(1+\tau)\rho_s+\tau(2-d_s^+-d_u^+-6\tau)+\sigma+\sigma_s}\varepsilon_0^{A'\tau},$$

with

(9.258)
$$A' := A+\rho_s-d_s^0+2-d_s^+-d_u^+-6\tau.$$

We want to show that the right-hand side of (9.257) is smaller than

(9.259)
$$B_0=\Big(\frac{x}{\varepsilon_0|P_u|}\Big)^{-\rho_0}\Big(\frac{|I|}{\varepsilon_0}\Big)^{2\sigma_0+\sigma_1}.$$

As $\rho_0>\rho_s$, it is sufficient to check this at $\widetilde{x}_{cr}$. After multiplying by $(\frac{\widetilde{x}_{cr}}{\varepsilon_0|P_u|})^{\rho_s}$, we have to show that

(9.260)
$$C\Big(\frac{\log|\widetilde{I}|^{-1}}{\log\varepsilon_0^{-1}}\Big)^{2\tau^{-1}}\Big(\frac{|\widetilde{I}|}{\varepsilon_0}\Big)^{(1+\tau)\rho_s+\tau(2-d_s^+-d_u^+-6\tau)+\sigma+\sigma_s}\varepsilon_0^{A'\tau}$$

is smaller than

(9.261)
$$\Big(\frac{|\widetilde{I}|}{\varepsilon_0}\Big)^{\frac{\sigma_0(\rho_s-\rho_0)}{\rho_0-\rho_1}+(1+\tau)(2\sigma_0+\sigma_1)}\varepsilon_0^{\tau(2\sigma_0+\sigma_1)}.$$

We will therefore ask that

$$\frac{\sigma_0(\rho_s-\rho_0)}{\rho_0-\rho_1}+(1+\tau)(2\sigma_0+\sigma_1)$$

(9.262)
$$\leqslant(1+\tau)\rho_s+\tau(2-d_s^+-d_u^+-6\tau)+\sigma+\sigma_s-\tau^2,$$

and

(9.263)
$$A+\rho_s-d_s^0+2-d_s^+-d_u^+-6\tau\geqslant 2\sigma_0+\sigma_1+1.$$

Then the required estimate is a consequence of

(9.264)
$$C\Big(\frac{\log|\widetilde{I}|^{-1}}{\log\varepsilon_0^{-1}}\Big)^{2\tau^{-1}}\Big(\frac{|\widetilde{I}|}{\varepsilon_0}\Big)^{\tau^2}\varepsilon_0^{\tau}\ll 1.$$

9.12.5. *Requirements on the exponents.* — Let us recall the various conditions imposed in this section on the various exponents. In Section 8, we introduced $d_s^* = d_s^0 + \varepsilon_0^{\frac{1}{5}d_s^0}$, $d_u^* = d_u^0 + \varepsilon_0^{\frac{1}{5}d_u^0}$, $d_s^+ = d_s^0 + C\eta\tau^{-1}$, $d_u^+ = d_u^0 + C\eta\tau^{-1}$.

- The exponent σ in (SR1) was defined in (9.49), Section 9.5, by

$$\sigma = \min(1 - d_u^+ - (1+\tau)d_s^* - \tau - c\eta\tau^{-1},$$
$$1 - d_s^+ - (1+\tau)d_u^* - \tau - c\eta\tau^{-1}). \tag{9.265}$$

- The exponents σ_s, σ_u in (SR2) were defined in (9.63), Section 9.6.1, by

$$\sigma_s := 1 - 3\tau - d_s^+, \qquad \sigma_u := 1 - 3\tau - d_u^+. \tag{9.266}$$

- The exponents ρ_s, ρ_u in (SR2) were asked to be close to d_s^0, d_u^0 respectively, to satisfy in (9.38), Section 9.4

$$\rho_u > d_u^0 + C\varepsilon_0, \qquad \rho_s > d_s^0 + C\varepsilon_0, \tag{9.267}$$

and later in (9.62) (Section 9.6.1) the stronger condition

$$\rho_u > d_u^* + \varepsilon_0^\tau, \qquad \rho_s > d_s^* + \varepsilon_0^\tau. \tag{9.268}$$

- The exponent ρ_0 in $(SR3)_s$ was required to be close to d_s^0. We asked also that $\rho_s > d_s^0 + C\varepsilon_0$ (cf. (9.38), Section 9.4), $\rho_0 > \rho_s + \varepsilon_0^\tau$ (cf. (9.62), Section 9.6.1), $\rho_0 \geqslant d_s^* + \varepsilon_0^{\tau^2/2}$ (cf. (9.98), Section 9.7.3), and finally (cf. (9.195), Section 9.10)

$$\rho_0 > d_s^* + C\tau^{-2}\frac{\log\log\varepsilon_0^{-1}}{\log\varepsilon_0^{-1}}. \tag{9.269}$$

- The exponent ρ_1 in $(SR3)_s$ was required to be close to $d_s^0(d_u^0 + 2d_s^0 - 1)\times (d_s^0 + d_u^0)^{-1}$.
- The exponents ρ_0', ρ_1' were defined by $\rho_0' = \frac{d_u^0}{d_s^0}\rho_0$, $\rho_1' = \frac{d_u^0}{d_s^0}\rho_1$.
- The exponents σ_0, σ_1 in (SR3) were required to be close to $1 - d_s^0$, $d_s^0 - d_u^0$ respectively. In (9.97), Section 9.7.3, we asked that

$$\sigma_0 + \sigma_1 \leqslant 1 - 3\tau - d_u^+, \tag{9.270}$$

and in (9.106), Section 9.7.4, that

$$\sigma_0 \leqslant 1 - 3\tau - d_s^+. \tag{9.271}$$

Finally, in (9.251), (9.254), (9.262) above, we required that

$$
\begin{aligned}
&2\sigma_0+\sigma_1 \leqslant 2-d_s^+-d_u^+-8\tau,\\
&-\frac{\rho_0\sigma_0}{\rho_0-\rho_1}+(1+\tau)(2\sigma_0+\sigma_1)\leqslant 2\sigma+\tau(2-d_s^+-d_u^+-6\tau)-\tau^2,\\
&\frac{\sigma_0(\rho_s-\rho_0)}{\rho_0-\rho_1}+(1+\tau)(2\sigma_0+\sigma_1)\\
&\qquad\leqslant (1+\tau)\rho_s+\tau(2-d_s^+-d_u^+-6\tau)+\sigma+\sigma_s-\tau^2.
\end{aligned}
\tag{9.272}
$$

– The exponent A in (SR3) was required above to satisfy

$$A-d_s^0-d_u^0+2-d_s^+-d_u^+-6\tau \geqslant 2\sigma_0+\sigma_1+1, \tag{9.273}$$

$$A+\rho_s-d_s^0+2-d_s^+-d_u^+-6\tau \geqslant 2\sigma_0+\sigma_1+1. \tag{9.274}$$

We have proved that, if all these conditions are satisfied, all candidates but a proportion not larger than $C|\widetilde{I}|^{\tau^2}$ satisfy $(SR3)_s$. Before checking that these requirements on the exponents are compatible, we need to review (briefly) the induction step for $(SR3)_u$ because new requirements will appear from it.

9.13. *The induction step for* $(SR3)_u$. — The proof that most candidates I in a strongly regular parent $\widetilde{I}$ satisfy $(SR3)_u$ follows the same plan than for $(SR3)_s$. However, condition (SR3) is not symmetric, hence we must check that the various steps work in the same way. We review briefly these steps below.

– *Very small values of x* (Section 9.7.1): we use now

$$\#Bi_-(I, I_\alpha, I_\omega; x) \leqslant Cx^{-d_u^*}, \tag{9.275}$$

for $x \leqslant x'_{\min} := |\widetilde{I}|^{C(\rho'_0-d_u^*)^{-1}}$.

– *Old and new elements* (Section 9.7.2) are defined as before.

– Observe that σ_0, σ_1 play the same role in $(SR3)_u$ than in $(SR3)_s$. Therefore, the results in Sections 9.7.3, 9.7.4, 9.7.5 work exactly in the same way, replacing $\widetilde{x}_{cr}$ in Proposition 42 by $\widetilde{x}'_{cr} := \varepsilon_0|Q_s|(\frac{|\widetilde{I}|}{\varepsilon_0})^{\frac{\sigma_0}{\rho'_0-\rho'_1}}$. We require as in (9.98) that $\rho'_0 \geqslant d_u^* + \varepsilon_0^{\tau^2/2}$, but this is actually a consequence of the definition of ρ'_0 and of (9.269) above.

– *Size of* Bi^{new} (Section 9.8). One proceeds as in Sections 9.8.1, 9.8.2, separating then the sum corresponding to (9.139) into two parts S'_l and S'_s. In the first sum (Section 9.8.3), the exponent ρ_1 must be replaced by $\rho'_1 = \rho_1\frac{d_u^0}{d_s^0} \leqslant \rho_1$. In particular the exponent $\sigma_0+\sigma_1-\rho'_1$ in (9.146) is still positive. The same considerations apply for S'_s. In the case $d_s^0 = d_u^0 = 1/2$, the exponent σ'_2 corresponding to σ_2 now depends on ω_s.

- Except for the obvious modifications in the definitions of the $b^+(P_\alpha, \ell)$, $b^-(Q_\omega, \ell)$, Section 9.9 is unchanged. In particular, the set L is defined in the same way; to estimate its cardinality, we now assume

$$\rho'_0 > d_u^* + C\tau^{-2}\frac{\log\log\varepsilon_0^{-1}}{\log\varepsilon_0^{-1}}, \tag{9.276}$$

but, in view of the definition of $\rho'_0 = \rho_0 d_u^0/d_s^0$, this is the same thing than (9.269), up to the value of the constant C.

- The quantities $\widehat{B}'_i$, $i = 1, 2, 3$ that appear instead of the $\widehat{B}_i$ of Proposition 46 are

$$\widehat{B}'_1 = (\#C_+(\widetilde{I}))(\#C_-(\widetilde{I}))(\log|\widetilde{I}|^{-1})^2(\rho'_0 - d_u^*)^{-2},$$

$$\widehat{B}'_2 = \left(\frac{\log|\widetilde{I}|^{-1}}{\log\varepsilon_0^{-1}}\right)^{2\tau^{-1}} (C|\widetilde{I}|^{1+\tau}x^{-1})^{\rho_u}(\#C_-(\widetilde{I}))\sum_{C_+(\widetilde{I})}\left(\max_{\widetilde{I}}|Q_\alpha|\right)^{\rho_u},$$

$$\widehat{B}'_3 = \varepsilon_0^{\rho'_1/3}|\widetilde{I}|^{-\varepsilon_0^{\rho'_1/3}}(\log|\widetilde{I}|^{-1})^{3\tau^{-1}}\widetilde{B}'.$$

- Dealing with $\widehat{B}'_3$ requires (9.251) as for $\widehat{B}_3$.
- Dealing with $\widehat{B}'_1$ requires (9.254), (9.255) as for $\widehat{B}_1$. Indeed, observe that in (9.254), we have

$$\frac{\rho_0}{\rho_0 - \rho_1} = \frac{\rho'_0}{\rho'_0 - \rho'_1} \tag{9.277}$$

in view of the proportionality of the exponents by the factor d_u^0/d_s^0.

- To control $\widehat{B}'_2$, we now require, corresponding to (9.262), (9.263)

$$\frac{\sigma_0(\rho_u - \rho'_0)}{\rho'_0 - \rho'_1} + (1+\tau)(2\sigma_0 + \sigma_1)$$
$$\leqslant (1+\tau)\rho_u + \tau(2 - d_s^+ - d_u^+ - 6\tau) + \sigma + \sigma_u - \tau^2, \tag{9.278}$$
$$A + \rho_u - d_u^0 + 2 - d_s^+ - d_u^+ - 6\tau \geqslant 2\sigma_0 + \sigma_1 + 1. \tag{9.279}$$

9.14. *Conclusion.*

Theorem **3.** — *Assume that the exponents in* (SR1), (SR2), (SR3) *satisfy the requirements of Section 9.12.5 and also (9.278), (9.279). Then, all candidates* I *in a strongly regular interval* $\widetilde{I}$ *are strongly regular, except for a proportion not larger than* $C|\widetilde{I}|^{\tau^2}$. *Moreover, it is possible to choose the exponents in order to satisfy these assumptions.*

Proof. — The first statement has been proved in Sections 9.5 through 9.13!

We choose the undefined exponents (σ, σ_s, σ_u are already defined) in the following order. First, take $\rho_u = d_u^* + 2\varepsilon_0^\tau$, $\rho_s = \frac{d_s^0}{d_u^0}\rho_u$. Then (9.268) is satisfied, and thus also (9.267). Next, choose

$$\rho_1 = d_s^0(d_u^0 + 2d_s^0 - 1)(d_s^0 + d_u^0)^{-1},$$

$$\rho_1' = d_u^0(d_u^0 + 2d_s^0 - 1)(d_s^0 + d_u^0)^{-1},$$

$$\rho_0 = d_s^0\left(1 + \tau^{-3}\frac{\log\log\varepsilon_0^{-1}}{\log\varepsilon_0^{-1}}\right),$$

$$\rho_0' = d_u^0\left(1 + \tau^{-3}\frac{\log\log\varepsilon_0^{-1}}{\log\varepsilon_0^{-1}}\right).$$

Then all the requirements on the ρ exponents are satisfied.

The exponents σ_0, σ_1 must satisfy (9.270), (9.271), the three relations in (9.272), and (9.278). Each of this six relations defines in the σ_0, σ_1-plane an affine half-plane whose boundary passes very close to the point $(1 - d_s^0, 1 - d_u^0)$ which has to be close to (σ_0, σ_1). Observe first that the coefficient of σ_1 in each of the six relations is positive. This implies that the intersection of the 6 *linear* half-planes parallel to these 6 affine half-planes contains an open cone at the origin. But then it is possible to find (σ_0, σ_1) close to $(1 - d_s^0, 1 - d_u^0)$ in the intersection of the affine half-planes.

Finally, the constraints on A are clearly satisfied if A is large enough. Actually, $A = 3$ works! □

Definition **13.** — *A strongly regular parameter (in* I_0*) is one which is the intersection of a decreasing sequence of strongly regular parameter intervals.*

Corollary **15.** — *Except for a set of relative measure* $\leqslant C\varepsilon_0^{\tau^2}$, *parameters in* I_0 *are strongly regular.*

Proof. — By Proposition 31, the starting interval is strongly regular. From the theorem, in a strongly regular parameter interval of level k, all points are contained in a strongly regular interval of the next level except for a set of relative Lebesgue measure $\leqslant \varepsilon_k^{\tau^2}$. As we have

$$\sum_{k\geqslant 0}\varepsilon_k^{\tau^2} \leqslant C\varepsilon_0^{\tau^2}, \tag{9.280}$$

the statement of the Corollary follows. □

10. The well-behaved part of the dynamics for strongly regular parameters

10.1. *Prime elements and prime decomposition.* — In the last two sections, we fix a strongly regular parameter, i.e. the intersection of a decreasing sequence $(I_m)_{m\geqslant 0}$ of strongly regular parameter intervals.

The sequence $\mathcal{R}(I_m)$ is increasing and we set

$$\mathcal{R} = \bigcup_{m\geqslant 0} \mathcal{R}(I_m). \tag{10.1}$$

Definition **14.** — *An element* $(P, Q, n) \in \mathcal{R}$ *is prime if* $n > 0$ *and it cannot be written as a simple composition of two shorter elements.*

Obviously, for any $(a, a') \in \mathcal{B}$, the element $(P_{aa'}, Q_{aa'}, 1)$ is prime. Such elements are called *trivial* primes. Non trivial primes are those of length bigger than 1.

There are only finitely many trivial primes. On the other hand, there are typically countably many non trivial ones.

Lemma **14.** — *Let* $(P, Q, n) \in \mathcal{R}$ *be an element which can be written as a simple composition* $(P, Q, n) = (P_1, Q_1, n_1) * (P_2, Q_2, n_2)$. *Let* $(\widetilde{P}_2, \widetilde{Q}_2, \widetilde{n}_2)$ *be the element such that* $\widetilde{P}_2$ *is the parent of* P_2, *and let* $(\widetilde{P}, \widetilde{Q}, \widetilde{n}) := (P_1, Q_1, n_1) * (\widetilde{P}_2, \widetilde{Q}_2, \widetilde{n}_2)$. *Then* $\widetilde{P}$ *is the parent of* P.

Proof. — If P_2 is a simple child, we have $n_2 = \widetilde{n}_2 + 1$, $n = \widetilde{n} + 1$ and P is a simple child of $\widetilde{P}$.

We now assume that P_2 is a non-simple child of $\widetilde{P}_2$. Let $(P', Q', n') \in \mathcal{R}$ the element such that P' is the child of $\widetilde{P}$ containing P. We will show that $P' = P$. We have $g_t^{\widetilde{n}}(P) \subset g_t^{\widetilde{n}_2}(P_2) \subset L_u$, hence P' also is a non-simple child.

Applying twice Proposition 7 (Section 6.2), we can write, in some $\mathcal{R}(I_m)$

$$(P_2, Q_2, n_2) \in (\widetilde{P}_2, \widetilde{Q}_2, \widetilde{n}_2) \,\square\, (P_3, Q_3, n_3),$$
$$(P', Q', n') \in (\widetilde{P}, \widetilde{Q}, \widetilde{n}) \,\square\, (P'_3, Q'_3, n'_3).$$

We have $\widetilde{Q} \subset \widetilde{Q}_2$ and also $P_3 \subset P'_3$ because $P \subseteq P'$. As $\widetilde{Q}_2 \pitchfork_{I_m} P_3$ and $\widetilde{Q} \pitchfork_{I_m} P'_3$ hold, concavity (Proposition 9, Section 6.3) imply that $\widetilde{Q}_2 \pitchfork_{I_m} P'_3$ also holds. As P_2 is a child of $\widetilde{P}_2$, we must have $P_3 = P'_3$ and $P' = P$. $\square$

Proposition **47.** — *Any element* $(P, Q, n) \in \mathcal{R}$ *with* $n > 0$ *can be uniquely written as a simple composition of a finite sequence of prime elements.*

Proof. — The existence of such a decomposition is clear. We have to show it is unique. Assume on the opposite that we can write

$$(\mathrm{P},\mathrm{Q},n) = (\mathrm{P}_1,\mathrm{Q}_1,n_1) * \cdots * (\mathrm{P}_r,\mathrm{Q}_r,n_r)$$
$$= (\mathrm{P}'_1,\mathrm{Q}'_1,n'_1) * \cdots * (\mathrm{P}'_s,\mathrm{Q}'_s,n'_s). \tag{10.2}$$

It is sufficient to show that $(\mathrm{P}_1,\mathrm{Q}_1,n_1) = (\mathrm{P}'_1,\mathrm{Q}'_1,n'_1)$. This is true if $n_1 = n'_1$. Assume for instance that $n_1 < n'_1$. Then we have $\mathrm{P} \subset \mathrm{P}'_1 \subset \mathrm{P}_1$ with $\mathrm{P}'_1 \neq \mathrm{P}_1$. But a repeated application of the lemma above shows that all rectangles in-between P and P_1 can be written as simple compositions of $(\mathrm{P}_1,\mathrm{Q}_1,n_1)$ with some other element. In particular, this is the case for $(\mathrm{P}'_1,\mathrm{Q}'_1,n'_1)$, which cannot be prime. □

Remark **14.** — In the prime decomposition

$$(\mathrm{P},\mathrm{Q},n) = (\mathrm{P}_1,\mathrm{Q}_1,n_1) * \cdots * (\mathrm{P}_r,\mathrm{Q}_r,n_r), \tag{10.3}$$

P_1 can be characterized as the thinnest prime rectangle containing P.

We will denote by $\mathcal{P}$ the set of prime elements of $\mathcal{R}$. We denote by $\mathcal{R}^*$ the set of elements of $\mathcal{R}$ of length > 0.

Let $(\mathrm{P},\mathrm{Q},n)$ be an element of $\mathcal{R}^*$ and let

$$(\mathrm{P},\mathrm{Q},n) = (\mathrm{P}_1,\mathrm{Q}_1,n_1) * \cdots * (\mathrm{P}_r,\mathrm{Q}_r,n_r), \tag{10.4}$$

be its prime decomposition. We define

$$\mathrm{T}^+((\mathrm{P},\mathrm{Q},n)) = (\mathrm{P}_2,\mathrm{Q}_2,n_2) * \cdots * (\mathrm{P}_r,\mathrm{Q}_r,n_r),$$
$$\mathrm{T}^-((\mathrm{P},\mathrm{Q},n)) = (\mathrm{P}_1,\mathrm{Q}_1,n_1) * \cdots * (\mathrm{P}_{r-1},\mathrm{Q}_{r-1},n_{r-1}), \tag{10.5}$$

if $r > 1$. When $(\mathrm{P},\mathrm{Q},n)$ is prime, with $\mathrm{P} \subset \mathrm{R}_a$ and $\mathrm{Q} \subset \mathrm{R}_{a'}$, we set

$$\mathrm{T}^+((\mathrm{P},\mathrm{Q},n)) = (\mathrm{R}_{a'},\mathrm{R}_{a'},0),$$
$$\mathrm{T}^-((\mathrm{P},\mathrm{Q},n)) = (\mathrm{R}_a,\mathrm{R}_a,0). \tag{10.6}$$

For $\mathrm{S} = (\mathrm{P},\mathrm{Q},n) \in \mathcal{R}$, we write $\mathrm{S} * \mathcal{R}$, resp. $\mathcal{R} * \mathrm{S}$, for the set of elements which can be written as $(\mathrm{P},\mathrm{Q},n) * (\mathrm{P}',\mathrm{Q}',n')$, resp. $(\mathrm{P}',\mathrm{Q}',n') * (\mathrm{P},\mathrm{Q},n)$, for some $(\mathrm{P}',\mathrm{Q}',n') \in \mathcal{R}$. We have partitions

$$\mathcal{R}^* = \bigsqcup_{\mathcal{P}} \mathrm{S} * \mathcal{R} = \bigsqcup_{\mathcal{P}} \mathcal{R} * \mathrm{S}. \tag{10.7}$$

10.2. *Number of factors in a prime decomposition.* — We write $r(\mathrm{S})$ for the number of factors in the prime decomposition of an element S of $\mathcal{R}$ (setting $r(\mathrm{S}) = 0$ *if* S has length 0). Let $(\mathrm{P},\mathrm{Q},n)$, $(\mathrm{P}',\mathrm{Q}',n')$ be elements of $\mathcal{R}$ such that P′ is a child of P. When P′ is a simple child, it is obtained by simple composition of P with an element of length 1 and we have

$$r(\mathrm{P}',\mathrm{Q}',n') = r(\mathrm{P},\mathrm{Q},n) + 1. \tag{10.8}$$

On the other hand, assume that P′ is a non-simple child of P. Let m be such that $(P', Q', n') \in \mathcal{R}(I_m)$. By Proposition 7 (Section 6.2), we can write

$$(P', Q', n') \in (P, Q, n) \,\Box\, (\widehat{P}, \widehat{Q}, \widehat{n}), \tag{10.9}$$

for some $(\widehat{P}, \widehat{Q}, \widehat{n}) \in \mathcal{R}(I_m)$ such that $Q \pitchfork_{I_m} \widehat{P}$ holds. Let

$$(P, Q, n) = (P_1, Q_1, n_1) * \cdots * (P_r, Q_r, n_r) \tag{10.10}$$

be the prime decomposition of (P, Q, n).

Define, for $1 \leqslant i \leqslant r$

$$\begin{aligned} (P^i, Q^i, n^i) &= (P_i, Q_i, n_i) * \cdots * (P_r, Q_r, n_r) \\ &= (T^+)^{i-1}(P, Q, n). \end{aligned} \tag{10.11}$$

We have an increasing sequence

$$Q = Q^1 \subset Q^2 \subset \cdots \subset Q^r = Q_r. \tag{10.12}$$

Let r' be the largest integer in $\{1, \ldots, r\}$ such that $Q^{r'}$ and $\widehat{P}$ are I_l-transverse for some $l \geqslant m$ (and then for all large enough l). Define $(\widetilde{P}, \widetilde{Q}, \widetilde{n}) \in \mathcal{R}$ by the condition $Q' \subset \widetilde{Q}$ and

$$(\widetilde{P}, \widetilde{Q}, \widetilde{n}) \in (P^{r'}, Q^{r'}, n^{r'}) \,\Box\, (\widehat{P}, \widehat{Q}, \widehat{n}). \tag{10.13}$$

Proposition **48.** — *The simple composition*

$$(P', Q', n') = (P_1, Q_1, n_1) * \cdots * (P_{r'-1}, Q_{r'-1}, n_{r'-1}) * (\widetilde{P}, \widetilde{Q}, \widetilde{n}). \tag{10.14}$$

is the decomposition of (P', Q', n') *in prime factors. In particular, we have*

$$r(P', Q', n') = r' \leqslant r(P, Q, n).$$

Proof. — It is sufficient to show that $(\widetilde{P}, \widetilde{Q}, \widetilde{n})$ is prime. Assume by contradiction that we can write

$$(\widetilde{P}, \widetilde{Q}, \widetilde{n}) = (\widetilde{P}_1, \widetilde{Q}_1, \widetilde{n}_1) * (\widetilde{P}_2, \widetilde{Q}_2, \widetilde{n}_2) \tag{10.15}$$

with $\widetilde{n}_1, \widetilde{n}_2 > 0$. Define

$$(P'_1, Q'_1, n'_1) = (P_1, Q_1, n_1) * \cdots * (P_{r'-1}, Q_{r'-1}, n_{r'-1}) * (\widetilde{P}_1, \widetilde{Q}_1, \widetilde{n}_1). \tag{10.16}$$

Claim. — *There exists* $r' \leqslant j \leqslant r$ *such that*

$$(P'_1, Q'_1, n'_1) = (P_1, Q_1, n_1) * \cdots * (P_j, Q_j, n_j). \tag{10.17}$$

Proof of claim. — We have $P' \subset P'_1$, $P' \neq P'_1$, hence $P \subset P'_1$, $n'_1 \leqslant n$. As $\widetilde{n}_1 > 0$, the smallest integer j such that

$$(\mathbf{10.18}) \qquad n'_1 \leqslant n_1 + \cdots + n_j,$$

satisfies $r' \leqslant j \leqslant r$. From

$$(\mathbf{10.19}) \qquad (P', Q', n') = (P'_1, Q'_1, n'_1) * (\widetilde{P}_2, \widetilde{Q}_2, \widetilde{n}_2)$$

and Lemma 14, we can write $(P_1, Q_1, n_1) * \cdots * (P_j, Q_j, n_j)$ as a simple composition

$$(\mathbf{10.20}) \qquad (P_1, Q_1, n_1) * \cdots * (P_j, Q_j, n_j) = (P'_1, Q'_1, n'_1) * (\overline{P}, \overline{Q}, \overline{n})$$

for some $(\overline{P}, \overline{Q}, \overline{n}) \in \mathcal{R}$. Still by Lemma 14, we can write

$$(\mathbf{10.21}) \qquad (P_j, Q_j, n_j) = (\overline{P}', \overline{Q}', \overline{n}') * (\overline{P}, \overline{Q}, \overline{n})$$

for some $(\overline{P}', \overline{Q}', \overline{n}') \in \mathcal{R}$. We have $\overline{n}' > 0$ by definition of j. As (P_j, Q_j, n_j) is prime, we have $\overline{n} = 0$, which proves the claim

We will now prove that, for l large enough, $Q^{j+1} \pitchfork_{I_l} \widehat{P}$ holds, a contradiction with the definition of r'. Indeed, from (10.19) we have $\widetilde{P}_2 \subset P^{j+1}$, $\widetilde{P}_2 \neq P^{j+1}$. Let $(P^*, Q^*, n^*) \in \mathcal{R}$ the element such that P^* is the child of P^{j+1} containing $\widetilde{P}_2$. By Proposition 7 (Section 6.2), we can write (with l large enough) $(P^*, Q^*, n^*) \in (P^{j+1}, Q^{j+1}, n^{j+1}) \,\square\, (\widehat{P}^*, \widehat{Q}^*, \widehat{n}^*)$, for some $(\widehat{P}^*, \widehat{Q}^*, \widehat{n}^*)$ with $\widehat{P}^* \supset \widehat{P}$. Then $Q^{j+1} \pitchfork_{I_l} \widehat{P}^*$ holds and $Q^{j+1} \pitchfork_{I_l} \widehat{P}$ also holds.

Thus, we obtain a contradiction and the proof of the proposition is complete. □

10.3. *A weighted estimate on the number of children.* — We present in this subsection a variation over the estimates in Section 8.3, which will be important in the definition of a transfer operator.

As mentioned already in Section 9.4, there exist from classical results of Bowen, Ruelle and Sinai an invariant probability measure m^+ on K such that the measure of an I_0-defined rectangle P is of the order $|P|^{d_s^t}$, where d_s^t is the transverse Hausdorff dimension of the stable foliation $W^s(K)$. Recall that d_s^t is a smooth, hence Lipschitz, function of t.

We fix a constant $\kappa \in (0, 1)$ close to 1, but independent of ε_0. Let also $d_s^- < d_s^t$, close to d_s^t.

For $S = (P, Q, n)$, we set

$$(\mathbf{10.22}) \qquad \|P\| = |P|^{d_s^-} \kappa^{r(S)}$$

(we will also write $r(P)$ instead of $r(S)$).

Proposition **49.** — *Assume that* $\log \kappa^{-1} < C^{-1}$ *and* $d_s^t - d_s^- < C^{-1} \log \kappa^{-1}$ *with* C *large enough. For any* $m \geqslant 1$, *any* $(P, Q, n) \in \mathcal{R}$, *we have*

$$\sum_{P'} \|P'\| \leqslant C\kappa^{\frac{m}{2}} \|P\|$$

where the sum in the left-hand side is over elements (P', Q', n') *such that* P′ *is a descendant of the* m^{th} *generation of* P.

We will first state a Lemma, then prove the proposition from the Lemma, and finally prove the Lemma.

Lemma **15.** — *Let* $\varepsilon_1 > 0$. *If* ε_0 *is small enough, we have*

$$\sum_{P'} \|P'\| \leqslant \varepsilon_1 \|P\|$$

for all $(P, Q, n) \in \mathcal{R}$, *where the sum in the left-hand side is over non-simple children of* P.

Proof of the Proposition. — Let $m_0 \geqslant 1$ be an integer to be determined later. Consider all chains

$$P = P^0 \supset P^1 \supset \cdots \supset P^{m_0} = P' \tag{10.23}$$

where P is given and P^{i+1} is a child of P^i.

First consider the case where P^{i+1} is, for each i, a simple child of P^i. One has then $r(P') = m_0 + r(P)$, and one can write $(P', Q', n') = (P, Q, n) * (\widehat{P}, \widehat{Q}, \widehat{n})$ with $(\widehat{P}, \widehat{Q}, \widehat{n}) \in \mathcal{R}(I_0)$. We have

$$\begin{aligned}
\|P'\| &\leqslant C\kappa^{m_0} \|P\| |\widehat{P}|^{d_s^-} \\
&\leqslant C\kappa^{m_0} \|P\| m^+(\widehat{P}) |\widehat{P}|^{d_s^- - d_s^t} \\
&\leqslant C\kappa^{m_0} \|P\| m^+(\widehat{P}) \exp(Cm_0(d_s^t - d_s^0)) \\
&\leqslant C\kappa^{\frac{2}{3}m_0} \|P\| m^+(\widehat{P}),
\end{aligned}$$

if $d_s^t - d_s^- < C^{-1} \log \kappa^{-1}$ with C large enough. Under this condition the part of the sum in Proposition 49 corresponding to simple descendants satisfies (as $\sum m^+(\widehat{P}) \leqslant 1$)

$$\sum \|P'\| \leqslant C\kappa^{\frac{2}{3}m_0} \|P\|. \tag{10.24}$$

We choose m_0 such that in (10.24) we have

$$C\kappa^{\frac{2}{3}m_0} \leqslant \frac{1}{2} \kappa^{\frac{m_0}{2}}. \tag{10.25}$$

On the other hand, from the lemma above, it follows that for every $\widetilde{P}$, we have

$$(\mathbf{10.26}) \qquad \sum \|\widetilde{P}'\| \leqslant C\|\widetilde{P}\|$$

where the sum is over all children of $\widetilde{P}$. Using the lemma again, when we sum over chains such that P_{i+1} is a non-simple child of P_i for at least one i, we obtain

$$(\mathbf{10.27}) \qquad \sum \|\widetilde{P}'\| \leqslant m_0 C^{m_0-1}\varepsilon_1\|P\|.$$

Taking ε_1 small enough, we obtain

$$(\mathbf{10.28}) \qquad \sum \|P'\| \leqslant \kappa^{\frac{m_0}{2}}\|P\|$$

where the sum is now over all chains. The proposition follows immediately from (10.28) and (10.26). □

Lemma **16.** — *Let* $(P_0, Q_0, n_0), (P'_0, Q'_0, n'_0), (P_1, Q_1, n_1) \in \mathcal{R}$ *with* $Q_0 \subset Q'_0$. *If* $Q_0 \pitchfork_{I_m} P_1$ *holds for m large enough and*

$$8|Q'_0|^{1-\eta} < \delta(Q_0, P_1),$$

then $Q'_0 \pitchfork_{I_m} P_1$ *holds for m large enough.*

Proof. — If $3|P_1|^{1-\eta} < \delta(Q_0, P_1)$, $Q'_0 \,\overline{\pitchfork}_{I_m}\, P_1$ holds for m large enough by direct verification of (T1), (T2), (T3) of Section 5.4. If on the other hand, $3|P_1|^{1-\eta} \geqslant \delta(Q_0, P_1)$, we can apply Proposition 21 in Section 8.1 to conclude that $Q'_0 \pitchfork_{I_m} P_1$ holds for m large enough. □

Proof of Lemma 15. — Let $(P, Q, n) \in \mathcal{R}$. Any non-simple child P' of P is obtained as

$$(\mathbf{10.29}) \qquad (P', Q', n') \in (P, Q, n) \,\square\, (P_1, Q_1, n_1)$$

and we denote by $\widetilde{P}_1$, the parent of P_1. One has

$$(\mathbf{10.30}) \qquad |P'| \leqslant C|P||P_1|\delta(Q, P_1)^{-\frac{1}{2}}.$$

Therefore, we will have

$$(\mathbf{10.31}) \qquad \|P\|^{-1}\sum \|P'\| \leqslant C\sum |P_1|^{d_s^-}\kappa^{r(P')-r(P)}\delta(Q, P_1)^{-\frac{1}{2}d_s^-}.$$

By Proposition 48, there is an increasing sequence

$$(\mathbf{10.32}) \qquad Q = Q^1 \subset Q^2 \subset \cdots \subset Q^{r(P)} = Q_{r(P)}$$

such that $r(P')$ is the largest integer r' for which $Q^{r'}$ and P_1 are I_m-transverse for large enough m. We claim that

$$r(P) - r(P') \leqslant C \log(\delta(Q, P_1))^{-1}. \tag{10.33}$$

Indeed, let $1 \leqslant r \leqslant r(P)$ such that

$$|Q^r| \leqslant \delta(Q, P_1)^2. \tag{10.34}$$

By Lemma 16, $Q^r \pitchfork_{I_m} P_1$ holds for m large enough, hence $r \leqslant r(P)$. On the other hand, there exists $\kappa^* \in (0, 1)$ such that

$$|Q^r| \leqslant \kappa^* |Q^{r+1}| \tag{10.35}$$

for $r < r(P)$. The claim follows.

Therefore, if κ is close enough to 1, we have

$$\kappa^{r(P')-r(P)} \leqslant (\delta(Q, P_1))^{-\frac{1}{6}d_s^-} \tag{10.36}$$

and the right-hand side of (10.31) is bounded by

$$C \sum |P_1|^{d_s^-} (\delta(Q, P_1))^{-\frac{2}{3}d_s^-}. \tag{10.37}$$

Using (R7) (Section 5.4), this is smaller than

$$C \sum |P_1|^{\frac{1}{3}d_s^-}. \tag{10.38}$$

To estimate this sum, we first fix the parent $\widetilde{P}_1$ and sum over children P_1; it follows from Proposition 26 (Section 8.2) that the corresponding sum is bounded by $C|\widetilde{P}_1|^{\frac{1}{3}d_s^-}$. Then (10.38) is not greater than

$$C \sum |\widetilde{P}_1|^{\frac{1}{3}d_s^-}. \tag{10.39}$$

For each integer m, let us count now how many $\widetilde{P}_1$ may satisfy

$$2^{-m} \geqslant |\widetilde{P}_1| \geqslant 2^{-m-1}. \tag{10.40}$$

As P' is a child of P, Q is transverse to P_1 but not to $\widetilde{P}_1$. From Lemma 16, we have therefore

$$\delta(Q, \widetilde{P}_1) \leqslant 8 |\widetilde{P}_1|^{1-\eta} \leqslant 8\, 2^{-m(1-\eta)} \tag{10.41}$$

which shows that there are no more than $C2^{m\eta}$ such $\widetilde{P}_1$'s. This implies that the sum (10.39) is at most of order $C\varepsilon_0^{\frac{1}{3}d_s^-}$, which yields the statement of the lemma. □

Remark **15.** — In Lemma 15, the value d_s^- that has been used to define $\|P\|$ is irrelevant. The assertion of the lemma is still true if we replace d_s^- by any positive number bounded away from 0.

Corollary **16.** — *Let* $\varepsilon_1 > 0$. *If* ε_0 *is small enough, we have*

$$\sum_{\mathcal{P}} n|P|^{d_s^-} < \varepsilon_1$$

where the sum in the left-hand side is over non-trivial primes (P, Q, n).

Proof. — Let (P, Q, n) be a non trivial prime. We have, by Proposition 12 in Section 6.6.2

$$n \leqslant \left(\log(C|P|^{-1})\right)^{\frac{\log 2}{\log 3/2}}. \tag{10.42}$$

Choose $\widehat{d_s^-} < d_s^-$ but so close to d_s^- that the hypothesis $d_s^t - \widehat{d_s^-} < C^{-1}\log\kappa^{-1}$ of Proposition 49 is still satisfied. If ε_0 is small enough, we have, for any non trivial prime P, as $|P| < \varepsilon_0$

$$n|P|^{d_s^-} \leqslant |P|^{\widehat{d_s^-}}. \tag{10.43}$$

Observe also that the thinnest $(\widetilde{P}, \widetilde{Q}, \widetilde{n}) \in \mathcal{R}(I_0)$ with $P \subset \widetilde{P}$ satisfies $\widetilde{Q} \subset Q_u$ hence $|\widetilde{P}| \leqslant \varepsilon_0^\alpha$ for some fixed positive α.

We apply the proposition using $\widehat{d_s^-}$ instead of d_s^-; we obtain, as $r(P) = 1$ for primes

$$\begin{aligned}\sum n|P|^{d_s^-} &\leqslant \sum |P|^{\widehat{d_s^-}}\\ &= \kappa^{-1}\sum \|P\|\\ &\leqslant C\kappa^{-1}(1-\kappa^{\frac{1}{2}})^{-1}\sum \|\widetilde{P}\|,\end{aligned} \tag{10.44}$$

where, in the last sum, $(\widetilde{P}, \widetilde{Q}, \widetilde{n})$ runs through the elements of $\mathcal{R}(I_0)$ with $|\widetilde{P}|$ of the order of ε_0^α. We have

$$\begin{aligned}\sum \|\widetilde{P}\| &\leqslant C\sum \varepsilon_0^{\alpha\widehat{d_s^-}}\kappa^{C^{-1}\log\varepsilon_0^{-1}}\\ &\leqslant C\varepsilon_0^{\alpha(\widehat{d_s^-}-d_s^t)+C^{-1}\log\kappa^{-1}},\end{aligned} \tag{10.45}$$

and the exponent is positive from the hypothesis of Proposition 49. Putting this into (10.44) yields Corollary 16. □

Remark **16.** — We do not know the range of d for which

$$\sum_{\mathcal{P}} |\mathrm{P}|^d$$

is convergent. The corollary shows at least that there are relatively few primes: the sum is convergent for $d > d_s^0 - \mathrm{C}^{-1}$, C large enough.

10.4. *Stable curves.* — Let $(\mathrm{P}_k, \mathrm{Q}_k, n_k)_{k\geqslant 0}$ be a sequence of elements of $\mathcal{R}$ such that P_{k+1} is strictly contained in P_k for $k \geqslant 0$. Let R_a be the rectangle of the Markov partition which contains P_0. The vertical part of the boundary of P_k is the union of two graphs $\{x_a = \varphi_k^{\pm}(y_a)\}$.

Proposition **50.** — *The intersection $\bigcap_{k\geqslant 0} \mathrm{P}_k$ is the graph $\{x_a = \varphi_\infty(y_a)\}$ of a $\mathrm{C}^{1+\mathrm{Lip}}$ function, whose $\mathrm{C}^{1+\mathrm{Lip}}$ norm of φ_∞ is bounded independently of the sequence $(\mathrm{P}_k)_{k\geqslant 0}$. Moreover, we have, for all $k \geqslant 0, y_a \in \mathrm{I}_a^u$*

$$|\varphi_k^{\pm}(y_a) - \varphi_\infty(y_a)| \leqslant \mathrm{C}|\mathrm{P}_k|,$$
$$|\mathrm{D}\varphi_k^{\pm}(y_a) - \mathrm{D}\varphi_\infty(y_a)| \leqslant \mathrm{C}|\mathrm{P}_k|.$$

Proof. — Let $(\mathrm{A}_k, \mathrm{B}_k)$ be the implicit representation associated to $(\mathrm{P}_k, \mathrm{Q}_k, n_k)$. We have

$$\varphi_k^{\pm}(y_a) = \mathrm{A}_k(y_a, x_b^{\pm}), \tag{10.46}$$

where $x_b^{\pm}$ are the endpoints of I_b^s and $\mathrm{Q}_k \subset \mathrm{R}_b$. The partial derivatives $\mathrm{A}_{k,y}$, $\mathrm{A}_{k,yy}$ are bounded by u_0^{-1}, $2\mathrm{D}_0$ respectively and we have

$$|\varphi_k^{+}(y_a) - \varphi_k^{-}(y_a)| \leqslant \mathrm{C}|\mathrm{P}_k|. \tag{10.47}$$

This implies the statement of the proposition, except for the last inequality.

To compare the derivatives of $\varphi_k^{\pm}$ and $\varphi_{k+1}^{\pm}$, we use (A.66) in Appendix A if P_{k+1} is a simple child of P_k, (A.86) if it is a non-simple child. We obtain that, for every y_a, x, there exists x^* such that

$$|\mathrm{A}_{k+1,y}(y_a, x) - \mathrm{A}_{k,y}(y_a, x^*)| \leqslant \mathrm{C}|\mathrm{P}_k||\mathrm{Q}_k|, \tag{10.48}$$

in the first case and

$$|\mathrm{A}_{k+1,y}(y_a, x) - \mathrm{A}_{k,y}(y_a, x^*)| \leqslant \mathrm{C}|\mathrm{P}_k||\mathrm{Q}_k|\delta^{-1/2}, \tag{10.49}$$

in the second case. Here, we have $\delta \ll |\mathrm{Q}_k|$ from (R7) in Section 5.4. Using that

$$|\mathrm{A}_{k,y}(y_a, x^*) - \mathrm{A}_{k,y}(y_a, x_b^{\pm})| \leqslant \mathrm{C}\|\mathrm{A}_{k,xy}\|_\infty \leqslant \mathrm{C}|\mathrm{P}_k|, \tag{10.50}$$

we conclude that in both cases we have

$$|D\varphi_k^{\pm}(y_a) - D\varphi_{k+1}^{\pm}(y_a)| \leqslant C|P_k|. \tag{10.51}$$

As there exists $\kappa^* \in (0,1)$ such that $|P_{k+1}| \leqslant \kappa^* |P_k|$ for all $k \geqslant 0$, we obtain, for all $l > k$

$$|D\varphi_k^{\pm}(y_a) - D\varphi_l^{\pm}(y_a)| \leqslant C|P_k|, \tag{10.52}$$

and the required inequality follows, letting l go to $+\infty$. □

Definition **15.**

1. *A stable curve is the intersection $\omega = \bigcap_{k\geqslant 0} P_k$ of a decreasing sequence of vertical-like rectangles as above. An unstable curve is the intersection $\omega' = \bigcap_{k\geqslant 0} Q'_k$ of a decreasing sequence of horizontal-like strips.*
2. *The set of stable curves, resp. unstable curves, is denoted by $\mathcal{R}_+^\infty$, resp. $\mathcal{R}_-^\infty$. The union of stable curves, resp. unstable curves, is denoted by $\widetilde{\mathcal{R}}_+^\infty$, resp. $\widetilde{\mathcal{R}}_-^\infty$.*
3. *Any stable curve $\omega \subset R_a$ has a canonical defining sequence characterized by the following conditions: $P_0 = R_a$ and, for each k, P_{k+1} is a child of P_k.*
4. *Two stable curves are equal or disjoint. Hence there is a canonical projection*

$$\pi : \widetilde{\mathcal{R}}_+^\infty \mapsto \mathcal{R}_+^\infty.$$

We will now define dynamics on a part of the sets $\mathcal{R}_+^\infty$, $\widetilde{\mathcal{R}}_+^\infty$.

Let $\mathcal{N}_+$ be the set of stable curves ω which are contained in infinitely many prime elements and let $\mathcal{D}_+$ be the complementary subset in $\mathcal{R}_+^\infty$. For $(P, Q, n) \in \mathcal{P}$, denote by $\mathcal{R}_+^\infty(P)$ the set of stable curves $\omega \in \mathcal{D}_+$ such that P is the thinnest prime containing ω.

We, thus, have partitions

$$\mathcal{R}_+^\infty = \mathcal{N}_+ \bigsqcup \mathcal{D}_+, \tag{10.53}$$

$$\mathcal{D}_+ = \bigsqcup_{\mathcal{P}} \mathcal{R}_+^\infty(P). \tag{10.54}$$

We denote by $\widetilde{\mathcal{N}}_+$, $\widetilde{\mathcal{D}}_+$, $\widetilde{\mathcal{R}}_+^\infty(P)$ the respective pre-images by π.

Let $(P, Q, n) \in \mathcal{P}$, $\omega \in \mathcal{R}_+^\infty(P)$. For any (P_k, Q_k, n_k) with $\omega \subset P_k \subset P$, we can write (cf. Remark after Proposition 47)

$$(P_k, Q_k, n_k) = (P, Q, n) * (P'_k, Q'_k, n'_k) \tag{10.55}$$

for some $(P'_k, Q'_k, n'_k) \in \mathcal{R}$; we have

$$T^+(P_k, Q_k, n_k) = (P'_k, Q'_k, n'_k) \tag{10.56}$$

and we define $\omega' = T^+(\omega)$ to be the stable curve obtained by the intersection of the P'_k when P_k decrease to ω. We have

$$(\mathbf{10.57}) \qquad g^n(P_k) \subset P'_k, \qquad g^n(\omega) \subset \omega'$$

and we also define

$$(\mathbf{10.58}) \qquad \widetilde{T}^+ / \widetilde{\mathcal{R}}_+^\infty(P) = g^n / \widetilde{\mathcal{R}}_+^\infty(P).$$

We, thus, have a commutative diagram

$$(\mathbf{10.59}) \qquad \begin{array}{ccc} \widetilde{\mathcal{D}}_+ & \xrightarrow{\widetilde{T}^+} & \widetilde{\mathcal{R}}_+^\infty \\ \downarrow \pi & & \downarrow \pi \\ \mathcal{D}_+ & \xrightarrow{T^+} & \mathcal{R}_+^\infty \end{array}$$

We observe that for $(P, Q, n) \in \mathcal{P}$ with $Q \subset R_a$, the image $T^+(\omega)$ of any $\omega \in \mathcal{R}_+^\infty(P)$ is contained in R_a.

Conversely, let $(P, Q, n) \in \mathcal{P}$ with $Q \subset R_a$ and let $\omega' \in \mathcal{R}_+^\infty$, $\omega' \subset R_a$. For any (P'_k, Q'_k, n'_k) with $\omega \subset P'_k$, we define (P_k, Q_k, n_k) by (10.55); the intersection $\omega = \bigcap_{k \geqslant 0} P_k$ is the unique stable curve in $\mathcal{R}_+^\infty(P)$ such that $T^+(\omega) = \omega'$.

Thus, T^+ induces a bijection from $\mathcal{R}_+^\infty(P)$ on the set $\mathcal{R}_+^\infty(a)$ of stable curves contained in R_a. For $\omega \in \mathcal{R}_+^\infty(P)$, we have

$$(\mathbf{10.60}) \qquad \widetilde{T}^+(\omega) = \omega' \cap Q.$$

10.5. *Topology and geometry of $\mathcal{R}_+^\infty$ and $\widetilde{\mathcal{R}}_+^\infty$.* — Each stable curve is a compact subset of $R = \bigcup R_a$. Therefore, $\mathcal{R}_+^\infty$ may be viewed as a subset of the set of non empty compact subsets of R endowed with the Hausdorff topology. The topology induced on $\mathcal{R}_+^\infty$ can also be viewed directly: for any $\omega = \bigcap P_k$ in $\mathcal{R}_+^\infty$, a basis of neighbourhoods of ω is obtained by considering for each k the set V_k of stable curves contained in P_k.

Equipped with this topology, $\mathcal{R}_+^\infty$ is a Cantor set. Each $\mathcal{R}_+^\infty(P)$, $P \in \mathcal{P}$, is a closed subset, and also a Cantor set. The restriction of T^+ to each $\mathcal{R}_+^\infty(P)$ is a homeomorphism onto $\mathcal{R}_+^\infty(a)$ (with $Q \subset R_a$).

However, the subset $\mathcal{N}_+$ may be dense and the map T^+ in general is not continuous on the whole of $\mathcal{D}_+$. We will see in the sequel that $\mathcal{N}_+$ is, in some appropriate sense, negligible.

For each $\omega \in \mathcal{R}_+^\infty(a)$, we denote by φ_ω the $C^{1+\mathrm{Lip}}$ map such that $\omega = \{x_a = \varphi_\omega(y_a)\}$; for each $a \in \mathcal{a}$, each $y_a^0 \in I_a^u$, the map

$$(\mathbf{10.61}) \qquad \begin{aligned} \phi_{y_a^0} : \mathcal{R}_+^\infty(a) &\mapsto I_a^s, \\ \omega &\mapsto \varphi_\omega(y_a^0) \end{aligned}$$

is a homeomorphism onto its image. Letting y_a^0 vary, we get an homeomorphism from $\mathcal{R}_+^\infty(a) \times I_a^u$ onto $\widetilde{\mathcal{R}}_+^\infty(a)$.

Regarding the transverse regularity of the partial foliation $\widetilde{\mathcal{R}}_+^\infty(a)$, we have the following result.

Proposition **51.** — *For all $a \in \mathfrak{a}$, all distinct $\omega, \omega' \in \mathcal{R}_+^\infty(a)$, all $y \in I_a^u$, we have*

$$\left| \frac{\partial}{\partial y} \log |\varphi_\omega - \varphi_{\omega'}| \right| \leqslant C.$$

In particular, the homeomorphisms $\phi_{y'} \circ \phi_y^{-1}$ are bi-Lipschitzian, uniformly in y, y'.

Proof. — For $(P, Q, n) \in \mathcal{R}$, let $y = \varphi_P^\pm(x)$ be the vertical-like curves bounding P. We will prove that, for $(P, Q, n), (P', Q', n') \in \mathcal{R}$ with P, P′ contained in the same R_a and $P \cap P' = \emptyset$, $\varepsilon, \varepsilon' \in \{\pm\}$, we have, for all $y \in I_a^u$

$$\left| \frac{\partial}{\partial y} \log |\varphi_P^\varepsilon - \varphi_{P'}^{\varepsilon'}| \right| \leqslant C. \tag{10.62}$$

This clearly implies the estimate of the proposition. To prove (10.62), we first observe that the inequality is obvious if we allowed the constant in the right-hand term to depend on P, P′. Let $N > 0$ and C(N) be the smallest constant such that (10.62) is satisfied when $(P, Q, n), (P', Q', n') \in \mathcal{R}$ are as required with $n, n' < N$. We will show that C(N) stays bounded.

Let $(P, Q, n), (P', Q', n') \in \mathcal{R}$ with P, P′ contained in the same R_a, $P \cap P' = \emptyset$ and $\max(n, n') = N$. Let $(\widetilde{P}, \widetilde{Q}, \widetilde{n}) \in \mathcal{R}$ be the element such that $\widetilde{P}$ is the smallest rectangle containing both P and P′. Let $\overline{P}, \overline{P}'$ be the children of $\widetilde{P}$ which contain P, P′ respectively. We distinguish two cases.

1. *At least one of $\overline{P}, \overline{P}'$ is a simple child.*

   In this case, we observe that, for all $y \in I_a^u$, we have

$$C^{-1}|\widetilde{P}| \leqslant |\varphi_P^\varepsilon(y) - \varphi_{P'}^{\varepsilon'}(y)| \leqslant C|\widetilde{P}|. \tag{10.63}$$

On the other hand, we have seen in the proof of Proposition 50 (cf. (10.52)) that

$$\begin{aligned} |D\varphi_P^\varepsilon(y) - D\varphi_{\widetilde{P}}^+(y)| &\leqslant C|\widetilde{P}|, \\ |D\varphi_{P'}^{\varepsilon'}(y) - D\varphi_{\widetilde{P}}^+(y)| &\leqslant C|\widetilde{P}|. \end{aligned} \tag{10.64}$$

We obtain (10.62) with some uniform constant C_0.

2. *Both* $\overline{\mathrm{P}}, \overline{\mathrm{P}}'$ *are non-simple children.*
Let $\bar{\varepsilon}, \bar{\varepsilon}'$ such that

$$|\varphi_{\mathrm{P}}^{\varepsilon}(y) - \varphi_{\mathrm{P}'}^{\varepsilon'}(y)| \geqslant |\varphi_{\overline{\mathrm{P}}}^{\bar{\varepsilon}}(y) - \varphi_{\overline{\mathrm{P}}'}^{\bar{\varepsilon}'}(y)|.$$

We apply the proposition in Appendix B, whose hypotheses are clearly satisfied. By definition of C(N), we obtain

$$\textbf{(10.65)} \qquad \left| \frac{\partial}{\partial y} \log |\varphi_{\overline{\mathrm{P}}}^{\bar{\varepsilon}} - \varphi_{\overline{\mathrm{P}}'}^{\bar{\varepsilon}'}| \right| \leqslant \mathrm{C}(\mathrm{C}(\mathrm{N})|\widetilde{\mathrm{Q}}|^{1/2} + 1).$$

From (10.52), we have

$$|\mathrm{D}\varphi_{\mathrm{P}}^{\varepsilon}(y) - \mathrm{D}\varphi_{\overline{\mathrm{P}}}^{\bar{\varepsilon}}(y)| \leqslant \mathrm{C}|\overline{\mathrm{P}}|,$$

$$|\mathrm{D}\varphi_{\mathrm{P}'}^{\varepsilon'}(y) - \mathrm{D}\varphi_{\overline{\mathrm{P}}'}^{\bar{\varepsilon}'}(y)| \leqslant \mathrm{C}|\overline{\mathrm{P}}'|$$

and from (MP7)

$$|\varphi_{\mathrm{P}}^{\varepsilon}(y) - \varphi_{\mathrm{P}'}^{\varepsilon'}(y)| \geqslant \mathrm{C}^{-1}(|\overline{\mathrm{P}}| + |\overline{\mathrm{P}}'|).$$

In view of this and (10.65), (10.66), we obtain (10.62) with a constant $\mathrm{C} = \mathrm{C}_1(1 + |\widetilde{\mathrm{Q}}|^{1/2}\mathrm{C}(\mathrm{N}))$ for some uniform constant C_1.

Observe that, when the second case occurs, N has to be large (at least of the order of $\log \varepsilon_0^{-1}$) and we must have $|\widetilde{\mathrm{Q}}| \ll \varepsilon_0$. Therefore, one has $\mathrm{C}(\mathrm{N}) \leqslant \mathrm{C}_0$ for $\mathrm{N} \ll \log \varepsilon_0^{-1}$ and

$$\textbf{(10.66)} \qquad \mathrm{C}(\mathrm{N}+1) \leqslant \max(\mathrm{C}_0, \mathrm{C}_1(1 + \mathrm{C}(\mathrm{N})\varepsilon_0^{1/2})).$$

This implies that C(N) is bounded (by $\max(\mathrm{C}_0, 2\mathrm{C}_1)$, as soon as $\varepsilon_0^{1/2}\max(\mathrm{C}_0, 2\mathrm{C}_1) < 1$) and ends the proof of the proposition. □

The result of Proposition 51 implies that the transverse Hausdorff dimension $d_s = d_s(g)$ of $\widetilde{\mathcal{R}}_+^\infty$ is well-defined, being equal to the Hausdorff dimension of $\phi_y(\mathcal{R}_+^\infty(a))$ for any $a \in \mathcal{A}, y \in \mathrm{I}_a^u$. We have just proved that it does not depend on y. That it does not depend on a is seen as follows: for $(a, a') \in \mathcal{B}$, g sends $\widetilde{\mathrm{R}}_+^\infty \cap \mathrm{P}_{aa'}$ into $\widetilde{\mathrm{R}}_+^\infty \cap \mathrm{R}_{a'}$; the transverse Hausdorff dimension of $\widetilde{\mathrm{R}}_+^\infty \cap \mathrm{R}_{a'}$ is therefore not smaller than that of $\widetilde{\mathrm{R}}_+^\infty \cap \mathrm{R}_a$; as this is true for all $(a, a') \in \mathcal{B}$, the conclusion follows.

We will also identify below in this section the transverse Hausdorff dimension d_s through a transfer operator in the classical manner of Bowen, Ruelle and Sinai.

Let $\omega, \widehat{\omega}$ be two stable curves and $j > 0$. Assume that ω and $\widehat{\omega}$ belong to the domain of $(\mathrm{T}^+)^j$. We say that ω and $\widehat{\omega}$ belong to the same component of the domain of $(\mathrm{T}^+)^j$ if, for each $0 \leqslant i < j$, there exists a prime P_i such that $(\mathrm{T}^+)^i(\omega)$ and $(\mathrm{T}^+)^i(\widehat{\omega})$ belong to $\mathcal{R}_+^\infty(\mathrm{P}_i)$.

Proposition **52.** — *There exists* $\theta_0 \in (0, 1)$ *such that, if* $\omega, \widehat{\omega}$ *belong to the same component of the domain of* $(T^+)^j$, *we have*

$$|\varphi_\omega(y) - \varphi_{\widehat{\omega}}(y)| \leqslant C\theta_0^j,$$
$$|D\varphi_\omega(y) - D\varphi_{\widehat{\omega}}(y)| \leqslant C\theta_0^j,$$

for all y.

Proof. — With $(T^+)^i(\omega) \in \mathcal{R}_+^\infty(P_i)$ for $0 \leqslant i < j$, let $(P, Q, n) = (P_0, Q_0, n_0) * \cdots * (P_{j-1}, Q_{j-1}, n_{j-1})$. We have $\omega, \widehat{\omega} \subset P$ and $|P| \leqslant C\theta_0^j$ for some fixed $\theta_0 \in (0, 1)$. The estimates are therefore a consequence of those in Proposition 50. □

10.6. *Transverse dilatation.* — This subsection is a preparation for the definition of a transfer operator in the next subsection. The weight function in this transfer operator is, up to a coboundary term, given by a transverse dilatation.

Let $(P, Q, n) \in \mathcal{P}$, $\omega = \{x = \varphi_\omega(y)\}$ a stable curve in $\mathcal{R}_+^\infty(P)$, $\omega' = T^+(\omega)$ its image. Let (A, B) be the implicit representation of (P, Q, n).

For $z = (\varphi_\omega(y), y) \in \omega$, let

$$v_\omega(z) = \frac{\partial}{\partial y} + D\varphi_\omega(y)\frac{\partial}{\partial x} \tag{10.67}$$

be the normalized tangent vector to ω at z.

The matrix of $D\widetilde{T}_+$ at z, computed in the bases $(\frac{\partial}{\partial x}, v_\omega(z))$ at z, $(\frac{\partial}{\partial x}, v_{\omega'}(z'))$ at $z' = \widetilde{T}^+(z)$, is lower triangular; the first diagonal coefficient is

$$A_x^{-1}(y, x')\left(1 - B_x(y, x')\, D\varphi_{\omega'}(y')\right). \tag{10.68}$$

We denote by $\widetilde{b}(z)$ the logarithm of the absolute value of this coefficient. As φ_ω is $C^{1+\mathrm{Lip}}$ uniformly in ω and $g^n : P \mapsto Q$ has bounded distortion, we have, for all $z, z^* \in \omega$

$$|\widetilde{b}(z) - \widetilde{b}(z^*)| \leqslant C|z - z^*|. \tag{10.69}$$

Let $j > 0$ and let $\omega, \widehat{\omega}$ be stable curves which belong to the same component of the domain of $(T^+)^j$. Let $z, \hat{z}$ be point of $\omega, \widehat{\omega}$, respectively, with the same y coordinate. It follows from Proposition 52 that one has

$$|\widetilde{b}(z) - \widetilde{b}(\hat{z})| \leqslant C\theta_0^j. \tag{10.70}$$

We also have, from the definition of $\widetilde{b}$:

$$|\widetilde{b}(z) + \log|P|| \leqslant C. \tag{10.71}$$

We want to get rid of the dependence of $\widetilde{b}$ on the y coordinate along ω by adding a coboundary term. We define

$$\mathcal{D}_+^\infty = \bigcap_{j\geqslant 0} \mathrm{Dom}(\mathrm{T}^+)^j = \bigcap_{j\geqslant 0} (\mathrm{T}^+)^{-j}(\mathcal{D}_+). \tag{10.72}$$

For each a, fix some $y_a^0 \in \mathrm{I}_a^u$. Then, for $\omega \in \mathcal{D}_+^\infty$, $\omega \subset \mathrm{R}_a$, $z \in \omega$, define

$$\Delta b(z) = \sum_{i\geqslant 0} \Big(\widetilde{b}((\widetilde{\mathrm{T}}^+)^i(z)) - \widetilde{b}((\widetilde{\mathrm{T}}^+)^i(z^0))\Big), \tag{10.73}$$

where z^0 is the point on ω with vertical coordinate equal to y_a^0.

From the cone condition, we have, for $i \geqslant 0$:

$$\|(\widetilde{\mathrm{T}}^+)^i(z) - (\widetilde{\mathrm{T}}^+)^i(z^0)\| \leqslant \mathrm{C}\lambda^{-i}. \tag{10.74}$$

The series defining Δb is uniformly convergent from (10.69), (10.74), and Δb is bounded on $\widetilde{\mathcal{D}}_+^\infty := \pi^{-1}(\mathcal{D}_+^\infty)$.

Write z^1 for the point on $\mathrm{T}_+(\omega) \in \mathrm{R}_{a'}$ with vertical coordinate $y_{a'}^0$. We have

$$\Delta b(z) - \Delta b(\widetilde{\mathrm{T}}^+(z)) = \widetilde{b}(z) - b(\omega), \tag{10.75}$$

with

$$b(\omega) = \widetilde{b}(z^0) + \sum_{i\geqslant 0} \Big[\widetilde{b}((\widetilde{\mathrm{T}}^+)^{i+1}(z^0)) - \widetilde{b}((\widetilde{\mathrm{T}}^+)^i(z^1))\Big]. \tag{10.76}$$

We call b the (logarithmic) mean transverse dilatation.

Proposition **53.** — *The mean transverse dilatation b, which differs from $\widetilde{b}$ on $\mathcal{D}_+^\infty$ by the coboundary of the bounded function Δb, satisfies*

$$|b(\omega) - b(\widehat{\omega})| \leqslant \mathrm{C}\theta_1^j \tag{10.77}$$

if ω, $\widehat{\omega}$ belong to the same component of the domain of $(\mathrm{T}^+)^j$. Here θ_1 is a fixed constant in $(0, 1)$ larger than θ_0.

Proof. — We have only to prove (10.77). Let $z^0 \in \omega$, $z^1 \in \mathrm{T}^+(\omega)$ as above and let $\hat{z}^0 \in \widehat{\omega}$, $\hat{z}^1 \in \mathrm{T}^+(\widehat{\omega})$ be similarly defined. We have, for $i \geqslant 0$, using (10.69) and (10.74):

$$|\widetilde{b}((\widetilde{\mathrm{T}}^+)^{i+1}(z^0)) - \widetilde{b}((\widetilde{\mathrm{T}}^+)^i(z^1))| \leqslant \mathrm{C}\lambda^{-i}, \tag{10.78}$$

$$|\widetilde{b}((\widetilde{\mathrm{T}}^+)^{i+1}(\hat{z}^0)) - \widetilde{b}((\widetilde{\mathrm{T}}^+)^i(\hat{z}^1))| \leqslant \mathrm{C}\lambda^{-i}. \tag{10.79}$$

From (10.70), we also have

$$|\widetilde{b}(z^0) - \widetilde{b}(\hat{z}^0)| \leqslant \mathrm{C}\theta_0^j. \tag{10.80}$$

For $0 \leqslant i < j/2$, we will compare $\widetilde{b}((\widetilde{T}^+)^{i+1}(z^0))$ and $\widetilde{b}((\widetilde{T}^+)^{i+1}(\hat{z}^0))$ as follows. Let $\widetilde{z}^{i+1}$ be the point on $T_+^{i+1}(\widehat{\omega})$ with the same y-coordinate as $(\widetilde{T}^+)^{i+1}(z^0)$. From (10.70), we have

$$(\mathbf{10.81}) \qquad |\widetilde{b}(\widetilde{z}^{i+1}) - \widetilde{b}((\widetilde{T}^+)^{i+1}(z^0)))| \leqslant C\theta_0^{j-i-1},$$

From Proposition 52, the cone condition, and (10.69), we also have

$$(\mathbf{10.82}) \qquad \begin{aligned} \|\widetilde{z}^{i+1} - (\widetilde{T}^+)^{i+1}(z^0))\| &\leqslant C\theta_0^{j-i-1}, \\ \|\widetilde{z}^{i+1} - (\widetilde{T}^+)^{i+1}(\hat{z}^0)\| &\leqslant C\theta_0^{j-i-1}, \\ |\widetilde{b}(\widetilde{z}^{i+1}) - \widetilde{b}((\widetilde{T}^+)^{i+1}(\hat{z}^0)))| &\leqslant C\theta_0^{j-i-1}. \end{aligned}$$

We therefore have

$$(\mathbf{10.83}) \qquad |\widetilde{b}((\widetilde{T}^+)^{i+1}(\hat{z}^0))) - \widetilde{b}((\widetilde{T}^+)^{i+1}(z^0)))| \leqslant C\theta_0^{j-i-1}.$$

Similarly, we get, for $0 \leqslant i < j/2$

$$(\mathbf{10.84}) \qquad |\widetilde{b}((\widetilde{T}^+)^{i}(\hat{z}^1))) - \widetilde{b}((\widetilde{T}^+)^{i}(z^1)))| \leqslant C\theta_0^{j-i}.$$

To estimate the difference $b(\omega) - b(\widehat{\omega})$, we use first (10.80), then (10.78), (10.79) to truncate the sums at $j/2$, and finally (10.81), (10.82) to estimate the difference of the remaining terms. We obtain (10.77) with

$$(\mathbf{10.85}) \qquad \theta_1 = \max(\theta_0^{1/2}, \lambda^{-1/2}). \qquad \square$$

10.7. *Definition of a transfer operator.* — As Δb is bounded, it follows from (10.71) that

$$(\mathbf{10.86}) \qquad |b(\omega) + \log|P|| \leqslant C$$

for all $(P, Q, n) \in \mathcal{P}$, $\omega \in \mathcal{R}_+^\infty(P)$.

It is then a consequence of Corollary 16 in Section 10.3 that the series

$$(\mathbf{10.87}) \qquad \sum_{T^+\omega'=\omega} \exp(-db(\omega'))$$

over pre-images ω' of a given stable curve ω is converging, uniformly in ω, for $d \geqslant d_s^-$. Here $d_s^- < d_s^i$ should satisfy the hypotheses of Proposition 49 and Corollary 16. We will, therefore, define a transfer operator L_d for $d \geqslant d_s^-$ as follows: for a bounded function h defined on $\mathcal{D}_+^\infty$, for $\omega \in \mathcal{D}_+^\infty$, we set

$$(\mathbf{10.88}) \qquad L_d h(\omega) = \sum_{T^+\omega'=\omega} \exp(-db(\omega'))h(\omega').$$

We can also view this sum over pre-images as a sum over inverse branches of T^+, which are in one-to-one correspondence with the primes (P, Q, n) such that Q and ω belong to the same rectangle R_a. Accordingly, we split the series in two parts: a finite sum corresponding to the trivial primes, (cf. Section 10.1), which we denote by L_d^0 and which is defined for all values of d, and a perturbative term which we denote by ΔL_d. The formula (10.88) defines a bounded operator from the space of bounded functions on $\mathcal{D}_+^\infty$ into itself, but to have nice spectral properties we need, as usual, to restrict to spaces of slightly more regular functions.

Let θ be a constant with

$$\theta_1 < \theta < 1 \tag{10.89}$$

where θ_1 comes from Proposition 53 and satisfies $\theta_1 > \lambda^{-1}$ (cf. (10.85)). Denote by E the space of bounded functions h on $\mathcal{D}_+^\infty$ which satisfy, for some constant $C > 0$,

$$|h(\omega) - h(\widehat{\omega})| \leqslant C\theta^j \tag{10.90}$$

whenever ω, $\widehat{\omega}$ belong to the same component of the domain of $(T^+)^j$. We denote by $\|h\|_\infty$ the usual norm on bounded functions, by $|h|_E$ the best possible C in (10.90), and set

$$\|h\|_E = \max(|h|_E, \|h\|_\infty). \tag{10.91}$$

It is clear that E is a Banach space.

Proposition **54.** — *For $d \geqslant d_s^-$, L_d restricts to a bounded operator on* E. *Moreover, the norm of the perturbative part ΔL_d is as small as we want if ε_0 is small enough.*

Proof. — Let $h \in E$, $\omega, \widehat{\omega} \in \mathcal{D}_+^\infty, j > 0$. Assume that ω, $\widehat{\omega}$ belong to the same component of the domain of $(T^+)^j$. Let (P, Q, n) be a prime such that Q, ω, $\widehat{\omega}$ belong to the same rectangle R_a, and let ω_1, $\widehat{\omega}_1$ be the inverse images of ω, $\widehat{\omega}$ by T^+ corresponding to this inverse branch. By the definition of $|\ |_E$, we have

$$|h(\omega_1) - h(\widehat{\omega}_1)| \leqslant |h|_E\theta^{j+1}. \tag{10.92}$$

From Proposition 53, we have

$$|b(\omega_1) - b(\widehat{\omega}_1)| \leqslant C\theta_1^{j+1}. \tag{10.93}$$

It follows from (10.86) that

$$|\exp(-db(\omega_1)) - \exp(-db(\widehat{\omega}_1))| \leqslant C(d)|P|^d\theta_1^{j+1}. \tag{10.94}$$

Putting together (10.92) and (10.94), we have

$$\begin{aligned} &|h(\omega_1)\exp(-db(\omega_1)) - h(\widehat{\omega}_1)\exp(-db(\widehat{\omega}_1))| \\ &\leqslant C(d)|P|^d(\theta^{j+1}|h|_E + \theta_1^{j+1}\|h\|_\infty). \end{aligned} \tag{10.95}$$

Summing over (non trivial) primes yields for $d \geqslant d_s^-$:

(**10.96**) $$|\Delta L_d h|_E < \varepsilon_1 \|h\|_E,$$

(**10.97**) $$|L_d h|_E < C\|h\|_E,$$

where ε_1 can be made arbitrarily small if ε_0 is small enough, according to Corollary 16. The same estimates (for $d \geqslant d_s^-$) for $\|\Delta L_d h\|_\infty$ and $\|L_d h\|_\infty$ are easier and can be seen directly. The proposition follows. □

10.8. *Spectral properties of the transfer operator.* — Let us denote by $\mathcal{R}_+^\infty(K)$ the set of stable curves ω which are intersections of a sequence of rectangles belonging to $\mathcal{R}(I_0)$; these stable curves are precisely those which meet the initial horseshoe K.

Observe that $\mathcal{R}_+^\infty(K) \subset \mathcal{D}_+^\infty$. Denote by E_K the space of bounded functions h on $\mathcal{R}_+^\infty(K)$ which satisfy

(**10.98**) $$|h(\omega) - h(\widehat{\omega})| \leqslant C\theta^j,$$

whenever ω, $\widehat{\omega}$ belong to the same component of the domain of $(T^+)^j$. Define $|h|_{E_K}$, $\|h\|_{E_K}$ as above, which makes E_K a Banach space.

Let $h \in E$; the restriction of h to $\mathcal{R}_+^\infty(K)$ belongs to E_K and we have

(**10.99**) $$\|h/\mathcal{R}_+^\infty(K)\|_{E_K} \leqslant \|h\|_E.$$

The formula for L_d^0 defines a bounded operator, still denoted by L_d^0, on E_K and we have a commutative diagram

(**10.100**) $$\begin{array}{ccc} E & \xrightarrow[L_d^0]{} & E \\ {\scriptstyle r}\downarrow & & \downarrow{\scriptstyle r} \\ E_K & \xrightarrow[L_d^0]{} & E_K \end{array}$$

where $r : E \mapsto E_K$ is the restriction operator. The bounded operator $L_d^0 : E_K \mapsto E_K$ is the subject of the classical theory by Bowen, Ruelle, Sinai for uniformly hyperbolic systems.

Let us recall some standard results of this theory.

(a) There is a direct sum invariant decomposition

(**10.101**) $$E_K = \mathbf{R}h'_d \oplus H'_d$$

depending analytically on the parameter d, such that h'_d is a positive eigenfunction, with associated eigenvalue $\lambda'_d > 0$, and such that

(**10.102**) $$sp\left(L_d^0/H'_d\right) \subset \{|z| < \lambda'_d\}.$$

(b) There exists a (unique) probability measure μ'_d on $\mathcal{R}_+^\infty(\mathrm{K})$ such that

$$\mathrm{H}'_d = \left\{ h \in \mathrm{E}_\mathrm{K}, \int h d\mu'_d = 0 \right\}. \tag{10.103}$$

One normalizes h'_d to have $\int h'_d d\mu'_d = 1$. Then, the probability measure $\nu'_d = h'_d \mu'_d$ is invariant under the restriction of T^+ to $\mathcal{R}_+^\infty(\mathrm{K})$ (observe that $\widetilde{\mathrm{T}}^+$ on $\widetilde{\mathrm{R}}_+^\infty(\mathrm{K})$ is just the restriction of g).

Let E^0 be the kernel of the restriction operator $r : \mathrm{E} \mapsto \mathrm{E}_\mathrm{K}$. It is invariant under L_d^0.

Lemma **17.** — *One has, for all $d \in \mathbf{R}$.*

$$sp\left(\mathrm{L}_d^0/\mathrm{E}^0\right) \subset \{|z| \leqslant \theta \lambda'_d\}.$$

Proof. — Let $h \in \mathrm{E}^0, j \geqslant 0$. We have

$$(\mathrm{L}_d^0)^j h(\omega) = \sum_{(\mathrm{T}^+)^j(\omega')=\omega}^{0} h(\omega') \exp(-db^{(j)}(\omega')), \tag{10.104}$$

where the symbol $\sum^0$ indicates that we only consider inverse branches of T^+ associated with trivial primes. The notation $b^{(j)}$ denotes the Birkhoff sum

$$b^{(j)}(\omega') = \sum_{0 \leqslant i < j} b((\mathrm{T}^+)^i(\omega')). \tag{10.105}$$

We observe that in the sum in (10.104), each ω' belongs to the same component of the domain of $(\mathrm{T}^+)^j$ as a stable curve in $\mathcal{R}_+^\infty(\mathrm{K})$. As h belongs to E^0, this implies that for such a ω' we have

$$|h(\omega')| \leqslant |h|_\mathrm{E} \theta^j. \tag{10.106}$$

On the other hand, we have

$$\sum^{0} \exp(-db^{(j)}(\omega')) \leqslant \mathrm{C} \lambda_d^{'j}, \tag{10.107}$$

and it follows that

$$\|(\mathrm{L}_d^0)^j h\|_\infty \leqslant \mathrm{C} \lambda_d^{'j} \theta^j \|h\|_\mathrm{E}. \tag{10.108}$$

Let $\widehat{\omega} \in \mathcal{R}_+^\infty$ belong to the same component of the domain of $(\mathrm{T}^+)^\ell$ as ω. Denote by $\widehat{\omega}'$ the inverse image of $\widehat{\omega}$ associated to the same sequence of trivial primes as ω'. We have

$$|h(\omega') - h(\widehat{\omega}')| \leqslant |h|_\mathrm{E} \theta^{j+\ell}, \tag{10.109}$$

and, from Proposition 53

$$(\mathbf{10.110}) \qquad |b^{(j)}(\omega') - b^{(j)}(\widehat{\omega}')| \leqslant C\theta_1^{\ell}.$$

Using also (10.106) and (10.107), we obtain

$$(\mathbf{10.111}) \qquad \left|(L_d^0)^j h(\omega) - (L_d^0)^j h(\widehat{\omega})\right| \leqslant C\theta^{j+\ell}\lambda_d'^j \|h\|_E,$$

which implies the statement of the Lemma. □

We deduce from Lemma 17 that there is a unique function in E, still denoted by h'_d, which restricts to h'_d on $\mathcal{R}_+^\infty(K)$ and satisfies

$$(\mathbf{10.112}) \qquad L_d^0(h'_d) = \lambda'_d h'_d.$$

Moreover, defining a supplementary hyperplane by

$$(\mathbf{10.113}) \qquad H''_d = r^{-1}(H'_d) \oplus E^0,$$

we have that H''_d is invariant under L_d^0 and

$$(\mathbf{10.114}) \qquad sp\left(L_d^0/H''_d\right) \subset \{|z| \leqslant \lambda''_d\},$$

where $\lambda''_d < \lambda'_d$ is independent of ε_0.

Using Proposition 54, we now consider L_d itself, assuming that ε_0 is small enough and $d \geqslant d_s^-$.

As the norm of the perturbation part ΔL_d is arbitrarily small, we conclude that L_d has a positive eigenfunction h_d, with associated eigenvalue λ_d arbitrarily close to λ'_d, and an invariant supplementary hyperplane H_d satisfying

$$(\mathbf{10.115}) \qquad sp\left(L_d/H_d\right) \subset \{|z| < \lambda_d\}.$$

Moreover, h_d, λ_d and H_d depend analytically on d for $d > d_s^-$ because L_d does. We check that

$$(\mathbf{10.116}) \qquad h_d \geqslant C^{-1} > 0.$$

Indeed, the sequence $h^{(n)} = \lambda_d^{-n} L_d^n(1)$ converges to a positive multiple of h_d. We have

$$(\mathbf{10.117}) \qquad h^{(n)}(\omega) = \lambda_d^{-n} \sum_{(T^+)^n(\omega')=\omega} \exp(-d b^{(n)}(\omega')).$$

Let ω, $\widehat{\omega}$ be elements of $\mathcal{D}_+^\infty$ in the same rectangle R_a; let ω', $\widehat{\omega}'$ be pre-images of ω, $\widehat{\omega}$ by $(T^+)^n$ associated with the same sequence of primes. We have (cf. (10.110))

$$(\mathbf{10.118}) \qquad |b^{(n)}(\omega') - b^{(n)}(\widehat{\omega}')| \leqslant C,$$

and it follows that

(**10.119**) $$C^{-1} \leqslant (h^{(n)}(\omega'))^{-1} h^{(n)}(\omega) \leqslant C.$$

This implies (10.116). One normalizes h_d in order to have

(**10.120**) $$h_d = \lim_{n \to +\infty} \lambda_d^{-n} L_d^n(1).$$

Denote then by μ_d the linear form on E with kernel H_d normalized by $\mu_d(h_d) = 1$. We have, for all $h \in E$

(**10.121**) $$\lim_{n \to \infty} \lambda_d^{-n} L_d^n h = \mu_d(h) h_d.$$

As L_d is a positive operator, μ_d is positive. Observe also that for all $(P, Q, n) \in \mathcal{R}$, the characteristic function χ_P (equal to 1 if $\omega \subset P$, 0 otherwise) belongs to E and satisfies $L^n \chi_P > 0$ everywhere for some $n > 0$. Therefore, there exists a unique probability measure on $\mathcal{R}_+^\infty$, still denoted by μ_d, which coincides with μ_d on the intersection of E with $C(\mathcal{R}_+^\infty)$.

10.9. *The Gibbs measure.* — From the defining property (10.121) of μ_d, we have, for all $h \in E$

(**10.122**) $$\mu_d(L_d h) = \lambda_d \mu_d(h).$$

We will now check the classical Jacobian property for μ_d.

Let $(P, Q, n) \in \mathcal{P}$, with $Q \subset R_a$. The application T^+ is a bijection T_P from the set $\mathcal{R}_+^\infty(P)$ onto $\mathcal{R}_+^\infty(a)$.

Let h be a function in E which vanishes outside $\mathcal{R}_+^\infty(P)$. Then, $L_d h$ vanishes outside $\mathcal{R}_+^\infty(a)$, and satisfies on $\mathcal{R}_+^\infty(a)$

$$L_d h(\omega) = h(T_P^{-1}\omega) \exp(-db(T_P^{-1}\omega)).$$

Plugging this into (10.122) gives

(**10.123**) $$\lambda_d \int h(\omega) d\mu_d(\omega) = \int_{\mathcal{R}_+^\infty(a)} h(T_P^{-1}\omega) \exp(-db(T_P^{-1}\omega)) d\mu_d(\omega).$$

This relation is the Jacobian property of the measure μ_d.

Consider in particular the case where h is the characteristic function of $\mathcal{R}_+^\infty(P)$. We then obtain

(**10.124**) $$\lambda_d \mu_d(\mathcal{R}_+^\infty(P)) = \int_{\mathcal{R}_+^\infty(a)} \exp(-db(T_P^{-1}\omega)) d\mu_d(\omega).$$

We now will specify the value of d by asking that

(**10.125**) $$\lambda_d = 1.$$

Indeed, we have the following

Proposition **55.** — *One has* $\frac{\partial}{\partial d}\lambda_d < 0$ *for* $d \geqslant d_s^-$, *and also* $\lambda_{d_s^-} > 1$, $\lim_{d\to+\infty}\lambda_d = 0$. *Therefore, there exists a unique* $d \geqslant d_s^-$ *with* $\lambda_d = 1$.

Proof. — We first prove that $\frac{\partial}{\partial d}\lambda_d < 0$. The Birkhoff sums of $\widetilde{b}$ for $\widetilde{T}^+$ grow at least linearly, and $b - \widetilde{b}$ is the coboundary of a bounded function. Therefore, there exists $j > 0$ such that the Birkhoff sum $b^{(j)}$ of b for T^+ is everywhere > 1. For $h \in E$, we have

$$\textbf{(10.126)} \qquad \frac{\partial}{\partial d}L_d^j(h)(\omega) = -\sum_{(T^+)^j(\omega')=\omega} h(\omega')b^{(j)}(\omega')\exp(-db^{(j)}(\omega'))$$

which is everywhere < 0 is $h > 0$. We will apply this with $h = h_d$. We differentiate with respect to d the relation $L_d^j(h_d) = \lambda_d^j h_d$ to get

$$\textbf{(10.127)} \qquad \frac{\partial}{\partial d}L_d^j(h_d) - \left(\frac{\partial}{\partial d}\lambda_d^j\right)h_d = (L_d^j - \lambda_d^j)\left(\frac{\partial}{\partial d}h_d\right).$$

The right-hand term belongs to H_d. Applying μ_d shows that $\frac{\partial}{\partial d}\lambda_d^j$ and then also $\frac{\partial}{\partial d}\lambda_d$ is < 0.

Next, d_s^- was chosen in order to be smaller than the transverse Hausdorff dimension of $W^s(K)$. This means that the eigenvalue $\lambda'_{d_s^-}$ for $L^0_{d_s^-}$ on E_K satisfies $\lambda'_{d_s^-} > 1$. As $\Delta L_{d_s^-}$ is also a nonnegative operator, we have $\lambda_{d_s^-} \geqslant \lambda'_{d_s^-} > 1$.

Finally, with j as above and $h \in E$, $h > 0$ we have

$$\textbf{(10.128)} \qquad L_d^j(h)(\omega) = \sum_{(T^+)^j(\omega')=\omega} h(\omega')\exp(-db^{(j)}(\omega')) \leqslant e^{d_s^- - d}L^j_{d_s^-}(h)(\omega)$$

which implies that $\lambda_d \leqslant e^{\frac{1}{j}(d_s^- - d)}\lambda_{d_s^-}$. □

We will denote by d_s the value of d such that $\lambda_d = 1$. We shall indeed see that d_s is the transverse Hausdorff dimension of $\widetilde{\mathcal{R}}_+^\infty$ which we were able to define in Section 10.5.

We just write μ for the measure μ_{d_s} and h^* for the eigenfunction h_{d_s}.

Proposition **56.** — *For any* $(P, Q, n) \in \mathcal{R}$, *we have*

$$C^{-1}|P|^{d_s} \leqslant \mu(\{\omega \subset P\}) \leqslant C|P|^{d_s}.$$

Proof. — Let

$$\textbf{(10.129)} \qquad (P, Q, n) = (P_1, Q_1, n_1) * \cdots * (P_r, Q_r, n_r)$$

be the prime decomposition of (P, Q, n). If $\omega \in \mathcal{D}_+^\infty$ satisfies $(T^+)^i(\omega) \in \mathcal{R}_+^\infty(P_{i+1})$ for $0 \leqslant i < r$, we claim that

$$\textbf{(10.130)} \qquad C^{-1}|P|^{d_s} \leqslant \exp(-d_s b^{(r)}(\omega)) \leqslant C|P|^{d_s}$$

(see the definition of $b^{(r)}$ in (10.105)). Indeed, let $z \in \omega$; denoting, by $\widetilde{b}^{(r)}$ the Birkhoff sum of $\widetilde{b}$ for $\widetilde{T}^+$, $\widetilde{b}^{(r)}(z)$ is the logarithm of the absolute value of the first diagonal coefficient of the matrix of $D(\widetilde{T}^+)^r$ at z, hence we have, by bounded distortion

$$\textbf{(10.131)} \qquad C^{-1}|P| \leqslant \exp(-\widetilde{b}^{(r)}(z)) \leqslant C|P|.$$

On the other hand, as $b - \widetilde{b}$ is the coboundary of a bounded function the difference $|\widetilde{b}^{(r)}(z) - b^{(r)}(\omega)|$ is bounded by C and the claim follows.

From the Jacobian property, we have

$$\begin{aligned} \mu(\{\omega \subset P\}) &\geqslant C^{-1}|P|^{d_s}\mu(\{\omega \in \mathcal{R}_+^\infty(a)\}) \\ \textbf{(10.132)} \qquad &\geqslant C^{-1}|P|^{d_s}, \end{aligned}$$

where R_a is the rectangle containing Q.

For the opposite inequality, we have also to take into account the other inverse branches of T_+^r when we estimate $L_{d_s}^r(\chi_P)$, where χ_P is the characteristic function of $\{\omega \subset P\}$. For $0 \leqslant i \leqslant r$, let

$$\textbf{(10.133)} \qquad (P^i, Q^i, n^i) = (P_{i+1}, Q_{i+1}, n_{i+1}) * \cdots * (P_r, Q_r, n_r)$$

(with $(P^r, Q^r, n^r) = (R_a, R_a, 0)$). We have

$$\textbf{(10.134)} \qquad L_{d_s}\chi_P = \chi_P^1 + \Delta\chi_P^1$$

where

$$\textbf{(10.135)} \qquad \chi_P^1(\omega^1) = \begin{cases} 0 & \text{if } \omega^1 \not\subset P^1, \\ \exp(-d_s b(\omega^0)) & \text{if } \omega^1 = T^+(\omega^0) \text{ for some } \omega^0 \in \mathcal{R}_+^\infty(P) \end{cases}$$

and

$$\textbf{(10.136)} \qquad \Delta\chi_P^1 \leqslant C\sum |P_1^*|^{d_s},$$

where the sum runs over prime elements (P_1^*, Q_1^*, n_1^*) with P_1^* contained in P and distinct from P (when $r = 1$). By Proposition 49 in Section 10.3, we obtain

$$\textbf{(10.137)} \qquad \mu(\Delta\chi_P^1) \leqslant C|P|^{d_s}\kappa^{\frac{r-1}{2}}.$$

If $r > 1$, we write similarly

$$\textbf{(10.138)} \qquad L_d\chi_P^1 = \chi_P^2 + \Delta\chi_P^2,$$

where χ_P^2 is associated with the inverse branch defined by the prime P_2 and vanishes outside P^2. The perturbative term satisfies

$$\textbf{(10.139)} \qquad \Delta\chi_P^2 \leqslant C|P_1|^{d_s}\sum |P_2^*|^{d_s},$$

where the sum now is over primes P_2^* contained in P^1 and distinct from P^1 (when $r=2$). Proposition 49 now gives

$$\mu(\Delta\chi_P^2) \leqslant C|P_1|^{d_s}|P^1|^{d_s}\kappa^{\frac{r-2}{2}}$$

(10.140)
$$\leqslant C|P|^{d_s}\kappa^{\frac{r-2}{2}}.$$

We iterate this process. At step i, we will have

(10.141)
$$\mu(\Delta\chi_P^i) \leqslant C|P|^{d_s}\kappa^{\frac{r-i}{2}}$$

where the constant C does not gets worse by the same argument used above to justify (10.130).

At the last step, we have from (10.130)

(10.142)
$$\mu(\chi_P^r) \leqslant C|P|^{d_s}.$$

The contribution of the perturbative terms is bounded by

(10.143)
$$\mu\left(\sum_1^{r-1}\Delta\chi_P^i\right) \leqslant C|P|^{d_s}\sum_1^r\kappa^{\frac{r-i}{2}} \leqslant C|P|^{d_s}.$$

□

Corollary **17.** — *The transverse Hausdorff dimension of* $\widetilde{\mathcal{R}}_+^\infty$ *is* $\leqslant d_s$. *More precisely, for any* C^1 *curve* γ *which is transverse to* $\widetilde{\mathcal{R}}_+^\infty$, *the Hausdorff measure in dimension* d_s *of the intersection of* γ *with* $\widetilde{\mathcal{R}}_+^\infty$ *is finite.*

We will see below that the transverse Hausdorff dimension is equal to d_s.

Proof. — Let $\delta > 0$, choose a finite collection of disjoint rectangles P_i with $|P_i| \leqslant \delta$ for each i and $\widetilde{\mathcal{R}}_+^\infty \subset \cup P_i$. We have

$$1 = \sum \mu(P_i) \geqslant C^{-1}\sum |P_i|^{d_s}$$

(10.144)
$$\geqslant C^{-1}\sum [\mathrm{diam}(\gamma\cap P_i)]^{d_s}$$

and the statement of the Corollary follows. □

The following statement shows that the dynamics T^+ is only undefined on a small set.

Proposition **57.** — *The transverse Hausdorff dimension of the set* $\widetilde{\mathcal{R}}_+^\infty - \widetilde{\mathcal{D}}_+^\infty$ *is* $\leqslant d_s^- < d_s$. *Moreover, we have*

$$\mu(\widetilde{\mathcal{R}}_+^\infty - \widetilde{\mathcal{D}}_+^\infty) = 0.$$

Proof. — We have

$$\widetilde{\mathcal{R}}_+^\infty - \widetilde{\mathcal{D}}_+^\infty = \bigcup_{n\geqslant 0} (\mathrm{T}^+)^{-n}(\mathcal{N}_+). \tag{10.145}$$

As each $(\mathrm{T}^+)^n$ has countably many inverse branches which are Lipschitzian, it is sufficient to prove that the transverse Hausdorff dimension of $\mathcal{N}_+$ is $\leqslant d_s^-$ and that $\mu(\mathcal{N}_+) = 0$. By the definition of $\mathcal{N}_+$, for any $\delta > 0$, the union of prime rectangles P with $|\mathrm{P}| < \delta$ contains $\mathcal{N}_+$. It then follows from Corollary 16 (in Section 10.3) that the Hausdorff dimension of $\mathcal{N}_+$ is $\leqslant d_s^-$ and (using also Proposition 56) that $\mu(\mathcal{N}_+) = 0$. □

10.10. *Transverse Hausdorff dimension of* $\widetilde{\mathcal{R}}_+^\infty$.

Theorem **4.** — *The transverse Hausdorff dimension of* $\widetilde{\mathcal{R}}_+^\infty$ *is the number* d_s *characterized by* $\lambda_{d_s} = 1$.

Remark **17.** — We have already seen that the Hausdorff measure in dimension d_s of the intersection of $\widetilde{\mathcal{R}}_+^\infty$ with a transverse curve is always finite. We do not know whether it is positive or always zero.

10.10.1. *Proof.* — Let γ be a horizontal segment in some R_a. We denote by $[\gamma]$ the set of stable curves which meet γ. We will show that, for all γ, we have

$$\mathrm{A}(\gamma) := \frac{\mu([\gamma])}{(\operatorname{diam}\gamma)^{d_s}} \leqslant \mathrm{C}\log^{\mathrm{C}_0}(\operatorname{diam}\gamma)^{-1}. \tag{10.146}$$

This clearly implies that the transverse Hausdorff dimension of $\widetilde{\mathcal{R}}_+^\infty$ is $\geqslant d_s$, which is sufficient to prove the theorem in view of Corollary 17.

10.10.2. We start with some preliminary work. In the rectangle R_{a_s} which contains L_s, we choose a horizontal line $\mathcal{J} = \{y_s = y_s^*\}$. We will use $\bar{x}$ to denote the x_s coordinate on $\mathcal{J}$ (we use a different notation because we will have in the same formulas points on $\mathcal{J}$ and points in R_{a_s}). Let J be the set of $\bar{x}$ such that $(\bar{x}, y_s^*) \in \mathrm{P}_s$, and J^∞ the set of $\bar{x} \in \mathrm{J}$ such that $(\bar{x}, y_s^*) \in \mathrm{P}_s \cap \widetilde{\mathcal{R}}_+^\infty$. For $\bar{x} \in \mathrm{J}^\infty$, let $x_s = \varphi(y_s, \bar{x})$ be the equation of the stable curve through $(\bar{x}, y_s^*)$ (thus, we have $\bar{x} = \varphi(y_s^*, \bar{x})$).

For each $\bar{x} \in \mathrm{J}^\infty$, φ is a $\mathrm{C}^{1+\mathrm{Lip}}$ function of y_s. Moreover, from (R4) in Section 5.3 and Proposition 50 in Section 10.4, we have

$$|\varphi_y(y_s, \bar{x})| \leqslant \mathrm{C}\varepsilon_0, \tag{10.147}$$

$$|\varphi_y(y_s, \bar{x}) - \varphi_y(y_s', \bar{x})| \leqslant \mathrm{C}\varepsilon_0 |y_s - y_s'|. \tag{10.148}$$

On the other hand, it follows from Proposition 51 in Section 10.5 that, for $\overline{x}, \overline{x}' \in J^{\infty}$ with $\overline{x}' > \overline{x}$, we have

$$C^{-1}(\overline{x}' - \overline{x}) \leqslant \varphi(y_s, \overline{x}') - \varphi(y_s, \overline{x}) \leqslant C(\overline{x}' - \overline{x}). \tag{10.149}$$

We extend the definition of φ letting $\overline{x}$ run in the whole interval J, to obtain an homeomorphism from $I^u_{a_s} \times J$ onto P_s and still having (10.147)–(10.149), with now $\overline{x}, \overline{x}' \in J$. This can be done for instance by linear interpolation in the $\overline{x}$ variable, for each fixed y_s.

The next step is to switch, via the diffeomorphism G_+ of Section 2.3, from the coordinates x_s, y_s to the coordinates x_s, w. We have (with the notations of Section 2.3) $y_s = Y_s(w, x_s)$; plugging this into φ gives a family of curves parametrized by $\overline{x}$

$$x_s = \varphi(Y_s(w, x_s), \overline{x}). \tag{10.150}$$

Lemma **18.** — *Each curve* $\{x_s = \varphi(Y_s(w, x_s), \overline{x})\}$ *is a graph* $\{x_s = \phi(w, \overline{x})\}$ *of a* $C^{1+\mathrm{Lip}}$ *function of* w. *The function* ϕ *satisfies the same relations* (10.147)–(10.149) *than* φ, *namely*

$$|\phi_w(w, \overline{x})| \leqslant C\varepsilon_0,$$
$$|\phi_w(w, \overline{x}) - \phi_w(w', \overline{x})| \leqslant C\varepsilon_0 |w - w'|,$$
$$C^{-1}(\overline{x}' - \overline{x}) \leqslant \phi(w, \overline{x}') - \phi(w, \overline{x}) \leqslant C(\overline{x}' - \overline{x}),$$

for all $w, w', \overline{x} < \overline{x}'$.

Proof. — In view of (10.147), the first statement follows from the implicit function theorem, which gives also

$$\phi_w = \varphi_y Y_{s,w}(1 - \varphi_y Y_{s,x})^{-1}. \tag{10.151}$$

The first two estimates of the lemma now follow from (10.147), (10.148) and the fact that the partial derivatives of Y_s of first and second order are bounded.

For the last inequality, let $x_s = \phi(w, \overline{x})$, $x'_s = \phi(w, \overline{x}')$. We have $x_s = \varphi(Y_s(w, x_s), \overline{x})$, $x'_s = \varphi(Y_s(w, x'_s), \overline{x}')$ and let $x^*_s = \varphi(Y_s(w, x_s), \overline{x}')$. Then, from (10.149), we get

$$C^{-1}(\overline{x}' - \overline{x}) \leqslant x^*_s - x_s \leqslant C(\overline{x}' - \overline{x}). \tag{10.152}$$

From (10.147), one obtains

$$|x'_s - x^*_s| \leqslant C\varepsilon_0(x'_s - x_s), \tag{10.153}$$

from which we deduce as required

$$C^{-1}(\overline{x}' - \overline{x}) \leqslant x'_s - x_s \leqslant C(\overline{x}' - \overline{x}). \tag{10.154}$$

□

10.10.3. In the rectangle R_{a_u} which contains L_u, consider a C^2 curve $\gamma = \{y_u = \psi(x_u)\}$ contained in Q_u, where ψ satisfies

$$(\mathbf{10.155}) \qquad |\psi_x| \leqslant C\varepsilon_0, \qquad |\psi_{xx}| \leqslant C\varepsilon_0.$$

Using the diffeomorphism G_- of Section 2.3 to switch to the coordinates w, y_u via $x_u = X_u(w, y_u)$, the curve is transformed into $\{y_u = \psi(X_u(w, y_u))\}$. By the implicit function theorem, this is still a graph $\{y_u = \Psi(w)\}$, with Ψ satisfying

$$(\mathbf{10.156}) \qquad |\Psi_w| \leqslant C\varepsilon_0, \qquad |\Psi_{ww}| \leqslant C\varepsilon_0.$$

In the spirit of Section 3.5, we now introduce

$$(\mathbf{10.157}) \qquad C(w, \overline{x}) := w^2 - \theta(\Psi(w), \phi(w, \overline{x})).$$

Observe that, for each $\overline{x}$, the zeros of $C(w, \overline{x})$ correspond to the points of intersection of the curve $G(\gamma \cap L_u)$ with the curve $x_s = \varphi(y_s, \overline{x})$.

Lemma **19.**

1. *For each* $\overline{x} \in J$, $C(w, \overline{x})$ *is a* C^{1+Lip} *function of* w, *satisfying*

$$|C_w(w, \overline{x}) - 2w| \leqslant C\varepsilon_0,$$
$$|C_w(w, \overline{x}) - C_w(w', \overline{x}) - 2(w - w')| \leqslant C\varepsilon_0 |w - w'|.$$

2. *For each* $\overline{x} \in J$, $C(w, \overline{x})$ *attains its minimum value at a unique point* w^*, *which is in the interior of the domain of definition of the* w *variable. Writing*

$$\delta(\overline{x}) := -\min_w C(w, \overline{x}),$$

one has, for $\overline{x} < \overline{x}'$ *in* $\mathcal{J}$

$$C^{-1}(\overline{x}' - \overline{x}) \leqslant |\delta(\overline{x}') - \delta(\overline{x})| \leqslant C(\overline{x}' - \overline{x}).$$

3. *Let* $\overline{x}_0, \overline{x}_1 \in J$ *such that* $\delta(\overline{x}_0) > \delta(\overline{x}_1) > 0$ *and let* w_0, w_1 *such that* $C(w_0, \overline{x}_0) = C(w_1, \overline{x}_1) = 0$. *We have*

$$\delta(\overline{x}_0)^{1/2} |w_0 - w_1| \geqslant C^{-1} |\overline{x}_0 - \overline{x}_1|.$$

Proof. — The first part of the lemma follows immediately from the corresponding properties of Ψ and ϕ, using that the partial derivatives of θ of order one and two are bounded. The first statement in the second part of the lemma is an immediate consequence of the properties of C_w stated in the first part. One has also, from the first part

$$(\mathbf{10.158}) \qquad |C_w(w, \overline{x}) - 2(w - w^*)| \leqslant C\varepsilon_0 |w - w^*|,$$

$$(\mathbf{10.159}) \qquad |C(w,\bar{x})+\delta(\bar{x})-(w-w^*)^2| \leqslant C\varepsilon_0|w-w^*|^2.$$

To prove the inequality in the second part of the lemma, the upper bound is just a consequence from the fact that, for each w, $C(w,\bar{x})$ is a Lipschitz function of $\bar{x}$. For the lower bound, we first recall that the partial derivatives θ_x, θ_y do not vanish and are bounded away from 0. Assume for instance that $\theta_x > C^{-1}$. Let w^* be the point where $C(w,\bar{x})$ attains its minimal value $-\delta(\bar{x})$. We have, from Lemma 18,

$$\phi(w^*,\bar{x}') \geqslant \phi(w^*,\bar{x}) + C^{-1}(\bar{x}'-\bar{x}),$$

hence

$$\begin{aligned} C(w^*,\bar{x}') &= (w^*)^2 - \theta(\Psi(w^*),\phi(w^*,\bar{x}')) \\ &\leqslant (w^*)^2 - \theta(\Psi(w^*),\phi(w^*,\bar{x})) - C^{-1}(\bar{x}'-\bar{x}) \\ &= -\delta(\bar{x}) - C^{-1}(\bar{x}'-\bar{x}), \end{aligned}$$

and the lower bound follows.

In the setting of the third part of the lemma, we have

$$(\mathbf{10.160}) \qquad C(w_0,\bar{x}_0) - C(w_1,\bar{x}_0) = C(w_1,\bar{x}_1) - C(w_1,\bar{x}_0),$$

where, by an argument just seen above,

$$(\mathbf{10.161}) \qquad |C(w_1,\bar{x}_1) - C(w_1,\bar{x}_0)| \geqslant C^{-1}|\bar{x}_0-\bar{x}_1|.$$

On the other hand, we have

$$(\mathbf{10.162}) \qquad |C(w_0,\bar{x}_0) - C(w_1,\bar{x}_0)| \leqslant C|w_0-w_1| \max_w |C_w(w,\bar{x}_0)|,$$

where the maximum is taken for w between w_0 and w_1. But it follows from the first part of the lemma (second inequality) that this maximum is taken at w_0 (because we have assumed that $\delta(\bar{x}_0) > \delta(\bar{x}_1) > 0$), and then from (10.158), (10.159) that $|C_w(w_0,\bar{x}_0)| \leqslant C\delta(\bar{x}_0)^{1/2}$. Plugging this above completes the proof of the third part of the lemma. □

10.10.4. We now come back to the proof of the estimate (10.146). Let γ_0 be a horizontal closed segment in some rectangle R_a. Clearly, we may assume that $\mu([\gamma_0]) > 0$. By shortening γ_0 if necessary, we can assume that there is a stable curve through each endpoint of γ_0. Let $(P_0, Q_0, n_0) \in \mathcal{R}$ the element such that P_0 is the thinnest rectangle containing any stable curve in $[\gamma_0]$. There are at least two children of P_0 which contain a stable curve in $[\gamma_0]$.

We say that γ_0 has complexity 0 if at least one of the following two conditions are satisfied:

- At least one stable curve in $[\gamma_0]$ is contained in a simple child of P_0.

- There is a child P'_0 of P_0 such that the set of stable curves in $[\gamma_0]$ contained in P'_0 has μ-measure $\geqslant \frac{1}{3}\mu([\gamma_0])$.

If none of these conditions, we say that γ_0 has complexity > 0.

When γ_0 has complexity 0, it is easy to obtain (10.146) (and even better). Assume first that the first condition is satisfied. Then, as γ_0 intersects at least two children of P_0, we have

$$\text{(10.163)} \qquad \operatorname{diam} \gamma_0 \geqslant C^{-1}|P_0|.$$

On the other hand, we have

$$\text{(10.164)} \qquad \mu([\gamma_0]) \leqslant \mu(\{\omega \subset P_0\}) \leqslant C|P_0|^{d_s}$$

by Proposition 56, which proves (10.146).

Assume now that the second condition is satisfied. Then, as γ_0 intersects at least two children of P_0, we have now

$$\text{(10.165)} \qquad \operatorname{diam} \gamma_0 \geqslant C^{-1}|P'_0|.$$

On the other hand, we have

$$\text{(10.166)} \qquad \mu([\gamma_0]) \leqslant 3\mu(\{\omega \subset P'_0\}) \leqslant C|P'_0|^{d_s}$$

by Proposition 56, which proves again (10.146).

We have shown that $A(\gamma_0)$ is bounded for segments of complexity 0.

10.10.5. We now assume that the complexity of γ_0 is > 0. If some stable curve through an endpoint of γ_0 is contained in a child P'_0 of P_0, but there is a stable curve contained in P'_0 not in $[\gamma_0]$, we shorten γ_0 to remove from $[\gamma_0]$ all stable curves contained in P'_0. We do this for both endpoints. The shortened curve, that we denote by γ'_0, satisfies $\mu([\gamma'_0]) \geqslant \frac{1}{3}\mu([\gamma_0])$ (because the second condition above is not satisfied). The negation of the second condition also implies that at least two children of P_0 contain a curve in $[\gamma'_0]$. We can still assume there is a stable curve through each endpoint of γ'_0. Any child P'_0 of P_0 which contains a curve in $[\gamma'_0]$ is non-simple, and all stable curves contained in P'_0 belong to $[\gamma'_0]$.

In particular, P_0 has non-simple children, hence Q_0 is contained in Q_u. By (R4), $g_t^{n_0}(\gamma'_0)$ is a graph $\{y_u = \psi(x_u)\}$ satisfying (10.155). As there is a stable curve through each endpoint of γ'_0, the images of these endpoints by $g_t^{n_0}$ are contained in L_u; but then (10.155) implies that $g_t^{n_0}(\gamma'_0)$ is contained in L_u. Let $J_1 \subset J$ be the compact interval image by the projection on the second coordinate $\overline{x}$ of the curve $G \circ g_t^{n_0}(\gamma'_0)$.

We define $\delta_0 := \max_{J_1} \delta(\overline{x})$, $\delta'_0 := \min_{J_1} \delta(\overline{x})$. From Lemma 19, part 3, we have

$$\text{(10.167)} \qquad \delta_0^{1/2} \operatorname{diam} G \circ g_t^{n_0}(\gamma'_0) \geqslant C^{-1}|J_1|,$$

from which it follows that

$$\delta_0^{1/2}\operatorname{diam}\gamma_0' \geqslant C^{-1}|P_0||J_1|. \tag{10.168}$$

We denote by $(P'_{0,i})_i$ the (non-simple) children of P_0 which contain a stable curve in $[\gamma_0]$. Each $P'_{0,i}$ is obtained from its parent P_0 by parabolic composition:

$$(P'_{0,i}, Q'_{0,i}, n'_{0,i}) \in (P_0, Q_0, n_0) \,\square\, (P_{0,i}, Q_{0,i}, n_{0,i}). \tag{10.169}$$

Observe that it is possible that two $P'_{0,i}$ (but no more than two) correspond to the same $P_{0,i}$. The widths are related through

$$C^{-1} \leqslant |P'_{0,i}||P_0|^{-1}|P_{0,i}|^{-1}\delta(Q_0, P_{0,i})^{\frac{1}{2}} \leqslant C. \tag{10.170}$$

Here, $\delta(Q_0, P_{0,i})$ is the quantity of Section 3.5.

For each i, we choose a stable curve $\{x_s = \varphi(y_s, \overline{x}_i)\}$ contained in $P_{0,i}$ and such that $\delta(\overline{x}_i)$ is not of the form $\delta_0 2^{-l}$. We have $\overline{x}_i \in J_1$ by construction of γ_0'. We also have (cf. Section 3.6.3)

$$|\delta(Q_0, P_{0,i}) - \delta(\overline{x}_i)| \leqslant C(|Q| + |P_{0,i}|), \tag{10.171}$$

and therefore, in view of (R7) in Section 5.4

$$\delta(\overline{x}_i) \gg (|Q| + |P_{0,i}|), \tag{10.172}$$

$$C^{-1} \leqslant |P'_{0,i}||P_0|^{-1}|P_{0,i}|^{-1}\delta(\overline{x}_i)^{\frac{1}{2}} \leqslant C. \tag{10.173}$$

We now distinguish two cases

- $\delta_0' < \frac{1}{2}\delta_0$.
  For any nonnegative integer l, let $J^*_{1,l}$ be the set of $\overline{x} \in J$ such that $\delta_0 2^{-l} \geqslant \delta(\overline{x}) \geqslant \delta_0 2^{-l-1}$. By Lemma 19, part 2, it is a compact interval satisfying

$$|J^*_{1,l}| \leqslant C\delta_0 2^{-l}. \tag{10.174}$$

  Let $\mathcal{L}$ be the set of nonnegative integers l such that $J^*_{1,l}$ contains some $\overline{x}_i$. It follows from (10.172) that $\mathcal{L}$ is finite. From (10.172), it also follows that it is possible to find, for $l \in \mathcal{L}$, an interval $J_{1,l}$ whose endpoints are at a distance $\ll \delta_0 2^{-l}$ from those of $J^*_{1,l}$ and which has the following property: if $\overline{x}_i \in J^*_{1,l}$, for any stable curve $\varphi(y_s, \overline{x})$ contained in $P_{0,i}$, one has $\overline{x} \in J_{1,l}$. One has still

$$|J_{1,l}| \leqslant C\delta_0 2^{-l}. \tag{10.175}$$

- $\delta_0' \geqslant \frac{1}{2}\delta_0$.
  In this case, we set $\mathcal{L} = \{0\}, J_{1,0} = J_1$.

In both cases, for $l \in \mathcal{L}$, let $\gamma_{1,l} \subset \mathcal{J}$ be the horizontal segment $J_{1,l} \times \{y_s^*\}$.

10.10.6. In the first case, we write

$$\mu([\gamma_0]) \leqslant 3\mu([\gamma_0']) \leqslant C\sum_i |P'_{0,i}|^{d_s} \quad \text{(from Proposition 56)}$$

$$\leqslant C|P_0|^{d_s}\sum_i |P_{0,i}|^{d_s}\delta(\overline{x}_i)^{-\frac{1}{2}d_s} \quad \text{(from (10.173))}$$

$$\leqslant C|P_0|^{d_s}\delta_0^{-\frac{1}{2}d_s}\sum_{\mathcal{L}} 2^{\frac{ld_s}{2}}\sum^{(l)} |P_{0,i}|^{d_s}$$

$$\leqslant C|P_0|^{d_s}\delta_0^{-\frac{1}{2}d_s}\sum_{\mathcal{L}} 2^{\frac{ld_s}{2}}\sum^{(l)} \mu(\{\omega\subset P_{0,i}\})$$

(from Proposition 56)

$$\leqslant C|P_0|^{d_s}\delta_0^{-\frac{1}{2}d_s}\sum_{\mathcal{L}} 2^{\frac{ld_s}{2}}\mu([\gamma_{1,l}]). \tag{10.176}$$

We have written $\sum^{(l)}$ for the partial sum over those $P_{0,i}$ satisfying $\overline{x}_i \in J^*_{1,l}$.

When $\delta_0' \geqslant \frac{1}{2}\delta_0$, a similar but simpler argument gives

$$\mu([\gamma_0]) \leqslant C|P_0|^{d_s}\delta_0^{-\frac{1}{2}d_s}\mu([\gamma_{1,0}]). \tag{10.177}$$

By definition of $A(\gamma)$, we have, for all l

$$\mu([\gamma_{1,l}]) = A(\gamma_{1,l})(\operatorname{diam}\gamma_{1,l})^{d_s}. \tag{10.178}$$

When $\delta_0' \geqslant \frac{1}{2}\delta_0$, we now use (10.168), (10.177) to conclude that

$$A(\gamma_0) \leqslant CA(\gamma_{1,0}). \tag{10.179}$$

When $\delta_0' < \frac{1}{2}\delta_0$, we use (10.176), (10.175), (10.168) to obtain

$$\mu([\gamma_0]) \leqslant C\max_{\mathcal{L}} A(\gamma_{1,l})|P_0|^{d_s}\delta_0^{\frac{1}{2}d_s}\sum_{\mathcal{L}} 2^{\frac{-ld_s}{2}}$$

$$\leqslant C\max_{\mathcal{L}} A(\gamma_{1,l})|P_0|^{d_s}\delta_0^{\frac{1}{2}d_s}$$

$$\leqslant C\max_{\mathcal{L}} A(\gamma_{1,l})(\delta_0|J_1|^{-1}\operatorname{diam}\gamma_0)^{d_s}. \tag{10.180}$$

But in this case, we have $|J_1| \geqslant C^{-1}\delta_0$ from Lemma 19, part 2, and we conclude that

$$A(\gamma_0) \leqslant C\max_{\mathcal{L}} A(\gamma_{1,l}). \tag{10.181}$$

To obtain (10.146), it is thus sufficient to define, for all horizontal segments γ_0 with $\mu([\gamma_0]) > 0$, a complexity index $c(\gamma_0) \in \mathbf{N}$ which vanishes as explained in Section 10.10.4 and satisfies otherwise

$$c(\gamma_0) \leqslant C \log\log |P_0|^{-1}, \tag{10.182}$$

$$c(\gamma_0) = 1 + \max_{\mathcal{L}} c(\gamma_{1,l}). \tag{10.183}$$

10.10.7. We want to use (10.183) to give an inductive definition of $c(\gamma_0)$. This will work if the $\gamma_{1,l}$ are in some sense "simpler" than γ_0. If all $\gamma_{1,l}$ have complexity 0, we just set $c(\gamma_0) = 1$. Assume therefore that some $\gamma_{1,l}$, that we just denote by γ_1, has complexity > 0 (according to Section 10.10.4). Observe that $J^*_{1,l}$ must contain at least two $\bar{x}_i$: otherwise, the unique $P_{0,i}$ such that $\bar{x}_i \in J^*_{1,l}$ would be the thinnest rectangle containing any stable curve in $[\gamma_1]$, and γ_1 would have complexity 0 by the first condition of Section 10.10.4.

Therefore, there exists an element $(P_1, Q_1, n_1) \in \mathcal{R}$ with the following properties:

- each $P_{0,i}$ with $\delta(\bar{x}_i) \in J^*_{1,l}$ is contained in some non-simple child of P_1;
- at least two non-simple children of P_1 contain some such $P_{0,i}$.

Lemma **20.** — *One has*

$$|P_1|^{1-\eta} \geqslant C^{-1}\delta_0 2^{-l}.$$

Moreover P_1 *is* I*-critical for any parameter interval* I *containing* t.

Proof. — Let i such that $P_{0,i} \subset P_1$. There exists a parameter interval I containing t such that Q_0 and $P_{0,i}$ are I-transverse. On the other hand, as $P'_{0,i}$ is a child of P_0, Q_0 and P_1 are I-critically related for every parameter interval I containing t. By Proposition 21 in Section 8.1, we must have $|P_1| > \frac{1}{3}|Q_0|$. If we had $|P_1|^{1-\eta} \ll \delta_0 2^{-l}$, we would also have $|Q_0|^{1-\eta} \ll \delta_0 2^{-l}$ and (T1), (T2), (T3) for $Q_0 \,\overline{\pitchfork}_I\, P_1$ would hold for a sufficiently small parameter interval I containing t. This contradiction proves the first statement. The second follow from Proposition 24 in Section 8.2. □

10.10.8. We assume in this subsection that P_0 also is I-critical for any I containing t.

Let I^* be the largest parameter interval for which we have

$$|P_0| > |I^*|^{\beta}. \tag{10.184}$$

Observe that P_0 is I^*-defined by Corollary 6 in Section 6.6.3. Then, (P_0, Q_0, n_0) cannot be I^*-bicritical as I^* is β-regular, hence Q_0 is I^*-transverse. This implies that there exists $P_{0,i}$ with $\delta(\bar{x}_i) \in J^*_{1,l}$ and P^* containing the stable curve $\{x_s = \varphi(y_s, \bar{x}_i)\}$ such that Q_0 and P^* are I^*-transverse. From Section 3.6 and Lemma 3 in Section 6.6.3, we have

$$\delta_0 2^{-l} \geqslant \delta(\bar{x}_i) \geqslant \delta(Q_0, P^*) \geqslant |I^*|. \tag{10.185}$$

On the other hand, it follows from the definition of I^* that

$$(\mathbf{10.186}) \qquad |I^*|^\beta \geqslant |P_0|^{1+\tau}.$$

Inequalities (10.185), (10.186), and Lemma 20 give

$$(\mathbf{10.187}) \qquad |P_0| < |P_1|^{\frac{1}{2}(1+\beta)}.$$

This estimate means that every P_1 that occur (for the various $\gamma_1 = \gamma_{1,l}$ such that the complexity of $\gamma_{1,l}$ is > 0) is indeed simpler than P_0. Moreover, by Lemma 20, the hypothesis of Section 10.10.8 is satisfied by any such P_1. We can therefore under the assumption of Section 10.10.8 use (10.183) to define inductively $c(\gamma_0)$. The inequality (10.182) follows then from (10.187).

10.10.9. Finally we deal with the general case. By Lemma 20, the index $c(\gamma_{1,l})$ is already defined for every $l \in \mathcal{L}$. We define $c(\gamma_0)$ by (10.183) and check (10.182). For every $l \in \mathcal{L}$, we have either $c(\gamma_{1,l}) = 0$ or $c(\gamma_{1,l}) > 0$ and

$$(\mathbf{10.188}) \qquad c(\gamma_{1,l}) \leqslant C \log\log |P_1|^{-1},$$

where P_1 is associated to $\gamma_{1,l}$ as in Section 10.10.7.

From Proposition 12 in Section 6.6.2, we have

$$(\mathbf{10.189}) \qquad \log\log |P_0|^{-1} \geqslant C^{-1} \log n_0.$$

On the other hand, we have, from Lemma 20 and (10.172)

$$(\mathbf{10.190}) \qquad C^{-1} \log |P_1|^{-1} \leqslant \log(\delta_0 2^{-l}) \leqslant \log |Q_0|^{-1} \leqslant C n_0.$$

From this, we obtain

$$(\mathbf{10.191}) \qquad \log\log |P_1|^{-1} \leqslant \log\log |P_0|^{-1} + C,$$

and (10.182) follows.

The proof of the theorem is now complete.

10.11. *Invariant measures.* — From the Gibbs measure μ, which is not invariant but has the Jacobian property, we define a measure ν on $\mathcal{R}_+^\infty$ by $d\nu = h^* d\mu$. Recall that $\mu(\mathcal{R}_+^\infty - \mathcal{D}_+^\infty) = 0$ and that h^* is normalized by $\mu(h^*) = 1$. This means that ν is a probability measure on $\mathcal{R}_+^\infty$ and that $\nu(\mathcal{D}_+^\infty) = 1$.

Proposition **58.** — *The probability measure $d\nu = h^* d\mu$ is T^+-invariant, ergodic. It satisfies, for all $(P, Q, n) \in \mathcal{R}$:*

$$C^{-1}|P|^{d_s} \leqslant \nu(\{\omega \subset P\}) \leqslant C|P|^{d_s}.$$

Proof. — As h^* is bounded and bounded away from 0 (cf. (10.116)), the estimate for $\nu(\{\omega \subset P\})$ follows from the same estimate for $\mu(\{\omega \subset P\})$ in Proposition 56.

Let us prove that ν is T_+-invariant. We first observe that, if $h_0, h_1 \in E$, the product $h_0 h_1$ also belongs to E; indeed we have

$$(\mathbf{10.192}) \qquad |h_0 h_1|_E \leqslant \|h_0\|_\infty |h_1|_E + |h_0|_E \|h_1\|_\infty .$$

In particular, for any $h \in E$, hh^* also belongs to E. Let $h \in E$. We write

$$\int h(T^+\omega) d\nu(\omega) = \int h(T^+\omega) h^*(\omega) d\mu(\omega)$$

$$(\mathbf{10.193}) \qquad = \sum_{\mathcal{P}} \int h(T^+\omega) h^*(\omega) \chi_P^*(\omega) d\mu(\omega),$$

where χ_P^* is the characteristic function of $\mathcal{R}_+^\infty(P)$. The Jacobian property (10.123) (Section 10.9) gives

$$\int h(T^+\omega) h^*(\omega) \chi_P^*(\omega) d\mu(\omega)$$

$$(\mathbf{10.194}) \qquad = \int_{\mathcal{R}_+^\infty(a)} h(\omega) h^*(T_P^{-1}(\omega)) \exp(-d_s b(T_P^{-1}(\omega))) d\mu(\omega)$$

where $Q \subset R_a$ and T_P is the restriction of T^+ to $\mathcal{R}_+^\infty(P)$. Summing over P and using that h^* is L_{d_s}-invariant gives

$$(\mathbf{10.195}) \qquad \int h(T^+\omega) d\nu(\omega) = \int h(\omega) d\nu(\omega).$$

But $E \cap C(\mathcal{R}_+^\infty)$ is dense in the space of continuous functions $C(\mathcal{R}_+^\infty)$; the invariance of ν follows.

Let us prove that the invariant measure ν is ergodic. Let $A \subset \mathcal{R}_+^\infty$ be a T^+-invariant Borel subset with $\nu(A) > 0$ and A^c its complement. Let $\varepsilon > 0$. We will prove that there exists $a \in \mathcal{A}$ such that

$$(\mathbf{10.196}) \qquad \nu(A \cap \mathcal{R}_+^\infty(a)) \geqslant (1 - \varepsilon) \nu(\mathcal{R}_+^\infty(a)).$$

As $\varepsilon > 0$ is arbitrary, this easily implies $\nu(A) = 1$.

As $\nu(A) > 0$, we can find (P, Q, n) such that

$$(\mathbf{10.197}) \qquad \nu(\{\omega \subset P\} \cap A^c) \leqslant \varepsilon' \nu(\{\omega \subset P\}),$$

where $\varepsilon' \varepsilon^{-1}$ is small. Let r be the number of factors in the prime decomposition of (P, Q, n). Up to a set of measure 0, we have

$$(\mathbf{10.198}) \qquad \{\omega \subset P\} = \bigcup_{0 \leqslant j \leqslant r} \bigcup_{P_j} (T^+)^{-j} (\mathcal{R}_+^\infty(P_j)) \bmod 0$$

where P_j runs through prime elements satisfying $P_j \subset (T^+)^j(P)$ and $(T^+)^{-j}$ is the inverse branch of $(T^+)^j$ whose image contains P. From (10.197), there exists $0 \leqslant j \leqslant r$ and P_j such that

$$\textbf{(10.199)} \qquad \nu(A^c \cap (T^+)^{-j}(\mathcal{R}_+^\infty(P_j))) \leqslant \varepsilon' \nu((T^+)^{-j}(\mathcal{R}_+^\infty(P_j))).$$

We apply the Jacobian property, taking (10.118) into account to get (10.196) with $\varepsilon = C\varepsilon'$. We have proved that ν is ergodic, and the proof of the proposition is complete. □

We will now lift ν to obtain a $\widetilde{T}^+$-invariant probability measure on $\widetilde{\mathcal{R}}_+^\infty$.

Proposition **59.** — *There exists a unique probability measure* $\widetilde{\nu}$ *on* $\widetilde{\mathcal{R}}_+^\infty$ *which is* $\widetilde{T}^+$*-invariant and projects onto* ν *under* π. *It is ergodic.*

Proof. — The arguments are standard.

Existence. — Denote by $\mathcal{M}(\nu)$ the set of probability measures on $\widetilde{\mathcal{R}}_+^\infty$ which project onto ν. This is a compact set for the weak topology, invariant under $\widetilde{T}^+$ because ν is T^+-invariant. One obtains a $\widetilde{T}^+$-invariant measure in $\mathcal{M}(\nu)$ by taking any $\widetilde{\nu}_0 \in \mathcal{M}(\nu)$ and choosing a weak limit of a subsequence of

$$\textbf{(10.200)} \qquad \frac{1}{n}\sum_0^{n-1}[(\widetilde{T}^+)^j]^*(\widetilde{\nu}_0).$$

Uniqueness. — The set of fixed points for the action of $\widetilde{T}^+$ on $\mathcal{M}(\nu)$ is thus non-empty. It is also compact and convex. If it has more than one point, it has at least two distinct extremal points $\widetilde{\nu}_0$, $\widetilde{\nu}_1$. As ν is ergodic, $\widetilde{\nu}_0$ and $\widetilde{\nu}_1$ are also ergodic. Still by the ergodicity of ν, some stable curve ω must meet the basins of both $\widetilde{\nu}_0$ and $\widetilde{\nu}_1$. But stable curves are contracted exponentially fast under positive iteration by T^+; we should thus have $\widetilde{\nu}_0 = \widetilde{\nu}_1$, a contradiction.

We have already said that $\widetilde{\nu}$ is ergodic. □

Finally, we want to "spread" the $\widetilde{T}^+$-invariant measure $\widetilde{\nu}$ in order to obtain a g-invariant measure σ. Let $\Lambda = \Lambda_g$ as in the Introduction (cf. Section 1.2).

We first observe that the support of $\widetilde{\nu}$ is contained into $\Lambda \cap \widetilde{\mathcal{R}}_+^\infty$: if $N \subset \widetilde{\mathcal{R}}_+^\infty$ is compact and disjoint from Λ, then N is disjoint from the image of $(\widetilde{T}^+)^j$ if j is large enough, hence $\widetilde{\nu}(N) = 0$.

Let now h be a continuous, and thus bounded, function on Λ. For $x \in \Lambda \cap \mathcal{D}_+^\infty$, we write

$$\textbf{(10.201)} \qquad \widetilde{T}^+(x) = g^{N(x)}(x),$$

where $N(x) = n$ if $x \in \widetilde{\mathcal{R}}_{+}^{\infty}(P)$ with $(P, Q, n) \in \mathcal{P}$. We define:

$$\text{(10.202)} \qquad Sh(x) = \sum_{0 \leqslant j < N(x)} h(g^j(x)).$$

The function Sh is defined $\widetilde{\nu}$-almost everywhere. It satisfies:

$$\text{(10.203)} \qquad |Sh(x)| \leqslant \|h\|_{\infty} N(x).$$

By Proposition 58 and Corollary 16 in Section 10.3, the function N is $\widetilde{\nu}$-integrable. We have therefore defined an operator

$$\text{(10.204)} \qquad S : C(\Lambda) \mapsto L^1(\widetilde{\nu}),$$

where $C(\Lambda)$ stands for the space of continuous functions on $\Lambda = \Lambda_g$.

We define a finite measure σ on Λ by

$$\text{(10.205)} \qquad \int h d\sigma = \int Sh d\widetilde{\nu},$$

for $h \in C(\Lambda)$. From the definition of Sh, we have

$$\text{(10.206)} \qquad S(h \circ g) = Sh + h \circ \widetilde{T}^{+} - h.$$

Thus, the $\widetilde{T}^{+}$-invariance of $\widetilde{\nu}$ implies that σ is g-invariant. It is ergodic. The Lyapunov exponents of $\widetilde{T}^{+}$ for $\widetilde{\nu}$ are non-zero because $\widetilde{T}^{+}$ is uniformly hyperbolic. To get the Lyapunov exponents of g for σ we have only to change time, which is possible since N is $\widetilde{\nu}$-integrable.

In the next and last section, we will see that in some appropriate geometric sense, the measure σ captures "most" of the dynamics on Λ, and therefore can be considered as a naturally defined geometric invariant measure on Λ.

We end this section by observing that everything that has been done for T^{+} and positive iteration in Section 10, can also be done for T^{-} and negative iteration, leading to another naturally defined geometric invariant measure σ^{-} on Λ.

11. Some further geometric properties of the invariant set

In this final section we pursue the geometric study of the invariant set $\Lambda = \Lambda_g$ in two directions. First, we will describe in a rather precise way, both from a dynamical and a geometrical point of view, the intersection of an unstable curve in $\mathcal{R}_{-}^{\infty}$, as defined in Section 10.4, with the invariant set Λ. In the second part of the section, we prove that Λ is a saddle-like invariant set in the measure-theoretical sense: both its stable and unstable sets have Lebesgue measure 0; thus, no attractors are present in Λ.

11.1. *One-dimensional analysis of the invariant set.* — Let $\omega^* \in \mathcal{R}_-^\infty$ be an unstable curve as defined in Section 10.4. Let $(P_k^*, Q_k^*, n_k^*)_{k\geqslant 0}$ be the canonical sequence associated to ω^* (cf. definition also in Section 10.4). We have

$$\omega^* = \bigcap_{k\geqslant 0} Q_k^*, \tag{11.1}$$

where Q_0^* is a rectangle R_a and Q_{k+1}^* is a child of Q_k^* for each $k \geqslant 0$. We want to analyze the intersection $\omega^* \cap \Lambda$. In Section 10, we have analyzed the set $\widetilde{\mathcal{R}}_+^\infty$ and we know, in particular, that $\omega^* \cap \Lambda$ contains the subset $\omega^* \cap \widetilde{\mathcal{R}}_+^\infty$; this last subset has Hausdorff dimension d_s characterized in terms of the transfer operator studied in Section 10; in particular, this dimension is independent of ω^*.

Let us summarize the results of our analysis in this section.

Theorem **5.** — *The intersection $\omega^* \cap \Lambda$ is the disjoint union of*

- *a, at most countable, family of Cantor sets $\Lambda_i(\omega^*)$,*
- *a, at most countable, set $Cr(\omega^*)$,*
- *an exceptional set $\mathcal{E}(\omega^*)$,*

with the following properties

(i) *For each i, there exists a piece $\omega^*(i)$ of ω^* containing $\Lambda_i(\omega^*)$, an unstable curve $\omega^*{}_i$ and an integer n_i such that*

$$g^{n_i}(\omega^*(i)) = \omega_i^*, \tag{11.2}$$

$$g^{n_i}(\Lambda_i(\omega^*)) = \omega_i^* \cap \widetilde{\mathcal{R}}_+^\infty. \tag{11.3}$$

In particular, there is a special index $i = 0$ for which $n_0 = 0$, $\omega^*(0) = \omega_0^* = \omega^*$, $\Lambda_0(\omega^*) = \omega^* \cap \widetilde{\mathcal{R}}_+^\infty$.

(ii) *For every point $c \in Cr(\omega^*)$, there exists a stable curve $\omega_+(c) \in \mathcal{R}_+^\infty$, an unstable curve $\omega_-(c) \in \mathcal{R}_-^\infty$, a positive integer $n(c)$ such that $g^{n(c)}(c)$ is a quadratic tangency point between $\omega^+(c)$ and $g^{N_0}(\omega^-(c) \cap L_u)$.*

(iii) *The Hausdorff dimension of $\mathcal{E}(\omega^*)$ is not greater than*

$$\left(\frac{1}{d_s^0} + \frac{1}{2(d_s^0 + d_u^0 - 1)}\right)^{-1} + o(1) \tag{11.4}$$

where the $o(1)$ term is small provided τ is small enough. Consequently, the Hausdorff dimension of $\omega^* \cap \Lambda$ is equal to d_s.

(iv) *Every point $x \in \mathcal{E}(\omega^*)$ is the intersection of a decreasing sequence of pieces $(\omega^*(i_n(x)))_{n\geqslant 0}$.*

Remark **18.**

1. The structure of $\omega^* \cap \Lambda$ will be made more precise in the next subsections. We have tried here to extract the most significant features of our analysis.
2. Even with $d_s^0 + d_u^0 > 1$, it may happen that Λ is a uniformly hyperbolic horseshoe; then, the family $(\Lambda_i(\omega^*))_i$ is finite, $Cr(\omega^*)$ and $\mathcal{E}(\omega^*)$ are empty. When Λ is not uniformly hyperbolic, the family $(\Lambda_i(\omega^*))_i$ is countable and $\mathcal{E}(\omega^*)$ is a Cantor set; it is not clear in this case if $Cr(\omega^*)$ can be empty.

11.2. *Parabolic cores.* — Let $(P, Q, n) \in \mathcal{R}$, $\mathcal{R}$ as in Section 10.1.

Definition **16.** — *The parabolic core of* P, *denoted by* $c(P)$, *is the set of points of* $W^s(\Lambda, \widehat{R})$ *which belong to* P *but not to any child of* P. *The parabolic core of* Q, *denoted by* $c(Q)$, *is the set of points of* $W^u(\Lambda, \widehat{R})$ *which belong to* Q *but not to any child of* Q.

We have partitions

$$R \cap W^s(\Lambda, \widehat{R}) = \bigsqcup_{\mathcal{R}} c(P) \sqcup \widetilde{\mathcal{R}}_+^\infty, \tag{11.5}$$

$$R \cap W^u(\Lambda, \widehat{R}) = \bigsqcup_{\mathcal{R}} c(Q) \sqcup \widetilde{\mathcal{R}}_-^\infty. \tag{11.6}$$

If R_a is the rectangle which contains ω^*, we also have

$$\omega^* \cap \Lambda = \bigsqcup_{P \subset R_a} (\omega^* \cap c(P)) \sqcup (\omega^* \cap \widetilde{\mathcal{R}}_+^\infty). \tag{11.7}$$

The parabolic core is empty if and only if P is I-decomposable for a small enough parameter interval containing the given strongly regular parameter value. In particular, $c(P)$ is empty if Q is I-transverse. Thus, the union in (11.5), (11.7) can be restricted to those $(P, Q, n) \in \mathcal{R}$ such that Q is I-critical for all I.

We will denote by $C(\omega^*)$ the set of elements $(P, Q, n) \in \mathcal{R}$ such that $c(P) \cap \omega^*$ is not empty. For any $(P, Q, n) \in C(\omega^*)$, Q is I-critical for all I.

11.3. *Decomposition of* $c(P) \cap \omega^*$. — Let $(P, Q, n) \in C(\omega^*)$. For $k \geqslant 0$, set

$$(P_k, Q_k, n_k) = (P_k^*, Q_k^*, n_k^*) * (P, Q, n), \tag{11.8}$$

$$\omega_P^* = \bigcap_{k \geqslant 0} Q_k. \tag{11.9}$$

The unstable curve ω_P^* is contained in Q and we have

$$g^n(\omega^* \cap c(P)) \subset \omega_P^* \cap L_u. \tag{11.10}$$

We define a tree $\mathcal{A}(\omega^*, P)$ as follows. The vertices are the rectangles $P' \subset P_s$ with the following property: for any parameter interval I (containing the given parameter value, say t), for any $Q_k \supset \omega^*_P$, Q_k and P' are not I-separated, and Q_k and the parent of P' are I-critically related.

We connect two vertices by an (oriented) edge if one is the parent of the other. We say that a vertex P' is *critical* if, for all I and $Q_k \supset \omega^*_P$, Q_k and P' are I-critically related. Otherwise, we say that P' is transverse. The parent of a vertex is always a critical vertex, except if this vertex is P_s, the *root* of the tree. When P' is a transverse vertex, the smallest integer k such that Q_k, P are I-transverse for I small enough is called the *level* of P'.

Let P' be a critical vertex; then, for every parameter interval $I \ni t$, P' is I-critical and, therefore, decomposable.

Let P' be a transverse vertex of level 0. We have $Q_0 = Q$. Therefore, the parabolic composition $(P, Q, n) \square (P', Q', n')$ is well defined and produces two children of P.

Let P' be a transverse vertex of level $k > 0$. For all $m \geqslant k$, the parabolic composition

$$(P_m, Q_m, n_m) \square (P', Q', n')$$

is well-defined and produces two elements $(P^\pm_m, Q^\pm_m, n^\pm_m)$. The formulas

$$\begin{aligned} \omega^*_{P,P',+} &:= \bigcap Q^+_m, \\ \omega^*_{P,P',-} &:= \bigcap Q^-_m, \end{aligned} \tag{11.11}$$

define unstable curves $\omega^*_{P,P',\pm}$ contained in Q'. We also define pieces $\omega^*(P, P', \pm)$ of ω^* through

$$g^{n_{P,P'}}(\omega^*(P, P', \pm)) = \omega^*_{P,P',\pm}, \tag{11.12}$$

$$n_{P,P'} := n + n' + N_0. \tag{11.13}$$

Lemma **21.** — *Let x be a point in $\omega^* \cap c(P)$, $y = g^{n+N_0}(x)$. Either y belongs to a transverse vertex of level > 0 or it belongs to an infinite decreasing sequence of critical vertices.*

Proof. — We have $g^n(x) \in L_u$ (cf. (11.10)), $y \in L_s \subset P_s$, and P_s is the root and a critical vertex of the tree $\mathcal{A}(\omega^*, P)$. We assume that the first possibility in the statement of the lemma does not hold and construct, starting with P_s, a sequence of critical vertices containing y.

Assume that y belongs to a critical vertex P'. As P' is indecomposable and $y \in W^s(\Lambda)$, y belongs to some child P'_1 of P'. This rectangle is a vertex of the tree: otherwise, Q_k and P'_1 would be I-separated if I and Q_k are thin enough, and then $g^{N_0}(g^n(\omega^*) \cap L_u) \cap P'_1$ (which contains y) would be empty. The vertex P'_1 cannot be transverse of level 0 because, as remarked above, the parabolic composition of (P, Q, n) and (P'_1, Q'_1, n'_1) would produce a child of P containing x, contradicting the hypothesis that $x \in c(P)$. Finally P'_1 cannot be transverse of level > 0 by hypothesis. It must be a critical vertex, and the induction step is complete. □

Proposition **60.** — *There is at most one point $x \in \omega^* \cap c(\mathrm{P})$ such that $y = g^{n+\mathrm{N}_0}(x)$ belongs to a decreasing sequence of critical vertices. When such a point exists, the intersection of this decreasing sequence of vertices is a stable curve which intersects $g^{\mathrm{N}_0}(\mathrm{L}_u \cap \omega_\mathrm{P}^*)$ at y as a quadratic tangency point.*

Proof. — Let x be a point in $\omega^* \cap c(\mathrm{P})$ such that $y = g^{n+\mathrm{N}_0}(x)$ belongs to a decreasing sequence $(\mathrm{P}'_\ell)_{\ell \geqslant 0}$ of critical vertices. Denote by ω_+ the stable curve which is the intersection of these critical vertices. For all parameter intervals I, all $k \geqslant 0$, $\ell \geqslant 0$, Q_k and P'_ℓ are I-critically related. This implies that

$$\lim_{\substack{k \to +\infty \\ \ell \to +\infty}} \delta(\mathrm{Q}_k, \mathrm{P}'_\ell) = 0. \tag{11.14}$$

For large k and ℓ, let γ_k (resp. (γ'_ℓ) be the image in Q_k (resp. the inverse image in P'_ℓ) of the intersection of P_k with an horizontal curve (resp. the intersection of Q'_ℓ with a vertical curve). By (11.14), the distance between the vertical-like curve γ'_ℓ and the tip of the parabolic-like curve $g^{\mathrm{N}_0}(\gamma_k)$ goes to zero as k, ℓ go to $+\infty$. Passing to the limit, we see that ω_+ has a tangency with $g^{\mathrm{N}_0}(\omega_\mathrm{P}^* \cap \mathrm{L}_u)$. This tangency is quadratic in the following sense (cf. also the remark after the end of the proof): First, $g^{\mathrm{N}_0}(\omega_\mathrm{P}^* \cap \mathrm{L}_u)$ is contained, with the exception of the tangency point, in one of the components of $\mathrm{P}_s - \omega_+$; moreover, the angle between the tangent lines to $\omega_+(x)$, $g^{\mathrm{N}_0}(\mathrm{L}_u \cap \omega_\mathrm{P}^*)$ at points on these curves at the same distance and on the same side of the tangency point is of the same order as this distance to the tangency point. This is a consequence of the uniform estimates (3.21), (3.22) in Section 3.5.

As ω_+ and $g^{\mathrm{N}_0}(\mathrm{L}_u \cap \omega_\mathrm{P}^*)$ meet at only one point, this point must be y. If x' is a point with the same property as x, and we construct ω'_+ in the same way as ω_+, we must have $\omega_+ = \omega'_+$ because otherwise $g^{\mathrm{N}_0}(\mathrm{L}_u \cap \omega_\mathrm{P}^*) \cap \omega_+$ or $g^{\mathrm{N}_0}(\mathrm{L}_u \cap \omega_\mathrm{P}^*) \cap \omega'_+$ is empty. But, then, we have $y' := g^{n+\mathrm{N}_0}(x') = y$ and $x' = x$. □

Remark **19.** — Calculations involving partial derivatives of higher order for the maps (A, B), which implicitly represent elements of $\mathcal{R}$, show that stable curves and unstable curves are actually of class C^∞, with uniform estimates in the C^k topology for all k. Then, quadratic tangency can be taken in the usual sense. However, the calculations involved, especially when considering parabolic composition, are quite long and not very interesting; we decided to stick to the $\mathrm{C}^{1+\mathrm{Lip}}$ regularity class, where the notion of "quadratic" tangency, as explained in the proof of Proposition 60, still makes sense.

It is easy to see exactly when a point $x \in \omega^* \cap c(\mathrm{P})$ with the property specified in Proposition 60 does exist: a necessary and sufficient condition is that the tree $\mathcal{A}(\omega^*, \mathrm{P})$ is infinite. In this case, the point x will be a point of the set $Cr(\omega^*)$ in the statement of Theorem 5 and the point $y = g^{n+\mathrm{N}_0}(x)$ is said to be *critical*.

Summarizing what we have established so far, two cases may happen:

(1) The tree $\mathcal{A}(\omega^*, \mathrm{P})$ is finite. Then, the intersection $\omega^* \cap c(\mathrm{P})$ is the finite disjoint union of the sets

$$\omega^*(\mathrm{P}, \mathrm{P}', \pm) \cap \Lambda \tag{11.15}$$

where P′ runs through the vertices of the tree which are transverse of level > 0. The image under $g^{n_{\mathrm{P},\mathrm{P}'}}$ of the set (11.15) is the intersection $\omega^*_{\mathrm{P},\mathrm{P}',\pm} \cap \Lambda$.

(2) The tree $\mathcal{A}(\omega^*, \mathrm{P})$ is infinite. Then, the intersection $\omega^* \cap c(\mathrm{P})$ is the countable disjoint union of the sets $\omega^*(\mathrm{P}, \mathrm{P}', \pm) \cap \Lambda$ as above and a single point $x \in Cr(\omega^*)$. The point $x = x_\mathrm{P}$ is the limit of the pieces $\omega^*(\mathrm{P}, \mathrm{P}', \pm)$ (whose diameters goes to 0 as $|\mathrm{P}'|$ goes to 0).

11.4. *The structure of* $\omega^* \cap \Lambda$. — We are now ready to prove all the statements in Theorem 5, mentioned above in Section 11.1, with the exception of (iii) (the estimate on the Hausdorff dimension of $\mathcal{E}(\omega^*)$).

The structure of $\omega^* \cap \Lambda$ that we are looking for, which is roughly described in Theorem 5, is obtained by iterating the partition (11.7) and the decomposition of $\omega^* \cap c(\mathrm{P})$ described in Section 11.3.

At the first step, we have partitioned $\omega^* \cap \Lambda$ into the following subsets:

- the intersection $\omega^* \cap \widetilde{\mathcal{R}}^\infty_+$; points in this set are said of type I;
- for each $(\mathrm{P}, \mathrm{Q}, n) \in \mathrm{C}(\omega^*)$ such that $\mathcal{A}(\mathrm{P}, \omega^*)$ is infinite, a point x_P such that $y_\mathrm{P} = g^{n+\mathrm{N}_0}(x_\mathrm{P})$ is critical; such points x_P are said of type II;
- for each $(\mathrm{P}, \mathrm{Q}, n) \in \mathrm{C}(\omega^*)$, each vertex $(\mathrm{P}', \mathrm{Q}', n')$ of $\mathcal{A}(\omega^*, \mathrm{P})$ which is transverse of level bigger than 0, each $\varepsilon \in \{+, -\}$, the intersection $\omega^*(\mathrm{P}, \mathrm{P}', \varepsilon) \cap \Lambda$; the image of this set under $g^{n_{\mathrm{P},\mathrm{P}'}}$ is the intersection $\omega^*_{\mathrm{P},\mathrm{P}',\varepsilon} \cap \Lambda$ of another unstable curve with Λ.

The intersection $\omega^*_{\mathrm{P},\mathrm{P}',\varepsilon} \cap \Lambda$ will be analyzed in the same way that $\omega^* \cap \Lambda$.

Consider a point $z_0 \in \omega^* \cap \Lambda$. If it is of type I, it belongs to the set $\Lambda_0(\omega^*) := \omega^* \cap \widetilde{\mathcal{R}}^\infty_+$ of the statement of Theorem 5. If it is of type II, it belongs to $Cr(\omega^*)$. Assume now that it is of type III. Then, it belongs to some $\omega^*(\mathrm{P}, \mathrm{P}', \varepsilon) \cap \Lambda$ as above. Define

$$z_1 = g^{n_{\mathrm{P},\mathrm{P}'}}(z_0), \tag{11.16}$$

which belongs to $\omega^*_{\mathrm{P},\mathrm{P}',\varepsilon} \cap \Lambda =: \omega^*_1$. This point may in turn be of type I, II, III with respect to ω^*_1. The process stops if z_1 is of type I or II; if z_1 is of type III, it belongs to some piece $\omega^*_1(\mathrm{P}_1, \mathrm{P}'_1, \varepsilon_1)$; we define

$$z_2 = g^{n_{\mathrm{P}_1,\mathrm{P}'_1}}(z_1), \tag{11.17}$$

which belongs to $\omega^*_2 \cap \Lambda$, with

$$\omega^*_2 := g^{n_{\mathrm{P}_1,\mathrm{P}'_1}}(\omega^*_1(\mathrm{P}_1, \mathrm{P}'_1, \varepsilon)). \tag{11.18}$$

Iterating this process lead to one of three possible outcomes:

(1) the z_k's are defined and of type III for all $k \geqslant 0$; the corresponding initial points z_0 form the set $\mathcal{E}(\omega^*)$.
(2) the z_k's are defined for $0 \leqslant k \leqslant \ell$ and z_ℓ is of type I, i.e. it belongs to $\widetilde{\mathcal{R}}_+^\infty$; let $(P_k, P'_k, \varepsilon_k)$ for $0 \leqslant k < \ell$ be the data involved in the definitions of the z_k's. We collect together the initial points z_0's with the same set of data; such a set form one of the Cantor sets $\Lambda_i(\omega^*)$ in Theorem 5.
(3) the z_k's are defined for $0 \leqslant k \leqslant \ell$ and z_ℓ is of type II. Then z_0 belongs to the set $Cr(\omega^*)$.

We have now completely defined the partition of $\omega^* \cap \Lambda$ described in Theorem 5. The properties (i), (ii), (iv) follow immediately from the definitions.

11.5. *Hausdorff dimension of the exceptional set $\mathcal{E}(\omega^*)$.*

11.5.1. The self-similar structure apparent in the definition of $\mathcal{E}(\omega^*)$ is the key to obtain an estimate of the dimension of this set. More specifically, we have

$$\mathcal{E}(\omega^*) = \bigsqcup_{(P,P',\varepsilon)} g^{-n_{P,P'}}(\mathcal{E}(\omega^*{}_{P,P',\varepsilon})), \tag{11.19}$$

where $\varepsilon \in \{+, -\}$, P runs through $C(\omega^*)$ and P′ through vertices of $\mathcal{A}(\omega^*, P)$ which are transverse of level > 0.

11.5.2. The estimate on the Hausdorff dimension will follow from a standard result that we formulate in a general setting.

Let $\alpha, d \in (0, 1)$. Let Ω be a set. For each $\omega \in \Omega$, we are given

- a subset $E(\omega) \subset [0, 1]$ which is the union of at most countably many disjoint compact subintervals of $[0, 1]$;
- a map $F_\omega = (g_\omega, f_\omega)$ from $E(\omega)$ into $[0, 1] \times \Omega$.

These data satisfy

- Each map f_ω is constant on each component of $E(\omega)$.
- The restriction of each map g_ω to each component of $E(\omega)$ is an uniformly expansive $C^{1+\alpha}$ diffeomorphism onto $[0, 1]$ with uniformly bounded distortion.

The second condition means that there exists $0 < \lambda < 1$, $C_0 > 0$, independent of ω, such that, writing h_J for the inverse of the restriction of g_ω to a component J of $E(\omega)$, we have, for $x, y \in [0, 1]$

$$|Dh_J(x)| \leqslant \lambda,$$
$$|\log|Dh_J(x)| - \log|Dh_J(y)|| \leqslant C_0|x - y|^\alpha.$$

Define $E := \{(x, \omega) \in [0, 1] \times \Omega, x \in E(\omega)\}$ and the map $F : E \to [0, 1] \times \Omega$ by $F(x, \omega) = F_\omega(x)$. Let $\mathcal{E} = \bigcap_{n \geqslant 0} F^{-n}(E)$. For $\omega \in \Omega$, let $\mathcal{E}(\omega) := \{x \in [0, 1], (x, \omega) \in \mathcal{E}\}$.

Proposition **61.** — *Assume that one has, for every $\omega \in \Omega$*

$$\sum_{J} |J|^{d} \leqslant \exp(-C_0 d(1-\lambda^{\alpha})^{-1}),$$

where the sum runs over the components J *of* $E(\omega)$. *Then the Hausdorff dimension of each set* $\mathcal{E}(\omega)$ *is at most* d.

Proof. — For $n > 0$, $\omega \in \Omega$, let $E_n := \bigcap_{0 \leqslant m < n} F^{-m}(E)$ be the domain of F^n, and $E_n(\omega) := \{x \in [0,1], (x,\omega) \in E_n\}$ be the fiber of E_n. Each $E_n(\omega)$ is the union of at most countably many disjoint compact intervals.

Let $n > 0$, $\omega \in \Omega$, J be a component of $E_n(\omega)$. There exists ω' and a $C^{1+\alpha}$ diffeomorphism h_J from $[0,1]$ onto J such that $F^n(x,\omega) = (h_J^{-1}(x), \omega')$ for $x \in J$. Moreover, we have, for $x, y \in [0,1]$

$$|Dh_J(x)| \leqslant \lambda^n,$$
$$|\log|Dh_J(x)| - \log|Dh_J(y)|| \leqslant C_0(1-\lambda^{\alpha})^{-1}|x-y|^{\alpha}.$$

For $n > 0$, $\omega \in \Omega$, define

$$S_n(\omega) = \sum_{J} |J|^{d},$$

where the sum runs over the components J of $E_n(\omega)$. Here, each interval has length $\leqslant \lambda^n$. Therefore the components of $E_n(\omega)$ form a covering of $\mathcal{E}(\omega)$ by intervals of small diameter. We have by hypothesis $S_1(\omega) \leqslant \exp(-C_0 d(1-\lambda^{\alpha})^{-1})$. We will show that $S_{n+1}(\omega) \leqslant S_n(\omega)$, which implies the conclusion of the proposition.

In the sum for $S_{n+1}(\omega)$, we first sum over components J^* of $E_{n+1}(\omega)$ which are contained in a fixed component J of $E_n(\omega)$. With ω' as above, such J^* are exactly the images by h_J of the components J' of $E(\omega')$. The lengths are related through the mean-value theorem by

$$|J| = |Dh_J(x)|, \qquad |J^*| = |Dh_J(y)||J'|,$$

for some $x \in [0,1]$, $y \in J'$. We therefore have

$$|J^*| \leqslant \exp(C_0(1-\lambda^{\alpha})^{-1})|J||J'|.$$

We finally obtain

$$S_{n+1}(\omega) = \sum_{J}\sum_{J'} |J^*|^{d}$$
$$\leqslant \exp(C_0 d(1-\lambda^{\alpha})^{-1}) \sum_{J}\Big(|J|^{d} \sum_{J'} |J'|^{d}\Big)$$

$$= \exp(C_0 d(1-\lambda^\alpha)^{-1}) \sum_J (|J|^d S_1(\omega'))$$

$$\leqslant \sum_J |J|^d = S_n(\omega).$$

□

11.5.3. We now come back to the setting of the theorem.

Lemma **22.** — *The maps*

$$g^{n_{P,P'}} : \omega^*(P, P', \varepsilon) \to \omega^*{}_{P,P',\varepsilon}$$

have uniformly bounded distortion.

Proof. — Let k be an integer larger than the level of the transverse vertex P′. Then, the parabolic composition of (P_k, Q_k, n_k) (cf. (11.8)) and (P', Q', n') is defined and produces an element (P'_k, Q'_k, n'_k) such that Q'_k contains $\omega^*_{P,P',\varepsilon}$. Let γ^*_k be an horizontal segment in P^*_k, γ_k its image under $g^{n^*_k}$, γ'_k the image of $\gamma^*_k \cap P'_k$ under $g^{n'_k}$.

The affine-like maps

$$g^{n^*_k} : P^*_k \to Q^*_k, \qquad g^{n'_k} : P'_k \to Q'_k, \tag{11.20}$$

have bounded distortion, hence the one-dimensional map

$$g^{n^*_k} \circ (g^{n'_k})^{-1} : \gamma'_k \to \gamma_k \tag{11.21}$$

have also uniformly bounded distortion. Letting k go to $+\infty$, γ'_k converge to $\omega^*_{P,P',\varepsilon}$ and γ_k to ω^* in the $C^{2-\varepsilon}$-topology for all $\varepsilon > 0$. The statement of the lemma follows. □

Lemma **23.** — *Let*

$$\delta(\omega^*_P, P') = \lim_{k\to+\infty} \delta(Q_k, P').$$

We have

$$C^{-1} \leqslant \frac{\operatorname{diam} \omega^*(P, P', \varepsilon)}{|P||P'|(\delta(\omega^*_P, P'))^{-\frac{1}{2}}} \leqslant C.$$

Proof. — As in the proof of Lemma 22, we write

$$g^{n_{P,P'}} = g^{n'_k} \circ (g^{n^*_k})^{-1}. \tag{11.22}$$

From the estimate (3.27) for parabolic composition in Section 3.5, we have

$$C^{-1} \leqslant \frac{\operatorname{diam}(\gamma^*_k \cap P'_k)}{|P_k|, |P'|(\delta(Q_k, P'))^{-\frac{1}{2}}} \leqslant C. \tag{11.23}$$

We also have, from the estimates on simple composition

$$(\mathbf{11.24})\qquad C^{-1} \leqslant \frac{\operatorname{diam}\omega^*(P, P', \varepsilon)|P_k^*|}{\operatorname{diam}\gamma_k^*} \leqslant C,$$

$$(\mathbf{11.25})\qquad C^{-1} \leqslant \frac{|P_k|}{|P||P_k^*|} \leqslant C.$$

Multiplying these three inequalities yields the Lemma. □

11.5.4. From (11.19), we see that $\mathcal{E}(\omega^*)$ is a set of the type considered in Proposition 61. The property of uniformly bounded distortion has been checked in Lemma 22, and the property of uniform expansion is clear from Lemma 23.

Let us introduce

$$(\mathbf{11.26})\qquad \chi(d) = \sum_{(P,P',\varepsilon)} [\operatorname{diam}\omega^*(P, P', \varepsilon)]^d.$$

If we are able, for some value of d, to show that the series defining χ is convergent and $\chi(d)$ is small enough, then by Proposition 61, we will deduce that the Hausdorff dimension of $\mathcal{E}(\omega^*)$ is $\leqslant d$.

In order to study χ, we will first fix P in $C(\omega^*)$ and sum over (P', ε). As ε takes only two values, and in view of Lemma 23, we define, for $P \in C(\omega^*)$:

$$(\mathbf{11.27})\qquad \chi_P(d) = \sum_{P'} |P'|^d \delta(\omega_P^*, P')^{-\frac{1}{2}d}.$$

We will then have

$$(\mathbf{11.28})\qquad \chi(d) \leqslant C \sum_P |P|^d \chi_P(d).$$

In the sum (11.27), P′ is a transverse vertex of level > 0. We claim that

$$(\mathbf{11.29})\qquad \delta(\omega_P^*, P') \leqslant \delta_{\max} := \min(\varepsilon_0, C|Q|^{1-\eta}).$$

The bound by ε_0 is clear. To show that $\delta(\omega_P^*, P') \leqslant C|Q|^{1-\eta}$, we recall that, for large k and small I containing t, Q_k and P′ are I-transverse, while Q and P′ are I-critically related for any I containing t. By Proposition 21 in Section 8.1, we must have $|Q| > \frac{1}{3}|P'|$. If we had $\delta(\omega_P^*, P') \gg |Q|^{1-\eta}$, we would also have $\delta(Q, P') \gg |Q|^{1-\eta}$, $\delta(Q, P') \gg |P'|^{1-\eta}$; conditions (T1), (T2), (T3) for $Q \pitchfork_I P'$ would be satisfied for I small enough, a contradiction. The claim is proved.

In the series (11.27), we first sum over those P′ such that

$$(\mathbf{11.30})\qquad 2^{-\ell}\delta_{\max} \geqslant \delta(\omega_P^*, P') \geqslant 2^{-\ell-1}\delta_{\max}$$

for some fixed $\ell \geqslant 0$. This allows us to write

$$\chi_{\mathrm{P}}(d) \leqslant \mathrm{C}\delta_{\max}^{-d/2} \sum_{\ell \geqslant 0} 2^{\frac{\ell d}{2}} \Big(\sum^{(\ell)} |\mathrm{P}'|^d \Big), \tag{11.31}$$

where $\sum^{(\ell)}$ means that P′ is constrained by (11.30). We divide $\sum^{(\ell)}$ into two parts.

11.5.5. In the first part, denoted by $\sum_1^{(\ell)}$, we consider only those P′ such that its parent $\widetilde{\mathrm{P}}'$ satisfies

$$|\widetilde{\mathrm{P}}'| \leqslant 2^{-\ell} \delta_{\max}. \tag{11.32}$$

To estimate $\sum_1^{(\ell)} |\mathrm{P}'|^d$, first observe that, with d bounded away from 0, it follows from Proposition 26 in Section 8.1 that the sum of $|\mathrm{P}'|^d$ over children of a fixed parent $\widetilde{\mathrm{P}}'$ is bounded by $\mathrm{C}|\widetilde{\mathrm{P}}'|^d$. We must therefore bound $\sum_1^{(\ell)} |\widetilde{\mathrm{P}}'|^d$.

Also, as $\widetilde{\mathrm{P}}'$ is a critical vertex, $\widetilde{\mathrm{P}}'$ cannot be very thin: we have $\delta(\omega_{\mathrm{P}}^*, \mathrm{P}') \leqslant 3|\mathrm{P}'|^{1-\eta}$. Indeed, otherwise, for large k and small I containing t, conditions (T1), (T2), (T3) for $\mathrm{Q}_k \pitchfork_{\mathrm{I}} \mathrm{P}'$ would be satisfied. We have therefore

$$|\widetilde{\mathrm{P}}'| \geqslant \mathrm{C}^{-1} [\delta(\omega_{\mathrm{P}}^*, \mathrm{P}')]^{(1-\eta)^{-1}} \geqslant \mathrm{C}^{-1} (\delta_{\max} 2^{-\ell})^{(1-\eta)^{-1}}. \tag{11.33}$$

Finally, the number of $\widetilde{\mathrm{P}}'$ with $|\widetilde{\mathrm{P}}'|$ of order $2^{-m-\ell}\delta_{\max}$ is at most $\mathrm{C}2^m$ and the integer m here is restricted by (11.33) to the range

$$1 \leqslant 2^m \leqslant \mathrm{C}(\delta_{\max} 2^{-\ell})^{-\eta(1-\eta)^{-1}}. \tag{11.34}$$

We, therefore, obtain for d bounded away from 0 and 1,

$$\begin{aligned} \sum_1^{(\ell)} |\mathrm{P}'|^d &\leqslant \mathrm{C} \sum_1^{(\ell)} |\widetilde{\mathrm{P}}'|^d \\ &\leqslant \mathrm{C}\delta_{\max}^d 2^{-\ell d} \sum_m 2^{m(1-d)} \\ &\leqslant \mathrm{C}(\delta_{\max} 2^{-\ell})^{d-\eta}. \end{aligned} \tag{11.35}$$

11.5.6. In the second part of $\sum^{(\ell)}$, denoted by $\sum_2^{(\ell)}$, we have on the opposite

$$|\widetilde{\mathrm{P}}'| > 2^{-\ell} \delta_{\max}. \tag{11.36}$$

We claim that, in this case, the number of possible $\widetilde{\mathrm{P}}'$ is bounded by C.

- If there is some $\widetilde{P}'$ with $|\widetilde{P}'| \gg 2^{-\ell}\delta_{\max}$, it will contain all P′ satisfying (11.30), because (according to (MP7)) there is a strip of width $\geqslant C^{-1}|\widetilde{P}'|$ along each vertical boundary of $\widetilde{P}'$ which does not intersect Λ and therefore does not contain any P′ satisfying (11.30). Two such $\widetilde{P}'$ cannot both be the parents of the same P′, hence there is at most one such $\widetilde{P}'$.
- The remaining $\widetilde{P}'$ satisfy $C2^{-\ell}\delta_{\max} > |\widetilde{P}'| > 2^{-\ell}\delta_{\max}$. We divide this range into a bounded number of shorter range, such that two $\widetilde{P}'$ with widths in the same range are disjoint. But there can be at most C disjoint $\widetilde{P}'$ with $|\widetilde{P}'| > 2^{-\ell}\delta_{\max}$ and containing some P′ satisfying (11.30).

This proves the claim.

As each P′ is a transverse vertex, we must have (by (R7))

(**11.37**) $$|P'| \leqslant C(2^{-\ell}\delta_{\max})^{(1-\eta)^{-1}}.$$

In particular, from (11.36), (11.37), P′ is a non-simple child of $\widetilde{P}'$. From Proposition 26 in Section 8.2, the number of P′ with |P′| of order $2^{-m}\varepsilon_0$ is at most $2^{c'm\eta}$.

We have

(**11.38**) $$\sum_2^{(\ell)} |P'|^d \leqslant \varepsilon_0^d \sum_m 2^{-m(d-c'\eta)} \leqslant C\varepsilon_0^{c'\eta}(2^{-\ell}\delta_{\max})^{d-c'\eta}.$$

11.5.7. Putting (11.35) and (11.38) together yields (replacing if necessary c' by $\max(c', 1)$)

(**11.39**) $$\sum^{(\ell)} |P'|^d \leqslant C(\delta_{\max}2^{-\ell})^{d-c'\eta}$$

and introducing this in (11.31) allows us to estimate χ_P:

(**11.40**) $$\chi_P(d) \leqslant C\delta_{\max}^{\frac{1}{2}d-c'\eta}.$$

Finally, we obtain

(**11.41**) $$\chi(d) \leqslant C \sum_{C(\omega^*)} |P|^d[\min(\varepsilon_0, |Q|)]^{\frac{1}{2}d-C\eta}.$$

11.5.8. We do not know exactly the set $C(\omega^*)$, but we know that if $(P, Q, n) \in C(\omega^*)$, the parabolic core $c(P)$ is non-empty and Q must be I-critical for all parameter intervals I containing the given parameter value.

We use Hölder's inequality to separate the P and Q in (11.41): for any $p, q > 1$ such that

(**11.42**) $$\frac{1}{p} + \frac{1}{q} = 1,$$

we have

$$(\mathbf{11.43}) \qquad \chi(d) \leqslant C\chi_+(d)^{\frac{1}{p}}\chi_-(d)^{\frac{1}{q}}$$

where

$$(\mathbf{11.44}) \qquad \chi_+(d) = \sum_{Q \text{ critical}} |P|^{dp},$$

$$(\mathbf{11.45}) \qquad \chi_-(d) = \sum_{Q \text{ critical}} \min(\varepsilon_0, |Q|)^{(\frac{1}{2}d - C\eta)q}.$$

We will choose d, p, q satisfying (11.42) in order to have $\chi_+(d)$ bounded and $\chi_-(d)$ small (when ε_0 is small). For such a choice, we can conclude that the Hausdorff dimension of $\mathcal{E}(\omega^*)$ is $\leqslant d$.

11.5.9. We first consider $\chi_-(d)$. We write $d^- = q(\frac{1}{2}d - C\eta)$. We divide the sum for $\chi_-(d)$ into three parts according to the width of Q. We recall that $|Q_u|$ is of the same order than ε_0.

- $|Q| \geqslant |Q_u|$.
  In this case, as Q is critical, it contains Q_u. The number of such Q is $\leqslant C\log\varepsilon_0^{-1}$, hence this part of the sum is bounded by $C\varepsilon_0^{d^-}\log\varepsilon_0^{-1}$ hence this part of the sum is bounded by $C\varepsilon_0^{d^-}\log\varepsilon_0^{-1}$.
- $|Q_u| > |Q| \geqslant C_0^{-1}\varepsilon_0^{1+\tau} = C_0^{-1}\varepsilon_1$.
  In this case, Q must be contained in Q_u. By Corollary 13 (Section 8.3), the number of such Q is at most $C\varepsilon_0^{-\tau d_u^*}$ and the corresponding part of the sum is bounded by $C\varepsilon_0^{d^- - \tau d_u^*}$.
- $C_0^{-1}\varepsilon_l > |Q| \geqslant C_0^{-1}\varepsilon_{l+1}$ for some $l > 0$.
  We first fix l and will afterwards sum over l.
  Let I be the parameter interval of length ε_{l-1} containing t. By Corollary 6 (Section 6.6.3), Q is I-defined. As Q is I-critical, it is I-special; by Proposition 18 (Section 7.7), we have $\max_I |Q| \leqslant \varepsilon_l$ (we choose C_0 in order to obtain this). By definition of $\mathcal{C}_-(I)$ (in Section 9.1), there exists $(P', Q', n') \in \mathcal{C}_-(I)$ such that $Q \subset Q'$. We have $\max_I |Q'| \leqslant \varepsilon_l$ and $|Q| \geqslant C_0^{-1}\varepsilon_{l+1}$; by Corollary 13, the number of Q for given Q' is $\leqslant C\varepsilon_l^{-\tau d_u^*}$. On the other hand, as I is strongly regular, we have from $(SR1)_u$ (Section 9.1) that

$$(\mathbf{11.46}) \qquad \#\mathcal{C}_-(I) \leqslant C\left(\frac{|I|}{\varepsilon_0}\right)^{\sigma}\varepsilon_0^{-\tau d_u^0},$$

where σ, defined in (9.49) (Section 9.4), is close to $1 - d_u^0 - d_s^0$.

The part of the sum for $\chi_-(d)$ corresponding to this fixed value of l is therefore bounded by

$$\textbf{(11.47)} \qquad C\varepsilon_0^{-\tau d_u^0}\left(\frac{\varepsilon_{l-1}}{\varepsilon_0}\right)^{\sigma}\varepsilon_l^{-\tau d_u^*}\varepsilon_l^{d^-} = C\varepsilon_0^{-\sigma-\tau d_u^0}\varepsilon_{l-1}^{\sigma+(1+\tau)(d^- -\tau d_u^*)}.$$

For the sum over l to converge, we ask that

$$\textbf{(11.48)} \qquad d^- \geqslant -\frac{\sigma}{1+\tau} + \tau d_u^* + \tau,$$

where the right-hand side is close to $d_s^0 + d_u^0 - 1$. When this is the case, the sum over l is, for ε_0 small enough, bounded by $C\varepsilon_0^{-\sigma-\tau d_u^0+\tau}$. From the definition of σ in (9.49), we have, as $1 - d_s^0 - d_u^0 \leqslant 0$, that $\sigma \leqslant -\tau$.

We conclude that, if (11.48) holds, we have $\chi_-(d) < \varepsilon_0^{\tau}$.

11.5.10. We now consider $\chi_+(d)$. We write $d^+ = pd$. We divide again the sum for $\chi_+(d)$ into three parts according to the width of Q.

- $|Q| \geqslant |Q_u|$.
  As Q must contain Q_u, this part of the sum is bounded by C (if d^+ is bounded away from 0, which will be the case).
- $|Q_u| > |Q| \geqslant (2\varepsilon_0)^{(1-\eta)^{-1}}$.
  In this case, Q must be contained in Q_u. By Corollary 13 (Section 8.3), the number of such Q is at most $C\varepsilon_0^{-C\eta}$. By Corollary 6 (Section 6.6.3), Q is I_0-defined, hence (P, Q, n) is the simple composition of some element in $\mathcal{R}(I_0)$ by (P_u, Q_u, n_u). In particular, we have $|P| \leqslant C|P_u|$ and this part of the sum is bounded by $\varepsilon_0^{-C\eta}|P_u|^{d^+} \leqslant C$ (with d^+ bounded away from 0).
- $(2\varepsilon_l)^{(1-\eta)^{-1}} > |Q| \geqslant (2\varepsilon_{l+1})^{(1-\eta)^{-1}}$ for some $l \geqslant 0$.
  We first fix l and will afterwards sum over l.
  Let I be the parameter interval of length ε_l containing t. By Corollary 6 (Section 6.6.3), Q is I-defined; as $(2\varepsilon_l)^{(1-\eta)^{-1}} > |Q|$, Q is thin I-critical. By definition of $\widehat{\mathcal{C}}_-(I)$ (Section 9.1), there exists $(P', Q', n') \in \widehat{\mathcal{C}}_-(I)$ such that $Q \subset Q'$. As I is strongly regular, condition $(SR2)'_u$ (Section 9.1) is satisfied.
  We take $d^+ = \rho_s$, where ρ_s is the exponent in $(SR2)'_u$, defined in Section 9.14 and close to d_s^0. However, to apply $(SR2)'_u$, we need the Q to be disjoint. To achieve this, we divide the range

$$(2\varepsilon_l)^{(1-\eta)^{-1}} > |Q| \geqslant (2\varepsilon_{l+1})^{(1-\eta)^{-1}}$$

  into several smaller ranges

$$(1 - C^{-1})^i(2\varepsilon_l)^{(1-\eta)^{-1}} > |Q| \geqslant (1 - C^{-1})^{i+1}(2\varepsilon_l)^{(1-\eta)^{-1}}, \quad i = 0, 1, \ldots$$

where C is large enough to insure that two Q with widths in the same range are disjoint. The number of smaller ranges is $\leqslant C\tau \log \varepsilon_l^{-1}$. The bound for $\sum |P|^{d^+}$ in each range given by $(SR2)'_u$ is $C|P_u|^{\rho_s}(\frac{\varepsilon_l}{\varepsilon_0})^{\sigma_s}$, where $\sigma_s = 1 - d_s^+ - 3\tau$ (cf. (9.63) in Section 9.6.1) is close to $1 - d_s^0$. Therefore the contribution to the sum for $\chi_+(d)$ coming for this value of l is bounded by

$$C\tau |P_u|^{\rho_s} \left(\frac{\varepsilon_l}{\varepsilon_0}\right)^{\sigma_s} \log \varepsilon_l^{-1}.$$

The sum over $l \geqslant 0$ is clearly converging and bounded by C (much better actually).

With $d^+ = \rho_s$, we obtain therefore $\chi_+(d) \leqslant C$.

11.5.11. We can now choose d, p, q to finish the proof. We take $\frac{1}{p} = \frac{d}{\rho_s}$, as already mentioned. In order to satisfy (11.48), we take

$$\frac{1}{q} = \frac{d - 2C\eta}{2(-\frac{\sigma}{1+\tau} + \tau d_u^* + \tau)}.$$

The relation $\frac{1}{p} + \frac{1}{q} = 1$ determines d. When $\tau \gg \eta \gg \varepsilon_0$ are small, we obtain a value for d close to

$$\left(\frac{1}{d_s^0} + \frac{1}{2(d_s^0 + d_u^0 - 1)}\right)^{-1} < d_s^0.$$

As explained above, the Hausdorff dimension of $\mathcal{E}(\omega^*)$ is $\leqslant d$ and the proof of the theorem is complete.

11.6. *The stable and unstable sets of* Λ. — Our goal at the end of this final section is to prove that the invariant set Λ is a saddle-like object in the following measure-theoretical sense:

Theorem **6.** — *For a strongly regular parameter, both the stable set* $W^s(\Lambda)$ *and the unstable set* $W^u(\Lambda)$ *have Lebesgue measure* 0.

The situation is symmetrical and we will deal with the stable set.

We have:

$$W^s(\Lambda) = \bigcup_{n \geqslant 0} g^{-n}(W^s(\Lambda, \widehat{R}) \cap R). \tag{11.49}$$

Therefore, it is sufficient to show that $W^s(\Lambda, \widehat{R}) \cap R$ has Lebesgue measure 0. We write

$$R \cap W^s(\Lambda, \widehat{R}) = \bigcup_{n \geqslant 0} \left(W^s(\Lambda, \widehat{R}) \cap R \cap g^{-n}(\widetilde{\mathcal{R}}_+^\infty)\right) \sqcup \mathcal{E}^+, \tag{11.50}$$

with

$$(\mathbf{11.51}) \qquad \mathcal{E}^+ = \{z \in W^s(\Lambda, \widehat{R}) \cap R, g^n(z) \notin \widetilde{\mathcal{R}}_+^\infty \text{ for all } n \geqslant 0\}.$$

We have seen in Section 10 that $\widetilde{\mathcal{R}}_+^\infty$ is Lipschitzian with transverse Hausdorff dimension d_s. Therefore, the Hausdorff dimension of $\widetilde{\mathcal{R}}_+^\infty$ is $1 + d_s$ and its Lebesgue measure is 0. The same is true of $g^{-n}(\widetilde{\mathcal{R}}_+^\infty)$. We have to prove that the Lebesgue measure of $\mathcal{E}^+$ is equal to 0.

11.7. *Decomposition of* $\mathcal{E}^+$. — By the definition of $\mathcal{E}^+$ and of the parabolic cores, we can write

$$(\mathbf{11.52}) \qquad \mathcal{E}^+ = \bigsqcup_{P_0} \mathcal{E}^+(P_0),$$

where

$$(\mathbf{11.53}) \qquad \mathcal{E}^+(P_0) = \mathcal{E}^+ \cap c(P_0)$$

and (P_0, Q_0, n_0) runs through the set $\mathcal{C}_-$ of elements of R with $c(P_0) \neq \emptyset$. In particular, Q_0 is I-critical for all I containing the given parameter value.

For any such P_0, we have

$$(\mathbf{11.54}) \qquad g^{n_0}(\mathcal{E}^+(P_0)) \subset Q_0 \cap L_u \cap \mathcal{E}^+,$$

$$(\mathbf{11.55}) \qquad g^{n_0+N_0}(\mathcal{E}^+(P_0)) \subset L_s \cap \mathcal{E}^+.$$

For $P_1 \in \mathcal{C}_-$, define

$$(\mathbf{11.56}) \qquad \mathcal{E}^+(P_0, P_1) = \{z \in \mathcal{E}^+(P_0), g^{n_0+N_0}(z) \in c(P_1)\}.$$

We have a partition

$$(\mathbf{11.57}) \qquad \mathcal{E}^+(P_0) = \bigsqcup_{P_1} \mathcal{E}^+(P_0, P_1).$$

At step k, we have a partition

$$(\mathbf{11.58}) \qquad \mathcal{E}^+ = \bigsqcup_{P_0,\dots,P_k} \mathcal{E}^+(P_0, \dots, P_k)$$

where the (P_i, Q_i, n_i) run through $\mathcal{C}_-$. We write

$$m_0 = n_0,$$

$$m_1 = n_0 + N_0 + n_1,$$

$$(\mathbf{11.59})\qquad \begin{aligned}&\vdots\\ m_j &= n_0 + \mathrm{N}_0 + n_1 + \mathrm{N}_0 + \cdots + n_{j-1} + \mathrm{N}_0 + n_j\\ &= m_{j-1} + \mathrm{N}_0 + n_j.\end{aligned}$$

For $0 \leqslant j \leqslant k$, we have

$$(\mathbf{11.60})\qquad g^{m_j}(\mathcal{E}^+(\mathrm{P}_0, \dots, \mathrm{P}_k)) \subset \mathrm{Q}_j \cap \mathrm{L}_u \cap \mathcal{E}^+,$$

$$(\mathbf{11.61})\qquad g^{m_j+\mathrm{N}_0}(\mathcal{E}^+(\mathrm{P}_0, \dots, \mathrm{P}_k)) \subset \mathrm{L}_s \cap \mathcal{E}^+.$$

We define, for $\mathrm{P}_{k+1} \in \mathcal{C}_-$

$$(\mathbf{11.62})\qquad \mathcal{E}^+(\mathrm{P}_0, \dots, \mathrm{P}_k, \mathrm{P}_{k+1}) = \{z \in \mathcal{E}^+(\mathrm{P}_0, \dots, \mathrm{P}_k), g^{m_k+\mathrm{N}_0}(z) \in c(\mathrm{P}_{k+1})\}$$

and we have

$$(\mathbf{11.63})\qquad \mathcal{E}^+(\mathrm{P}_0, \dots, \mathrm{P}_k) = \bigsqcup_{\mathrm{P}_{k+1}} \mathcal{E}^+(\mathrm{P}_0, \dots, \mathrm{P}_k, \mathrm{P}_{k+1}).$$

However, in order to have $\mathcal{E}^+(\mathrm{P}_0, \dots, \mathrm{P}_k) \neq \emptyset$ strong restrictions on the P_i must take place. We have already mentioned that $(\mathrm{P}_i, \mathrm{Q}_j, n_i) \in \mathcal{C}_-$. This is the only restriction on $(\mathrm{P}_0, \mathrm{Q}_0, n_0)$. But, from (11.55), P_1 must meet P_s and we also know that Q_1 is critical. As the parameter is regular, we must have

$$(\mathbf{11.64})\qquad \max(|\mathrm{P}_1|, |\mathrm{Q}_1|) \leqslant \varepsilon_0^{\beta}.$$

Lemma **24.** — *Let $k \geqslant 1$. Assume that $\mathcal{E}^+(\mathrm{P}_0, \dots, \mathrm{P}_{k+1})$ is not-empty. We have*

$$\max(|\mathrm{P}_{k+1}|, |\mathrm{Q}_{k+1}|) \leqslant \mathrm{C}|\mathrm{Q}_k|^{\widetilde{\beta}},$$

with $\widetilde{\beta} = \beta(1-\eta)(1+\tau)^{-1}$.

Proof. — Let I be the largest parameter interval containing t such that

$$(\mathbf{11.65})\qquad \max(|\mathrm{P}_{k+1}|, |\mathrm{Q}_{k+1}|) > |\mathrm{I}|^{\beta}.$$

We first observe that, as $\mathrm{P}_{k+1} \subset \mathrm{P}_s$ and $\mathrm{Q}_{k+1} \subset \mathrm{Q}_u$, we have $\max(|\mathrm{P}_{k+1}|, |\mathrm{Q}_{k+1}|) \leqslant \varepsilon_0^{\beta}$ and I is not the starting interval I_0. Therefore, we have, by definition of I

$$(\mathbf{11.66})\qquad \max(|\mathrm{P}_{k+1}|, |\mathrm{Q}_{k+1}|) \leqslant |\mathrm{I}|^{\frac{\beta}{1+\tau}}.$$

Let $(\widetilde{\mathrm{P}}_{k+1}, \widetilde{\mathrm{Q}}_{k+1}, \widetilde{n}_{k+1}) \in \mathcal{R}(\mathrm{I})$ be the element such that $\widetilde{\mathrm{P}}_{k+1}$ is the thinnest I-defined rectangle containing P_{k+1}. We claim that $\widetilde{\mathrm{P}}_{k+1}$ is I-transverse. Indeed

- If $(\widetilde{\mathrm{P}}_{k+1}, \widetilde{\mathrm{Q}}_{k+1}, \widetilde{n}_{k+1}) = (\mathrm{P}_{k+1}, \mathrm{Q}_{k+1}, n_{k+1})$, this follows from (11.65), as Q_{k+1} is I-critical and I is β-regular.

– If $(\widetilde{P}_{k+1}, \widetilde{Q}_{k+1}, \widetilde{n}_{k+1}) \neq (P_{k+1}, Q_{k+1}, n_{k+1})$, from Corollary 5 (Section 6.6.3), we have that $|Q_{k+1}| \leqslant |I|^{\beta+\frac{1}{3}}$ and therefore $|\widetilde{P}_{k+1}| > |P_{k+1}| > |I|^{\beta}$. As $\widetilde{Q}_{k+1}$ is I-critical (by the structure theorem of Section 6.5), $\widetilde{P}_{k+1}$ must again be I-transverse.

The claim is proved. Next, we observe that Q_k and $\widetilde{P}_{k+1}$ cannot be I-separated, because $G(Q_k \cap L_u) \cap c(P_{k+1})$ contains $g^{m_k+N_0}(\mathcal{E}^+(P_0, \ldots, P_{k+1}))$. On the other hand, there cannot exist an element $(P^*, Q^*, n^*) \in \mathcal{R}(I)$ with $Q^* \supset Q_k$ and $Q^* \pitchfork_I P_{k+1}$: we would have $Q_k \pitchfork_{I'} P_{k+1}$ for small I′ and the corresponding non-simple descendants of P_k would contain $g^{m_{k-1}+N_0}(\mathcal{E}^+(P_0, \ldots, P_{k+1}))$, in contradiction with the definition of $c(P_k)$.

Therefore, the I-transversality of $\widetilde{P}_{k+1}$ implies the existence of $(P', Q', n') \in \mathcal{R}(I)$ with $Q' \subset Q_k$ and $Q' \pitchfork_I \widetilde{P}_{k+1}$. By coherence, Q_k is I-defined. By Proposition 10 (Section 6.4), as Q_k and $\widetilde{P}_{k+1}$ are not I-transverse, we must have $2|Q_k|^{1-\eta} > |I|$.

The estimate of the lemma follows from this and (11.66). □

Taking $\widehat{\beta} < \widetilde{\beta}$ but close to β and ε_0 sufficiently small, the estimate of the lemma and (11.64) give

$$\max(|P_j|, |Q_j|) \leqslant \varepsilon_0^{\widehat{\beta}'}. \quad \textbf{(11.67)}$$

11.8. *Size and area of parabolic cores.*

Proposition **62.** — *Let* $(P, Q, n) \in \mathcal{C}_-$. *With* Leb *standing for Lebesgue measure, we have*

$$\operatorname{diam}(g^n(c(P))) \leqslant C|Q|^{\frac{1}{2}(1-\eta)}, \quad \textbf{(11.68)}$$

$$\operatorname{Leb}(g^n(c(P))) \leqslant C|Q|^{\frac{3}{2}-\frac{1}{2}\eta}, \quad \textbf{(11.69)}$$

$$\operatorname{Leb}(c(P)) \leqslant C|P||Q|^{\frac{1}{2}(1-\eta)}. \quad \textbf{(11.70)}$$

Remark **20.** — A posteriori, $c(P)$, which is contained in $W^s(\Lambda, \widehat{R})$, will have zero Lebesgue measure. However, we estimate here the diameter and Lebesgue measure of a larger set, as will be apparent in the proof.

Proof. — We start with a general observation on an affine-map with implicit representation (A, B). The Jacobian of the map is the product $A_x^{-1}B_y$. The distortion of Lebesgue measure under the map, which is produced by the oscillation of the logarithm of the Jacobian, is, therefore, controlled by the distortion of the affine-like map in the sense of Section 3.2. In particular, the distortion of Lebesgue measure by the restriction of iterates corresponding to the elements of $\mathcal{R}$ is uniformly bounded.

Thus, the third inequality (11.70) in the proposition is a consequence of the second. On the other hand, as $g^n(c(P)) \subset Q$, the second inequality (11.69) is an obvious consequence of the first. We have, therefore, only to prove (11.68). Set

$$Z = g^{n+N_0}(c(P)). \quad \textbf{(11.71)}$$

This set is contained in $g^{N_0}(Q \cap L_u) \cap W^s(\Lambda, \widehat{R})$, and, a fortiori, in P_s. we have to show that $\operatorname{diam} Z \leqslant C|Q|^{\frac{1}{2}(1-\eta)}$.

In order to do this, we will use the machinery introduced in the proof of Theorem 4, Sections 10.10.2 and 10.10.3. We extended the lamination of P_s by stable curves into a foliation by C^{1+Lip} vertical-like curves $\{x_s = \varphi(y_s, \overline{x})\}$ with Lipschitzian holonomy. Then, given a horizontal-like curve $\gamma = \{y_u = \psi(x_u\}$ in Q_u satisfying (10.155), we introduced a function $C(w, \overline{x})$ (vanishing at the points of intersection of $G_0 \circ G_-(\gamma)$ with $G_+(\{x_s = \varphi(y_s, \overline{x})\}))$ and a function $\delta(\overline{x}) = -\min_w C(w, \overline{x})$, whose properties are given in Lemma 19 in Section 10.10.3.

We choose for γ one of the two horizontal-like boundary curves of Q, more precisely the one which gives the greater values of δ (corresponding to δ_L, δ_{LR} in the context of Section 3.6.

Lemma **25.** — *If a curve $\{x_s = \varphi(y_s, \overline{x})\}$ contains a point of Z, then we have*

$$0 \leqslant \delta(\overline{x}) \leqslant C|Q|^{1-\eta}.$$

Proof. — That $\delta(\overline{x}) \geqslant 0$ follows from the choice of γ and the fact that $Z \subset G(Q \cap L_u)$. We prove the other inequality by contradiction, assuming that there is a point $z' = g^{n+N_0}(z) = (\varphi(y_s, \overline{x}), y_s) \in Z$ with $\delta(\overline{x}) \geqslant C_0|Q|^{1-\eta}$, C_0 large.

Let $(P', Q', n') \in \mathcal{R}$ be an element such that $z' \in P'$. We claim that $|P'| \geqslant 3|Q'|$. Indeed, if we had $|P'| < 3|Q'|$, the inequality $\delta(\overline{x}) \geqslant C_0|Q|^{1-\eta}$ with C_0 large enough would directly imply (T1), (T2), (T3) for $Q \,\overline{\pitchfork}_I\, P'$, I small enough. But then z would belong to a descendant of P, in contradiction with the definition of $c(P)$. The claim is proved.

It follows from the claim that there exists a thinnest rectangle P' containing z'. As z' belongs to $W_s(\Lambda)$ but not to any child of P', P' is not I-decomposable for any parameter interval I containing t; therefore P' is I-transverse for I small enough.

As $z' \in G(Q \cap L_u) \cap P'$, Q and P' are not I-separated. We have also already seen that they cannot be I-transverse. Therefore, as P' is I-transverse, there exists $(P^*, Q^*, n^*) \in \mathcal{R}$ with $Q \supset Q^*$ such that Q^* and P' are I-transverse. But then, by Proposition 21 (Section 8.1), it follows from $|P'| \geqslant 3|Q'|$ that $Q \pitchfork_{I'} P'$ holds for I′ small enough, a contradiction. □

We have shown that Z is contained in the lenticular region bounded on one side by $G(\gamma \cap L_u)$ and on the other by the curve $\{x_s = \varphi(y_s, \overline{x}^*)\}$, with $\delta(\overline{x}^*) = C|Q|^{1-\eta}$. The quadratic geometry of C, δ given by Lemma 19 guarantees that this lenticular region has diameter $\leqslant C|Q|^{\frac{1}{2}(1-\eta)}$. □

11.9. *Proof of Theorem 6.* — We will estimate first the Lebesgue measure of each domain $\mathcal{E}^+(P_0, \dots, P_k)$. We have

$$g^{m_{k-1}+N_0}(\mathcal{E}^+(P_0, \dots, P_k)) \subset c(P_k). \tag{11.72}$$

We now use that both the fixed map g^{N_0} and the affine-like iterates $g^{n_j} : P_j \to Q_j$ (for $0 \leqslant j < k$) have uniformly bounded distortion with respect to Lebesgue measure. We are, therefore, able to deduce from (11.69) in Proposition 62 that

$$\text{(11.73)} \qquad \mathrm{Leb}(\mathcal{E}^+(P_0, \dots, P_k)) \leqslant C^{k+1} |Q_k|^{\frac{3}{2} - \frac{1}{2}\eta} \prod_0^k \frac{|P_j|}{|Q_j|}.$$

By Lemma 24, we have $|P_{j+1}| \ll |Q_j|$ for $j > 0$ and it is easy to check that this still holds for $j = 0$ (using (11.64) if $|Q_0| \geqslant \varepsilon_0$; if $Q_0 \subset Q_u$, the argument of Lemma 24 applies). It then follows from (11.73) that we have (for $k > 0$)

$$\text{(11.74)} \qquad \mathrm{Leb}(\mathcal{E}^+(P_0, \dots, P_k)) \ll |P_0||Q_k|^{\frac{1}{2}(1-\eta)}.$$

To obtain the estimate for $\mathcal{E}^+$, we have to sum over sequences $(P_0, \dots, P_k)$. We first estimate, when (P_0, Q_0, n_0) and (P_k, Q_k, n_k) are fixed, how many admissible sequences have these two extremities.

The element $(P_{k-1}, Q_{k-1}, n_{k-1})$ must satisfy

$$\text{(11.75)} \qquad |Q_{k-1}| \geqslant C^{-1} \max(|P_k||Q_k|)^{1/\widetilde{\beta}}.$$

On the other hand, as Q_{k-1}, P_k are neither separated nor transverse (for every parameter interval containing t), we must have, as $|Q_{k-1}| \gg |P_k|$

$$\text{(11.76)} \qquad 0 \leqslant \delta_{LR}(Q_{k-1}, P_k) \leqslant C|Q_{k-1}|^{1-\eta}.$$

For every scale $2^{-l} \geqslant C^{-1} \max(|P_k||Q_k|)^{1/\widetilde{\beta}}$, this will give at most $C2^{l\eta}$ possibilities for Q_{k-1} with $2^{-l} \geqslant |Q_{k-1}| > 2^{-l-1}$. Summing over l gives at most $C|Q_k|^{-\eta/\widetilde{\beta}}$ total possibilities for Q_{k-1}.

We repeat this, with Q_{k-1} now fixed, for $Q_{k-2}, \dots$. We obtain a number of possible admissible sequence not greater than

$$\text{(11.77)} \qquad C^l |Q_k|^{-\eta(\widetilde{\beta}^{-1} + \dots + \widetilde{\beta}^{-l})}$$

with $l \leqslant C \log\log |Q_k|^{-1}$. We conclude that the total number of admissible sequences with fixed extremity Q_k is bounded by $C|Q_k|^{-C\eta}$.

Therefore, we obtain

$$\text{(11.78)} \qquad \mathrm{Leb}(\mathcal{E}^+) \leqslant \sum_{P_0, Q_k} |P_0||Q_k|^{\frac{1}{2} - C\eta}.$$

We first claim that the sum $\sum |P_0|$ is bounded. Actually, this is true even if we do not restrict $(P_0, Q_0, n_0) \in \mathcal{R}$ by the condition that Q_0 is critical. Indeed, when we sum over a given generation (with a fixed number of ascendants), the P_0 are disjoint hence the sum is bounded. But the sum of the widths of the children is always smaller by a definite factor than the width of the parent; therefore, the sum over the various generations is bounded.

Finally, we claim that the sum $\sum_{Q_k} |Q_k|^{\frac{1}{2}-C\eta}$ is arbitrarily small (when k is large). Indeed, it has been shown in Section 11.5.9 that the sum $\sum_{Q \text{ critical}} |Q|^{d^-}$ is convergent provided (cf. (11.48))

$$(\mathbf{11.79}) \qquad d^- \geqslant -\frac{\sigma}{1+\tau} + \tau d_u^* + \tau.$$

Here, the right-hand side is close to $d_s^0 + d_u^0 - 1$ (when $\tau \gg \eta \gg \varepsilon_0$ are small). On the other hand, under condition (H4), the maximum value of $d_s^0 + d_u^0 - 1$ is $1/5$. Therefore, the sum $\sum_{Q \text{ critical}} |Q|^{\frac{1}{2}-C\eta}$ is convergent. But we have

$$(\mathbf{11.80}) \qquad |Q_k| \leqslant \varepsilon_0^{\widetilde{\beta}^k}.$$

We deduce that

$$(\mathbf{11.81}) \qquad \lim_{k\to+\infty} \sum_{Q_k} |Q_k|^{\frac{1}{2}-C\eta} = 0,$$

and this concludes the proof of Theorem 6.

We can sum up the results in Sections 10 and 11 by rephrasing our main result as follows:

Theorem **7.** — *Assume* (H1)–(H4). *Then, for most* $g \in \mathcal{U}_+$, $\Lambda_g \subset W^s(\Lambda_g)$ *and* $\Lambda_g \subset W^u(\Lambda_g)$ *carry geometric invariant measures, à la Sinai-Ruelle-Bowen* [Si], [Ru], [BR], *with non-zero Lyapunov exponents. Both* $W^s(\Lambda_g)$ *and* $W^u(\Lambda_g)$ *have Lebesgue measure zero and thus* Λ_g *carries no attractors nor repellors.*

Appendix A: Composition formulas for affine-like maps

We mostly recall in this appendix the formulas for the simple and parabolic compositions of implicitly defined affine-like maps.

We follow closely [PY2]. The main difference with [PY2] is that we consider maps depending on a parameter t, and we are interested also in some partial derivatives with respect to the parameter.

A.1 *Formulas for simple composition.* — Here we consider a map $F_t : (x_0, y_0) \mapsto (x_1, y_1)$ implicitly defined by

$$(\mathbf{A.1}) \qquad \begin{cases} x_0 = A(y_0, x_1, t), \\ y_1 = B(y_0, x_1, t), \end{cases}$$

and a map $F'_t : (x_1, y_1) \mapsto (x_2, y_2)$ implicitly defined by

$$(\mathbf{A.2}) \qquad \begin{cases} x_1 = A'(y_1, x_2, t), \\ y_2 = B'(y_1, x_2, t). \end{cases}$$

The composition $F''_t = F'_t \circ F_t$ is implicitly defined by

$$(\mathbf{A.3}) \qquad \begin{cases} x_0 = A''(y_0, x_2, t), \\ y_2 = B''(y_0, x_2, t) \end{cases}$$

and we want to relate the partial derivatives of A'', B'' to those of A, B, A', B'. Set

$$(\mathbf{A.4}) \qquad \Delta := 1 - A'_y(y_1, x_2, t) B_x(y_0, x_1, t).$$

When we solve the system (A.1), (A.2) for x_1, y_1, we obtain

$$(\mathbf{A.5}) \qquad \begin{cases} x_1 = X(y_0, x_2, t), \\ y_1 = Y(y_0, x_2, t), \end{cases}$$

where the partial derivatives of X, Y are given by

$$(\mathbf{A.6}) \qquad \begin{cases} X_x = A'_x \Delta^{-1}, \\ X_y = A'_y B_y \Delta^{-1}, \\ X_t = (A'_t + A'_y B_t) \Delta^{-1}, \\ Y_x = A'_x B_x \Delta^{-1}, \\ Y_y = B_y \Delta^{-1}, \\ Y_t = (B_t + A'_t B_x) \Delta^{-1}. \end{cases}$$

We have

$$(\mathbf{A.7}) \qquad \begin{cases} A''(y_0, x_2, t) = A(y_0, X, t), \\ B''(y_0, x_2, t) = B'(Y, x_2, t), \end{cases}$$

which gives

$$(\mathbf{A.8}) \qquad \begin{cases} A''_x = A_x A'_x \Delta^{-1}, \\ B''_y = B'_y B_y \Delta^{-1}, \end{cases}$$

$$(\mathbf{A.9}) \qquad \begin{cases} A''_y = A_y + A_x X_y, \\ B''_x = B'_x + B'_y Y_x, \end{cases}$$

$$\textbf{(A.10)} \qquad \begin{cases} A''_t = A_t + A_x X_t, \\ B''_t = B'_t + B'_y Y_t. \end{cases}$$

Next, from (A.4), we have

$$\textbf{(A.11)} \qquad \begin{cases} -\Delta_x = B_{xx} X_x A'_y + B_x A'_{xy} + B_x A'_{yy} Y_x, \\ -\Delta_y = A'_{yy} Y_y B_x + A'_y B_{xy} + A'_y B_{xx} X_y, \\ -\Delta_t = B_{xt} A'_y + B_{xx} X_t A'_y + B_x A'_{yt} + B_x A'_{yy} Y_t. \end{cases}$$

Taking logarithmic derivatives in (A.8) gives

$$\textbf{(A.12)} \qquad \partial_x \log |A''_x| = \partial_x \log |A'_x| + Y_x \partial_y \log |A'_x| + X_x \partial_x \log |A_x| - \Delta_x \Delta^{-1},$$

$$\textbf{(A.13)} \qquad \partial_y \log |A''_x| = \partial_y \log |A_x| + X_y \partial_x \log |A_x| + Y_y \partial_y \log |A'_x| - \Delta_y \Delta^{-1},$$

$$\textbf{(A.14)} \qquad \begin{aligned} \partial_t \log |A''_x| = \partial_t \log |A_x| + \partial_t \log |A'_x| + X_t \partial_x \log |A_x| \\ + Y_t \partial_y \log |A'_x| - \Delta_t \Delta^{-1}, \end{aligned}$$

$$\textbf{(A.15)} \qquad \partial_y \log |B''_y| = \partial_y \log |B_y| + X_y \partial_x \log |B_y| + Y_y \partial_y \log |B'_y| - \Delta_y \Delta^{-1},$$

$$\textbf{(A.16)} \qquad \partial_x \log |B''_y| = \partial_x \log |B'_y| + Y_x \partial_y \log |B'_y| + X_x \partial_x \log |B_y| - \Delta_x \Delta^{-1},$$

$$\textbf{(A.17)} \qquad \begin{aligned} \partial_t \log |B''_y| = \partial_t \log |B'_y| + \partial_t \log |B_y| + Y_t \partial_y \log |B'_y| \\ + X_t \partial_x \log |B_y| - \Delta_t \Delta^{-1}. \end{aligned}$$

Taking derivatives in (A.9) gives

$$\textbf{(A.18)} \qquad \begin{cases} A''_{yy} = A_{yy} + 2A_{xy} X_y + A_{xx} X_y^2 + A_x X_{yy}, \\ B''_{xx} = B'_{xx} + 2B'_{xy} Y_x + B'_{yy} Y_x^2 + B'_y Y_{xx}, \end{cases}$$

$$\textbf{(A.19)} \qquad \begin{cases} A''_{yt} = A_{yt} + X_t A_{xy} + X_y A_{xt} + X_t X_y A_{xx} + A_x X_{yt}, \\ B''_{xt} = B'_{xt} + Y_t B'_{xy} + Y_x B'_{yt} + Y_t Y_x B'_{yy} + B'_y Y_{xt}, \end{cases}$$

where the partial derivatives of X, Y are obtained from (A.6):

$$\textbf{(A.20)} \qquad X_{yy} = B_y \Delta^{-1} (A'_{yy} Y_y + A'_y \partial_y \log |B_y| + A'_y X_y \partial_x \log |B_y| - A'_y \Delta_y \Delta^{-1}),$$

$$\textbf{(A.21)} \qquad Y_{xx} = A'_x \Delta^{-1} (B_{xx} X_x + B_x \partial_x \log |A'_x| + B_x Y_x \partial_y \log |A'_x| - B_x \Delta_x \Delta^{-1}),$$

$$\textbf{(A.22)} \qquad X_{yt} = B_y \Delta^{-1} (A'_{yy} Y_t + A'_{yt} + A'_y \partial_t \log |B_y| + A'_y X_t \partial_x \log |B_y| - A'_y \Delta_t \Delta^{-1}),$$

$$\textbf{(A.23)} \qquad Y_{xt} = A'_x \Delta^{-1} (B_{xx} X_t + B_{xt} + B_x \partial_t \log |A'_x| + B_x Y_t \partial_y \log |A'_x| - B_x \Delta_t \Delta^{-1}).$$

A.2 *Formulas for parabolic composition.* — We have now a fold map $G_t = G_+ \circ G_0 \circ G_-$:

$$(x_u, y_u) \xrightarrow{G_-} (w, y_u) \xrightarrow{G_0} (x_s, w) \xrightarrow{G_+} (x_s, y_s),$$

with

$$\text{(A.24)} \qquad \begin{cases} y_s = Y_s(w, x_s, t), \\ x_u = X_u(w, y_u, t), \end{cases}$$

$$\text{(A.25)} \qquad w^2 = \theta(y_u, x_s, t).$$

We also have an affine like map $F_0 : (x_0, y_0) \mapsto (x_u, y_u)$ implicitly defined by

$$\text{(A.26)} \qquad \begin{cases} x_0 = A_0(y_0, x_u, t), \\ y_u = B_0(y_0, x_u, t), \end{cases}$$

and another affine-like map $F_1 : (x_1, y_1) \mapsto (x_1, y_1)$ implicitly defined by

$$\text{(A.27)} \qquad \begin{cases} x_s = A_1(y_s, x_1, t), \\ y_1 = B_1(y_s, x_1, t). \end{cases}$$

We assume that (PC1), (PC2) in Section 3.5 are satisfied. As we have seen in [PY2] and Section 3.5, the first step is to write

$$\text{(A.28)} \qquad \begin{cases} x_u = X(w, y_0, t), \\ y_s = Y(w, x_1, t), \end{cases}$$

where the partial derivatives of X, Y are given by

$$\text{(A.29)} \qquad \begin{cases} X_w = X_{u,w} \Delta_0^{-1}, \\ X_y = X_{u,y} B_{0,y} \Delta_0^{-1}, \\ X_t = (X_{u,t} + X_{u,y} B_{0,t}) \Delta_0^{-1}, \\ Y_w = Y_{s,w} \Delta_1^{-1}, \\ Y_x = Y_{s,x} A_{1,x} \Delta_1^{-1}, \\ Y_t = (Y_{s,t} + Y_{s,x} A_{1,t}) \Delta_1^{-1}, \end{cases}$$

$$\text{(A.30)} \qquad \begin{cases} \Delta_0 = 1 - X_{u,y} B_{0,x}, \\ \Delta_1 = 1 - Y_{s,x} A_{1,y}. \end{cases}$$

We set

$$\text{(A.31)}\qquad \begin{cases} \overline{Y}(w, y_0, t) := B_0(y_0, X, t), \\ \overline{X}(w, x_1, t) := A_1(Y, x_1, t), \end{cases}$$

$$\text{(A.32)}\qquad C(w, y_0, x_1, t) := w^2 - \theta(\overline{X}, \overline{Y}, t).$$

The partial derivatives are given by

$$\text{(A.33)}\qquad \begin{cases} \overline{Y}_w = B_{0,x}X_w, \\ \overline{Y}_y = B_{0,y} + B_{0,x}X_y = B_{0,y}\Delta_0^{-1}, \\ \overline{Y}_t = B_{0,t} + B_{0,x}X_t = (B_{0,t} + B_{0,x}X_{u,t})\Delta_0^{-1}, \end{cases}$$

$$\text{(A.34)}\qquad \begin{cases} \overline{X}_w = A_{1,y}Y_w, \\ \overline{X}_x = A_{1,x} + A_{1,y}Y_x = A_{1,x}\Delta_1^{-1}, \\ \overline{X}_t = A_{1,t} + A_{1,y}Y_t = (A_{1,t} + A_{1,y}Y_{s,t})\Delta_1^{-1}, \end{cases}$$

$$\text{(A.35)}\qquad \begin{cases} -C_w = -2w + \theta_x\overline{X}_w + \theta_y\overline{Y}_w, \\ -C_x = \theta_x\overline{X}_x, \\ -C_y = \theta_y\overline{Y}_y, \\ -C_t = \theta_x\overline{X}_t + \theta_y\overline{Y}_t + \theta_t. \end{cases}$$

We solve

$$\text{(A.36)}\qquad C(w, y_0, x_1, t) = 0$$

to define

$$\text{(A.37)}\qquad w = W(y_0, x_1, t)$$

(there are two solutions W^+ and W^-).

The corresponding branch of the parabolic composition is implicitly defined by

$$\text{(A.38)}\qquad \begin{cases} x_0 = A_0(y_0, X(W, y_0, t), t) =: A(y_0, x_1, t), \\ y_1 = B_1(Y(W, x_1, t), x_1, t) =: B(y_0, x_1, t). \end{cases}$$

The partial derivatives of A, B, W are given by

$$\text{(A.39)}\qquad \begin{cases} A_x = A_{0,x}X_wW_x, \\ A_y = A_{0,y} + A_{0,x}(X_y + X_wW_y), \\ A_t = A_{0,t} + A_{0,x}(X_t + X_wW_t), \end{cases}$$

$$\text{(A.40)}\qquad \begin{cases} B_y = B_{1,y} Y_w W_y, \\ B_x = B_{1,x} + B_{1,y}(Y_x + Y_w W_x), \\ B_t = B_{1,t} + B_{1,y}(Y_t + Y_w W_t), \end{cases}$$

$$\text{(A.41)}\qquad \begin{cases} W_x = -C_x C_w^{-1}, \\ W_y = -C_y C_w^{-1}, \\ W_t = -C_t C_w^{-1}. \end{cases}$$

Substituting (A.29), (A.41), (A.35), (A.34) in the formulas (A.39)–(A.40) leads to

$$\text{(A.42)}\qquad \begin{cases} A_x = A_{0,x} A_{1,x} C_w^{-1} \theta_x X_{u,w} \Delta_0^{-1} \Delta_1^{-1}, \\ B_y = B_{1,y} B_{0,y} C_w^{-1} \theta_y Y_{s,w} \Delta_0^{-1} \Delta_1^{-1}, \end{cases}$$

$$\text{(A.43)}\qquad \begin{cases} A_y = A_{0,y} + A_{0,x} B_{0,y} \Delta_0^{-1} (X_{u,y} + X_{u,w} \theta_y \Delta_0^{-1} C_w^{-1}), \\ B_x = B_{1,x} + B_{1,y} A_{1,x} \Delta_1^{-1} (Y_{s,x} + Y_{s,w} \theta_x \Delta_1^{-1} C_w^{-1}), \end{cases}$$

$$\text{(A.44)}\qquad \begin{cases} A_t = A_{0,t} + A_{0,x} \Delta_0^{-1} [X_{u,t} + X_{u,y} B_{0,t} + X_{u,w} C_w^{-1} (\theta_t + \theta_x \overline{X}_t + \theta_y \overline{Y}_t)], \\ B_t = B_{1,t} + B_{1,y} \Delta_1^{-1} [Y_{s,t} + Y_{s,x} A_{1,t} + Y_{s,w} C_w^{-1} (\theta_t + \theta_x \overline{X}_t + \theta_y \overline{Y}_t)]. \end{cases}$$

Taking the logarithmic derivatives in the first formula of (A.39), we obtain

$$\text{(A.45)}\qquad \partial_x \log|A_x| = W_x X_w \partial_x \log|A_{0,x}| + W_x \partial_w \log|X_w| + \partial_x \log|W_x|,$$

$$\text{(A.46)}\qquad \begin{aligned} \partial_y \log|A_x| = {} & \partial_y \log|A_{0,x}| + \partial_x \log|A_{0,x}|(X_y + X_w W_y) \\ & + \partial_y \log|X_w| + W_y \partial_w \log|X_w| + \partial_y \log|W_x|, \end{aligned}$$

$$\text{(A.47)}\qquad \begin{aligned} \partial_t \log|A_x| = {} & \partial_t \log|A_{0,x}| + \partial_x \log|A_{0,x}|(X_t + X_w W_t) \\ & + \partial_t \log|X_w| + W_t \partial_w \log|X_w| + \partial_t \log|W_x|. \end{aligned}$$

From the second formula in (A.39), one gets

$$\text{(A.48)}\qquad \begin{aligned} A_{yy} = {} & A_{0,yy} + 2A_{0,xy}(X_y + X_w W_y) + A_{0,xx}(X_y + X_w W_y)^2 \\ & + A_{0,x}(X_{yy} + 2X_{wy} W_y + X_{ww} W_y^2 + X_w W_{yy}), \end{aligned}$$

$$\text{(A.49)}\qquad \begin{aligned} A_{yt} = {} & A_{0,yt} + A_{0,xy}(X_y + X_w W_t) + A_{0,xt}(X_y + X_w W_y) \\ & + A_{0,xx}(X_t + X_w W_t)(X_y + X_w W_y) \\ & + A_{0,x}(X_{yt} + X_{wy} W_t + X_{wt} W_y + X_{ww} W_y W_t + X_w W_{yt}). \end{aligned}$$

The symmetric formulas for B are

$$\text{(A.50)}\qquad \partial_y \log|B_y| = W_y Y_w \partial_y \log|B_{1,y}| + \partial_y \log|W_y| + W_y \partial_w \log|Y_w|,$$

$$\partial_x \log |B_y| = \partial_x \log |B_{1,y}| + \partial_x \log |Y_w|$$

(A.51) $$+ \partial_x \log |W_y| + \partial_y \log |B_{1,y}|(Y_x + Y_w W_x) + W_x \partial_w \log |Y_w|,$$

$$\partial_t \log |B_y| = \partial_t \log |B_{1,y}| + \partial_y \log |B_{1,y}|(Y_t + Y_w W_t)$$

(A.52) $$+ \partial_t \log |Y_w| + W_t \partial_w \log |Y_w| + \partial_t \log |W_y|,$$

$$B_{xx} = B_{1,xx} + 2B_{1,xy}(Y_x + Y_w W_x) + B_{1,yy}(Y_x + Y_w W_x)^2$$

(A.53) $$+ B_{1,y}(Y_{xx} + 2Y_{wx} W_x + Y_{ww} W_x^2 + Y_w W_{xx}),$$

$$B_{xt} = B_{1,xt} + B_{1,xy}(Y_t + Y_w W_t) + B_{1,yt}(Y_x + Y_w W_x)$$

$$+ B_{1,yy}(Y_x + Y_w W_x)(Y_t + Y_w W_t)$$

(A.54) $$+ B_{1,y}(Y_{xt} + Y_{wx} W_t + Y_{wt} W_x + Y_{ww} W_x W_t + Y_w W_{xt}).$$

In formulas (A.45)–(A.54), the partial derivatives of order 2 of W are obtained from (A.41):

(A.55) $$\begin{cases} W_{xx} = -C_w^{-1}(C_{ww} W_x^2 + 2C_{wx} W_x + C_{xx}), \\ W_{xy} = -C_w^{-1}(C_{ww} W_x W_y + C_{wx} W_y + C_{wy} W_x + C_{xy}), \\ W_{yy} = -C_w^{-1}(C_{ww} W_y^2 + 2C_{wy} W_y + C_{yy}), \\ W_{xt} = -C_w^{-1}(C_{ww} W_x W_t + C_{wx} W_t + C_{wt} W_x + C_{xt}), \\ W_{yt} = -C_w^{-1}(C_{ww} W_y W_t + C_{wy} W_t + C_{wt} W_y + C_{yt}). \end{cases}$$

The partial derivatives of order 2 of C are obtained from (A.35):

(A.56) $$-C_{ww} = -2 + \theta_x \overline{X}_{ww} + \theta_y \overline{Y}_{ww} + \theta_{xx} \overline{X}_w^2 + 2\theta_{xy} \overline{X}_w \overline{Y}_w + \theta_{yy} \overline{Y}_w^2,$$

(A.57) $$\begin{cases} -C_{wx} = \theta_{xx} \overline{X}_w \overline{X}_x + \theta_{xy} \overline{Y}_w \overline{X}_x + \theta_x \overline{X}_{wx}, \\ -C_{wy} = \theta_{yy} \overline{Y}_w \overline{Y}_y + \theta_{xy} \overline{X}_w \overline{Y}_y + \theta_y \overline{Y}_{wy}, \end{cases}$$

$$-C_{wt} = \theta_{xt} \overline{X}_w + \theta_{xx} \overline{X}_w \overline{X}_t + \theta_{xy}(\overline{X}_w \overline{Y}_t + \overline{Y}_w \overline{X}_t) + \theta_{yy} \overline{Y}_w \overline{Y}_t$$

(A.58) $$+ \theta_{yt} \overline{Y}_w + \theta_x \overline{X}_{wt} + \theta_y \overline{Y}_{wt},$$

(A.59) $$\begin{cases} -C_{xx} = \theta_{xx} \overline{X}_x^2 + \theta_x \overline{X}_{xx}, \\ -C_{xy} = \theta_{xy} \overline{X}_x \overline{Y}_y, \\ -C_{yy} = \theta_{yy} \overline{Y}_y^2 + \theta_y \overline{Y}_{yy}, \end{cases}$$

(A.60) $$\begin{cases} -C_{xt} = \theta_{xt} \overline{X}_x + \theta_{xx} \overline{X}_x \overline{X}_t + \theta_{xy} \overline{X}_x \overline{Y}_t + \theta_x \overline{X}_{xt}, \\ -C_{yt} = \theta_{yt} \overline{Y}_y + \theta_{yy} \overline{Y}_y \overline{Y}_t + \theta_{xy} \overline{Y}_y \overline{X}_t + \theta_y \overline{Y}_{yt}. \end{cases}$$

The partial derivatives of order 2 of $\overline{X}$, $\overline{Y}$ are obtained from (A.34):

$$(\mathbf{A.61})\qquad \begin{cases} \overline{X}_{ww} = A_{1,yy}Y_w^2 + A_{1,y}Y_{ww}, \\ \overline{X}_{wx} = A_{1,xy}Y_w + A_{1,yy}Y_wY_x + A_{1,y}Y_{wx}, \\ \overline{X}_{wt} = A_{1,yt}Y_w + A_{1,yy}Y_wY_t + A_{1,y}Y_{wt}, \\ \overline{X}_{xx} = A_{1,xx} + 2A_{1,xy}Y_x + A_{1,yy}Y_x^2 + A_{1,y}Y_{xx}, \\ \overline{X}_{xt} = A_{1,xt} + A_{1,xy}Y_t + A_{1,yt}Y_x + A_{1,yy}Y_xY_t + A_{1,y}Y_{xt}, \end{cases}$$

$$(\mathbf{A.62})\qquad \begin{cases} \overline{Y}_{ww} = B_{0,xx}X_w^2 + B_{0,x}X_{ww}, \\ \overline{Y}_{wy} = B_{0,xy}X_w + B_{0,xx}X_wX_y + B_{0,x}X_{wy}, \\ \overline{Y}_{wt} = B_{0,xt}X_w + B_{0,xx}X_wX_t + B_{0,x}X_{wt}, \\ \overline{Y}_{yy} = B_{0,yy} + 2B_{0,xy}X_y + B_{0,xx}X_y^2 + B_{0,x}X_{yy}, \\ \overline{Y}_{yt} = B_{0,yt} + B_{0,xy}X_t + B_{0,xt}X_y + B_{0,xx}X_yX_t + B_{0,x}X_{yt}. \end{cases}$$

Finally, from (A.29), we obtain

$$(\mathbf{A.63})\qquad \begin{cases} X_{ww} = \Delta_0^{-1}(X_{u,ww} + 2X_{u,wy}\overline{Y}_w + X_{u,yy}\overline{Y}_w^2 + X_{u,y}B_{0,xx}X_w^2), \\ X_{wy} = \Delta_0^{-1}(X_{u,wy}\overline{Y}_y + X_{u,yy}\overline{Y}_w\overline{Y}_y + X_{u,y}X_w(B_{0,xy} + B_{0,xx}X_y)), \\ X_{yy} = \Delta_0^{-1}(X_{u,yy}\overline{Y}_y^2 + X_{u,y}(B_{0,yy} + 2B_{0,xy}X_y + B_{0,xx}X_y^2)), \\ X_{wt} = \Delta_0^{-1}[X_{u,wt} + X_{u,wy}\overline{Y}_t + X_{u,yy}\overline{Y}_w\overline{Y}_t + X_{u,yt}\overline{Y}_w \\ \qquad + X_{u,y}X_w(B_{0,xt} + B_{0,xx}X_t)], \\ X_{yt} = \Delta_0^{-1}[X_{u,yt}\overline{Y}_y + X_{u,yy}\overline{Y}_y\overline{Y}_t \\ \qquad + X_{u,y}(B_{0,yt} + B_{0,xy}X_t + B_{0,xt}X_y + B_{0,xx}X_yX_t)], \end{cases}$$

$$(\mathbf{A.64})\qquad \begin{cases} Y_{ww} = \Delta_1^{-1}(Y_{s,ww} + 2Y_{s,wx}\overline{X}_w + Y_{s,xx}\overline{X}_w^2 + Y_{s,x}A_{1,yy}Y_w^2), \\ Y_{wx} = \Delta_1^{-1}(Y_{s,wx}\overline{X}_x + Y_{s,xx}\overline{X}_w\overline{X}_x + Y_{s,x}Y_w(A_{1,xy} + A_{1,yy}Y_x)), \\ Y_{xx} = \Delta_1^{-1}(Y_{s,xx}\overline{X}_x^2 + Y_{s,x}(A_{1,xx} + 2A_{1,xy}Y_x + A_{1,yy}Y_x^2)), \\ Y_{wt} = \Delta_1^{-1}[Y_{s,wt} + Y_{s,wx}\overline{X}_t + Y_{s,xx}\overline{X}_w\overline{X}_t + Y_{s,xt}\overline{X}_w \\ \qquad + Y_{s,x}Y_w(A_{1,yt} + A_{1,yy}Y_t)], \\ Y_{xt} = \Delta_1^{-1}[Y_{s,xt}\overline{X}_x + Y_{s,xx}\overline{X}_x\overline{X}_t \\ \qquad + Y_{s,x}(A_{1,xt} + A_{1,xy}Y_t + A_{1,yt}Y_x + A_{1,yy}Y_xY_t)]. \end{cases}$$

A.3 *Estimates for simple composition.* — We keep the notations of Section A.1 above and use also the notations $|P|$, $|Q|$, $|P'|$, $|Q'|$ introduced in Section 3.2. Constants depending only on the cone condition satisfied by F_t, F'_t are denoted by C_0, those depending also on the distortions of F_t, F'_t by C.

From (A.6), we get

$$(\mathbf{A.65})\qquad \begin{cases} |X_x| \leqslant C_0|P'|, \\ |X_y| \leqslant C_0|Q|, \\ |Y_x| \leqslant C_0|P'|, \\ |Y_y| \leqslant C_0|Q|, \\ |X_t| \leqslant C_0(|A'_t| + |B_t|), \\ |Y_t| \leqslant C_0(|A'_t| + |B_t|). \end{cases}$$

Then (A.9) gives

$$(\mathbf{A.66})\qquad \begin{cases} |A''_y - A_y| \leqslant C_0|P||Q|, \\ |B''_x - B'_x| \leqslant C_0|P'||Q'|, \end{cases}$$

and (A.10) gives

$$(\mathbf{A.67})\qquad \begin{cases} |A''_t - A_t| \leqslant C_0|P|(|A'_t| + |B_t|), \\ |B''_t - B'_t| \leqslant C_0|Q'|(|A'_t| + |B_t|). \end{cases}$$

Next, (A.11) gives

$$(\mathbf{A.68})\qquad \begin{cases} |\Delta_x| \leqslant C_0|P'|(|B_{xx}| + |A'_{yy}| + |\partial_y \log |A'_x||) \leqslant C|P'|, \\ |\Delta_y| \leqslant C_0|Q|(|B_{xx}| + |A'_{yy}| + |\partial_x \log |B_y||) \leqslant C|Q|, \\ |\Delta_t| \leqslant C(|B_{xt}| + |A'_{yt}| + |A'_t| + |B_t|). \end{cases}$$

From (A.20)–(A.23) we obtain

$$(\mathbf{A.69})\qquad \begin{cases} |X_{yy}| \leqslant C_0|Q|(|A'_{yy}| + |B_{xx}| + |\partial_x \log |B_y|| + |\partial_y \log |B_y||) \leqslant C|Q|, \\ |Y_{xx}| \leqslant C_0|P'|(|B_{xx}| + |A'_{yy}| + |\partial_y \log |A'_x|| + |\partial_x \log |A'_x||) \leqslant C|P'|, \\ |X_{yt}| \leqslant C|Q|(|B_{xt}| + |A'_{yt}| + |A'_t| + |B_t| + |\partial_t \log |B_y||), \\ |Y_{xt}| \leqslant C|P'|(|B_{xt}| + |A'_{yt}| + |A'_t| + |B_t| + |\partial_t \log |A'_x||). \end{cases}$$

Then, from (A.12), (A.13), (A.15), (A.16), we get

$$(\mathbf{A.70})\qquad \begin{cases} |\partial_x \log |A''_x| - \partial_x \log |A'_x|| \leqslant C_0|P'|(D(F) + D(F')), \\ |\partial_y \log |A''_x| - \partial_y \log |A_x|| \leqslant C_0|Q|(D(F) + D(F')), \\ |\partial_x \log |B''_y| - \partial_x \log |B'_y|| \leqslant C_0|P'|(D(F) + D(F')), \\ |\partial_y \log |B''_y| - \partial_y \log |B_y|| \leqslant C_0|Q|(D(F) + D(F')), \end{cases}$$

while (A.18) gives

$$(\mathbf{A.71})\qquad \begin{cases} |A''_{yy} - A_{yy}| \leqslant C_0|P||Q|(D(F) + D(F')), \\ |B''_{xx} - B'_{xx}| \leqslant C_0|P'||Q'|(D(F) + D(F')). \end{cases}$$

Formulas (A.70) and (A.71) give the formula (3.13) for the distortion of a simple composition.

From (A.14), (A.17), we get

$$\textbf{(A.72)}\qquad \begin{cases} |\partial_t \log |A''_x| - \partial_t \log |A'_x| - \partial_t \log |A_x|| \leqslant C(|B_{xt}| + |A'_{yt}| + |A'_t| + |B_t|), \\ |\partial_t \log |B''_y| - \partial_t \log |B'_y| - \partial_t \log |B_y|| \leqslant C(|B_{xt}| + |A'_{yt}| + |A'_t| + |B_t|), \end{cases}$$

while (A.19) gives

$$\textbf{(A.73)}\qquad \begin{cases} |A''_{yt} - A_{yt}| \leqslant C|P|[|A'_t| + |B_t| + |Q|(|B_{xt}| + |A'_{yt}| + \partial_t \log |A_x| + \partial_t \log |B_y|)], \\ |B''_{xt} - B'_{xt}| \leqslant C|Q'|[|A'_t| + |B_t| + |P'|(|B_{xt}| + |A'_{yt}| + \partial_t \log |A'_x| + \partial_t \log |B'_y|)]. \end{cases}$$

Formulas (A.72), (A.73) are used in Section 7.4.

A.4 *Estimates for parabolic composition.* — The context and notations are those of Section A.2. We derive estimates from the formulas in this subsection, assuming that the maps F_0, F_1 satisfy (see (R4) in Section 5.3)

$$\textbf{(A.74)}\qquad \begin{cases} |A_{1,y}| < C\varepsilon_0, & |A_{1,yy}| < C\varepsilon_0, \\ |B_{0,x}| < C\varepsilon_0, & |B_{0,xx}| < C\varepsilon_0. \end{cases}$$

We write δ for $\delta(Q_0, P_1)$. We assume that (see (R7) in Section 5.4)

$$\textbf{(A.75)}\qquad \delta(Q_0, P_1) \geqslant C^{-1}\left(|P_1|^{1-\eta} + |Q_0|^{1-\eta}\right).$$

We first deal with the partial derivatives not involving time. From (A.29), we get

$$\textbf{(A.76)}\qquad \begin{cases} C^{-1} \leqslant |X_w| \leqslant C, \\ C^{-1} \leqslant |Y_w| \leqslant C, \\ |X_y| \leqslant C|Q_0|, \\ |Y_x| \leqslant C|P_1|. \end{cases}$$

Then, from (A.33), (A.34), we obtain

$$\textbf{(A.77)}\qquad \begin{cases} C^{-1}|P_1| \leqslant |\overline{X}_x| \leqslant C|P_1|, \\ C^{-1}|Q_0| \leqslant |\overline{Y}_y| \leqslant C|Q_0|, \\ |\overline{X}_w| \leqslant C\varepsilon_0, \\ |\overline{Y}_w| \leqslant C\varepsilon_0. \end{cases}$$

Putting this into (A.35) gives

$$\textbf{(A.78)}\qquad \begin{cases} C^{-1}|P_1| \leqslant |C_x| \leqslant C|P_1|, \\ C^{-1}|Q_0| \leqslant |C_y| \leqslant C|Q_0|. \end{cases}$$

From (A.63), (A.64), one gets

$$\textbf{(A.79)} \qquad \begin{cases} |X_{ww}| \leqslant C, \\ |Y_{ww}| \leqslant C, \end{cases}$$

and then, from (A.61), (A.62)

$$\textbf{(A.80)} \qquad \begin{cases} |\overline{X}_{ww}| \leqslant C\varepsilon_0, \\ |\overline{Y}_{ww}| \leqslant C\varepsilon_0. \end{cases}$$

Taking (A.77), (A.80) into (A.56) gives

$$\textbf{(A.81)} \qquad |C_{ww} - 2| \leqslant C\varepsilon_0.$$

Fix y_0, x_1 and denote by w^* the point where C takes its minimal value $\overline{C}(y_0, x_1)$. We have

$$\textbf{(A.82)} \qquad \begin{cases} |C_w(w, y_0, x_1) - 2(w - w^*)| \leqslant C\varepsilon_0|w - w^*|, \\ |C(w, y_0, x_1) - (w - w^*)^2 - \overline{C}(y_0, x_1)| \leqslant C\varepsilon_0|w - w^*|^2. \end{cases}$$

This gives, as $C(W(y_0, x_1), y_0, x_1) = 0$,

$$\textbf{(A.83)} \qquad ||C_w(W(y_0, x_1), y_0, x_1)| - 2|\overline{C}(y_0, x_1)|^{1/2}| \leqslant C\varepsilon_0|\overline{C}(y_0, x_1)|^{1/2}.$$

From (A.75), (A.78) and the definition of $\delta = \min -C$, we have thus

$$\textbf{(A.84)} \qquad C^{-1}\delta^{1/2} \leqslant |C_w| \leqslant C\delta^{1/2}.$$

Formulas (3.27), (3.28) of Section 3.5 are then a consequence of (A.42).

From (A.41), we now get

$$\textbf{(A.85)} \qquad \begin{cases} C^{-1}|P_1|\delta^{-1/2} \leqslant |W_x| \leqslant C|P_1|\delta^{-1/2}, \\ C^{-1}|Q_0|\delta^{-1/2} \leqslant |W_y| \leqslant C|Q_0|\delta^{-1/2}. \end{cases}$$

Plugging into (A.39), (A.40) gives

$$\textbf{(A.86)} \qquad \begin{cases} |A_y - A_{0,y}| \leqslant C|P_0||Q_0|\delta^{-1/2}, \\ |B_x - B_{1,x}| \leqslant C|P_1||Q_1|\delta^{-1/2}. \end{cases}$$

Next, from (A.63), (A.64), we have

$$\textbf{(A.87)} \qquad \begin{cases} |X_{wy}| \leqslant C|Q_0|, \\ |X_{yy}| \leqslant C|Q_0|, \\ |Y_{wx}| \leqslant C|P_1|, \\ |Y_{xx}| \leqslant C|P_1|, \end{cases}$$

and then, from (A.61), (A.62)

$$(\mathbf{A.88})\qquad \begin{cases} |\overline{X}_{wx}| \leqslant C|P_1|, \\ |\overline{X}_{xx}| \leqslant C|P_1|, \\ |\overline{Y}_{wy}| \leqslant C|Q_0|, \\ |\overline{Y}_{yy}| \leqslant C|Q_0|. \end{cases}$$

Plugging into (A.57), (A.59) gives

$$(\mathbf{A.89})\qquad \begin{cases} |C_{wy}| \leqslant C|Q_0|, \\ |C_{yy}| \leqslant C|Q_0|, \\ |C_{wx}| \leqslant C|P_1|, \\ |C_{xx}| \leqslant C|P_1|, \\ |C_{xy}| \leqslant C|P_1||Q_0|. \end{cases}$$

We can now estimate the second partial derivatives of W from (A.55)

$$(\mathbf{A.90})\qquad \begin{cases} |W_{yy}| \leqslant C|Q_0|\delta^{-1/2}, \\ |W_{xx}| \leqslant C|P_1|\delta^{-1/2}, \\ |W_{xy}| \leqslant C|P_1||Q_0|\delta^{-1/2}, \end{cases}$$

which finally gives from (A.48), (A.53)

$$(\mathbf{A.91})\qquad \begin{cases} |A_{yy} - A_{0,yy}| \leqslant C|P_0||Q_0|\delta^{-1/2}, \\ |B_{xx} - B_{1,xx}| \leqslant C|P_1||Q_1|\delta^{-1/2}. \end{cases}$$

These estimates are used in Section 7.5. We now turn to derivatives involving the parameter. First, from (A.29), we get

$$(\mathbf{A.92})\qquad \begin{cases} |X_t| \leqslant C(1 + |B_{0,t}|), \\ |Y_t| \leqslant C(1 + |A_{1,t}|), \end{cases}$$

and then, from (A.33), (A.34)

$$(\mathbf{A.93})\qquad \begin{cases} |\overline{X}_t| \leqslant C(\varepsilon_0 + |B_{0,t}|), \\ |\overline{Y}_t| \leqslant C(\varepsilon_0 + |A_{1,t}|). \end{cases}$$

Plugging this into (A.35) gives

$$(\mathbf{A.94})\qquad |C_t| \leqslant C(1 + |B_{0,t}| + |A_{1,t}|),$$

and then, from (A.41)

$$(\mathbf{A.95}) \qquad |W_t| \leqslant C\delta^{-1/2}(1 + |B_{0,t}| + |A_{1,t}|).$$

From (A.44), one then concludes that

$$(\mathbf{A.96}) \qquad \begin{cases} |A_t - A_{0,t}| \leqslant C|P_0|\delta^{-1/2}(1 + |B_{0,t}| + |A_{1,t}|), \\ |B_t - B_{1,t}| \leqslant C|Q_1|\delta^{-1/2}(1 + |B_{0,t}| + |A_{1,t}|). \end{cases}$$

From now on, we assume that the estimates of Proposition 17 in Section 7.6 are satisfied, i.e.

$$(\mathbf{A.97}) \qquad \begin{cases} |A_{1,t}| \leqslant C\varepsilon_0^{1/2}, \\ |B_{0,t}| \leqslant C\varepsilon_0^{1/2}. \end{cases}$$

Observe then that in (A.92), (A.95), we get the estimates

$$(\mathbf{A.98}) \qquad \begin{cases} |X_t| \leqslant C, \\ |Y_t| \leqslant C, \\ |\overline{X}_t| \leqslant C\varepsilon_0^{1/2}, \\ |\overline{Y}_t| \leqslant C\varepsilon_0^{1/2}, \\ |C_t| \leqslant C, \\ |W_t| \leqslant C\delta^{-1/2}. \end{cases}$$

From (A.63), (A.64), we obtain

$$(\mathbf{A.99}) \qquad \begin{cases} |X_{wt}| \leqslant C(1 + |B_{0,xt}|), \\ |X_{yt}| \leqslant C|Q_0|(1 + |B_{0,xt}| + |\partial_t \log|B_{0,y}|), \\ |Y_{wt}| \leqslant C(1 + |A_{1,yt}|), \\ |Y_{xt}| \leqslant C|P_1|(1 + |A_{1,yt}| + |\partial_t \log|A_{1,x}|), \end{cases}$$

and then, from (A.61), (A.62)

$$(\mathbf{A.100}) \qquad \begin{cases} |\overline{Y}_{wt}| \leqslant C(\varepsilon_0 + |B_{0,xt}|), \\ |\overline{Y}_{yt}| \leqslant C|Q_0|(1 + |B_{0,xt}| + |\partial_t \log|B_{0,y}|), \\ |\overline{X}_{wt}| \leqslant C(\varepsilon_0 + |A_{1,yt}|), \\ |\overline{X}_{xt}| \leqslant C|P_1|(1 + |A_{1,yt}| + |\partial_t \log|A_{1,x}|). \end{cases}$$

Taking this into (A.58), (A.60) gives

$$(\mathbf{A.101}) \qquad \begin{cases} |C_{wt}| \leqslant C(\varepsilon_0 + |B_{0,xt}| + |A_{1,yt}|), \\ |C_{yt}| \leqslant C|Q_0|(1 + |B_{0,xt}| + |\partial_t \log|B_{0,y}|), \\ |C_{xt}| \leqslant C|P_1|(1 + |A_{1,yt}| + |\partial_t \log|A_{1,x}|). \end{cases}$$

We can now estimate the partial derivatives of W from (A.55)

$$\textbf{(A.102)}\qquad \begin{cases} |W_{xt}| \leqslant C|P_1|\delta^{-1/2}[\delta^{-1}+\delta^{-1/2}(|B_{0,xt}|+|A_{1,yt}|)+|\partial_t \log|A_{1,x}||], \\ |W_{yt}| \leqslant C|Q_0|\delta^{-1/2}[\delta^{-1}+\delta^{-1/2}(|B_{0,xt}|+|A_{1,yt}|)+|\partial_t \log|B_{0,y}||]. \end{cases}$$

Finally, we obtain from (A.49), (A.54)

$$\textbf{(A.103)}\qquad \begin{cases} |A_{yt}-A_{0,yt}| \leqslant C|P_0|\delta^{-1/2}+C|P_0||Q_0|\delta^{-1/2}[\delta^{-1}+\delta^{-1/2}(|B_{0,xt}|+|A_{1,yt}|) \\ \qquad\qquad +|\partial_t \log|B_{0,y}|+|\partial_t \log|A_{0,x}||], \\ |B_{xt}-B_{1,xt}| \leqslant C|Q_1|\delta^{-1/2}+C|P_1||Q_1|\delta^{-1/2}[\delta^{-1}+\delta^{-1/2}(|B_{0,xt}|+|A_{1,yt}|) \\ \qquad\qquad +|\partial_t \log|B_{1,y}|+|\partial_t \log|A_{1,x}||], \end{cases}$$

and from (A.47), (A.52)

$$\textbf{(A.104)}\qquad \begin{cases} |\partial_t \log|A_x|-\partial_t \log|A_{0,x}|| \leqslant C[\delta^{-1}+\delta^{-1/2}(|B_{0,xt}|+|A_{1,yt}|)+|\partial_t \log|A_{1,x}||], \\ |\partial_t \log|B_y|-\partial_t \log|B_{1,y}|| \leqslant C[\delta^{-1}+\delta^{-1/2}(|B_{0,xt}|+|A_{1,yt}|)+|\partial_t \log|B_{0,y}||]. \end{cases}$$

These formulas are used in Section 7.7.

Appendix B: On the Lipschitz regularity of $\widetilde{R}_+^\infty$

The goal of this Appendix is to perform some calculations that are used in Section 10.5, Proposition 51.

We recall the setting We have rectangles $R_0 = I_0^s \times I_0^u$, $R_{a_s} = I_u^s \times I_u^u$, $R_{a_u} = I_s^s \times I_s^u$ with respective coordinates (x_0, y_0), (x_u, y_u), (x_s, y_s).

We have an affine-like map F with domain $P \subset R_0$, image $Q \subset Q_u \subset R_{a_u}$, and implicit representation (A, B). We assume as usual that the cone condition is satisfied and that the distortion is bounded by C. We also have a folding map G with domain $L_u \subset I_u^s \times I_u^u$, image $L_s \subset I_s^s \times I_s^u$. The map G is implicitly defined by the system

$$\textbf{(B.1)}\qquad \begin{cases} y_s = Y_s(w, x_s), \\ x_u = X_u(w, y_u), \\ w^2 = \theta(y_u, x_s) \end{cases}$$

with Y_s, X_u, θ as in Sections 2.3 and 3.5. One has (cf. (R4) in Section 5.3):

$$\textbf{(B.2)}\qquad |B_x| < C\varepsilon_0, \qquad |B_{xx}| < C\varepsilon_0.$$

Consider a vertical-like C^2 curve $\omega = \{x_s = \varphi(y_s)\} \subset P_s \subset R_{a_s}$ intersecting L_s and satisfying

$$\textbf{(B.3)}\qquad \left|\frac{\partial\varphi}{\partial y}\right| < C\varepsilon_0, \qquad \left|\frac{\partial^2\varphi}{\partial y^2}\right| < C\varepsilon_0.$$

As in Section 3.5 (formulas (3.14), (3.15)) we eliminate y_u in the system $x_u = X_u(w, y_u)$, $y_u = B(y_0, x_u)$ to write

(**B.4**) $$x_u = X(w, y_0).$$

Similarly, in the equation

(**B.5**) $$y_s = Y_s(w, \varphi(y_s)),$$

we solve for y_s to write

(**B.6**) $$y_s = Y(w).$$

We set

(**B.7**) $$C(w, y_0) = w^2 - \theta(B(y_0, X(w, y_0)), \varphi(Y(w))).$$

We assume that, for all $y_0 \in I_0^u$

(**B.8**) $$\overline{C}(y_0) := \min_w C(w, y_0) \leqslant -C^{-1}|Q|^{1-\eta}.$$

Suppose now that we have two vertical-like C^2 curves $\omega_i = \{x_s = \varphi_i(y_s)\}$, $i = 0, 1$ as above satisfying (B.3), (B.8). We also assume that $\varphi_0(y_s) \neq \varphi_1(y_s)$ for all y_s and that we have for some $T > 0$, everywhere on I_s^u

(**B.9**) $$\left| \frac{\partial}{\partial y} \log |\varphi_1 - \varphi_0| \right| \leqslant T.$$

For $i = 0, 1$, let $\Omega_i^{\pm}$ be the connected components of $F^{-1}(Q \cap G^{-1}(\omega_i \cap L_s))$.

Proposition. — The curves $\Omega_i^{\pm}$ are graphs

(**B.10**) $$\Omega_i^{\pm}(s) = \{x_0 = \Phi_i^{\pm}(y_0)\}$$

which satisfy, everywhere on I_0^u

(**B.11**) $$\left| \frac{\partial}{\partial y} \log |\Phi_1^+ - \Phi_0^+| \right| \leqslant T',$$

(**B.12**) $$\left| \frac{\partial}{\partial y} \log |\Phi_1^- - \Phi_0^-| \right| \leqslant T',$$

(**B.13**) $$\left| \frac{\partial}{\partial y} \log |\Phi_1^+ - \Phi_0^-| \right| \leqslant T'',$$

(**B.14**) $$\left| \frac{\partial}{\partial y} \log |\Phi_1^- - \Phi_0^+| \right| \leqslant T'',$$

with $T' = C(1 + |Q|^{\frac{1}{2}} T)$, $T'' = C$.

Remark. — In (B.13), we allow $\omega_0 = \omega_1$.

Proof. — Let C_i, $i = 0, 1$, be the function defined by (B.7) for ω_i. One solves $C_i = 0$ to get

$$w = W_i^{\pm}(y_0), \tag{B.15}$$

$$\Phi_i^{\pm}(y_0) = A(y_0, X(W_i^{\pm}(y_0), y_0)). \tag{B.16}$$

Let $\varepsilon, \varepsilon' \in \{+, -\}$, $W_0 = W_0^{\varepsilon}$, $W_1 = W_1^{\varepsilon'}$, $\Phi_0 = \Phi_0^{\varepsilon}$, $\Phi_1 = \Phi_1^{\varepsilon'}$. Define, for $s \in [0, 1]$

$$W_s = sW_1 + (1 - s)W_0. \tag{B.17}$$

One has

$$\begin{aligned}(\Phi_1 - \Phi_0)(y_0) \\ = (W_1 - W_0)(y_0) \int_0^1 A_x(y_0, X(W_s(y_0), y_0)) X_w(W_s(y_0), y_0) ds.\end{aligned} \tag{B.18}$$

Set

$$a_s(y_0) = A_x(y_0, X(W_s(y_0), y_0)) X_w(W_s(y_0), y_0). \tag{B.19}$$

From (A.76), we have, for all s, y_0

$$C^{-1}|P| \leqslant |a_s(y_0)| \leqslant C|P|. \tag{B.20}$$

One has

$$\begin{aligned}\frac{\partial}{\partial y} \log |a_s| = \partial_y \log |A_x| + \partial_x \log |A_x| \left(X_y + X_w \frac{\partial W_s}{\partial y} \right) \\ + X_w^{-1} \left(X_{wy} + X_{ww} \frac{\partial W_s}{\partial y} \right).\end{aligned} \tag{B.21}$$

From (A.76), (A.79), (A.87) we get

$$\left| \frac{\partial}{\partial y} \log |a_s| \right| \leqslant C \left(1 + \left| \frac{\partial W_s}{\partial y} \right| \right). \tag{B.22}$$

But we see as for (A.85) that we have, for $i = 0, 1$

$$\left| \frac{\partial W_i}{\partial y} \right| \leqslant C|Q|\delta_i^{-1/2}, \tag{B.23}$$

where $\delta_i := -\max_{y_0} \min_w C(w, y_0)$. We have assumed that

$$\delta_i \geqslant C^{-1}|Q|^{1-\eta}. \tag{B.24}$$

It then follows from (B.22) that the logarithmic derivative of $|a_s|$ is bounded by C. From (B.20), the same is true for the logarithmic derivative of $|\int_0^1 a_s ds|$. To prove the estimates of the proposition, in view of (B.18), it remains to see that the logarithmic derivative of $|W_1 - W_0|$ is bounded by $C(1 + |Q|^{\frac{1}{2}}T)$ or C (depending whether ε and ε' are equal or not).

1. We assume first that $\varepsilon \neq \varepsilon'$.

In this case, from (B.24) and the quadratic behavior of C_i (cf. (A.82)), we have

(B.25) $$|W_1 - W_0| \geqslant C^{-1}|Q|^{\frac{1}{2}(1-\eta)}.$$

Combining this with (B.23), (B.24) shows that the logarithmic derivative of $|W_1 - W_0|$ is bounded by C.

2. We assume now that $\varepsilon = \varepsilon'$.

We may assume that

(B.26) $$|W_1 - W_0| \ll |Q|^{\frac{1}{2}(1-\eta)},$$

otherwise we conclude as the first case. We interpolate between the two curves ω_0, ω_1 defining

(B.27) $$\varphi(y_s, s) = s\varphi_1(y_s) + (1-s)\varphi_0(y_s).$$

We have

(B.28) $$\left|\frac{\partial}{\partial y}\log\left|\frac{\partial \varphi}{\partial s}\right|\right| \leqslant T.$$

In the equation

(B.29) $$y_s = Y_s(w, \varphi(y_s, s)),$$

we solve for y_s to write

(B.30) $$y_s = Y(w, s).$$

We then set

(B.31) $$\overline{X}(w, s) := \varphi(Y(w, s), s),$$

(B.32) $$C(w, y_0, s) := w^2 - \theta(\overline{Y}(w, y_0), \overline{X}(w, s)).$$

Then, one solves $C = 0$ to get

(B.33) $$w = W^{\varepsilon}(y_0, s).$$

One has

(B.34)
$$\frac{\partial W^{\varepsilon}}{\partial s} = -\frac{\partial C}{\partial s} C_w^{-1} = C_w^{-1}\theta_x\Big(\frac{\partial \varphi}{\partial s} + \frac{\partial \varphi}{\partial y}\frac{\partial Y}{\partial s}\Big).$$

From (B.29), (B.30), one obtains

(B.35)
$$\frac{\partial Y}{\partial s} = Y_{s,x}\frac{\partial \varphi}{\partial s}\Big(1 - Y_{s,x}\frac{\partial \varphi}{\partial y}\Big)^{-1},$$

where, here and below $Y_{s,x}$ stands for the partial derivative of Y_s with respect to x (and not the second partial derivative of Y!). Plugging this into (B.34), one obtains

(B.36)
$$\frac{\partial W^{\varepsilon}}{\partial s} = C_w^{-1}\theta_x\frac{\partial \varphi}{\partial s}\Big(1 - Y_{s,x}\frac{\partial \varphi}{\partial y}\Big)^{-1}.$$

We now take the logarithmic derivative of this product of four terms with relation to y_0. We obtain a sum of four terms:

(B.37)
$$Z_1 = -C_w^{-1}[C_{ww}W_y + C_{wy}],$$

(B.38)
$$Z_2 = \partial_y \log|\theta_x|(\overline{Y}_y + \overline{Y}_w W_y) + \partial_x \log|\theta_x|\overline{X}_w W_y,$$

(B.39)
$$Z_3 = \Big(\frac{\partial}{\partial y}\log\Big|\frac{\partial \varphi}{\partial s}\Big|\Big)Y_w W_y,$$

(B.40)
$$Z_4 = \Big(1 - Y_{s,x}\frac{\partial \varphi}{\partial y}\Big)^{-1}\Big\{Y_{s,x}\frac{\partial^2 \varphi}{\partial y^2}Y_w W_y + \frac{\partial \varphi}{\partial y}(Y_{s,xw}W_y + Y_{s,xx}\overline{X}_w W_y)\Big\}.$$

The partial derivatives of X have been estimated in Appendix A (formulas (A.76), (A.79), (A.87))

(B.41)
$$|X_w| \leqslant C, \qquad |X_y| \leqslant C|Q|, \qquad |X_{ww}| \leqslant C, \qquad |X_{wy}| \leqslant C|Q|.$$

The partial derivatives of Y are estimated by

(B.42)
$$|Y_w| \leqslant C, \qquad |Y_{ww}| \leqslant C.$$

Then we get

(B.43)
$$|\overline{X}_w| \leqslant C\varepsilon_0, \qquad |\overline{Y}_w| \leqslant C\varepsilon_0, \qquad |\overline{Y}_y| \leqslant C|Q|,$$

(B.44)
$$|\overline{X}_{ww}| \leqslant C\varepsilon_0, \qquad |\overline{Y}_{ww}| \leqslant C\varepsilon_0, \qquad |\overline{Y}_{wy}| \leqslant C|Q|,$$

(B.45)
$$|C_{wy}| \leqslant C|Q|, \qquad |C_{ww} - 2| \leqslant C\varepsilon_0,$$

(B.46)
$$|W_y| \leqslant C|Q||C_w^{-1}|.$$

From (B.26), we have $|\delta_0 - \delta_1| \ll \delta_0$, and therefore we have

$$\textbf{(B.47)} \qquad |C_w^{-1}(W^\varepsilon(y_0, s), y_0, s)| \leqslant C\delta_0^{-1/2},$$

for all y_0, s. This gives, as $\delta_0 \geqslant C^{-1}|Q|^{1-\eta}$

$$\textbf{(B.48)} \qquad |Z_i| \leqslant C \quad \text{for } i = 1, 2, 4;$$

$$\textbf{(B.49)} \qquad |Z_3| \leqslant C|Q|^{1/2}T.$$

This proves that

$$\textbf{(B.50)} \qquad \left|\frac{\partial}{\partial y}\log\left|\frac{\partial W^\varepsilon}{\partial s}\right|\right| \leqslant C(1 + |Q|^{1/2}T).$$

But then, we have

$$\begin{aligned}
\left|\frac{\partial}{\partial y}(W_1^\varepsilon - W_0^\varepsilon)\right| &= \left|\frac{\partial}{\partial y}\int_0^1 \frac{\partial W^\varepsilon}{\partial s} ds\right| \\
&\leqslant C(1 + |Q|^{1/2}T)\int_0^1 \left|\frac{\partial W^\varepsilon}{\partial s}\right| ds \\
&= C(1 + |Q|^{1/2}T)|W_1^\varepsilon - W_0^\varepsilon|,
\end{aligned}$$

as $\frac{\partial W^\varepsilon}{\partial s}$ has constant sign.

The required estimate on the logarithmic derivative of $|W_1 - W_0|$ has been obtained in both cases. The proof of the proposition is complete. □

Appendix C: A toy model for the transversality relation

C.1 Our goal in this appendix is to explain why the complicated definition of the transversality relation in Section 5.4, is in some way "natural", if we require some useful properties for the proof of our Main Theorem. The toy model that we are considering is an abstract one. It is much simpler than the real situation of Section 5 because the sets in which the relation takes place are well defined to begin with: in Section 5, we need to know the transversality relation in order to construct the classes $\mathcal{R}(I)$.

C.2 A partially ordered set X is a *forest* if, for any $x_0 \in X$, the set $\{x \geqslant x_0\}$ is finite and totally ordered. A *tree* is a forest with a single maximal element. Let $X_1, \ldots, X_n$ be forests and let A be a subset of $X = X_1 \times \cdots \times X_n$ (one should think of A as the graph of an *n-ary* relation). We say that A is *hereditary* if whenever $x = (x_1, \ldots, x_n), y = (y_1, \ldots, y_n)$ are such that and $y_i \leqslant x_i$ for all $1 \leqslant i \leqslant n$ (abbreviated as $y \leqslant x$), then $y \in A$ if $x \in A$.

Two points $x = (x_1, \ldots, x_n)$, $y = (y_1, \ldots, y_n)$ of X are *coordinate-wise comparable* (*c-comparable* for short) if for each $i \in \{1, \ldots, n\}$, we have $x_i \geqslant y_i$ or $x_i \leqslant y_i$. In this case, we set

$$x \vee y = (\max(x_i, y_i))_{1 \leqslant i \leqslant n}. \tag{C.1}$$

The set A is concave if, whenever $x, y \in A$ are *c*-comparable, then the point $x \vee y$ also belongs to A.

The intersection of hereditary, resp. concave, subsets of X is hereditary, resp. concave. It follows that any subset $A \subset X$ is contained in a smallest concave hereditary subset, called the *c.h*-envelope and denoted by $\widehat{A}$.

Example. — When the number of factors $n = 1$, any subset is concave: the *c.h*-envelope of $A \subset X_1$ is the set of $x \in X_1$ such that $x \leqslant y$ for some $y \in A$.

C.3 We construct the *c.h*-envelope when $n = 2$.

Proposition. — *Let* X_1, X_2 *be forests and* A *be a subset of* $X = X_1 \times X_2$. *Let* A_1 *be the set of* $x \in X$ *such that* $x = y \vee z$ *for some c-comparable* $y, z \in A$. *The c.h-envelope of* A *is equal to the set* A_2 *of* $t = (t_1, t_2)$ *such that* $t_1 \leqslant x_1$, $t_2 \leqslant x_2$ *for some* $x = (x_1, x_2) \in A_1$.

Proof. — It is clear that the set A_2 defined in the proposition is hereditary and it is contained in the *c.h*-envelope $\widehat{A}$ of A. We have to prove that A_2 is concave. We first prove the □

Lemma. — *If* $y, z \in A_1$ *are c-comparable,* $y \vee z$ *also belongs to* A_1.

Proof. — By the definition of A_1, we can write $y = y' \vee y''$, $z = z' \vee z''$ with y', y'', z', z'' in A, y' and y'' *c*-comparable, z' and z'' *c*-comparable. We may assume that $y_1 \leqslant z_1$ and $y_2 \geqslant z_2$, and also $z_1 = z'_1, y_2 = y'_2$; then, we have $z'_1 = z_1 \geqslant y_1 \geqslant y'_1$ and $y'_2 = y_2 \geqslant z_2 \geqslant z'_2$, hence y', z' are *c*-comparable with $y' \vee z' = y \vee z$. □

End of the Proof of the Proposition. — Let $t', t'' \in A_2$ be *c*-comparable and let $x', x'' \in A_1$ be such that $x'_i \geqslant t'_i$, $x''_i \geqslant t''_i$ for $i = 1, 2$. As X_1 and X_2 are forests, x' and x'' are *c*-comparable. From the lemma, $x' \vee x''$ belongs to A_1; then $t' \vee t''$ belongs to A_2. □

C.4 For $n \geqslant 3$, the situation is more complicated, as the two examples below indicate.

Example. — Let X_1, X_2, X_3 be forests and let x, y, z be three points of $X = X_1 \times X_2 \times X_3$ such that

$$x_1 \geqslant y_1 \geqslant z_1, \tag{C.2}$$

(**C.3**) $y_2 \geqslant x_2, \quad y_2 \geqslant z_2,$

(**C.4**) $z_3 \geqslant x_3, \quad z_3 \geqslant y_3.$

Let $A = \{x, y, z\} \subset X$. If we define, as in the proposition above,

(**C.5**) $A_1 = \{u \vee v,\ u, v \in A,\ u, v$ c-comparable$\}$

and if we assume that x_2, z_2 are *not* comparable and x_3, y_3 are *not* comparable, then we have

(**C.6**) $A_1 = \{x, y, z, y \vee z = (y_1, y_2, z_3)\}.$

On the other hand, the point $w = (x_1, y_2, z_3) = x \vee (y \vee z)$ certainly belongs to the $c.h$-envelope of A, but does not satisfy $w_i \leqslant u_i$ $(i = 1, 2, 3)$ for any $u \in A_1$. This example shows that the analogue of the proposition above is false for $n = 3$.

Example. — Let X_1, X_2, X_3 forests and let $x, y, z \in X = X_1 \times X_2 \times X_3$ such that

(**C.7**) $x_1 \geqslant y_1, \quad x_1 \geqslant z_1,$

(**C.8**) $y_2 \geqslant x_2, \quad y_2 \geqslant z_2,$

(**C.9**) $z_3 \geqslant x_3, \quad z_3 \geqslant y_3,$

but none of the pairs (y_1, z_1), (x_2, z_2), (x_3, y_3) is made of comparable elements. Let $A = \{x, y, z\}$. The sets $\{u \leqslant x\}$, $\{v \leqslant y\}$, $\{w \leqslant z\}$ are disjoint and their union is the $c.h$-envelope of A: any $u \leqslant x$, $v \leqslant y$ cannot be c-comparable; otherwise, as X_3 is a forest and x, y are larger than $u \wedge v = (\min(u_i, v_i))$, x_3 and y_3 would be comparable. On the other hand, if $u' \leqslant x$, $u'' \leqslant x$ and u', u'' c-comparable, then $u' \vee u'' \leqslant x$.

C.5 We have the following partial result:

Proposition. — *Let* $X_1, \ldots, X_n$ *be forests and let* A *be a subset of* $X = X_1 \times \cdots \times X_n$. *Let* A_1 *be the set of elements* $x \in X$ *for which there exists* $x^1, x^2, \ldots, x^n$ *in* A *with* $x_j = x^j_j \geqslant x^i_j$ *for all* $1 \leqslant i, j \leqslant n$. *Let* A_2 *be the set of elements* $y \in X$ *such that* $y \leqslant x$ *for some* x *in* A_1. *Then* A_2 *is hereditary and concave.*

Remark. — Example 2 above shows that A_2 can be strictly larger than the $c.h$-envelope. Example 1 shows that the straightforward generalization of the case $n = 2$ does not work.

Proof of the Proposition. — It is very similar to the proof of the proposition in Section C.3 above and left to the reader. □

C.6 We will now see how the definition of the transversality relation in Section 5.4 is a natural consequence of the proposition above. As observed earlier, an essential difference with the toy model is that the transversality relation is used to construct the classes $\mathcal{R}(I)$. So, let us just try to define the relation for the starting class $\mathcal{R}(I_0)$ associated to the initial horseshoe K. We would have:

(**C.10**) $$X_1 = \{(P, Q, n) \in \mathcal{R}(I_0), Q \subset Q_u\},$$

(**C.11**) $$X_2 = \{(P', Q', n') \in \mathcal{R}(I_0), P \subset P_s\},$$

and X_3 is the set of parameter intervals. All sets are partially ordered by inclusion (of the Q's for X_1, of the P's for X_2), and are obviously *trees*, with respective roots (P_u, Q_u, n_u), (P_s, Q_s, n_s), I_0.

We start from an intuitive definition of transversality: for $(P, Q, n) \in X_1$, $(P', Q', n') \in X_2$, $I \in X_3$, we write

$$Q \,\widehat{\pitchfork}_I\, P'$$

if for all $t \in I$ we have

(**C.12**) $$\delta(Q, P') \geqslant 2\max(I, |Q|^{1-\eta}, |P'|^{1-\eta}).$$

(The number η in the exponent is necessary in order to keep the distortions under control.)

The corresponding subset of $X_1 \times X_2 \times X_3$ is

(**C.13**) $$A = \{(Q, P', I), Q \,\widehat{\pitchfork}_I\, P'\}.$$

This set is hereditary but it is not, a priori, concave. The concavity property (Proposition 9 in Section 6.3) is very useful in many places. So, we wish to replace A by a larger set which is hereditary and concave. If we apply the recipe of the proposition in Section C.5, we are led first to define a set A_1 and then a set A_2. According to the proposition, A_1 should be the set of (Q, P', I) for which there exist $Q_2, Q_3 \subset Q$, $P'_1, P'_3 \subset P'$, $I_1, I_2 \subset I$ satisfying

(**C.14**) $$Q \,\widehat{\pitchfork}_{I_1}\, P'_1, \qquad Q_2 \,\widehat{\pitchfork}_{I_2}\, P', \qquad Q_3 \,\widehat{\pitchfork}_I\, P'_3.$$

As I_1, I_2 can be chosen arbitrarily small, we can replace them with single values $t_1, t_2 \in I$ of the parameter; the three conditions in (C.14) become:

- there exists $P'_1 \subset P'$, $t_1 \in I$ such that

(**C.15**) $$\delta(Q, P'_1) \geqslant 2\max(|Q|^{1-\eta}, |P'_1|^{1-\eta})$$

- there exists $Q_2 \subset Q$, $t_2 \in I$ such that

(**C.16**) $$\delta(Q_2, P') \geqslant 2\max(|Q_2|^{1-\eta}, |P'|^{1-\eta})$$

– there exists $Q_3 \subset Q$, $P'_3 \subset P'$ such that

$$\delta(Q_3, P'_3) \geqslant 2\max(|I|, |Q_3|^{1-\eta}, |P'_3|^{1-\eta}) \tag{C.17}$$

for all $t \in I$.

As P'_1, Q_2, Q_3, P'_3 may be chosen arbitrarily thin, it is natural to replace (C.15), (C.16), (C.17) by

$$\delta(Q, P'_1) \geqslant 2|Q|^{1-\eta} \quad \text{for } t = t_1; \tag{C.15)'}$$

$$\delta(Q_2, P') \geqslant 2|P'|^{1-\eta} \quad \text{for } t = t_2; \tag{C.16)'}$$

$$\delta(Q_3, P'_3) \geqslant 2|I| \quad \text{for all } t \in I. \tag{C.17)'}$$

Finally, the largest t value of $\delta(Q, P'_1)$ that one can hope for (by choosing $P'_1 \subset P'$ appropriately) is $\delta_R(Q, P')$; similarly, the largest value of $\delta(Q_2, P')$ that one can hope for is $\delta_L(Q, P')$ and the largest value of $\delta(Q_3, P'_3)$ that one can hope for is $\delta_{LR}(Q, P')$. Notice that we need anyway to eliminate P'_1, Q_2, P'_3, Q_3 from the definition because in $\mathcal{R}(I)$ (instead of $\mathcal{R}(I_0)$), elements are constructed inductively and thinner rectangles are constructed at the end. Replacing $\delta(Q, P'_1)$ by $\delta_R(Q, P')$, $\delta(Q_2, P')$ by $\delta_L(Q, P')$ and $\delta(Q_3, P'_3)$ by $\delta_{LR}(Q, P')$, we obtain the three conditions, (T1), (T2), (T3) in Section 5.4. This defines $\overline{⫛}$.

The last step is to go from A_1 to A_2, taking the hereditary envelope of A_1, which corresponds exactly to the transition from $\overline{⫛}$ to ⫛ in Section 5.4.

REFERENCES

[BC] M. Benedicks and L. Carleson, The dynamics of the Hénon map, *Ann. Math.*, **133** (1991), 73–169.

[BDV] C. Bonatti, L. Diaz, and M. Viana, *Dynamics Beyond Uniform Hyperbolicity*, Encyclopedia of Math. Sciences, vol. 102, Springer, Berlin, 2004.

[BR] R. Bowen and D. Ruelle, The ergodic theory of Axiom A flows, *Invent. Math.*, **29** (1975), 181–202.

[C] E. Colli, Infinitely many coexisting strange attractors, *Ann. Inst. Henri Poincaré Anal. Non Linéaire*, **15** (1998), 539–579.

[CL] M. L. Cartwright and J. E. Littlewood, On nonlinear differential equations of the second order I, *J. Lond. Math. Soc.*, **29** (1945), 180–189.

[L] M. Levi, Qualitative analysis of the periodically forced relaxation oscillations, *Mem. Am. Math. Soc.*, **32** (1981), 244.

[MV] L. Mora and M. Viana, Abundance of strange attractors, *Acta Math.*, **171** (1993), 1–71.

[MPV] C. G. Moreira, J. Palis, and M. Viana, Homoclinic tangencies and fractal invariants in arbitrary dimension, *C. R. Acad. Sci. Paris Sér. I Math.*, **333** (2001), 475–480.

[MY] C. G. Moreira and J.-C. Yoccoz, Stable intersections of regular Cantor sets with large Hausdorff dimensions, *Ann. Math.*, **154** (2001), 45–96. Corrigendum in *Ann. Math.*, **154** (2001), 527.

[N] S. Newhouse, The abundance of wild hyperbolic sets and nonsmooth stable sets for diffeomorphisms, *Publ. Math. I.H.E.S.*, **50** (1979), 101–151.

[NP] S. Newhouse and J. Palis, Cycles and bifurcation theory, *Astérisque*, **31** (1976), 44–140.

[P1] J. Palis, A global view of Dynamics and a conjecture on the denseness of finitude of attractors, *Astérisque*, **261** (2000), 335–347. Géométrie complexe et systèmes dynamiques (Orsay, 1995).

[P2] J. Palis, A global perspective for non-conservative dynamics, *Ann. Inst. Henri Poincaré Anal. Non Linéaire*, **22** (2005), 485–507.

[P3] J. Palis, Open questions leading to a global perspective in dynamics, *Nonlinearity*, **21** (2008), 37–43.

[PT] J. Palis and F. Takens, Hyperbolic and the creation of homoclinic orbits, *Ann. Math.*, **125** (1987), 337–374.

[PY1] J. Palis and J.-C. Yoccoz, Homoclinic tangencies for hyperbolic sets of large Hausdorff dimension, *Acta Math.*, **172** (1994), 91–136.

[PY2] J. Palis and J.-C. Yoccoz, *Implicit Formalism for Affine-like Map and Parabolic Composition*, Global Analysis of Dynamical Systems, pp. 67–87, Institut of Phys., IOP, London, 2001.

[PY3] J. Palis and J.-C. Yoccoz, Fers à cheval non uniformément hyperboliques engendrés par une bifurcation homocline et densité nulle des attracteurs, *C. R. Acad. Sci. Paris*, **333** (2001), 867–871.

[Po] H. Poincaré, *Les méthodes nouvelles de la mécanique céleste*, vol. III, pp. 189, Gauthier-Villars, Paris, 1899.

[Ru] D. Ruelle, A measure associated with Axiom A attractors, *Am. J. Math.*, **98** (1976), 619–654.

[S] S. Smale, Differentiable dynamical systems, *Bull. Am. Math. Soc.*, **73** (1967), 747–817.

[Si] Ya. Sinai, Gibbs measures in ergodic theory, *Russ. Math. Surv.*, **27** (1972), 21–69.

J. P.
Instituto Nacional de Matemática Pura e Aplicada,
Estrada Dona Castorina, 110 Jardim Botânico,
22460-320, Rio de Janeiro RJ, Brazil
jpalis@impa.br

J.-C. Y.
Collège de France,
3 Rue d'Ulm,
75005 Paris, France
jean-c.yoccoz@college-de-france.fr

Manuscrit reçu le 27 avril 2006
publié en ligne le 15 octobre 2009.

List of Publications of Jacob Palis

1968

[1] On Morse-Smale Diffeomorphisms, Bulletin American Math. Society, 741, (985-988).

[2] Local Structure of Hyperbolic Fixed Points in Banach Space, Anais da Academia Brasileira de Ciencias, 40, (263-266).

1969

[3] On Morse-Smale Dynamical Systems, Topology, 19, (385-405).

1970

[4] Structural Stability Theorems, with S.Smale, Proceedings of the Institute on Global Analysis, American Math. Society, Vol. XIV, (223-232).

[5] A Note on Ω-Stability, Proceedings of the Institute on Global Analysis, American Mathematical Society, Vol. XIV, (220-222).

[6] Neighborhoods of Hyperbolic Sets, with M.Hirsch, C. Pugh and M. Shub, Inventiones Mathematicae, 9, (212-234).

[7] Ω-Explosions, Bulletin of the Brazillian Mathematical Society, 1, (55-57).

1971

[8] Ω-Stability and Explosions, Lecture Notes in Mathematics, 1206, Springer-Verlag, (40-42).

[9] Ω-Explosions for Flows, Proceeding s American Math. Society, 27, (85-90).

[10] Sistemas Dinamicos, Seminar Notes, IMPA.

1973

[11] Hyperbolic Nonwandering Sets on Two-Dimensional Manifolds, with S. Newhouse, in Dynamical Systems, Academic Press, (293-302).

[12] Bifurcations of Morse-Smale Dynamical Systems, with S. Newhouse, in Dynamical Systems, Academic Press, (303-366.

1974

[13] Vector Fields Generate Few Diffeomorphisms, Bulletin American Mathematical Society, 80, (503-505).

1975

[14] Non Differentiability of Invariant Foliations, with C. Pugh and C. Robinson, Lecture Notes in Mathematics, 468, Springer-Verlag, (234-241).

[15] Genericity Theorems in Topological Dynamics, with C. Pugh, M. Shub and D. Sullivan, Lecture Notes in Mathematics, 468, Springer-Verlag, (241-251).

[16] Fifty Problems in Dynamical Systems, with C. Pugh, Lecture Notes in Mathematics, 468, Springer-Verlag, (345-353).

[17] Arcs of Dynamical Systems: Bifurcations and Stability, Lecture Notes in Mathematics, 468, Springer-Verlag, (48-53).

1976

[18] Cycles and Bifurcations Theory, with S. Newhouse, Astérisque, 31, (44-140).

[19] Stable Arcs of Diffeomorphisms, with S. Newhouse and F. Takens, Bulletin American Mathematical Society, 82, (499-502).

[20] La Topologie du Feuilletage d'un Champ de Vecteurs Holomorphe près d'une Singularitè, with C. Camacho and N. Kuiper, C.R. Acad. Sc. Paris, 282, (959-961).

1977

[21] Topological Equivalence of Normally Hyperbolic Vector Fields, with F. Takens, Topology, 16, (335-345).

[22] Some Developments on Stability and Bifurcations of Dynamical Systems, Lecture Notes in Mathematics, 597, Springer-Verlag, (495-509).

[23] Geometry and Topology, editor, with M. do Carmo, Proc. of the III Latin American Mathematical School, Lecture Notes in Mathematics, 597, Springer-Verlag.

1978

[24] The Topology of Holomorphic Flows near a Singularity, with C. Camacho and N. Kuiper, Publications Math. Institut Hautes Études Scientifiques, 48, (5-38).

[25] Introdução aos Sistemas Dinamicos, with W.de Melo, Notes of the X Bazilian Mathematical Colloquium, 1975, and book in Projeto Euclides, IMPA-CNPq.

[26] Centralizeres of Diffeomorphisms and Stability of Suspended Foliations, Lecture Notes in Mathematics, 652, Springer-Verlag, (114-121).

[27] Invariantes de Conjugação e Módulos de Estabilidade dos Sistemas Dinâmicos, Proceedings of the XI Brazilian Mathematical Colloquium.

[28] A Differentiable Invariant of Topological Conjugacies and Moduli of Stability, Astérisque, 51, (335-346).

[29] Moduli of Stability and Bifurcation Theory, Proceedings of the International Congress of Mathematicians, Helsinki, (835-839).

1980

[30] Moduli of Stability for Diffeomorphisms, with W. de Melo, Proc. Symp. Dyn. Systems, Lecture Notes in Mathematics, 819, Springer Verlag, (318-339).

1981

[31] Characterization of the Modulus of Stability for a Class of Diffeomorphisms, with W. de Melo and S.Van Strien, Lecture Notes in Mathematics, 898, Springer-Verlag, (266-285).

[32] Families of Vector Fields with Finite Moduli of Stability, with I. P. Malta, Lecture Notes in Mathematics, 898, Springer-Verlag, (212-229).

1982

[33] Geometric Theory of Dynamical Systems, with W. de Melo, Springer-Verlag. Translated into Russian and Chinese.

1983

[34] Bifurcations and Stability of Families of Diffeomorphisms, with S. Newhouse and F. Takens, Publications Math. Institut Hautes Études Scientifiques, 57, (5-72).

[35] Geometric Dynamics, Proc. Int. Symposium Dynamical Systems, IMPA-1981. Lecture Notes in Mathematics, 1007, Springer-Verlag.

[36] Stability of Parameterized Families of Gradient Vector Fields, with F. Takens, Annals of Mathematics, 118, (383-421).

[37] A Note on the Inclination Lemma and Feigenbaum's Rate of Approach, Lecture Notes in Mathematics, 1007, Springer-Verlag (630-636).

1985

[38] The Dynamics of a Diffeomorphism and Rigidity of its Centralizer, Singularities and Dynamical Systems, North Holland, (15-21).

[39] Topological Invariants as Translation Number, with R. Roussarie, in Dynamical Systems and Bifurcations, Lecture Notes in Mathematics, 1125, Springer-Verlag, (64-86).

[40] Cycles and Measure of Bifurcation Sets for Two-Dimensional Diffeomorphisms, with F. Takens, Inventiones Mathematicae, 82, (397-422).

1987

[41] Homoclinic Orbits, Hyperbolic Dynamics and Fractional Dimension of Cantor Sets, Contemporary Mathematics, 58, (203-216).

[42] Dimensões Fracionárias de Conjuntos de Cantor e Dinâmica Hiperbólica, Proceedings of the XV Brazillian Mathematical Colloquium, (341-353).

[43] Hyperbolicity and Creation of Homoclinic Orbits, with F.Takens, Annals of Mathematics, 125, (337-374).

1988

[44] On the Solution of the Stability Conjecture (Mañé) and the -Stability Conjecture, Proceedings of the XVI Brazillian Mathematical Colloquium, (599-606).

[45] On the Continuity of Hausdorff Dimension and Limit Capacity for Horseshoes, with M.Viana, Dynamical Systems, Lecture Notes in Mathematics, 1331, Springer-Verlag, (150-160).

[46] Homoclinic Bifurcations and Hyperbolic Dynamics, with F. Takens, Lecture Notes, 66, Proceedings of the XVI Brazillian Mathematical Colloquium, (10-15).

[47] Topics in Dynamical Systems, Editor with R. Bamón e R. Labarca, Lecture Notes in Mathematics, 1331, Springer-Verlag.

[48] On the Ω-Stability Conjecture, Publications Math. Institut Hautes Etudes Scientifiques, 66, (210-215).

1989

[49] On the Solution of the Stability Conjecture and the Ω-Stability Conjecture, IX International Congress on Mathematical Physics Adam Higher (469-471)

[50] Rigidity of Centralizeres of Diffeomorphisms, with J.C.Yoccoz, Ann. Scient. Ècole Normale Superièure, 22, (81-98).

[51] Centralizers of Anosov Diffeomorphisms on Tori, with J.C.Yoccoz, Ann. Scient. Ecole Normale Superieure, 22, (99-108).

[52] Homoclinic Bifurcations and Fractional Dimensions, Publicaciones Matematicas del Uruguay, 1, (55-66).

[53] Gradient Flows, Stability Theory and Related Topics in Dynamical Systems, World Scientific, (142-145).

[54] Bifurcations and Global Stability Families of Gradients, with M. J. Carneiro, Publications Mathematiques Institut Hautes Etudes Scientifiques, 70, (103-168).

1990

[55] Centralizers of Diffeormorphisms, Pitman Research Notes in Math., 221, (19-23).

[56] Chaotic or Turbulent Systems, Attractors and Homoclinic Bifurcations, Revista Matematica Universitaria, Brazilian Mathematical Society.

[57] Differentiable Conjugacies of Morse-Smale Diffeomorphisms, with J.C. Yoccoz, Bulletin Brazilian Mathematical Society, New Series, v.2, (25-48).

1991

[58] Homoclinic Bifurcations, Sensitive-Chaotic Dynamics and Strange Attractors, Dyn. Systems and Related Topics, World Scientific, (466-473).

1992

[59] A Glimpse at Dynamical Systems: the Long Trajectory from the Sixties to Present Developments, Prize Talk at the Third World Academy of Sciences, Proceedings of the TWAS.

[60] New Developments in Dynamics: Hyperbolicity and Chaotic Dynamics, Chaos, Resonance and Collective Dynamical Phenomena in the Solar Systems, International Astron. Union, Kluwer Acad. Publ., 1992 (363-369).

1993

[61] Dynamical Systems, with R. Bamon, R. Labarca e J. Lewowicz, Pitman Research Notes in Math. 285.

[62] Hyperbolicity and Sensitive-Chaotic Dynamics at Homoclinic Bifurcations, Fractal Dimensions and Infinitely Many Attractors, with F. Takens, book, Cambridge University Press.

[63] On the Contribution of Smale to Dynamical Systems, From Topology to Computation, volume in honour of Stephen Smale, Springer-Verlag, (165-178).

1994

[64] Homoclinic Tangencies for Hyperbolic Sets of Large Hausdorff Dimension, with J.C. Yoccoz, Acta Mathematica, 172, (91-136).

[65] High Dimension Diffeomorphisms Displaying Infinitely Many Sinks, with M. Viana, Annals of Mathematics, 140, (207 - 250).

[66] A View on Chaotic Dynamical Systems, Brazilian Journal of Physics, 24, (926-930).

1995

[67] Chaotic and Complex Systems, Science International, 58, November (27-31).

[68] A Global View and Conjectures on Chaotic Dynamical Systems, Dynamical Systems and Chaos, World Scientific, 1, (217-225).

1996

[69] From Dynamical Stability and Hyperbolicity to Finitude of Ergodic Attractors, Proceedings of the Third World Academy of Sciences, 11th General Conference, Italy.

1997

[70] On the Arithmetic Sum of Regular Cantor Sets, with J. C. Yoccoz, Annales de l'Inst. Henri Poincarè, Analyse Non Lineaire, 14, (439-456).

1999

[71] Chaotic and Complex Systems, Caos e Complexidade, Editora UFRJ, (27-38).

[72] Uncertainty-Chaos in Dynamics. A Global view, Medal Lecture Third World Academy of Sciences 1998, Proceedings of 10th General Meeting, (33-38).

2000

[73] A Global View of Dynamics and a Conjecture on the Denseness of Finitude of Attractors, Astérisque, 261, (339-351).

[74] Homoclinic bifurcations: from Poincaré to present time, The Mathematical Sciences, After the Year 2000, World Scientific, (123-134).

2001

[75] Nonuniformily Hyperbolic Horseshoes Unleashed by Homoclinic Bifurcations and Zero Density of Attractors, with J.C.Yoccoz, C.R. Ac.Sc. Paris, 333, (1-5).

[76] Homoclinic tangencies and fractal invariants in arbitrary dimension, with C. Moreira and M. Viana, in C.R. Ac.Sc. Paris, 333 (5).

[77] Implicit formalism for affine-like map and parabolic composition, with J.C. Yoccoz, Global Analysis of Dynamical systems, Inst. of Phys., IOP-London, (67-87).

[78] Wonders and Frontiers of Sciences – CNPq 45 Years, Jacob Palis and José Galizia Tundisi Eds. Publication of MCT-CNPq.

2002

[79] Chaotic and Complex Systems, Current Science, 82 (4), (403-406).

2005

[80] A Global Perspective for Non-Conservative Dynamics, Ann. Inst. Henri Poincaré, Analyse Non Linéaire, 22, (485-507).

2008

[81] Open Questions Leading to A Global Perspective In Dynamics, Invited paper Nonlinearity 21 (2008) T37-T43.

2009

[82] Non-Uniformly Hyperbolic Horseshoes Arising from Bifurcations of Poincaré, Heteroclinic Cycles with J.C. Yoccoz , Publications Mathematiques Institut Hautes Études Scientifiques, No. 110, 1—217.

2010

[83] Dynamical Systems, Chaotic Behaviour - Uncertainty, Interdisciplimary Forum, Fondazione Internazionale Premio E. Balzan Milano, Italia.

2011

[84] On Floris Takens and our joint mathematical work. Indagationes Mathematicae 22, 144-146.

2013

[85] On the Finiteness of Attractors for One-Dimensional Maps with Discontinuities, http://arxiv.org/abs/1401.0232

List of Ph.D. Students of Jacob Palis at IMPA

1. Welington de Melo (1972)
2. Pedro Mendes (1972)
3. Ricardo Mañé (1973)
4. Geovan Tavares dos Santos (1976)
5. Paulo Sad (1977)
6. Artur Oscar Lopes (1977)
7. Iaci P. Malta (1978)
8. Maria Izabel T. Camacho (1978)
9. Genesio dos Reis (1978)
10. Luiz Fernando da Rocha (1978)
11. Jorge A. Beloqui (1981)
12. Rafael Labarca (1985)
13. Marcelo Viana (1990)
14. Jorge Manuel Rocha (1990)
15. Leonardo Mora (1991)
16. Sergio Plaza (1988)
17. Jaques Gheiner (1989)
18. Roberto Markarian (1990)
19. Pedro Miguel Rosa Dias (1993)
20. Lorenzo Diaz (1990)
21. Jaime Alberto Vera (1990)
22. Neptali Romero (1992)
23. Raul Ures (1993)
24. Carlos Gustavo de Araujo Moreira (1993)
25. Carlos Arnoldo Morales (1994)
26. Enrique Pujals (1996)
27. Eleonora Catsigeras (1995)
28. Bernardo San Martin (1995)
29. Eduardo Colli (1996)
30. Stefano Luzzatto (1995)
31. Martin Sambarino (1998)
32. Roger Metzger (1998)
33. Fernando Sanchez-Salas (1998)
34. José Carlos Martin-Rivas (1992)
35. Vilton Pinheiro (2000)
36. Frederico Rodriguez-Hertz (2001)
37. Ali Tahzibi (2002)
38. Germán Aubin Arroyo Camacho (2002)
39. Carlos Vasquez (2003)
40. Luis Bladismir Leal (2005)
41. Paulo Brandão (2013)

Acknowledgments

The Brazilian Mathematical Society (SBM), the Institute for Pure and Applied Mathematics (IMPA) and Springer-Verlag, express their gratitude to the following copyright holders for granting permission to reprint these contributions:

American Mathematical Society

Palis, J.; Smale, S.: Structural stability theorems. Global Analysis (Proc. Sympos. Pure Math., Vol. XIV, Berkeley, Calif.,1968) (1970), 223-231, Amer. Math. Soc., Providence, R.I.

Palis, J.: A note on Ω-stability. Global Analysis (Proc. Sympos. Pure Math., Vol. XIV, Berkeley, Calif., 1968) (1970), 221-222, Amer. Math. Soc., Providence, R.I.

Elsevier

Palis, J.: On Morse-Smale dynamical systems. Topology 8 (1969), 385-404, Pergamon Press.

Newhouse, S.; Palis, J.: Hyperbolic nonwandering sets on two-dimensional manifolds. Dynamical systems (Proc. Sympos., Univ. Bahia, Salvador, 1971), 293-301. Academic Press, New York, 1973.

Palis, J.; Takens, F.: Topological equivalence of normally hyperbolic dynamical systems. Topology 16 (1977), no. 4, 335-345, Pergamon Press.

Palis, J.-; Yoccoz, J.-C.: Centralizers of Anosov diffeomorphisms on tori. Ann. Sci. École Norm. Sup. (4) 22 (1989), no. 1, 99-108, Gauthier-Villars.

Palis, J.-; Yoccoz, J.-C.: On the arithmetic sum of regular Cantor sets. Ann. Inst. H. Poincaré Anal. Non Linèaire 14 (1997), no. 4, 439-456, Gauthier-Villars.

Palis, J.: A global perspective for non-conservative dynamics. Ann. Inst. H. Poincaré Anal. Non Linèaire 22 (2005), no. 4, 485-507.

Finnish Academy of Sciences and Letters

Palis, J.: Moduli of stability and bifurcation theory. Proceedings of the International Congress of Mathematicians (Helsinki, 1978), 835-839, Acad. Sci. Fennica, Helsinki, 1980.

French Academy of Sciences and Elsevier Masson SAS

Palis, J.; Yoccoz, J-C.: Fers à cheval non uniformément hyperboliques engendrés par une bifurcation homocline et densité nulle des attracteurs. (French) [Non-uniformly hyperbolic horseshoes generated by homoclinic bifurcations and zero density of attractors] C. R. Acad. Sci. Paris Sér. I Math. 333 (2001), no. 9, 867-871.

Moreira, C.G. ; Palis, J.; Viana, M.: Homoclinic tangencies and fractal invariants in arbitrary dimension. C. R. Acad. Sci. Paris Sér. I Math. 333 (2001), no. 5, 475-480.

Institute des Hautes Études Scientifiques

Camacho, César; Kuiper, Nicolaas H.; Palis, Jacob: The topology of holomorphic flows with singularity. Publ. Math. Inst. Hautes Études Sci. No. 48 (1978), 5-38.

Newhouse, S.; Palis, J.; Takens, F.: Bifurcations and stability of families of diffeomorphisms. Publ. Math. Inst. Hautes Études Sci. No. 57 (1983), 5-71.

Palis, J.: On the C^1 Ω-stability conjecture. Publ. Math. Inst. Hautes Études Sci. No. 66 (1988), 211-215.

Palis, J.; Yoccoz, J-C.: Non-uniformly hyperbolic horseshoes arising from bifurcations of Poincaré heteroclinic cycles. Publ. Math. Inst. Hautes Études Sci. No. 110 (2009), 1-217.

Princeton University and Institute for Advanced Study

Palis, J.; Takens, F.: Stability of parametrized families of gradient vector fields. Ann. of Math. (2) 118 (1983), no. 3, 383-421.

Palis, J.; Takens, F.: Hyperbolicity and the creation of homoclinic orbits. Ann. of Math. (2) 125 (1987), no. 2, 337-374.

Palis, J.; Viana, M.: High dimension diffeomorphisms displaying infinitely many periodic attractors. Ann. of Math. (2) 140 (1994), no. 1, 207-250.

Societé Mathematique de France, Paris

Palis, J.: A global view of dynamics and a conjecture on the denseness of finitude of attractors. Géométrie complexe et systemes dynamiques (Orsay, 1995). Astérisque No. 261 (2000), 335-347.

Kolibri 360 GmbH
Ferdinand-Jühlke-Straße 7
99095 Erfurt, Deutschland
produktsicherheit@kolibri360.de